Biomathematical and Biomechanical Modeling of the Circulatory and Ventilatory Systems

Volume 3

For further volumes:
http://www.springer.com/series/10155

Marc Thiriet

Signaling at the Cell Surface in the Circulatory and Ventilatory Systems

Springer

Marc Thiriet
Project-team INRIA-UPMC-CNRS REO
Laboratoire Jacques-Louis Lions, CNRS UMR 7598
Université Pierre et Marie Curie
Place Jussieu 4
75252 Paris Cedex 05
France

ISSN 2193-1682
ISBN 978-1-4614-1990-7 e-ISBN 978-1-4614-1991-4
DOI 10.1007/978-1-4614-1991-4
Springer New York Dordrecht Heidelberg London

Library of Congress Control Number: 2011943886

Printed on acid-free paper

Springer is part of Springer Science+Business Media (www.springer.com)

Contents

Introduction

"... *sunt quaedam corpora quorum concursus,*
motus, ordo, positua, figurae
efficiunt ignis, mutatoque ordine mutant
naturam ...
[... some materials exist, of which interaction, motion, order, position, configuration produce fire, and if order differs, a different pattern...] " (Lucretius) [1]

Volume 3 of the book series Biomathematical and Biomechanical Modeling of the Circulatory and Ventilatory Systems aims at presenting major sets of signaling receptors mainly located at the plasma membrane, [1] in a modeling framework rather than biological perspective. Collecting signaling effectors, their main interactions, and major properties are the first tasks required for any modeling of cell signaling processes.

The major objective is to comprehend the complexity of natural phenomena to model these events. In other words, to yield the maximal amount of known information (i.e., to briefly describe the huge number of signaling mediators and their main known features) enables the depiction of any signaling cascade using a suitable data set with a minimal content in required quantities. Handling of complex signaling networks and their complex behavior leads to a compulsory preview to distinguish primary from secondary elements. In physics, a similar strategy is carried out during phenomenological analysis and scaling. Knowledge of all possible known interactors is mandatory to avoid forgetting any important contributor and to understand the complex behavior of signaling cascades. Once all of the mediators of the pathway of interest are identified, major contributors are selected as parameters of modeling equation set and minor are rejected.

1. Signaling receptors involved in cell adhesions are described in Vol. 1 (Chap. 7. Plasma Membrane). Intracellular receptors sense steroid and thyroid hormones, vitamin-A and -D, metabolites (e.g., fatty and bile acids and sterols), and xenobiotics.

> "*The thirty spokes merge in the center to form a wheel; but it is on the empty central space of the axle that the usefulness of the wheel depends. Clay is shaped into pots; but it is on their empty hollowness that our utilization relies. The door and windows are created in walls for a living space; but it is on the empty space that makes it livable.*" (Attributed to Lao Tzu: Tao te Ching [The classic of the way of virtue]; 6th century B.C.E.)

The molecule selection stage is not an obvious task as: (1) the number of known molecular participants of any signaling cascade is often quite large; (2) involved effectors possess many names; (3) some effector aliases designate different types of molecules; (4) most mediators interact with many partners; (5) crosstalk exists with other signaling axes; (6) the finely tuned intracellular cascade of reactions has a complex functioning; and, last but not least, (7) some regulators and mediator properties as well as values of kinetic coefficients remain unknown.

Presentation of biochemical properties of involved molecules goes beyond the scope of this book. However, structural motifs are sometimes given to understand the binding of a signaling mediator or conformational changes that contribute to activate or inactivate an enzyme involved in the next step of the signaling cascade of chemical reactions. Similarly, amino acid residues specifically targeted during post-translational modifications are often given. For example, phosphorylation of a given target amino acid triggers activation (activatory site of a signaling mediator), whereas that of another residue (inhibitory site of a signaling mediator) primes deactivation.

> "*On ne pourra bien dessiner le simple qu'après une étude approfondie du complexe.*
> [The simple (model) will be adequately designed only after a deep investigation of the complex (reality).] " (G. Bachelard)

The set of books devoted to Circulatory and Ventilatory Systems in the framework of Biomathematical and Biomechanical Modeling aims at providing basic knowledge and state of the art on the biology and the mechanics of blood and air flows. The cardiovascular and respiratory systems are tightly coupled, as their primary function is the supply of oxygen (O_2) to and removal of carbon dioxide (CO_2) from the body's cells. Oxygen is not only a nutrient that is used in cellular respiration, but also a component of structural molecules of living organisms, such as carbohydrates, proteins, and lipids. Carbon dioxide is produced during cell respiration. It is an acidic oxide that, in an aqueous solution, converts into anhydride of carbonic acid (H_2CO_3). It is then carried in blood mostly as bicarbonate ions (HCO_3^-) owing to carbonic anhydrase in erythrocytes, but also small fractions that are either dissolved in the plasma or bound to hemoglobin as carbamino compounds. Carbon dioxide is one of the mediators of autoregulation of local blood supply. It also influences blood pH via bicarbonate ions. Last, but not least, it participates in the regulation of air and blood flows by the nervous system.

Explorations of blood and air flows in the cardiovascular and respiratory systems will require development of models that couple different length and time scales. Like physiology, biomechanics cope with the macroscopic scale. Like molecular biology, biomathematics can deal with the nano- and microscopic scales that must be coupled to macroscopic events to take into account the fundamental features of living cells and tissues that sense, react, and adapt to applied loadings. Therefore, the three basic natural sciences - biology, chemistry, and physics - interact with mathematics to explain the functioning of physiological flows, such as those experienced by the circulatory and ventilatory systems. In the near future, biomechanical models should be coupled to biomathematical models of cell signaling and tissue adaptation to better describe the reality, although its complexity still necessitates abstraction. The main objective of the present 8-volume publication is to present data that will be used as inputs for multiscale models.

Biological systems (from molecular level to physiological apparatus) [2] are characterized by their complicated structure, variable nature, and complex behavior. Processing of signals that control the activity of transcription factors and the expression of genes to direct cell decision (differentiation, growth, proliferation, or death), organization of metabolism, cell communication for coordinated action in a tissue, all rely on non-linear dynamics that control spatial distribution and clustering of molecular species at a given time. Fast protein modifications that result from protein interactions in the cytoplasm propagate signals and lead to either relatively slow transcription and translation or direct release of stored substances.

Complexity arises from the large number of involved quantities that are related by non-linear relationships. Therefore, kinetic and transport equations with associated rates, kinetics coefficients, and transport coefficients that govern cell signaling and tissue remodeling are strongly coupled.

Tier architecture of any living system is characterized by its communication means and regulation procedures. It allow integration of environmental changes to adapt. Multiple molecules interact to create the adaptable activity of the cells, tissues, organs, and body. Any integrative model then incorporates a set of models developed at distinct length scales that also includes response characteristic times to efficiently describe the structure–function relationships of the explored physiological system.

Cells communicate with: (1) themselves internally or by secreting regulators (intra- and autocrine signaling, respectively); (2) neighboring cells by direct contact (juxtacrine signaling) or over short distances (paracrine signaling, when signals target a similar or different cell type in the immediate vicinity through a tiny space of extracellular medium); and (3) remote cells, i.e., over large distances (endocrine signaling). Endocrine signals that are transmitted by endocrine cells are called hormones. They circulate in blood to reach their targets. Neurotransmitters represent an example of paracrine signals. Some signaling molecules can function as both a hormone and a neurotransmitter, such as adrenaline and noradrenaline, whether they

---

2. In physiology, apparatus ([Latin] apparatus: preparation, planning; apparo/apparere: to prepare [or adparo (ad: toward, paro/parare: to make ready)]) refers to group of organs that collectively carry out a specific task.

are released from cells of the adrenal medulla (inner region of the adrenal gland) [3] or neurons. Certain messengers such as estrogens are released by the ovary and operate as hormones on other organs such as the uterus or act locally as auto- or paracrine regulators. Cells receive information from themselves and their environment via proteic receptors of the cell surface.

Cells appropriately respond in a controlled or coherent manner to external stimuli (adaptation robustness). Specific responses characterized by intracellular biochemical reaction cascades can be generated over a wide range of parameter variation. Signals are transduced by information processing networks that are characterized by signal transduction complexity and between-pathway connectivity.

Signaling initiation and first steps of molecular interactions and transformations of most pathways occur at the cell membrane and cortex. The dynamics of a biochemical process can be represented by a set of equations that link the time variations of concentrations of implicated substances to production and consumption rates, which depends on concentrations of interacting molecules and spatial coordinates within the cell and possibly the extracellular compartment. The cell is a heterogeneous medium, even inside the cytosol and organelles.

Any complicated physiological system can be analyzed by decomposition into simple parts with identified functions. The combination of these functions allows us to deduce system functioning due to linear interactions. Deconstruction into parts of physiological systems is necessary to understand part behavior as well as to determine between-part interactions. The cell is a complex system constituted of many components. The features of complex systems are adaptation, self-organization, and emergence. Cells self-organize to operate with optimal performance. The behavior of a complex system is not necessarily predictible from the properties of its elementary constituents, which can non-linearly interact with feedback loops, contributing to system bulk behavior. The organization and bulk behavior of a complex system not only results from the simultaneous activities of its constituents, but also emerges from the sum of the interactions among its constituents. A complex system adapts by changing its organization and possibly its structure to environmental stimuli. Yet, a predictive model requires a theory, or at least a framework, that involves relationships.

Input data for integrative investigation of the complex dynamic cardiovascular and respiratory system include knowledge accumulated at various length scales, from molecular biology to physiology on the one hand, and histology to anatomy on the other hand. Tier architecture of living systems is characterized by its communication means and regulation procedures that integrate environmental changes to adapt. Multiple molecules interact to create the adaptable activities of cells, tissues, organs, and body. A huge quantity of these molecules forms a complex reaction set with feedback loops and a hierarchical organization. Studies from molecular cascades primed by mechanical stresses to cell, then to tissues and organs need to be combined to study living systems with complex dynamics; but future investigations are still needed to

3. Adrenal medullary cells that are grouped around blood vessels operate as postganglionic neurons of the sympathetic nervous system.

mimic more accurately system functioning and interaction with the environment, using multiscale modeling. An integrative model also incorporates behavior at various time scales, including response characteristic times, cardiac cycle (s), and diurnal periodicities (h), to efficiently describe the structure–function relationships of the explored physiological system.

Models of biochemical reaction cascades are generally described by mass action equations based on involved molecule concentrations and chemical kinetic coefficients for each elementary reaction:

$$\text{substrate} + \text{enzyme} \rightleftharpoons \text{complex} \rightleftharpoons \text{enzyme} + \text{product}.$$

According to the mass-action law, the rate of change in concentration of a chemical species in chemical elementary, slow reactions at local equilibrium in ideal gases and dilute solutions is proportional to the product of the concentrations of the different involved reagents, raised to given powers. Complex reactions are often considered as a succession of elementary reactions. Generalized power law models have been proposed for complex processes.

Chemical reaction cascades, in which the product of a reaction enters another reaction and different species can be recycled, represent complicated processes. In addition, in biochemical reactions, enzymes, after catalyzing a chemical reactiont, are generally released in free forms ready to enter another reaction cycle. However, enzyme can be sequestered or degraded. Therefore, recycling is not complete.

Models of cell response to environmental stimuli that treat metabolic and signaling networks can have a good predictive potential owing to the limited number of possible states, as cells optimally function in a bounded parameter space (experienced states are defined by a given set of physical and chemical parameters that evolve in known value ranges with identified relationships among them).

The text has been split into a book set according to the length scale. Volume 1 of this book series introduces cells (microscopic scale) involved not only in the architecture of the cardiovascular and respiratory systems as well as those convected by blood to ensure body homeostasis and defense against pathogens, but also cells that regulate blood circulation and the body's respiration to adapt air and blood flows to the body's need.

The remote control as well as major events of the cell life are detailed in Volume 2. Cells of the cardiovascular and ventilatory apparatus are actually strongly regulated by themselves, their neighboring cells, as well as remote cells of the endocrine and nervous systems to adapt to local conditions as well as regional and general stimuli.

Volumes 3 and 4 are aimed at describing components of cascades of chemical reactions (nanoscopic scale) that enable cellular responses to environmental stimuli, in particular mechanical stresses. Extracellular messengers (locally released agents for auto- and paracrine regulation, hormones, growth factors, cytokines, chemokines, and constituents of the extracellular matrix, as well as mechanical stresses) activate various types of receptors to initiate signaling cascades (Vol. 3). Intracellular signaling pathways are usually composed of multiple nodes and hubs that correspond to major mediators (Vol. 4). Signaling pathways trigger release of substances from intracellular stores and gene expression to produce effectors for intra- or autocrine regulation as well as close or remote control. Cell activities can be modeled using systems of equations to predict outcomes.

Volume 5 deals with tissues (mesoscopic scale) of the cardiovascular (heart, blood and lymph vessels, as well as blood and lymph) and respiratory apparatus (airways and lungs), including interactions between adjoining cells. Cell activities involved in adaptation (mechanotransduction, stress-induced tissue remodeling in response to acute or chronic loadings, angiogenesis, blood coagulation, as well as inflammation and healing) can be described using mathematical models.

The wetted surface of any segment or organ of the cardiovascular system is covered by the endothelium, which constitutes the interface between the flowing blood and the deformable solid wall. The endothelium is a layer of connected and anchorage-dependent cells. The endothelium has several functions. It controls molecule exchange between the blood and the vessel wall and perfused tissues. It regulates flowing cell adhesion on the blood vessel wall and extravasation, especially for immune defense. It controls coagulation and thrombolysis. It regulates the vasomotor tone and proliferation of vascular smooth muscle cells via the release of several compounds. It is required in angiogenesis. Endothelial cells detect hemodynamic stresses via mechanosensors.

The blood vessel wall is a living tissue that quickly reacts to loads applied on it by the flowing blood. In any segment of a blood vessel, the endothelial and smooth muscle cells sense the large-amplitude space and time variations in small-magnitude wall shear stress and wall stretch generated by the large-magnitude blood pressure. These cells respond with a short time scale (from seconds to hours) to adapt the vessel caliber according to the loading, especially when changes exceed the limits of the usual stress range. This regulatory mechanism is much quicker than the nervous and hormonal control. The mechanotransduction pathways determine the local vasomotor tone and subsequently the lumen bore of the reacting blood vessel.

Volume 6 focuses on the functioning of the cardiovascular and respiratory apparatus (macroscopic scale) and diseases associated with blood and air flows. Local flow disturbances can trigger pathophysiological processes and/or result from diseases of conduit walls or their environment. Volume 6 contains chapters on anatomy and physiology of the cardiovascular and respiratory systems as well as medical signals and images, and functional tests. In addition, it presents pathologies of the fluid convection duct network (i.e., heart, blood vessels, and respiratory tract) and their treatment that are targeted by biomechanical studies. Nowadays, the development of medical devices and techniques incorporates a stage of numerical tests in addition to experimental procedures. Moreover, a huge number of research teams develop tools for computer-aided diagnosis, therapy (e.g., treatment planning and navigation hard- and softwares), and prognosis.

Volume 7 copes with mechanics of air and blood flows (macroscopic scale), taking into account rheology of blood and deformable walls of respiratory conduits and blood vessels and providing insights in numerical simulations of these types of flows.

Volume 8 presents a set of glossaries to rapidly get a basic knowledge on coonstituents, structures, and parameters used in models, as this information arises from diverse scientific disciplines. Specific vocabulary used in each of study field can indeed limit easy access to a field by researchers of other disciplines. Moreover, same words can be used in different knowledge fields, but with distinct meanings.

The present Volume is composed of 10 chapters. Transduction of mechanical stresses applied both on the wetted surface of physiological conduits and within the wall by cells involves plasmalemmal ion carriers and receptors, among other molecules of the cell surface. The signal then propagates within the cell using specific pathways. Chapter 1 describes signaling pathways that allow the cell to transduce received signals to determine its functioning for given missions. It give insights on effector stage of the initiated signaling pathways that is achieved by an activation cascade, with assembling of molecular complexes and reversible protein modifications, especially phosphorylation. Chapters 2 to 5 cope with ion and molecule carriers as well as receptors of the cell–matrix mass transfer. Ion fluxes particularly generate action potential in the heart wall that then travels through conduction (nodal) tissue and myocardium to trigger contraction of the cardiac pump. Heart contraction requires calcium influx and energy, whereas cardiomyocyte relaxation needs calcium efflux. Some ion channels are also primed by mechanical loads. Both ion and molecule carriers participate in cell signaling and exchange between cells and their environment. Chapter 6 introduces cell-surface receptors to various ligands, details intracellular receptors and their translational partners, reports on some sets of plasmalemmal receptors (guanylate cyclase receptors for natriuretic peptides and nitric oxide, adenylate cyclase sensors, renin and prorenin receptors, and receptors of the plasminogen–plasmin cascade), and provides examples with the set of adipokine receptors. Chapters 7 to 11 describe various families of receptors, starting with the largest group of plasmalemmal receptors, i.e., G-protein-coupled receptors (Chap. 7). Then follow receptor protein tyrosine and serine/threonine kinases (Chap. 8), receptor protein tyrosine phosphatases (Chap. 9), receptors implicated in morphogenesis (Chap. 10; receptors of the Notch, Hedgehog, and Wnt pathways, as well as glycoprotein EpCAM, plexins [semaphorin receptors] and Robo), to terminate on blood cell receptors, i.e., receptors of the immune system (Chap. 11).

Volume 4 is composed of 10 chapters. This volume is mainly aimed at giving the major features of intracellular signaling mediators, including their various names, to handle the complexity of molecular biology. Once the analysis step is completed, the synthesis step that retains the major signaling components enable mathematical modeling. Chapters 1 to 9 enumerate the main families of signaling mediators and their members. Volume 4 begins by components of lipid signaling (Chap. 1) and a preambule to protein kinases (Chap. 2) to give a survey on cytoplasmic protein tyrosine (Chap. 3), serine/threonine (Chap. 4), mitogen-activated protein kinase modules (Chap. 5), and dual-specificity (Chap. 6) kinases as well as protein phosphatases (Chap. 7). Then follow heterotrimeric and monomeric guanosine triphosphatases (Chap. 8). Chapter 9 presents signaling gas as well as major transcription factors and coregulators. Chapter 10 gives examples of signaling pathways, such as those involving cyclic adenosine and guanosine monophosphates, adhesion and matrix molecules, and calcium, as well as those involved in oxygen sensing, insulin stimulation, angiogenesis, and mechanotransduction.

Once collected, the set of mediators involved in cell signaling is sorted and selected signaling components constitute the set of model variables. Primary mediators are indeed kept in modeling of regulated cellular processes, whereas multiple secondary signaling components are discarded to handle simple, representative modeling and eventually manage their inverse problems. As mathematics deals with abstraction, modeling based on the transport equation may be preferred to that governed by the mass-action law used by chemists. The major drawback of the latter is the large number of kinetic coefficients, the values of which often still remain unknown even after efficient data mining.

Common abbreviations such as "a.k.a." and "w.r.t." that stands for "also known as" and "with respect to", respectively, are used throughout the text to lighten sentences. Latin-derived shortened expressions are also widely utilized (but not italicized despite their latin origin): "e.g." (exempli gratia) and "i.e." (id est) mean "for example" and "in other words", respectively. Rules adopted for substance aliases as well as alias meaning and other notations are given at the end of this book.

Acknowledgments

These books result from lectures given at Université Pierre et Marie Curie in the framework of prerequisite training of Master "Mathematical Modeling", part of Master of "Mathematics and Applications", Centre de Recherches Mathématiques, [4] and Taida Institute for Mathematical Sciences, [5] the latter two in the framework of agreements with the French National Institute for Research in Computer Science and Control. [6] These lectures mainly aim at introducing students in mathematics to basic knowledge in biology, medicine, rheology, and fluid mechanics in order to conceive, design, implement, and optimize appropriate models of biological systems at various length scales in normal and pathological conditions. These books may also support the elaboration of proposals following suitable calls of granting agencies, in particular ICT calls "Virtual Physiological Human" of the European Commission. The author takes the opportunity to thank the members of ERCIM office (European Consortium of Public Research Institutes) and all of the participant teams of the working group "IM2IM" that yields a proper framework for such proposals. These books have been strongly supported by Springer staff members. The author thanks especially S.K. Heukerott and D. Packer for their help and comments.

The author, an investigator from the French National Center for Scientific Research [7] wishes to acknowledge members of the INRIA-UPMC-CNRS team "REO", [8] and Laboratoire Jacques-Louis Lions, [9] as well as CRM (Y. Bourgault, M. Delfour, A. Fortin, and A. Garon), being a staff member in these research units, and

4. CRM: www.crm.umontreal.ca.
5. TIMS: www.tims.ntu.edu.tw.
6. Institut National de la Recherche en Informatique et Automatique (INRIA; www.inria.fr).
7. Centre National de la Recherche Scientifique (CNRS; www.cnrs.fr
8. www-roc.inria.fr/reo
9. www.ann.jussieu.fr

TIMS (I.L. Chern, C.S. Lin, and T.W.H. Sheu), as well as members of the Dept. of Bioengineering and Robotics from Tohoku University (Japan) led by T. Yamaguchi for joint PhD experience and research. The author also acknowledges the patience of his family (Anne, Maud, Julien, Jean, Raphaëlle, Alrik, Matthieu [Matthew], Alexandre, Joanna, Damien, and Frédéric [Frydsek]). This book is dedicated to the author's father and grandfathers.

1

Signal Transduction

In any multicellular organism, the body's cell communicate. Cell signaling enables the coordination of cell activities, from basic life to tissue development and remodeling. These basic cellular activities include uptake of nutrients, excretion of wastes and toxins, and interactions with neighboring or remote cells of the organism, in addition to processing of possible occurrence of developmental abnormalities (cancers) and invasion of pathogens.

Any cell emits, receives, transmits, stores, and processes information. Cell signaling involve, a single or many given signals. Signaling events include signal synthesis in and release from the sending cell, transmission, reception by the receiving cell, and response of the latter. Transport of signaling molecules can be carried out over short or long distance (i.e., within the producing cell or between remote cells), and slowly or rapidly whatever the type of communication (i.e., from intra- to endocrine signals), whether the signaling molecule is synthesized or stored in a given subcellular compartment, and whether the transmission path is the nerve or blood stream. In addition, exocytosis from and possible recycling for further molecule processing within the sending cell on the one hand, surface signaling, signaling during endocytosis into the receiving cell and receptor recycling for further delayed signaling are crucial steps of cell signaling. In polarized cells, the cell side from which the produced signaling molecule is secreted influences the transport mode, range, and speed, as well as the type of receptors in apposed receiving cells.

Various types of signals encompass chemical, electrochemical, physical, and mechanical stimuli. Signal transduction that regulates cell fate (differentiation, growth, division, migration, transformation, and death) as well as tissular homeostasis and immune defense against invading pathogens relies on: (1) plasmalemmal (Chaps. 7 to 11) and intracellular (Chap. 6) receptors, and incorporated adaptors; as well as (2) ion carriers (Chaps. 2 and 3);[1] (3) cell surface transporters (Chaps. 4 and 5); (4) cell adhesion molecules that links any cell to apposed cells and/or its extracellular matrix (Vol. 1 – Chap. 7. Plasma Membrane); and (5) specialized plasmalem-

1. In particular, in neurons and myocytes, electrochemical signaling is controlled and modulated by activity of ion carriers (ion channels, pumps, exchangers, etc.).

mal nanodomains, such as membrane rafts, tetraspanin-enriched nanodomains, and calveolae (Vol. 1 – Chap. 9. Intracellular Transport).

Cell signaling relies on cascades of molecular interactions and chemical reactions in response to extra- or intracellular stimulus. Mechanosomes are constituted of cell adhesion molecules, such as cadherins and integrins, kinases and phosphatases, and transcription factors that translocates to the nucleus to acheive their mission, i.e., gene transcription fot protein synthesis.

1.1 Main Signaling Features

Signaling pathways that originate at the plasma membrane use a set of proteic and lipidic mediators (Table 1.1). These mediators either possess a catalytic function or participate in the regulation of the activity or localization of other effectors of the signal-transduction cascade.

The structure of cell signaling mediators is often modular. It contains distinct binding, regulatory, and possible catalytic domains. Catalysis is regulated by interactions between enzymes, substrates, possible cofactors, and messengers that allosterically modulate enzyme activity. Post-translational modifications of signaling effectors influence binding and activity of these molecules.

Signaling pathways and their kinetics can be explored via mutant signaling proteins, [2] by preventing connections between interacting signaling mediators, as well as using direct and allosteric inhibitors of effectors [2]. The latter induces conformational changes that prevent ligand binding, hence activation of signaling effectors. Rewiring cell signaling circuits and constructing novel signaling sensors, such as receptors activated solely by synthetic ligands (RASSL), have potential applications in medicine and biotechnology [3].

1.1.1 Types of Cell Communications

Any cell communicates with itself (intracellular communication [intracrine regulation] and autocrine signaling) and others (intercellular communication) using various types of messengers. Chemical intercellular communications occur via direct contact (*juxtacrine signaling*) as well as over short (*auto- and paracrine signaling*) and long (*endocrine signaling*) distances.

In juxtacrine signaling, plasma membranes of interacting cells come into contact. Ligand–receptor interactions at this interface trigger intracellular signaling. Plasmalemmal receptors are organized into nanodomains at the interface between 2 apposed cells. Between-protein interactions depend on protein density and local membrane curvature. In juxtacrine signaling, intermembrane protein binding has a cooperative effect for other pairs of proteins, as the plasma membrane can bend to

2. Following gain-of-function mutations, hyperactive mutant proteins can be inhibited by molecules that target the active site. Conversely, loss-of-function mutations can be rescued by agents that allow signaling from mutant protein.

accommodate short and long intermembrane pairs. These pairs can be segregated according to their size and intermembrane spacing to guide signaling. Binding of pairs that creates different sizes of intermembrane spacing is segregated to minimize membrane bending [4]. High membrane bending is actually unfavorable.

1.1.2 Phases of Cell Communications

Signaling pathways can be defined by an *initiation stage* and *effector stage*. The initiation phase is triggered by activation of a receptor and/or ion carrier (among others) that is associated with a conformational change of the receptor after ligand binding or membrane deformation that results from a mechanical stress field in the case of a mechanosensitive plasmalemmal molecule. This chemical event generates

Table 1.1. Cell signaling circuit. Cells sense developmental and environmental cues and process received information using intracellular signaling networks that cause various types of response programs and outputs, such as gene transcription, molecule secretion, and cytoskeleton reorganization. Signaling activators can either be components of preformed complexes (fully scaffolded complexes) or undergo an active recruitment (partially scaffolded complexes). Signaling cascades of chemical reactions achieve diverse but selective responses via signal integration and discrimination, especially at signaling hubs and between-pathway crosstalk. Temporal activation kinetics and spatial distribution of signaling effectors, i.e., spatial organization of between-mediator interactions, control cell decisions. Signaling circuits can generate switches, pulses, and oscillations. Feedback loops regulate signaling duration and intensity. Between-pathway crosstalk contributes to robustness of cellular responses. Signaling modules are shared by signaling pathways, but connections are enabled or disabled according to the context for specific outcomes.

Step	Element
Input signal	Chemical cues (neurotransmitters, hormones, growth factors, cytokines, chemoattractants, nucleotides, ions, etc.) Mechanical cues (stretch) Physical cues (temperature, light, osmolarity, electric field, magnetic field, etc.)
Sensor	Receptors, ion carriers, adhesion molecules, other plasmalemmal enzymes
Signal processing	Chemical reaction cascade
Output (cell response)	Protein synthesis, stored molecule exocytosis, cytoskeleton remodeling
Effect	Cell growth, proliferation, differentiation, morphology, contraction, polarization, migration, survival, senescence, death Intracellular molecular transport

the effector stage with more or less numerous chemical reactions that lead to specific cell responses. Signaling complexes are often ephemeral.

1.1.3 Main Signaling Mediators

The *Guide to Receptors and Channels* of the British Pharmacological Society presents the main signaling mediators at the plasma membrane and in the cytoplasm [5]: (1) guanine nucleotide-binding (G)-protein-coupled receptors (or 7-transmembrane receptors; Chap. 7); (2) catalytic receptors, such as receptor kinases (Chap. 8); (3) nuclear receptors (Sect. 6.3); (4) ligand-gated ion channels (Sect. 2.5);[3] (5) ion channels (Chap. 3),[4] (6) transporters (Chap. 4); and (7) enzymes (Vol 4 – Chaps. 1. Signaling Lipids to 9. Other Major Signaling Mediators).

Integrins are major plasmalemmal receptors that are responsible for dynamical interactions between cells and their environment. Cell-surface integrins recognize and bind extracellular matrix proteins and control the organization of the cytoskeleton, in addition to initiating signaling pathways that regulate the cell behavior.

1.1.4 Signaling Cascade

Many cell stimuli induce signaling cascades that terminate by protein import into the nucleus to activate transcription of target genes. Most of these proteins contain a domain that binds to importins with which they translocate into the nucleus through the nuclear pores (Vol. 1 – Chap. 9. Intracellular Transport). In the nucleus, control of gene transcription results from combination of many factors, such as transcription factors, epigenetic mechanisms, and interactions with small RNA molecules (Vol. 1 – Chap. 5. Protein Synthesis and Sect. 1.2.8).

Blood coagulation yields an example of a signaling cascade (Fig. 1.1; Vol. 5 – Chap. 9. Endothelium). Blood clotting involves a cascade of tens of chemical reactions. Competition between formation of enzymes at each step and diffusive mixing controls the progression of the whole signaling cascade. Therefore, efficiency of the process relies on the spatial organization of involved mediators. Spacing between reactants is more relevant than the overall amount of reactants.

3. Ligand-gated ion channels are opened by: (1) neurotransmitters serotonin (ionotropic $5HT_3$), acetylcholine (nicotinic acetylcholine receptors [nAChR]), γ-aminobutyric acid ($GABA_A$ receptors, i.e., Cl^- channels), glutamate (ionotropic glutamate receptors, i.e., AMPA-type [GluR1–GluR4], kainate-type [GluK1–GluK5], and NMDA-type [GluN1, GluN2a–2d, GluN3a–3b] receptors), and glycine (glycine channels–receptors GlyR); (2) nucleotides (P2X receptors); and (3) zinc (zinc-activated channels).

4. In the above-mentioned guide, ion channels comprise acid-sensing (proton-gated) ion channels (ASIC), aquaporins, voltage-gated Ca^{++}, CatSper, and Cl^- channels, connexins and pannexins; cyclic nucleotide-gated, epithelial Na^+, and hyperpolarization-activated, cyclic nucleotide-gated channels (HCN), IP_3 receptors, K^+ channels, ryanodine receptors, sodium leak, voltage-gated Na^+, and transient receptor potential cation (TRP) channels.

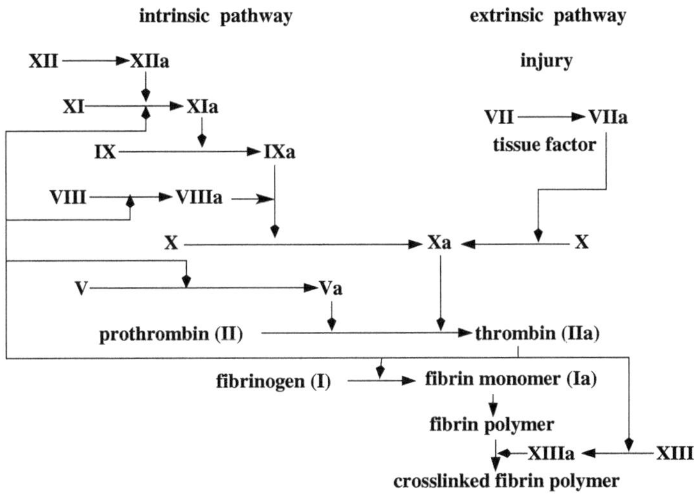

Figure 1.1. Sketch of pathways involved in blood coagulation cascade. See Table 1.2 for the definition of clotting factors.

Table 1.2. Coagulation factors.

Coagulation factor	Molecule
Factor I	Fibrinogen
Factor II	Prothrombin
Factor III	Tissue thromboplastin
Factor IV	Ca^{++}
Factor V	Proaccelerin
Factor VI	Accelerin
Factor VII	Proconvertin
Factor VIII	Antihemophilic factor A
Factor IX	Plasma thromboplastic component
Factor X	Prothrombin-converting enzyme
Factor XI	Plasma thromboplastin antecedent
Factor XII	Hageman or contact factor
Factor XIII	Fibrin-stabalizing factor

1.1.5 Features of Signaling Cascades

Signaling cascades are characterized by: (1) *robustness*, i.e., ability to cope with unpredictable variations;[5] (2) *modularity*; and (3) *evolvability*. Hundreds of mechanical, physical, and chemical stimuli control the cell function using a limited directory of signaling pathways that prime distinct cell responses. Loss in function of a sin-

5. Biological control networks are robust against perturbations, as they are able to maintain the concentration of a mediator at steady state within a narrow range, thereby allowing sustained function, regardless of variations in amounts of other network components.

gle component is tolerated owing to compensatory signaling that allows the cell to achieve regulatory pathway robustness.

Controllability of complex, self-organized, signaling networks relies on the network structure, especially nodes with a low-interaction degree that can determine the network behavior and performance (given signaling delay and duration) [6].

1.1.5.1 First, Second, and Third Messengers

Extracellular *first messengers* modify intracellular concentrations of *second messengers* and, in turn, the latter change the concentrations of and activate via post-translational modifications and conformational variations their signaling effectors.[6]

First messenger can work both on the cell surface and inside the cell, i.e., can be considered as both an *intra- and extracellular first messenger*). Most often, it binds to a plasmalemmal receptor and initiate a signaling from the cell surface. Secondarily, the liganded receptor is internalized and can trigger a new type of signaling from the surface of endosomes, with which it travels within the cell. In addition, some first messengers such as members of a major category of hormones — hydrophobic steroid hormones — cross the plasma membrane to tether to their cognate intracellular receptors. The resulting complex serves as a transcription factor. This complex can be considered as an *intracellular second messenger*. In both cases, the first messenger primes signaling when it resides inside the cell.

The prototypical second messengers are (3′-5′)-cyclic adenosine monophosphate (cAMP; Vol. 4 – Chap. 10. Signaling Pathways) and clacium ion (Ca^{++}). Intracellular cAMP is generated owing to a membrane-bound enzyme — adenylate cyclase — upon activation of any cell-surface receptor coupled to a stimulatory subunit of guanine nucleotide-binding (G) protein (Chap. 7), in particular, by hydrophilic peptidic and proteic hormones. Once synthesized within the cytosol, this second messenger provokes a cascade of activations of protein kinases that ultimately causes a cell response. Among these activated kinases, an enzyme that represents a unique, major hub in the signaling cascade may be called an *intracellular third-order messenger*. Mediator cAMP via enzymatic changes activates the transcription factor cAMP response element-binding protein (CREB) that can be considered as a *intracellular fourth-order messenger*. Once bound to its response element (5′ TGACGTCA 3′) in promoters of target genes, it elicits the gene transcription. Inside the cell, cAMP is catabolyzed by a phosphodiesterase, thereby terminating signaling.

Agent cAMP can also be actively transported through the plasma membrane to the extracellular space via an ATP-binding cassette effluxer (Sect. 4.18), an adenosine triphosphatase (ATPase), which catalyzes the transformation of adenosine triphosphate (ATP), a major source of chemical energy, into adenosine diphosphate (ADP). Extracellular cAMP then can be sequentially processed, first by a type-2 transmembrane ectonucleotide pyrophosphatase (ENPP)–ectophosphodiesterase to

6. The second messenger concept was proposed in 1958 by T. Rall and E. Sutherland who identified a mediator of the intracellular actions of hormones glucagon and adrenaline on glycogen metabolism in the liver [7].

adenosine monophosphate, and then by ecto-5′-nucleotidase to adenosine. Adenosine can then act as a auto- (the cell signals to itself), juxta- (the cell signals to apposed cells), and paracrine (the cell signals to adjoining cells) messenger by binding its cognate P1 receptors. Therefore, this (first signaling wave-induced) *secondary extracellular messenger* derived after exocytosis from processing of the intracellular second messenger (hence a metabolite formed upon stimulation by the *primary extracellular messenger*), can launch a second, delayed signaling wave, which differ from the initial signaling wave.

In addition, cAMP can be exported to and circulates in blood. It can then operate as a prohormone, i.e., as a precursor of the endocrine regulator — adenosine — that functions upon remote processing of cAMP mediator. In particular, extrusion of cAMP can continue after termination of a pulse of stimulations from flowing blood cells.

Furthermore, an intracellular second messenger can act secondarily as an extracellular signal. Free calcium ions, another type of *intracellular second messenger*, can also act as an (exocytozed, non-metabolized) *extracellularly operating second messenger* with a staggered activity, which initiates a supplementary distinct signaling. After activation of suitable receptors by Ca^{++}-mobilizing agonist and resulting formation of inositol trisphosphate, the liberation of stored Ca^{++} ions through IP_3 receptor channels augments the intracellular Ca^{++} pool. A fraction of cytosolic Ca^{++} ions can be exported through the plasma membrane Ca^{++} ATPase, hence elevating the extracellular Ca^{++} concentration. On the other hand, store emptying also triggers Ca^{++} influx via store-operated Ca^{++} channels of the plasma membrane. However, these 2 exchange processes with Ca^{++} influx and efflux across the plasma membrane may create temporally and spatially segregated extracellular zones of instability of the Ca^{++} content. The buffering capacity for Ca^{++} ions in the extracellular medium is actually much lower than that inside the cell. Marked fluctuations in free Ca^{++} ions in the extracellular space may trigger signaling. Local Ca^{++} fluctuations outside the cell may indeed influence Ca^{++}-sensitive channels and receptors (e.g., G-protein-coupled calcium-sensing receptor, Ca^{++}-gated K^+ and non-selective cation channels, etc.) on the cell surface in an auto-, juxta-, and paracrine manner.

Cyclic guanosine monophosphate (cGMP) serves as the second messenger for first messengers natriuretic peptides and nitric oxide. Agent cGMP operates via cGMP-dependent protein kinase-G that phosphorylates substrate proteins in the cytoplasm. Other second messengers that relay signals received at receptors on the cell surface include inositol (1,4,5)-trisphosphate (IP_3) and diacylglycerol (DAG). Peptide and protein hormones, such as angiotensin-2 and vasopressin, and neurotransmitters such as γ-aminobutyric acid (GABA), bind to their cognate G-protein-coupled receptors to activate phospholipase-C. This intracellular enzyme hydrolyzes phosphatidylinositol (4,5)-bisphosphate (PIP_2) in the inner layer of the plasma membrane to produce both DAG and IP_3 second messengers. The former stimulates calcium-dependent protein kinase-C; the latter binds to IP_3 receptors (IP_3R) on the endoplasmic reticulum to release of calcium ions into the cytosol.

Signal transduction relies on a cascade of chemical reactions during which a set of effectors are successively activated. Among these effectors, some can be defined as *i*th-order messengers (to avoid confusion with the usual terminology).

The reaction cascade reaches the level of transcription factors that, once activated, prime transcription of early immediate-early genes, or primary response genes. These genes synthesize another groups of transcription factors (e.g., Fos, Jun, and early growth response protein Egr1) that are defined as *third messengers*. These transcription factors initiate a transcriptional cascades that target delayed early genes to cause changes in cell phenotypes. Other stimulated immediate-early genes encode paracrine mediators of cellular communication. Their products (e.g., prostaglandin synthase-2, inducible nitric oxide synthase, cytokines, and chemokines) modulate the behavior of neighboring cells.

1.1.5.2 Outside–In vs. Inside–Out Signaling

Cells usually experience *outside–in signaling*. Extracellular first messengers, such as hormones, neurotransmitters, nutrients, growth factors, and other agents, modify intracellular concentrations of the so-called second messengers and, in turn, those of their signaling effectors.

Cells can also trigger *inside–out signaling*. Cytoplasmic events are transmitted to external ligand-binding domains that enable bidirectional communication between apposing cells. These connected cells have ligands and complementary receptors on their respective plasma membranes.

Integrins signal in the 2 directions with different consequences. During *integrin inside–out signaling*, an intracellular activator, such as talin or kindlins, binds to integrin and causes a conformational change that increases its affinity for extracellular ligands. Integrin inside–out signaling enables strong interactions between integrins and extracellular matrix proteins, thereby controling adhesion strength and permitting integrins to transmit forces for matrix remodeling and cell migration. Integrins that are activated directly by extracellular factors are able to transmit information into cells from their environment. During *integrin outside–in signaling*, the binding of integrins to their extracellular ligands changes the integrin conformation and promotes integrin clustering. The 2 unidirectional signalings, in fact, are often closely linked, as ligand binding associated with outside–in signaling stimulates integrins and, conversely, integrin activation increases ligand binding for inside–out signaling.

1.1.5.3 Modularity of Intracellular Signaling Cascades

Signaling networks contain diverse functional modules (subnetworks) that can be rapidly connected to shift from one subcellular localization and function to others. Signaling mediators possess different regions, such as regulatory, binding, and eventually catalytic domains. The modular organization of signaling mediators and networks creates behaviors that fit a constantly changing environment (cell suitability).

The modular architecture of signaling networks has been used to develop simpler representations of its constituents, such as amplification and adaptation modules. The *amplification module* not only detects changes in environmental conditions and generates an intracellular signal, but also amplifies signals over a given dynamical range. The *adaptation module* maintains the intracellular signal at a steady state, whatever the ambient concentration of ligand, which can fluctuate. However, as the behavior of signaling pathways is complex, the nature of the coupling between constituents must be determined.

For example, the 3-tiered MAPK module combined with negative feedback constitutes a negative feedback amplifier that confer robustness, output stabilization, and linearization of non-linear signal amplification. A negative feedback amplifier [8]: (1) converts intrinsic switch-like activation kinetics into graded linear responses; (2) conveys robustness to changes in rates of reactions within the negative feedback amplifier module; and (3) stabilizes outputs in response to drug-induced perturbations of the amplifier.

Static changes in signal intensity (step experiments) enable the testing of receptor sensitivity. Signaling pathways can be sensitive to time changes in ligand concentrations [9]. Time-varying signals, such as exponential monotonic variation (increase or decrease) and exponentially varying oscillatory (sine wave) inputs, can be used to explore the gradient sensitivity of the pathway and its frequency response (i.e., amplitude and phase of the response to exponential sine waves) that yields the frequency band over which the pathway can adequately process the input signal.

Dynamical stimuli allow the testing of the adaptation kinetics near the steady state. Negative feedback loops yield oscillations and switch-like responses. In response to an unsteady input, a negative feedback signal must exactly cancel the change in input signal to support a steady state, i.e., the maintenance of a constant output. Conversely, large-amplitude fluctuations result from attenuated feedback due to enzyme saturation. The time-gradient sensitivity depends on the coupling mode between the amplification and adaptation modules [9].

During organ development, signaling modules, each with a given amount of proteins, are integrated into complex networks with a given spatiotemporal organization [10]. For example, 4 networks of cardiac development proteins have been assigned to the morphogenesis of atrial septum, atrioventricular valves (and their precursors, the endocardial cushions), myocardial trabeculae, and outflow tract. The functional modules can be recycled during cardiogenesis. [7]

1.1.5.4 Signal Integration

Cell signaling pathways can use common components. Mechanisms for signaling insulation or, conversely, collaboration are thus necessary. In any case, signal

7. Functional modules include signaling mediators of BMP/TGFβ, cell cycle regulators, FGF–PDGF, focal adhesions, HER, Notch, retinoic acid, semaphorins, Wnt pathways, in addition to transcription factors. Some transcriptional regulators are only active during a brief period, whereas others, such as GATA4, NKx2-5, and TBx5, are continuously expressed. Yet, the latter type of regulators participate in modules that vary in behavior and composition.

processing efficiency and precision rely on signal discrimination and integration. Signaling integration occurs down to the level of gene regulation via the assembly and disassembly of transcriptional complexes.

1.1.5.5 Spatial Organization of Signaling

The spatial organization of regulatory components govers the signaling efficiency. The spatial organization starts from structuring of signaling complexes at the plasma membrane and continues with various regulatory nodes and hubs at specific subcellular locations, in the cytosol and at endomembranes. In particular, the spatial organization of signaling complexes at cell–matrix and –cell adhesion sites are modulated by mechanical forces.

Reversible post-translational modifications of regulatory factors influence their localization. The segregation and compartmentation of proteins generate spatial gradients. Mono- and trimeric guanine nucleotide-binding proteins (binding of GTP or GDP), kinases, calcium-responsive proteins undergo reaction cycles that enable their coupling to and decoupling from signaling cascades. In particular, the steady state distribution and rate of exchange of guanine nucleotides of small GTPases in different subcellular compartments diversify the features of GTPase-dependent signal transduction.

Intracellular situation then depends on the residence sites of modifying and demodifying enzymes. In addition to cellular organelles bounded by membranes, the evolving, adaptive cytoskeleton contributes to spatial heterogeneity. Flux of signaling effectors inside the cell relies on the cell transfer machinery, in particular Rab GTPases (Vols. 1 – Chap. 9. Intracellular Transport and 4 – Chap. 8. Guanosine Triphosphatases and Their Regulators). Numerous isoforms of Rab GTPases are involved in vesicular transport between cell compartments.

1.1.5.6 Transport and Reaction of Signaling Substances

Diffusion of molecules eliminates spatial heterogeneity. Yet, when the diffusivities of interacting chemical species are different, diffusion can generate spatial subdomains, even in an initially uniform medium. Regulatory (activatory or inhibitory) domains depend on effective diffusion coefficients and, hence, concentrations of activator, substrate, and deactivator (production and consumption rates of involved chemical species) as well as reaction kinetics. In addition, free, inactive molecules are transported along cytoskeletal tracks.

Intracellular gradients of concentration of a signaling mediator can result from the spatial separation of its activator and deactivator. Protein phosphorylation by a kinase at the plasma membrane in response to an external stimulus and dephosphorylation by a cytosolic phosphatase generate a concentration gradient of the modified protein with high and low concentrations of phosphorylated protein close to the plasma membrane and in the cell interior, respectively. On the other hand, scaffold proteins can form complexes with antagonist enzymes, hence confining the active and inactive form of a mediator at a given site.

1.1.5.7 Modalities of Signal Transmission

Modifications of post-translational state, enzymatic activity, or evolving concentration of mediators can act as molecular signals that are relayed and interpreted to control cell function. In addition, endocytosis of activated receptors modulates activity of signaling pathways, as it attenuates or enhances signal magnitude [11] (Vol. 1 – Chap. 9. Intracellular Transport). [8]

Signaling during Endocytosis

Endosomes not only serves for recycling or degradation of plasmalemmal receptors, but also as signaling mediators. Activated receptors can indeed accumulate in endosomes. Signaling can then be initiated and terminated from endosomal receptors. Moreover, certain signaling components are exclusively located in endosomes. Receptors actually can continue to transmit signals from endosomes that differ from those sent from the plasma membrane. Receptor signaling from endosome membranes is regulated by ligand availability, receptor coupling to signaling effectors, and subcellular location of signaling messengers [12].

1.1.5.8 Regulation of Signaling Cascades

Signaling pathways are regulated via: (1) *cytosolic feedback loops* at the receptor and effector levels, among other strategies; (2) *nuclear feedback loops* that control gene transcription; (3) *post-translational modifications* of proteic mediators; and (4) post-transcriptional supervision by *microRNAs* that provide robustness to signaling programs.

1.1.5.9 Magnitude and Duration of Stimuli

Signal duration (transient vs. sustained), magnitude, and location are important features of signal transduction. Cells are able to perceive extracellular signal intensity and duration and couple this information with the activation of distinct programs. In some circumstances, only the highest dose of messengers can provoke a cell response. A gradient of extracellular messenger can be translated into a graded intracellular effector signaling, thus inducing a range of cell responses among cells of a stimulated tissue. In addition, cells interpret extracellular signals according to history and environment to optimize the repertoire of instructions that fits the present needs.

Signaling-termination enzymes incorporated in a given pathway render transient the active state of pathway effector. Signaling termination often depends on integration of phosphorylation or dephosphorylation and ubiquitination or deubiquitination of pathway members.

---

8. Endocytosis of Notch fragment and Wnt receptors (Chap. 10) promotes signaling.

1.1.5.10 Retroactivity

Reverse signaling can incorporate sequestration of given mediator (a given cascade node i that receives an input signal I_i and yields an output cue O_i, which serves as input I_{i+1} for the next node in the cascade of reversible reactions) in a signaling cascade by a downstream component $i+k$ that complexes with substrate i. Owing to the existence of reversible modifications, an effector can inhibit directly or not an affector, as it can transmit a retroactive signal that activates one of the antagonist converter enzymes that direct the fate of the substrate i, such as kinase and phosphatase as well as ubiquitinase and deubiquitinase.

1.1.5.11 Signaling Dynamics

The *dynamic range of activation* of a given signaling molecule (i.e., responsiveness of cell outcome to incremental activation of a signaling molecule) can be a better predictor of functional response than signal strength, which most often corresponds to basal or hyperstimulated signaling state [13].

Molecular signals can change their meaning and relative strength according to subcellular location and time duration, as well as signaling context (i.e., relative time of occurrence or simultaneous committment of other signals). Signal transduction is optimized for dynamical range over which signaling occurs rather than signal strength to maximize cell response to diverse ranges of stimuli.

Signaling pathways relay information about changes in cell environment with a given frequency according to their *bandwidth*, i.e., information amount transmitted by the pathway per unit time [14]. The larger the bandwidth of a signaling pathway, the shorter the response time. When stimulation frequency is greater than a given threshold, the cell does not respond. When stimulation frequencies match the cell pathway bandwidth, the cell follows with fidelity imposed changes. Chemical reactions of the activated pathway are coupled to stimulations. In the case of physical stresses such as osmotic variations of the extracellular medium beyond the limit, mechanical response is decoupled from chemical reactions. Whereas physical response evolves with the environmental fluctuations, the reaction cascade cannot be activated between successive stimulation cycles. Signaling pathways that rely on at least 2 branches can integrate fast signals as well as record slow signals.

1.1.5.12 Intercellular Variability in Cell Response to Signaling

Cells of the same population respond differently to identical, external, uniform stimuli according to synthesis level and activation state of mediators (*intercellular variability* in event probability and timing). For example, the percentage of responders to apoptosis inducers (Vol. 2 – Chap. 4. Cell Survival and Death) depends on an intrinsic random factor, the natural difference in gene expression level in human cell lines. [15].[9]

9. Variability in time to death is determined by differences in the reaction rate for activation of pro-apoptotic BID protein. (Activated initiator caspases convert BID into a truncated active

1.1.5.13 Cell Response to Relative Signal Intensity

Signal-transduction cascades in a cell population mask variability between individual cells. However, cells perceive stimuli on the basis of stimulus magnitude relative to the background level. Signaling pathways indeed respond to relative elevation in mediators. Therefore, cells behave like physiological sensory systems, such as vision and hearing, that display a response proportional to the amplitude change in the stimulus relative to the background (*Weber's law*) [16]. [10]

Cells can generate signaling from reliable relative changes more easily than from reliable absolute changes. Cells exposed to Wnt morphogen respond to concentration changes in β-catenin between post- and preWnt stimulation states rather than the absolute level of β-catenins [17]. Timing of extracellular signal-regulated kinase ERK2 signaling dynamics is more precise between cells than amplitude. Cells respond to the relative level of stimulated ERK2 with respect to background level. [11]

The *Michaelian response* is described by: (1) a synthesis rate of a transcript that increases when the corresponding transcription activator binds to the gene promoter and (2) a degradation rate that is proportional to transcript level.

Negative cooperativity means that the transcription of a given gene is controlled by 2 transcription factor binding sites, the binding of the first being antagonized by that of the second. This phenomenon enables a response according to the Weber's law over a wider range of stimuli. Adaptation to a stimulus requires a return to the basal state and can cause the same response to the same reapplied input, whatever the background status.

The incoherent feedforward loop in transcription networks can behave as a concentration-change detector [16]. A stimulus triggers a response using 2 pathways with a more or less large time lag, where one pathway raises the output and the other decreases it. For example, an activator can control the expression of a target gene as well as that of a repressor of this gene. The magnitude and duration of gene expression depends mainly on relative changes in input, not on its absolute level. The incoherent feedforward loop then achieves not only pulse generation and signal

---

form.) Active protein BID induces assembly of 2 pore-forming proteins BAX and BAK that prime mitochondrial outer-membrane permeabilization. Variations in the expression of several proteins control the rate of cell death. In the case of apoptosis mediated by TNFSF10, or tumor-necrosis factor-related apoptosis-inducing ligand (TRAIL), some cells in a clonal population die while others survive. Furthermore, among cells that die, the time between TNFSF10 exposure and caspase activation is highly variable.

10. Weber's Law states that a whisper is audible in a quiet room, but in a noisy environment, intense sound can be unheard. Similarly, a light source can remain unseen in a bright room, whereas it is easily detected in a dark space.

11. Extracellular signal-regulated kinase ERK2 rapidly translocates in the nucleus after stimulation by epidermal growth factor. The ratio of stimulated to unstimulated nuclear ERK2 is similar among cells, but basal levels of nuclear ERK2 vary widely among cells [18]. Peak nuclear accumulation of ERK2 that occurs about 10 mn after stimulation is proportional to basal level. Concentration of ERK2 returns to original basal level in each cell, when stimulation disappears in about 30 mn. Yet, some cells exhibit a second peak of ERK2 translocation at about 40 mn.

adaptation, but also fold-change detection, unlike a transcriptional negative feedback loop [16].

1.1.5.14 Signaling Specificity

The spatial organization and segregation of parallel signaling pathways triggered by a given signal relies on the compartmentalization of signaling enzymes that possess a broad spectrum to ensure the specificity of responses of a given signal [24]. [12]

1.1.5.15 Cell Response to Biological Cycles

Cells also manage many cyclic processes, such as circadian (Vol. 2 – Chap. 5. Circadian Clock) and cell division cycles (Vol. 2 – Chap. 2. Cell Growth and Proliferation), that must coordinate with each other to optimize performance. Cell cycle progression that is independent of the cell cycle phase slows down during a specific circadian interval, according to a model of coupled biological oscillators [19].

1.1.5.16 Behavior of Isolated Cells and Cell Populations

Cells detect, analyze, and process received information using biochemical strategies and decide. Cooperative responses are promoted in a multicellular organism. According to probabilistic approaches, [13] cells must first estimate and infer signals from noise, then estimate the most likely state of their environment and eventually anticipate future state of their environment (Bayesian inference) based on sensed extracellular signals (e.g., local concentrations of metabolites, hormones, growth factors, etc.) and sequential updating from their short-term history [20]. [14] Given the most probable states, cells must weigh the costs (expenditure of cellular energy) and benefits (e.g., proliferation rate) of each potential response and select a course of action, taking into account competitive or cooperative actions undertaken by other cells.

Cell signaling occurs in a population of heterogeneous cells, but the overall tissue response is controlled to avoid fluctuations via paracrine signaling (signaling robustness in cell populations). Cells are characterized by their synchrony under the control

---

12. Estrogen hormones diffuse across the plasma membrane and meet estrogen receptors in a given subcellular pool. The localization of nuclear receptors ERα and ERβ (NR3a1 and NR3a2) at the plasma membrane, in the cytosol, or nucleus influences the type of estrogen signaling.

13. Statistical inference deals with deduction of a quantity that cannot be observed directly from an observable quantity that is correlated with the quantity of interest. Decision theory is based on the search for an optimal response given uncertain information by weighing advantages and drawbacks of possible response. Evolutionary theory considers scenarios in which decisions result from competing options.

14. The response to a stimulus can be associated with the response to another signal, even in the absence of this signal due to the inference from the usual context during which the stimulus occurs (Pavlovian-like event). Such anticipation has been learned over an evolutionary time scale [20].

of the circadian clock due to a coordination of cells that operate with various phase lags via a feedback exerted by the temporal heterogeneity of paracrine signaling that entrains the dynamical response of the cell population. Similarly, the heterogeneity of electrophysiological behavior of cardiomyocytes across the heart wall enables damping of delays of occurrence of action potential from the nodal tissue for a coordinated contraction.

Signaling mediated by NFκB transcription factor is aimed at regulating individual cell decisions toward cell proliferation or death as well as the overall population response needed in immunity and inflammation. Upon TNFα stimulation, NFκB commutes between the cytoplasm and nucleus, running back and forth (robust oscillations). Although the initial response may be initially synchronized, later oscillations can be asynchronous between cells [21].

In NFκB signaling, the precise timing of negative feedbacks that rely on delayed transcription of inhibitors IκBα and IκBε is optimized for heterogeneous timing of NFκB oscillations of individual cells, thereby reducing sensitivity to perturbations [21]. Maximization of the heterogeneity of phases of NFκB oscillations between individual cells can be balanced by the temporal heterogeneity of secretion of paracrine signals by individual cells that minimizes fluctuations within the cell population. Cell heterogeneity is associated with an overall coordination and stability of cell population responses by decreasing temporal fluctuations via a suitable paracrine signaling.

1.1.5.17 Drug Design

Last, but not least, knowledge accumulated in signal transduction in cells allows identification of appropriate targets for the development of new more or less specific drugs that do not inflict detrimental effects on other cell functions as well as in other cell types.

1.1.5.18 Modeling

Temporal dynamics of signaling networks can be analyzed using ordinary differential equations that express time derivatives of concentrations of involved chemical species with respect to production and consumption rates, without considering space (hence assuming a homogeneous reaction medium). However, these simple models are able to display complex temporal dynamics. Nonetheless, ordinary differential equations can roughly account for spatial dynamics using coarse compartmentation (e.g., cytoplasm and nucleus as well-mixed subcellular domains).

Classical chemical kinetic equations rely on the assumption that all involved chemical species are fully mixed and, therefore, randomly distributed in solution. Yet, signaling cascades should be modeled by equations that take into account spatial positions of reactants.

Partial differential equations express both temporal and spatial derivatives of concentrations of involved chemical species. The so-called reaction-transport equations describe the concentration differences that result from production and consumption as well as transport of these species that can travel from the plasma membrane to the nucleus.

1.2 Signal Processing

Cell communications with other cells and the extracellular matrix involve cell-surface proteins, such as receptors, cell adhesion molecules, ion carriers, and plasmalemmal constituents of vesicular transport. Signaling pathways consist of cascades of proteins where each protein activates the next effector. A signaling cascade is triggered when a receptor is activated by its ligand or sustains a proper conformation change caused by mechanical stress.

Signal propagation can be achieved by proteic complex assemblages, protein activation cascades, protein turnovers, and post-translational modifications of proteins, such as phosphorylation and dephosphorylation, glycosylation, ubiquitination, and acetylation, [15] which reversibly modify the properties of involved proteic effectors.

1.2.1 Transducers

Transducers can involve: (1) mechanosensitive ion channels such as large conductance, large pore, low ion selectivity carriers; (2) conformational changes of molecules; and (3) molecular switches in the plasma membrane or cytosol. Signaling pathways contain protein interactions that occur in nodes [16] of signaling pathways. Similarly to lumped parameter models of biofluid flow in a conduit network, pathway modeling can focus on effector interactions and signaling networks can be considered to mimic electrical analogs, the effectors being represented as nodes linked via electrical resistances. Mathematical modeling of cell behavior is, indeed, based on molecular interactions inside and between cells involved in cell signaling.

1.2.2 Molecule Translocation

Protein sorting (targeting) refers to the mechanism by which cells transport proteins to the appropriate positions inside (organelle lumen or membrane, cell cortex, or plasma membrane) or outside the cell. Targeting signals enable the cellular transport machinery to correctly position a given protein.

Receptor excitation triggers the displacement of cytosolic adaptors and enzymes to the plasma membrane. These *translocations* correspond to effective control mechanisms aimed at switching on the signaling pathway. A signaling complex with scaffold proteins, and probably membrane ion channels and cytoskeletal components,

---

15. Acetylation affects the functions of various proteins, such as histones, many transcription factors, enzymes (histone acetyltransferases and deacetylases, kinases, phosphatases, etc.), as well as cytoskeleton-associated proteins.

16. A pathway node represents an effector, i.e., a single protein, proteic complex, or gene, with a given task.

is assembled. Receptor complexes recruit cytoplasmic signaling components, which initiate a biochemical cascade that finally causes changes in gene expression.

Vesicular transport operates in signal transduction via degradation of transmembrane receptors and control by vesicle-associated regulators. Vesicle-associated signal transducers serve as staging areas for regulators to enhance effector interaction and activation.

1.2.3 Proteic Interactions – Interactomes

Cells are permanently subjected to signals and stresses. Proteic interaction networks depend on environmental conditions. Between-protein interactions enable signal propagation most often after post-translational modifications. The human interactome contains approximately 130,000 binary interactions. Interactomes with kinases, phosphatases, and transcription factors, as well as DNA repair factors, are dynamical structures that change with modifying conditions, e.g., in the absence or presence of DNA-damaging agent [22]. Functional relations between mediators reorganize.

Signal transduction commonly starts at the cell surface as a result of signaling that emanates from endo-, para-, juxta-, and autocrine regulation (including the nervous control), but neither from intracrine regulation nor hormonal cues that target intracellular receptors. At the plasma membrane, which is intimately connected to the cortical cytoskeleton, extracellular messengers trigger clustering of signaling proteins or activate pre-existing clusters to transmit a message. Some membrane nanodomains promote interproteic interactions.

Many plasmalemmal molecules form nanoclusters (e.g., receptor Tyr kinases with integrins) and collaborate in signal transmission that requires interactions between receptors, signaling effectors, and possibly adhesion molecules.

Association and dissociation between interacting molecules in the plasma membrane or cytoplasm occur continuously. Molecular interactions that involve specific proteins can induce similar substrate activation, but cause distinct effects. In other words, a given signaling component can contribute to numerous processes via subtle variations in signaling processing, such as recruitment to specific subcellular loci, duration of local recruitment, and mode of activation, that enables transmission of specific signals. In addition, signal-dependent formation of proteic complexes can create local foci of amplified enzyme activity.

Protein activity depends on local interactions between amino acids as well as linkage between functional distant domains. Protein activity can in fact be regulated by ligand binding to a distant domain. Proteic complexes are conserved among species, but interproteic interactions can change among them.

1.2.3.1 Transducisome

Ephemeral or long-lasting *transducisomes* can include receptors, adaptors, enzymatic mediators, and ion channels. Mechanosensitive ion channels can form clusters with their associated molecules, such as extracellular matrix and cytoskeletal (actin and tubulin) proteins.

1.2.3.2 Protein Domains and Protein Interactions

Between-protein interactions rely on protein domains. The shortest domains such as *zinc finger domains* are stabilized by metal ions or disulfide bridges. Signaling calmodulin and troponin-C contain calcium-binding *EF hand domain*. *Immunoglobulin-like domains* (IgV, IgC1, IgC2, and IgI) are involved in protein–protein interactions that occur for cell adhesion and between-cell molecular recognition. *Pleckstrin homology domain* (PH) binds phosphoinositides. *Phosphotyrosine-binding domain* (PTB) binds phosphorylated Tyr residues that are targeted during signal transduction. *Src homology-2 domain* (SH2) recognizes specific amino acid sequences of proteins containing Tyr^P residues and *Src homology-3 domain* (SH3) targets proline-rich sequences.

1.2.3.3 Allostery

Allostery that governs almost all metabolisms and gene regulation results from events that occur at given signaling effectors. Macromolecules, mostly proteins, can, indeed, transmit regulatory effects induced by binding to an allosteric ligand at one effector site to a different, often distant, functional site that is allosterically excited, ready to be stimulated by a signaling regulator. Allosteric signal transduction relies on multiple conformations that characterize signaling mediators.

1.2.3.4 Transcriptional Coregulators

The activation of gene transcription by transcription factors is achieved via the recruitment of coregulators. Such proteic interactions not only connect transcription factors to the transcription machinery, but also allow chromatin remodeling that favors the assembly of the transcription initiation complex.

1.2.3.5 Measurements

Molecular interaction forces of ligand–receptor complexes can be measured using atomic force microscopy. The binding kinetics between plasmalemmal receptors and their extracellular ligands as well as the distribution of receptors, can also be assessed by mechanical imaging [23]. [17]

1.2.3.6 Scaffold Proteins

Scaffold proteins are binding molecules that not only serve as assembly platforms, but can have an active role in effector enzyme activation. Scaffold proteins

17. The heterogeneous distribution of vascular endothelial growth factor receptor-2 (Sect. 8.2.5.8) in microvascular endothelial cells, as well as linkages between VEGFR2 and the cortical cytoskeleton, has thus been demonstrated.

are managers of signaling organization, as they simultaneously bind at least 2 signaling mediators. Signaling scaffold actually recruits and links at least an enzyme and its substrate to form a molecular complex. Scaffold proteins either form fully scaffolded, pre-assembled signaling complexes or target another scaffold protein that has constructed a partially scaffolded complex.

Scaffold proteins participate in the regulation of signaling cascades that control cell functioning. Scaffold proteins can allow the efficient relay of signals to successive enzymes involved in a given signaling cascade that are pre-assembled on the same complex by these scaffolds [24]. Moreover, scaffold proteins can anchor proteic complexes to given subcellular sites, insulate signaling modules, and enable immediate feedback.

Scaffold proteins influence the localization of signaling molecules as well as thresholds and dynamics of signaling reactions. They can amplify signal transduction, control signaling duration, and coordinate positive and negative feedback loops that can induce oscillatory signaling pattern.

Among multiple proteic scaffolds, MAPK enzymes can act as pivotal scaffolds that allow corresponding signaling pathways to acquire new interaction partners [25]. Scaffold A-kinase anchoring proteins often bind activators (affectors), kinases, phosphatases, and effectors. These signaling platforms thus couple activation–deactivation cycles.

Mathematical modeling is able to enhance the knowledge of scaffold protein function. Modeling indeed shows that scaffolds operate as: (1) platforms on which signaling molecules assemble; (2) attractors for optimal localization of signaling proteins; (3) coordinators of feedback loops; (4) modifiers of signaling pathways that can protect activated signaling effectors from inactivation; (5) designers of signaling thresholds; and (6) framers of signaling behaviors, as they can create transient or sustained signaling.

Scaffold proteins enhance signaling pathway efficiency by assembling their components and specificity by preventing spurious interactions between mediators. Scaffold proteins facilitate the conversion of a distributive process that requires multiple independent interactions to a processive mechanism that initiates a single interaction for many actions. Catalytic scaffolds enhance or inhibit effector activation.

Scaffold proteins enable translocation of signaling effectors in sites of optimized reaction. They can amplify an initially weak stimulus. Conversely, by simultaneously binding several signaling mediators, they also restrict the ability of a kinase to phosphorylate many targets. In addition, scaffold proteins can protect activated signaling molecules from inactivation and degradation for prolonged signaling. Conversely, they can impede signal transduction by limiting the number of downstream mediators that must undergo activation. They can intervene in signaling termination that relies on inactivation of all signaling components (Table 1.3).

1.2.3.7 Adaptor Proteins

Adaptor proteins connect proteins in order to form multimolecular complexes involved in intracellular transmission of signals. Adaptors are accessory proteins with

Table 1.3. Adaptors, dockers, and scaffolds in cell signaling.

Type	Role
Adaptor protein	Signaling molecule interactions
Docking protein	Attachment of substances to molecular complexes Signaling repressor or activator
Scaffold protein	Assembling of signaling complexes Maintenance of pre-assembled signaling complexes Decyphering signals by proper chemical reactions Prevention of irrelevant interactions Spatial regulation of signaling steps (signaling specificity) (control of signaling inactivation) Temporal regulation (up/downloading signaling mediators for signal transduction or termination (adjustment of activation kinetics)

respect to main effectors of a signaling pathway. They do not have enzymatic activity, but mediate specific between-protein interactions that are mandatory for the functioning of the reaction cascade. The specificity of signal transduction relies on the recruitment of signaling components, such as small guanosine triphosphatases and protein kinases, that is induced by adaptors.

Adaptors usually contain several binding domains for interactions with several specific proteins. Adaptor proteins differentially localize signaling effectors to specific subcellular sites, thereby differentially directing effector activity to different substrates.

Various signaling molecules that interact to propagate a signal can assemble into a complex within a few minutes after ligand–receptor binding. An example is given by interaction of member TNFRSF5 (or CD40) of the tumor-necrosis factor receptor superfamily with its ligand and adaptors TRAF2 and TRAF3, ubiquitin conjugase UbC13, cellular inhibitor of apoptosis proteins IAP1 and -2, IκB kinase (IKK) regulatory subunit IKKγ (NFκB essential modulator [NEMo]), and mitogen-activated protein kinase kinase kinase (MAP3K) [26]. After 30 mn, the complex has moved to the cytoplasm and this translocation drives the activation of signaling mediator MAP3K upon IAp1/2–induced degradation of TRAF3. This type of cytokine signaling is thus initiated by the translocation of ligand-activated proteic complex and internal reactions that control spatial and temporal separation of MAPK and IKK signaling.

1.2.3.8 Compartmental Organization

Signaling has a compartmental organization, the successive events rapidly occurring in spatially restricted domains. Signal transduction usually starts with interactions between multiple types of proteins at the cell surface.

Besides, cells connected by gap junctions communicate rapidly with each other, either by ionic current or diffusion of low-molecular-weight second messengers (calcium ions, cyclic nucleotides cAMP and cGMP, diacylglycerol, and inositol (1,4,5)-trisphosphate). Cells then use chemical communication, releasing stimulating molecules, which diffuse to a target cell with suitable receptors.

Cells can differentially respond to a given substance according to the subcellular location and kinetics of ligand binding and subsequent clustering of molecules. Many biological processes are compartmentalized within the cell. The location of proteins and other signaling molecules is spatially restricted. The spatial organization of these compartments is often bounded by cellular membranes. Non-membrane-bound intracellular compartments can be defined from their composition, a missing component preventing their functioning. All the subcellular compartments can facilitate the control of complex signal transduction.

Pathway cycles are formed by convertible types of signaling proteins, resulting from activity of either kinase or phosphatase on phosphorylated proteins, or activity of either guanine nucleotide-exchange factor or GTPase-activating protein on small GTPases. Multisite phosphorylation increases the sensitivity of the signaling dynamics. The activity of a cycle with 2 opposing enzymes depends on the location of the involved molecules within the cell. This location can depend on the time and space concentration gradient of the target protein, hence on the protein diffusivity and the signaling dynamics with its given response rates.

Signaling molecules commonly do not spread by diffusion. They can instead use endocytic vesicles (signaling endosomes) or other cell transport means driven by given molecular motors [27]. Endocytosis indeed is not only aimed at ending signaling via degradation of activated receptor complexes after internalization from the plasma membrane. The signal transduction can exploit the compartmentation of endocytosis. Signaling factors (e.g., SHC, GRB2, and SOS) can be found on early endosomes [28]. Endocytosis provides signaling temporal regulation according to the transport kinetics. Endocytosis can either lead to quick signal transduction, or, inversely, slow down the process.

1.2.3.9 Signaling Nanodomains in the Plasma Membrane

Spatial organization is a common feature of cell signaling. Intracellular signaling cascades operate far from reaction equilibrium and homogeneity. Constraints on molecular movement and assembly reflect on cell signaling.

The plasma membrane is a heterogeneous fluid mixture of lipids, proteins, and other molecules that provides the environment in which nearly all signal transductions start. Plasmalemmal cholesterol and other lipidic components contribute to membrane organization, as they are organized into separated miscibility phases. Numerous receptors and signaling molecules form clusters in the plasma membrane. Therefore, many signaling cascades originate from plasmalemmal clusters.

The cortical actin cytoskeleton controls the dynamical spatial organization of plasmalemmal signaling clusters, as it carries molecules and can interact with the components of signaling cascades. contributes to cluster assembly.

The prototypical cluster is composed of receptor Tyr kinase (Chap. 8), phospholipids (Vols. 1 – Chap. 7. Plasma Membrane and 4 – Chap. 1. Signaling Lipids), and small GTPase Ras (Vol. 4 – Chap. 8. Guanosine Triphosphatases and their Regulators). Once activated, Ras^{GTP} recruits GRB2 adaptor and SOS guanine nucleotide-exchange factor that acts on Ras. Small GTPase Ras is an allosteric activator of SOS, the catalytic module of which is intrinsically inactive. It thus primes a positive Ras–SOS feedback loop. Tethering of Ras to the plasma membrane and localization in nanoclusters heightens SOS catalytic activity up to 500-fold with respect to reactions between SOS and Ras in a simple solution [29]. [18] Clusters of receptors, such as those of the HER and ephrin receptor family, as well as those of signaling mediators such as small GTPase Ras have been detected.

1.2.3.10 Signaling Nodes and Hubs

Between-protein interactions act as simple *nodes* for cue transmission or as *hubs*. Nodes are characterized by a single interaction. On the other hand, multiple signaling mediators localize to hubs. Hubs are then involved in multiple connections. Hubs correspond to essential proteins. Scaffolded huddles at hubs can coordinate adequate responses in space and time, support suitable conformational changes of effectors, and carry out necessary post-translational modifications.

Both nodes and hubs ensure connectivity of the signaling network. A node neighborhood is essential as proteins tend to cluster in densely connected modules enriched in essential proteins [30]. Some scaffold proteins can serve as platforms for the spatial grouping of successive signaling enzymes, thereby allowing the linear transfer of a signal from one enzyme to the next in a signaling cascade.

Signaling pathways can share common mediators. However, these mediators are not necessarily common nodes of signal transduction between pathways, as they can connect to different scaffolds and complexes.

Two independent chemical signals, such as second messengers cAMP and Ca^{++}, can converge at a given signaling hub of a pathway to elicit a given event. [19] Therefore, these intracellular signals, must be adequately processed. Synchronization and integration of these signals can initiate a common outcome that increases signal speed and precision [24]. Some scaffold proteins can integrate distinct cues to coordinate the cell response.

---

18. The uncontrolled activation of SOS by Ras is prevented by SOS regulatory domains that block the allosteric Ras binding site. This inhibition is relieved by the interaction with PIP_2 or phosphatidic acid in the membrane.

19. In cardiomyocytes, phosphorylated protein kinase-D is released from A-kinase anchor protein AKAP13 and translocates into the nucleus, where it phosphorylates histone deacetylase HDAC5. Phosphorylated HDAC5 then exits the nucleus, then DNA can relax and enable the transcription of genes involved in hypertrophy. Calcium and diacylglycerol on the one hand and cAMP on the other promote the activation of protein kinase-A and -C, respectively. These kinases act synergistically in PKD phosphorylation [24].

1.2.4 Lipidic Interactions

Lipids are essential for cells to transduce signals. Phosphoinositides control diverse cellular processes. Plasmalemmal phosphatidylserine (at the inner leaflet of the lipid bilayer) initiates multiple processes such as blood coagulation and clearance of apoptotic bodies. Anionic phosphatidylserine promotes the recruitment of positively charged, polycationic proteins.

Lipid interactions with lipids and proteins are also required to form signaling complexes. For example, phospholipase-D2 hydrolyzes plasmalemmal phosphatidylcholine into phosphatidic acid, which recruits the guanine nucleotide-exchange factor Son of sevenless. The SOS mediator activates Ras during epidermal growth factor signaling [31] (Sect. 8.2.5.2).[20]

1.2.5 Protein Modifications

Protein modifications involve attachment or detachment, reversible or not, of small chemical groups, such as phosphate, acetyl, or methyl groups, as well as entire proteins, glucids, or lipids. A protein can thus be modified by the covalent attachment of another, usually smaller, protein, such as ubiquitin that is synthesized as an inactive precursor or ubiquitin-related modifiers. Assigned enzymes confer selectivity to the reaction that attaches the modifier. Protein modifications consume energy.

Acetyl groups and ubiquitins are most often attached to Lys (K), methyl groups to Arg (R), and phosphate groups to Ser (S), Thr (T), and Tyr (Y) residues. Acetyl and phosphate groups are usually attached as single, specific moieties (functional groups). On the other hand, 2 to 3 methyl groups can be transferred to the same acceptor site. Successive conjugations of ubiquitin on the same target produce polyubiquitin chains. Ubiquitin can also be attached at distinct loci of a given substrate.

Protein modifications modulate their properties. Acetylation and phosphorylation influence the substrate properties, as they change its surface charge (direct effect). On the other hand, post-translational modifications can operate via modified target by effectors (indirect effect). Acetylation, methylation, phosphorylation, and ubiquitination of histones-H3 and -H4 affect their control of gene expression.

Some protein post-translational modifications are stable and important for maturation and proper folding of newly synthesized proteins, such as glycosylation, lipidation, and disulfide bridge formation. Others, such as phosphorylation, are more transient and essential in cell signaling. A common outcome of protein post-translational modification is the creation of a high-affinity binding site for the selective interaction with a specific binding protein.

Interplay between different protein modifications regulates protein function. Stress-induced phosphorylation of transcription factor P53 regulates its acetylation.

---

20. The GRB2–SOS interaction promotes SOS recruitment to activated receptors. Phosphatidic acid can be converted to diacylglycerol by phosphatidic acid phosphatase that recruits RasGRP1.

Acetylation of SMAD7 inhibits its ubiquitination. Trans- (by TβR2) and autophosphorylation of TβR1 is required for its sumoylation that improves SMAD3 phosphorylation. Sumoylation of transcription regulators, such as heat shock proteins and transcription factors GATA1 and MEF2, depends on phosphorylation. Protein modifications cause conformational changes of signaling mediators, thereby influencing protein interactions, especially enzymatic activity and nucleic acid binding.

1.2.5.1 Phosphorylation and Dephosphorylation

Phosphorylation (i.e., addition of a phosphate group to serine [Ser or simply S], threonine [Thr or T], or tyrosine [Tyr or Y], which yields phosphoSer [Ser^P], phosphoThr [Thr^P], and phosphoTyr [Tyr^P]) that relies on ATP supply participates in the control of signaling tasks. Phosphate groups are added to and removed from molecules, then recycled.

Protein Activation by Phosphorylation or Dephosphorylation

Phosphates carry a distinctive molecular signal with their strong negative charge. The presence or absence of a phosphate can switch a protein from an active or inactive state and conversely according to the position of target residues and generated conformational change. Phosphates interact strongly with proteic chains, so an added phosphate can cause a protein to fold in a specific manner or stabilize a proteic complex.

Cell Signaling Based on Phosphorylation and Dephosphorylation

Inside any cell, a large collection of kinases and phosphatases control protein phosphorylation and dephosphorylation, respectively. Cell response to stimuli relies on integration of a network of coordinated signaling pathways that are actually controlled, at least partly, by cycles of phosphorylation and dephosphorylation for a particular signaling outcome. Signaling can be regulated by phosphorylation–dephosphorylation cycles of one or several specific sites of sequential messengers. In addition to protein kinases and phosphatases, protein phosphorylation state depends on activators, inhibitors, and scaffolds. Alterations in protein kinase activity is frequently implicated in disease development.

Protein serine, threonine, or tyrosine kinases and Ser^P, Thr^P, and Tyr^P phosphatases are characterized by their substrate specificity.[21] Kinases and phosphatases operate in signaling cascade, from stimulus perception by receptor and sensor proteins, fine-tuning by scaffold proteins, pathway crosstalk by connector proteins, to signal integration via allosteric effects.[22]

21. Some docking surfaces can allow several low-affinity bindings that confer overall binding specificity.

22. In prokaryotes (bacteria and archaea), signaling proteins can be phosphorylated not only at serine, threonine, or tyrosine residues by kinases, but also at aspartate and histidine by histidine kinases, as well as arginine.

Sequential Phosphorylation of an Amino Acid

Primary phosphorylation of a given suitable amino acid residue aims at stabilizing the activation segment in a conformation suitable for substrate binding. Eventual *secondary phosphorylation* of the activation segment has a variable function, such as substrate recruitment or full kinase activity, according to the kinase type.

Sequential Phosphorylation of a Set of Kinases

Sequential activation of a kinase cascade can culminate in activation of mitogen-activated protein kinases (MAPK) via double phosphorylation. Translocation of MAPK to the nucleus then enables direct phopshorylation of transcription factors that activate transcription of immediate-early genes, such as AP1 components Jun and Fos. Immediate-early genes regulate a second wave of transcription of delayed-early genes that encode a large range of proteins, including negative regulators.

Autophosphorylation

Autophosphorylation of protein kinases such as receptor Tyr kinase dimers outside or inside of the activation segment follows changes in conformation so that local reorganization enables arrangement of target residues in positions for phosphoryl transfer. Cytoplasmic kinases can produce oligomers. Calmodulin-dependent protein kinase CamK2 forms dodecamers that autoactivate rapidly in response to high concentration of Ca^{++}–calmodulin complexes. Increase in local kinase concentration and subsequent self-association (dimerization) facilitate autoactivation. Active kinases can be defined by dimer with phosphorylation sites of activation segments close to the catalytic site of the interacting protomer [32].

Multisite Phosphorylation

Different phospho-sites in a given signaling mediator can coexist. They can be regulated by different processes according to the signaling type, hence the mode of between-protein interaction. In general, phosphorylation is regulated differently on different sites within the same protein [33].

The number of phosphorylated sites on a given protein can be very high ($O[100]$). Phosphorylation of different sites of a target protein can typically serve different functions. In other words, multisite phosphorylation generates many phosphorylated forms that can have distinct effects, as both the position and number of phosphates influence the biological outcome. When kinases and antagonist phosphatases act on a substrate that possesses many phosphorylated sites, the regulatory network that processes information is flexible to respond to various types of stimuli.

Cell-Specific Effects

Both kinases and antikinases have cell-specific effects. In particuler, kinases that control cell proliferation and survival differ among cell lines, especially those that are rate-limiting for these processes [34]. [23] In addition, cells in culture develop different sensitivities to kinase loss.

An Example: α-Dystroglycan

α-Dystroglycan is a plasmalemmal receptor that anchors the basal lamina to the plasma membrane of myocytes, as it binds proteins of the extracellular matrix that contain laminin-G domains. [24] Post-translational modification is required to bind laminin. Mature, laminin-binding α-dystroglycan, indeed, undergoes ^{O}mannosyl phosphorylation on its mucin-like domain and post-phosphoryl modification that is mediated by the like-acetyl glucosaminyl transferase protein [35].

Phosphorylation–Dephosphorylation Cycle

The phosphorylation–dephosphorylation cycle corresponds to protein oscillations between a basal and a phosphorylated state, with possible intermediate states. In the usual phosphorylation of a substrate (effector E) by a kinase K, 3 intermediate states occur to reach the phosphorylated active effector (E^P): (1) the kinase binds its substrate (K–E); (2) the substrate undergoes a first phosphorylation (E^P), which is irreversible in the presence of ATP, and separates immediately from the kinase; (3) the kinase binds the phosphorylated substrate (K–E^P); and (4) the substrate undergoes a second phosphorylation (E^{PP}). In addition, molecules diffuse in the cytosol with a diffusion characteristic time ($L^{\star 2}/\mathcal{D}$ of about 10 s ($L^\star$: diffusion distance of the order of 10 μm; $\mathcal{D}$: diffusivity of the order of 10 μm^2/s).

Double phosphorylation of effector E by kinase K comprises 3 intermediate states: (1) binding of E to K (K–E complex); (2) first irreversible phosphorylation and separation of phosphorylated effector E^P from K; (3) binding of E^P to K (K–E^P complex); (4) second phosphorylation (E^{PP}).

$$E + K \underset{k'_1}{\overset{k_1}{\rightleftarrows}} E - -K \xrightarrow{k_2\ (ATP)} E^P + K \underset{k'_3}{\overset{k_3}{\rightleftarrows}} E^P - -K$$

$$E^P - -K \xrightarrow{k_4\ (ATP)} E^{PP} + K \xrightarrow{\kappa} E$$

Double phosphorylation can be modeled by the following set of differential equations:

23. Tested cells include multiple sets of 2 types of primary cells from the same tissue (e.g., keratinocytes and fibroblasts simultaneously isolated from the same tissue), but possibly from different individuals, essentially isogenic cells that vary only by the introduction and expression of a single gene; immortalized, but not transformed cells; and numerous tumor cell lines derived from similar cell origins as well as various tissues.

24. Defective binding with laminin causes congenital muscular dystrophy.

$$d[E]/dt = -k_1[K][E] + k_1'[K-E] + \kappa[E^{PP}] \tag{1.1}$$
$$d[E-K]/dt = k_1[K][E] - (k_1' + k_2)[K-E] \tag{1.2}$$
$$d[E^{P}]/dt = k_2[K-E] + k_3'[K--E^{P}] - k_3[E^{P}][K] \tag{1.3}$$
$$d[E^{P}-K]/dt = k_3[E^{P}][K] - (k_3' + k_4)[K-E^{P}] \tag{1.4}$$
$$d[E^{PP}]/dt = k_4[K-E^{P}] - \kappa[E^{PP}] \tag{1.5}$$

Most often, a steady regime and negligible intermediate states are assumed. The double phosphorylation then obeys to the simple equation of a single phosphorylation [36].

$$[E] + [K] \underset{\mathtt{k}'}{\overset{\mathtt{k}}{\rightleftarrows}} [E^{P}].$$

The matter conservation states that

$$[E]_{tot} = [E] + [E^{P}],$$

where the concentration $[E]_{tot}$ is the total value of $[E]$.

Each effector activated by a kinase follows a reaction-diffusion equation which is much simpler when the kinase is located at the plasma membrane in the steady state, which is fully defined with the boundary conditions at the plasma membrane:

$$\mathcal{D}\nabla^2[E^{P}] = k'[E^{P}],$$

where ∇^2 is the Laplace operator. The resolution of the diffusion-reaction equation describes the regulation of the transduction cascade [36].

Modifications of proteins such as the dual phosphorylation can occur at distinct cell sites. Enzymes that modify proteins act processively or distributively. In a *processive mechanism*, the enzyme remains bound to its substrate during the entire processing, from first to last protein modification. A *distributive mechanism* means that enzymes release their substrates between modifications of different sites. This process can generate an ultrasensitive response, as it can amplify weak signals using positive feedback, and lead to bistability via enzyme sequestration. Bistability can lead to a all-or-none response that buffers the cell against fluctuations of an input signal and lock the cell in a given state. Slow ADP release by kinases can protect against rapid rebindings. Rapid enzyme rebinding to substrate can transform a distributive mechanism into a quasi-processive mechanism. Rapid rebinding of the enzyme to its substrate after modification of the first site reduces response time, i.e., determines the cell decision that relies on signal duration, but leads to a loss in sensitivity and bistability. Signal transduction cascades can thus be defined by a set of features that encompasses the gain (input–output relationship), signal-to-noise ratio, effect of spatiotemporal fluctuations, response time, and eventual bistability.

1.2.5.2 Acetylation

Acetylation is the post-translational addition of acetyl groups to given residues of proteins. Lysine acetylation, the transfer of acetyl groups from acetyl coenzyme-A as acetyl group donor onto lysine residues, is associated with charge neutralization and influences properties of proteins.

In the nucleus, acetylation and deacetylation of lysine residues of chromatin-associated proteins, such as histones H2a, H2b, H3, and H4, as well as nuclear transcription regulators, contributes to the control of gene transcription. Histone acetylation contributes to the activation of gene transcription due to the weakening of histone–DNA interaction.

A series of post-translational modifications on histones initiated by phosphorylation of histone-3 allows the recruitment of histone acetyltransferase MOF that acetylates histone-3 (Lys16). Histone-3 then acts as a platform that recruits positive transcription elongation factor PTEFb via bromodomain-containing protein BRD4 [37].

In addition, acetylation of transcription factors, such as P53 and E2F1, increases binding to DNA. Acetylation stabilizes E2F1 factor. The bromodomain of various transcription factors and coactivators recognizes acetylated lysines to participate in the formation of transcriptional and chromatin-remodeling complexes [37]. Acetylation of SMAD7 by P300 transcriptional coactivator that is a histone acetyltransferase as well as E4 polyubiquitin ligase competes with ubiquitination on the same lysine residue, hence avoiding degradation.

Acetylation yields an additional regulatory mechanism to control the activity of metabolic enzymes, therefore responding to changes in metabolic demands. These metabolic enzymes can be involved in glycolysis, gluconeogenesis, fatty acid and glycogen metabolism, and the tricarboxylic acid and urea cycles. In human livers, acetylation that leads to either activation or inhibition of these enzymes targets a large proportion of enzymes in various metabolic pathways [38]. Acetylation also controls the stability of some enzymes. [25]

The retained N-terminal methionine residue of a nascent protein is often N-terminally acetylated. The removal of N-terminal Met by Met-aminopeptidases frequently leads to N-terminal acetylation of resulting N-terminal alanine, valine, serine, threonine, and cysteine residues. More than 80% of human proteins are N-terminally acetylated. N-terminally acetylated residue may act as a degradation signal (acetylated N-terminal degron [$_{acN}$degron]) [39].

Acetylation also influences interproteic interactions. For example, acetylation of heat shock protein HSP90 increases its interaction with the glucocorticoid receptor.

1.2.5.3 Glycosylation

Protein glycosylation is another modification that regulates protein's final destination within or outside the cell. Glycosylation also determines half-life of proteins that travel through the body. Glycosylation tags proteins with linear and/or branched moieties. Glycans intervene in cell adhesion, self-recognition, protein transport, and

---

25. Acetylation stimulates enoyl-coenzyme-A hydratase and 3-hydroxyacyl coenzyme-A dehydrogenase in fatty acid oxidation and malate dehydrogenase in the tricarboxylic acid cycle. On the other hand, acetylation inhibits argininosuccinate lyase in the urea cycle and destabilizes phosphoenolpyruvate carboxykinase in gluconeogenesis [38].

receptor activation. Aberrant glycosylation alters ion channel activity and electrical signaling in neurons and myocytes.

Glycosyl-Phosphatidylinositol Anchorage

Multiple proteins are post-translationally modified by a glycosyl-phosphatidylinositol (GPI) at their C-termini and anchored to the plasma membrane via lipid segments. Glycosyl-phosphatidylinositol-anchored proteins (gpiAP)[26] are mainly associated with plasmalemmal nanodomains, the so-called membrane rafts (10–200 nm),[27] that are enriched in sphingolipids and cholesterol. These proteins act as enzymes, receptors, regulators, adhesion molecules, and immunological mediators [40].

Among GPI-anchored heparan sulfate proteoglycans, glypicans regulate morphogens, such as secreted ligands Wnt and Hedgehog (Chap. 10). In humans, Notum (Wingful) pectinacetylesterase ortholog cleaves and releases glypicans and other gpiAPs from the plasma membrane. Released glypicans impede the interaction between Wnt (especially Wnt11) and its receptor. In addition, angiotensin-converting enzyme not only cleaves angiotensin-1 and bradykinin, but also GPI-anchored proteins.

Sialylation and Desialylation

In the heart, certain ^{N}glycans produced by cardiomyocytes allows the identification of diverse cell types, as these glycosylated proteins can direct cardiomyocyte excitability. Voltage-gated ion channels, such as Na_V and K_V channels, are heavily glycosylated. Sialylation degree (from absence to complete sialylation) influences gating of $Na_V1.5$ channel [41]. Polysialyltransferase ST8sia2 that is expressed only in the neonatal atrium is required for normal action potential waveforms [42]. Less sialylated atrial Na_V channels have more depolarized gating potentials and faster recovery rate with respect to adequately processed channels. Therefore, cardiomyocyte excitability depends not only on expression and distribution of ion channel isoforms, but also on their post-translational modifications.

Sialidases remove sialic acids from glycoproteins and glycolipids. Sialidase Neu3 controls transmembrane signaling for cell differentiation, growth, and apoptosis via protein interactions with other signaling molecules [43]. Enzyme Neu3 at the plasma membrane tethers to caveolin-1. In response to growth factors such as epidermal growth factor, Neu3 translocates to membrane ruffles with Rac1 (Vol. 4 – Chap. 8. Guanosine Triphosphatases and their Regulators) to enhance cell motion.

---

26. Glycosyl-phosphatidylinositol-anchored protein has a phosphatidylinositol backbone (with ethanolamine phosphate, mannose, and glucosamine) and at least a glycan branch. Precursors of gpiAPs synthesized in the endoplasmic reticulum have a GPI-attachment signal peptide. GpiAPs are transported by vesicles from the endoplasmic reticulum to the Golgi body, where they are remodeled.

27. Lipid-modified proteins of membrane rafts are either located in the outer leaflet such as gpiAPs or in the inner leaflet such as Src protein Tyr kinases.

In addition, Neu3 participates in neurite outgrowth. Moreover, overexpressed Neu3 that is linked to the adaptor growth factor receptor-binding protein GRB2 decreases insulin-stimulated phosphorylation of insulin receptor and insulin receptor substrate IRS1. In response to insulin, Neu3 is phosphorylated and links to GRB2 adaptor. Sialidase Neu3 is able to interact with caveolin-1, GRB2, Rac1, β_4-integrin, and epidermal epidermal growth factor receptor. Furthermore, Neu3 activates EGFR, kinases FAK and ILK, adaptor SHC, and β4-integrin. Enzyme Neu3 can modulate the activity of plasmalemmal effectors of signaling pathways.

Acetylglucosamination

Protein acetylglucosamination is an ubiquitous, reversible protein modification that targets serine and threonine residues of proteins. It serves as a nutrient and stress sensor to modulate the activity of many nuclear and cytoplasmic proteins. Protein acetylglucosamination of nuclear and cytoplasmic proteins by O-linked attachment of monosaccharide β^{N}acetylglucosamine (^{O}Glc^{NAc}) is involved in the regulation of many cell processes, such as nuclear transport, translation and transcription, signal transduction, cytoskeletal reorganization, proteasomal degradation, and apoptosis. Acute protein acetylglucosamination corresponds to a stress response to enhance cell survival. In the cardiovascular system, protein acetylglucosamination promotes cell protection, but also contributes to insulin resistance [44].

The addition of a single O-linked ^{N}acetylglucosamine to Ser and Thr residues is catalyzed by O-linked ^{N}acetylglucosaminyltransferase [28] and removed by β^{N}acetylglucosaminidase.

Different nucleoplasmic, cytosolic, and mitochondrial isoforms of O-linked ^{N}acetylglucosaminyltransferases exist. These enzymes can complex with catalytic subunits of protein phosphatase-1. β^{N}Acetylglucosaminidase is encoded by a single gene, but has 2 splice variants.

Acetylglucosamination–deacetylglucosamination cycles most often target phosphorylation sites, although acetylglucosamination sites can be distant from phosphorylation sites. Consequently, these post-translational protein modifications are additional sources of crosstalk between pathways to regulate protein function and signal transduction. Attachment of ^{O}Glc^{NAc} competes with that of phosphate and disturbs signaling, especially in immunocytes [45].

In hepatocytes, phosphorylation of cAMP-response element-binding coactivator CRTC2 promotes its interaction with 14-3-3 proteins and subsequent cytosolic retention, whereas O-glycosylation at the same residues primes CRTC2 translocation into the nucleus to regulate the expression of genes of gluconeogenesis.

^{N}Acetylglucosaminyltransferases particularly target phosphatidylinositol 3-kinase, protein kinase-B, glycogen synthase, endothelial nitric oxide synthase, insulin

28. O-Linked ^{N}acetylglucosaminyltransferase is ubiquitous. Nonetheless, it particularly abounds in the pancreas, adipose tissue, brain, and skeletal and cardiac muscles. This enzyme acts as a dimer. It is able to bind phosphatidylinositol (3,4,5)-trisphosphate for recruitment at the plasma membrane. It is phosphorylated (activated) by calcium–calmodulin-dependent protein kinase-4 in response to depolarization.

receptor substrate-1, transcription factor P53, nuclear factor-κB, nuclear factor of activated T cells, etc.

Acetylglucosamination and phosphorylation regulates spindle assembly and cytokinesis during the mitosis, as many acetylglucosamination sites are either identical to phosphorylation sites or close to them [46]. Overexpression of ${}^{O}\text{Glc}^{\text{NAc}}$ transferase (OGT) increases the phosphorylation of cyclin-dependent kinase CDK1, as it decreases the abundance of Polo-like kinase PLK1 and augments and reduces activity of kinase and phosphatase of CDK1, i.e., PKMyt1 and CDC25c, respectively (Vol. 2 – Chap. 2. Cell Growth and Proliferation).

Unlike extracellular glycans, acetylglucosamination is rapidly done (at a rate comparable to that of phosphorylation) on regulatory proteins in response to metabolic and environmental signals. When phosphorylation is impeded (e.g., glycogen synthase kinase-3 inhibition), acetylglucosamination can also decay. However, when acetylglucosamination increases, phosphorylation loci can change. Elevated acetylglucosamination lowers phosphorylation at some sites and heightens phosphorylation at others [47]. Interplay between acetylglucosamination and phosphorylation can result from diverse mechanisms: (1) competition for occupancy on same residues; (2) competition via hindrance by residence on proximal sites; (3) phosphorylation of ${}^{O}\text{Glc}^{\text{NAc}}$ transferase (or ^{O}glcnacase); and (4) acetylglucosamination of kinases or phosphatases.

Polyadpribosylation

Adpribosylation is a post-translational modification of proteins catalyzed by ADPribosyl transferases (ART) using nicotinamide adenine dinucleotide (NAD^+) as a substrate. Monoribosylation can be extended into linear or branched polymers of ^{ADP}ribose (PAR) [29] by poly^{ADP}ribosylpolymerases (PARP). Conversely, poly^{ADP}ribosylglycosidases (PARG) [30] and ^{ADP}ribosylhydrolase like-2 enzyme (ADPRHl2). [31] catalyze deadpribosylation of mono- or polyadpribosylated proteins.

Polyadpribosylation is an element of the response to DNA damage, as DNA-strand breakage primes the synthesis of poly^{ADP}ribose by PARP1. [32] In addition,

29. The synthesis of poly^{ADP}ribose relies on 3 types of transferase activities: (1) initiation, i.e., transfer of mono^{ADP}ribose to an acceptor amino acid of the corresponding substrate by mono- and polyADPribosyl transferases; (2) elongation of ^{ADP}ribose and poly^{ADP}ribose by poly^{ADP}ribosyltransferases (PARP1–PARP4, PARP5a–PARP5b, PARP6, PARP8–PARP12, and PARP14–PARP16); and (3) branching of the polymer by PARP1 and PARP2.

30. Poly^{ADP}ribosylglycohydrolase is encoded by a single gene, but different isoforms exist with various subcellular locations. The full-length isoform (PARG110) mainly localizes to the nucleus; a small isoform (mtPARG) may function in mitochondria. 2 other isoforms (PARG103 and PARG99) reside in the cytosol.

31. A.k.a. ^{ADP}ribosylhydrolase ARH3. Three ^{ADP}ribosylhydrolases exist (ARH1–ARH3).

32. Polymerases PARP1 and PARP2 are stimulated by DNA breaks. However, PARP1 may also be activated by some types of DNA structures, such as hairpins, cruciforms, and supercoiled DNAs. Activation of PARP1 by tolerable amounts of DNA breaks favors DNA repair. Conversely, severe DNA injury that overactivates PARP1 triggers cell apoptosis via apoptosis-inducing factor.

PARP1 controls chromatin architecture, nucleosome removal, and transcription activation in response to corresponding stimulus [48]. $Poly^{ADP}$riboses interact not only with many repair factors, but also with checkpoint proteins to control cell division. Furthermore, PARP1, PARP2, PARP9, and PARP14 act as transcription cofactors, as they influence the formation and/or activity of various transcription factor complexes, without necessarily synthesizing $poly^{ADP}$riboses. $Poly^{ADP}$riboses regulate gene expression by targeting insulators that limit heterochromatin–euchromatin borders and prevent inappropriate cross-activation of neighboring genes. $Poly^{ADP}$riboses inhibit DNA methyltransferase DNMT1 for chromatin insulation.

Polyadpribosylation also represents a post-translational modification of proteins. $Poly^{ADP}$ribosylpolymerases [33] catalyze the polymerization of ^{ADP}ribose units from NAD^+ that leads to ramified $poly^{ADP}$riboses [48]. $Poly^{ADP}$riboses can covalently link to acceptor proteins (mainly via glutamic residues).

$Poly^{ADP}$riboses can be degraded either by $poly^{ADP}$ribosylglycohydrolase and ADPRHl2. The last moiety may be removed by a $mono^{ADP}$ribosyl–protein hydrolase or lyase. $Poly^{ADP}$ribose degradation occurs only a few minutes after its synthesis.

Polyadpribosylation also regulates protein function via non-covalent binding as in proteic interactions with proteins or nucleic acids. Non-covalent $poly^{ADP}$ribose binding may attract proteins to specific subcellular sites. In addition, a given protein can both non-covalently and covalently bind to $poly^{ADP}$ribose for its recruitment to a specific site, where the recruited protein can be modified [48]. Diverse functional outcomes can arise from different sizes and branching types of $poly^{ADP}$riboses.

$Poly^{ADP}$ribose can serve as a local ATP source [48]. AMP-activated protein kinase, which is activated when the ATP concentration decays, phosphorylates PARP1. $Poly^{ADP}$ribose is involved in the sequential activation of calpain and pro-apoptotic factor B-cell lymphoma-2-associated X protein that leads to translocation of apoptosis-inducing factor from mitochondria to the cell nucleus and cell apoptosis (Vol. 2 – Chap. 4. Cell Survival and Death). ^{ADP}Ribose is also a second messenger that activates transient receptor potential melastatin channel TRPM2 (Sect. 2.3.4.3) for Ca^{++} influx in response to oxidative stress.

1.2.5.4 Hydroxylation

The addition of hydroxyl groups to proline residues happens in secreted proteins and components of the cellular oxygen sensor. Hydroxylation of Pro residues

33. $Poly^{ADP}$ribosylpolymerases catalyze successive addition of ^{ADP}ribose units to either ^{ADP}ribose or proteins. The $poly^{ADP}$ribosylpolymerase family contains 17 identified members. Certain members may act as $mono^{ADP}$ribosyltransferases. Activity of PARP1 can be modulated by post-translational modifications. Phosphorylation of PARP1 by extracellular signal-regulated kinases ERK1 and ERK2 yields a maximal activation after DNA damage [48]. Acetylation of PARP1 by histone acetyltransferases P300 and cAMP-responsive element-binding protein-binding protein (CBP) enhances the synergistic activity of PARP1, P300, and CBP in inflammation.

is catalyzed by prolyl hydroxylases.[34] Hydroxylated hypoxia-inducible factor-α is recognized by von Hippel-Lindau ubiquitin ligase.

1.2.5.5 Methylation

Methylation is the conjugation of a methyl group to amino acid residues, such as lysine and arginine. Various degrees of methylation (mono-, di-, or trimethylation) can occur on the same residue. Two groups of enzymes — methyltransferases and demethylases — regulate the methylation status of target proteins.

Methylation occurs not only on nucleic bases in DNA,[35] but also on various amino acids in proteins.[36] Histone methylation allows interactions with binding partners that, in turn, regulate a DNA-dependent process, such as transcription, replication, or repair.

Methylation of histones is one of the epigenetic modifications that influence chromatin structure and recruitment of non-histone proteins to chromatin.[37] Non-histone proteins, such as transcription factor P53 and small RNA-binding PIWI proteins, are also regulated by methylation, either positively or negatively according to the site and degree of methylation.

Many classes of protein carboxyl methyltransferases are distinguished according to methyl acceptor. Methyltransferases act in transcriptional repression, as they promote the formation of heterochromatin by methylating histone tails.[38] Histone lysine methyltransferases indeed cause gene silencing. ^{L}Isoaspartate (^{D}aspartate) O-methyltransferase (or protein carboxyl methyltransferase PCMT1) catalyzes the transfer of a methyl group from ^{S}adenosyl^{L}methionine to the free carboxyl groups of ^{D}aspartyl and ^{L}isoaspartyl residues. It then converts abnormal ^{L}isoaspartyl and ^{D}aspartyl residues to normal ^{L}aspartyl residues to maintain proper protein conformation. Leucine carboxyl methyltransferases methylate the carboxyl group of Leu residues. ^{S}Adenosyl^{L}methionine-dependent Arg ^{N}methyltransferases modulate both intra- and intermolecular interactions of target proteins and regulate their function. Arginine methylation is thus a post-translational modification caused by arginine methylase that is used in signaling pathways.

Three main forms of methylarginine have been identified: $^{N^G}$monomethylarginine (MMA) and $^{N^G N^G}$ asymmetrical (aDMA) and symmetrical (sDMA) dimethylargi-

34. However, removal of hydroxyl groups has not yet been reported.

35. DNA methylation on cytosine residues serves to regulate gene expression. Methylation can also serve to protect DNA from enzymatic cleavage.

36. Methionine methyltransferase has 2 substrates (^{S}adenosylmethionine and ^{L}methionine) to produce ^{S}adenosylhomocysteine and ^{S}methyl^{L}methionine. 5-Methyltetrahydrofolate-homocysteine methyltransferase is also called methionine synthase, as it produces methionine from homocysteine. Glycine methyltransferase catalyzes the chemical reaction from ^{S}adenosyl^{L}methionine and glycine to ^{S}adenosyl^{L}homocysteine and sarcosine.

37. Methylated histones include histone 3 (Lys4, Lys9, Lys27, Lys36, Lys79, and Arg2) and -4 (Lys20) [37].

38. Distinct histone acetyl- and methyl-transferases modify histone tails at different positions and yield a combinatorial interplay between these modification types.

nines. Arginine methylation modifies transcription factors, histones, and splicing and ribosomal proteins that interact with RNAs [49]. The majority of nuclear asymmetrical dimethylarginine residues is found in heterogeneous nuclear ribonucleoproteins that operate in pre-mRNA processing and nucleocytoplasmic RNA transport. RNA-binding proteins that are methylated on arginine residues are monomethylated or asymmetrically dimethylated rather than symmetrically dimethylated. However, certain spliceosomal small nuclear ribonucleoproteins (Vol. 1 – Chap. 5. Protein Synthesis) are symmetrically dimethylated. Myelin basic protein that belongs to the first group of detected arginine-methylated proteins contains monomethyl arginine and symmetrically dimethylated arginine residues.

The family of protein arginine methyltransferases encompasses many members (PRMT1–PRMT8) [49, 608].[39] Methyltransferases have been classified as type-1 and -2 enzymes whether, after monomethylating arginine, further dimethylation is asymmetrical or symmetrical, respectively. Most PRMTs belong to the type-1 enzyme set, whereas PRMT5[40] is a type-2 methyltransferase. Arginine methyltransferases are involved in transcription, nucleocytoplasmic transport, protein sorting, and cell signaling. They interact with interferon receptors, Janus kinases, transcription factors STAT, and the P160 family of transcriptional coactivators of nuclear hormone receptors.

The activity of the transcription factor P53 (vol. 4 – Chap. 9. Other Major Signaling Mediators) that triggers either cell cycle arrest or apoptosis is regulated by numerous post-translational modifications[41] in addition to accessory proteins and cofactor binding. Transcription factor P53 bears lysine methylation that can be reversed by demethylation. Methyl lysine provides docking sites for effector proteins. Methylation can also serve to prevent alternate post-translational modifications on the same lysine residue. Lysine can be mono-, di-, or trimethylated. Methylation level is correlated with distinct genomic locations and functions [51]. Protein Arg methyltransferase PRMT5 targets P53 that then forms a defense complex with stress-responsive activator of P300 (StRAP), or Ser/Thr kinase receptor-associated protein, and junction-mediating and regulatory protein (JMY) in response to DNA damage to promote cell cycle arrest rather than death [52]. Transcription cofactor JMY of P53 binds to transcription coactivator and histone acetyltransferases cAMP responsive element-binding protein-binding protein (CBP) and P300 to activate P53-dependent transcription. Upon DNA damage, P53 arginine methylation stimulates the Cki1A gene to prime G1 arrest. Furthermore, arginine methylation of P53 regulates its activity by modulating its subcellular location (predominately nuclear or cytoplasmic) and oligomerization (PRMT5 decreases the formation of P53 dimers and tetramers).

Leucine carboxyl methyltransferase LCMT1 causes reversible carboxy methylation of C-terminal leucine of the catalytic subunit of protein Ser/Thr PP2 phos-

39. Protein arginine methyltransferase PRMT4 is mainly known as coactivator-associated arginine methyltransferase CARM1, in addition to histone-arginine methyltransferase [608].

40. A.k.a. Janus kinase-binding protein-1.

41. Phosphorylation of transcription factor P53 by related kinases ATMK (ataxia telangiectasia mutated) and ATRK (ataxia telangiectasia and Rad3-related) is a response to DNA damage.

phatase. Methyltransferase LCMT2 is involved in post-translational modification of phenylalanine tRNA.

Nuclear receptor-binding SET domain-containing histone-3 methyltransferase (or lysine methylase) NSD1 [42] methylates the transcription factor $P65_{NF\kappa B}$ [43] (Lys37 monomethylation) that participates in inflammation, immunity, as well as cell proliferation and apoptosis [53]. Conversely, lysine demethylase F-box and leucine-rich repeat protein FBxL11 inhibits NFκB activity. Methylated $P65_{NF\kappa B}$ is confined to the nucleus. Monomethylation promotes $P65_{NF\kappa B}$ binding to promoters of a subset of NFκB-regulated genes. [44] Other histone methylases include NSD2 and NSD3.

1.2.5.6 Myristoylation

Myristoylation is an irreversible protein modification that happens either post-translationally or during translation (co-translational modification). A myristoyl group derived from myristic acid is attached to N-terminal amino acid (often glycine) of a nascent polypeptide. Myristoylation is catalyzed by ^{N}myristoyltransferase.

1.2.5.7 Nitrosylation

Nitric oxide-mediated, redox-based S-nitrosylation is a post-translational modification that changes protein function. [45] It is commonly used in signal transduction. Nitric oxide, possibly in its ^{S}nitrosothiol form, can actually target cysteine residue for S-nitrosylation to modulate enzyme activitity.

Fluid flow interacts with conduit walls, as it transmits normal and tangential frictional forces to the wall constituents. In endothelial cells of blood vessels, transmitted mechanical stresses can activate endothelial nitric oxide synthase (NOS3) via the PI3K–PKB pathway, thereby increasing nitric oxide production.

Numerous proteins with various functions can experience S-nitrosylation to adapt to flow-applied stress field [55]. More than 100 proteins, such as tropomyosin (Cys170) and vimentin (Cys328), undergo S-nitrosylation in endothelial cells. [46] The extent of S-nitrosylation in vascular endothelia depends on mechanical force magnitude.

---

42. A.k.a. SETD7, SET7, and SET9.
43. A.k.a. RelA component of the NFκB heterodimer.
44. Activity of NFκB transcription factor also depends on other modifications, such as ubiquitination, phosphorylation, nitrosylation, and acetylation. Acetylation of P65 at Lys218 and Lys221 inhibits IκB binding and enhances DNA binding, whereas its acetylation at Lys122 and Lys123 inhibits its transcriptional activity [54].
45. S-nitrosylation is the covalent attachment of nitric oxide to the thiol side chain of Cys residues.
46. Nitrosylated proteins include protein disulfide isomerase, ubiquitin-specific isopeptidase in early endosomes, transitional endoplasmic reticulum ATPase, mitochondrial HSP70, heat shock transcription factor HSF2 that modulates thioredoxin transcription under oxidative stress, etc. [55].

Protein denitrosylation can be triggered by the stimulation of cell-surface receptors, such as G-protein-coupled receptors, receptor Tyr kinases, and members of the tumor-necrosis factor superfamily of receptors.

Protein denitrosylation by denitrosylases, such as ^{S}nitrosoglutathione reductase and thioredoxin kernel, serves to regulate S-nitrosylation and protect the cell against nitrosative stress. The *thioredoxin kernel* includes thioredoxin and thioredoxin reductase as well as the *glutathione kernel* with glutathione and ^{S}nitrosoglutathione reductase [56]. [47]

Regulated denitrosylation can promote activation of enzymes, such as caspase-3, nitric oxide synthase NOS3, and IκB kinase-β [56]. It can affect between-protein interactions via scaffold proteins such as β-arrestins.

1.2.5.8 Palmitoylation

Palmitoylation (or palmitylation) consists of reversible addition of palmitate to a cysteine residue of membrane proteins. Palmitoylation–depalmitoylation cycles enables the cell to regulate the location of specific proteins. Palmitoylation enhances the hydrophobicity of proteins and contributes to membrane linkage. Palmitoylation also intervenes in cellular transport of proteins between membrane compartments and modulates between-protein interactions.

Palmitoylation is experienced by many signaling mediators and affectors, such as hRas, $G\alpha_s$ subunit of G proteins, β2-adrenergic receptor, and endothelial nitric oxide synthase.

1.2.5.9 Prenylation

Post-translational modification with isoprenoids modulates between-protein and protein–lipid interactions. Protein prenylation means that either a farnesyl or a geranylgeranyl moiety from a soluble isoprenoid pyrophosphate is attached to target proteins.

Most of the prenylated proteins are subunits of heterotrimeric G proteins and monomeric Ras GTPases that act in signaling pathways. Protein prenylation is carried out by 3 heterodimeric enzymes: farnesyltransferase, geranylgeranyltransferase-1, and Rab geranylgeranyltransferase [57]. Unlike other protein prenyltransferases, Rab geranylgeranyltransferase does not recognize its protein substrate directly but operates in coordination with Rab escort protein.

1.2.5.10 Sulfation

Tyrosine sulfation is a post-translational modification in which a sulfate group is added to a Tyr residue of a protein. Sulfation contributes to strengthening of between-protein interactions. In humans, proteins that experience Tyr sulfation include ad-

---

47. A.k.a. glutathione-dependent formaldehyde dehydrogenase and class-3 alcohol dehydrogenase ADH5.

hesion molecules, G-protein-coupled receptors, coagulation factors, serine protease inhibitors (serpins), extracellular matrix proteins, and hormones.

Tyrosylprotein sulfotransferase (TPST) catalyzes the transfer of sulfate from 3′-phosphoadenosine 5′-phosphosulfate (PAPS) to the hydroxyl group of Tyr residues to form a Tyr sulfate ester [58]. Two types of tyrosylprotein sulfotransferases have been identified (TPST1–TPST2). In the trans-Golgi network, protein Tyr sulfation is catalyzed by TPST1 enzyme. This post-translational modification is restricted to proteins of the secretory pathway.

Sulfation can also refer to the generation of sulfated glycosaminoglycans. The sulfate group is added either via oxygen (O-sulfation) or nitrogen (N-sulfation).

1.2.5.11 Ubiquitination

Ubiquitination remodels proteins and affects their location and interactions with other proteins. Limited ubiquitination modifies the activity of some proteins and can be reversible. Ubiquitination not only regulates proteolysis, but also contributes to many proteasome-independent activities (non-degradative ubiquitin signaling).

Ubiquitination starts with the ATP-dependent activation and adenylation of the C-terminal glycine of ubiquitin (Ub), followed by the transfer of Ub to a lysine residue of a substrate via a series of thioester intermediates owing to the set of E1 to E3 enzymes, i.e., Ub activase, conjugase, and ligase, respectively.

Ubiquitin Types

Dedicated enzymes confer selectivity to the reaction. Ubiquitin is encoded by 4 polyubiquitin genes: UBA52,[48] UBA80,[49] UBB, and UBC that encode ubiquitin A52 residue ribosomal protein fusion product-1, ubiquitin carboxyl extension protein-80, ubiquitin-B, and polyubiquitin-C, respectively [59].

Ubiquitin is transcribed and translated as a complex of multiple ubiquitins or ubiquitin linked to N-terminus of 2 ribosomal proteins: 40S ribosomal protein L40 and 60S ribosomal protein S27a. Free ubiquitin is created from these precursors by deubiquitinases. Linear ubiquitin chains not only are ubiquitin precursors, but also contribute to signaling such as in the nuclear factor-κB pathway.

Ubiquitination Reaction

Attachment of ubiquitin consumes energy. ATP-Dependent ubiquitin-activase attaches ubiquitin. Ubiquitin is then transferred to ubiquitin carrier enzyme, or Ub conjugase. Ubiquitin ligase attaches ubiquitin to the proteic substrate (ubiquitin–protein ligation).

The human genome express a single Ub-activating, about 60 Ub-conjugating, and about 600 Ub-ligating enzymes to ensure the specificity of substrate selection.

48. A.k.a. RPL40, CEP52, and HUBCEP52.
49. A.k.a. RPS27A, UBCEP80, UBCEP1, CEP80, and HUBCEP80.

Table 1.4. Protein ubiquitination modes and effects. Non-proteolytic functions of ubiquitins often involve mono- or polyubiquitination via Lys63-linked chains. All of Lys residues of ubiquitin can be used for further ubiquitination. Ubiquitination thus produces structurally discernable signals on the modified substrate. Ubiquitin signals in both monomeric and polymeric forms are often transmitted via ubiquitin receptors that recognize the modifier. Some ubiquitin receptors are highly selective for a particular type of polyubiquitin linkage. They can discriminate between Lys63-linked and linear chains that are structurally similar.

Mode	Function
Monoubiquitination	Receptor endocytosis, DNA-damage repair
(Lys63)	Signaling (e.g., via TNFR and DDR)
Multimonoubiquitination	
Lys48-linked polyubiquitin chain	Protein degradation
Lys63-linked polyubiquitin chain	Non-degradative signaling

The class of Ub ligases comprises 2 main groups: (1) HECT (homologous to E6AP C-terminus) and (2) RING (really interesting new gene). Other types encompass α, F-box, and U-box Ub ligases. U-box ligases accelerate polyubiquitination of ubiquitinated proteins.

Non-proteolytic functions of ubiquitin involve specific Ub ligases. In other words, ubiquitination via Lys48 or another Lys relies on the specificity of Ub ligases, because no specific signals exist in substrate proteins to determine the type of lysine ubiquitination.

Modes of Ubiquitination

Ubiquitous ubiquitins can be conjugated to target proteins either as monomers or chains. *Monoubiquitination*, i.e., protein tagged with a single ubiquitin at a single site regulates DNA repair and receptor endocytosis (Table 1.4). *Multimonoubiquitination* refers to the connection of a single ubiquitin at multiple sites. *Polyubiquitination*, i.e., protein labeled by a chain of ubiquitins, serves for protein degradation by several proteolytic enzymes. In fact, Ubiquitins can be subsequently added to a protein-linked ubiquitin. Polyubiquitination leads to recycling of ubiquitins after substrate proteolysis.

Ubiquitin Chains

Ubiquitination generates chains that are assembled between C-terminal Gly and any one of 7 Lys residues of ubiquitin (Lys6, Lys11, Lys27, Lys29, Lys33, Lys48, or Lys63), whereas N-terminus can serve to extend ubiquitin chains to form linear chains. Ubiquitin chains linked by lysine residues at position 48 (Lys48) tag proteins for proteasomal degradation. Ubiquitination can use other lysine linkers [50] of ubiquitin chains for other purposes such as DNA damage.

50. I.e., Lys6, Lys11, Lys27, Lys29, Lys33, and Lys63.

Different types of ubiquitin chains exist: (1) *homotypic* when the same ubiquitin Lys is used for conjugation; (2) *mixed linkage* if several distinct Lys connect ubiquitins; and (3) *heterologous linkage* in the case of ubiquitin connected to ubiquitin-like modifiers (SUMo–Ub chains).

Proteolytic Function of Ubiquitins

The *ubiquitin–proteasome system* (UPS) degrades more than 70% of intracellular proteins. The classical intracellular function of ubiquitin is the regulation of protein turnover and protection against damaged or misfolded proteins.

The ubiquitin–proteasome system controls almost all cellular processes, as it degrades signaling regulators. Protein destruction starts by its tagging with ubiquitins by a coordinated action of an enzyme chain that comprises an Ub activase, conjugase, and ligase.

Multicatalytic 26S proteasome selectively degrades ubiquitin-conjugated proteins in a non-lysosomal, ATP-dependent process.[51] 26S Proteasome (prosome, macropain) non-ATPase regulator subunit Psmd4[52] binds to polyubiquitin chains.

Extrinsic ubiquitin receptors (e.g., repair protein RAD23b and ubiquilin Ubqln2) cooperate in ubiquitinated protein recognition. Ubiquitin-receptor proteins have a ubiquitin-like- (UBL) and at least one ubiquitin-associated (UBA) domain. Domain UBL is recognized by the 19S proteasome and UBA domain binds ubiquitin. Ubiquitin-receptor proteins inhibit deubiquitination.

On the other hand, the integral plasmalemmal glycoprotein that serves as a proteasomal ubiquitin receptor, adhesion regulating molecule AdRM1[53] binds ubiquitin to dock ubiquitin conjugates at the proteasome as well as deubiquitinating enzyme [60].

Non-Degradative Functions of Ubiquitins

Non-proteolytic functions of ubiquitin encompass: (1) DNA damage and repair; (2) cell signaling; (3) intracellular molecular transfer; and (4) transcriptional control.

Ubiquitination modulates endocytosis via association with caveolin or clathrin for *signaling* from endosome or *proteolysis* by the proteasome [61]. Signals transmitted by receptors from the endosome may differ from those initiated at the plasma membrane.

---

51. 26S Proteasome is composed of at least 33 different subunits arranged in 2 subcomplexes: a catalytic 20S core and 1 or 2 19S regulators. The 20S core is composed of 2 rings with 7 α subunits and 2 rings with 7 β subunits. The 19S regulator contains a base with 6 ATPase subunits and 2 non-ATPase subunits and a lid with up to 10 non-ATPase subunits. The regulatory component recognizes ubiquitinated substrates, deconjugates ubiquitin chains, and unfolds substrates before their translocation into the core component.
52. A.k.a. proteasomal regulatory particle non-ATPase RPN10.
53. A.k.a. RPN13.

Substrate Activation

Unanchored polyubiquitin chains can activate protein kinases. Ubiquitin ligase TRAF6 contributes to the activation of nuclear factor-κB (Vol. 4 – Chap. 9. Other Major Signaling Mediators) and mitogen-activated protein kinases (Vol. 4 – Chap. 5. Mitogen-Activated Protein Kinase Modules) in several signaling pathways, especially those emanating from Toll-like and interleukin-1 receptors (Chap. 11).

The TRAF6 ligase operates with the ubiquitin conjugase complex that consists of UbC13 [54] and UEv1a [55] to catalyze Lys63-linked polyubiquitination that can stimulate MAP3K7 kinase, which subsequently activates IκB kinase and NFκB. Lys63-Linked polyubiquitin chains bind to MAP3K7IP2 [56] and causes MAP3K7 autophosphorylation (activation) [62]. Furthermore, polyubiquitin chains synthesized by Ub conjugase UbE2d3 [57] and TRAF6 activate the IKK complex.

Ubiquitins in DNA Repair

The DNA-damage processing relies on 3 complementary, cooperating, controlled mechanisms: (1) checkpoints for DNA-damage sensing that initiate an appropriate response such as phosphorylation cascade primed by ataxia telangiectasia mutated kinase (ATMK) and ataxia telangiectasia and Rad3-related kinase (ATRK); (2) DNA repair of interstrand crosslinks using the Fanconi anemia pathway; and (3) DNA-damage tolerance to bypass lesions during DNA replication (delayed repair) to avoid interference with DNA synthesis. [58] DNA-damage processing is supported by suitable, reversible, post-translational modification of proteins by ubiquitins [63].

A DNA double-strand break primes a signaling mediated by a ubiquitination cascade that involves [63]: (1) ubiquitin ligases, such as RING finger proteins RNF8 and RNF168, [59] HECT domain and RCC1-like domain-containing protein HERC2, [60] delayed repair protein Radiation sensitivity protein Rad18 homolog, and BrCa1–BARD1 heterodimer (i.e., a functional Ub ligase); [61] (2) modification of histone

---

54. A.k.a. UbE2n.

55. A.k.a. UbE2v1.

56. A.k.a. TGFβ-activated kinase (TAK1)-binding protein TAB2.

57. A.k.a. ubiquitin-conjugase homolog UbCH5c.

58. Monoubiquitination and Lys63-linked polyubiquitination of the replication clamp protein Proliferating cell nuclear antigen (PCNA) controls replicative lesion bypass by translesion synthesis and an error-free, recombination-like mechanism, respectively [63].

59. Ubiquitin ligase RNF168 recognizes ubiquitinated histone cluster-2 proteins $H2a^U$ and $H2ax^U$ [63]. It works together with ubiquitin receptor-associated protein RAP80. The latter mediates BrCa1 recruitment to damaged chromatin.

60. Ubiquitin ligase HERC2 interacts with RNF8. It promotes interaction of RNF8 with UbC13 enzyme [63]. The latter is the partner of both RNF8 and RNF168. It cooperates with ubiquitin-conjugase variants UbE2V1 (or UEV1) and UbE2V2 (or UEV2) for the synthesis of Lys63-linked polyubiquitin chains.

61. Ubiquitin ligase Breast and ovarian cancer susceptibility protein BrCa1 is recruited for the G2–M checkpoint arrest and initiation of a break repair by homologous recom-

cluster-2 proteins H2a that are involved in the structure of chromatin;[62] and (3) formation of Lys63-linked polyubiquitin chains as well as monoubiquitin signals on target proteins.[63]

Extracellular Ubiquitin

In addition to mediating signaling inside the cell, ubiquitin can operate as a *paracrine messenger* for cell fate. Extracellular ubiquitin has pleiotropic activities, such as regulation of immune response and inflammation, neuroprotection, and control of cell growth and apoptosis. Extracellular ubiquitin secreted from cells and incorporated into hematopoietic cells can suppress cell growth and promote apoptosis via proteasome-dependent degradation of selective cellular proteins such as signal transducer and activator of transduction STAT3 [65].

On the other hand, once released in the extracellular medium and following its reuptake, ubiquitin can prevent apoptosis of adult rat ventriculomyocytes initiated by β-adrenergic receptors (βAR) [64]. Cell survival signaling is mediated by the PI3K–PKB pathway that inhibits glycogen synthase kinase-3β and Jun N-terminal kinase, which promote the activation of the mitochondrial pathway of apoptosis.[64] Therefore, βAR stimulation heightens the extracellular ubiquitin concentration in ventriculomyocytes (at least those in rat ventricles). Extracellular ubiquitin counteracts βAR-stimulated apoptosis by inhibiting GSK3β and JNKs, the latter act-

---

bination [63]. The BrCa1–BARD1 (BrCa1-associated RING domain protein) heterodimer forms independent complexes. A BrCa1–BARD1-based complex contains scaffold abraxas, receptor-associated protein RAP80, BrCa1–BrCa2-containing complex subunits BrCC36 and BrCC45, and BrCa1a complex subunit MERIT40 [63]. This complex is required for BrCa1 recruitment.

62. Histone H2ax21 is one of the earliest subtrates of ATMK on DNA double-strand breaks. Phosphorylated H2ax labels the affected chromatin domain to recruit various effectors of the damage response [63].

63. Processing of DNA interstrand crosslinks relies on monoubiquitination of complementation group-D2 and -I Fanconi anemia proteins (FancD2 and FancI) that causes their recruitment to chromatin and initiation of recombination-mediated repair [63]. The Fanconi anemia DNA repair network involves specific factors encoded by 13 genes, mutations of which cause Fanconi anemia. The FancM–FAAP24 recognition complex (FancM: Fanconi anemia group-M protein; FAAP24: Fanconi anemia-associated protein-24) together with DNA-binding histone-fold proteins FAAP10 and FAAP16 binds to DNA interstrand crosslinks and recruits the core complex (FancA–FancB, FancE–FancG, and FAAP100) and the catalytic subunit FancL via FancG protein. The core complex monoubiquitinates the FancD2–FancI complex and causes its retention in chromatin foci. The FancD1–FancJ–FancN repair complex contributes to the BrCa axis that controls homologous recombination. The same molecule can act in the Fanconi anemia DNA repair network and DNA double-strand break pathway, as FancD1 is another alias for BrCa2, FancJ is BrCa1-interacting protein C-terminal helicase BACH1, and FancN corresponds to Partner and localizer of BrCA2 protein PALB2.

64. Ubiquitin reuptake may be coupled to specific Ub ligases that avoid Lys48 to activate the survival pathways. Ubiquitination of PKB1 and PKB2 by Ub ligase TRAF6 promotes their recruitment to the plasma membrane, where protein kinase-B is activated by PI3K enzyme.

ing downstream from activated GSK3β, as well as impeding elevation in cytosolic cytochrome-C, a marker of mitochondrial pathway of apoptosis.

1.2.5.12 Deubiquitination

Deubiquitinases (DUb) [65] form a class of isopeptidases that disassemble ubiquitin chains, process ubiquitin precursors, and remove ubiquitin from labeled proteins, hence conferring reversibility to ubiquitination. Both their activity and specificity are controlled by regulators.

Most deubiquitinases are cysteine proteases, some are metalloproteases. The human genome encodes a large number of DUbs that are classified into 5 families [66]: (1) ubiquitin C-terminal hydrolases (UCH) that are Cys proteases; (2) ubiquitin-specific Cys proteases (USP); [66] (3) ovarian tumor superfamily Cys proteases (OTU), such as tumor-necrosis factor-α-induced protein TNFαIP3; [67] (4) Machado-Joseph neurodegenerative disease protein domain-containing Cys proteases (MJD or Josephin), such as ataxin-3; and (5) JAB1/MPN/Mov34 zinc metalloenzymes (JAMM MPN^+ DUbs) that commonly associate with proteic complexes. [68]

Deubiquitinases have 3 main types of activities [66]: (1) generation of free ubiquitin from precursors, i.e., polyubiquitins transcribed from several genes or ubiquitin complexes with ribosomal proteins (Sect. 1.2.5.11) as well as unanchored ubiquitin chains; (2) removal of ubiquitin chains from ubiquinated proteins to rescue these proteins from degradation or remove a non-degradative ubiquitin signal; and (3) recycling of ubiquitin from degraded proteins, hence preventing degradation of ubiquitin with its substrate. In addition, some DUbs might exchange ubiquitin signal types. Many deubiquitinases link to Ub ligases that have an intrinsic tendency to self-ubiquitinate. [69] Some deubiquitinases such as TNFαIP3 combine DUb and

65. A.k.a. deubiquitinating or deubiquitylating (iso)peptidases as well as ubiquitin proteases, hydrolyases, and isopeptidases.

66. Deubiquitinase USP44 targets CDC20, a coactivator of the ubiquitin ligase anaphase-promoting complex or cyclosome that promotes the degradation of many cell cycle regulators (Vol. 2 – Chap. 2. Cell Growth and Proliferation). Deubiquitinase USP44 participates in the regulation of the spindle checkpoint and correct execution of chromosome segregation. It prevents inappropriate ubiquitination of CDC20 by APC/C after a precocious release from MAD2, hence premature checkpoint inactivation.

67. A.k.a. A20 and TNFαIP2.

68. Deubiquitinases JAMM MPN^+ are components of the 26S proteasome as non-ATPase subunit Psmd14 (a.k.a. Poh1 and Pad1), with COP9 signalosome as subunit CSN5, endosomal sorting complex required for transport (ESCRT) as associated DUb STAM-binding protein (STAMBP), and both DNA-damage response-involved BrCa1 (breast and ovarian cancer susceptibility protein-1) A complex and BRISC isopeptidase complex as Lys63-specific deubiquitinase BRCC36.

69. Deubiquitinase USP19 stabilizes RING finger Ub ligase RNF123 for cyclin-dependent kinase inhibitor CKI1b. Deubiquitinase USP8 maintains Ub ligase RNF41 (a.k.a. neuregulin receptor degradation protein NRDP1) that inhibits receptor Tyr kinase HER3 and can auto-ubiquitinate [66].

ligase activity to modulate the ubiquitination status of mediators of signaling cascades.

Deubiquitinases distinguish between ubiquitin-like molecules, isopeptides, and linear peptides as well as between different types of ubiquitin linkage and chain structures [66]. 26S Proteasome-associated ubiquitin-specific protease USP14 is specific for Lys48-linked ubiquitin chains. Deubiquitinase USPL2 [70] hydrolyzes Lys63-linked and linear chains.

Multiple ubiquitin-binding domains (UBD) serve in between-protein interactions: zinc finger ubiquitin-specific protease domain (ZnFUbP), ubiquitin-interacting motif (UIM), ubiquitin-associated sequence (UbA) that binds to monoubiquitin with low affinity, and at least one ubiquitin-like region (UbL) without C-terminal Gly–Gly motif for cleavage by deubiquitinases [66]. [71]

Deubiquitinase activity depends on regulatory mechanisms, such as post-translational modifications, allosteric interactions, and subcellular localization [66]. Subcellular localization determines substrate availability.

Phosphorylation and oxidation can cause inactivation of deubiquitinases. [72] Autoproteolysis or cleavage by proteases eliminates deubiquitinase activity. [73] Molecular interactions can change catalysis rate (allosteric activation). [74]

The SAGA (Spt–Ada–GCN5 acetyltransferase) complex is a transcriptional coactivator that functions as a histone acetylase and deubiquitinase. It is recruited to gene promoter regions by activators. It contains 21 proteins. The SAGA deubiquitinating tetramer binds to monoubiquitinated histones H2b in nucleosomes [67]. Ubiquitin-specific protease UbP8 tethers to Sgf11, Sus1, and Sgf73. The SAGA-deubiquitinating subcomplex domains are interconnected and stabilized by zinc atoms.

---

70. This ubiquitin specific processing C-terminal protease is also called cylindromatosis gene product (Cyld).

71. UIM Domain of ataxin-3 confers selectivity for Lys63.

72. Phosphorylation of deubiquitinase USPL2 prevents deubiquitination of tumor-necrosis factor receptor-associated factor TRAF2 and TNF-induced NFκB response to inflammatory stimuli. Oxidation by hydrogen peroxide of the DUb Cezanne impedes deubiquitination of receptor-interacting protein kinase RIPK1 [66].

73. Ubiquitin-specific protease USP1 is inactivated by autoproteolysis. Protein TNFαIP3 undergoes cleavage by mucosa-associated lymphoid tissue lymphoma translocation paracaspase MALT1 [66].

74. Endosomal deubiquitinases STAMBP (a.k.a. associated molecule with the SH3 domain of STAM [signal transducing adaptor molecule; AMSH]) and USP8 are activated by binding to UIM-containing signal transducing adaptor molecule STAM2. The TNFαIP3 deubiquitinase (or A20) binds to ubiquitin-binding proteins transiently expressed axonal glycoprotein-1-binding protein TAX1BP1 and A20-binding inhibitors of NFκB (ABIN) that may affect substrate targeting or catalysis rate. The USP1-associated factor UAF1 (a.k.a. WD repeat endosomal protein WDR48) heterodimerizes with USP1, USP12, and USP46 to increase their catalytic rate [66].

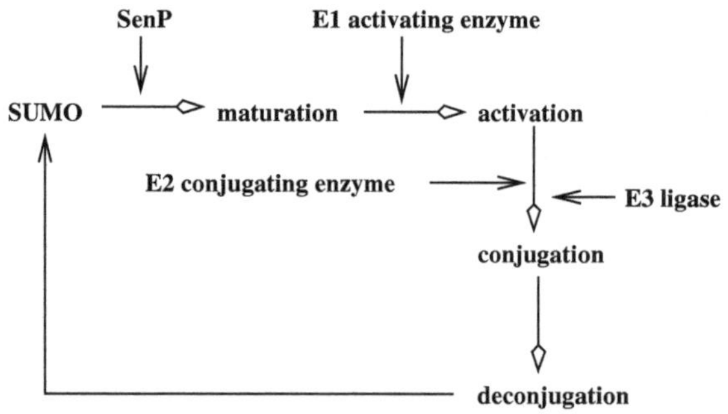

Figure 1.2. Sumoylation and sumodeconjugation cycle (Source: [68]).

1.2.5.13 Sumoylation

Small ubiquitin-related modifier (SUMo) couples to target proteins and reversibly changes inter- or intramolecular interactions of modified proteins. Sumoylation regulates the location, activity, and stability of substrates.

Sumoylation Reaction

Similarly to ubiquitination, but with different enzymes, sumoylation requires an energy-consuming conjugation cascade of consecutive enzymatic steps (Fig. 1.2). Attachment of SUMo to a protein involves SUMo (E1) activase (the AOS1–UbA2 heterodimer forms a SUMo–adenylate conjugate), SUMo (E2) conjugase UbC9, and usually one among several SUMo (E3) ligases that facilitates the transfer of SUMo to the substrate. SUMo Ligases bind UbC9 and SUMo, as well as their targets.

Sumoylation and Desumoylation Cycle

Many proteins undergo rapid cycles of SUMo attachment and deconjugation. SUMo-specific proteases (SenP1–SenP3 and SenP5–SenP7) remove SUMo from substrates, with distinct activities, targets, and locations. In particular, SenP3 and SenP5 preferentially deconjugate SUMo2 and SUMo3 from substrates. Sumoylation and desumoylation cycles sometimes occur only at certain times (e.g., at a certain stage of the cell cycle) and places (e.g., the nucleus) [68].

SUMo Types

The SUMo family includes 4 distinct proteins (SUMo1–SUMo4). Whereas the 3 first subtypes are ubiquitous, SUMo4 is expressed mainly in the kidney, lymph nodes, and spleen [69]. The SUMo1 protein and the twins SUMo2 and SUMo3 are the main SUMo proteins that are expressed at all developmental stages [68]. The SUMo2 and SUMo3 proteins exist at much higher concentrations than SUMo1. The

SUMo1 to SUMo3 proteins use the same basic conjugation machinery, but they conjugate to different targets, are differentially affected by isopeptidases, respond differently to stress, and can be distinguished by their ability (SUMo2 and SUMo3) or inability (SUMo1) to form SUMo chains.

Proteins SUMos are synthesized as immature precursors. These precursors undergo proteolysis by SUMo-specific isopeptidases, sentrin-specific proteases (SenP). Many targets have clear preference for a given SUMo type (e.g., RanGAP1 is exclusively sumoylated by SUMo1). A target can experience either mono- or polysumoylation.

Sumoylation and Protein Interactions

Change in interaction between the target and its partner due to sumoylation results from conformational changes in the target that reveal, mask, or destroy existing binding sites, or addition of interfaces for a partner. A small percentage of most targets is generally sumoylated at a given time, but low-level sumoylation can be efficient [69]. Localization of components (enzymes, substrates, and partners) controls the process.

Sumoylation can interact with other types of post-translational modifications. Phosphorylation can support or impede sumoylation. Acetylation, methylation, and ubiquitination can compete for modification of the same amino acid acceptor as that targeted by sumoylation.

When sumoylation generates a binding site that allows high-affinity interactions, sumoylated proteins undergo specific interaction with effectors that have low affinity for both the non-sumoylated form of interacting proteins and free SUMo. A SUMo-interacting motif mediates low-affinity, non-covalent interactions between SUMo-interacting motif-containing proteins and free SUMo [68]. The SUMo-interacting motif pertains to enzymes of the SUMo conjugation machinery, SUMo target proteins, and proteins involved in SUMo-dependent repression of gene transcription.

Sumoylation Effects

Reversible sumoylation contributes to chromatin structure, DNA repair, gene transcription, cell cycle progression, homologous recombination (rearrangement of DNA strands during cell division), metabolism, transport,[75] ion flux,[76] mitochondrial fission and fusion, cell apoptosis, and signal transduction [68, 69].

Small ubiquitin-like modifier is added to many nuclear proteins to regulate transcription and nuclear transport. Transmembrane proteins (e.g., ion channel $K_V1.5$, and TGFβ receptor-1) can also be sumoylated to regulate their stability, activity, internalization, and/or recycling.

---

75. E.g., sumoylated, translocated RanGAP1 interacts with nuclear-pore protein RanBP2 and sumoylated glutamate receptor GluR6 undergoes endocytosis.

76. E.g., reduced activity of sumoylated potassium-leak channel $K_{2P}1$ and voltage-gated potassium channel $K_V1.5$).

1.2.5.14 Attachment of Other Ubiquitin-like Proteins

In addition to ubiquitination and sumoylation (conjugation of small ubiquitin-related modifiers [SUMo]), other ubiquitin-like proteins can be attached to proteins. Isgylation is the ligation of interferon-stimulated gene product ISG15; neddylation the ligation of neural precursor cell expressed, developmentally downregulated protein NEDD8; and urmylation the ligation of ubiquitin-related modifier-1 homolog (Urm1).

Neddylation

Neddylation competes with other post-translational modifications of proteins (Vol. 1 – Chap. 5. Protein Synthesis). It yields a binding surface to recruit new partners. Neural precursor cell expressed developmentally downregulated protein NEDD8 is attached to a lysine of proteic subtrates. Initially, NEDD8 is activated by an activase $E1^{NEDD8}$. Activated NEDD8 is then transferred by a conjugase $E2^{NEDD8}$ to a ligase $E3^{NEDD8}$ for a specific conjugation of NEDD8 to its substrate.

Urmylation

Ubiquitin-related modifier-1 homolog (Urm1) is used to modify both proteins and RNAs (Vol. 1 – Chap. 5. Protein Synthesis).[77] The human genome contains a single Uba4-like E1 enzyme encoded by the MOCS3 gene and at least 2 ubiquitin-like proteins encoded by 2 genes URM1 and MOCS2. Modifier Urm1 functions as a sulfur carrier in tRNA thiolation via the formation of a thiocarboxylate at the C-terminal glycine residue of Urm1. Ubiquitous Urm1 is conjugated to lysine residues of substrate proteins [70].[78] Protein modifier Urm1 is appended 3 components of its own pathway (MOCS3, ATPBD3, and CTU2). In addition, Urm1 is conjugated to the nucleocytoplasmic transport factor, cellular apoptosis susceptibility protein.

Oxidative stress impedes sumoylation. On the other hand, it favors protein urmylation [70].[79] Protein modifier Urm1 can then serve as sensor to alarm the cell or a marker for proteolysis.

---

77. Modifier Urm1 is thiocarboxylated and acts as a sulfur donor in tRNA thiolation.

78. Similarly to ubiquitination, urmylation involves a thioester intermediate and causes the formation of a covalent peptide bond between Urm1 and its substrates. Yet, conjugation of Urm1 involves the transformation of the C-terminal glycine of Urm1 to a thiocarboxylate by the addition of sulfur. The sulfur atom from cysteine is transferred by a series of reactions to MOCS3 sulfurtransferase. Next, MOCS3 adenylates the Urm1 C-terminus before the transfer of sulfur to the C-terminal glycine. Then, Urm1 associates with the thiouridylases CTU1 and CTU2.

79. Oxidative stress occurs when the concentrations of oxidizing radicals such as hydrogen peroxide exceed the capacity of the cell to eliminate these species. Usually, cells protect against or repair damages caused to DNA, lipids, and proteins by oxidants during cellular metabolism. One technique relies on reversible or irreversible modifications of cysteine residues, such as glutathionylation and nitrosylation. Reduced glutathione scavenges oxidants, as it forms oxidized glutathione to protect proteins and other molecules against damage from oxidation,

Table 1.5. Protein cleavage and signaling (Sources: [71, 1124, 1322]). For example, proteolysis by matrix metalloproteinases (MMP) creates space for cell migration, produces substrate fragments with biological activity, releases ECM-bound growth factors, regulates tissue architecture, activates or deactivates the activity of signaling effectors. Substrates of MMPs include growth factors, receptor Tyr kinases, cell adhesion molecules, cytokines and chemokines, other MMPs, and proteases. In particular, MMPs release stored VEGF for angiogenesis and also cleave VEGF. Truncated VEGF increases vessel diameter; uncleaved VEGF supports vessel sprouting.

Substrate	Role, cleaved fragment effect
Cleavage of structural matrix proteins	
Collagen-1	Cell migration
Collagen-2	Bone morphogenetic protein antagonist fragment
Collagen-4	Anti-angiogenic fragment
Collagen-18	Anti-angiogenic fragment
Cleavage of chemokines	
CCL2	Decrease in chemokine activity
CXCL8 (interleukin-8)	Increase in chemokine activity
Cleavage of signaling molecules	
Insulin growth factor-binding protein	Release IGF
Latent transforming growth factor-β	TGFβ
Vascular endothelial growth factor	Truncated VEGF
Cleavage of transmembrane proteins	
Notch	HRT, HES3, sHH, STAT3 expression
β-amyloid precursor protein	EGFR expression

1.2.5.15 Proteolysis

Regulated *proteolysis* is required in many cellular processes, especially when the cell must react quickly. Gene expression and signaling particularly undergo quality control. Irreversible proteolysis is initiated after a specific post-translational cleavage using specific adaptors. Proteases cleave their substrates of the cytoplasm, plasma membrane, and extracellular matrix, and therefore release reactants, convert structural matrix proteins to signaling molecules, regulate growth factors, and modify signaling (Table 1.5).

Proteolysis at the cell surface includes 2 kinds of enzymes: (1) membrane-tethered proteases, which are soluble enzymes; and (2) integral membrane proteins responsible for the regulated intramembrane proteolysis. Transmembrane serine proteases are involved in cell signaling and interaction with its environment. Proteases of the ADAM and BACE families cleave plasmalemmal proteins. [80]

80. Adamlysin ADAM10 targets E-cadherin; BACE1 processes P-selectin ligand-1, neuregulin-1, and voltage-gated sodium channels; γ-secretase cleaves Notch receptor; and SPPL2 degrades tumor-necrosis factor-α.

Intramembrane-cleaving proteases (ICliP) form a family of enzymes that cleave hydrophobic substrates within the lipid bilayer. This family includes: (1) the site-2 protease zinc metalloproteases, presenilin and presenilin-like aspartyl proteases; and (2) the rhomboid set of serine proteases. Presenilin is the catalytic component of γ-secretase complex.

1.2.6 Reversible Oxidation of Kinases and Phosphatases

Signal transduction regulation can rely on the reduction–oxidation process. Protein Tyr kinases and phosphatases can undergo reversible oxidation by reactive oxygen species, such as hydrogen peroxide and superoxide, that can oxidize free sulfhydryl groups of proteins into sulfenic, sulfinic, or sulfonic acids.[81] However, reversible oxidation that can regulate the enzyme activity requires that a cysteine residue is not oxidized further than sulfenic acid (S–OH), as higher oxidation degree to sulfinic ($S–O_2H$) or sulfonic ($S–O_3H$) acid is usually irreversible.

Reversible Cys oxidation inactivates fibroblast growth factor receptors (RTKs) and Src kinases (NRTKs) [72]. Reduced kinase Src is fully active, whereas oxidized Src only retain 8 to 25% of its full activity. Oxidation of a specific Cys residue (Cys277)[82] causes Src homodimerization by a disulfide bridge. Protein Tyr kinases that lack a cysteine residue at the corresponding position are not regulated by a similar reduction–oxidation process. However, Janus kinase-2 that is also reversibly inactivated by oxidation does not contain a Cys residue in the Gly loop of its catalytic domain.

On the other hand, all protein Tyr phosphatases (PTP) and dual specificity phosphatases contain a Cys residue in their catalytic site. Reversible oxidation of this residue inactivates PTP activity [73].[83]

---

81. Reactive oxygen species not only generate oxidative stress, but also can serve as signaling mediators that are recruited by certain pathways activated by growth factors and cell adhesion molecules. For example, NADPH oxidase catalyzes the single electron reduction of molecular oxygen to produce superoxide (O_2^-) that is rapidly converted to hydrogen peroxide (H_2O_2).

82. This cysteine residue located in the Gly loop of the catalytic domain is conserved only in 8 human PTKs (3 among SRC superfamily kinases, all 4 kinases of the FGFR family, and TNK1 of the ACK family).

83. B-cell antigen receptor (Sect. 11.2.1) oligomerizes upon antigen stimulation and is phosphorylated by SRCB family kinase Lyn to recruit and activate Syk kinase. This event triggers a burst of phosphorylation (activation) of effectors. Timing and magnitude of BCR-generated signal is determined by a feedback loop between BCR-induced Ca^{++} release and ROS production [74]. The ROS mediators regulate the level of Tyr phosphorylation. Effectors Ca^{++} and ROS act cooperatively to amplify the early signal generated by regulating the concentration of produced oxidants that determine pulses of phosphatase inactivation. Two waves of BCR-induced phosphorylation are associated with 2 phases of Ca^{++} influx from cellular stores and then extracellular space. Intensity of the first phosphorylation signal has a crucial impact on the extent and rate of signal progression. Calcium release from the endoplasmic reticulum that is enhanced by ROS directs the proportion of activated kinases Lyn that is kept by ROS by inhibiting PTPs, probably PTPn6.

1.2.7 Receptor Endocytosis

The endosomal shuttle with its various compartments is a dynamical hub where protein sorting directs protein fluxes between the Golgi body, plasma membrane, and lysosomes. Endosomal cargos are assigned for a selected endosomal compartment whether endocytosis aims at recycling, degrading, or signaling.

Cis- and trans-acting factors (i.e., intra- and intermolecular agents that act from the same molecule and a different molecule, respectively) ensure that endosomal sorting is executed properly to avoid inappropriate lysosomal degradation of plasmalemmal receptors and carriers that prevents suitable environmental sensing and nutrient uptake. Ubiquitin acts as a cis-acting signal, whereas endosomal sorting complex required for transport (ESCRT) operates as a trans-acting factor for multivesicular body sorting.[84]

Signaling receptors undergo a rapid endocytosis following activation to move in a series of endosomal compartments with sorting to recycling endosomes or multivesicular bodies and lysosomes. Endocytosis thus prevents excessive ligand-induced activation triggered from the plasma membrane by decreasing the number of receptors available to extracellular ligands as well as plasmalemmal effectors. However, recycling leads to reactivation.

Moreover, many types of receptor remain active in endosomes and interact with downstream effectors, thereby allowing endosome-specific signaling after internalization.

Certain mediators can regulate the residence time of signaling components in early endocytic vesicles. In any case, fusion between a multivesicular body and lysosome deliver intralumenal vesicles of the multivesicular body to the hydrolytic medium of the lysosome. In summary, signal transduction is coupled to endocytosis by a set of mediators with dual activities [11] (Table 1.6).

1.2.8 Gene Expression

Transcriptional regulators, costimulators, and corepressors, control the activity of specific genes according to the presence of signaling mediators as well as metal ions, metabolites, and possible drugs that change the folding, hence the affinity of transcription factors for cognate DNA regulatory site.

Allosteric regulation results from communication between proteins. The connection of a first protein influences the affinity with which a second protein that can adopt 2 alternative conformations, i.e., inactive low- and functional high-affinity for its target (with possible intermediary states), binds to a third molecule. Regulated transcriptional regulators enable cells to respond to changes in environmental conditions.

Environmental signals can be efficiently processed by the cell using a pathway that coordinates the activity of multiple functionally related messenger RNAs.

---

84. Multivesicular body, or endosome, results from cargo partitioning into intralumenal vesicles within endosome.

Table 1.6. Endocytosis and signaling crosstalk (Source: [11]). Many proteins have a dual function in signaling and endocytosis. Adaptor proteins can represent hubs between endocytosis and signal transduction because they interact with 2 alternative types of effectors. β-Arrestins mediate GPCR endocytosis by binding to clathrin and adaptor complex AP2 and participate in signaling by linking components of the MAPK pathway. Similarly, GRB2 operates in endocytosis by interacting with the ubiquitin ligase CBL as well as in signaling by connecting to SOS that stimulates small GTPase Ras. Kinases that have a large set of substrates, such as protein kinase-C and Src kinases, are involved in various signaling cascades and modulate endocytosis of receptors (APPL: adaptor protein, phosphotyrosine interaction, PH domain and leucine zipper-containing; CDC: cell division cycle protein; EGFR: epidermal growth factor receptor; ERK: extracellular signal-regulated kinase; Fz: Frizzled; GPCR: G-protein-coupled receptor; GRB: growth factor receptor-bound protein; MAPK: mitogen-activated protein kinase; MP: MAP2K1 partner; NTRK: nerve growth factor receptor; PKB: protein kinase-B; RTK: receptor Tyr kinase; Ub: ubiquitin).

Protein	Endocytosis	Signaling
APPL1/2	NTRK1 endocytosis	PKB and ERK1/2 activation
β-arrestin	GPCR internalization	GPCR desensitization at cell surface
		Scaffold for MAPK modules
Epsin	Ub–cargo recruitment	CDC42 activation
GRB2	RTK Internalization,	Ras activation
	Ub-Dependent degradation	
MP1–p14–p18	Late endosome genesis	ERK1/2 activation
	Endosome transport	
P38MAPK	EGFR endocytosis	Responses to stress
	Rab5 regulation	Inflammation
Intersectin	EGFR internalization	CDC42 activation
Disheveled	Fz internalization	Wnt signaling

Gene expression, indeed, depends on coordinated transcription, translational activation, and mRNA stability, as well as post-transcriptional events controlled by RNA-binding proteins (RNABP) and small non-coding RNAs [75]. The components of the ribonucleoproteic complexes (RNA-binding proteins, non-coding RNAs, and metabolites) interact with mRNA regulatory elements, the untranslated sequence elements for regulation (USER). Such RNA operons and/or regulons guarantee the coregulated expression of a set of proteins acting in a given pathway. RNA Operons and regulons can be activated or repressed by RNABP phosphorylation.

1.3 Signaling Triggered by Ligand-Bound Receptor

The cell signaling pathways, once the receptor is activated by its ligand, involve: (1) molecular transformations with a given reaction rate; (2) intermolecular associations due to changing binding affinity; and (3) intracellular translocations for optimized reaction compartmentation (Fig. 1.3).

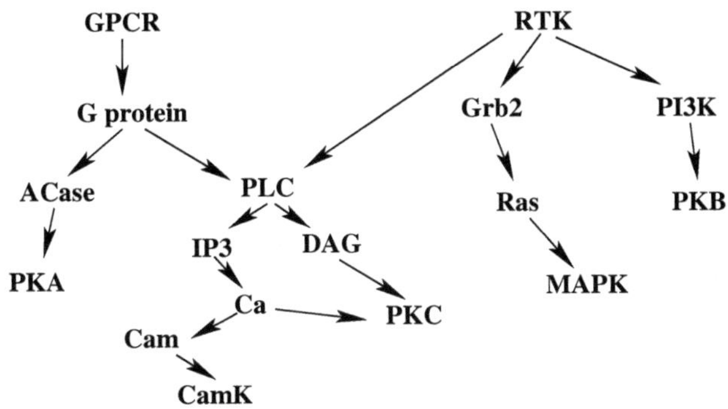

Figure 1.3. Examples of pathways activated by G-protein-coupled receptors (GPCR) and receptor Tyr kinases (RTK). Several types of reaction cascades can be triggered by each type of receptor. The pathways are selected according to the context. The same signaling modules can be used by different receptors, specific responses being achieved by different processes (Source: [76]).

Signaling pathways can lead to the phosphorylation of transcriptions factors that, in turn, control the expression of target genes. Furthermore, protein kinase-A and most of the mitogen-activated protein kinases can link to target genes, either at the promoter, or at the transcribed regions of genes [77].

Plasmalemmal, ligand-bound receptors are removed from the cell membrane and incorporated into endocytic vesicles that fuse with early endosomes. Ubiquitin supports the vesicle incorporation. Early endosomes contain phosphatidylinositol 3-phosphate, early endosome antigen-1, and Rab5 small GTPase. The latter is replaced by Rab7 in late endosome.

Endosomal sorting complexes required for transport (ESCRT0–ESCRT3; Vol. 1 – Chap. 9. Intracellular Transport) are used for transport of ubiquinated proteins from endosomes to lysosomes via multivesicular bodies. Proteins ESCRTs interact between them as well as with clathrin and 3-phosphoinosides. The ESCRT0 agent recruits ESCRT1, which connects to ESCRT2; the latter links to ESCRT3 [78]. In opposition to other ESCRTs, ESCRT3 does not recognize ubiquitin; instead it recruits deubiquitinases to remove ubiquitin before incorporation into multivesicular bodies.

1.3.1 Signaling Initiation

Plasmalemmal receptor types include: (1) receptors coupled to *G proteins* that activate or inactivate enzymes to modulate the formation of second messengers (Table 1.7);[85] (2) receptors, the cytoplasmic domain of which is activated when the

85. Subfamilies of Gα subunits include Gs and Gi that respectively activate and inhibit adenylate cyclase to regulate the intracellular cAMP level. Members of the Gq family stimulate enzymes that activate phospholipases to release second messengers derived from mem-

Table 1.7. Examples of second messengers and corresponding receptor ligands. Vasopressin (antidiuretic hormone) receptors interact with specific kinases. Atrial, brain, and C-type natriuretic peptides produced by endothelial cells and monocytes are vasorelaxants and inhibit the proliferation of vascular smooth muscle cells, via guanylate cyclases coupled to the plasmalemmal natriuretic peptide receptors. Adrenoceptors-β1 and -β2, as well as D_1 to D_5 dopamine receptors interact with Gs and Gi subunits of G proteins coupled to adenylate cyclase, whereas α1- and α2-adrenergic receptors function with Gq coupled to phospholipase-C and Gi, respectively. Erythropoietin regulates the proliferation, differentiation, and maturation of erythroid cells via Epo receptor, a member of class-1 cytokine receptors, which uses the JaK–STAT pathway. The parathyroid hormone receptor is a G-protein-coupled receptor leading to cAMP pathway. The calcitonin G-protein-coupled receptor is also associated with adenylate cyclase. Whereas the insulin receptor is a receptor Tyr kinase, the glucagon receptor belongs to the superclass of G-protein-coupled receptors that activates adenylate cyclase. Somatostatin acts on insulin and glucagon secretion via 2 somatostatin-receptor subtypes. Somatostatin produced by several body's tissues inhibits growth hormone release (negative feedback loop). Growth hormone receptor is a receptor Tyr kinase that activates the receptor-associated Tyr kinase JaK2. The prolactin receptor belongs to the cytokine receptor superclass. The oxytocin receptor is a G-protein-coupled receptor that requires Mg^{++}.

Second messenger	Activators
cAMP	Catecholamines (α2, β), dopamine (D_1–D_5), glucagon, calcitonin, oxytocin, adrenomedullin, parathyroid hormone
cGMP	Natriuretic peptide, nitric oxide
Protein kinase	Insulin, growth hormone, prolactin, erythropoietin, growth factors, antidiuretic hormone, angiopoietins, ephrins
Phosphoinositides and calcium	Catecholamines (α1), angiotensin-2, vasopressin

receptor is linked to its ligand and activates one or more specific enzymes to simultaneously stimulate multiple signaling pathways (Table 1.8);[86] and (3) receptors linked directly or indirectly to ion channels.

Plasmalemmal receptors with a catalytic activity (recepzymes) have cytoplasmic domains that either have an intrinsic enzymatic activity or are associated with an intracellular enzyme (enzyme-linked receptors [extrinsic enzymatic activity]). In both cases, enzymatic activity is stimulated by ligand binding to the receptor. Transmem-

brane lipids. Activated phospholipase-C releases IP_3 and DAG, and phospholipase-A2 and -D initiate arachidonic and phosphatidic acid pathways.

86. Once the cytoplasmic domain of the receptor is activated by phosphorylation, the receptor connects to effectors, such as cytosolic Tyr kinases, guanylate cyclase, phospholipase-C, and stimulators or inhibitors of small GTPases involved in the cytoskeleton arrangement and endothelial permeability (RHO family). For example, VEGFR2 acts via the PLC and Ras pathways.

Table 1.8. Main receptors and effectors (GPCR: G-protein-coupled receptor; RTK: receptor Tyr kinase; RSTK: receptor Ser/Thr kinase; cAMP: cyclic adenosine monophosphate; cGMP: cyclic guanosine monophosphate; PK(A,B,C): protein kinase-A, -B, -C; PL(A,C,D): phospholipase-A, -C, -D; MAPK: mitogen-activated protein kinase; PI3K: phosphatidylinositol 3-kinase; AA: arachidonic acid; Lkt: leukotriene; PG: prostaglandin; GRK: GPCR kinase).

Receptor	Effectors and pathways
GPCR	cAMP–PKA
	PLC–PKC
	PLA, PLD
	MAPK
RTK	PI3K–PKB
	PI3K–PLC
	PLA–AA–Lkt, PG
	GTPases–MAPK
RSTK	GRK
	cAMP, cGMP
	PLC–PKC–Ca^{++}

brane receptors with intrinsic enzymatic activity include adenylate- and guanylate cyclases, Tyr phosphatases, Tyr and Ser/Thr kinases, and plexins.

1.3.2 Molecule Transformations and Multicomponent Complexes

Signal transduction enables the cell to adapt to the changing environment using various procedures at the molecular scale. Information is often transmitted via protein modifications, especially reversible protein phosphorylation. The reversible phosphorylation targets not only signaling receptors and effectors, including many transcriptional regulators, but also accessory mediators of the signaling pathway (e.g., ubiquitin ligases and guanine nucleotide-exchange factors).

Whatever the signaling substance, the ligand-bound receptor undergoes a conformational change. The receptor is then usually phosphorylated by an associated protein kinase. The majority of proteins contain multiple phosphorylation loci with different kinetics. Phosphorylation sites depend on the stimulus type, time dynamics, and subcellular location. Owing to several phosphorylation loci, proteins can serve as hubs for signal integration.

Proteins participating in a shared cellular process, especially in cellular transport and signaling pathways, can transiently interact between them or form more or less stable proteic complexes. Furthermore, additional protein interactions can occur within proteic complexes, between complexes and between free proteins and complexes.

The location of signaling complexes at plasmalemmal nanoenvironments modulates signaling outputs. Membrane rafts (Vols. 1 – Chaps. 7. Plasma Membrane and 9. Intracellular Transport, and 2 – Chap. 6. Cell Motility) that contain a cluster of proteins can change their size and composition in response to stimuli for protein interactions [79]. Specific anchors enable orientation of non-specific enzymes toward specific targets and remove them from undesirable substrates.

Most regulated processes in the cell, characterized by a great sensitivity and a high specificity, especially the signaling pathways, with its triggers, switches, transducers (activators and inhibitors), and adaptors, require the assembly and disassembly of multicomponent, functional complexes made of proteins and other molecules. For example, weak interactions between growth factors and certain receptor domains are enhanced by cooperative interactions with additional receptor domains and/or other compounds, such as heparan sulfate proteoglycans, to form specific complexes, the weak interactions being replaced by stronger and specific dockings [80]. The assembling dynamics start with non-specific, transient associations. The first assembling stages involve electrostatic interactions between encounters. The non-specific complexes quickly separate. Partners of specific complexes reorient and, after desolvation (water molecules move away from protein surfaces), form much stronger links between them.

1.3.3 Coupled Pathways

Signal transduction is usually characterized by a non-linear, interconnected nature (Fig. 1.4). The number and variety of plasmalemmal receptors demonstrate that cells receive a large amount of information from their environments. G-protein-coupled receptors, receptor Tyr kinases, and integrins are 3 major receptor classes. Sensed information that can often be random, noisy, or even contradictory is then integrated to make non-trivial decisions [81]. Plasmalemmal receptor stimulation can activate a network of tens or even hundreds of cytoplasmic proteins. These networks are not necessarily receptor-specific because different receptors often activate common sets of proteins, but they generate distinct responses.

Many signaling effectors contain a catalytic (e.g., kinase or phosphatase) and at least one regulatory domain, such as the auto-inhibitory motif that represses the constitutively active, catalytic domains. The modular structure can yield complex behavior, such as input-dependent gating, non-linear functioning, and multiple signal integration.

The conversion during the evolution from linear to complex network architecture confers: (1) *robustness* that enhances *reproducibility* (despite input variations and signal processing hazards, as distinct inputs always trigger the same unique outputs) and *reliability* of signal transfer; and (2) *evolvability* with increasing control required of signaling pathways from unicellular to multicellular organisms. The *modular structure* comprises several modules that are partially redundant. Module redundancy allow compensatory functioning in the case of component failure.

Cells coherently respond to external stimuli (adaptation robustness). Specific responses characterized by appropriate cascades of intracellular chemical reactions can be generated over a wide range of parameter variations. Signals are transduced by information processing that is characterized by signal transduction *complexity* and between-pathway *connectivity*.

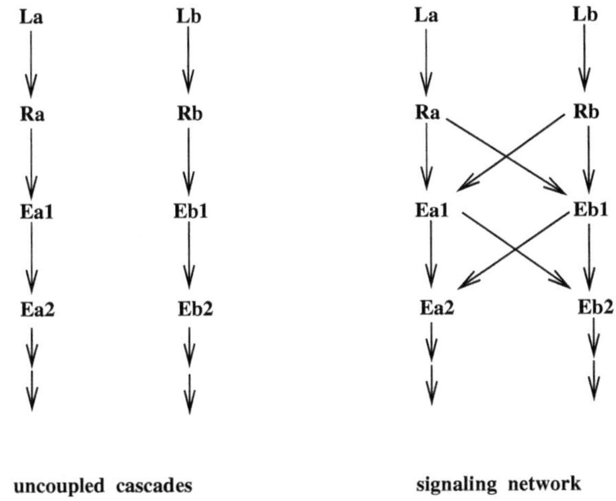

Figure 1.4. Uncoupled cascades (**left**) and signaling networks (**right**). Signaling networks with interacting parallel cascades achieve a storeyed network that imparts robustness and guarantees output reproducibility. Interconnected pathways of biochemical signal transduction networks resemble parallel, distributed processing networks. The finely tuned information processor is able to make decisions in a complex context. The network is able to treat different combinations of a huge number of environmental stimuli to generate a small number of appropriate responses (pattern recognition system).

The architecture of chemical reaction cascades with their storied configuration, branching, and modularity, as well as functional characteristics with feedback loops, serve as information relay networks that avoid failures in the transmission of internal and external signals.

Feedback loops yield the main functional feature that accounts for signaling robustness. Positive feedback loops enhance the amplitude and prolong the active state even for a transient stimulus. Negative feedback loops stop complex chemical reaction cascades at different stories of the reaction cascade, hence determining the signaling duration (activation window).

1.3.4 Feedback Loops

Positive and negative feedback loops are common regulators in cell signaling. Feedback loops connect output signals back to inputs. Multiple feedback loops nested in time and space generate desired signaling responses. Receptor-triggered calcium signaling in non-excitable cells (e.g., histamine-triggered cytosolic Ca^{++} signals in epithelial cells) primes various kinds of responses, such as spikes, oscillations, and plateaus. Periodic oscillations can serve as a clock.[87]

87. Oscillation period must remain a precise number of time units despite slight variations in the individual components of a clock (robustness). In the case of environmental change, the system must be retuned to adjust (tunability) [82].

Oscillations can be generated by a negative feedback with a delay inherent in the system due to processes involved in gene expression. Both robustness and tunability are important features of an oscillatory system. [88]

Negative feedback can: (1) stabilize basal signaling levels (*basal homeostat*); [89] (2) limit maximal signaling output (*output limiter*); [90] (3) enable adaptive responses (*adaptor*); [91] and (4) create transient signal responses (*transient generator*) [83]. [92]

Combination of 2 negative feedback loops can independently stabilize basal signaling and limit maximal signal output. [93]

Positive feedback can: (1) amplify signaling responses (*amplifier*); [94] (2) alter kinetics (*accelerator* in saturating conditions or *delay* in non-saturating conditions); [95] and (3) create *bistable switches*. [96]

Dual fast and slow positive feedback loops can exhibit transient or persistent bistable states if the fast or both positive feedbacks are committed. A first set of stimuli with sufficiently high magnitude can trigger the fast reversible positive feedback and following activators can commit the slow positive feedback and trigger a

---

88. The repressilator is defined by a network of 3 genes that sequentially repress one another. The 3 repressive interactions leads to a bulk negative feedback with a delay. The repressilator does oscillate, but oscillations are not robust, as oscillations are variable. Moreover, the repressilator is not tunable, as reaction rate constantly changes and generally abolishes oscillation rather than changing its frequency [82].

89. Small-amplitude negative feedback loop stabilizes the basal signaling state without preventing strong input signals from triggering maximal pathway activation. Small deviations of an input signal are thus suppressed, whereas large changes can initiate signaling.

90. Upon stimulation, the output signal rapidly increases, but it is attenuated once it passes a threshold. For example, receptor-triggered increase in Ca^{++} concentration in the cytosol are stabilized by a rapid negative feedback resulting from Ca^{++} uptake by mitochondria.

91. Adaptive signaling responds to changes in input rather than magnitude of input signal by using partial deactivation and subsequent internalization of activated plasmalemmal receptors to prevent saturation and allow subsequent signaling when additional stimuli occur.

92. Strong negative feedback loops that are triggered after a delay convert a constant input into transient output signals with amplitudes that depend on input magnitude. Delayed activation (after 10–60 s) of the calcium–calmodulin-regulated plasma membrane Ca^{++} pump following cytosolic Ca^{++} level increase augments PMCA activity for Ca^{++} efflux.

93. Combined negative feedbacks that switch Ca^{++} influx from intracellular store to extracellular space (first negative feedback, basal homeostat) and cause mitochondrial Ca^{++} uptake (second negative feedback, limiter) operate independently.

94. Positive feedback yields absolute and relative (sigmoidal output–input curve) amplification. Activation of inositol trisphosphate receptors on endoplasmic reticulum surface by Ca^{++} binding primes an initial Ca^{++} release from its store that, in turn, triggers a positive feedback loop due to linking of multiple released Ca^{++} to fully activates IP_3Rs.

95. Positive feedback can change the timing of the signaling response to more rapidly reach saturation response or prolong the time required to reach a higher steady state.

96. Upon reception of input below a critical threshold, the output remains near its basal state, whereas for inputs above the threshold, the output increases to a high, active state. Dose–response curve hysteresis allows a bistable system to remain in the active state, as the stimulus required to keep the system in the active state is lower than the input required to initiate the transition from basal to active state.

persistent output increase. In the case of Ca^{++} signaling, Ca^{++}-activated IP_3Rs can provide rapid and localized positive feedback and Ca^{++}-stimulated phospholipase-C that yields a second slow positive feedback produces rapidly diffusible IP_3, therefore generating global robust bistability.

Interactions between both negative and positive feedback loops ensure robust, tunable oscillations in gene expression. [97] Coupled positive and negative feedback loops are usual in cell signaling. Mixtures of single positive and negative feedback can indeed create single pulses or oscillatory outputs [83]. Negative feedback that terminates positive feedback is induced only at high concentrations of the output or after a time delay. Delayed negative feedback can force a bistable system back to the inactive state and create a pulse output with a fixed amplitude and duration. After a pulse output, a new positive feedback cycle can be triggered to generate periodic cycling between high and low outputs in response to a steady input (*oscillator*).

Positive and negative feedback loops can be active in a restricted region of the cell rather than the whole cell. Mixed positive and negative feedbacks then trigger local signals, self-propagating waves, or cell polarization [83]. [98] Adding a negative feedback loop to an oscillator (positive and negative feedback loops) can cause sharper spikes, increased input range over which oscillations occur, and increased output frequency range (*robust oscillator*). [99]

Feedback circuits can use pre-existing components that undergo post-translational modifications to enable immediate tuning of the output. They can also utilize newly synthesized components that efficiently characterize responding pathways by the duration of signaling activity (Fig. 1.5) according to the target component type (already existing or not) and target storey of the signaling pathway (mRNA, transcription factor, or upstream effector).

---

97. Supplementing a core delayed negative feedback circuit with a positive feedback loop yields robust and tunable oscillations [83–85].

98. In the case of Ca^{++} signaling under weak stimulation, local positive feedback that activates IP_3R quickly inactivates, and diffusing calcium ions are restrained to a tiny distance from the source, thereby creating local Ca^{++} pulses. When the local signal is strong enough, new local pulses triggered at adjoining sites keep the signal and allow a self-propagating signal to move away from its source (traveling wave). Neutrophil migration is regulated by a positive feedback loop, as guanine-exchange activator of small guanosine triphosphatase Rac (RacGEF) DOCK activates lamellipodium extension and is recruited to sites of lamellipodium extension. Once the positive feedback-driven local recruitment has been initiated, DOCK concentration decays in the entire cell. This subsequent negative feedback can prevent the formation of a second cell leading edge.

99. Stromal interaction molecule STIM2 keeps basal Ca^{++} concentration in the cytosol and lumen of the endoplasmic reticulum at about 50 nmol and 400 μmol, respectively. Agent STIM2 triggers influx of extracellular Ca^{++} at endoplasmic reticulum–plasma membrane junctions in response to reduced Ca^{++} concentration in the lumen of the endoplasmic reticulum. Calcium–calmodulin-dependent protein kinase-2-mediated inhibitory phosphorylation of inositol trisphosphate receptors serves as a second negative feedback loop that prolongs the time between Ca^{++} spikes and hence extends the output frequency range.

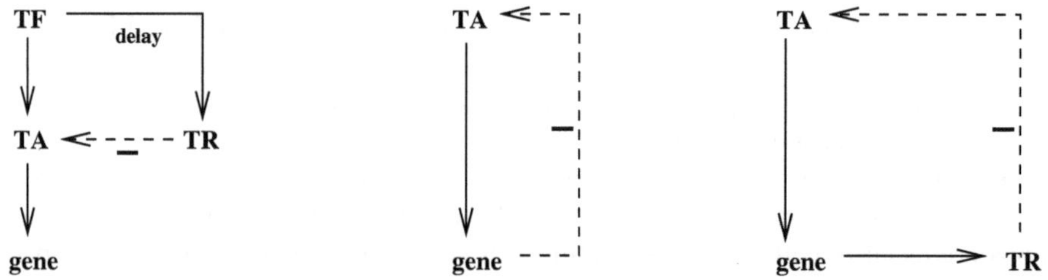

Figure 1.5. Various types of feedback loops define the signaling duration. (**Left**) Type 1 with quick production of a transcription activator and slow generation of a transcription repressor (pulsed activation). (**Middle**) Type 2 based on the lag between transcription and translation (fast inhibition); (**Right**) Type 3 made of a transcription activator that regulates its own transcription repressor.

1.3.5 Cell Type Specificity

Signaling pathways are generally made of modules with a variable node number, each node being defined by molecular interaction or assembling of molecular complexes composed of involved enzymes, adaptors, and other types of transducers.

Different cell types can use a common transducer node or module but distinct kinases and other effectors at other storeys to generate diverse responses according to the cell type (cell type specificity) [87]. Moreover, cell type-specific output can be induced by a given stimulus, when signal activation, combination of activation events, signal processing, signaling dynamics, or responsiveness is cell specific.

1.3.6 Signal Specificity

Signal transduction networks allow cells to sense and respond to disparate extracellular signals. The location of signaling molecules allows *pathway specificity*.

Scaffold proteins strongly contribute to the specificity by allocating adequate sites within the cell for activation of following storeys of the reaction cascade. Scaffolds organize signaling complexes to target pathway effectors to specific substrates and facilitate communication with other pathways. Adaptor proteins regulate the kinetics, amplitude, and location of signaling nodes, providing an efficient mechanism that enables an individual extracellular stimulus to elicit a specific biological response. In the case of the Ras–Raf–MAPK–ERK cascade activated by different growth factors, scaffolds include kinase suppressor of Ras (KSR), β-arrestin, MAP2K partner-1, Similar expression to Fgf genes (SEF),[100] and IQGAP1 [88]. The spatial organization of the various components of the cAMP pathway triggered by different neuromediators and hormones segregates suitable responses among all possible cell functional reactions. Phosphodiesterases (Vol. 4 – Chap. 10. Signaling Pathways) hydrolyze cAMP, and hence prevent cAMP diffusion in the whole cytosol and inadequate target activation in cAMP signaling in cardiomyocytes [89].

100. A.k.a. interleukin-17 receptor-D (IL17Rd), not SAA3 enhancer factor (also SEF), or transcription factor TFCP2 (a.k.a. LBP1c, LSF1d, and LSF).

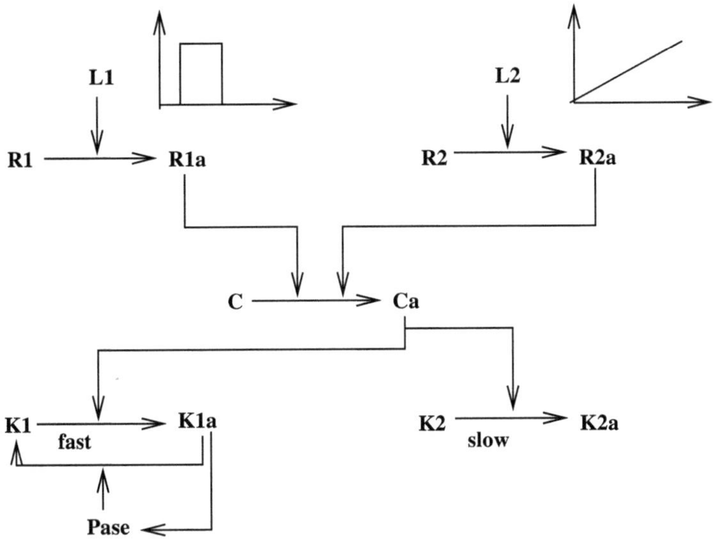

Figure 1.6. Signaling discrimination among transient and slowly varying cues (Source: [90]). Temporal profiles of propagated signals that share the same signaling module can be associated with pathway specificity according to distinct chemical kinetics and pathway architectures: (1) fast, transient activation of the MAPK module followed by rapid deactivation and (2) slow activation of the MAPK module with the maximum amplitude occurring after a long duration (e.g., $O[1\,\text{h}]$) and persistent activity over a significant time interval. Signaling by pathways 1 and 2 associated with fast and slow dynamics is initiated when ligands L1 and L2 bind to cognate receptors R1 and R2 (R1a and R2a) that prime reaction cascades down to a common pathway component C, which once activated (Ca) activates (directly or not) specific kinases K1 or K2 (without noticeable activation of the other), respectively. Activation of pathways 1 and 2 generates a fast, transient and slow, persistent rise in activated C (Ca), respectively. Kinase K1 remains inactive when the signal varies slowly. Enzyme K1 has a rapid activation kinetics relative to the time scale of the transient signal, thereby being excited by rapidly and transiently activated common component C to produce a strong response. Moreover, K1 activates a phosphatase (Pase) that deactivates K1 (negative feedback). Enzyme K1 then belongs to signaling species with fast activation and deactivation kinetics. Enzyme K2 has slow activation kinetics, hence fast, transient signal from pathway 1 has no effect on K2.

Dynamical response (*kinetic insulation*) achieves signal specificity: pathway temporal organization downstream of a common component (especially MAPK modules) can separate signals according to activation and deactivation time scales and signaling duration downstream of common components, such as MAP3K [90] (Fig. 1.6).

Distinct chemical kinetics of biochemical cascades can prevent crosstalk and hence encode information into temporal patterns via an unique biochemical module that can lead to a single cell response. Quick and transient activation (with rapid deactivation) can activate a given type of MAPK, whereas slow and sustained activation triggers another type of MAPK, both MAPK types being stimulated by the same

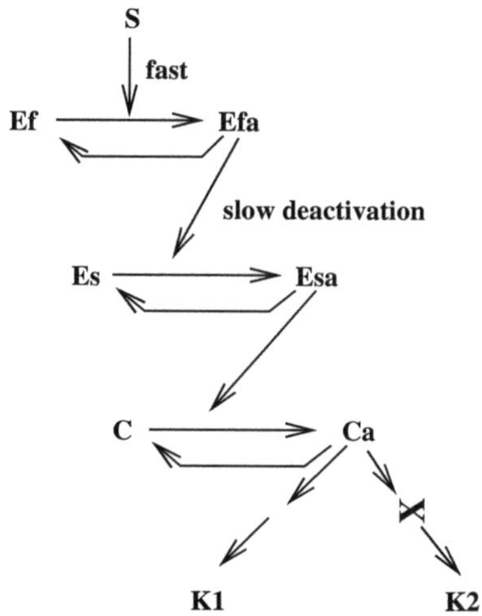

Figure 1.7. Signaling pathways can transmit multiple signals through a single module for specific responses using several procedures, among which temporal patterns, scaffold proteins, cross-inhibition, etc. Dose and duration of stimuli influence the response pattern. The pathway can respond to both transient, fast (starting-stopping cue) and sustained, slowly varying (ramp) stimuli owing to a multistorey, adaptive pathway module (Source: [90]). Pathway response is achieved by signaling effectors according to their activation and deactivation kinetics. Any effector with fast activation and deactivation kinetics is a priori insensitive to slowly increasing cues. The pathway module with an upstream fast-reacting effector (node Ef) with slow deactivation kinetics followed by an adaptive species transmits inputs, once a threshold is reached, into a slow output. Node Ef is rapidly activated by the transient stimulus to form stable Efa. Persistent Efa then slowly activates downstream slow-reacting effector (node Es) that generates the slow response by activating a corresponding cognate type of kinases and not others via a common pathway component C. Kinetic insulation activates one subpathway without neutralizing the other, as long as the shared component is not saturated.

MAP3K. Certain pathways require transient signals to produce a response, whereas others need slowly increasing signals. However, some pathways can respond to both transient, fast and sustained, slowly varying (ramp) stimuli owing to an adaptive pathway module (Fig. 1.7).

1.3.7 Pathway Complexity

Signaling pathways can be represented as linear or complex cascades (Fig. 1.8). In linear cascades, a ligand binds to the receptor, which activates an effector. This effector either carries out the action or stimulates another single downstream effector, which in turn does the same activity until the effector leads to the signaling response, whether the pathway requires a single stage or is multitiered. Complicated pathways are composed of multiple steps characterized by linear interactions.

Complex networks are characterized by nodes and hubs formed by effectors. [101] They usually are constituted by protein isoforms. These related proteins cause a pathway divergence with multiple routes. The node is positively and negatively regulated. Furthermore, the node can be the biochemical site for crosstalk with other signaling networks.

The concept of stimulus-specific interconnected signaling pathways, which includes timing, amplitude, and duration of signaling responses for coordinated behavior in time and space, as well as subcellular locations and specific protein recruitments, relate extracellular signals to the expression of nuclear transcription factors of specific genes more efficiently than linear pathways [91]. Signaling networks have various kinetics, activation being either transient or sustained [92]. Brief stimuli can cause sustained kinase activation at low concentrations of corresponding phosphatase (bistable behavior) [93]. This state can be eliminated by kinase-induced increase in related phosphatase. The signaling network can then respond proportionally (monostable state).

A single transduction pathway can be rather complex due to its numerous inputs, outputs, and regulatory nodes. Moreover, different signaling pathways interact. For example, synergistic activity results from costimulation with ligands that mobilize calcium ions and from ligands that trigger cAMP production. Costimulation of receptors causes a peculiar pattern of production and suppression of members of a set of targeted molecules, whereas a single receptor stimulation can either favor or prevent the production of the entire set of molecules. Input ligands can modulate signaling pathways rather than control a given output.

Coupled positive and negative feedback loops induce complex dynamics, favoring the occurrence of instabilities and oscillatory behavior. Positive feedback amplifies the signal, whereas negative feedback attenuates it. A given effector can not only stimulate its activator, but also inactivate its inhibitors. Signaling cascade steps with positive feedbacks are endowed with bistability, characterized by 2 stable activities, basal and stimulated. Negative feedback can stabilize the cascade outputs (monostable state) and adapt the pathway to parameter variations.

Information processing by cells requires the coordinated activity of a network of signaling pathways. Crosstalk between pathways allows combinations of stimuli. The architecture of the signaling network in which the majority of inputs modulates signaling (only a few inputs independently control cellular outputs) has been proposed in macrophages, selecting activation of receptor Tyr kinases and G-protein-coupled, Toll-like, and cytokine receptors [94]. Cells can distinguish between adaptive and maladaptive signals that share the same pathway.

Spatial and temporal signaling dynamics after receptor stimulation determine the specificity of cellular responses via activation of kinase and phosphatase cas-

101. These nodes and hubs act as mesh nodes in spatial discretizations of any explored organ used to solve differential equations that govern a biological process. The nodal values of involved physical quantities depend on values at adjoining nodes. They also operate similarly to nodes of a communication network, receiving and providing information to and from neighboring nodes.

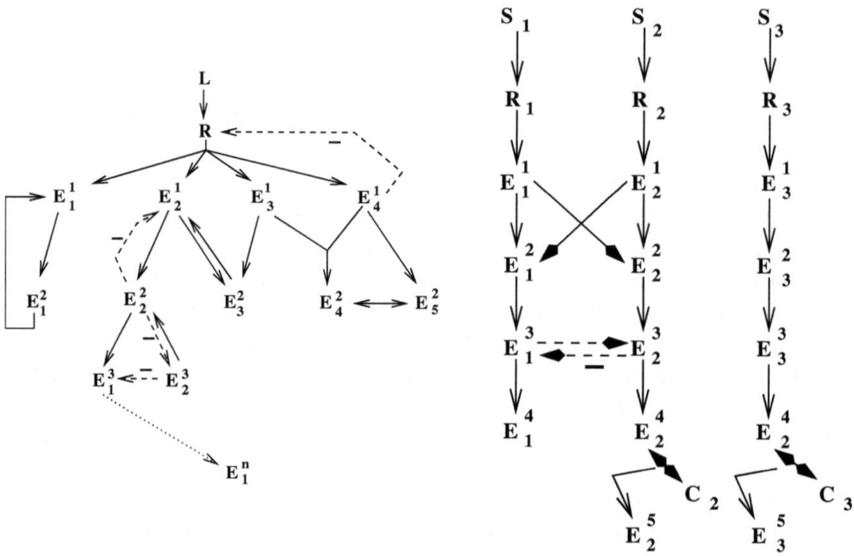

Figure 1.8. (**Left**) Complex pathway with non-linear interactions between effectors (L: ligand; R: receptor; E_i^s: effector i in pathway stage s). Diagrams are useful for representing reactions of a signaling process. The binding of a ligand to its receptor triggers the activation of signaling pathways through effectors and intracellular second messengers, which finally lead to biological effects. Complex signaling pathways are characterized by cooperation, reciprocal effects, interaction, crosstalk, and positive and negative feedback. (**Right**) Parallel interacting signaling pathways activated by their respective signals (S_1, S_2, and S_3), with targeted components (effectors E_1, E_2, and E_3) are activated when they simultaneously receive their signals. Specificity is lost when both pathways are activated, although they are exposed to a single signal due to mutual activation or shared components. They achieve different responses when they are exposed to their own signals and either cross-inhibition or insulation. Cross-activation between pathways triggered by S_1 and S_2 downstream from tier 1 is associated with mutual inhibition downstream from tier 2 (downstream from the cross-activation node, here between E_1^3 and E_2^3) to eliminate unwanted interactions between pathways. Insulation between pathways triggered by S_2 and S_3 is achieved when shared component E_2^4 is incorporated into different molecular complexes (C_2 or C_3, made of different adaptor and scaffold proteins), each one being specific for each signal that must be processed.

cades. Distinct spatiotemporal activation of a given signaling pathway, according to temporal control and spatial distribution of activated transducers, causes different responses. The magnitude of kinase activation depends on the stimulus amount (activation differs from an all-or-none response) [95]; but gene expression, which depends on the nature of the inductive signal, can be independent of the amplitude of kinase activation.

The mitogen-activated protein kinase pathway (Vol. 4 – Chap. 5. Mitogen-Activated Protein Kinase Modules) governs cell fate by transducing and processing multiple growth factor signalings. The Raf–MAP2K–ERK cascade, a module of 3 nodes activated by many external growth factor receptors, is controlled by Ras GTPase to particularly regulate cell proliferation and differentiation. Small GTPase Ras recruits Raf MAP3K from the cytosol to the plasma membrane for activation. In turn, MAP2K, ERK, and scaffold proteins translocate to the plasma membrane to be also activated.

The signal sensitivity of the module is determined by its subcellular location [96]. The activation threshold is low at the plasma membrane and high in the cytosol. The MAPK module with a greater complex lifetime at the plasma membrane than in the cytosol transmits signals with a larger efficiency. Dephosphorylation of MAP2K and ERK occurs about 2 and 5 times faster, respectively, in the cytosol than at the plasma membrane. Cells can hence respond appropriately to physiological stimuli with maximal outputs for low inputs.

The Raf–MAP2K–ERK cascade is activated by many external cues bound to growth factor receptors and, hence, must coordinate these signals and maintain specific responses. Different effectors or modules, upstream or downstream from the Raf–MAP2K–ERK cascade, or their intracellular location can explain distinct regulatory mechanisms. Epidermal growth factor uses the Raf1–MAP2K1/2–ERK1/2 pathway, as well as neuronal growth factor, but they have different effects. On the one hand, EGF transiently activates ERK, leading to cell proliferation. On the other NGF generates sustained ERK activation, causing cell differentiation [97]. Stimulation by EGF promotes a negative feedback loop from ERK to Raf1, whereas NGF elicits a positive feedback loop from ERK to Raf1, with a persistent response (Fig. 1.9). Another short negative feedback from ERK1/2 to MEK1/2 exists for integrin signaling. The cell response can be altered by pathway changes. Blockage of protein kinase-C in NGF-stimulated cells transiently activates ERK (NGF-stimulated cells proliferate rather than differentiate). Therefore, PKCδ is involved in the NGF-induced positive feedback. Conversely, concomitant PKC activation in EGF-stimulated cells yields sustained ERK signaling, transforming negative feedback from ERK into positive feedback (PC12 cells differentiate rather than proliferate). Moreover, the response feature varies between EGF and NGF. Augmentation in EGF level (negative feedback) gradually increases ERK phosphorylation, whereas NGF (positive feedback) drives all-or-nothing ERK activation (switch-like, bistable dynamics).

1.3.8 Modeling and Simulation

Mathematical modeling is aimed at describing the macroscopic reactions of cells and tissues from the nanoscale mechanisms. Mathematical modeling and simulation of signaling pathways can identify novel regulatory elements and crosstalk between pathways. Furthermore, they can correct known regulatory networks and plan new experimental strategies for better designing reaction cascades.

Computational models of regulatory networks relies on several methodologies. Main mathematical modelings of regulated signal transduction include [98]: (1) *continuous models* based on differential equations such as the *mass action kinetics* with its set of kinetic coefficients of molecular interactions; and (2) *Boolean models* determined by individual input–output relationships at each protein node of the molecular

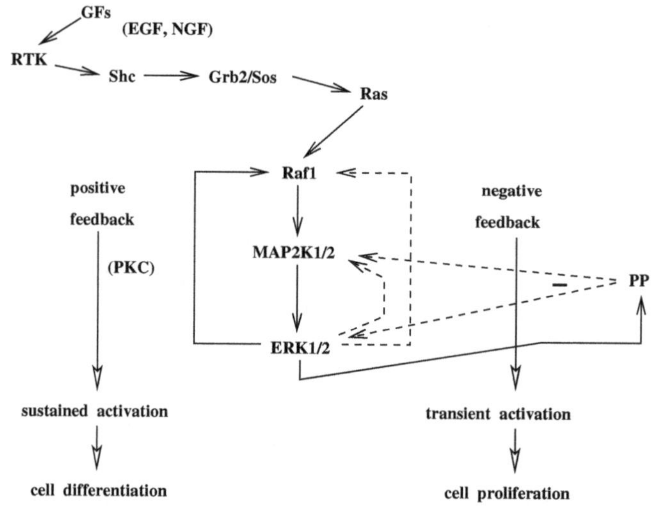

Figure 1.9. The Raf–MAP2K–ERK cascade can be activated by multiple ligand-bound receptors (Source: [97]). Different receptors activate this common 3-node module of the corresponding signaling pathways that generates distinct outputs. Pathway specificity can be determined according to the type of feedback loops. A positive feedback loop from ERK to Raf1 induces persistent activation leading to cell differentiation (PC12 cells subjected to NGF stimulation via NTRK1 (or TRKa) receptor differentiate). A negative feedback loop from ERK to Raf1 explains the occurrence of transient activation generating cell proliferation (PC12 cells subjected to EGF stimulation via EGF receptors proliferate).

cascade of events. Continuous models are associated with real-value parameters that evolve over a continuous timescale. However, quantitative data almost always lack and several parameters must be estimated. Discrete-state models ignore kinetic properties of the process. Deterministic models based on chemical rate equations ignore inherent molecular fluctuations around the average concentration, as molecules are finite in number. When these fluctuations become important, statistical models can be required.

Modeling of behavioral networks is mostly described by differential equations of rise or decay in concentration of biochemical species of interest. The equation parameters describing molecule synthesis, decay, transformation, and transport (rates of reaction and diffusion, activation thresholds, etc.) must be identified using available data. [102] Hybrid systems that incorporate both continuous and discrete dynamics model both the logic of interactions and dynamics of biochemical changes, simplifying the parameter identification problem.

102. Parameter identification can be done by solving the model differential equations, starting from a guess for the numerical values of the parameters, and solving the equations using suitable numerical algorithms. The simulation results are then compared with experimental data. If the results do not match, a new guess is made for a new simulation. In the absence of a suitable parameter set, the model must be modified.

A cascade model of 2-state *reaction-diffusion equations* aimed at representing morphogen gradient has been developed [99]. A cascade of extracellular proteolysis reactions generates a concentration field (ligand gradient). Diffusible extracellular ligand (ligand mobile state) interrupts diffusion (ligand immobile state) when it binds to its cognate receptor and activates its signaling effector, which moves within the cell (with an intracellular mediator gradient) until it reaches the cell nucleus (mediator immobile state). Ligand can afterward dissociate from its receptor or undergo endocytosis (new ligand mobile state). Diffusible intracellular mediator travels in (mediator mobile state) and out (new mediator mobile state) of the nucleus. Each stage of a cascade then deals with a diffusible particle (ligand and mediator) that reversibly binds to immobile traps (plasmalemmal receptors and cell nuclei). Immobile particles (extracellular ligands) in the first stage initiate the production of mobile, active particles (intracellular mediators) from a passive form in the second stage. The output (i.e., spatial distribution of immobile particles) of a stage yields an input to the next stage. The cascade is initiated by injection of mobile particles.

Simulations combined with experimental techniques for assessment of kinetic coefficients of molecular interactions can be aimed at understanding biological complexity at the molecular levels. The chemical reaction theory based on the mass action kinetics [103] and existence of steady states has been extended to complex chemical reaction sets that are split into submodules, which are often analyzed separately.

The huge number of data in signal transduction pathways as well as the complexity both in storeyed architecture rich in interconnections (hubs in addition to simple pathway protein nodes) and functioning with feedback loops leads to a reductionist strategy that designs meaningful signaling networks.

The temporal dynamics of signaling networks are commonly described by chemical kinetics equations (currently ordinary differential equations [ODE]), one for each time-dependent molecule involved in the chemical transformation. The concentration depends on production and consumption, both having given rates. Given initial concentrations $c_i(0)$ for each chemical species i and the kinetic coefficients k_j (forward reaction) and k'_j (backward reaction) for each reaction j, the time changes in concentration are determined by solving a system of ordinary differential equations derived from the studied biochemical reaction cascade. A set of coupled ordinary differential equations for the ensemble of elementary reactions with forward and reverse rate constants usually givcs thc rates of production and consumption of individual chemical species, using mass action kinetics. Time integration of equations yields the concentration of each implicated species at given times.

Differential equation-based models are difficult to parameterize realistically. Logical models of regulatory networks that are the simplest type are then proposed. They represent the local state of each involved entity (molecules and genes) at any time as a discrete level. A logic table that represents an instruction set is assigned to

103. The mass action law states that reaction rates are proportional to the concentrations of the reacting species. More precisely, chemical reaction rates are defined as the product of a rate constant and the concentration of the reactants. The calibration stage is aimed at adjusting the model parameters to experimental data.

each node (entity). The system evolution is assumed to occur in discrete time steps during which entity levels are updated. The global state of the model corresponds to the combination of entity states. A sequence of consecutive global states represents a trajectory. In a *Boolean model*, an entity can attain 2 alternative levels: either expressed or active (1), or unexpressed or inactive (0). The level of each entity is updated according to the levels of other entities using a specific Boolean function. Boolean networks can be used to analyze the relationship between regulation functions and network stability. Regulatory relationships between network components are usually derived from experimental data. However, the analysis can yield a large number of regulation functions that are equally consistent with experimental data.

A qualitative, discrete Boolean model of signal transduction can be elaborated to avoid the problem of parameterizing a quantitative model [81]. The logic at each node that is expressed by on–off law gives the main drawback of the model. However, entities that exhibit complex regulation can be represented by multiple nodes, each corresponding to a distinct state of these entities. The strong downside of Boolean models of biochemical reaction cascades is due to data size: for any node with N inputs, there are $2N$ combinations of those inputs in each logic table. Combinations that cannot be derived in a straightforward manner from the literature must be deduced indirectly. Dose–response curves are aimed at demonstrating agreements between qualitative results of the Boolean model and the observed facts.

Probabilistic Boolean models are developed to express uncertainty in the regulatory logic as several candidate regulatory functions can be associated with a given entity. Each regulation function has a probability depending on its compatibility with prior data. At each time step, each entity is combined with a regulation function that is randomly selected according to the defined probabilities. A probabilistic Boolean model of regulated networks generates a sequence of global states that constitutes a *Markov chain* [98]. [104]

Another type of nanoscopic scale modeling focuses on protein motions. Two main approaches carry out simulations of protein motions. In molecular dynamics, molecules interact at a given temperature following the laws of classical mechanics. In normal mode analysis, the simple harmonic motions of the molecule about a local energy minimum are calculated. Other commonly used techniques are continuum electrostatic calculations and Brownian dynamics, a variant of molecular dynamics.

Physicochemical models of cell signaling pathways of short-term vascular regulation and long-term vascular adaptation can be based on deterministic or stochastic forms of ordinary and partial differential equations. These models not only describe the set of implicated elementary reactions, following a reductionist strategy (after a selection of the degree of details), but also changes in location, [105] as well as other non-enzymatic changes of state, such as assembling into multicomponent complexes.

104. A Markov chain is a stochastic process in which future states for a given present state are independent of past states, as the present state yields the complete information that influences the evolution of the process.

105. Between-compartment transport is done with a given rate, whereas the transport is assumed to be instantaneous within a compartment. Otherwise, species concentrations must be described with respect to space using partial differential equations.

1.4 MicroRNAs in Cell Signaling

MicroRNAs (miRs) are 20- to 25-nucleotide-long, non-coding RNAs (Vol. 1 – Chap. 5. Protein Synthesis).[106] Most MIR genes are located at intergenic regions, whereas other miRs (25–30%) are embedded within introns of coding genes. Because microRNAs participate in the post-transcriptional control of gene expression, they influence cell responsiveness to extracellular signals that activate genes to produce signaling mediators, including plasmalemmal receptors (Table 1.9).[107]

MicroRNAs repress some signaling pathways in the absence of stimulation, especially those that must be expressed at low levels. *Default repression* ensures that gene expression is activated exclusively in the presence of signaling effector, but repressed in its absence or in the presence of inputs that are too weak or too transient [100]. MicroRNAs can buffer the effect of signaling mediators, as they can raise the activation threshold to restrict the response to appropriate zones of competence. This damping effect occurs early, before the synthesis and translocation of messengers, i.e., at the earliest stage of the deployment of the signaling program. On the other hand, microRNAs can operate in *default activation*. The suppression of signaling mediators can actually relieve an inhibition and cause the gene transcription. Some miRs target inhibitors of signaling cascades downstream from receptor Tyr kinases (Chap. 8). The specific microRNA milieu of each cell type contributes to suit cell responses.

Coherent regulation means that, upon stimulation, the synthesis of a mediator is triggered, once the repression by a miR is relieved, as the miR inhibitor is repressed (coherent feed-forward loop). *Incoherent regulation* refers to simultaneous stimulation of a mediator and its miR inhibitor (incoherent feed-forward loop). *Reciprocal inhibition* (bistable behavior) corresponds to the activation of a transcription factor that inhibits the expression of a miR, which, in turn, prevents the expression of this transcription factor. Reciprocal inhibition between transcription factors zinc finger

---

106. MicroRNAs primary transcripts (pri-miR) are long precursors produced by RNA polymerase-2 that fold on themselves to form one or more hairpin structures. Pri-miRs are then processed in the nucleus into precursor miRs (pre-miR) by the Drosha endonuclease complex that recognizes hairpin structures. Pre-miRs translocate to the cytoplasm using exportin-5. In the cytosol, pre-miRs are detected by Dicer ribonuclease and the RISC-loading complex subunit transactivation responsive RNA (TAR)-binding protein (TRBP or TARBP2) and processed into about 20-nucleotide mature miRs. Generally, a single strand is selected as the bioactive mature miR and the other one is degraded (strand-selection stage). Mature miRs then complex with Argonaute proteins to form RNA-induced silencing complexes (RISC). Once incorporated in RISC, single-stranded miRs are unwound by Dicer to prevent translation from target mRNAs.

107. During embryo- and fetogenesis, multiple morphogens (transforming growth factor-β, Wnt, Hedgehog, and Notch) as well as Hippo, or STK3 and STK4 kinases, and signaling cascades primed by receptor Tyr kinases (epidermal, vascular endothelial, and platelet-derived growth factor receptor) synergistically regulate tissue patterning and growth. Peptidylprolyl isomerase-D (or cyclophilin-D) is a Nodal-related TGFβ morphogen in zebrafish that is also controlled by miR430.

Table 1.9. MicroRNAs in cell signaling (Source: [100]; FrmD: FERM domain-containing protein; GPCR: G-protein-coupled receptor; LaTS: large tumor suppressor; LEF1: lymphoid enhancer-binding factor-1; Lefty: left–right determination factor; MAPK: mitogen-activated protein kinase; PI3K: phosphatidylinositol 3-kinase; GY-box: GUCUUCC motif; Brd-box: AGCUUUA motif; K-box: cUGUGAUa motif). Sprouty-related protein with an EVH1 domain SpRED1 is an inhibitor of the Ras–Raf interaction and MAPK signaling. Nodal, a potent morphogen of the TGFβ superfamily that specifies the body axis, has an asymmetrical activity due to miR15 and miR16 that target Nodal activin receptor AcvR2a (at least in Xenopus laevis).

Target	MicroRNAs
	Transforming growth factor-β signaling
AcvR2a	MiR15, miR16
SMAD3/4/5	MiR23b
Lefty1/2	MiR430
	Wnt signaling
T-cell factor	MiR8
LEF1	MiR203
GPCR177	MiR8
	Hedgehog signaling
Smoothened	MiR324-5p, miR125b, miR326
Gli1	MiR324-5p
	Notch signaling
HES1	GY-box, Brd-box, and K-box-targeting miRs
	Hippo signaling
LaTS1/2	MiR372, miR373
FrmD6	MiR278
	Transcription factors
P53	MiR125b
P63	MiR203
	Kinases and phosphatases
PI3KR2	MiR126
PTen	MiR21, miR26a, miR216a, miR217
Sprouty-1	MiR21
SpRED1	MiR126
	Small (monomeric) GTPases
K/N-Ras	Let7

E-box-binding homeobox proteins ZEB1 and ZEB2 and miR200 family members regulates the switch between epithelial and mesenchymal states. [108]

108. Epithelial to mesenchymal transition relies on a gene expression program with loss of cell adhesion. This program is thus activated during embryogenesis and adult tissue remod-

MicroRNAs can modify tissue responsiveness over time.[109] In addition, microRNAs can sharpen messenger gradients and act as signaling amplifiers. MicroRNA clusters can concomitantly target several mediators. Despite their intrinsic weak effect, their simultaneous action on a set of regulators can amplify their effect.[110] MicroRNA126 is the most abundant miR in endothelial cells, in which it sustains signaling from vascular endothelial growth factor by impeding the production of its natural repressors Spred1 and PIK3R2 mRNAs, hence promoting angiogenesis.

MicroRNAs operate as signaling coordinators that allow crosstalk between signaling cascades, as miRs can connect distinct signal transduction axes. A single miR can act simultaneously on 2 signaling cascades to coordinate their activities.[111] MicroRNAs can be regulated by a signaling pathway and, in turn, target a component of another pathway to coordinate its activity.

MicroRNAs can dampen environmental and genetic fluctuations to confer robustness (i.e., generate an invariant reaction) to the cell response to extracellular signals. MicroRNAs can actually operate as signaling balancers and buffers. They can target both an activator and inhibitor of a single pathway [100].

On the other hand, some signaling mediators can contribute to miR processing [100]. Signaling initiated by transforming growth factor-β and bone morphogenetic proteins stimulates SMAD effectors that promote a rapid increase in the concentration of mature miR21. The latter mediates the TGFβ-induced change of phenotype of vascular smooth muscle cells to contractile status.

---

eling. Members of the miR200 family specify the epithelial phenotype, as they prevent the expression of transcriptional repressors of E-cadherin, zinc finger E-box-binding homeobox proteins ZEB1 and ZEB2. The latter repress miR200 primary transcript.

109. Self-renewal of progenitor cells must be restrained when cells need to differentiate. Agent Let7 regulates progenitor cell phenotype, i.e., the transition from undifferentiated to proliferative progenitors. Expression of Let7 is low in progenitor cells, but increases during cell differentiation to progressively preclude signaling from small GTPase Ras [100]. This elevation in Let7 concentration results from an increase in its transcription rate. The marker of undifferentiated human embryonic stem cells Lin28 indeed blocks the cleavage of Let7 pri-miR and pre-miR by Drosha and Dicer, respectively.

110. MicroRNA21 targets both PTen phosphatase and Sprouty, thereby enhancing signaling triggered by receptor Tyr kinases [100].

111. MicroRNA15 and -16 connect Wnt and transforming growth factor-β signaling [100]. MicroRNA203 favors skin regeneration and self-renewal, as it antagonizes both Wnt signaling via the transcriptional cofactor lymphoid enhancer-binding factor LEF1 and P63 activity [100]. In glomerular mesangial cells, TGFβ provokes miR192 expression. MicroRNA192 hinders the transcription factor zinc finger E-box-binding homeobox ZEB2, thereby relieving the repression of miR216a and miR217 that can then inhibit phosphatase and tensin homolog deleted on chromosome 10 (PTen). This inhibition favors the activation of protein kinase-B [100].

1.5 Adenosine Triphosphate

Adenosine triphosphate is a major cell molecule that supplies chemical energy and a phosphate group. Therefore, ATP is permanently produced and consumed inside cells at high rates. In addition to being an energy source for intracellular functions, ATP also functions as a regulator of both intra- and extracellular processes. A steep concentration gradient ($\mathcal{O}$[1 mmol] in the cytosol vs. $\mathcal{O}$[1 nmol] in the extracellular medium) and the transmembrane potential favor ATP release. Most ATP molecules indeed exist in anionic forms at physiological pH.

The ATP messenger furnishes a suitable introduction to signaling agents, as it is released from cells by many types of carriers (vesicles, ion channels, and molecule transporters), and signals to cells using different types of receptors (iono- and metabotropic receptors).

1.5.1 ATP Messenger and Neurotransmitter

Once expelled from the cell, ATP operates as an extracellular signaling molecule and neurotransmitter. In addition to outside–in and inside–out signaling, ATP release via plasmalemmal carriers and exocytosis represent a third type of communication used for both inside–out and outside–in signaling, and, in the latter case, for both auto-, juxta-, and paracrine signaling.

Extracellular ATP messenger and other nucleotides, such as ADP, uridine triphosphate (UTP), and related molecules such as uridine diphosphate sugars, are ubiquitous modulators of cell function. They target nucleotide-activated P2 receptors (Sect. 7.13.3). In addition, ATP can signal via its metabolite adenosine. Extracellular ATP liberated during hypoxia or inflammation can indeed signal directly to purinergic receptors or, after proceesing, can activate adenosine receptors (Sect. 7.13.2).

1.5.2 Basal and Stimulated ATP Release

Basal ATP release can be increased by physical (e.g., osmotic cell swelling or shrinking, stretching, etc.) or chemical stimuli. Non-excitatory cells can release ATP by exocytosis in response to biochemical and mechanical stimuli, in addition to devoted carriers [101]. Exocytotic vesicular release of ATP has been demonstrated for various cell types (vascular endothelial and urothelial cells, osteoblasts, fibroblasts, chromaffin cells, and astrocytes) [102].

Cells release ATP that activates ligand-gated ion channels P2X and G-protein-coupled P2Y receptors for auto- or paracrine regulation. This process establishes the basal level of activation (*set point*) for various signal transduction pathways implicated in tissue perfusion, ion transport, cell volume regulation, neuronal signaling, and immunity [103].

The ATP messenger influences intracellular concentrations of second messengers, such as Ca^{++} and cAMP. It also tunes the activation of protein kinases, such as cAMP-dependent protein kinase-A, protein kinase-C, and Ca^{++}–calmodulin-dependent protein kinases (Vol. 4 – Chap. 2. Preamble to Protein Kinases). The ATP agent thus contributes to ambient levels of activation of signaling mediators that can interfere with other signal transduction pathways [103]. Receptors of ATP can trigger additive or antagonistic actions to those launched by other agonists. It can also desensitize other receptor types. Moreover, released ATP modulates the intracellular concentration of Ca^{++} ions via gap junctions [103].

1.5.3 Cell Volume Control and Molecular Exchanges

Extracellular ATP intervenes on ion transport in numerous cell types, in particular when their osmotic environment changes to maintain their volume. Osmotic stress elicits ATP liberation, hence activation of P2 receptors. Subsequently, a Ca^{++}-dependent control of ion transfer across the plasma membrane takes place, in addition to Ca^{++}-independent purinergic signaling [103].

In erythrocytes, P2 receptors potentiate the regulatory volume decrease primed by cell swelling, as they stimulate Ca^{++}-dependent K^+ efflux as well as extracellular signal-regulated kinases ERK1 and ERK2 [103].

In airway epithelial cells, ATP contributes to the control of the airway surface fluid layer via an autocrine regulation of cystic fibrosis transmembrane conductance regulators (CFTR Cl^- channels) [103] (Sects. 3.5.9 and 4.18.8.2).

In renal epithelial cells, ATP modulates Na^+ transport via its $P2Y_2$ receptors. Activated basolateral $P2Y_2$ inhibits water transport primed by vasopressin in the medullary collecting duct [103].

1.5.4 Cellular Processes for ATP Release

The ATP messenger can be released from cells via: (1) exocytosis of ATP-containing vesicles; (2) transport via connexin and pannexin hemichannels (Vol. 1 – Chap. 7. Plasma Membrane); (3) transfer through nucleoside transporters; and (4) transport through ATP-binding cassette proteins.

The ATP molecule exists mostly as an anion (ATP^{4-} and $^{Mg}ATP^{2-}$) at physiological pH. It can thus be electrogenically translocated through anion channels. Cystic fibrosis transmembrane regulator can regulate ATP release [103]. Chloride channel CFTR is able to bind ATP and possesses an ATP-anion channel activity.

Among volume-activated (VAAC) or -regulated anion (VRAC) channels, volume expansion-sensing outwardly rectifying anion channel (VSOR) is most prominently activated and ubiquitously expressed [104].[112] The unitary conductance, voltage dependency, anion selectivity, pH dependency, and pharmacology of the VSOR anion channel are distinct from those of volume-sensitive ClC2 channel (Sect. 3.5.1).

112. Volume-sensitive outwardly rectifying anion channels are particularly used in pathological circumstances. For example, they serve for glutamate release from astrocytes under ischemia and inflammatory stimuli such as bradykinin via B_2 receptor and reactive oxygen species. Activity of VSORs is modulated by numerous substances, such as kinases EGFR, PI3K, RoCK, and Src, as well as phospholipase-Cγ. Hence, activated AT_1 receptors trigger several signaling axes via EGFR (Ras–Raf–MAP2K–ERK, PI3K–NOx–H_2O_2, and Src–PLCγ–Ca^{++}), in addition to the Rho–RoCK–MLCK pathway to enhance the VSOR activity [104].

Nucleotide-sensitive Cl^- channel ClNS1a (Sect. 3.5.8) may contribute to VSOR current. The VSOR channel participates not only in cell-volume regulation, but also cell proliferation, differentiation, migration, and apoptosis.[113] However, neither CFTR nor VSOR channels are involved in swelling-induced ATP release from human epithelial intestine cells [105].

Swelling-activated Maxi-anion channel is characterized by a voltage-dependent inactivation. It is inhibited by intracellular arachidonic acid. Its wide pore (at the nanoscopic scale) is indispensable to convey ATP^{4-} and $ATP^{Mg^{--}}$ anions [105]. Maxi-anion channel can then function as an ATP-conductive carrier for intercellular purinergic communications. Its molecular identity remains to be fully determined.

The $P2X_7$ receptor is unusual among members of the P2X receptor family. In addition to the opening of the usual cation channel with low ATP concentrations, high concentrations of ATP enable opening of large pores permeable to molecules (up to 900 Da) [102].

1.5.5 Neuroregulator ATP

Adenosine triphosphate provokes various effects in the central nervous system. It also potentiates acetylcholine (ACh) action in the neuromuscular junction (skeletal muscle). In the absence of classical neurotransmitters acetylcholine and noradrenaline, ATP intervenes as a neurotransmitter in many physiological systems, such as the digestive and urinary tracts, vas deferens, and arteries (purinergic neurotransmission to smooth muscle) [106].

Non-synaptic, non-vesicular release of the neurotransmitter adenosine triphosphate from axons can occur through volume-activated anion channels (VAAC) gated by depolarization-induced microscopic axon swelling during action potential firing [107]. This non-vesicular, non-synaptic communication between axons and astrocytes yields an additional, local signaling axis.

1.5.5.1 Neurocotransmitter ATP

In the nervous system, ATP can act as an excitatory cotransmitter. The ATP modulator operates directly and via its metabolites on neuronal, neuroendocrine, and glial cells, especially in sensory transduction (retina, olfactory epithelium, taste buds, and cochlea). In neurons, exocytotic vesicles contain up to 1000 mmol of nucleotides [106].

In purinergic neuromuscular junctions, ATP is synthesized and stored in vesicles in nerve varicosities. The ATP transmitter is released by exocytosis and acts on

113. Cell division and migration is accompanied by cell volume changes. Cells regulate their volume using regulatory volume decrease that is achieved mainly by the release of K^+ and Cl^- ions and osmotically driven water flux. Except in erythrocytes and myocytes, the K^+ conductance is much larger than that of Cl^- anion. Therefore, a marked increase in Cl^- conductance has greater consequence for swollen cells. Osmotic cell swelling activates or upregulates many anion channels, such as Maxi-anion channel, ClC2, and bestrophin oligomeric Cl^- channels (Sect. 3.5.3).

postjunctional P2 purinoceptors on smooth muscle cells. In the synaptic cleft, ATP is catabolized by ATPases and nucleotidase into adenosine. The latter is taken up by nerve varicosities for ATP resynthesis. In addition, adenosine acts on prejunctional P1 purinoceptors of nerve varicosities to modulate transmitter release. Moreover, adenosine can be converted into inosine by adenosine deaminase and removed by blood circulation.

On the other hand, ATP catabolite – adenosine – produced by ectoenzymatic breakdown of ATP attaches to presynaptic P1 receptors to inhibit the release of excitatory neurotransmitters in both the peripheral and central nervous systems.

Purinergic Transmission

Non-adrenergic, non-cholinergic neurotransmission can be ensured by ATP. Neurotransmitter ATP is responsible for *inhibitory junction potentials* at junctions between nerves and smooth muscles. This inhibitory neurotransmitter rapidly causes hyperpolarizations and associated smooth muscle relaxations [106]. In fact, ATP operates as a cotransmitter with vasoactive intestinal polypeptide and nitric oxide in non-adrenergic, non-cholinergic, inhibitory nerves [106]. [114]

Purinergic Cotransmission

Nerves can release several neurotransmitters in proportions that vary in different tissue types. The ATP modulator can be liberated together with one of the classical neurotransmitters acetylcholine and noradrenaline.

ATP–NAd Cotransmission

The ATP coneurotransmitter is released from some sympathetic nerves, in addition to non-adrenergic, non-cholinergic, inhibitory nerves. Noradrenaline (NAd) and ATP are coneurotransmitters in sympathetic nerves that provoke *excitatory junction potentials* [106]. In the hypothalamus, ATP and NAd act synergistically to release vasopressin and oxytocin.

Coneurotransmitters ATP and NAd can be simultaneously released from terminal varicosities of sympathetic nerves. Noradrenaline targets postjunctional $\alpha 1$-adrenoceptor that causes Ca^{++} influx (within the cytosol) via inositol triphosphate. Adenosine triphosphate binds to the $P2X_1$ ion channel that also primes Ca^{++} influx. Cytosolic accumulation of Ca^{++} ions elicits smooth muscle contraction.

ATP–ACh and Other Cotransmissions

Cotransmitters ACh and ATP are released from parasympathetic nerves that supply the bladder. Subpopulations of sensory nerves employ ATP in addition to substance-P and calcitonin gene-related peptide [106].

---

114. Hence, purinergic transmission refers to the absence of cotransmission with the classical neurotransmitters acetylcholine and noradrenaline.

The ATP cotransmitter can also be released with dopamine, γ-amino butyric acid (GABA), glutamate, and serotonin (or 5-hydroxytryptamine) in the central nervous system [106]. In addition to glutamate, ATP is involved in long-term potentiation in hippocampal CA1 neurons for learning and memory.

1.5.5.2 Neuromorphogen ATP

Moreover, extracellular ATP regulates neuron proliferation and differentiation via multiple types of P2 receptors of both the P2X and P2Y classes [106]. Released ATP in particular from dorsal root ganglion neurons activates $P2Y_2$ receptor to serve as a coactivator of neuronal differentiation via the high-affinity neurotrophin NTRK1 receptor (Sect. 8.2.12). The ATP morphogen also promotes cell proliferation of some neuron types. Multiple P2 receptor types intervene in neuronal differentiation and proliferation, such as members of the P2X family. On the other hand, $P2Y_1$ receptor precludes the differentiation of neural progenitor cells [103].

1.5.5.3 Neuroremodeler ATP

In the central nervous system, nucleotide-based signaling is involved in nervous tissue remodeling caused by trauma, stroke, ischemia, or neurodegenerative disorders. In combination with epidermal, fibroblast, and platelet-derived growth factor, ATP can stimulate astrocyte proliferation [106].

Extracellular signaling mediator ATP links neurons to glial cells. Once released from astrocytes, it may trigger cellular responses to injury, as it causes reactive astrogliosis and glial scar formation. It operates via multiple P2X and P2Y receptor types on astrocytes, oligodendrocytes, and immune microglial cells [106].

1.5.5.4 Neuroregulator ATP of Physiological Functions

Interactions between purinergic and nitrergic neurotransmitters participate in the regulation of hormone secretion and body temperature in the hypothalus as well as cardiovascular and respiratory control in the brainstem [106].

1.5.6 ATP Release by Endothelial Cells

The ATP vasoregulator released from endothelial cells that experience hypoxia and abnormal shear stress act on endothelial P2 receptors (autocrine signaling). Subsequently, nitric oxide is liberated and targets adjoining smooth muscle cells that can then relax.

Its derived molecule adenosine contributes to functional and reactive hyperemia, i.e., the increase in blood flow to perfused tissues with an augmented activity and after occlusion, respectively. Functional hyperemia results from vasodilatation by relaxation of arteriolar smooth muscle cells and opening of precapillary sphincters.

Constitutive ATP release from the wetted membrane of endothelial cells modulates the vascular tone via P2X and P2Y receptors. In addition to basal release, increased ATP release from endothelial cells in response to abnormal shear magnitude, hypoxia, or ischemia promotes P2-mediated vasodilation. Arterioles release ATP that is antagonized by ADP that binds to $P2Y_{13}$ receptors on endothelial cells. This mechanism inhibits ATP liberation from erythrocytes as well as further release from the endothelium to prevent an excessive response [103].

Agent ATP that targets the vascular endothelium can be released from both endothelial cells and perivascular nerves. Cotransmitter ATP liberated with noradrenaline from perivascular sympathetic nerves at the adventitia–media border generates a vasoconstriction via $P2X_1$ receptors on smooth muscle cells [103, 106].

In addition, adenosine produced rapidly from ATP by ectoenzymes causes vasodilation. It connects to P1 receptors on smooth muscle cells as well as perivascular nerve terminal varicosities to inhibit transmitter release [106].

Both ATP and adenosine are involved in the control of cell migration, proliferation, and death that are all required during angiogenesis [106].

Compound ADP is also released from aggregated platelets. In addition, it acts on endothelial P2Y receptors to stimulate vasodilation for enhanced blood supply in repair materials.

1.5.7 ATP Release by Thrombocytes

Activated platelets constitute an important source of extracellular adenine nucleotides (ATP, ADP, and AMP) upon degranulation and liberation of dense granule content (ATP and ADP) during inflammation and vessel injury. Extracellular ATP then regulates platelet reactivity via platelet nucleotide receptors or via its metabolite ADP.

Subsequent catabolism of ATP and ADP to AMP and adenosine inhibits platelet aggregation. This inhibition is required in normal conditions. Both endothelial, membrane-bound, ectoATP/ADPase ENTPD1 and plasma nucleotidases regulate concentrations of circulating adenine nucleotides. Ectonucleoside triphosphate diphosphohydrolases ENTPD1 hence controls hemostasis, as it promotes ADP formation from released ATP [108]. Nonetheless, a plasma nucleotidase hydrolyzes ATP directly to AMP, thereby preventing ADP accumulation. Moreover, generated adenosine is a potent, local inhibitor of platelet reactivity that avoid thrombus formation.

Depolarization of endothelial cells causes the endothelial production of O_2^- that hinders the activity of endothelial ectonucleotidases [109]. Superoxide dismutase antagonizes this process. Increased transmural pressure also causes endothelial depolarization. Depolarization of the endothelial interface favors platelet aggregation.

1.5.8 ATP Release by Leukocytes

During inflammation, activated neutrophils release soluble secretagog adenosine monophosphate (AMP), in addition to ATP, that primes electrogenic chloride secretion (and water transport) in epithelial cells [110].

Agent AMP is converted into adenosine by the plasmalemmal enzyme adenosine ecto-5′-nucleotidase. Adenosine blocks AMP activity, but activates its cognate receptors for para- and autocrine effects. On the other hand, ectonucleoside triphosphate diphosphohydrolase that lodges on the endothelium hydrolyzes extracellular ATP and ADP into AMP. All ectonucleoside triphosphate diphosphohydrolases (ENTPD1–ENTPD3 and ENTPD5–ENTPD6) are widely distributed.

1.5.9 ATP Release by Erythrocytes

Erythrocytes release ATP into blood in response to small changes in pH, oxygen concentration, or osmotic pressure. Erythrocytes participate in the regulation of blood pressure by releasing ATP that acts as a vasodilator in response to increased shear stress [111]. Mechanical stress-induced ATP release is triggered by deformation of the plasma membrane. The dynamics of deformation-induced ATP release are characterized by the following features [112]: (1) a significant time delay occurs between the onset of increased shear stress and ATP release; (2) a rising RBC rigidity decreases the amount of ATP released, without modifying the delay; and (3) the critical duration of increased shear stress needed to activate ATP release correlates with a membrane relaxation time.

Release of ATP is mechanically coupled to the retraction of the cell membrane and not only on its deformation. Therefore, the retraction of the spectrin–actin cytoskeleton triggers the mechanosensitive ATP release and a shear-dependent membrane viscosity controls the rate of release. Release of ATP by mechanotransduction relies on 2 distinct time scales. The activation time is estimated to be 3 to 6 ms. Deformations thus need to occur over longer time to induce ATP release. The time required for ATP release with respect to the onset of increased shear stress varies from 25 to 75 ms according to the shear stress magnitude. It remains insensitive to the plasma membrane rheology. This time constant matches the regulation requirements when the flow velocity is slow enough so that the release occurs in a vascular segment close to the high shear region.

1.5.10 Target Receptors of Extracellular ATP

Plasmalemmal receptors activated by extracellular ATP belong to 2 categories: P2X ion channel (Sect. 2.5.7) and G-protein-coupled P2Y receptors (Sect. 7.13.3).[115] Some receptors mediate fast, transient signaling for secretion, neurotransmission, and neuromodulation. Others mediate slow, sustained signaling for cell proliferation, differentiation, motility, and apoptosis. Messenger ATP preferentially activates $P2Y_2$ and $P2Y_{11}$ [103]. Nevertheless, ATP and its derived products interact with other P2Y receptors. Furthermore, ATP activates all of the 7 P2X homo- or heteromeric receptors.

115. Receptor subtypes for purines and pyrimidines encompass: 4 P1 adenosine receptors (A_1–A_3), 7 homo- and heteromultimeric P2X ionotropic receptors ($P2X_1$–$P2X_7$), and 8 P2Y metabotropic receptors ($P2Y_1$, $P2Y_2$, $P2Y_4$, $P2Y_6$, and $P2Y_{11}$–$P2Y_{14}$).

1.5.11 Extracellular Metabolism of Nucleotides

Extracellular ATP is hydrolyzed by membrane ectonucleotidases. These enzymes constitute several families: (1) ectonucleoside triphosphate diphosphohydrolases (ENTPD); (2) ectonucleotide pyrophosphatases (ENPP); (3) alkaline phosphatases (AlP); and (4) ecto-5′-nucleotidase (NT5E).[116] They thus control extracellular nucleotide concentrations and the activation of type-2 purinergic receptors (P2). On the other hand, ectonucleoside diphosphokinase (ENDPK) converts diphosphate nucleotides into triphosphate nucleotides.

Ectonucleotideases hydrolyze the 4 nucleotides ATP, ADP, UTP, and UDP with similar values of Michaelis coefficient (k_M) and maximum velocity (V_{max} [113]. Conversely, both ADP and UDP are effective substrates of ectonucleoside diphosphokinase that gives rise to ATP and UTP. The rate of extracellular phosphorylation by ectonucleoside diphosphokinases exceeds (up to 20-fold) that of nucleotide hydrolysis by ectonucleotidases [113].

Adenylate compounds tether to metabotropic P2Y receptors. Adenosine, the final product of ectoATPases, activates its cognate P1 receptors. Both P1 or P2Y can hetero-oligomerize with ectonucleotidases on cell surface.

The concentrations of extracellular nucleotides, and thus activation of P2 receptor, are regulated by ectonucleotidases that are membrane-bound enzymes with an extracellular catalytic site.

Members of a first family of ectonucleotideases, *ectonucleoside triphosphate diphosphohydrolases*[117] hydrolyze nucleotide di- and triphosphates into nucleotide monophosphates. In humans, 5 isoenzymes exist: ENTPD1,[118] ENTPD2,[119] ENTPD3,[120] ENTPD5,[121] and ENTPD6.[122]

The ENTPD1 enzyme synthesized in vascular endothelial and smooth muscle cells hydrolyzes ATP directly into AMP as well as ADP to AMP and UTP into UDP; ENTPD2, possibly localized to the surface of adventitial fibroblasts, converts ATP to ADP. Nucleotides UTP and UDP that are equipotent vasoconstrictors are released from endothelial cells and activated platelets.

Five members of a second family of ectonucleotidases — the *ectonucleotide pyrophosphatases and phosphodiesterases* (ENPP) — convert nucleotide triphosphates into nucleotide monophosphates and extracellular inorganic pyrophosphate as well as cyclic adenosine monophosphate into adenosine (Sect. 7.13.2).

Ecto-alkaline phosphatases, AlP^I (intestinal), AlP^L (tissue non-specific, liver in particular), AlP^{Pl} (placental), and AlP^{PlL2} (placental-like-2), that pertain to a third class of ectonucleotideases convert AMP into adenosine.

116. A.k.a. 5′-ribonucleotide phosphohydrolase or CD73.
117. A.k.a. NTPDases, ATPDases, as well as CD39 and CD39-like molecules CD39L1–CD39L4; Vol. 4 – Chap. 1. Signaling Lipids.
118. A.k.a. NTPDase1 and CD39.
119. A.k.a. CD39L1.
120. A.k.a. CD39L3.
121. A.k.a. CD39L4.
122. A.k.a. CD39L2.

Ecto-5′-*nucleotidase* is anchored to the plasma membrane by a glycosyl-phosphatidylinositol linkage. At neutral pH, it catalyzes the conversion of purine 5-′mononucleotides to nucleosides. Agent AMP is the preferred substrate of NT5E that transforms AMP into adenosine.

Nucleoside diphosphate kinases (NDPK) contribute to the maintenance of the intracellular pool of deoxynucleotide triphosphates (dNTP) and nucleotide triphosphates via the transfer of phosphate from NTP donor to NDP acceptor. They accept γ-phosphate from both ATP and GTP providers. Some species of these enzymes have functions unrelated to their catalytic activity. Isoenzyme NDPKb is a DNA-binding protein that acts as a transcriptional activator of the human MYC gene. Nucleoside diphosphate kinases are typically intracellular enzymes. However, ectoNDPKs reside on the cell surface.

Receptors of the apical surface of human airway epithelial cells can be targeted by nucleotides such as UTP delivered from UDP-containing aerosols to elicit Cl^- secretion in individuals with cystic fibrosis. In the lumen of the respiratory tract, ectonucleoside diphosphokinase promotes the conversion of UDP into UTP in the presence of ATP [114]. The UTP ligand selectively activates the $P2Y_4$ receptor. It is also the most potent agonist for $P2Y_2$ receptors. In human nasal epithelial cells, UDP targets a UDP receptor that differs from the $P2Y_2$ receptor [114]. [123]

123. The $P2Y_6$ receptor is potently and selectively activated by UDP. It resides in several tissues, such as the heart, lung, kidney, digestive tract, and spleen, as well as vascular smooth muscle cells. In the absence of UTP, UDP remains inactive on $P2Y_2$ and $P2Y_4$ receptors [114].

2

Ion Carriers

Any cell is enclosed by a plasma membrane impermeable to many life-sustaining substances. Inside the cell, organelles are wrapped in membranes (Vol. 1 – Chap. 4. Cell Structure and Function); mitochondria and the nucleus are surrounded by an envelope, i.e., a double membrane. Cell membranes are barriers to ion flux. Cellular functioning results from the interplay between membrane proteins, such as cognate receptors of numerous compounds, substance and ion carriers on the one hand, and signaling molecules, nutrient and waste chemicals, and ions on the other. Activity of these membrane proteins, carriers and receptors is partly modulated by exocytosis (Vol. 1 – Chap. 9. Intracellular Transport).[1] In particular, the ionic composition of the cytosol and intracellular organelles is regulated by the transfer of ion carriers.

2.1 Connexins and Pannexins

Connexins and pannexins constitute 2 families of proteins involved in intercellular communication. Connexins mainly form intercellular channels, the so-called gap junctions (Vol. 1 – Chap. 7. Plasma Membrane), that bridge adjoining cells across the extracellular space. Pannexins generate non-junctional channels that operate as paracrine regulators, as they release adenosine triphosphate (ATP), thereby modulating the range of intercellular Ca^{++} transmission between cells such as astrocytes.

2.1.1 Connexins

Connexins are encoded by 21 genes.[2] Connexins are commonly named according to their molecular weights (kDa). Six connexins form hemichannels, or connex-

1. Late stages of membrane protein processing such as appropriate folding and assembling as well as post-translational modifications happen in the endoplasmic reticulum and Golgi body. Delivery of vesicular cargos to the suitable membrane is ensured by exocytosis that needs a cascade of proteic interactions for proper sorting.

2. Connexins include Cx23, Cx25, Cx26, Cx30.1 to Cx30.3, Cx31, Cx31.1, Cx31.9, Cx32, Cx36, Cx37, Cx40, Cx40.1, Cx43, Cx45 to Cx47, Cx50, Cx59, and Cx62.

ons, that assemble to build a gap junction. Connexins combine into both homo- and heteromers with distinct functional properties.

Gap junction is a channel between interiors of coupled cells (inner diameter of ~1.4 nm) composed of 2 apposed connexons. It enables the diffusion of molecules (weight up to 1 kDa), such as some nutrients, metabolites (ATP, glucose, glutamate, and lactate), second messengers (Ca^{++}, inositol triphosphate [I(1,4,5)P_3], and cyclic adenosine monophosphate [cAMP]), as well as ions.

Gap junctions are involved in many processes, such as myocyte contraction, regulation of neuronal excitability, and control of epithelial electrolyte transport. Different connexin compositions explain differences in conductance, selectivity, and/or asymmetric voltage gating.

Gating of gap junctions is voltage sensitive [115]. Maximal junctional conductance occurs at zero transjunctional voltage and declines in response to hyperpolarization or depolarization of either coupled cell.

Each apposed hemichannel has 2 mechanisms of gating in response to transjunctional voltage: the fast and slow (or loop) gate. In addition, gap junctions can be gated by intracellular H^+, Ca^{++}, post-translational modifications, and some chemical agents. Changes in intracellular pH modify significantly electrical and metabolic intercellular communications through gap junctions.[3] Modulation of the voltage-gating properties can explain H^+-dependent change in junctional conductance.

2.1.2 Pannexins

Pannexins (Px1–Px3) contribute to gap junction structure in humans [5]. In addition, they form channels to release ATP in erythrocytes and taste receptor cells as well as epoxyeicosatrienoic acids in erythrocytes.[4] They are involved in early stages of innate immunity, as they interact with $P2X_7$ purinergic receptor (Sect. 2.5.7) to liberate interleukin-1β [118] (Vol. 2 – Chap. 3. Growth Factors).

Pannexins Px1 and Px2 are widely expressed in the brain. Non-junctional channels, i.e., pannexons and some connexons, such as Cx46 and Cx50, open by membrane depolarization [115]. However, their activation threshold depends on channel type.

Erythrocyte release ATP through pannexin-1 in response to hypoxia as well as activated I prostanoid receptor using the AC–cAMP–PKA pathway [119].

---

3. Junctional conductance and transjunctional voltage dependence of junctional conductance are influenced by intracellular pH, at least in connexin-45 homotypic and Cx45–Cx43 heterotypic gap junctions [116].

4. Erythrocytes, or red blood cells, are reservoirs of vasodilatory, anti-aggregatory, and anti-inflammatory epoxyeicosatrienoic acids (EETs). These lipids are produced by epoxidation of arachidonic acid using hemoglobin. Synthesis and release of EETs from erythrocytes are mediated by ATP stimulation of $P2X_7$ receptors coupled to ATP transporters, pannexin-1, and CFTR channels [117].

2.2 Ion Carriers

Ion carriers, i.e., ion channels, transporters, exchangers, and pumps, convey ions across cellular membranes. In this book set, *carrier* defines any transmembrane transport process; otherwise, the terminology is given by the Nomenclature Committee of the International Union of Biochemistry and Molecular Biology.

2.2.1 Ion Carriers in Cell Signaling

Ions are ubiquitously involved in cell signaling, but must cross cellular membranes to exert their effects. Therefore, ion carriers that reside in cellular membranes act as signaling mediators. Activity of ion carriers is influenced by phospholipid-mediated activation, protein kinase-induced phosphorylation, and ion-based modulation. Ion carriers can be connected to or indirectly interact with membrane receptors.[5] Multiple generated ionic currents contribute to the diversity in signaling axes that are activated by G-protein-coupled receptors or receptor Tyr kinases. Plasmalemmal receptors can use ion channels as effectors. Different receptors can exhibit signaling convergence when they target the same ion channel.

Among ions, divalent cations, especially Ca^{++}, serve as messengers. Intracellular calcium ions actually participate in the regulation of nearly every aspect of cell life. Calcium ions can be delivered to Ca^{++}-sensitive effectors by influx through plasmalemmal ion channels as well as Ca^{++} release from intracellular store. In many non-excitable cells, activation of phospholipase-C-linked receptors causes a biphasic increase in cytosolic Ca^{++} concentration: (1) an initial, transient elevation owing to the release of Ca^{++} from the endoplasmic reticulum (receptor-operated cytosolic Ca^{++} influx) that is followed by (2) a much smaller, sustained augmentation, as Ca^{++} depletion in cell stores primes opening of store-operated Ca^{++} channels in the plasma membrane. Both receptor- and store-operated calcium entry rely, at least partly, on the TRPC superfamily of ion channels. In addition, voltage-gated Ca^{++} channels are implicated in excitation–contraction coupling in both cardiomyocytes and smooth muscle cells. Voltage-gated potassium channels contribute to changes in transmembrane voltage (membrane potential) of excitable cells, such as neurons and myocytes, and more precisely, cell repolarization, interspike membrane potential, and action potential frequency, i.e., signal transmission.

2.2.2 Types of Ion Carriers

Membrane transport is done via multiple proteic carrier types (Table 2.1). Coupled transporters (antiporters and symporters) switch on the flux of one ion species against its electrochemical gradient with the reverse motion of another. An *antiporter* moves 2 ions in opposite directions across the membrane. An *exporter* functions in the outward direction; an *importer* inwardly. A *pore* enables a non-specific passage

5. Ion carriers can be directly linked to complexes of G-protein-coupled receptors, G proteins, and ion channels.

Table 2.1. Membrane carrier proteins bind a substance and transport it across the membrane, undergoing conformational changes. Active transport proteins — ion pumps and exchangers — that carry ions against their concentration gradients constitute a single category of ion carriers. Ion pumps are sometimes called primary active carriers to differentiate from secondary active carriers, such as cotransporters, or symporters, and countertransporters, or exchangers. Ion pumps bind ATP directly (direct active transport). An ion pump that conveys a single or several ions is also called an ion transporter; an ion pump can be a uniporter or an antiporter. Some ion pumps exchange ions. Ion transporters (antiport and symport pumps) use the energy created by ion pumps (indirect active transport).

	Passive ion transport
Channel	Water-filled proteic pore (for a specific ion or a small molecule)
	Active ion transport
Pump	Energy-consuming carrier (ATPase for ions or small molecules)
Transporter	Specific active carrier
Exporter	Cellular efflux
Importer	Cellular influx
	Direct active ion transport
Uniporter	Transport of a single molecule
	Indirect active ion transport
Antiporter	Coupled transport in the opposite direction (ions or small molecules, simultaneously or sequentially)
Symporter	Coupled transport in the same direction

of solutes of different sizes. A *symporter* simultaneously carries 2 chemical species in the same direction, at least one of them being ionic and driven by its electrochemical potential gradient. A *Uniporter* displaces a single chemical species across the cellular membrane.

2.2.3 Transmembrane Transporters

A *transporter* is a membrane protein for a specific chemical species. In humans, about 40 families of secondary transporters participate in many functions (e.g., uptake of nutrients in intestine, transport of ions in kidneys such as Na^+–Cl^- symporter in distal convoluted tubule, and removal of neurotransmitters from synaptic clefts such as serotonin transporter).

A transmembrane transport protein can be defined by its mechanism of action, set of substrates, polarity of transport, regulation, subcellular location, and possible energy source. Although transporters of a given family can have several modes of action or use a mechanism distinct from that of other family members, most often family members share similar function and action mechanism.

Transporters can: (1) act by diffusion without energy input; (2) couple electrochemical potential gradients of ions to solute ingress (secondary active transport); or (3) use chemical reactions (primary active transport). Primary active transporters uses the energy released from light, redox reactions, or adenosine triphosphate hydrolysis to translocate substrates across the membrane. Secondary active ion gradient-coupled transporters are proteins that use the free energy stored in an ion gradient for substrate transport to allow the passage of specific solutes, such as nutrients and neurotransmitters, across the cell membrane by transferring them in association with one or more cations, often sodium ions. Transporters carry solutes against their chemical or electrochemical potential gradient.

A alternating access mechanism (i.e., alternative opening and closing to extra- and intracellular media) arises after substrate binding at a single site.[6] The binding site opens alternately to the one and, then, the other membrane side as a result of reciprocal opening and closing of cavities connecting the binding site to either side of the membrane.

Two kinds of inhibitors of secondary transporters exist [120]: competitive and non-competitive. In competitive inhibition, the substrate-binding site and inhibitor site overlap. Competitive inhibitors stabilize an opening of the extracellular side of the transporter. Non-competitive inhibitors stabilize the occluded state of secondary transporters and prevent further conformational change that is needed for the transport cycle.

According to the transporter classification approved by the International Union of Biochemistry and Molecular Biology (www.tcdb.org) and Human Genome Nomenclature Committee, transporters are designated by a 5-character code [121] (Tables 2.2 to 2.12): (1) a first number N1 gives transporter class (channel, carrier, primary active transporter, group translocator, or transmembrane electron flow carrier); (2) a letter L1 yields transporter subclass; (3) a second number N2 corresponds to transporter superfamily; (4) a third number N3 to transporter family; and (5) a fourth number N4 to transporter type.

Ion pumps are also called primary pumps. Secondary pumps such as *cotransporters* (countertransporters or exchangers) exploit the energy stored in electrochemical gradient of ions (currently Na^+) by coupling movement of these ions to adverse transport of another substrate.

2.2.4 Ion Carrier Features

The motion of ions across cell membranes ensures various physiological processes, such as action-potential generation and propagation, hormone secretion, muscle contraction, regulation of blood volume and pressure, control of electrolyte and

6. This binding site is located at the transporter center and surrounded by inner helices. It is flanked by 2 gates that control access to the outside and inside of the cell. Only one of these gates is able to open at a given time, allowing substrate or ion to reach the binding site.

Table 2.2. Transporter superfamilies (Source: Transporter Classification Database; www.tcdb.org; *i*: number [N2.N3.N4]). Transporters are designated by a 5-character code: (1) a first number N1 gives transporter class (channel, carrier, primary active transporter, group translocator, or transmembrane electron flow carrier); (2) a letter L1 yields transporter subclass; (3) a second number N2 corresponds to transporter superfamily; (4) a third number N3 to transporter family; and (5) a fourth number N4 to transporter type.

Superfamily	Code
Aerolysin	1.C.*i*
APC	2.A.*i*
BART	2.A.*i*
Cecropin	1.C.*i*
CPA	2.A.*i* and 3.B.*i*
Defensin	1.C.*i*
Huwentoxin	8.B.*i*
IT	2.A.*i* and 9.B.*i*
LysE	2.A.*i*
Melittin	1.C.*i*
MerTCFH	9.A.*i*
MFS	2.A.*i*
OmpG	1.B.*i*
Opr	1.B.*i*
PTS-AG	4.A.*i*
RTX-toxin	1.C.*i*
VIC	1.A.*i*
YaeT/TpsB	1.B.*i*

water balance, as well as hydrogen ion control (or acidobasic equilibrium), fertilization, immune responses, and cell proliferation (cell division) and apoptosis.[7]

The opening and closing of ion channels and pumps can be disturbed by gene mutations, leading to alterations in ion conductance through the central pore and possible ion leakage (*channelopathies*).

Differences in ion concentrations or electrical potential between both sides of the plasma membrane, created by proteic ion carriers, produce electrical signals in excitable cells.[8] Changes in membrane electrical polarization occur when ion channels and pumps open.

Ion carriers are characterized by size, charge, and number of transported ions. Ion current depends on the sign of ionic charge, magnitude, and direction of their transmembrane concentration gradient, and strength and sign of the transmembrane electric potential difference. Controlled ion motions into and out of the cell as well

7. Other activities associated with ion fluxes also comprise sensory transduction, learning, memory, etc.

8. Electrical potential difference exists across the plasma membrane, the negative side being on the cytosolic face of the membrane. They are expressed in mV, with a negative sign.

Table 2.3. Transporters in humans: annexins and ATP-binding cassette transporters (**Part 1**; Source: Transporter Classification Database; www.tcdb.org).

Set	Code
	Annexins
ANXA1–A11 and A13	1.A.31.1.1
	ATP-binding cassette transporters
	Subfamily A
ABCA1–12	3.A.1.211.1 and -3 (ABCA2 and -9)
	Subfamily B (MDR/TAP)
ABCB1, -4–10, -10P, 11	3.A.1.201.x, 3.A.1.209.1, 3.A.1.210.x, 3.A.1.212.1
	Subfamily C (CFTR/MRP)
ABCC1–6, -8–13	3.A.1.208.x
	Subfamily D (ALD)
ABCD1–4, 1P1–P4	3.A.1.203.x
	Subfamily E (OABP)
ABCE1	3.A.1.120.4
	Subfamily F (GCN20)
ABCF1–3	3.A.1.120.x
	Subfamily G (WHITE)
ABCG1–5, -8	3.A.1.204.x

as cell organelles use 2 main types of carriers — ion channels and pumps — that are responsible for regulated passive and active transport, respectively.

According to ion types, ion flux and its associated transmembrane electric current change the membrane potential and additionally initiates a signaling cascade as the ion can serve as a messenger (e.g., Ca^{++}). A current of Cl^- usually aims to stabilize membrane potentials or to facilitate transmembrane flux of electrolytes and water.

2.2.5 Ion Channels and Pumps

Ion channels and pumps differ by their structural and functional features [208]. (1) An ion pump has at least 2 gates that must open and close alternately to provide access to its binding sites from only one side of the membrane at a given instant, whereas an ion channel may contain a single gate. However, ion channels can have an additional gate. Voltage-gated K^+, Na^+, and Ca^{++} channels contain an activation gate connected to a voltage sensor and a separate inactivation gate. Closure of a single gate impedes ion flux.[9] (2) The region of cation channel that recognizes and selects ions to be transported is separated from the activation gate, whereas the ion-binding locus is remodeled at each alternation of access in cation pumps. However, slight differences distinguish anion channels and pumps. (3) Selected ions diffuse rapidly

9. Homodimeric Cl^-–H^+ pump is a double-barrelled protein that comprises fast gates (timescale $O[10\,\text{ms}]$) at each barrel end and a common slow gate (timescale $O[1\,\text{s}]$).

Table 2.4. Transporters in humans: pumps or ATPases (**Part 2**; Source: Transporter Classification Database; www.tcdb.org).

Set	Code
Class 1	
Type 8A, aminophospholipid transporter (APLT)	
ATP8A1–A2	3.A.3.8.1
Type 8B	
ATP8B1–8B5	3.A.3.8.1
Class 2	
ATP9A–9B	3.A.3.8.x
Class 5	
ATP10A, -10B, -10C	3.A.3.8.1
Class 6	
ATP11A–11C	3.A.3.8.1
Na^+–K^+ transporting (α and β)	
ATP1A1–1A4	3.A.3.1.1
ATP1B1–1B4, -1B3P1, -1BL1	3.A.3.1.x
H^+–K^+ transporting (α and β)	
ATP12A	3.A.3.1.1
H^+–K^+ exchanging	
ATP4A–4B	3.A.3.1.2
Ca^{++} transporting	
ATP2A1–2A3	3.A.3.2.7
Plasma membrane	
ATP2B1–2B4	3.A.3.2.1
Type 2C	
ATP2C1	3.A.3.2.5
Mg^{++} transporting	
ATP3	

through open ion channels with a speed that can approach that of the diffusion limit, whereas they slowly move through ion pumps.

A given family of ion carriers can be constituted by different types of carriers [208].[10] Certain ion carriers can simultaneously display both pump and chan-

10. The ClC set includes both Cl^- channels and Cl^-–H^+ pumps. Cystic fibrosis transmembrane conductance regulator is the single member of the set of ATP-binding cassette proteins that functions as an ion channel (for Cl^-), whereas the others are mostly ATP-dependent transporters. Binding of ATP opens the pore of cystic fibrosis transmembrane conductance regulator, whereas it closes ATP-binding cassette proteins.

Table 2.5. Transporters in humans: ion channels (**Part 3**; Source: Transporter Classification Database; www.tcdb.org).

Set	Code
ATP synthase, H^+ transporting, mitochondrial	
ATP5A1–5A2, -5AL1–L2, -5AP1–P3,	3.A.2.1.3
ATP5B, -5BL1–L2, ATP5C1–C2, ATP5D,	3.A.2.1.3
ATP5E, -5EP1, ATPF1, ATP5G1–5G3, -5GP1–P4,	3.A.2.1.3
ATP5H, ATP5I, ATP5J, -5J2, -5J2LP, -5J2P2–P3,	3.A.2.1.3
ATP5O, ATP5S	3.A.2.1.3
H^+ transporting, lysosomal	
ATP6AP1–AP2	
ATP6V0A1–6V0A2, -6V0A4,	3.A.2.2.3
ATP6V0B–C, -6V0D1–D2, -6V0E	3.A.2.2.3
ATP6V1A, -6V1B1–B2, -6V1C1–C2,	3.A.2.2.3
ATP6V1D, -6V1E1–E2, -EL1, -EP1–EP2,	3.A.2.2.3
ATP6V1F, -6V1G1–G3, -GP1, -6V1H	3.A.2.2.3
Cu^{++} transporting	
ATP7A–7B	3.A.3.5.x
ATP synthase mitochondrial F1 complex assembly factor	
ATPAF1–AF2	
ATPase inhibitory factor	
ATPIF1	

nel function when ions travel a part of the pore using separate paths [208]. Hybrid behavior involves transport of different ions through a single molecule.

2.2.5.1 Ion Channels

Ion channel is a gated pore that allows passive ion flux according to concentration and electrical potential gradients. An ion channel has 2 conformational states, open and closed, determined by a conformational change that displaces a gate. This gate prevents ion movement through the pore in the prohibitive conformation, but not in the permissive state. Ion channels are involved in specific transport of ions or uncharged molecules.

Ion channels are not confined to the plasma membrane. Intracellular ion channels also lodge in membranes of intracellular organelles, such as exo- and endocytotic vesicles, synaptic vesicles, lysozymes, mitochondria, and the endoplasmic reticulum.

Structural and Functional Features

Ion channels have a typical caliber of about 8 nm, with a central pore of about 2 nm, and a length of about 12 nm, hence exceeding the lipid bilayer thickness. The central pore has a narrow part.

Table 2.6. Transporters in humans: ion channels (**Part 4**; Source: Transporter Classification Database; www.tcdb.org).

Set	Code
Calcium channels	
CACNA1A–1I, -1S, CACNA2D1–2D4	1.A.1.11.x
CACNB1–B4,	
CACNG1–G8	8.A.16.1.x or 1.C.12.1.3 (G8)
Sodium channels	
Voltage-gated	
SCN1A–1B, SCN2A1–2A2, SCN2B,	1.A.1.10.1, 8.A.17.1.1 (1B), and 8.A.18.1.1 (2B)
SCN3A–3B, SCN4A–4B,	1.A.1.10.1, 8.A.17.1.2 (3B), and 8.A.18.1.1 (4B)
SCN5A, SCN7A, SCN8A, SCN9A	1.A.1.10.1
Non-voltage-gated	
SCNN1A–N1B, -N1D, -N1G	1.A.6.1.1
Sodium channel modifier-1	
SCNM1	

Hydrophilic charged elements cross the hydrophobic cell lipid bilayer using ion channels through transmembrane hydrophilic pores. Ion channel controls the electrical state of the cell, as it enables certain ions to traverse the membrane and excludes others. [11] Relative selective permeability describes the ability of an ion channel to permit the passage of a single ion type or a set of ions.

Ion channels are composed of one or more pore-forming subunits, often in association with accessory subunits. [12] The basic structure contains 4 (tetrameric channel) or 5 (pentameric channel) transmembrane helices fitted together, which edge the central pore. For example, the 4 subunits of Na^+ and K^+ tetrameric channels surround a central pore with a selectivity filter formed by P loops [13] from each of the subunits [123]. However, ion channels such as certain voltage-gated channels can use a single unit with various domains (monomeric channel). Others are dimers, with similar (homodimeric channel) or different (heterodimeric channel) subunits.

Ion channels are characterized by: (1) gating (pore opening and closing) kinetics; (2) possible modulation by intracellular ions and molecules; and (3) channel conductance. [14]

11. For instance, channels selective for sodium ions allow Na^+ to enter the cell. Channels selective for potassium ions enable K^+ to exit the cell, thus generating membrane hyperpolarization and reducing the cell's electrical excitability.

12. Accessory subunits contribute to the diversity of ion channels.

13. Each subunit of the ion channel can have 2 membrane-spanning helices, outer TM1 and inner TM2 helix, with an intervening P-loop.

14. The channel conductance is a measure of the current crossing a single open channel at a given membrane potential.

Table 2.7. Transporters in humans: solute carriers (**Part 5**; Source: Transporter Classification Database; www.tcdb.org).

Set	Code
Potassium channels	
Subfamily K	
KCNK1–K7, -K9–10, -K12–13, -K15–16	1.A.1.8.x and 1.A.1.9.1
Subfamily T	
KCNT1–T2	1.A.1.3.1
Subfamily U	
KCNU1	
Subfamily V	
KCNV1–V2	1.A.1.2.1
Voltage-gated channels	
KCNA1–A7, -A10, -AB1–3,	1.A.1.2.2 and 8.A.5.1.1 (-ABx)
KCNB1–B2, KCNC1–C4, KCND1–D3,	1.A.1.2.x
KCNE1–E4, -E1L, KCNF1, KCNG1–G4,	8.A.10.1.1 (-E), 1.A.1.2.1
KCNH1–H8	1.A.1.5.2
K_V channel-interacting proteins	
KCNIP1–P2, -P4	
Inwardly rectifying channels	
Large conductance calcium-activated	
KCNMA1, MB1–B4, -B3L	1.A.1.3.2 (MA1) and 8.A.14.1.1
Intermediate/small conductance calcium-activated	
KCNN1–N4	1.A.1.16.x
Delayed rectifiers	
KCNS1–S3	1.A.1.2.1
Voltage-gated, KQT-like subfamily	
KCNQ1–Q5, -Q1OT1	1.A.1.15.x
Potassium channel regulator	
KCNRG	1.A.1.2.2

Three functional states of ion channels can be defined: resting (closed), open (conducting), and inactivated. Ion channels share similar features, relative selective permeability and gating (Table 2.13). Gating refers to the channel opening, depending on the presence of external signals, such as a ligand (ligand-gated channels) or an ionic flux (voltage-gated channels). Gating is characterized by transitions between conductive and non-conductive conformations via structural rearrangements of the flexible selectivity filter.[15] In addition to voltage, opening and closing of ion channels can also be regulated by changes in temperature or mechanical stress. The gating kinetics deal with the onset and duration of the conductive state. Gating can be modulated by biochemical reactions such as phosphorylation.

15. Channel opening can be due to a rotation of a cytoplasmic domain of the membrane protein into its cytoplasmic end. The rotation can be rfesult from a change in transmembrane

Table 2.8. Transporters in humans: solute carriers (**Part 6**; Source: Transporter Classification Database; www.tcdb.org).

Set	Code
Family 1: glutamate transporter	
SLC1A1–1A7	2.A.23.2.x
Family 2: facilitated glucose transporter	
SLC2A1–2A14	2.A.1.1.x
Family 3: cystine, dibasic, and neutral amino acid transporters, activator of cystine, dibasic and neutral amino acid transport	
SLC3A1–3A2	8.A.9.1.x
Family 4: anion exchangers, sodium bicarbonate cotransporters, and transporter-like proteins	
SLC4A1–4A5, -4A7–A11	2.A.31.x.x
Family 5: sodium–glucose cotransporter, inositol and iodide transporters, sodium iodide symporter, and sodium-dependent vitamin transporter	
SLC5A1–5A8	2.A.21.x.x
Family 6: neurotransmitter transporters	
SLC6A1–6A16	2.A.22.x.x
Family 7: cationic and neutral amino acid transporters	
SLC7A1–7A11, -7A13	2.A.3.x.x
Family 8: sodium–calcium exchangers	
SLC8A1–8A3	2.A.19.3.1
Family 9: sodium–hydrogen exchangers	
SLC9A1–9A9, -9A3P, 9A3P2, 9A3R1–R2	2.A.36.1.x

The ion flux through the cell membrane across the corresponding channels depends mainly on 3 factors: (1) the difference in ion concentrations on both sides of the membrane; (2) voltage across the membrane; and (3) membrane permeability that depends on ion channel activity. The membrane permeability depends on: (1) the number of plasmalemmal channels; (2) opening time duration; and (3) channel conductance.

Ion channels can be either specific for a given ion or non-selective, such as serotonin-gated ion channel. Non-selective stretch-activated cation channels let calcium, sodium, and potassium cross, whereas others are selective for potassium. Nicotinic acetylcholine receptors are ion channels in nerves and myocytes that are per-

potential. In voltage-dependent channels, a voltage-sensing domain, linked to the pore gate, is able to transduce the energy provided by the transmembrane voltage into gate motion.

Table 2.9. Transporters in humans: solute carriers (**Part 7**; Source: Transporter Classification Database; www.tcdb.org).

Set	Code
Family 10: sodium–bile acid cotransporters	
SLC10A1–10A5	2.A.28.1.x
Family 11: proton-coupled divalent metal ion transporters	
SLC11A1–11A2	2.A.55.2.x
Family 12: sodium–potassium–chloride transporters	
SLC12A1–12A9	2.A.30.x.x
Family 13: sodium–sulfate symporters and sodium-dependent dicarboxylate or citrate transporter	
SLC13A1–13A5	2.A.47.1.x
Family 14: urea transporters	
SLC14A1–14A2	1.A.28.1.x
Family 15	
SLC15A1–15A4	2.A.17.x.x
Family 16: monocarboxylic acid transporters	
SLC16A1–16A8, -16A10–16A12	2.A.1.13.x
Family 17: sodium phosphate, anion–sugar transporter and sodium-dependent inorganic phosphate cotransporter	
SLC17A1–17A8	2.A.1.14.x
Family 18: vesicular monoamine and acetylcholine	
SLC18A1–18A3	2.A.1.2.x
Family 19: folate and thiamine transporters	
SLC19A1–19A3	2.A.48.x.1

meable to all cations, including sodium, potassium, and calcium ions.[16] Nucleotide P2X receptor channels, sensitive to ATP, are sodium and calcium channels. Certain voltage-activated cation channels in the cardiac pacemaker and Purkinje fibers are permeable to both sodium and potassium ions. Voltage-dependent anion channels in the mitochondrial outer membrane are slightly selective for anions. Anion channels usually are less selective than cation channels.

16. Nicotinic acetylcholine receptor (nAChR) is a member of the set of pentameric neurotransmitter-gated ion channels. Helices are arranged around a central aqueous pore: TM2 α-helix lines the cation-selective permeation channel; TM1 and TM3 transmembrane α-helices are packed against the pore-lining TM2 α-helix, thus shielding TM2 from the surrounding lipids; and the most lipid-exposed, outer TM4 helix. The transient open-channel conformation ($\sim$50 ms) of the nicotinic acetylcholine receptor is due to pore dilation that results from rapid rearrangement of TM1/2/3 transmembrane helices [124]. The quick closing and opening kinetics allows fast postsynaptic response upon neurotransmitter binding.

Table 2.10. Transporters in humans: solute carriers (**Part 8**; Source: Transporter Classification Database; www.tcdb.org).

Set	Code
Family 20: phosphate transporters	
SLC20A1–20A2, -20A1P1	2.A.20.2.3
Family 22: organic cation and anion/cation transporters	
SLC22A1–22A17, -22A1L, -22A1LS	2.A.1.x.x
Family 23: nucleobase transporters	
SLC23A1–23A4	2.A.40.6.1
Family 24: sodium–potassium–calcium exchanger	
SLC24A1–24A5	2.A.19.4.x
Family 25: mitochondrial carriers	
SLC25A1–25A6, -25A10–A22, -25A26–A28	2.A.29.x.x
Family 26: sulfate transporter and other members	
SLC26A1–26A4, -26A6–A11	2.A.53.x.x
Family 27: fatty acid transporters	
SLC27A1–27A6	9.B.17.1.x
Family 28; sodium-coupled nucleoside transporters	
SLC28A1–28A3	2.A.41.2.x
Family 29: nucleoside transporters	
SLC29A1–29A4	2.A.57.1.x

Mechanism of Activation

Ion channels include: (1) electrically gated channels (activation by change in membrane voltage or electric field); (2) chemically gated channels (excitation by ligand binding); and (3) mechanically gated channels (stimulation by stress or strain).

Intracellular signaling proteins can modulate the ion channel activity. Ion channels are regulated by various mechanisms, such as phosphorylation by protein kinases (Vol. 4 – Chap. 2. Preamble to Protein Kinases). Ion channels can be substrates not only for protein kinases, but also phosphoprotein phosphatases (Vol. 4 – Chap. 7. Cytosolic Protein Phosphatases). Hence, phosphorylation and dephosphorylation modulate channel gating. Kinases and phosphatases can especially bind to voltage- and calcium-gated ion channels, either directly or via anchoring proteins such as A-kinase anchoring proteins (Vol. 4 – Chap. 9. Other Major Signaling Mediators) to form regulatory proteic complexes. Calmodulin, a calcium sensor, can also bind to ion channels in the absence of calcium. The calcium–calmodulin complex modulates the channel activity.

Table 2.11. Transporters in humans: solute carriers (**Part 9**; Source: Transporter Classification Database; www.tcdb.org). Glycans-modifying proteins and lipids modulate protein interactions and mediate protein folding and quality control, protein targeting and clearance, cell adhesion and signaling, and cell–matrix interactions. The conjugation of sialic acid to terminal positions of glycans is catalyzed by different substrate- and linkage-specific sialyltransferases. The addition of sialic acid to glycoproteins and glycolipids at the Golgi body requires not only sialyltransferases, but also glycoconjugate substrates and nucleotide sugar donor, ^{CMP}sialic acid. ^{CMP}Sialic acid is transported into the lumen of the Golgi body through the CMP-sialic acid antiporter, which also carries the nucleotide cytidine monophosphate (CMP) into the cytosol. ^{UDPN}acetylglucosamine 2-epimerase catalyzes the first steps of sialic acid synthesis (UDP: nucleotide uridine diphosphate).

Set	Code
Family 30: zinc transporters	
SLC30A1–30A9	2.A.4.x.x
Family 31: copper transporters	
SLC31A1–31A2, -31A1P	9.A.12.1.2
Family 32: GABA vesicular transporter	
SLC32A1	
Family 33: acetylCoA transporter	
SLC33A1	2.A.1.25.1
Family 34: sodium–phosphate transporters	
SLC34A1–34A3	2.A.58.1.x
Family 35: ^{CMP}sialic acid, ^{UDP}galactose, ^{UDPN}acetylglucosamine transporter, among other members	
SLC35A1–35A5, SLC35B1–35B4, SLC35C1–35C2,	2.A.7.x.x
SLC35D1–35D3, SLC35E1–35E4, SLC35F1–35F5	2.A.7.x.x

Voltage-Gated Ion Channels

Voltage-gated (-activated, -dependent, or -sensitive) ion channels are activated at a given membrane polarity threshold and create electrical impulses in excitable cells.[17] Voltage-gated K_V and Na_V channels produce nerve impulses. Voltage-gated Ca_V channels initiate muscle contraction. In addition, voltage-gated ion channels generate and propagate the action potential for excitation–contraction coupling in the heart.

Voltage sensing is not restricted to 6-transmembrane segment, voltage-gated ion channels. In addition voltage-gated ion channels, other proteins, such as phosphatase (e.g., Ciona intestinalis voltage sensor-containing phosphatase) and proton-permeating voltage sensor only protein (VSOP; human ortholog HV1) with 4 trans-

17. Certain cyclic nucleotide-gated channels are similar to voltage-gated ion channels. Some enzymes also have voltage sensors that allow the plasmalemmal voltage to regulate their catalytic activity.

Table 2.12. Transporters in humans: solute carriers (**Part 10**; Source: Transporter Classification Database; www.tcdb.org).

Set	Code
Family 36: proton–amino acid symporter	
SLC36A1–36A4	2.A.18.8.x
Family 37: glycerol 3-phosphate transporters	
SLC37A1–37A4	2.A.1.4.6
Family 38	
SLC38A1–38A6	2.A.18.6.x
Family 39: zinc transporters	
SLC39A1–39A14	2.A.5.x.x
Family 40: iron-regulated transporter	
SLC40A1	9.A.23.1.1
Family 41	
SLC41A1	9.A.19.4.1
Family 43	
SLC43A1	2.A.1.44.1
Organic anion transporters	
SLCO1A2, SLCO1B1/3, SLCO1C1, SLCO2A1/B1, SLCO3A1, SLCO4A1/C1, SLCO5A1, SLCO6A1	2.A.60.1.x

membrane segments, but with neither pore domain nor phosphatase motif, contain a voltage sensor domain (VSD) [125]. [18]

Voltage sensors change the channel conformation in response to voltage variations across cell membranes. The closed channel corresponds to a distribution of conformations according to the magnitude of membrane polarization. On the other hand, channel opening results from a specific conformation. The transition from open to closed state can occur over a very small voltage change (a few hundredths of volt). Conformational modifications arise from the movement of positively charged amino acids — arginine (mostly) or lysine — and their interaction with negatively charged amino acids, i.e., transmembrane transfer of gating charges of voltage-sensors [126].

Gating also has a component sensitive to the osmolarity of channel surroundings. Hydrophobic cavities are structural motifs in voltage-gated K^+, Na^+, and Ca^{++} channels. Hydrophobic gating may be a common component in the overall gating. Ions and copermeating water molecules can move in single file through the selectivity filter. A dewetting transition of the hydrophobic pore cavity causes the open pore to collapse into a closed conformation [127].

18. In several proton-conducting proteins, hydrogen ions are transferred by hopping via hydrogen-bound chains rather than by ion diffusion through an aqueous pore.

Table 2.13. Classification of ion channels.

Channel set	Examples
Voltage-gated channels	Voltage-gated sodium channel Voltage-gated potassium channel Voltage-gated sodium–potassium channel Voltage-gated proton channel Voltage-gated calcium channel Voltage-gated chloride channel (ClC) Mitochondrial voltage-gated anion channel
Extracellular ligand-gated channels	GABA-gated Cl^- channel Glycine-gated Cl^- channel Glutamate-gated Ca^{++} channel P2X Nicotine acetylcholine receptor Serotonin-gated cation channel ATP-gated outward rectifying Cl^- channel
Intracellular ligand-gated channels	CFTR Cl^- channels ATP-binding cassette proteins (ABC) IP3R–TRP TRPM7 Nucleotide receptor ATP-sensitive inward rectifying K^+ channel (ROMK, K_{IR}, IRK/K_{IR}2, BIR, etc.) Calcium-activated Cl^- channel (ClCa) Calcium-activated K^+ channels (SK, BK) ATP-gated K^+ channel (K(ATP)) cGMP-gated ion channels (CNG) G-protein-activated inward rectifying K^+ channel (GIRK/K_{IR}3) Calcium-release Ca^{++} channels Ryanodine (Ca^{++} channel) receptor Epithelial Na^+ channel (ENaC)
Mechanosensory channels	TRP Stretch-activated ion channels
Volume-regulated channels channels	VRAC
Miscellaneous	Gap junctions Peptide ion channels

Voltage-gated, tetrameric ion channels have 6 transmembrane segments (TM1–TM6). Subunits TM1 to TM4 constitute the voltage-sensor domain, TM5 and TM6 form the hydrophilic pore. The voltage sensor domain is aimed at sensing transmembrane potential. In most voltage-gated channels, when the membrane is depolarized from a resting hyperpolarized state, the voltage sensor undergoes conformational changes leading to pore opening. Sensor TM4 has periodically aligned positively charged amino acids. The voltage-sensor domain, without the pore domain, generates depolarization-induced outward currents and inward tail currents during repolarization.

In voltage-gated proton channel (H_V), the voltage-sensor domain is characterized by a pH-dependent gating, [19] zinc ion sensitivity, and hydrogen ion selectivity [128]. The H_V channel localizes to blood cells, particularly granulocytes and monocytes, and alveolar epithelium cells, among others. [20] The voltage-sensor domain works not only for voltage sensing, but also constitutes a voltage-gated proton channel. Proton efflux through these H_V channels leads to membrane depolarization.

Ion channels can complex between them such as large-conductance Ca_V channels and Ca^{++}-activated voltage-gated K^+ channels ($K_{Ca}1.1$ or BK) for fast and localized control of neuronal firing and release of hormones and neurotransmitters in the central nervous system [129].

Voltage-dependent anion channel (VDAC), or mitochondrial porin, the most abundant protein in the mitochondrial outer membrane, functions in cell apoptosis, as it releases apoptotic factors. These porins are also responsible for the energy maintenance of the cytosol.

Ligand-Gated Ion Channels

Ligand-gated ion channels open when they are bound with their corresponding ligand. Ion channels can be gated either directly by ligand binding or indirectly via a cascade of molecular reactions leading to channel opening. They can be activated by G-protein-coupled receptors.

Extracellular and intracellular ligands can activate ion channels. Usual intracellular ligands include Ca^{++}, ATP, cAMP and cGMP, and phosphatidylinositol. Stored calcium is the main source of Ca^{++} influx in non-excitable cells. The cytosolic influx can require calcium release-activated calcium channels. Intracellular ligand-gated ion channels modify the intracellular concentration of second messengers and can indirectly act on coupling proteins and cofactors, especially those involved in long-lasting processes in smooth muscle cells. Glutamate ion channels, which are activated by glutamate, do not share sequence similarity with any other ligand-gated ion channels.

Ligand-gated ion channels can be opened by both full and partial agonists. Partial agonists induce a maximal response, which is a fraction of that of a full agonist. Partial agonists of nicotinic acetylcholine and glycine receptors drive an open–shut reaction similar to that of full agonists, but the response to partial agonists is limited by an earlier conformation change that happens when the channel is still shut [130].

19. The voltage-triggered opening strongly depends on both the intra- and extracellular pH. Decrease in extracellular pH causes a shift of the current-voltage relationship in the positive direction, whereas intracellular pH changes generate a shift in the opposite direction.

20. The H_V channel resides also in neurons and skeletal myocytes.

Mechanical Stress-Gated Ion Channels

Mechanotransduction is the conversion of mechanical force into chemical signals that generate biological responses. Mechanical stimuli drive morphogenesis, proprioception, touch and pain sensation and hearing, lung growth, bone and muscle adaptation, and the regulation of the vasomotor tone and, hence, blood pressure. *Mechanically gated ion channels* (MGIC) act as force sensors and react in less than 5 ms, whereas *mechanosensitive ion channels* (MSIC) are activated by second mediators downstream from force sensors.

Mechanogated cation channel lodge in many cell types. Two molecules are necessary and sufficient for the functioning of mechanogated channels: Piezo-1 and -2 [131].[21] These proteins are large, multipass transmembrane (24–36 TMs) molecules. Piezo-1 resides at both the plasma membrane and endoplasmic reticulum. Piezo-1 and -2 are strongly detected in the lung and bladder [131].[22] Piezo proteins may serve as non-conducting subunits of ion channels required for proper activity of mechanogated channels or modulators of these channels.

Mechanosensitive ion channels have been classified into: (1) *volume-activated ion channels* (VAIC) that respond to an increase in cell volume and (2) *stretch-activated ion channels* (SAIC) that detect and respond to forces undergone by the cell as well as cell deformation in the absence of cell volume change. Stretch-activated ion channels thus open and close in response to mechanical stimuli. Whereas VAICs respond to cell swelling after a lag time of at least 1 mn, SAICs are promptly activated.

Cells are made of about 80% water. The osmotic pressure, fundamental to cell survival, induces stretch force on the cell membrane. Two major sets of ubiquitous stretch-activated channels exist [132]: (1) ion channels that require force transmission from the cytoskeleton and (2) ion channels that respond to stress mediated via the lipid bilayer.

Both SAICs and VAICs have been further categorized into subsets owing to their ion selectivity (cation non-selective [$S(V)ACC_{NS}$], K^+-selective [$S(V)AC_K$], or Cl^--selective [$S(V)AC_{Cl}$] channels [133].[23] Stretch-activated channel is mostly either $SACC_{NS}$ or SAC_K.

21. A.k.a. Fam38a and -b (πιεσω: to strongly press; πιεζω: to squeeze, press, pinch; πιεσις: squeezing, compression; πιεσιμος: pressing, as some ion channels can be activated by pressure). The name Piezo may also originates from the method used for its discovery. Forces were applied via a piezo-electrically driven glass probe to the cell surface of several rodent cell lines that express mechanogated channels [131].

22. Piezo-1 is also observed at high levels in the skin (Piezo-2 at a low levels) and Piezo-2 in sensory neurons of the dorsal root ganglia (Piezo-1 at a low levels); Piezo-1 and -2 are synthesized at mid levels in the colon, and Piezo-1 in the kidney (Piezo-2 at a low levels); Piezo-1 and -2 are identified to a much lesser extent in the central nervous system, heart, skeletal muscles, and digestive tract.

23. Activation of potassium-selective channels generally causes membrane repolarization and hyperpolarization, whereas activation of cation-non-selective channels tends to depolarize resting cells and to repolarize cells at more positive membrane potentials during the action potential [133].

Mechanosensitive ion channels heve been detected in a wide range of species and tissues. They contribute to the regulation of cell volume and contractile activity of myocytes, proprioception (the sense of the relative position of the body's parts) and senses (e.g., touch and hearing). Mechanical stimuli usually increase their open probability, although stretch-inactivated channels have been observed. Therefore, they translate mechanical signals (stress and strain, as well as membrane curvature or thickness) into an electrochemical signal.

Two models of channel gating by mechanical stimuli exist [134]. In the *membrane model*, the channel undergoes surface tension or bending of the lipid bilayer caused by a mechanical force that results in a hydrophobic mismatch, which primes channel opening. The *tether model* supposes that accessory proteins (elements of the cytoskeleton and/or extracellular matrix) are bound to ion channels and that tether deformation directly or indirectly (via a second mediator) induces a conformational change. Heteromultimeric channels that participate in mechanotransduction can be composed of pore-forming and force-sensing subunits.

Volume-regulated anion channels (VRAC) are found in epithelial cells. Stretch-activated calcium and inward-rectifying potassium channels drive an ion influx leading to cell depolarization [135].

Several mechanosensitive ion channels belong to the transient receptor potential (TRP) proteins [136]. Channels TRPs are polymodal, as they are activated by several types of stimuli. Channels TRPs can have enzymatic activity.

Members of the TRP channel superfamily (Sect. 2.3; TRPC3, TRPC6, TRPV2, and TRPM4) are mediators of the tone of vascular smooth muscle cells. Transient receptor potential melastatin-like protein TRPM7 is an ion channel for Ca^{++} and Mg^{++} ions. It is involved in mechanosensation in vascular smooth muscle cells at high pressure. In addition, TRPM7 is required for thymic maturation of T lymphocytes. The TRPV4 channel is implicated in osmoregulation. In osmosensitive neurons of the circumventricular organs, [24] it primes the release of antidiuretic hormone (or vasopressin) and causes water reabsorption in kidneys and colon. It can be activated by a second mediator stimulated by hypo- and/or hyperosmotic stress (possible phosphorylation by Src kinase or activation by (5,6)-epoxyeicosatrienoic acid).

2.2.5.2 Ion Pumps

Ion pumps use the energy from ATP hydrolysis (ATPases) to transport ions against their electrochemical gradient. P-Type ATPases are phosphorylated (hence their name P-type) using ATP once the carried ion has entered into the binding pocket through the open cytoplasmic entry, whereas the exit remains closed. Pump phosphorylation shuts the cytoplasmic side gate, temporarily sequestrating bound ion before a conformational change opens the extracellular side gate and lowers affinity for the transported ions. Ion release into the extracellular medium is followed by binding of a countertransported ion (e.g., K^+ in the case of Na^+–K^+ and H^+–K^+ ATPases, or H^+ in the case of Ca^{++}–H^+ sarco(endo)plasmic reticulum ATPase). Closure of

24. I.e., organum vasculosum lamina terminalis and subfornical organ.

the extracellular side gate that is triggered by counter-ion binding leads to pump dephosphorylation and stabilization of an occluded state with counter-ion sequestration. Another rearrangement reopens the cytoplasmic side gate, thereby restoring the conformation with lowered affinity for the countertransported ions, but elevated affinity for the transported ions.

2.3 Superfamily of Transient Receptor Potential Channels

Transient receptor potential channels (TRP) constitute a superfamily of ubiquitous proteins. Many TRP channels reside mainly in the plasma membrane, where they transduce diverse stimuli. Numerous identified TRP channel genes are subdivided into 7 families.

2.3.1 Classification of TRP Channels

Transient receptor potential channels have a diverse permeability to ions, although in general, they are non-selective cation channels. However, certain TRPs are highly selective for calcium ions. Transient receptor potential cation channels are indeed responsible for calcium influx into non-excitable cells.

The 7 major TRP families, i.e., the TRPA (ankyrin-like), TRPC (canonical), TRPM (melastatin), TRPML (mucolipin), TRPN (or no mechanoreceptor potential-C [NoMPC]), TRPP (polycystin), and TRPV (vanilloid receptor) families, differ in selectivity and activation mechanisms. In mammals, TRPs are also subdivided into 2 groups: group 1 with TRPA, TRPC, TRPM, and TRPV families and group 2 with TRPML and TRPP.

2.3.1.1 Canonical TRP Channels

Seven receptor-activated, non-selective cation channels (TRPC1–TRPC7) constitute the family of canonical transient receptor potential channels. However, TRPC2 is a pseudogene in humans (but not in other mammalian species.[25]

Two TRPC epifamilies have been defined: epifamily 1 with TRPC1, -4, and -5 and -2 with TRPC3, -6, and -7. Based on sequence homology and functional similarities, 7 members of the TRPC family are subdivided into 4 groups: group 1 with TRPC1; group 2 with TRPC2; group 3 with TRPC3, -6, and -7; and group 4

25. A pseudogene is a dysfunctional DNA component related to a known gene without protein-coding ability due to truncation and mismatches of base pairs in comparison with the functional gene. A pseudogene is thus related to a functional gene, but defective because of some genetic lesions. Pseudogenes are characterized by: (1) sequence homology between the DNA sequences of the pseudogene and those of the parent gene (sufficient number of identical base pairs) and (2) non-functionality that results from various genetic disablements associated with termination codons and reading frame shifts caused by insertions or deletions of nucleotides.

with TRPC4 and -5 [138]. Therefore, the TRPC epifamily 1 is composed of group-1 and -4 TRPCs. Channel TRPC1 is indeed closely related to TRPC4 and -5. Moreover, TRPC1 can tether to TRPC4 and -5 to form heteromultimeric channels. [26] Channels TRPC3, TRPC6, and TRPC7 can also form functional homo- and heteromeric ion channels.

These channels are activated by PLC-coupled receptors. They are able to assemble between them. Store-operated TRPC1 channel abounds in endothelial cells, among others. Channel TRPC2 is tissue specific. Channel TRPC3 interacts with IP_3R.

Channels TRPC3, TRPC6, and TRPC7 are closely related in molecular structure and function. These channels are activated by diacylglycerol independently of protein kinase-C. Receptor-activated non-selective cation channel TRPC6 is highly expressed in manifold cell types, particularly vascular smooth muscle cells. It is involved in the depolarization of smooth muscle cells. Mechanosensitive TRPC6 channel that is activated by PLC-coupled receptors sense the membrane stretch independently of phospholipase-C activity [139]. It can thus regulate the myogenic tone according to the applied pressure in blood vessels. Channels TRPC4 and TRPC5 are store-independent, non-specific cation channels.

Several transient receptor potential canonical channels are plasmalemmal store-operated channels activated by receptor stimulation. Calcium sensor stromal interacting molecule-1 binds TRPC1, TRPC4, and TRPC5 [140]. STIM1 does not bind TRPC3, TRPC6, and TRPC7. However, it intervenes in heteromultimerization of TRPC3 with TRPC1 and of TRPC6 with TRPC4.

2.3.1.2 Families of TRP-Related Channels

The TRP-related set is associated with the olfaction and osmolarity transduction pathways as well as specific genetic diseases. In addition to the canonical family, the TRP-related set is subdivided into the following families: (1) the vanilloid family (TRPV), with TRPV5, or epithelial calcium channel (ECaC), [27] TRPV6, or calcium transporter-1 (CaT1), [28] the osmoreceptor TRPV3, [29] and mammalian vanilloid receptor-1 (VR1) [30] and its homolog VRL1; (2) the polycystin family (TRPP) with polycystin-1 and -2; [31] (3) the no mechanoreceptor potential-C family (TRPN),

26. Electrophysiological properties of TRPC1–TRPC5 heterotetramers differ from that of TRPC5 homotetramers [138].

27. Epithelial calcium channel (ECaC1, or CaT2) is highly selective for Ca^{++}.

28. The CaT1 channel (a.k.a. ECaC2) is permeable to Na^+, K^+, and Ca^{++}.

29. The osmoreceptor TRPV3, or OTRPC4, is synthesized in the kidney (distal nephron), heart, and liver. It is a non-selective cation channel.

30. Vanilloid receptor-1 is a heat-gated ion channel.

31. Polycystin-1 and -2 are encoded by PKD1 and PKD2, respectively. A non-selective cation channel results from coassembly of polycystin-1 and -2. Polycystin-L (PCL) is a calcium-regulated cation channel permeable to Na^+, K^+, and Ca^{++}.

not detected in mammals; (4) the mucolipin family (TRPML) with mucolipin-1;[32] (5) the melastatin family (TRPM)[33] that has a potential role in the regulation of the cell cycle and calcium influx in lymphocytes, particularly with chanzymes that possess both ion channel and enzymatic activity, such as TRPM2[34] and TRPM7;[35] and (6) the ankyrin-like family (TRPA).

2.3.2 Structure of TRP Channels

The TRP channels are homo- or heterotetramers that contain 6 transmembrane segments per subunit with a pore region between TM5 and TM6, and intracellular amino (N)- and carboxy (C)-termini. Most TRP channels are non-selective cation channels with low-voltage dependence.

2.3.3 TRP Channel Activity

The TRP channels mediate cell responses to growth factors, pheromones, odorants, mechanical and physical factors, chemical agents, and metabolic stresses [141]. They are thus involved in mechano-, chemo-, and thermosensation, as well as vision, olfaction, and taste. In addition, TRPs intervene in Ca^{++} and Mg^{++} homeostasis and various Ca^{++}-dependent signalings that regulate cell growth, proliferation, migration, and death. In particular, TRPs act as store-operated Ca^{++} channels that prime Ca^{++} entry upon Ca^{++} release from intracellular stores.

Channel activity is influenced by: (1) several physical parameters, such as temperature, voltage, osmolarity, and mechanical forces, and (2) chemical agents, such as ions (mainly Ca^{++} and Mg^{++}, in addition to H^+), as well as interactions with external ligands or cellular proteins.

Heteromultimeric calcium store-operated TRPs and store depletion-independent IP_3- and DAG-activated TRP-like channels (TRPL), with calmodulin-binding sites, are light-activated channels. Heat sensor transient receptor-potential vanilloid TRPV1 is activated by low pH and inflammatory molecules. Osmosensor transient receptor potential vanilloid TRPV4,[36] a non-selective cation channel with a slight preference for calcium ions, is highly expressed in the kidney, inner-ear hair cells, and paraventricular regions of the central nervous system. Like all polymodal TRPs, TRPV4 can

32. Mucolipin-1, encoded by the MCOLN1 gene, is also called mucolipidin. The founding member TRPML1 of this family is mutated in the lysosomal storage disease mucolipidosis-4

33. The TRPM1 channel is a Ca^{++}-permeable channel of vesicles that influence melanin synthesis. Melanin is synthesized within melanosomes that contain large amounts of calcium. The expression of TRPM1 is reduced in melanomas.

34. Cation channel TRPM2, or LTRPC2, is activated by ^{ADP}ribose and oxidants such as hydrogen peroxide. It acts as a specific adenosine diphosphate–ribose pyrophosphatase (or adenosine diphosphoribose hydrolase). It is permeable to both monovalent and divalent cations.

35. Intracellular ligand-gated, ATP^{Mg}-regulated TRPM7 is also called LTRPC7 or TRPP-LIK (transient receptor potential phospholipase-C-interacting kinase). It is permeable to K^+, Na^+, and Ca^{++} (in decreasing order of permeability). It also permeates Mg^{++}.

36. A.k.a. vanilloid receptor-related osmotically activated channel (VROAC).

Table 2.14. Thermosensitive TRP channels (Sources: [143, 144]). ThermoTRP channels are homotetramers. Channels TRPV1 to TRPV4 are activated by relatively high temperature, whereas TRPA1 and TRPM8 are activated by relatively low temperature. Membrane voltage can change the temperature threshold for channel opening. Channels TRPV1, -2, and -4 also transduce osmotic and mechanical stimuli.

Receptor	Temperature threshold (C)	Expression	Other stimuli
TRPA1	<17	Sensory neurons, Inner ear	Mechanical stimulus
TRPV1	≤43	Sensory neurons, Epithelial cells	Protons (pH<6), Lipids Weak voltage dependency
TRPV2	≤52	Sensory, motor neurons, Heart, lung, etc.	Mechanical stress
TRPV3	≤32	Sensory neurons, Keratinocytes	
TRPV4	≤27	Sensory neurons, Skin, lung, kidney, etc.	Hypo-osmolarity, Mechanical stress, Epoxyeicosatrienoic acids
TRPM4	15–35	Ubiquitous	Calcium ions
TRPM5	15–35	Taste cells, Pancreas	Calcium ions Phospholipase-C
TRPM8	∼25	Sensory neurons	Menthol Weak voltage dependency

also be activated by heat and amino acids. Ubiquitination and subsequent endocytosis control TRPV4 density at the plasma membrane [142].

2.3.3.1 Thermosensitive TRP Channels

Thermal sensors of afferent nerves that can detect changes in ambient temperature with a relatively large temperature spectrum (from warm to very hot as well as from cool to intense cold) are associated with temperature-sensitive members of the TRP superfamily (Table 2.14). [37]

37. The transduction of heat stimuli by C-fiber mechanoheat nociceptors occurs at different skin depths [143]. Thermosensitive ion channels contribute to temperature detection by cutaneous nociceptors. Heat-gated ion channels (TRPA1, TRPV1–TRPV4, TRPM4–TRPM5, and TRPM8) indeed reside in a subset of primary afferent neurons. The temperature threshold from which TRPs are activated ranges from relatively warm (TRPV3–TRPV4 and TRPM4–TRPM5) to extremely hot (TRPV2), whereas TRPA1 and TRPM8 are activated by cold. The heat threshold of C-fiber mechanoheat nociceptors depends on the absolute temperature rather than the rate of temperature change.

2.3.3.2 TRP Channels, Phosphoinositides, and Calcium Influx

Channels TRPs are activated by the phosphoinositide cascade. The transduction cascade is mediated by G-protein-activated phospholipase-C and produced second messengers diacylglycerol and inositol trisphosphate.

Deactivation of channel activity is regulated by protein kinase-C. Enzymes PLC and PKC complex with TRPs and scaffold proteins. Channels TRPs drive inositide-mediated calcium influx channels that include both store-operated (SOC) owing to a coupling between TRP and inositol triphosphate receptor IP_3R and store-independent channels via PLC activation.

Cytosolic calcium concentration that governs multiple cellular functions rises quickly in response to calcium-mobilizing messengers by opening of Ca^{++}-permeable channels of the plasma membrane as well as membranes of intracellular calcium stores, particularly the endoplasmic reticulum.

In the endoplasmic reticulum, in addition to inositol trisphosphate- and ryanodine-sensitive Ca^{++} channels that release Ca^{++} in the cytosol in response to inositol trisphosphate and cyclic adenosine diphosphate–ribose, respectively, TRPP2 and TRPV1 can also contribute to Ca^{++} influx (in the cytosol).

Lysosomes can also serve as calcium stores. Calcium is released from lysosomes in response to another calcium-mobilizing messenger, nicotinic acid adenine dinucleotide phosphate (NAADP), that activates both TRPM2 and TRPML1 [145]. Mucolipin-type TRPs are located on intracellular membranes of the endosomal and lysosomal network. Calcium-permeable melastatin-related TRPM1 and TRPM2 also reside in vesicles and lysosomes, respectively.

Homologs of TRP channels are involved in the control of vascular smooth muscle tone via calcium influx. Channel TRPC1 is involved in store-operated activity. Channel TRPC6 is a component of the α1-adrenoreceptor activator calcium-permeable, non-selective cation channel. Channel TRPC4 expressed in vascular endothelial cells mediates vasorelaxation of blood vessels by synthesis and release of vasoactive effectors.

Calcium influx through plasmalemmal TRP channels can trigger cell apoptosis. In addition, TRPP2 [38] of endoplasmic reticulum membrane counteracts the activity of sarco(endo)plasmic Ca^{++} ATPase that pumps Ca^{++} into its store, as it lowers Ca^{++} concentration in the endoplasmic reticulum [146]. It then protects cells against apoptosis because Ca^{++} content in the endoplasmic reticulum determines the amplitude of mitochondrial Ca^{++} signal that activates permeability transition pores, thereby causing mitochondrial depolarization and cytochrome-C release. [39] In addition, TRPP2 homotrimer and PKD1 form a complex on the cell surface [147].

---

38. Channel TRPP2 (previously termed polycystin-2) resides in the plasma membrane, primary cilium, and endoplasmic reticulum. Transport of TRPP2 between the endoplasmic reticulum and plasma membrane is controlled by phosphorylation-dependent interaction with the adaptors phosphofurin acidic cluster sorting protein PACS1 and PACS2.

39. In normal conditions, Ca^{++} continuously cycles between the endoplasmic reticulum and mitochondria. Calcium ions are stored in the endoplasmic reticulum by ATPases (SERCA) and released by IP_3-gated channels (IP_3R) and TRPP2. Calcium ions can enter into and exit from

Mutations in the PKD1 (Trpp1) and PKD2 (Trpp2) genes cause autosomal dominant polycystic kidney disease.

The TRP channels can link to phosphoinositides via specific binding sites. Positively charged TRPs can also interact electrostatically with negatively charged phosphatidylinositol phosphates. Moreover, phosphatidylinositol (4,5)-biphosphate can regulate TRPs such as TRPV1 via the accessory protein phosphoinositide-interacting regulator of transient receptor potential channels PIRT [148].

Gating of TRPs depends on plasmalemmal phosphatidylinositol phosphates, particularly $PI(4,5)P_2$ of the cytoplasmic leaflet of the plasma membrane [148].[40]

Several mechanisms could cause functional changes in TRP channel activity: (1) direct binding that generate a conformational change, hence a new functional state; (2) competitive binding to the same site on TRP between phosphoinositides and activators or inhibitors such as calmodulin modulator; (3) indirect interaction via phosphoinositide-binding accessory proteins (activators or inhibitors); and (4) interaction with an agonist-delivering enzyme that leads to the formation of a signalplex such as $PLC\gamma1$–$PI(4,5)P_2$ complex [148].

Interactions between PIs and TRPs intervene in: (1) signal desensitization (intensity of a constant stimulus decreases with time);[41] (2) modulation of cell excitability;[42] (3) TRP transport to the plasma membrane; and (4) feedback control of Ca^{++} and Mg^{++} apical influx and basolateral extrusion in transepithelial transport using TRP channels.[43]

Interactions between PIs and TRPs can also have pathological consequences because of TRP channel mutations or dysregulation by lack in phosphoinositide availability due to defects in enzymes involved in phosphoinositide metabolism [148].

TRPs in the Vasculature

Transient receptor potential channels intervene in many functions of vascular endothelial and smooth muscle cells, such as arterial tone regulation, angiogenesis, and wall permeability [149] (Table 2.15). Vascular smooth muscle cells contract upon stimulation by numerous agents (endothelin, histamine, noradrenaline, serotonin, etc.) that bind to G-protein-coupled or protein Tyr kinase receptors to activate

---

mitochondria through Ca^{++} uniporter and Na^+–Ca^{++} exchanger, respectively. The passive Ca^{++} egress raises via TRPP2 and reduces the sensitivity to apoptosis.

40. The $PI(4,5)P_2$ agent also modulates the activity of various other ion channels and transporters. In addition, phospholipase-C targets $PI(4,5)P_2$ to form signaling molecules inositol trisphosphate and diacylglycerol. Lipid $PI(4,5)P_2$ also serves as substrate for PI3K to synthesize another signaling molecule, phosphatidylinositol (3,4,5)-triphosphate.

41. Desensitization of TRPs following stimulation by muscarinic receptors can be used for muscarinic fine-tuning of smooth muscle contraction.

42. Voltage-dependent, non-selective cation channels operate in neuron excitability.

43. The TRPV5 and TRPV6 channels involved in transepithelial transport of Ca^{++} are inactivated by reduced concentration of $PI(4,5)P_2$ that depends on Ca^{++} and PLC. Both TRPM6 and TRPM7 that carry Mg^{++} are inactivated by intracellular Mg^{++} concentration via Mg^{++} interaction with $PI(4,5)P_2$.

Table 2.15. Transient receptor potential channels detected in arteries, particularly endothelial (EC) and smooth muscle (SMC) cells and their effects (Source: [149]). All TRPC isoforms have been detected in ECs and SMCs, in which they can be involved in store-operated Ca^{++} entry (SOC; capacitive Ca^{++} entry) in the cytosol as well as receptor-operated Ca^{++} influx (ROC) following the activation by receptor ligands (e.g., ATP, bradykinin, and thrombin) used for activity of endothelial vasoregulators or SMC contraction. Channels TRPC1, TRPC3, and TRPC4 are examples of store-operated Ca^{++} channels in ECs and SMCs. Uridine trisphosphate causes depolarization and activation of Ca_V1 channels as well as Ca^{++} influx via TRPC3 for contraction of vascular smooth muscle cells. The TRPV4 channel is involved in vasodilation induced by epoxyeicosatrienoic acids such as (5,6)EET. It forms a Ca^{++} signaling complex with ryanodine receptors and Ca^{++}-sensitive, large conductance K^+ channels (BK) to elicit smooth muscle hyperpolarization due to spontaneous transient outward currents and, subsequently, vasodilation. Small resistance arteries constrict in response to rising intraluminal pressure (vascular autoregulatory myogenic response) via activation of mechanosensitive TRPC6, TRPM4, and TRPV2, as well as Ca_V channels. Activation of thrombin receptors or proteinase-activated receptor-1 in pulmonary endothelial cells induces a calcium influx through TRPC4 and an increase in endothelial capillary permeability.

Channel	Location	Function
TRPC1	EC, SMC	Store-operated calcium influx
TRPC2	EC, SMC	
TRPC3	EC, SMC	SOC, ROC, UTP-induced depolarization, oxidative stress response
TRPC4	EC, SMC	SOC, ROC, endothelium-dependent vasodilation, endothelial permeability (thrombin), hypoxia-induced EC proliferation
TRPC5	EC, SMC	
TRPC6	EC, SMC	Pressure-induced vasoconstriction
TRPC7	EC, SMC	
TRPM1		
TRPM2		
TRPM3		
TRPM4	EC, SMC	Pressure-induced vasoconstriction
TRPM5		
TRPM6		
TRPM7		
TRPM8		
TRPV1		
TRPV2	EC, SMC	Pressure-induced vasoconstriction
TRPV3		
TRPV4	EC, SMC	EET-mediated vasodilation, thermosensitivity

phospholipase-C, thereby raising cytosolic Ca^{++} level with an early, IP_3-mediated, transient phase followed by a secondary, TRP-induced, sustained phase.

In the heart, the 2 TRPC groups function in coordinated complexes [150]. Inhibition of TRPCs in cardiomyocytes reduces the activity in the PP3–NFAT (nuclear factor of activated T cell) pathway that is involved in maladaptive hypertrophy.

2.3.4 Families of Transient Receptor Potential Channels

2.3.4.1 Ankyrin-like Transient Receptor Potential Channel

The TRPA1 channel [44] is the single mammalian member of the ankyrin-like transient receptor potential channel family. It is permeable to cations with a similar affinity for mono- and divalent cations ($P_{Ca}/P_{Na} = 0.8$).

The TRPA1 channel is frequently coexpressed with the TRPV1 channel. This outward rectifier inactivates in response to prolonged cooling, but is activated by arachidonic acid upon stimulation of phospholipase-C as well as intracellular Zn^{++} and elevated intracellular Ca^{++} level [5].

The TRPA1 channel is produced in a subset of peripheral sensory neurons that express the nociceptive markers calcitonin gene-related peptide and substance-P [143]. It is synthesized in unmyelinated and thinly myelinated sensory neurons (in approximately 50% of TRPV1+ peptidergic neurons). [45]

It can be activated by numerous substances. Endogenous agonists include products of oxidative stress and lipid peroxidation pathways, e.g., 4-hydroxynonenal and 4-oxononenal [151]. Exogenous agonists comprise garlic (allicin).

The TRPA1 channel participates in chemosensation, i.e., response to chemical irritants by subsets of sensory neurons in the lung, eye, and mucous membranes, and pain transduction [152]. Endocannabinoids such as anandamide, a polyunsaturated fatty acid amide, not only activate G-protein-coupled cannabinoid receptors CB_1 and CB_2 (Sect. 7.13.13), [46] but also modulate activity of TRPA1, which can be considered as an ionotropic cannabinoid receptor. The TRPA1 channel may also contribute to cold sensing by temperature-sensitive neurons.

In the cardiovascular system, TRPA1 can cause a vasodilation [151]. Agonists of TRPA1 can stimulate CGRP-dependent, cyclooxygenase-independent relaxation of rat mesenteric arteries as well as that of cerebral arteries via endothelial $K_{Ca}3.1$ channels. Agonists of TRPA1 can also induce a peripheral vasodilation, thereby rising blood flow in the skin. The relaxation of smooth muscle cells results from endothelial-dependent (via nitric oxide) and -independent mechanisms. Agonists of TRPA1 can act on peripheral nerves and cause a local sensory neurogenic vasodilation and bradycardia associated with bradypnea. The TRPA1 channel interacts with

44. A.k.a. ankyrin repeat and transmembrane domain-containing protein AnkTM1.

45. It has been detected on murine post-ganglionic sympathetic neurons and, in humans, on motoneurons and neurons of the intestinal myenteric plexus, as well as on basal keratinocytes and rat endothelial cells [151].

46. The CB_1 receptor resides in the central nervous system. It is coupled to Gi protein, hence inhibiting cAMP and Ca^{++} channels, but stimulating K^+ channels. The CB_2 receptor is expressed in immunocytes.

TRPV1; they both localize to sensory nerves. Agonists of TRPA1 may provoke vasodilation owing to the release of sensory nerve-derived vasodilatory neuropeptides such as calcitonin gene-related peptide and substance-P. However, systemic circulatory and pulmonary effects of TRPA1 excitation may mainly be associated with the Bezold-Jarisch reflex initiated by activated mechano- and chemoreceptors and cardiopulmonary afferents, as vagal fibers innervate both the heart and lungs. The vasovagal reflex with an increase in parasympathetic vagal output to the heart can be associated with postganglionic sympathetic activation and noradrenaline release, which explains the relatively sustained hypertensive response that follows a transient hypotensive response [151]. In addition, TRPA1 may also be activated in the nucleus tractus solitarius.

2.3.4.2 Canonical Transient Receptor Potential Channels

Canonical transient receptor potential channels can be activated by 3 distinct mechanisms according to the subunit composition of TRPC homo- and heterotetramers or the signaling environment [138] (Table 2.16): (1) phospholipase-C-mediated activation (activation of the TRPC epifamily-2 [also TRPC group-3]), as PLC increases diacylglycerol level and lowers phosphatidylinositol (4,5)-bisphosphate content in the plasma membrane;[47] (2) stimulation of exocytosis of TRPC channels sequestered in vesicles;[48] and (3) depletion of intracellular Ca^{++} stores (hence their name: store-operated or capacitative calcium entry channels).[49] Functioning of TRPC as store-operated channels activated by depletion of intracellular calcium stores or receptor-operated channels insensitive to store depletion is highly controversial [5]. Channels TRPC are linked to signaling molecules via adaptors to generate stable signaling complexes (Table 2.17).

47. Channels TRPC4 and -5 are inhibited by PIP_2 as well as protein kinase-C. Degradation of PIP_2 mediated by PLC relieves this inhibition. On the other hand, PKC activated by IP_3 produced by activated PLC limits resulting channel activation duration. In some circumstances, TRPC activation mode via PLC stimulation by a G-protein-coupled receptor requires Src kinase.

48. Growth factors such as EGF promote the insertion of TRPC5 (but not TRPC1–TRPC5 heterotetramers) from a cell's peripheral pool of vesicles into the plasma membrane. Translocation of TRPC can involve phosphoinositide 3-kinase, Rac1 GTPase, phosphatidylinositol 4-kinase, and phosphatidylinositol 4-phosphate 5-kinase [138]. Rapid vesicular insertion of TRP corresponds to the fast incorporation of channel from a subplasmalemmal pool of vesicles to the cell surface [153]. Nerve, brain-derived, and insulin growth factors promote the rapid exocytosis of TRPC5 into the neuronal membrane.

49. The sensor of intracellular Ca^{++} store depletion is the Ca^{++}-binding membrane protein STIM1. The STIM1 protein redistributes and aggregates into nanodomains at junctions between the endoplasmic reticulum and plasma membrane (plasmerosomes), where it associates with and activates ion channels, such as Ca^{++} release-activated Ca^{++} channel Orai1 and store-operated TRPC1 channel. Store-operated Ca^{++} entry activated by depletion of endoplasmic reticulum Ca^{++} stores through TRPC1 is gated by TRPC1–STIM1 clustering. The latter results from TRPC1 dissociation from plasmalemmal scaffold caveolin-1 [154].

Table 2.16. Activators and functional characteristics of TRPC channels (Source: [5]; X_e: extracellular compound X; GTPγS: guanosine 5′-O-[γ-thio]triphosphate, a G-protein-activating analog of guanosine triphosphate. LPC: lysophosphatidylcholine).

Type	Activators	Functional characteristics
TRPC1	Gq/11, PLCγ, stretch, nitrosylation (NO)	Mono- and divalent cations, monovalent cation flux blocked by Ca_e^{++}
TRPC3	Gq/11, PLCγ, $PI(4,5)P_2$	Mono- and divalent cations ($P_{Ca}/P_{Na} = 1.6$), monovalent cation flux blocked by Ca_e^{++}, dual (inward and outward) rectification, relieved of inhibition by Ca^{++}–calmodulin, by IP_3Rs, inhibition by PKG
TRPC4	Gq/11, GTPγS with Ca_e^{++}, nitrosylation (NO)	Dual rectification, inhibition by $PI(4,5)P_2$, potentiation by H_e^+
TRPC5	Gq/11, IP_3, LPC, $\uparrow[Ca^{++}]_e$	Mono- and divalent cations ($P_{Ca}/P_{Na} = 1.8$–9.5), potentiation by H_e^+, dual rectification (homomer), but outward rectifier (TRPC5–TRPC1 or –TRPC4)
TRPC6	Gq/11, 20HETE, stretch, Src kinases	Mono- and divalent cations ($P_{Ca}/P_{Na} = 4.5$–5.0), monovalent cation flux blocked by Ca_e^{++} and Mg_e^{++}, dual rectification
TRPC7	Gq/11	Non-selective mono- and divalent cations ($P_{Ca}/P_{Na} = 5.9$), inhibited by Ca^{++}–calmodulin, monovalent cation flux blocked by Ca_e^{++} and Mg_e^{++}, modest outward rectification

TRPC1

The full-length TRPC1 polypeptide is associated with several splice variants. It is widely expressed in different cell types. [50] In cells, it resides in the plasma membrane and in the endoplasmic reticulum and Golgi body [156]. It heterotetramerizes with other TRPC polypeptides.

50. E.g., the brain, heart, liver, salivary glands, testis, and ovaries, and, in particular, in neurons and smooth muscle and endothelial cells.

Table 2.17. Partners of TRPC channels and resulting effects (Source: [155]; CaBP: Ca^{++}-binding protein that is specifically expressed in neurons; NCX: Na^{+}–Ca^{++} exchanger; NHERF: Na^{+}–H^{+} exchanger regulatory factor; PLC: phospholipase-C; VAMP: vesicle-associated membrane protein; ZO: zonula occludens).

Partner	Effect
CaBP1	TRPC5 inactivation
Calmodulin	TRPC inactivation
Caveolin-1	TRPC1/3 transport and anchoring
Diacylglycerol	TRPC3/6/7 activation
	(activation by DAG faster than phosphorylation by PKC),
	relieves PKC-induced inhibition of TRPC4/5
	(inhibition by PKC quicker than activation by DAG)
NHERF	TRPC4/5 plasmalemmal organization
Homer	TRPC1/2/5 activation and inactivation
	TRPC–IP_3R linking
Immunophilin	Assembly of TRPC–FKBP12 or -52 complex
IP_3R	TRPC activation
Junctate	TRPC–IP_3R linking
NCX1	TRPC3 coupling for Ca^{++} entry
PLC	TRPC3 functioning
RhoA	TRPC1 functioning
Stathmin	TRPC5 microtubular transport
VAMP2	TRPC3 vesicular trafficking
ZO1	TRPC4 scaffolding in astrocytes

The TRPC1 channel interacts with calmodulin, caveolin-1, IP_3 receptor, Homer, phospholipase-C and several other proteins [156].[51] Cation entry through TRPC1 is inhibited by Ca^{++} through interaction with calmodulin (negative feedback).

The TRPC1 channel may be serve as store-operated Ca^{++} channel (SOC) by direct interaction with IP_3R as well as non-SOC, plasmalemmal, Ca^{++}-permeable, channel activated by direct interaction with IP_3R or stretch [156].

The TRPC1 channel participates in smooth muscle contraction, stem cell differentiation, endothelium-induced vasoconstriction, endothelial cell permeability, salivary gland secretion, glutamate-mediated neurotransmission, growth cone movement, neuroprotection, neuronal differentiation and the regulation of liver cell volume [156].

TRPC3

Lipid-regulated homomeric TRPC3 is permeable to Ca^{++} and, to a much lower extent, Na^{+} cation. In the heart, TRPCs are affectors of the Ca^{++}–PP3–NFAT path-

51. E.g., $G\alpha_{q/11}$, glutamate metabotropic receptor-1α, fibroblast growth factor receptor-1, RhoA GTPase, ATPases PMCA and SERCA, PKCα, β-tubulin, antiviral dynamin family member myxovirus resistance Mx1, and TRPP2.

way that controls the transcription of cardiac genes involved in the hypertrophic remodeling of the heart (Vol. 5 – Chap. 5. Cardiomyocytes). In murine atriomyocytes, protein kinase C modulates TRPC3 activity, thereby switching between TRPC3-mediated transcriptional activation and inactivation [157].

With-no-lysine kinase WNK4 regulates diverse ion transporters in the kidney and vasculature to control blood pressure. In vascular smooth muscle cells, WNK4 restricts Ca^{++} influx through TRPC3 channel [158]. Therefore, WNK4 controls arterial resistance to avoid hypertension.

TRPC4

Ion channels formed by TRPC4 are permeable to Ca^{++}, but with a variable Ca^{++} selectivity according to the channel subtype. Among numerous splice variants of TRPC4, TRPC4α and TRPC4β are the most abundant isoforms [159]. Its binding partners include protein 4.1, regulator SLC9a3R1,[52] calmodulin, and IP_3 receptors. The TRPC4 channel supports Ca^{++} influx needed for the activity of vascular endothelial cells as well as for neurotransmitter release from thalamic interneuronal dendrites [159].

TRPC5

Homomultimeric TRPC5 and heteromultimeric TRPC1–TRPC5 channels can be activated by reduced extracellular levels of thioredoxin. Thioredoxin is an intracellular redox protein that can be secreted. It breaks a disulphide bridge in the extracellular loop adjacent to the ion-selectivity filter [160].

TRPC6

Regional alveolar hypoxia causes local vasoconstriction in the pulmonary arterial bed to shift blood flow to well-ventilated regions and, hence, preserve the ventilation–perfusion ratio. The TRPC6 channel regulates acute hypoxic pulmonary vasoconstriction in smooth muscle cells of pulmonary arterioles [161]. Yet, pulmonary hypertension caused by chronic hypoxia does not depend on TRPC6 activity.

TRPC7

Like TRPC3 and TRPC6, TRPC7 is directly activated by diacylglycerol. It has a constitutive activity. It is inhibited by extracellular Ca^{++} ion [162]. In humans, it resides in endothelial cells, but not in smooth muscle cells.

---

52. A.k.a. ezrin–radixin–moesin-binding phosphoprotein EBP50 and Na^+–H^+ exchange regulatory cofactor NHERF1.

2.3.4.3 Melastatin-Related Transient Receptor Potential Channels

The family of melastatin-related transient receptor potential channels encompasses 8 members that can be classified into 4 groups on the basis of sequence homology: group 1 with TRPM1 and -3; group 2 with TRPM2 and -8; group 3 with TRPM4 and -5; and group 4 with TRPM6 and -7 (Table 2.18). Channels TRPM1 and TRPM2 localize to intracellular membranes. They are involved in melanin synthesis and oxidative stress-induced cell death, respectively. On the other hand, TRPM4 and TRPM5 are impermeable to Ca^{++} ions.

TRPM1 Channel

Melastatin-related transient receptor potential channel TRPM1 is downregulated in melanoma cell lines. Channel TRPM1 may exist as 5 splice variants. It is involved in normal melanocyte pigmentation [5].

TRPM2 Channel

Melastatin-related transient receptor potential channel TRPM2 [53] is a Ca^{++}-permeable, non-selective cation channel. It is involved in oxidative stress-induced cell death and inflammation. The TRPM2 channel is an ^{ADP}ribose hydrolase that may act as a redox sensor. It is predominantly expressed in the brain, but also in the heart, lung, liver, spleen, and bone marrow.

The TRPM2 channel can be activated by oxidative stress as well as ^{ADP}ribose (ADPR), cyclic ^{ADP}ribose (cADPR), NAD^+, and nicotinic acid adenine dinucleotide phosphate (NAADP), at least in vitro. Intracellular Ca^{++} can activate TRPM2 and its alternatively spliced isoforms [54] independently of ^{ADP}ribose [163]. On the other hand, ^{ADP}ribose that can activate TRPM2 (but not its splice variants) requires both adequate intracellular Ca^{++} concentration and functional calmodulin-binding motif.

TRPM3 Channel

Changes of the extracellular osmolarity provoke cell swelling or shrinkage that results from increase or decrease in intracellular water content in association with ionic concentration gradient. Cells modulate the activity of their ion carrier, hence intracellular ion concentrations, to adapt to environmental conditions.

Transient receptor potential channel melastatin TRPM3 as well as vanilloid-like TRPV4 channel are volume-regulated cation channels [164]. The TRPM3 channel

53. A.k.a. transient receptor potential cation channel TRPC7, estrogen responsive element-associated gene EREG1, and long transient receptor potential channel LTRPC2.

54. Splice variants of TRPM2 include shorter form ($TRPM2_S$), C-terminal ($TRPM2_{\Delta C}$) and N-terminal ($TRPM2_{\Delta N}$) truncated fragments. Isoform $TRPM2_S$ prevents activation of full-length $TRPM2_L$ by ROS (H_2O_2).

Table 2.18. Activators and functional characteristics of TRPM channels (Source: [5]; $X_{e(i)}$: extracellular (intracellular) compound X; ADPR: ^{ADP}ribose; cADPR: cyclic ADPR; RNS, ROS: reactive nitrogen, oxygen species).

Type	Activators	Functional characteristics
TRPM1	Constitutively active outward rectification	Mono- and divalent cations,
TRPM2	$ADPR_e$, cADPR, RNS, ROS, Ca^{++}–calmodulin, heat (∼35 C)	Mono- and divalent cations ($P_{Ca}/P_{Na} = 0.6$–0.7), inactivation at negative potentials, potentiation by arachidonic acid, no rectification
TRPM3	Small constitutive activity, strong depolarization, cell swelling	TRPM3α1: monovalent cations TRPM3α2: mono- and divalent cations
TRPM4	Ca_i^{++}, $PI(4,5)P_2$, depolarization, heat	Monovalent cations, strong outward rectification, slow activation at positive potentials, rapid deactivation at negative potentials
TRPM5	Gq/11, Ca_i^{++}, IP_3, $PI(4,5)P_2$, depolarization, heat	Monovalent cations ($P_{Ca}/P_{Na} = 0.05$), strong outward rectification, transient activation and subsequent desensitization by Ca_i^{++}, slow activation at positive potentials, rapid deactivation at negative potentials
TRPM6	Constitutively active, $\downarrow Mg_i^{++}$	Mono- and divalent cations ($Mg^{++} > Ca^{++}$; $P_{Ca}/P_{Na} = 6.9$), potentiation by H_e^+, strong outward rectification, inhibition by Mg_i^{++} and ATP_i
TRPM7	Gs, $PI(4,5)P_2$	Mono- and divalent cations ($P_{Ca}/P_{Na} = 0.34$), potentiation by ATP_i, H_e^+, inhibition by intracellular divalent cations, inhibition by Gi, outward rectification
TRPM8	Depolarization, cooling (<22–26C) $PI(4,5)P_2$	Mono- and divalent cations ($P_{Ca}/P_{Na} = 1.0$–3.3), modulation by LPC, outward rectification

exists as multiple splice variants that differ in their biophysical properties. It is characterized by variability in transcripts due to alternative start positions and different C-terminal ends. Variants of TRPM3 have, in particular, different activation features.

In the kidney, TRPM3 is mainly located in collecting duct as well as, to a lesser extent, proximal convoluted tubule. The TRPM3 channel is also detected in the central nervous system,[55] pancreas, and ovary [164]. It is activated by sphingosine and its precursor dihydrosphingosine, but not its metabolites ceramides and sphingosine 1-phosphate [164].

TRPM4 Channel

The TRPM4 channel is a Ca^{++}-activated, voltage-gated, non-selective cation channel. It is synthesized in many tissues and cell types. Its expression reaches its highest levels in heart, pancreas, colon, prostate, and placenta [165]. Two splice variants exist. Full-length TRPM4b carries monovalent cations. The TRPM4b variant and TRPM5 can act as calcium-activated cation channels.

The TRPM4 channel regulates Ca^{++} entry into mastocytes and dendritic cells during migration [5].

The activity of TRPM4 is attenuated by intracellular ATP and other adenine nucleotides, but, unlike TRPM5, is insensitive to acid pH. A weak voltage dependence modulates channel gating by intracellular Ca^{++} as well as $PI(4,5)P_2$-mediated regulation. Several positively charged residues in the transmembrane segment 4 may act as voltage sensors.

The TRPM4 channel is a temperature-sensitive, heat-activated ion channels. Inward currents induced by TRPM4 rise abruptly at temperatures between 288 and 308 K (15 and 35 °C). In addition, TRPM4 contributes to dampen the cytokine secretion by T lymphocytes[56] and constriction of smooth muscle cells of cerebral arteries [165]. Following activation, Ca^{++} sensitivity of TRPM4 decreases. However, intracellular ATP^{Mg} at a concentration that is able to activate lipid kinases and $PI(4,5)P_2$ restores TRPM4 currents.

TRPM5 Channel

The TRPM5 channel is activated directly by an elevated Ca^{++} concentration, especially upon activation of G-protein-coupled taste receptors, transducing sweet

55. I.e., cerebellum, choroid plexi, locus coerulus, posterior hypothalamus, and substantia nigra.

56. Activated T-cell receptors prime the PLC pathway that leads to depletion of intracellular Ca^{++} stores and subsequent Ca^{++} entry through Ca^{++}-release-activated Ca^{++} channels. This Ca^{++} influx activates TRPM4 and Ca^{++}-activated K^+ channels. Sodium influx through TRPM4 depolarizes the cell membrane that promotes the opening of voltage-gated K^+ channels and impedes further Ca^{++} entry. This set of interacting channels thus causes Ca^{++} oscillations in response to stimulated T-cell receptors [165].

and bitter stimuli [5].[57] It is detected at high levels not only in the tongue, but also in the small intestine and stomach, as well as, to a lesser extent, other tissues [165].

Like TRPM4, TRPM5 is a monovalent-specific ion channel [143]. In the presence of a stimulating intracellular Ca^{++} concentration, cell membrane depolarization raises the opening probabilility of the TRPM5 channel. Like TRPM4, Ca^{++} sensitivity of TRPM5 lowers once TRPM5 has been activated [165]. The TRPM5 channel is inhibited by reduced pH (≤ 6), but, unlike TRPM4, is insensitive to ATP and other adenine nucleotides. Like TRPM4, TRPM5 is a heat-activated ion channel. Elevated temperature (288–308 K, i.e., 15–35 °C) enhances the gustatory response to sweet compounds [143].

TRPM6 Chanzyme

Transient receptor potential channel melastatin TRPM6 as well as its closest TRP family member TRPM7 possesses a kinase domain that is absent in other TRP channels.

TRPM7 Chanzyme

Melastatin-related transient receptor potential channel TRPM7 has several funtions in vascular endothelial cells. The TRPM7 channel is a non-selective cation channel with predominant permeability for Ca^{++} and Mg^{++} ions.

Channel TRPM7 hinders the growth and proliferation of endothelial cells [166]. It lowers the activity of extracellular signal-regulated kinase. In addition, TRPM7 impedes nitric oxide production by nitric oxide synthase.

TRPM8 Channel

Melastatin-related transient receptor potential channel TRPM8[58] is a non-selective cation channel with relatively high Ca^{++} permeability [143]:

$$\mathcal{P}_{Ca}/\mathcal{P}_{Na} = 3.3 .$$

Activation of TRPM8 by depolarization depends strongly on temperature, as the channel closing rate lowers with decreasing temperature.

The TRPM8 channel is expressed in a subpopulation of sensory neurons as well as in the respiratory and urinary tract and vasculature. It is a cold sensor associated

57. Upon ligand-binding, G-protein-coupled taste receptor triggers the Gβγ–PLCβ2 axis that leads to messengers DAG and IP_3. The IP_3 mediator activates its cognate receptors to release Ca^{++} from intracellular stores. Elevation in intracellular Ca^{++} level opens TRPM5 channels and the resulting Na^+ influx depolarizes the taste cell membrane and releases transmitter.

58. A.k.a. cold- and menthol-sensitive receptor CMR1. Menthol that is found in mint provokes a cooling sensation. The TRPM8 channel can also be activated by other cooling compounds, such as menthone, eucalyptol, and icilin, albeit with lower efficacy than menthol.

with polymodal nociceptors that respond to noxious cold, heat, and pinching and connect to Aδ- and C-sensory nerves of a subset of neurons of the dorsal root and trigeminal sensory ganglia, where it is coexpressed with receptor Tyr kinase NTRK1 (but neither NTRK2 nor NTRK3) or temperature-sensitive TRPA1 channel [168]. Phosphatidylinositol (4,5)-bisphosphate modulates the capacity of TRPM8 to respond to cold.

In arterial smooth muscle cells, according to the vasomotor tone, activated TRPM8 causes a vasoconstriction or a vasodilation [167]. Hydrolysis of PIP_2 by activated phospholipase-C hinders TRPM8 function. Protein kinase-C that operates upstream from TRPM8 and does not phosphorylate TRPM8 lowers TRPM8 activity. Increased intracellular pH also precludes TRPM8 functioning.

2.3.4.4 Mucolipin-Related Transient Potential Channels

Transient receptor potential mucolipin channels are located in the plasma and intracellular membranes. The TRPML family consists of 3 mammalian members (TRPML1–TRPML3).

The inward rectifier TRPML1 that is encoded by the gene MCOLN1 resides in late endosomes and lysosomes. It contributes to sorting and transport of late endosomes as well as fusion between late endosomes and lysosomes.

The other members of the TRPML family localize to specific endosomes and lysosomes as well as the plasma membrane. The TRPML channels can function as calcium-permeable channels.

The TRPML3 channel is activated by extracellular Na^+ and membrane depolarization [5]. This outward rectifier carries Na^+ better than K^+, in addition to Ca^{++} (P_{Ca}/P_K ~350).

2.3.4.5 Polycystin-Related Transient Potential Channels

In humans, the polycystin family contains 8 widespread members. These proteins can be subdivided according to their structure into groups 1 of TRPP1-like and 2 of TRPP2-like proteins. In addition to TRPP1,[59] the TRPP1 subfamily includes polycystic kidney disease (PKD; or polycystin) and receptor for egg jelly (REJ) homolog PKDREJ, and related PKD1L1 to PKD1L3 protein. The TRPP2 subfamily contains TRPP2 (or PKD2), TRPP3 (or PKD2L1), and TRPP5 (or PKD2L2) [5].[60]

Polycystin-related transient receptor potential channel TRPP1 does not produce ionic current, but constitutively activates G proteins and subsequently Jun N-terminal

59. A.k.a. polycystic kidney disease PKD1 protein (not protein kinase-D1 [PKD1] and polycystin-1 (PC1).

60. The Trpp1 (PKD1) and Trpp2 (PKD2) genes can bear loss-of-function mutations that provoke autosomal dominant *polycystic kidney disease* (ADPKD). This nephropathy is characterized by defects in polarization and function of epithelial cells of the nephron that cause abnormal tubular cell growth and fluid-filled cysts. This disease is often associated with hypertension, intracranial aneurysms, and mitral valve prolapse.

kinase. Proteins TRPP1 and TRPP2 couple to act as a signaling complex. This complex suppresses stimulation of G protein that results from the constitutive activity of TRPP1 and TRPP2.

TRPP1

Channel TRPP1 can undergo a proteolytic cleavage (Sect. 7.3.2). The released C-terminus then translocates to the nucleus and activates the AP1 transcription factor. Channel TRPP1 can also activate signaling primed by G proteins, such as the phospholipase-C and adenylate cyclase axes, eventually in collaboration with TRPP2 [169]. Although TRPP1 distribution among cell types and within the cell differs from that of TRPP2, thereby having independent functions, TRPP2 and TRRP1 channels can operate as interacting partners once they heteromultimerize by coassembly of their C-termini.

The TRPP1–TRPP2 heteromer promotes TRPP2 translocation to and retention in the plasma membrane. It serves as non-selective cation channel with permeability and sensitivity to di- and trivalent cations similar to those of the TRPP2 homomer. In the TRPP1–TRPP2 complex, both TRPP types have reciprocal stabilizing effects on each other's activity [169]. Channel TRPP1 represses TRPP2 constitutive activity to avoid uncontrolled, detrimental function. Conversely, TRPP2 precludes TRPP1 constitutive activation of G proteins. The TRPP1–TRPP2 heteromer thus not only controls TRPP2 activity, but also that of G proteins. Channels TRPP1 and TRPP2 colocalize in the primary cilium[61] of renal epithelial cells, where they might sense fluid-applied stress.

TRPP2

The TRPP2 channel is implicated in Ca^{++}-dependent mechanosensation associated with cell proliferation, among other functions. It links to cortactin and cadherin via TRPP1. Moreover, its activity increases when it connects to α-actinin.

Channels related to TRPP2 have large conductance for mono- and divalent cations, such as Na^+, K^+, Ba^{++}, and Ca^{++} [169].

Channel TRPP2 is expressed in various tissues (e.g., adrenal gland and ovary) and cell types, such as epithelial and vascular smooth muscle cells as well as cardiomyocytes. It causes Ca^{++} influx at both the plasma (particularly at its basolateral segment) and primary ciliary membranes, where it can heteromultimerize with TRPP1, especially in primary cilium, as well as possibly other members of the TRP channel superfamily, such as TRPC1 and TRPV4 [169]. In addition, TRPP2 acts as a Ca^{++}-release channel in the endoplasmic reticulum that is directly activated by Ca^{++}, where it may interact with IP_3Rs to amplify Ca^{++} transients. It has also been detected in mitotic spindles during the cell division cycle.

61. The primary cilium is a single, immotile, microtubule-based, cortical structure that arises from the centriole and projects into the lumen of the tubule. Proteins that contribute to the assembly or function of primary cilia include cystin, polaris, inversin, and kinesin-2.

Phosphorylation of TRPP2 (Ser812) by casein kinase-2 markedly increases TRPP2 sensitivity to Ca^{++}.[62] However, TRPP2 dephosphorylation is required for its translocation to the plasma membrane. Channel TRPP2 actually travels between the endoplasmic reticulum, Golgi body, and plasma membrane. Accumulation of TRPP2 at the plasma membrane is prevented by casein kinase-2 that phosphorylates TRPP2, thereby promoting the binding of phosphofurin acidic cluster sorting adaptors PACS1 and PACS2 [169]. 14 kDa Polycystin-2 interactor, Golgi body and endoplasmic reticulum-associated protein PIGEA14 also interacts with the TRPP2 C-terminus and causes the redistribution of TRPP2 from the endoplasmic reticulum to the trans-Golgi network.

TRPP3

The TRPP3 channel is characterized by a low constitutive activity that is enhanced by membrane depolarization. Changes in cell volume affect its voltage-dependent gating. Furthermore, it is activated and subsequently inactivated by intracellular Ca^{++} ions. The TRPP3 channel is more permeable to Ca^{++} than to monovalent cations ($\mathcal{P}_{Ca}/\mathcal{P}_{Na} = 4.0$–$4.3$), whereas TRPP2 has a lower selectivity ($\mathcal{P}_{Ca}/\mathcal{P}_{Na} = 1$–$3$).

Unlike TRPP2 that is retained in the endoplasmic reticulum by phosphofurin acidic cluster sorting proteins (PACS), TRPP3 as well as TRPP5 lack the PACS-binding in their C-termini. Like TRPP2, TRPP3 is activated by Ca^{++} ions.

2.3.4.6 Vanilloid Receptor-Related Transient Potential Channels

Vanilloid receptor-related transient potential channels are sensitive to temperature with various temperature sensitivities, except TRPV5 and TRPV6: warm (TRPV3 and TRPV4; TRPV4 temperature activation threshold 298–306 K [25–33 °C]), hot (TRPV1), and very hot (TRPV2).

Channels TRPV1 to TRPV4 differ from TRPV5 and TRPV6 not only in their sequences, but also in their biophysical properties (Table 2.19). Based on structure and function, 4 groups of tetrameric TRPV channels can be defined : group 1 with TRPV1 and -2; 2 with TRPV3; 3 with TRPV4; and 4 with TRPV5 and -6.

TRPV1

Pain and heat perception relies on neural signals that originate from terminals of nociceptors and propagate into neurons of the spinal cord and brainstem to be transmitted to specific brain regions. Activated ion channels on nociceptor terminals depolarize corresponding neurons. Among these ion channels, members of the TRP superfamily, especially TRPV1, play a crucial role [170]. The concentration of TRPV1 reaches its highest level in sensory neurons, especially in those of the dorsal

62. The maximum open probability of TRPP2 indeed occurs for a Ca^{++} concentration at which the non-phosphorylated form remains closed.

Table 2.19. Activators and functional characteristics of TRPV channels (Source: [5]).

Type	Activators	Functional characteristics
TRPV1	Depolarization, nitrosylation (NO), heat (>43C at pH 7.4), H_e^+, some eicosanoids,	Mono- and divalent cations ($P_{Ca}/P_{Na} = 9.6$), outward rectification, inhibition by Ca^{++}–calmodulin, inhibition by cooling
TRPV2	Cannabidiol	Mono- and divalent cations ($P_{Ca}/P_{Na} = 0.9$–2.9), dual rectification, translocation to plasma membrane in response to IGF1 and stretch
TRPV3	Heat (23–39C), nitrosylation (NO)	Mono- and divalent cations, outward rectification, potentiation by arachidonic acid
TRPV4	Constitutively active, nitrosylation (NO), heat (>24–32C), cell swelling, epoxyeicosatrieonic acids,	Mono- and divalent cations ($P_{Ca}/P_{Na} = 6$–10), dual rectification, potentiation by Ca^{++}–calmodulin, Src, sensitization by PKC, inhibition by $\uparrow Ca_i^{++}$
TRPV5	Constitutively active,	Mono- and divalent cations ($P_{Ca}/P_{Na} > 107$), inward rectification, feedback inhibition by Ca^{++}, Inhibition by acidosis
TRPV6	Constitutively active,	Mono- and divalent cations ($P_{Ca}/P_{Na} > 130$), inward rectification, inhibition by Ca_i^{++}, Mg_i^{++}, slow inactivation by Ca^{++}–calmodulin, phosphorylation by PKC impedes Ca^{++}-calmodulin binding

root and trigeminal sensory ganglia that may correspond to unmyelinated, polymodal C-fibers involved in nociception [143].

The activation of TRPV1 by depolarization depends on temperature. Channel opening rate increases with rising temperature. The TRPV1 channel can be activated by heat above a temperature threshold of 316 K (43 °C). [63] Heat may initiate a painful response.

63. The TRPV1 channel may alternatively be activated at usual ambiant temperature when the hydrogen ion concentration heightens (pH < 6.0) [143].

A TRPV1 variant in magnocellular neurons of the supraoptic and paraventricular nuclei of the hypothalamus as well as osmotically sensitive circumventricular organ OVLT (organum vasculosum lamina terminalis) participate in response to hypertonicity, such as vasopressin secretion [171]. Some splice variants of TRPV1 inhibit coexpressed TRPV1 [5].

The TRPV1 channel primes influx of calcium and sodium ions, with a high Ca^{++} permeability ($\mathcal{P}_{Ca}/\mathcal{P}_{Na} = 9.6$) [143]). It exists as homo- and heteromers that are linked to regulatory proteins.

Many TRPV1 motifs modulate channel function, as they serve in [170]: (1) multimerization; (2) channel permeability and ion selectivity; (3) voltage, hydrogen ion, [64] and heat activation; [65] (4) sensitization by protein kinases; (5) desensitization by Ca^{++}-calmodulin binding to TRPV1 C-terminus; and (6) modulation by lipids. [66]

The TRPV1 channel is phosphorylated by protein kinase-A [67] and -C, [68] especially PKCϵ and PKCμ isoforms, as well as calmodulin-dependent kinase CamK2 [69] and Src kinase. Phosphorylation of TRPV1 by PKA sensitizes TRPV1 to heat. Heat-activated TRPV1 is sensitized by protein kinase-Cϵ. Phosphorylation of TRPV1 by PKC potentiates hydrogen ion-mediated response and reduces the temperature threshold for TRPV1 activation.

TRPV1-sensitizing agents, such as adenosine, extracellular ATP, bradykinin, histamine, prostaglandins (PGE2 and PGI2), serotonin, trypsin, and tryptase, bind to their cognate G-protein-coupled receptors to stimulate protein kinase-A or -C and inhibit phosphatidylinositol (4,5)-bisphosphate, as well as possibly involve 12-hydroperoxyeicosatetraenoic acid [143, 170]. [70]

Eicosanoids and endocannabinoids produced by lipoxygenase activate TRPV1 channel. Fatty acid amide hydrolase targets the cannabinoid anandamide [71] to produce arachidonic acid and ethanolamide [170]. Arachidonic acid is processed by lipoxygenase into TRPV1 ligands, such as leukotriene-B4, 5-hydroxyeicosatetraenoic acid (HETE), and 12- and 15-hydroperoxyeicosatetraenoic acids (HPETE). Anandamide processed by lipoxygenase yields HETE ethanolamides (HETEE) and HPETE ethanolamides (HPETEE) that can also be TRPV1 agonists. Products of anandamide after lipoxygenase processing are potent inhibitors of fatty acid amide

64. Pain sensation is heightened by acidic extracellular pH during ischemia and inflammation.

65. Like TRPM8 and TRPA1, TRPV1 to TRPV4 are thermosensitive.

66. The TRPV1 channel is inactivated by PIP_2 binding. Phospholipase-C-mediated PIP_2 hydrolysis relieves this inhibition. Anandamide and some lipoxygenase products activate TRPV1 channel.

67. At Ser116, Thr144, Thr370, and Ser502 residues.

68. At Ser502, Thr704, and Ser800 residues.

69. At Ser502 and Thr704 residues.

70. These inflammatory mediators potentiate hydrogen ion action that results from inflammation-induced tissue acidification and lower the temperature threshold for heat activation.

71. Anandamide is synthesized from phosphodiesterase-mediated cleavage of the phospholipid precursor ^{N}arachidonoyl-phosphatidylethanolamine in a calcium-dependent fashion.

hydrolase. In addition, protein kinase-C that directly activates TRPV1 also sensitizes TRPV1 to its ligands. Moreover, anandamide directly activates protein kinase-C.

Elevated arterial pressure is able to provoke the generation of 20-hydroxy-eicosatetraenoic acid that activates TRPV1 in C-fibers, thereby leading to nerve depolarization and vasoactive neuropeptide release [171].

TRPV2

The ligand-operated, voltage-independent, non-selective Ca^{++} channel TRPV2 possesses multiple phosphorylation sites for protein kinase-A (residues 98 and 326) and -C (residues 97, 111, 322, and 734), and protein Tyr kinase (residues 106, 223, and 330).

This ligand-operated, voltage-independent, non-selective, calcium-permeable, cation channel abounds in the central nervous system, lung, and liver. It is also synthesized in the heart, blood vessels, kidneys, gastrointestinal tract, pancreas, skeletal muscle, and adipose tissues.

Channel TRPV2 is activated by high temperatures (threshold >325 K [52 °C]), but not moderate heat, vanilloids, and hydrogen ions. It is then associated with type-1 Aδ mechano- and heat-sensitive nociceptors. It has a relatively high Ca^{++} permeability ($\mathcal{P}_{Ca}/\mathcal{P}_{Na} = 2.9$). It resides in the dorsal root sensory ganglia. In the spinal cord, it lodges in lamina-1 and inner lamina-2 related to myelinated nociceptors as well as lamina-3 and -4 that are targeted by non-nociceptive Aβ fibers [143].

In arterial smooth muscle cells that produce the TRPV2 channel is d irespond to hypotonic stimulus by allowing calcium influx [171]. In addition, calcium entry is required for cell-cycle progression. Insulin-like growth factor-1 is a progression factor that causes TRPV2 translocation to the plasma membrane [172].

The TRPV2 channel corresponds to the growth factor-regulated, Ca^{++}-permeable cation channel (GRC). Insulin-like growth factor-1 stimulates calcium entry for cell cycle progression independently of PI3K and the Ras pathway through this growth factor-regulated channel [605].

The TRPV2 channel is also regulated by insulin for its autocrine action on pancreatic β-cells [606]. In addition, Head activator that localizes to hypothalamus and intestine at high concentrations in humans activates and causes translocation of the growth factor-regulated Ca^{++}-permeable channel TRPV2 [604].

TRPV3

The TRPV3 channel is activated by warm temperatures (307–311 K [34–38 °C]). It is produced in sensory and hypothalamic neurons and keratinocytes [143]. It can form with TRPV1 a functional hetero-oligomer.

TRPV4

The TRPV4 channel [72] preferentially forms homomers. Some TRPV4 splice variants do not oligomerize and are retained intracellularly. The TRPV4 channel transfers Ca^{++} and Na^{+} ions, with a relatively high Ca^{++} permeability (permeability ratio $\mathcal{P}_{Ca}/\mathcal{P}_{Na} = 6$).

The TRPV4 channel is widely expressed. It is detected in the brain, heart, kidney, liver, skin, etc. More precisely, it localizes to para- and sympathetic nerves, endothelia, respiratory epithelium, and airway and arterial smooth muscles [73]

The TRPV4 channel is implicated in heat sensing, pain, and vascular regulation. In rat vascular smooth muscle cells, angiotensin provokes the transient association of TRPV4 with β-arrestin. [74] Angiotensin G-protein-coupled receptor AT_{1A} signals via both G proteins and β-arrestin to prime different types of cell responses. Adaptor β-arrestin-1 is recruited to the plasmalemmal AT_{1A}–TRPV4 complex [173]. Channel TRPV4 is then ubiquitinated by ubiquitin ligase Itch that interacts with β-arrestin-1, thereby avoiding excessive Ca^{++} influx.

TRPV4 and Physical Stimuli

The TRPV4 channel is highly sensitive to extracellular osmolarity changes. This mechano- and osmosensor maintains cellular osmotic homeostasis. Extracellular hypo-osmolarity activates TRPV4 and causes Ca^{++} influx, whereas hyperosmolarity above 300 mosm/l impedes Ca^{++} flux [174]. Hypotonic stress augments phosphorylation of TRPV4 by members of the SRC family kinases (Vol. 4 – Chap. 3. Cytosolic Protein Tyr Kinases) such as Lyn kinase (Tyr253).

The TRPV4 channel regulates volume change, i.e., causes regulatory volume decrease and increase in response to hypo- and hypertonicity, respectively. For example, it mediates Ca^{++} influx into bronchial epithelial cells that triggers regulatory volume decrease via Ca^{++}-dependent potassium channels [171]. Moreover, TRPV4 interacts with aquaporin-5 in regulatory volume decrease of salivary gland epithelium.

In addition, TRPV4 is a temperature-sensitive channel that is activated by warm temperature (300–308 K [27–35 °C]) [143]. Gating of TRPV4 is augmented by simultaneous hypotonicity and heat. The TRPV4 channel can be activated bya hyperviscous, isotonic medium. It also responds to mechanical stress, acidic pH, and adequate ligands.

---

72. A.k.a. OSM9-like TRP channel OTRPC4, vanilloid receptor-related osmotically activated channel (VROAC), vanilloid receptor-like channel VRL2, and TRP12.

73. The highest TRPV4 expression is detected in kidneys, but restricted to epithelial cells of the water-impermeable part of the nephron, i.e., the thin and thick ascending limbs of the Henle's loop, and distal and connecting tubules.

74. β-Arrestins desensitize various receptor types (G-protein-coupled receptors, receptor Tyr kinases, and receptor Ser/Thr kinases. They also support GPCR endocytosis and G-protein-independent signaling.

TRPV4 and Chemical Stimuli

The TRPV4 channel can be activated by products of metabolism of arachidonic acid and cannabinoid anadamide, such as epoxyeicosatrienoic acids (5,6)- and (8,9)-EETs produced by cytochrome-P450 epoxygenase [174]. Outward and inward fluxes through TRPV4 are strongly reduced by loss of extracellular Ca^{++} as well as reduction in intracellular Ca^{++} concentration.

The TRPV4 channel has a calmodulin-binding site in its intracellular C-terminus. Calmodulin influences the channel's kinetics, but not its activation or inactivation. Upon Ca^{++} entry, Ca^{++}–calmodulin interacts with the C-terminal binding domain to raise both the speed and amplitude of TRPV4 activity (positive feedback) [174]. Yet, Ca^{++} can limit Ca^{++} ingress through TRPV4 and other Ca^{++} channels, such as TRPV1 and TRPV6 (negative feedback). The activity of TRPV4 can also be inhibited by calcium–calmodulin.

Phosphorylation (Tyr253) by SRC family kinases such as Lyn kinase stimulates TRPV4 activity [164]. In addition, microtubule-associated protein MAP7 favors TRPV4 translocation to the plasma membrane.

TRPV4 and Mechanical Stimuli

Hemodynamic stress (without TRPV4 phosphorylation) primes the activity of mechanosensitive Ca^{++}-permeable TRPV4, hence Ca^{++} accumulates in the cytosol. Among its functions, arterial endothelium controls the contractile state of medial smooth muscle cells (Vol. 5 – Chaps. 8. Smooth Muscle Cells and 9. Endothelium). Endothelial cells regulate local arterial lumen bore via vasoactive agents such as the so-called endothelium-derived hyperpolarizing factor. This electrochemical agent as well as the synthesis of some vasodilators, such as nitric oxide and prostacyclin, depend on Ca^{++} [175]. In small arteries, TRPV4 activity initiates endothelial hyperpolarization to cause vasodilation.

Calcium entry through TRPV4 targets Ca^{++}-activated $K_{Ca}2.3$ and $K_{CaCa}3.1$ channels, thereby generating endothelial hyperpolarization that propagates to adjoining smooth muscle cells via myoendothelial gap junctions. In smooth muscle cells, this hyperpolarization inhibits Ca_V1 channels, thereby favoring relaxation. In large arteries, the endothelium-derived hyperpolarizing factor contributes much less than nitric oxide to endothelium-mediated vasodilation. Upon Ca^{++} influx through TRPV4, Ca^{++}–calmodulin activates endothelial NOS3 nitric oxide synthase. Last, the Ca^{++}-dependent release of arachidonic acid by phospholipase-A2 launches prostacyclin synthesis by cyclooxgenase-1.

TRPV4 and Ciliary Beat Frequency

The ciliary beat frequency of ciliated cells such as bronchial epithelial cells is influenced by TRPV4 gating [171]. The ciliated respiratory epithelium conveys the airway lining fluid and mucus layer with entrapped particles (Vol. 5 – Chap. 12. Airway Surface Liquid and Respiratory Mucus). The rate of mucociliary clearance in airways depends on ciliary beat frequency. Among the members of the TRP superfamily, TRPC1, TRPC4, and TRPC6, as well as TRPV2 and TRPV4 reside in the respiratory epithelium.

Both mechanical and chemical paracrine stimulation of epithelial ciliated cells initiate an increase in intracellular Ca^{++} concentration, then ciliary beat frequency. Several intracellular signals influence the ciliary beat frequency, such as cAMP, cGMP, and nitric oxide. G-Protein-coupled P2Y receptors increase the ciliary beat frequency. Agent ATP actually promotes both the release of calcium ions from IP_3-sensitive intracellular stores and entry across the plasma membrane. [75]

The TRPV4 mechanosensor activated by arachidonic acid generated by phospholipase-A2 provokes a Ca^{++} entry and raises the ciliary beat frequency [177]. Cation channel TRPV4 mediates Ca^{++} influx in response to mechanical and osmotic stimuli in ciliated epithelial cells.

The TRPV4 channel contributes to an ATP-induced increase in ciliary beat frequency. It acts in receptor-operated, but not store-operated Ca^{++} entry [176]. [76] Nonetheless, the *autoregulation of ciliary beat frequency* in the presence of a high viscosity medium is preserved in the absence of TRPV4 despite a reduced Ca^{++} signal. [77]

TRPV5

Channels TRPV5 and TRPV6 that function as homo- and heterotetramers are Ca^{++} selective. They are involved in Ca^{++} absorption and reabsorption across intestinal and renal tubule epithelia [5]. However, in the absence of extracellular Ca^{++}, TRPV5 carries monovalent cations. In the presence of extracellular Ca^{++}, Ca^{++}-induced inactivation occurs at hyperpolarized potentials. Blockade of TRPV5 and TRPV6 by extracellular Mg^{++} depends on membrane potential. Channels TRPV5 and TRPV6 differ in their kinetics of Ca^{++}-dependent inactivation and recovery from inactivation.

---

75. Airway epithelial cells release ATP in response to various stimuli, including mechanical stresses during tidal breathing. Basal concentration of ATP in airway surface liquid can increase from the nanomolar to micromolar range in response to some stimuli. Cell response primed by ATP is mainly mediated by $P2Y_2$ receptors (with minor contributions of other receptors). Low ATP concentrations typically generate oscillatory Ca^{++} signals in ciliated cells associated with a store-operated process. A micromolar ATP concentration provokes a peak Ca^{++} release from IP_3-responsive stores and a more sustained Ca^{++} influx through TRPV4 channels [176].

76. Calcium influx can arise once calcium stores are depleted from plasmalemmal Ca^{++} channels, i.e., store-operated STIM–Orai-mediated calcium entry, and/or via phospholipase-C-dependent receptor-operated calcium entry. Different TRP channels respond to one or several signaling molecules that result from stimulation of PLC activity, such as diacylglycerol, inositol triphosphate, or arachidonic acid.

77. Airway ciliated cells are able to maintain their beating frequency in high viscosity conditions. This mechanism that is termed ciliary beat frequency autoregulation depends on Ca^{++} entry and subsequent activation of cilia. Contribution of TRPV4 to this autoregulation is small [176].

TRPV6

The TRPV6 channel exhibits a biphasic inactivation kinetics with a fast component (timescale O[1 ms]) and a following slow, Ca^{++}–calmodulin-dependent component. Intracellular Mg^{++} exerts an inhibition that is alleviated by hyperpolarization and contributes to the time-dependent activation and deactivation of weak monovalent cation flux.

2.3.5 TRP Channels in the Cardiovascular Apparatus

Transient receptor potential channels are activated or modulated by multiple agents. They can prime Ca^{++} influx and then contribute to acute hemodynamic control and long-term remodeling of the cardiovascular apparatus (Tables 2.20 and 2.21).

The expression pattern of TRPs depends on the animal species and location in the vasculature. In vascular smooth muscle cells, the most commonly expressed TRPC isoforms are TRPC1, TRPC4, and TRPC6. Production of TRPC4 stimulates the growth of pulmonary arterial smooth muscle cells. In hearts, the order of relative density of TRPC isoforms is [178]:

$$\text{TRPC1} > \text{TRPC4} \sim \text{TRPC6} > \text{TRPC5} \gg \text{TRPC3}.$$

The TRPC1 channel is involved in neointimal hyperplasia after vascular injury via store (depletion)-operated Ca^{++} entry. The TRPC6 channel contributes to receptor-operated and mechanosensitive Ca^{++} mobilization that causes vasoconstriction and myogenic response as well as pulmonary arterial wall cell proliferation.

The TRPM7 channel serves as a pro-proliferative vascular Mg^{++} entry channel. Polycystin-related TRP channels (TRPP1 and TRPP2) cause Ca^{++} entry, thereby participating in the maintenance of vascular integrity.

The TRPV1 channel participates in blood flow and pressure regulation via vasoactive neuropeptide release. The TRPV2 channel may correspond to a Ca^{++}-overloading factor associated with dystrophic cardiomyopathy. The TRPV4 channel is a mediator of endothelium-dependent hyperpolarization.

Among TRPs, some are activated in response to: (1) receptor stimulation, the so-called *receptor-operated Ca^{++}-entry channels* (ROCC) or *non-capacitative Ca^{++} entry* (NCCE); (2) depletion of intracellular Ca^{++} store, the so-called *store-operated Ca^{++}-entry channels* (SOC) or *capacitative Ca^{++} entry* (CCE); and (3) mechanical stress imposed on the plasma membrane, the so-called *mechanosensitive Ca^{++}-entry channels* (MSCC).

Other stimulators comprise various chemical and physical agents, such as lipids, acids, oxidative stress, heat or cold exposure, and hypo-osmolarity. Stimuli such as growth factors, angiotensin-2, and endothelin, as well as mechanical stresses applied by the flowing blood activate at least 3 TRP channels (TRPC1, TRPC6, TRPM7) to provoke the growth and proliferation of vascular smooth muscle cells.

The vasomotor tone is regulated by neurotransmitters released from perivascular nerves, circulating hormones, and locally released factors. At least 5 TRP subtypes

Table 2.20. Major TRP channels in the cardiovascular apparatus (**Part 1**; Source: [178]). Channel types TRPC5, TRPC7, TRPM2, TRPM3, TRPM5, TRPM6, and TRPM8 are not listed because of poorly identified function (12LOx: 12-lipooxygenase; AA: arachidonic acid; Cam: calmodulin; CamK: calmodulin-dependent kinase; CSN: cardiac perivascular sensory nerve; GF: growth factor; IP_3R: IP_3 receptor; MSCC: mechanosensitive calcium channel (Ca^{++} entry); NCX: Na^+–Ca^{++} exchanger; NGF: nerve growth factor; PI3K: phosphatidylinositol 3-kinase; PIP2: phosphatidylinositol (4,5)-bisphosphate; PKA, protein kinase-A; PKG: protein kinase G; PP3: phosphatase calcineurin; PSN: perivascular sensory nerve; RyR: ryanodine receptor; SOC: store-operated channel (Ca^{++} entry); VEC: vascular endothelial cell; VSMC: vascular smooth muscle cell). The TRP channels operate as non-voltage-gated, non-selective cationic channels. The myogenic response (Bayliss effect) corresponds to the reflex vasoconstriction induced by elevated intravascular pressure.

Type	Distribution	Regulator	Effect
TRPC1	Ubiquitous	IP_3R, Ca^{++}-Cam, Homer, caveolin-1	Vasoconstriction, VSMC proliferation, SOC, MSCC
TRPC3	Heart, Artery, vein	GPCR, IP_3R, RyR, NCX1, Ca^{++}-Cam, PKG, DAG,	Vasoconstriction, SOC, MSCC
TRPC4	Heart, Artery	GPCR, GF, Ca^{++}-Cam	NO production, microvascular permeability, SOC, MSCC
TRPC6	Ubiquitous	GPCR, GF, IP_3R, DAG, Ca^{++}-Cam, HETE, CamK2, PKC	Vasoconstriction, VSMC proliferation SOC, MSCC
TRPM4	Heart, Artery	Ca^{++} influx, PIP2, ATP, ADP, membrane potential	Myogenic response, Delayed afterdepolarization
TRPM7	Heart, Artery	Mg^{++} loss, pH, PIP2, cAMP, PKA, annexin-1, mechanical stress	VSMC proliferation, Mg^{++} homeostasis

(TRPC1, TRPC6, TRPM4, TRPV1, and TRPV4) are involved in vasomotor tone regulation [178].[78]

Vascular smooth muscle cells can change from a contractile phenotype to a migratory and proliferative state during vessel wall inflammation. Several TRP subtypes contribute to neointimal hyperplasia after vascular injury and pulmonary arterial hypertension (TRPC1, TRPC3, TRPC6, and TRPM7) [178]. In addition, in patients with essential hypertension, expression of TRPC3 and TRPC5 rises significantly.

In cardiomyocytes, TRPC3 is tightly coupled to Na^+–Ca^{++} exchanger NCX1. Abnormal sustained Na^+ entry through TRPC3 channels may cause Ca^{++} over-

78. In rat cerebral artery, TRPC3 channel is implicated in membrane depolarization and vasoconstriction generated by endothelium-released vasoconstrictor UTP [178].

Table 2.21. Major TRP channels in the cardiovascular apparatus (**Part 2**; Source: [178]; 2APB: 2-aminoethoxydiphenyl borate, an inhibitor of IP_3 receptors and TRP channels, but activator at higher concentrations of TRPV1 to TRPV3). Channel type TRPV3 is not listed because of poorly identified function.

Type	Distribution	Regulator	Effect
TRPP2	Heart, Artery	Ca^{++}, mechanical stress	Vascular integrity MSCC Contribution
TRPV1	CSN, PSN, VEC	12-LOx metabolites, endocannabinoid, 20-HETE, bradykinin, ATP, NGF, PKA, PKC, PI3K, CamK2, PP3, PIP2, 2APB	Vasodilation (CGRP release), myogenic response (substance-P release,) heat and acid sensing
TRPV2	Heart, Artery	GF, cell stretch	Heat sensing, MSCC contribution
TRPV4	Heart, Artery	Hypo-osmolarity, heat, endocannabinoid, AA, EETs, PKC, Ca^{++}-Cam, shear stress	EDHF-dependent vasorelaxation, MSCC contribution

load via NCX1 and delayed afterdepolarizations [178]). Channels TRPM4 and TRPM7 that are expressed at relatively high level in hearts may function as Ca^{++}-activated non-selective cation channels and constitutively activated cation channels, respectively, and may generate arrhythmogenic delayed afterdepolarizations. Growth factor-regulated Ca^{++} channel TRPV2 can cause sustained Ca^{++} overload and dystrophic cardiomyopathy [178]. Channels TRPP2 to TRPP4 may be implicated in pressure overload-induced cardiomyopathies.

2.3.6 Cyclic Nucleotide-Gated Channels

Cyclic nucleotide-gated (CNG) channels are responsible for signaling in primary sensory cells, such as retinal photoreceptors, olfactory neurons, and pinealocytes in epiphysis (or pineal gland).

The CNG proteins are tetrameric, voltage-independent, cation channels. Each subunit has 6 transmembrane (TM) domains with a pore-forming domain between TM5 and TM6 [5]. Subunits CNGα1 to CNGα3 [79] form homomeric functional channels (Table 2.22). Auxiliary subunits CNGα4, CNGβ1, and CNGβ3, [80] do not build functional channels.

79. These subunits are encoded by the CNGA1 to CNGA3 genes, respectively.
80. These auxiliary subunits are encoded by the CNGA4, CNGB1, and CNGB3 genes, respectively.

Table 2.22. Cyclic nucleotide-gated (CNG) channels (Source: [5]).

Type	Activators
CNGα1	cGMP ≫ cAMP
CNGα2	cGMP ∼ cAMP
CNGα3	cGMP ≫ cAMP

2.4 Hyperpolarization-Activated Cyclic Nucleotide-Gated Channels

Hyperpolarization-activated, cyclic nucleotide-gated, non-selective cation channels (HCN) are permeable to monovalent (Na^+ and K^+) as well as divalent (Ca^{++}) cations. Cyclic nucleotides cAMP and cGMP activate HCN channels. They shift HCN activation to more positive voltages.

The HCN channels generate pacemaker currents I_f [81] in many excitable cells, such as cardiac nodal cells and neurons. In hearts, hyperpolarization-activated cation inward current initiates and modulates the rhythm of cardiac pacemaker cells and neonatal cardiocytes. The HCN channels not only generate cardiac pacemaker depolarizations, but also control neuronal excitability and synaptic remodeling (plasticity), as well as sleep–wake cycle.

Channels HCNs are regulated by cAMP. [82] Their activity increases when cellular cAMP level heightens, as during sympathetic stimulation, and decays when it lowers, as during vagal excitation.

Hyperpolarization-activated, cyclic nucleotide-gated channels that generate the I_f current in sinoatrial cells are not implicated in heart rate increase. They are required to maintain a stable cardiac rhythm, especially after sympathetic stimulation via β-adrenoceptors of sinoatrial cells [179]. Cardiac, cAMP-regulated HCNs determine basal and maximal firing rates, as they generate a voltage-dependent depolarizing offset current [180]. Other ion carriers (e.g., voltage-gated Ca^{++} channel, ryanodine receptor, and Na^+–Ca^{++} exchanger) as well as distinct mediators (protein kinase-A and calmodulin-dependent protein kinase) are responsible for heart rate adaptation during physical activity.

2.4.1 Molecule Diversity

Each HCN subunit consists of 6 transmembrane domains (TM), a pore region between TM5 and TM6, and a binding domain to cyclic nucleotides in its cytoplasmic C-terminus. Subunits of HCN assemble into functional homo- or heterotetramers.

Four identified subtypes of tetrameric HCN channels exist (HCN1–HCN4; Table 2.23). Isochannels HCNs have similar structure, except in their cytoplasmic N-

81. A.k.a. I_{HCN}, I_q, and I_h.

82. Regulator cAMP binds to the cyclic nucleotide-binding domain of HCNs and elicits a positive shift in the voltage dependence of activation.

Table 2.23. Types of hyperpolarization-activated, cyclic nucleotide-gated, non-selective cation channels (HCN; Source: [5]). The HCN channels are permeable to both Na^+ and K^+ ions (Na^+/K^+ permeability ratio ~ 0.2).

Type	Activators
HCN1	cAMP > cGMP (both weak)
HCN2	cAMP > cGMP
HCN3	
HCN4	cAMP > cGMP

and C-termini. They have distinct physical properties (e.g., activation rate), expression patterns, and cAMP-modulation extent. Isoform HCN1 is the fastest channel, then HCN2, HCN3, and HCN4. In addition, HCN channels can form heteromers (e.g., the HCN1–HCN4 complex).

2.4.2 Cellular Distribution

Expression of HCN1 is restricted to the brain [181]. [83] The channel HCN1 is weakly modulated by cAMP. It exhibits a faster time course of activation than HCN2 and HCN3 channels.

The HCN2 channel abounds in the central nervous system. [84] It is the dominant isoform in atria and ventricles. [85] It is much more sensitive to cytosolic cAMP than HCN1 channel.

The HCN3 channel is strongly expressed in the brain. [86] The HCN4 channel is the most highly produced HCN isoform in the early sinoatrial node.

Major cardiac isoforms HCN2 and HCN4 contribute to the automatism of the sinoatrial node among several other ion carriers. The pacemaker activity of the sinoatrial node is controlled by the autonomic nervous system. Cholinergic and β-adrenergic stimulation slows and accelerates spontaneous firing of sinoatrial node, respectively.

2.5 Ligand-Gated Ion Channels

The nervous system transmits electrochemical signals down to synapses where they are converted into chemical signals, as neurotransmitters are released into synaptic gaps. In the post-synaptic membrane, ligand-gated ion channels convert chemical signals back into an electrochemical signals.

83. I.e., cerebral cortex, hippocampus (hippocampal CA1 neurons), cerebellum, and facial motor nucleus.

84. I.e., thalamus, mamillary bodies, pontine nucleus, ventral cochlear nucleus, and nucleus of the trapezoid body.

85. Lack in HCN2 leads to sinoatrial dysrhythmia [182].

86. Especially in the hypothalamus (supraoptic nucleus) and area postrema.

Ligand-gated ion channels (LGIC) are integral membrane proteins that contain a pore to enable fluxes of selected ions across the plasma membrane. Upon ligand binding, these channels undergo a conformational change that opens the channel.

Ligand-gated ion channels constitute a set that include: (1) ionotropic receptors of neurotransmitters, such as (i) excitatory, cation-selective, nicotinic acetylcholine; (ii) inhibitory, anion-selective, γ-aminobutyric acid $GABA_A$; (iii) excitatory, non-selective cation, ionotropic glutamate; (iv) inhibitory, anion-selective, glycine; and (v) excitatory, cation-selective, serotonin $5HT_3$ receptors that mediate fast synaptic transmission (time scale $O[1\,ms]$); (2) nucleotide P2X receptors; and (3) zinc-activated channels.

Nicotinic acetylcholine, $5HT_3$, $GABA_A$, glycine receptors, and zinc-activated channels are pentamers, whereas ionotropic glutamate and P2X receptors are tetramers and trimers, respectively.

Most of LGIC agonists bind to distinct types of receptors that act in completely different ways, either as ion channels, or classical receptors. Acetylcholine indeed binds to nicotinic and muscarinic G-protein-coupled acetylcholine receptors (Sect. 7.13.1); serotonin to $5HT_3$ and metabotropic G-protein-coupled receptors (Sect. 7.13.56); glutamate to ionotropic and metabotropic G-protein-coupled receptors (Sect. 7.13.27); and nucleotides to ionotropic P2X and G-protein-coupled P2Y receptors (Sect. 7.13.3).

2.5.1 Superfamily of Cys-Loop Ligand-Gated Ion Channels

Fast inhibitory neurotransmission (time scale of ms) that modulates both the magnitude and duration of neuronal signaling results from the tethering of inhibitory neurotransmitters to their specific receptors of the Cys-loop family in the plasma membranes of neurons. These receptors operate as chloride channels. Chloride ions repolarize the cell.

The superfamily of Cys-loop ligand-gated ion channels includes nicotinic acetylcholine, $GABA_A$, glycine, and serotonin $5HT_3$ receptors. These pentamers are usually composed of 2 α and 3 other subunits among β, γ, and δ subunits, but some subtypes can be made of 5 α subunits. All subunits contain an extracellular N-terminus, 3 transmembrane domains; a cytoplasmic loop, and an extracellular C-terminus [183]. They possess a characteristic loop formed by a disulfide bond between α subunits. Each subunit has 4 membrane-spanning α helices (TM1–TM4); TM2 of the 2 αsubunits forms the central pore and contains regions responsible for channel gating and ion selectivity. The intracellular segment connects to the cytoskeleton.

N-Glycosylation of Cys-loop receptors enables their plasmalemmal insertion. Subunits of $GABA_A$ are glycosylated. Glycans regulate the stability, assembly, and localization of $GABA_A$ channels, according to the glycosylation site [184]. [87]

---

87. Subunit $\beta 2$ contains 3 N-linked glycosylation sites (N32 and N104 in loop L3, and N173 in the Cys-loop). Site N173 is used for stability of $\beta 2$ subunits; N104 for $\alpha 1 \beta 2$ receptor assembly and/or stability in the endoplasmic reticulum and cell surface expression.

Table 2.24. Nicotinic acetylcholine receptors (Source: [5]). Subunits of nAChRs are encoded by 17 genes. In brains, 9 and 3 genes encode α- and β subunits, respectively. Various combinations of nAChR subunits differ in their pharmacological and kinetic properties as well as cellular and subcellular localization.

Subtype	Other name	Features
α1	Muscle-type	$(\alpha1)_2\beta\gamma\delta$: $P_{Ca}/P_{Na} = 0.16–0.20$
		$(\alpha1)_2\beta\epsilon\delta P_{Ca}/P_{Na} = 0.65–1.38$
α2		α2β2: $P_{Ca}/P_{Na} \sim 1.5$
α3	Autonomic,	α3β2: $P_{Ca}/P_{Na} = 1.5$
	ganglionic	α3β4: $P_{Ca}/P_{Na} = 0.78–1.1$
α4	Neuronal	α-bungarotoxin-insensitive
		α4β2: $P_{Ca}/P_{Na} = 1.65$
α6		
α7		α-bungarotoxin-sensitive
		$P_{Ca}/P_{Na} = 6.6–20$
α9		$(\alpha9)_5$: $P_{Ca}/P_{Na} = 9$
		α9α10: $P_{Ca}/P_{Na} = 9$

2.5.2 Nicotinic Acetylcholine Receptor Channel

Among neurotransmitter receptors that are involved in drug addiction, nicotinic acetylcholine receptor oligomers (nAChR) contributes to nicotine addiction caused by tobacco smoking [185].[88]

Nicotinic acetylcholine receptors are homo- and heteropentamers formed from members of a set of 17 subunits (α1–α10, β1–β4, δ, ϵ, and γ) [5] (Table 2.24).[89] Abbreviations of nAChR subtypes are often based on the predominant α subunit. Pentamers nAChRs are distributed throughout the central and peripheral nervous systems.

Plasmalemmal nAChR channels respond to ACh and nicotinic agonists by fast opening (μs–ms) to Na^+, K^+ and, in some cases, Ca^{++} cations. In the peripheral nervous system, nAChRs are reponsible for the rapid, phasic effects of local, short-lasting, high-magnitude ACh release [185].

In the central nervous system, nAChRs are involved in few types of fast transmission. Yet, nAChRs are targeted by tonic ACh release at lower, modulatory concentrations [185]. Two main nAChR types in the brain are α7 homo- and α4β2 hetero-oligomers. Whereas β2-containing and α7 nAChRs are widespread in the brain,

88. Every year, more than 5 million people worldwide die from consequences of smoking such as lung cancer. When compulsion is established and access to the drug is prevented, somatic and affective symptoms arise (motivational withdrawal syndrome). Nicotine withdrawal syndromes mobilize brain circuits characterized by the presence α2 and α5 to α7, as well as β4 nAChR subunits that regulate the expression of somatic symptoms, in addition to α6 and β2 subunits that contribute to affective symptoms [185].

89. All subunits except avian α8 subunit have been identified in mammals. Pentamers $(\alpha1)_2\beta1\epsilon\delta$ resides at the neuromuscular junction. Other nicotinic receptors combine mainly α2 to α6 and β2 to β4 subunits [5].

other nAChR subunits have a more restricted expression. High-frequency, high-burst firing require the activation of β2, but not α7 nAChRs [185].

The α7 homopentamer is characterized by a fast activation, low affinity, and high Ca^{++} permeability [185]. The α4β2 heteropentamer is typified by a high affinity and slow desensitization.

Various nicotinic acetylcholine receptors, such as α4-, α6-, β2-, and possibly β3-containing nAChRs control cognition (e.g., exploration, spatial memory, and navigation) and locomotion [185]. [90]

2.5.3 γ-Aminobutyric Acid Receptor Channel

The main inhibitory neurotransmitter receptor in the brain is the type-A γ-aminobutyric acid (GABA) receptor of the Cys-loop receptor superfamily of ligand-gated ion channels. Fast-responding ionotropic $GABA_A$ receptors are ligand-gated, Cl^--selective, pentameric channels, whereas $GABA_B$ receptors are G-protein-coupled receptors. Chloride flux prevents firing of new action potentials.

---

90. Subunit β3 can assemble with α4, α6, β2 subunits in nigrostriatal dopaminergic pathways [185]. Subunits α2, α4, α7, and β2 contribute to the effect of nicotine on hippocampal synaptic adaptivity (plasticity). Nicotinic agonists enhance cognition and psychomotricity. Conversely, nicotinic antagonists alter cognitive performance.

Nicotine targets nAChRs on both dopaminergic and gabaergic neurons in the ventral tegmental area (VTA), or nucleus A10, of the midbrain. Nicotine modulate firing of dopaminergic VTA neurons that mediates nicotine reinforcing effects. Various nAChR oligomers are expressed on: (1) γ-aminobutyric acid (GABA)-containing interneurons in the ventral tegmental area; (2) excitatory or inhibitory inputs to the ventral tegmental area; and (3) dopaminergic VTA neurons. Subunits α4 and β2, but not α6, are involved in the transition from tonic to phasic firing of dopaminergic neurons in the ventral tegmental area that intervenes in reinforcement [185]. Acute exposure to nicotine activates α4β2 and α7 nAChRs and a nervous circuit that includes the striatum, hippocampus, and amygdala and is controlled by the prefrontal cortex. Both α7 and β2 nAChRs are detected in the basal and lateral amygdala nuclei.

Chronic exposure to ACh or nicotinic agonists increases dopamine release in the nucleus accumbens. Moreover, it causes a gradual decrease in the activation rate (100 ms–mn) and desensitized and closed state, as well as additional long-term changes in receptor properties [185]. It also provokes an augmented number of high-affinity receptors in the central nervous system.

During long-term exposure to nicotine, the sensitization (i.e., increased sensitivity to subsequent nicotine exposures) of dopaminergic neurons in the ventral tegmental area relies on: (1) increased nAChR production; (2) presynaptic compensation to opposing action of some nAChR subtypes and decreased contribution of β2 nAChRs by cholinergic transmission via α7 subunit; and (3) elevated bursting activity of VTA dopaminergic neurons.

The ventral tegmental area project to the prefrontal cortex as well as limbic and striatal regions, in particular the brain stress- and habenulo-interpeduncular system, as well as hippocampus, amygdala, and nucleus accumbens. In the nucleus accumbens, α4- and α6 subunits are required for dopamine release [185]. The prefrontal cortex, amygdala, and hippocampus provide glutamatergic input to neurons in the nucleus accumbens that are also regulated by dopaminergic VTA neurons.

$GABA_A$ glycoproteins comprise 6 α, 3 β, 3 γ, 1 δ, 1 ϵ, 1 θ, 1 π (restricted to reproductive tissue), and 3 ρ subunits [5]. Many subtypes of $GABA_A$ heteropentamers contain α, β, and γ subunits with the possible stoichiometry $(\alpha)_2(\beta)_2\gamma$. In the central nervous system, the α1β2γ2 hetero-oligomer constitutes the largest population of $GABA_A$ receptors, followed by α2β3γ2 and α3β3γ2 isoforms. Subunits α and β contribute to the GABA-binding site. Homo- or hetero-oligomers formed by ρ subunits are also termed $GABA_C$ receptors.

2.5.4 Glutamate Receptor Channels

Ionotropic glutamate receptors (iGluR) are ligand-gated, non-selective cation channels that carry K^+, Na^+, and Ca^{++} in response to glutamate binding. They cause depolarizing, excitatory current that triggers an action potential in postsynaptic neurons. Ion flux speed and duration depend on receptor subtype.

Ionotropic glutamate receptor channels include 3 main receptor categories originally named according to their preferred synthetic agonist: (1) ^{N}methyl ^{D}aspartic acid (NMDA) receptors (NMDAR or ^{NMDA}Glu receptor);[91] (2) kainate receptors (KaR);[92] and (3) α-amino 3-hydroxy 5-methyl 4-isoxazolepropionic acid (AMPA) receptors (AMPAR or ^{AMPA}Glu receptor)[93] (Table 2.25). The receptor heterogeneity within each category arises from homo- and hetero-oligomeric assembly of distinct subunit types into cation-selective tetramers.

Ionotropic glutamate receptors are tetramers or rather dimer of dimers. All glutamate receptor subunits have an extracellular N-terminus with a ligand-binding motif, membrane-spanning region with 3 transmembrane domains (M1, M3, and M4) and a channel lining re-entrant P-loop (M2) between M1 and M3, and an intracellular C-terminus.

2.5.4.1 Ionotropic AMPA-Type Glutamate Receptors

Ionotropic α-amino 3-hydroxy 5-methyl 4-isoxazolepropionic acid receptor connects to glutamate for fast synaptic transmission in the central nervous system. AMPA-Type glutamate receptor mediates most of the fast excitatory transmission in the central nervous system. It contributes to synaptic transmission, strength, and remodeling such as long-term potentiation that is responsible for a sustained increase in amplitude of excitatory postsynaptic potential at postsynaptic membranes.[94]

---

91. ^{N}methyl ^{D}aspartic acid is an amino acid derivative that acts as a specific ligand of NMDA receptor. It mimics the action of glutamate on NMDA receptor. Unlike glutamate, NMDA binds to NMDA receptors only.

92. Kainate is a drug that has been isolated from red alga Digenea simplex.

93. α-Amino 3-hydroxy 5-methyl 4-isoxazolepropionic acid is a manufactured glutamate analog.

94. Excitatory postsynaptic potential corresponds to a transient depolarization of postsynaptic membrane potential caused by flux of cations into postsynaptic cells through ligand-gated ion channels, i.e., excitatory postsynaptic current. Brief periods of high activity can sustain

Table 2.25. Ionotropic glutamate receptor subunits (EAA: excitatory amino acid receptor). Activation of NMDA receptor requires both glycine and glutamate. Alternative splicing leads to functionally distinct subunits from each corresponding gene. Subunits of the category of ionotropic glutamate receptor-δ are considered orphan subunits, as they do not bind glutamate and related agonists (but have binding partners and antagonists).

Subunit	Other aliases	Gene
	Ionotropic, AMPA-type glutamate receptors	
GluR1	GluA1, GluRa, GluRk1, GluH1	GRIA1
GluR2	GluA2, GluRb, GluRk2	GRIA2
GluR3	GluA3, GluRc, GluRk3	GRIA3
GluR4	GluA4, GluRd	GRIA4
	Ionotropic, kainate-type glutamate receptors	
GluK1	GluR5, GluK5, EAA3, GluRβ1	GRIK1
GluK2	GluR6, GluK6, EAA4, GluRβ2	GRIK2
GluK3	GluR7, GluK7, EAA5	GRIK3
GluK4	KA1, EAA1, GluRγ1	GRIK4
GluK5	KA2, EAA2, GluRγ2	GRIK5
	Ionotropic, NMDA-type Glutamate receptors	
GluN1	NR1, NMDAR1,	GRIN1
GluN2a	NR2a, NMDAR2a, GluRε1	GRIN2A
GluN2b	NR2b, NMDAR2b, GluRε2	GRIN2B
GluN2c	NR2c, NMDAR2c, GluRε3	GRIN2C
GluN2d	NR2d, NMDAR2d, GluRε4	GRIN2D
GluN3a	NR3a, NMDAR3a	GRIN3A
GluN3b	NR3b, NMDAR3b	GRIN3B
	Glutamate receptor, ionotropic, δ	
GluD1	GluRδ1	GRID1
GluD2	GluRδ2	GRID2

Four types of AMPAR subunits (GluR1–GluR4) tetramerize. Most AMPARs are heterotetramers of GluR2[95] and either GluR1,[96] GluR3,[97] or GluR4[98] subunits.[99] Subunits GluR1 to GluR4 can exist as 2 alternative spliced variants (flip and flop) that differ in their desensitization kinetics. Numerous processes influence receptor

---

synaptic transmission between neurons (*long-term potentiation*), especially for learning and memory. Channels AMPARs at postsynaptic membranes are primarily responsible for long-term potentiation.

95. A.k.a. GluA2, GluRb, and GluRk2.
96. A.k.a. GluA1, GluRa, and GluRk1.
97. A.k.a. GluA3, GluRc, and GluRk3.
98. A.k.a. GluA4 and GluRd.
99. The GluR1–GluR2 heterotetramer prevails at postsynaptic sites under basal conditions. In the hippocampus, GluR1–GluR2 and GluR2–GluR3 heterotetramers account for the majority of AMPAR channels.

affinity for glutamate, ion selectivity, and conductance, as well as gating, in addition to their location and stability within the cell membrane.

The function of homo- or heterotetrameric ^{AMPA}Glu receptors is regulated by their subunit composition, post-transcriptional (alternative mRNA splicing and RNA editing), post-translational modifications (phosphorylation, glycosylation, and palmitoylation), and protein interactions.

Several interactors affect ^{AMPA}Glu receptor subcellular localization, synaptic stabilization, and kinetics. Auxiliary proteins thus control the AMPAR activity. Transmembrane AMPAR regulatory proteins (TARPγ2–TARPγ5 and TARPγ7–TARPγ8) act as auxiliary subunits that influence receptor trafficking and function. Cornichon proteins (Cnih2–Cnih3) constitute another category of AMPAR regulatory proteins.

Subunits of AMPAR channels differ mainly in their C-termini that determine their interactions with scaffold proteins. Subunit GluR1 binds to Disc large homolog DLg1 [100] Subunit GluR2 connects to adaptor protein kinase-Cα-binding protein PICK1 and glutamate receptor-interacting protein GRIP1. [101] The latter impedes AMPAR recycling to the synaptic membrane [186]. ^{N}Ethylmaleimide-sensitive fusion protein (NSF) also interacts with GluR2 to limit receptor endocytosis.

Type-1 *transmembrane AMPAR regulatory proteins* (TARP; i.e., γ2 to γ4 and γ8) that act as auxiliary subunits to AMPA receptors influence AMPAR transport, conductance, and gating. Type-2 TARP γ5 subunit interacts selectively with AMPARs that contain RNA-edited GluA2 subunits. [102] The TARP proteins [187]: (1) increase the surface expression of AMPA receptors; (2) slow receptor deactivation, as they delay channel closure after glutamate removal; (3) decrease receptor desensitization, as they reduce the decline in response in the permanent presence of glutamate; and (4) augment channel conductance.

The transmembrane 44-kDa, cystine-knot, AMPAR modulating protein (CKAMP44) reaches its highest level in the hippocampus. [103] Unlike TARPs, CKAMP44 reduces AMPAR responses, but does not affect NMDAR-mediated currents [188]. It does not affect AMPAR surface density. It causes stronger and faster desensitization and slows recovery from desensitization, thereby attenuating facilitation of the postsynaptic response. Protein CKAMP44 thus modulates short-term remodeling at specific excitatory synapses.

Phosphorylation of AMPARs regulates channel location, conductance, and gating. Subunit GluR1 has 4 known phosphorylation sites (Ser818, Ser831, Thr840, and Ser845) that are targeted by: (1) protein kinase-A (Ser845) for elevated peak response open probability [189]; (2) protein kinase-C (Ser818) for AMPAR remodeling-driven synaptic incorporation and long-term potentiation [190]; and (3) calmodulin-dependent kinase CamK2 (Ser831) for augmented delivery to the synapse and conductance [191, 192].

---

100. A.k.a. synapse-associated protein SAP97.
101. A.k.a. AMPAR-binding protein-interacting protein (ABP).
102. The GRIA2 mRNA can undergo RNA editing, in which the codon that encodes a P-loop glutamine residue is converted to 1 encoding arginine.
103. The Ckamp44 gene is specifically expressed in the brain.

Channels AMPARs open and close quickly. They thus contribute to fast excitatory synaptic transmission in the central nervous system. Their permeability to Ca^{++}, K^+ and Na^+ is governed by GluR2 subunit. Subunit GluR2 renders the channel impermeable to Ca^{++} ions. Moreover, in the absence of GluR2 subunits, AMPAR favors inward flux (inward rectifier) rather than outward current. A sequence prior to the C-terminus of transmembrane domain 4 in all AMPAR subunits determines the speed of desensitization and resensitization [193].

Central noradrenergic signaling intervenes in arousal and learning. The GluR1 subunit complexes with β2-adrenergic receptor as well as Gs subunit, adenylate cyclase, and protein kinase-A to prime a localized cAMP signaling in neurons via γ subunit (stargazin) of Ca_V1 channel and scaffold or anchor protein Disc large homolog DLg4 that clusters receptors, ion channels, and associated signaling proteins at postsynaptic sites [194]. Only GluR1 linked to β2AR is phosphorylated by protein kinase-A on β2AR stimulation. In addition, GluR1 connects to protein kinase-A and the counteracting phosphatase PP3 [104] to optimize the efficacy of its regulation via A-kinase anchor protein AKAP5 and DLg1, or alternatively DLg4. Phosphorylation by PKA increases plasmalemmal GluR1 density (enhanced exocytosis and/or inhibited endocytosis) at postsynaptic sites, hence amplitude of excitatory postsynaptic currents.

Acquisition and consolidation of memories of physiological or psychological disturbances is modulated by glucocorticoids released by adrenal glands via activation of mineralocorticoid and glucocorticoid receptors [195]. [105] Glucocorticoids (e.g., cortisol) regulate the exo- and endocytosis of AMPARs involved in rapid and persistent synaptic transmission and remodeling (plasticity) that support long-term, adaptive responses to stress. In addition to their genomic effect, glucocorticoids influence synaptic transmission via rapid or slowly developing action on AMPARs according to the type of targeted subunits.

2.5.4.2 Ionotropic Kainate-Type Glutamate Receptors

Ionotropic kainate receptors contribute to excitatory neurotransmission, as they activate postsynaptic receptors, as well as inhibitory neurotransmission, as they modulate GABA release. Kainate receptor channels are mainly permeable to Na^+ and K^+ ions. Their permeability to Ca^{++} is usually very low. Kainate receptors participate in synaptic remodeling.

---

104. A.k.a. calcineurin and PP2b.

105. Stress rapidly triggers signaling by the autonomic nervous system. Adrenaline and noradrenaline are secreted into the blood circulation. The hypothalamus–pituitary–adrenal axis is activated and augments the time-evolving (circadian rhythm-dependent) baseline release of glucocorticoid hormones by adrenal glands. In the brain (more precisely in the hippocampus, amygdala, and the prefrontal cortex), glucocorticoids bind to its cognate nuclear receptors, mineralocorticoid (MR, or NR3c2) and glucocorticoid (GR, or NR3c1) receptors. Via these receptors and in interaction with other neuromodulators, such as noradrenaline, corticotropin-releasing hormone, and endocannabinoids, corticosteroid hormones promote memory consolidation.

Table 2.26. Selective ligands of NMDA receptor channels with potency orders (Source: [5]).

Ligands	Potency order
Selective ligands of GluN1 (glycine site)	
Glycine	GluN2d > GluN2c > GluN2b > GluN2a
^{D}Serine	GluN2d > GluN2c > GluN2b > GluN2a
Selective ligands of GluN2 (glutamate site)	
NMDA	GluN2d > GluN2c > GluN2b > GluN2a
^{L}Aspartate	GluN2d = GluN2b > GluN2c = GluN2a
^{D}Aspartate	GluN2d > GluN2c = GluN2b > GluN2a

These tetramers are made of subunits among 5 types (GluK1–GluK5). Subunits GluK1, [106] GluK2, [107] and GluK3 [108] can form homo- and heteromers, whereas GluK4 [109] and GluK5 [110] only form functional receptors with subunits GluK1 to GluK3. The GRIK1 and GRIK2 RNAs, but not other kainate receptor subunit RNAs, can undergo RNA editing.

2.5.4.3 Ionotropic NMDA-Type Glutamate Receptors

Ionotropic ^{N}methyl ^{D}aspartic acid-type receptors for glutamate and glycine are ligand-gated, voltage-dependent ion channels. They require the coactivation by 2 ligands – glutamate and glycine – to allow influx of Na^+ and small amounts of Ca^{++} ions and outflux of K^+ ions.

These heterotetramers are composed of GluN1, [111] GluN2a, [112] GluN2b, [113] GluN2c, [114] GluN2d, [115] GluN3a, [116] and GluN3b [117] subunits [5]. Alternative splicing can generate 8 GluN1 isoforms with different features. Various splice variants of GluN2b to GluN2d and GluN3a exist. Subunits of the GluN1 and GluN2 subfamilies bind neurotransmitters glycine and glutamate, respectively (Table 2.26). Subunits encoded by the GRIN3 gene family (GRIN3A–GRIN3B) also possess glycine-binding modules; they lower NMDAR conductance.

Functional NMDA receptors correspond at least to dimer of heterodimers made of GluN1 and GluN2. However, heterotrimers can incorporate multiple subtypes of

106. A.k.a. GluR5 and GluK5.
107. A.k.a. GluR6 and GluK6.
108. A.k.a. GluR7 and GluK7.
109. A.k.a. KA1.
110. A.k.a. KA2.
111. A.k.a. NMDAR1, NR1, and GluR1.
112. A.k.a. NMDAR2a, NR2a, and GluRε1.
113. A.k.a. NMDAR2b, NR2b, and GluRε2.
114. A.k.a. NMDAR2c, NR2c, and GluRε3.
115. A.k.a. NMDAR2d, NR2d, and GluRε4.
116. A.k.a. NR3a and NMDAR3a.
117. A.k.a. NR3b and NMDAR3b.

GluN2 and GluN3 subunits. NMDA-Type glutamate receptor channels (^{NMDA}Glu receptor) commonly have a relatively high permeability to Ca^{++}. ^{NMDA}Gly receptors that contain both GluN1 and GluN3 subunits (GluN1–GluN3 complexes) are activated by glycine alone [196]. [118]

^{NMDA}Glu receptors have an internal binding site for an Mg^{++} ion that creates a voltage-dependant block. Ion Mg^{++} can also potentiate NMDA activity at positive membrane potentials. Since the Mg^{++} block must be removed by outward current, NMDAs rely on AMPAR-mediated excitatory postsynaptic potential to open. [119] NMDA-Type glutamate receptors control synaptic remodeling (plasticity). The most explored forms of adaptivity at excitatory synapses are actually NMDAR-dependent long-term potentiation (LTP) and depression (LTD). Changes in synaptic connectivity support learning and memory.

Inhibitory sites exist not only for Mg^{++}, but also Zn^{++} and H^+ ions. Each NMDAR subunit has a cytoplasmic domain that is targeted by protein kinases and phosphatases as well as adaptors and scaffold proteins. Src kinases enhance NMDAR function [197]. In addition, NMDARs are regulated by endogenous neuroactive steroids [5].

> " *Life is pleasant. Death is peaceful. It's the transition that's troublesome.* " *(I. Asimov).*

Receptors NMDARs can both promote neuronal health and death, according to the receptor location. Stimulation of synaptic NMDARs, particularly trans-synaptical activation, that operate primarily via nuclear Ca^{++}, ensures neuroprotection, whereas stimulation of extrasynaptic NMDARs fosters neuron death [198]. Synaptic NMDARs control a nuclear, Ca^{++}-regulated, gene program that protects against excitotoxic and apoptotic stimuli. Extrasynaptic signals are enhanced by ischemia-induced activation of death-associated protein kinase (DAPK). [120]

118. The GluN1–GluN3 complex can lodge in myelin sheath such as that surrounding the optic nerve, where it can prevent Ca^{++} overload, but not in the oligodendrocyte soma. The receptors are activated by ^{D}serine. Subunits GluN2 that also reside in myelin do not influence the response to ^{D}serine.

119. On ligand binding, AMPARs prime a Na^+ flux into the postsynaptic cell. The resulting depolarization removes Mg^{++} out into the extracellular space, allowing NMDAR to open, hence carrying both Na^+ and Ca^{++} ions. Calcium ions stimulate calmodulin-dependent kinase CamK2 that enhances AMPAR conductance.

120. Extrasynaptic NMDARs lodge in the soma, dendritic shaft, neck of dendritic spine, and close to the postsynaptic density, i.e., perisynaptic NMDARs localize to a region within 100 nm of postsynaptic densities. They are not activated during low-frequency synaptic events. They trigger dephosphorylation (inactivation) of the prosurvival transcription factor CREB, nuclear import of the prodeath transcription factor FOxO, inactivation of extracellular signal-regulated kinases ERK1 and ERK2, cleavage by calpain of PTPn5 phosphatase, and activation of P38MAPK.

2.5.4.4 Ionotropic Glutamate Receptor-δ Channels

Subunits δ1 [121] and δ2 [122] of excitatory glutamate receptor ligand-gated ion channel (iGluR) are encoded by the GRID1 and GRID2 genes, respectively. Ion channels of the ionotropic glutamate receptor family allow fast excitatory synaptic transmission in the central nervous system. In addition, they support synaptic adaptivity (plasticity). Subunits GluD*i* are homologous to AMPA, kainate, and NMDA subunits.

Glutamate receptor subunit δ2 is expressed in cerebellar Purkinje cells. It interacts with ionotropic glutamate receptor subunits AMPAR1 (or GluR1) and AMPAR2 (or GluR2), as well as kainate-type subunits GluK2 and GluK5, in addition to GRID2 gene product-interacting protein-1 (GRID2IP1 or delphilin), PDZ coiled-coil domain-binding partner for the GTPase RhoJ (GOPC), and non-receptor protein Tyr phosphatase PTPn4 [59].

The GluD2 channel subunit contributes to the normal development of cerebellar circuits. Once synthesized by postsynaptic neurons, GluD2 is necessary and sufficient to prime the formation of new synapses between cerebellar neurons and their maintenance [199].

Cerebellin-1 (Cbln1) [123] is a member of the C1q and tumor-necrosis factor superfamily (C1qTNFSF) [124] and a soluble protein secreted from cerebellar granule cells (i.e., presynaptically synthesized molecule). It binds to the extracellular N terminus of GluD2 on Purkinje cells. The Cbln1–GluD2 *synapse-organizer complex* causes: [125] (1) presynaptic differentiation and (2) postsynaptic clustering of synapse-specific molecules. Both processes are needed for synapse formation between heterologous granule cells and Purkinje cells [199].

---

121. A.k.a. GluD1 and GluRδ1.

122. A.k.a. GluD2 and GluRδ2.

123. Cerebellin-1 is also produced in various brain regions, such as the olfactory bulb, entorhinal cortex, and certain thalamic nuclei. In these cerebral regions, GluD2 channel is not expressed, but GluD1 is highly synthesized, especially during development [199].

124. C1q is the recognition protein of the classical complement pathway. This protein connects innate to acquired immunity. Agent C1q can engage many ligands via its globular domain (gC1q). This gC1q domain is also found in many non-complement proteins. It has a fold similar to that of ligands of the tumor-necrosis factor superfamily. The members of the C1qTNFSF are involved in immune defense, inflammation, apoptosis, auto-immunity, cell differentiation, and organogenesis [200].

125. Other synapse-organizing complexes comprise: neuroligin–neurexin, synaptic cell adhesion molecule SynCAM–SynCAM (a.k.a. cell adhesion molecule-1 CADM1–CADM1, immunoglobulin superclass member-4 IGSF4a–IGSF4a, and nectin-like protein-2 NecL2–NecL2), ephrin-B–EPHb, FGFR2b receptor, as well as the pentraxin complex that includes neuronal activity-regulated pentraxin (NARP), neuronal pentraxin-1 (NP1 or NPtx1), and AMPAR receptors, and the NGL3–PTPRf complex (composed of the DLg4-interacting postsynaptic adhesion molecule netrin-G ligand-3 and receptor protein Tyr phosphatase PTPRf [199].

Table 2.27. Selective ligands of homo-oligomeric glycine receptor channels with potency order (Source: [5]). Four known isoforms of the α and a single β subunit exist. Heteropentameric inhibitory GlyR has an invariant (3:2) stoichiometry; the main type being $(\alpha1)_3(\beta)_2$ receptor. α Subunits can also form functional homopentameric receptors.

Subunit	Potency order
GlyRα1	Glycine > β-alanine > taurine
GlyRα2	Glycine > β-alanine > taurine
GlyRα3	Glycine > β-alanine > taurine
GlyRα4	

2.5.5 Glycine Receptor Channels

Glycine receptors (GlyR) are Cl^- channels made mainly of 5 subunits that surround a central pore. These heteropentamers are constituted of α (encoded by the GLRA1–GLRA4 genes) and β subunits (encoded by the GLRB gene). Further diversity arises from alternative splicing of the primary gene transcripts for α1 ($\alpha1^{ins}$ and $\alpha1^{del}$), α2 (α2a and α2b), α3 ($\alpha3_S$ and $\alpha3_L$) and β(β7) subunits as well as mRNA editing of α3 subunit [5]. In addition, the α4 subunit has not been detected in adult humans. Subunit β of glycine receptor contributes to ligand binding, anchors the receptor to the cytoskeletal attachment protein gephyrin, [126] and reduces channel conductance.

Gβγ subunit of guanine nucleotide-binding protein (G proteins; Vol. 4 – Chap. 8. Guanosine Triphosphatases and their Regulators) enhances the open state probability of glycine receptors (Table 2.27). Intracellular Cl^- concentration modulates the kinetics of glycine receptors. Intracellular Ca^{++} increases the affinity of glycine receptors, produces fast stimulation of their activity, and prolonges channel opening. Cation Zn^{++} acts via distinct binding sites of high- and low-affinity to enhance channel function at low concentrations (<10 μmol) and inhibit at higher concentrations in a subunit-selective manner [5].

2.5.6 Serotonin Receptor Channels

Serotonin (5-hydroxytryptamine) $5HT_3$ receptor is composed of 5 subunits arranged around a central ion-conducting pore that is permeable to Na^+, K^+, and Ca^{++} ions. [127] Whereas other 5HT (metabotropic) receptors are G-protein-coupled receptors, $5HT_3$ operates as an inwardly rectifying channel.

---

126. Gephyrin is a component of the postsynaptic protein network of inhibitory synapses associated with glycine and/or $GABA_A$ receptors. Gephyrin binds to numerous subsynaptic proteins and clusters and anchors receptors to the synapse [5].

127. The pore of the $5HT_3$ receptor is predominantly sodium and potassium selective [201]. Its opening causes a rapidly activating and then desensitizing inward current. A large loop between membrane-spanning α helices TM3 and TM4 forms the intracellular domain involved in channel conductance and modulation.

Five subunits are encoded by the HTR3A to HTR3E genes. All functional $5HT_3$ receptors must contain at least one A subunit. Functional channels consist of homo- and heteropentamers [5]. Therefore, only $5HT_{3A}$ subunits form functional homopentameric channels; all other subunit types form functional channels with the $5HT_{3A}$ subunit.

The diversity of the $5HT_3$ channel is increased by alternatively spliced variants of the $5HT_{3A}$ and $5HT_{3E}$ subunits. The relative permeability to divalent cations is reduced by coexpression of the $5HT_{3B}$ subunit.

Functional receptors are glycosylated (Asn109, Asn175, and Asn191) [201]. Activity of $5HT_3$ channel can be modified by phosphorylation.

Serotonin receptor channels are located throughout the central and peripheral nervous systems. Serotonin binding leads to an excitatory response in neurons. Postsynaptic receptors $5HT_{3A}$ can mediate fast excitatory synaptic transmission, whereas presynaptic $5HT_{3A}$ receptors on nerve terminals modulate neurotransmitter release.

The $5HT_3$ receptor interacts various ligands [201]. Dopamine may be a weak partial agonist in vivo. Compounds can target other sites than the usual binding site on the molecule to modulate the agonist-induced response (e.g., steroids and cannabinoids). The $5HT_{3A}$ subunit can also link in vitro with subunits from related receptors such as the nicotinic acetylcholine receptor $\alpha 4$ subunit as well as the light chain of microtubule-associated protein-1B [201].

2.5.7 Ionotropic Nucleotide Receptors

Ionotropic nucleotide receptors [128] are cation-permeable, ligand-gated ion channels that bind extracellular adenosine triphosphate. [129] Binding of extracellular ATP to trimeric P2X receptors triggers opening of a transmembrane pore, thereby allowing sodium, potassium, and calcium ions to move down their electrochemical gradients, change membrane voltage, and activate intracellular signaling cascades.

Some P2X receptors are also activated by uracil nucleotides such as UTP that stimulates cardiomyocytes, as well as diadenosine polyphosphates. [130] Several P2X receptors can also inhibit ATPase activity [5].

128. Purinergic P2 receptor, or purinoceptor, is a misnomer, as both purine and pyrimidine species are ligands. The P2 receptors are better defined as nucleotide receptors (Sects. 7.13.2 and 7.13.3).

129. Intra- and extracellular ATP concentrations equal to about 10 mmol and 10 nmol, respectively. Among other functions, ATP acts as a cotransmitter in perivascular sympathetic nerves. Extracellular ATP can have many sources, such as nerve terminals, endothelial and smooth muscle cells, erythrocytes, activated platelets, inflammatory cells, and myocytes [202]. The preferred expelled anion is $^{Mg}ATP^{2-}$. Extracellular ATP is rapidly degraded by ectonucleotidases.

130. Ubiquitous adenosine polyphosphates produced and released from platelets or chromaffin cells form a group of adenosine dinucleotides that consist of 2 adenosine molecules bridged by a variable number of phosphates. They inhibit ATP-activated K^+ channels and stimulate ryanodine receptors (Ca^{++}-induced Ca^{++} release) in cardiomyocytes.

Ionotropic P2X receptors are involved in synaptic transmission, inflammation, and sensing of taste and pain. In the central nervous system, presynaptic neurons that express P2X receptors enhance the release of neurotransmitters, such as glutamate and γ-aminobutyric acid. On the other hand, postsynaptic neurons need P2X to transmit an ATP-induced postsynaptic signal. In the peripheral nervous system, afferent neurons with P2X receptors sense various stimuli such as distension.

In pancreatic islets, extracellular ATP targets $P2X_3$ receptors to enhance human β-cell sensitivity and responsiveness to glucose fluctuations [203]. Activation of $P2X_3$ by released ATP raises the intracellular calcium concentration and amplifies glucose-induced insulin secretion by β cells that can respond to relatively modest changes in glucose concentration under physiological conditions. Released ATP creates a positive autocrine feedback loop.

2.5.7.1 Molecule Diversity

Purinoceptors P2Xs are trimers. Seven distinct genes encode P2X receptors ($P2X_1$–$P2X_7$). They are homo- (e.g., $P2X_1$ in smooth muscle cells as well as $P2X_2$ to $P2X_4$ and $P2X_7$) or heteropolymers (e.g., $P2X_2$–$P2X_3$ and $P2X_1$–$P2X_5$) [5]. The hemichannel pannexin-1 can be implicated in the pore formation induced by $P2X_7$. Homomers $P2X_2$ and $P2X_3$ are able to combine and form an unique heteromeric, functional $P2X_2$–$P2X_3$ channel, as in sensory neurons. Ionotropic receptors $P2X_4$ and $P2X_7$ do not heteromerize, whereas $P2X_6$ needs to heteromerize with $P2X_1$, $P2X_2$, $P2X_4$, or $P2X_5$ to form an active channel.

2.5.7.2 Cellular Distribution

The P2X receptors are expressed in brain, heart, etc., as well as on leukocytes and thrombocytes. The $P2X_1$ receptor is prevalent in smooth muscle cells, whereas $P2X_2$ is prominent in the autonomic nervous system.

Adenosine triphosphate released through pannexin hemichannels from CD4+ helper T cells after stimulation of the T-cell receptor is an autocrine regulator that binds to P2X receptors and activates the mitogen-activated protein kinase module. Autocrine messenger ATP secreted by CD4+, CD25+, FoxP3+ regulatory T cells also links to $P2X_7$ on their cell surface to impede their immunosuppressive action and causes their conversion into IL17-secreting helper T_{H17} effector cells [204].

2.5.7.3 Functional Features

Channel opening duration depends on the receptor type. Upon continuous ATP binding, $P2X_1$ and $P2X_3$ desensitize rapidly (a few hundred ms), whereas $P2X_2$ remains open. Furthermore, P2X sensitivity to ATP is modulated by changes in extracellular pH and heavy metals (e.g., zinc and cadmium; allosteric modulation). ATP sensitivity of $P2X_1$, $P2X_3$, and $P2X_4$ is attenuated when extracellular pH drops below 7, whereas that of $P2X_2$ rises. Zinc potentiates ATP-gated currents through

Table 2.28. Ionotropic P2X nucleotide receptors of the heart (Source: [202]). Cardiomyocytes exhibit both auto- and paracrine nucleotide regulations. Ionotropic receptors $P2X_1$ to $P2X_7$ as well as metabotropic $P2Y_1$, $P2Y_2$, $P2Y_4$, $P2Y_6$, and $P2Y_{11}$ exist in cardiomyocytes. The $P2X_3$ receptor is also found in human fetal heart where it represents the most abundant type. Receptors $P2X_1$, $P2X_2$, and $P2X_4$ also localize to smooth muscle cells of coronary vessels. Messenger ATP induces non-specific cationic and Cl^- fluxes that depolarize the cells. Extracellular ATP inhibits Ca_V1 channels via Gi protein and activated protein kinase-G and activates inward-rectifying (ACh- and ATP-activated K^+ fluxes) and outward K^+ currents. Nucleotide receptors P2Y via Gs protein activate adenylate cyclase-5 that produces cAMP, thereby reducing intracellular ATP level, which is then less available for channel functioning and counteracting enhancement of $i_{K_{ATP}}$ ($K_{IR}6.2$) by extracellular ATP. They also target the protein Tyr kinase Fyn via Gq protein to activate phospholipase-Cγ that generates inositol trisphosphate and stimulates Cl^-–HCO_3^- exchangers, which induce a large transient acidosis. Phospholipase-A2 and -D are also activated via protein kinase PKC-α or other PKC isoforms acting via MAPK activation on the one hand and PKCζ on the other.

Type	Carried ions	Desensitization	Modulation
$P2X_1$	Na^+, K^+, Ca^{++}	Fast	$H^+ \downarrow$
$P2X_2$	Na^+, K^+	Slow	$H^+ \uparrow$; $Ca^{++} \downarrow$
$P2X_3$	Na^+, K^+, Ca^{++}	Fast	$H^+ \downarrow$; $Ca^{++} \downarrow, \uparrow$
$P2X_4$	Na^+, K^+	Slow	$H^+ \downarrow$
$P2X_5$	Na^+, K^+, Ca^{++}	Slow	$Ca^{++} \downarrow$
$P2X_6$	Na^+, K^+, Ca^{++}	Slow	
$P2X_7$		Pore	
$P2X_{2--3}$	Na^+, K^+, Ca^{++}	Slow	$H^+ \downarrow, \uparrow$
$P2X_{1--5}$		Biphasic	$H^+ \downarrow$
	$Ca^{++} \downarrow$		
$P2X_{2--6}$		Slow	

$P2X_2$, $P2X_3$, and $P2X_4$, and represses cation flux through $P2X_1$ receptor. Moreover, Mg^{++} reduces $P2X_2$ affinity to ATP.

The P2X receptors modulate synaptic transmission in pre- and postsynaptic nerve terminals. They mediate nociception [205]. They also initiate contraction in striated myocytes, including cardiomyocytes (Table 2.28), and smooth muscle cells. They temper the cardiac rhythm and contractility, as well as the vascular tone.

2.5.8 Zinc-Activated Channel

Zinc-activated channel (ZAC) likely exists as a homopentamer. It forms a cation-selective channel that operates as an outwardly rectifying channel. It displays a constitutive activity that is excited by Zn^{++} [5]. It has been detected in humans. [131]

131. Also in chimpanzees, dogs, and cows, but not in rodents.

2.6 Chanzymes

Certain ion channels have an enzymatic activity and are thus called chanzymes. The cation channels TRPM6 and TRPM7 are associated with an α-kinase. The TRPM7 channel is involved in cellular Mg^{++} homeostasis. It is highly permeable to Ca^{++}. It is activated by phospholipase-C. It also serves as a receptor for signal transduction. It indeed modulates the function of the actomyosin cytoskeleton. It phosphorylates the myosin-2A heavy chain. [132] Channel TRPM7, activated by bradykinin that targets a Gq-coupled receptor and primes the PLC axis, initiates a Ca^{++}-dependent phosphorylation of myosin heavy chains and disassembly of myosin-2A at the cell cortex, promoting cell spreading and augmented adhesion to the extracellular matrix [206].

2.7 Ion Carriers and Regulation of H^+ Concentration

In general, the cytoplasm acidifies because of protonation that must be corrected by deprotonation. Cytosolic acidification results from catabolism, especially ATP production in the cytoplasm by glycolysis that generates lactate (anaerobic component) and oxidative phosphorylation in mitochondria that fabricates CO_2 (aerobic component), and negative membrane potential that is created by accumulation of H^+ imported through cation channels and loss of basic HCO_3^- exported through anion channels [207].

Proton buffers, i.e., intracellular weak acids and bases, such as phosphate groups and side chains of amino acids, and bicarbonates have a finite capacity, [133] whereas plasmalemmal transporters, such as Na^+–H^+ exchangers and HCO_3^- transporters control pH at an optimal level for protein interactions (Table 2.29).

Hydrogen ion extrusion and entry that are finely regulated are responsible for pH homeostasis in the cytosol as well as inside organelles, including the nucleus, endoplasmic reticulum, Golgi body, vesicles of endo- and exocytosis, and peroxisomes (Table 2.30).

Numerous proteins implicated in signal transduction, intercellular communication, cytoskeletal dynamics, vesicular transport, and membrane excitability, are

132. Myosin-2 is the major motor protein responsible for cell contractility (Vol. 1 – Chap. 6. Cell Cytoskeleton). Cell contractility associated with myosin-2 is regulated by Rho GTPases and their effectors, myosin light-chain kinases and phosphatases, as well as inhibitory myosin heavy-chain kinases. Myosin-2 assembles into bipolar thick filaments and generates a tension of the cell cortex by pulling together oppositely oriented actin filaments.

133. Cells constantly produce chemical energy, hence CO_2 that with water, generates H_2CO_3 that can be converted into HCO_3^- and H^+ ions. Inter-conversion of CO2 and HCO_3^- occurs spontaneously. Nevertheless, this reaction is greatly accelerated by carbonic anhydrase. Among 11 known isoforms, CA1 to CA3, CA7, and CA8 reside in the cytosol; CA5 in mitochondria; and others at the plasma membrane. Plasmalemmal CA4 is anchored to the extracellular surface by a glycosyl phosphatidylinositol anchor, whereas CA9, CA12, and CA14 possess a transmembrane segment. Isoform CA6 is secreted from the cell [207].

Table 2.29. Ion carriers that regulate cytoplasmic pH (Source: [207]). These ion carriers use the energy stored in the inwardly directed electrochemical Na^+ gradient established by Na^+–K^+ ATPases. In most mammalian cells, hydrogen ions are extruded from cytosol against their electrochemical gradient by coupling to other substances through exchangers, antiporters, and cotransporters, or symporters.

Type	Alias
Plasmalemmal alkalinizing carriers	
Na^+–H^+ exchangers	NHE
Na^+–HCO_3^- cotransporters	NBC
Na^+–H^+ antiporter	NHA
Na^+-dependent Cl^-–HCO_3^- exchangers (some cell types)	NDCBE
Monocarboxylate–proton cotransporter	MCT
Vacuolar H^+ ATPase (some epithelial cells and osteoclasts)	vATPase
Plasmalemmal acidificating carriers	
Cl^-–HCO_3^- exchanger	AE1
Anion exchangers	AE
Plasma membrane Ca^{++} ATPase	PMCA
K^+–H^+ exchanging ATPase (acid-secreting gastric and nephron distal tubule cells)	ATP4α–ATP4β

Table 2.30. Values of pH in different cellular compartments (Source: [207]).

Organelle	pH
Cytosol	7.3–7.4
Nucleoplasm	~7.2
Endoplasmic reticulum	~7.2
cis-Golgi network	~6.7
trans-Golgi network	~6.0
Secretory vesicle	~5.5
Mitochondrium	~8.0 (matrix)
Peroxisomes	~7.0
Early endosome	~6.3
Recycling endosome	~6.5
Late endosome	~5.5 (bulk luminal fluid)
Lysosome	~4.7

highly sensitive to small changes in pH of their surrounding milieu, such as certain cation channels, connexins, actin-binding proteins (cofilin, gelsolin, and villin), neurotransmitter-gated receptors, H^+-sensing G-protein-coupled receptors, and subtype-A2 of vATPase (ATP6V0a2).

Alkali cation–proton exchangers are homodimers that couple H^+ transfer across membranes to the countertransport of monovalent cations, such as Na^+ or K^+:

Table 2.31. Coupling stoichiometry of sodium–bicarbonate cotransporters. (Source: [207]; AE: anion exchanger; NBCe: Na^+–HCO_3^- cotransporter electrogenic isoform; NBCn: Na^+–HCO_3^- cotransporter electroneutral isoform; SBC: sodium–bicarbonate cotransporter; SLC: solute carrier family). Isoform NBCe1 carries Na^+ and HCO_3^- into the blood across the basolateral surface of renal proximal tubule cells.

Type	Other aliases	Stoichiometry
NBCe1	SLC4a4, NBC1, NBC2	1 Na^+:2 HCO_3^-
NBCe2	SLC4a5	1 Na^+:3 HCO_3^-
NBCn1	SLC4a7	1 Na^+:1 HCO_3^-
SBC5	SLC4a9, AE4	1 Na^+:1 HCO_3^-
NBCn2	SLC4a10	1 Na^+:1 HCO_3^-

Na^+–H^+ exchangers NHE1 (or SLC9a1) to NHE9 (or SLC9a9); Na^+–H^+ antiporter NHA1 [134] and NHA2. [135] Alkali cation–proton exchangers reside in specific subdomains of the plasma membrane, such as apical or basolateral surfaces of epithelia cells, intercalated discs and T-tubules of cardiomyocytes, lamellipodia of migrating fibroblasts, as well as vesicles of exo- and endocytosis [207].

Acid-extruding *bicarbonate transporters* include 6 plasmalemmal Na^+–HCO_3^- cotransporters (NBC) with different coupling stoichiometry (Table 2.31).

Monocarboxylate–proton cotransporters (MCT; Sect. 4.4.2) mediate the transport of monocarboxylic acids, essentially lactate, that accumulates during anaerobic metabolism, with hydrogen ions, especially in myocytes [207]. *Acid-importing transporters* are anion exchangers that exchange Cl^- for HCO_3^-. They comprise AE1 (SLC4a1) to AE3 (aSLC4a3)) and 5 electrogenic or electroneutral Cl^-–HCO_3^- exchangers of the SLC26a family. Inward Cl^- transfer yields the driving force for HCO_3^- efflux. *Cystic fibrosis transmembrane conductance regulator* (CFTR) has a HCO_3^- permeability that is about one-third of that of Cl^- anion. [136]

Vacuolar ATPases are composed of 2 domains: an integral membrane complex with 6 subunits V0 and a peripheral complex of 8 subunits V1. The latter converts the chemical energy stored in ATP for hydrogen ion displacement, whereas V0 yields the path for hydrogen ion flux across the membrane. Electrogenic vATPases generate a transmembrane voltage. The regulator of H^+ ATPases of the vacuolar and endosomal membranes (RAVE) regulate pump assembly in vesicles.

Vesicles of endocytic and secretory routes are characterized by luminal relative acidity owing to the concerted action of vacuolar H^+ ATPases and Cl^-–H^+, K^+–H^+, and Na^+–H^+ exchangers. Progressive acidification of exocytic vesicles allows proper post-translational modifications, sorting, and transport of newly synthesized

134. A.k.a. sodium–hydrogen exchanger-like domain-containing protein NHEDC1 in humans.

135. A.k.a. NHEDC2.

136. It contributes significantly to pancreatic base secretion [207].

proteins. Likewise, graded acidification of endocytic vesicles allocate a route to internalized cargos to recycling or degradation.

In endosomes, electrogenic 2 Cl^-:1 H^+ exchangers balance charge influx by vATPases, as they import anions. Other H^+ efflux from endosomal lumen is caused by sarco(endo)plasmic reticulum Ca^{++} ATPases (SERCA) and secretory pathway Ca^{++} ATPases (SPCA), as well as non-selective alkali cation–H^+ exchangers, i.e., Na^+–H^+ exchangers NHE6 to NHE9, and a Zn^{++}-inhibitable, voltage-sensitive H^+ channel [207].

In cell organelles, H^+ entry causes neutralizing flux of counter-ions, such as import of anions or export of cations. Several anion carriers exist in endomembranes, such as stretch-gated Mid1-related chloride channel (MClC), [137] Golgi anion channel (GolAC), and members of the Cl^- channel and transporter ClC set. Voltage-dependent chloride channels ClC3 to ClC7 (Sect. 3.5.1) may mediate Cl^- entry into endosomes and lysosomes. Cation channels and transporters also operate in secretory vesicles, in which K^+, Na^+, Mg^{++}, and Ca^{++} act as counter-ions in the vesicular lumen.

The mitochondrial matrix is alkaline (pH∼8.0) because of H^+ extrusion through the inner membrane by components of the electron transport chain. Inner membrane H^+–ATP synthase (F1F0-ATPase) then generates ATP from ADP and inorganic phosphate. The H^+ gradient regulates Na^+, K^+, and Ca^{++} homeostasis through the synergistical actions of alkali cation–proton and sodium–calcium exchanges as well as Ca^{++} uniport. Calcium ions can also exit the matrix on transient opening of the permeability transition pore.

137. A.k.a. ClIC-like chloride channel ClCC1.

3

Main Sets of Ion Channels and Pumps

Ion fluxes across cell membranes are mainly controlled by ion channels and pumps [208]. Fluxes of Na^+ and K^+ cations change membrane potential that can act as physical signals. Calcium current not only serves as a physical signal, but also as a chemical cue, as it operates as a secong messenger in cell signal transduction. Flux of Cl^- anion aims at stabilizing membrane potentials and facilitating transmembrane electrolytes and water movement.

3.1 Introduction

Ion carriers are transmembrane proteins that enable transmembrane transport of anions and cations. Transmembrane ion-transport proteins encompass ion pumps, exchangers, and channels. Active transport proteins — ion pumps and exchangers — that carry ions against their concentration gradients are usually grouped in a single category.

Ion carriers cycle between conformations for which either an ion can bind to the extra- or intracellular side of the membrane, or cross the membrane. An ion carrier either transports only one or a few ions per conformational cycle, [1] or flux of ions (a single ion channel may carry up to thousands of ions).

The traffic of ions across cell membranes is mainly controlled by ion channels and pumps. Open ion channels are water-filled proteic channels. They facilitate the fast diffusion (passive transport) of a huge amount of selected ions down their electrical and concentration gradients.

Catalytic ion pumps used for the active transport couple the energy-consuming ion movement across a membrane to ATP hydrolysis. Ion pumps generate gradients across cell membranes, which are then be used by ion channels. An ion pump that conveys a single or several ions is also called an ion transporter; an ion pump can be a uniporter or an antiporter.

1. Uniporter facilitates ion diffusion. Antiporter exchange a species for another across a membrane. Symporter is a cotransporter. In particular, adenine nucleotide translocase is an ADP^{3-}–ATP^{4-} exchanger (1:1 stoichiometry) across the inner mitochondrial membrane.

Ion carriers are transported to specific membrane domains to fullfill their functions. Ankyrins link ion carriers, such as voltage-gated Na^+ and ATP-sensitive K^+ channels [2] Na^+–K^+ ATPase, and anion and Na^+–Ca^{++} exchangers, to the actin cytoskeleton. Ankyrins can serve as membrane scaffolds to tether plasmalemmal ion carriers to the spectrin–actin cytoskeleton and factors in membrane delivery.

3.1.1 Ion Channels

Ions are passively transported through *ion channels* down electrical and concentration gradients, at speeds that can approach the diffusion limit. Because ion flows through channels dissipate their driving gradients, channels contain gates that are regulated to open only when needed.

Ion channels are pore-forming, integral membrane, multisubunit proteins that contribute to establishment of a small transmembrane voltage gradient. In some ion channels, ion passage through the pore is gated, as these channels open (or close) by conformational changes when bound to specific chemicals (agonists) or subjected to physical agents (electrical signals, temperature, mechanical force, etc.). These electrical, chemical, physical, and/or mechanical agents are detected by a sensor associated with the channel gate.

Ion channels have 3 essential structural and functional elements: (1) a central tunnel or pore through which ions move down their electrochemical gradient; (2) a selectivity filter that dictates ion type(s) that can cross the pore; and (3) a gate that controls the switch between open and closed conformations. A single conformational change can allow permeation flux of more than 100 ions per second. Although an ion channel needs a single gate, many channels have activation and inactivation gates that open simultaneously for ion passage.

3.1.1.1 Neuronal Voltage-Gated Ion channels

Voltage-gated sodium and potassium channels can generate and propagate action potentials in the nervous system. Voltage-gated calcium channels in nerve terminals causes calcium influx and neurotransmitter release. In dendrites, hyperpolarization-activated cyclic nucleotide-gated cation channels (HCN) and K_V channels control back-propagation and time course and magnitude of passive spread of synaptic potentials [209]. [3] Density in dendritic Ca_V channels is higher in the proximal segment.

Density and spatial distribution in axons and dendrites of neuronal voltage-gated ion channels, in addition to their physicochemical properties, determine the signaling features of polarized neurons. In general, voltage-gated sodium [4] and potassium

2. I.e., K_{ATP} or $K_{IR}6.2$.

3. Back-propagation of action potentials can direct long-term potentiation or depression according to their timing with respect to synaptic input.

4. In axons, Na_V channels cause action potential propagation along the unmyelinated axon as well as, in myelinated axons, its initiation at the axon initial segment and nodes of Ranvier, in addition to back-propagation in dendrites [209].

channels K_V1 and K_V7 reside in axons,[5] whereas K_V2 and K_V4[6] as well as HCN channels are located in dendrites [209]. However, different types of voltage-gated Ca_V and K_V3 channels localize in axons and dendrites.

3.1.2 Ion Pumps

Electrochemical gradients are generated by *ion pumps*, or ion adenosine triphosphatases (ATPases), that hydrolyze ATP, hence consuming chemical energy (active transport). These transmembrane proteins use energy from ATP to move ions across cellular membranes against their concentration gradient.

The chemical energy (ATP) can be used directly or indirectly. Ion pumps bind ATP directly (*direct active transport*). Ion transporters (antiport and symport pumps) use the energy created by ion pumps (*indirect active transport*).

Ion pumps are sometimes called primary active carriers to differentiate from secondary active carriers that couple downhill motions of ions to uphill fluxes of other substrate, such as *co-* and *countertransporters*, also called *exchangers*.

Ion exchangers use the energy stored in the electrochemical gradient of a first ion species to move it down its gradient across the membrane in exchange for countertransport of a second ion species. Some ion pumps exchange ions (e.g., H^+–K^+ ATPases).

Ion pumps generate large electrical and chemical gradients that drive ionotropic signaling through ion channels as well as the uptake and extrusion of other solutes through secondary transporters and facilitators. Ion pumps can receive their energy by favorable redox mechanisms (e.g., complex I–III of the mitochondrial membrane) or ATP hydrolysis (e.g., ATP-binding cassette transporters in addition to ATPases).

An ion pump needs at least 2 gates that must open and close in strict alternation to provide access to its binding sites from only one side of the membrane at a time, hence slowing down ion motion [208]. Autophosphorylation of a pump that exports a given ion from the cytosol closes the cytoplasmic gate, temporarily entrapping the bound ion before a conformational change opens the extracytoplasmic gate to the extracellular space or organelle lumen whether the pump resides in the plasma or organelle membrane. The smaller affinity for the transported ion in that conformation leads to their release to the extracytoplasmic compartment and binding of the countertransported ions [208]. Closure of the extracytoplasmic gate that is triggered by counter-ion binding leads to dephosphorylation and stabilization of another occluded state that entraps the counter-ion. A second large rearrangement reopens the cytoplasmic gate, restoring the conformation with lowered affinity for the counter-ion, but higher affinity for the primary ion.

---

5. In the case of myelinated axons, they reside in the axon initial segment, nodes of Ranvier, juxtaparanodes, and presynaptic terminals.

6. Clusters of $K_V2.1$ channels are detected in the soma and proximal dendrites, whereas $K_V4.2$ channels are found more prominently in distal dendrites [209].

ATP-dependent ion pumps are categorized according to the structure and transport mechanism. The 4 major superclasses of ATPases include E-,[7] F-,[8] P-,[9] and V-type ATPases.[10]

3.1.2.1 P-type ATPases

Many ion ATPases belong to the superclass of phosphorylated type of cation pumps, or P-type ATPases. Five main classes of P-type ATPases exist (P1–P5): (1) heavy metal ATPases, (2) Ca^{++} ATPases, (3) H^+ ATPases, (4) amino–phospholipid ATPases, and (5) a set with unknown specificity. These classes of P-type ATPases can be further decomposed into subclasses.

In particular, Na^+–K^+ ATPase, a member of the superclass of P-type ATPases,[11] exports 3 Na^+ successively at different speeds from the cytosol, then imports 2 K^+ during ATP hydrolysis.

P-Type ATPases constitute a family of pumps that carry cations inside or outside the cell. Six-transmembrane helix-containing P-type ATPases possess a cytoplasmic core with phosphorylation, nucleotide-binding, and actuator domains that enable autophosphorylation and ATP hydrolysis.

They undergo conformational changes that create 2 enzymatic states — E1 and E2 — with high- and low-affinity for the primary transported ions, respectively. Ion transport through P-type ATPases relies on changes between these 2 conformations E1 and E2 linked to the autophosphorylation–dephosphorylation cycle. This cycle allows coupling of ion transport and countertransport, as it controls opening and closing of cytosolic and extracytosolic gates [339].

3.2 Calcium Carriers

Calcium is an ubiquitous second messenger (Vol. 4 – Chap. 10. Signaling Pathways) that regulates many biological processes, such as cell signaling and transport, membrane excitability, substance secretion, cell motility, myocyte contraction, synaptic transmission, learning, memory, and fertilization, as well as embryo- and fetogenesis (cell proliferation and differentiation). Divalent cation calcium controls both fast responses, such as secretion and contraction, and long time scale events such as transcription, growth, and cell division.

---

7. E-Type ATPases are cell-surface nucleoside triphosphatases for Ca^{++}), Cu^{++}, and Mg^{++}, in addition to inorganic ion transport.

8. The prototypical F-type ATPase is the mitochondrial F1F0 ATPase for H^+ transport.

9. P-Type ATPases lodge in the membrane of the cell and its organelles. They comprise Ca^{++} ATPases, H^+–K^+ ATPases, and Na^+–K^+ ATPases, among others.

10. V-Tupe ATPases, or V1VO ATPases, primarily reside in vesicles. These H^+ carriers possess a proton-conducting segment and an extramembrane catalytic sector.

11. Sodium–potassium pump belongs to the P2 class with sarco(endo)plamic reticulum Ca^{++} pumps.

Calcium concentration is relatively high in the extracellular medium (millimolar range), whereas it is much lower in the cytosol of cells at rest (O[100 nmol]). Furthermore, an additional Ca^{++} source exists in the intracellular medium. The endoplasmic reticulum (sarcoplasmic reticulum in striated myocytes; Vol. 1 – Chap. 4. Cell Structure and Function) serves as the main specialized Ca^{++} storage organelle (estimated free Ca^{++} concentration ~1 mmol).

Neurotransmitters, hormones, growth factors, ion fluxes, and other regulators provoke a release of Ca^{++} from intracellular Ca^{++}-storing organelles. Calcium release mainly from the endoplasmic reticulum can rapidly yield small, localized or large, global Ca^{++} signals. Therefore, Ca^{++} entry into the cytosol (Ca^{++} influx) across the membrane of the cell surface or endoplasmic reticulum can trigger signaling. In the cell, Ca^{++} ions regulate between-protein interactions, gene transcription, nuclear-pore transport regulation, exo- and endocytosis (Vol. 1 – Chap. 9. Intracellular Transport) [12] and apoptosis (calcium being potentially cytotoxic). [13]

Calcium carriers (channels, exchangers, and pumps) span cellular membranes and respond to many types of physical or chemical stimuli. Among membrane proteins that are permeable to calcium, certain form calcium-selective pores. There are multiple types of calcium channels, including certain members of the transient receptor potential channel (TRP) family.

Intracellular calcium concentration is regulated by various types of Ca^{++} channels, pumps, and exchangers. The intracellular free calcium concentration can fluctuate between 10 nmol and 1 μmol, whereas the extracellular free calcium concentration approximately equals 1 mmol. Within a single contraction–relaxation cycle, the free calcium level in the cardiomyocyte cytosol increases and decreases 100-fold.

Two kinds of plasmalemmal ion channels control calcium influx. One channel allows transient calcium entry and the other slower, more sustained flux. An acute rise in intracellular Ca^{++} concentration by either a release from intracellular stores or entry across the plasma membrane is used in particular by T lymphocytes, endothelial and smooth muscle cells, and cardiomyocytes.

Calcium influx into the cytosol from intracellular stores is regulated by second messengers inositol (1,4,5)-trisphosphate, cyclic ^{ADP}ribose (cADPR), and nicotinic acid adenine dinucleotide phosphate (NAADP). Inositol trisphosphate and cyclic ^{ADP}ribose prime Ca^{++} release from endoplasmic reticulum stores through IP_3 and ryanodine receptors. Nicotinic acid adenine dinucleotide phosphate mobilizes Ca^{++}

12. Intracellular transport is composed of calcium-dependent and -independent steps. Calcium can act as a cofactor in the regulation of certain membrane fusion such as fusion of secretory granules and synaptic vesicles with the plasma membrane.

13. Intracellular calcium compartmentation and transport between the endoplasmic reticulum and mitochondria are modulated by BCL2 family proteins. Moreover, Ca^{++}-regulated processes are also involved in clearance of apoptotic cells and cell debris by phagocytes [210]. Voltage-independent B-type Ca^{++} channels, which interact with plasma membrane Ca^{++} ATPase and are functionally coupled to mitochondria via Ca^{++}, modulate apoptosis of cardiomyocytes [211, 212]. In addition, Na^{+}–Ca^{++} exchanger NCX3 is cleaved by Ca^{++}-dependent calpains, leading to calcium overload and cell death [213]. Conversely, upregulation of NCX1 and NCX3 protect against ischemia [214].

from acidic vesicles through NAADP receptors that are 2-pore channels of endosomal and lysosomal membranes [215].

In T lymphocytes, in addition to IP_3 and cADPR, NAADP is involved in initiation and propagation of Ca^{++} signaling from ryanodine receptors RyR1 for T-cell activation on formation of immunological synapses [216].

In non-excitable cells, plasmalemmal calcium influx channels are classified into 2 categories: (1) *store-operated channels* (SOC) [14] that are responsible for calcium release-activated calcium flux and (2) *store-independent channels*.

The activation of both channel types requires phospholipase-C. Store-operated channels are activated by the reduction in stored calcium ions. Store-independent channels are stimulated by either elevation in the cytosolic calcium level or diacylglycerol. [15]

To conserve cell surface mass and area, exo- and endocytosis are coupled. *Vesicular calcium channels* in the plasma membrane are constituted by clusters of Ca^{++} channels at sites of vesicle fusion. Channels $Ca_V2.1$ form clusters at the presynaptic membrane as well as axoplasm and axolemma of developing axons, where they provoke Ca^{++} influx, hence triggering vesicle fusion [217].

Mitochondria can transport and buffer huge amounts of Ca^{++} across the inner membrane, especially near the endoplasmic reticulum and plasma membrane. Consequently, the mitochondrial Ca^{++} uptake contributes to the control of cytosolic Ca^{++} signals and subsequent regulation of diverse processes (e.g., rapid hormone secretion and prolonged cell differentiation). Mitochondrial Ca^{++} uptake can stimulate dehydrogenases of the tricarboxylic acid cycle. [16] *Mitochondrial calcium uptake protein*-1 (MiCU1) localizes to the mitochondrial inner membrane [218]. This high-capacity uniporter depends on membrane potential. It is active when the extramitochondrial Ca^{++} concentration is in the micromolar range. It couples cytosolic calcium transients to the activation of matrix dehydrogenases. This calcium sensor can be considered as the founding member of a set of mitochondrial calcium carriers [218].

3.2.1 Calcium Release-Activated Ca^{++} Channels

In many non-excitable cells, activated phospholipase-C-linked receptors cause a biphasic augmentation in cytosolic Ca^{++} concentration. The initial, transient increase that results from Ca^{++} release from the endoplasmic reticulum is followed by a much smaller, but sustained elevation. Calcium release-activated Ca^{++} channels

14. Persistent Ca^{++} influx is triggered by a sensor that monitors the Ca^{++} concentration in the lumen of the endoplasmic reticulum and directly gates the opening of plasmalemmal Ca^{++} channels when the endoplasmic reticulum Ca^{++} concentration drops. This process is called *store-operated Ca^{++} entry* (SOCE).

15. Diacylglycerol can be recycled to PIP_2 by the phosphatidylinositol cycle or converted to phosphatidic acid by DAG kinase and ^{CDP}DAG by cytidine diphosphate–diacylglycerol (CD) synthase (CDS).

16. Calcium signals ATP consumption in the cytosol and stimulates ATP production in mitochondria (feedforward mechanism). However, excessive Ca^{++} uptake can prime the appearance of mitochondrial permeability transition pores and apoptosis.

(CRAC) actually allow sustained Ca^{++} signaling in cells after Ca^{++} release from the endoplasmic reticulum.

Store-operated CRAC channels mediate Ca^{++} influx across the plasma membrane when intracellular stores of calcium are depleted. [17]

In excitable cells such as smooth muscle cells, sustained Ca^{++} entry is not only due to *receptor-operated* Ca^{++} channels, i.e., voltage-gated Ca_V channels, [18] but also to Ca^{++} release-activated Ca^{++} channels. [19] The relative contribution of each source depends on the smooth muscle cell type. Store-operated Ca^{++} channels have a greater importance in tonic smooth muscle cells such as those of pulmonary artery wall. On the other hand, sarco(endo)plasmic reticulum Ca^{++} ATPase (SERCA) (Sect. 3.2.7) controls Ca^{++} stores. [20]

The endoplasmic reticulum informs the plasma membrane of Ca^{++} depletion. Two protein types, transmembrane stromal interaction molecule (StIM) and Orai of the membrane of the endoplasmic reticulum and cell surface, respectively, explain CRAC channel activity. The StIM1 sensor detects depletion of calcium from endoplasmic reticulum, oligomerizes, and then recruits plasmalemmal calcium channel Orai1 and interacts directly with it. [21] This interaction enables the opening of Orai1 channels [220]. [22] In fact, 2 StIM (StIM1–StIM2) and 3 Orai isoforms (Orai1–Orai3) exist in humans.

---

17. One of the main Ca^{++}-regulated transcription factors is cytoplasmic nuclear factor of activated T cells (NFAT). Sustained Ca^{++} influx activates calmodulin-dependent protein phosphatase PP3, which dephosphorylates NFAT, leading to NFAT translocation to the nucleus. The stromal interaction molecule (StIM) is required with plasmalemmal Orai1 for the activation of Ca^{++} release-activated Ca^{++} channels [219]. When calcium is depleted from its stores, StIM1 translocates to vesicular structures near the plasma membrane to activate calcium release-activated calcium channels.

18. Release of Ca^{++} from its intracellular stores in response to IP_3, possibly amplified by Ca^{++}-induced Ca^{++} release from ryanodine-sensitive Ca^{++} channels, activates Ca^{++}-dependent chloride channels. In response to Ca^{++} influx, opening of Ca^{++}-activated chloride channels causes an initial, transient current that depolarizes the cell membrane and provokes opening of voltage-gated Ca^{++} channels.

19. Sustained Ca^{++} entry results from non-selective cation channels. The NO–cGMP pathway inhibits Ca^{++} release-activated Ca^{++} channels that are activated by endothelin in arterial smooth muscle cells. In addition, cGMP can also favor refilling of the endoplasmic reticulum.

20. In human tracheal smooth muscle cells, SERCA2b isoform replenishes the endoplasmic reticulum.

21. As soon as calcium store empties, StIM1 sensor that is distributed throughout the endoplasmic reticulum moves and creates clusters close to the plasma membrane (within 25 nm of the plasma membrane), immediately before the activation of CRAC currents. Currents CRAC follow Orai1 displacements, the coordinated redistribution of both StIM1 and Orai1 in the endoplasmic reticulum and plasma membrane, respectively, being required for their interaction.

22. Orai1 channel is also called CRACM1. The Orai1 channel is mainly a plasmalemmal dimer at rest [221]. It multimerizes and ensures the selectivity for calcium ions. Upon interaction with StIM, Orai1 dimers indeed dimerize. Tetramers Orai1 constitute Ca^{++}-selective pores.

In many cell types, Ca^{++} is released from its intracellular store , mainly the endoplasmic reticulum, by the inositol (1,4,5)-trisphosphate receptor attached to its IP_3 ligand. A very small number of IP_3Rs exists in the plasma membrane (especially of B lymphocytes) [222]. The IP_3R and Orai1 channels have large and small conductance, and are poorly and highly selective for calcium, respectively.

Calcium release from the endoplasmic reticulum leads to aggregation of Ca^{++}-sensing stromal interaction molecule-1 in areas of the endoplasmic reticulum close to the plasma membrane, where they can interact with plasmalemmal Orai1 (Fig. 3.1). The coupling between Orai1 and StIM1 explains the store-operated Ca^{++} release-activated Ca^{++} influx [223]. Activity of StIM1 and Orai1 are coordinated. [23]

Excitable and non-excitable cells differ by their ability to increase the concentration of cytosolic calcium. Non-excitable cells, such as lymphocytes, dendritic cells, and mastocytes, have a low density of voltage-gated Ca^{++} channels, but have Ca^{++} release-activated Ca^{++} (CRAC) channels. On the other hand, excitable cells, such as neurons and myocytes, have numerous voltage-gated Ca^{++} channels that are activated by depolarization, but a small pool of store-operated Ca^{++} channels.

In addition to activation of store-operated, plasmalemmal Ca^{++} channels Orai, StIM1 represses voltage-gated $Ca_V1.2$ channel in various cell types (e.g., vascular smooth muscle cells and neurons). The StIM1–$Ca_V1.2$ interaction in *plasmerosomes* [24] prevents channel opening (acute inhibition) and removes the channel from the cell surface (slow inhibition by endocytosis) [225, 226]. [25]

3.2.1.1 Family of Plasmalemmal Orai Channels

Channels that are activated by StIM belong to canonical transient receptor potential (TRPC1–TRPC6) and Orai channel families. Channels Orai1 and Orai3 are expressed in lymphoid organs, skeletal muscle, liver, and skin, whereas Orai2 is predominantly found in human lungs, kidneys, and spleen. Heteromeric Orai1–Orai3 channels have a more reduced Ca^{++} selectivity and inactivation rate than those of their homomeric forms [227]. The Ca^{++}-dependent inactivation rate rises from Orai1, Orai2, to Orai3 (low, moderate, and very fast, respectively). The fast Ca^{++}-dependent inactivation that limits Ca^{++} influx relies on Orai C-terminus as well as StIM1 [228]. The StIM1 sensor binds to TRPCs to heteromultimerize them as well as allow their opening by electrostatic interaction.

---

23. Overexpression of either protein fails to significantly increase the calcium influx. Overexpression of both proteins greatly amplifies store-operated currents.

24. The plasmerosome is the unit made by the plasmalemmal nanodomain and the adjoining junctional sarco(endo)plasmic reticulum, separated by a tiny (12–20 nm) cytosolic junctional space, with its associated molecules. Substance content in a segment of the plasma membrane and that of the apposed membrane of the endoplasmic reticulum (ER) constitute the plasmerosome.

25. The cytosolic CRAC-activating domain (CAD) of StIM1 recruits both Orai and $Ca_V1.2$ to plasmerosomes and activates Orai, but inhibits $Ca_V1.2$.

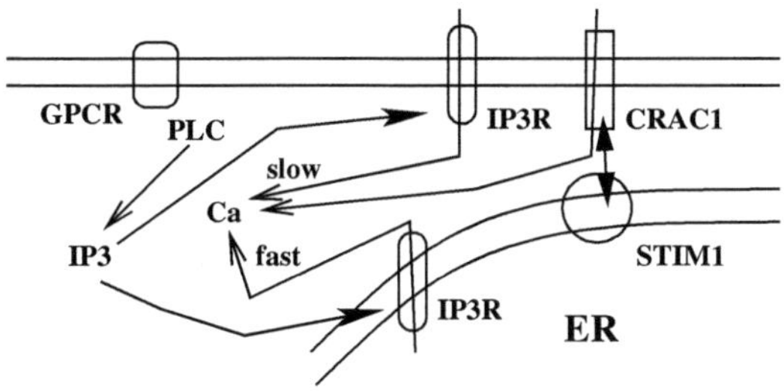

Figure 3.1. Receptor-mediated calcium transports in the cytosol (Sources: [223, 224]). Ligand-bound G-protein-coupled receptors stimulate phospholipase-C, which generates inositol (1,4,5)-trisphosphate (IP_3). The latter interacts with IP_3 receptors (IP_3R) of the endoplasmic reticulum membrane. The IP_3R channels, quickly gated by IP_3, release Ca^{++} from its store, leading to a fast initial rise in cytosolic calcium concentration and a rapid calcium store depletion. In many cell types, the decrease in stored calcium triggers store-operated channels in the plasma membrane, leading to the additional Ca^{++} release-activated Ca^{++} (CRAC) current. Long-term response involves a slower and more persistent calcium entry through a small amount of plasmalemmal IP_3Rs, slowly gated by IP_3, and a large quantity of Orai1 (or CRACM1). Channel Orai1 couples to calcium sensors StIM1 for store-operated Ca^{++} release-activated Ca^{++} influx. When endoplasmic reticulum Ca^{++} stores are filled, monomeric and dimeric StIM colocalizes with microtubules. Binding of Ca^{++} to StIM1 impedes its multimerization. When these stores are depleted, StIM oligomerization occurs within seconds and triggers StIM translocation toward the plasma membrane, close enough to the cell surface to interact with plasmalemmal proteins (100–200 nm from the cell surface). Oligomers StIM1, but not monomers or dimers, accumulate near the plasma membrane. Here, StIM couples to Orai channels and causes their tetramerization, aggregation, and gating, hence concentrating Ca^{++} entry at StIM–Orai cluster sites. Therefore, depletion of intracellular calcium stores following IP_3-induced Ca^{++} release activates store-operated Ca^{++} channels in the plasma membrane for Ca^{++} entry in the cytosol and subsequent refilling of Ca^{++} stores. Store-operated channels possess various levels of cation selectivity and permeability. One store-operated Ca^{++} flux is the Ca^{++} release-activated Ca^{++} current through Orai channels activated by StIM1 sensor. Store-operated and Ca^{++} release-activated Ca^{++} channels include members of canonical transient receptor potentials (TRPC1–TRPC6) and Orai (Orai1–Orai3) channels. StIM1 also interacts with TRPC channels to provoke their heteromultimerization and activation.

3.2.1.2 Family of StIM Sensors

Upon store depletion, both StIM1 and StIM2 on endoplasmic reticulum aggregate, redistribute to junctions adjacent to the plasma membrane, organize Orai channels into clusters, and activate SOC (CRAC) channels in the plasma membrane.

The StIM2 sensor has a lower effective affinity for luminal endoplasmic reticulum Ca^{++} than StIM1 (400 and 200 μmol, respectively) [224]. During modest depletion of endoplasmic reticulum Ca^{++} stores, StIM1 can remain distributed along the endoplasmic reticulum, whereas StIM2 already forms clusters near the plasma membrane.

3.2.1.3 SOCs in Vascular Smooth Muscle Cells

Stromal interaction molecule StIM1 initiates opening of store-operated Ca^{++} channels in vascular smooth muscle cells to regulate their contraction, proliferation, and migration. Effect of StIM1 in smooth muscle cells varies according to the cell type, i.e., the vascular bed. In human coronary artery smooth muscle cell, StIM1 causes store-operated calcium influx, thereby provoking cell proliferation [229]. On the other hand, in human saphenous vein smooth muscle cells, StIM1 that associates with TRPC1 favors cell migration, but not proliferation [230].

3.2.1.4 Mitochondrial Stores

Stimulated Ca^{++} release can load mitochondria with a fraction of cytosolic Ca^{++}. Delivery of Ca^{++} from mitochondria by Na^{+}–Ca^{++} exchanger generates a gradual increase in cytoplasmic Ca^{++} concentration that provokes a second Ca^{++} release from the endoplasmic reticulum [231]. Therefore, a second peak in Ca^{++} cytosolic concentration occurs that allows partial reloading of mitochondria (Vol. 4 – Chap. 10. Signaling Pathways).

3.2.1.5 SOC Modulation

Calcium fluxes from intracellular stores via Ca^{++} release channels are associated with countermotion of ions across the endoplasmic reticulum membrane to balance the transient negative potential generated by Ca^{++} release. Trimeric intracellular cation channels (TRIC) function in synchronization with Ca^{++} influx from intracellular stores [232]. Channel TRICa [26] is distributed throughout the endoplasmic reticulum of excitable cells (nervous and muscular tissues), whereas TRICb [27] is present in the endoplasmic reticulum of most tissues.

Calcium influx through CRACs can be suppressed by an increase in cytosolic Ca^{++} concentration via a negative feedback, when the intracellular Ca^{++} concentration remains high during prolonged periods. Both calmodulin and StIM1 contribute to inactivation of store-operated Ca^{++} release-activated Ca^{++} channels. [28]

In T lymphocytes, Ca^{++}-binding CRAC regulator-2A (CRACR2a) regulates store-operated Ca^{++} influx. At low intracellular Ca^{++} concentration, CRACR2a promotes the binding of StIM sensors to Orai channels and stabilizes this interaction, thereby enhancing CRAC channel activity [234]. The CRACR2a protein forms the Orai–StIM–CRACR2a ternary complex with a 11- to 14-nm protrusion into the

26. A.k.a. transmembrane protein TMEM38a.

27. A.k.a. transmembrane protein TMEM38b.

28. Ca^{++}-Dependent CRAC inactivation depends on a short motif (residues 475–483) of Orai1-binding CRAC activation domain of StIM1 sensor [233]. Neutralization of aspartate or glutamate pairs in this sequence reduces or enhances Ca^{++}-dependent inactivation, whereas the combined neutralization of 6 acidic residues supresses inactivation entirely. Calmodulin-binding site of Orai1 also contains a sequence (residues 68–91) that participates in Ca^{++}-dependent inactivation.

cytoplasm. As intracellular Ca^{++} level rises, CRACR2a binds Ca^{++} and dissociates the Orai–StIM complex. The CRACR2b homolog has a relatively low sequence identity with CRACR2a and may have different functions.

3.2.2 Calcium Channel-Induced Ca^{++} Release

Calcium channel-induced (receptor-operated) Ca^{++} release (CCICR) occurs in the absence of Ca^{++} influx from the extracellular medium by activation of Ca_V1 channel that provokes Ca^{++} release from the endoplasmic reticulum and contraction of smooth muscle cells of basilar arteries of rats and coronary arteries of diverse mammalians (porcs, rats, and humans) via the G-protein–PLC–IP_3 cascade [235].

Calcium channel-induced Ca^{++} release can be primed by chemical agents such as ATP or small depolarization without opening of Ca^{++} channels. Hypoxia by opening ATP-regulated K^+ channels due to elevated intracellular ATP level induces membrane hyperpolarization. Hypoxia also closes voltage-dependent Ca^{++} channel activity in most vascular beds, thereby inhibiting CCICR and causing a vasodilation independently of K_{ATP} channels, inhibition of Ca^{++} import through voltage-dependent Ca^{++} channels, and RoCK kinase.

3.2.3 Voltage-Gated Calcium Channels

Voltage-gated Ca^{++} channels reside in membranes of most excitable cells. Voltage-gated calcium channels (Ca_V) are classified into 3 main types (Tables 3.1 and 3.2: (1) high-voltage-activated, dihydropyridine-sensitive channels (Ca_V1.x, i.e., $Ca_V1.1$–$Ca_V1.4$); [29] (2) high-voltage activated, dihydropyridine-insensitive channels (Ca_V2.x) that comprises $Ca_V2.1$, [30] $Ca_V2.2$; [31] and $Ca_V2.3$ channels; [32] and (3) low-voltage-activated, more or less rapidly inactivating, Ca^{++} channels (Ca_V3.x, i.e., $Ca_V3.1$–$Ca_V3.3$). [33]

3.2.3.1 Excitation–Contraction Coupling

Excitation–contraction coupling corresponds to muscle contraction triggered by an electrochemical signal. Sarcolemmal depolarization activates Ca_V channels that stimulate intracellular Ca^{++} ryanodine channels (sarcoplasmic reticulum Ca^{++} release channels), thereby releasing large amounts of calcium to initiate muscle contraction. In the myocardium, $Ca_V1.2$ is involved in excitation–contraction coupling. Slow calcium channels induce the slow depolarization necessary for the normal functioning of heart nodal tissue. Calcium release channels cause calcium delivery from the sarcoplasmic reticulum to the cytosol of cardiomyocytes of smooth muscle cells.

29. A.k.a. L-type (long-lasting) Ca^{++} channels.
30. A.k.a. P-type (Purkinje cells) and Q-type Ca^{++} channels.
31. A.k.a. N-type (related to noradrenaline) Ca^{++} channels.
32. A.k.a. R-type Ca^{++} channels.
33. A.k.a. T-type (transient) Ca^{++} channels.

Table 3.1. Main feature of voltage-gated Ca^{++} channels (Source: [5]; CMC: cardiomyocyte; SkM: skeletal myocyte; SMC: smooth muscle cell). Voltage-gated calcium channels exist in the plasma membrane as heteromers with an $\alpha 1$ subunit that forms the pore and associates with auxiliary subunits β and $\alpha 2\delta$. The $\alpha 2\delta$ dimer that has many isotypes ($\alpha 2\delta 1$–$\alpha 2\delta 4$) influences the transfer and kinetic and voltage-dependent properties of $Ca_V 1$ and $Ca_V 2$ heteromeric channels.

Type	Other names	Functional characteristics
$Ca_V 1.1$	L-type (SkM), α1s	High voltage-activated channel, slow inactivation
$Ca_V 1.2$	L-type (CMC/SMC), α1c	High voltage-activated channel, slow inactivation
$Ca_V 1.3$	L-type, α1d	Low to moderate voltage-activated channel, slow inactivation
$Ca_V 1.4$	L-type, α1f	Moderate voltage-activated channel, slow inactivation
$Ca_V 2.1$	P-, Q-type, α1a	Moderate voltage-activated channel, moderate inactivation
$Ca_V 2.2$	N-type, α1b	High voltage-activated channel, moderate inactivation
$Ca_V 2.3$	R-type, α1e	Moderate voltage-activated channel, fast inactivation
$Ca_V 3.1$	T-type, α1g	Low voltage-activated channel, fast inactivation
$Ca_V 3.2$	T-type, α1h	Low voltage-activated channel, fast inactivation
$Ca_V 3.3$	T-type, α1i	Low voltage-activated channel, moderate inactivation

In addition to its effects on contractility of smooth muscle cells,[34] Ca^{++} influences its own flux, as it modulates the activity of both voltage-dependent and -independent ion channels and controls gene transcription. Ca^{++} influx through $Ca_V 1$ can lead to phosphorylation (activation) of cAMP response element-binding protein.

In fish skeletal muscle cells, sarcolemmal depolarization activates the sarcolemmal $Ca_V 1.1$ for sarcomere contraction. Simultaneous with the excitation–contraction coupling, a small and slowly activating Ca^{++} inward current happens through pore-forming Ca_VV1.1α1s non-Ca^{++} conductive channels. These channels act as voltage sensors to trigger opening of ryanodine receptors [237].

34. Although Ca^{++} influx is necessary for effective SMC contraction, voltage-dependent Ca^{++} entry across the plasma membrane can initiate contraction, as membrane potential also determines smooth muscle tone.

Table 3.2. Voltage-gated calcium channel isoforms and genes. Calcium channels form hetero-oligomers. Four different genes encode the β subunit of Ca_V channels. Each gene product can be alternatively spliced, giving rise to 2 to 4 splice variants. In cardiomyocytes, β2 subunit predominates, whereas in smooth muscle cells at least 3 β subunit types have been identified (β1b, -2, and -3). The α2–δ dimer localizes to the extracellular edge of the cell, connected to the plasma membrane by a glycosyl phosphatidylinositol anchor [236].

Gene	Channel
Subunit α	
CACNA1S	$Ca_V1.1$
CACNA1C	$Ca_V1.2$
CACNA1D	$Ca_V1.3$
CACNA1F	$Ca_V1.4$
CACNA1A	$Ca_V2.1$
CACNA1B	$Ca_V2.2$
CACNA1E	$Ca_V2.3$
CACNA1G	$Ca_V3.1$
CACNA1H	$Ca_V3.2$
CACNA1I	$Ca_V3.3$
Subunit β	
CACNB1	β1
CACNB2	β2
CACNB3	β3
CACNB4	β4
Subunit α2δ	
CACNA2D1	α2δ1
CACNA2D2	α2δ2
CACNA2D3	α2δ3
CACNA2D4	α2δ4
Subunit γ	
CACNG1	γ1
CACNG2	γ2
CACNG3	γ3
CACNG4	γ4
CACNG5	γ5
CACNG6	γ6
CACNG7	γ7
CACNG8	γ8

3.2.3.2 Calcium–Calmodulin Binding

Calcium and calmodulin binding to Ca_V1 and Ca_V2 provokes distinct channel regulations. Calcium can bind to calcium sensors of the channel; a first and second sensor are sensitive to local Ca^{++} level and weaker distant Ca^{++} sources, respectively. Spatial Ca^{++} selectivity of remote calcium sensor can be switched to local calcium sensing owing to an additional Ca^{++}–Cam binding site (NSCaTE),[35] particularly in $Ca_V1.2$ and $Ca_V1.3$, which naturally contain NSCaTE [238]. Channel Ca_V2 lacks NSCaTE. The existence or absence of NSCaTE explains differences in calcium sensing according to calcium location (nearby or remotely), especially between Ca_V1 and Ca_V2 channels.

3.2.3.3 Ca_V1 Channels

High-voltage-gated calcium channels include Ca_V1 and Ca_V2 channels. The regulation of Ca_V channels is governed by a set of feedbacks. In particular, the Ca_V function is modulated by interactions with calmodulin. Calcium–calmodulin plays a major role, as elevated cytoplasmic calcium either fosters channel closing or opening (calcium-dependent inactivation or facilitation), whether a single or 2 Ca^{++}–calmodulin are tethered to $Ca_V1.2$, respectively [239]. Two Ca^{++}–calmodulin can indeed bind simultaneously to one monomeric $Ca_V1.2$ channel [239]. In addition, according to the type of calmodulin-binding domain of $Ca_V1.2$, the connection of Ca^{++}–calmodulin is labile or not.

When a subject is frightened, the sympathetic nervous system trigger a *fight-or-flight response* by the release of noradrenaline that activates the Gs-coupled β-adrenoceptors, in particular to yield positive chronotropy and inotropy. Activated β-adrenergic receptors stimulate adenylate cyclase that launches a phosphorylation program by protein kinase-A to increase contractility of skeletal and cardiac muscles. One of the main PKA substrates is Ca_V1 channels. The basal activity of $Ca_V1.2$, which is regulated by phosphorylation of serine and threonine residues at the border of regulatory domains, is multiplied at least 3 times upon activation of the βAR–AC–cAMP–PKA pathway [240]. In addition, the regulation of the cardiac frequency requires a proper matching between $Ca_V1.2$ and A-kinase anchoring proteins.

3.2.3.4 $Ca_V1.2$ Isoforms

Alternative splicing of $Ca_V1.2$ generates tissue-specific variants. Three isoforms are present in smooth muscle cells. The $Ca_V1.2a$ and $Ca_V1.2b$ channels correspond to long and short isoforms, respectively.

3.2.3.5 $Ca_V1.2l$ in the Cardiovascular System

Voltage-gated $Ca_V1.2$ channel is involved in excitation–contraction coupling in both cardiomyocytes ($Ca_V1.2a$) and smooth muscle cells ($Ca_V1.2b$). The cardiac and smooth muscle α1 subunits are splice variants of the same gene.

---

35. NSCaTE stands for N-terminal spatial calcium-transforming element.

$Ca_V 1.2$ in Smooth Muscle Cells

In smooth muscle cells of resistance arteries, such as cerebral arterioles (bore 100–200 μm), $Ca_V 1.2$ channels are the primary Ca^{++} entry agents [241]. Predominance of certain exons in smooth muscle cell $Ca_V 1.2$ channels causes a hyperpolarizing shift in the voltage sensitivity of currents toward the arterial voltage range.

$Ca_V 1.2$ in Cardiomyocytes

In cardiomyocytes, the magnitude of the intracellular Ca^{++} transients relies on that of Ca^{++} currents and release from intracellular stores. Influx of Ca^{++} through $Ca_V 1.2$ can be facilitated or restrained according to subject's activity. The availability of $Ca_V 1.2$ is regulated by the transmembrane potential, protein kinases and phosphatases, and Ca^{++}-binding proteins.

A train of repetitive depolarizations or a strong depolarizing prepulse shifts $Ca_V 1.2$ from a resting gating pattern into a frequency- or voltage-dependent mode of gating, respectively, characterized by long-duration opening and high-open probability. This positive feedback mechanism, the so-called $Ca_V 1.2$ *facilitation*, relies on elevated intracellular Ca^{++} concentration in the vicinity of the channel (calcium-dependent facilitation). On the other hand, because excessive Ca^{++} influx is toxic, *voltage-* and *calcium-dependent inactivation* balance Ca^{++}-dependent facilitation. Therefore, augmented intracellular Ca^{++} concentration is responsible for 2 opposing signals transmitted to $Ca_V 1.2$: (1) enhancement, or facilitation, and (2) inactivation to limit Ca^{++} accumulation in the cytosol.

Voltage- and frequency-dependent (Ca^{++}-dependent) activation of $Ca_V 1.2$, as well as Ca^{++}-dependent inactivation, depend on high-affinity binding of calmodulin and/or phosphorylation (Ser1512 and Ser1570 of $Ca_V 1.2$ $\alpha 1$ subunit in murine hearts) by calmodulin kinase CamK2 [242]. Facilitation of $Ca_V 1.2$ activity may be triggered by phosphorylation of different subunits that are expressed in a tissue-specific manner. Like $Ca_V 1.2$ $\alpha 1$ subunit that exists in neurons, smooth muscle cells, and cardiomyocytes, phosphorylation (Thr498 in rabbits) of the neuronal $\beta 2a$ subunit of calcium channel (not detected in murine hearts) also causes facilitation. In any case, other processes contribute to Ca^{++}-dependent facilitation.

Voltage-dependent inactivation and recovery from inactivation, i.e., the number of $Ca_V 1.2$ channels available for activation, also depend on CamK2 activity. Calmodulin-dependent kinase-2 enhances the activity of $Ca_V 1.2$, as it accelerates recovery from inactivation and positively shifts the voltage dependence of steady-state inactivation [242].

In cardiomyocytes, 20-hydroxyeicosatetraenoic acid (20HETE; concentration 10–100 nmol) stimulates $Ca_V 1.2$ channel via protein kinase-C and superoxide produced by NADPH oxidase [243].

$Ca_V 1.2$ Regulation by Kinases

The $Ca_V 1.2$ channel is regulated by various kinases, such as cAMP-dependent (PKA) and cGMP-dependent (PKG) protein kinase, as well as protein kinase-C, in addition to protein Tyr kinases and calmodulin-dependent kinase, which thus modulate the cell contractility [244].[36] Conversely, protein phosphatases PP1 and PP2 dephosphorylates $Ca_V1.2$ channel.

3.2.3.6 $Ca_V2.2$ Isoforms

Many $Ca_V2.2$ isoforms are ubiquitous. Channels $Ca_V2.2$ are inhibited by G-protein-coupled receptors in either a voltage-dependent[37] or -independent manner.[38]

3.2.4 Two-Pore Calcium Channels

Nicotinic acid adenine dinucleotide phosphate (NAADP) is one of the most potent messengers with inositol (1,4,5)-trisphosphate and cyclic ^{ADP}ribose (cADPR) that is produced from nicotinamide adenine dinucleotide (NAD^+) by ADP–ribosyl cyclases. It actually triggers calcium signaling, as it causes Ca^{++} release from intracellular stores. Nicotinic acid adenine dinucleotide phosphate mobilizes Ca^{++} from acidic vesicles through NAADP receptors that are 2-pore channels of endosomal membranes [215]. Calcium release from vesicle stores is subsequently amplified by Ca^{++}-induced Ca^{++} release from the endoplasmic reticulum through IP_3 receptors.

In addition, NAADP targets on lysosomes the Ca^{++} channel transient receptor potential mucolipin-1 (TRPML1) that operates either as a H^+ channel or iron release channel as well as on the endoplasmic reticulum and nuclear envelope the Ca^{++} channel RyR1 ryanodine receptor.[39]

Two-pore channels are closely related to voltage-gated cation channels of sperm and, more distantly, to TRP channels. Two-pore, cation-selective channels are intracellular members of the category of voltage-gated ion channels. Unlike other calcium channels that possess 4 domains with 6 transmembrane segments, TPC1 only contains 2 domains, also with 6 transmembrane segments. Two-pore channel TPC1 encoded by the TPCN1 gene resides on endosomal membranes, whereas TPC2 encoded

---

36. Protein kinase-G phosphorylates (inactivates) $Ca_V1.2a$ (Ser533) and $Ca_V1.2b$ (Ser528). It can also activate a protein phosphatase that then dephosphorylates $Ca_V1.2$ channel. In addition, cGMP can stimulate phosphodiesterase PDE2 that reduces cAMP level, hence inhibiting $Ca_V1.2$ channel. Protein kinase-A phosphorylates (activates) $Ca_V1.2a$ (Ser1928) and $Ca_V1.2b$ (Ser1923). Protein kinase-C phosphorylates (inactivates) $Ca_V1.2a$ (Thr27 and Thr31), but can also activate $Ca_V1.2$ channel.

37. Inhibition by binding of $G\beta\gamma$ subunits, released from $G\alpha_{i/o}$ subunits, to $Ca_V2.2$ decreases during depolarization, probably due to dissociation of $G\beta\gamma$.

38. Inhibition is mediated by several mechanisms. The first one involves Gq and phospholipase-C, the second one Gi/o acting via Tyr kinase.

39. In T lymphocytes upon formation of immunological synapses, in addition to inositol trisphosphate and cyclic ^{ADP}ribose, nicotinic acid adenine dinucleotide phosphate is involved in initiation and propagation of Ca^{++} signaling from RyR1 ryanodine receptors, whereas mucolipin-1 plays an accessory role [216]. Agent NAADP primes Ca^{++} release within seconds, whereas IP_3 and cADPR trigger Ca^{++} liberation from intracellular stores within minutes and tens of minutes, respectively.

by the TPCN2 gene is located on lysosomal membranes [215]. The TPC2 channel is expressed in most human tissues, with higher levels in the liver and kidneys.

3.2.5 Inositol Trisphosphate-Sensitive Calcium-Release Channels

Mammalian cells have 2 main channels responsible for Ca^{++} influx due to release from intracellular stores, i.e., efflux from endoplasmic reticulum: (1) inositol trisphosphate-sensitive Ca^{++} release channels that are designated as inositol trisphosphate receptors (IP_3R) and (2) ryanodine-sensitive Ca^{++} release channels that are labeled ryanodine receptors (RyR). Inositol trisphosphate and ryanodine receptors are gated by Ca^{++} and pyridine nucleotide cyclic ^{ADP}ribose.

Calcium puffs occur with short latencies of 100 to 200 ms after release of IP_3 messenger. They involve pre-established, stable clusters of at least 4 IP_3Rs in the endoplasmic reticulum membrane [245].

Ubiquitous IP_3R is a membrane glycoproteic complex that serves as Ca^{++} channel activated by inositol (1,4,5)-triphosphate.[40] Most of the IP_3Rs in the cell resides on Ca^{++}-storage endoplasmic reticulum. Calcium-release channels IP_3Rs are actually responsible for Ca^{++} influx into the cytosol from cellular Ca^{++} store. Local and global Ca^{++} signals then regulate numerous cell processes.

3.2.5.1 Molecular Diversity

Three IP_3R isoforms (IP_3R1–IP_3R3) are encoded by 3 different genes (ITPR1–ITPR3; Table 3.3). They have distinct and overlapping expression patterns. Most cells outside the central nervous system produce more than one IP_3R type.[41] Expression level of IP_3R isoforms can change during cell differentiation in response to various stimuli, and after ubiquitination for proteasomal degradation [246].

Permeation, gating, and regulation of IP_3R channels by ligands (IP_3, Ca^{++}, ATP, and H^+) and IP_3R-interacting proteins (caldendrin, BCL2-related proteins, Na^+–

40. Upon ligand binding, activated receptor Tyr kinases and G-protein-coupled receptors activate phospholipases PLCβ and PLCγ that hydrolyze plasmalemmal lipid phosphatidylinositol (4,5)-bisphosphate to generate inositol (1,4,5)-trisphosphate. The latter diffuses in the cytoplasm and binds to its receptor on the plasma membrane and, most often, on the endoplasmic reticulum, i.e., in contact with extracellular calcium reservoir and intracellular calcium store. In resting condition, estimated cytosolic Ca^{++} concentration ranges from 50 to 100 nmol. This low concentration is maintained by Ca^{++} carriers located in endoplasmic reticulum and plasma membranes. In particular, Ca^{++} ATPases of endoplasmic reticulum membrane accumulate Ca^{++} in the endoplasmic reticulum lumen that contains high concentrations of Ca^{++}-binding proteins. Total Ca^{++} amount in the lumen can be greater than 1 mmol; concentration of free Ca^{++} can span between 100 and 700 µmol [246]. Calcium can move in the cytoplasm by diffusion or bind mobile and immobile Ca^{++}-binding proteins that can serve as buffers (Vol. 4 – Chap. 10. Signaling Pathways). Calcium ions can thus rapidly dissipate near IP_3Rs. Distribution and concentration of Ca^{++}-handling proteins enable spatial and temporal regulation to provide specific Ca^{++} signals.

41. For example, all 3 IP_3Rs reside in ovarian granulosa cells.

Table 3.3. Inositol (1,4,5)-trisphosphate receptors–channels (Source: [5]). These tetrameric, ligand-gated Ca^{++} channels release Ca^{++} from intracellular stores such as the endoplasmic reticulum. Channels IP_3R associate with some proteins, such as calmodulin, FKBP, and PP3 via FKBP. They are phosphorylated by protein kinase-A, -C, and -G, as well as calmodulin-dependent kinase CamK2.

Type	Activators
IP_3R1	IP_3 (nmol–µmol), cytosolic Ca^{++} (<750µmol), cytosolic ATP (< mmol)
IP_3R2	IP_3 (nmol–µmol), cytosolic Ca^{++} (nmol)
IP_3R3	IP_3 (nmol–µmol), cytosolic Ca^{++} (nmol)

K^+ ATPase α subunit, A-kinase anchoring protein AKAP9, etc.) confer the specificity to channel subtypes. Permeation properties of IP_3R isoforms are similar. According to gating properties of homotetrameric IP_3Rs, IP_3R1 isoform induces irregular $[Ca^{++}]_i$ oscillations;[42] IP_3R2 subtype causes long-lasting regular $[Ca^{++}]_i$ oscillations; IP_3R3 isochannel generates monophasic $[Ca^{++}]_i$ responses [246]. However, channel density and spatial distribution as well as many other agents influence functioning of IP_3R isoform.

Additional IP_3R diversity results from splice variants. Channel IP_3R1 can include sequences encoded by 3 alternatively spliced exons (SE1–SE3). Adult cerebellum expresses 13 splice IP_3R1 forms. The long IP_3R1 form is distinguished by inclusion of SE2 exon, whereas short variant corresponds to SE2− IP_3R1.[43] Isoform IP_3R2 can also have a splice variant. Glycoprotein IP_3R3 does not have known alternatively spliced variant. Last, IP_3R isoforms can form homo- and hetero-oligomers.

3.2.5.2 Cellular IP_3R Distribution

Receptor IP_3R1 is the most widely expressed in all tissue. Both IP_3R2 and IP_3R3 are less widespread, but can be prominent in some cell types. Cerebellar Purkinje neurons predominantly synthesize IP_3R1, cardiomyocytes IP_3R2 (2-fold higher abundance than that of IP_3R1 and IP_3R3 [248]), hepatocytes IP_3R2, and insulin-secreting pancreatic β cells IP_3R3. In the gastrointestinal tract, IP_3R3 level is comparable to that of IP_3R1. Olfactory tissue produces the short S2−[44] IP_3R1 variant and IP_3R3. The latter is also synthesized by taste bud cells. Lung cells express moderatly more IP_3R2 than IP_3R1 and IP_3R3 (39, 31, and 30%, respectively [248]).

42. In cerebellar granule cells, IP_3R1 expression is regulated by Ca^{++} influx through Ca_V1 channels or [N]methyl [D]aspartate receptors.

43. Cerebellar granule cells express SE1+, SE2+, SE3+ IP_3R isoforms, whereas Purkinje neurons predominantly express SE1−, SE2+, SE3− variants [247].

44. Exclusion of S2 exon that characterizes the neuronal, long form of IP_3R1.

Table 3.4. Relative abundance of IP_3R subtypes in the heart and lungs (Source: [248]).

Organ	IP_3R1	IP_3R2	IP_3R3
Heart	25	48	27
Lung	31	39	30

Heart

Many hormones ignite IP_3 production in cardiomyocytes. Subsequent Ca^{++} release through IP_3Rs induces inotropic, chronotropic, and arrhythmogenic effects. Ventriculomyocytes express IP_3R, but to a lower (6-fold) extent than atriomyocytes [249].

In atrial myocytes, the predominant IP_3R isoform IP_3R2 colocalizes with the junctional ryanodine receptor RyR2 in the subsarcolemmal space. Subtype IP_3R2 abolishes the positive inotropic effect of endothelin-1 and protects from its atrial arrhythmogenic effect [250].

Ventriculomyocytes synthesize much lower levels of IP_3Rs than atriomyocytes. Both IP_3R2 and IP_3R3 are expressed in ventricular myocytes [251]. The predominant IP_3R isoform IP_3R2 (Table 3.4) operates in the hypertrophic response downstream from prohypertrophic G-protein-coupled receptors [252].

Owing to Ca^{++} liberation from sarcoplasmic reticulum (Ca^{++} *puffs*), ventriculomyocyte IP_3Rs facilitate Ca^{++} release through ryanodine receptor clusters (Ca^{++} *sparks*) to ensure positive inotropic effect. However, activated IP_3R can elevate diastolic intracellular Ca^{++} concentration and increase the propensity for spontaneous, arrhythmogenic Ca^{++} release events.

In cardiomyocytes, IP_3 receptors are expressed at low level, but their density is higher in nodal pacemaker and conducting myocytes than in contractile myocytes (Vol. 5 – Chap. 5. Cardiomyocytes). Subtype IP_3R differs in these 2 types of myocytes. Receptors IP_3R1 and IP_3R2 reside in conducting and contractile atriomyocytes, respectively. They are located mainly on the nuclear membrane, although a small number localizes on endoplasmic reticulum close to ryanodine receptors. The IP_3 messenger binds its IP_3R1 receptor on endoplasmic reticulum to generate Ca^{++} signals that subsequently activate protein kinases and/or contribute to Ca^{++}-induced Ca^{++} release through ryanodine receptors to increase Ca^{++} cue. In atrial myocytes, IP_3R2 activation contributes to excitation–contraction coupling, as it enhances Ca^{++} transients. However, this process can initiate inappropriate Ca^{++} transients that can generate ectopic beats [253]. Any IP_3R2 close to the sarcolemma can disturb functioning of sarcolemmal ion channels and exchangers. They can indeed interfere with Ca_V channels to shorten action potential duration or activate Na^+–Ca^{++} exchangers to enhance Na^+ entry.

Calcium influx in mammalian hearts is mainly generated by ryanodine and inositol (1,4,5)-trisphosphate receptors in adults and embryos, respectively. Embryonic and neonatal cardiomyocytes lack T-tubules. Juvenile cardiomyocytes have immature T-tubules. In the fetal heart, the sarcoplasmic reticulum is sparse and its Ca^{++}-

storing capacity is low compared with that in adults. In neonatal (days 1–2) and juvenile (days 8–10) rat cardiomyocytes, both ryanodine and IP_3 receptors contribute to the Ca^{++}-induced Ca^{++} release for cardiac contraction [254].

Blood Vessels

In blood vessels, the control of the vasomotor tone depends on extra- and intracellular calcium. Two major types of calcium channels in vascular smooth muscle cells include plasmalemmal Ca_V channels and endoplasmic reticulum IP_3Rs, as ryanodine receptor amount is relatively low in vascular smooth muscle.

Subtype IP_3R1 is expressed in relatively high amounts in smooth muscle cells. Its subcellular localization depends on the SMC type.[45] The other IP_3R isoforms are also produced, but their expression depends on the developmental status.[46]

Upon activation by IP_3, IP_3Rs launch smooth muscle contraction. Receptor IP_3R1 is predominantly expressed in rat aorta and basilar and mesenteric arteries [256]. Subtype IP_3R1 is detected in both smooth muscle and endothelial cells, whereas IP_3R2 and IP_3R3 are restricted to the endothelium. In rats, cerebral artery smooth muscle cells produce all 3 IP_3R isoforms, IP_3R1 being the most abundant isoform (82% of total IP_3R) [257].

In endothelial cells, caveolin-1 regulates Ca^{++} store release-induced Ca^{++} influx, as it interacts with both TRPC1 and IP_3R3 (these 2 receptors also connects each other) [258].

Lung

Receptors IP_3R1 to IP_3R3 in the lung are produced at approximately the same level. In airway smooth muscle cells, ATP that is secreted by airway epithelial cells (in addition to be a spasmogen released by mastocytes), activates $P2Y_2$ or $P2Y_4$ receptors that stimulate phospholipase-C for Ca^{++} influx through IP_3Rs [259].

In airway epithelial cells, calcium influx results from auto- and/or paracrine activation at the apical or basolateral membrane of G-protein-coupled receptors linked to phospholipase-C. The 3 IP_3R isoforms can be found in human nasal epithelial as well as in epithelium that coats intrathoracic airways from trachea to bronchioles [260]. In human tracheal gland cells, Ca^{++} response chiefly depends on the presence and activity of IP_3Rs.

45. In many cell types, the density of IP_3R is highest in the perinuclear endoplasmic reticulum [255]. In phasic smooth muscle cells, IP_3R1 is predominantly located in the peripheral smooth endoplasmic reticulum with calsequestrin. In tonic smooth muscle cells, the endoplasmic reticulum is centrally located. In aorta, IP_3R1 is thus predominantly centrally situated.

46. In neonatal smooth muscle cells, IP_3R3 the predominant isoform but its level decreases during development, whereas that of IP_3R1 increases [255]. Subtype IP_3R1 localizes throughout the cytoplasm, IP_3R2 mainly near the plasma membrane and nucleus, and IP_3R3 in the perinuclear region.

Pancreas

Parasympathetic stimulation of pancreatic islets augments glucose-stimulated insulin secretion via Ca^{++} release through IP_3Rs. Ankyrin-B that abounds in pancreatic β cells binds to IP_3Rs [261]. Ankyrin-B contributes to the potentiation of insulin secretion by muscarinic agonists as well as IP_3R density in pancreatic β cells.

3.2.5.3 Structural and Functional Features

Structurally, IP_3R contains a cytoplasmic N-terminus, a hydrophobic region with 6 transmembrane helices (TM1–TM6) bound by linker regions, which determine a pore-forming region with a selectivity filter, and a C-terminus with a suppressor domain-interacting region, which is involved in ion permeation.[47]

Concentrations of free ATP and ATP^{Mg} in the cytoplasm range from 3 to 8 mmol and 400 to 600 μmol, respectively [246]. According to ATP affinity of IP_3R1 and IP_3R3, IP_3R responds to change in free ATP concentration. Activity of IP_3R can then be tuned by the metabolic state of the cell. The endoplasmic reticulum is tightly coupled to mitochondria. The latter releases free ATP and rapidly takes up Ca^{++}, particulalrly in nanodomains of close endoplasmic reticulum–mitochondria apposition. Besides, the number of ATP-binding sites varies according to IP_3R isoform. Other nucleotides, such as ATP^{Mg}, ADP, AMP, and GTP, as well as adenine and adenosine can potentiate IP_3R activity with various degrees of influence according to ligand and isoform types. In addition, ATP binding to IP_3R regulates its phosphorylation by protein kinase-A.

The IP_3R channels are phosphorylated by numerous kinases, such as cAMP-dependent protein kinase PKA, cGMP-dependent protein kinase PKG, calmodulin-dependent protein kinase CamK2, protein kinase-C, and various protein Tyr kinases. Phosphorylation by PKA enhances IP_3R sensitivity to IP_3. Kinase PKA, protein phosphatases PP1 and PP2, and IP_3R1 can complex together via AKAP9 adaptor. Yet, PP1α is able to bind directly IP_3R1. Isoform IP_3R1 is more sensitive to PKA than IP_3R2 and IP_3R3.

47. Functionally, the N-terminus can be decomposed into an IP_3-binding motif with a IP_3-binding suppressor sequence and a coupling domain with functional loci that connect the N-terminus to transmembrane region. Between IP_3-binding domain and pore, a segment contains phosphorylation sites as well as nucleotide- (ATP-) and protein-binding motifs, including proteolysis sequences. The transmembrane domain possesses an oligomerizing region.

Kinase PKG inhibits IP_3R in smooth muscle via phosphorylated accessory protein IP_3R-associated cGMP kinase substrate (IRAG).[48] Protein kinase PKG1β can complex with IP_3R and IRAG to prevent Ca^{++} release.[49]

Kinase CamK2 also precludes IP_3R activity. Kinase PKB modulates protein interactions with IP_3R. Cytosolic protein Tyr kinases Src and Fyn raise IP_3-binding affinity. The CcnB–CDK1 complex phosphorylates both IP_3R1 (Ser421 and Thr799) and IP_3R3 (Ser795) to enhance IP_3-mediated Ca^{++} release.

The IP_3R channels interact with numerous proteins [246]: calmodulin, Ca^{++}-binding proteins (CaBP) in the brain, calcium- and integrin-binding protein CIB1,[50] adaptor RACK1, dimeric Gβγ subunit of guanine nucleotide-binding (G) protein,[51] phosphorylated IP_3R-binding protein released with IP_3 (IRBIT),[52] chromogranins,[53] endoplasmic reticulum lumen-specific protein ERP44,[54] FK506-binding protein FKBP12,[55] glyceraldehyde 3-phosphate dehydrogenase (GAPDH),[56] membrane-associated IP_3R-interacting protein (ITPRIP),[57] anti-apoptotic proteins BCL2 and BCLxL,[58] pro-apoptotic mediator cytochrome-C,[59] Huntingtin–Huntingtin-associated protein HAP1 complex,[60] caspase-3 and calpain involved in IP_3R degrada-

---

48. A.k.a. murine retrovirus integration site 1 homolog (MRVI1). Protein IRAG abounds in platelets. Enzyme PKG1 is a mediator of the NO–cGMP pathway that prevents platelet activation and aggregation in normal conditions, hence thrombosis. Protein IRAG complexes with PKG1β and IP_3R1 to be phosphorylated as well as IP_3R1 [262]. Among the 2 splice variants of PKG1 — PKG1α and PKG1β—, PKG1β, but not PKG1α, binds to the general transcriptional factor GTF2-1 and IRAG [263].

49. Calcium release is precluded by IRAG and PKG1β in the presence of cGMP, whereas bradykinin-stimulated calcium release remains unaffected by the expression of either IRAG or PKG1β [264]. Activated PKG1 decreases IP_3-stimulated elevation in intracellular calcium, induces smooth muscle relaxation, and contributes to the antiproliferative and pro-apoptotic effects of nitric oxide (NO–cGMP–PKG1 pathway).

50. A.k.a. calmyrin or KIP. Phosphorylation and other protein modifications and interactions may regulate the affinity of CaBP1 and CIB1 with IP_3Rs.

51. Adaptor RACK1 increases the affinity of IP_3R for IP_3. G-protein subunit Gβγ can activate IP_3R independently of PLC and IP_3.

52. Protein IRBIT binds to IP_3R and reduces IP_3 sensitivity.

53. Chromogranins are high-capacity, low-affinity Ca^{++}-binding proteins in secretory granules. Chromogranin–IP_3R interaction may prime Ca^{++} release associated with exocytosis.

54. Protein ERP44 hinders IP_3R activity.

55. Binding of FKBP12 to phosphatase PP3 enables its recruitment to IP_3R.

56. Enzyme GAPDH may couple channel activity to cell metabolic state.

57. A.k.a. protein Danger. It enhances Ca^{++}-mediated inhibition of IP_3R [265].

58. Calcium ion released by IP_3R can be taken up by mitochondria to stimulate oxidative phosphorylation and enhance ATP production. On the other hand, Ca^{++} uptake that coincides with apoptotic stimuli can trigger swelling and rupture of the outer mitochondrial membrane. Apoptosis-protective BCL2 and BCLxL sensitizes IP_3R to low IP_3 level.

59. Cytochrome-C binding to IP_3R relieves high Ca^{++} inhibition of Ca^{++} release.

60. Protein HAP1 enhances IP_3 sensitivity of IP_3R.

tion, adaptors, such as neuronal protein 4.1N,[61] ankyrin,[62] and possibly Homer,[63] ion carriers, such as transient receptor potential channels TRPC1, TRPC3, TRPC4, and TRPC6, and Na^+–K^+ ATPase.[64]

Divalent cations enter into IP_3R pore more readily than monovalent cations. Glycoprotein IP_3R is a divalent cation-selective channel with a selectivity for Ca^{++} and Mg^{++} over K^+ of about 8 and 5, respectively [246]. Whereas IP_3 and Ca^{++} are essential IP_3R ligands, other ligands such as ATP are not required for IP_3R activation, but enhance IP_3R sensitivity to Ca^{++} for biphasic stimulation (*positive feedback*).

However, Ca^{++} can also inhibits IP_3R (*negative feedback*). Inhibition effect of Ca^{++} vary among IP_3R isoforms [246]. Calcium ion is influenced by cytosolic ATP content as well as the presence or absence of interacting proteins. Ligand IP_3 regulates IP_3R channel activity by modulating its sensitivity to Ca^{++} inhibition. At very low IP_3 concentrations, Ca^{++}-sensitive IP_3R begins to be inhibited by Ca^{++} before full activation. On the other hand, IP_3R exposed to higher IP_3 concentration has lower susceptibility to Ca^{++} inhibition. Inhibition by Ca^{++} thus enables a graded Ca^{++} release according to the stimulus intensity. This inhibition is associated with a distinct Ca^{++}-binding site in IP_3R (*IP_3-independent inhibitory Ca^{++} sensor*). Furthermore, IP_3 regulates IP_3R activity by modifying the functional affinity and nature of *IP_3-dependent Ca^{++} sensor*, as it can transform inhibitory to excitatory Ca^{++}-binding site and conversely. Therefore, combined Ca^{++} excitation and inhibition of IP_3R clusters that modulate IP_3 effect sow tuned signals instead of all-or-nothing cue. Binding of ATP to IP_3R can modify the properties of both Ca^{++} activation and inhibition IP_3R sites.

3.2.5.4 Mechanism of Action

Permeation and gating of IP_3Rs depend on structural IP_3R features, redox and phosphorylation status, ion permeation mechanism, and channel gating regulation by IP_3, Ca^{++}, and IP_3R-interacting proteins. Tetrameric channel IP_3R binds IP_3 with high affinity (Kd 10–100 nmol) [246]. Gating of IP_3R results from many factors, such as channel activation, inhibition, inactivation, stochastic attrition, and sequestration, that depend on ligand sensitivities and concentrations.

Behavior of ligand-gated IP_3Rs depends not only on IP_3 type and concentration, but also on recruitmtent of IP_3Rs that can arrange in clusters to form discrete release sites. In the temporal domain, IP_3-mediated Ca^{++} spikes or oscillations are characterized by their frequency and amplitude. In the spatial domain, Ca^{++} signals originated at specific locations remain highly localized or propagate. In any case,

61. A.k.a. erythrocyte membrane protein band 4.1-like EPB41L1. It translocates IP_3R1.

62. Ankyrin-B attenuates IP_3 sensitivity. Ankyrin can also recruit IP_3Rs as well as Na^+–K^+ ATPase and Na^+–Ca^{++} exchanger into membrane rafts.

63. Scaffold Homer binds to receptors mGluR1α, transient receptor potential channels TRPC1, and ryanodine receptors, among others.

64. Sodium–potassium ATPase can associate with Src kinase, PLCγ1, and all 3 IP_3R isoforms.

Ca^{++} can generate Ca^{++} waves with velocity of a few tens of μm/s. Spatially restricted stimuli originate from appropriate colocalization of Ca^{++}-handling proteins. According to the amount of IP_3 and excited IP_3Rs, Ca^{++} signals include: (1) *blip* (the smallest cue) generated by a weak activation (small $[IP_3]$) of a single or few concerted IP_3Rs triggered by intermediate $[IP_3]$; (2) *puff* by coordinated action of multiple clustered IP_3Rs; and (3) *wave*, i.e., a global propagating Ca^{++} signal by high $[IP_3]$ and large amounts of aggregated IP_3Rs. Concerted IP_3R opening is triggered by Ca^{++} release from one activating channel that stimulates gating of nearby channels (*Ca^{++}-induced Ca^{++} release*). Spatial organization of IP_3Rs into a collection of Ca^{++}-release sites and distribution of IP_3R clusters together with the positive feedback of Ca^{++} on these Ca^{++}-activated IP_3Rs thus enable long-range Ca^{++} signals. Other mechanisms grade Ca^{++} release with stimulus intensity and terminate Ca^{++} release.

Protein IP_3R is a calcium-selective channel whose gating is regulated not only by IP_3, but also by other ligands as well, in particular cytoplasmic Ca^{++}. Calcium ion signals initiated by IP_3 require local interactions between IP_3 receptors. As IP_3R is characterized by a rapid stimulation and slow inhibition by cytosolic Ca^{++}, IP_3Rs that are initially randomly distributed (separation of ~1 μm) can be recruited to rapidly form clusters of IP_3Rs [266]. These reversibly aggregated IP_3Rs open together to yield a local Ca^{++} puff. When these bursts become more frequent, they prime Ca^{++} waves. At resting cytosolic $[Ca^{++}]$, clustered IP_3Rs open independently with a lower opening probability, shorter open time, and less sensitivity than lone IP_3Rs. Elevated cytosolic $[Ca^{++}]$ changes behavior of aggregated IP_3Rs. Gating of a collection of IP_3Rs becomes coupled, and the opening duration is prolonged.

3.2.6 Ryanodine-Sensitive Calcium-Release Channels

Ryanodine receptors (RyR1–RyR3) that are calcium-release channels sensitive to ryanodine[65] reside on intracellular Ca^{++}-storage organelles to release Ca^{++} in the cytosol. The endoplasmic reticulum in non-myocytes and smooth muscle cells and the sarcoplasmic reticulum of striated myocytes are the main Ca^{++}-storage organelles. The endo- and sarcoplasmic reticulum store Ca^{++} bound to calreticulin and calsequestrin, respectively. Calcium ion is released from intracellular stores through both ubiquitous, large, high-conductance ryanodine- and IP_3-sensitive Ca^{++}-release channels.

Activity of ryanodine receptors–channels is influenced by its associated proteins, such as tacrolimus (FK506)-binding protein, calmodulin, triadin, calsequestrin, junctin, and sorcin, as well as protein kinases and phosphatases. Ryanodine receptors participate in various Ca^{++}-induced cell functions, such as neurotransmission, secretion, contraction, etc.

Tetrameric ryanodine receptors share structural homology and function similarly to tetrameric IP_3-sensitive Ca^{++}-release channels (IP_3 receptors; Sect. 3.2.5). However, distinct functional features between IP_3Rs and RyRs explain spatiotemporal

65. Plant (Ryania speciosa) alkaloid ryanodine is a ligand for the activated calcium channel RyR that can block Ca^{++} release.

Table 3.5. Endogenous activators of ryanodine receptors (Source: [5]).

Type	Activators
RyR1	Cytosolic Ca^{++} (μmol), luminal Ca^{++}, calmodulin at low cytosolic Ca^{++}, protein kinase-A, calmodulin-dependent kinase, cytosolic ATP (mmol), depolarization via Ca_V1 channel
RyR2	Cytosolic Ca^{++} (μmol), luminal Ca^{++}, cytosolic ATP (mmol), protein kinase-A, calmodulin-dependent kinase
RyR3	Cytosolic Ca^{++} (μmol), cytosolic ATP (mmol), calmodulin at low cytosolic Ca^{++}

properties of intracellular Ca^{++} signaling. Calcium release from a ryanodine receptor cluster produces a spatiotemporally restricted rise in cytosolic Ca^{++}, the so-called *calcium spark*. Localized, hurried Ca^{++} influx in the cytosol can lead to *calcium wave* that results from possibly combined ligand-mediated activation of RyRs, such as CICR feedback.

In arterial smooth muscle cells, Ca^{++} sparks occur transiently near the plasma membrane and act to deliver locally high levels (μmol) of Ca^{++} ions to plasmalemmal Ca^{++}-sensitive ion channels, mainly Ca^{++}-activated Cl^- channels (Cl_{Ca}; Sect. 3.5.6) and large-conductance Ca^{++}-activated K^+ channels (BK; Sect. 3.4.6), without changing the global cytosolic Ca^{++} concentration [267]. Activation of BK channels by Ca^{++} sparks contributes to the regulation of arterial lumen caliber. Interaction between Ca^{++} and BK channels can be influenced by membrane potential and modulatory β subunits of BK channels. On the other hand, Cl_{Ca} channels are not present in all smooth muscle cell types. In addition, Ca^{++} sparks can also influence the activity of Ca^{++}-dependent transcription factors, such as Fos.

3.2.6.1 Molecular Diversity

Three ryanodine receptor–calcium release channel isoforms are encoded by 3 genes (RYR1–RYR3): (1) RyR1 is primarily expressed in skeletal muscle (skeletal RyR isoform); (2) RyR2 in cardiomyocytes (cardiac RyR isoform); and (3) RyR3 in non-muscle cells (Table 3.5). Therefore, in mammalian striated muscles, the expression of RyR isoforms is tissue specific. Channel RyR3 is also observed in mammalian striated myocytes, but at relatively low level (i.e., <5% of the overall RyR population in diaphragm [268]).

Non-muscular tissues generally contain different RyR isoforms (as well as several IP_3R isoforms). All 3 RyR isoforms are identified in the porcine central nervous system [66] as well as esophagus [269]. Isoforms RyR2 and RyR3 are detected in myocardium.

RyR1

In addition to skeletal myocytes, RyR1 is synthesized in neurons, [67] smooth muscle cells, and some immunocyte types (dendritic cells and B lymphocytes) [270].

In skeletal myocytes, RyR1 channels are arranged in highly ordered arrays on the junctional face of the sarcoplasmic reticulum close (~10 nm) to the T-tubule membrane that contains Ca_V1 channels. Tetramers $Ca_V1.1$ and RyR1 are connected and aligned so that the signal transmission between these 2 channel types can occur during the very brief (~2 ms) action potential.

The amount of uncoupled RyR1s differs among skeletal muscles. [68] The RyR channels in all striated myocyte types serve as hubs to form large complex with components of 3 compartments: extracellular medium, cytosol, and sarcoplasmic reticulum lumen. These components are linked by transmembrane proteins of sarcolemma and sarcoplasmic reticulum membrane, such as Ca^{++} channels. Consequently, RyR is able to sense its environment related to these 3 compartments to appropriately regulate Ca^{++} influx (into the cytosol).

Ryanodine receptors cooperate with FK506-binding protein (FKBP). Ubiquitous FK506-binding protein (FKBP) is a peptidyl-prolyl rotamase (cis–trans isomerase) that is specifically inhibited by T-cell inhibitor FK506 [271]. Channel RyR1 of sarcoplasmic reticulum in skeletal muscle cells contains 4 tightly associated FK506-binding proteins FKBP12 [272], whereas RyR2 links to FKBP12-6.

Isotype RyR1 is permeable to Ca^{++} and other divalent cations as well as as monovalent cations (K^+ and Na^+) and Cl^- under certain physiological situations [270].

Unlike in cardiomyocytes, Ca^{++} release in skeletal muscle does not depend on Ca^{++} entry. However, during prolonged trains of action potentials, Ca^{++} influx through store-operated and depolarization-gated channels contributes to the sustained Ca^{++} transients [270].

Activity of RyR1 is enhanced in vitro by oxygen, superoxide anion, hydrogen peroxide, nitric oxide, hydroxyl radical, reduced glutathione and oxidized glutathione (or glutathione disulfide), and ^{S}nitrosoglutathione [270]. Oxidizing and reducing agents impact RyR function by modifying Mg^{++} sensitivity. In addition, RyR1 is S-nitrosylated (Cys315, Cys811, Cys906, and Cys3635).

66. I.e., parietal, frontal, and temporal lobes of cerebrum; thalamus and hypothalamus; cerebellum; and brainstem.

67. I.e., cerebral cortex, cerebellum, hippocampus, and brainstem. Subtype RyR1 is more highly expressed in the stratum oriens, stratum pyramidale and stratum radiatum of the CA1 and CA3 subfields of the hippocampus. The highest density of RyR1 is detected in somata of pyramidal cells within the stratum pyramidale and parts of their apical dendrites in the stratum radiatum [270].

68. Slow-twitch skeletal muscles can have at least 3 uncoupled RyR1 for each $Ca_V1.1$–RyR1 complex [268].

The effect of Mg^{++} [69] antagonizes that of Ca^{++}, as it prevents Ca^{++} release. On the other hand, ATP [70] activates RyR1 [270]. In the presence of physiological levels of Mg^{++} and ATP, RyR1 requires less Ca^{++} to activate than RyR2 or RyR3 channel. Subtype RyR1 is more sensitive to Mg^{++} than RyR2 channel. ATP^{Mg} improves the gating transition to the open state of RyR1 [270].

RyR2

In cardiomyocytes, about 1 $Ca_V1.1$ channel exists for every 5 to 10 RyR2 subtypes. The $Ca_V1.1$ channel launches Ca^{++} influx that activates adjoining RyR2 (depolarization-induced Ca^{++} release [DICR]). In cardiomyocytes, $Ca_V1.1$ and RyR2 are not aligned. Because cardiac action potential is relatively long (~100 ms), diffusible messenger Ca^{++} slows $Ca_V1.1$–RyR2 signaling, but yields time to regulate the channel coupling. In addition, activity of RyR1 and RyR2 is influenced by Ca^{++} currents of other origins that can particularly regulate $Ca_V1.1$-uncoupled RyRs (CICR, but not DICR).

Isoform RyR2 binds several molecules, such as calmodulin, protein kinase-A, protein phosphatase-1 and -2, triadin, and calsequestrin. Subtype IP_3R2 releases calcium at a much smaller rate and extent than RyR2. Both RyR2 and IP_3R2 are inhibited by high calcium level and calmodulin and activated by protein kinases PKA and PKC and calmodulin-dependent protein kinase-2.

RyR3

Subtype RyR3 is activated by Ca^{++} at about 100 nmol and inactivated at about 10 mmol [273]. Calmodulin activates RyR3 at low Ca^{++} concentrations, but inactivates it at high Ca^{++} concentrations. Channel RyR3 is activated by ATP, inhibited by Mg^{++}, and not influenced by cyclic ^{ADP}ribose.

Sympathetic nerve efferents from the superior cervical ganglion protect the cerebral vasculature during periods of acute hypertension. Activity of these nerves depends on Ca^{++} release through RyRs. In adult rats, RyR2 and RyR3 are predominant isoforms in these nerves. Nitric oxide synthesized from neuronal nitric oxide synthase NOS1 in neurons that innervate the superior cervical ganglion and cerebral vessels modulates cADPR level, hence RyR activity. During aging, expression of RyR3, its modulator NOS1, as well as SERCA Ca^{++} ATPases declines [274].

Tissue-specific expression of splice variants of RyRs can explain different pharmacological and functional properties according to tissue type. Seven alternatively spliced variants of RyR3 can be detected [282]. Splice variant $RyR3_{(AS-8a)}$ that lacks a 29-amino acid fragment (His4406–Lys4434) abounds in smooth muscle cells, but not in striated myocytes (skeletal muscles and heart) and brain cells. It forms functional heteromeric channels with RyR3. It can also build a complex with RyR2, thereby impeding RyR2 activity.

---

69. Intracellular free Mg^{++} concentration ranges 600 µmol to 1 mmol.
70. Intracellular free ATP concentration equals about 5 mmol.

3.2.6.2 Structural and Functional Features

Tetramer RyR is a huge protein (~1/10 ribosome size). The RyR proteins share significant homology with IP_3R Ca^{++} channels. The highest density is observed in striated myocytes. The ability of RyR protein to carry Ca^{++} depends on its permeation (conductance and ion selectivity) and gating properties (durations of open and closed configurations and opening and closure patterns) as well as kinetics of RyR activity (rate constants). Channel RyR is a poorly selective Ca^{++} channel (selectivity $Ca^{++}/K^{+} \sim 6$) with very high conductance. Other permeable cations include Mg^{++}, Na^{+}, and K^{+}.

The RyR channel possesses a membrane-spanning domain between N- and C-termini. The large cytoplasmic region (~29×29×12 nm [268]) contains several ligand-binding sites, especially Ca^{++} sensors, and phosphorylation motifs. Channel discrimination between ion types happens in the pore that has given caliber, length, charge, etc. Channel pore also determines transport rate. Structural heterogeneity exists among RyR isoforms that can contribute to isoform-specific functional attributes. However, permeation properties of RyR isoforms is similar.

3.2.6.3 Regulation Overview

The RyR channel is regulated by Ca^{++}, Mg^{++}, and ATP. Sarco(endo)plasmic reticulum Ca^{++} ATPase uses the energy stored in ATP to pump calcium back into the sarcoplasmic reticulum. In resting condition, RyR cytosolic and sarcoplasmic reticulum parts bathe in media with a free Ca^{++} concentration of nearly 100 nmol and 1 mmol, respectively. The cytosol contains other minerals and nucleotides (K^{+}: ~140 mmol; Mg^{++}: ~1 mmol; ATP: 5–10 mmol, etc.) [268]). The cytosol also contains some proteins that interact with RyR channels.

Post-Translational Modifications

Post-translational modifications of RyR impact its channel function, especially its sensitivity to Mg^{++}. Channel RyR can undergo oxidation that can prevent regulator binding. Rabbit skeletal RyRs in lipid bilayers can be activated or inhibited by nitric oxide according to its concentration, membrane potential, and the presence of channel agonists. Nitric oxide modulates the redox status of RyRs; S-nitrosylation and/or oxidation by NO change RyR activity. Initial NO release activates RyRs, but strong concentration inhibits RyR activity [275].

Phosphorylation–dephosphorylation cycle regulates RyR action, as it sensitizes and desensitizes RyR to Mg^{++} and ATP. Protein kinase PKA and phosphatase PP1 activates and inactivates RyR1, respectively [276]. Calmodulin-dependent kinase and protein kinase-A can target RyR2. Increased phosphorylation of RyR2 by PKA (Ser2808) caused by persistent excitation of the sympathetic system (β-adrenergic stimulation) increases Ca^{++} release by heightening RyR2 sensitivity to diastolic

Ca^{++}, but does not markedly alter the cellular function [277].[71] Once phosphorylated by protein kinase-A, RyR2 dissociates from its regulator FKBP12-6[72] that controls RyR opening probability.[73] Phosphorylation of RyR by protein kinase-C decreases RyR Ca^{++} sensitivity.

Cyclic ^{ADP}Ribose

Intracellular messenger cyclic adenosine diphosphate–ribose (cADPR), a metabolite of nicotinic acid adenine dinucleotide phosphate (NADP) by adenosine diphosphate–ribosyl cyclase, is a powerful Ca^{++}-releasing agent that acts on RyRs.[74]

The effect of cADPR on RyR depends on calmodulin. Agent cADPR activates RyR by binding to RyR accessory protein FKBP12-6. In coronary arterial smooth muscle cells, the sarcoplasmic reticulum can produce superoxide via nicotinamide adenine dinucleotide phosphate oxidases (NAD(P)H oxidases [NOx]) via the cADPR–RyR–Ca^{++} pathway; IP_3Rs are not involved [279].

In renal vasculature subjected to angiotensin-2 and endothelin-1 stimulation, the cADPR–RyR axis generates large Ca^{++} influx to hasten contraction of preglomerular arterioles [280].

The ADPR cyclase and RyR contribute to the Ca^{++} response initiated by IP_3R (Ca^{++}-induced Ca^{++} release). In addition, cyclic ^{ADP}ribose also activates SERCA Ca^{++} ATPase that increases $[Ca^{++}]_{SR}$ [281].

Calcium Ion

As IP_3Rs, Ca^{++} action on RyRs is not banal. In addition to being carried through RyRs, Ca^{++} switches these channels on and off. The RyR channels are activated and inhibited by low (1–10 μmol) and high (1–10 mmol) Ca^{++} cytosolic concentration, respectively.

Although the Ca^{++}-induced Ca^{++} release process governs the activity of most the RyR types, one RyR type is coupled to a voltage sensor in the plasma membrane. Various other partners interact with and regulate RyRs such as Homer dimer. In heart and pancreas, messenger cyclic ^{ADP}ribose contributes to RyR activation.

Striated Myocytes – Cardiomyocytes

Functional heterogeneity exists among RyR isoforms (RyR1–RyR3) as well as for a given RyR isoform, as all RyR1s do not have the same response kinetics to

71. However, a hyperactive RyR2 with increasing opening probability upon phosphorylation by PKA or Ca^{++}–calmodulin-dependent kinase can cause arrhythmogenesis, whereas a regular PKA-dependent phosphorylation leads to a positive inotropic effect.

72. A.k.a. calstabine-2.

73. In heart failure, RyR2 is hyperphosphorylated. Hyperphosphorylation of RyR2 dissociates the channel-stabilizing protein FKBP12-6, hence triggering a Ca^{++} leak.

74. In pancreatic β cells in response to elevated ATP level after glucose stimulation, cADPR mobilizes Ca^{++} from the endoplasmic reticulum through RyRs, independently of ATP-sensitive K^+ channels to trigger insulin secretion [278].

Ca^{++} signal [268].[75] Moreover, Ca^{++} binding to the luminal RyR part that contains both activating and inactivating sites regulates the channel. The sensitivity of RyRs to certain cytosolic agonists rises at high Ca^{++} concentration in the sarcoplasmic reticulum lumen. Luminal Ca^{++} may act, at least partly, via partners, such as calsequestrin and membrane-associated proteins triadin and junctin.

In striated myocytes, RyRs represent the primary source for Ca^{++} release during excitation–contraction coupling that follows depolarization-induced trans-sarcolemmal Ca^{++} influx through voltage-gated Ca^{++} channels (*depolarization-induced* Ca^{++} *release* [DICR]). The initial relatively fast component of the trans-sarcolemmal Ca^{++} flux through $Ca_V1.1$[76] of the sarcolemma, especially on transverse tubules, triggers Ca^{++} release from the sarcoplasmic reticulum. The subsequent slow component that can correspond to non-inactivating, voltage-gated Ca^{++} channels can load the sarcoplasmic reticulum with an amount of Ca^{++} available for release during the next cardiac beat [283]. Calcium influx not only primes Ca^{++}-*induced* Ca^{++} *release* (CICR; *positive feedback*), but also inactivates this process, as large increase in intracellular free Ca^{++} inhibits further release (*negative feedback*). During the following relaxation, Ca^{++} reaccumulates in its intracellular stores mainly through sarcoplasmic reticulum Ca^{++} ATPases and exits from the cell through sarcolemmal Ca^{++} ATPases and Na^+–Ca^{++} exchangers.

Cytosolic Ca^{++} is able to establish a positive feedback, as a small amount of Ca^{++} in the cytosol close to RyR primes additional Ca^{++} release. This calcium-induced calcium release (CICR) is sensitive to both the speed and amplitude of applied Ca^{++} trigger [284]. Elevated concentration in free Ca^{++} primes Ca^{++} release from sarcoplasmic reticulum up to an optimum. A supraoptimal increase in free Ca^{++} inactivates the calcium-induced calcium release (negative feedback). Furthermore, augmentation of free Ca^{++} level below that necessary for activation of CICR can cause partial inactivation. In addition, amplitude of Ca^{++} release induced by a given increase of free cytosolic Ca^{++} concentration decays when the rate of this increase diminishes. Therefore, inactivating Ca^{++}-binding site of endoplasmic reticulum Ca^{++}-sensitive Ca^{++} channels has a higher affinity for Ca^{++} than activating Ca^{++}-binding site, but a lower rate constant. The removal of Ca^{++}-triggered inactivation of Ca^{++} release corresponds to the slowest stage of the process to ensure a refractory period.

Rapid changes in cytosolic Ca^{++} concentration are needed for signaling. These changes are achieved via RyRs that open and close quickly. Ca^{++} Sparks involve nearly simultaneous opening of many RyRs followed by synchronized closing. Depletion of sarcoplasmic reticulum luminal Ca^{++} sensed by calsequestrin causes channel closing in cardiomyocytes. Skeletal myocytes display a low rate of depletion from a much greater Ca^{++} reservoir. Sarcoplasmic reticulum buffer yields a kinetically distinct Ca^{++} compartment close to Ca^{++}-release channels. Delayed de-

75. Channels RyR2 and RyR3 respond more homogeneously to cytosolic Ca^{++}.

76. More precisely α1s-subunit of these L-type (long-lasting) calcium channel, a.k.a. as heteropentameric dihydropyridine receptor (DHPR).

pletion corresponds to flux from the sarcoplasmic reticulum lumen to reload Ca^{++} lost from the buffer during a spark [286].

Smooth Myocytes

Arterial smooth muscle cells either contract or relax according to the magnitude of Ca^{++} signals. Large global intracellular Ca^{++} signals activate the contractile cytoskeleton, whereas small localized Ca^{++} sparks from activated ryanodine receptors of the sarcoplasmic reticulum activate Ca^{++}-dependent K^{+} channels in the plasma membrane that elicit hyperpolarization and vascular smooth muscle cell relaxation (*vasodilation*) [285].

Other Small Molecules

The cytosol of most cells contains ATP (~5 mmol total ATP; ~300 µmol free ATP) and free Mg^{++} (~1 mmol). Most ATP is bound to Mg^{++} ion. Cytosolic free ATP is an effective RyR activator, whereas cytosolic Mg^{++} is a potent RyR inhibitor. Free ATP is a much more effective on RyR1 than RyR2 and RyR3 [268]. Mg^{++} competes with Ca^{++} at the Ca^{++} activation site as well as high Ca^{++} inhibition site. Channel RyR1 is more sensitive to Mg^{++} action on high Ca^{++} inhibition site than RyR2 and RyR3. In the presence of normal levels of Mg^{++} and ATP, RyR1 requires less Ca^{++} for its activation than RyR2 and RyR3.

Proteic Regulators

Protein RyR actually comprises binding sites for interacting peptides, such as annexin, calmodulin, calsequestrin, FKBPs, kinases, phosphatases, triadin, caffeine, imperatoxin,[77] and ryanodine, among others.

Ryanodine receptors of the sarcoplasmic reticulum are activated by a positive rising cell potential. Calcium-release activated calcium channel (CRAC) is strongly selective after stimulation by depletion of intracellular calcium levels and stores. Store-operated calcium channels are found in particuler in arteriolar smooth muscle cells.

FK506-Binding Protein

In skeletal muscles, ryanodine receptor is tightly associated with FK506-binding protein (FKBP) in a 1:4 stoichiometric ratio [287].[78] Proline isomerases FKBPs stabilize and coordinate the activity of RyR subunits. They ensure the coupled gating between neighboring RyR1, causing simultaneous opening and closing of adjacent

77. Imperatoxin is a peptide toxin from the venom of the African scorpion Pandinus imperator. Activator imperatoxin-A and inhibitor imperatoxin-I enhances and represses Ca^{++} influx of Ca^{++} into the cytosol from the sarcoplasmic reticulum.

78. I.e., 4 12-kDa FK506-binding protein (FKBP12) molecules per tetrameric RyR channel.

channels. Protein FKBP12 also renders RyRs into inward-rectifying channels, favoring lumen to cytoplasm currents and blocking the reverse [270].

In fact, FKBP12 and FKBP12-6 associate with all RyR subtypes (RyR1–RyR3). Isoforms RyR1 and RyR3 bind both FKBP12 and FKBP12-6, whereas RyR2 preferentially binds to FKBP12-6, but also FKBP12. Calstabin-2, or FKBP12-6, is a subunit that stabilizes the closed state of the ryanodine receptor, avoiding diastolic Ca^{++} leak from the sarcoplasmic reticulum. Regulator FKBP12 couples RyR1 and $Ca_V1.1$, hence enabling voltage sensors to potently activate RyRs [287].

Calmodulins and Other Calcium-Binding Proteins

Ubiquitous calcium-binding calmodulin [79] operates as a resident regulatory subunit of RyRs, as it modulates RyR activity in response to change in $[Ca^{++}]_i$. Half-times of calmodulin binding and dissociation are 30 ± 10 mn and 50 ± 20 s, respectively [288].

In humans, calmodulins and calmodulin-like proteins (CamL) form a family of nearly 20 members. [80] Calmodulin-binding proteins (CamBP) interact with calmodulin in a calcium-dependent and -independent fashion. [81]

Calmodulin activates or inhibits RyR1 and RyR3 at low (e.g., 0.2 μmol) and high Ca^{++} level (micro–millimolar range), respectively [268]. [82] Calcium binding to calmodulin determines the effect of calmodulin on both RyR1 and RyR2. [83] For submicromolar Ca^{++} concentration, apocalmodulin binding causes partial RyR1 activation, whereas for micromolar Ca^{++} concentration, Ca^{++}–Cam binding promotes RyR1 inhibition [290]. Calcium–calmodulin impedes RyR2 functioning, whatever

79. A portmanteau for calcium-modulated protein (alias Cam).

80. These members include signal transducers calmodulins (Cam1–Cam3 [phosphorylase kinases δ]) that are encoded by 3 genes (CALM1–CALM3), and calmodulin-like proteins (CamL1, CamL3–CamL6) that are encoded by corresponding genes (CALML1, CALML3–CALML6), as well as guanylate cyclase-activating proteins, calcium and integrin-binding protein-1 (calmyrin), calmodulin-related protein NB1, neurocalcin-δ, neuronal calcium sensor-1, recoverin, and troponin-C.

81. Calcium–calmodulin can bind to various proteins, such as: (1) kinases (e.g., myosin light-chain kinase, calcium–calmodulin-dependent protein kinase, and inositol-trisphosphate 3-kinase-A); (2) phosphatases (e.g., calcium–calmodulin-dependent protein kinase phosphatase and phosphatase-3); (3) signaling proteins (e.g.,adenylate cyclase-1 and nitric-oxide synthase-1); and (4) cytoskeletal proteins (e.g., caldesmon, dystrophin, α1-syntrophin, and spectrin). Calcium-free *apocalmodulin* (ApoCam) can also bind to proteins, such as: (1) neuroproteins (e.g., glutamate decarboxylase, neuromodulin, and neurogranin) and (2) cytoskeletal proteins (e.g., myosin-1A, protein kinase-G1α, and Ras GTPase-activating-like protein IQGAP1), to elicit calcium-independent cell response.

82. Calcium–calmodulin also exerts 2 opposing effects on $Ca_V2.1$, first promoting then inhibiting channel opening, whether Ca^{++} binds to C- or N-terminus of calmodulin, respectively.

83. The affinity of calmodulin for RyR2 lowers in failing hearts. Agent FKBP12-6 associated with RyR2 enables calmodulin binding to RyR2. Moreover, normal interdomain interaction between N-terminal and central domains of RyR2 enables Cam–RyR2 binding [289].

the Ca^{++} concentration (100 nM–1 mmol). The activation of RyR1 by calmodulin is associated with an approximately 6-fold increase in Ca^{++} sensitivity, whereas Ca^{++} sensitivity of RyR2 remains similar in the absence and presence of calmodulin [291]. Calmodulin inhibits calcium release in the absence of ATP by reducing channel opening duration without effect on conductance [292].

The S100 calcium-binding protein family member S100a1 binds to RyR and promotes Ca^{++} release. The Ca^{++}–S100a1 complex competes with Ca^{++}–calmodulin for the same binding site on RyR1 [293]. The Ca^{++}–S100a1 complex stimulates RyR channel.

Calmodulin-Dependent Kinase

Calmodulin-dependent kinase phosphorylates RyR to maintain the force–frequency relationship in the heart.[84] Myocardial contractility rises proportionally with increasing heart rate to enhance the cardiac output during exercise. Systolic Ca^{++} transient amplitude is elevated at faster pacing frequencies. Heart rate-dependent phosphorylation of ryanodine receptor RyR2 (Ser2814) by the dodecameric holoenzyme calcium–calmodulin-dependent kinase CamK2δ explain this force–frequency relationship [294]. This effect actually raises systolic Ca^{++} transients and contributes to enhance cardiac contractility at faster heart rates.

This kinase is activated by Ca^{++}–calmodulin when Ca^{++} level is high. When it is activated, it autophosphorylates (Thr287) and remains active even when Ca^{++} concentration decays. The CamK2δ kinase cycles between a phosphorylated active state during systole and a dephosphorylated resting state during diastole. When the cardiac frequency increases, CamK2δ does not have sufficient time to dephosphorylate before the onset of the next systole. Activated CamK2δ then progressively accumulates [294].

Activated CamK2δ phosphorylates multiple Ca^{++}-handling proteins, such as $Ca_V1.2$ (Thr498) and phospholamban, in addition to RyR2, to increase the open probability of both channels and reduce the inhibition of phospholamban on the sarcoplasmic reticulum Ca^{++} ATPase SERCA2a [294].

Calsequestrin, Junctin, and Triadin

The calcium-storage protein calsequestrin (Csq) encoded by the CASQ gene is a low-affinity, high-capacity Ca^{++} buffer that resides inside the sarcoplasmic reticulum lumen near RyRs. It enables the increase in the total sarcoplasmic reticulum Ca^{++} content to an amount that can exceed 20 mmol and simultaneously maintains luminal free Ca^{++} concentration of 1 mmol. Calsequestrin's properties are similar in the cardiac and skeletal myocytes.

Two Csq isoforms are expressed in striated myocytes. Isoform Csq1 is the single isoform expressed in fast-twitch skeletal muscle and the major isoform in slow-twitch skeletal muscle. Isoform Csq2 is the sole isoform expressed in cardiomyocytes

84. The force–frequency relationship is also termed Bowditch and Treppe effect (Vol. 6 – Chap. 3. Cardiovascular Physiology).

and a minor transcript in slow-twitch skeletal muscle. Calsequestrins Csq1 and Csq2 are luminal Ca^{++} sensors for RyR1 and RyR2. Calsequestrin enhances Ca^{++} release in cardiomyocytes, whereas it depresses Ca^{++} in skeletal myocytes [295].

Activity of Csq1 depends on the Ca^{++}-dependent polymerization–depolymerization cycle. In general, Csq1 polymers inhibit RyR1, whereas Csq1 monomers activate RyR1. Luminal Ca^{++} concentration influences Csq1 binding to small anchoring proteins junctin and triadin that are embedded in the sarcoplasmic reticulum membrane.[85] In humans, junctin[86] is present both in cardiac and skeletal myocytes [298]. A human junctin isoform-1 has been identified. Both junctin and triadin bind to Csq and RyRs; they anchor Csq close to the RyR.

Once bound to triadin and junctin, Csq1 forms linear polymers near the luminal aperture of RyRs. Calcium ions favor Csq polymerization and stabilization. They are adsorbed on these polymers rather than bound, as Ca^{++} is able to diffuse laterally and remains deliverable to open RyRs (proximate source of releasable Ca^{++} [286]). At rest, Csq polymers provide maximum Ca^{++} binding. Calcium release lowers local $[Ca^{++}]_{SR}$. The Csq polymer then disassembles, and Ca^{++} dissociates and causes intraluminal transient. When Ca^{++} release continues, Csq fully depolymerizes. Intraluminal transients help sustain the Ca^{++} concentration gradient that drives flux, thereby delaying the effect of depletion under prolonged Ca^{++} release. Depletion becomes greater as Ca^{++} replenishes the proximate store and Csq polymer reassembles.

At low Ca^{++} concentration (below resting level, i.e., <1 mmol), monomeric Csq1 strongly binds to triadin and junctin. At high Ca^{++} concentration (≥ 3 mmol), Csq1 polymerizes and dissociates from junctin and triadin over a period of 2 to 3 mn [295]. When luminal Ca^{++} is lower than 100 μmol, Csq1 depolymerizes and its oligomers (but not monomers) dissociate from triadin and junctin, again over a period of few minutes. The Csq1 monomers activate RyR1, whereas polymerized Csq1 does not inhibit RyR1 activity when luminal Ca^{++} ranges from 3 to 5 mmol. Calsequestrin thus operates as a dynamical store, as Csq1 inhibits RyR1 only when: (1) Csq1 binds to junctin; (2) luminal Ca^{++} concentration is moderately low (1 mmol); and (3) Csq1 is in a polymer form [295]. Inhibition by Csq1 of RyR1 is maintained only briefly with a $[Ca^{++}]$ of about 1 mmol. Inhibition of RyR1 is relieved: (1) upon exposure to lower Ca^{++} concentration (≤ 100 μmol) during few mn (2–3 mn) that primes Csq1 depolymerization and dissociation of its oligomers from anchoring proteins (only monomers remain attached to anchoring proteins); and (2) upon exposure to higher Ca^{++} concentration (≥ 3 mmol) during several mn (3–5 mn) that causes Csq1 polymerization and dissociation from junctin and triadin. Isoform Csq1 conserves Ca^{++}

85. Triadin, a 95-kDa membrane glycoprotein, abounds in the junctional sarcoplasmic reticulum. Junctional sarcoplasmic reticulum is the sarcoplasmic reticulum part in close contact with the sarcolemma. Triadin contains a single transmembrane domain that separates the protein into N-terminal cytoplasmic and C-terminal luminal domains [296, 297].

86. Junctin is a 26-kDa protein encoded by the ASPH (aspartate β-hydroxylase) gene. It possesses a short, cytoplasmic N-terminal domain, a single membrane-spanning sequence, and a longer, highly charged C-terminus.

in the sarcoplasmic reticulum lumen when intraluminal Ca^{++} briefly falls. Activity of RyR1 falls after a sudden decrease in luminal Ca^{++} concentration.

Effect of Csq2 in the heart can strongly differ from that of Csq1 in skeletal muscle. Ca^{++}-binding capacity of Csq1 is greater than that of Csq2 [299].[87] At resting level ($[Ca^{++}]_{SR} = 1$ mmol), Csq2 is a monomer, whereas Csq1 is a polymer. The RyR2 channels are activated by Csq2 when $[Ca^{++}]_{SR} \leq 1$ mmol [295]. In response to an action potential, Csq2 ensures a maximal Ca^{++} release, whereas Csq1 reduces RyR1 activity and impedes Ca^{++} release.

In cardiomyocytes, calsequestrin Csq2 and triadin are located in junctional sarcoplasmic reticulum. Calsequestrin-2–triadin–junctin–RyR2 complex contributes to Ca^{++} storage into and release from sarcoplasmic reticulum during excitation–contraction coupling, avoiding premature Ca^{++} release [300]. Moreover, Csq2 and triadin participate in organization of junctional sarcoplasmic reticulum.

In skeletal muscle, junctin and triadin exert independent effects on RyRs; triadin alone modifies excitation–contraction coupling; junctin alone supports interactions between calsequestrin and ryanodine receptors.

Ryanodine receptors are similarly activated by triadin or junctin that acts on independent sites on the RyR luminal side [301]. Nonetheless, triadin and junctin do not similarly transmit information between calsequestrin and ryanodine receptors. Calsequestrin inhibits junctin–triadin–RyR and junctin–RyR complexes, but not triadin–RyR aggregates. Like RyR–triadin complex, activity of RyR–triadin–calsequestrin complex increases when luminal Ca^{++} falls, whereas that of RyR–junctin–calsequestrin complex is further curtailed when luminal Ca^{++} decays (from 1 mmol to less than 100 μmol). In addition, overexpressed triadin self-aggregates and removes its excitatory influence [295].

Junctophilins

Junctophilins (JP) interact with membrane lipid and stabilize close apposition of the junctional endoplasmic reticulum and the plasma membrane. Subtype JP1 is the major isoform expressed in skeletal myocytes. Neural isoforms JP3 and JP4 couples ^{N}methyl ^{D}aspartate receptors, RyRs, and activation of small conductance, Ca^{++}-activated K^{+} channels [270].

Sorcin

Ubiquitous Ca^{++}-binding sorcin interacts with the C-terminal endoproteolytic fragment of presenilin-2. This interaction can modulate the release of Ca^{++} by RyR1 [270].

87. In normal conditions, Csq has high capacity and low affinity for Ca^{++}, as it binds about 40 to 80 ions per molecule with a binding constant of about 1 mmol. An increase in Ca^{++} binding is correlated with protein oligomerization. Isoforms Csq1 and Csq2 with net charges of about -80 and -69, respectively, binds approximately 80 and 60 Ca^{++} per molecule. Deletion of the negatively charged, disordered C-terminal (27 amino acids) of Csq2 causes a 50% reduction of its calcium-binding capacity and a loss of Ca^{++}-dependent tetramerization.

Homer

Homer proteins with its subtypes encoded by different genes and its splice variants modulate RyR activity. Homer-2B activates RyR1 and significantly increases Ca^{++} responsiveness [270]. Both short and long forms of Homer bind to RyR1 to regulate channel function in a biphasic manner. Long and short forms act in combination to yield an additive effect.

Selenoprotein-N

Selenoprotein-N (SelN, SelN1, or SePn), a member of the selenocysteine (^{Se}Cys)-containing protein family, tethers to RyR for full RyR activity [270]. [88]

Nicotine

Nicotine, a psychoactive substance of cigarette smoke, [89] provokes a long-lasting strengthening of Ca^{++} signaling via pre- and postsynaptic nicotinic acetylcholine receptors (Sect. 2.5.2), [90] Ca^{++} influx, neuronal depolarization, and Ca^{++}-dependent changes in gene transcription, thereby modulating neuron activity. In mice, nicotine upregulates the production of RyR2 on the endoplasmic reticulum in neurons of numerous brain regions associated with cognition and addiction, especially the cortex and ventral midbrain [302].

Nicotine operates via the cAMP-responsive element-binding transcription factor to reinforce Ca^{++}-induced Ca^{++} release and insert new Ca^{++} channels into postsynaptic membranes. [91] In addition, RyR2 is involved in a positive feedback loop for long-term phosphorylation of CREB [302].

---

88. Selenium is a nutrient in humans. The set of functional selenoproteins includes: 5 glutathione peroxidases (GPx1–GPx4 and GPx6); 3 thioredoxin reductases (TxnRd1–TxnRd3); 3 iodothyronine deiodinases (DIo1–DIo3); selenophosphate synthase SPS2; and selenoproteins (SelH, SelI, SelK, SelM–SelP, SelR–SelT, SelV, SelW, and Sel15). Selenoprotein-P is the most common selenoprotein in the plasma. Selenoprotein-N may be an endoplasmic reticulum glycoprotein, the loss of which causes muscular dystrophy. Two alternatively spliced transcript variants generate distinct SelN isoforms.

89. Nicotine enhances cognition, contributes to neuroprotection, and causes addiction. Persistent changes in synaptic responses are involved in memory, learning, and addiction. Amplification of synaptic glutamatergic activity in short-term nicotine signaling does not depend on gene transcription and does not involve significantly RyRs, whereas long-term signaling relies on RyR2 [302].

90. Subtype $\alpha 4\beta 2$ nAChR in both pre- and postsynaptic elements of the cortex and midbrain, which modulate the neuronal activity via Na^{+}-dependent depolarization, are especially activated [302].

91. Postsynaptic stimulation can initiate a program of gene transcription for dendritic and synaptic remodeling together with an upregulation of receptors–channels such as ionotropic glutamate receptors in the mesolimbic dopamine pathway implicated in the development of nicotine addiction. Upregulation of RyR2 in glutamatergic presynaptic terminals can be actually associated with an amplified CICR in the dopaminergic postsynaptic neurons [302].

3.2.7 Sarco(endo)plasmic Reticulum Calcium ATPase

Sarco(endo)plasmic reticulum Ca^{++} ATPase (SERCA) is a member of the superclass of phosphorylated-type (P-type) ATPase, with a relatively large ATP-binding domain. P-Type ATPases actively transport cytoplasmic Ca^{++} with a countertransport of luminal H^+ to the cytoplasm. Agent ATP is not only a catalytic substrate, but also a stimulatory cofactor. The SERCA pumps that carry cytosolic Ca^{++} into the endoplasmic reticulum maintain a steep calcium concentration difference between the lumen of the endoplasmic reticulum and the cytosol so that fast Ca^{++}-mediated signaling can occur. This intracellular membrane pump that sequesters calcium strongly contributes to the recovery phase following excitation of myocytes and neurons.

Activity of SERCA, especially myocyte relaxation, is controlled by addition and removal of a phosphate group. A P-type pump adopts 2 main conformations (E1 and E2) [303].[92] The SERCA pump cycles between an *E1 conformation* with high-affinity Ca^{++}-binding sites facing the cytoplasm and *E2 conformation* with low-affinity Ca^{++}-binding sites facing the lumen of the endoplasmic reticulum. In E1 state, SERCA binds 2 Ca^{++} ions, whereas in E2 state, cation-binding sites target H^+ ions.[93] Access to ion-binding sites alternates between the 2 sides of the membrane. Transition states associated with phosphorylation $E1^P$ and $E2^P$ cause channel occlusion and trapping of bound calcium and hydrogen ions, respectively. The SERCA pump releases 2 Ca^{++} ions sequentially. Ions can move against an ion gradient (from a low- to a high-concentration medium) without backflux.

3.2.7.1 SERCA Isoforms

Three main isoforms (SERCA1–SERCA3) are encoded by 3 major paralog genes (ATP2A1–ATP2A3). In addition, alternative splicing of the SERCA messenger RNA yields splice variants. The SERCA1a pump is mainly expressed in fast skeletal muscle; SERCA1b in fetal and neonatal tissues; SERCA2a in heart and slow skeletal muscle, as well as in smooth muscle and endothelial cells; and SERCA2b in smooth muscle cells (Table 3.6). Other variants are observed in non-myocytes and neuronal cells. The SERCA2 pump is found in all brain regions. Isoforms SERCA3a and SERCA3b are synthesized in endothelial cells, among other cells.

92. P-Type ATPases are also called E1–E2 ATPases.

93. The SERCA pump in E1 state binds Ca^{++} ions coming from the cytosol. Calcium binding promotes pump phosphorylation using ATP ($E1^P$ state with 2 trapped Ca^{++} ions). The SERCA pump then releases temporarily bound ADP, changes its configuration that leads to Ca^{++} decaying affinity and translocation ($E2^P$ state with separation of TM1–TM2, TM3–TM4, and TM5–TM6 segments). Calcium efflux (pump-forward direction corresponding to flux toward the extracytosolic side) hence occurs. Influx of H^+ (in the reverse direction, i.e., countertransported H^+ from outside of the cytosol) starts with exposed pump binding sites in the $E2^P$ state that link and trap H^+ ions. Hydrogen ion binding triggers pump dephosphorylation (E2 state). Agent ADP favors dephosphorylation and ATP regeneration. The ion channel reopens to the cytosol (E1 state) freeing H^+ and ready to accept Ca^{++}. The steps of the ion-transfer cycle are reversible.

Table 3.6. Types and locations of sarco(endo)plasmic reticulum Ca^{++} ATPase (SERCA). Activity of SERCAs varies, as their affinity for calcium ions adjusts. Among all SERCA isoforms, SERCA2b possesses the highest Ca^{++} affinity. In the heart, Ca^{++} affinity of SERCA2a is regulated mainly by the reversible interaction with phospholamban. Phospholamban as well as sarcolipin interact with SERCA1a pump. Phospholamban is produced only in cardiac, slow-twitch, and smooth myocytes, whereas sarcolipin is highly expressed in fast-twitch and, to a lesser extent, in cardiomyocytes.

Type	Locations
	SERCA1 Isoform (ATP2A1 gene)
SERCA1a	Adult fast-twitch skeletal muscle
SERCA1b	Neonatal fast-twitch skeletal muscle
	SERCA2 Isoform (ATP2A2 gene)
SERCA2a	Myocardium, slow-twitch skeletal muscle Smooth muscle cells
SERCA2b	Ubiquitous (vascular SMC)
	SERCA3 Isoform (ATP2A3 gene)
SERCA3a	Platelet, lymphocytes, endothelial cells, Purkinje neurons Pancreas, salivary glands
SERCA3b	Endothelial cells Kidney, pancreas
SERCA3c	Kidney, pancreas

SERCA1

Fast-twitch skeletal-muscle isoform SERCA1a is composed of 10 transmembrane helices (TM1–TM10) and 3 cytosolic domains: (1) actuator that controls the opening and closing of the protein lumen; (2) nucleotide binding site; and (3) phosphorylation motif. The SERCA1a pump is able to generate a transmembrane Ca^{++} concentration ratio that ranges from micromolar to millimolar level with countertransport of H^+ ions. Sarcolipin, expressed in fast-twitch skeletal muscle and atrial myocardium inhibits SERCA1a and SERCA2a pumps.

SERCA2

In the cardiomyocyte, reuptake of calcium ions in the sarcoplasmic reticulum is done through Ca^{++}–Mg^{++} ATPase SERCA2a, which has a high affinity, but a low transport capacity. The SERCA2a pump is able to lower the calcium concentration to about 70%. It transports 2 Ca^{++} per ATP. It is connected to its regulatory protein phospholamban. Phospholamban binding to SERCA hampers calcium flux.

A high Ca^{++} concentration in the endoplasmic reticulum is mandatory for the maintenance of a proper functioning of the endoplasmic reticulum. In the liver,

SERCA2b is required to avoid endoplasmic reticulum stress and enable glucose and lipid homeostasis [304].

3.2.8 Plasma Membrane Calcium ATPase

Plasma membrane Ca^{++} ATPase extrudes calcium ions from the cytosol. It is hence associated with the recovery phase after cell activities resulting from Ca^{++} influx. It belongs to the P-type superclass of ATPases. Calcium–calmodulin is a major PMCA regulator.

3.2.8.1 PMCA Structure and Function

Plasma membrane Ca^{++} ATPase pumps (PMCA) and Na^{+}–Ca^{++} exchangers (antiporter NCX) extrude calcium ions. The set of Na^{+}–Ca^{++} exchangers comprises the cardiac isoform NCX1, NCX2, which is preferentially expressed in the brain, and NCX3 in skeletal muscles. Na^{+}–Ca^{++} exchangers can lead to Ca^{++} efflux against Ca^{++} electrochemical gradient without additional energy consumption, because Na^{+} entry along the Na^{+} gradient yields the necessary energy.

The active pump is a homodimer. Each subunit possesses 10 transmembrane segments, 2 large intracellular loops, and cytosolic C- and N-termini. The PMCA pump contains an auto-inhibitory domain that binds to Ca^{++}–calmodulin for activation. Calcium–calmodulin-dependent PMCA has a high affinity, but a low capacity for Ca^{++} ions. It is involved in the spatial and temporal control of Ca^{++} levels. It is dedicated to Ca^{++} removal from the intra- to extracellular space. The PMCA pump transports one Ca^{++}, whereas SERCA carries 2 Ca^{++} ions.

Plasma membrane Ca^{++} ATPase is the single carrier for calcium extrusion in the majority of cells. On the other hand, in cardiomyocytes, PMCAs play only a minor role in calcium level, being mainly involved in signaling [305]. The PMCA pump, indeed, does not significantly modify CMC contractile performance.

3.2.8.2 PMCA Subtypes

Plasma membrane Ca^{++} ATPase (PMCA1–PMCA4) are encoded by 4 genes (ATP2B1–ATP2B4). Various gene products and splice variants have different patterns of expression. In adults, PMCA1 and PMCA4 are found in all tissues, whereas PMCA2 and PMCA3 are primarily expressed in excitable cells of the nervous system and muscles [306]. Different PMCA isoforms contribute to control slow, tonic Ca^{++} signals in some cells and rapid Ca^{++} extrusion in others.

Alternatively spliced PMCA pumps yield dynamical elements to respond to a changing need of Ca^{++} regulation. Alternative splicing targets 2 PMCA regulatory domains in the first intracellular loop and C-terminus. [94]

94. The first cytosolic loop contains a putative G-protein-binding sequence and phospholipid-sensitivity site. The C-terminus comprises interaction motifs for calmodulin binding, phosphorylation, and coupling to PDZ domain-containing anchoring and signaling proteins.

In cardiomyocytes, the slow sarcolemmal Ca^{++} ATPase PMCA contributes to about 1 to 2% of the calcium efflux. Its activity is much more important in smooth and non-muscle cells. The sarcolemmal Ca^{++} pump transports one Ca^{++} per ATP. All isoforms are expressed in cardiomyocytes, PMCA1 being the main cardiac type. Activity of PMCA depends on calmodulin binding and phosphorylation by PKA and/or PKC.

Two types of vascular smooth muscle cells exist in canine arteries and veins: (1) type-1 cells located in the media that express muscle-specific proteins and (2) type-2 cells located in both media and adventitia that do not express muscle specific proteins. Both cell types synthesize plasma membrane Ca^{++} ATPases. Whereas type-1 cells produce PMCA1b, PMCA4a, and PMCA4b, type-2 cells express PMCA1b and PMCA4b [307]. Carotid arterial and saphenous venous smooth muscle cells exhibit a time-dependent upregulation of PMCA4a that depends on PI3K to favor cell proliferation. In vascular smooth muscle cells, both PMCA1 and PMCA4 are detected. Isoform PMCA1 represses cell proliferation due to the inhibition of transcription factor MyB.

Overexpression of nitric oxide synthase-1-associated isoform PMCA4b in arterial smooth muscle cells of mice increases the blood pressure [308], the pump being mainly involved in signal transduction rather than muscle relaxation. The PMCA pump localizes to caveolae, nodes of cell signaling. Calcium efflux by PMCA regulates calcium–calmodulin-dependent NOS1 activity (Vol. 4 – Chap. 9. Other Major Signaling Mediators); NOS1 is also located in caveolae.

The PMCA4 pump is ubiquitous, but its function is tissue specific, due to its interactions with other molecules [212]. Many PMCA-interacting proteins exist, such as Ras-associated factor-1, Ca^{++}–calmodulin-dependent protein phosphatase-3, and cytoskeletal syntrophin. The PMCA pumps are tethered to dystrophin complex via syntrophin. Certain PMCA partners recruit PMCA to complexes involved in the nitric oxide pathway, pre- and postsynaptic Ca^{++} signaling, and actin-cytoskeleton remodeling.

3.2.9 Secretory Pathway Calcium ATPase

Secretory pathway calcium ATPases (SPCA1–SPCA2) enable accumulation of Ca^{++} and Mn^{++} ions in the Golgi body, hence preventing manganese toxicity. Several proteins of the secretory pathway require divalent ions, particularly Ca^{++} or Mn^{++}, that act as metal cofactors, such as aminopeptidase-P, proprotein convertases, and sulfotransferases. The SPCA pumps are P-type ATPases that are encoded by the genes ATP2C1 and ATP2C2.

3.2.10 Sodium–Calcium Exchangers

Sodium–calcium exchangers (NCX) are expressed at high levels in cardiomyocytes (Vol. 6 – Chap. 5. Cardiomyocytes). They exchange (extrude or import) 1 Ca^{++} ion for 3 Na^{+} ions (entering or leaving the cell, according to the functioning mode, i.e., forward or reverse mode, respectively). In the heart, NCX is the main source of Ca^{++} extrusion.

The Na^{+}–Ca^{++} exchanger is voltage sensitive and contributes to Ca^{++} influx during cell activation. It has a relatively low affinity, but a high capacity for Ca^{++} ion.

3.2.10.1 NCX Family

Mammalian sodium–calcium exchangers comprises not only NCX1 to NCX3, but also NCKX1 to NCKX4 and NCLX proteins. Members of the NCX subfamily have a stoichiometry of 3 Na^{+}:1 Ca^{++} and those of the NCKX subfamily a stoichiometry of 4 Na^{+}:1 Ca^{++}:1 K^{+} [309]. The NCLX carrier is a K^{+}-independent lithium–calcium exchanger that does not convey Ba^{++} or Zn^{++} cations. Cations Na^{+} and Li^{+} influence similarly Ca^{++} transport rate. Activity of NCLX is strongly inhibited by zinc cation. It is a mitochondrial inner membrane protein (localization to cristae).

NCX1

Carrier NCX1 of the SLC8 family eliminates calcium ions from cardiomyocytes during diastole. It extrudes less than 30% of calcium out of the cytosol of cardiomyocytes. It corrects Na^{+} gradient determined by Na^{+}–K^{+} ATPase, which is phosphorylated by protein kinases PKA and/or PKC (Vol. 4 – Chap. 4. Cytosolic Protein Ser/Thr Kinases).

In the absence of NCX1, cardiomyocytes limit Ca^{++} influx through Ca_V1 by 50% [310]. Cardiomyocyte adaptation hence does not increase other Ca^{++} transfer mechanisms. The NCX1 exchanger is activated by intracellular Ca^{++} binding that relieves Na^{+}-dependent inactivation. The calcium-binding domain of NCX is able to detect fast changes in intracellular Ca^{++} level during the cardiomyocyte functioning [311]. Elevated concentrations of cytosolic Na^{+} either inactivate or activate NCX according to phosphatidyl (4,5)-bisphosphate levels. In the heart, NCX transports 10 to 15 times more calcium than PMCA pump.

NCKX Subfamily

Plasma membrane Na^{+}–Ca^{++}–K^{+} exchangers mainly mediate Ca^{++} extrusion together with K^{+} at the expense of the Na^{+} electrochemical gradient. Calcium extrusion corresponds to the forward mode of exchange of NCKX. However, the ion transfer through NCKX has a reverse mode (Ca^{++} import). The transport of Ca^{++} in NCKX is coupled to the symport of K^{+} at a 1:1 stoichiometry and antiport of Na^{+} at a stoichiometry of 4 Na^{+} for 1 Ca^{++}. The key feature of both NCX and NCKX proteins is the absolute selectivity for Na^{+} ions over all other alkali cations to initiate Ca^{++} transport (not Li^{+}).

The NCKX subfamily, the most numerous Ca^{++} extrusion protein family in terms of distinct genes, comprises 5 known members (NCKX1–NCKX5) that belong to the solute carrier family SLC24 of sodium–potassium–calcium exchanger (SLC24a1–SLC24a5).[95]

Subtype NCKX1 of the retinal rod outer segment mediates extrusion of Ca^{++} previously entered through the light-sensitive cyclic nucleotide-gated channel. A major difference between NCXs and NCKXs is the required Ca^{++} and K^{+}+ cotransport for NCKX proteins, but not for NCX carriers. Yet, a K^{+}-independent component of Ca^{++} transport exist for both NCKX1 and NCKX2 under certain conditions [312].

Subtype NCKX2 is widely distributed in rat brain. Isotype NCKX3 also abounds particularly in the brain. In addition, it is synthesized in smooth muscle cells, particularly in arteries, and in the lung, as well as intestine and uterus. Isoform NCKX4 is uniformly expressed in the brain; it abounds in the aorta, lung, and thymus, and, to a lesser extent, in heart, digestive tract, spleen, lymph node, skeletal muscle, kidney and adrenal gland [312].

On the other hand, NCKX5 is expressed intracellularly rather than in the plasma membrane [312]. It may thus function as a Ca^{++} regulator in intracellular compartments.

NCLX Subfamily

Ubiquitous K^{+}-dependent Na^{+}–Ca^{++} exchanger NCLX (or NCKX6) participates in the maintenance of the cellular Ca^{++} homeostasis in diverse cell types.[96]

In addition to their metabolic role (e.g.,oxidative phosphorylation), mitochondria serve as Ca^{++} stores. Owing to a steep mitochondrial membrane potential, Ca^{++} enters into mitochondrion through an uniporter and is extruded by Na^{+}–Ca^{++} and Li^{+}–Ca^{++} exchanger [313]. Mitochondrial Ca^{++} efflux coupled to mitochondrial Na^{+} influx through NCLX depends strictly on Na^{+}, but not on K^{+}.[97]

3.2.11 Calcium Channel Expression during the Cell Cycle

Quiescent, hypertrophic, and proliferating cells use different types of calcium signaling [314]. In quiescent cells, calcium signals mostly comprise elementary calcium events, such as sparks and puffs, produced by localized Ca^{++} release through clustered IP_3R and RyR receptors. This type of calcium signal promotes activation of transcription factor cAMP response element-binding protein for cell cycle arrest. Cell proliferation is characterized by a sustained increase in cytosolic calcium due to: (1) enhanced IP_3R excitation by IP_3; (2) improved activity of store-operated Ca^{++} channels and Ca_V3 channels; (3) attenuated cytosolic Ca^{++} removal due to inhibition of plasma membrane and sarco(endo)plasmic reticulum Ca^{++} ATPases. This type of

95. Protein NCKX6 (SLC24a6), in fact, corresponds to NCLX.

96. The NCLX protein abounds in skeletal muscle, pancreas, and stomach.

97. The H^{+}–Ca^{++} exchanger that is primarily observed in non-excitable cells do not have a clear role in Ca^{++} efflux.

calcium signal elicits activation of transcription factor nuclear factor of activated T lymphocytes.

In healthy, quiescent, contractile, vascular smooth muscle cells, Ca^{++} signaling that regulates contraction involves plasmalemmal Ca_V and sarcoplasmic reticulum RyR channels. On the other hand, proliferative and motile smooth muscle cells are characterized by a loss of Ca_V1 and gain in Ca_V3 and transient receptor potential channels (e.g., TRPC1, TRPC3, TRPC4, and TRPC6) [315]. Proliferative smooth muscle cells also lose RyR3 and sarco(endo)plasmic reticulum Ca^{++} ATPase SERCA2a, but Na^+–Ca^{++} exchangers are overexpressed. In addition, a 6-fold increase in expression of IP_3R1 happens when smooth muscle cells progress from phase G0 to G1–S transition of the cell division cycle owing to cell cycle-associated transcription factor MyB that regulates IP_3R1 transcription [316].

Expression of ion channels devoted to other ions than Ca^{++} also changes. Calcium-activated K^+ channels $K_{Ca}1.1$ and $K_{Ca}3.1$ are down- and upregulated, respectively, when smooth muscle cells switch from a contractile to a proliferative phenotype [315]. In addition, $Na_V1.7$ channel[98] that is not synthesized in human coronary arterial smooth muscle cells as well as large voltage-gated Na^+ channels are produced in motile cells. Inwardly rectifying K^+ channels (K_{IR}) have an augmented expression in proliferative cells. Voltage-gated K^+ channels K_V1 and $K_V3.4$ (that is characterized by a rapid inactivation) are main contributors to K_V currents in human uterine contractile and proliferating smooth muscle cells, respectively. Voltage-gated Cl^- channel ClC3 is implicated in smooth muscle cell proliferation.

Last, but not least, cell proliferation influences synthesis of regulators of ion channels. Ubiquitous Ca^{++}–calmodulin-dependent protein kinase-2 is a multifunctional kinase that mediates many effects of Ca^{++} signaling (Vol. 4 – Chap. 4. Cytosolic Protein Ser/Thr Kinases). In proliferating and motile smooth muscle cells, CamK2δ2 and CamK2γ are over- and underexpressed, respectively [315]. Distinct isoforms of a given kinase, indeed, translate frequency and amplitude content of Ca^{++} signals as well as their spatiotemporal distributions and target different effectors for different patterns of gene expression to generate different outcomes.

3.3 Sodium Carriers

Sodium channels deactivate quickly compared with calcium channels. Fast Na^+ channels play a key role for rapid communication between cells. In the nephron, sodium channels control the natriuresis, and subsequently the blood volume. Voltage-gated sodium channels yield the action potential of neurons, and voltage-gated potassium channels drive the repolarization. During the propagation of the action potential in the heart, sodium channels also first activate. Although voltage-sensitive sodium channels play a central role in electrogenesis in excitable cells, these channels exist in non-excitable cells within (e.g., glial cells) and outside the nervous system.

98. A.k.a. tetrodotoxin (TTx)-sensitive voltage-gated Na^+ channel, peripheral sodium channel PN1, and human neuroendocrine channel (hNE).

Sodium channels have a cytoplasmic activation gate and an inactivation component that diffuses toward the corresponding receptor in the channel to inactivate the sodium channel. The inactivation component is inhibited when the activation gate is open.

The ENaC family of sodium-selective channels that controls Na^+ transport across epithelia and thus fluid clearance is composed of many proteins. The ENaC channel is regulated by an oxygen-sensitive mechanism. Acid-sensing ion channels (ASIC) are gated by a decrease in extracellular pH.

3.3.1 Epithelial Sodium Channel

Epithelial sodium channels (ENaC)[99] are permeable to H^+, Li^+, and especially Na^+. They can cooperate for Na^+ transport with Na^+–K^+ ATPase. They control the body's fluid homeostasis.

Epithelial sodium channels have substantial sequence homology with degenerins (Deg), and, hence, constitute the ENAC–DEG superfamily.[100] In fact, the ENAC–DEG superfamily is composed of 3 sets: (1) Epithelial Na^+ channels; (2) brain Na^+ channels (BNaC); and (3) amiloride-sensitive cation channel 5 (ACCn5), or intestinal Na^+ channel (INaC) that is similar to the rodent channel brain–liver–intestine Na^+ channel (BLINaC). Both ENAC and BNAC genes generate different splice variants. The human genome gives rise to 11 proteins of the ENAC–DEG superfamily (ENaCα–ENaCδ, ASIC1–ASIC4, and 2 or 3 proteins related to ACCn5 (INaC) [317].

Heterotrimeric (α–β–γ trimer) and heterotetrameric ($[\alpha]_2$–β–γ tetramer) ENaCs comprise 3 different subunits that are encoded by the genes SCNN1A, SCNN1B, SCNN1D, and SCNN1G (Table 3.7). Each subunit contains 2 transmembrane helices, an extracellular loop, and cytosolic N- and C-termini. In addition, δ subunit encoded by gene SCNN1D can form functional ion channels in pancreas, testis, and ovary with β and γ subunits. Subunit α, β, and γ abound in small and mid-size airways [317].

Baroreceptor nerve endings detect acute fluctuations in arterial pressure. Members of the ENAC–DEG superfamily of cation channels contribute to the baroreceptor activity. Both β and γ subunits of ENaC channel are produced in rat aortic and

99. α Non-voltage-gated epithelial sodium channel is also named sodium channel non-neuronal-1 (SCNN1) or amiloride-sensitive sodium channel (ASSC). Sodium reabsorption is suppressed by diuretic amiloride.

100. The name degenerin is related to selective degeneration of sensory neurons involved in touch and proprioception that result from mutations of the DEG1 gene and other related gene. Degenerins that localize mainly to the central and peripheral nervous system are also called brain Na^+ channels (BNaC). As they are activated by extracellular H^+ ions, these channels were called acid-sensing ion channels (ASIC). The ASIC family comprises ASIC1a (or BNaC2), ASIC1b (or ASICβ), splice variants of ASIC1, ASIC2a (a.k.a. Deg, Deg1, BNaC1, and BNC1), ASIC2b (Deg2), ASIC3 (a.k.a. DRASIC and TNaC1), and ASIC4 (or SPASIC) [317]. However, both Deg and ENaC proteins differ from ASIC proteins.

Table 3.7. Structure of heterotrimeric (α–β–γ trimer) and heterotetrameric ($[\alpha]_2$–β–γ tetramer) epithelial sodium channels (ENaC).

Gene	Subunit
SCNN1A	Subunit α
SCNN1B	Subunit β
SCNN1D	Subunit δ
SCNN1G	Subunit γ

carotid baroreceptor neurons and in their terminals. In particular, γ subunit localizes to the site of mechanotransduction in baroreceptor nerve terminals innervating the aortic arch and carotid sinus [318]. In addition, ASIC2, a mechanosensitive constituent of the heterotrimeric channel of the ASIC family of the ENAC–DEG superfamily on aortic baroreceptor neurons in the nodose ganglia and their terminals also participates in the arterial mechano- and chemosensory reflex [319].

The ENaC channel colocalizes with caveolin-1 and inducible form of heme oxygenase, i.e., isoenzyme heme oxygenase-1 (HOx1). [101] Heme associates with and dissociates from substrates similarly to other hemoproteins (e.g., hemoglobin, myoglobin, cytochromes, and nitric oxide synthases) and heme-binding proteins. Heme metabolism by heme oxygenase operates as an O_2 sensor for ENaC activity regulation [321].

The ENaC channel is activated by serine peptidases, such as furin, [102] serine protease-8 (PrsS8), [103] plasmin, and elastase [5]. The activation of ENaCs by proteases is repressed by palate, lung, and nasal epithelium carcinoma-associated protein PLuNC [104] that is secreted by airway cells and binds specifically to ENaC to prevent its cleavage.

Phosphatidylinositides, such as $PI(4,5)P_2$ and $PI(3,4,5)P_3$, stabilize ENaC gating, as they bind to β and γ subunits, respectively. Moreover, ENaC phosphorylation by extracellular signal-regulated protein kinases ERK1 and -2 prevent ENaC withdrawal

101. Heme oxygenase catalyzes the degradation of heme to produce biliverdin, iron, and carbon monoxide. Three known isoforms of heme oxygenases exist: (1) ubiquitous, inducible HOx1 that is synthesized from the gene HMOX1 in response to oxidative stress, hypoxia, heavy metals, cytokines, etc.; (2) ubiquitous, constitutive HOx2; and (3) oxygen sensor HOx3 that is not catalytically active. Heme oxygenase-2 serves as an oxygen sensor in the carotid body and regulates activity of high-conductance Ca^{++}-activated K^+ channels (BK; Sect. 3.4.6).

102. A.k.a. paired basic amino acid-cleaving enzyme (PACE). It is encoded by the gene FUR (FES upstream region). This proprotein convertase processes precursor proteins.

103. A.k.a. prostasin.

104. A.k.a. secretory protein in upper respiratory tract (SPURT); short palate, lung, and nasal epithelium clone-1 (SPLuNC1); lung-specific X protein (LUNX); nasopharyngeal epithelium-specific gene product (NaSG); and long LPLUNC3. Proteins of the human PLUNC family that intervene in innate immunity in airways can be subdivided into short (SPLuNC1–SPLuNC3) and long (LPLuNC1–LPLuNC4 and LPLuNC6) proteins [320].

from apical membrane [5]. This process may contribute to cAMP-mediated increase in Na^+ conductance.

Epithelial sodium channel and cystic fibrosis transmembrane conductance regulator are colocalized at the apical surface of respiratory epithelia, where CFTR can regulate ENaC activity. Members of the 70-kDa heat shock protein family contribute to appropriate folding and transport in the cell of newly synthesized proteins. The HSP73 protein [105] is expressed constitutively, whereas HSP70 is produced on stress and heat shock. The former is a cofactor for ubiquitination and promotes lysosomal degradation of intracellular proteins. It also catalyzes ATP-dependent uncoating of clathrin-coated pits. The HSP73 protein can reduce ENaC surface expression, whereas HSP70 has the opposite seffect [322].

3.3.1.1 ENaC Channels in Kidneys and Respiratory Tract

The ENaC channel localizes to apical membrane of polarized epithelial cells, particularly in the kidney, respiratory ducts, lung alveoli, and descending colon, where they mediate Na^+ entry. In the distal part of nephron tubule, Na^+ reabsorption, which is regulated by aldosterone, glucocorticoids, and vasopressin, participates in the control of the sodium balance and, hence, blood volume and pressure. [106] Production of ENaC is required for the lung fluid balance. Glucocorticoids regulate ENaC activity in the respiratory tract. This control is potentiated by thyroid hormones.

Hypoxia affects Na^+ absorption in the distal nephron and respiratory epithelia. Oxygen-dependent regulation of epithelial Na^+ transport is based on ENaC channels. Hypoxia actually reduces epithelial Na^+ transport to compensate attenuated Na^+–K^+ ATPase during cellular stress. Activity of ENaC is sensitive to heme protein that is synthesized on the matrix side of the inner mitochondrial membrane and released after hypoxia, and catabolized by heme oxygenase. Heme decreases channel open probability. Conversely, NADPH that supports heme oxygenase activity and heme oxygenase in normoxia increases channel activity. Carbon monoxide produced by heme oxygenase activates ENaC when it cooperates with NADPH in normoxia. In hypoxia, heme precludes ENaC action even in the presence of NADPH. Heme oxygenase substrate and product inhibits and stimulates ENaC functioning, respectively.

3.3.1.2 ENaC Channel in the Nephron

In the nephron, the hormonal control of ion transport regulates blood volume and electrolyte concentration (Fig. 3.2). Hormonal control of transepithelial sodium transport mainly through ENaC uses the PI3K (Vol. 4 – Chap. 1. Signaling Lipids) and Raf–ERK1/2 (Vol. 4 – Chap. 5. Mitogen-Activated Protein Kinase Modules) pathways. The ENaC regulatory complex that modulates ENaC plasmalemmal density comprises: (1) inhibitors Raf1 and ubiquitin ligase NEDD4-2 as well as (2) aldosterone-stimulated activators serum- and glucocorticoid-induced kinase

105. A.k.a. HSPa8 and Hsc70.

106. The activity of ENaC in colon is also modulated by aldosterone.

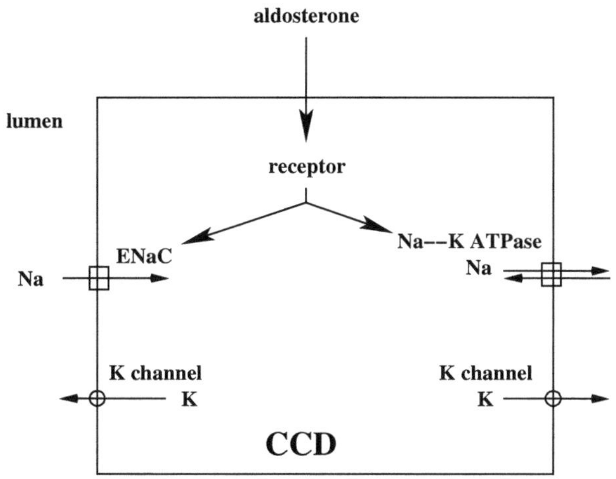

Figure 3.2. Two-step transepithelial Na^+ transport in a principal cell of the connecting tubule or cortical collecting duct (CCD; Source: [317]). Epithelial sodium channel (ENaC) maintains the balance between Na^+ intake and excretion by the nephron. The ENaC channel localizes to the apical membrane of polarized epithelial cells to cause Na^+ entry from the tubule lumen. In CCD, ENaC colocalizes with aquaporin-2; both are stimulated by vasopressin. Sodium–potassium ATPase extrudes Na^+ at the basolateral side. Potassium channels reside on basolateral and apical membranes. Apical K^+ channels secrete K^+ into the tubular lumen. Sodium reabsorption increase in response to aldosterone or vasopressin on dehydration and/or salt deprivation to maintain sodium and fluid balance. Aldosterone binds to intracellular receptors that translocate to the nucleus to regulate the expression and location of ENaC and Na^+–K^+ ATPase.

SGK1 and Raf1-interacting glucocorticoid-induced leucine zipper protein GILZ1. Activators SGK1 and GILZ1 synergistically stimulate ENaC expression [323]. Kinase SGK1 phosphorylates (inhibits) NEDD4-2 that promotes ENaC degradation. Protein GILZ1 antagonizes ERK1 and -2 that leads to ENaC phosphorylation for NEDD4-mediated ENaC destruction, whereas PKC activates the Raf1–ERK1/2 module. In addition, GILZ1 interacts with and inhibits the ENaC–Raf1–NEDD4-2 complex and recruits SGK1 to it.

ENaC Channels in the Respiratory Tract

In the respiratory tract, the periciliary fluid layer is a thin, lubricating film for the mucus coating that entraps foreign particles. This double-decker layer lines the epithelial wetted surface to protect airways from infection by an efficient mucociliary clearance (Vol. 5 – Chap. 12. Airway Surface Liquid and Respiratory Mucus). Excess airway surface liquid is eliminated by airway surface liquid absorption associated with Na^+ flux through ENaC channels. Diluted or concentrated airway surface liquid can be sensed by specific controllers that can adapt ion transport rates to absorb or secrete airway surface liquid, respectively.

Sensors in the respiratory epithelium that regulate the volume of the lumen-lining liquid control the activity of ENaC, as they check hydration levels of respiratory conduits. Activity of ENaC is modulated by membrane-anchored proteases and soluble protease inhibitors. Epithelial Na^+ channel, indeed, must be cleaved (activated) by extracellular channel-activating proteases such as prostasin and/or intracellular furin-type proteases to convey Na^+ ions. The ENaC channel can also be cleaved by serine proteases such as trypsin that are antagonized by protease inhibitors.

Secretory protein in upper respiratory tract (SPURT) [107] can operate as a soluble ENaC regulator and airway-coating liquid-volume sensor [324]. Once bound to ENaC, SPURT prevents ENaC cleavage and activation by serine proteases, although its does not strongly affect the activity of extracellular serine proteases. Secreted SPURTs exist as short and long isoforms. In normal airways, SPURT is highly, moderately, and poorly expressed in submucosal glands and epithelium of proximal and distal airways, respectively.

3.3.2 Hydrogen-Gated Sodium Channels – Acid-Sensing Ion Channel

Hydrogen-gated Na^+ channels, or acid-sensing ion channels (ASIC), [108] are voltage-independent, proton-activated, [109] sodium-selective, trimeric channels that belong to the superfamily of epithelial sodium channels and degenerins [110]

---

107. A.k.a. soft palate, lung, and nasal epithelial carcinoma-associated protein (PLuNC or SPLuNC1).

108. The word "acid-sensing ion channel" is a misnomer, as these channels do not directly deal with acids, although hydrogen ion is related to acids and medium acidification. Yet, it is kept in the present text to enable the current usage of aliases ASIC used to designate the family members. Hydrogen ions in water correspond to H^+ or hydronium (H_3O^+). Besides, pH is a measure of the acidity or basicity of an aqueous solution. In a solution, pH is defined as a negative decimal logarithm (base 10) of the hydrogen ion activity (in units of molar concentration: mol/l):

$$\text{pH} = -\log_{10}(a_{\text{H}^+}) = \log_{10}(1/a_{\text{H}^+}).$$

The hydrogen ion activity is the product of the concentration and an activity coefficient (in a [non-highly] concentrated solution, its value equals about 1). Water exposed to air absorbs carbon dioxide that is then converted into carbonic acid, which dissociates to liberate bicarbonate anion and hydrogen cations:

$$H_2O + CO_2 \leftrightarrow H_2CO_3 \leftrightarrow HCO_3^- + H^+.$$

The transfer of hydrogen ions is a fundamental process in acid–base reactions.

109. Proton (symbol $^1H^+$; πρωτον: first [πρωτονεως: to sail for the first time]); i.e., ion of protium, the much more common hydrogen isotope, that does not possess neutron in its nucleus, is the traditional word for the hydrogen cation (H^+). Hydron is the general name for cations of hydrogen regardless of their isotopic composition, as it refers collectively to proton, deuteron ($^2H^+$; [δευτερος: second]; i.e., the nucleus of deuterium, a stable isotope of hydrogen, the nucleus of which contains one proton and one neutron), and triton ($^3H^+$; [τριτος: third]; i.e., the nucleus of tritium composed of 1 proton and 2 neutrons).

110. Ion channel families of the epithelial sodium channel–degenerin superfamily differ among invertebrates and mammals. Degenerins in Caenorhabditis elegans participate in

The nervous system generates rapid autonomic and behavioral responses to adapt to a changing environment. In the central nervous system, more precisely in the temporal lobe of the brain, the amygdala coordinates fear responses to a variety of threats. Proton-gated sodium channels in the amygdala mediate fear responses associated with CO_2 inhalation. Amygdala cells indeed possess numerous acid-sensing ASIC1a subtype that can act as chemosensors for the CO_2 concentration [325]. Dissolved carbon dioxide causes blood and extracellular acidosis that triggers cationic current through ASIC1a channels.

Proton-gated Na^+ channels (ASIC1–ASIC4) form functional homo- and heterotrimers with various pH sensitivities and desensitization patterns. Splice variants of ASIC1 include ASIC1a, [111] ASIC1b1, [112] ASIC1b2, [113] and those of ASIC2 ASIC2a, [114] and ASIC2b [5]. [115] The ASIC1a isoform is also permeable to Ca^{++} ions. The ASIC4 channel represses the expression of ASIC1a and ASIC3 channels.

The ASIC genes especially encode H^+-gated cation channels in both the central and peripheral nervous system, where they participate in neuronal sensitivity to acidosis.

Channels ASICs have also been detected in taste receptor cells (ASIC1–ASIC3), photoreceptors and retinal cells (ASIC1–ASIC3), cochlear hair cells (ASIC1b), testis (ASIC3), respiratory epithelial cells (ASIC1a and ASIC3), vascular smooth muscle cells (ASIC1–ASIC3), immunocytes (ASIC1, -3, and -4) and bone (ASIC1–ASIC3) [5].

Channels ASIC1a, ASIC2a, and ASIC2b have a similar widespread distribution pattern in the central nervous system [317]. [116] The ASIC2a channel intervenes in neurons undergoing acidosis during cerebral ischemia.

The ASIC3 channel is produced in sympathetic cardiac nerves and can contribute to pain sensation following tissue acidosis during cardiac ischemia. The ASIC4 channel is coproduced with other family members in many areas of the central nervous system. In addition, ASIC4 is strongly expressed in the pituitary gland. In the peripheral nervous system, ASICs localize predominantly to small sensory neurons involved in pain sensation.

---

mechanosensing in neurons, as they convert mechanical forces into electrochemical signals. Phe–Met–Arg–Phe (FMRF)-amide-gated ion channel (FANaC) was originally isolated from the nervous system of the snail Helix aspersa and later found in other mollusks, but no mammalian FANaC homologs have been identified [317]. The Drosophila family are present exclusively in fly ovary and testis.

111. A.k.a. ASICα and BNaC2α.
112. A.k.a. ASICβ1 and BNaC2β.
113. A.k.a. ASICβ2.
114. A.k.a. BNaC1α, BNC1a, and MDeg1.
115. A.k.a. BNaC1β and MDeg2.
116. The highest concentrations are detected in the hippocampus, cerebellum, neo- and allocortical regions, main olfactory bulb, habenula, and basolateral amygdaloid nuclei.

3.3.3 Sodium–Hydrogen Exchangers

Sodium–proton antiporters (NHA) or Na^+–H^+ exchangers (NHE) catalyze the electroneutral exchange of Na^+ and H^+ down their respective concentration gradients. The human genome contains 11 known genes that encode Na^+–H^+ exchangers that are members of the solute carrier family SLC9A (SLC9A1–SLC9A11, i.e., NHE1–NHE11). These isoforms differ by their tissue distribution, membrane localization, inhibitor sensitivity, and regulation mode. Yet, they have structural similarities.

3.3.3.1 NHE1

The sodium–hydrogen antiporter-1, or exchanger isoform NHE1, is an ubiquitous, integral membrane enzyme involved in volume- and pH-regulation of cells. It regulates intracellular pH by removing one intracellular H^+ in exchange for one extracellular Na^+.

In epithelial cells, NHE1 is mainly present at the basolateral surface, where it can link to ezrin, radixin, and moesin proteins, which are cell cytoskeleton crosslinkers, to regulate protein kinase-B-dependent cell survival.

In fibroblasts, NHE1 also interacts with ERM^P proteins. The phosphoERM content depends on the activity of several kinases such as Rho-associated, coiled-coil-containing protein kinase, an effector of Rho GTPase (Vol. 4 – Chap. 8. Guanosine Triphosphatases and Their Regulators).

Left ventriculomyocytes are sensitive to a small decrease in intracellular pH that can occur during mild ischemia. They respond by increasing activity of Na^+–H^+ exchangers (NHE1). In addition, cardiac sarcolemmal Na^+–H^+ exchangers intervene in the development of cardiac hypertrophy. Augmented NHE1 activity actually activates the protein kinase-B pathway [326].

In rat left ventriculomyocytes under basal conditions, activated ezrin, radixin, and moesin proteins are predominantly located at intercalated discs, whereas inactive ERM proteins are retained in the cytoplasm. Activation requires binding of ERM proteins to phosphatidylinositol (4,5)-biphosphate followed by phosphorylation that selects ERM localization to actin-rich membrane regions. In response to cell acidification, activated ERM redistribute to transverse tubules with a marked increase in ERM^P as well as PKB^P and $GSK3\beta^P$. Activated ERM proteins mediate the effects of acid-induced activation of sarcolemmal Na^+–H^+ exchanger that prime the NHE1–ERM–PKB–GSK3β axis [326].

Sodium–hydrogen exchanger isoform NHE1 facilitates differentiation of embryonic stem cells into cardiomyocytes [327].

Sodium–hydrogen exchanger isoform NHE1 regulates pH in arterial endothelial and smooth muscle cells. Insulin inhibits Na^+–H^+ exchanger NHE1 in vascular smooth muscle and endothelial cells of mouse mesenteric arteries via hydrogen peroxide-mediated inhibition of Tyr phosphatase PTPn11 [328]. Insulin induces intracellular acidification and increases intracellular concentration of reactive oxygen species.

Table 3.8. The sodium–proton exchanger family (Source: [329]; DT: digestive tract; SkM: skeletal muscle).

Gene	Name	Other alias	Substrate	Tissue distribution
SLC9A1	NHE1	APNH	Na^+, H^+, Li^+, $NH4^+$	Ubiquitous Plasma membrane (basolateral)
SLC9A2	NHE2		Na^+, H^+, Li^+, $NH4^+$	DT>SkM≫kidney, brain≫ heart, lung Plasma membrane (apical)
SLC9A3	NHE3		Na^+, H^+, Li^+, $NH4^+$	DT>kidney≫brain Recycling endosome
SLC9A4	NHE4		Na^+, H^+, Li^+	DT≫kidney, brain Plasma membrane (basolateral)
SLC9A5	NHE5		Na^+, H^+, Li^+	Neurons Plasma membrane, recycling endosomes, synaptic vesicles
SLC9A6	NHE6			Ubiquitous Recycling endosome
SLC9A7	NHE7		Na^+, K^+, H^+, Li^+	Ubiquitous Trans-Golgi network, endosomes
SLC9A8	NHE8			Ubiquitous Plasma membrane

3.3.3.2 Other NHE Subtypes

Sodium–proton exchangers differ in their tissue distribution, cell localization (plasmalemmal or organellar NHE), kinetic properties, sensitivity to antagonists, and regulation by hormonal and mechanical stimuli [329] (Table 3.8).

Renal NHEs contribute to the maintenance of the hydrogen ion control (acid–base balance), volume regulation, and blood pressure control. In the kidney, all NHE isoforms are expressed at various levels. The NHE3 subtype localizes to the apical membrane of the proximal convoluted tubule and thick ascending limb of Henle's loop (ansa nephroni) and, to a lesser extent, in the upper portion of long descending thin limbs. Other isoforms operate in the proximal tubule.

3.3.3.3 NHE3

The Na^+–H^+ exchanger NHE3 (or SLC9a3) is a 12-transmembrane domain protein of the apical membrane (brush border) [117] of Na^+-absorptive cells of the renal proximal tubule and intestinal epithelium [330]. In the kidney, NHE3 is responsible for 50 to 60% of Na^+-dependent reabsorption of bicarbonate and fluid in the

117. A brush border, or striated border, is the luminal surface of an epithelium covered by microvilli (caliber ~100 nm; length 100–2,000 nm). Brush border cells reside in the small intestine, where nutrient absorption and terminal carbohydrate digestion happen as well as in the kidney, more precisely in the proximal tubule of the nephron.

proximal tubule. The short-term regulation of NHE3 density relies on exocytosis and turnover rates.

3.3.4 Voltage-Insensitive, Non-Selective, Sodium Leak Channel

Voltage-insensitive, non-selective, sodium leak channel $Na_{Vi}2.1$ is encoded by the gene NALCN. Structurally, it is a member of the family of voltage-gated Na^+ channels, but it possesses distinct ion selectivity and properties [5]. It is widely distributed within the central nervous system. It is also synthesized in heart and pancreas. It is activated by SRC family kinases (Vol. 4 – Chap. 3. Cytosolic Protein Tyr Kinases; Table 3.9).

3.3.5 Voltage-Gated Sodium Channels

Voltage-gated sodium channels (Na_V) are involved in the generation and propagation of electrochemical impulses in neurons and myocytes, secretion of hormones and neurotransmitters, and regulation of contraction in cardiomyocytes, skeletal myocytes, and vascular smooth muscle cells [331]. Their modulation participates in the regulation of the nervous and muscular functions. Voltage-gated sodium channels are also synthesized at low concentrations in non-excitable cells,

Voltage-gated Na^+ channels are encoded by a single gene. Nine types of Na_V channels have been identified ($Na_V1.1$–$Na_V1.9$; Table 3.9); $Na_V1.5$ channel is the cardiac isoform. A related protein has also been detected [331].

Voltage-gated sodium channels are made of one pore-forming α subunit associated with 1 or 2 β subunits. Subunit α contains 4 domains (D1–D4). Each domain has 6 transmembrane segments (TM1–TM6). These segments form 2 basic units [332]: (1) a central pore that is constituted by the set of TM5 and TM6 segments of the 4 domains and allows a rapid permeation of ions across the cell membrane (> 100 ions/s); and (2) 4 similar peripheral voltage sensors formed by TM1 to TM4 of each domain that also serve as binding sites for some drugs and toxins. Each of the 4 voltage sensors activate in response to changes in voltage. Voltage sensors of D1 to D3 prime channel opening, whereas that of D4 triggers fast inactivation. Some properties of the 9 $Na_V\alpha$ subunits are given in Table 3.10.

3.3.5.1 $Na_V1.1$ to $Na_V1.3$

In rodents, $Na_V1.3$ is highly expressed in the fetal nervous system. On the other hand, $Na_V1.1$, $Na_V1.2$, and $Na_V1.6$ abound in the adult central nervous system [333].

Both $Na_V1.1$ and $Na_V1.6$ are produced in the peripheral nervous system. However, the most abundant voltage-gated sodium channels in the peripheral nervous system are $Na_V1.7$ to $Na_V1.9_V$ channels.

In general, $Na_V1.1$ and $Na_V1.3$ localize to the soma of the neuron, where they may integrate synaptic inputs and contribute to the setting of the threshold of action potential initiation and propagation to the axon and dendrites [333].

Table 3.9. Voltage-gated sodium channel isoforms and genes (Sources: [5, 331]; Na_S: rabbit Schwann cell sodium channel; PN: peripheral nerve sodium channel; SNS: sensory neuron-specific sodium channel). Voltage-gated sodium channels are the main carriers of ion fluxes responsible for the fast upstroke of action potentials in excitable cells. Voltage-gated sodium channels possess a pore-forming α subunit that can associate with either 1 or 2 β auxiliary subunits. In the adult central nervous system and heart, they contain β1 to β4 subunits, whereas in adult skeletal muscle, they have only the β1 subunit. The pore-forming α subunit is sufficient for functioning, but the localization, kinetics, voltage dependence of channel gating, and interactions with cell adhesion molecules, extracellular matrix, and cytoskeleton depend on the auxiliary β subunit.

Gene	Channel	Alternative names	Features
		$Na_V\alpha$ subunits	
		Type-1 isoforms	
SCN1A	$Na_V1.1$	Brain type 1	Fast inactivation (0.7 ms)
SCN2A	$Na_V1.2$	Brain type 2	Fast inactivation (0.8 ms)
SCN3A	$Na_V1.3$	Brain type 3	Fast inactivation (0.8 ms)
SCN4A	$Na_V1.4$	Skeletal muscle type 1	Fast inactivation (0.6 ms)
SCN5A	$Na_V1.5$	Skeletal muscle type 2, Cardiac type	Fast inactivation (1 ms)
SCN8A	$Na_V1.6$	PN4, NaCH6	Fast inactivation (1 ms)
SCN9A	$Na_V1.7$	PN1, Na_S	Fast inactivation (0.5 ms)
SCN10A	$Na_V1.8$	PN3, SNS1	Slow inactivation (6 ms)
SCN11A	$Na_V1.9$	NaN, SNS2	Slow inactivation (16 ms)
		Type-2 isoforms	
SCN6A	$Na_V2.1$		Voltage-insensitive, non-selective Activation by SRC family kinases
SCN7A	$Na_V2.2$		
		$Na_V\beta$ subunits	
SCN1B	$Na_V\beta1$		
SCN2B	$Na_V\beta2$		
SCN3B	$Na_V\beta3$		
SCN4B	$Na_V\beta4$		

The $Na_V1.2$ channel is expressed in unmyelinated axons, where it conducts the action potential [333]. The $Na_V1.6$ channel replaces $Na_V1.2$ in maturing nodes of Ranvier.

3.3.5.2 $Na_V1.4$

The $Na_V1.4$ channel is produced in skeletal muscles. Calmodulin binding to the skeletal myocyte $Na_V1.4$ regulate its conductance, but not the inactivation kinetics [334].

Table 3.10. Properties of $Na_V\alpha$ subunits (Source: [331]; CNS: central nervous system; DRG: dorsal root ganglion; GDNF: glial cell-derived growth factor; NGF: nerve growth factor; PNS: peripheral nervous system).

Type	Properties
$Na_V1.1$	Central neurons (primarily soma), cardiomyocytes Association with subunits β1 to β4 Ion selectivity: $Na^+ > K^+ > Ca^{++}$
$Na_V1.2$	Central neurons (primarily unmyelinated and premyelinated axons) Association with subunits β1 to β4 Ion selectivity: $Na^+ > K^+ > Ca^{++}$
$Na_V1.3$	Central neurons (primarily embryonic and early prenatal life), cardiomyocytes Association with subunits β1 and β3 Ion selectivity: $Na^+ > K^+ > Ca^{++}$
$Na_V1.4$	Skeletal muscle (highly in adults) Association with subunits β1 Ion selectivity: $Na^+ > K^+$
$Na_V1.5$	Cardiomyocytes, immature skeletal myocytes Association with subunits β1 to β4 Ion selectivity: $Na^+ > K^+ > Ca^{++}$
$Na_V1.6$	Output neurons of the cerebellum, cerebral cortex, and hippocampus; cerebellar Purkinje cells; brainstem; spinal cord; astrocytes; Schwann cells; DRG; nodes of Ranvier of PNS (sensory and motor axons) and CNS Association with subunits β1 and β2 Ion selectivity: Na^+
$Na_V1.7$	DRG and sympathetic neurons, Schwann cells, neuroendocrine cells Association with subunits β1 and β2 Ion selectivity: Na^+
$Na_V1.8$	Small and med-size DRG neurons Ion selectivity: Na^+ Synthesis regulated by NGF and GDNF
$Na_V1.9$	C-type DRG and trigeminal neurons Ion selectivity: Na^+ Synthesis regulated by GDNF

3.3.5.3 $Na_V1.5$

In excitable cells, voltage-gated sodium channels initiate action potential. In cardiomyocytes, $Na_V1.5$ is the primary voltage-gated Na^+ channel pore-forming α subunit that can associate with accessory $Na_V\beta1$ to β4 subunits. In excitable ventriculomyocytes, $Na_V1.5$ complexes mainly with β1 and β2 subunits that regulate channel plasmalemmal density and gating, in addition to operating as scaffolds for ankyrins. Numerous $Na_V1.5$-binding partners are tightly connected to microtubules and the actin cytoskeleton, such as ankyrin-G, syntrophin, and dystrophin. In ventriculomyocytes, tubulin polymerization attenuates sarcolemmal density of $Na_V1.5$–$\beta1$ and $Na_V1.5$ alone, especially at intercalated discs, but also in membrane nanodomains of transverse tubules and other regions of the sarcolemma, where $Na_V1.5$ can colocalize with caveolin-3, possibly by endocytosis [335]. Furthermore, tubulin polymerization accelerates the final phase of fast inactivation of $Na_V1.5$–$\beta1$ channel and slow inactivation of $Na_V1.5$ without $\beta1$ subunit [335]. [118]

3.3.5.4 $Na_V1.6$

Voltage-gated sodium channel $Na_V1.6$ is the prominent somatodendritic Na_V channel subunit in hippocampal pyramidal cells. It localizes not only to nodes of Ranvier and axon initial segments, but also, albeit to a much lower extent, in dendrites of CA1 pyramidal cells [336]. [119]

3.3.5.5 $Na_V1.7$

Whereas cultured human aortic smooth muscle cells strongly expressed the gene SCN9A that encodes $Na_V1.7$, this channel is not markedly expressed in aorta, except after balloon injury [337]. It then participates in cell migration as well as endo- and exocytosis.

Mutations in the gene SCN9A that encodes $Na_V1.7$ cause a congenital inability to experience pain anosmia in humans [338].

The $Na_V1.7$ is widespread in the peripheral nervous system. It localizes to axons, where it may assist in the initiation and conduction of the action potential [333].

3.3.5.6 $Na_V1.8$ and $Na_V1.9$

In the peripheral nervous system, the $Na_V1.8$ and $Na_V1.9$ channels have a more restricted expression pattern than that of $Na_V1.7$. They are observed in small sensory neurons of the dorsal root and trigeminal ganglia. They intervene in the pain perception [333].

3.3.5.7 Na_V Channels in Neurons

Neurons integrate their synaptic inputs and generate their output signals in particular from voltage-dependent ion channels that reside on their surface (axosomatodendritic membrane). Channels $Na_V1.1$, $Na_V1.2$, and $Na_V1.6$ are synthesized in adult brains.

118. Microtubules also interact also with $Ca_V1.2$, K_V7, and $K_V11.2$. They influence $K_V11.2$ activation time and $Ca_V1.2$ gating, but do not affect their plasmalemmal density.

119. Density of Na_V (per μm^2 plasma membrane) is estimated to be equal to 5 in somata and proximal apical dendrites, 3 in proximal oblique dendrites, and 2 in distal apical dendrites [336].

Table 3.11. Sodium–potassium ATPase subunits (Source: [5]). The active Na^+–K^+ ATPase is a heteromultimer with large, catalytic α subunits and small, glycoproteic β subunits; the third γ subunit, an ion transport regulator, modulates its activity. However, the sodium–potassium pump does not depend on the γ subunit to be functional.

Subunit	Gene
α1	ATP1A1
α2	ATP1A2
α3	ATP1A3
α4	ATP1A4
β1	ATP1B1
β2	ATP1B2
β3	ATP1B3
β4	ATP1B4
γ	FXYD2

Synaptic adaptivity (plasticity) in cortical pyramidal cells ensures cellular learning and training. Input synapses that distribute over a huge surface that covers the large dendritic tree, must be capable of sensing the precise timing of the output signal. Fast sodium-based, back-propagating action potentials initiated in the axon initial segment yield information to the dendritic tree and influence dendritic synapses that can remodel.

Voltage-gated sodium currents in dendrites of hippocampal and neocortical pyramidal cells support back-propagation of action potentials. They also underlie nonlinear synaptic integration and dendritic sodium spike generation [336]. In addition, axonal and somatodendritic Na_V currents differ in their activation and inactivation properties due to different compositions in Na_V subunits or distinct post-translational modifications.

3.3.6 Sodium–Potassium Pump

Ubiquitous sodium pumps (Na^+–K^+ ATPase) use energy derived from ATP hydrolysis to pump ions across membranes, hence maintaining gradients of Na^+ and K^+ across the plasma membrane and contributing to membrane potential, particularly in cardiomyocytes. Sodium–potassium ATPase is involved in electrochemical signaling and regulation of cell volume and intracellular pH.

3.3.6.1 Structure

The Na^+–K^+ ATPase have additional transmembrane helices (i.e., P-type ATPase TM1–TM6 and TM7–TM10) [339]. The cell-surface Na^+–K^+ ATPase is made of 3 subunits (α–γ) with multiple isoforms (α1–α4 and β1–β4; Table 3.11). Its specific β and γ subunits are associated with transmembrane helices TM7 to TM10 and TM9, respectively.

Ion-binding sites of the Na^+–K^+ ATPase are involved in cation selection. Ion-binding site 1 and- 2 bind either Na^+ or K^+; site 3 only Na^+ in a hydrated form [340].

Its γ subunit heightens its affinity for ATP, but lowers it for both Na^+ and K^+ ions. γ Subunit serves as a ion transport regulator of the pump, especially in renal cells, where mutations of its FXYD2 gene cause dominant renal hypomagnesemia. Subunit γ is not required to build a functional ion pump.

In humans, $\beta 1$ subunit is ubiquitous; $\beta 2$ subunit resides predominantly in neurons; and $\beta 3$ subunit in testes [341]. Subunit β enables a proper transfer of the $\alpha\beta$ complex to the plasma membrane [341]. This subunit influences K^+ affinity of the pump. Its main interaction motif point between the β and α subunit also operates for E2 stabilization with bound K^+ ion. β Subunit interact with members of a family of Na^+–K^+ ATPase-interacting proteins (NKAIP). Similarly, $\alpha 3$ subunit binds to extracellular matrix agrin [341].

In humans, $\alpha 1$ subunit is the predominant, ubiquitous isoform; $\alpha 2$ subunit is mainly produced in skeletal, heart, and smooth myocytes, as well as in the brain, lung, and adipose tissue; $\alpha 3$ subunit in neurons and cardiomyocytes; and $\alpha 4$ subunit in testes [341].

Subunit α contains binding sites for Na^+, K^+, and ATP. Isoforms α confer different plasmalemmal distributions and kinetic properties. Plasmalemmal $\alpha 2$ pumps are confined to *plasmerosomes*, whereas $\alpha 1$ pumps are more uniformly distributed [342]. As sodium ion affinity depends on α isoform, the ionic composition of cytoplasmic nanodomains close to different sodium pumps can differ. Besides, $\alpha 2$ pumps colocalize with Na^+–Ca^{++} exchangers to couple their activities. Cardiac $\alpha 2$ pumps increase calcium transients during the contractile cycle, whereas $\alpha 1$ Na^+–K^+ ATPase have opposite effects [343].

Access to ion-binding sites of Na^+–K^+ ATPase alternates from one side of the membrane to the other to avoid dissipation in electrochemical gradient [344]. [120]

3.3.6.2 Partners

Most of the FXYD proteins connect to Na^+–K^+ ATPase and regulate its activity. Small, regulatory, single-span membrane, FXYD proteins possess a cytoplasmic C-terminus and an extracellular FXYD sequence. They influence Na^+ and K^+ affinities as well as the maximal transport rate of the pump [341]. α-Subunit TM9 contains the main interaction motif for FXYD2 protein, or γ subunit.

In vascular smooth muscle cells, Na^+–K^+ ATPase can affect the regulation of intracellular Ca^{++} concentration, as it couples to the Na^+–Ca^{++} exchanger. Major Ca^{++} clearance proteins that prime the relaxation of the arterial media include sarco(endo)plasmic reticulum (SERCA) and plasma membrane (PMCA) Ca^{++} ATPases. Na^+–K^+ ATPase in cooperation with Na^+–Ca^{++} exchanger is a minor contributor of the Ca^{++} clearance [345].

120. When ion access alternation is disrupted and binding sites are simultaneously accessible from both sides of the membrane, Na^+–K^+ pump is transformed into an ion channel.

Na^+–K^+ ATPase is the single known receptor for cardiotonic steroids, or cardiac glycosides (e.g., ouabain, digitalis, and digoxin), that hamper pump activity. [121] The cardiac glycoside-binding site [122] belongs to the catalytic subunit isoform α2 of Na^+–K^+ ATPases that localizes in brain, heart, vasculature, and other tissues involved in adaptation to exercise. In skeletal striated myocytes, this binding site regulates resting and contraction-induced Na^+–K^+ ATPase transport according to binding and unbinding of endogenous Na^+–K^+ ATPase ligands. Cardiac glycoside-insensitive or -unbound isoform lowers resting, but enhances contraction-induced Na^+–K^+ ATPase transport and consequently exercise performance [347]. This site also participates in blood pressure regulation mediated by adrenocorticotropic hormone [348]. [123]

3.3.6.3 Signaling Mediator

The sodium pump is not only an ion carrier, but also a signaling mediator. Receptor Na^+–K^+ ATPase generates positive inotropic effects and ensures cardioprotection, at least in rodent cardiomyocytes. Na^+–K^+ ATPase localizes to caveolar signalosome, where they can interact with IP_3Rs. This interaction, together with Src-dependent phosphorylation (activation) of phospholipase-Cα1, stimulate IP_3 synthesis and subsequent Ca^{++} release from the sarco(endo)plasmic reticulum. Upon ouabain binding to the sodium pump, activated Src kinase phosphorylates the epidermal growth factor receptor and stimulates the Ras–Raf–MAP2K–ERK1/2 pathway [349]. Na^+–K^+ ATPase also binds to phosphoinositide 3-kinase and can then trigger the PI3K–PKB axis. Kinases ERK1 and ERK2, PI3K, and PKB are prosurvival proteins in the heart.

121. Cardiotonic steroids were introduced as positive inotropic drugs for the treatment of congestive heart failure. However, they possess toxic effects (cardiac arrhythmias, neuronal disturbances, etc.). Positive inotropic effect of cardiotonic steroids results from the coupling between Na^+–K^+ ATPase and Na^+–Ca^{++} exchanger. Inhibition of the Na^+–K^+ ATPase by cardiotonic steroids causes a decreased Na^+ gradient and a subsequently reduced activity of NCX (forward mode inhibition) and increased cytosolic Ca^{++} concentration. At the opposite, nanomolar concentrations of cardiotonic steroids can activate Na^+–K^+ ATPase [346]. In plasmerosomes, inhibition of Na^+–K^+ ATPase can provoke a local, transient increase of the subplasmalemmal Na^+ concentration that activates NCX (reverse mode), thereby causing a small, local elevation of Ca^{++} concentration within the plasmerosome, which stimulates the release of Ca^{++} from its intracellular store [346].

122. Cardiac glycoside-binding site of Na^+–K^+ ATPase that is highly conserved in all α isoforms binds to endogenous cardiotonic steroids as well as cardiac glycosides, such as ouabain, digoxin, digitoxin, and their antagonists.

123. Chronic administration of ACTH causes hypertension in wild-type mice, but not in mice with an ouabain-resistant α2 isoform of Na^+–K^+ ATPase. Conversely, mice with an ouabain-sensitive α1 isoform such as in humans develop ACTH-induced hypertension to a greater extent than wild-type mice.

3.3.7 Sodium Symporters

Solute–sodium symporters couple sodium outflux with nutrient import. They do not share amino acid sequence, but can have structural resemblance with neurotransmitter sodium symporters. Mutations in glucose and iodide symporters cause metabolic disorders.

3.4 Potassium Carriers

Potassium channels constitute the most widely distributed set of ion channels. They have diverse structures and functions, such as the maintenance of the resting cell membrane potential and repolarization in excitable cells. They determine the cell excitability and frequency, duration, and shape of action potentials (Vol. 5 – Chap. 5. Cardiomyocytes). Potassium channels also regulate the secretion of hormones and neurotransmitters. Their activity can be regulated by voltage, calcium, and neurotransmitters, as well as their signaling pathways.

Potassium channels are tetramers, like many other ion channels.[124] The 4 subunits favor cation binding near the center of the membrane. Pore domains of potassium channels are characterized by a high degree of functional and structural similarities.

About 70 different genes encode K^+ channel α subunits in the human genome. These pore-forming α subunits tether to auxiliary regulatory subunits. The pore loop contains the TVGYG sequence that confers K^+ selectivity to all known K^+ channels. Each subunit can contain identity tags for connections between channel subunits that can restrict subunit assemblies. Two domains are crucial for channel assembly: K_T- and NAB (N-terminal A and B box) domains that specify subunit types, which can heteromerize.[125]

Some types of K^+ channels are able to respond to a specific stimulus, such as change in transmembrane electrical potential difference or ligand binding (Table 3.12), owing to proteic domains attached to the pore that confer gating features. Stimulus-induced conformational changes are transmitted to the pore via extensions of inner helices.

Potassium channels are cation-selective channels that have an equilibrium potential close to the cell resting potential. Potassium channels have been classified according to their structural and functional properties. The potassium channel superclass can be organized into 4 main classes: (1) calcium-activated; (2) voltage-gated; (3) inwardly rectifying that carries K^+ more easily in the inward direction; and (4) tandem pore domain K^+ channels that are constitutively open and, when they are widely open, carry K^+ across the membrane almost as fast as in water.

---

124. E.g., calcium, sodium, non-selective cation, cyclic nucleotide-gated, glutamate-gated, transient receptor potential channels, etc.

125. The NAB domains; such as $NAB_{K_{V1}}$ to $NAB_{K_{V4}}$, mediate the assembly of both subunit α–α and α–β dimers.

Table 3.12. The superclass of potassium channel and generated currents with usual abbreviations. Outward- and inward-rectifying K^+ channels generate fluxes more easily into the out- and inward direction (export and import), respectively. A proper sequence of cardiac electrochemical activity requires regional differences in magnitude of ionic currents within the myocardium. In the left ventricular free wall, the density of the transient outward K^+ current is larger in the outer zone than that in the inner region. Transmural i_{Kto} gradient causes ventricular repolarization to run from the epicardial to endocardial wall layer. A small and large fraction of i_{Kto} current is generated by $K_V1.4$ and K_V4, respectively. In atriomyocyte, early and later repolarizations rely on both i_{Kto} and i_{Kur} and both i_{Kur} and i_{K1}, respectively.

K^+ current type	Current notation
Voltage-gated channels	
Transient outward	i_{Kto}
Rapid transient outward	$i_{\text{Kto(r)}}$
Slow transient outward	$i_{\text{Kto(s)}}$
Ultrarapid delayed outward rectifier	i_{Kur}
Rapid delayed outward rectifier	i_{Kr}
Slow delayed outward rectifier	i_{Ks}
Inward rectifier	i_{K1}
Ligand-gated channels	
ATP-sensitive	$i_{\text{K}_{\text{ATP}}}$
Acetylcholine-activated	i_{ACh}
Calcium-activated	
Sodium-activated	
Arachidonic acid-activated	
Cyclic nucleotide-gated	
Adenosine-sensitive	

3.4.1 Ligand-Gated Potassium Channels

Calcium-activated potassium channels (K_{Ca}) have been discovered in erythrocytes (or red blood cells).

Acetylcholine-activated K^+ channels ($i_{\text{K}_{\text{ACh}}}$) are activated by muscarinic receptors. These heterotetramers are composed of 2 GIRK1 and 2 GIRK4 subunits (GIRK: Gβγ-regulated inwardly rectifying K^+ channel). Acetylcholine-activated K^+ channels slow the heart frequency.

Cyclic nucleotide-gated channels (CNG) link changes in intracellular cAMP and cGMP to excitability. Conductance increase with respect to steady-state opening of some K^+ channel type is induced by protons acting on the intracellular side [350]. Gating is modulated by transmembrane voltage. Inactivation depends on external K^+ concentration.

Table 3.13. Groups and classes within the superclass of potassium channels and their main functions (**Part 1**; Source: Wikipedia; 2TM: two transmembrane-containing channel subunits; 1P: one pore-domain channel). The class of inwardly rectifying potassium channels (K_{IR}) is constituted of 7 families ($K_{IR}1$–$K_{IR}7$). Some of these families correspond to the GIRK ($K_{IR}3$ family), IRK ($K_{IR}2$ family), ROMK ($K_{IR}1.1$ type), and ATP-sensitive K^+ channel ($K_{IR}6$ family) families.

Category	Subclasses Families	Function
Inwardly rectifying 2TM and 1P	ROMK ($K_{IR}1.1$)	Renal K^+ secretion or reuptake
	$K_{IR}2.i$	Cardiomyocyte repolarization
	$K_{IR}2.1/2$	i_{K1} current
	$K_{IR}2.1$ (SMC)	NOS1/3 regulation
	GPCR regulated ($K_{IR}3.i$)	GPCR inhibitory effect (neurons and heart)
	$K_{IR}3.1$–4	Cardiac i_{ACh} current
	ATP-sensitive ($K_{IR}6.i$)	Insulin secretion (pancreatic β cells)
	$K_{IR}6.1/2$	Upregulation of cardiac NOS3
	$K_{IR}6.2$	Cardiac $i_{K_{ATP}}$ current Myocardial ischemia preconditioning

3.4.2 Potassium Channel Structure and Groups

The elementary potassium channel subunit contains 2 transmembrane segments (TM) with a pore loop (P). Potassium channel subunits can contain 2, 4, or 6 to 7 transmembrane segments. According to the number of transmembrane domains, potassium channels are classified into 3 main groups: 2TM, 4TM, and 6TM (Tables 3.13 and 3.14).

Subunits of K^+ channels with 2 or 6-to-7 transmembrane domains form a single pore-forming domain-containing channel (K_{1P}). These K_{1P} channels are homo- or heterotetramers.

Subunits of K^+ channels with 4 transmembrane domains consist of 2 pore-forming domains (K_{2P}) arranged in tandem that form dimers. Potassium channels ($K_{2P}i.1$ encoded by KCNKi, $i = 1$–7, 9–10, 12–13, 15–18) that cause a background current are strongly regulated.

The 2TM group of K^+ channels, i.e., K^+ channels formed by 4 (2TM–1P) subunits, generally functions as inward rectifier K^+ channels that include several subsets ($K_{IR}1.i$–$K_{IR}7.i$; Table 3.15). They conduct a substantial current near the resting potential, but carry few or no ions at depolarized potentials. $K_{IR}2.i$ are strong inward-rectifier K^+ channels in hearts (inwardly rectifying current i_{K1}; Vol. 5 – Chap. 5. Cardiomyocytes). The $K_{IR}3.i$ channels are G-protein-activated inward rectifier K^+ (GIRK) channels responsible for i_{ACh} current. ATP-sensitive inward rectifiers $K_{IR}6.i$ are ligand-gated channels responsible for $i_{K_{ATP}}$ current in the myocardium. Pore-

Table 3.14. Groups and classes within the superclass of potassium channels and their main functions (**Part 2**; Source: Wikipedia; 4 (6)TM: four (six-to-seven) transmembrane-containing channel subunits; 1 (2P): one (two) pore-domain channel). Two pore-domain potassium channels comprise 15 members that are regulated by several factors, such as oxygen, hydrogen ion, mechanical stretch, and G proteins. Voltage-gated potassium channels are constituted by α subunits that form the conductance pore and auxiliary β subunits that modulate the channel activity. Subunits α can be categorized into 12 families (K_V1–K_V12). Calcium-activated potassium channels are composed of BK, IK, and SK channels according to conductance (big, intermediate, and small). Subunits $K_{Ca}\alpha$ with 6 transmembrane segments form homo- and heterotetramers. Activity of $K_{Ca}1$ channel depends on both Ca^{++} and voltage. Channels $K_{Ca}4$ and $K_{Ca}5$ are responsive to intracellular Na^+ and Cl^- instead of Ca^{++}.

Category	Subclasses families	Function
Tandem pore domain 4TM and 2P	TWIK TRAAK TREK TASK	Resting potential
Voltage-gated 6TM and 1P	HERG ($K_V11.1$) K_VLQT1 ($K_V7.1$)	Repolarization Action potential frequency limitor
Calcium-activated 6TM and 1P	BK ($K_{Ca}1.1$) IK ($K_{Ca}5.1$) SK ($K_{Ca}2/3/4.i$)	Inhibition following stimuli with calcium influx

forming α subunits form tetramers. Heteromeric channels may be formed within families (e.g., $K_{IR}3.2$–$K_{IR}3.3$ complex).

The 4TM group is made of K^+ channels responsible for background currents, with subsets TWIK, [126] TREK, TASK, TALK, THIK, and TRESK (Tables 3.16 and 3.17). Potassium channels of the 4TM group are open at all voltages. They are regulated by numerous mediators and neurotransmitters. The pore-forming α subunit contains 2-pore domains (K_{2P} channels). They can form functional dimers rather than usual tetramers. In addition, they can form heterodimers (e.g., $K_{2P}3.1$–$K_{2P}9.1$ complex). The usual division into subfamilies is based on similarities in both structural and functional properties.

126. TWIK stands for 2-pore domain weak inward-rectifying-related K^+ channel that can be activated by both force and osmolarity, as well as lysophosphatidylcholine, phosphatidylinositol (4,5)-bisphosphate, and phosphatidic acid.

Table 3.15. The 2TM group of K^+ channels, i.e., inward rectifier K^+ channels (Source: [5]).

Family	Component(s)	Activators	Functional characteristics
$K_{IR}1$	$K_{IR}1.1$ (ROMK1)		Inward rectifier
$K_{IR}2$	$K_{IR}2.1$–2.4 (IRK1–4)		Strong inward rectifier
$K_{IR}3$	$K_{IR}3.1$–3.4 (GIRK1–4) (i_{ACh})	PIP2, $G\beta\gamma$	G-protein-activated inward rectifier
$K_{IR}4$	$K_{IR}4.1$–4.2		Inward rectifier
$K_{IR}5$	$K_{IR}5.1$		Inward-rectifier
$K_{IR}6$	$K_{IR}6.1$–6.2 ($i_{K_{ATP}}$)		ATP-sensitive inward rectifier
$K_{IR}7$	$K_{IR}7.1$		Inward rectifier

Table 3.16. Two pore-domain potassium channels form dimers composed of 2 inward rectifier α subunits (TALK: TWIK-related alkaline-pH-activated K^+ channel; TASK: TWIK-related acid-sensitive K^+ channel; THIK: tandem-pore-domain halothane-inhibited K^+ channel; TRAAK: TWIK-related arachidonic acid-stimulated K^+ channel; TREK: TWIK-related K^+ channel; TRESK: TWIK-related spinal cord K^+ channel; TWIK: tandem of P-domains in a weak inwardly rectifying K^+ channel).

Gene	Channel	Alias
KCNK1	$K_{2P}1.1$	TWIK1
KCNK2	$K_{2P}2.1$	TREK1
KCNK3	$K_{2P}3.1$	TASK1
KCNK4	$K_{2P}4.1$	TRAAK1
KCNK5	$K_{2P}5.1$	TASK2
KCNK6	$K_{2P}6.1$	TWIK2
KCNK7	$K_{2P}7.1$	TWIK3
KCNK8	$K_{2P}8.1$	TWIK4
KCNK9	$K_{2P}9.1$	TASK3
KCNK10	$K_{2P}10.1$	TREK2
KCNK12	$K_{2P}12.1$	THIK2
KCNK13	$K_{2P}13.1$	THIK1
KCNK14	$K_{2P}14.1$	TASK4
KCNK15	$K_{2P}15.1$	TASK5
KCNK16	$K_{2P}16.1$	TALK1
KCNK17	$K_{2P}17.1$	TALK2
KCNK18	$K_{2P}18.1$	TRESK

The `6TM` group encompasses voltage-gated K_V channels with several families (Tables 3.18 and 3.19): (1) the KCNQ family with cardiac types K_V7,[127] (2) EAG

127. Decrease in $K_V7.1$ that are encoded by the KCNQ1 gene are involved in myocardial slow repolarization (i_{Ks} current, in opposition to rapid phase of cardiac repolarization cur-

Table 3.17. Group 4TM of K^+ channels (Source: [5]). Many potassium-selective pore-forming subunits with 4 transmembrane regions and 2 pore domains per subunit constitute the K_{2P} group. Members of this group can be categorized into subclasses on the basis of structural and functional properties: TWIK- (TWIK1–TWIK3), TASK- (TASK1, -3, and -5), TALK- (TALK1–TALK2 [or TASK4] and TASK2), TREK- (TREK1–TREK2, and TRAAK), and THIK subclass (THIK1–THIK2).

Subclass	Subtypes
TWIK	$K_{2P}1.1$ (TWIK1), $K_{2P}6.1$ (TWIK2), $K_{2P}7.1$ (TWIK3)
TREK	$K_{2P}2.1$ (TREK1), $K_{2P}10.1$ (TREK2), $K_{2P}4.1$ (TRAAK)
TASK	$K_{2P}3.1$ (TASK1), $K_{2P}9.1$ (TASK3), $K_{2P}15.1$ (TASK5)
TALK	$K_{2P}5.1$ (TASK2), $K_{2P}16.1$ (TALK1), $K_{2P}17.1$ (TALK2)
THIK	$K_{2P}13.1$ (THIK1), $K_{2P}12.1$ (THIK2)
TRESK	$K_{2P}18.1$ (TRESK)

channels that include cardiac delayed rectifying outward K^+ channels that generate fast-activating (i_{Kr}) ventricular repolarization currents,[128] and (3) the class of Ca^{++}-activated K^+ channels with BK and SK channel subfamilies. The pore-forming α subunits form tetramers. Heteromeric channels can be formed between-family members (e.g., $K_V1.1$–$K_V1.2$ and $K_V7.2$–$K_V7.3$ complexes).

In addition to the Slo1 subunit of tetrameric BK channels, the 6TM-group Slo family also comprises SLACK (sequence-like a calcium-activated K^+ channel; or Slo2.2) and SLICK (sequence-like an intermediate conductance K^+ channel; or Slo2.1) high-conductance, Na^+-activated K^+ channels (K_{Na}).

3.4.3 Gating Modes

Many potassium channels possess a time-dependent activity that can switch abruptly from high to low open probability. The selectivity filter and adjacent channel components are characterized by a conformational flexibility, with partial and fully-open states. Conformational fluctuations can occur at the selectivity filter of the conductive channel.

Therefore, K^+ channels can control the magnitude and duration of ionic fluxes using different modes of gating. The gating modes rely on specific interactions between the structural elements that surround the selectivity filter. These interactions, in cooperation with ion occupancy, influence the conformational state of the pore [352].

Potassium channels undergo a gating cycle with conductive and non-conductive states as well as various conductance levels that arise from conformational changes

---

rent i_{Kr}) and are associated with cardiac arrhythmias [351]. Delayed cardiac repolarization (long QT interval in ECG) and prolonged cardiac action potentials increase the risk of early after-depolarization that can give rise to torsades de pointes. On the other hand, premature repolarizations (short QT interval) can lead to sudden cardiac death.

128. The rapid phase of cardiac repolarization is ensured by K^+ channels encoded by human ether-a-go-go related gene (HERG or $K_V11.1$).

Table 3.18. Genes of the 6TM-group K^+ channels (α subunits for K_V). Groups of voltage-gated channels K_V5, K_V6, K_V7, and K_V9 do not form functional channels as homotetramers, but can heterotetramerize with K_V2 family members to form conductive channels. Auxiliary regulatory β subunits ($\beta1$–$\beta4$) connect to pore-forming Slo1 α subunits to form large-conductance, voltage- and Ca^{++}-activated K^+ channels that are commonly termed BK channels. Sodium- and chloride-gated, ATP-sensitive, K^+ channel Slick and Slack are ligand-gated, high conductance K^+ channels (K_{Na}) that are weakly sensitive to Ca^{++} and membrane voltage.

Family members	Genes
$K_V1.1$–$K_V1.8$	KCNA1–KCNA7, KCNA10
$K_V2.1$–$K_V2.2$	KCNB1–KCNB2
$K_V3.1$–$K_V3.4$	KCNC1–KCNC4
$K_V4.1$–$K_V4.3$	KCND1–KCNC3
$K_V7.1$–$K_V7.5$	KCNQ1–KCNQ5
$K_V10.1$–$K_V10.2$ (EAG1–EAG2)	KCNH1, KCNH5
$K_V11.1$–$K_V11.3$ (ERG1–ERG3)	KCNH2, KCNH6, KCNH7
$K_V12.1$–$K_V12.3$ (ELK1–ELK3)	KCNH8, KCNH3, KCNH4
$K_{Ca}1.1$ (MaxiK, Slo1, BK)	KCNMA1
$K_{Ca}4.1$ (Slack, Slo2.2)	KCNT1
$K_{Ca}4.2$ (Slick, Slo2.1)	KCNT2
$K_{Ca}5.1$ (Slo3)	KCNMA3, KCNMC1

at the selectivity filter with a given duration. The fully-open conformation of the activation gate results from a saturating stimulus. On the other hand, non-conductive states with lifetimes ranging from ms to s are associated with a stationary gating. Transitions between these non-conductive and conductive states define gating mode shifts. These transitions can result from post-translational modifications as well as conformational fluctuations at or near the filter [352]. The pore, selectivity filter, and external vestibule are dynamical structures where small, local conformational changes, which include motions of the carbonyl oxygens, small fluctuations of the filter backbone, or changes in the configuration and occupancy of water molecules behind the filter, can strongly influence the channel gating pattern. These structural changes indeed modulate the interactions between ions and the channel filter.

Table 3.19. Group 6TM of voltage-gated K^+ channels (Source: [5]). The first voltage-gated K^+ channel (Shaker) was cloned from Drosophila. Afterward, 4 Drosophila subfamilies were described (Shaker, Shab, Shal, Shaw). The first cloned mammalian K^+ channels are related to these subfamilies.

Family	Members (associated subunits) [Functional characteristics]
K_V1 (Shaker)	$K_V1.1$–1.8 ($K_V\beta1$–2)
K_V2 (Shab)	$K_V2.1$–2.2 ($K_V5.1$, $K_V6.1$–6.3, $K_V8.1$, $K_V9.1$–9.3)
K_V3 (Shaw)	$K_V3.1$–3.4 (MiRP2 or $K_V3.4$)
K_V4 (Shal)	$K_V4.1$–4.3 (KChIP, KChAP)
K_V7 (KCNQ)	$K_V7.1$–7.5 (MinK; MiRP2: $K_V.7.1$) [$K_V7.1$: cardiac IK_S; $K_V7.2$–7.3: M current]
K_V10, K_V11, K_V12 (EAG)	$K_V10.1$–10.2, $K_V11.1$–11.3, $K_V12.1$–12.3 (MinK β subunit [KCNE1], MiRP1 [KCNE2]: $K_V11.1$) [$K_V11.1$: cardiac IK_R]
$K_{Ca}1$, $K_{Ca}4$, $K_{Ca}5$ (BK)	$K_{Ca}1.1$, $K_{Ca}4.1$–4.2, $K_{Ca}5.1$, Slack, Slick [MaxiK: $K_{Ca}1.1$; K_{Na}: Slack, Slick]
$K_{Ca}2$, $K_{Ca}3$ (IK, SK)	$K_{Ca}2.1$–2.3, $K_{Ca}3.1$ (SK4, IK) [SK_{Ca}: $K_{Ca}2.1$–2.3; IK_{Ca}: $K_{Ca}3.1$]

3.4.4 Inwardly Rectifying Potassium Channels

Inward rectifier potassium channels (K_{IR} or IRK) constitute a class of K^+-selective ion channels ($K_{IR}1$–$K_{IR}7$) that lack voltage-dependent gating. They conduct K^+ ions most efficiently in the inward direction, i.e., into the cell. [129]

Inwardly rectifying potassium channels control the resting membrane potential, and hence the functioning of excitable cells. They possess ion-binding sites along

129. The term "inward rectification" refers to a change in slope of the current–voltage relationship of the channel at the reversal potential, i.e., the zero current level that occurs at the equilibrium potential for K^+ ions. At membrane potentials much greater than the equilibrium potential for K^+, the outward current is very small compared to the inward current at potentials much smaller than the equilibrium potential. This rectification is due to occlusion

the ion-conduction pore that are both conductive and inhibitory [353]. These channels, indeed, carry K^+ ions into the cell at internal negative membrane voltages. At internal positive membrane voltages, they are blocked by intracellular multivalent ions.

3.4.4.1 K_{IR} Class

Inwardly rectifying potassium channels (Table 3.20) that set the resting membrane potential and modulate membrane excitability are gated by plasmalemmal phosphatidylinositol (4,5)-bisphosphate.

Plasmalemmal density and conductance of $K_{IR}2.3$ channel rises when $K_{IR}2.3$ links to scaffold Disc large homolog DLg1 [354]. [130]

Among K_{IR} channels, atrial K_{ACh} channels are heterotetramers composed of $K_{IR}3.1$ (GIRK1) and $K_{IR}3.4$ (GIRK4) subunits. [131]

Unlike most K_{IR} channels, $K_{IR}3$ (e.g., heteromers $K_{IR}3.1$–$K_{IR}3.2$, and $K_{IR}3.1$–$K_{IR}3.4$, as well as homomer $K_{IR}3.2$) are activated by $PI(4,5)P_2$ only in the presence of an additional gating molecule, such as subunit $G\beta\gamma$ of G proteins or intracellular sodium. [132] Both $K_{IR}3$ and $K_{IR}5.1$ require sodium for channel gating. [133] Sodium triggers a structural switch that allows interactions between Na^+-sensitive K_{IR} channels and $PI(4,5)P_2$ [355]. Moreover, $K_{IR}5.1$ confers Na^+ sensitivity, as well as CO_2 and H^+ sensitivities, on channels with which it makes heteromers.

3.4.4.2 G-Protein-Coupled Inwardly Rectifying Channels K^+ (GIRK)

G-protein-gated inwardly rectifying potassium channels (GIRK) are heteromers composed of 4 different α subunits (GIRK1–GIRK4, or $K_{IR}3.1$–$K_{IR}3.4$, encoded by the genes KCNJ3, KCNJ6, KCNJ9, and KCNJ5, respectively). The production of GIRK4 is low. Different splice variants exist. For example, at least 3 splice variants of the GIRK2 subunit have been detected (GIRK2a–GIRK2c).

The intracellular N- and C-termini of GIRK channels contain interacting domains with the $G\alpha$ and $G\beta\gamma$ subunits of G proteins that regulate the channel opening.

Intracellular transfer and function of GIRKs depend on the composition of channel subunits. Subunits GIRK1 and GIRK3 that lack an endoplasmic reticulum export signal are sequestered in the endoplasmic reticulum. To be conveyed to the plasma membrane, they must assemble with GIRK2 or GIRK4 subunits. The latter have transport signal motifs as well as a VL sequence for endocytosis [356]. The GIRK3

---

of the central pore by intracellular Mg^{++} and polyamines at potentials above the equilibrium potential for K^+ cations.

130. A.k.a. synapse-associated protein SAP97.

131. Homotetramers generally show little or no activity, except in the case of mutations.

132. Intracellular sodium also activates sodium-activated potassium channels (K_{Na}) and represses the activity of epithelial Na^+ channels.

133. The $K_{IR}5.1$ channel is expressed by renal tubular epithelial cells involved in Na^+ reabsorption and in brainstem neurons that act as CO_2 chemoreceptors in the cardiorespiratory nuclei.

Table 3.20. Inwardly rectifing potassium channels (Source: Wikipedia). Each group of K_{IR} channels ($K_{IR}1$–$K_{IR}7$) has multiple members ($K_{IR}ij$). The K_{IR} channels are either homo- (each subunit is made of 2 membrane-spanning α helices TM1 and TM2) or heterotetramers (SUR2: sulfonyl urea receptor, or ABCc9, an ATP-binding cassette protein that serves as a subunit linked to pore-forming inward rectifiers $K_{IR}6.1$ and $K_{IR}6.2$).

Gene	Channel	Alias
KCNJ1	$K_{IR}1.1$	ROMK1
KCNJ2	$K_{IR}2.1$	IRK1
KCNJ12	$K_{IR}2.2$	IRK2
KCNJ4	$K_{IR}2.3$	IRK3
KCNJ14	$K_{IR}2.4$	IRK4
KCNJ3	$K_{IR}3.1$	GIRK1, KGA
KCNJ6	$K_{IR}3.2$	GIRK2
KCNJ9	$K_{IR}3.3$	GIRK3
KCNJ5	$K_{IR}3.4$	GIRK4
KCNJ10	$K_{IR}4.1$	$K_{IR}1.2$
KCNJ15	$K_{IR}4.2$	$K_{IR}1.3$
KCNJ16	$K_{IR}5.1$	BIR9
KCNJ8	$K_{IR}6.1$	K_{ATP} (SUR2b)
KCNJ11	$K_{IR}6.2$	K_{ATP} (SUR1, SUR2a, and SUR2b)
KCNJ13	$K_{IR}7.1$	$K_{IR}1.4$

subunit contains a lysosome-targeting motif that limits its surface accumulation, even in assembly with GIRK2 or GIRK4.

G-protein-coupled inward rectifiers are bound to and modified by integrins. They are inactivated by stretch. G-protein–activated inward rectifier K^+ channels are found in atriomyocytes, but not in the sinoatrial node.

Neuronal GIRK Channels

The basal activity of GIRKs contributes to the resting potential of neurons. Under physiological conditions, the resting membrane potential of a neuron is positive with respect to the equilibrium potential for K^+, and the small outward K^+ current through GIRKs decreases the neuronal excitability. Channels GIRKs hyperpolarize neurons in response to the activation of many types of G-protein-coupled receptors, thus reducing membrane excitability.

In the central nervous system, 3 primary GIRK subunits form homotetramers (e.g., GIRK2–GIRK2) and heterotetramers (GIRK1–GIRK2, GIRK1–GIRK3, and GIRK2–GIRK3) that have different biophysical properties [357]. [134] Most neurons express GIRK1, GIRK2, and GIRK3. The predominant form of GIRKs in the brain is a GIRK1–GIRK2 heterotetramer.

134. The GIRK2 homotetramer has short open times (average <0.5 ms), whereas heterotetramers have longer open times (average 1–2 ms).

Three GIRK2 splice variants that are expressed in the brain differ in the length of the C-terminus: GIRK2c [135] contains a PDZ-binding motif that is absent in GIRK2a [136] and GIRK2b. Splice variants also exist for GIRK1, but not GIRK3 and GIRK4.

Various neurotransmitters, such as acetylcholine, adenosine, dopamine, γ-amino-butyric acid, opioids, serotonin, and somatostatin, activate GIRKs, once they have stimulated their cognate G-protein-coupled receptors. The Gβγ dimers released from G proteins (separated from Gα subunits Gi and Go) bind directly to GIRKs and cause their opening [357].

Once activated by neurotransmitters, such as A_1 adenosine, D_2 dopamine, type-B γ-aminobutyric acid ($GABA_B$), and somatostatin Sst_5 receptors, Gi/o-protein-coupled receptors inhibits [357]: (1) emitting neurons (GIRK-mediated neuronal self-inhibition); (2) connecting neurons (slow inhibitory postsynaptic potential); and (3) neuronal networks, when a neurotransmitter is released from many neurons (large-scale effect of neuromodulators mediated by a moderate activation of GIRKs called volume transmission). [137] Upon activation by adenosine, GIRK channels contribute to reversal of long-term potentiation of glutamatergic transmission (depotentiation).

The GIRK channels interact with multiple intracellular proteins that modulate their functioning. They are inactive in the absence of bound G-protein subunits. Binding of Gα primes GIRK activity in the absence of receptor signaling (basal activity). Activation of Gi or Go by Gi/o-coupled receptors, such as γ-aminobutyric acid $GABA_B$, dopamine, and adenosine A_1 receptors, stimulates GIRKs by direct binding of Gβγ to their intracellular domains [356]. On the other hand, activation of Gq and G11 by Gq/11-coupled receptors precludes both basal and Gβγ-stimulated GIRK activity via stimulation of protein kinase-C. Phosphatidylinositol (4,5)-bisphosphate heightens GIRK activity.

Activity of GIRK channels participates in the modulation of memory as well as pain perception by endogenous substances, such as endorphins and endocannabinoids [357]. Endorphins (endogenous opioid polypeptides, i.e., endogenous morphines) produced by the pituitary gland and hypothalamus act as pain relievers. Endocannabinoids are lipid messengers (hence hydrophobic molecules with restricted zones of influence) synthesized by the nervous and immune systems.

The GIRK channels as well as SK channels hyperpolarize the membrane and limit the neuronal excitability once activated. Both channel types belong to proteic complexes that contain protein kinases and phosphatases. These agents modulate the channel activity. Channel partners include protein kinase-A, protein phosphatases PP1 and PP2, and receptor for activated-C kinase RACK1. [138] Protein kinase-A and

135. A.k.a. $K_{ATP}2$, GIRK2a1, and BIR1.

136. A.k.a. GIRK2 and GIRK2-1.

137. Dopamine released from a single site can diffuse and encounter extrasynaptic dopamine autoreceptors on dopamine-containing axons as well as parareceptors on neighboring cells. It can thus activate many dopamine receptors on several neurons and influence thousands of synapses simultaneously.

138. A.k.a. GNB2L1.

phosphatase-2 facilitate and repress the channel activation, respectively [356]. In addition, Na^+ ions stimulate GIRKs [357].

Activated phospholipase-C reduces the concentration of membrane phosphatidylinositol (4,5)-bisphosphate, thereby stabilizing the channel in its open configuration. Yet, it also activates protein kinase-C that phosphorylates GIRK, thus impeding channel opening [356]. Therefore, phosphorylations by PKA and PKC rise and reduce GIRK activity, respectively.

Excitatory transmission that provokes long-term potentiation via ^{N}methyl ^{D}aspartate receptors and calcium–calmodulin-dependent kinase CamK2 also facilitates GIRK inhibitory current to possibly narrow the window of time of excitatory input detection [356].

Various regulators of G-protein signaling proteins (RGS; Sect. 7.8 and Vol. 4 – Chap. 8. Guanosine Triphosphatases and Their Regulators) also modulate GIRK activity. They all contain a GTPase-activating protein domain that inactivates heterotrimeric G protein. In addition, RGS proteins influence interactions with GIRK partners.

The GIRK channels form signaling complexes. Muscarinic M_2 receptors that can dimerize with metabotropic $GABA_B$ subunit-2 can tether to GIRKs. In excitatory synapses, the GIRK channels are coupled to $GABA_B$, but not ionotropic $GABA_A$ receptors (Cl^- channels). In gabaergic interneurons, complexes formed by $GABA_B$ receptors, G proteins, and GIRKs, but not RGS2, activate $GABA_B$ and open GIRK channels [356]. In dopaminergic neurons, RGS2 weakens the coupling between $GABA_B$ receptors and GIRK channels that contain GIRK2 and/or GIRK3 subunits, hence relieving inhibition of dopaminergic neurons.

3.4.4.3 Renal Outer Medullary Potassium Channel (ROMK)

Potassium inwardly rectifying channel ($K_{IR}1.1$), or the renal outer medullary potassium channel (ROMK), is an ATP-dependent K^+ channel. It exports K^+ ions out of cells. It recycles K^+ in the thick ascending limb of the Henle's loop and allows its secretion in the cortical collecting duct of the nephron. The ROMK channel excretes K^+ that enters cells through Na^+–K^+ ATPase at the basolateral membrane under normal dietary K^+ intake. Both ROMK and Ca^{++}-activated, big-conductance K^+ channel (BK) contribute to K^+ elimination when K^+ diet increases.

With no lysine kinase WNK4 inhibits ROMK, as WNK4 stimulates clathrin-dependent endocytosis of ROMK. The WNK1 kinase also prevents ROMK functioning in the cortical collecting duct. Hyperkalemia increases the expression of a kidney specific splice variant of WNK1 (ksWNK1) that antagonizes WNK1 inhibition. On the other hand, serum- and glucocorticoid-regulated kinase SGK1 phosphorylates (inactivates; Ser1169) WNK4 in the connecting tubule and cortical collecting duct.

The activity of the ROMK channel rises in the apical membrane of distal nephron and promotes K^+ excretion during hyperkalemia, but not blood volume depletion. Hyperkalemia increases aldosterone release and SGK1 production to enhance renal K^+ removal. In addition, SGK1 stimulates Na^+–K^+ ATPase and epithelial sodium channel. Conversely, Src kinase prevents SGK1 from relieving the inhibition of

ROMK by WNK4, hence restoring the WNK4-mediated inhibition of ROMK and repressing K^+ excretion [358]. Activity of Src is suppressed and promoted by hyper- and hypokalemia, respectively. Kinases of the SRC family phosphorylate ROMK (Tyr337), thereby facilitating internalization of ROMK channels. They then preserve ROMK inhibition in response to volume depletion.

3.4.4.4 Inwardly Rectifying Potassium Channels (IRK)

Strong inwardly rectifying potassium channels (IRK) convey K^+ anions more easily in the inward direction, i.e., favor K^+ influx, to set the membrane potential back to its resting value. The 4 members of the IRK family (IRK1–IRK4 or $K_{IR}2.1$–$K_{IR}2.4$) form homo- and heterotetramers.

The IRK1 channel can link to $K_{IR}2.2$ and $K_{IR}4.1$ as well as Disc large homologs DLg1 and DLg4 (members of the membrane-associated guanylate kinase [MAGUK] family involved in postsynaptic densities) and AKAP5 (member of the A-kinase anchoring protein family).

The IRK2 channel can connect to $K_{IR}2.1$ and $K_{IR}2.3$ as well as DLg1, DLg4, and mammalian abnormal cell lineage-seven (Lin7) homolog proteins MALS1 and MALS3.[139]

The IRK3 channel can heteromerize with $K_{IR}2.1$ and $K_{IR}2.3$ as well as DLg2[140] and DLg4 adaptors. The IRK4 channel can form heteromeric channels with IRK1.

3.4.4.5 ATP-Sensitive Potassium Channels

ATP-sensitive potassium channels (K_{ATP}), members of the class of K_{IR} inward rectifiers, reduce the action potential duration.

K_{ATP} and Cell Metabolism

Channels K_{ATP} are electrochemical transducers of the metabolic state of excitable cells, such as neurons, cardiac and skeletal myocytes, smooth muscle cells, as well as some renal and pancreatic cells. ATP-sensitive K^+ channel links electrical excitability to the cellular metabolic state, or cellular energetics, particularly during ischemia.[141]

In fact, decreased metabolism (reduced ATP-to-ADP ratio) opens K_{ATP}, hence causing K^+ efflux and membrane hyperpolarization. Conversely, increased metabolism closes K_{ATP}, thereby generating membrane depolarization and subsequently

139. A.k.a. Lin7 homolog-A and -C as well as vertebrate homologs of Caenorhabditis elegans Lin7 VeLi1 and VeLi3.

140. A.k.a. PSD93 and 110-kDa channel-associated protein of synapses (chapsyn-110).

141. The K_{ATP} channels are closed in basal conditions due to inhibition by intracellular ATP, but open during metabolic stress such as hypoxia. They thus contribute to adaptive response to metabolic stress. In vascular smooth muscle, K_{ATP} channels regulate basal vasomotor tone and promote vasodilation in response to severe hypoxia.

triggering various cell responses according to cell type, such as regulation of the vascular tone, contraction of striated myocytes, release of hormones and neurotransmitters, and synaptic transmission.

In striated myocytes (heart and skeletal muscle), sarcolemmal ATP-sensitive K^+ channels couple membrane excitability with cellular metabolism and control energy expenditure. Excess energy provided by nutrition is stored using glycogen or lipid deposition that can be afterward transformed into ATP. The K_{ATP} channels govern energy economy in myocytes, because they support glycogen and lipid depots. Impaired K_{ATP} function, even with adequate substrate availability, causes additional energy cost, reduces the body's weight, and limits physical endurance [359].

Structure

The K_{ATP} channels are constituted by 2 different proteins: pore-forming $K_{IR}6$ ($K_{IR}6.1$ or $K_{IR}6.2$) [142] and regulatory sulfonylurea receptor (SUR1, SUR2a, and SUR2b) [143] of the superclass of ATP-binding cassette transporters [360]. [144]

Whereas in vascular smooth muscle cells K_{ATP} channel results from a combination of $K_{IR}6.1$ and SUR2b, in pancreatic insulin-secreting β cells, K_{ATP} channel is composed of $K_{IR}6.2$ and SUR1. The $K_{IR}6.1$ subunit contributes to the maintenance of vascular tone. The $K_{IR}6.2$ subunit participates in glucose homeostasis. Four $K_{IR}6.2$ subunits generate the channel pore, each $K_{IR}6.2$ subunit being linked to one SUR1 protein to form an hetero-octamer. The SUR protein contains one nucleotide-binding site.

Partners and Regulators

The K_{ATP} channel is unique among K^+ channels, as its activity depends strongly on cytosolic ATP concentration. It is gated by 2 mechanisms: (1) fast, ligand-independent and (2) slow, ligand-dependent gating. It is inhibited by ATP binding and activated by Mg^{++}-bound nucleotides. [145]

The K_{ATP} channel can be formed from $K_{IR}6.2$ subunits with or without SUR subunits. The $K_{IR}6.2$ subunit has binding sites for ATP (free or ATP^{Mg}) and PIP_2 (mutual exclusion) that stabilize closure and opening of the channel, respectively. Magnesium-dependent ATP hydrolysis on SUR subunit overcomes the inhibitory

142. Subunit $K_{IR}6.1$ is the dominant pore-forming subunit of K_{ATP} in vascular smooth muscle cells. Subunit $K_{IR}6.2$ is predominant in pancreatic β cells, cardiomyocytes, and non-vascular smooth muscles.

143. Subunit sulfonylurea receptor (SUR) can intervene differently according to the localization of K_{ATP} in membrane compartments.

144. Proteins SUR1 and SUR2 correspond to the ABC family C subtypes 8 (ABCc8) and 9 (ABCc9).

145. Intracellular nucleotide ATP, in the absence of Mg^{++}, binds directly to and represses $K_{IR}6$ protein. On the other hand, both ATP^{Mg} and ADP^{Mg} bind to SUR and activate the channel.

effect of ATP on $K_{IR}6.2$ subunit. However, channel opening by ADP^{Mg} is more efficient. The state of K_{ATP} thus depends on the metabolic activity of the cell.

Channels K_{ATP} reside within dynamical complexes that include enzymes and cytoskeletal and adaptor proteins, in addition to its subunits. Partner enzymes comprise creatine kinase, adenylate kinase, and lactate dehydrogenase, among others. In addition, syntaxin-1A, a soluble ^{N}ethylmaleimide-sensitive factor attachment protein receptor (SNARE) protein, and cAMP-binding guanine nucleotide-exchange factor RapGEF4 bind to SUR1 [361].

The function of K_{ATP} depends also on actin binding in myocytes. The actin cytoskeleton controls both the channel exocytosis and its ATP sensitivity. The C-terminus of $K_{IR}6.2$ possesses an ankyrin-B-binding motif that serves for membrane targeting during intracellular transfer as well as ATP sensitivity, hence K_{ATP} gating [362]. Ankyrin-B connects to $K_{IR}6.2$ C-terminus, but not $K_{IR}6.1$ that is coproduced with $K_{IR}6.2$ in many excitable cells.

In addition to intracellular nucleotides, K_{ATP} activity is regulated by numerous factors, such as phospholipids (e.g., phosphatidylinositol (4,5)-bisphosphate), acyl-CoA, H^+, as well as by the cytoskeleton, in addition to phosphorylation (e.g., by PKA and PKC). In particular, PIP_2 binds to $K_{IR}6$ and activates K_{ATP}. On the other hand, PIP_2 hydrolysis by phospholipase-C inhibits K_{ATP} channel.

Inhibition mediated by ATP and activation induced by PIP_2 of $K_{IR}6$–SUR channels relies on a negative heterotropic cooperativity [363]. The allosteric ligand ATP interacts with K_{ATP} in both the open and closed states [364]. Regulators ATP and PIP_2 stabilize the closed and open states, respectively. Subunit $K_{IR}6.2$ regulates ATP sensitivity. Subunits SUR control channel kinetics.

Two long closed states, locked by ATP (C_{ATP}) and unlocked (C_0) are used to model the activity of $K_{IR}6.2$–SUR channels [364]. Binding of ATP to the channel in its C_0 state initiates C_{ATP} state. Channel behavior can be explained by models with the following assumptions [363]: (1) the channel undergoes ATP-insensitive transitions from the open state to a short closed state (C_s) and a longer unlocked closed state (C_l); (2) the C_l state is destabilized in the presence of SUR1; (3) ATP locks the C_l state into C_{ATP} state; and (4) PIP_2 stabilizes the open state (O), hence reducing ATP sensitivity. The ATP level can modulate the equilibrium between C_{ATP} and C_l states, thereby prolonging closed state duration.

$$C_{ATP} \underset{k_{Cl-ATP}}{\overset{k_{ATP-Cl}}{\rightleftarrows}} C_l \underset{k_{O-Cl}}{\overset{k_{Cl-O}}{\rightleftarrows}} O \underset{k_{Cs-O}}{\overset{k_{O-Cs}}{\rightleftarrows}} C_s.$$

Subunit SUR and Subcellular Localization

In cardiomyocytes, the majority of plasmalemmal K_{ATP} channels are $K_{IR}6.2$–SUR2a dimers located in caveolin-3-enriched nanodomains [365]. [146] The pore-forming subunit of $K_{IR}6.2$ frequently associates with caveolin-3 as well as adenosine receptors A_1 that are coupled to K_{ATP} via inhibitory G protein (Gi; Sect. 7.6.2 and Vol. 4 – Chap. 8. Guanosine Triphosphatases and their Regulators) and protein

146. The $K_{IR}6.1$–SUR1 complex is also detected.

kinase-C. [147] Millimolar concentrations of intracellular ATP during ischemia suffice to open K_{ATP} channel, whereas micromolar concentration of ATP is required for its inhibition. Attenuated ATP sensitivity results from signaling such as that triggered by adenosine receptors. [148] In vascular smooth muscle cells, $K_{IR}6.1$–SUR2b channel also lodges in caveolae.

In addition to the binding of Mg^{++}–adenosine nucleotides (ATP^{Mg} and ADP^{Mg}), SUR subunit binds to drugs of the sulfonylurea class of K_{ATP} blockers (hence its name). At least in cardiomyocytes, SUR1 subunit tethered to $K_{IR}6.2$ targets the channel to the plasma membrane, [149] whereas subunits SUR2a and SUR2b tag the channel for the endosomal–lysosomal compartment, thereby representing an available pool for turnover and recycling of plasmalemmal channels [366]. The intracellular pool of SUR2-containing K_{ATP} derived by endocytosis from the plasma membrane can play a cardioprotective role by serving as a reservoir to modulate the cell surface K_{ATP} density under stress, such as ischemia.

Hydrogen Sulfide, an Opener of K_{ATP} channel

In vascular smooth muscle cells, hydrogen sulfide activates K_{ATP} to cause vasodilation [367]. This opener of K_{ATP} does not enhance channel conductance. In pancreatic β cells, H_2S stimulates also K_{ATP}, thereby contributing to the regulation of glucose-induced insulin secretion [368].

In vitro, hydrogen sulfide enhances angiogenic potential of endothelial cells, as it improves cell proliferation, migration, and tube formation [369]. Hydrogen sulfide operates via K_{ATP} and increases phosphorylation of PKB, ERK, and P38MAPK kinases in endothelial cells.

3.4.5 Voltage-Gated K_V Channels

In excitable cells, Voltage-gated K^+ channels (K_V; Tables 3.21 and 3.22) open in response to changes in voltage difference between fluids separated by the plasma membrane. The K_V channels open under membrane depolarization. They control the repolarization of neurons and other excitable cells.

Subunits α ($K_V\alpha1$–$K_V\alpha12$; Table 3.21) that form the conductance pore are grouped into 6 groups: (1) delayed rectifier, (2) A-type, (3) outward rectifying, (4) inward rectifying, (5) slowly activating, and (6) modifier–silencer type.

---

147. Adenosine receptor activates K_{ATP} owing to caveolin that also influences other modulators of K_{ATP} channels. Gi Protein and protein kinase-C localize to caveolae. Caveolin-3, but not caveolin-1, plays an important role in the regulation of K_{ATP} by adenosine receptors in cardiomyocytes. Prolonged stimulation (15–30 mn) causes K_{ATP} internalization in cardiomyocytes [365].

148. The K_{ATP} channels are also regulated by numerous other activators released during ischemia, such as nitric oxide, noradrenaline, protein kinase-C, and phosphatidylinositol (4,5)-bisphosphate.

149. The SUR1 subunit links to several proteins involved in exocytosis.

Table 3.21. Subunits α of K_V channels (Sources: [370], Wikipedia; EAG: Ether-a-go-go; ELK: Ether-a-go-go-like K^+ channel; ERG: Ether-a-go-go related gene; HERG or ERG1: human Ether-a-go-go-related gene). Each gene encodes a subunit. A functional channel is constituted by 4 subunits. The K_V channels are homo- or heterotetramers. Modifier subunits do not produce functional channels, but form heterotetramers with K_V2 family subunits.

Gene	Type
	Delayed rectifier
KCNA1–KCNA7, KCNA10	$K_V1.1$–$K_V1.8$ (Shaker-related family)
KCNB1–KCNB2	$K_V2.1$–$K_V2.2$ (Shab-related family)
KCNC1–KCNC2	$K_V3.1$–$K_V3.2$ (Shaw-related family)
KCNQ1–KCNQ5	$K_V7.1$–$K_V7.5$ (Shal-related family)
	($K_V7.1$: K_VLQT; $K_V7.2$: KQT2)
KCNH1	$K_V10.1$ (EAG1)
	A-type potassium channel
KCNC3–KCNC4	$K_V3.3$–$K_V3.4$ (Shaw-related family)
KCND1–KCND3	$K_V4.1$–$K_V4.3$ (Shal-related family)
	Outward-rectifying
KCNH5	$K_V10.2$ (EAG2)
	Inward-rectifying
KCNH2, KCNH6–KCNH7	$K_V11.1$–$K_V11.3$ (ERG1–ERG3)
	Slowly activating
KCNH3, KCNH4, KCNH8	$K_V12.2$, $K_V12.3$, $K_V12.1$ (ELK2, ELK3, ELK1)
	Modifier/silencer
KCNF1	$K_V5.1$
KCNG1–KCNG4	$K_V6.1$–$K_V6.4$
KCNV1–KCNV2	$K_V8.1$–$K_V8.2$
KCNS1–KCNS3	$K_V9.1$–$K_V9.3$

Voltage-gated potassium channels are used for cell repolarization, such as in the myocardium. Transient outward K^+ channels rapidly activate and inactivate. Delayed rectifier K^+ channels have slower activity than transient-outward K^+ channels. Inward rectifiers K_{IR} that lack steep voltage-gating, modulate the cell excitability.

3.4.5.1 Structure and Function

The pore domain of K_V channels contains all molecular elements necessary for ion conduction and activation and inactivation gating. Pore-forming α subunits are sufficient to generate functional K^+ channels. Yet, they can connect to function-modifying, cytoplasmic β subunits.

The $K_V\beta1.1$ subunit and its splice variants quickly inactivate delayed rectifiers such as $K_V1.1$. On the other hand, cytoplasmic $K_V\beta2.1$ subunit does not induce fast inactivation, but causes kinetic changes. Other β subunits, such as MinK and Slo1 subunits of $BK_{V,Ca}$ channels, are transmembrane proteins.

Voltage-sensor domains are structurally and functionally conserved proteic modules that consist of 4 transmembrane segments (TM1–TM4) with charged residues.

Table 3.22. Subunits β of K_V channels (Sources: [370], Wikipedia). The family of calcium-binding, voltage-gated potassium channel-interacting proteins (KChIP) that are encoded by the KCNIP genes belong to the recoverin branch of the EF-hand superfamily. The KCNIP3 gene (a.k.a. calsenilin) is a calcium-regulated transcriptional repressor and interacts with presenilins. Other accessory proteins that connect to K_V tetramers and modify their properties include calmodulin, which binds to K_V10.

Gene	Type	Associated K_V	Cardiac expression
KCNAB1	$K_V\beta1$	K_V1, K_V2	+
KCNAB1	$K_V\beta2$	K_V1, K_V2	+
KCNAB1	$K_V\beta3$	K_V1, K_V2	
KCNE1	MinK	K_V11	+
KCNE2	MiRP1		+
KCNE3	MiRP2		+
KCNE4	MiRP3		
KCNE5	MiRP4		
KCNE1L	KCNE1-like		
KCNIP1	KChIP1	K_V4	
KCNIP2	KChIP2		
KCNIP3	KChIP3		
KCNIP4	KChIP4		
PIAS	KChAP		+
FREQ	NCS1		+

These modules confer voltage sensitivity to many ion channels. The TM4 segment represents the major component of the voltage sensor for gating, as it contains, in most channels, a large part of the gating charge of the voltage sensor, although negative charges in TM2 and TM3 segments also contribute. Depolarization of the plasma membrane causes a movement of TM4 that then induces further conformational changes to open the channel.

Some homo- and heterotetrameric channels formed by $K_V\alpha$ subunits constitute non-functional complexes. However, these non-functional tetramers can modify the activity of other $K_V\alpha$ subunits. In particular, $K_V\alpha9.1$ and -9.2 specifically modulate $K_V\alpha2.1$ and -2.2 function, as they change channel kinetics and expression level [371]. Modulatory subunits $K_V5.1$ and $K_V6.1$ are involved in the regulation of functional $K_V2.1$ and $K_V2.2$ subunits, as they can inhibit, change channel kinetics, and/or provoke shifts in voltage dependency of channel inactivation.

Voltage-gated *potassium channel-interacting proteins* (KChIP) operate as β subunits specific for K_V4 to regulate their functioning [372]. Four K^+ channel-interacting proteins exist (KChIP1–KChIP4) with several spliced isoforms. They enhance the surface expression of these channels and modulate their kinetic behavior, like neuronal Ca^{++} sensor NCS1 (a.k.a. frequenin) [373]. [150]

On the other hand, *potassium channel-associated protein* (KChAP) serves as a chaperone for $K_V1.3$, $K_V2.1$, and $K_V4.3$ channels.

150. Both neuronal Ca^{++}-binding proteins and KChIPs are produced in smooth muscle cells.

Post-translational modifications such as palmitoylation that assist in membrane association and interaction with other proteins can modulate voltage sensing and kinetics via protein–membrane interactions (e.g., $K_V1.1$ [374]) or enhance the surface density (e.g., $K_V1.5$ [375]) of K_V channels. Palmitoylation of the KChIP auxiliary subunits also controls plasmalemmal residence of $K_V4.3$ channels [376].[151]

3.4.5.2 Activation and Inactivation

The functional behavior of ion channels can be defined by activation and inactivation gating. In K^+ channels, inactivation is coupled to activation owing to side-chain rearrangements [377]. The selectivity filter aims at optimally selecting ions, i.e. impeding the passage of most ions, but allowing a fast K^+ translocation. It can be involved in both fast and long-lasting inactivation gating.

Certain voltage-gated, Ca^{++}-independent K^+ channels can cause slow delayed rectifying currents that develop after the Na^+ depolarizing current. Their delayed onset of activation is followed by slow inactivation.

Delayed rectifying outward potassium channels that generate fast- (i_{Kr}) and slow-activating (i_{Ks}) ventricular repolarization currents control the duration of cardiac action potentials. Cell repolarization is mainly carried out by activated fast-delayed, outwardly rectifying K^+ channels. Slow-delayed, K^+ outward rectifiers serve as a repolarization reserve.

A-Type Function

Channels K_V that provoke rapid A-type currents are characterized by fast activation and inactivation O[100 ms]) [373]. Channels $K_V1.4$, $K_V3.4$, and $K_V4.1$ to $K_V4.3$ can generate A-type currents. The A-type current is regulated by Ca^{++}–calmodulin-dependent kinase CamK2 that reduces the rate of inactivation and protein phosphatases PP1 and PP3 that raise the rate of inactivation of A-type currents.

The A-type currents are observed in neurons, atrial and ventricular myocytes (transient outward K^+ current), and all smooth muscle cell types. The A-type currents in smooth muscle cells may be related to the maintenance of membrane potential and regulation of excitability [373].

Fast-inactivating A-type K^+ currents inhibit back-propagation of dendritic action potentials. They are also implicated in synaptic adaptivity (plasticity). They regulate heart excitability.

N-Type Function

The rate of inactivation of K_V channels determines the firing rate of neurons. Some K_V close rapidly during continued membrane depolarization and are responsible for N-type inactivation,[152] whereas other K_V channels show little or no in-

151. Among the 3 splice variants of rat KChIP2 with variable N-terminus, the 2 longer isoforms that can be palmitoylated promote $K_V4.3$ accumulation in the plasma membrane.

152. N-type inactivation of K_V channels is a process in which the protein N-terminus binds to the open-channel surface and occludes its pore.

activation. Fast N-type inactivation that happens after an initial activation despite a sustained depolarization can be explained by an auto-inhibitory binding of the N-terminus of K_V to the open pore that occludes it.

C-Type Function

Activation is classically associated with conformational rearrangements at the inner helix bundle. Slow, incomplete C-type inactivation (time scale ms–s) originates from transitions at the selectivity filter i n almost all K^+ channels. Pore constriction develops with much slower kinetics. It is highly modulated by permeant ions and pore blockers.

Most K^+ channels open through coupled gates at both ends of the permeation path. A minimal kinetic cycle coupling these 2 dynamical gates is assumed to involve transitions between 4 states [377]: (1) closed inner gate with open-conductive filter (C/O); (2) open inner gate with open-conductive filter (O/O); (3) open inner gate with inactivated filter (O/I); and (4) closed inner gate with inactivated filter (C/I). Transition occurs between a closed-conductive conformation in which the inner gate is closed, but the selectivity filter is conductive, to an open-inactivated one in which the inner gate is fully open, but the selectivity filter is inactivated [378].

3.4.5.3 Influence of the Membrane

The membrane environment influences the ion channel function and kinetics. Membrane lipids can convert A-type channels into delayed rectifiers and conversely [379]. Phosphoinositides remove N-type inactivation from A-type channels by immobilizing the inactivation domains.

Amphiphilic arachidonic acid endows delayed rectifiers with rapid voltage-dependent inactivation. It may modulate properties of the channel by: (1) changing the physical characteristics of the lipid bilayer; (2) modifying the hydrophobic interaction between the membrane protein and lipid bilayer; and (3) binding to a specific site involved in gating or permeation of the channel.

Phosphatidylinositol (4,5)-bisphosphate and intracellular ATP^{Mg} control the activity of the $K_V7.1$–MinK channel responsible for the delayed rectifying K^+ current (i_{Ks}) that controls the duration of the action potential of human hearts [380] (Vol. 5 – Chap. 5. Cardiomyocytes). The former stabilizes the channel in the open state. Moreover, PIP_2 slows the deactivation kinetics and shifts the voltage dependency of the channel activation toward negative potentials.

Certain voltage-gated potassium channels have a component at the external face of the channel with positively charged amino acids that senses the change in voltage across the cell membrane. In non-excitable cells, voltage-gated K^+ channels can be activated by changes in lipid charges of the adjoining plasma membrane [381]. Gating of K_V needs interaction with several species of phospholipids [382]. Voltage-gated potassium channel opening is indeed ensured by the presence of negatively charged phosphate groups in suitable plasmalemmal phospholipids [383]. Voltage changes generate sensor motions within the lipid bilayer, leading to opening of the channel central pore.

3.4.5.4 Channel Density

Activity of voltage-gated potassium channels depends not only on their opening rate, which determine ion flux intensity, but also on their plasmalemmal density. The density of K_V channels is correlated to: (1) synthesis rate (with adequate folding to avoid degradation) and (2) exo- and endocytosis rates.

Ion channels are assembled in the endoplasmic reticulum and transported to the cell membrane. Voltage-gated potassium channels can form functional homotetramers. However, many K_V link to ancillary β subunits. Channels K_V are then internalized and subsequently degraded via NEDD4-2-dependent ubiquitination (half-life 1–35 h). Endocytosis of the K_V7.1–MinK complex (but not homomeric K_V7.1) is, at least partly, mediated by MinK [384]. The MinK subunit undergoes dynamin-2- and clathrin-dependent endocytosis that augments K_V7.1 homomer density at the plasma membrane. In other words, K_V7.1 subunits outnumber MinK subunits (ratio 5:1) in human ventriculomyocyte sarcolemma. Nevertheless, new ancillary subunits MinK or members of the MIRP (MinK-related peptide) family such as MiRP2 can then form new complexes with K_V7.1. The MiRP2–K_V7.1 complex causes current of similar amplitude, but with much faster kinetics.

3.4.5.5 K_V1 Family

Members of the K_V1 family can assemble with other members of this family to form heteromultimers (Tables 3.23 and 3.24), but not with members of other K_V families.

K_V1.1

The K_V1.1 channel maintains the membrane potential and modulates the excitability in neurons and myocytes. Abundant, highly unsaturated fatty acids, such as arachidonic acid, docosahexaenoic acid, and anandamide block conduction of K_V1 channels.[153] Once K_V1 has been opened by depolarization, these lipids interact with hydrophobic residues that line the inner cavity of the pore, hence occluding the permeation pathway [385].

K_V1.2

Actin-binding protein cortactin[154] regulates the activity of K_V1.2 [386]. The actin cytoskeleton mediates intracellular dynamin-dependent transport. Cortactin is required for normal plasmalemmal K_V1.2 expression and inhibition of its elimination

153. In neurons, various signals cause the release of arachidonic acid, docosahexaenoic acid, and anandamide (^{N}arachidonylethanolamide, AEA) from the cell membrane into the cytosol.
154. Cortactin is a crosslinker for actin filament and an activator of the ARP2–ARP3 complex.

Table 3.23. Subunits of the $K_V\alpha1$ family (**Part 1**; Source: [370]; SNAP: synaptosomal-associated protein). Potassium channel-associated protein (KChAP) modulates the expression of K_V channels. Cell recognition molecule contactin-associated protein CntnAP2 (a.k.a. CAsPr2 and neurexin homolog Nrxn4), that operates as a cell-adhesion molecule and receptor is located at juxtaparanodes of myelinated axons, where it links to K^+ channels. Disc large homolog-4 (DLg4; a.k.a. postsynaptic density adaptor PSD95 and synapse-associated protein SAP90), is a member of the membrane-associated guanylate kinase (MAGUK) that heteromultimerizes with DLg2 (or PSD93). The latter is recruited into the same NMDA receptor and K^+ channel clusters. Disc large homolog-1 (DLg1; a.k.a. SAP97) is another MAGUK family member that is involved in the transfer of ionotropic receptors from the endoplasmic reticulum to the plasma membrane. The SNAP25 protein is a component of the SNARE complex that execute fusion between synaptic vesicles and the plasma membrane by forming a complex between these 2 cell constituents. The $K_V\alpha1$ subunit can associate with subunits of $K_V\beta1$ to $K_V\beta3$ families, whereas $K_V\alpha2$ subunit links to $K_V\beta4$ subunits.

Channel	Partners	Distribution
$K_V1.1$	$K_V\beta1/\beta2$, DLg1/4, SNAP25, $K_V1.i$	Neurons, pancreatic islet cells, cardiomyocytes, skeletal myocytes
$K_V1.2$	$K_V\beta1/\beta2$, DLg1/4, SNAP25, CntnAP2, RhoA, $K_V1.i$	Neurons, Schwann cells, myocytes
$K_V1.3$	$K_V\beta$, KChAP, β_1-integrin, DLg, $K_V1.i$	Microglia, oligodendrocytes, platelets, lymphocytes, macrophages, fibroblasts, pancreatic islet cells, osteoclasts
$K_V1.4$	$K_V\beta$, KChAP, DLg1/4, SNAP25, α-actinin-2, $K_V1.i$	Neurons, cardiomyocytes, skeletal myocytes, pancreatic islet cells

from the cell membrane. Cortactin–$K_V1.2$ interaction decreases [155] after Tyr phosphorylation that leads to $K_V1.2$ endocytosis to affect membrane excitability. Removal of $K_V1.2$ can be generated by muscarinic receptor M_1.

K_V1.3

Voltage-gated delayed rectifier $K_V1.3$ regulates membrane potential and calcium signaling, especially in lymphocytes and oligodendrocytes. Production of $K_V1.3$ rises in effector memory T cells.

155. This interaction is not suppressed when endocytosis occurs.

Table 3.24. $K_V\alpha1$ subunits (**Part 2**).

Channel	Partners	Distribution
$K_V1.5$	$K_V\beta1$, $K_V\beta2$, KCNA3B, SAP97, Src, Fyn, α-actinin-2, caveolin, $K_V1.i$	Neurons, oligodendrocytes, Schwann cells, microglia, smooth muscle cells, cardiomyocytes
$K_V1.6$	$K_V\beta1$, $K_V\beta2$, CntnAP2, $K_V1.i$	Astrocytes, oligodendrocytes, neurons, cardiomyocytes, smooth muscle cells
$K_V1.7$	$K_V1.i$	Cardiomyocytes, skeletal myocytes, smooth muscle cells
$K_V1.8$	KCNa4b $K_V1.i$	Kidney, heart, skeletal muscle, brain, adrenal gland

K_V1.4

Activity of voltage-gated delayed rectifier $K_V1.4$ is regulated by phosphorylation and dephosphorylation by calmodulin-dependent kinase CamK2 and calcineurin (PP3), respectively. It is involved in neuronal afterhypolarization.

K_V1.5

The $K_V1.5$ channel generates the ultrarapid repolarizing current (i_{Kur}) in atriomyocytes and regulates the vascular tone with oxygen. It modulates neuron excitability.

In human hearts, $K_V1.5$ is phosphorylated by Src kinase. In addition, $K_V1.5$ is targeted by ubiquitin-like SUMo (SUMo1–SUMo3) and SUMo conjugasee UbC9 for adaptive tuning of the electrical excitability of cells [387]. This interaction reversibly modulates $K_V1.5$ activity, because $K_V1.5$ is liberated from SUMo by SUMo-specific proteases SENP2. Sumoylation of $K_V1.5$ can change the excitability of atriomyocytes and vascular smooth muscle cells, in particular in the pulmonary vasculature according to oxygen level.

Glucocorticoids promote $K_V1.5$ expression. Cholesterol modulates the turnover of voltage-gated potassium channels, hence the atrial repolarizing current through $K_V1.5$ channels. Cholesterol actually regulates plasmalemmal density of $K_V1.5$, as it participates in its exocytosis via Rab11-associated recycling endosome [388].[156] In adult atriomyocytes, recycling endosomes constitute a storage compartment of $K_V1.5$ that can be recruited to the plasma membrane upon cholesterol depletion. However, cholesterol depletion also decreases K_V4, hence fast component of i_{Kto} current.

156. Small GTPases Rab regulate sorting, tethering, and docking of vesicles (Vol. 4 – Chap. 8. Guanosine Triphosphatases and Their Regulators). Small GTPase Rab11 is linked to recycling endosomes.

Table 3.25. Family of K_V2 channels (Source: [370]).

Channel	Partners	Distribution
$K_V2.1$	$K_V5.1$, $K_V6.1$–6.3, $K_V8.1$, $K_V9.1$–9.3, KChAP, Fyn kinase	Neurons, Schwann cells, cardiomyocytes, smooth muscle cells, skeletal myocytes
$K_V2.2$	$K_V8.1$, K_V9, $K_V\beta4$, KChAP	Neurons, cardiomyocytes, smooth muscle cells

Table 3.26. Family of K_V3 channels (Source: [370]).

Channel	Partners	Distribution
$K_V3.1$		Neurons, skeletal myocytes, airway smooth muscle cells
$K_V3.2$	$K_V3.1$	Neurons, Schwann cells, pancreatic islets
$K_V3.3$		Neurons, smooth muscle cells
$K_V3.4$	MiRP2	Neurons, skeletal myocytes, pancreatic acinar cells, parathyroid

3.4.5.6 K_V2 Family

Voltage-gated $K_V2.1$ delayed rectifier operates in oxygen-sensitive pulmonary artery as well as in neurons (particularly in hippocampal and globus pallidus neurons; Table 3.25).

Angiotensin-2 type-1 receptor mediates inhibition of $K_V2.2$ in brainstem and hypothalamic neurons [370].

3.4.5.7 K_V3 Family

Voltage-gated $K_V3.1$ delayed rectifier is involved in high-firing frequency of auditory and fast-spiking gabaergic interneurons (Table 3.26). It also regulates action potential duration in presynaptic terminals.

The $K_V3.2$ channel contributes to fast-spiking gabaergic interneurons of the neocortex, hippocampus, and caudate nucleus as well as GABA release via the regulation of action potential duration in presynaptic terminals [370]. It has a fast deactivation rate. Its activity is modulated by protein kinase-A.

Both $K_V3.3$ and $K_V3.4$ belongs to the group of voltage-gated A-type potassium channels. It forms with MirP2 low-voltage-activating K^+ channels that regulate skeletal muscle resting potential.

$K_V3.4$ and the Cell Cycle in Smooth Muscle Cells

Unlike fully-differentiated striated myocytes, smooth muscle cells are able to modify their phenotype in response to environmental changes. Upon exposure to

Table 3.27. Family of K_V4 channels (Source: [370]; KChIP: K_V channel-interacting protein).

Channel	Partners	Distribution
$K_V4.1$	KChIP1	Brain, heart, lung, thyroid, kidney digestive tract, liver, pancreas, pulmonary artery
$K_V4.2$	KChIP1, KChIP2, KChIP4, Frequenin, PSD95	Neurons, cardiomyocytes
$K_V4.3$	KChIP1, KChIP2, KChIP4,	Neurons, cardiomyocytes, smooth muscle cells

physiological (e.g., hormones and growth factors) or pathophysiological (e.g. hypoxia and acidosis) stimuli, vascular smooth muscle cells can switch from a usual contractile state to a proliferative and migratory behavior to remodel blood vessels.

The switch from a contractile to a proliferative state is characterized by transient changes in ion transport in distinct phases of the cell division cycle. Cchannels K_V, such as $K_V10.1$ and $K_V11.1$, as well as other types of K^+ channels, such as $K_{Ca}1.1$, $K_{Ca}3.1$, and $K_{2P}9.1$ (or TASK3), intervene during the cell cycle, particularly in the G1–S transition.

The $K_V3.4$ channel controls the cell cycle of cultured human smooth muscle cells of uterine arteries [389]. Hyperpolarization caused by this K^+ channel promotes entry into the cell cycle and progression through the G1 phase.

Accessory protein MinK-related protein MiRP2 modulates the gating, kinetics, unitary conductance, and drug sensitivity of $K_V3.4$ channel. Like $K_V3.4$, MiRP2 expression undergoes cell cycle-dependent variations. The increased expression of $K_V3.4$ as the cell cycle progresses from the G0 to G1 stage and in the G2–M transition is associated with a progressive decline in MiRP2, hence in $K_V3.4$–MiRP2 complexes.

3.4.5.8 K_V4 Family

In excitable cells, widespread A-type, low-voltage-activated K_V4 channels control action potential firing, dendritic activity, and synaptic signal integration. The K_V4 channels are complexes that contain K^+ channel-interacting proteins (KChIP), which possess calcium-binding domains to confer calcium dependence (Table 3.27).

Accessory protein KChIP1 increases $K_V4.1$ density, accelerates inactivation rate, and shifts steady-state inactivation to more depolarized potentials. It raises ionic flux across $K_V4.2$, slows onset of inactivation, and accelerates recovery from inactivation.

Frequenin, a calcium-binding protein, elevates $K_V4.2$ current amplitude, lowers inactivation rate, and accelerates recovery from inactivation.

Scaffold Disc large homolog DLg4 is involved in $K_V4.2$ trafficking. Pore-forming proteins $K_V4.2$ and/or $K_V4.3$ complex with auxiliary KChIP2 subunits.

Heteromultimers of $K_V4.2$ and $K_V4.3$ subunits and KChIP are responsible for the transient outward i_{Kto} current in the heart. The expression of i_{Kto} current-inducing

Table 3.28. Family of K_V5 modifiers (Source: [370]).

Channel	Partners	Distribution
$K_V5.1$	$K_V2.1$–$K_V2.2$	Brain, heart, skeletal muscle, liver, kidney, pancreas

subunits differs between animal species and myocardial regions across the ventricular wall [390]. Whereas $K_V4.2$ level correlates with i_{Kto} magnitude across left ventricular walls of rodents, its production is very low in that of large animals and humans. Subunit KChIP2 is expressed according a transmural gradient in canine hearts.

The K_V4 subunit can complex with KChIP3 and $Ca_V3.2$ or $Ca_V3.3$ channels [391]. These low-voltage-activated Ca^{++}-regulated complexes permit subthreshold calcium signals to set neuronal excitability and operate at membrane potentials below those required to trigger an action potential.

$K_V4.3$

Heart failure is characterized by a reduction of transient outward current (i_{Kto}) and excessive activation of Ca^{++}–calmodulin-dependent kinase CamK2 in ventriculomyocytes.

The $K_V4.3$ channel responsible for the i_{Kto} current is coupled to CamK2 kinase in cardiomyocyte sarcolemma [392]. Dissociation of CamK2 from the $K_V4.3$–CamK2 complex increases CamK2 autophosphorylation and fosters Ca_V1 activity (hence L-type calcium current). On the other hand, $K_V4.3$ precludes CamK2 activation by binding to the calmodulin-binding site. Enzyme CamK2 can complex with $K_V4.3$ via the Disc large homolog DLg1 adaptor.

3.4.5.9 K_V5 Family

The $K_V5.1$ modifier channel operates only as a modulator of K_V2 channels (Table 3.28).

3.4.5.10 K_V6 Family

Alone, $K_V6.1$ subunit is unable to elicit any current, but can impede $K_V2.1$ current, although less effectively than $K_V5.1$. It also can suppress, to a lesser extent, $K_V2.2$ (Table 3.29).

3.4.5.11 K_V7 Family

Slow-delayed, outwardly rectifying potassium channels are composed of a pore-forming α subunit $K_V7.1$ and auxiliary β subunit MinK encoded by genes KCNQ1

Table 3.29. Family of K_V6 proteins (Source: [370]).

Channel	Partners	Distribution
$K_V6.1$	K_V2	Skeletal muscle, brain, bone, skin, kidney, pancreas, uterus, ovary, placenta, germ cell, prostate, testis
$K_V6.2$	K_V2	Myocardium, fetal brain, germinal center B cells
$K_V6.3$	$K_V2.1$	Brain, spinal cord, pituitary, testis, small intestine, thymus, adrenal gland
$K_V6.4$	$K_V2.1$	Brain, liver, small intestine, colon

Table 3.30. Family of K_V7 channels (Source: [370]).

Channel	Partners	Distribution
$K_V7.1$	MinK, MiRP2 lung, pancreas, cochlea, placenta	Heart, kidney, rectum, ear, germ,
$K_V7.2$	$K_V7.3$, MiRP1	Brain, sympathetic ganglia, eye, Heart, lung, breast, small intestine, germ cell, placenta, testis
$K_V7.3$	$K_V7.2$, $K_V7.5$	Brain, retina, colon, testis
$K_V7.4$	$K_V7.3$	Brainstem auditory nuclei, cochlea, vestibular organs, placenta
$K_V7.5$	$K_V7.3$	Brain, sympathetic ganglia, skeletal muscle

Table 3.31. Family of K_V8 channels (Source: [370]).

Channel	Partners	Distribution
$K_V8.1$	K_V2	Brain, kidney
$K_V8.2$	K_V2	Lung, kidney, liver, pancreas, spleen, colon, thymus, prostate, testis, ovary

and KCNE1, respectively (Table 3.30). The $K_V7.1$ channel intervenes in repolarization of cardiomyocytes. It also determines subthreshold excitability of neurons. The MinK subunit slows $K_V7.1$ activation (5–10-fold), impedes its inactivation, and increases its conductance (4 fold) [384]. The $K_V7.1$–MinK complex hence produces larger current amplitude with slower kinetics than $K_V7.1$ homomer.

3.4.5.12 K_V8 Family

The K_V8 channels regulate membrane potential and action potential frequency and modulate the activity of K_V2 channels (Table 3.31).

Table 3.32. Family of K_V9 proteins (Source: [370]).

Channel	Partners	Distribution
$K_V9.1$	K_V2	Brain
$K_V9.2$	K_V2	Brain, retina, spinal cord, pulmonary artery
$K_V9.3$	K_V2	Brain, eye, heart, kidney, lung, muscle, skin, stomach, colon, breast, uterus, testis, pulmonary artery smooth muscle cells

Table 3.33. Family of K_V10 channels (Source: [370]; KCR: K^+ channel regulator). Auxiliary hyperkinetic β subunit of K^+ channels regulates their firing properties and K^+ current. Ion channel auxiliary protein slowpoke channel-binding protein (SloB) also interacts with Ca^{++}-dependent K^+ channel (BK). Epsins contribute to membrane deformations that occur during endocytosis. Calmodulin binds to and regulates numerous proteins and serves as a calcium sensor and signal transducer.

Channel	Partners	Distribution
$K_V10.1$	$K_V\beta$	Brain
	Calmodulin, SloB, epsin	
$K_V10.2$	$K_V\beta$	Brain
	Calmodulin, SloB, epsin	

The $K_V8.1$ channel changes $K_V2.1$ and $K_V2.2$ expression level as well as their kinetics and half-inactivation potential toward hyperpolarization. The $K_V8.2$ channel modifies $K_V2.1$ activation threshold and kinetics.

3.4.5.13 K_V9 Family

Channels $K_V9.1$ and $K_V9.2$ colocalize with $K_V2.1$ and/or $K_V2.2$ (Table 3.32). They modulate the activity of $K_V2.1$ and $K_V2.2$ channels, as they change their expression rate and kinetics as well as shift the half-inactivation potential to more polarized values. The $K_V9.1$ channel enhances the conductance of $K_V2.1$ channel.

3.4.5.14 K_V10 Family

The $K_V10.1$ and $K_V10.2$ channels can heteromerize and connect to several partners (Table 3.33). The $K_V10.1$ channel controls the cell cycle.

3.4.5.15 K_V11 Family

Voltage-gated $K_V11.1$ channel [157] can homomerize (Table 3.34. It is expressed in multiple cell types such as cardiomyocytes. The Herg gene encodes the pore-forming

Table 3.34. Family of K_V11 channels (Source: [370]).

Channel	Partners	Distribution
$K_V11.1$	MinK, MiRP1 $K_V11.2$, $K_V11.3$	Heart, blood cells, brain, kidney, lung, liver, pancreas, small intestine, tonsil, ovary, uterus, testis, prostate
$K_V11.2$	$K_V11.1$, $K_V11.3$	Brain, uterus
$K_V11.3$	$K_V11.1$, $K_V11.2$	Brain, sympathetic ganglia

α subunit that co-assemble to create rapidly activating delayed rectifier K^+ current in the heart.

In the heart, the $K_V11.1$ channel mainly intervenes in repolarization (action potential phase 3), although it open early upon membrane depolarization and is rapidly inactivated. The $K_V11.1$ channel generates its maximal repolarizing current during phases 2 and 3 of the action potential to restore the resting potential, hence to govern the duration of the QT interval on ECG.

The intensity of K^+ flux depends on plasmalemmal $K_V11.1$ density. Protein kinase-A, upon sustained activity of cAMP, promotes a 2- to 4-fold production of adequately folded $K_V11.1$ channel [393]. [158]

Mutations in human ether-a-go-go-related gene-1 are responsible for type-2 long-QT syndrome (LQT2). Diverse mutation types can cause [394]: (1) attenuated transfer to the plasma membrane, as transcripts are retained in the endoplasmic reticulum; (2) channel gating at more negative voltages; and/or (3) rapid deactivation kinetics.

3.4.5.16 K_V12 Family

The N- and C-termini of K_V12 channels (Table 3.35) possess a light oxygen voltage (LOV) and cyclic nucleotide-binding (CNB) domain, respectively.

3.4.5.17 K_V Channels in Vascular Smooth Muscle Cells

In arterial smooth muscle cells, K_V channels abounds. They modulate the vasomotor tone, as they influence the activity of voltage-gated Ca^{++} channels. Activated K_V channels cause membrane hyperpolarization and vasodilation, whereas their inhibition leads to membrane depolarization and vasoconstriction. They are inhibited by vasoconstrictors, such as angiotensin-2, endothelin-1, serotonin, and thromboxane-A2. These vasoconstrictors bind to their cognate Gq/11-coupled receptors (AT_1, ET_A,

157. A.k.a. human ether-a-go-go-related gene product (HERG).

158. In cardiomyocytes, the G protein–ACase–cAMP–PKA pathway influences the activity of numerous ion channels and transporters, such as $K_V7.1$-MinK complex (slowly activating, delayed rectifying potassium current), ryanodine receptors that dissociate from stabilizing FKBP12.6, and $K_V11.1$ associated with its ancillary subunits MinK and MiRP1.

Table 3.35. Family of K_V12 channels (Source: [370]).

Channel	Distribution
$K_V12.1$	Brain, sympathetic ganglia, lung, colon, testis, uterus
$K_V12.2$	Brain, nerve, eye, lung Lymphocytes
$K_V12.3$	Brain, cerebellum, pituitary gland, esophagus

$5HT_{2A}$, and TP, respectively; Chap. 7). These receptors activate phospholipase-C to produce inositol trisphosphate and diacylglycerol with consequent Ca^{++} release from intracellular stores and activation of protein kinase-C. Protein kinase-C inhibits arterial K_V channels that colocalize with caveolin, possibly via caveolin-dependent endocytosis. Endothelin-1 and angiotensin-2 operate via PKCα and PKCϵ, respectively, using different PKC-interacting proteins rather than separate populations of K_V channels [395]. Angiotensin-2 also inhibits ATP-sensitive K^+ channels in arterial smooth muscle cells via PKCϵ.

3.4.6 Calcium-Gated Potassium Channels BK, IK, and SK

Potassium channels BK, IK, and SK ($K_{Ca}i$) constitute a class of 8 members most often activated by intracellular Ca^{++}. However, some ($K_{Ca}4$ and $K_{Ca}5$) are responsive to intracellular Na^+ and Cl^-. Furthermore, $K_{Ca}1$ is activated by both Ca^{++} and voltage.

Calcium-activated small (SK) and intermediate (IK) conductance channels in endothelial cells and large (BK) conductance channels in vascular smooth muscle cells participate in endothelium-derived hyperpolarization responsible for arteriolar dilations.

Both SK and BK channels localize to caveolae. Caveolin-1 interacts with and inhibits the BK channel, thereby limiting the contribution of this channel to the vasodilation resulting from endothelium-derived hyperpolarizing factor on microvascular smooth muscle cells [396].

3.4.6.1 Large-Conductance, Calcium- and Voltage-Gated Potassium Channels

Large-conductance, voltage-gated,, Ca^{++}-activated potassium channel owing to large single-channel currents is also called big potassium channel.[159] Ubiquitous BK channel actually has the largest single-channel conductance of all K^+-selective channels.

Potassium BK channel can be activated synergistically by both voltage and intracellular Ca^{++} concentration. It intervenes in the regulation of the vascular smooth muscle tone and neurotransmitter release.

159. Alias BK, BK_{Ca}, or $BK_{V,Ca}$, as well as $K_{Ca}1$, MaxiK, and Slo1 channel.

The BK channel is synthesized in many cell types. It particularly abounds in neurons and smooth muscle cells. Membrane depolarization and increased intracellular Ca^{++} concentration cause BK channels to open. This event provokes membrane hyperpolarization and closes voltage-dependent channels and reduces Ca^{++} influx.

In addition to the resulting negative feedback regulation of Ca^{++} signaling, BK channel acts as a negative and positive feedback regulator in excitable cells whether they conduct repolarizing or hyperpolarizing outward currents. It controls firing patterns in neurons and contributes to the modulation of the vasomotor tone of blood vessels.

Structure

The BK channel is a tetramer. Each BK channel protomer is itself a dimer with a pore-forming, voltage-sensing α subunit (Slo1) and one of the 4 modulatory β subunits. The α subunit is encoded by the gene KCNMA1 (SLO1) in the human genome, whereas modulatory β subunits are synthesized from a small set of genes (KCNMB1–KCNMB4 and KCNMB3L). Channel BK isoforms then arise from many sources, especially from distinct β subunits that confer variable Ca^{++} sensitivity and inactivation rate.

As in other voltage-gated K^+ channels, 4 pore-forming α subunits form a functional BK channel. Each α subunit possesses an extracellular N-terminus, 7 transmembrane segments (TM0–TM6), [160] and a large intracellular C-terminus with regulator of conductance for K^+ domains (RCK1 and RCK2) that confer sensitivity to Ca^{++} and other intracellular ligands. [161]

The TM1–TM4 transmembrane region of BK channels forms a voltage-sensing domain. The large intracellular C-terminus contains 3 Ca^{++}-binding sites [399]: a Ca^{++} bowl and 2 other Ca^{++} and (or) Mg^{++}-binding sites.

The BK channel has a structure similar to that of voltage-gated K^+ channels. Modifications in membrane potential displace the voltage sensor. The resulting conformational change then drives the channel opening. However, unlike typical K_V channels, more gating charges of voltage-sensor domains (TM1–TM4) of voltage-sensitive BK channels pertain to transmembrane segments other than TM4 segment [397]. Segments TM2 and TM4 possess distinct voltage dependence, but interact functionally. Activation of TM2 or TM4 segment when the other segment has been activated produces more charge movement. When one of these 2 segments is neutralized, the charge motion is reduced when its intact partner activates. According

160. Transmembrane segments TM1 to TM4 contain the voltage sensor with a high density of charged residues (Asp153 and Arg167 in TM2, Asp186 in TM3, and Arg213 in TM4 [397]), whereas TM5 and TM6 segments contribute to the K^+-selective pore. The large cytoplasmic domain has multiple divalent cation sensors. Therefore, the pore-forming α subunit contains both voltage and Ca^{++} sensors.

161. Four cytoplasmic regions, more precisely 2 tandem C-terminal RCK domains from each of 4 channel subunits, form a gating ring at the intracellular membrane surface with 4 Ca^{++}-binding sites on the outer perimeter [398]. The Ca^{++}-gating ring that regulates the pore directly may also modulate the function of voltage sensor.

to the dynamical field focusing theory, segment activation causes aqueous crevices to form, focusing the field. As a result, in the TM2–TM4 doubly activated state, TM2-dependent gating may occur in BK channels.

Alternative Splicing

Alternative splicing of KCNMA1 transcripts create channels that differ in conductance and gating properties. Subunits α of BK channel can have an alternatively spliced insert — the stress-regulated exon (StREx) — that interacts with the plasma membrane via palmitoylation of cysteine residues of cytosolic StREx C terminus. The StREx insert yields a cysteine-rich domain in the intracellular linker (palmitoylation of which serves as an anchor to the plasma membrane) between 2 RCK domains. Phosphorylation of serine residue immediately near palmitoylated cysteine residues in StREx by protein kinase-A of a single α subunit dissociates the C terminus from the plasma membrane, thus inhibiting BK activity [400]. Palmitoylation is required for BK regulation by protein kinase-A.

Post-translational Modifications

Phosphorylation

The BK channel is regulated by several Ser/Thr protein kinases, such as PKA, PKC, and PKG. Numerous hormones, neurotransmitters, and drugs bind to G-protein-coupled receptors that then signal via activated protein kinases PKA, PKC, and PKG.

In smooth muscle cells, PKA and PKG activate BK channels, as they increase the voltage- and Ca^{++} sensitivity of BK channels. Hence, PKA or PKG contribute to smooth muscle relaxation. Protein kinase-C exerts opposite effects, thereby favoring vasoconstriction. Phosphorylation (Ser695 between RCK1 and -2 domains) by protein kinase-C actually inhibits BK channel [401]. In addition, phosphorylation (Ser695) by PKC renders BK channel insensitive to stimulation by protein kinases PKA or PKG.

Phosphorylation (Ser695; inactivation) by PKC depends on a preceding phosphorylation (Ser1151) in the C-terminus of the channel α subunit. This sequential phosphorylation of 2 distinct Ser residues in the BK channel C-terminus of the channel protein controls the regulation of BK channels by PKA and PKG in tracheal smooth muscle cells. [162]

Phosphorylation of only one α subunit within the pore (Ser1151 and Ser695) suffices to inhibit BK channel activity. Protein phosphatase-1 that associates with the channel constantly counteracts Ser695 phosphorylation.

162. Phosphorylation (Ser1151) by PKC also influences stimulation of BK channel activity by protein kinase-A and -G.

Myristoylation

Protein myristoylation enables cells to anchor proteins into the internal leaflet of the plasma membrane (Sect. 1.2.5.6).[163] The BK channel is internally myristoylated, myristic acid being connected to intracellular loops (serine and threonine residues of loop 1 or 3) of the molecule [402]. Myristoylation participates in the regulation of the channel density on the cell surface.

Influence Agents

The BK channel is sensitive to different ions (Ca^{++}, Na^+, and H^+), which modulate BK function, and second messengers activated by G-protein-coupled receptors. Because the BK channel contains both voltage- and ligand-gating domains, it is activated by both depolarization and elevation in intracellular Ca^{++} concentration. Calcium influx that activates the BK channel and provokes local hyperpolarization of the plasma membrane may facilitate endocytosis.

Intracellular H^+ stimulates BK, thus linking membrane excitability to cell metabolism owing to a common motif that mediates stimulation by both H^+ and Ca^{++}.[164] Low intracellular pH facilitates channel opening.[165]

BK Channel in the Carotid Body

Constitutive heme oxygenase-2 is an oxygen sensor for the BK channel [404]. Modulation of BK actvity by oxygen is indeed crucial in the carotid body. Heme oxygenase-2 forms a complex with the BK channel to enhance the channel activity in normoxia. Inhibition of the BK channel by hypoxia augments by HO2 activity.

Carbon monoxide is not only an odourless and tasteless poison gas released during the incomplete combustion of carbon-based fuels, but also a cellular messenger and regulator of ion channels, such as calcium-activated K^+ (BK), voltage-activated $K_V2.1$ and Ca_V1, ligand-gated ionotropic P2X (e.g., $P2X_2$ and $P2X_4$), TREK1, and epithelial Na^+ (ENaC) channels. It can be produced from the catalysis of heme by heme oxygenase (HO) [405, 406].[166] Carbon monoxide can either activates, inhibits, or both activates and inhibits according to the circumstances the channel functioning.

163. Usually, myristoylation targets an N-terminal glycine (irreversible N-terminal myristoylation) and rarely an internal amino acid residue (internal myristoylation).

164. Histidine residues of RCK1 domain serve as H^+ sensors and participate in Ca^{++} activation [403].

165. Activation of BK channel by H^+ requires electrostatic interactions between histidine residues and an adjacent negatively charged residue in Ca^{++}-sensing segment.

166. Heme is a moiety (part of a molecule) of different hemoproteins, such as hemoglobin, cytochrome-C, and soluble guanylate cyclase. Its turnover is then an essential feature of organism life. Heme oxygenase catalyzes heme degradation to CO, ferrous Fe^{++} cation, and biliverdin. Biliverdin is converted to bilirubin by biliverdin reductase; these molecules are potent cellular anti-oxidants. Three paralogs of HO exist (HO1–HO3): 2 active isoforms and a probable pseudogene (HO3). Ubiquitous isoform HO1 is inducible (stress response protein). The other active isoform, HO2, is highly expressed in endothelial and smooth muscle cells as

BK Channel in Neurons

Under normal conditions, BK activation requires adequate Ca^{++} concentrations, hence Ca^{++} sources must be in close proximity. In neurons, the BK channel can complex with $Ca_V1.2$, $Ca_V2.1$, and $Ca_V2.2$ [129]. Owing to the BK–Ca_V complexes, Ca^{++} influx through Ca_V channels quickly (submillisecond kinetics) activates BK channels. The BK–Ca_V complexes enable BK-mediated membrane hyperpolarization that controls neuronal firing pattern and the release of hormones and transmitters in the central nervous system.

BK Channel in Smooth Muscle Cells

Plasmalemmal BK channels of smooth muscle cells are activated by localized calcium sparks produced by ryanodine receptors. Tetrameric BK channels regulate the vascular smooth muscle tone. It is known for sustaining cerebral and coronary arterial tone and for its linkage to vasodilatory β-adrenergic receptors as well as counterbalancing vasoconstrictory receptors.

In vascular smooth muscle cells, Ca^{++}-permeable transient receptor potential canonical TRPC1 forms a signaling complex with BK that causes membrane hyperpolarization [407]. In contractile smooth muscle cells, hyperpolarization counterbalances depolarization, thereby limiting Ca^{++} influx through Ca_V1 and preventing excessive contraction of smooth muscle cells by contractile agents.

Vascular smooth muscle cells with a contractile phenotype express $K_{Ca}1.1$ (BK) channels. In contractile smooth muscle cells of rat basilar artery, epidermal growth factor stimulates $K_{Ca}1.1$ and causes a membrane hyperpolarization required for cell proliferation. Among EGFR ligands, epidermal growth factor, heparin-binding EGF-like growth factor, and transforming growth factor-α provoke a 20% increase in $K_{Ca}1.1$ current [408]. Signaling by EGFR in contractile vascular smooth muscle cells is mediated by adenylate cyclase AC5 and protein kinase-A.

When smooth muscle cells switch from contractile to proliferative phenotype, the plasmalemmal density of BK as well as that of Ca_V1 decays, whereas that of intermediate-conductance, Ca^{++}-activated K^+ channels (IK_{Ca} or $K_{Ca}3.1$) rises [409]. Repressor element RE1-silencing transcription factor (REST) [167] that hampers the KCNN4 gene, hence $K_{Ca}3.1$ synthesis, indeed, declines during cell proliferation. Unlike BK channels, IK channels that do not depend on depolarization can open at more hyperpolarized membrane potentials. They enhance the electrochemical gradient for Ca^{++} entry through voltage-independent Ca^{++} channels.

Channel $K_{Ca}1.1$ can connect to vasoconstrictory thromboxane-A2 receptors (TP). Liganded TP receptor trans-inhibits BK independently of G-protein activation to support vasoconstriction. [410].

---

well as astrocytes (but not HO1). Subtype HO2 is constitutively expressed in many (but not all) cell types (neurons, glial cells, renal and airway epithelial cells, and type-1 cells of the carotid body.

167. A.k.a. neural-restrictive silencer factor (NRSF).

3.4.6.2 Intermediate-Conductance Calcium-Gated Potassium Channels

Intermediate-conductance, Ca^{++}-activated K^+ channel $K_{Ca}3.1$ controls K^+ efflux and maintains erythrocyte water balance [411]. The IK channel also operates in T-lymphocyte activation. It also resides in B lymphocytes, endothelial cells, fibroblasts, macrophages, and dedifferentiated smooth muscle cells. The IK channel contributes to proliferation of B and T lymphocytes, fibroblasts, macrophages, and vascular endothelial and smooth muscle cells. It also intervenes in the migration of smooth muscle cells and macrophages as well as platelet aggregation. It causes a hyperpolarization upon excitation by acetylcholine and endothelium-derived hyperpolarizing factor [412].

In the blood vessel wall, $K_{Ca}3.1$ channel operates in vasodilation as well as intimal hyperplasia (Vol. 6 – Chap. 7. Vascular Diseases). Endothelial cells produce abundantly $K_{Ca}3.1$, whereas smooth muscle cells and fibroblasts often produce a tiny amount. [168]

In endothelial cells, fibroblast growth factor-2 and vascular endothelial growth factor upregulate $K_{Ca}3.1$ production [412]. Channel $K_{Ca}3.1$ participates with channel SK ($K_{Ca}2$) [169] in non-nitric oxide-, non-prostacyclin-mediated, endothelium-dependent vasodilation (Vol. 5 – Chaps. 8. Smooth Muscle Cells and 9. Endothelium), in which several possible factors intervene (e.g., K^+ flux, epoxyeicosatrienoic acids, cAMP, cytochrome-P450-2c, and hydrogen peroxide). Whereas, aorta contains a large amount of nitric oxide synthase NOS3, distal mesenteric arteries have a relatively poor NOS3 content. However, $K_{Ca}3.1$ is also needed for NO-mediated vasodilation.

Activated epidermal growth factor receptor (EGFR; Sect. 8.2.5.2) induces a sustained increase in intermediate-conductance $K_{Ca}3.1$ channels for cell proliferation [408].

In smooth muscle cells, $K_{Ca}3.1$ can directly inhibit $K_{Ca}1.1$. Calcium influx through Ca_V1 channels upregulate smooth muscle-specific genes via the RoCK–myocardin–serum response factor pathway; this influx is repressed by K^+ efflux through $K_{Ca}3.1$ and subsequent hyperpolarization (Fig. 3.3). Channel $K_{Ca}3.1$ is also necessary for growth factor-induced smooth muscle phenotype modulation mediated by Activator protein-1 [412]. Growth factors bind to their receptors to initiate a signaling cascade in which Activator protein-1, in concert with suppression of inhibition caused by Repressor element-1 silencing transcription factor, upregulates IK channel.

3.4.6.3 Small-Conductance Calcium-Gated Potassium Channels

Small-conductance, Ca^{++}-activated, voltage-independent K^+ channels (SK or $K_{Ca}2$) mediate afterhyperpolarizations in neurons and repolarization in cardiomyocytes. The SK channel is gated solely by intracellular Ca^{++} in the submicromolar

168. Differentiated, quiescent smooth muscle cells express predominantly $K_{Ca}1.1$ (BK) channel.

169. Both channels $K_{Ca}2$ and $K_{Ca}3.1$ are located with connexins in myoendothelial junctions.

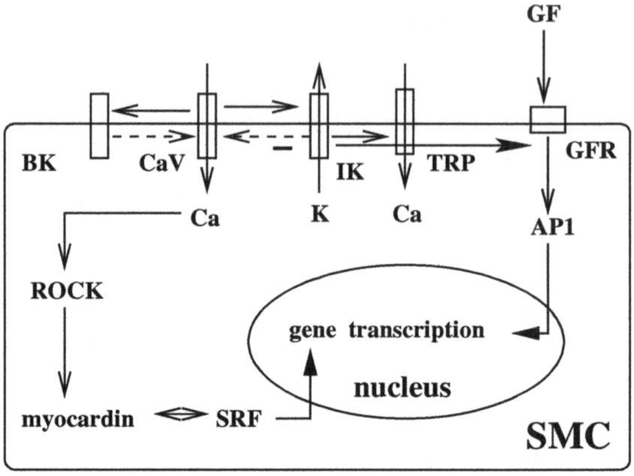

Figure 3.3. Potassium channel IK in smooth muscle cell (SMC; Source: [412]). Intermediate-conductance, calcium-gated potassium channel impedes the activity of Ca_V1 channel, but favors that of transient receptor channel (TRP). Like IK channel, large-conductance calcium-gated potassium channel (BK) that is stimulated by the opening of Ca_V precludes Ca_V function. Calcium entry triggers the activation of RoCK kinase that targets the regulatory cofactor myocardin and transcription factor serum response factor (SRF), which are needed for SMC differentiation. Potassium IK channel also intervenes in growth factor (GF)-induced modulation of SMC phenotype mediated by transcription factor Activator protein-1 (AP1). Upon stimulation by ligand-bound growth factor receptor (GFR), AP1 undergoes nuclear import to activite its specific genes, particularly IK channel gene.

Table 3.36. Potassium intermediate–small conductance calcium-activated channel subunits (IKCA: intermediate conductance calcium-activated potassium channel protein).

Subunit	Aliases	Gene	Partner
SK1	$K_{Ca}2.1$	KCNN1	Calmodulin
SK2	$K_{Ca}2.2$	KCNN2	Calmodulin
SK3	$K_{Ca}2.3$	KCNN3	Calmodulin
SK4	$K_{Ca}3.1$, IKCa1	KCNN4	Calmodulin

range. Like many ion channels, it forms molecular complexes not only with pore-forming and auxiliary subunits, but also scaffold, cytoskeletal, and regulatory proteins.

The SK channels are tetramers of α subunits (Table 3.36). Each α subunit contains 6 transmembrane domains (TM1–TM6), a pore-forming loop (P), and intracellular N- and C-termini. Calmodulin and protein phosphatase-2A tether to adjoining sites of the C-terminus.

The SK1 subunit is encoded by the KCNN1 gene. The SK2 and SK3 subunits encoded by the genes KCNN2 and KCNN3, respectively, form homo- and heteromeric channels in the brain [356]. An SK2 isoform has an extended N-terminus.

The SK channel senses intracellular Ca^{++} concentration via calmodulin. The binding of calcium–calmodulin complex to the SK channel provokes a rapid channel opening. The SK channel is gated mainly by intracellular Ca^{++} ions. It actually has a high selectivity for Ca^{++} over other divalent ions.[170] Channel activation is followed by a slow deactivation.

The SK channel is constitutively assembled with CK2 casein kinase and PP2 phosphatase. It is regulated not only by its gating sensor calmodulin, but also by CK2 kinase and PP2 phosphatase that modulate its activity. The CK2 kinase actually does not phosphorylate SK, but SK-bound calmodulin only when SK channel is closed, hence accelerating SK deactivation and reducing its Ca^{++} sensitivity [356]. Conversely, PP2 dephosphorylates SK-bound calmodulin and raises the Ca^{++} sensitivity of SK channel.

Adequate ion channel exocytosis and recycling endocytosis via early endosome back to the cell surface is responsible for precise channel localization and density in the plasma membrane. Plasmalemmal insertion of $K_{Ca}2.2$ depends on ^{F}actin crosslinker α-actinin-2, like Ca_V1 channel [414]. Interaction with cytoskeletal α-actinin-2 allows endocytosis-mediated recycling toward the plasma membrane.

SK Channel in the Nervous System

Calcium influx through voltage-gated R-type Ca^{++} channels and, in dendritic spines, ^{N}methyl ^{D}aspartate receptors (NMDAR)[171] increases SK activity. Upon long-term potentiation, spine SK channels are phosphorylated by protein kinase-A to rapidly cause endocytosis. On the other hand, SK repolarizing activity attenuates α-amino 3-hydroxy 5-methyl 4-isoxazole propionic acid (AMPAR)-mediated spine depolarization as well as depolarization-induced unblocking of NMDARs by external Mg^{++} [356].

Small conductance Ca^{++}-activated potassium channels lodge in glutamatergic synapses, where they limit synaptic transmission and remodeling. Activated SK channels reduce excitatory postsynaptic potentials. In the basolateral amygdala, activation of β adrenoceptors by noradrenaline targets SK channels via activated protein kinase-A for enhanced long-term potentiation at glutamatergic synapses and emotional memory (fear learning) [415]. Channel SK is constitutively recycled from the postsynaptic membrane. Activated β adrenoceptors remove SK from excitatory synapses, thereby enhancing synaptic transmission.

Synaptic Transmission Modulation and Synaptic Remodeling

In neurons, both activated GIRK and SK channels hyperpolarize the membrane, hence inhibiting neuronal excitability. These voltage-independent ion channels signal via molecular complexes in membrane nanodomains. G-protein-coupled inwardly rectifying and small-conductance Ca^{++}-activated K^{+} channels are located in

170. Trivalent ions, such as terbium (Tb^{3+}) and europium (Eu^{3+}), can be used as alternative ligands for Ca^{++} to study Ca^{++}-binding proteins. Nanomolar concentrations of Ca^{++} substitute lanthanide ion that has an apparent affinity more than 100-fold higher than Ca^{++} binds to the same site as Ca^{++} to activate the channel [413].

171. The SK2 channel colocalizes with ^{NMDA}Glu receptors in postsynaptic densities.

postsynaptic densities of dendritic branches, spines, and shafts in many central neurons. Their activities depend on their interactions with multiple co-assembled partners. Like α-amino 3-hydroxy 5-methyl 4-isoxazole propionic acid receptors (AMPAR), GIRK, and SK channels undergo rapid transport (GIRK and SK are inhibitory channels, whereas AMPARs are excitatory channels).

3.4.7 Sodium-Activated Potassium Channels

Slack (sequence-like a calcium-activated K^+ channel or Slo2.2) and Slick (sequence like an intermediate conductance K^+ channel or Slo2.1) high-conductance, Na^+-activated K^+ channels (K_{Na}) have been originally detected in cardiomyocytes, in which they may provide protection against ischemia that leads to an increase in intracellular Na^+ concentration.

Sodium-activated potassium channels are also produced at high levels in neurons of the medial nucleus of the trapezoid body in the auditory brainstem, where they contribute to timing at high frequencies [416].[172] In addition, Slack and Slick channels contribute to setting and stabilization of the resting membrane potential in certain dorsal root ganglion neurons [417].

Channels Slack and Slick are gated by intracellular Na^+ and Cl^- rather than Ca^{++} ions. Moreover, Slack and Slick channels are activated and inhibited by neuromodulators via Gq-protein-coupled receptors, such as M_1 muscarinic acetylcholine and $mGlu_1$ metabotropic glutamate receptor, respectively [418].

Fragile X mental retardation protein-1 (FMR1 or FMRP) that binds to RNA and cytosolic proteins involved in protein synthesis also targets Slack channel [419]. This channel maintains neuronal activity during high-frequency stimulation. Protein FMRP thus allows a sustained channel activity during repetitive neuronal stimulation.

3.4.8 Hyperpolarization-Activated Cyclic Nucleotide-Gated Potassium Channels

Among members of the family of cyclic nucleotide-gated channels (CNG), hyperpolarization-activated, cyclic nucleotide-gated K^+ channels (HCN) comprise 4 known isotypes (HCN1–HCN4; Sect. 2.4).

These 2 families of channels are activated by ligands such as cyclic adenosine (cAMP) or guanosine (cGMP) monophosphate. They can then serve as switches that transduce changes in intracellular concentrations of cyclic nucleotides to initiate variation in membrane potential and calcium concentration.

They can also be gated by plasmalemmal voltage variations. Both families actually belong to the class of voltage-dependent K^+ channels that contain 6 transmembrane segments (TM1–TM6) with a positively charged TM4 segment and an ion pore between TM5 and TM6. However, they are not selective for K^+ ions.

---

172. A given neuron exhibits regular spiking, burst firing, adaptation, or high-frequency firing according to signaling context and activity of specific ion channels synthesized by this neuron

A tetrameric, cGMP-gated, K^+-selective channel (tetraKCNG) that is made of 4 KCNG domains exists in sperm [420]. Each KCNG domain has 6 transmembrane segments, an ion pore with K^+ selectivity (GYGD), and a cyclic nucleotide-binding domain (CNBD).

3.4.9 Potassium Channels of the TWIK Subclass

Background potassium channels are composed of K_{2P} channel subunits, also called tandem of P domains in a weak inwardly rectifying K^+ channel (TWIK) subunits, that are encoded by the KCNK*i* genes.

Dimeric 2-pore-domain potassium channels have unusual gating properties, enabling background activity and membrane-stretch sensitivity. The K_{2P} channel subunits are subdivided into 6 main structural subclasses [421]: subclass 1 with TWIKs; 2 with TWIK-related K^+ (TREK) and TWIK-related arachidonic acid-stimulated (TRAAK); 3 with TWIK-related acid-sensitive K^+ (TASK); 4 with tandem-pore-domain halothane-inhibited K^+ (THIK); 5 with TWIK-related alkaline-pH-activated K^+ (TALK); and 6 with TWIK-related spinal cord K^+ (TRESK) channels.

Whereas the TWIK1 channel generates weakly inwardly rectifying K^+ currents, the TREK1 channel produces outwardly rectifying K^+ currents [422]. Both channels are open at the resting potential and control the resting membrane potential near the K^+ equilibrium potential in a numerous cell types. The TASK1 channel is the first cloned mammalian voltage-insensitive channel that possesses all the characteristics of background conductance and serves as a sensor of external pH variations [422].

Small ubiquitin modifier SUMo1 regulates dimeric $K_{2P}1$ channels. Agent SUMo1 and SUMo-activating- SAE1 and -conjugating enzyme UbC9 colocalize with $K_{2P}1$ channels in the plasma membrane. The $K_{2P}1$ channel is activated by SUMo isopeptidase sentrin-specific protease SenP1 and inhibited by SUMo1 [423].

In the central nervous system, the neuronal network responsible for the respiratory rhythm comprises several groups of respiratory neurons that spread from the caudal ventrolateral medulla to the dorsolateral pons. The activity of this network adjusts to variations of O_2, CO_2, and H^+ during diverse conditions (e.g., sleep, exercise, or high altitude). Changes in arterial CO_2 and pH as well as acute O_2 variations are detected by peripheral chemoreceptors, mainly carotid bodies; long periods of hypoxemia as well as variations in CO_2 and pH by regions of the brainstem, such as the raphe, parafacial respiratory group (pFRG), and retrotrapezoid nucleus [424]. Channels K_{2P} — TASK1 to -3 — that are encoded by the genes KCNK3, KCNK5, and KCNK9 produce background K^+ currents that are inhibited by hypoxia and acidification. Oxygen-sensitive background K^+ currents in carotid-body type-1 cells are generated by TASK1 and TASK3 channels. Channels TASK1 and TASK3 are also expressed in multiple populations of chemosensitive neurons (medullary raphe, retrotrapezoid nucleus, pre-Bötzinger and Bötzinger complexes, lateral reticular nucleus, hypoglossal motoneurons, and locus coeruleus) [424].

Acid- and volume-sensitive TASK2 K^+ channel abounds in renal proximal tubules and papillary collecting ducts, where it intervenes in bicarbonate reabsorption, as well as in salivary glands and colon. In nephron proximal tubules, luminal

Na^+–H^+ exchanger and H^+ ATPase, carbonic anhydrases-2 and -4, and basolateral HCO_3^- transporter are functionally coupled to transport HCO_3^- across the epithelium [425]. Several K^+ channels reside in proximal tubules, such as $K_V7.1$–MinK, $K_V1.8$, TASK2 ($K_{2P}5.1$), TREK2b (a splice variant of TREK2 or $K_{2P}10.1$), TWIK1 ($K_{2P}1.1$), $K_{IR}2.1$, $K_{IR}4.2$, $K_{IR}5.1$), and $K_{IR}7.1$.

In the central nervous system, TASK2 is restricted to a few brainstem nuclei, including the ventral medullary surface, where it acts as a central H^+, O_2, and CO_2 sensor for respiratory adaptation to hypoxia and hypercapnia [424]. Reactive oxygen species that are generated during hypoxia activate TASK2 channels.

The TREK proteins are 2-pore-domain K_{2P} channels, which produce outwardly rectifying currents. They are stimulated by mechanical stretch, intracellular acidification, temperature, polyunsaturated fatty acids (arachidonic acid, lysophospholipids, phosphatidylinositol (4,5)-bisphosphate) [426]. They are inhibited by G-protein-coupled receptors activated by neurotransmitters and hormones.

Voltage-independent TREK1 with 4 transmembrane elements is activated by intracellular H^+.[173] A-kinase-anchoring protein AKAP5, a constituent of TREK1 channel,[174] transforms low-activity outwardly rectifying current into robust leak conductance insensitive to stimulators but hampered by GPCRs.[175]

Background potassium channel TREK1 of endothelial cells in small mesenteric arteries and skin vessels is regulated by multiple chemical (hormones and neurotransmitters acting via GPCRs, such as serotonin and intracellular pH) and physical stimuli [421].[176]

In endothelial cells, vasodilators acetylcholine and bradykinin provoke the production of nitric oxide via TREK1 [427]. Nitric oxide acts on underlying smooth muscle cells, inducing a relaxation.

173. Among the large set of distinct types of K^+ channels, only a few are activated by intracellular H^+. In addition, most K^+ channels are inhibited by H^+.

174. The TREK1 channel encoded by the KCNK2 gene is highly produced in the brain, particularly in γ-aminobutyric acid-containing neurons of the caudate nucleus and putamen. It is also found in the prefrontal cortex, hippocampus, hypothalamus, midbrain serotonergic neurons of the dorsal raphé nucleus, and sensory neurons of the dorsal root ganglia. It also lodges in peripheral tissues.

175. Inhibition of the TREK1–AKAP5 complex by Gs-coupled receptors is faster, but has the same effect as that of TREK1 alone. Inhibition of TREK1–AKAP5 by Gq-coupled receptors is much more reduced than that of TREK1.

176. The TREK1 channel is activated by intracellular acidosis, depolarization, high temperature, and stretch. It is closed by the PKA and PKC pathways. It is inhibited by stimulated GPCRs (Gs and Gq). Its activity is modulated by polyunsaturated fatty acids such as arachidonic acid, which reversibly open TREK1 channels. Hydrolysis of $PI(4,5)P_2$ by PLC shifts TREK1 voltage-dependent activation toward more depolarized voltages. Mechanical and lipid activations of TREK1 can be functionally linked. The C-terminal domain of TREK1 regulates the voltage- and time-dependent gating. A-kinase anchoring protein AKAP5 interacts with TREK1 and modifies its regulation. Adaptor AKAP5 augments the inhibition of TREK1 by Gs-coupled receptors and reverses the downmodulation of TREK1 by Gq-coupled receptors [421].

3.5 Chloride Carriers

Chloride channels contribute to membrane excitability (especially in neurons and smooth muscle cells), transepithelial transport, regulation of cell volume, pH control of intracellular organelles, cell cycle, and apoptosis. Chloride flux is also required for olfactory perception, regulation of vascular tone, and epithelial electrolyte and fluid secretion.

Chloride channels constitute many families: (1) γ-aminobutyric acid and glycine receptors–channels; (2) Ca^{++}-activated channels anoctamins; (3) voltage-sensitive channels (ClC1–ClC9); (4) volume-regulated channels; (5) high (maxi)-conductance channels; (6) bestrophins (Best1–Best4); and (7) cystic fibrosis transmembrane conductance regulator (CFTR). In the heart, CFTR mediates protein kinase-A-stimulated Cl^- currents.

According to the activation mode, 3 main types of chloride channels exist: volume-regulated, voltage-gated, and ligand-gated.[177] Chloride channels are thus also classified as: (1) voltage-gated Cl^- channels (ClC1–ClC7 and ClCk1–ClCk2); (2) calcium-activated Cl^- channels (CaCC); (3) high-conductance Cl^- channels (maxiCl); (4) cystic fibrosis transmembrane conductance regulator (CFTR); and (5) volume-regulated anion channels (VRAC).

Voltage-gated chloride channels ClCk1 and ClCk2 as well as calcium-activated Cl^- channels abound in the kidney. Intracellular chloride channels (ClIC) are located in the nuclear membrane. These channels could be involved in cell cycle regulation. Eicosanoid-modulated chloride channels are directly gated by epoxyeicosatrienoic acids.

3.5.1 Voltage-Gated Chloride Channels

Voltage-gated ClC channels belong to the largest chloride channel class. Nine identified ClC channels are encoded by the CLCN family genes (CLCN1–CLCN7 and CLCNKA–CLCNKB). They can be subdivided into 3 groups [5] (Table 3.37): group 1 with plasmalemmal channels ClC1 and ClC2 as well as human ClCk1 and ClCk2 (or ClCka and ClCkb); 2 with ClC3 to ClC5 that are predominantly intracellular; and 3 with ClC6 and ClC7 that are also intracellular. Alternative splicing increases the structural diversity.

Group-2 and -3 ClC channels operate as Cl^-–H^+ antiporters, or exchangers, rather than classical Cl^- channels (2:1 stoichiometry). Therefore, the ClC class includes both Cl^- channels and Cl^-–H^+ antiporters that can be grouped into 2 functional sets: (1) channels that enable a passive diffusion down the Cl^- electrochemical gradient and (2) secondary active transporters that couple Cl^- motion in one

177. γ-Aminobutyric acid (GABA) and glycine channels in the postsynaptic neuron and skeletal muscles belong to the family of chloride channels. Mediator GABA is synthezised from glutamate. Glycine channels are mainly located in the brain. Other types include cyclic adenosine monophosphate-activated, calcium-activated, and calcium-dependent ATP/UTP-activated chloride channels.

Table 3.37. Voltage-gated ClC channels (Source: [5]; AA: arachidonic acid). Different ClC class members act as chloride ion channels or as chloride–proton exchangers that are responsible for passive and active ion movement, respectively.

Type	Activators	Functional characteristics
		Group 1 (plasmalemmal)
ClC1	Constitutively active	Depolarization gating; inwardly rectifying; incomplete deactivation upon repolarization
ClC2	AA, amidation	Hyperpolarization gating; inward rectification; activation by cell swelling, PKA, and weak extracellular acidosis; potentiation by SGK1; inhibition by CDK1–cyclin-B
		(Kidney subgroup
ClCk1	Constitutively active (barttin)	Inhibition by extracellular acidosis; potentiation by extracellular Ca^{++}
ClCk2	Constitutively active (barttin)	Bidirectional rectification; inhibition by extracellular acidosis; potentiation by extracellular Ca^{++}
		Group 2 (predominantly intracellular)
ClC3		Cl^-–H^+ antiporter; outward rectification; potentiation by CamK3; inhibition by I(3,4,5,6)P4 and external acidosis
ClC4		Cl^-–H^+ antiporter; outward rectification; inhibition by extracellular acidosis;
ClC5		Cl^-–H^+ antiporter; outward rectification; potentiation by intracellular acidosis; inhibition by extracellular acidosis
		Group 1 (intracellular)
ClC6		Cl^-–H^+ antiporter
ClC7		Cl^-–H^+ antiporter

direction (downhill movement) to H^+ displacement in the opposite direction (uphill movement).[178] Nevertheless, owing to the conservation of specific amino acids, these functionally distinct ClCs share the same basic architecture and can then be members of the same class of ion carriers.

The ClC channels participate in the maintenance of the electrical stability of the resting myocyte sarcolemma, regulation of transepithelial fluid flow through renal cells, acidification of endosomal and lysosomal content. They are activated by strong hyperpolarization or cell volume changes, but not ligand binding.

The ClC proteins are homodimers with a separate ion track within each subunit. Each ClC subunit with its cytoplasmic C-terminus and transmembrane region

178. In ClC exchangers, the Cl^- gradient generates uphill H^+ transport; the H^+ gradient creates a Cl^- transport with a coupling ratio of 2 Cl^- for 1 H^+ under normal conditions of transmembrane voltage and ion concentrations.

actually contributes to a single pore and dimeric ClC channel thus contains 2 independently gated pores, although they can function simultaneously [5].

Movements of a single acidic side chain of a gating glutamate enable the ion transport through the aqueous pore in the center of the protein and the coupling between H^+ and Cl^- in Cl^-–H^+ antiporter. The cytoplasmic regulatory domains interact with the transmembrane domains. Conformational changes can thus be transmitted to the ion pore. Yet, a minimal conformational change can ensure the ion transfer. The side chain of the gating glutamate indeed alternately allows access to one side of the membrane or the other through the ion permeation channel [428].

Slightly outward rectifying chloride channel ClCk1 is a kidney-specific channel involved in transepithelial chloride transport in the inner medullary thin ascending limb of Henle's loop (Vols. 2 – Chap. 1. Remote Control Cells and 5 – Chap. 3. Cardiovascular Physiology).

3.5.1.1 ClC0 Carrier

Gating (opening and closing) of the ClC0 channel involves the transmembrane flux of protons, so that this channel operates as an active transporter [429].

3.5.1.2 ClC1 Channel

The ClC1 channel regulates the electric excitability of the skeletal myocyte membrane. It accounts for 75% of the membrane conductance at rest in skeletal myocyte [5]. It is also expressed in the sarco(endo)plasmic reticulum of skeletal myocytes.

3.5.1.3 ClC2 Channel

The ClC2 channel is ubiquitously synthesized, particularly in pulmonary epithelia. It can be activated by strong hyperpolarization and cell swelling [5]. Nonetheless, it does not correspond to volume-regulated anion channel.

3.5.1.4 ClCk Channels

Chloride channels ClCk1 and ClCk2 in renal and inner ear epithelia are crucial for renal Cl^- reabsorption and inner ear K^+ secretion [430]. They require the accessory Cl^- channel β subunit *barttin* for adequate tubular salt reabsorption.[179] The

179. The Bartter's syndrome results from an autosomal recessive salt-losing nephropathy. The renal salt loss is caused by an impaired transepithelial reabsorption of sodium chloride in the nephron. Sodium chloride is taken up apically by the combined activity of Na^+–K^+–$2Cl^-$ cotransporters NKCC2 and K^+ channels ROMK. Chloride ions exit from the cell through basolateral ClCk2 channels. Mutations in the 3 corresponding genes provoke Bartter's syndrome. Bartter's syndrome associated with sensorineural deafness (BSND) is caused by mutations in barttin.

ClCk2 channel is non-functional without barttin. On the other hand, ClCk1 homomers are active without barttin, but with a reduced conductance and altered voltage-dependent activation. Barttin modulates not only the exocytosis of ClCk channels, but also their function [431]. Coexpression of barttin enhances plasma membrane insertion. Furthermore, it heightens conductance and gating of ClCk channels.

3.5.1.5 ClC3 Transporter

The ClC3 carrier, or Cl^-–H^+ exchange transporter-3, exists with 2 alternatively spliced isoforms. The ClC3a isoform localizes in late endosomes, where it acts as an anion shunt during acidification. The ClC3b isoform resides in the Golgi body, where it colocalizes with cystic fibrosis transmembrane conductance regulator [432]. Both channel types bind to the Golgi body via Golgi-associated PDZ and coiled-coil motif-containing protein (GoPC). Both channels also connect to Na^+–H^+ exchange regulatory cofactor NHERF1 [180] NHERF3. [181]

3.5.1.6 ClC4 Carrier

The ClC4 transporter, or Cl^-–H^+ exchanger-4, can operate in 2 transport modes: (1) a slippage mode with which ClC4 behaves as an ion channel and (2) an exchanger mode in which unitary transport rate is 10-fold lower [5].

3.5.1.7 ClC5 Carrier

Mutations of the CLCN5 gene cause X-linked familial renal tubular disorder Dent disease that is characterized by low-molecular weight proteinuria, hypercalciuria, and nephrocalcinosis (nephrolithiasis). The ClC5 exchanger is expressed in renal proximal tubule cells. Its highest density is observed below the brush border where endocytotic vesicles accumulate. In this region, ClC5 colocalizes with H^+ ATPase [433]. The ClC5 carrier may contribute to protein endocytosis in proximal tubules.

Endosomal ClC5 carrier exploits the proton gradient generated by a proton pump (H^+ ATPase) to concentrate chloride ions inside endosomes [434]. Chloride concentration rises in exchange for protons through $2\,Cl^-$–H^+ exchanger, hence limiting acidification in endosomes of proximal tubular and intercalated cells.

3.5.1.8 ClC6 Carrier

The ClC6 exchanger is encoded by the CLCN6 gene that generates different alternatively spliced transcript variants. Among several truncated isoforms that result from alternative splicing (ClC6a–ClC6d), ClC6a has a relatively broad expression pattern (brain, kidney, skeletal muscle, pancreas, thymus, and testis), whereas ClC6c is restricted to kidney [435].

180. A.k.a. ERM-binding phosphoprotein EBP50 and SLC9a3r1.
181. A.k.a. SLC9a3r2 and PDZK1.

3.5.1.9 ClC7 Carrier

The ClC7 exchanger links to the β subunit osteopetrosis-associated transmembrane protein OsTM1 that increases its stability [5]. In bone-resorbing osteoclasts that remodel bones together with bone-synthesizing osteoblasts, microphthalmia transcription factor (MiTF) [182] regulates both the CLCN7 and Ostm1 genes during osteoclast maturation [436].

Lysosomal ClC7 carrier allows chloride entry into the organelle lumen toward a positive electrical potential that a proton pump has generated across the membrane [437]. This influx of chloride dissipates the change in potential and limits vesicle acidification.

3.5.2 Chloride Channels of the Anoctamin Family

The anoctamin channel family that comprises 10 detected mammalian members and multiple splice variants is the second largest of the 5 identified Cl^--channel families. [183] In bronchial epithelial cells that experience an intracellular accumulation of calcium ions, Cl^- current can result from the activity of calcium-activated chloride channel TMEM16A (or anoctamin-1) [438]. The TMEM16 channel produces Ca^{++}-activated Cl^- currents with kinetics similar to those of calcium-activated chloride channels [5].

3.5.3 Bestrophins

Bestrophins that are encoded by the genes BEST1 to BEST4 form chloride channels that are activated by Ca^{++} ions [5]. The BEST1 gene [184] is regulated by microphthalmia transcription factor. Bestrophin-1 [185] can interact with protein phosphatase-2 [439].

Bestrophin-2 is mainly synthesized in the colon as well as basolateral membrane of retinal pigment epithelial cells. It contribute to the transfer of water, ions, and metabolites between photoreceptors and choroid.

In vascular smooth muscle cells, fast, strongly inward rectifier bestrophin-3 [186] causes cGMP-dependent Ca^{++}-activated Cl^- current [440]. Bestrophin-4 yields the highest Cl^- flux among the 4 human bestrophins [441].

3.5.4 Maxi and Tweety Homologs

Chloride Maxi channel (pore radius $\sim$ 1.3 nm; Sect. 3.5.4) is a high-conductance, volume- and voltage-dependent, anion-selective, ATP-conductive, large-conductance

182. A.k.a.class-E basic helix-loop-helix protein bHLHe32.
183. Anoctamin channels can serve as biomarkers, as they abound in tumors.
184. The BEST1 gene is mutated in Best macular dystrophy.
185. A.k.a. vitelliform macular dystrophy protein VDM2.
186. A.k.a. vitelliform macular dystrophy-2-like protein-3.

channel. It can also transfer large organic anions, such as aspartate (Asp^-) and glutamate (Glu^-; cut-off radius ~ 0.35 nm), as well as ATP^{4-} and $^{Mg}ATP^{2-}$ (cut-off radius 0.58–0.65 nm).

Anion Maxi channel then serves as a path for regulated ATP release in numerous cell types. In particular, they participate in swelling-induced release of ATP. Astrocytes release gliotransmitters, such as glutamate and ATP. In cultured astrocytes, the massive ATP release happens mainly through gadolinium-sensitive Maxi anion channels, as exocytosis vesicles, gap junction hemi-channels (connexins Cx32, Cx37, and Cx43, as well as pannexin-1), ABC transporters, such as ABCb1 and ABCc1, $P2X_7$ receptor, and anion channels, such as cystic fibrosis transmembrane conductance regulator and volume-sensitive outwardly rectifying chloride channels do not significantly contribute to the process [442].

Anion Maxi channel is widely distributed (skeletal and cardiomyocytes, neurons, glial cells, lymphocytes, secreting and absorbing epithelial cells, renal macula densa cells, and human placenta syncytiotrophoblasts) [5].

In rat cardiomyocytes, Maxi channel mainly resides at openings of transverse tubules, [187] like the ATP-sensitive $K_{IR}6$ channel, and along Z-lines [443]. It coexists with many types of ionotropic P2X and metabotropic P2Y receptors. Its activation produces repolarizing outward current associated with Cl^- influx to maintain the T-tubule membrane potential.

Chloride Maxi channel may be homologous to the plasmalemmal, volume- and voltage-dependent, ATP-conductive, large-conductance, anion channel (VDACL; not the porin ion channel VDAC located on the outer mitochondrial membrane). [188]

In renal macula densa cells, salt stress activates Maxi anion channel and induces massive release of ATP (tubuloglomerular feedback [444]. Anion Maxi channel is stimulated by hypotonic or ischemic stress. It is activated by G-protein-coupled receptors. It is also regulated by annexin-6 [5] as well as cAMP, actin, protein kinase-C, and Ser/Thr protein phosphatases (PP1–PP6, but not Mg^{++}-dependent PPM1 and Mg^{++}- and Ca^{++}-dependent PP7) [444]. Cytosolic ATP^{Mg} prevents the activation of Maxi channel as channel phosphorylation impedes its opening. Anion Maxi channel is activated by protein Tyr phosphatase PTPRb [444].

Tweety Homolog Channels

Human Ca^{++}-independent, volume-sensitive, large-conductance Cl^- channels — the tweety homologs (TtyH1–TtyH3) — constitute a family of Maxi Cl^- channels. Tweety homolog channels are characterized by 5 transmembrane domains, an extracellular N-terminus, and an intracellular C-terminus.

187. Transverse tubules are invaginations of the sarcolemma at the Z-line. Many of the proteins involved in excitation–contraction coupling are concentrated at T-tubules, such as $Ca_V1.2a$ channels.

188. Mitochondrial porin, a voltage-dependent anion channel, can also localize to the plasma membrane. Deletion of all the 3 genes that encode VDAC isoforms do not influence the activity of Maxi anion channel [442].

N-glycosylation can be required for exocytosis to the plasma membrane. Nevertheless, N-glycosylation is not the determining factor for TtyH2 transfer to the plasma membrane [445]. Ubiquitin ligase NEDD4-2 binds to TtyH2 and TtyYH3, but not TtyH1, to regulate their cell surface density [446]. [189]

3.5.5 Volume-Regulated Chloride Channels

Volume-regulated (activated) chloride channel (VRClC), [190] participates in regulatory volume decrease in response to cell swelling. It also contributes to the regulation of membrane excitability, transcellular Cl^- transport, angiogenesis, cell proliferation and death, as well as glutamate release from astrocytes [5].

The VRClC channels (pore radius ~ 0.63 nm) carry monovalent anions. They are activated by stretch of β_1-integrins. On hypotonic stimuli, activated VRACs cause efflux of organic osmolytes, such as amino acids and polyols. Their rate of swelling-induced activation is modulated by intracellular ATP concentration. They are inactivated at positive membrane potentials. They are inhibited by increased intracellular free Mg^{++} concentration. Kinase PKCα is required for their optimal activity. The VRAC protein may not represent a single type of molecule, but various channels.

Cell swelling results from exposure to hyposmotic external medium. This event can stimulate Cl^- intracellular channel ClIC2 (Sect. 3.5.7) that is not a VRAC, Maxi Cl^- channel (Sect. 3.5.4), and nucleotide-sensitive, swelling-dependent Cl^- channels (Sect. 3.5.8).

Cell swelling-activated, intracellular ATP-sensitive, outwardly rectifying Cl^- current $i_{Cl,swell}$ has been observed in both atrial and ventricular myocytes and nodal cells. β_1-Integrin stretch also causes $i_{Cl,swell}$. Cell volume-sensitive Cl^- current contributes to cell volume regulation and modulates cardiac electrical activity and apoptosis.

Signaling mediators comprise angiotensin-2, epidermal growth factor receptor, phosphoinositide 3-kinase, SRC family kinases, focal adhesion kinase, protein Tyr kinase, protein kinase-C, and reactive oxygen species [447]. Sphingosine kinase mediates PI3K activation primed by angiotensin-2 and EGFR hyperactivity in vascular smooth muscle cells. Ceramide elicits ROS production via NADPH oxidase. Alterations in sphingolipid metabolism occur in cardiovascular diseases, such as atherosclerosis, ischemia–reperfusion injury, and congestive heart failure. Among sphingolipids, ceramide and sphingosine are active molecules that accumulate in cardiovascular diseases and cause apoptosis. In rabbit ventriculomyocytes, ceramide elicits persistent, outwardly rectifying Cl^- current [447].

---

189. The NEDD4-2 Ub ligase regulates several ion channels, such as epithelial Na^+ channel, Cl^- channel ClC5, voltage-gated Na^+ channels $Na_V1.2$, $Na_V1.5$, $Na_V1.7$, and $Na_V1.8$, as well as voltage-gated K^+ heteromeric channels $K_V7.2$–$K_V7.3$ and $K_V7.3$–$K_V7.5$ [446].

190. A.k.a. volume-regulated anion channel (VRAC), volume expansion-sensing outwardly rectifying anion channel (VSOR), and volume-sensitive organic osmolyte/anion channel (VSOAC).

Voltage-Dependent Anion Channels

Voltage-dependent anion channels (VDAC) are pore-forming proteins, called porins. Mitochondrial voltage-dependent anion channels ($VDAC^{MT}$) transport various metabolites, such as ATP and ADP, across the mitochondrial outer membrane. The plasmalemmal splice variant ($VDAC1^{(PM)}$) may achieve the same task.

In cultured nasal and tracheal epithelial cells, $VDAC1^{(PM)}$ participates in regulatory volume decrease and ATP release with other carriers [448]. Extracellular ATP released by airway epithelial cells regulates the mucociliary clearance. In airway epithelia, released ATP acts through the $P2Y_2$ receptor to activate phospholipase-C, thereby causing Ca^{++} influx and stimulation of protein kinase-C. These processes elicit Cl^- secretion via Ca^{++}-activated Cl^- and CFTR channels. In addition, they inhibit epithelial sodium channel (ENaC). [191] Therefore, ciliary beat frequency and mucin release increases.

Extracellular ATP is catabolized into adenosine (Sect. 1.5.11). Adenosine, in turn, binds to its cognate A_{2B} receptors (Sect. 7.13.2) to promote cAMP-regulated CFTR activity. Hence, adenosine synergyzes the effect of ATP.

3.5.6 Calcium-Activated Chloride (Pseudo)Channels

Calcium-activated chloride channels (ClCa or CaCC), or calcium-activated chloride channel regulators, are encoded by the genes CLCA1 to CLCA4. They reside in many cell types, where they mediate epithelial secretion, sensory signal transduction, and smooth muscle contraction.

Electrolyte transport by nasal epithelia controls the quantity and composition of the nasal fluid. Calcium-activated Cl^- channels (ClCa1–ClCa3), as well as Na^+–HCO_3^+ (NBC) and K^+–Cl^- cotransporters KCC1 and KCC4, are synthesized in nasal mucosa [449]. The ClCa channels probably operate as cell adhesion proteins or are secreted rather than forming channels.

The ClCa1 channel resides in the intestine [450]. The ClCa2 channel localizes mainly to larynx and tracheobronchial tree [451]. It is also produced in endothelia of small pulmonary arteries and arterioles, as well as subpleural and interlobular venules [452]. It colocalizes with β_4-integrin and may operate in basal cell attachment in stratified epithelia [453]. The β_4Itg–ClCa2 complex serves as lung-specific endothelial cell adhesion molecule (LuECAM1) for blood-borne cancer cells that cause lung metastasis [452]. Therefore, ClCa2 Channel has a dual function, as it carries chloride ions and acts in intercellular adhesion.

The ClCa3 channel that is a structurally divergent member of the CLCA family is expressed in the respiratory tract, spleen, thymus, and mammary gland [454].

191. ENaC channel constitutes the rate-limiting step for sodium reabsorption in epithelial cells that line the distal part of the renal tubule, distal colon, and ducts of several exocrine glands, in addition to airway epithelium. Its activity is stimulated by vasopressin and aldosterone.

The ClCa4 channel is primarily synthesized in the digestive tract, mainly in colon, but also detected in smaller amounts in brain, salivary and mammary glands, and urogenital organs.

Calcium-activated chloride channel subunits include transmembrane protein with unknown function TMem16a and TMem16b among the 10 members of the mammalian TMEM16 family. Transmembrane protein TMem16a localizes to the apical membranes of epithelial cells in exocrine glands and trachea, as well as smooth muscle cells of airway wall and reproductive tracts and interstitial cells of Cajal [192] in the gastrointestinal tract [455]. In the trachea, TMem16a promotes UTP receptor-regulated transepithelial transport and adequate mucus motion. Airway smooth muscle cells have an endogenous pacemaker driven by a Ca^{++} oscillator. [193] The TMem16b protein can be also detected in photoreceptor terminals and olfactory neuron cilia, where TMem16b-containing ClCa channels fulfill the negative and positive feedback regulation in these sensory neurons, respectively.

3.5.7 Chloride Intracellular (Pseudo)Channels

Chloride intracellular channels (ClIC1–ClIC6) contribute to stabilize cell membrane potential as well as to transepithelial transport, maintenance of intracellular pH, and regulation of cell volume. In addition to the chloride channel activity, ubiquitous CLIC family members participate in the cytoskeleton activity, cell cycle control, and cell differentiation. Both ClIC1 and ClIC4 exist in both soluble and membrane-bound configurations.

Many ClICs connect to cytoskeletal proteins. Channels ClIC1 and ClIC5, but not ClIC4, are strongly and reversibly inactivated by ^{F}actin [456]. Therefore, actin-regulated ClICs can modify ion transport during cellular events associated with actin cytoskeleton restructuring.

The ClIC channels bypass the classical secretory pathway and auto-insert directly into membranes. Since intracellular ClICs lack the membrane-spanning domains that characterize ion channels, they may be activators of chloride channels rather than real chloride channels. In fact, several alternative ways in which CLICs can display intracellular ion channel activity exist. As they are channel-forming proteins, they operate as ion channel components or ion channels.

Scaffold proteins serve to localize signaling mediators to specific areas to facilitate their precise organization and transmission of intracellular signaling events. A-Kinase-anchoring proteins anchor protein kinase-A to specific subcellular locations and recruit other signaling proteins. All CLIC family members are able to bind

---

192. Interstitial cells of Cajal of the digestive tract are pacemaker cells that control SMC contraction, as they generate rhythmic slow waves in the electrochemically coupled smooth muscle cells.

193. Neurotransmitters and hormones, such as acetylcholine and noradrenaline, release Ca^{++} from its internal stores. Calcium sparks can then activate CaCC, thereby causing membrane depolarization and further calcium import through voltage-gated calcium channels.

centrosomal giant scaffold AKAP9 [194] located at the centrosome and on the Golgi body.

ClIC1

The ClIC1 channel [195] localizes principally to the cell nucleus. Nevertheless, it carries chloride ion across both nuclear and plasma membranes. It interacts with the transfer protein particle complex TraPPC2 that is involved in targeting and fusion of endoplasmic reticulum–Golgi body transfer vesicles, and A-kinase anchoring protein AKAP9 [59].

ClIC2

The ClIC2 channel is a monomer that has an enzymatic activity (low glutathione peroxidase activity). It is widely distributed in tissues. It impedes activity of cardiac ryanodine receptor, hence Ca^{++} release from cardiac sarcoplasmic reticulum [457].

ClIC3

The ClIC3 channel is predominantly localized in the nucleus. It interacts with mitogen-activated protein kinase MAPK15 [458].

ClIC4

Ubiquitous chloride intracellular channel-4 (ClIC4) is targeted by P53 and MyC transcription factors. This multifunctional protein shuttles between the cytoplasm and nucleus. Cytoplasmic ClIC4 translocates to the nucleus during metabolic stress, DNA damage, growth arrest, and apoptosis.

The ClIC4 channel interacts with transcription factor Schnurri-2 homolog [196] that operates in the bone morphogenetic protein signaling. Transforming growth factor-β promotes the expression of ClIC4 and Schnurri-2 as well as their association in the cytoplasm and translocation to the nucleus [459].

Nuclear ClIC4 tethers to $SMAD2^P$ and $SMAD3^P$ and protects them from dephosphorylation by nuclear phosphatases such as magnesium-dependent protein phosphatase PPM1a [197] Therefore, ClIC4 and Schnurri-2 pertain to the group of nuclear SMAD regulators for precise TGFβ signaling control.

The ClIC4 channel colocalizes with scaffold AKAP9. It resides in mitochondria, cortical actin structures, and nuclear matrix. It can also accumulate in intercellular

194. A.k.a. 350-kDa centrosome and Golgi body-located protein kinase-N-associated protein (CGNAP).

195. A.k.a. nuclear chloride ion channel NCC27.

196. A.k.a. MHC-binding protein MBP2, ZNF40b, and human immunodeficiency virus type-1 enhancer-binding protein HIVEP2.

197. A.k.a. PP2cα.

junctions. In the central nervous system, ClIC4 colocalizes with caveolin and can form a complex with dynamin-1, α tubulin, β actin, creatine kinase, and some 14-3-3 isoforms [460].

ClIC5

The ClIC5 channel is almost equally permeable to Na^+, K^+, and Cl^- ions. The CLIC5 gene encodes 2 isoforms following alternative splicing (ClIC5a, or ClIC5, and ClIC5b). Scaffold AKAP9 links specifically to ClIC5b at the Golgi body as well as, to a lesser extent, centrosome [461, 462].

The ClIC5a channel is a component of an ezrin-containing cytoskeletal complex formed during actin polymerization in apical microvilli. It can contribute to assembly or maintenance of ^{F}actin-based cortical structures, in addition to operating as a chloride channel restricted to intracellular membranes [463].

ClIC6

The ClIC6 channel interacts with dopamine receptors D_2 to D_4 as well as multiple PDZ domain-containing protein MuPP1 [198] and cytoskeletal radixin that links actin to the plasma membrane [464].

3.5.8 Nucleotide-Sensitive Chloride Channels

Two types (ClNS1a–ClNS1b) of nucleotide-sensitive, swelling-dependent, Cl^- channels are encoded by 2 genes (CLNS1A–CLNS1B). Integrin-binding proteins encompass 3 types of molecules: (1) cytoskeletal proteins (e.g., talin and α-actinin); (2) cytosolic proteins (e.g., coregulator integrin-β_3-binding protein [ItgB3BP]); [199] and (3) membrane proteins. The ClNS1a channel can bind to $\alpha_{2B}\beta_3$-integrin in platelets [465].

3.5.9 Cystic Fibrosis Transmembrane Conductance Regulator

Among the set of ATP-binding cassette transporter (ABC), cystic fibrosis transmembrane conductance regulator (Sect. 4.18.8.2) is a cAMP-regulated, Cl^- anion channel of epithelial cell membranes, involved in fluid transport across various epithelia.

Ion channel CFTR also functions as a regulator of other ion channels, in particular Cl^-–HCO_3^- exchanger. In addition, CFTR regulator: (1) inhibits epithelial Na^+ channel (ENaC), Ca^{++}-activated Cl^- channels (ClCa), and volume-regulated anion channel (VRAC); (2) activates outwardly rectifying chloride channel (ORCC

198. Protein MuPP1, or MPDZ, also connects to claudins and serotonin receptors $5HT_{2C}$. It is able to form a complex with synaptic GTPase-activating protein SynGAP (or RasA1) and Ca^{++}–calmodulin-dependent kinase CamK2.
199. A.k.a. nuclear receptor-interacting factor NRIF3 and β3-endonexin.

such as ClC3); (3) enhances sensitivity of the renal outer medullary K^+ channel ROMK2; and (4) regulates TRPV4 that yields the Ca^{++} signal for regulatory volume decrease in airway epithelia [5]. Moreover, CFTR links to chloride–bicarbonate exchangers SLC26a3 and SLC26a6 to mutually enhance their activities. Phosphorylation of CFTR by protein kinase-A facilitates this interaction.

The CFTR channel also transports ATP, sodium and bicarbonate ions, and water. It is activated by protein kinase-A, cAMP, and ATP. It is synthesized in endothelial cells, cardiomyocytes, erythrocytes, and respiratory epithelial cells (for mucus formation), among others. Its permeability depends on animal species. It provokes a slight outward rectification.

Its phosphorylation by protein kinase-A is necessary for its activation by ATP. Protein kinases PKC as well as PKG2 in intestinal epithelial cells stimulate CFTR activity. The CFTR channel is also regulated by several interacting proteins, such as syntaxin-1A, syntaxin-binding protein StxBP1, and PDZ domain-containing proteins (e.g., Na^+–H^+ exchanger regulatory factor NHERF1 [200] and 70-kDa CFTR-associated protein CAP70). [201]

The CFTR channel is not only gated by ligands, but also membrane stretch, independently from cytosolic factors [466]. Its activation by stretch causes chloride transport in human airway epithelial cells. The stretch-mediated activation lowers in the presence of cAMP, another CFTR activator.

3.6 Proton Carriers

Protons, or hydrogen ions, exist in solution almost entirely in hydrated form as hydronium ions (H_3O^+). [202] However, all water-filled pores and proton-selective channels conduct protons as H^+ rather than short-lifetime H_3O^+ ions [467].

Protons have a 5-fold higher conductivity in water than other cations such as K^+. The prototropic transfer [203] corresponds to proton motion via hydrogen-bound chains that account for the high mobility of protons in aqueous solution, as they can hop from 1 water molecule or chain of ionizable amino acids to another, or even use the movement of associated molecules. Water is an ideal medium for prototropic conduction because of its propensity to form hydrogen bonds [467]. Proton transfer via transient protonation of side chains of constituent residue may account for several H^+ transport proteins such H^+ ATPases.

200. A.k.a. sodium–hydrogen antiporter-3 regulator and ERM-binding protein EBP50.

201. Both NHERF1 and CAP70 bind to inducible NOS2 nitric oxide synthase.

202. A.k.a. oxonium and hydroxonium ions. Ionic proton concentration in solution ($[H_3O^+] \sim 40$ nmol) is tiny w.r.t. total hydrogen concentration in water (110 mol) [467]. The average H_3O^+ lifetime O[1 ps]) in liquid water at ambient temperature. Other larger ionic species might exist, as free pairs of electrons on one water molecule and might be able to exert sufficient force on a hydrogen held by a pair of electrons on another water molecule to bind the 2 molecules together. The 2 main larger species are the so-called Zundel cation $H_5O_2^+$ and Eigen cation $H_9O_2^+$.

203. A.k.a. Grotthuss mechanism.

Voltage-gated proton channels represent a subset of proton channels that have voltage-dependent gating. They differ from most ion channels in their extraordinarily high selectivity and tiny conductance.

3.6.1 Voltage-Gated Proton Channels

Voltage-gated, H^+-gradient-sensitive, proton-selective channels (H_V) open with depolarization, according to pH value. The H_V channel opens only when the electrochemical gradient is outward, hence enabling H^+ outflux.

The H_V channels are observed in respiratory epithelia, including lung alveoli, airway glands, kidney, skeletal muscles, and microglia, as well as granulocytes (basophil, eosinophil, and neutrophil), mastocytes, B-and T lymphocytes, and monocytes and macrophages [467].

The H_V channel forms dimers, but the dimerization is not necessary as single subunits can act as proton channels [468]. The H_V dimer has 2 pores similarly to voltage-gated chloride ClC channels.

Each of the 6 transmembrane-spanning channel protomer that constitutes voltage- and ligand-gated cation channels contains a voltage-sensor domain (VSD) that serves to regulate the opening and closing of the central ion-conducting pore. All proton channels conduct protons by a hydrogen-bound chain mechanism in which the proton hops from one water or titratable group to the next (prototropic transfer or Grotthuss mechanism). Yet, transmembrane H^+ flux through voltage-gated H^+ channels is faster than H_3O^+ diffusion rate in water.

Voltage-gated cation channels typically achieve rapid, selective transmembrane ion flux by coordinating the dehydrated permeant ion in a tetrameric selectivity filter structure. The H_V1 channel most likely forms an internal aqueous wire for selective proton transfer.[204] Transfer of H^+ ions occurs in a water wire within the central crevice of the voltage-sensor domain [469].

Arachidonate-activable, NADPH oxidase-associated, voltage-gated proton channel is gated not only by depolarization, but also both external and internal pH. Its external and internal sensing sites are inhibited by extracellular polyvalent cations. Gating can be blocked by Zn^{++}.

On the other hand, elevated cytosolic Ca^{++} concentration increases moderately H^+ flux with an efficiency that depends on intracellular pH (but Ca^{++} does not activate H^+ channel) [467]. Furthermore, H^+ channel activation is enhanced by phosphorylation.

3.6.1.1 Alveolar Epithelium

Alveolar epithelium possesses a high density of H_V channels that can engage in elimination of CO_2 across the thin alveolocapillary barrier. The alveolocapillary membrane separates the blood that flows unidirectionally in the capillary mesh

204. A transient aqueous wire can occur if, due to thermal fluctuations, a chain of water molecules align across the membrane or a channel segment.

around the alveolar sac from the alveolar gas mixture. The composition of alveolar gas relies on diffusion in gaseous phase governed by the Stefan-Maxwell equations (Vols. 1 and 7). Gas diffusion is coupled at the exit of the pulmonary acinus to oscillatory convection described by the Navier-Stokes equations.

Carbon dioxide leaves the blood and crosses the capillary endothelium to reach the alveolar epithelium. In alveolar epithelial cells, carbonic anhydrase-2 catalyzes the conversion of CO_2 and H_2O into H^+ and HCO_3^- ions [467]. These ions diffuse across the cell, H^+ bound to mobile buffer. Hydrogen cation leave the cell by voltage-gated H^+ channels of the apical membrane to enter into the surfactant. Bicarbonate anions exit the cell through Cl^-–HCO_3^- exchangers. Extruded HCO_3^- and H^+ can then recombine to form CO_2 and H_2O in the aqueous subphase. Water is reabsorbed, whereas CO_2 penetrates into the gaseous phase to be exhaled.

3.6.1.2 Hypoxic Pulmonary Vasoconstriction

Pulmonary and systemic arteries differ in their behavior when they experience hypoxia. Pulmonary arteries constrict, whereas systemic arteries relax. Voltage-gated H^+ channel on vascular smooth muscle cells contributes to this process, whereas NADPH oxidase acts as an O_2 sensor in airway chemoreceptors of neuroepithelial bodies [467].

3.6.1.3 Phagocytosis

Phagocytosis of pathogen invaders is initiated by the respiratory burst, during which nicotinamide adenine dinucleotide phosphate (NADPH) oxidase produces large amounts of superoxide anion. When microbes are engulfed by phagocytes (eosinophils, neutrophils, and macrophages), NADPH oxidases assemble in the membrane and begins to produce reactive oxygen species to partake in pathogen killing.

NADPH Oxidase is electrogenic, moving electrons across the membrane. Proton channels open to allow proton flux to compensate electron flux. Oxidation of NADPH releases one proton into the cytoplasm, and regeneration of NADPH produces a second proton via the hexose monophosphate shunt. However, NADPH oxidase operates optimally at neutral pH. Produced protons must then be extruded by proton channels to ensure the synthesis of reactive oxygen species during phagocytosis.

During phagocytosis, intracellular pH evolution is triphasic: (1) rapid, strong acidification; (2) rapid, partial or entire recovery; and (3) slow acidification that may reflect the onset of apoptosis.

Voltage-gated proton channels are the first carriers to respond [470]. Afterward, recovery requires the additional contribution of Na^+–H^+ antiporter, whereas H^+ ATPase does not participate markedly in the cell response.

3.6.2 Proton Pump

Proton pump transports protons across the plasma membrane as well as membranes of cell organelles such as mitochondria. Proton ATPases are classified into type `F` (chloroplast, mitochondria, and bacteria), `V` (vacuolar, or more broadly organelles), and `P` (plants, fungi, bacteria) proton pumps.

3.6.2.1 Subunit F_0 of F_0F_1 ATP Synthase

The F_0F_1 ATP synthase consists of 2 main parts: a subunit F_0 that is embedded in the membrane and an ATP-binding F_1 subunit. The F_0 subunit serves as a proton channel.

In the cell respiration, proton pumps grab H^+ ions from the mitochondrial matrix to release them into the intermembrane space. Multisubunit mitochondrial ATP synthase catalyzes ATP synthesis using an electrochemical gradient of protons across the inner membrane during oxidative phosphorylation.

3.6.2.2 Vacuolar ATPase

Vacuolar ATPase (vATPase) is a multisubunit enzyme that acidifies intracellular organelles, such as endosomes, lysosomes, and secretory vesicles. Vacuolar ATPases also intervene in synaptic transmission from neurons. In addition to pumps, other types of transmembrane ATPases include cotransporters and exchangers (e.g., H^+–K^+ ATPase).

3.6.2.3 Gastric Proton Pump

Gastric proton pump, or hydrogen–potassium ATPase, acidifies stomach content. Heterodimeric H^+–K^+ ATPase contains $\alpha 1$ and β subunits encoded by the ATP4A and ATP4B genes. The α subunit has 10 transmembrane domains with catalytic and pore functions, whereas the β subunit has a single transmembrane domain that serves for intracellular transfer and to stabilize the α subunit.

3.6.2.4 H^+–K^+ ATPase (Non-gastric)

Non-gastric (ouabain-sensitive) H^+–K^+ ATPase has a catalytic α subunit encoded by the ATP12A gene that belongs to the superclass of P-type cation transport ATPases and contributes to K^+ influx. Subunit $\alpha 2$ (ATP12a) can be expressed with $\beta 1$ subunit (encoded by the ATP1B1 gene) of Na^+–K^+ ATPase [5].

Table 3.38. Aminophospholipid-transporting ATPases (Source: [5]; BRIC: benign recurrent intrahepatic cholestasis; FIC: familial intrahepatic cholestasis; PFIC: progressive familial intrahepatic cholestasis [severe form]).

Type	Gene	Other name
ATPase1a	ATP8A1	Chromaffin granule ATPase-2
ATPase1b	ATP8A2	ML1
ATPase1c	ATP8B1	BRIC, FIC1, PFIC1
ATPase1d	ATP8B2	
ATPase1k	ATP8B3	
ATPase1m	ATP8B4	
ATPase2a	ATP9A	
ATPase2b	ATP9B	
ATPase5a	ATP10A	Aminophospholipid translocase-5a, ATPase5.10a, ATPase10c, ATPVA, ATPVC
ATPase5b	ATP10B	ATPase5.10b
ATPase5d	ATP10D	ATPase5.10d
ATPase1h	ATP11A	ATPase6.11a, ATPase1s
ATPase1f	ATP11B	ATPase6.11b, ATPase1r
ATPase1g	ATP11C	ATPase6.11c, ATPase1g, ATPase1q

3.7 Other Types of ATPases

3.7.1 Copper-Transporting ATPases

Copper-transporting ATPases (Cu^{++} ATPase-1 and -2)[205] export Cu^{++} ions across the plasma membrane and import Cu^{++} across intracellular membranes. The ATP7a and ATP7b carriers belong to the superclass of P-type ATPases.

Copper-transporting ATPases contain 8 transmembrane domains. They associate with multiple copper chaperone proteins. Both isoforms interact with anti-oxidant protein AtOx1 and glutaredoxin Glrx1 [59].[206] Copper-transporting ATPase-1 is found in most tissues, but not the liver. In cells, it usually resides in the Golgi body. Copper-transporting ATPase-2 (ATP7b) is a monomer that, in particular, causes efflux of hepatic copper into the bile.

205. A.k.a. ATP7a and ATP7b, respectively, as well as Cu^{++}-transporting α and β polypeptides. They are encoded by the genes ATP7A and ATP7B, mutations of which cause Wilson and Menkes diseases that are inherited disorders of copper metabolism.

206. Protein Glrx1 catalyzes the reduction of disulfide bridges and reverses glutathionylation of proteins to control their activities.

3.7.2 Phospholipid-Translocating Mg^{++} ATPases

Phospholipid-transporting ATPases[207] carry phosphatidylserine and phosphatidylethanolamine from one side of the phospholipid bilayer to the other. They constitute a class of aminophospholipid-transporting ATPases within the superclass of P-type cation-transport ATPases (Table 3.38).

207. A.k.a. Mg^{++} ATPases, aminophospholipid-transporting ATPases, ATP phosphohydrolases, ATP-dependent aminophospholipid translocases, and flippases.

4

Transmembrane Compound Carriers

The majority of biological solutes are charged organic or inorganic molecules. Many hydrophobic molecules bind to plasma proteins, such as albumin and globulins, to be transported throughout the body (Vol. 5 – Chap. 1. Blood). In plasma, lipoproteins carry fatty acid and sterol derivatives. Chaperone proteins, such as fatty acid-binding proteins, allow both trans- and intracellular transport of some hydrophobic molecules.

Any molecule that enters or leaves a cell or, within a cell, an organelle must penetrate membranes that constitute impermeable barriers. With the exception of diatomic gas (oxygen and nitrogen), ions, small molecules, and even some macromolecules must cross membranes via specialized carrying transmembrane proteins. Because molecules that are acquired by cells from their environment or secreted from cells are diverse, many different carriers are encoded by the genome. In addition, membrane carriers regulate absorption, distribution within the body, and excretion of drugs among other substrates.

When a substrate in the extracellular medium connects to the outward-facing binding site, the transporter undergoes a set of conformational changes to create a passage to be crossed until the inward-facing conformation allows substrate release into the cytosol (alternating access).[1]

Nutrient transporter-related receptors, the so-called *transceptors*, are carrier-like proteins with a receptor function, but without transport capacity. They then function as nutrient sensors. Binding of a cognate ligand provokes a conformational change that triggers signaling. In particular, solute carrier family SLC38a2 member, an amino acid transporter, may operate as a transceptor [471].

On the other hand, nutrient transporters can have an additional nutrient signaling function. For example, in astrocytes, glutamate transporter EAAT1 stimulates the ex-

1. The occluded conformation corresponds to the state in which the substrate is shielded from both the extra- and intracellular media. Transporters switch from an outward- to an inward-facing conformation with substrate-binding site oriented toward the extra- and intracellular fluid, respectively.

tracellular signal-regulated kinase signaling in response to its substrate [471] (Vol. 4 – Chap. 5. Mitogen-Activated Protein Kinase Modules).

Pharmacokinetics (PK) [2] is the study of the fate of substances (nutrients, metabolites, hormones and growth factors, as well as toxins, and mainly drugs) administered externally to a living organism. Investigations focus on mechanisms of absorption and tissue distribution of an administered drug, the delay of action and duration of effect, hepatic uptake, processing, and export, other chemical changes experienced by the substance in the body, and effects and renal, biliary, and intestinal routes of excretion of drug catabolites. Compounds transporters (or carriers) of the *ATP-binding cassette* (ABC) and *solute carrier* superclasses convey drugs through biological membranes in both directions (influx and efflux; Table 4.1). Whereas enzymes operate mainly in the compound metabolism, carriers intervene in their absorption, distribution in the body, metabolism, and excretion. In particular, members of the class of solute carrier organic anion transport (SLCO), i.e., of the traditional SLC21 class, or organic anion transporting polypeptide (OATP), carry pharmaceuticals, such as HMGCoA reductase inhibitors (statins), angiotensin-converting enzyme inhibitors, angiotensin receptor antagonists, and cardiac glycosides [472].

4.1 Superclass of Solute Carriers

The solute carrier (SLC) superclass of membrane transporters includes 47 classes and at least 360 members in humans. The SLC nomenclature is based on genes that encode these transporters. Names of SLC members have SLC`ixj` format, with the integer `i`, single letter `x`, and integer `j` that denote the class, family, and subtype, respectively (Tables 4.2 to 4.4).

Transporters from the same SLC gene family share at least 20% amino acid sequence identity. On the other hand, several SLC families consist of members with great functional diversity. [3]

4.2 Class of Solute Carrier Organic Anion Transporters (SLCO)

Human uptake and efflux carriers of the class of solute carrier organic anion transporters (SLCO) include, at least, several members of the subclass 1 (SLCO1a2, SLCO1b1, SLCO1b3, and SLCO1c1), few members of the subclass 2 (SLCO2a1 and SLCO2b1), as well as a representative of classes 3 to 6 (SLCO3a1, SLCO4c1, SLCO5a1, and SLCO6a1) [472]. Apical or basolateral localization of SLCO transporters depends on the mammalian species and, in a given species, on the cell type. In

2. φαρμακον: drug and κινεω: to move, set in motion, disturb, or stir; i.e., "what the body does to the drug".

3. The SLC5 class of Na^+–glucose transporter encompasses not only Na^+-coupled glucose transporters, but also transporters for iodide, choline, vitamins, and short-chain fatty acids.

Table 4.1. Transmembrane transporters of compounds that pertain to the ATP-binding cassette (ABC) and solute carrier superclasses (M: mitochondrium; PM: plasma membrane; V: vesicle). Substrates of exporters, or effluxers, of the ABC superclass include ions, amino acids, peptides, sugars, and other hydrophilic molecules. The ABC superclass consists of primary active transporters. However, some members are subunits of ion channels or constitute ion channels, such as sulfonylurea receptors and cystic fibrosis transmembrane regulator, for which ATP hydrolysis enables the activation of the ion channel represented by the ABC protein itself or its partners. The ABCE and ABCF classes that consist of 1 ABCe and 3 ABCf proteins are sometimes regrouped (the number of ABC classes then equals 6). The SLC superclass comprises facilitative transporters (downhill solute transfer according to the electrochemical gradient) and secondary active transporters (uphill solute flux against the electrochemical gradient by coupling to the downhill cotransport). The class of solute carrier organic anion transport (SLCO), i.e., the traditional SLC21 class, or organic anion transporting polypeptide (OATP), is isolated for a better identification. The SLCO carriers serve in the transmembrane transfer of amphipathic endo- and exogenous organic compounds (bile acids, steroid conjugates, xenobiotics, and drugs). Transport through SLCO carriers is independent of sodium, chloride, and potassium gradients, membrane potential, and ATP level. The electroneutral exchange may result from the coupling of cellular uptake of organic anions to the efflux of neutralizing anions (e.g., bicarbonate, glutathione, or glutathione conjugates). The expression pattern of ABC, SLC, and SLCO transporters varies according to the cell type.

Set	Number of members	Number of subsets	Sites	Effects
ABC	49	7 (A to G)	PM	Export (in humans) Active transport
SLC	316	51 (1 to 51)	PM M,V	Export, import Exchange Passive and secondary active (coupled) transport
SLCO	36	6 (1 to 6)		Uptake, excretion

humans, the tissue distribution of SLCO transporters is either specific to a biological tissue or wide to ubiquitous (Table 4.5).

The SLCO carriers convey endogenous substances (e.g., hormones [thyroxine] and their conjugates [estradiol 17β-glucuronide],[4] steroid conjugates, bile acids, bilirubin, auto- and paracrine eicosanoid lipid messengers [leukotrienes and prostaglandins]), in addition to drugs [472] (Table 4.6). Various SLCO carriers are used for a bidirectional transport. Numerous SLCO carriers transport their substrates more efficiently at low extracellular pH. The material transfer is countered by bicarbonate efflux.

4. Estradiol is conjugated in the liver using sulfate and glucuronide to be afterward excreted via the kidney.

Table 4.2. Solute carrier (SLC) superclass (**Part 1**; Source: Wikipedia).

Name	Transporter type and known isoforms
SLC1	High-affinity glutamate and neutral amino acid transporters (SLC1a1–SLC1a7)
SLC2	Facilitative GLUT transporters (SLC2a1–SLC2a14)
SLC3	Heavy subunits of heterodimeric amino acid transporters (SLC3a1–SLC3a2)
SLC4	Bicarbonate transporters (SLC4a–SLC4a11)
SLC5	Sodium glucose cotransporters (SLC5a1–SLC5a12)
SLC6	Sodium- and chloride- dependent sodium:neurotransmitter symporters (SLC6a1–SLC6a20)
SLC7	Cationic amino acid transporter/glycoprotein-associated (SLC7a1–SLC7a14)
SLC8	Na^+–Ca^{++} exchangers (SLC8a1–SLC8a3)
SLC9	Na^+–H^+ exchangers (SLC9a1–SLC9a11)
SLC10	Sodium bile-salt cotransporters (SLC10a1–SLC10a7)
SLC11	Proton-coupled metal-ion transporters (SLC11a1–SLC11a2)
SLC12	Electroneutral cation-Cl^- cotransporters (SLC12a1–SLC12a9)
SLC13	Na^+-sulfate/carboxylate cotransporters (SLC13a1–SLC13a5)
SLC14	Urea transporters (SLC14a1–SLC14a2)
SLC15	Proton oligopeptide cotransporters (SLC15a1–SLC15a4)

4.3 Amino Acid Transporters

Amino acid transporters are plasmalemmal proteins that carry amino acids through cell membrane. Many belongs to the solute carrier superclass. Solutes transported by SLC superclass members consist of both charged and uncharged organic molecules as well as inorganic ions. The SLC superclass comprises facilitative and secondary active transporters that carry solutes downward and against electrochemical gradient, respectively.

4.3.1 Members of Solute Carrier Superclass

Amino acid transporters of the solute carrier superclass include: (1) high-affinity glutamate and neutral amino acid transporter; (2) heavy subunits of heteromeric

Table 4.3. Solute carrier (SLC) superclass (**Part 2**; Source: Wikipedia).

Name	Transporter type and known isoforms
SLC16	Monocarboxylate transporters (SLC16a1–SLC16a14)
SLC17	Vesicular glutamate transporters (SLC17a1–SLC17a9)
SLC18	Vesicular amine transporters (SLC18a1–SLC18a3)
SLC19	Folate/thiamine transporters (SLC19a1–SLC19a3)
SLC20	Type-3 Na^+–phosphate cotransporters (SLC20a1–SLC20a2)
SLC21	Organic anion transporters (SLCO1a2/b1/b3–b4/c1; SLCO2a1–b1; SLCO3a1; SLCO4a1/c1; SLCO5a1; SLCO6a1)
SLC22	Organic cation/anion/zwitterion transporters (SLC22a1–SLC22a20)
SLC23	Na^+-dependent ascorbic acid transporters (SLC23a1–SLC23a4)
SLC24	Na^+–(Ca^{++}/K^+) exchangers (SLC24a1–SLC24a6)
SLC25	Mitochondrial carriers (SLC25a1–SLC25a46)
SLC26	Multifunctional anion exchangers (SLC26a1–SLC26a11)
SLC27	Fatty acid transport proteins (SLC27a1–SLC27a6)
SLC28	Na^+-coupled nucleoside transporters (SLC28a1–SLC28a3)
SLC29	Facilitative nucleoside transporters (SLC29a1–SLC29a4)
SLC30	Zinc efflux proteins (SLC30a1–SLC30a10)

amino acid transporters; (3) cationic amino acid–glycoprotein-associated transporters; (4) proton oligopeptide cotransporter; (5) vesicular glutamate and amine transporters; (6) multifunctional anion exchanger; (7) vesicular inhibitory amino acid transporter; (8) proton-coupled amino acid transporter; and (9) sodium-coupled neutral amino acid transporter.

The class 1 includes 5 high-affinity glutamate transporters (SLC1a1–SLC1a3, SLC1a6, and SLC1a7) and 2 neutral amino acid transporters (SLC1a4 and SLC1a5). These transporters carry ^{L}isomer of glutamic acid (^{L}Glu), ^{L}Asp, and ^{D}Asp, cotransport 3 Na^+ and 1 H^+, and countertransport 1 K^+ [473]. Neutral amino acid transporters carry small neutral amino acids, such as Ala, Ser, Cys, and Thr in exchange of Na^+ ions.

Table 4.4. Solute carrier (SLC) superclass (**Part 3**; Source: Wikipedia).

Name	Transporter type and known isoforms
SLC31	Copper transporters (SLC31a1)
SLC32	Vesicular inhibitory amino acid transporter (SLC32a1)
SLC33	Acetyl-CoA transporter (SLC33a1)
SLC34	Type-2 Na^+-phosphate cotransporters (SLC34a1–SLC34a3)
SLC35	Nucleoside-sugar transporters (SLC35a1–SLC35a5; SLC35b1– SLC35b4; SLC35c1–SLC35c2; SLC35d1–SLC35d3; SLC35e1–SLC35e4)
SLC36	Proton-coupled amino acid transporters (SLC36a1–SLC36a4)
SLC37	Sugar-phosphate/phosphate exchangers (SLC37a1–SLC37a4)
SLC38	Sodium-coupled neutral amino acid transporters (SLC38a1–SLC38a6)
SLC39	Metal ion transporters (SLC39a1–SLC39a14)
SLC40	Basolateral iron transporter (SLC40a1)
SLC41	MgtE-like magnesium transporters (SLC41a1–SLC41a3)
SLC42	Rh ammonium transporters (RhAG, RhBG, RhCG)
SLC43	Na^+-independent, system-L-like amino acid transporters (SLC43a1–SLC43a3)
SLC44	Choline-like transporters (SLC44a1–SLC44a5)
SLC45	Putative sugar transporters (SLC45a1–SLC45a4)
SLC46	Heme transporters (SLC46a1–SLC46a2)
SLC47	Multidrug and toxin extrusion proteins (SLC47a1–SLC47a2)

Heterodimeric amino acid transporters (HAAT) are composed of light and heavy subunits linked by a disulfide bridge. The heavy and light glycoproteic subunits belong to the SLC3 and SLC7 classes of amino acid transporters, respectively. The HAAT transporters carry small and large neutral amino acids as well as anionic and cationic amino acids [474].

The SLC class 7 is split into 2 families [475]: (1) cationic amino acid transporters (CAAT; SLC7a1–SLC7a4) and (2) glycoprotein-associated amino acid transporters

Table 4.5. Subcellular localization and tissue distribution of some human uptake OATP transporters (Source: [472]; ND: not determined).

Type	Localization	Distribution
SLCO1a2	Apical	Liver (cholangiocytes)
		Small Intestine (enterocytes [villus tip])
	Kidney (distal tubule)	
	ND	Brain (capillary endothelial cells)
SLCO1b1	Basolateral	Liver (hepatocytes)
SLCO1b3	Basolateral	Liver (hepatocytes [centrilobular])
SLCO1c1	ND	Testes (Leydig cells)
SLCO2b1	Basolateral	Liver (hepatocytes)
		Placenta (syncytiotrophoblasts)
	Apical	Small intestine (enterocytes)
	ND	Heart (vascular endothelium)
SLCO4a1	Apical	Placenta (syncytiotrophoblasts)

Table 4.6. Examples of substrates of certain SLCO transporters (Source: [472]).

Type	Substrates
SLCO1a2	Cholic acid, prostaglandin-E2, thyroxine, triiodothyronine
SLCO1b1	Bilirubin and its conjugates, cholic acid, estradiol conjugates, leukotriene-C4, prostaglandin-E2, thromboxane-B2, thyroxine, triiodothyronine
SLCO1b3	Bilirubin conjugates, cholecystokinin-8, leukotriene-C4, thyroxine, triiodothyronine

(GPAAAT; SLC7a5–SLC7a11), which are light, catalytic chains of the heterodimeric amino acid transporters.

Members of the SLC class 15 (SLC15a1–SLC15a4) are electrogenic transporters that use the proton-motive force for uphill transport of short peptides (e.g., free histidine and di- and tripeptides) [476].

Members of the SLC class 17, initially defined as phosphate transporters, carry organic anions [477].

Members of the SLC class 18 convey positively charged amines against an electrochemical gradient across the vesicular membrane by proton pumping into the vesicle owing to an ATPase, at a ratio of one translocated amine per 2 displaced protons [478]. Vesicular amine transporters (VAT) are integral membrane proteins of secretory vesicles in neurons and endocrine cells. Monoamine transporters SLC18a1 and SLC18a2 and acetylcholine transporter SLC18a3 transport their respective cargos into secretory vesicles to discharge them into the extracellular space.

Members of the SLC class 26 transport multiple mono- and divalent anions, such as chloride, sulfate, bicarbonate, formate, oxalate, and hydroxyl ions [479]. Anions are carried by SLC26 class members with variable specificity and electrogenic Cl^-–HCO_3^- and Cl^-–OH^- exchange. [5]

The SLC class 32 contains a single member: the H^+-coupled vesicular inhibitory amino acid transporter or vesicular GABA and glycine transporter [480]. On synaptic vesicles of gabaergic and glycinergic neurons, as well as some endocrine cells, it ensures the H^+ ATPase-driven uptake and subsequent exocytosis of inhibitory amino acids.

Proton-coupled amino acid transporters (PAAT) of the SLC class 36 in endomembranes and the plasma membrane of specialized cells carry neutral amino acids, such as glycine, alanine, and proline in exchange for H^+ [481].

The sodium-coupled neutral amino acid transporters (SNAAT1–SNAAT5) of the SLC class 38 countertransport H^+ in exchange for small, aliphatic amino acids, especially glutamine [482].

4.3.2 Cysteine and Cystine Transporters

Cysteine and cystine, the free thiol- and disulfide-bound forms of the amino acid, respectively, exist in an equilibrium defined by redox conditions in the intra- and extracellular spaces. Oxidizing and reducing environment favor cystine and cysteine, respectively. Cystine and cysteine uptake necessitates distinct transporters. However, cystathionase produces cysteine from methionine.

Cysteine belongs to the set of amino acids (e.g., tryptophan) the level of which is regulated by antigen-presenting cells to participate in the control of T-cell activation [483]. Dendritic cells control the proliferation of T lymphocytes, as they regulate cysteine availability. Moreover, interaction between macrophages, another type of antigen-presenting cell, and T lymphocytes primes generation of cysteine by macrophages. Cysteine is then released for T-cell proliferation. [6]

4.4 Symporters or Secondary Active Transporters

Secondary active transporters (e.g., sodium–leucine symporter [LeuT] and sodium–glucose/galactose symporter SGlT) have alternating structures that achieve alternate opening and closing of inward- and outward-facing cavities from substrate-bound configuration occluded to both sides of the membrane for alternating access to either side of the membrane. Outward- and inward-facing cavities are symmetrically arranged on opposite sides of the membrane. Reciprocal opening and closing of these cavities is synchronized by inverted repeat transmembrane helices 3 and 8 among the helix set [484].

5. Stoichiometry of Cl^-–HCO_3^- exchange differ between SLC26 paralogs. The SLC26a3 transporter carries at least 2 Cl^- per HCO_3^-, whereas SLC26A6 transports at least 2 HCO_3^- per Cl^- ion.

6. Lymphocytes lack both cystine transporter and cystathionase.

The site of the cation uptake and release is also controlled by the conformational changes, because the ion-binding site is located between transmembrane helices TM1 and TM8. The cation-binding sites depend on the transporter type, which can be located on the surface of the outward- or inward-facing cavity.

4.4.1 Sodium–Taurocholate Cotransporter, A SLC10 Symporter

Sodium–taurocholate cotransporting polypeptide NTCP, or SLC10a1, that is encoded by the Slc10A1 gene is a hepatic Na^+–bile acid symporter of the SLC10 class. Its substrates are restricted to unconjugated bile acids, glycine- and taurine-conjugated bile acids, steroid sulfate conjugates, and iodothyronines [485]. It resides in the basolateral (sinusoidal) membrane of hepatocytes, where it extracts conjugated bile acids from portal blood. It is also expressed at a low level on the luminal membrane of pancreatic acinar cells.

4.4.2 Monocarboxylate Transporters, SLC16 Members

The transport of energy-rich compounds, such as lactate, pyruvate, and other monocarboxylates, can contribute to the supply of energy. In fact, the transport of lactic acid across the plasma membrane is necessary for the cell metabolism and pH regulation. Lactic acid produced by glycolysis must be removed; lactic acid must be pick up by cells that utilize it for gluconeogenesis (cells of the liver and kidney) or as a respiratory fuel (myocytes of the heart and muscle).

Monocarboxylates include, in particular lactate, pyruvate, and ketone bodies acetoacetate and β-hydroxybutyrate. The transmembrane transfer of monocarboxylates is ensured by different isoforms of the monocarboxylate transporter (MCT) family of carriers. The monocarboxylate cotransporter family comprises 14 detected members (Table 4.7). Both MCT1 and MCT4 require the auxiliary protein basigin [7] to operate as a functional protein at the plasma membrane. The MCT carriers cotransport a proton with a lactate anion, or with other monocarboxylates (pyruvate, hydroxybutyrate, or acetacetate) in an electroneutral mode. The efflux of 1 H^+ ion with 1 lactate releases cells from the acid load generated by glucose metabolism. Conversly, proton-dependent uptake of lactate at physiological concentrations decreases the cytosolic pH. These transporters may differ in their affinity for certain substrates and stereoselectivity (^{L}lactate vs. ^{D}lactate) [486].

Mitochondria also possess H^+–monocarboxylate transporters that are adapted to their specific roles [486]. In addition, distinct Na^+–monocarboxylate cotransporters on the luminal surface of intestinal and kidney epithelia enable the active uptake of lactate, pyruvate, and ketone bodies.

Monocarboxylic acid transporter-1 (MCT1), or SLC16a1, is expressed in almost all cell types. In particular, MCT1 is found in heart, skeletal muscle, and intestine.

7. A.k.a. CD147, collagenase stimulatory factor, extracellular matrix metallopeptidase inducer (emmprin), and OK blood group antigen.

Table 4.7. Monocarboxylic acid transporters (MCT; TAT: T-type amino-acid transporter).

Type	Other alias
SLC16a1	MCT1
SLC16a2	MCT8
SLC16a3	MCT4
SLC16a4	MCT5
SLC16a5	MCT6
SLC16a6	MCT7
SLC16a7	MCT2
SLC16a8	MCT3
SLC16a9	MCT9
SLC16a10	MCT10, TAT1
SLC16a11	MCT11
SLC16a12	MCT12
SLC16a13	MCT13
SLC16a14	MCT14

It resides exclusively in the plasma membrane. In polarized cells, MCT1 usually localizes to the basolateral membrane. In the nervous system, the electroneutral MCT1 cotransporter may be involved in the shuttling of energy-rich substrates between astrocytes and neurons.

The MCT2 transporter is synthesized predominantly in the brain, heart, kidney, liver, and testis [487]. The MCT3 transporter abounds in muscles and placenta. Both MCT4 and MCT5 are predominant in the kidney and placenta. Transporters MCT6 and MCT7 are almost ubiquitous [487].

The MCT1 transporter warrants the proton-linked efflux of lactates that accumulate during anaerobic glycolysis, especially in myocytes. It is not glycosylated, but connects to transmembrane glycoproteins basigin and embigin [488]. In addition to lactate, MCT1 conveys the reversible transport of various monocarboxylic acids, such as pyruvate, ^{D}- and ^{L}3-hydroxybutyrate, acetoacetate, α-oxoisohexanoate, and α-oxoisovalerate, but not of dicarboxylic and tricarboxylic acids [487].

Monocarboxylate transporters and Na^+–HCO_3^- cotransporters (NBC) [8] coexist in many epithelial and glial cells, where they carry acid and base equivalents. The lactate–H^+ cotransport via MCT1 stimulates NBC activity [489]. Moreover, the lactate–H^+ cotransport through MCT1 augments about 2-fold when MCT1 is co-expressed with NBC carrier. The facilitation of MCT1 activity may result from the apparent buffer capacity due to NBC cotransporters. [9]

8. The sodium-bicarbonate cotransporter carries 1 Na^+ ion with 2 HCO_3^- ions. It operate in epithelial cells and, in the brain, in all types of macroglial cells (astrocytes, oligodendrocytes, Schwann cells, and retinal Müller cells).

9. The bicarbonate shuttling through the NBC cotransporter attenuates the intracellular pH changes caused by in- or outwardly cotransported lactate and protons through the MCT1 carrier, and hence counteracts the dissipation of the proton gradient across the cell membrane, which may reduce MCT1 activity.

4.5 Ion Transporters

Ions are vital elements for living cells that are conveyed from environment via various types of transporters, in addition to channels and pumps.

4.5.1 Copper Exporters and Importers

Copper [10] is a ubiquitous nutrient for cells. This cofactor for several enzymes can accept and donate electrons as it changes its oxidation state between mono- (Cu^+) and divalent Cu^{++} cation.

Copper is widely used in reactions for iron oxidation and mobilization, cell respiration, anti-oxidant defense, neurotransmitter synthesis (e.g., catecholamine), neuropeptide amidation, neuronal myelination, connective tissue formation, and angiogenesis.

Cuproproteins include cytochrome-C oxidase and NADH dehydrogenase-2. Secretory cuproenzymes encompass dopamine-β-hydroxylase, tyrosinase, ceruloplasmin, lysyl oxidase, and amine oxidases [490]. Copper is passed to cuproproteins from copper metallochaperones that operate as intracellular Cu^+ shuttles.

Copper is less abundant in cells than other metallic cations, such as iron and zinc. Copper is a potent stabilizer that induces a higher relative stability of complexes than that formed by most divalent metal ions (*Irving-Williams series*):

$$Mg^{++} \sim Ca^{++} < Mn^{++} < Fe^{++} < Co^{++} < Ni^{++} < Cu^{++} < Zn^{++}.$$

Copper cation also has a high ligand-field stabilization energy [490].

Copper participates in the formation of reactive oxygen species when unbound to proteins within the cell. Cells use copper toxicity to eliminate pathogens. Activated macrophages and neutrophils release ROS using NADPH oxidase and copper for storage in phagosomes.

Plasmalemmal and intracellular copper transporters as well as soluble copper chaperones carry this mineral cation to its destinations inside the cell. Copper is needed within mitochondria to supply Cu_A and intramembrane Cu_B sites of cytochrome oxidase, within the trans-Golgi network to supply secreted cuproproteins, as well as within the cytosol to supply SOD1 superoxide dismutase [490].

Copper-binding metallochaperones, such as anti-oxidant protein AtOx1, transport copper to different intracellular sites for storage or assembly of cupro-enzymes. The intracellular copper concentration is controlled by the balance between the activity of copper-efflux ATPases (ATP7a and ATP7b effluxers; Sect. 3.7.1) and copper permeases (CTr1 and CTr2 importers).

Copper is absorbed through the intestine. Dietary intake of copper salts generally exceeds tissue demands. The Cu ATPase ATP7a, a copper efflux pump of the basolateral plasma membrane, accepts copper from AtOx1 and facilitates copper export from enterocytes to blood. Copper is then transported to the liver, into which it enters using copper transporter CTr1 to be stored [491]. Copper is then used in the liver for

10. Cu: cuprum.

many purposes such as the synthesis of ceruloplasmin, a secreted copper ferroxidase. Excess copper is exported from hepatocytes by apical cell-surface ATP7b enzyme. In fact, a small amount is then mobilized into the blood circulation to supply cells. Copper is exported via the bile using ATP7b carrier to avoid liver and brain copper overload.

4.5.1.1 Copper Deficiency

Copper deficiency causes cardiovascular, hematological, and neurological disorders as well as connective tissue deficits. In the heart, copper deficiency leads to cardiac hypertrophy and cardiomyopathy. In cardiomyocytes, the lack in copper transporter CTr1 and cardiac copper deficiency elicit a signal that determines an increase in: (1) concentration of the ATP7a copper-exporting ATPase in liver and intestine and copper chaperone for superoxide dismutase (CCS) [11] as well as (2) copper plasma level subsequent to elevated copper uptake and mobilization into the blood circulation [493].

4.5.1.2 Handling of Copper Toxicity

Copper is an essential trace element for cells, but it has toxic side effects. Therefore, the intracellular concentration of free copper must be restricted and copper transfer between proteins must minimize toxicity. Free copper is then nearly unavailable in cells.

Copper enters the cell as Cu^+ through high-affinity plasmalemmal copper transporters or low-affinity permeases. All copper ions are then exchanged between cuproproteins and copper ligands such as reduced glutathione. Copper chaperones, such as cytoplasmic copper chaperone for superoxide dismutase, cytochrome-C oxidase copper chaperone COx17, [12] cytosolic copper carrier anti-oxidant protein AtOx1, [13] then bind and transport Cu^+ to cytosolic enzymes or membrane transporters. [14]

Intracellular copper proteins involved in redox catalysis, copper transport between cells as well as different cellular subcompartments, and copper storage possess various Cu^+-binding affinities. Fast copper transfer require metal-mediated between-protein interactions supported by protein-specific recognition [494].

In synergy with $^{(Cu,Zn)}$SOD1, metallothioneins that bind both biological (copper, selenium, and zinc) and xenobiotic (arsenic, cadmium, mercury, and silver) heavy metals have the highest affinity for Cu^+ cation. They may contribute to the regulation

11. Various superoxide dismutases (SOD) exist. These anti-oxidant enzymes employ a copper, manganese, iron, or nickel cofactor. Copper SODs (SOD1) obtain copper via interaction with CCS copper chaperone [492]. Subpopulations of copper–zinc superoxide dismutases localize to the cytosol, mitochondria, secretory system, and nucleus.

12. A.k.a. cytochrome-C oxidase assembly homolog.

13. A.k.a. human anti-oxidant protein-1 (AtX1) homolog (HAH1).

14. Copper chaperone COx17 and its partners Synthesis of cytochrome-C oxidase SCO1 and SCO2 are all involved in the assembly of the Cu_A site of cytochrome-C oxidase (CcO) within the mitochondrial intermembrane space.

of cellular copper distribution [494]. Metallothionein-2 (MT2) localizes in both the cytoplasm and mitochondrial intermembrane space.

4.5.2 Iron Transporters

Iron is a component of oxidative metabolism and a cofactor for various enzymes. It contributes to the function of many hemoproteins and iron–sulfur proteins. Iron can serve both as an electron donor and acceptor. However, it can be toxic in high concentrations. Intracellular iron concentration is thus regulated by iron transport and storage. Numerous proteins, such as transferrin, transferrin receptor, and ferritin, control iron transport, uptake, storage, and export.

In plasma, transferrin (Tf) binds to extracellular ferric ion. Fe^{3+} Transferrin binds to its cognate TfR1 receptor. The Tf–TfR1 complex is internalized via clathrin-coated pits. Ferric ion is then released and reduced into Fe^{++} ions. Ferrous ion (Fe^{++}) is transported out of endosomes by divalent metal transporter-1 (DMT1). [15] Iron regulatory proteins (IRP) regulate the production of the transporter DMT1 that enhances iron influx. However, iron must not accumulate inside the cell to avoid subsequent mitochondrial dysfunction and oxidative stress.

Iron importer mitoferrin-1 (or SLC25a37) of the mitochondrial inner membrane carries iron to synthesize mitochondrial heme [16] and iron–sulfur proteins, especially in erythroblasts [495]. Mitoferrin-1 indeed abounds in both fetal and adult erythropoietic cells. The ABCb10 transporter of the mitochondrial inner membrane interacts with mitoferrin-1 (but not mitoferrin-2) to enhance its stability and promote mitoferrin-1-dependent mitochondrial iron import [495].

4.5.3 Magnesium Transporters

Magnesium ion is the most abundant intracellular divalent cation in cells. The majority of Mg^{++} is bound to proteins, phospholipids, nucleic acids, and ATP. Magnesium ion serves as a cofactor of ATP (ATP^{Mg}: >90% of cellular ATP). In addition, intracellular Mg^{++} modulates the activity of many ion channels.

Free and total intracellular Mg^{++} concentrations range approximately from 0.5 to 1 mmol and 17 to 20 mmol, respectively. Serum Mg^{++} concentration equals 1 to 2 mmol. In the cytosol, free Mg^{++} concentration fluctuates much less than that of Ca^{++} ions.

In humans, SLC41a1 and SLC41a2 proteins may complement Mg^{++} transport rather than carry Mg^{++}, as they do not convey significant Mg^{++} flux. In mammals, Mg^{++} transporters MagT1 and tumor suppressor candidate TuSC3 [17] allows Mg^{++}

15. A.k.a. natural resistance-associated macrophage protein NRAMP2.

16. Imported iron is incorporated to protoporphyrin-9 by ferrochelatase to produce heme.

17. A.k.a. protein M33, N33, MRT7, and human ortholog of the oligosaccharyltransferase complex subunit OST3a (Saccharomyces cerevisae). It is a subunit of the oligosaccharyltransferase complex that catalyzes the transfer of an oligosaccharide chain on nascent proteins (N-glycosylation) in the endoplamic reticulum lumen. The other subunit OST3b corresponds to

influx [496]. In humans, MagT1 is ubiquitous and TuSC3 is produced in most non-lymphoid tissues. The production of the Mg^{++} transporter is upregulated when the extracellular Mg^{++} concentration is low.

4.6 Cation–Chloride Cotransporters

Three types of cation–chloride cotransporters are encoded by distinct genes: K^+–Cl^- (KCC), Na^+–Cl^- (NCC), and Na^+–K^+–$2Cl^-$ (NKCC) cotransporters. All 3 co-transporters require Cl^- as well as at least one cation, either Na^+ and/or K^+. All 3 cotransport types are electrically neutral. The corresponding genes constitute the cation–chloride cotransporter (CCC) gene family.

4.6.1 K^+–Cl^- Cotransporters

Potassium–chloride cotransporter (KCC) is a membrane protein that mediates a secondary active, coupled, electroneutral transport of K^+ and Cl^- across the plasma membrane. It is activated by dephosphorylation. The KCC isoforms are similar to other members of the electroneutral cation-chloride cotransporter set (i.e., NKCCs and NCCs).

Four K^+–Cl^- cotransporter isoforms (KCC1–KCC4) that are encoded by their own genes (Slc12A4–Slc12A7). The KCC family can be subdivided into 2 groups: (1) KCC2 and KCC4 on the one hand and (2) KCC1 and KCC3 on the other. Alternative splicing and diverse promoter usage generate further molecular heterogeneity.

The KCC2 subtype is a neuron-specific cotransporters, whereas the other 3 KCCs are widely distributed. In particular, KCC1 is detected in glomerular mesangial and endothelial cells of renal vessels. The KCC3 isoform that has been initially cloned from vascular endothelial cells is expressed in the brain, heart, kidney, lung, and muscle. The KCC4 isotype abounds in the heart and kidney.

Potassium–chloride cotransporters contribute to several cellular function, such as cell volume regulation, epithelial ion transport, and osmotic homeostasis.

Insulin-like growth factor IGF1 increases KCC activity by activating phosphatidylinositol 3-kinase and mitogen-activated protein kinase cascades, hence stimulating protein kinase-D and extracellular signal-regulated kinase ERK1 and ERK2, respectively [497].

4.6.2 Na^+–Cl^- Cotransporters

Sodium–chloride cotransporter (NCC), or symporter, primarily removes Na^+ and Cl^- from the distal convoluted tubule, at the beginning of the aldosterone-sensitive nephron [498]. This electroneutral cation-coupled chloride cotransporter is regulated by WNK kinases (Vol. 4 – Chap. 4. Cytosolic Protein Ser/Thr Kinases); Table 4.8). It belongs to the solute carrier class SLC12; it is encoded by the Slc12A3 gene.

---

the magnesium transporter MagT1. Alternative splicing produces 2 TuSC3 isoforms, TuSC3-1

Table 4.8. Effect of kinases WNK1 and WNK4 on transport in distal tubule of nephron (Source: [498]; ENaC: epithelial Na^+ channel; NKCC: apical Na^+–K^+–$2Cl^-$ cotransporter; ROMK: renal outer medullary K^+ channel).

Protein	WNK1	WNK4
NCC		$\downarrow$
NKCC		$\downarrow$
ENaC	$\uparrow$	$\downarrow$
ROMK	$\downarrow$	$\downarrow$
Claudin-4	$\uparrow$	$\uparrow$

The distal convoluted tubule (DCT) is responsible for the reabsorption of 5 to 10% of the glomerular filtrate. It is divided into upstream (DCT1) and downstream (DCT2) segments, respectively. The major salt reabsorption carrier in the distal convoluted tubule is the Na^+–Cl^- cotransporter (NCC). In DCT1, salt is carried exclusively by NCC, whereas in DCT2, NCC operates in synergy with the epithelial sodium channel (ENaC). The sodium gradient that drives transport from the lumen to the interstitium is generated and maintained by the Na^+–K^+ ATPase in the basolateral membrane. A part of the potassium ions that enter into cell at the basolateral membrane is secreted at the luminal membrane via K^+ channels and K^+–Cl^- cotransporter [498]. In addition, NCC modulates magnesium reabsorption in parallel with sodium reabsorption as well as calcium reabsorption inversely with sodium reabsorption. Because NCC is located after the macula densa, the salt reabsorption rate is not subjected to the tubuloglomerular feedback, hence influencing extracellular fluid balance and arterial blood pressure.

4.6.2.1 NCC and WNK Kinases

Angiotensin-2 acts via its receptor AT1R to raise NCC activity, as it inhibits NCC inhibitor WNK4 [498]. Angiotensin-2 also leads to elevated phosphorylation of Ste20-related Pro/Ala-rich kinase (SPAK or STK39; Vol. 4 – Chap. 4. Cytosolic Protein Ser/Thr Kinases). The ATn2–WNK4–SPAK–NCC axis thus participates in the renal response to intravascular volume depletion.

Two WNK1 isoforms are synthesized from alternative splicing in the nephron: (1) the full long form ($WNK1_L$) expressed along the entire nephron and (2) a shorter transcript ($WNK1_S$) only located in the distal convoluted and connecting tubule. The $WNK1_S$ isoform interacts with $WNK1_L$ and relieves the $WNK1_L$-induced inhibition of WNK4, thereby allowing WNK4 to inhibit NCC [498]. The $WNK1_S$ isotype interacts also with $WNK1_L$ to regulate the activity of renal outer medullary K^+ channels. The $WNK1_L$ isoenzyme reduces ROMK function. The $WNK1_S$ subtype prevents $WNK1_L$ activity on ROMK channel. Moreover, $WNK1_L$ modulates ENaC activity

---

and TuSC3-2. Proteins MagT1 and TuSC3 may function cooperatively for Mg^{++} uptake, at least in some cells.

via SGK that phosphorylates (inhibits) neural precursor cell expressed, developmentally downregulated protein NEDD4 that lowers ENaC activity, as it promotes ENaC clathrin-dependent endocytosis.

As a possible sensor of cell volume and/or intracellular Cl^- concentration, WNK3 regulates all the members of the SLC12 class. It is detected in many tissues. In kidneys, it is located along the entire nephron. It activates NCC, NKCC1, and NKCC2 cotransporters, but inhibits K^+–Cl^- cotransporters that lower intracellular chloride concentration (KCC1–KCC4). Therefore, WNK3 activates chloride influx and inactivates Cl^- efflux. In addition, WNK3 may interact with and regulate WNK4, and conversely [498]. Whereas WNK3 is a NCC activator, WNK4 is a NCC inhibitor.

4.6.3 Na^+–K^+–$2Cl^-$ Cotransporters

Sodium–potassium–chloride cotransporter (NKCC), or symporter, as it carries ions in the same direction in an electrically neutral manner, contributes to the active transport of sodium, potassium, and chloride ions (in- and outflux). The NKCC cotransporter contributes to the maintenance of cell volume. The NKCC cotransporter maintains intracellular Cl^- concentration at high levels to promote net salt transport in epithelial cells and to set synaptic potentials in neurons [499].

These cotransporters lodges in various cell types, particularly reabsorptive and secretory epithelial cells, myocytes, neurons, endothelial cells, fibroblasts, and blood cells. These cotransporters participates in salt secretion by intestinal, salivary gland, and respiratory epithelial cells, among other secretory epithelia.

Ion translocation through Na^+–K^+–$2Cl^-$ cotransporters requires that all 3 involved ion types be simultaneously present on the same side of the plasma membrane. Overall transport of Na^+, K^+, and Cl^- through NKCC is not driven by the transmembrane voltage. Moreover, ion transport does not change the transmembrane voltage.

In most cells, Na^+ and K^+ gradients are established by the Na^+ pump such that the outwardly directed concentration gradient for K^+ exceeds the inwardly directed concentration gradient for Na^+. Therefore, the transmembrane Cl^- gradient determines whether the net NKCC flux favors import or export. Unlike Na^+ and K^+, the intracellular Cl^- concentration can vary over a wide range (from few to 50–60 mmol) according to activity of other Cl^- carriers (Table 4.9).

Two NKCC subtypes exist (NKCC1–NKCC2 or SLC12a2–SLC12a1). Isoform NKCC1 is widely distributed, especially in cells of exocrine glands that secrete fluids, whereas NKCC2 resides specifically in the kidney [500]. More precisely, NKCC2 is specifically detected in the apical membrane of cells of the thick ascending limb of the loop of Henle of the nephron, where it extracts Na^+, K^+, and Cl^- from urine. Isoform NKCC2 also lodges in the renal juxtaglomerular apparatus. Subtype NKCC1 mainly localizes to basolateral membrane close to blood capillaries, hence transporting Na^+, K^+, and Cl^- from blood into cell.

The NKCC2 isoform is smaller than the NKCC1 cotransporter. It has at least 3 alternatively spliced variants [499]. Different NKCC2 isoforms with distinct Na^+

Table 4.9. Typical concentration (mmol) of Cl^-, K^+, and Na^+ in extra- and intracellular media (Source: [499]). Differences in concentration of ions on opposite sides of the plasma membrane produce a voltage change (transmembrane potential). The largest contributions are usually yielded by sodium (Na^+) and chloride (Cl^-) ions, which have high concentrations in the extracellular medium, and potassium (K^+) ions, which, together with proteic anions, have high concentrations in the intracellular medium. Calcium ion serves as a second messenger.

	Intracellular space	Extracellular space
Cl^-		110
K^+	140	4.5
Na^+	15	140

and K^+ affinities may explain regional differences in Na^+ reabsorption along the thick ascending limb of the loop of Henle. In addition, NKCC2b splice variant is found in macula densa cells that are involved in tubuloglomerular feedback and renin secretion.

Ion transport across the cell membrane requires energy that is provided by Na^+ electrochemical gradient established by Na^+–K^+ ATPase. Symporter NKCC is thus a secondary active cotransporter, as it takes its energy from the combined chemical gradients of the transported ions [500].

Agent ATP intervenes in NKCC activity. Cotransporter NKCC actually undergoes regulatory phosphorylation–dephosphorylation cycles, hence its dependence on ATP [499]. The NKCC cotransporter must be phosphorylated to cause cotransport. Protein kinase-A and -C may phosphorylate NKCC cotransporter. The regulation of Na^+–K^+–$2Cl^-$ cotransporter is mediated, at least in some cases, by Mg^{++}-dependent phosphorylation by protein kinase-A [500]. However, the set of protein kinases and phosphatases that regulate NKCC function remains to be determined.

In addition, intracellular Cl^- concentration in the physiological range and above it strongly inhibits NKCC cotransporter. However, intracellular Cl^- is neither the only, nor the most effective anion that inhibits NKCC cotransporter [499].[18] The Cl^--associated regulatory motif differs from Cl^--transport sites. Intracellular Ca^{++} concentration may also activate NKCC cotransporter [499]. Last, but not least, like other ion carriers, the cytoskeleton participates in the regulation of NKCC activity.

Water and salt secretion involves the coordinated action of multiple ion carriers. Chloride efflux across the apical membrane uses multiple Cl^- channel types [500]. Conversely, in absorptive thick ascending limb epithelium, apical Cl^- influx is done by Na^+–K^+–$2Cl^-$ or Na^+–Cl^- cotransporters with K^+ recycling through apical K^+ channels and Cl^- efflux via basolateral Cl^- channels.

Secretory epithelia such as respiratory epithelia secrete a Cl^--rich aqueous solution. Secretion requires the transepithelial transport of Cl^- ions, together with attendant cations and osmotic motion of water, from blood to gland lumen, i.e., from

18. Anions I^- and Br^- are more potent inhibitors than Cl^- [499]. Anions nitrate (NO_3^-) and sulfate (SO_4^{2-}) have much less inhibitory potency.

the basolateral through NKCC to apical membrane through Cl^- channels [499]. Basolateral NKCCs are functionally coupled to apical Cl^- channels. Moreover, NKCC activity rises when intracellular Cl^- concentration lowers.

4.7 Ion-Coupled Solute Transporters

Energy can be yielded to active solute transport across cellular membranes by 2 main processes: (1) ATP hydrolysis (ATP-dependent transporters) or (2) electrochemical ion gradients, i.e., cotransport of ions, such as H^+, Na^+, Cl^-, and/or countertransport of K^+ or OH^- [501].

Most mammalian ion-coupled organic solute transporters can be grouped into several transporter categories: (1) sodium-coupled transporter category that includes Na^+–glucose cotransporters (SGlT1–SGlT3; Sect. 4.4) [19] and inositol transporter SLC5a3; [20] (2) Na^+- and K^+-coupled transporter category that comprises neurotransmitter transporters of adrenaline, dopamine, γ-aminobutyric acid, glycine, noradrenaline, proline, and serotonin (Sect. 4.8), as well as transporters of β-amino acids; (3) Na^+- and K^+-dependent glutamate–neurotransmitter category that consists of high-affinity glutamate transporters (EAAT1–EAAT4; Sect. 4.8.2.2) and glutamate–neutral amino acid transporters SLC1a4 (or ASCT1; ASC: Ala–Ser–Cys); and (4) H^+-coupled oligopeptide transporter category that encompasses H^+-coupled transporter of oligopeptides and peptide-derived antibiotics PepT1 and PepT2 in the intestine and kidney.

4.8 Neurotransmitter Transporters

4.8.1 Choline and Acetylcholine Transporters

High-affinity choline transporter (ChT1) is encoded by the Slc5A7 gene. In mammals, the solute carrier family SLC55A of sodium-dependent transporters includes Na^+–substrate cotransporters for glucose, myoinositol, [21] and iodide [5]. It is expressed mainly in cholinergic neurons, on nerve cell terminals and synaptic vesicles.

Choline is generated by hydrolysis of acetylcholine by acetylcholinesterase. The ChT1 transporter helps maintain acetylcholine synthesis in presynaptic terminals of

19. A.k.a. SLC5a1, SLC5a2, and SLC5a4 respectively. Active absorption across the luminal (apical) membrane of epithelial cells of small intestine and nephron proximal tubule of each glucose molecule through SGlT1 and SGlT2 is linked to the cotransport of 2 Na^+ and 1 Na^+ ion, respectively. Na^+–K^+ ATPase in the basolateral membrane maintains the required inward Na^+ flux and electrochemical gradient. Glucose is then conveyed across the basal membrane into the blood stream through glucose transporters, such as GluT1 and GluT2.

20. A.k.a. sodium–myoinositol cotransporter (MIT).

21. Myo-inositol, formerly called meso-inositol, the most prominent natural stereoisomer of inositol, the structural basis for secondary lipidic messengers, such as inositol phosphates, phosphatidylinositols, and phosphatidylinositol phosphates.

cholinergic neurons in the central nervous system, including the brainstem and spinal cord, the sympathetic and parasympathetic autonomic ganglia, and target tissues such as neuromuscular junctions.

Vesicular acetylcholine transporter (VAChT) is a neurotransmitter transporter that loads acetylcholine into secretory vesicles in neurons and neuroendocrine cells. It is a member of the set of neurotransmitter transporters that also includes *vesicular monoamine transporters* VMAT1 and VMAT2 (Sect. 4.8.3). The VAChT transporter is assigned exclusively to small synaptic vesicles in all cholinergic neurons. In some neuronal cell lines, VAChT resides on small synaptic vesicles that do not contain VMAT2 transporter [502].

4.8.2 Sodium- and Chloride-Dependent Neurotransmitter Transporters

Sodium- and chloride-dependent neurotransmitter receptor transporters of the solute carrier class 6 include γ-aminobutyric acid, glycine, and monoamine transporters. They recycle neurotransmitter for reuse by neurons and maintain low extracellular concentrations of neurotransmitters.

In addition, the vesicular transporter SLC32a1 [22] concentrate GABA as well as glycine within synaptic vesicles [5].

Three vesicular glutamate transporters (VGluT1–VGluT3) of the SLC17 class of sodium-dependent inorganic phosphate cotransporters (SLC17a7, SLC17a6, and SLC17a8, respectively) accumulate glutamate in synaptic vesicles [5].

4.8.2.1 GABA Transporters

Plasmalemmal Na^+–Cl^--dependent γ-aminobutyric acid transporters are members of the SLC6A family (GAT1–GAT3 and BGT1; Table 4.10). The GABA transporters reside on both neurons and glial cells. These neurotransmitter transporters retrieve released GABA from the synaptic cleft into the neuron. They use Na^+ gradient for GABA uptake. In addition, several electrogenic ion movements can occur through GAT1. The GAT1 transporter undergoes regulated endo- and exocytosis according to the context, i.e., interactions with numerous proteins, especially protein kinase-C and syntaxin-1A [504].

4.8.2.2 Glutamate Transporters

Glutamate transport plays an important role in cellular amino acid metabolism and in fast excitatory synaptic transmission in the central nervous system. In small intestine and proximal tubule of the nephron, glutamate transporters are involved in transepithelial glutamate transport. In the central nervous system, they contribute to rapid removal of released glutamate from the synaptic cleft, hence termination of postsynaptic action of glutamate. Glutamate transporters localize in pre- or postsynaptic membranes as well as the plasma membrane of glial cells.

22. A.k.a. vesicular GABA transporter (VGAT) and vesicular inhibitory amino acid transporter (VIAAT).

Table 4.10. The GABA transporters and their endogenous substrates (Source: [5]; BGT: betaine–GABA transporter; GAT: GABA transporter). β-Alanine, or 3-aminopropanoic acid, is an endogenous amino acid with the amine group at the β position (2 atoms [rather than 1] away) from the carboxylate group and without chiral center, unlike its usual counterpart.

Type	Other names	Endogenous substrates
GAT1	SLC6a1, mGAT1	GABA
GAT2	SLC6a16, mGAT3	GABA, β-alanine
GAT3	SLC6a11, mGAT4, GATb	GABA, β-alanine
BGT1	SLC6a12, mGAT2	GABA, betaine

Table 4.11. Glutamate transporters of the SLC1 family (AAAT: adipocyte amino acid transporter; ASCT: alanine–serine–cysteine–threonine transporter; EAAC: excitatory amino acid carrier [electrogenic, Na^+-dependent, Cl^-independent, high-affinity, ^{L}glutamate and ^{D} and ^{L}aspartate cotransporter]; EAAT: excitatory amino acid transporter; GlAsT: (glial, high-affinity) glutamate aspartate transporter; GLT; glial glutamate transporter; SATT: serine–alanine–threonine transporter). Glutamate serves as a metabolic fuel in cells. In particular, in the microvilli of enterocytes, the EEAT3 transporter is responsible for the uptake of glutamate by enterocytes. Glutamate modulates neuronal migration, outgrowth of neuronal processes, and synapse elimination during the brain development. It is also the main excitatory neurotransmitter in the brain. The EEAT transporters that convey Cl^- ions are gated by the joint binding of Na^+ and glutamate. The EAAT transporters reduce the extracellular concentration of glutamate by rapid uptake into nerve terminals and glial cells (astrocytes, microglial cells, and oligodendrocytes). Glial glutamate transporters EEAT1 and EEAT2 actually represent the main mechanism of glutamate removal from the extracellular fluid to avoid toxicity.

Member	Other alias	Distribution
SLC1a1	EEAT3, EAAC1	Neurons, Purkinje cells, glial cells, epithelial cells (small intestine and kidney)
SLC1a2	EEAT2, GLT1	Glial and endothelial cells
SLC1a3	EEAT1, GlAsT1	Neurons, glial and endothelial cells
SLC1a4	ASCT1, SATT	Neurons, cerebellar Purkinje cells, glial cells
SLC1a5	ASCT2, AAAT	Glial cells
SLC1a6	EEAT4	Cerebellar Purkinje cells
SLC1a7	EAAT5	Retinal cells

Glutamate–Aspartate Transporters (EAATs)

Plasmalemmal Na^+-dependent glutamate–aspartate transporters, or excitatory amino acid transporters (EAAT1–EAAT5; Table 4.12) are members of the solute carrier class 1 (Table 4.11). Glutamate transporters are trimers. However, each monomer is a functional unit capable of substrate permeation [5]. Na^+–K^+ ATPase coassembles with EAAT1 and EAAT2 transporters.

Glutamate transporters of the EAAT group include glial glutamate transporters EAAT1 and EAAT2, neuronal and epithelial glutamate transporter EAAT3, and cerebellar glutamate transporter EAAT4. In the nervous system, glutamate transporters are indeed located on neurons (predominantly EAAT3 to EAAT5) and glial cells (mainly EAAT1 and EAAT2). They regulate excitatory neurotransmission, maintain low extracellular concentrations of glutamate, and provide glutamate for the glutamate-glutamine cycle [5].

The EAAT1 transporter (SLC1a3) carries glutamic and aspartic acid coupled to the cotransport of 3 Na^+ cations and 1 H^+ cation and countertransport of 1 K^+ cation. It abounds in glial cells.[23] It is also synthesized in numerous other cells such as cardiomyocytes.

The EAAT2 transporter is responsible for the majority of glutamate clearance from the synaptic cleft in the brain. Splice variants of EAAT2 (SLC1a2) homo- and heteromerize [5].

Both EEAT2 and EEAT3 carriers enable the cysteine transport. However, in renal cells, only the cysteine transport by the EEAT3 carrier regulates the intracellular glutathione (GSH) content [506].[24] Protein kinase-C is a translocator and activator of EAAT3 transporter. The EAAT3 partner ARL6IP5 interacts with EAAT3 at the plasma membrane and prevents the cellular GSH production, thereby enhancing the resistance to oxidative stress.

The high-affinity glutamate transporter EAAT3 (or SLC1a1) is strongly expressed in the central nervous system (glutamatergic as well as some non-glutamatergic neurons, such as gabaergic cerebellar Purkinje cells, where glutamate can serve as a GABA precursor, and cholinergic α-motoneurons of the spinal cord) as well as in the intestine, and, to a lesser extent, heart, kidney, and liver [501]. In the kidney, EAAT3 in the apical membrane of proximal tubular cells is responsible for glutamate reabsorption [5]. The EAAT3 transporter has a polarized location, as it resides in the apical surface of epithelial cells in the kidney and gut.

Glutamate uptake by EAAT3 is coupled to the cotransport of 2 Na^+ and countertransport of 1 K^+ and 1 OH^- ion. Transport through EAAT3 depends on extracellular Na^+, but not Cl^- anion, unlike GABA transporter GAT1 of the Na^+- and Cl^--coupled neurotransmitter transporter group [501].

The glutamate transporter EAAT3-interacting protein GTRAP3-18, or adpribosylation factor-like ARL6-interacting protein ARL6IP5,[25] is a microtubule-associated protein that is involved in the restructuring of the actin cytoskeleton and activation of the MAPK module [507]. It intervenes in cell migration via ERK effectors focal adhesion kinase and cyclooxygenase-2.

---

23. In glial cells, the glutamate transport through EEAT3 is electrogenic and coupled to the cotransport of 2 Na^+ ions and the countertransport of 1 K^+ and either the countertransport of 1 OH– or the cotransport of 1 H^+ ion [505].

24. Glutathione is synthesized from glutamate, glycine, and cysteine. Astrocytes and immature neurons take up cystine that is then converted to cysteine. Mature neurons, which are unable to ingest cystine, pick up cysteine. Cysteine is the rate-limiting factor for GSH synthesis. Glutathione aids maintain sulfhydryl groups of proteins in the reduced state and the iron heme in the ferrous state. It also serves as a reducing agent for glutaredoxin, DNA, reactive oxygen species, and free radicals.

25. A.k.a. addicsin, cytoskeleton-related vitamin-A-responsive protein, prenylated Rab acceptor protein PRAP2, and JWA protein.

The EAAT4 transporter reaches its highest concentrations in the cerebellum. Two proteins, the glutamate transporter EAAT4-associated proteins GTrAP41 [26] and GTrAP48 [27] interact with the EAAT4 transporter.

The retina-specific EAAT5 transporter localizes to both cone and rod photoreceptor terminals and axon terminals of rod bipolar cells [508]. It acts as an inhibitory presynaptic receptor on rod bipolar cell axon terminals.

Glutamate Neutral Amino Acid Transporters (ASCTs)

Sodium-dependent neurotransmitter transporters of the SLC1 class also includes glutamate and neutral amino acid (hence their alias ASCT for alanine–serine–cysteine–threonine transporter) [28] transporters SLC1a4 [29] and SLC1a5 [30] Transporters in this family contain an associated Cl^- channel that opens on linkage of a transportable amino acid, but Cl^- flux is uncoupled from the amino acid import. Transporters SLC1a4 and SLC1a5 carry an overlapping, but non-identical set of neutral amino acids. Transporter SLC1a5 only conveys glutamine.

The ASCT1 transporter localizes to both glial cells and neurons, at least in rats [509]. It is expressed at similar concentrations in both the developing and adult brains. It can be detected in hippocampal pyramidal and dentate granule neurons, cerebellar Purkinje cells, and, in the cerebral cortex, neuron somas in all lamina and astrocytes.

Neutral amino acid transporter ASCT1 not only mediates the efflux of glutamate from the neuron into the synaptic junction via a calcium-independent release, but also the efflux of ^{L}serine from glial cells and its uptake by neurons [510]. [31] The ASCT1 carrier has a sodium-dependent amino acid transport activity. It conveys primarily zwitterionic amino acids at physiological pH. [32] When the extracellular pH is lower than 7.4, the ASCT1 carrier shifts its substrate specificity to anionic amino acids.

26. A.k.a. spectrin chain-β, non-erythrocytic (brain) type 2 (SptβN2).

27. A.k.a. Rho guanine nucleotide-exchange factor (GEF)-11 (RhoGEF11).

28. Neutral amino acid transporters can carry neutral amino acids alanine, serine, cysteine and threonin. ^{L}Cysteine is also conveyed in a pH-independent manner by the glutamate EAAT3 transporter.

29. A.k.a. alanine–serine–cysteine–threonine transporter ASCT1 and serine–alanine–threonine transporter (SATT).

30. A.k.a. ASCT2 and adipocyte amino acid transporter (AAAT). Transporter SLC1a5 also serves as a retrovirus receptor on N-deglycosylation of its C-terminus. Hence, SLC1a5 has been previously named baboon M7- (M7V1) and feline RD114 virus receptor (RDRC).

31. ^{L}Serine is used for the synthesis of coagonists of ^{NMDA}Glu receptor, ^{D}serine and glycine. The ASCT1 carrier predominantly contributes to the uptake of ^{L}serine in neurons.

32. An amino acid has both a basic amine (NH_2) and an acidic carboxylic (COOH) group. Amino acids are zwitterions, as a hydrogen ion can be transferred from the carboxylic group to the amine group to create an ion with separate parts containing a negative (COO^-) and a positive (NH_3^+) charge, but without overall electrical charge. Any solution of an amino acid can contain this zwitterion. The pH of the solution can be changed with the addition of either acids or alkalis. When the pH of a solution of an amino acid increases (e.g., when hydroxide ions are added), the hydrogen ion is removed from the NH_3^+ group to generate an anion with

Table 4.12. Glutamate transporters and their endogenous substrates (Source: [5]; ASCT: alanine–serine–cysteine–threonine transporter; EAAT: excitatory amino acid transporter). Transported small neutral amino acids by ASCT carriers include ^{L}alanine, ^{L}serine, ^{L}cysteine, and ^{L}threonin. All glutamate transporters are chloride carriers during the transmembrane amino acid transfer; the magnitude and direction of the substrate-activated chloride flux do not influence amino acid transport (thermodynamically uncoupled flux).

Type	SLC1A code	Endogenous substrates
	Glutamate aspartate transporters	
EAAT1	SLC1a3	^{L}glutamate, ^{L}aspartate
EAAT2	SLC1a2	^{L}glutamate, ^{L}aspartate
EAAT3	SLC1a1	^{L}glutamate, ^{L}aspartate
EAAT4	SLC1a6	^{L}glutamate, ^{L}aspartate
EAAT5	SLC1a7	^{L}glutamate, ^{L}aspartate
Glutamate and neutral amino acid transporters		
ASCT1	SLC1a4	^{L}alanine
ASCT2	SLC1a5	Glutamate

The ASCT1 transporter functions as an electroneutral exchanger with chloride channel activity [511].[33] Like other glutamate EAAT transporters, ASCT1 requires extracellular Na^+ ions. An anion-conducting state associates with an amino acid-bound state. However, the stoichiometry used by ASCT1, i.e., the interaction of 1 Na^+ ion with the ASCT1 transporter, differs from that of EAAT transporters that involve the cotransport of multiple Na^+ ions with glutamate. In addition, unlike other glutamate transporters, which complete the transport by a countertransport of K^+ ions and exchange a much smaller amount of amino acid, ASCT1 is insensitive to K^+ ions and carries only electroneutral, one-for-one, homo- and heteroexchange of neutral amino acids and sodium, the current occurring during transport being a thermodynamically uncoupled chloride flux. Therefore, the resulting amino acid flux is negligible, but a net transport of a particular substrate can be achieved when a transmembrane concentration gradient for another one of its substrates exists.

The ASCT2 transporter is a mediator of glutamine efflux from astrocytes owing to an exchange with extracellular amino acids. Extracellular alanine, serine, cysteine, threonine, glutamine, or leucine causes a rapid release of glutamine through the ASCT2 carrier [512]. The ASCT2 transporter thus operates in the glutamine-glutamate cycle between neurons and glial cells, as it facilitates the efflux of glutamine from glial cells.

a COO^- group. When the pH a solution of an amino acid decays, the COO^- group picks up a hydrogen ion to reform a COOH group and the amino acid becomes a cation due to its NH_3^+ group.

33. The chloride conductance is activated during glutamate transport. An anion selectivity sequence exists in ASCT1 and others glutamate EAAT transporters [511]: SCN^- (thiocyanate ion) > $NO3^-$ (nitrate ion) > I^- (iodide ion) > Cl^-.

Table 4.13. Glycine transporters and their endogenous substrate (Source: [5]).

Type	Other names	Endogenous substrates
GlyT1	SLC6a9	Glycine
GlyT2	SLC6a5	Glycine

4.8.2.3 Glycine Transporters

In the central nervous system, the action of glycine terminates due to its rapid uptake into nerve terminals and adjacent glial cells via high-affinity glycine transporters. Plasmalemmal, sodium- and chloride-dependent glycine transporters are members of the SLC6 class. They are encoded by 2 genes. Transporters GlyT1 and GlyT2 (Table 4.13) are predominantly located on glial cells and neurons, respectively [5]. Variants of GlyT1 (GlyT1a–GlyT1e) and GlyT2 (GlyT2a–GlyT2c) have been identified.

The GlyT1 isoform in glial cells that surrounds glutamatergic synapses regulate synaptic glycine concentrations and influences NMDA-type glutamate receptor (GluN)-mediated neurotransmission (Sect. 2.5.4).

The GlyT2 transporter localizes to axons and boutons of glycinergic neurons. It contributes to transmitter loading of synaptic vesicles and termination of inhibitory neurotransmission.

4.8.2.4 Monoamine Transporters

Plasmalemmal, sodium- and chloride-dependent monoamine neurotransmitter transporters (Table 4.14) carry the hormone–transmitters adrenaline, noradrenaline, dopamine, and serotonin. They are members of the SLC6 class. They include dopamine (DAT1), noradrenaline (NAT1), serotonin (SerT1), and vesicular monoamine (VMAT1–VMAT2) transporters.[34] They remove neurotransmitters from the synaptic cleft, thus terminating signaling.

The dopamine active transporter is a symporter that couples dopamine movement across the cell membrane to ion fluxes. It actually links the sequential cotransport of 2 Na^+ and 1 Cl^- ion to ingress of one dopamine molecule. The energy for dopamine reuptake is supplied by plasmalemmal Na^+–K^+ ATPase. Noradrenaline transporter can also transport dopamine.

Serotonin Transporter

The SerT1 transporter localizes to presynaptic membranes of 5HT-secreting neurons. It also serves to transport serotonin through the plasma membrane of platelets

34. The SLC6 class of Na^+- and Cl^--dependent neurotransmitter transporters also include γ-aminobutyric acid (GAT1) and glycine (GlyT1–GlyT2) transporters.

Table 4.14. Monoamine transporters and their endogenous substrates (Source: [5]; DAT: dopamine active transporter; NAT: noradrenaline transporter; NET: norepinephrine transporter; SerT: serotonin transporter; 5HTT: 5-hydroxytryptamine transporter).

Type	Other names	Endogenous substrates
DAT1	SLC6a3	Dopamine, adrenaline, noradrenaline
NAT1	SLC6a2, NET	Noradrenaline, adrenaline, dopamine
SerT1	SLC6a4, 5HTT	Serotonin (5-hydroxytryptamine)

and epithelial cells of gastrointestinal and pulmonary tracts and placenta [503].[35] Once in the cytosol, serotonin can be used as a substrate by transglutaminases. "Serotonylation" of proteins contributes to the regulation of cytoskeleton dynamics and granule secretion.

The SerT1 transporter catalyzes the 5HT influx together with Na^+ and Cl^- (stoichiometry 1:1:1) and coupled efflux of one K^+ or H^+ in the absence of K^+ cation [503]. Splice variants of the SerT1 transporter have been detected.[36]

Transport capacity of SerT1 depends on both the plasmalemmal transporter density and its intrinsic activity, which is influenced by various regulators (Table 4.15). In platelets, activated histamine receptor causes a PKG-mediated elevation in 5HT transport. Adenosine receptor stimulation has a similar effect. Cytokine receptor of tumor-necrosis factor-α and interleukin-1β as well as fibrinogen receptor $\alpha_{2B}\beta_3$-integrin increase plasmalemmal SerT1 density, whereas α2-adrenergic receptor reduces SerT1 activity in synaptosomes [503].

Serotonin Transporter in Pulmonary Arterial Hypertension

Pulmonary arterial hypertension (PAH) is characterized by an increase in pulmonary vascular resistance, vascular remodeling, and right ventricle failure. Human pulmonary arterial smooth muscle cells from idiopathic pulmonary arterial hypertension display an increased expression of serotonin transporter with elevated serotonin-induced proliferation [513]. The $5HT_{1B}$ receptor in pulmonary arteries cooperates with the SerT receptor to launch the ERK pathway and generate vasoconstriction of pulmonary arteries and proliferation of smooth muscle cells.

35. The SerT1 transporter is expressed in the central (CNS) and enteric (ENS) nervous system, adrenomedullary neuroendocrine serotonin system (NSS), hypothalamo–pituitary–adrenocortical axis (HPA), and peripheral serotonin system (PSS; i.e., heart, blood vessels, lungs, and pancreas as well as platelets.

36. Whereas SerT1a is more abundant in the central nervous system and adrenal medulla, variants SerT1a and SerT1b are equally expressed in the stomach and heart, and SerT1c is only expressed in the gut and, to a lesser extent, in heart [503].

Table 4.15. Regulators of the SerT1 transporter (Source: [503]; MAPK: mitogen-activated protein kinase; MARCKS: myristoylated alanine-rich C kinase substrate; NOS: nitric oxide synthase; PKC, PKG: protein kinase-C, -G; SCAMP: secretory carrier membrane protein). Protein kinase-C induces a biphasic attenuation of 5HT uptake, as it first inhibits transport and then lowers plasmalemmal SerT1 density. Protein phosphatase-2 (PP2) antagonizes PKC activity. Protein kinase-G increases plasmalemmal SerT1 density and transport rate. Activated P38MAPK enhances 5HT transport via elevated plasmalemmal SerT1 density. Some plasmalemmal receptors modulate SerT1 density and activity via its phosphorylation and dephosphorylation. Syntaxin-1A, a component of the synaptic vesicle docking and fusion, has an affinity for SerT1 that is controlled by calmodulin-dependent kinase-2. Phosphorylation by PKC of MARCKS-related protein MARCKSL1 (or MacMARCKS) reduces its membrane attachment and affinity for cytoskeletal proteins, thus facilitating SerT1 recycling (PKC-induced endocytosis).

Process	Agents
Phosphorylation	PKC, PKG, P38MAPK
Dephosphorylation	PP2
Exocytosis	Syntaxin-1A, SCAMP2, NOS1
Endocytosis	MARCKSL1, Hic5

4.8.3 Vesicular Monoamine Transporters

Vesicular monoamine and acetylcholine transporters (VAChT; Sect. 4.8.1) constitute a subclass (SLC8a). The family of vesicular monoamine transporters includes VMAT1 [37] and VMAT2. [38]

Vesicular monoamine transporter VMAT1 carries biogenic amines, except histamine, whereas VMAT2 transports adrenaline, dopamine, histamine, noradrenaline, and serotonin [514]. Movement of biogenic amines across membranes of secretory vesicles is powered by a transmembrane electrochemical H^+ gradient generated by H^+ ATPase.

The VMAT1 ransporter is detected mainly on neurohormone-containing, large dense-core vesicles in various endocrine and neuroendocrine cells, but not exclusively [514]. Large dense-core vesicles store biogenic amines (120–160 nmol) as well as neural peptide hormones. In chromaffin neuroendocrine cells in the adrenal gland medulla that secrete adrenaline, noradrenaline, enkephalin, and endorphins into the blood stream, catecholamine–H^+ exchanger VMAT1 transports newly synthesized adrenaline from the cytosol into chromaffin granules, where adrenaline is stored in preparation for release.

In noradrenergic neurons of the central nervous system, VMAT2 resides preferentially on large dense-core vesicles, but also on small synaptic vesicles (45–50 nmol) [514]. In dopaminergic neurons of the substantia nigra, VMAT2 is found primarily on small synaptic vesicles in axons and dendrites. In peripheral neurons, monoamines are stored in small dense-core vesicles.

37. A.k.a. SLC18a1 and VAT1.
38. A.k.a. SLC18a2.

Transporter VMAT2 has approximately a 2- to 4-fold greater affinity for biogenic amines (adrenaline, dopamine, noradrenaline, and serotonin) and substituted aromatic amines such as methylenedioxymethamphetamine, than VMAT1 [514]. Its affinity for serotonin is about 10-fold higher than that for histamine. Moreover, VMAT2 possesses a 10- to 20-fold larger affinity for unsubstituted aromatic amines, such as amphetamine, phenylethylamine, and histamine (i.e., like that for serotonin) than that of the VMAT1 transporter.

4.9 Adenine Nucleotide Transporters

Adenine nucleotide transporters (ANT) are integral membrane proteins responsible for the specific exchange of adenine nucleotides across the inner mitochondrial membrane. The exchange is specific for ADP and ATP, as AMP is released only after its conversion to ADP. The transport is electrogenic because ATP has 4 negative charges (ATP^{4-}) and ADP 3 (ADP^{3-}). The ANT transporters can function as monomers or dimers.

The ANT carrier function is influenced by the metabolic state of mitochondria. It is thus regulated by the concentrations of ADP and ATP (ADP/ATP ratio), the mitochondrial inner membrane potential, fatty acids, and temperature. Apoptosis protector BCL2 and pro-apoptotic BAX assists and reduces ADP–ATP exchange, respectively. Calcium ions, lipid messengers, such as ceramide and ganglioside GD3, reactive oxygen species, pro-apoptotic B-cell lymphoma protein-2 family members (e.g., BAX and BID), and nitric oxide can convert adenine nucleotide translocator into a large channel. In humans, 3 different isoforms of adenine nucleotide translocator (ANT1–ANT3) exist.

The ANT1 isoform is preferentially expressed in skeletal muscle, heart, and brain. It has 2 functions, as it can act as [515]: (1) a specific ADP/ATP translocator and (2) a non-specific pore. The ANT1 subtype interacts with constitutive mitochondrial proteins, such as ATP synthase, cyclophilin-F, a matrix protein at the inner membrane, and voltage-dependent anion channel (VDAC), an outer membrane protein at contact site of mitochondrial membranes. Mitochondrial glutathione S-transferase may stabilize ANT1. In astrocytes, transforming growth factor-β1 promotes ANT1 expression via SMADs and SP1 transcription factors.

The ANT2 subtype is preferentially expressed in the lung, kidney, liver, and spleen. Like ANT1, ANT2 acts as a specific ADP/ATP translocator and a non-specific pore that contributes to the apoptotic mitochondrial membrane permeabilization. Like ANT1, ANT2 interacts with ATP synthase, cyclophilin-F, VDAC, and glutathione S-transferase [516].

4.10 Nucleoside Transporters

Cells obtain nucleic acid precursors via synthesis of nucleotides and salvage of exogenous nucleobases and nucleosides. Nucleoside transporters are subdivided

Table 4.16. Nucleoside transporters and their endogenous substrates (Source: [5]; CNT: concentrative nucleoside transporter; ENT: equilibrative nucleoside transporter; PMAT: plasma membrane monoamine transporter).

Type	Other names	Endogenous substrates
CNT1	SLC28a1, N2/cit	Uridine, cytidine, thymidine, adenosine
CNT2	SLC28a2, N1/cif, SPNT	Adenosine, guanosine, inosine, thymidine
CNT3	SLC28a3, N3/cib	Uridine, cytidine, thymidine, adenosine, Guanosine, inosine
ENT1	SLC29a1	Adenosine, guanosine, inosine, uridine, Thymidine, cytidine
ENT2	SLC29a2, DER12, HNP36	Adenosine, guanosine, inosine, uridine, Thymidine, hypoxanthine
ENT3	SLC29a3	Nucleosides, nucleobases
ENT4	SLC29a4, PMAT	Adenosine

into 2 subclasses: (1) the class-28 sodium-dependent solute carriers, or concentrative nucleoside transporters (SLC28ai or CNTi) with 3 members (CNT1–CNT3; Table 4.16) and (2) the class-29 equilibrative solute carriers, or equilibrative nucleoside transporters (SLC29ai or ENTi) with 4 members (ENT1–ENT4) [5]. Nucleoside transporters also modulate extra- and intracellular concentrations of adenosine.

Production of concentrative nucleoside transporters, particularly CNT2, is associated with cell differentiation, whereas the concentration of equilibrative nucleoside transporters, mostly ENT1, rises during cell proliferation, i.e., wound healing and stimulation by growth factors (e.g., EGF and related TGFα) [517]. The ENT1 and ENT2 transporters have similar broad permeant selectivities for purine and pyrimidine nucleosides [518]. The ENT2 transporter also efficiently transports nucleobases. The ENT3 transporter that carries nucleosides and nucleobases operates in intracellular membranes. The ENT4 transporter is selective for adenosine, but also conveys various organic cations and biogenic amines (e.g., dopamine and serotonin). Both ENT3 and ENT4 are sensitive to pH; they act optimally under acidic conditions.

In polarized cells, ENT1 and ENT2 transporters reside in the basolateral membrane and, in collaboration with concentrative nucleoside transporters that localize to the apical membrane, enable the transepithelial transport of nucleosides and various nucleoside-derived drugs. The ENT1 transporter is a transmembrane glycoprotein that lodges not only in the plasma membrane, but also in mitochondrial membranes. The ENT2 transporter particularly abounds in skeletal muscle [519]. The intracellular nucleoside ENT3 transporter may act in lysosomal transport of nucleosides. The ENT4 transporter is situated in the plasma membrane. In adult mice, cardiomyocytes express predominantly ENT1, an essential agent for purine nucleoside uptake, besides adenosine receptors ENT2, ENT3, and ENT4 [520].

4.11 Nucleobase–Ascorbate Transporters

Nucleobase–ascorbate transporters constitute 2 families. Members of the nucleobase–cation symporter-1 (NCS1) family take up nucleobases.[39] Members of the nucleobase–cation symporter-2 family (NSC2), or nucleobase–ascorbate transporters (NAT),[40] are ubiquitous transmembrane proteins responsible for the uptake of nucleobase or ascorbate (secondary active transport).

In mammals, the transfer of vitamin-C is ensured by the sodium-dependent vitamin-C transporters SVCT1 (SLC23a1) and SVCT2 (SLC23a2) of the NAT/NCS2 family.[41]

4.12 Fatty Acid-Binding Proteins

Fatty acid-binding proteins transport many agents in the intra- and extracellular media (Table 4.17). They are chaperones for long-chain fatty acids, eicosanoids, retinols, retinoic acids, and related metabolites. In the cell, they interact with nuclear receptors, mainly peroxisome proliferator-activated and retinoic acid receptors (Sect. 6.3.6.5), as well as enzymes.

4.13 Retinoid-Binding Proteins

Retinoids are small, hydrophobic, liposoluble, active vitamin-A metabolites: retinol (the most usable form of vitamin-A [vitamin A alcohol]), retinal (aldehyde form), retinoic acid, and retinyl ester.

They participate in multiple processes, such as embryo- and fetogenesis and differentiation of stem cells, hematopoiesis, functioning of mucus-secreting epithelial cells, vision (11-cis-retinal),[42] maintenance of immune cell types (retroretinoids), and glycoprotein synthesis. All-trans-[43] and 9-cis-retinoic acids are potent regulators of gene transcription [522]. Biological effects of retinoids are mediated by 2 categories of proteins: (1) retinoid receptors that are transcription factors of the class of nuclear hormone receptors and (2) retinoid-binding proteins (Table 4.18).

---

39. Nucleobases comprise cytosine, guanine, adenine in both DNA and RNA, thymine in DNA, and uracil in RNA. Adenine and guanine are purines; cytosine, thymine, and uracil are pyrimidines.

40. The NAT family can be subdivided into 3 subfamilies according to the substrate types [521]: (1) transporters specific for oxidized purines xanthine and/or uric acid in bacteria, fungi, and plants; (2) transporters specific for uracil only in bacteria; and (3) transporters specific for ^{L}ascorbic acid in vertebrates. The mammalian NAT proteins are sodium symporters (but bacterial homologs are proton symporters; e.g., Escherichia coli uracil–H^+ symporter UraA).

41. The mammalian nucleobase–ascorbate transporters — SVCT1 and SVCT2 — recognise specifically ^{L}ascorbic acid, but not nucleobases purines or pyrimidines.

42. 11-cis-Retinal serves as chromophore for visual pigment rhodopsin.

43. The words retinol, retinal, and retinoic acid designate the respective all-trans-isomers.

Table 4.17. Fatty acid-binding proteins (Source: [5]).

Type	Other names
FABP1	Liver FABP (lFABP)
FABP2	Intestinal FABP (iFABP or FABPi)
FABP3	Heart FABP (hFABP), muscle FABP (muFABP), mammary-derived growth inhibitor
FABP4	Adipocyte FABP (aFABP), adipocyte lipid-binding protein (ALBP)
FABP5	Epidermal FABP (eFABP), keratinocyte FABP (kFABP), psoriasis-associated FABP (paFABP)
FABP6	Ileal FABP, gastrotropin, ileal bile acid transporter (IBABP)
FABP7	Brain FABP (bFABP), brain lipid-binding protein (BLBP), mammary-derived growth inhibitor-related protein
FABP8	Myelin FABP (myFABP), peripheral myelin protein-2 (PMP2 or MP2)
FABP9	Testis FABP (tFABP), PERF

Retinoid-binding proteins are transport proteins that solubilize their labile ligands in aqueous spaces. Moreover, they regulate localization, metabolism, and activities of retinoids. All known retinoid-metabolizing enzymes can use free retinoids. Retinoid-binding proteins can protect retinoids from some enzymes, whereas they can favor the action of others [522].

Soluble retinol-binding proteins constitute a family of carriers that bind retinol in intra- and extracellular (RBP3) spaces, including plasma (RBP4). Inside cells, all-trans-retinols and their oxidation products, all-trans-retinals, are associated with retinol-binding proteins RBP1 or RBP2 [522]. Retinol-binding protein RBP4 that may enable the secretion of retinol from the liver belongs to the lipocalin family with epididymal retinoic acid-binding protein and β-lactoglobulin.[44] Like many lipocalins that interact with accessory proteins, RBP4 connects to transthyretin,[45] a carrier of thyroxine (T_4) and retinol in blood and cerebrospinal fluid, to prevent its filtration by renal glomeruli. Protein RBP4 is mainly synthesized in liver (also the major vitamin-A storage site), but also other tissues [522].[46]

Cellular retinol-binding proteins (cRBP1, -2, -5, and -7) and retinoic acid-binding proteins (cRABP1–cRABP2) belong to the category of intracellular lipid-binding proteins (iLBP) that includes the fatty acid-binding proteins [522] (Sect. 4.12). Protein RBP1 is expressed in fetuses and adults. In adults, it particularly abounds in the lung, liver, kidney, and reproductive organs, but also in choroid plexi and retinal pigment epithelium [522]. On the other hand, in adults, RBP2 is restricted to enterocytes of the small intestine.

44. Lipocalins constitute a family of proteins that carry small hydrophobic molecules, such as steroids, bilins (bile pigments or bilanes), retinoids, and lipids. They are involved in inflammation and detoxification. They also intervene in prostaglandin synthesis and retinoid binding. They are respiratory allergens of animals.

45. A portmanteau for `trans`porter of `thy`roxine and `ret`in`ol`.

46. I.e., the adipose tissue, kidney, lung, heart, skeletal muscle, spleen, eye, and testis.

The retinol-binding protein RBP1 may carry retinol from blood into cells. Binding of cRBP1 to all-trans retinol permits the presentation of retinol to lecithin–retinol acyltransferase (LRAT)[47] to form retinyl ester as well as the conversion of retinol to retinal by the retinol dehydrogenase isoenzymes.

The ratio of apo- to holocRBP determines the flux of retinol.[48] Agent holocRBP1 favors retinol esterification by lecithin–retinol acyltransferase, whereas apocRBP1 precludes retinol esterification, but stimulates retinyl ester hydrolysis [522]. Protein RBP2 may impede reoxidation of retinol to retinal and help retinol esterification by lecithin–retinol acyltransferase.

Cellular retinoic acid-binding proteins have different distribution patterns. In adults, cRABP1 is produced almost ubiquitously, whereas cRABP2 is only expressed in skin, uterus, and ovary [522]. Both cRABP1 and -2 can reside in the nucleus, where they can deliver retinoic acid to retinoic-acid receptors. Protein cRABP1 enhances the conversion of retinoic acid into polar metabolites [522].

Retinaldehyde-binding protein (RAlBP) is detected in the retinal pigment epithelium and neural retina, as well as in the pineal gland, ciliary body, cornea, optic nerve, and brain [522]. Protein cRAlBP enables isomerization of all-trans-retinol to 11-cis-retinol and promotes oxidation of 11-cis-retinol to 11-cis-retinal [522].

4.14 Flavonoid Transporter

Flavonoids[49] in fruits and vegetables are ketone-containing compounds that lower the risk of cardiovascular diseases. Flavonoid-rich diets, indeed, have anti-oxidant, anti-inflammatory, and vasodilatory effects. Dietary flavonoid uptake by vascular endothelial cells is mediated by a plasmalemmal bilirubin-specific transporter, the bilitranslocase [523]. In endothelial cells, flavonoids target various substrates, such as nitric oxide synthase, NADPH oxidase, adenosine deaminase, and 5′-nucleotidase.

47. Another enzyme, acylCoA–retinol acyltransferase (ARAT) catalyzes retinyl ester synthesis.

48. Agent holocRBP represents the most abundant form of unesterified retinol. It is a substrate for NADP-dependent retinol dehydrogenase. Produced retinal from holocRBP is converted to retinoic acid in the presence of apocRBP. Agent apocRBP stimulates the hydrolysis of membrane-bound retinyl esters, serves as a carrier for the released retinol, and is involved in the mobilization of membrane-bound retinyl esters.

49. A.k.a. vitamin-P and citrin.

Table 4.18. Retinoid-binding proteins (Sources: [5, 59, 522]; cRBP: cellular retinol-binding protein; cRABP: cellular retinoic acid-binding protein; cRAlBP: cellular retinaldehyde-binding protein; iRBP: interphotoreceptor retinoid-binding protein or interstitial retinol-binding protein; pRBP: plasma retinol-binding protein; RBP: retinol-binding protein; RBPc: retinol-binding protein, cellular; RlBP: retinaldehyde-binding protein). Proteins cRAlBP and iRBP reside specifically in retina and nervous system.

Type	Other names	Major ligands Interactions
		Retinol-binding protein
RBP1	cRBP1, RBPc1	All-trans-retinol, all-trans-retinal Low-density lipoprotein-related protein-2
RBP2	cRBP2, RBPc2	All-trans-retinol, all-trans-retinal
RBP3	iRBP	All-trans-retinol, 11-cis-retinal, Long-chain fatty acids
RBP4	pRBP	All-trans-retinol Pre-albumin, amyloidosis type 1
RBP5	cRBP3, hRBPiso	All-trans-retinol
RBP7	cRBP4	All-trans-retinol
		Retinaldehyde-binding protein
RlBP1	cRAlBP	11-cis-retinol, 11-cis-retinal Short-chain dehydrogenase–reductase SDR9c5
		Retinoic acid-binding proteins
cRABP1		All-trans-retinoic acid
cRABP2	RBP6	All-trans-retinoic acid Cyclin-D3

4.15 Citrate and Succinate Transporters

Mitochondrial tricarboxylate transport protein SLC25a1 [50] serves as citrate transport protein. [51] Prostatic citrate carrier is an isoform of SLC25a1 mitochondrial transporter [524]. [52]

Plasmalemmal citrate transporters of the SLC13 class include sodium-dependent citrate transporters SLC13a2, SLC13a3, and SLC13a5 (or NaCT). Mammalian low- and high-affinity sodium–dicarboxylate cotransporters, such as SLC13a2 [53] and

50. A.k.a. SLC20a3 and mitochondrial citrate carrier (MCiC).

51. Multiple carrier proteins that are involved in energy transfer reside in the inner membranes of mitochondria. They include: ADP–ATP translocase, 2-oxoglutarate–malate carrier, phosphate carrier, tricarboxylate transport protein, etc.

52. The prostate produces and secretes large amounts (up to 180 mmol) of citrate into the prostatic fluid. Potassium-dependent released prostatic citrate acts mainly as an energy substrate for sperm that can then raise their ATP production. In prostatic epithelial cells, mitochondrial aconitase synthesizes citrate. This rate-limiting synthase is regulated by Zn^{++} and hormones.

53. A.k.a. sodium-dependent dicarboxylate transporter NaDC1.

SLC13a3,[54] carry succinate and other Krebs-cycle intermediates and operate in the handling of citrate by kidneys.

4.16 Aquaporins

Although, in most cells, water crosses the lipid bilayer by osmosis, additional mechanisms of water transport explains high water permeability of some cells. Integral membrane channel aquaporins (Aqp) that form pores in the cell membrane regulate the water flux across the plasma membrane. Aquaporins are entangled in secretory and absorptive functions, especially for fluid production and reabsorption, in the lung, kidneys, eyes, salivary and sweat glands, gastrointestinal tract, etc.

4.16.1 Aquaporin Family

The aquaporin family include 13 types identified in mammals (Aqp0–Aqp12; Tables 4.19 and 4.20). They are divided into 3 subfamilies based on the primary sequences: (1) *water-selective* Aqps (Aqp0–Aqp2, Aqp4–Aqp6, and Aqp8); (2) *aquaglyceroporins* (Aqp3, -7, -9, and -10); and (3) *superaquaporins* (Aqp11 and Aqp12). Several aquaporin types can be detected in a given tissue, as tissues contain many different cell types.[55]

Aquaporins (Table 4.21) are synthesized by many cell types from the AQP gene family. These water channels carry water molecules in and out of the cell, but prevent the passage of ions and other small solutes. However, *aquaglyceroporins* also transport small uncharged solutes, such as glycerol, CO_2, ammonia, and urea according to pore size. Yet, Aqp6 that is characterized by a significant basal water permeability, conducts anions. Seven aquaporin types (i.e., Aqp0, -1, -4, -5, -8, -9, and -11) have been identified in the liver and biliary tree, where they contribute to bile secretion [527].

Aquaporin-1 not only serves in water flux, but also is a cGMP-gated ion channel. Water channel Aqp1, like certain RH family members, is permeable to gas, such as CO_2 and NH_3. The Aqp1 tetramer of the erythrocyte membrane carries CO_2 delivered by cells [530]. It is also permeable to nitric oxide. Isotransporters Aqp1 (human erythrocyte), Aqp3 (rat astrocytic endfeet at the blood–brain barrier), Aqp5 (rat alveolar type-1 pneumocytes), Aqp8, and Aqp9 are permeable to NH_3.

4.16.2 Water-Selective Aquaporins

Trans- and paracellular (through tight junction) transports of water and electrolytes maintain homeostasis at a given time (owing to biological cycles). Claudin-7

54. A.k.a. NaDC3.

55. For example, in developing orofacial tissues, Aqp1, Aqpn3, and Aqpn4 are identified in developing tooth, Meckel's cartilage, submandibular gland, and masseter and infrahyoid muscles (Aqp3 also in gingiva); Aqpn10 in developing tooth, gingiva, and masseter and infrahyoid muscles; Aqp8 and Aqp9 in masseter muscle [525].

Table 4.19. Aquaporin family (**Part 1**; Source: [526]). In humans, aquaporins supplement ammonia transport of Rhesus proteins (Aqp3, -7, -8, -9, and possibly -10) and urea transporters (Aqp7, -9, and possibly -3). In the kidney, at least 7 aquaporins are synthesized at distinct sites. The kidney's basic structural and functional unit is the nephron that regulates concentrations of water and soluble substances like ions by filtering blood, reabsorbing a part, and excreting the rest as urine, thereby controlling blood volume and pressure, levels of electrolytes and metabolites, and regulates blood pH (ADH: antidiuretic hormone [vasopressin]; CCT: cortical collecting tubule; ICT: initial collecting tubule MCD: medullary collecting duct (I/O: inner/outer segment); PCT: proximal convoluted tubule; PST: proximal straight tubule; tDLH: thin descending limb of loop of Henle; NH_3: ammonia gas).

Type	Location	Function
Aqp0	Lens	Water channel, junctional protein
Aqp1	Renal nephron	Water reabsorption
	(PCT, PST, tDLH)	NH_3 transport
	(apical membrane)	
	Erythrocyte	Water and CO_2 transport
	Multiple organs	
Aqp2	Renal nephron	Water reabsorption in response to ADH
	(ICT, CCT, I/OMCD)	
	(apical membrane)	
	Testis	
Aqp3	Renal nephron	Water reabsorption
	(MCD)	Glycerol, ammonia, and urea transport
	(basolateral membrane)	
	Astrocyte	
	Multiple organs	
Aqp4	Renal nephron	Water reabsorption
	(MCD)	CO_2 transport
	(basolateral membrane)	
	Brain and other organs	
Aqp5	Salivary acinar cell	
	Alveolar type-1 pneumocyte	Water and CO_2 transport
	Eye	
Aqp6	Intracellular vesicles	Water and anion transport
	(CT)	

and -3 and occludin regulate the permeability of tight junctions, whereas aquaporin-5 is the major transcellular water transporter in salivary acinar cells. Aquaporin-5 in salivary gland is required for water transport not only across the plasma membrane, but also through tight junctions [535]. Aquaporins, claudins, and occludins are involved in the crosstalk between both trans- and paracellular transports that are coordinated rather than compensatory.

Table 4.20. Aquaporin family (**Part 2**; Source: [526]).

Type	Location	Function
Aqp7	Renal nephron (PST)	Water transport
	Spermatocyte, adipocyte	Glycerol, ammonia, and urea transport
Aqp8	Proximal tubule, collecting duct (intracellular)	Water and NH_3 transport (glycerol impermeable)
	Heart, pancreas, liver, colon, testis, epididymis, placenta	
Aqp9	Leukocytes	Water transport
	Lung, brain, liver, spleen, testis, epididymis	Glycerol, NH_3, and urea transport
Aqp10	Intestinal epithelium (apically)	Water, glycerol, and NH_3 transport
Aqp11	Proximal tubule, urothelium, hepatobiliary system	Development of the proximal tubule
	Endoplasmic reticulum	
Aqp12	Pancreatic acinus	Glycerol and small solutes transport
		Digestive enzyme secretion

Table 4.21. Aquaporin features (Source: [5]).

Type	Activators	Inhibitors	Permeability
Aqp0		Hg^{++}	Water (low)
Aqp1	cGMP	Hg^{++}, Ag^{+}	Water (high)
Aqp2		Hg^{++}	Water (high)
Aqp3		Hg^{++}, H^{+}	Water (high), glycerol
Aqp4		PKC	Water (high)
Aqp5		Hg^{++}	Water (high)
Aqp6	H^{+}	Hg^{++}	Water (low), anions
Aqp7		Hg^{++}	Water (high), glycerol
Aqp8		Hg^{++}	Water (high)
Aqp9		Hg^{++}	Water (low), glycerol
Aqp10		Hg^{++}	Water (low), glycerol

4.16.3 Aquaglyceroporins

Aquaporin-3, as well as Aqp8, but not Aqp1, support the uptake of hydrogen peroxide (H_2O_2) produced by plasmalemmal NADPH oxidase to trigger redox signaling for cell growth, differentiation, and migration, but avoiding non-specific oxidative stress [528]. Intracellular accumulation of H_2O_2 can indeed be regulated by Aqp3.

Aquaporin-7, the expression of which is inhibited by insulin, also carries glycerol[56] and thus regulates fat accumulation. Aquaporin-7 is the main cardiac glycerol uptake channel [534]. In healthy cardiomyocytes, ATP, the source of chemical energy for all energy-consuming reactions, is mainly produced by fatty acid oxidation in mitochondria, whereas the remaining ATP is yielded by carbohydrate oxidation using primarily glucose and ketone bodies. The heart thus produces chemical energy from different substrates and can shift between sources, according to supply availability. Under high ATP demand conditions, glycolysis and phosphotransferase reactions provide alternative sources of ATP synthesis. In cardiomyocytes, glycerol that is phosphorylated to glycerol 3-phosphate represents another metabolic substrate. It contributes not only to phospholipid synthesis and lipid storage as triacylglycerol, but also energy production via oxidation, especially during stress. Under low fatty acid oxidation conditions, heart favors glycerol over glucose consumption.

4.16.4 Structural and Functional Features

Functional aquaporins (at least Aqp1–Aqp5) form tetramers; each monomer acts as a water channel. However, Aqp4 can form larger oligomers in the plasma membrane. Aquaporins are constituted by 6 transmembrane α-helices with 5 interhelical extra- and intracellular loops and N- and C- termini on the inner surface of the membrane [526]. Structural features determine Aqp localization owing to apical or basolateral sorting motif as well as post-translational glycosylation.

The aromatic–arginine selectivity filter is constituted by a group of amino acids that serves to bind to water molecules and exclude other molecules. Ammonia-permeable aquaporins differ from other aquaporins by a less restrictive aromatic–arginine domain.

The unit of water permeability of Aqp1, i.e., conductance per monomer, is very high ($\sim 3\times10^9$/s). Water channel gating depends on local chemical and physical conditions. Lowered cytosolic pH and cAMP stimuli preclude water permeability elevation induced by vasopressin. At pH lower than 5.5, water permeability and anion conductance through Aqp6 augment rapidly and reversibly.

4.16.5 Aquaporins in the Respiratory Epithelium

Low-magnitude mechanical stresses activate the non-selective cation channel transient receptor potential vanilloid TRPV4 and, subsequently, voltage-gated calcium Ca_V1 channel. The resulting increase in intracellular calcium triggers endocytosis and degradation of apical aquaporin-5. Modulation of paracellular permeability of the epithelial barrier and water flux in respiratory epithelia by aerodynamic stresses transmitted to the airway surface liquid and epithelium relies on aquaporin-5 [531].

56. Glycerol binds 3 fatty acids to form triglycerides. Glycerol released from adipocytes into the blood is taken up mainly by the liver and converted to glucose.

In addition, aerodynamic stresses also provoke actin and desmoplakin redistribution [531]. Desmoplakin, a component of desmosomes (Vol. 1 – Chap. 7. Plasma Membrane), connects to intermediate filaments. Aquaporin-0 and -2 bind to intermediate filament proteins and actin, respectively.

4.16.6 Aquaporins in the Nephron

At least 6 aquaporin types are located in the kidney. Aquaporin-1 abounds intensely in the proximal tubule and descending thin limb of Henle's loop, where it causes water reabsorption, but is absent in water-impermeable thin and thick ascending limbs and distal tubule. Because it does not reside in the collecting duct, it is not entailed in vasopressin regulation.

Aquaporin-2 is exclusively expressed in principal cells of the connecting tubule and collecting duct. It is the predominant water channel regulated by vasopressin [57] that activates the cAMP–PKA pathway. [58] Acute regulation of water reabsorption accelerates Aqp2 vesicular transport between an intracellular reservoir and apical plasma membrane. [59] Channel Aqp2 is expressed in nephron cells that can experience potentially hypertonic environment and osmotic stress. [60] Cell survival in a hypertonic environment is ensured by the *hypertonic stress response*. [61] Channel Aqp2 is

---

57. Upon vasopressin binding to Gs-coupled receptors V_2 (Sect. 7.13.63) that stimulates adenylate cyclase, intracellular concentration of cyclic adenosine monophosphate rises. Messenger cAMP then activates protein kinase-A. Prostaglandin PGe2 prevents vasopressin-induced water permeability by reducing cAMP level as well as removing Aqp2 from the plasma membrane. In addition, hypercalcemia and hypokalemia lower Aqp2 expression. Hypercalcemia also represses Na^+–Ka^+–$2Cl^-$ cotransporter NKCC2 (kidney-specific Na^+–Ka^+–Cl^- symporter SLC12a1 or bumetanide-sensitive sodium-potassium-chloride cotransporter BSC1). During pregnancy, extracellular fluid volume rises (30–50%): the renin-angiotensin-aldosterone system is activated, the threshold of vasopressin secretion is reset, and Aqp2 synthesis increases. Steroid female hormones can influence plasma vasopressin concentration.

58. In kidneys, many water and ion transport are mediated via cyclic adenosine monophosphate that activates protein kinase-A (PKA) as well as Rap guanine nucleotide-exchange factors RapGEF3 and RapGEF4 (a.k.a. exchange proteins directly activated by cAMP EPAC1 and -2; alternative pathways to PKA) [529]. Activators RapGEF3 and RapGEF4 are expressed at the brush border of proximal tubule cells, thick ascending limbs of Henle's loop (mainly apically), and distal convoluted tubule and connecting tubule (RapGEF3 apically). In the collecting duct, RapGEF3 expression is dispersed in intercalated cells only of the cortical segment, principal and intercalated cells of outer medullary segment, and mainly Aqp2– cells in the inner medullary segment, whereas RapGEF4 is produced in cortical principal cells, dispersed and apical in outer medullary segment, and all cells of inner medullary segment.

59. Protein kinase-A (Vol. 4 – Chap. 4. Cytosolic Protein Ser/Thr Kinases) phosphorylates Aqp2 monomers (Ser256) to enhance Aqp2 transport that entails microtubules and actin–myosin filaments (Vol. I, Part A – Chaps. 6. Cell Cytoskeleton and 9. Intracellular Transport).

60. High level of hypertonicity can happen during periods of antidiuresis to allow for efficient water reabsorption.

61. Gene expression that allows accumulation of compatible osmolytes within the cell is primed by the transcription factor nuclear factor of activated T cells NFAT5 (or tonicity-

also involved in long-term adaptation of body water balance, i.e., prolonged changes in body hydration status, via its production level.

Isotypes Aqp3 and Aqp4 are both present in basolateral plasma membrane of principal cells of collecting duct and represent passages for water reabsorbed apically via Aqp2.

Channel Aqp6 resides in intracellular vesicles in type-A intercalated cells of the collecting duct. Channel Aqp7 abounds in the brush border of proximal tubule cells.

Channel Aqp8 lodges intracellularly at small levels in principal cells of the proximal tubule and collecting duct. Isotypes Aqp9 and Aqp10 are also expressed in kidneys.

In kidneys, aquaporin-2 contributes to water absorption and urine concentration. Like any carrier, its density at the apical plasma membrane of epithelial cells of the collecting duct of the nephron results from the balance between exo- and endocytosis, the latter yielding both recycling and degradation routes. These processes depend on Aqp2 C-terminus phosphorylation and subsequent between-protein interactions. Many proteins, indeed, bind to aquaporin-2, such as actin, clathrin, dynamin, annexin-2, AP2 adaptor complex, myelin and lymphocyte protein (MAL), Rap1 GTPase-activating signal-induced proliferation-associated protein SIPA1, LIP5 that participates in multivesicular body formation and sorting, phosphatase PP1c, HSP70, HSP73 [62] and polyubiquitin-C (for polyubiquitin chain) [59, 533]. Exocytosis of Aqp2 relies on activated adenylate cyclase that stimulates protein kinase-A via cAMP.

Vasopressin regulates phosphorylation of Aqp2 (Ser256, Ser261, Ser264, and Ser269) [533]. In fact, Aqp2 endocytosis rate depends on its phosphorylation on 2 residues (Ser256 and Ser269). [63]

Nitric oxide and atrial natriuretic peptide that increase the cGMP concentration also elicit apical Aqp2 accumulation. On the other hand, prostaglandin-E2 and dopamine can provoke Aqp2 endocytosis.

---

responsive enhancer binding protein [TonEBP]). Expression of Aqp2 results from hypertonic stress response not only via NFAT5, but also and independently via calcium- and calcineurin-dependent, cytoplasmic nuclear factor of activated T-cell pathway [532].

62. A.k.a. Hsc70 and HSPa8. It is expressed constitutively, whereas HSP70 expression is induced by stress and heat shock. Both are members of the cytoplasmic 70-kDa heat shock protein family. These members are involved in appropriate folding and exocytosis of newly synthesized proteins. Agents HSP70 and HSP73 can have antagonistic effects.

63. Residue Ser256 is necessary for Aqp2 exocytosis, whereas phosphorylation of both Ser256 and Ser269 can modulate Aqp2 endocytosis. Aquaporin-2(Ser256^{P}) is distributed throughout the cell, whereas Aqp2$^{S269^{P}}$ is detected only in the apical plasma membrane. On the other hand, neither Ser261 nor Ser264 significantly influences cAMP-induced Aqp2 transfer.

4.17 Glucose Carriers

Two groups of glucose transporters exist: (1) the GLUT group, i.e., the SLC2A family, of facilitated diffusion glucose transporters that are uniporters and (2) the SGLT group of sodium–glucose carriers that are cotransporters or symporters.

4.17.1 Sodium–Glucose Cotransporters – Active Transport

The SGlT1 carrier, or SLC5a1,[64] is involved in the active transport of glucose and galactose. Transport of glucose across the apical membrane of intestinal and renal epithelial cells depends on secondary active Na^+–glucose symporters — SGlT1 and SGlT2 — that use the energy provided by Na^+ cotransport down the electrochemical gradient.

The SLC5a1 carrier is strongly produced in the small intestine and, to a lesser extent, in kidneys, lungs, and liver [501]. The production of SLC5a2 (or SGlT2) is restricted to the kidney cortex.

The low-affinity Na^+–glucose cotransporter SGlT3, or SLC5a4,[65] resides in the intestine, spleen, liver, and muscles, as well as, to a lower level, kidneys. In fact, glucose carrier-like SGlT3 protein is unable to transport glucose, but may act as a glucose sensor (transceptor).

4.17.2 Glucose Transporters – Passive Transport

Glucose transporters (GluT) of the SLC2 class are responsible for the uptake of several monosaccharides, such as glucose, fructose, mannose, galactose, and glucosamine. Many glucose transporter isoforms exist with a given tissue distribution, substrate specificity, transport kinetics, and regulated expression. The GluT (SLC2) class indeed include 13 known members that are decomposed into 3 families on the basis of their structural characteristics (sequence homology) [536]: family 1 of glucose transporters (GluT1–GluT4); 2 of fructose transporters (GluT5, -7, -9, and -11); and 3 (GluT6, -8, -10, and -12, as well as H^+–myoinositol transporter HMIT1 encoded by the SLC2A13 gene.

Facilitative glucose transporters share common structural features. Glucose transporters contain 12 transmembrane helices with a large cytoplasmic loop between helice 6 and -7 and cytoplasmic N- and C-termini. The cytoplasmic N-terminus of GluT4 comprises a domain that can mediate intracellular sequestration. Glucose transporters are also characterized by their sugar transporter motif that corresponds to a N-linked glycosylation site.

4.17.2.1 Glucose Transporter GluT1

The GluT1 transporter is expressed in most tissues, especially in insulin-responsive sites for glucose disposal, i.e., myocytes and adipocytes. It is highly produced

64. A.k.a. sodium–glucose transporter (NaGT).
65. A.k.a. solute carrier neutral amino acid transporter SAAT1 and SLC54.

in endothelial cells and erythrocytes. It localizes predominantly to the plasma membrane, even under basal conditions.

4.17.2.2 Glucose Transporter GluT2

The GluT2 transporter is a low-affinity isoform (high Michaelis-Menten constant) that resides in the liver, intestine, and kidney, as well as pancreatic β cells [536]. This transporter is a component of glucose sensor pancreatic β cells. It also allows glucose absorption by intestinal epithelial cells.

4.17.2.3 Glucose Transporter GluT3

The GluT3 transporter is expressed primarily in neurons. Glucose hence crosses the blood–brain barrier (Vol. 5 – Chap. 7. Vessel Wall) to enter neurons through endothelial GluT1 and neuronal GluT3 transporters.

4.17.2.4 Glucose Transporter GluT4

The GluT4 transporter is the SLC2a4 protein. The major substrate of GluT4 is ^{D}glucose. Other sugars, such as ^{D}mannose and 2-deoxyglucose, as well as, to a lesser extent, ^{D}galactose and ^{D}fructose, are also carried.

The GluT4 transporter is synthesized in many tissues and abounds in insulin-sensitive tissues: white and brown adipose tissues, skeletal muscles, and myocardium. Unlike most other GluT isoforms, it is sequestered in intracellular storage vesicles in basal conditions.

GluT4 in Cells at Rest

At rest, GluT4 indeed mainly resides in small vesicles scattered throughout the cytoplasm, especially recycling endosomes in the perinuclear region. The GluT4 transporter is linked to a specialized compartment of the recycling endosomal network. Insulin-responsive amino-peptidase (IRAP) [66] and GluT4 are sequestered in the same recycling endosomal compartment [537]. [67]

---

66. A.k.a. cystinyl aminopeptidase, leucyl–cystinyl aminopeptidase (LNPep), placental leucine aminopeptidase (PLAP), and oxytocinase.

67. Membrane amino-peptidases process regulatory peptides. Peptidase IRAP that colocalizes with GluT4 in insulin-responsive storage vesicles cleaves vasopressin, oxytocin, kallidin (or ^{Lys}bradykinin; released from low-molecular weight kininogen by tissue kallikrein), ^{Met}enkaphalin (or opioid growth factor [OGF], an opioid peptide neurotransmitter, the other form of enkephalin being ^{Leu}enkephalin), dynorphin, and angiotensin-3 and -4 [536]. It also serves as angiotensin-4 receptor.

Stimulation by Insulin

The PI3K (Vol. 4 – Chap. 1. Signaling Lipids) and RhoQ (Vol. 4 – Chap. 8. GTPases and Their Regulators) pathways are triggered by insulin to translocate GluT4 to the plasma membrane, the exocytosis relying on the actin cytoskeleton. Filamentous actin, the so-called *caveolin-associated* ^{F}*actin* (Cav–actin), localizes around caveolae.

Insulin causes actin rearrangements not only in the cell cortex, but also perinuclear regions [536]. Insulin-stimulated GluT4 translocation also requires myosin-1C action. The concentration and composition of phosphoinositides that modulate actin dynamics and membrane trafficking are regulated by both kinases (e.g., PI3K1a) and phosphatases (e.g., PTen and SHIPs). Guanine nucleotide-exchange factor RacGEFs, such as Vav1 and TIAM1, interact with $PI(3,4,5)P_3$ to be activated and stimulate GTPases that control the actin cytoskeleton. Phosphoinositide $PI(3,4,5)P_3$ is also targeted by ArfGEFs and ArfGAPs that control activity of ARF1 and ARF6, which are involved in actin reorganization. On the other hand, $PI(4,5)P_2$ also contributes to the regulation of the actin structuring, as it modulates activities of multiple actin-binding proteins.[68] In addition, phosphatidylinositol 4-phosphate 5-kinase-α provokes actin nucleation from sphingolipid–cholesterol-rich rafts.

The microtubule cytoskeleton as well as intermediate filaments also participate in insulin-induced GluT4 translocation via their respective components α-tubulin and vimentin as well as microtubule nanomotors dynein and kinesin.

In atriomyocytes, GluT4 is detected in secretory granules that contain atrial natriuretic factor. In response to insulin or exercise, GluT4 rapidly translocates to the plasma membrane. Conversely, upon excitation arrest, GluT4 endocytosis uses both clathrin-dependent and -independent pathways [537].

The expression of the GLUT4 gene is regulated by several transcription factors, such as myocyte enhancer factor MEF2a and GLUT4 enhancer factor (GEF) that cooperate in GLUT4 transcription. Krüppel-like factor KlF15, thyroid hormone receptor THRα1, myogenic differentiation factor, peroxisome proliferator-activated receptor-γ coactivator PGC1, nuclear receptor NR4a1, sterol response element-binding protein SREBP1c, liver X receptor LXRα, and CCAAT/enhancer-binding protein C/EBPα increase GLUT4 transcription [537]. On the other hand, nuclear factor NF1, histone deacetylase HDAC5, early B-cell factor-1,[69] and peroxisome proliferator-activated receptor-γ repress GLUT4 transcription.

Insulin regulates glucose homeostasis by increasing the glucose uptake rate, especially into myocytes and adipocytes, using glucose transporter GluT4. During exercise, GluT4 also undergoes a rapid redistribution within the cell to raise the sur-

68. Phosphoinositide $PI(4,5)P_2$ inhibits actin-severing proteins such as the actin depolymerizing factor (ADF)–cofilin complex, capping proteins such as gelsolin, and profilin. Conversely, it activates actin crosslinking proteins such as α-actinin and links actin to the plasma membrane via vinculin, talin, and ezrin–radixin–moesin proteins. It also stimulates WASP family proteins in conjunction with CDC42 and RhoQ GTPases for ARP2/3-induced actin nucleation.

69. A.k.a. olfactory neuron protein Olf1.

face GluT4 density. Translocation of GluT4 involves the PI3K–PKB pathway upon insulin stimulation and AMP-activated protein kinase in the case of exercise. Vesicular transport leads to tethering, docking, and fusion of GluT4-containing vesicles at the plasma membrane. Kinases AMPK and PKB phosphorylate GTPase activating proteins that target monomeric GTPases Rab8, -10, and -14 that are implicated in GluT4 transfer [537].

The adipokine retinol-binding protein-4 (RBP4) is secreted by adipose tissues in the absence of glucose uptake by GluT4. It restricts glucose uptake in the liver and skeletal muscle, inhibiting insulin signaling. It increases glucose output by the liver, stimulating the gluconeogenic enzyme phosphoenolpyruvate carboxykinase in the liver [538].

In adipocytes, atypical protein kinase CDK5 that is stimulated by insulin regulates GluT4-mediated glucose uptake [539]. Activated CDK5 phosphorylates calcium-binding, integral vesicular membrane protein synaptotagmin-1. The latter then connects to GluT4 transporter.

4.17.2.5 Glucose Transporter GluT8

The GluT8 transporter is synthesized mainly in testis and in smaller amounts in most other tissues. In cells, it is predominantly located in intracellular microsomes and moves to the plasma membrane in response to IGF1 and insulin.

The GluT8 transporter can be a multifunctional sugar transporter needed during embryogenesis [540]. In adipocytes, its expression rises markedly when these cells undergo differentiation. Moreover, it is sensitive to prolonged hypoxia and glucose deprivation. In liver, GluT8 is produced in perivenous hepatocytes, where it regulates glycolytic flux.

4.17.2.6 Glucose Transporter GluT9

The GluT9 transporter is devoted to hexose transport across membranes, but with low-affinity glucose and fructose transport [541]. In humans, 2 splice variant transcripts differ only in their cytoplasmic N-terminus: short GluT9$^{\Delta N}$ (short N-terminus) and long GluT9^{WT} (wild type) forms. Both forms are present mainly in the liver and kidney; otherwise, they are differentially expressed in various other organs. Highest GLUT9^{WT} levels are observed in the kidney, liver, and lung. In polarized cells, GluT9$^{\Delta N}$ and GluT9 (GluT9^{WT}) localize to the apical and basolateral membranes, respectively.

The GluT9 transporter located in the basolateral membrane of hepatocytes and both apical and basolateral membranes of the distal nephron cells also serves as urate transporter [542].[70] Urate–anion exchanger SLC22a12 (or Urat1)[71] enables urate reabsorption and secretion across the apical membrane of epithelial cells of the

70. Urate is the end product of purine degradation.

71. A.k.a. organic anion transporter 4-like protein (OATL4) and renal-specific transporter (RST).

proximal convoluted tubule. Organic anion transporter SLC22a11 [72] in apical membrane can also transport urate. Voltage-dependent GluT9 of the basolateral membrane of proximal tubular epithelial cells allows urate reabsorption, whereas organic anion transporters SLC22a6 (or OAT1) and SLC22a8 (or OAT3) of the basolateral membrane mainly cause urate excretion.

4.17.2.7 Glucose Transporter GluT10

The GluT10 subtype is longer than other known members of the GLUT superfamily. In humans, GluT10 is detected in the heart, lung, brain, kidney, liver, skeletal muscle, pancreas, and placenta [543]. ^{D}Glucose and ^{D}galactose compete with 2-deoxy ^{D}glucose (carrying Michaelis constant ~ 0.3 mmol).

In most cells, mitochondria are the major source of reactive oxygen species. Mitochondria incorporate and recycle vitamin-C. Transporter GLuT10 localizes predominantly to mitochondria of smooth muscle cells and insulin-stimulated adipocytes. It facilitates the cellular uptake and transport of the oxidized form of vitamin C — ^{L}dehydroascorbic acid (DHA) — into mitochondria that protects cells against oxidative stress [544]. [73]

4.18 Superclass of ATP-Binding Cassette Transporters

Adenosine triphosphate-binding cassette transporters, also called traffic ATPases, constitute a superclass of ubiquitous transmembrane proteins involved in ATP-powered cellular export or import of various substances, such as ions, phosphate esters, inorganic phosphate and sulfate, glucids, lipids, cyclic nucleotides, iron-chelator complexes, steroids, vitamins, peptides, and proteins, as well as rare elements and drugs (Tables 4.22 and 4.23). The ABC transporters convert chemical energy from ATP hydrolysis to mechanical energy for the transmembrane transfer of substrates.

4.18.1 Classification of ABC transporters

In humans, the superclass of ABC transporters that are encoded by 49 known human ABC genes is one of the largest sets of proteins. The ABC transporter superclass is decomposed into 7 classes [545]: (1) class A (ABC1 with 12 known members); (2) class B, or class of multiple drug resistance and transporter associated with antigen processing (MDR–TAP with 11 identified members); (3) class C, or class of cystic fibrosis transmembrane conductance regulator and multidrug resistance-like proteins (CFTR–MRP with 13 detected members); (4) class D, or class of adrenoleukodystrophy gene products (ALD with 4 known members); (5) class E, or

72. A.k.a. organic anion transporter OAT4.

73. Mutations in the glucose transporter GLUT10 gene alter angiogenesis and cause arterial tortuosity syndrome.

Table 4.22. Superclass of ATP-binding cassette transporters (**Part 1**; Source: [5]).

Type	Other names
ABCa1	Cholesterol efflux regulatory protein (CERP)
ABCa2	
ABCa3	
ABCa4	Retinal-specific ATP-binding cassette transporter
ABCa5	
ABCa6	
ABCa7	
ABCa8	
ABCa9	
ABCa10	
ABCa12	
ABCa13	
ABCb1	Multidrug resistance protein-1, P-glycoprotein-1, CD243
ABCb2	Antigen peptide transporter APT1, peptide supply factor PSF1, Peptide transporter involved in antigen processing TAP1
ABCb3	Antigen peptide transporter APT2, peptide supply factor PSF2, Peptide transporter involved in antigen processing TAP2
ABCb4	Multidrug resistance protein-3, P-glycoprotein-3
ABCb5	
ABCb6	Mitochondrial ABC transporter MTABC3
ABCb7	
ABCb8	Mitochondrial ABC transporter MABC1
ABCb9	TAP-like protein (TAPL)
ABCb10	Mitochondrial ABC transporter MTABC2
ABCb11	Bile salt export pump (BSEP), P-glycoprotein-4, sister of P-glycoprotein (SPGP), Progressive familial intrahepatic cholestasis protein PFIC2

class of oligoadenylate-binding proteins (OABP), [74] with one member; (6) class F, or class of general control proteins (GCN20 with 3 known members); [75] and (7) class G, or class of White proteins (with 5 known members). [76]

74. A.k.a. ribonuclease-4 inhibitor (RNaseLI).

75. General amino acid-based control (GCN) is a regulatory mechanism that augments transcription of numerous genes in the case of purines or amino acid starvation. The GCN2 kinase, or eukaryotic translation initiation factor-2α kinase-4 (eIF2αK4) in humans, phosphorylates the α subunit of translation initiation factor eIF2 that delivers initiator tRNA to ribosome 40S subunit, thereby stimulating translation initiation, especially that of transcriptional activator GCN4 [547]. Protein GCN20 is a member of the ATP-binding cassette superclass of proteins. Complex components GCN1 and GCN20 couples the kinase activity of GCN2 to the availability of amino acids.

76. In Drosophila melanogaster, half transporter members also comprise ABC transporter protein expressed in trachea (ATET) as well as Brown and Scarlet). ABC Transporter White is an eye pigmentation marker. With its binding partners Brown and Scarlet, it is involved in

Table 4.23. Superclass of ATP-binding cassette transporters (**Part 2**; Source: [5]).

Type	Main names
ABCc1	Multidrug resistance-associated protein MRP1, Leukotriene C4 transporter
ABCc2	MRP2, canalicular multispecific organic anion transporter cMOAT1
ABCc3	MRP3, multispecific organic anion transporter MOATd (cMOAT2)
ABCc4	MRP4, multispecific organic anion transporter MOATb
ABCc5	MRP5, multispecific organic anion transporter MOATc
ABCc6	MRP6, multispecific organic anion transporter MOATe
ABCc7	Cystic fibrosis transmembrane conductance regulator (CFTR)
ABCc8	Sulfonylurea receptor SUR1
ABCc9	Sulfonylurea receptor SUR2
ABCc10	MRP7
ABCc11	MRP8
ABCc12	MRP9
ABCc13	
ABCd1	Adrenoleukodystrophy protein (ALDP)
ABCd2	Adrenoleukodystrophy-related protein (ALDR or ALDL1)
ABCd3	70-kDa peroxisomal membrane protein (PMP70 or PxMP1)
ABCd4	PMP69, PXMP1L
ABCe1	Oligoadenylate-binding protein (OABP), ribonuclease 4 inhibitor
ABCf1	TNFα-stimulated ABC protein
ABCf2	Iron-inhibited ABC transporter-2
ABCf3	
ABCg1	White protein homolog-1
ABCg2	Placenta-specific ATP-binding cassette transporter (ABCP), Breast cancer resistance protein (BCRP), Mitoxantrone resistance-associated protein (MXR)
ABCg4	White protein homolog-2
ABCg5	White protein homolog-3, sterolin-1
ABCg8	Sterolin-2

Another classification proposes 8 sets [546]: in addition to ABC1, MDR–TAP, MRP–CFTR, ALD, OABP, GCN20, and White classes, it includes the ASNA group, i.e., arsenite [77] and antimonite [78] transporters or ^{L}asparaginases (ASnA1–ASnA2). In humans, arsenical pump-driving ATPase, or arsenite-stimulated ATPase, ASnA1 is the homolog of the bacterial ArsA protein, a catalytic component of the multisubunit oxyanion pump that is responsible for resistance to arsenicals and antimonials.

loading pigment granules in eyes. It may also contribute to the transport of biogenic amines, such as dopamine, histamine, and serotonin [550].

77. Chemical compound with an arsenic (As) oxyanion ($As_iO_j^{k-}$).
78. An antimonite refers to salts of antimony (Sb).

4.18.2 Structure of ABC Transporters

Many functional ATP-binding cassette transporters are homo- or heterodimers. Most ATP-binding cassette importers and exporters are constituted at least of 4 domains: 2 ABC domains with 2 nucleotide-binding sites (NBD) devoted to ATP hydrolysis and 2 groups of transmembrane domains (TMD).

Additional domains can complement these core components to confer a regulatory function. Transmembrane domains that are usually composed of 6 transmembrane (TM) helices serve as substrate-binding elements and carriers across cell membranes.

Exporters have a set of 12 TM helices; importers 10 to 20 TM helices according to the mass and chemical nature of their substrate. Nucleotide ATP binds at the interface between 2 ABC domains. The cycle of ATP binding and hydrolysis switches ABC transporters from ATP-binding, closed state to ATP-free, open conformation.

All ABC transporters contain 2 cytosolic nucleotide-binding domains in 2 transmembrane segments. A large number of ABC proteins are active pumps, i.e., they carry their substrate against the concentration gradient of the substrate.

However, exceptions exist. In humans, ABC50[79] is an ABC dimer without TMDs that can be a pseudotransporter. It has nuclear targeting signals and might be involved in aminoacyl–tRNA binding. The cystic fibrosis transmembrane conductance regulator (ABC35 or ABCc7) is a chloride channel. The sulfonylurea receptors (SUR1 and SUR2, i.e., ABCc8 and ABCc9) are intracellular ATP sensors that regulate the permeability of some types of potassium channels ($K_{IR}6.1$ and $K_{IR}6.2$, i.e., K_{ATP}).

In addition, a third transmembrane domain (TMD0) that is composed of 5 TM helices can be attached to the N-terminus of several class-C members.

4.18.3 ABC Exporters and Importers

The ABC transporters that are both importers and exporters have not yet been detected. They function as *importers* or *exporters* in various types of organisms. They experience a series of nucleotide- and substrate-dependent conformational changes that alternate access to intra- and extracellular medium (inward- and outward-facing conformations).

Most exporters are homodimers that consist of 2 half transporters with a TMD fused to a nucleotide-binding domain (NBD). The ABC importers have additional regulatory elements that recruit high-affinity binding proteins. The ABC importers, but not the ABC exporters, can actually require a binding proteic partner to carry their substrate.

For example, maltose transporter of the ABC superclass interacts with substrate-loaded maltose-binding protein (MBP) in the periplasm[80] induce a partial closure

79. A.k.a. GCN20 homolog ABCf1.

80. The periplasm, or periplasmic space, is the space between the inner cytoplasmic bilipd layer and outer membrane of bacteria.

of the maltose–maltodextrin transporter MalK dimer in the cytoplasm. ATP binding to this conformation then promotes progression to the outward-facing state [548]. Binding of both MBP and ATP are required to form the outward-facing conformation.

In certain types of ABC transporters with 2 ATP-binding sites, [81] ATP binding seems to cause unidirectional transport because it promotes an outward-facing conformation, whereas dissociation of ATP-hydrolysis products elicits an inward-facing state [549]. The exporter cleft, only accessible from the cytoplasm, is separated from the extracellular medium by a closed gate. Binding of 2 ATP molecules on ATP-binding sites closes the gap. In ATP-free ABC importers, the attached binding protein appears to play the gating role.

4.18.4 Full and Half ABC Transporters

The majority of functional ABC transporters are full transporters with their $(TMD\text{-}ABC)_2$ arrangement, such as some transporters of the classes A, B, and C (e.g., ABCa1, ABCb1b, and ABCc7 [CFTR]). Half transporter members of the ABC superclass include ABCb2, ABCb3, ABCb7, ABCd1 to ABCd3, ABCg1, ABCg2, ABCg5 and ABCg8 proteins.

Full transporters are mostly located in the plasma membrane. Half transporters in functional homo- or heterodimers localize to membranes of cell organelles, such as the endoplasmic reticulum, peroxisome, or mitochondria.

4.18.5 Role of ABC Transporters

The ABC transporters are involved in lipid transport, multidrug resistance, antigen presentation, mitochondrial iron homeostasis, and ATP-dependent regulation of ion channels. Thisd latter function is particularly ensured by cystic fibrosis transmembrane conductance regulator (Sect. 3.5.9) and sulfonylurea receptors, subunits of inward-rectifier potassium channels ($K_{IR}6$; Sect. 3.4.4). The ABC transporters also participate in the creation and maintenance of electrochemical gradients across the plasma membrane.

The ABC transporters export lipid-modified peptides that are essential for cell migration, especially during embryo- and fetogenesis. Enzyme 3-hydroxy 3-methylglutaryl coenzyme-A reductase (HMGCoAR) adds geranylgeranyl moieties to C-termini of peptides. Germ cell attractant that is geranyl modified can be exported by ABC transporters [551].

The ABC transporters can be targeted for apical or basolateral plasma membrane in human, polarized, ciliated, bronchial cells. Different localization patterns suggest specific roles for ABC transporters in basal and ciliated cells [552].

81. Vitamin-B12 transporter has a nucleotide-free conformation.

Table 4.24. Aliases of class-A ABC Transporters.

Type	Aliases
ABCa1	ABC1, CERP
ABCa2	ABC2
ABCa3	ABC3, ABCc, SMDP3
ABCa4	ABC10, ABCR, ARMD2, CoRD3, RmP, StgD1
ABCa5	ABC13
ABCa6	ARA
ABCa7	ABCx, SSn, macrophage ABC transporter
ABCa8	
ABCa9	
ABCa10	
ABCa11	(pseudogene)
ABCa12	ABC12, ICR2B, ICR5, LI2
ABCa13	

4.18.5.1 Sphingosine 1-Phosphate Transporter

Sphingosine 1-phosphate in blood is supplied mainly by erythrocytes. Platelets are another sources of S1P in plasma. Sphingosine 1-phosphate operates in lymphocyte egress from lymphoid organs (thymus and secondary lymphoid organs). Lipidic signal S1P transmitter is exported from erythrocytes and thrombocytes through ABC transporters. The rate of S1P transport increases with S1P concentration (apparent Michaelis constant ∼21 μmol) [553]. Sphingosine 1-phosphate transporter is implicated in migration of myocardial precursors [554]. [82]

4.18.6 Class-A ABC Transporters

Prototypic full ATP-binding cassette transporter of the class A ABCa1 (Table 4.24) is a facilitator of cellular cholesterol and phospholipid export. Transporters ABCa3 and ABCa4 intervene in lung surfactant and retinaldehyde processing, respectively.

4.18.6.1 ABCa1 Transporter

At the subcellular location level, transporter ABCa1 [83] localizes to the plasma membrane as well as on intracellular organelles, such as the Golgi body, early and late endosomes, and lysosomes. At the cell distribution level, ABCa1 resides prominently in macrophages and hepatocytes. At the organ distribution level, ABCa1 is synthesized in kidneys, adrenal glands, intestine, and central nervous system. The

82. Lipid mediator sphingosine 1-phosphate (Vol. 2 – Chap. 3. Growth Factors) operates in vascular development, particularly during embryo- and fetogenesis.

83. A.k.a. ABC1 and cholesterol efflux regulatory protein (CERP).

highest expression levels may be detected not only in the liver, but also the lung, adrenal glands, and placenta, in addition to various fetal tissues [546].

The ABCa1 transporter supports the transfer of lipids between intracellular sites and the extracellular medium. In the presence of apolipoprotein-A1, ABCa1 promotes cellular efflux of phosphatidylcholine, sphingomyelin, and cholesterol to apolipoprotein-1. Therefore, ABCa1 is a major regulator of cellular cholesterol and phospholipid homeostasis. Cholesterol is a component of cell membranes and a precursor in some metabolisms, such as steroid hormone and bile acid synthesis.

In humans, the production of ABCa1 in monocytes and macrophages depends on sterols. It is inversely regulated by cholesterol import and export. The basal concentration of ABCa1 in macrophages is low. It rises upon cholesterol influx owing to the oxysterol-dependent activation of the LXR–RXR heterodimeric transcription factor (Sect. 6.3.6) and lowers upon cholesterol efflux.

In macrophages, ABCa1 excretes cholesterol that accumulates with the uptake of oxidized cholesterol-carrying lipoproteins [555]. Polyunsaturated fatty acids increase the already rapid turnover of ABCa1 in macrophages, thereby lowering ABCa1 density at the cell surface [556]. Conversely, lipid-poor apolipoproteins can stabilize ABCa1 protein. The ABCa1 transporter interacts with different lipid-poor apolipoproteins, such as ApoA1, ApoA2, ApoE, and likely ApoA4 [556].

In hepatocytes, ABCa1 contributes to the production of the precursor forms of high-density lipoproteins [555]. In addition, at least in mice, ABCa1 is involved in the engulfment of apoptotic cells by macrophages.

The maintenance of the membrane cholesterol-to-phospholipid ratio relies on the balance between the cellular input of cholesterol from endogenous synthesis and uptake from plasma lipoproteins and the output from cells. Cholesterol egress is controlled by the same genes that regulate cholesterol ingress, i.e., the genes of membrane-bound transcriptional activators, the sterol regulatory element-binding proteins SREBP1 and SREBP2. The SREBP transcription factors regulate cholesterol level. In the nucleus, SREBPs enhance the transcription of cholesterol-synthesizing genes, such as those that encode 3-hydroxy 3-methylglutaryl coenzyme-A (HMGCoA) synthase and HMGCoA reductase [557].

Besides, hepatocytes produce 2 alternatively spliced transcripts of SREBP1 (SREBP1a and SREBP1c) [555]. The SREBP1a isoform predominates in most non-hepatic tissues. In hepatocytes, SREBP1c transcription is enhanced by insulin in synergy with liver X receptors. Agent SREBP1 preferentially activates the synthesis of fatty acids that are incorporated into triglycerides [555]. The SREBP2 factor favors the activation of the synthesis and uptake of cholesterol.

Small non-coding RNA miR33 participates in cholesterol homeostasis. The MIR33B and MIR33A genes are embedded within the Srebp1 and Srebp2 genes. Both microRNA-33 isoforms preclude the production of ABCa1 that regulates the synthesis of high-density lipoprotein, thereby blocking cholesterol egress to apolipoprotein-A1 [557, 558]. In mouse macrophages, miR33 targets ABCg1, hence reducing cholesterol efflux to nascent high-density lipoproteins. In the liver, miR33 regulates HDL genesis.

4.18.6.2 ABCa2 Transporter

The ABCa2 transporter [84] is predominantly synthesized in the nervous system, where it can operate in lipid export. In human macrophages, ABCa2 is produced during cholesterol import from a cholesterol-responsive gene [559].

4.18.6.3 ABCa3 Transporter

The ABCa3 transporter [85] is particularly involved in the transport of lipids. Its activity is primed by membrane lipids. The ABCa3 ATPase is N-glycosylated in the endoplasmic reticulum and can be cleaved in the post-Golgi compartment [560].

The ABCa3 transporter lodges predominantly in the lung alveolar type-2 cells, where it localizes to lamellar bodies that are lysosomes specialized in surfactant storage. It is also synthesized, albeit to a lower extent, in brain, heart, kidney, and pancreas [560].

The ABCa3 transporter intervenes in the metabolism of the lung surfactant, especially that of phosphatidylcholine and phosphatidylglycerol, as well as lamellar body genesis. Mature surfactant proteins SftPb and SftPc localize within lamellar bodies and associate with surfactant phospholipids.

In the lung, the production of ABCa3 is triggered by glucocorticoids, once the glucocorticoid receptor is bound to the AbcA3 promoter [560]. Sterol response element-binding protein SREBP1c and thyroid transcription factor TTF1 upregulate ABCa3 synthesis. Interleukin-6 that activates signal transducer and activator of transcription STAT3 also enhances ABCa3 expression.

4.18.6.4 Other Class-A ABC Transporters

Retinal photoreceptor-specific transporter ABCa4 [86] localizes to the rim of the rod outer segment discs. It is involved in the transport of retinoid compounds across the outer segment disc membrane following the photoactivation of rhodopsin. ^{N}Retinylidene phosphatidylethanolamine is its preferred retinoid substrate in the absence of ATP [561]. [87]

The full ABCa5 transporter (or ABC13) is synthesized in brain, heart, lung, thyroid gland, and prostate. It is located in lysosomes and late endosomes.

---

84. A.k.a. ATP-binding cassette transporter-2 (ABC2).

85. A.k.a. ATP-binding cassette transporter-3 (ABC3) and ABCc transporter.

86. A.k.a. ABC10, ABC transporter of the rod outer segment (ABCR), age-related macular degeneration protein ARMD2, cone-rod dystrophy protein CoRD3, photoreceptor Rim protein (RmP), and Stargardt disease protein StgD1.

87. ^{N}retinyl phosphatidylethanolamine, the reduced form of ^{N}retinylidene phosphatidylethanolamine that does not activate the ATPase activity of ABCa4, binds to ABCa4 in stoichiometric amounts. ^{N}RetinylPE displaces ^{N}retinylidenePE from ABCa4. In addition, ATP binding to ABCa4 and hydrolysis dissociates the retinoids ^{N}retinylPE and ^{N}retinylidenePE from ABCa4.

The ABCa6 transporter [88] is produced in hematopoietic precursors. It is ubiquitous with highest levels in heart, liver, lung, and brain. In human macrophages, it is regulated by cholesterol.

The full ABCa7 transporter [89] is detected predominantly in peripheral leukocytes as well as in the thymus, spleen, and bone marrow. Alternative splicing of the AbcA7 gene creates 2 transcript variants. The ABCa7 transporter, like ABCa1, is devoted to cellular phosphatidylcholine and sphingomyelin efflux via apolipoprotein-1, but, unlike ABCa1, has no effect on cholesterol efflux [562].

Transporters ABCa8, a drug transporter, and ABCa9 have a relatively large amino acid sequence homology (~70%). The ABCA9 transporter is ubiquitous, with the highest levels in heart and brain. In human macrophages, 3 truncated forms may exist [563]. It may play a role in monocyte differentiation as well as lipid homeostasis in macrophages.

Cholesterol-regulated ABCa10 transporter is ubiquitous with its highest levels in the heart, brain, and gastrointestinal tract [564]. Various truncated forms of ABCa10 have been detected in human macrophages. The ABCa10 synthesis is suppressed by cholesterol import into macrophages (cholesterol-responsive AbcA10 gene).

The AbcA11P gene corresponds to a pseudogene. The ABCa12 transporter [90] is synthesized upon stimulation by the activators peroxisome proliferator-activated receptors PPARβ (or nuclear receptor NR1c2) and PPARγ (or NR1c3) as well as liver X receptor (NR1h2/3) in cultured human keratinocytes [565].

The ABCa13 transporter is a susceptibility factor for both schizophrenia and bipolar disorder. It is the largest known ABC transporter. It contains a hydrophobic and large hydrophilic region.

4.18.7 Class-B ABC Transporters (MDR–TAP)

Human class-B ABC proteins (Table 4.25) are either full transporters, such as ABCb1, ABCb4, and ABCb11, or half transporters, such as ABCb2 and ABCb3, as well as mitochondrial transporters, such as ABCb6 to ABCb8, and ABCb10.

4.18.7.1 Multidrug Resistance

Multidrug resistance of cancer cells is caused by plasmalemmal ABCb1 protein on the apical surface of polarized epithelial cells, such as the brush border membrane of intestinal cells, the biliary canalicular membrane of hepatocytes, and the luminal membrane of epithelial cells of renal proximal tubules, as well as the blood–brain barrier and at choroid plexi [546].

MDR1 transporter can extrude numerous structurally unrelated, hydrophobic, toxics from the cell, including antibiotics and anticancerous agents. Therefore, it

88. A.k.a. anthracycline resistance-associated protein (ARA).
89. A.k.a. ABCx, autoantigen SSn, and macrophage ABC transporter.
90. A.k.a. ATP-binding cassette ABC12, lamellar ichthyosis LI2, and ichthyosis congenita ICR2b and ICR5.

Table 4.25. Aliases of class-B ABC Transporters (MDR: multiple drug resistance; TAP: transporter associated with antigen processing).

Type	Aliases
ABCb1	MDR1, PGP, PGP1
ABCb2	ABC17, TAP1, RING4, APT1, PSF1
ABCb3	ABC18, TAP2, RING11, APT2, PSF2
ABCb4	ABC21, MDR2, MDR3, GBD1, PFIC3, PGP3
ABCb5	
ABCb6	ABC14, MtABC3, PRP, UMAT
ABCb7	ABC7, ATM1P, AsAT
ABCb8	MABC1 (MtABC1)
ABCb9	ABC9, TAPL
ABCb10	MtABC2
ABCb11	ABC16, SPGP, BSEP, BRIC2,

reduces the concentration of these compounds to subtoxic levels in target cells and causes drug resistance in infectious microorganisms and tumors.

In fact, multidrug transporters belong to 4 categories: (1) the major facilitator superfamily (MFS) of secondary active transporters; (2) the small multidrug resistance (SMR); (3) the resistance, nodulation and cell division (RND); and (4) the ATP-binding cassette (ABC) family.

4.18.7.2 Transporter Associated with Antigen-Processing Complex

The major histocompatibility complex (MHC) encodes the class-1 and -2 glycoproteins that present peptides for immunorecognition by cytotoxic and helper T lymphocytes, respectively.[91] Plasmalemmal major histocompatibility complexes are inspected by CD8+ cytotoxic T lymphocytes that discern virus-infected or malignant cells.

Transporters associated with antigen processing (TAP) mediate immune recognition of virally or malignantly transformed cells via MHC class-1 antigen presentation. They allow translocation and proteasomal degradation of products down to the endoplasmic reticulum lumen for loading onto MHC class-1 molecules.

The TAP1 and TAP2 subunits are half ABC transporters. In fact, the TAP molecule is a heterodimer composed of ABCb2[92] and ABCb3 half transporters.[93]

Each half transporter consists of a transmembrane domain with 6 transmembrane helices connected by loops and cytosolic nucleotide-binding domain for ATP

91. Class-2 MHC molecules associate mainly with peptides derived from endocytosed extracellular proteins. Class-1 MHC molecules bind peptides generated by intracellularly degraded proteins.

92. A.k.a. ABC17, transporter of antigenic peptides TAP1, Really interesting new gene-4 protein (RING4), antigen peptide transporter APT1, and peptide supply factor PSF1.

93. A.k.a. ABC18, TAP2, RING11, APT2, and PSF2.

hydrolysis [566]. The N-terminal domain of each subunit serves for assembly of a MHC class-1 peptide-loading complex with tapasin, protein disulfide isomerase PDIa3,[94] calreticulin, MHC heavy chain, and antigen-presenting β2-microglobulin, a MHC class-1 molecule on all nucleated cells. The transport is composed of ATP-independent peptide-binding and ATP-dependent translocation steps.

4.18.7.3 ABCb1 Transporter

In humans, ABC exporters are required for lipid export. The ABCb1 transporter[95] of the MDR–TAP class maintains the cholesterol distribution across the leaflets of the plasma membrane. It also conveys other lipophilic compounds.

The ABCb1 ATPase is a drug efflux pump for xenobiotics with a wide substrate spectrum. It lowers drug accumulation in cells. It then often supports the development of resistance to anticancer drugs.

The ABCb1 transporter abounds at the blood–brain barrier. Its transport activity in cerebral capillaries is rapidly reduced by inflammatory cytokines, such as tumor-necrosis factor-α that acts via the TNFRSF1a receptor, as well as endothelin-1 via its ET_B receptor, nitric oxide synthase, and protein kinase-Cβ1 [567].

4.18.7.4 Other Class-B Transporters

The ABCb4 full transporter[96] is a member of the P-glycoprotein set. This hepatobiliary phospholipid transporter targets phosphatidylcholine for secretion from hepatocytes. Both ABCb4 and ABCb11 are involved in different types of progressive familiar intrahepatic cholestasis. This phospholipid (phosphatidylcholine) translocator is not implicated in multidrug resistance [546].

The ABCb5 transporter is also a member of the P-glycoprotein set. It is preferentially expressed by cells of the melanocytic lineage. Two isoforms exist in melanocytes: ABCb5α and ABCb5β.

The ABCb6 half transporter[97] is located in the outer membrane of mitochondria, where it carries porphyrins that will be converted to heme.

Half transporters ABCb7[98] and ABCb8[99] transport heme from mitochondria to cytosol. Proteins ABCb7 and ABCb8 form functional dimers. Substrates of ABCb7 transporter comprise iron–sulfur cluster precursors.

---

94. A.k.a. endoplasmic reticulum-resident protein ERP57, ERP60, and ERP61, as well as 58-kDa glucose-regulated protein GRP58 (or GRP57) and 58-kDa microsomal protein.

95. A.k.a. permeability glycoprotein (PGP or P-glycoprotein-1) as well as ABC20, CD243, GP170, and multiple drug resistance (MDR or MDR1).

96. A.k.a. ABC21, GBD1, multidrug resistance protein MDR2 and MDR3, progressive familial intrahepatic cholestasis protein PFIC3, and P-glycoprotein-3.

97. A.k.a. ABC14, mitochondrial transporter MtABC3, P-glycoprotein-related protein (PRP), and ubiquitous mammalian ABC half transporter (UMAT).

98. A.k.a. ABC7, ABC half transporter of the mitochondrial inner membrane ATM1P, and aspartate aminotransferase (AsAT).

99. A.k.a. mitochondrial ATP-binding cassette MABC1.

Another half transporter ABCb9 [100] may operate in lysosomes, as it colocalizes with the lysosomal-associated membrane proteins LAMP1 and LAMP2.

The ABCb10 transporter [101] is a mitochondrial inner membrane erythroid transporter involved in heme biosynthesis.

The ABCb11 transporter [102] carries taurocholate and other cholate conjugates from hepatocytes to the bile. It is the major canalicular bile salt export pump for active transport of bile salts across the canalicular membrane of hepatocytes into bile. Its activity is necessary for phosphatidylcholine secretion via ABCb4 (or PGY3).

4.18.8 Class-C ABC Transporters (MRP–CFTR)

The majority of non-ABCb1-mediated multidrug resistance is due to the overexpression of ABCc1 protein or human multidrug resistance-associated protein (hMRP1).

In ciliated polarized epithelial cells, MRPs are restricted to the basolateral surface, whereas other ABC superclass members, such as cystic fibrosis transmembrane conductance regulators and ABCb1, are located at the apical surface. In basal cells without tight junctions, MRPs are detected along the entire plasma membrane, whereas CFTR and ABCb1 are not identified.

Proteins of the ABC class C (Table 4.26), such as ABCc1 to ABCc3, ABCc6, ABCc8, and ABCc9, but not ABCc4, ABCc5, and ABCc7, possess the N-terminal TMD0 domain [546].

4.18.8.1 Multidrug Resistance-Associated Protein

Multidrug resistance-associated protein (MRP) transmembrane transporter conveys various lipophilic substrates. This ATP-dependent pump carries glutathione, glucuronate, and sulfate conjugates, especially exporting glutathione conjugate leukotriene-C4. It participates in cell detoxification by efflux of glutathione conjugates. In particular, MRP1 transports both hydrophobic anticancer agents and anionic glutathione drug conjugates.

4.18.8.2 Cystic Fibrosis Transmembrane Conductance Regulator

Cystic fibrosis transmembrane conductance regulator (Sect. 3.5.9) is the full transporter ABCc7. It is regulated by phosphorylation and intracellular nucleotides. It functions at apical plasma membrane of epithelial cells as a cAMP-dependent chloride channel that can regulate other ion channels and protein secretion. This transmembrane protein actually also acts as a conductance regulator that controls

100. A.k.a. ABC9 and TAP-like protein (TAPL).

101. A.k.a. mitochondrial transporter MtABC2.

102. A.k.a. ABC16, bile salt export pump (BSEP), sister of P-glycoprotein (SPGP), benign recurrent- BRIC2 and progressive familial intrahepatic cholestasis [severe form] protein PFIC2, as well as PGY4.

Table 4.26. Aliases of class-C ABC Transporters (++, +, +/−: high, intermediate, and low production; ARA: anthracycline resistance-associated protein; CMOAT: canalicular multispecific organic anion transporter; MRP: multidrug resistance-associated protein). Protein short type of multidrug resistance homolog (SMRP) is a partial sequence of ABCc5 transporter.

Type	Aliases	Location
ABCc1	MRP1	Ubiquitous
ABCc2	MRP2, CMOAT1	Predominantly in liver and kidney
ABCc3	MRP3, CMOAT2, MOATd	Hepatocyte, enterocyte
ABCc4	MRP4, MOATb	Widely distributed (not in liver)
ABCc5	MRP5, MOATc, (SMRP)	Ubiquitous (skeletal muscle [++], kidney, heart, brain, and testis [+], liver [+/−])
ABCc6	MRP6, MOATe, ARA	Predominantly in liver and kidney
ABCc7	CFTR	Epithelia (lung, kidney, intestine, pancreas, liver, and sweat glands)
ABCc8	SUR1	Smooth and striated myocytes
ABCc9	SUR2	
ABCc10	MRP7	
ABCc11	MRP8	Predominantly in testis and breast
ABCc12	MRP9	Ubiquitous
ABCc13	PRED6	Bone marrow, leukocytes

the activity of various other ion carriers [568]. This cAMP-activated Cl^- channel is defective in cystic fibrosis (or mucoviscidosis).

An alternatively spliced form of CFTR that is detected in the heart can generate robust cAMP-activated chloride currents [546]. In men with congenital bilateral absence of vas deferens, another alternative splicing mRNA variant gives rise to 2 transcripts, one normal with intact exon 9 and the other with in-frame deletion of exon 9.

The CFTR channel regulates certain transmembrane proteins, such as Cl^-–HCO_3^- exchangers, epithelial Na^+ (ENaC) [569], and outwardly rectifying Cl^- (ORCC) channels [570]. The CFTR channel colocalizes with ENaC and ORCC channels. It directly inhibits ENaC channel. Furthermore, it impedes ENaC activation by protein kinase-A. The CFTR channel activated by PKA also causes ATP efflux. Once released, ATP binds to nucleotide receptors and causes outwardly rectifying Cl^- currents.

The CFTR channel can also activate swelling-activated and Ca^{++}-gated Cl^- currents [568]. It controls not only the activity of other Cl^- channels (ClCa and ORCC), but also cAMP-activated, voltage-gated delayed rectifier $K_V7.1$ channels (MinK–K_VLQT1). The $K_V7.1$ subunit can heteromerize with cell swelling-activated Iswell Cl^- channel. In addition, multimers constituted by sulfonylurea receptors (of the superclass of ABC transporters) and K^+ channel subunits, such as ATP-sensitive K^+ channels ROMK2 ($K_{IR}1.2$) as well as $K_{IR}1.1a$ and $K_{IR}6.1$, the activity of which is precluded by sulfonylurea, have an enhanced sensitivity in the presence of CFTR.

The CFTR channel participates in the regulation of airway surface liquid pH that influences activity of ion channels, contact of mucus to membranes, and attachment of bacteria and viruses to mucus. Channel CFTR controls the activity of 2 antiporters that regulate cytosolic pH: Na^+–H^+ and Cl^-–HCO_3^- exchangers [568]. Channel CFTR is able to bind to NHERF regulator of Na^+–H^+ exchanger NHE3. Airway surface liquid pH is regulated by H^+–K^+ ATPase and Cl^-–HCO_3^- exchanger. In addition, CFTR located in intracellular membrane of endosomal vesicles can counterbalance the activity of H^+ ATPase that acidifies the endosomal compartment. The performance of enzymes that intervene in protein sialylation in the endoplasmic reticulum depends on H^+ ion.

The CFTR channel can activate aquaporin, hence forming a functional controller that contributes to transepithelial ion and water transport. Besides, CFTR interacts with cortical actin and syntaxin-1A as well as nucleotides ATP and GTP for hydrolysis. Nucleoside diphosphate kinases that transfer phosphates from ATP to GDP to generate GTP are expressed in the airway epithelium, where they are involved in the control of ion secretion [568].

4.18.8.3 Other Class-C Transporters

Organic anion transporter ABCc1 (or multidrug resistance protein MRP1) carries oxidized glutathione, cysteinyl leukotrienes, glucuronides, and sulfate conjugates of steroid hormones and bile salts, and activated aflatoxin-B1. [103] It also transports drugs and other hydrophobic compounds in the presence of glutathione.

Detoxification of many endo- and exogenous lipophilic compounds in the liver is achieved by conjugation with glutathione, glucuronide, or sulfate, thereby creating negatively charged, amphiphilic compounds that are secreted into bile or urine. Hepatobiliary excretion of these conjugates is mediated by ABCc class members such as ABCc2 located exclusively in the canalicular membrane of the hepatocyte [546].

The ABCc2 transporter localizes to the canalicular (apical) surface of hepatocytes, where it acts in biliary transport of mainly anionic conjugates with glutathione, sulfate, or glucuronosyl (e.g., glucuronosyl bilirubin). [104] Other substrates encompass anticancer drugs such as vinblastine. The ABCc2 ATPase thus contributes to drug resistance.

The ABCc3 transporter may intervene in biliary and intestinal excretion of organic anions such as bile salts. It is expressed in the liver, kidney, and colon.

The ABCc4 transporter contributes to prostaglandin transport. This organic anion transporter also transfers cyclic nucleotides and some nucleoside monophosphate analogs such as nucleoside-based antiviral drugs. It is widespread, with high levels in the prostate, but not in the liver.

The ABCc5 transporter exports cyclic nucleotides. It is expressed at its highest levels in skeletal muscle, at intermediate levels in the brain, heart, kidney, and testis,

---

103. Aflatoxins are mycotoxins produced by many species of aspergilli (notably Aspergillus flavus and parasiticus). Aflatoxin-B1 is the most toxic species.

104. Mutations of the AbcC2 gene cause the Dubin-Johnson syndrome characterized by a defective secretion of amphiphilic, anionic conjugates from hepatocytes into the bile.

and, at low levels, in liver. A short type of multidrug resistance protein homolog (SMRP) is a partial sequence of ABCc5 transporter.

The ABCc6 transporter lodges primarily in the liver and kidney. Like ABCc4, this organic anion transporter transports cyclic nucleotides and some nucleoside monophosphate analogs including nucleoside-based antiviral drugs. It can be co-expressed with ABCc1 protein.

The ABCc8 transporter, or sulfonylurea receptor SUR1, is a modulator of ATP-sensitive potassium channels and insulin release. Sulfonylurea receptor may sense changes in ATP and ADP concentration. Once activated by sulfonylurea, the sulfonylurea receptor of pancreatic β cells inhibits $K_{IR}6.2$ [546]. The ABCc8 protein also forms ATP-sensitive potassium channels in cardiac, skeletal, and vascular and non-vascular smooth myocytes.

The AbcC9 gene gives rise to sulfonylurea receptor splice variants SUR2a and SUR2b that are subunits of the inward rectifier K^+ channels $K_{IR}6$ to create i_{ATP} current. ATP sensitivity and pharmacological properties of $K_{IR}6$ channels associated with SUR2 differ from those of channels tethered to SUR1 subunit.

Full transporters ABCc10 and ABCc11 are lipophilic anion pumps. Proteins ABCc10 and ABCc7 (CFTR) transport estradiol-17β glucuronide and, to a lesser extent, leukotriene-C4.

The ABCc11 ATPase transfers bile acids, conjugated steroids, and cyclic nucleotides. It is expressed at low levels in many tissues, except kidney, spleen, and colon. The long ABCc11 transcript is specifically expressed in mammary gland; the short transcript in testis.

The ABCc12 protein, like ABCc11, possesses several alternatively spliced variants. It is produced at low levels in testis, prostate, and ovary.

The ABCc13 transporter abounds in the fetal liver and bone marrow, as well as circulating leukocytes.

4.18.9 Class-D of ABC Transporters (ALD)

Four half transporters (ABCd1–ABCd4) localize to the peroxisomal membrane. Their mutant forms are involved in different peroxisomal disorders.[105] All known peroxisomal ABC transporters are half transporters that require another half transporter partner to form functional homo- or heterodimeric transporters.

4.18.9.1 Peroxisomal Transport

Peroxisomes are organelles implicated in the metabolism of fatty acids and many other metabolites. They contain enzymes for certain oxidative reactions, such as cata-

105. Adrenoleukodystrophy (ALD) is a severe X-linked neurodegenerative disorder. This disease is characterized by the accumulation of unbranched, saturated fatty acids connected to cholesterol esters in the brain white matter and adrenal cortex and some sphingolipids of the brain because of an impaired degradation of very-long-chain fatty acids in peroxisomes. This accumulation alters myelin formation in the central nervous system and steroidogenesis in the adrenal gland.

Table 4.27. Aliases of class-D ABC transporters (ALD, ALDP: adrenoleukodystrophy protein; ALDL(R,RP): adrenoleukodystrophy protein-like (-related) protein; Pex: peroxin; PMP: peroxisomal membrane protein; PxMP1L: PMP1-like protein).

Type	Aliases
ABCd1	ABC42, ALD, ALDP, Pex1
ABCd2	ABC39, ALDL1, ALDR, ALDRP
ABCd3	ABC43, PMP70, PxMP1
ABCd4	ABC41, PMP69, PxMP1L, P70R, P79R

lase, ^{D}amino acid oxidase, and uric acid oxidase. Peroxisomes degrade fatty acids (β-oxidation) to generate acetyl coenzyme-A. AcetylCoA can enter the citric acid cycle or be used in the synthesis of acetylcholine by choline acetyltransferase.

Peroxin receptors Pex5 and Pex7 accompany cargos that contain a peroxisomal targeting signal PTS1 and PTS2, respectively, into peroxisomes, where they release their ligands (import in peroxisomes) and then return to the cytosol for recycling.

4.18.9.2 ABCd Class Proteins

The ABCd1 half transporter is involved in import of fatty acids and/or fatty acyl-coenzyme-A in peroxisomes (Table 4.27). This peroxisomal membrane protein is oriented toward the cytosol. The ABCd1 protein can interact with Pex6 and Pex26 peroxisomal proteins.

Half transporters ABCd2 to ABCd4 are also involved in peroxisomal import of fatty acids and/or fatty acylCoAs. [106] Proteins ABCd2 and ABCd4 are potential partners of ABCd1 and ABCd3 to form heterodimers.

4.18.10 Class-E of ABC Transporters (OABP)

The ABCe1 transporter is a soluble protein, unlike other ATP-binding cassette proteins (Table 4.28). It possesses 2 ABC domains, but no transmembrane regions. It is involved in transcription initiation, ribosome genesis, and inhibition of interferon-induced ribonuclease-L. [107]

The 2-5A antiviral pathway is a RNA degradation axis stimulated by interferon. Interferon activates the genes that encode several double-stranded RNA (dsRNA)-dependent synthases. These synthases transiently generate unstable, 5′-triphosphorylated, (2′,5′)-phosphodiester-linked oligoadenylates (2-5A) from ATP. Agent 2-5A activates the endoribonuclease RNaseL to destroy single-stranded RNA [571]. [108]

106. Mutations of the AbcD3 gene cause the Zellweger cerebro-hepato-renal syndrome.

107. Activation of ribonuclease-L impedes protein synthesis during the antiviral action of interferon that relies on (2′,5′)-oligoisoadenylate synthetase-dependent) ribonuclease-L, or (2,5A)-dependent ribonuclease.

108. At least 3 (2′,5′)-phosphodiester-linked oligoadenylates and a single 5′-phosphoryl group are required for maximal activation of RNaseL.

Table 4.28. Aliases of class-E, -F, and -G ABC transporters (ABCP: ABC protein expressed in placenta; BCRP: breast cancer resistance protein; HuSSY: human sequence similar to yeast; MABC: mitochondrial ABC transporter; MXR: mitoxantrone resistance; OABP: oligoadenylate-binding protein; RLI, RNaseL1, and RNaseLI: interferon-induced ribonuclease-L inhibitor; RNs4I: ribonuclease-4 inhibitor; STSL: sitosterolemia protein; Wht: White protein homolog).

Type	Aliases
ABCe1	ABC38, RLI, RNaseL1, RNs4I, OABP
ABCf1	ABC27, ABC50
ABCf2	ABC28, MABC1, HuSSY18
ABCg1	Wht1, ABC8
ABCg2	ABC15, ABCP, BCRP, MXR, CD338
ABCg4	Wht2
ABCg5	Sterolin-1, STSL
ABCg8	Sterolin-2, STSL

The 2-5A antiviral pathway is composed of 3 enzyme types: interferon-inducible 2-5A synthases, 2-5A phosphodiesterase, and 2-5A-dependent endoribonuclease L (RNaseL). 2-5A Synthases are activated by double-stranded RNA. They convert ATP into the unusual series of 2-5A oligomers. The RNaseL enzyme is activated by a subnanomolar concentration of 2-5A molecules. It leads to the inhibition of protein synthesis by mRNA cleavage. The ABCe1 protein, i.e., the RNaseL inhibitor, prevents the binding of 2-5A to RNaseL enzyme.

4.18.11 Class-F ABC Transporters (GCN20)

The ABCf1 transporter [109] (homolog of yeast GCN20) [110] that lacks a transmembrane domain is more hydrophilic than other members of the ABC superclass [546]. It is rather a pseudotransporter that has 3 nuclear localization signals.

The ABCf1 transporter is regulated by tumor-necrosis factor-α to participate in inflammation caused by this cytokine (Table 4.28). [111] It is implicated in the enhancement of protein synthesis, in addition to inflammation. It indeed tethers to eukaryotic initiation factor eIF2 and ribosomes [572].

The ABCf2 half transporter, or iron-inhibited ABC transporter-2, localizes to mitochondrial membranes.

109. A.k.a. ABC27 and ABC50.

110. Agent GCN20, a member of the ABC superclass, is a positive effector of the mediator GCN2 of the general amino acid-based control response that phosphorylates the translation initiation factor eIF2α to stimulate the translation of Gcn4 mRNA in amino acid-starved cells.

111. The ABCf1 transporter is hence termed TNFα-stimulated ABC protein.

4.18.12 Class-G ABC Transporters (WHITE Class)

The ABC transporters of the ABCG (WHITE) class, such as ABCg1, ABCg5, and ABCg8 (Table 4.28), participate in the regulation of lipid transfer in macrophages, hepatocytes, and intestinal mucosa cells. [112]

The ABCg1 half transporter contributes to cholesterol and phospholipid efflux [573]. It plays an important role in tryptophan and guanine uptake [546]. [113]

Proteins ABCg1 and ABCa1 suppress leukocytosis [114] that favors atherosclerosis. Both transporters protect against atherosclerosis, as they promote cholesterol efflux from foam cells to apolipoprotein-A1 and high-density lipoproteins (cholesterol acceptors) that transport cholesterol from peripheral tissues to the liver [574]. Transcription factor liver X receptor primes the synthesis of both ABCg1 and ABCa1 proteins.

Furthermore, ABC transporters as well as lipoprotein receptors are highly expressed in hematopoietic cells. High-density lipoproteins are engaged in the regulation of stem cell proliferation in the bone marrow. [115] In hematopoietic progenitor stem cells such as granulocyte-monocyte progenitors that give rise to neutrophils or monocytes, growth factor receptors are organized in membrane rafts to promote cell proliferation and migration. These membrane rafts contain a high content of cholesterol and glycolipids. Excess cholesterol is removed by ABC transporters. Conversely, elevated formation of membrane rafts due to augmented cholesterol concentration helps to cluster IL3 and CSF2 receptors.

The ABCg2 half transporter functions as a xenobiotic transporter involved in multidrug resistance. It also conveys organic anions, cholesterol, and steroids (estradiol, progesterone, and testosterone).

The ABCg4 transporter removes cellular cholesterol and generates cholesterol-rich high-density lipoproteins.

Two half transporters ABCg5 and ABCg8 operate as heterodimers. Proteins ABCg5 and ABCg8 limit intestinal absorption and promote biliary excretion of neutral sterols [575]. Agent ABCg5 localizes to the apical membranes of mouse enterocytes and hepatocytes. It heterodimerizes with ABCg1, ABCg2, and ABCg4. However, only the ABCg5–ABCg8 dimer is functional and promotes sterol excretion into bile.

112. The gene AbcG3 is not found in humans. In mice, ABCg3 abounds in the thymus and spleen.

113. The former is the precursor of serotonin.

114. Elevated number of leukocytes, or white blood cells, in blood, especially neutrophils and monocytes.

115. In mice with hematopoietic cells deficient in both transporters, the number of Lin−, SCA1+, SCFR+ (LSK) pluripotent hematopoietic stem and multipotent progenitor cells (common myeloid- and granulocyte–monocyte progenitors) rises in the bone marrow due to increased responsiveness to interleukin-3 and granulocyte–macrophage colony-stimulating factor (CSF2) [574]. In addition, atherosclerosis occurs earlier.

4.18.13 Arsenite Transporters

The arsenite transporter is composed of 2 regulatory and 3 structural subunits that are encoded by distinct genes. The human ArsA homolog AsnA1 is a monomeric ABC ATPase. The arsenical pump component ArsA possesses 2 ATP-binding sites and an arsenic dioxide (AsO_2)-binding domain. The other homolog Ans2, a membrane translocase, like Ans1, has a single ABC domain. The ArsA–ArsB complex carries arsenite and antimonite.

4.19 Gas Transporters

Gas transport through membranes supports nutritive and scavenging transport, oxidative metabolism, and signaling. Gas channel SLC42a1, or RhAG (type-A Rhesus blood group (Rh)-associated glycoproteic ammonium transporter)[116] transport ammonia (NH_3) in erythrocytes. Both SLC42a1 and Aqp1 allow erythrocytes to convey NH_3 from tissues where NH_3 gradient fosters NH_3 uptake to liver where NH_3 gradient elicits NH_3 egress for NH_3 detoxification. Aquaporin Aqp1 also carries hydrophobic NO and O_2 molecules.

Like ion channels, gas channels exhibit selectivity. Among Aqp1 of human erythrocyte, M23 variant of Aqp4 of rat astrocytic endfeet at the blood–brain barrier, Aqp5 of rat alveolar type-1 pneumocyte, and RhAG of human erythrocyte, the relative index of CO_2/NH_3 permeability varies widely [530]:

$$\text{Aqp4} \sim \text{Aqp5} \gg \text{Aqp1} > \text{RhAG},$$

all 4 channels being permeable to CO_2 and NH_3, except Aqp4 and Aqp5 that are impermeable to NH_3. Selectivity of transit molecules does not rely on size, but other features in relation with those of the channel pore. Hydrophilic NH_3 that has the same electronic configuration with H_2O probably also moves through monomeric aquapores, whereas less hydrophilic CO_2 can move through the hydrophobic central pores of multimeric channels.

116. A.k.a. CD241, erythrocyte plasma membrane 50-kDa glycoprotein Rh50GP, Rh50, and Rh50a.

5

Receptors of Cell–Matrix Mass Transfer

The plasma membrane of most cell types contains proteins that are devoted to endo- and exocytosis. In addition, at least some of them can serve as signaling mediators. Two major categories of endocytic receptors — low-density lipoprotein and scavenger receptors — exist. Members of these receptor categories bind to their cognate ligands for their intracellular transfer as well as transmit possible signaling associated with these molecules.

Cell signaling is integrated from the extracellular medium. Plasmalemmal and cell surface-anchored mediators contribute to signal modulation and integration of different signaling pathways. Signaling axes indeed display intracellular links as well as extracellular crosstalks. Interactions between ligands, their receptors, and eventual antagonists can occur outside the cell. In addition, these ligands are internalized after binding to their plasmalemmal receptors into intracellular endosomes, from which they can signal during their motion to another cell compartment. Moreover, endocytic receptors regulate the concentration of their specific ligands in the extracellular space as well as their own density at the cell surface, as these liganded receptors, once internalized, can undergo lysosomal degradation or be recycled back to the plasma membrane. Inside the cytoplasm, liganded receptors can also be unloaded. Subsequently, ligands can travel to lysosomes and receptor be recycled. Receptor regeneration at the cell surface enables the action of a second wave of ligands.

5.1 Endocytosis-Devoted Low-Density Lipoprotein Receptors

The family of low-density lipoprotein receptors (LDLR) includes many members [576] (Table 5.1): low-density lipoprotein receptor (LDLR), LDLR-related proteins (LRP1–LRP6 and LRP9), as well as the reelin receptors apolipoprotein-E receptor-2 (ApoER2 or LRP8) and very-low-density lipoprotein receptor (VLDLR).

Members of the LDLR family are transmembrane glycoproteins characterised by a large extracellular domain and comparatively short intracellular region. The cytoplasmic tail contains docking sites for adaptor proteins that possess Tyr^P-binding domains.

Table 5.1. Members of the family of low-density lipoprotein receptors and their aliases.

Main alias	Other alias(es)
LDLR	FH (FHC)
LRP1 (LRP1a)	ApoER (ApoER1), α2MR, IGFBP3R, TGFBR5, CD91
LRP1b	LRPDIT
LRP2	Megalin, GP330
LRP4	LRP10, synLRP, MEGF7
LRP5	LRP7, LR3, BMND1, EVR1, EVR4, HBM, OPPG, OPS, OPTA1, VBCH2
LRP6	ADCAD2
LRP3	LRP105
LRP9	LRP10, LR11, LRP11, SorL1, SorLa, SorLa1
LRP12	FAM4A1, HELG, RAY1, SEN4, ST7, TSG7
ApoER2	LRP8
VLDLR	CARMQ1, CHRMQ1

Members of the LDLR family binds numerous ligands (Table 5.2). They serve in ligand endocytosis. In particular, most members, such as LDLR, VLDLR, and LRP1, internalize lipoproteins. In addition, they operate as signal transducers or modulators of several signal transduction pathways, such as those triggered by BMP, Hedgehog, PDGF, reelin, TGFβ, and Wnt messengers. Numerous lipoprotein receptors are substrates of secretases that release their intracellular domains.

Receptors of the LDLR family have some overlapping ligand specificity as well as patterns of expression [577] (Table 5.3). The LDL receptor is widespread, but is prominent in the liver. Like LDLR, LRP1 is also widely synthesized, but mainly in the central nervous system and liver. Member LRP2 lodges predominantly in specialized epithelia of the central nervous system, lung, and kidney. The VLDL receptor is produced in the central nervous system, heart, muscle, and adipose tissue. The ApoE receptor-2 resides in the central nervous system and placenta.

5.1.1 Low-Density Lipoprotein Receptor

Low-density lipoprotein receptor [1] regulates cholesterol homeostasis via receptor-mediated endocytosis of lipoproteins. Hepatic uptake of plasma cholesteryl esters from LDL uses LDL receptors. Uptake of VLDL and scavenger receptors, at a greater extent than LDLR, contribute to foam cell formation from macrophages during atherogenesis [578].

The LDLR synthesis is promoted by depletion of cellular sterols. The pathway relies on diacylglycerol- and phospholipid-regulated protein kinase-Cε, but neither

1. A.k.a. familial hypercholesterolaemia protein (FH or FHC). Familial hypercholesterolemia is an autosomal dominant disease that results from mutations of the Ldlr gene and causes to a diminished catabolism and elevated concentration of LDL–cholesterol.

Table 5.2. Ligands of members of the family of low-density lipoprotein receptors (Main sources: [576, 577]; ApoER: apolipoprotein-E receptor; APP: amyloid precursor protein; CamK: Ca^{++}–calmodulin-dependent protein kinase; CK: casein kinase; DLg: Discs large homolog; GSK: glycogen synthase kinase; LDLR: low-density lipoprotein receptor; LR-PAP: LRP-associated protein; LRP: LDLR-related protein; NMDA: ^{N}methyl ^{D}aspartate-type glutamate receptor subunit; PAI: plasminogene activator inhibitor; PDGF: platelet-derived growth factor; SostDC: sclerostin domain-containing protein; TFPI: tissue factor pathway inhibitor; TGF: transforming growth factor; VLDLR: very-low-density lipoprotein receptor; 25(OH)D_3: 25-hydroxyvitamin-D3). Serpin-A1, -C1, -D1, and -E1 are also called α1-antitrypsin, antithrombin-3, heparin cofactor-2, and plasminogen activator inhibitor-1, respectively. Plasmatic α2-macroglobulin is able to inactivate numerous types of serine, cysteine, and aspartic peptidases, as well as metallopeptidases. Amyloid precursor protein (APP) is an integral membrane protein of many cell types that concentrates in neuronal synapses. Soluble oligomeric amyloid β pools are toxic for both neurons and blood vessel wall cells.

Receptor	Ligands
ApoER2	ApoE, serpin-D1–thrombin, reelin
LDLR	ApoA/B/E (LDL), lipoprotein lipase, (LRPAP1)
LRP1	ApoE (LDL), chylomicron remnants, , Lipoprotein lipase, hepatic lipase, α2-macroglobulin, APP, PDGF, TGFβ, tPA, pro-uPA, serpin-E1–tPA, serpin-E1–uPA, Serpin-C1–thrombin, serpin-A1–trypsin, α2-macroglobulin–peptidase, Soluble APP, TFPI, thrombospondins-1/2, LRPAP1, IGFBP3, TGFβ1
LRP1b	Synaptotagmin, laminin receptor precursors uPA–uPAR–PAI1 complex
LRP2	ApoB/E/H/J, lipoprotein lipase, hepatic lipase, Albumin, cubilin, retinol-binding protein, Vitamin-D-binding protein, sonic Hedgehog, BMP4 Thrombospondin-1, uPA–PAI1, pro-uPA, LRPAP1
LRP3	ApoB100, plasma peptidases, albumin, 25(OH)D_3, cubilin, LRP2BP
LRP4	Dickkopf-1, sclerostin, SostDC1, NMDA, DLg1/4, CamK2
LRP5	Frizzled, GSK3, CK1, axin, sclerostin
LRP6	Frizzled, GSK3, CK1, axin
LRP9	ApoE (HDL, VLDL), neurotensin, head activator, APP, GGA, PACS1, LRPAP1
VLDLR	ApoA/E (VLDL), lipoprotein lipase, uPA–PAI1, pro-uPA, Thrombospondin-1, TFPI, reelin, LRPAP1

PKCα, PKCγ, PKCδ, nor PKCζ [579]. Ubiquitin ligase Inducible degrader of LDLR

Table 5.3. Distribution and function of LDL receptor family members (Main source: [576]; BMP: bone morphogenetic protein; LMW: low-molecular weight; PDGFR: platelet-derived growth factor receptor; PTH: parathyroid hormone; SHh: sonic Hedgehog; VD–VDBP: vitamin-D–vitamin-D-binding protein complex; VSMC: vascular smooth muscle cell). Thyroglobulin is the thyroid prohormone that produces thyroxine (T_4) and tri-iodothyronine (T_3).

Receptor	Distribution	Effect
ApoER2	Central nervous system, testis	Neuron migration, synaptic transmission
LDLR	Ubiquitous	Cholesterol homeostasis
LRP1	Broad distribution (neurons, hepatocytes, macrophages, VSMC, etc.)	Ligand endocytosis, regulation of PDGF and TGFβ signaling, regulation of calcium flux, phagocytosis of apoptotic cells, embryo/fetogenesis, cell growth inhibition
LRP1b	Central nervous system; VSMC	Inhibition of cell migration
LRP2	Apical plasma membrane of absorptive and secretory epithelia, neuroectoderm, thyroid, parathyroid, type-2 pneumocyte	Supply of vitamins and nutrients, calcium homeostasis, reabsorption of excreted LMW proteins and VD–VDBP, PTH endocytosis, thyroglobulin uptake, BMP4 and SHh signaling
LRP3	Wide distribution (skeletal muscle, ovary)	
LRP4	Central nervous system, osteoblasts	Bone growth and turnover, tooth development, BMP, SHh, and Wnt signaling
LRP5	All tissues except brain	Wnt and BMP signaling, bone density regulation
LRP6	Osteoblasts, SMC	Wnt and BMP signalings, bone density regulation
LRP9	Nervous system; SMC (intracellular)	Intracellular transport, blood vessel wall remodeling, proliferation of neuronal precursor and neuroendocrine cells
LRP11	Wide distribution	
LRP12	Wide distribution	
VLDLR	Central nervous system, heart, adipose tissue, endothelial cells	Neuron migration, synaptic transmission

(IDOL) causes LDLR degradation. On the other hand, miR33 promotes LDLR expression and cholesterol efflux to ApoA1.

5.1.1.1 LDLRAP1

Low-density lipoprotein receptor adaptor protein LDLRAP1 [2] serves as an adaptor for LDLR endocytosis in hepatocytes. Whereas Dab2 adaptor operates in internalization of LDLRs (but not transferrin receptors) independently of LDLRAP1 and clathrin adaptor complex AP2, when it is is absent, LDLRAP1 triggers LDLR endocytosis in association with AP2 [580]. In addition, Dab2, but not LDLRAP1, promotes the efficient clustering of LDLR into clathrin-coated pits. Therefore, Dab2 in synergy with phospholipids and clathrin sorts LDLR to clathrin-coated pits, whereas LDLRAP1 may accelerate later stages in LDLR endocytosis in cooperation with AP2 [580].

5.1.2 Low-Density Lipoprotein Receptor-Related Proteins

Serpin–peptidase complexes, such as serpin-C1– (antithrombin-3)–thrombin, serpin-D1– (heparin cofactor-2)–thrombin, as well as serpin-A1– (α1-antitrypsin)–trypsin and –elastase, are removed from the blood circulation by receptors of the serpin–peptidase complex in the liver [577]. This *clearance receptor* (more precisely serpin–peptidase-complex clearance receptor) is the low-density lipoprotein receptor-related protein (LRP). [3]

5.1.2.1 LRP1 Clearance Receptor

Clearance receptor LRP1 (or LRP) [4] is a promiscuous and ubiquitous receptor that binds to more than 40 ligand types (lipoproteins, particularly certain apolipoprotein-E- and lipoprotein lipase-enriched lipoproteins, peptidases, peptidase–antipeptidase complexes [e.g., α2-macroglobulin– and serpin–peptidase complexes], growth factors, cytokines, extracellular matrix constituents, and viruses) [577].

The LRP1 protein is composed of a cytoplasmic domain, a transmembrane sequence, and a large extracellular region. It is synthesized as a precursor that is cleaved in the trans-Golgi network by furin into a heavy α and light β chain. The large subunit with ligand-binding sites tethers to the small subunit that contains the transmembrane domain to function at the cell surface.

---

2. A.k.a. autosomal recessive hypercholesterolemia protein (ARH, ARH1, or ARH2).

3. The alias LRP is also used to designate a category of the class of lipid phosphate phosphatases (LPP): LPP-related proteins (better alias LPPRP), or plasticity-related gene products (PRG). Enzymes LPPs are involved in regeneration processes and attenuation of effects induced by lysophosphatidic acid. Signaling by LPA is implicated in wound healing, brain development, vascular remodeling, and tumor progression. Three LPPs and a splice variant deactivate lysophosphatidic acid [581].

4. A.k.a. apolipoprotein-E receptor (ApoER or ApoER1), α2-macroglobulin receptor (α2MR), type-5 transforming growth factor-β receptor (TβR5), insulin-like growth factor-binding protein-3 receptor (IGFBP3R), laminin receptor, and cluster of differentiation CD91.

LRP1 and Peptidase–Antipeptidase Complexes

The affinity of LRP1 for peptidase–antipeptidase complexes varies. Complexes of tissue plasminogen activator or urokinase with plasminogen activator inhibitor PAI1 have the highest affinity for LRP. On the other hand, α1-antitrypsin–trypsin, heparin cofactor-2–thrombin, and antithrombin-3–thrombin complexes have a markedly reduced affinity for LRP w.r.t. PAI1–tPA and PAI1–uPA complexes. Agent LRP1 binds to these peptidase–antipeptidase complexes and mediates their endocytosis for lysosomal degradation in the liver.

LRP1 and Cell Fate

The LRP1 protein is required in the inhibition of cell growth by insulin-like growth factor-binding protein IGFBP3 and transforming growth factor TGFβ1 [582]. Agent IGFBP3 is a regulator of endocrine effects of insulin-like growth factors and P53-regulated apoptotic factors. Hence, IGFBP3R can serve as a cell-death receptor.

LRP1 in the Nervous System

In neurons, LRP1 colocalizes with postsynaptic proteins such as Disc large homolog DLg4 [5] and NMDA-type glutamate receptor subunits [576]. Agent LRP1 mediates the uptake of certain neural peptidases such as neuroserpin. Cholesterol is imported into neurons with apolipoprotein-E4 via LRP1 receptors.

In transmissible spongiform encephalopathies, the normal cellular form of prion protein cPrP undergoes a conformational change at the plasma membrane or in transfer vesicles into a misfolded, infectious form PrP^{Sc} (pathological isoform scrappy) that partly resists to peptidases [583]. The cPrP protein is transported along the secretory route to the cell surface, where it attaches to the outer leaflet of the plasma membrane. This glycosylphosphatidylinositol-anchored molecule localizes in cholesterol- and glycosphingolipid-rich plasmalemmal rafts of neurons. It can then be internalized by clathrin-dependent, Cu^{++}-mediated endocytosis [584]. Laminin receptor, i.e., LRP1, but not LRP1b, acts as a plasmalemmal receptor for cPrP [584]. It is not only involved in the cPrP metabolism, but also the prion propagation [583]. Heparan sulfate proteoglycans function as cofactors or coreceptors for the binding of cPrP to LRP1 receptor. Amyloid precursor protein (APP), another LRP1 ligand, does not modulate cPrP endocytosis [584].

LRP1 in Lungs

The LRP1 receptor modulates the response of alveolar macrophages to foreign bodies and damaged cells when these materials are opsonized by surfactant proteins [576]. Aggregated surfactant molecules bind to the calreticulin–LRP1 complex

5. A.k.a. postsynaptic density protein PDS95.

for phagocytosis. [6] Moreover, they promote the production of pro-inflammatory cytokines.

LRP1 in Blood Vessel Walls

Oxidation of low-density lipoprotein is an early event during the development of atherosclerosis. Low-density lipoproteins bind to plasmalemmal LRP1 on macrophages. Afterward, LDLs are oxidized by 15-lipoxygenase, which is highly expressed in macrophages [585]. [7]

The LRP1 receptor is involved in the modulation and integration of PDGF and TGFβ signals in vascular smooth muscle cells. Loss of LRP1 in smooth muscle cells leads to an hypertrophic media and disrupted elastic lamina as well as increased susceptibility to atherosclerosis with hyperactive PDGF signaling, augmented density of PDGFRβ and ERK1/2 phosphorylation [576]. The LRP1 protein binds directly to the PDGFbb dimer. Phosphorylated LRP1 interacts with SHC adaptor.

Thrombospondin induces the disassembly of focal adhesions via the calreticulin–LRP1 complex and activation of ERK kinases as well as PI3K [576]. This process contributes to the regulation of cell migration.

Stimulation of nuclear factor-κB and subsequent upregulation of E-selectin that promotes leukocyte adhesion by the chemokine CXCL4 depends partly on LRP1 [576].

LRP1 Antagonist LRPAP1

Low-density lipoprotein receptor-related protein (LRP)-associated protein-1 (LRPAP1) [8] connects to both LRP1 and LRP2 on the cell surface in the presence of calcium ions. It prevents ligand binding to both LRP1 and LRP2 proteins. This glycoprotein can also bind to heparin and heparan sulfate proteoglycans.

In addition to its location at the apical surface of epithelial cells, LRPAP1 abounds in the endoplasmic reticulum lumen, where it may act as a chaperone for receptor (LRPs, VLDLR, etc.) folding and/or transfer from the endoplasmic reticulum and Golgi body to the plasma membrane [587].

---

6. Calreticulin is also called calregulin, CRP55, calcium-binding protein CaBP3, calsequestrin-like protein, and endoplasmic reticulum resident protein ERp60. It binds to misfolded proteins and prevents their export from the endoplasmic reticulum to the Golgi body.

7. On the basis of its product derived from arachidonic acid, this type of lipoxygenase is called 15-lipoxygenase (15LOx) in humans and rabbits and 12LOx in pigs, rats, and mice. 15-Lipoxygenase is a non-heme, iron-containing dioxygenase that synthesizes 12-hydroperoxy eicosatetraenoic acids (12HPETE) and 15-HPETE [585]. It also oxidizes esterified fatty acids in lipoproteins (cholesteryl esters) and phospholipids. It can also produce 13-hydroperoxy octadecadienoic acid (13HPODE) from linoleic acid. 15-Lipoxygenase regulates the expression of pro-inflammatory cytokines and chemokines, hence inflammation in insulin-sensitive tissues such as adipose tissue and insulin resistance [586].

8. A.k.a. α2-macroglobulin receptor-associated protein (MRAP and α2MRAP), receptor-associated protein (RAP), and heparin-binding protein HBP44.

5.1.2.2 LRP1b

Low-density lipoprotein-related protein-1B [9] is detected in medial and intimal smooth muscle cells near atheromatous plaques [588]. The LRP1b protein favors the catabolism of membrane receptors, such as urokinase-type plasminogen activator receptor (uPAR) and platelet-derived growth factor receptor PDGFβ, hence affecting SMC migration. [10]

The structural similarity between LRP1 and LRP1b suggests similar sets of ligands and overlapping functions. Receptor LRP1b can bind and internalize 2 components of the plasminogen activation system: single chain urokinase and plasminogen activator inhibitor PAI1 [589].

Urokinase, or urokinase-type plasminogen activator (uPA), binds to uPAR receptor. [11] The uPA–uPAR complex that displays an enhanced peptidase activity catalyzes the conversion of plasminogen to plasmin. The latter targets matrix proteins to facilitate cell migration. On the other hand, PAI1 binding to uPA–PAI1 complex not only inactivates uPA, but also provokes LRP1 binding and formation of a quaternary complex. The uPA–PAI1–uPAR–LRP1 huddle is rapidly recruited to a clathrin-coated pit owing to LRP dominant endocytosis signals. Subsequent endocytosis and transport into early endosome is followed by sorting of LRP and uPAR into recycling vesicle. Once internalized, uPAR and LRP1 are indeed recycled back to the plasma membrane, thereby regenerating uPAR, whereas uPA and PAI1 are degraded in lysosomes. The regeneration of unoccupied uPAR at the cell surface maintains plasminogen activation and facilitates cell migration.

Like LRP1, LRP1b binds to uPAR in the presence of uPA–PAI1 complexes. Agent LRP1b can also mediate the delivery of uPA–PAI1 complex to the lysosome for degradation. However, LRP1 and LRP1b function differently in uPAR endocytosis. Receptor LRP1 enhances unoccupied uPAR recycling on the plasma membrane, whereas LRP1b retains the uPA–uPAR complex on the cell surface, as the uPA–PAI1–uPAR–LRP1b aggregate moves very slowly to a clathrin-coated pit, thereby precluding rapid endocytosis, recycling, and regeneration of unoccupied uPAR on the cell surface, and consequently cell migration [589]. [12] Receptor LRP1b limit the effectiveness of LRP1- or caveolae-mediated uPA–PAI1–uPAR clearance.

---

9. A.k.a. LDLR-related protein-deleted in tumor (LRPDIT).

10. Protein LRP1b augments PDGFRβ degradation and lowers PDGFR-mediated phosphorylation levels of ERK1 and ERK2 kinases. Therefore, it hampers the migration of smooth muscle cells in the presence of PDGFbb dimers [588].

11. Activity of uPA is regulated by glycosylphosphatidylinositol-anchored uPA receptor and plasminogen activator inhibitor PAI1. Binding of uPA to uPAR soars its catalytic rate. Active uPA bound to uPAR remains at the cell surface. On the other hand, the uPA–uPAR–PAI1 complex is rapidly internalized and degraded.

12. The cytoplasmic domain of LRP1b and LRP contain the same endocytosis motifs. However, the tail of LRP1b possesses an additional sequence w.r.t. that of LRP1. This insert may be responsible for the slower endocytosis rate of LRP1b by binding cytosolic adaptors that may mask endocytosis motifs.

5.1.2.3 LRP2

Low-density lipoprotein receptor LRP2 [13] has an extracellular region that contains 17 EGF-like, 36 LDLR type-A, and 37 LDLR type-B domains. Its cytoplasmic C-terminus possesses apical sorting motifs as well as sequences needed for endocytosis and signaling.

The extracellular domain of LRP2 binds to apoE-enriched β-VLDL, lipoprotein lipase-enriched VLDL, and apolipoprotein-J. It mediates LDL endocytosis and degradation via interaction with apolipoprotein-B100 [590]. In addition to lipoproteins, LRP2 links to vitamin-binding and carrier proteins, hormones, enzymes, as well as signaling molecules [576]. In the thyroid gland, endocytosis of thyroglobulin via LRP2 involves heparan sulfate proteoglycans.

The LRP2 protein binds to many, but not all, LRP1 ligands. The intracellular domain of LRP2 contains binding motifs for adaptors and regulators of endocytosis, such as LDLRAP1, LRP2BP, Disabled homolog Dab2, and Disc large homolog DLg4 [576].

The LRP2 receptor has been detected in the nephron, gallbladder, epididymal, and mammary epithelial cells, parathyroid cells, thyroid follicular cells, placental cytotrophoblasts, as well as yolk sacs, ciliary body of eyes, intestinal brush border, male reproductive tract, uterus and oviduct [591, 592]. It also localizes to type-2 pneumocytes, to which it supplies vitamin-E.

In the liver, the LRP2 receptor operates in the clearance of remnants lipoproteins and plasma peptidases bound to their inhibitors.

Liganded and Unliganded LRP2, LRPAP1, and Endocytosis

The ligand–LRP2 complex is transported via early endosomes. In the absence of ligands, the intracellular transfer relies on the receptor-associated protein LRPAP1 that impedes LPR2–ligand binding. The LRP2–LRPAP1 complex travels in late endosomes [591]. The LRP2 receptor returns to the plasma membrane, whereas LRPAP1 is degraded in lysosomes. Chaperone-like protein LRPAP1 functions in the LRP2 distribution between cellular compartments.

LRP2 in Kidneys

The LRP2 receptor localizes to clathrin-coated pits of several types of epithelia, especially those of renal glomerulus and proximal tubules of nephrons. In proximal tubules, it intervenes in reabsorption of filtered proteins. In particular, it causes albumin internalization.

In the proximal tubule, LRP2 links to the sodium–hydrogen exchanger. In addition to reabsorption of calcium, LRP2 is involved in the reuptake of low-molecular weight proteins, such as retinol (vitamin-A), vitamin-B12 and -D, vitamin-D-binding

13. A.k.a. megalin and glycoprotein GP330.

protein complexed to vitamin-D, and angiotensin-2 in the proximal tubule from the glomerular filtrate [591].

Carriers facilitate steroid uptake by endocytosis of steroid–carrier complexes w.r.t. passive diffusion. Internalization is followed by intracellular release of transported steroids. The LRP2 protein abounds in proximal tubules of kidneys. 25-Hydroxyvitamin-D3 ($25(OH)D_3$) [14] is a prehormone produced in the liver by hydroxylation of vitamin-D3 (cholecalciferol) by cholecalciferol 25-hydroxylase. It tethers to its plasmalemmal carrier, vitamin-D-binding protein (VDBP), which enters the proximal tubule by receptor-mediated endocytosis. The LRP2 protein mediates the uptake of the $25(OH)D_3$–VDBP complex filtered in renal glomeruli. In kidneys, calcifediol is converted by 25(OH)D-1α-hydroxylase into calcitriol (or $(1,25)(OH)_2D_3$).

LRP2 in Lungs

The LRP2 receptor operates as a coreceptor for sonic Hedgehog that can compete with its inhibitor BMP4 for LRP2 binding. During pulmonary development, sonic Hedgehog morphogen and LRP2 colocalize to tracheal and bronchial epithelial cells [591]. Sonic Hedgehog can bind to heparan sulfate proteoglycans that can then facilitate the binding to LRP2 receptor. The LRP2–SHh complex is endocytozed to regulate SHh availability to Patched as well as for lysosomal degradation.

In adult humans, LRP2 resides in type-2 lung epithelial cells. Agent LRP2 may participate in the regulation of peptidase activity and the supply of nutrients and vitamins such as vitamin-E [576].

Cubilin Coreceptor

The extracellular domain of LPR2 interacts with cubilin for the uptake of vitamins-B12 and -D [576]. Cubilin [15] is a membrane-associated endocytic coreceptor that colocalizes with LRP2 receptor. It is attached to the extracellular face of the epithelial cell membrane. It promotes endocytosis, as it sequesters the steroid–carrier complexes on the cell surface before LRP2-mediated internalization of cubilin-bound ligand [593]. Transforming growth factor-β1 lowers LRP2–cubilin-mediated endocytosis via SMAD2 and SMAD3 [594].

LRP2BP

The partner low-density lipoprotein receptor-related protein 2-binding protein (LRP2BP) is mainly produced in small intestine, colon, and testis, as well as leukocytes [595]. It localizes both in the cytoplasm and nucleus.

---

14. A.k.a. calcifediol, calcidiol, and 25-hydroxycholecalciferol.
15. A.k.a. intrinsic factor cobalamin receptor (IFCR), glycoprotein GP280, intestinal intrinsic factor receptor, and intrinsic factor vitamin-B12 receptor.

5.1.2.4 LRP3

Low-density lipoprotein receptor LRP3 [16] is a type-1 transmembrane protein with a cytoplasmic region that contains an internalization signal. It is widely distributed, with the highest expression in skeletal muscle and ovary [596].

The LRP3 receptor, but not LRP1, binds to nucleobindin-1 (NucB1), or calnuc, owing to an arginine (R)-rich region in the cytoplasmic tail. Nucleobindin-1 is an ubiquitous, multifunctional, multicompartmental Ca^{++}-binding protein. It resides in the cytosol, where it binds to different Gα subunits of guanine nucleotide-binding (G) protein, as well as in the Golgi lumen, where it constitutes a major Ca^{++} storage pool, and outside the cell.

5.1.2.5 LRP4

Low-density lipoprotein receptor LRP4 [17] localizes to dendrites and soma of neurons [597]. This endocytic, multiligand receptor interacts with postsynaptic scaffold proteins such as Disc large homologs DLg1 and DLg4. It associates with ^{N}methyl ^{D}aspartate (NMDA)-type glutamate receptor subunit. Once phosphorylated by Ca^{++}–calmodulin-dependent protein kinase CamK2 (Ser1887 and Ser1900), it cannot connect to Disc large homologs.

LRP4 in Bones and Teeth

Protein LRP4 is expressed in osteoblasts. It connects to many ligands, such as Dickkopf-1, sclerostin, and sclerostin domain-containing SostDC1 [598]. Dickkopf-1 is another potent soluble Wnt inhibitor regulated by bone morphogenetic proteins. Sclerostin is a potent osteocyte-secreted inhibitor of bone formation. Sclerostin domain-containing protein SostDC1 [18] is a secreted Wnt modulator that behaves as a context-dependent activator and inhibitor of Wnt signaling as well as a BMP antagonist.

Agent LRP4 contributes to bone mineral content (BMC) and density (BMD) and bone growth and turnover, as an integrator of Wnt and BMP bone morphogenetic protein) signalings. Extracellular modulator SostDC1 secreted from mesenchymal cells binds to BMPs as well as LRP4 on epithelial cells [599]. The LRP4–SostDC1 complex thus acts as an extracellular integrator of epithelial–mesenchymal cell communication based on the integration of Wnt and BMP pathways that uses SHh mediator.

The LRP4 receptor participates in BMP and Wnt signaling during tooth morphogenesis, as it binds to the BMP antagonist SostDC1 [599]. The negative Wnt coreceptor LRP4 antagonizes the canonical Wnt pathway, as it displaces homologous LRP5 and LRP6 from the receptor complex formed with Frizzled proteins.

---

16. A.k.a. 105-kDa LRP (LRP105).

17. A.k.a. multiple epidermal growth factor-like domain-containing protein-7 (MEGF7), synaptic LDLR-related protein (synLRP), and LRP10.

18. A.k.a. Wise homolog, uterine sensitization-associated gene product-1 (USAG1), and ectodermal BMP inhibitor (ectodin).

5.1.2.6 LRP5 and LRP6

Low-density lipoprotein receptors LRP5 [19] and LRP6 [20] connect to Frizzled receptors, thus participating in Wnt signaling. Binding of a Wnt ligand to the Frizzled–LRP5–LRP6 receptor complexes inhibits phosphorylation and degradation of its transcriptional coactivators β-catenins (Sect. 10.3).

LRP5 and LRP6 in Bones

Receptors LRP5 and LRP6 binds to sclerostin encoded by the SOST gene. [21] Sclerostin is produced and secreted by osteoclasts and osteocytes. It antagonizes Wnt messenger and some bone morphogenetic proteins (BMP6–BMP7) [600]. Sclerostin is a potent inhibitor of bone formation that directly binds to LRP5 and LRP6 coreceptors. In addition to its inhibition of BMPs, sclerostin exerts an anti-osteogenic effect, as it prevents the Wnt signaling via LRP5 and LRP6 in osteoblasts by disrupting Wnt-primed formation of Frizzled–LRP complexes.

Receptors LRP5 and LRP6 link to Dickkopf-1, another soluble inhibitor of the Wnt–β-catenin signaling. The Dkk1 production is regulated by bone morphogenetic proteins. Dickkopf-1 can remove sclerostin from the LRP5–sclerostin complex [598]. The Dkk1 mediator is involved in head and limb development during embryo- and fetogenesis. It also regulates postnatal bone accretion and maintenance of bone mass mainly by binding to LRP5 and LRP6 and using kringle-containing transmembrane proteins Kremen-1 or- 2 or independently of these Dickkopf receptors [598].

5.1.2.7 LRP9

Low-density lipoprotein receptors LRP9 [22] is synthesized as a proreceptor that is processed to the mature form by a furin-like propeptidase.

The LRP9 receptor is a type-1 membrane protein that constitutes with LRP3 and LRP12 a subfamily of the LDLR family. Members of this subfamily possess several specific structural features that distinguish them form other LDLR class members, such as a relatively large cytoplasmic tail and ligand-binding CUB domains in the extracellular region.

---

19. A.k.a. LRP7, LR3, as it possesses 3 ligand binding domains, osteoporosis-pseudoglioma syndrome protein (OPPG or OPS), osteopetrosis autosomal dominant disease type 1 (OptA1), bone mineral density variation BMND1, high bone mass protein (HBM), exudative vitreoretinopathy protein EVR1 or EVR4, and van Buchem disease type 2 (VBCH2).

20. A.k.a. autosomal dominant coronary artery disease type 2 protein (ADCAD2).

21. Mutations in the SOST gene cause sclerosteosis characterized by massive bone overgrowth. Van Buchem's disease is a generalized hyperostosis that results from a deletion downstream from the SOST gene that removes a SOST-specific regulatory element.

22. A.k.a. sorting protein- or sortilin-related receptor containing LDLR type-A repeats (SorLa, SorLa1, or SorL1) and LDLR relative with 11 ligand-binding repeats (LR11).

LRP9 in the Trans-Golgi Network and Endosomes

The LRP9 receptor localizes to the trans-Golgi network and endosomes rather than the plasma membrane. It interacts with clathrin adaptors involved in the transport between the trans-Golgi network and endosomes. Like mannose 6-phosphate receptor (or IGF2R), it carries ligands between these 2 intracellular compartments.

This sorting receptor regulates transfer and processing of amyloid precursor protein. It acts as a retention factor for APP in the trans-Golgi network, where it prevents APP release into regular processing pathways [601]. It interacts with the adaptors Golgi-localized, γ adaptin ear-containing, ARF-binding proteins (GGA) and phosphofurin acidic cluster sorting protein PACS1 that are involved in protein transport to and from the trans-Golgi network.

LRP9 in Neurons

Neurosecretases are involved in the release of fibril-forming neuropeptides and other products. Secretase-α, or metallopeptidases of the ADAM family, and -β,[23] and the γ-secretase multimer (γSC) support intramembranal cleavage.

Receptor LRP9 localizes to recycling endosomes, where it binds the retromer coat complex. The retromer complex has domains for cargo specification. Other adaptors linked to secretory endosomes include homohexameric AAA ATPase soluble ^{N}ethylmaleimide-sensitive factors (NSF), soluble NSF-attachment proteins, syntaxins, synaptobrevins, amyloid-β A4 precursor protein-binding adaptor APBA1, reticulons Rtn3 and Rtn4, and the Rtn4 receptor. The LRP9 protein also links to Golgi-localized γ-ear-containing adpribosylation factors-binding coat proteins (GGA) for protein transport to endosomes. The LRP9 receptor is required in the tranfer of amyloid precursor protein from membranes that possess non-amyloidogenic processing by α-secretase to the β–γ-secretase processor.

LRP9 in Smooth Muscle Cells

The LRP9 protein is markedly expressed in smooth muscle cells of the remodeled intima in animal atherosclerosis models [602]. It binds apolipoprotein-E-rich lipoproteins, especially β-very-low-density lipoproteins and ApoE-enriched HDL, with a high affinity that is similar to that of LDLRs and VLDLRs [602].

LRP9 Partners

The LRP9 receptor binds also LRPAP1 as well as neurotensin [603]. It also links to neuropeptide Head activator that stimulates cell proliferation of neuronal precursor

23. A.k.a. β(-site) amyloid precursor protein (APP)-converting (cleaving) enzyme (BACE), aspartyl peptidase-2, and membrane-associated aspartic peptidase-2 (memapsin-2).

and neuroendocrine cells. [24] The signaling cascade relies on Ca^{++} influx through TRPV2 that is a growth factor-regulated, Ca^{++}-permeable cation channel. [25]

The LRP9 protein that cycles between the trans-Golgi network and endosomes interacts directly with NucB1 [607]. Nucleobindin-1 and LRP9 colocalize at the surface of the trans-Golgi network and early endosomes. Depletion in NucB1 missorts LRP9 in the late endosome–lysosome compartment and enhances its lysosomal degradation.

5.1.2.8 LRP11

Low-density lipoprotein receptor-related protein-11 (LRP11) [26] is widespread. [27] According to [609], the lipoprotein receptor LRP11 is also termed sortilin-related receptor type-A repeats-containing protein, hence corresponding to LRP9. However, the Information Hyperlinked over Proteins website [608] mentions 2 different proteins: LRP9 (FLJ21930 or FLJ39258) and LRP11 (FLJ14735).

Protein LRP11 integrates components of pre- and post-Golgi membranes and clathrin-coated vesicles implicated in the transfer of cargos in association with lipoproteins [609].

5.1.2.9 LRP12

Low-density lipoprotein receptor-related protein-12 (LRP12) [28] is widespread. [29] It possesses a cytoplasmic tail that contains several motifs implicated in endocytosis and signal transduction. Its cytoplasmic domain indeed interacts at least with 3 mediators of signaling pathways [611]: (1) receptor for activated protein C kinase RACK1; (2) β_1-integrin-binding protein Itgβ1BP3; [30] and (3) SMAD anchor for receptor activation (SARA). [31]

---

24. Head activator localizes to hypothalamus and intestine at high concentrations in humans. This mitogen acts in the G2–M transition of the cell cycle. It operates via Gi/o subunit of G protein, Ca^{++} influx, and hyperpolarization of the transmembrane potential. Head activator causes activation and translocation of Ca^{++}-permeable TRPV2 channel [604].

25. Insulin and insulin-like growth factor-1 stimulate calcium entry through growth factor-regulated TRPV2 channel [605, 606].

26. A.k.a. MANSC3 [608].

27. I.e., connective tissue, heart, kidney, bone marrow, brain, liver, lung, pharynx, among other organs. In the central nervous system, LRP11 abounds in hippocampus, dentate gyrus, and cerebral cortex [609]. It colocalizes with β-amyloid precursor protein on Golgi membranes.

28. A.k.a. suppressor of tumorigenicity ST7, tumor suppressor gene product TSG7, as well as ETS7q, FAM4A1, HELG, RAY1, and SEN4.

29. It reaches its highest levels in the heart, liver, and pancreas [610]. It also lodges in the lung and, at high concentrations, in the digestive tract (mouth, intestine, and colon) as well as reproductive female system. It is detected in epithelial cells of endo- and exocrine glands.

30. A.k.a. muscle integrin-binding protein (MIBP).

31. A.k.a. zinc finger, FYVE domain-containing protein ZFYVE9 and mothers against decapentaplegic homolog-interacting protein (MADHIP).

Peroxisome proliferator-activated receptor-α-associated proteins that may affect signal transduction and PPARα (or NR1c1 nuclear receptor) activity include ribosomal protein-L11 (RPl11), a component of the large 60S subunit and regulator of P53, hence of the cell cycle, [32] and LRP12 [612].

5.1.3 ApoER2 (LRP8) and VLDLR

Very-low-density lipoprotein receptor (VLDLR) interacts with ApoE-containing lipoproteins, such as VLDL, but not LDL. It also binds to the peptidase–antipeptidase complexes. Like VLDLR, apolipoprotein-E receptor ApoER2 [33] binds to ApoE-rich lipoproteins, but not ApoB-containing lipoproteins. The Apoer2 gene gives rise to several alternatively spliced transcripts with different spatial and temporal expression [576].

The *reelin receptors* — ApoER2 and VLDLR — localize to both neurons and glial cells (astrocytes, oligodendrocytes, Purkinje cells, retinal pigment epithelial cells), where they can form clusters. The reelin receptors lodge also on endothelial and smooth muscle cells, macrophages, platelets, adipocytes, osteoblasts, Sertoli cells, placenta Hofbauer cells, and trophoblasts [600].

Reelin stimulates progenitor cells to differentiate into radial glia, as the producton of brain lipid-binding protein (BLBP), or brain-type fatty acid-binding protein FABP7, is promoted via the Notch1 cascade. Reelin is a serine peptidase that can influence cell adhesion and migration. Once bound to VLDLR and ApoER2, reelin is internalized.

Receptors ApoER2 and VLDLR participate in neuronal development. They transduce reelin signals. The latter contributes to the regulation of neuron migration and positioning in the developing brain. In adults, reelin is engaged in synaptic remodeling for long-term potentiation, [34] dendrite and dendritic spine development, as well as the regulation of migration of neuroblasts.

The reelin receptors recruit Dab1 adaptor. The latter is both a substrate and an activator for SRC family kinases. Phosphorylation of Dab1 leads to its ubiquitination and subsequent degradation (negative feedback). The reelin receptors are phosphorylated by Src and Fyn kinases to trigger the actin cytoskeleton reorganization. The reelin receptors also interact with CRK family proteins that function in the relocalization and activation of small GTPases by guanine nucleotide-exchange factors. Binding of CRK adaptors to reelin receptors results from CRK phosphorylation and activates Rap1 GTPase following phosphorylation of RapGEF1 [576].

32. Inhibition of gene expression by RPl11 results from a blockade of binding to the PPAR-response element (PPRE) of genes.

33. A.k.a. LDLR-related protein LRP8.

34. Reelin operates in long-term potentiation via the ApoER2 interaction with NMDA receptor.

5.2 Scavenger Receptors

Scavenger receptors constitute a superclass of receptors that recognize modified low-density lipoproteins by oxidation or acetylation. Scavenger receptors function in lipid transport, both in endo- and exocytosis. They, at least, contribute to the transfer of lipids to or from the plasma membrane. They can also escort their substrates during endocytosis (Vol. 1 – Chap. 9. Intracellular Transport). In particular, they lead to LDL clearance (scavenging) by macrophages.

Plasma lipoproteins (Vol. 5 – Chap. 1. Blood) carry and deliver lipids, such as triglycerides, cholesterol, and certain lipid-soluble vitamins, to the body's cells. Chylomicrons and VLDL particles carry dietary and endogenous lipids, respectively, to supply the energy providers, the fatty acids, to cells using triglyceride hydrolysis by endothelium-anchored lipoprotein lipase. The latter generates cholesterol-enriched residual lipoproteins, i.e., chylomicron remnants, intermediate (IDL)- and low (LDL)-density lipoproteins. These lipoproteins are removed from blood by receptors on cells to process these lipoproteins and store or secrete lipids. High-density lipoproteins receive cholesterol from cell membranes. Cholesterol is then esterified by lecithin–cholesterol acyltransferase. High-density lipoproteins then participate in cholesteryl ester delivery especially to steroidogenic cells and in cholesterol transport from tissues to hepatocytes for excretion out of the body. Cholesterol esters can be exchanged by cholesterol ester transfer protein from large CE-rich HDL3s for triacylglycerols and transported in TG-rich lipoproteins.

In addition to lipoproteins, scavenger receptors are able to link to endo- and exogenous molecules, in particular pathogen fragments, as well as endogenous neoantigens. They thus have been incorporated into the class of pattern recognition receptors [613] (Chap. 11).

Scavenger receptors are categorized into 8 classes (A–H) according to their structure [613] (Table 5.4). Class-A scavenger receptors are mainly expressed in macrophages. These trimers are composed of cytosolic, transmembrane, spacer, α-helical coiled-coil, collagen-like, and cysteine-rich domains. Class-B scavenger receptors with their prototypic member CD36 (ScaRb3) have 2 transmembrane regions. Class-C Drosophila DSRc does not have a human homolog.

Class-D scavenger receptor may contribute to the processing of oxidized LDL (oxLDL) in the endolysosomal compartment. Class-E scavenger receptor is expressed by vascular cells (endothelial and smooth muscle cells and macrophages). Class-F scavenger receptors are expressed by endothelial cells. Class-G scavenger receptor corresponds to membrane-bound chemokine CXCL16.

Class-H scavenger receptors fasciclin, EGF-like, laminin-type EGF-like and link domain-containing scavenger receptor FEEL1 [35] and FEEL2 [36] are primarily involved in bacterial binding.

35. A.k.a. Fas1 EGF-like and X-link domain-containing adhesion molecule FEX1 and stabilin-1 (Stab1).

36. A.k.a. FEX2, stabilin-2 (Stab2), and hyaluronan receptor for endocytosis (HaRE).

Table 5.4. Types of scavenger receptors and their aliases. Scavenger receptors constitute a superclass with 8 classes (LOX1: lectin-like oxidized LDL receptor; MaRCo: macrophage receptor with collagenous structure, class-A scavenger receptor family member [SCARAF]; MSR: macrophage scavenger receptor; SREC: scavenger receptor expressed by endothelial cells; SRPSOx: scavenger receptor phosphatidylserine and oxidized lipoprotein). They share the ability to tether to modified forms of low-density lipoproteins (LDL). They are also able to bind and clear (scavenge) a broad range of other modified self- and non-self ligands (apoptotic cells, anionic phospholipids, and amyloid and pathogen components).

	Class A
ScaRa1a/b/c	MSR1, CD204
ScaRa2	MaRCo
ScaRa3	CSR, CSR1, MSRL1, APC7
ScaRa4	SRLC, Colec12, ClP1, NSR2
ScaRa5	
	Class B
ScaRb1	CD36L1, Cla1
ScaRb2	CD36L2, LIMP2, LGP85
ScaRb3	CD36, GP3b, GP4, FAT, PAS4
	Class C
DSRc	(Drosophila)
	Class D
ScaRd1	CD68, GP110, macrosialin
	Class E
ScaRe1	OLR1, LOx1, CLec8a
	Class F
ScaRf1	SREC1
ScaRf2	SREC2, SREPCR, SRECRP1, NSR1
	Class G
CXCL16	SRPSOx, SCyB16
	Class H
FEEL1	FEX1, Stab1
FEEL2	FEX2, Stab2, HaRE
	Class I
HbSR	CD163

Class-I scavenger receptor CD163, or hemoglobin–haptoglobin scavenger receptor (HbSR) is involved in hemoglobin binding. Glucocorticoids and interleukin-10, but neither IL4 nor IL13, upregulate CD163 in cultured human monocytes.

5.2.1 Class-A Scavenger Receptors

Class-A scavenger receptors include 5 homotrimers that share common collagen-like domains for ligand binding. Class-A scavenger receptors preferentially bind to modified low-density lipoproteins, i.e., acetylated (acLDL) and oxidized (oxLDL) low-density lipoproteins.

Liganded ScaRa activate intracellular signaling mediated by phospholipase-Cγ1, phosphoinositide 3-kinase, and protein kinase-C. The ScaRa receptor causes endocytosis and clearance of bacteria, environmental particles, and DNA oligonucleotides [614].

The mechanism that regulates the transport of the macrophage ScaRa during ligand uptake (acetylated and oxidized low-density lipoproteins, but not LDL) relies on receptor phosphorylation. Phosphorylation occurs during endocytosis of acLDLs and oxLDLs by macrophages. Class-A scavenger receptors possess 3 phosphorylation sites in their cytoplasmic domains (Ser21, Thr30, and Ser49). [37]

5.2.1.1 Scavenger Receptor-A1

Class-A, type 1 scavenger receptors (ScaRa1), [38] i.e., the 3 alternatively spliced variants, have a cysteine-rich domain. The Msr1 gene indeed generates 3 differentially spliced mRNAs that code for type-1 transmembrane receptors ScaRa1a to ScaRa1c. The third variant, ScaRa1c, is retained in the endoplasmic reticulum, thereby being not accessible to extracellular ligands.

In atheroma, ScaRa1a and ScaRa1b are synthesized in vascular endothelial and smooth muscle cells, macrophages, and foam cells [613]. They recognize modified apolipoprotein-B in LDL particles. Both ScaRa1a and ScaRa1b have a lower affinity for oxidized LDLs than acetylated LDL particles.

The higher the density of class-A macrophage scavenger receptors, the greater the plasmatic HDL phospholipid and apolipoprotein-E levels [616]. In resident macrophages, acetylated low-density lipoproteins enhance cholesterol esterification. On the other hand, in activated macrophages, cholesterol esterification can be lower in the presence of acLDLs, but to a higher extent with β-very-low-density lipoproteins.

The expression of ScaRa1 and ScaRb3 increases during differentiation of monocytes into macrophages. Adipocyte-produced adiponectin suppresses ScaRa1 expression, but not that of ScaRb3 [617]. It also lowers synthesis of lipoprotein lipase, but not that of apolipoprotein-E. In addition, adiponectin precludes differentiation of macrophages into foam cells. Peroxisome proliferator-activated receptor-γ increases the production of adiponectin in adipocytes owing to the presence of a PPAR-responsive element in human adiponectin promoter [617]. The PPARγ–RXR heterodimer, i.e., the nuclear receptor NR1c3–NR2b complex, binds to PPRE to elicit the promoter activity. Similarly, LRH1 (NR5a2) connects to its LRHRE responsive

37. The Ser21 residue controls surface receptor expression. The $Ser49^P$ residue accelerates receptor internalization [615].

38. A.k.a. macrophage scavenger receptor MSR1, macrophage acetylated LDL receptor-1 and -2, and CD204.

element on adiponectin promoter to augment PPARγ-induced activation of adiponectin promoter [617].

5.2.1.2 Scavenger Receptor-A2

The ScaRa2 receptor[39] contains an extracellular collagen-like motif and a C-terminal cysteine-rich domain. In normal mice, its expression is restricted to macrophages in the spleen marginal zone and lymph nodes, where it intervenes in pathogen clearance [613].

Inhaled nanoparticles can reach lung alveoli, where alveolar macrophages reside. These cells constitute the first line of defense against various environmental particles and microorganisms. The ScaRa2 receptor mediates the ingestion of unopsonized environmental particles. It also clears nanoparticles (size 20–200 nm) that can deposit in alveoli [619].

5.2.1.3 Scavenger Receptor-A3

The ScaRa3 receptor[40] intervenes in clearance, phagocytosis, pathogen recognition, and inflammation.

The ScaRa3 receptor aims at scavenging oxidative molecules or products of oxidation during oxidative stress. In normal fibroblasts, its synthesis augments significantly upon exposure to UV radiation or hydrogen peroxide [620].

5.2.1.4 Scavenger Receptor-A4

The ScaRa4 receptor[41] is a membrane-type collectin from placenta. It is produced in vascular endothelial cells, but not in macrophages [621]. It reacts with oxLDLs, but not acLDLs.

Collectins constitute a family of C-type lectins (Vol. 1 – Chap. 7. Plasma Membrane) that have collagen-like sequences and carbohydrate recognition domains. They are involved in host defense, as they are able to bind to carbohydrate antigens of pathogens.[42] Like other collectins, ScaRa4 can bind and phagocytose not

39. A.k.a. macrophage receptor with collagenous structure (MARCO).

40. A.k.a. cellular stress response gene protein (CSR and CSR1), macrophage scavenger receptor-like MSRL1, or MSLR1, and APC7.

41. A.k.a. scavenger receptor with C-type lectin [SRCL], collectin family member Colec12, collectin placenta protein ClP1, and nurse cell scavenger receptor NSR2.

42. In addition to ScaRa4, 4 subfamilies of collectins exist: (1) subfamily of mannan-binding protein (MBPa and MBPc); (2) subfamily of surfactant protein-A (SPa); (3) subfamily of surfactant protein-D (SPd); and (4) subfamily of collectin liver-1 (ClL1). Collectins play an important role in innate immunity [621]. Mannan-binding proteins can destroy bacteria through activation of the complement pathway or opsonization via collectin receptors. Mannan-binding proteins and conglutinin of the SPd group target influenza A viruses. Surfactant protein-A amplifies the phagocytosis of bacteria by macrophages and opsonizes herpes simplex virus. Surfactant protein-D agglutinates bacteria and exerts an hemagglutination inhibition of influenza A virus.

only bacteria (e.g., Escherichia coli and Staphylococcus aureus), but also yeast (Saccharomyces cerevisiae) [621].

5.2.1.5 Scavenger Receptor-A5

Expression of ScaRa5 is restricted to the plasma membrane of epithelial cells. Inhaled viruses invade the respiratory epithelium. Airway epithelial cells then initiate innate and subsequent adaptive immune responses. Recognition of intra- or extracellular double-stranded RNA (dsRNA) that signals viral infection to the host immune system triggers an early and potent reaction to activate innate and adaptive immunity. Extracellular dsRNA is recognized and internalized by class-A scavenger receptors [614]. Once dsRNA is bound to its cognate surface receptors, it is internalized to activate suitable pathways. Epithelial cells then release inflammatory cytokines, such as tumor-necrosis factor-α and interleukin-6 and -8, as well as CCL5 chemokine. Double-stranded RNA-activated pathways are mediated by eukaryotic translation initiation factor-2α kinase-2, or protein kinase-R (Vol. 4 – Chap. 4. Cytosolic Protein Ser/Thr Kinases) and mitogen-activated protein kinase (Vol. 4 – Chap. 5. Mitogen-Activated Protein Kinase Modules)

5.2.2 Class-B Scavenger Receptors

Class-B scavenger receptors are also identified as members of the oxLDL receptor CD36 family. This family includes: ScaRb3 [43] as well as 2 alternatively spliced variants ScaRb1 [44] and ScaRb2 [45]

Unlike the SCARA family, CD36 is a type-3 (multiple transmembrane domain) receptor that traverses the membrane twice to form a heavily glycosylated extracellular loop with 2 short intracellular tails.

Scavenger receptors ScaRb1 and ScaRb3 bind to many common ligands, such as HDL, LDL, VLDL, acetylated and oxidized LDLs, and anionic phospholipid vesicles [622]. Nevertheless, these 2 receptors have specific lipid transport functions: ScaRb1 intervenes in the transport of cholesterol and cholesteryl ester from HDL particles as a major HDL receptor, whereas ScaRb3 promotes uptake of long-chain fatty acids.

5.2.2.1 Scavenger Receptor ScaRb1

Scavenger receptor ScaRb1 has 2 transmembrane and 2 cytoplasmic domains as well as a large extracellular domain with sites for N-linked glycosylation. Receptor

43. A.k.a. leukocyte differentiation antigen CD36, platelet glycoprotein GP3b or GP4, fatty acid translocase (FAT), periodic acid Schiff-positive glycoprotein PAS4 (staining behavior), platelet collagen receptor, and thrombospondin receptor.

44. A.k.a. CD36 antigen-like protein 1 (CD36L1), thrombospondin receptor-like protein-1, and collagen receptor type 1 (Cla1).

45. A.k.a. CD36L2, lysosomal integral membrane protein LIMP2, and 85-kDa lysosomal membrane sialoglycoprotein LGP85.

ScaRb1 may concentrate in caveolae (plasmalemmal nanodomains) [622]. However, ScaRb1 is not necessarily colocated with caveolin-1. Anyway, it clusters and forms patches in the plasma membrane, in particular in plasmalemmal microvillar domains.

The ScaRb1 glycoprotein is expressed in many organs, such as brain, intestine, and placenta, as well as cell types, such as endothelial and smooth muscle cells, platelets, macrophages, adipocytes, and keratinocytes [622]. However, it has its highest expression levels in tissues that use HDL free cholesterol and cholesteryl ester for bile acid synthesis (liver) and steroidogenesis (adrenal glands, ovary, and testis), as well as macrophages.

Scavenger receptor ScaRb1 is an HDL receptor for the selective uptake of cholesteryl esters from HDL into cells. It is then required for cholesterol delivery to the liver. It also facilitates cholesterol egress from cells to HDLs after direct interaction between HDLs and scavenger receptors, especially in the arterial wall (bidirectional flux). The ScaRb1 receptor is implicated with the ATP-binding cassette transporter ABCa1 in cholesterol efflux, the former as an HDL receptor, the latter as a lipid transporter. It thus participates in cellular transport at both ends of cholesterol convection in blood flow, i.e., the body's cell efflux (entry in blood stream) and liver influx (exit from blood stream).

The ScaRb1 receptor is also identified as an oxidized LDL receptor, as it connects not only to normal HDLs and LDLs, but also oxLDLs. Unlike the LDL receptor (LDLR) that causes endocytosis of the entire lipoprotein via clathrin-coated pits and lysosomes for degradation by an acidic cholesteryl ester hydrolase, ScaRb1 binds HDL without HDL degradation. Uptake by ScaRb1 of the core cholesteryl ester for cell delivery occurs without concomitant HDL internalization [622].

Cell delivery of HDL materials is a 2-step mechanism [622]: (1) lipoprotein binding to ScaRb1 in a metabolically active membrane domain and (2) selective transfer of cholesteryl ester by ScaRb1 to the plasma membrane, once HDL cholesteryl esters heve been removed from membrane-bound HDLs by a neutral cholesteryl ester hydrolase. This process is much more efficient than that using ScaRb3. On the other hand, ScaRb1 permits the delivery of free cholesterol for esterification to HDL, also using a metabolically active plasmalemmal platform [622].

The ScaRb1 receptor can also be involved in uptake of cholesteryl esters from apolipoprotein-B-containing LDLs, albeit with a lower efficiency than that for HDLs. Moreover, the LDL–ScaRb1 binding depends on the composition of LDLs in apolipoproteins (ApoA, ApoB, and ApoE) and sphingomyelin content. Differences in lipoprotein composition can explain targeting of either ScaRb or LDLR receptors [622].

In addition, ScaRb1 recognizes apoptotic cells, advanced glycation end-products, anionic phospholipids, serum amyloid-A, and β-amyloid [613].

The regulation of ScaRb1 depends on the cell type. In the liver and intestinal epithelium, but neither in adrenal glands nor macrophages, ScaRb1 interacts with PDZ domain-containing adaptor PDZK1.[46] The latter contributes to the regulation of

46. A.k.a. 70-kDa CFTR-associated protein CAP70, Na^+-dependent inorganic phosphate cotransporter C-terminal-associated protein NaPiCAP1, Na^+–H^+ exchange regulatory cofactor NHERF3, C-terminal linking and modulating protein [CLAMP], and Diphor-1.

ScaRb1 expression on the cell surface of hepatocytes [623]. The PDZK1-interacting domain of ScaRb1 is actually required for its plasmalemmal insertion. The PDZK1 adaptor and other PDZ domain-containing proteins may thus regulate ScaRb1 density in the plasma membrane, hence cholesterol transport. In the liver, Small PDZK1-associated protein (SPAP) lowers PDZK1 production, hence that of ScaRb1.

In addition, ScaRb1 stimulates nitric oxide production in endothelial cells (Vol. 5 – Chap. 9. Endothelium). Hence, HDLs prevent the inhibition of nitric oxide synthase NOS3 by oxidized LDLs via ScaRb1, calcium mobilization, increase in ceramide level, and possibly the PI3K–PKB pathway [622]. Yet, a significant contribution of induced vasodilation results from the interaction of HDL-associated lysophospholipids, such as sphingosyl phosphorylcholine, lysosulfatide (galactosylsphingosine sulfate), and sphingosine 1-phosphate that targets its receptor $S1P_3$ (Chap. 7). HDL-Associated estradiol provokes the same effect.

The processing mode used by ScaRb1 depends on its ligand type: either at the cell surface (e.g., cholesteryl ester) or inside the cell, often for degradation. The ScaRb1 receptor allows binding, endocytosis, and cellular accumulation of both monomeric and HDL-associated lipopolysaccharides [622].

5.2.2.2 Scavenger Receptor ScaRb2

In humans, scavenger receptor ScaRb2, a high-density lipoprotein receptor, is a type-3 double-transmembrane glycoprotein with cytoplasmic N- and C-termini. It localizes primarily to membranes of late endosomes and lysosomes.[47] It binds to thrombospondin-1.

The ScaRb2 receptor is encoded by an alternatively spliced mRNA from the SCARB1 gene. It differs from ScaRb1 only by its C-terminus. Like ScaRb1, ScaRb2 localizes to caveolae. It operates as an endocytic receptor that mediates HDL endocytosis via a clathrin-dependent, caveolae-independent route [624].

The ScaRb2 receptor acts in both cholesterol ingress and egress. In addition, cytoplasmic ScaRb2, but not cytoplasmic ScaRb1, can binds to phospholipase-$C\gamma 1$ [625].

Lysosomal Membrane and Its Proteic Constituents

Lysosomal membrane contains a characteristic set of 5 highly glycosylated transmembrane proteins, among which are the lysosomal-associated membrane proteins (LAMP).[48] This category includes LAMP1,[49] LAMP2,[50] and LAMP3.[51] In addi-

47. Late endosomes and lysosomes are in dynamical equilibrium, as they undergo multiple cycles of fusion and fission.

48. A.k.a. lysosomal membrane sialoglycoproteins (LGP).

49. A.k.a. LAMPa and LGP120, as well as CD107a.

50. A.k.a. LAMPb, LGP96, LGP110, LGPb, and CD107b.

51. A.k.a. dendritic cell lysosome-associated membrane glycoprotein (DCLAMP), TSC403, and CD208.

tion, lysosomal membrane is constituted of lysosomal integral membrane proteins LIMP1,[52] LIMP2,[53] and lysosomal acid phosphatase.

Proteins LAMP1 and LAMP2 are type-1 membrane protein, whereas ScaRb2 is a type-3 membrane protein. The LAMP1 protein is expressed on activated platelets and neutrophils, as well as T lymphocytes and endothelial cells. The LAMP2 protein is produced on activated platelets and weakly on activated neutrophils and endothelial cells.

The LIMP1 protein is synthesized in activated macrophages, platelets, and splenocytes, as well as monocytes, macrophages, mastcytes, fibroblasts, endothelial and smooth muscle cells, neurons, hepatocytes, enterocytes, osteoclasts, synoviocytes, and Langerhans cells.

Overexpression of ScaRb2, in synergy with small GTPase Rab5b, causes an enlargement of early and late endosomes and lysosomes, but neither LAMP1 nor LAMP2 [626]. Moreover, ScaRb2 overexpression impairs endocytosis function of these enlarged compartments.

5.2.2.3 Scavenger Receptor ScaRb3 (CD36)

Scavenger receptor ScaRb3, a palmitoylated, plasmalemmal glycoprotein, resides on monocytes, macrophages, platelets, erythrocytes, splenocytes, adipocytes, cardiomyocytes, hepatocytes, steroidogenic cells of adrenal glands, testis, and ovary, bronchiolar and mammary epithelial cells, as well as certain types of endothelial cells, such as those of the microvasculature, especially in skin and coronary capillaries. It is also detected in epithelia of the small intestine, kidney, sebaceous glands, and circumvallate papillae of the tongue, at least in rats [627].[54] Preferential residences of ScaRb3 correpond to tissues with active fatty acid metabolism, such as skeletal muscle, heart, and adipose tissue.

In humans, ScaRb3 possesses a high affinity for different types of lipoproteins: HDL, LDL, and VLDL [631]. Binding of HDL to ScaRb3 does not require the presence of apolipoprotein-E. The ScaRb3 receptor interacts also with numerous other ligands, such as collagen-1 and -4, thrombospondin-1, 37-kDa platelet-agglutinating protein PAP37, oxidized and acetylated low-density lipoproteins, long-chain fatty acids, β-amyloid, anionic phospholipids, as well as apoptotic cells, advanced glycation end products, and pathogen-derived ligands (e.g., Plasmodium falciparum peptides and bacterial lipopeptides) [613]. It plays a major role in the clearance of oxLDLs.[55] Unlike ScaRa1 and ScaRa2, ScaRb3 does not bind acetylated or oxi-

---

52. A.k.a. CD63, tetraspanin Tspan30, granulophysin, GP55, platelet glycoprotein GP40, or PltGP40, neuroglandular antigen (NGA), melanoma-associated antigen MLA1, or ME491, and ocular melanoma-associated antigen OMA81H.

53. A.k.a. LGP85 or ScaRb2.

54. Protein CD36 was originally identified as platelet receptor glycoprotein-4 that binds thrombospondin and Plasmodium falciparum-parasitized erythrocytes [628, 629]. Later, it was recognized as a macrophage receptor for moderately oxidized LDLs [630].

55. Oxidized phosphatidylcholine derived from oxidized 1-palmitoyl 2-arachidonoyl sn-glycero 3-phosphocholine (PAPC) and 1-palmitoyl 2-linoleoyl sn-glycero 3-phosphocholine

dized LDLs with high affinity, but links to lipoproteins (LDL, HDL, and VLDL). In addition, ScaRb3 can promote pro-inflammatory signaling in response to thrombospondin, amyloid peptides (β-amyloid, fibrillar ApoC2), and pathogen-derived ligands [613].

Expression of ScaRb3 in macrophages is upregulated by oxLDL products such as 4-hydroxy 2-nonenal (HNE) via PPARγ and nuclear respiratory factor NRF2 [632]. Besides, macrophages release lower amounts of pro-inflammatory cytokines than monocytes in coronary artery diseases. The higher the PPARγ expression, the lower the cytokine release [633].

Hyperglycemia fosters the transformation of macrophages and vascular smooth muscle cells into foam cells. This process relies on imbalanced lipid flux by increased ingress of modified low-density lipoproteins via ScaRb3 and reduced egress via ABCg1 transporter, as well as oxidative stress and activated nuclear factor-κB [634].

The ScaRb3 receptor can trigger several signaling pathways. Particularly in macrophages and microglial cells, as well as endothelial cells, ScaRb3 can associate with SRC family kinases, such as Fyn, Lyn, and Yes that are reversibly coupled to the inner leaflet of the plasma membrane. These kinases phosphorylate members of the mitogen-activated protein kinase modules, thereby leading to activation of ERK1 and ERK2, P38MAPK, and JNK [613]. Therefore, ScaRb3 causes diverse cellular responses according to the cell type, such as cell adhesion, migration, or death, as well as inflammatory gene expression.

Furthermore, ScaRb3 can interact with different coreceptors to regulate signal transduction axes. The ScaRb3 receptor cooperates with members of the Toll-like receptor family, such as TLR2 and TLR6 (Sect. 11.2.4), and members of the integrin class, such as $\alpha_3\beta_1$- and $\alpha_6\beta_1$-integrins [613]. [56]

5.2.3 Other Types of Scavenger Receptors

5.2.3.1 Class-D Scavenger Receptor ScaRd1 (CD68)

Scavenger receptor ScaRd1 [57] is a heavily glycosylated type-1 transmembrane glycoprotein that binds to low-density lipoprotein. It has a unique N-terminal mucin-like domain.

The ScaRd1 receptor is expressed on monocytes and macrophages. It is thus used as a marker of macrophages as well as preadipocytes. [58] Within macrophages,

---

(PLPC) in oxidized lipoproteins that abound in atherosclerotic lesions are high-affinity ScaRb3 ligands [613]. In atheroma, ScaRb3 mediates lipid uptake and apoptotic cell clearance.

56. The ScaRb3–TLR2/6 cooperation activates the MyD88–IRAK–TRAF6–IKK cascade to activate nuclear factor-κB. On macrophages, ScaRb3 dimerizes with $\alpha_V\beta_3$-integrin to form a non-opsonic receptor involved in phagocytosis.

57. A.k.a. cluster of differentiation CD68, GP110, and macrosialin homolog.

58. In subcutaneous adipose tissue, the proportion of stromovascular cells such as macrophages that are committed to pre-adipocytes lowers with obesity due to a greater recruitment of pre-adipocytes during previous adipogenesis or greater pre-adipocyte apoptosis [635]. On

ScaRd1 is predominantly expressed in late endosomes and lysosomes [613]. The ScaRd1 receptor is identified as an oxLDL-binding protein that may contribute to oxLDL endolysosomal processing.

5.2.3.2 Class-E Scavenger Receptor C-type Lectin ScaRe1

Scavenger receptor ScaRe1 [59] is a type-2 transmembrane protein. It has been associated more with the C-type lectin subclass than the scavenger receptor superclass [636].

The ScaRe1 receptor lodges on vascular endothelial and smooth muscle cells as well as macrophages, especially in arterial walls, in addition to platelets. Its production is triggered by inflammatory stimuli. Vascular endothelial cells internalize and degrade pro-atherogenic oxLDLs via this oxLDL receptor. The ScaRe1 receptor also binds to other ligands, such as phosphatidylserine, advanced glycation end products, apoptotic cells, platelets, and bacteria [613].

Among mechanisms of receptor–ligand internalization (clathrin-, caveolae-, and membrane raft-mediated uptake) that lead to distinct ligand fate (recycling or degradation), ScaRe1 uses an endocytosis pathway for delivery to the endosome–lysosome compartment. The ScaRe1 receptor indeed causes oxLDL internalization using large GTPase dynamin-2 [637].

5.2.3.3 Class-F Scavenger Receptors ScaRf1 and ScaRf2

Class-F scavenger receptors ScaRf1 and ScaRf2 [60] are type-1 transmembrane receptors with N-terminal epidermal growth factor-like domains, a transmembrane domain, and a long cytoplasmic tail.

These 2 receptors bind to modified LDLs. However, only ScaRf1 [61] internalizes its ligand for degradation [613]. In macrophages, ScaRf1 accounts for only 6% of total acetylated LDL degradation.

5.2.3.4 Class-G Scavenger Receptor CXCL16

Class-G scavenger receptor [62] is the CXC motif-containing chemokine CXCL16. It can bind to oxLDLs. It resides on endothelial and smooth muscle cells and macrophages, especially in atheromatous plaques [613].

---

the other hand, the fraction of stromovascular cells that are macrophages is greater in obese women.

59. A.k.a. lectin-like oxidized LDL receptor-1 (LOx1), oxidized low-density lipoprotein (lectin-like) receptor OLR1, and C-type lectin CLec8a.

60. A.k.a. scavenger receptors expressed by endothelial cells SREC1 and SREC2, respectively.

61. A.k.a. acetyl LDL receptor.

62. A.k.a. scavenger receptor phosphatidylserine and oxidized lipoprotein (SRPSOx) as well as small-inducible cytokine SCyB16.

6

Receptors

In biology, the word "receptor"[1] commonly refers to: (1) specialized proteins of cells that receive and respond to proteic or lipidic messengers as well as physical or mechanical signals, thereby being defined as chemical-responsive,[2] physical agent-responsive, and mechanical stress-responsive receptors;[3] (2) sensors that receive stimuli and produce an informative nerve impulse; and (3) immune receptors that reside on the surface of specialized cells of the immune system ("immunocytes") and bind to antigens.

1. Latin recipio, recipere: to receive, record, recover, etc.

2. The nouns "sensors" and "receptors" can be applied to different component types of the body according to the length scale, i.e., the cell and the molecule. The present context focuses on chemosensory molecules. In the narrowest sense, a chemoreceptor, or chemosensor, refers to a chemosensory cell of the nervous system that transduces a chemical signal into an action potential such as the chemoreceptors of the vasculature that regulate the circulatory and respiratory functions. These chemosensory cells actually relay the received information by triggering electrochemical impulses in coupled nerves to feed the cardiac, respiratory, and vasomotor centers of the medulla oblongata of the brainstem. The chemoreceptors of the carotid and aortic sensory corpuscles detect changes primarily in arterial O_2 partial pressure and also, to a lesser extent, in CO_2 partial pressure and pH. On the other hand, once bound and activated by a specific ligand, a chemosensory molecule initiates a cascade of chemical reactions within the receiving cell, which can then react to adapt, stimulate nearby cells, or send messages to remote targets. In any case, the response triggered by chemosensory molecules is delayed w.r.t. that primed by chemosensory cells (chemoreceptors). Obviously, a chemoreceptor possesses a set of chemosensory molecules (Sect. 6.10 and Vol. 2 – Chap. 1. Remote Control Cells).

3. Some chemosensory molecules that sense cell's chemical environment can also respond to a mechanical stress.

6.1 Introduction

A receptor is a molecule on the surface of a cell, or inside a cell, that transmits a received signal inside the cell, possibly down to the nucleus for gene transcription. [4] *Plasmalemmal receptors* encompass various types of cell-surface molecules with diverse functions, whereas *intracellular receptors* are transcription factors.

A receptor that sense chemical cues is bound to its ligands, i.e., specific chemical species (agonists and antagonists) that operate as signal transmitters, or *primary messengers*, emitted from an external source (endo-, para-, and juxtacrine messengers) or the same cell after secretion (autocrine messengers). Both agonists and antagonists can fix to receptors with or without effect, respectively. Nevertheless, antagonists block agonist fixation.

Primary messengers can be: (1) neurotransmitters released by nerve endings (nervous control); (2) hormones secreted by specialized cells and transported by the flowing blood (endocrine regulation; Vol. 2 – Chap. 1. Remote Control); and (3) other types of regulatory substances produced locally (intra-, auto-, juxta-, and paracrine regulators).

At least one, more or less, cognate type of signaling molecules can attach a given receptor. Ligand binding usually modifies and then stabilizes the receptor conformation in an activated state. Attachment of a ligand on its corresponding receptor induces a reaction cascade within the cell.

Receptor concentration as well as signal processing shape signal transmission. In the cell, signaling effectors can bind activated receptors either directly or indirectly via a binding protein that serves as an intermediary (adaptor, docker, or scaffold protein). Response magnitude following direct binding depends on receptor concentration at the plasma membrane. On the other hand, indirect binding controls sensitivity.

6.1.1 Catalytic and Non-Catalytic Receptors

Catalytic receptors involve, directly or not, a catalysis by enzymes. Receptors currently refer to plasmalemmal molecules with associated intrinsic or extrinsic catalytic activity that aims at post-translationally modifying signaling effectors. They include plasmalemmal receptor enzymes ("recepzymes") and enzyme-linked receptors.

Non-catalytic receptors include other types of plasmalemmal substances that transmit signals to the cell, i.e., adhesion molecules (Vol. 1 – Chap. 7. Plasma Membrane). They recruit signal transmitters as well as ligand-activated compounds that facilitate the transmembrane transfer of signaling mediators or carry messengers through the plasma membrane. Both types of receptors coexist on the cell surface.

Among adhesion molecules that transduce physical and chemical signals from their immediate environments, integrin receptors and their associated molecular clusters — integrin adhesomes — connect to components of the extracellular matrix and

4. A receptor in the plasma membrane is called a plasmalemmal receptor; a receptor within the cytoplasm and/or nucleus of the target cell an intracellular and, most often in the literature, a "nuclear" receptor (NR).

cell cytoskeleton. They then act as receptors and sensors that are able to trigger signaling.[5]

6.1.2 Cell-Surface and Intracellular Receptors

Many different kinds of receptors reside at the cell surface and bind to specific ligands, so the cell can respond to manifold environmental signals. On the other hand, thyroid hormones, gluco- and mineralocorticoids, as well as steroids diffuse across the plasmalemma to reach their intracellular receptors, most of them belonging to the nuclear envelope (Sect. 6.3).

Except for the receptors of these hormones, receptors of most messengers are located in the plasma membrane. In particular, a G-protein-coupled receptor that responds to estrogen (GPER)[6] localizes to intracellular membranes. It responds to progesterone analogs independently of G proteins [5].

Growth factors (Vol. 2 – Chap. 3. Growth Factors) control cell growth and survival, as, at least partly, they regulate cellular access to extracellular nutrients by maintaining an appropriate density in transporters and receptors, hence avoiding a decline in the mitochondrial membrane potential, altered mitochondrial permeability, and release of pro-apoptotic factors (Vol. 2 – Chap. 2. Cell Growth and Proliferation).[7]

6.1.3 Catalytic Receptor-Initiated Signaling

Once initiated by ligand binding to cognate receptor, signal transduction pathways generally activate a cascade of chemical reactions to launch the release of stored or newly synthesized messengers either within the cell (autocrine signaling) or in the cell neighborhood (Chap. 1). Messengers are indeed previously produced and stored in an inactive form in appropriate intracellular pools such as vesicles, thereby ready to rapidly prime signaling, or need to be synthesized, i.e., require gene transcription and tranlation (Vol. 1 – Chap. 5. Protein Synthesis), hence triggering a delayed response.

Growth factor receptors transmit signals emitted by growth factor binding. Intracellular signal transduction is governed by receptor autophosphorylation that is followed by a cascade of phosphorylation of membrane-anchored and cytosolic protein kinases and their substrates. Therefore, growth factor stimulation is characterized by a set of effectors with a given distribution of phosphorylation sites (Tyr^P, Thr^P, and Ser^P [or pTyr, pThr, and pSer]) and temporal dynamics.[8] Among phosphorylated signaling mediators, a majority is associated with the cytoskeleton or localizes

---

5. Integrin-$\alpha_{2B}\beta_3$ can interact with $G\alpha_{13}$ subunit of guanine nucleotide-binding proteins that are signaling mediators to regulate cell spreading [638].

6. A.k.a. GPR30.

7. Growth factors modulate the activity of the PI3K–PKB–TOR pathway.

8. Distribution of phosphoproteins with Tyr^P, Thr^P, and Ser^P is approximately 2%, 12%, and 86%, respectively [33].

to the nucleus [33]. Phosphorylation of signaling mediators can be followed by dephosphorylation by phosphatases, translocation to another subcellular location, and degradation via ubiquitination.

Coordinated behavior of cells require intercellular communication via secreted regulators and their receptors. Appropriate signals must be received at correct time. Among signaling molecules and receptors, bone morphogenetic proteins, fibroblast growth factors, Hedgehog, transforming growth factor-β, and Wnt morphogens regulate development, maintenance, and regeneration of most tissues.

6.1.4 Organization of Receptors at the Plasma Membrane

The spatial organization of receptors in lipid nanodomains ensures the transmission of weak signals as well as diversification of signaling [639]. In addition, homo- and heterotypic receptor clustering by direct interactions or via adaptors or lipids enables signal amplification by a simultaneous activation of multiple receptors. Receptor cooperativity integrate multiple signals to achieve a coordinated response.

Receptor clustering determines signal specificity and sensitivity to external stimuli. During chemotaxis, signaling sensitivity depends on the structuring of receptor–adaptor–kinase complexes at the plasma membrane.

The lipid composition of the plasma membrane and protein–lipid interactions influence ligand-induced receptor clustering. Conversely, activated receptors can provoke the hydrolysis of plasmalemmal phosphatidylcholine by phospholipase-D2 to produce phosphatidic acid.

Heterotypic receptor clustering occurs when coreceptors are requird for a proper receptor functioning (e.g., semaphorin–neuropilin complex).

6.1.5 Chemosensors

Strictly, a chemosensor is a sensory receptor that transduces a chemical signal into an action potential. Chemosensors include olfactory receptors, the vomeronasal organ that detect pheromones, taste buds of the oropharyngeal cavity (tongue, soft palate, epiglottis, and upper esophagus; Sect. 6.10), and chemosensors of the control of breathing and blood circulation.

In the respiratory tract, the *diffuse chemosensory apparatus* incorporates *solitary chemosensors* and *chemosensory clusters*.[9]

9. Chemosensory clusters are smaller than taste buds. They localize to the larynx and lower airways in contact with sensory endings of trigeminal and vagal nerve fibers, respectively. Solitary chemosensory cells of the upper respiratory apparatus, in particular the nose for bitter taste sensing, express molecules of the chemosensory cascade (e.g. taste receptors, G-protein subunit α-gustducin and enzyme PLCβ2 of bitter and sweet taste transduction, and ion channels IP_3R3 and sweet and bitter taste-sensing (and amino acid siganling) transient receptor potential cation channel TRPM5). They respond to taste stimuli and trigger a trigeminal reflex. They sense bacterial bitter compounds to avoid further inhalation of bacteria. At least in the mouse trachea, chemosensory cells of the lower respiratory apparatus are endowed with

Chemosensors that control the breathing frequency encompass: (1) central chemosensors of the medulla oblongata that can detect changes in pH of the cerebrospinal fluid and (2) peripheral chemoreceptors of the carotid and aortic bodies that detect changes in blood gas content, primarily the partial pressure of oxygen, but also the partial pressure of carbon dioxide and arterial pH. Aortic body cells sense changes in blood O_2 and CO_2, but not pH; carotid body cells detect modifications of all 3 types of agents. The latter do not desensitize, but the former can.

Furthermore, at least in the mouse trachea, solitary chemosensory cells, which differ from solitary neuroendocrine cells of the respiratory epithelium, monitor the composition of the airway lining fluid, especially bitter compounds. They can reduce the breathing frequency [640]. These brush cells of the respiratory epithelium that have direct contacts with sensory nerve fibers are characterized by an apical tuft of microvilli, which contain villin and fimbrin. At least 2 types of sensory nerve fibers close to tracheal cholinergic chemosensory cells exist: peptidergic (CGRP+) and non-peptidergic nerves. A subtype of each type synthesizes nicotinic acetylcholine receptors. [10] Therefore, the chemical composition of the luminal content of the tracheobronchial tree participates to the regulation of the body's respiration via cholinergic sensors. This reflex differs from purely local responses initiated by human airway ciliated cells that are able to react to bitter stimuli and increases the intracellular concentration of Ca^{++} ions and the ciliary beat frequency [641]. [11]

In addition, peripheral chemosensors transmit signals to the vagal center to regulate the heart beat frequency. Afferent fibers from mechano- and chemosensors respond to mechanical deformations of heart chambers and chemical species (e.g., bradykinin and prostaglandins), respectively. Atrial and ventricular ganglionated plexi contain heterogeneous populations of neurons that exert a control over broad and overlapping cardiac regions. In dogs, few (3%) atrial and ventricular neurons respond solely to regional mechanical deformation (mechanotransducing neurons), whereas the remaining atrial and ventricular neurons transduce multimodal (mechanical and chemical) stimuli (multimodal sensory neurons) [643]. The intrinsic cardiac nervous system represents the final relay station for the coordination of cardiac regions. [12] It comprises afferent sensory and adrenergic and cholinergic efferent motor neurons as well as interconnections at atrial and ventricular ganglionated plexi,

---

the synthesis and packaging machinery for acetylcholine, in addition to type-2 taste receptors (Tas2R105 and Tas2R108) and downstream effectors [640]. Vagal sensory nerve fibers possess nicotinic acetylcholine receptors.

10. The majority of cholinergic chemosensory cells do not receive direct innervation [640]. In particular, tracheal TRPM5+ cells are less innervated than those of the nasal cavity. Acetylcholine released from non-innervated chemosensory cells (chemoreceptors rather than strictly chemosensors) may exert paracrine effects.

11. Chemoreceptors of occupational dusts, such as dusts of calcium oxide, sodium borate, and calcium sulfate (in decreasing potency order) provoke secretion and stimulate the mucociliary transport [642].

12. E.g., right atrial ganglionated plexus (RAGP) near the sinoatrial node and inferior vena caval–inferior atrial ganglionated plexus (IVCIAGP) close to the atrioventricular node. The intrinsic cardiac nervous system also includes the dorsal atrial (DAGP); left atrial (LAGP);

that communicate with components of intrathoracic extracardiac ganglia under the control of adrenergic sympathetic and cholinergic parasympathetic efferent preganglionic neurons of the central nervous command (dual efferent parasympathetic inhibitor and sympathetic augmentor) and circulating catecholamines [644]. Cardiac chemosensitive endings are stimulated by numerous substances, especially those produced by a stressed myocardium such as an ischemic myocardium (e.g., bradykinin, prostaglandins, reactive oxygen species) and support reflex adjustments [645].

6.2 Plasmalemmal Receptors

Plasmalemmal receptors are mostly integral membrane proteins. They have 3 basic domains: (1) an extracellular segment with the ligand-binding domain; (2) a transmembrane domain that anchors the receptor in the membrane; and (3) a cytoplasmic domain that primes the signaling cascade by activating the first signaling effector. In some cases, other membrane proteins interact with the receptor to modulate its activity. Certain receptors cluster together in the membrane after ligand binding.

Receptor ligands include hormones, neurotransmitters (Table 6.1), growth factors, immunoglobulins, complement fragments, lipoproteins, components of the extracellular matrix, etc.

Messengers generally have specific receptors. However, a given messenger can bind to several receptor types. Adrenaline targets α1, α2, and β1 to β3 receptors (Sect. 7.13.5). Acetylcholine has nicotinic, or ionotropic receptors (nAChRs, i.e., ion channels) and muscarinic, or metabotropic receptors (mAChRs, i.e., G-protein-coupled receptors).

6.2.1 Main Families of Catalytic Plasmalemmal Receptors

The plasmalemmal receptors can be regrouped in several main families (Table 6.2): (1) receptor tyrosine kinases (e.g., insulin and growth factor receptors; Chap. 8); (2) receptor tyrosine phosphatases (e.g., PTPRc, or cluster determinant protein CD45 of T lymphocytes and macrophages; Chap. 9); (3) receptor serine/threonine kinases (e.g., receptors of the TGFβ superfamily; Chap. 8); (4) receptor serine/threonine phosphatases; and (5) G-protein-coupled receptors (Chap. 7).

Binding of an extracellular ligand to a G-protein-coupled receptor promotes receptor activation via its coupling to a trimeric G protein of the intracellular side of the plasma membrane to trigger a signaling cascade. On the other hand, cell growth and differentiation are regulated by the opposing activities of protein tyrosine kinases (PTK) and phosphatases (PTP). These enzyme sets contain families of transmembrane or receptor kinases and phosphatases. Receptor Ser/Thr kinases (RSTK) include several (12) type-1 (7) and -2 (5) receptors for the TGFβ superfamily ligands. Receptor Tyr kinases (RTK) encompass multiple species with isoforms generated by

---

right ventricular (RVGP); ventral septal ventricular (VSVGP); and cranial medial ventricular (CMVGP) ganglionated plexi [644].

Table 6.1. Examples of receptors of the autonomic nervous system. A given category of receptors can comprise ionotropic and metabotropic receptors, i.e., ligand-gated ion channels and catalytic receptors. The former signals via ion fluxes, the latter triggers a cascade of chemical reactions. These categories of receptors include acetylcholine receptors (nicotinic [nAChRs] and muscarinic [mAChRs]), γ-aminobutyric acid receptors (ionotropic [$GABA_A$] and metabotropic [$GABA_B$]), glutamate receptors (ionotropic [GluR1–GluR4, GluK1–GluK5, and GluN1–GluN3b] and metabotropic [$mGlu_1$–$mGlu_8$]), and serotonin receptors (ionotropic [$5HT_3$] and metabotropic [$5HT_1$–$5HT_2$ and $5HT_4$–$5HT_7$]). Nucleotide receptors also include ionotropic receptors ($P2X_1$–$P2X_7$) one the one hand and P1 receptors (A_1–A_3) and metabotropic ($P2Y_1$ –> $P2Y_{14}$) receptors.

Type	Location
	Adrenergic receptors
α1	Vascular smooth muscle cell
	(skin and splanchnic regions)
α2	Presynaptic nerve endings
	Platelets, adipocytes
β1	Nodal tissue, myocardium
β2	Vascular and bronchial smooth muscle cells
	Digestive tract, skeletal muscle
	Cholinergic receptors
Nicotinic	Autonomic ganglia
(ionotropic)	(sympathetic and parasympathetic)
	Adrenal medulla
	Neuromuscular junction
Muscarinic	Sinoatrial node
(metabotropic)	Non-vascular smooth muscle cell
	Glands
Dopamine receptor	Smooth muscle cells, neurons
Glutamate receptor	Astrocyte

distinct genes or alternative splicing ($>$ 50). G-Protein-coupled receptors constitute a huge set of molecules ($\sim$800).

6.2.1.1 G-Protein-Coupled Receptors

G-Protein-coupled receptors (GPCR; Table 6.3; Chap. 7) are also called 7-transmembrane receptors (7TMR). These receptors are coupled to heterotrimeric guanine nucleotide-binding (G) proteins (Vol. 4 – Chap. 8. Guanosine Triphosphatases and their Regulators). They represent the largest superfamily of plasmalemmal receptors.

A collection of distinct GDP- (inactive) or GTP-bound (active) Gα subunits confer pathway specificity. Activation of Gα subunit reduces its affinity for G$\beta\gamma$ dimer with respect to the inactive $G\alpha^{GDP}$ form. The active $G\alpha^{GTP}$ form then dissociates

Table 6.2. Receptor families. Heterotrimeric G proteins (Gαβγ) coupled to receptors transduce signals in various pathways. Both receptor tyrosine kinases (RTK) and receptor protein tyrosine phosphatases (RPTP) include several subsets. There are also receptor serine/threonine kinases (RSTK). Steroid receptors reside within the nucleus; glucocorticoid receptor localizes to the cytosol until it binds its ligand. T-Cell receptor is targeted by a combination of antigen fragment with a glycoprotein of the major histocompatibility complex. Toll-Like receptors are members of pattern-recognition receptors.

Receptor type	Features
GPCRs	Activation of GTPases and protein kinases
	Ion transport
Frizzled receptors	Wnt signaling
	Activation of cytosolic Disheveled
	β-Catenin nuclear translocation
Nucleotide receptors	Flow-stress sensing
RTKs	Cell growth and differentiation
	Activation of GTPases and protein kinases
RPTPs	Opposition to RTK
RSTKs	Cell growth and differentiation
	TGFβ superfamily
Nitric oxide receptors	Guanylate cyclases
	Effectors: protein kinase-B and -G
Notch receptors	Gene transcription
	Jagged/Serrate and Delta ligands
	NFκB activation
Cytokine receptors	GTPase activation
	NRTK activation
T-cell receptor (for antigen)	Dimeric receptor
	Rise in intracellular calcium level
	Activation of calcineurin
Toll-like receptors	Detection of pathogen molecules
	Activation of immune cells
Nuclear receptors	Location in nucleus or cytosol
	Ligands: gluco- and mineralocorticoids, steroids

from the plasmalemmal complex and binds effectors to trigger the corresponding signaling pathway. Intrinsic GTPase activity of Gα inactivates it and enables its rebinding to Gβγ (G-protein cycle).

Several modes of signal propagation via heterotrimeric G proteins exist whether both Gα and Gβγ subunits interact with signaling effectors (signal transmission

Table 6.3. Examples of G-protein-coupled receptors.

Ligand	Activity
Acetylcholine	Neurotransmission
Cathecholamine	Neurotransmission
Glutamate	Neurotransmission
(metabotropic)	Voltage-sensitive (some)
Adrenomedullin	Water and salt balance
Angiotensin-2	Blood pressure regulation
Endothelin-1	Vasomotor tone
Calcium	Hematopoiesis
Leukotriene	Chemotaxis, inflammation
Vasopressin	Water and salt balance

mode 1), only Gα (mode 2) or Gβγ subunit (mode 3) [13] is used to transmit the signal. In addition, distinct magnitude and rate of heterotrimer dissociation can explain regulation by some types of G-protein subunits [646]. [14]

Several neurotransmitters act via G-protein-coupled receptors to trigger slow excitation [15] of neurons, such as acetylcholine, noradrenaline, neurotensin, and substance-P. [16]

6.2.1.2 Conventional Catalytic Receptors

The majority of receptor enzymes and enzyme-linked receptors are protein kinases and connected to protein kinases, respectively. Various types of plasmalemmal enzymes exist, such as receptor Tyr kinases, Tyr kinase-associated receptors; receptor Ser/Thr kinases; receptor-like Tyr phosphatases; and receptor guanylate cyclases. In addition, G-protein-coupled receptors are associated with G-protein heterotrimers that have a GTPase activity upon activation by receptors. These heterotrimers should

13. In excitable cells, Gβγ dimers that are derived specifically from Gi/o-associated heterotrimers activate inwardly rectifying potassium (GIRK; or $K_{IR}3$) channels [646]. These channels slow synaptic potentials in neurons and cause cholinergic inhibition in atriomyocytes.

14. The Gα subunits confer signaling specificity, even though Gβγ dimer is responsible for GIRK activation. This specificity relies on pre-assembled signaling complexes as well as different capabilities to release free Gβγ dimers.

15. Unlike fast excitation of neurotransmitters, slow excitation is mainly attributed to suppressor K^+ flux and/or activated non-selective cation channel.

16. Neurotensin and substance-P activate a proteic complex made of neuronal cation channel that causes a small background leak Na^+ flux (sodium leak channel, non-selective [NaLCN]) at the resting membrane potential and protein Uncoordinated-80 homolog [647]. Substance-P binds to its G-protein-coupled tachykinin (neurokinin) receptor NK_1 (or TacR1) and then targets Src kinases (Part B – Chap. Cytosolic Protein Tyr Kinases), without requiring G-protein activation.

not be confused with small, monomeric GTPases that can be recruited by receptors to trigger a signal transduction pathway.

As G-protein-coupled receptors are classified in a superfamily, given their huge number, receptor Tyr kinases constitute the largest population of the set of catalytic receptors that comprises both receptor enzymes and enzyme-linked receptors. The majority of these molecules are receptors for growth factors, cytokines, and hormones.

Receptor enzymes that possess an intracellular catalytic domain are usually single-pass transmembrane receptors. Enzyme-linked receptors have an intracellular region that is connected more or less transiently to cytosolic enzymes that they activate, once liganded.

Most recepzymes dimerize once bound to their ligands to activate further signaling effectors. Ligand binding can lead to auto- as well as trans-phosphorylation. Phosphorylated amino acid residues then attract signaling mediators. They can serve as platforms that can trigger a multireaction process at the cell cortex.

Intracellular protein Tyr phosphatases control the phosphorylation status of activated receptor Tyr kinases. Receptor Tyr phosphatases can regulate their receptor Tyr kinase counterparts when both types of receptor enzymes are partners of plasmalemmal clusters.

6.2.1.3 Cytokine Receptors

Cytokine receptors (Table 6.4; Chap. 11) localize especially to blood and lymph cells, which originate from bone marrow (Vols. 1 – Chap. 2. Cells of the Blood Circulation and 5 – Chap. 1. Blood). They are classified into several groups: (1) immunoglobulin superfamily (IGSF) receptors (e.g., interleukin-1 receptor); (2) hematopoietic receptors (class 1; e.g., high-affinity, trimeric interleukin-2 receptor, hematopoietic growth factor receptors); (3) interferon receptors (class 2; e.g., receptors for Ifn1s, Ifn2s, and tissue factor); (4) tumor-necrosis factor receptors (class 3); and (5) G-protein-coupled receptors. Prolactin receptors (PrlR) and growth hormone receptors (GHR) pertain to the superfamily of cytokine receptors.

Type-1 cytokine receptors include: (1) type-1 interleukin receptors (IL2R–IL9R, IL11R–IL13R, IL15R, IL21R, ILR23R, and IL27R); (2) colony stimulating factor receptors (erythropoietin, gCSF [CSF3R], and gmCSF [CSF2R] receptors); and (3) others (growth hormone, prolactin, oncostatin-M, and leukemia inhibitory factor receptors). Type-2 cytokine receptors comprise: (1) interferon receptors (interferon-α/β and -γ receptors) and (2) type-2 interleukin receptors (IL10R, IL20R, IL22R, and IL28R). Cytokine receptors also consist of members of the immunoglobulin superfamily (IL1R, IL18R, CSF1R, and SCFR) and tumor-necrosis factor receptor family (e.g., TNFRSF1a and -b, TNFRSF3, TNFRSF5, TNFRSF7, and TNFRSF8) as well as TGFβ receptors and chemokine G-protein-coupled receptors.

Cytokines cause long-term transcription-dependent changes in the expression of receptors, coreceptors, signaling mediators, and adaptors (Chap. 1). Cytokines also coregulate the promoter activity of target genes directly or via accessibility to other transcription factors. Cytokines not only trigger specific gene expression programs,

Table 6.4. Types of cytokine receptors. Class-1 cytokine receptors are hematopoietin family receptors, class-2 cytokine receptors correspond to interferon and interleukin IL10 family receptors. The other main classes encompass TNF receptor family and IL1 and IL17 receptor families.

Type	Ligand
Class 1	
gmCSF subset	Granulocyte–monocyte colony-stimulating factor (CSF2)
	Interleukins IL3, IL5
IL6 subset	Interleukins IL6, IL11, IL12
IL2 subset	Interleukins IL2, IL4, IL7, IL9, IL15
Hormone receptors	Growth hormone, prolactin
Class 2	Interferons IFNα, IFNβ, IFNγ
TNF receptors	Tumor-necrosis factor superfamily members
Immunoglobulin superfamily	Macrophage colony-stimulating factor (CSF1) Interleukins IL1

but also control transcription-independent communications between cytokine receptors and other receptors coupled to common or distinct signaling mediators [648]. Cytokine signaling pathways can actually interact with other signaling pathways from inter-receptor crosstalk in plasmalemmal domains to collaborative action of transcription factors at gene promoters. [17] However, a signaling adaptor can be reused in an alternative pathway rather than acting as a hub for cross-talk between signaling pathways.

Cytokine receptors stimulated by at least 30 cytokines use 4 known Janus kinases (Part B – Chap. Cytosolic Protein Tyr Kinases) and 7 identified signal transducers and activators of transcription to launch transcriptional programs specified by signal-

17. Receptors (e.g., T-cell and B-cell antigen receptors as well as Fc receptors that bind immunoglobulins and immune complexes) that signal via transmembrane adaptors (e.g., FcRγ and DNAX activation protein DAP12), which contain immunoreceptor tyrosine-based activation motifs (ITAM), modulate signaling by heterologous receptors (i.e., receptors with distinct structure and function, such as Toll-like receptors, tumor-necrosis factor receptors, and cytokine receptors). They then heighten or dampen signaling initiated by other receptors. Conversely, Toll-like and cytokine receptors modulate signaling mediated by ITAM-containing protein via β2-integrins coupled to ITAM-containing adaptors (signaling cross-talk with bidirectional regulation) [649]. Interaction of ITAM-mediated signaling into signaling cascades with other signaling pathways tunes cellular responses to various extracellular stimuli. ITAM-Associated and Toll-like receptors activate distinct signaling pathways that converge on common signaling effectors such as nuclear factor-κB (Part B – Chap. Other Signaling Mediators) and mitogen-activated protein kinases (Part B – Chap. Mitogen-Activated Protein Kinase Modules). Death receptors (DR1–DR6) transmit signals by apical proteic complexes that are nucleated by adaptors Fas receptor-associated death domain (FADD) and tumor-necrosis factor receptor-associated death domain (TRADD). These adaptors also participate in complexes assembled by other immune cell receptors, such as pattern-recognition receptors [650].

ing effectors. Nevertheless, a given cytokine receptors can prime cell type-specific responses despite shared usage of a small set of JaK–STAT axes. Effective interpathway insulation possibly via different scaffolds avoids spurious response to a given stimulus.

6.2.2 Ionotropic Receptors – Ligand-Gated Ion Channels

Ligand-gated ion channels are cation or anion channels that pertain to the superfamily of multipass transmembrane proteins. They prime fast signalings, as, once liganded, they serve as open channels for the passage of small ionic second messengers, in addition to change in transmembrane potential (Sect. 2.5).

6.3 Intracellular or Nuclear Receptors

Nuclear receptors are intracellular proteins that sense hormones and other molecule types. These receptors constitute a class of specialized *transcription factors* with structural similarities that directly bind as homo- or heterodimers to specific sequences of DNA, the so-called response elements, in promoters of target genes. Once bound to their ligands, they promote or repress gene expression in association with other proteins. Resulting changes in receptor conformation actually allow recruitment or dissociation of proteic partners.

In humans, 48 nuclear receptors have been identified. They include *nuclear hormone* (NHR) and *orphan nuclear* (ONR) receptors (Tables 6.5 to 6.8). Aliases of nuclear receptors (NR) are defined according to the classification (subclass i [numeral], family X [letter], member j [numeral]). Whereas nuclear hormone receptors possess known hormonal ligands, agonists of orphan receptors have not been identified at least in the discovery period. Lipophilic hormones can traverse the plasma membrane to penetrate into the cytoplam, where NHRs transduce signals from these regulators. Several orphan receptors respond to xenobiotics. They then mediate transcription of detoxifying enzymes of the cytochrome-P450 superfamily.

6.3.1 Ligands

Ligands of nuclear hormone receptors include lipophilic hormones, such as gluco- and mineralocorticoids, sex steroids (estrogens, progesterone, and androgens), and thyroid hormones that cross the plasma membrane, as well as vitamin-A and -D. Some nuclear receptors such as farnesoid X receptor (FXR), liver X receptor (LXR), and peroxisome proliferator-activated receptors (PPAR) bind various metabolic intermediates, such as fatty acids, bile acids, and/or sterols, with relatively low affinity. Other nuclear receptors, such as constitutive androstane receptor (CAR) and pregnane X receptor (PXR), upregulate the expression of cytochrome-P450 enzymes.

Table 6.5. Nuclear receptors. (**Part 1**) Subclass 1 (NR1) – thyroid hormone receptor-like factors (Sources: [5] and Wikipedia; CAR: constitutive androstane receptor; EAR: viral oncogene vErbA-related gene product; FXR: farnesoid X receptor; LXR: liver X receptor-like factor; PAR: pregnane-activated receptor; PPAR: peroxisome proliferator-activated receptor; PXR: pregnane X receptor; RAR: retinoic acid receptor; ROR: RAR-related orphan receptor; SXR: steroid and xenobiotic receptor; THR: thyroid hormone receptor; VDR: vitamin-D receptor).

Type	Aliases	Ligands
	Family A: thyroid hormone receptors	
NR1a1	THRα, TRα, THRa ErbAα, ErbA1, EAR7	Thyroid hormones ($T_3 > T_4$)
NR1a2	THRβ, TRβ, THRb ErbAβ, ErbA2	
	Family B: retinoic acid receptors	
NR1b1	RARα	Vitamin-A and related compounds
NR1b2	RARβ	
NR1b3	RARγ	
	Family C: peroxisome proliferator-activated receptors	
NR1c1	PPARα	Free fatty acids,
NR1c2	PPARβ	Prostanoids,
NR1c3	PPARγ	Eicosanoids
	Family D: RevErbA factors	
NR1d1	RevErbAα EAR1, hRev	Heme
NR1d2	RevErbAβ EAR1β, RVR, BD73	
	Family F: RAR-related orphan receptors	
NR1f1	RORα	Cholesterol,
NR1f2	RORβ	Tretinoin (acid form of vitamin-A)
NR1f3	RORγ	
	Family H: liver X receptor-like factors	
NR1h2	LXRβ	Oxysterol
NR1h3	LXRα	Oxysterol
NR1h4	FXRα	Bile acid
NR1h5	FXRβ	Bile acid, lanosterol
	Family I: vitamin-D receptor-like factors	
NR1i1	VDR	Vitamin-D
NR1i2	PXR, SXR, PAR1	Xenobiotics
NR1i3	CAR, MB67	Xenobiotics, androstane, bile acids (high concentration)

6.3.2 Structure and Function

Nuclear receptors share common structural features. They are constituted by 6 domains (A–F; Table 6.9). The central DNA-binding domain (DBD; region C)

Table 6.6. Nuclear receptors. (**Part 2**) Subclass-2 (NR2) retinoid X receptor-like factors (Sources: [5] and Wikipedia; COUPTF: chicken ovalbumin upstream promoter-transcription factor; EAR: vErbA-related gene product; HNF: hepatocyte nuclear factor; MODY: maturity onset diabetes of the young; PNR: photoreceptor cell (retina)-specific nuclear receptor; RXR: retinoid X receptor; TCF: transcription factor; TLL: tailless; TLX: tailless homolog; TR: testicular receptor).

Type	Aliases	Ligands
	Family A: hepatocyte nuclear factors-4	
NR2a1	HNF4α	
	MODY1, TCF14	
NR2a2	HNF4β	
	HNF4γ	
	Family B: retinoid X receptors	
NR2b1	RXRα	9-cis retinoic acid
NR2b2	RXRβ	
NR2b3	RXRγ	
	Family C: testicular receptors	
NR2c1	TR2	
NR2c2	TR4	
	Family E: TLX–PNR factors	
NR2e1	TLL, TLX	
NR2e3	PNR	
	Family F: COUP–EAR factors	
NR2f1	COUPTF1,	
	COUPα, EAR3, SVP44	
NR2f2	COUPTF2,	
	COUPβ, ARP1, SVP40	
NR2f6	EAR2	

targets to specific DNA sequences with a response element. The ligand-binding domain (LBD; region-E) resides in the C-terminal half of the receptor. It recognizes specific hormonal and non-hormonal ligands. A hinge sequence (region D) of variable length is situated between the DBD and LBD domains. These domains contain dimerization motifs. The N-terminus of variable length (regions A and B) has an activation function. The C-terminus (region F) also has a variable length.

The function of nuclear receptors is determined by *activation function* domains AF1 and AF2 that reside in the N-terminus and C-terminal ligand-binding domain, respectively. Ligand-dependent recruitment of coactivators relies on the AF2 domain [653]. Nonetheless, a distinct set of coactivators is associated with the AF1 domain.

Nuclear receptors can exist as homo- or heterodimers. Each partner binds to a specific hormone response element (HRE) on DNA. These HRE regions consist of

Table 6.7. Nuclear receptors. (**Part 3**) Subclass 3 (NR3) steroid hormone receptors (Sources: [5, 651] and Wikipedia; AR: androgen receptor; ER: estrogen receptor; ERR: estrogen-related receptor; EsRL: estrogen receptor-like factor; GR: glucocorticoid receptor; MR: mineralocorticoid receptor; PR: progesterone receptor). The transcriptional activity of ERRs is enhanced by members of the steroid receptor coactivator (SRC) family (SRC1–SRC3), PGC1α and -β, proline-rich nuclear receptor coactivator PNRC1 and PNRC2, and transducin-like enhancer of split homolog TLE1, but reduced by the corepressor nuclear receptor-interacting protein NRIP1 and orphan nuclear receptor NR0b2 (or small heterodimer partner SHP). Natural ligands of ERRs remains unknown. The NR3B family function in a ligand-independent manner,

Type	Aliases	Ligands
	Family A: estrogen receptors	
NR3a1	ERα	Estrogens
NR3a2	ERβ	Estrogens
	Family B: estrogen-related receptors	
NR3b1	ERRα, ERR1, EsRL1	
NR3b2	ERRβ, ERR2, EsRL2	
NR3b3	ERRγ, ERR3, EsRL3	
	Family C: 3-ketosteroid receptors	
NR3c1	GR	Glucocorticoids: cortisol, corticosterone
NR3c2	MR	Mineralocorticoids: aldosterone
NR3c3	PR (PGR)	Progesterone
NR3c4	AR	Androgens: testosterone, 5α-dihydrotestosterone

2 half-sites separated by a nucleotide (DNA) spacer of variable length [652]. The second half-site has a sequence inverted from the first, hence the names "direct" and "inverted half-site repeat" for the first and second sequence, respectively.

Nuclear receptors complex with coregulatory proteins and nuclear receptors. Regulatory partners with enzymatic or scaffolding function form multicomponent assemblies with the nuclear receptors. These complexes can serve as coactivators or corepressors. In general, ligand binding to nuclear receptors causes an exchange of coactivators for corepressors to facilitate transcription. These transcriptional complexes influence chromatin remodeling via post-translational modifications, such as histone acetylation, deacetylation, and methylation, among others.

6.3.3 Classification

Nuclear receptors are categorized into different classes according to their mechanism of action and subcellular distribution in the absence of ligand. Nuclear receptors are located in the cytosol (distribution type 1) or nucleus (distribution type 2).

Nuclear receptors are subdivided into 4 mechanistic classes, according to their structure and DNA-binding mode [652]. Mechanistic class 1 receptors such as steroid

Table 6.8. Nuclear receptors. (**Part 4**) Subclasses 4 to 6 and 0 (Sources: [5] and Wikipedia; Ad4BP: adrenal 4-binding protein; DAX: dosage-sensitive sex reversal, adrenal hypoplasia critical region on chromosome X; GCNF: germ cell nuclear factor; LRH: liver receptor homolog; NOR: neuron-derived orphan receptor (mitogen-induced nuclear orphan receptor); NuRR: nuclear receptor-related factor; RTR: retinoid receptor-related testis-specific receptor; SF: steroidogenic factor (steroid hormone receptor); SHP: small heterodimer partner; TINuR: transcriptionally inducible nuclear receptor; 2DBDNR: 2-DNA-binding domain-containing nuclear receptors).

Type	Aliases	Ligands
	Class 4: NR4 – nerve growth factor-1B-like factors	
NR4a1	NGF1b, NuR77, NAK1, ST59, TR3, HMR	
NR4a2	NuRR1, NOT, TINuR, RNR1	
NR4a3	NOR1	
	Class 5: NR5 – steroidogenic factor-like factors	
NR5a1	SF1, Ad4BP	Phospholipids (phosphatidylinositols)
NR5a2	LRH1	Phospholipids (phosphatidylinositols)
	Class 6: NR6 – germ cell nuclear factor-like factor	
NR6a1	GCNF, RTR	
	Class 0: NR0 – miscellaneous)	
NR0b1	DAX1, AHCH	
NR0b2	SHP	
NR0c1	2DBDNR	

Table 6.9. Structure and function of nuclear receptors (Source: [652]; DBD: DNA-binding domain; LBD: ligand-binding domain).

Region	Domain	Function
A–B		Activation
C	DBD	DNA binding
D		Dimerization
E	LBD	Ligand binding
F		

hormone receptors comprise cytosolic nuclear receptors that, upon ligand binding, dissociate from heat shock protein chaperones. They then form homodimers, translocate into the nucleus, bind to hormone response elements, and recruit other proteins for DNA transcription. These homodimers bind to half-site response element inverted repeats. Mechanistic class 2 receptors are sequestered in the nucleus and form complexes with corepressors. Ligand-bound nuclear receptors dissociate from these receptors from corepressors and recruit coactivators. They bind as heterodimers with RXR receptors to DNA. Mechanistic class 3 receptors bind as homodimers to direct repeat instead of inverted repeat hormone response elements as class 1. Mechanistic class 4 receptors bind either as monomers or dimers to single site hormone response elements. Mechanistic class-3 and -4 nuclear receptors correspond to orphan receptors.

Members of the steroid and thyroid hormone receptors can be decomposed into 2 categories. Type-1 receptors include steroid hormone receptors of the class 3 (NR3 class), i.e., androgen (AR or NR3C4), estrogen (ERα–ERβ or NR3a1–NR3a2), progesterone (PR or NR3c3), glucocorticoid (GR or NR3c1), and mineralocorticoid (MR or NR3c2) receptors. Type-2 receptors encompass vitamin-A, calcitriol (active form of vitamin-D), retinoid (RARs and RXRs), and thyroid hormone receptors.

6.3.4 Transcriptional Regulation

Gene transcription, or mRNA synthesis, is the first stage of protein synthesis (Vol. 1 – Chap. 5. Protein Synthesis). To initiate transcription, RNA polymerase binds to core promoters (promoter sequences within that trigger transcription) in the presence of various specific transcription factors. Exchange of corepressors for coactivators is the basic switch from gene repression to activation.

6.3.4.1 Unliganded Nuclear Hormone Receptors – Corepressors

In the absence of hormone, nuclear hormone receptors act as transcriptional corepressors. Unliganded intracellular receptors can indeed bind to their target genes and complex with silencing mediators. These coregulatory proteins have a broad spectrum of association with many different nuclear receptors and transcription factors.

Transcription repression is ensured by a group of proteins, such as the nuclear receptor corepressors NCoR1 [18] and NCoR2. [19]

Nuclear receptors can inhibit the transcriptional activity of other types of transcription factors by trans-repression. In opposition to agonists, a set of antagonists, more precisely inverse agonists, fosters the interaction with corepressors. In addition, nuclear receptor-interacting protein NRIP2 [20] represses transactivation by a set of nuclear receptors.

The NCoR corepressors participate in the formation of several distinct corepressor complexes with histone deacetylases, such as HDAC1 and HDAC2 and transcription regulator Sin3 homolog-A, a component of a histone deacetylase complex [653].

---

18. A.k.a. thyroid, retinoic-acid receptor-associated corepressor TRAC1.
19. A.k.a. silencing mediator of retinoic acid and thyroid hormone receptor SMRT, SMRTe, and SMRTeτ, SMRT-associated polypeptide SMAP270, CTG repeat protein CTG26, T3 receptor-associating factor, TRAC1 (also!), and TNRC14.
20. This neuronal-specific corepressor is also called neuronal-interacting factor-X (NIX1).

Several combinations of these components support specific functions of nuclear receptor, e.g., the absence of Sin3a or HDAC and the presence of additional components. A first complex contains HDAC1, HDAC2, and Sin3a, among other components. A second complex comprises Swi–SNF-related matrix-associated actin-dependent regulator of chromatin SMARCa4, SMARCb1, SMARCc1, SMARCc2, and the nuclear corepressor zinc finger suppressor and Krüppel-associated box (KRAB)-associated protein KAP1. A third complex is constituted by transducin-β-like protein TBL1 and HDAC3.

Corepressors can be targetd by some signaling axes. For example, upon activation of mitogen-activated protein kinase, NCoR2 translocates from the nucleus to the cytoplasm, predominantly the perinuclear region [653]. In addition, Seven in absentia homolog ubiquitin ligase interacts with NCoR1 for proteosomal degradation.

Certain regulators can function as corepressors or coactivators according to the context. Small mothers against decapentaplegic homolog SMAD3 of the TGFβ pathway prevents transcription mediated by androgen receptor. On the other hand, it can act as a coactivator for liganded vitamin-D receptor, once complexed with a member of the NOAC family in the nucleus.

6.3.4.2 Liganded Nuclear Hormone Receptors – Coactivators

Liganded nuclear receptors interact directly with the so-called basal (or general) transcription factors, such as transcription initiation factor GTF2b and GTF2d [21] to influence the activity of basal transcription machinery.

Nuclear receptors bound to *hormone response elements* recruit several transcription coregulators which enhance or prevent transcription of target gene. These coregulators operate in chromatin remodeling that regulates gene accessibility to transcription or serve as bridges to stabilize binding of other coregulators. Coactivators often have an intrinsic histone acetyltransferase activity to weaken the association of histones to DNA. On the other hand, corepressors recruit histone deacetylases to strengthen the histone–DNA binding.

Coactivator Superfamily

The coactivator superfamily includes many families: the NCOA, P300–CBP, Mediator, PGC, and NCOA6 family (Table 6.10). [22] Coactivators of the NCOA family possess several binding domains, among which one for LBD of liganded nuclear receptor, a second for histone acetyltransferases of the CBP–P300 family, and a third for protein Arg methyltransferases of the PRMT family, in addition to the site devoted to members of the class 7 of bHLH-PAS transcription factors [653].

Coactivators assemble to form a complex between protein and subcomplexes, i.e., with the following elements: NCoA1–NCoA2–NCoA3, CARM1, CBP–P300,

21. A.k.a. TATA box-binding protein (TBP).

22. Coactivators contain at least an LXXLL (L: leucine; X: any amino acid) motif that is necessary and sufficient for ligand-dependent interactions with the activation function-2 (AF2) domain in the ligand-binding domain (LBD) of the nuclear receptor C-terminus (region E).

Table 6.10. Families of coactivators of nuclear receptors (CBP: cAMP-responsive element-binding protein (CREB)-binding protein; Med*i*: mediator complex subunit-*i*; NCoA: nuclear receptor coactivator). Coactivators, such as members of the CBP–P300, NOAC, and Mediator (DRIP–TRAP) families, are general coactivators for both nuclear hormone receptors (NHR) and non-NHR transcription factors.

Family	Members
NCOA	NCoA1, NCoA2, NCoA3
CBP–P300	CBP, P300
Mediator	Med1/4/6, Med12, Med14–Med17, Med21–Med28, Med30–Med31
	NCoA6

NCoA6, TGS1, and Med1–Med23 components. Some components operate as scaffold or co-integrator-associated proteins for the coactivator complex, others act as enzymes that, in particular, methylate (such as CARM1 and PIMT) or acetylate (such as CBP–P300) histones to reach target DNA segments.

Acetylation- and Methylation-Dependent Chromatin Remodeling

Coactivators recruit the chromatin-remodeling Switch–Sucrose non-fermentable (Swi–SNF) complex that can destabilize silencing histone–DNA interactions, owing to its ATPase activity. Mammalian Swi–SNF homologs form large proteic complexes. The Swi–SNF complex facilitates the binding of transcription factors to nucleosomal DNA and can cause ATP-dependent, local changes in chromatin structure.

Many coactivators possess a histone acetyltransferase activity, such as members of the CBP–P300 and NCOA family. Moreover, they can associate with K (lysine) acetyltransferase KAT2b. To ensure a stronger efficiency, coactivator-associated arginine (R) methyltransferase-1 (CARM1 or PRMT4), a member of the family of protein Arg methyltransferases (PRMT), catalyzes the methylation of Arg residues particularly in histones-3.[23] Therefore, coactivators are or recruit histone acetyltransferase to the transcription start site to relieve the inhibition by tightly packed chromatin and enable the RNA polymerase-2 complex to bind and initiate transcription.

6.3.4.3 Nuclear Receptor Coactivators

Nuclear receptor-mediated transcriptional initiation involves multiple cycles of coactivator assembly and disassembly. The NCOA family[24] comprises 3 known members: nuclear receptor coactivators (NCoA1–NCoA3), or steroid receptor coactivators (SRC1–SRC3; Table 6.11).

23. The PRMT family includes 9 known members (PRMT1–PRMT9). The PRMT9 protein corresponds to F-box protein FBxO11. The F-box proteins constitute one of the 4 subunits of ubiquitin ligase complex SKP1–cullin–F-box that operates in phosphorylation-dependent ubiquitination.

24. A.k.a. SRC1 or P160 family.

Table 6.11. Aliases of nuclear receptor coactivators (NCoA) of the NCOA (also P160 or SRC1) family.

Member	Aliases
NCoA1	SRC1, BHLHe74, KAT13A, Hin2, NY-REN52, RIP160
NCoA2	SRC2, BHLHe75, KAT13C, GRIP1, TIF2
NCoA3	SRC3, BHLHe42, KAT13B, TRAM1, ACTR, AIB1, CAGH16, CTG26, CIP, RAC3, TNRC14, TNRC16

Members of the NCOA family interact not only with liganded nuclear receptors, but also other transcriptional regulators, such as coactivator-associated arginine methyltransferase CARM1, CREB-binding protein (CBP), and K (lysine) acetyltransferase KAT2b.[25] In addition to their enzymatic activity as histone acetyltransferases, they serve as assembling platforms for coactivator complexes.

NCoA1

The NCoA1 coactivator is also called steroid receptor coactivator SRC1, class-E basic helix-loop-helix protein BHLHe74, K (lysine) acetyltransferase KAT13a, human homolog of "high in normal protein" HIN2,[26] renal carcinoma antigen NY-REN52, and retinoid X receptor-interacting protein RIP160.

NCoA2

The NCoA2 coactivator is also designated as SRC2, BHLHe75, glucocorticoid receptor-interacting protein GRIP1, KAT13c, and transcriptional intermediary factor TIF2.

NCoA3

The NCoA3 coactivator[27] regulates the assembly of the coactivator complex. Members of the NCOA family serve as substrates for coactivator-associated argi-

25. Histone acetyltransferase KAT2b is also called P300–CBP-associated factor (PCAF or P/CAF), and general control of amino acid synthesis GCN5 or GNC5-like protein (GCN5L or GCN5L1).

26. High in normal protein HIN1 is a secreted protein highly expressed in normal human breast epithelium. High in normal protein HIN2 is similar to uteroglobin-related protein UGRP1 that is specifically expressed in the lung. Protein UGRP1 is synthesized by Clara-like cells in the bronchial epithelium. It is upregulated in cystic fibrosis. It targets its scavenger receptor ScaRa2 (or macrophage scavenger receptor with collagenous structure [MaRCo]) during inflammation for pathogen clearance in the lung [654].

27. A.k.a. SRC3, another coactivator for nuclear thyroid hormone receptors ACTR, amplified in breast cancer protein-1 (AIB1), BHLHe42, co-integrator-associated protein CAGH16, CBP-interacting protein (pCIP), CTG26, KAT13b, receptor-associated coactivator RAC3, and thyroid hormone nuclear receptor activator molecule TRAM1, and thyroid hormone nuclear receptor coactivator TNRC14 and TNRC16.

nine (R) methyltransferase CARM1 [655]. The latter is a dual-function coactivator that not only activates transcription by modifying histones, but also terminates hormone signaling by disassembly of the coactivator complex. Methylation of NCoA3 that promotes dissociation of the NCoA3–CARM1 coactivator complex is regulated by estrogens.

Alternative splicing of NCOA3 transcript produces 2 proteins: a full-length protein ($NCoA3_L$) and $NCoA3_{\Delta 4}$. Whereas SRC3 acts in the nucleus to regulate the transcription of genes involved in growth and development, $NCoA3_{\Delta 4}$ localizes to the plasma membrane and operates as a cytoplasmic signaling coactivator. [28]

The latter actually couples epidermal growth factor receptor to focal adhesion kinase for cell migration [657]. Adaptor $NCoA3_{\Delta 4}$, upon linking EGFR and FAK, promotes EGF-induced phosphorylations of FAK and Src kinases. [29] Non-receptor Ser/Thr protein P21-activated kinase PAK1 is also activated, as it is recruited to EGFR by NCK1 adaptor. At the cell membrane, PAK1 phosphorylates $NCoA3_{\Delta 4}$ that then localizes to and functions at the plasma membrane [657].

6.3.4.4 P300–CBP Family of Histone Acetyltransferases

Histone acetyltransferases cAMP-responsive element-binding protein (CREB)-binding protein (CBP) and P300 constitute the P300–CBP family that can be enlarged as the NCOA (P160)–CBP–P300 family. [30] Both CBP and P300 can operate as histone acetyltransferases, polyubiquitin ligases (E4), and adaptors. The hormone response needs the recruitment of members of the NCOA family of receptor coactivator and CBP–P300 transcriptional coactivators. Coactivators NCoAs recruit histone acetyltransferases CBP and P300 for suitable transmission of the hormonal signal to the transcriptional machinery.

6.3.4.5 Mediator Complex

The *Mediator complex* with 20 to 30 subunits participates in transcription of numerous genes as a coactivator. It connects to the C-terminus of RNA polymerase-2, thereby bridging this holoenzyme [31] and transcription factors.

---

28. Overexpression of $NCoA3_{\Delta 4}$ promotes breast tumor metastasis. In addition, NCoA3 adaptor that bridges EGFR to FAK promotes also ovarian cancer cell migration, independently of its role in estrogen receptor signaling [656]. The NCoA3 coactivator thus also enhances estrogen-dependent growth of cancer cells.

29. Enzyme FAK is autophosphorylated (activated; Tyr397) when it binds to integrin complexes. Kinase Src is then recruited to the FAK complex, as it connects to $Tyr397^P$. Enzyme Src then phosphorylates FAK (multiple Tyr sites such as Tyr925). Phosphorylation at Tyr925 modifies the interaction of FAK with its partners, such as GRB2 and paxillin, to promote cell migration.

30. Histone acetyltransferase P300 is also called E1A-binding protein.

31. Active form of cofactor-activated apoenzyme.

Table 6.12. Aliases of some subunits of the coactivator Mediator (Med) of the DRIP–TRAP family (DRIP: vitamin-D receptor-interacting protein; TRAP: thyroid hormone receptor-associated protein).

Subunit	Aliases
Med1	DRIP205, DRIP230, TRAP220, ARC205, CRSP1, CRSP200, PBP, PPARBP, PPARGBP, RB18a, TRIP2
Med16	DRIP92, TRAP95
Med23	DRIP130, ARC130, CRSP3, CRSP130, and CRSP133, SuR2
Med24	TRAP100

Mediator is a proteic complex made of components of the DRIP–TRAP family [32] The DRIP–TRAP family is constituted, in particular, by Mediator of RNA polymerase-2-mediated transcription subunit-1 (Med1), [33] the general nuclear receptor coactivator NCoA6, and PPARγ coactivators PGC1α and PGC1β.

The Med23 subunit [34] forms a subcomplex of the Mediator complex with 2 other subunits: Med16 [35] and Med24 [658] (Table 6.12). [36]

NCoA6IP or Trimethylguanosine Synthase

Nuclear receptor coactivator-interacting protein NCoA6IP [37] is a strong enhancer of Med1 coactivator activity [659]. It also improves the action of NCoA6 coactivator. However, NCoA6IP represses the function of CBP and P300 coactivators.

32. DRIP: vitamin-D receptor-interacting protein; TRAP: thyroid hormone receptor-associated protein.

33. A.k.a. peroxisome proliferator-activated receptor-binding protein PPARBP, PPARγ-binding protein (PPARγBP), PPAR-binding protein (PBP), 205-kDa activator-recruited cofactor ARC205, cofactor required for SP1 transcriptional activation subunit CRSP1 and CRSP200, vitamin-D receptor-interacting proteic complex component DRIP205 and DRIP230, P53 regulatory protein RB18a, 220-kDa thyroid hormone receptor-associated proteic complex component TRAP220, and thyroid hormone receptor-interacting protein TRIP2.

34. A.k.a. 130-kDa activator-recruited cofactor component ARC130, CRSP3, CRSP130, and CRSP133, DRIP130, and human homolog of suppressor of activated Lethal protein Let60 Ras (SuR2). In Caenorhabditis elegans, the signal Lin3 (epidermal growth factor; Lin: abnormal cell lineage) stimulates the receptor Tyr kinase Lethal Let23 (EGFR), Lethal protein Let60 Ras (or Lin34), and the MAPK pathway. Protein SuR2 interacts with Lin1 transcrition factor.

35. A.k.a. 95-kDa thyroid hormone receptor-associated protein TRAP95 and 92-kDa vitamin-D3 receptor-interacting protein complex component DRIP92.

36. A.k.a. TRAP100.

37. A.k.a. trimethylguanosine synthase homolog TGS1, cap-specific guanine-N2 methyltransferase, CLL-associated antigen KW2, hepatocellular carcinoma-associated antigen HCA137, nuclear receptor coactivator-6-interacting protein NCoA6IP, and PRIP-interacting protein with methyltransferase motif (PIMT or PIPMT).

6.3.4.6 NCoA6

Nuclear receptor coactivator NCoA6 of the DRIP–TRAP family is also designated to as nuclear receptor coactivator (NRC). [38]

6.3.4.7 Ligand-Dependent Nuclear Receptor Corepressor

Ligand-dependent nuclear receptor corepressor (LCoR) interacts with class-1 and -2 agonist-bound nuclear receptors. This coregulator of gene transcription links (directly or not) to thyroid hormone (TRα, or NR1a1), retinoic acid (RARα–RARγ, or NR1b1–NR1b3), vitamin-D (VDR, or NR1i1), retinoid X (RXRα, or NR2b1), estrogen (ERα–ERβ, or NR3a1–NR3a2), glucocorticoid (GR, or NR3c1), and progesterone (PR, or NR3c3) receptors [660]. It can also interact with histone deacetylase-3 and -6. It can act as a corepressor of ligand-mediated transactivation of, at least, some nuclear receptors. In fact, LCoR may have a dual role as a coactivator and a corepressor according to cell and gene types and context.

In the presence of hormone, LCoR is recruited to the promoters of several NR3A1 and NR3C3 target genes with a peak recruitment from 30 to 45 mn after hormone stimulation [660]. In fact, LCoR binds to its cofactors progesterone (PR) and estrogen ERα receptor, and the corepressors C-terminal-binding protein CTBP1 (or CTBP) and histone deacetylase HDAC6 [660].

Via the CTBP corepressor complex that can bind to the growth hormone promoter, LCoR also connects to lysine (K)-specific demethylase LSD1. [39] In addition, LCoR acts as a scaffold that links to CTBP1, CTBP-interacting protein (CTIP), [40] and Polycomb RING finger protein BMI1 [41] to form complexes similar to the Polycomb group repressor complexes.

6.3.4.8 PPARγ Coactivators (PGC1)

The family of peroxisome proliferator-activated receptor (PPAR)γ (or nuclear receptor NR1c3) coactivators-1 include the tissue-specific and inducible transcriptional coactivator PGC1α [42] and the second member PGC1β. [43]

---

38. As well as activating signal co-integrator ASC2, amplified in breast cancer protein AIB3, cancer-amplified transcriptional coactivator, 250-kDa nuclear receptor-activating protein RAP250, peroxisome proliferator-activated receptor-interacting protein (PRIP), and thyroid hormone receptor-binding protein (TRBP).

39. The demethylase LSD1 is a member of the Krüppel-like zinc finger E-box-binding homeobox ZEB1–LSD1 repressor element-1 silencing transcription factor corepressor (CoREST) and CtBP corepressor complex [660].

40. A.k.a. retinoblastoma-binding protein RBBP8.

41. A.k.a. Polycomb group RING finger protein PCGF4 and RNF51.

42. A.k.a. ligand effect modulator LEM6.

43. A.k.a. PGC1-related estrogen receptor-α coactivator PERC and estrogen receptor-related receptor ligand ERRL1. The PGC1 proteins are human homologs of yeast DNA repair and TF2h regulator MMS19 and P68 RNA helicase.

In adipocytes of brown adipose tissue, PGC1 is activated by thermal stimuli. It then acts as a coactivator with CBP and NCoA factors for PPARγ and T_3R-mediated transcriptional activation [653].

In response to certain metabolic signals, both mitochondria and peroxisomes execute fatty acid degradation (oxidation) and detoxification of reactive oxygen species.[44] These 2 types of organelles can also undergo a rapid remodeling and divide. In response to cold exposure, transcriptional coactivator PGC1α controls the remodeling and genesis of mitochondria. When the energy demand rises, cells trigger a rapid and coordinated response to restructure their organelles (adaptation of enzymatic composition of organelles and organelle density) and maximize the use of fatty acids. In this context, PGC1α also regulates the remodeling and genesis of peroxisomes specialized in β-oxidation in brown adipocytes and hepatocytes [661].

The transcriptional coactivator PGC1α interacts with and regulates the activities of several nuclear receptors, in addition to PPARγ, such as estrogen, glucocorticoid, retinoic acid, and thyroid hormone receptors, as well as transcription factors cAMP response element-binding protein (CREB) and nuclear respiratory factors (NRF). It connects also to RNA polymerase-2, RNA-processing factors, and the Mediator complex.

The PGC1α coactivator operates in energy metabolism.[45] It is predominantly expressed in heart, brown adipose tissue, kidney, skeletal muscle, and liver. Activity of PGC1α is potentiated by Arg methylation by PRMT1 methyltransferase [662].

6.3.4.9 The PRMT Family of Protein Arginine Methyltransferases

Protein arginine methyltransferases (PRMT) catalyze the addition of 1 or 2 methyl groups to Arg residues. The mammalian PRMT family contains 11 members. According to the methylated form of Arg residues of their substrates, PRMTs are classified into 2 categories: (1) type-1 PRMTs (PRMT1, -3, -4, -6, and -8) that catalyze asymmetrical dimethylation of Arg and (2) type-2 PRMTs (PRMT5, -7, and -9) that catalyze symmetrical dimethylation of Arg residues.

Most methyltransferases use the methyl donor ^{S}adenosyl ^{L}methionine (AdoMet) as a cofactor. Protein Arg methylation contributes to the regulation of transcriptional regulation and signal transduction, among other tasks. The PRMT enzymes are recruited to gene promoters by transcription factors and methylated histones, transcription coactivators such as CBP and P300, as well as transcription elongation factors.

44. Peroxisomal oxidation enzymes, such as acylCoA oxidase ACOx1, enoylCoA hydratase–3-hydroxyacylCoA dehydrogenase (EHHADH; a.k.a. peroxisomal bifunctional enzyme [PBFE]), and acetylCoA acyltransferase ACAA1 (a.k.a. 3-ketoacylCoA thiolase and peroxisomal 3-oxoacyl-CoA thiolase (PThio) reduce long-chain fatty acids (carbon chains >22 components). On the other hand, mitochondrial oxidation enzymes only oxidize short-chain fatty acids (carbon chains <22 components).

45. Mitochondrial genesis and stimulation of oxidative phosphorylation particularly require the cooperation of PGC1α with nuclear respiratory factors NRF1 and NRF2 and estrogen-related receptor ERRα.

The PRMT enzymatic activity can be modulated by their regulated expression level, formation of complexes and inter-protein interactions, and post-translational modifications and modified accessibility to substrate. In addition, members of each PRMT class can differ by their enzymatic properties, with distinct specificities for substrates, and subcellular localization. In particular, PRMT1 and PRMT3 reside predominantly in the nucleus and cytoplasm, respectively.

Protein Arg Methyltransferase-1

Protein Arg methyltransferase-1 (PRMT1) is the predominant member of the PMRT family in mammalian cells. Isoforms PRMT1s with alternatively spliced N-terminus differ in intracellular localization. The PRMT1 oligomer interacts with nuclear receptor coactivator NCoA2. It methylates histones-4 [663].

The PRMT1 enzyme is a coactivator of several nuclear receptors, such as thyroid hormone, estrogen, and androgen receptors. It also links to hepatocyte nuclear factor HNF4 (NR2a) that regulates the expression of genes involved in glucose and lipid metabolism, as well as that of apolipoprotein-A1 [664].[46] It interacts with TR3 receptor (NR4a1) via a feedback loop [665]. It delays NR4a1 protein degradation, but methylates NR4a1 only in the case of direct interaction.

Shuttling of PRMT1 between the nucleus and cytoplasm depends on the methylation status of substrates and yields a dynamic mechanism for the regulation of substrate methylation [666]. Splicing variant PRMT1v2 contains a nuclear export signal that leads to a predominantly cytoplasmic localization.

Protein Arg Methyltransferase-2

Protein Arg methyltransferase-2 (PRMT2) is a coactivator for estrogen receptor ERα (NR3a1) [667].[47] It also binds to other types of nuclear receptors, such as estrogen ERβ (NR3a2), progesterone PR (or NR3c3), and thyroid hormone TRβ (NR1a2) receptors, as well as PPARγ (NR1c3), RARα (NR1b1), and RXRα (NR2b1). In addition, PRMT2 links to 2 other coactivators – NCoA1 and NCoA6 –, but without synergistic enhancement of NR3a1 transcriptional activity.

Protein Arg Methyltransferase-4 (CARM1)

Protein Arg ^{N}methyltransferase PRMT4, or coactivator-associated arginine methyltransferase CARM1, enhances the activity of coactivators, as it methylates not only histones-3, but also coactivators and histone acetyltransferases CBP and P300,

46. Hepatocyte nuclear factor HNF4 contributes to the regulation of tissue development, especially in the liver and intestine. In adult mammals, HNF4 is expressed in the liver, intestine, kidney, and pancreas.

47. It interacts directly with 3 regions of liganded ERα (AF1 and DNA- and hormone-binding domains.

thereby increasing nuclear receptor-dependent gene transcription.[48] Like PRMT1, it interacts with nuclear receptor coactivator NCoA2. Unlike PRMT1 that targets histone-4, PRMT4 methylates histone-3. Like PRMT1 and PRMT2, PRMT4 is involved in estrogen receptor-mediated transcriptional activation.

6.3.4.10 CAPER

Coactivator of activator protein-1 and estrogen receptors CAPER selectively binds Jun (activator protein AP1 component) and estradiol-bound ligand-binding domains of ERα and ERβ (NR3a1 and NR3a2) [668].

6.3.4.11 Multistage Activation of Gene Transcription

The activation of gene transcription is a multistep process. Once the nuclear receptor dimer such as the PPAR–RXR heterodimer is bound to the response element of DNA such as PPRE (Sect. 6.3.6.4), the NCoA1–NCoA2–NCoA3 subcomplex is recruited. This initial signaling platform, in turn, recruits histone acetyltransferases CBP and P300 as well as methyltransferase PRMT4. Proteins CBP and P300 acetylate histones, whereas PRMT4 methylates histones as well as CBP and P300 [659].

The NCoA–CBP–P300–PRMT4 platform recruits NCoA6IP, NCoA6, and subunits of Mediator. The latter is anchored to the coactivator complex by Med1 subunit. The NCoA6 component and its interactor NCoA6IP bridges the initial histone acetyltransferase-histone methyltransferase (NCoA–CBP–P300–PRMT4) complex to Mediator [659]. Mediator then facilitates interaction with RNA polymerase-2 complexes (basal transcriptional machinery).

6.3.5 Intracellular Hormone Receptors

Receptors of steroid (androgen, estrogen, glucocorticoid, mineralocorticoid, and progesterone) and thyroid hormones (Vol. 2 – Chap. 1. Remote Control Cells – Sect. Endocrine System and Hormones) reside either in the cytoplasm or nucleus. Intracellular hormone receptors are indeed exported from the nucleus to capture ligands into the cytosol up to the plasma membrane and to transfer these signaling molecules to the nucleus. The cytoplasmic hormone–receptor complex translocates to the nucleus (active transport) to operate as transcription factors. The 48 known members of the human intracellular hormone receptor family contribute to tissue development, cell growth, metabolism, and inflammation.

The so-called nuclear receptors tether to lipophilic ligands that include: (1) steroid hormones (estradiol, progesterone, testosterone, cortisol, and aldosterol); (2) thyroid hormones; (3) retinoic acid; and (4) seco-steroid (1α,25)-dihydroxyvitamin-D3.

48. The PRMT4 enzyme methylates the KIX domain of CBP and P300. KIX domain-methylated CBP and P300 fail to activate CREB-dependent genes, but are able to support gene transcription by nuclear receptors. Therefore, PRMT4 polarizes the activity of CBP and P300 toward nuclear receptor-targeted genes.

The nuclear hormone–receptor complex binds to specific *hormone response elements* (HRE) of DNA for gene transcription.[49] These hormone receptors thus are bifunctional, as they bind hormone and directly activates gene transcription, hence behaving like ligand-activated transcription factors for responsive genes. Steroid and thyroid hormone receptors are actually made of distinct domains. The single polypeptidic chain possesses transcription-activator, DNA-binding, and hormone-binding domains.

Thyroid hormone receptors are nuclear receptors that bind DNA in the absence of hormone, usually repressing gene transcription. Once bound to thyroid hormone, the cognate receptor behaves as a transcriptional activator.

Steroid hormone receptors exist as dimers coupled with chaperones, such as heat shock proteins HSP90 and HSP65, that are shed on binding to steroid hormone. They also affect transcription by interactions with other transcription factors, such as Activator protein-1 and nuclear factor-κB.

Splice variants of each of these receptors can form functional or non-functional monomers that can dimerize to form functional or non-functional receptors. Alternative splicing of PR mRNA produces monomer-A and -B that combine to produce functional AA, AB, and BB dimeric receptors with distinct characteristics [5].

Distribution of intracellular receptors, although dynamical, varies according to the receptor type: in about equal amount in both compartments (e.g., mineralocorticoid and vitamin-D receptors), predominantly nuclear (e.g., PPARs), or mostly cytoplasmic (e.g., androgen and glucocorticoid receptors). The presence of hydrophobic ligands that cross the plasma membrane changes the receptor localization.

Features, especially topological aspects, of the intracellular receptor-based signaling can be identified using generic mathematical models [669]. An optimal response relies on shuttling of intracellular receptors between the cytoplasm and nucleus with a preferential nuclear import of liganded receptor and export of free receptor.[50] Moreover, intracellular receptor may contribute more to ligand movement if it can directly collect the ligand from the plasma membrane, where 5 to 10% of the total cellular concentration can indeed localize. A theoretical optimal ratio of nuclear to cytoplasmic fractions of a given intracellular receptor depends on ligand properties and transcription activation requirements.

6.3.5.1 Sex Steroid Hormone Receptors

Sex steroid hormone receptors can regulate other nuclear receptors, such as peroxisome proliferator-activated receptor PPARα and liver X receptors, which govern metabolic pathways [670]. Nuclear steroid receptor coactivators that enhance the

---

49. Steroid hormones bind and activate their nuclear hormone receptors that then dimerize and bind to specific DNA response elements, interacting with coactivators and corepressors. Hormone response elements are DNA sequences of promoters of hormone-responsive genes.

50. When all intracellular receptors are constitutively bound to the DNA, the transcriptional response is very low. A high concentration of free intracellular receptors in the nucleus improves the sensitivity, but the response remains slow.

transcription initiation mediated by nuclear receptors and other transcription factors, shuttle between the cytoplasm and the nucleus.

Steroid receptor coactivator-interacting proteins (SIP) sequester steroid receptor coactivators in the cytoplasma [671]. Extracellular stimuli induce SIP phosphorylation by casein kinase-2 (Vol. 4 – Chap. 4. Cytosolic Protein Ser/Thr Kinases), thereby dissociating from SIPs the steroid receptor coactivators that can then move to the nucleus.

Intracellular Androgen Receptors

Androgen receptor (NR3c4) possesses 2 endogenous agonists – testosterone and its more active metabolite 5α-dihydrotestosterone – that regulate sexual differentiation and maturation. It contains 4 functional domains [5]: an N-terminal regulatory domain, DNA-binding motif, small hinge region, and ligand-binding site. The latter undergoes structural rearrangement once agonists, antagonists, or specific nuclear receptor modulators are bound.

Intracellular Estrogen Receptors

In addition to G-protein-coupled estrogen receptor [51] (Sect. 7.13.20), nuclear estrogen receptors are composed of 2 types (ERα–ERβ or NR3a1–NR3a2) that are encoded by distinct genes. Receptors ERα and ERβ differ in structure and tissue expression pattern. [52] Each type can homo- or heterodimerize. Ligand-bound estrogen receptors can also interact directly with other transcription factors, such as Fos and Jun [5]. In addition, ERβ has isoforms (ERβcx and ERβ2) [672]. These isoforms preferentially heterodimerize and inhibit ERα.

Ligand binding to estrogen receptors creates a conformational change that enables the recruitment and tethering of cofactors to regulate gene transcription according to the context. Estrogen receptors bind to estrogen response elements (ERE).

Steroid hormone 17β-estradiol (or estradiol [E_2]) contributes to the regulation of cell growth and differentiation, as well as function in various tissues, such as male and female reproductive tracts, mammary gland, skeletal muscle, and cardiovascular system [672]. The predominant effects of E_2 are mediated by intracellular receptors ERα and ERβ that interact with identical DNA response elements.

Estradiol operates via 4 signaling mechanisms [672]: (1) classical pathway that relies on the E_2–NR3a complexes, which targets EREs in target promoters; (2) ligand-independent axis stimulated by growth factors or cyclic adenosine monophosphate that trigger a kinase activation cascade down to mitogen-activated protein kinases that phosphorylate NR3a1/2s; (3) ERE-independent, E_2–NR3a complex-based route that uses other transcription factors, such as Fos and Jun, and targets

51. Receptor GPER is also called GPR30.

52. The ERα receptor is the predominant subtype expressed in mammary glands, uterus, cervix, vagina, and several other target organs. Distribution of ERβ is more restricted, as it is primarily detected in ovary, prostate, testis, spleen, lung, hypothalamus, and thymus [672].

AP1-responsive elements; and (4) plasmalemmal receptor-initiated signaling that rapidly generates the cell response.

Nuclear estrogen receptors can be activated not only by ligands, but also by ligand-independent mechanisms. The NR3a1 transcription factor is indeed activated by numerous other extracellular signals in the absence of ligands, such as cAMP and growth factors, e.g., epidermal growth factor and insulin-like growth factor-1 [667].

Estrogen receptor activated by E_2 enhances glucose stimulation of the insulin promoter and, hence, insulin synthesis, independently of the estrogen response element [673]. Extranuclear ERα interacts with Src kinase that activates extracellular signal-regulated kinases ERK1 and ERK2 to assist by phosphorylation the binding of the transcription factor neurogenic differentiation factor NeuroD1 to the insulin promoter.

Protein kinase-A activated by cAMP phosphorylates NR3a1, but this phosphorylation is dispensable for cAMP-induced NR3a1 activation. On the other hand, PKA also phosphorylates PRMT4 methyltransferase [674]. This phosphorylation is indispensable for cAMP-induced NR3a1 activation and PRMT4–NR3a1 interaction.

Segregation of Estrogen Signaling by Spatial Organization

Palmitoylation and subsequent interaction with caveolin-1 promote the translocation of ERα and ERβ to the plasma membrane [24]. This transfer supports the rapid activation of phosphoinositide 3-kinase and the Raf–MAP2K–ERK axis to promote cell survival and motility. In addition, crosstalk between cortical estrogen receptors and epidermal (EGF) or insulin-like growth factor-1 (IGF1) activates extracellular signal-regulated protein kinases. On the other hand, the nuclear localization is needed for developmental programs.

Sex Steroid Hormone Receptors in the Cardiovascular Apparatus

Sex steroid hormone receptors regulate the gene expression in vascular cells. Cardiovascular coregulators include NCoA3 coactivator and the myocardial androgen receptor coactivator four and a half LIM domain-containing protein FHL2 [670].[53] Estrogen and androgen receptors, as well as 2 progesterone receptor isoforms (PRa and PRb) are expressed in the vasculature. Estrogen and progesterone receptors in vascular cells can be activated in the absence of ligand by growth factors. Sex steroid hormone receptors, particularly estrogen receptors ERα and ERβ exist in cardiomyocytes, as well as aromatase.

A diet with (but not too rich in) red wine polyphenols generates an endothelium-dependent vasodilation (the "French paradox"). This vasorelaxation is mediated by estrogen receptor NR3a1 [675]. Delphinidin and NR3a1 agonist 17β-estradiol can

53. Coactivator NCoA3 of the NCOA (or SRC) family of nuclear hormone receptor coactivators recruit protein acetyltransferases, such as cAMP response element-binding protein (CBP) and P300, its associated factor P/CAF, and protein arginine methytransferases PRMT1 and PRMT4 to nuclear receptor target gene promoters to modify chromatin structure and/or assemble the transcription initiation complex.

cause vasodilation by activation of the nitric oxide pathway in endothelial cells, which involves caveolin-1, Src kinase, extracellular signal-regulated protein kinases ERK1 and ERK2, and NOS3.

Sex steroid hormone receptors stimulate endothelial nitric oxide synthase and cause vasodilation via activated estrogen receptors. Estrogen receptor-α has protective effects on injured blood vessels by promoting re-endothelialization and inhibiting smooth muscle cell proliferation. Estradiol acts on NR3a1 to upregulate the production of atheroprotective prostacyclin by activation of cyclooxygenase-2 [676].[54] In males, aromatase converts testosterone to estrogen, thereby allowing the maintenance of vascular relaxation. Testosterone thus is able to activate both androgen and estrogen receptors. Progesterone lowers the blood pressure. Estrogen receptor-β is required for vasodilation and normal blood pressure in both males and females [670].

Nuclear receptor coactivator-3 also targets transcription factor myocardin in vascular smooth muscle cells [678]. Myocardin, a serum response factor cofactor exclusively expressed in cardiac and smooth muscle cell lineages, is involved in SMC differentiation to the non-proliferative, contractile phenotype.[55] Estrogens, by acting on NCoA3, can limit or prevent abnormal proliferation of vascular smooth muscle cells.

6.3.5.2 Glucocorticoid Receptors

Glucocorticoid receptor (GR or GCR), or nuclear receptor NR3c1, is encoded by the NR3C1 gene. It is specifically targeted by cortisol and other glucocorticoids (Vol. 2 – Chap. 1. Remote Control Cells – Sect. Endocrine System and Hormones).[56]

The glucocorticoid receptor–hormone complex translocates into the cell nucleus, where it binds to *glucocorticoid response elements* (GRE) in the promoter of target genes (transactivation). On the other hand, activated glucocorticoid receptor can interact with specific transcription factors and prevent the transcription of targeted genes (transrepression).

In the absence of hormone, the glucocorticoid receptor resides in the cytosol, bound to cytosolic chaperone proteins, such as heat shock proteins HSP90 and HSP70 as well as FK506-binding protein FKBP52. Glucocorticoid hormones move into the cytoplasm and bind to the glucocorticoid receptor. They then cause a conformation change of the receptor, a release of chaperone proteins, and, then, the translo-

---

54. Estradiol effect is mediated by 2 nuclear estrogen receptors ERα and ERβ that are encoded by genes ESR1 and ESR2, respectively. Isotype ERα, but not ERβ, is necessary and sufficient to mediate most of the vascular effects of estradiol. Estrogen receptor-α stimulates target gene transcription owing to 2 activation functions AF1 and -2 located in 2 distinct domain of the protein. Activation function AF1 is not needed for vasculoprotection of 17β-estradiol [677]. AF1-Associated region of ERα is targeted by several protein kinases, such as MAPK, glycogen synthase kinase-3β, the CcnA–CDK2 complex, and CDK7.

55. Nuclear receptor coactivators regulate signaling by transcription factors such as AP1, NFκB, HNF1α, MEF2C, E2F1, and TEF.

56. The word glucocorticoid derives from the role of these hormones in the regulation of glucose metabolism, their synthesis in the adrenal cortex, and their steroidal structure.

cation of GR homodimers into the nucleus. Cortisol (or hydrocortisone) contributes to the regulation of metabolic (stimulation of gluconeogenesis as well as synthesis of regulators of lipid and protein metabolism), respiratory (production of surfactant), cardiovascular, and immunological functions. Activated glucocorticoid receptor promotes and impedes the production of anti-inflammatory and pro-inflammatory proteins, respectively.

In humans, 2 glucocorticoid receptor splice variants exist: the canonical GR, or GRα subtype (NR3c1α) and GRβ (NR3c1β). Whereas GRα acts as a ligand-activated transcription factor, GRβ does not bind glucocorticoids and lacks transactivational activity, but acts as an inhibitor of GRαisoform.

The transcription factors glucocorticoid receptor (NR3c1) and peroxisome proliferator-activated receptor-α (NR1b1) modulate the immune response. Simultaneous activation of PPARα and GR enhances repression of transcription factor NFκB and cytokine production; PPARα can then act as a GR potentiator for transrepression on NF-κB [679]. On the other hand, activated PPARα impedes the GRα stimulation of glucocorticoid response element; PPARα can thus serve as a GR antagonist for GR transactivation.

Growth arrest-specific, non-coding, single-stranded RNA GAS5 accumulates in cells that lack nutrients or growth factors and experience growth arrest. Agent GAS5 binds to the DNA-binding domain of the glucocorticoid receptor, thereby precluding the linkage of glucocorticoid receptor to glucocorticoid response element [680].

6.3.5.3 Mineralocorticoid Receptor

Mineralocorticoid receptor is also called aldosterone receptor or nuclear receptor NR3c2. It is expressed in many tissues, such as the heart, central nervous system, kidney, brown adipose tissue, colon, etc. Its activation leads to the synthesis of proteins that regulate ionic and water transport, mainly epithelial sodium channel and Na^+–K^+ pump, as well as serum- and glucocorticoid-induced kinase SGK1. It thus provokes sodium reabsorption and an increase in extracellular volume, as well as a potassium excretion to maintain a normal salt concentration in the organism. Mineralocorticoid receptor is targeted not only by mineralocorticoids such as aldosterone, but also glucocorticoids, such as cortisol and cortisone. Activated mineralocorticoid receptor homodimerizes, translocates to the cell nucleus, and binds to hormone response elements in the promoter of target genes.

6.3.5.4 Thyroid Hormone Receptors

Thyroid hormone receptors (THR) [57] belong to the subclass 1 of nuclear hormone receptors (NR1a1–NR1a2). They participate in the regulation of nutrient

57. Alias THR avoids confusion with testicular receptors TR2 and TR4. The testicular receptors are also members of the nuclear receptor family of intracellular transcription factors. They are encoded by 2 genes NR2C1 and NR2C2, respectively.

metabolism, cognition, and cardiovascular homeostasis. They are activated by triiodothyronine (T_3) and thyroxine (T_4). Once activated by a ligand, these transcription factors act as monomers and homo- or heterodimers with members of the retinoid X receptor family. Thyroid hormone receptors might localize to the plasma membrane by interacting with $\alpha_V\beta_3$-integrin [5].

Two THRα and 2 THRβ receptor splice variants are encoded by the THRA and THRB genes, respectively. The ThRα1 isoform is widespread with high levels in cardiac and skeletal muscles; ThRα2 is also broadly spread, but unable to bind thyroid hormones; ThRβ1 is predominantly synthesized in the brain, liver, and kidneys; ThRβ2 is restricted to the hypothalamus and pituitary gland.

6.3.6 Other Nuclear Receptors

Cells need to handle messages transmitted by hormones produced by the body for a requested task and then cleared as well as to respond to numerous exogenous chemicals, or xenobiotics, such as ingested food components, some vitamin types,[58] environmental pollutants, drugs, and carcinogens, that must be processed. Among xenobiotics, some species have both endo- and exogenous sources such as cholesterol. The body must then ensure their homeostasis to avoid their presence at a much higher concentration that can be deleterious. Nuclear receptors are involved in these processes.

6.3.6.1 Estrogen-Related Receptors

Three known estrogen-related receptors exist (ERRα–ERRγ), i.e., nuclear receptors NR3b1 to NR3b3 that are encoded by the genes ESRRA, ESRRB, and ESRRG.

The ERRα factor that has a wide tissue distribution interacts with transcription factor TF2b as well as lysosomal trafficking regulator and nuclear receptor coactivator-3 (NCoA3).

The ERRβ factor interacts with NCoA2. The ERRγ factor associates with NCoA1 and hepatocyte growth factor-regulated Tyr kinase substrate [59].

6.3.6.2 Farnesoid X Receptors

Farnesoid X receptor (FXR), a farnesol-activated protein, or nuclear receptor NR1h4, potently binds chenodeoxycholic acid as well as other bile acids, such as lithocholic and deoxycholic acids, in cells that express bile acid transporters. The production of FXR is restricted to the enterohepatic apparatus, kidneys, and adrenals.

In addition to the control of bile acid synthesis, FXR regulates the transport of bile acids to prevent their overaccumulation in hepatocytes. Two main transporters

58. Vitamin-C, or ascorbic acid, is a water-soluble sugar acid with anti-oxidant properties. It is a vitamin for humans, but not for many animal species that synthesize it. Vitamin-B7 is a cofactor for several carboxylases in the metabolism of fatty acids, catabolism of some amino acids, and gluconeogenesis. It is required in diet only in certain circumstances.

exist in hepatocytes: (1) SLC10a1 [59] for bile acid uptake, and (2) ABCb11 [60] for bile acid efflux into the bile. The FXR receptor prevents the synthesis of SLC10a1, but promotes that of ABCb11 carrier.

In enterocytes, bile acids upregulate transcription of *ileal bile acid-binding protein* (IBABP) involved in the absorption of conjugated bile acids via FXR receptor. [61] The IBABP protein is the single known cytosolic protein that binds and carries bile acids. Bile acids activate their FXR receptor to heighten IBABP transcription.

Bile acid sensor FXR participates in overall sterol metabolism, as it regulates the transcription of genes involved in bile acid metabolism [652]. [62]

Activated farnesoid X receptor translocates to the nucleus and forms heterodimers that associate with hormone response elements on DNA. Farnesoid X and liver X receptors are members of a steroid analog-activated nuclear receptor set (subclass 1 [NR1]) that heterodimerize with members of the retinoid X receptor family (subclass 2 [NR2]).

6.3.6.3 Liver X Receptors

Liver X receptors [63] act as oxysterol receptors and cholesterol sensors. They participate in the regulation of concentrations of cholesterol, hydroxycholesterols (OHC), fatty acids, and glucose. They maintain the cholesterol level, as they control the transcription of proteins that govern cholesterol catabolism, storage, absorption, and transport.

Two identified isoforms of LXR exist (LXRα–LXRβor NR1h3–NR1h2). The LXRβ receptor is ubiquitous. On the other hand, LXRα production is restricted to cells strongly involved in lipid metabolism [681].

Liver X receptors heterodimerize with retinoid X receptor. The RXR–LXR heterodimer preferentially binds to LXRE response elements. Liver X receptors target the genes that encode sterol regulatory element-binding protein SREBP1 and ATP-binding cassette transporters that facilitate efflux of cholesterol from cells (Tables 6.13 and 6.14). In enterocytes, ABC transporters limit cholesterol absorption.

In cells, the cholesterol concentration results from a balance among uptake, efflux, and synthesis. Sterol-responsive nuclear receptor LXR contributes to the maintenance of cholesterol homeostasis, as it promotes cholesterol efflux and suppres-

59. A.k.a. Na^+–taurocholate cotransporter polypeptide (NTCP) and liver bile acid transporter (LBAT).

60. A.k.a. bile salt export pump (BSEP).

61. Another transcript $IBABP_L$ (long form) exist. Its transcription is controlled by nuclear factor-κB rather than farnesoid X receptor [682].

62. Bile acid homeostasis is governed by a negative feedback of synthesis. Bile acids upregulate genes that export bile acids out of cells and repress genes responsible for bile acid synthesis and uptake. When the bile acid pool rises, CyP7a1 transcription is precluded. Bile acid-response element of the CYP7A1 promoter is targeted by the NR0b2 repressor during the bile acid-mediated inhibition of CyP7a1 triggered by FXR receptor [681]. The NR0b2 protein can form an inactivating heterodimer with NR5a2 to suppress CyP7a1 production.

63. The name comes from their initial isolation from a hepatic molecule search.

Table 6.13. Regulation of cholesterol (Cs) metabolism by nuclear receptors (**Part 1**; BA: bile acid; EL: endothelial lipase; FA: fatty acid; LPase: lipoprotein lipase; NPC1L1: Niemann-Pick protein-C1-like-1). In enterocytes, hepatocytes, and macrophages, liver X receptors (LXR) regulate storage and transport of cholesterol to reduce intracellular cholesterol level. In hepatocytes, LXRs contribute to cholesterol catabolism into bile acids, hence excretion of excess cholesterol into bile and feces. Excess cholesterol indeed generates oxysterols that stimulate the transcription of cholesterol 7α-hydroxylase of the cytochrome P450 superfamily (CyP7a1) under the control of NR1h3 (LXRα), NR5a1 (SF1; inhibition), and NR5a2 (LRH1; stimulation) due to the existence of LX- and bile acid-response element in the CYP7A1 promoter. In enterocytes, LXRs favor cholesterol excretion via ABC transporters. Bile acids upregulate the transcription of ileal bile acid-binding protein (IBABP). In other cell types, LXRs regulate the esterification and storage of cholesterol via sterol regulatory element-binding protein SREBP1c. Once cleaved from the membrane, SREBP1c elicits transcription of fatty acid-synthesizing enzymes. Cellular efflux of free cholesterol is achieved by ATP-binding cassette transporters (ABC) that deliver cholesterol to high-density lipoproteins, which transport cholesterol to the liver. The LXR receptors also favor the production of HDLs, as they upregulate ApoE. In HDLs, free cholesterol is esterified by lecithin–cholesterol acyltransferase (LCAT), the synthesis of which is promoted by LXRs.

Cell type	Carriers, enzymes	Function	Regulator
Enterocyte	ABCa1/g5/g8	Excretion	LXR (NR1h2/3; +)
	NPC1L1	Absorption	NR1c2 (PPARδ +)
	IBABP	Absorption	FXR (NR1h4; +)
	CyP7a1	Synthesis	NR5a2 (+)
Hepatocyte	ABCg5/g8	Cs Export	LXR–RXR (+)
	CyP7a1	BA Synthesis	NR0b2–NR5a2 (−) (FXR induction)
	SREBP1c	FA Synthesis	LXR (+)
	SREBP2	LDLR Expression	
		VLDL Production	FXR (−)
	BSEP (ABCb11)	BA Efflux	FXR (+)
	NTCP (SLC10a1)	BA Influx	FXR (−)
	ScaRb1	CE Uptake	LXR (+)
	LDLR	CE Uptake	LXR (+)
	ApoC2	LPase Cofactor	FXR (+)
	ApoC3	LPase Inhibitor	FXR (−)
	ApoA1	HDL Formation	FXR (−) NR1c2
	LPase	CE Uptake	LXR (+)

ses low-density lipoprotein uptake [683]. The LXR receptor elicits transcription of the ubiquitin ligase inducible degrader of LDL receptor (IDOL) that causes LDLR degradation, hence impeding LDL uptake.

Sterol regulatory element-binding protein SREBP2 and LXR control antagonistic transcriptional programs that stimulate cholesterol uptake and synthesis as well

Table 6.14. Regulation of cholesterol (Cs) metabolism by nuclear receptors (**Part 2**; BA: bile acid; EL: endothelial lipase; FA: fatty acid; LPase: lipoprotein lipase; NPC1L1: Niemann-Pick protein-C1-like-1). In macrophages, liganded RXR–LXR heterodimer stimulates synthesis of ABCa1 and ABCg1 transporters for cholesterol efflux. Scavenger receptor-B3 (ScaRb3 or CD36) binds many ligands, such as collagen, thrombospondin, oxidized LDLs, oxidized phospholipids, and long-chain fatty acids. Reduced ScaRb3 expression in macrophages attenuates atherosclerosis.

Cell type, tissue	Carrier, enzyme	Function	Regulator
Macrophage	ABCa1/g1	Cs Efflux Cs Deliver to HDLs	LXR–RXR (+)
	ApoA1	Cs Efflux	NR1c2 (+)
	ApoE	HDL Formation	LXR–RXR (+)
	ScaRa	ox/acLDL Endocytosis	
	ScaRb1	Cs Efflux	LXR–RXR (+)
	ScaRb2	CE Ingress, egress	
	ScaRb3	CE Uptake	NR2c2 (TR4; +) NR1c3 (PPARγ +) (PPAR–RXR)
	ScaRd1	oxLDL Endocytosis	
	LDLR	CE Uptake	LXR (−)
	LPase	CE Uptake	LXR (+)
Endothelial cell	EL	CE Uptake	
Plasma Lipoprotein	CETP		NR5a2 (+)

as cholesterol efflux, respectively. The SREBP2 protein is synthesized together with miR33, as they are cotranscribed. Agent SREBP2 favors production of LDL receptor in hepatocytes as well as LXR-regulated sterol transporters ABCa1 and ABCg1 [684]. MiR33 promotes LDLR expression and cholesterol efflux to ApoA1, but represses that of ABCa1 and ABCg1 (Table 6.15).

Since LXR is activated by specific oxysterols, but apparently not cholesterol, induction of ABCa1 expression following cellular uptake of cholesterol-rich lipoproteins requires the conversion of cholesterol into specific oxysterols, such as 22-OH-, 27-OH-, or (24,25)-epoxy-cholesterol [556].

6.3.6.4 Peroxisome Proliferator-Activated Receptors

Peroxisome proliferator-activated receptors are ligand-dependent transcription factors and nuclear hormone receptors of the NR1c family (NR1c1–NR1c3). [64] Three

64. The word "peroxisome proliferator-activated receptor" derives from the initial cloning of PPARα isoform as a target of various xenobiotics that cause proliferation of peroxisomes in the liver.

Table 6.15. Cholesterol homeostasis is coordinately controlled by sterol regulatory element-binding protein SREBP2, its coexpressed miR33 microRNA, and liver X receptor (LXR or NR1h2/3), among other agents (Source: [684]). The SREBP2 and LXR pathways that respond to changes in cellular sterols are associated via Ub ligase inducible degrader of LDLR (IDOL1) and miR33. MiR33 promotes synthesis and uptake of sterols and minimizes sterol loss, as it represses ABC transporters. 3-Hydroxy 3-methylglutaryl CoA reductase (HMGCR) is the rate-limiting enzyme of cholesterol synthesis.

Target	Stimulator	Inhibitor
HMGCR	SREBP2	Low cholesterol content
LDLR	SREBP2	LXR (IDOL1)
ABCa1/g1	SREBP2, LXR	miR33

Table 6.16. Aliases of peroxisome proliferator-activated receptors (CIMT: carotid intimal medial thickness; FAAR: fatty acid-activated receptor; Glm1: susceptibility to glioma type 1). The PPARα factor is highly expressed in organs in which high rates of fatty acid β-oxidation happens.

Type	Aliases	Distribution
NR1c1	PPARα	Heart, kidney, liver, intestine, brown adipose tissue
NR1c2	PPARδ, PPARβ, FAAR	Ubiquitous
NR1c3	PPARγ, CIMT1, GLM1	Adipose tissue (mainly)

PPAR isoforms (PPARα, PPARγ, and PPARδ) have distinct tissue distribution (Table 6.16).[65]

PPAR Function

The direct interaction between a ligand and the C-terminal helix (AF2 helix) in the ligand-binding domain (LBD) that constitutes the activation function AF2, provokes the receptor activation, as it forms binding interfaces with members of the steroid receptor coactivator family.[66]

Activated PPARs actually heterodimerize with RXRs and recruit cofactors. Like RAR–RXR complexes, these PPAR–RXR heterodimers specify the transcriptional response. Liganded PPARs bind preferentially to PPAR response elements (PPRE). They can also complex with other transcription factors such as CCAAT/enhancer-binding protein [652].

65. The transcription factor PPARδ is also called PPARβ and fatty acid-activated receptor (FAAR).

66. Partial agonists do not always interact with AF2 helix to activate PPARγ. For example, 15-oxo-eicosatetraenoic acid (15oxoETE) tethers to a site adjacent to AF2 helix. On the other hand, full agonists connect to the AF2 helix.

Many cell types express more than one PPAR isoform. Response specificity relies on selective coactivators and endogenous ligand types of both PPAR and RXR isoforms.

Peroxisome proliferator-activated receptors participate in the control of cell proliferation and differentiation, embryo- and fetogenesis, lipid and glucose metabolism, adipogenesis, nutrient metabolism (as lipid sensors), insulin sensitivity, energy balance, and the immune response.

The nuclear receptor PPARγ is a ligand-dependent transcription factor that coordinates gene expression related to glucose homeostasis and insulin sensitization.

PPARα

In vascular endothelial cells subjected to flow, Jun, a component of the transcription factor Activator protein AP1, upregulates the small, non-coding, single-stranded RNA — miR21 — that precludes the production of Pparα mRNA by binding to the 3′-untranslated region [686]. Conversely, nuclear PPARα lowers the pri-miR21 expression activated by AP1 (negative feedback loop). In addition, miR21 supports the adhesion of circulating monocytes to vascular endothelial cells, as it promotes the activation of AP1 and synthesis of the adhesion molecule VCAM1 (vascular cell adhesion molecule-1) and CCL2 chemokine.[67] Furthermore, the relieved inhibition of AP1 by PPARα heightens the production of miR21 that further represses PPARα (positive feedback loop).

PPARγ

Two variants of peroxisome proliferator-activated receptor-γ exist due to alternative promoter usage and alternative splicing: a long PPARγ2 and short PPARγ1 type [685]. Many tissues express a low PPARγ1 level. In particular, macrophages express PPARγ1, which is involved in lipid metabolism and cytokine production. The PPARγ2 isotype is selectively produced at very high levels in adipocytes (adipocyte-specific PPARγ2 isoform).

PPARγ in Adipogenesis

The PPARγ factor operates on most adipocyte-specific genes. It cooperates with the pro-adipogenic protein CCAAT/enhancer-binding protein C/EBPα.[68] Preadipocytes that begin to differentiate express C//EBPβ and -δ that prime the production of PPARγ and C/EBPα [685].

67. A.k.a. monocyte chemoattractant protein-1.

68. CCAAT/enhancer-binding proteins C/EBPα, -β, and -δare strongly involved in adipogenic differentiation.

Table 6.17. Ligands of PPARγ (Source: [5, 685]).

Eicosanoids (PGI2, 15-deoxy $\Delta^{(12,14)}$PGJ2, 13-HODE, 15-HETE, LTB4)
Fatty acids
Lysophosphatidic acid
Serotonin

PPARγ and Insulin Sensitivity

In adipocytes, PPARγ increases the concentration of insulin-stimulated GluT4 transporter. It may promote transfer of triglycerides and fatty acids from muscle and liver and their storage in adipose tissue [685]. Moreover, activated PPARγ reduces levels of cytokines tumor-necrosis factor-α and interleukin-6 in adipocytes. In obese subjects, these cytokines are involved in the development of the insulin resistance (Vol. 6 – Chap. 7. Vascular Diseases).

PPARγ in Inflammation

The PPARγ factor has pro-inflammatory ligand-independent and anti-inflammatory ligand-dependent effects in the regulation of the activity of the transcription factor nuclear factor-κB [687]. On the one hand, unliganded PPARγ connects to class-2 NFκB protein RelA (or P65). The formation of the PPARγ–RelA complex is mandatory to support the binding of RelA to promoters of genes that produce pro-inflammatory agents such as CCL2 and CCL5 chemokines, at least in rat glomerular mesangial cells, in response to pro-inflammatory signals.[69] On the other hand, liganded PPARγ prevents the formation of the PPARγ–RelA complex. Consequently, RelA cannot interact with its cognate promoters.

PPAR Ligands

The PPAR factors have many endogenous agonists, such as prostaglandin-J2 metabolite 15-deoxy $\Delta^{(12,14)}$PGJ2,[70] prostacyclin (PGI2), leukotriene-B4, lysophosphatidic acid, 13-hydroxy octadecadienoic acid (13-HODE), 15-hydroxy eicosatetraenoic acid (15-HETE), and many fatty acids [5] (Table 6.17).

Peroxisome proliferator-activated receptors are able to upregulate vascular endothelial growth factor in colon carcinoma cells [688]. Subtype PPARδ thus favors angiogenesis (Vol. 5 – Chap. 10. Vasculature Growth) and carcinogenesis (Vol. 2 – Chap. 2. Cell Growth and Proliferation). Moreover, PPARδ inhibits epithelial tumor

69. The pro-inflammatory cytokine tumor-necrosis factor-α actually fosters the phosphorylation and nuclear translocation of RelA and RelA interaction with PPARγ.

70. Prostaglandin-J2 (PGJ2) and its metabolites Δ^{12}PGJ2 and 15-deoxy $\Delta^{(12,14)}$PGJ2 are derivatives of prostaglandin-D2 (sequential conversion from PGD2). In activated macrophages, 15-deoxy $\Delta^{(12,14)}$PGJ2 represses the Nos2 and Tnfα genes. This inhibition at least partly depends on PPARγ.

cell apoptosis via VEGF autocrine positive signaling loop and the PI3K–PKB axis. The PPARδ isoform is stimulated by prostaglandin-E2 and the Wnt–βCtn pathway.

In response to cellular stresses, PPARα, PPARγ, and PPARδ upregulate cytoprotective heme oxygenase-1.[71] The PPARδ receptor acts via PPARγ coactivator PGC1α in vascular endothelial cells exposed to oxidative stresses [689]. Moreover, heme oxygenase-1 promotes PPARδ synthesis (positive feedback loop). In addition, PPARδ upregulates endothelial nitric oxide synthase NOS3 in endothelial cells. Therefore, PPARδ stimulates endothelial anti-oxidant and anti-inflammatory functions.

In macrophages, several factors control cholesterol uptake and egress, such as cytokines and activators of nuclear receptors, particularly PPAR receptors. These ligand-activated transcription factors regulate glucose and lipid homeostasis, especially during inflammation. They are activated by several fatty acids and their derived products. Cholesterol efflux mediated by PPARα and PPARγ involves liver X receptor LXRα (NR1h3) that causes the synthesis of ATP-binding cassette transporters. Enhanced production of transporters ABCa1 and ABCg1 results from the activation of PPARγ, once bound to both ScaRb3 and ghrelin receptor GHSR1a. In macrophages, the PPAR–LXRα–ABCa1 pathway promotes cholesterol efflux to apolipoprotein-A1 and high-density lipoproteins. Growth hormone-releasing peptide can provoke the activation of extracellular signal-regulated kinases ERK1 and ERK2 [690]. These kinases heighten the expression of cyclooxygenase-2 that converts arachidonic acid to bioactive lipids such as prostaglandins, in particular the natural PPARγ ligand 15-deoxy $\Delta^{(12,14)}$-prostaglandin-J2. The latter can activate not only PPARγ, but also PPARα. Oxidized LDLs also stimulate PPARα and -γ in macrophages via 15-deoxy $\Delta^{(12,14)}$PGJ2 production using the ERK1/2–COx2 pathway.

The nuclear receptor PPARγ recognizes various endo- and exogenous ligands. Fatty-acid metabolites activate PPARγ. In fact, several polyunsaturated fatty-acid metabolites produced via the cyclooxygenase- or lipoxygenase-mediated pathways are potent endogenous ligands for PPARγ in adipocytes and macrophages. Serotonin metabolites[72] bind directly to PPARγ to regulate macrophage function and adipogenesis [691]. A serotonin metabolite and a fatty-acid metabolite can simultaneously bind to distinct PPARγ sites that are adjoining to the AF2 locus; each metabolite type uses specific coregulator and/or heterodimer interfaces. In other words, ligand binding to a given subsite of the ligand-binding domain causes conformational changes that enable interactions with specific coregulators and heterodimer partners. Furthermore, these 2 ligand types jointly provoke a stronger PPARγ activation than that caused by a single ligand type.

---

71. Heme oxygenase-1 is the rate-limiting enzyme in the oxidative degradation of heme into biliverdin that releases free iron and carbon monoxide. Its end-products have anti-apoptotic, anti-oxidant, and anti-inflammatory, as well as pro-angiogenic effects.

72. Extracellular serotonin activates its cognate plasmalemmal receptors. It is subsequently incorporated into the cell by its specific transporter SerT. It is subsequently metabolized into 5-hydroxy indole and 5-methoxy indole acetates by several enzymes such as monoamine oxidase.

6.3.6.5 Retinoic Acid Receptors

Retinoids, i.e., retinoic acid (or vitamin-A) and its derivatives (all-trans-retinoic acid [ATRA] and 9-cis-retinoic acid), bind to retinoic acid receptors (RAR). Three detected RAR subtypes (RARα–RARγ or NR1b1–NR1b3) are encoded by distinct genes (RARA, RARB, and RARG, respectively).

These nuclear hormone receptors heterodimerize with retinoid X receptors. Cytoplasmic retinoid-binding proteins-1 to -4 may control the concentrations of intracellular retinoids available for interaction with their receptors [5].

In the absence of ligand, some nuclear receptors such as retinoic acid receptor act as transcriptional repressors that recruit nuclear receptor corepressor complexes to target genes. The RAR–RXR heterodimer binds to corepressor NCoR2 and hormone response elements. Ligand binding dissociates the corepressor and recruits a coactivator to promote transcription of target genes. Agonists of RARs facilitate the exchange between NCoR and NCoA complexes, as they destabilize NCoR– and stabilize NCoA–RAR complexes.

Two distinct types of retinoid antagonists exist: *neutral antagonist* and *inverse agonist* due to their ability to induce distinct coregulator interaction patterns [692]. [73]

Retinoic acids and RAR receptors support the synthesis of surfactant-B peptide by respiratory epithelial cells in cooperation with NCoA1, NCoA2, CBP, and NK2 homeobox-containing transcription factor-1 (NKx2-1) [694]. [74]

6.3.6.6 Retinoic X Receptors

Retinoid X receptors (RXR) are retinoid-responsive transcription factors of the NR2B family, particularly activated by 9-cis retinoic acid. They form RXR–RXR homodimers as well as heterodimers with members of the subclass-1 nuclear receptors, such as RXR–RAR and RXR–PPAR complexes. They enhance the DNA-binding activity of RARs and thyroid hormone receptors. Three RXR subtypes (RXRα–RXRγ, or NR2b1–NR2b3) are encoded by distinct genes (RXRA, RXRB, and RXRG, respectively).

---

73. Inverse agonist BMS493 strongly enhances NCoR interaction. On the other hand, neutral antagonist BMS614 antagonizes agonist-induced NCoA recruitment. This pattern has been confirmed [693]: (1) BMS614 antagonist of RAR efficiently prevents the recruitment of coactivators, whereas BMS493 is less efficient for this task and (2) BMS614 has little effect on the association with corepressors (it moderately decreases NCoR2 binding to RAR), but BMS493 strongly supports corepressor interaction.

74. A.k.a. thyroid-specific enhancer-binding protein (TEBP or T/EBP), thyroid nuclear factor-1, and thyroid transcription factor-1 (TTF1). It is only detectable in lungs and thyroid. It is an important regulatory transcription factor in lung development and maintenance of cells of terminal respiratory units, such as alveolar pneumocytes and Clara cells.

6.3.6.7 Retinoic Acid-Related Orphan Receptors

Retinoic acid-related orphan receptors (RORα–RORγ) [75] belong to the subclass-1 nuclear receptors (NR1f1–NR1f3). Isotype RORα may be synthesized with an agonist such as cholesterol [5].

Receptor RORα possesses neuroprotective and anti-inflammatory properties. In astrocytes, it regulates interleukin-6 production [695]. [76]

Two isoforms are produced from the same RORC gene: RORγ, or RORγ1, and RORγt, or RORγ2, with a shorter N-terminus. The former is expressed in many tissues. The latter is restricted to the thymus, where it is expressed exclusively in immature CD4+ and CD8+ thymocytes as well as lymphoid tissue inducer cells (LTi).

6.3.6.8 Subclass-1, Family-D, Nuclear Receptors

All members of the NR1D family of nuclear receptors that are related to the viral oncogene vErbA (RevErbA) bind heme. The NR1d1 isoform (or RevErbAα) is a transcriptional repressor. It is highly synthesized in the brain, liver, skeletal muscle, and adipose tissue. It participates in their development.

Heme receptor RevErbAα is a core negative component of the circadian network that controls circadian oscillation of several clock genes such as the Bmal1 gene (Vol. 2 – Chap. 5. Circadian Clock). It also regulates the transcriptional coactivator PGC1, thereby impacting glucose, lipid, and bile acid metabolism. Nuclear receptor RevErbAα interacts with corepressor NCoR and participates in the regulation of the circadian clock [696].

In humans, 2 promoters of the gene generate a full-length and short isoforms. The NR1d1 isoform binds to target response elements (RevRE). It recruits its nuclear receptor corepressor NCoR that activates HDAC3 histone deacetylases. The interaction between NR1d1 and the NCoR-HDAC3 complex is enhanced by heme.

It can be phosphorylated by glycogen synthase kinase GSK3β that contributes to its stability. In addition, 2 Ub ligases, HUWE1 (HECT, UBA, and WWE domain-containing enzyme) [77] and Myc-binding protein MycBP2 [78] ubiquitinate NR1d1 for its degradation [697].

The NR1d2 isoform (RevErbAβ) is another member of the REVERBA family of transcription factors. It is involved in lipid and energy homoeostasis in skeletal myocytes. Its gene is regulated by the circadian rhythm.

---

75. A.k.a. RZRαto RZRγ.

76. In normal conditions, RORα activates the synthesis of neuroprotective IL6 at basal concentrations. In inflammation, RORα is upregulated and represses IL6 production to impede IL6 toxicity that can arise in some circumstances.

77. A.k.a. ARF-binding protein ArfBP1.

78. A.k.a. protein associated with Myc (PAM).

6.3.6.9 Subclass-5, Family-A, Nuclear Receptors

The NR5A family of nuclear receptors includes 2 identified members. Both NR5a1 and NR5a2 are able to bind phospholipids [698]. Nuclear receptors NR5a1 and NR5a2 bind to identical DNA consensus sequences and share common ligands, but have distinct tissue expression patterns.

NR5a1

Nuclear receptor NR5a1 [79] is confined to steroidogenic cells and their central command. In particular, NR5a1 abounds in adrenals. It is involved in the development of the hypothalamic–pituitary–adrenal and –gonadal axes [681]. It is a competence factor for transcription of steroidogenic cytochrome P450 enzymes.

Glucocorticoid synthesis is regulated by the transcriptional control of steroidogenic enzymes of the cytochrome-P450 gene (CYP) family, especially by NR5a1 and NR5a2 receptors. Adrenals are the major source of glucocorticoids. Yet, an extra-adrenal glucocorticoid synthesis exist. Intestinal epithelium becomes an important source of glucocorticoids upon immunological stimuli [698].

NR5a2

Monomeric NR5a2 receptor [80] is the enterohepatic paralog of NR5a1 factor; its expression is limited to the liver, exocrine pancreas, intestine, and ovary [681].

The NR5a2 receptor intervenes in tissue development (where it is synthesized), reverse cholesterol transport, bile acid homeostasis, and steroidogenesis. In particular, it may serve as a competence factor for bile acid synthesis. Its target genes include NR0B2 and those that encode cholesterol ester transfer protein, and bile acid-synthesizing enzymes.

In the intestinal mucosa, NR5a2 is predominantly expressed by crypt cells, where it regulates the expression of cyclin-D1 and -E1 [698]. It thus promotes epithelial cell proliferation and crypt cell renewal. The synthesis of NR5a2 in the intestine results from immune stimulation. It then provokes the synthesis of steroidogenic enzymes CyP11a1 and CyP11b1, hence that of corticosterone [698].

6.3.6.10 Subclass-0, Family-B, Nuclear Receptors

Both NR0b1 [81] and NR0b2 [82] lacks a DNA-binding domain. Receptors NR0b1 and NR0b2 inhibit NR5a1 and NR5a2, respectively.

---

79. A.k.a. steroidogenic factor SF1 and adrenal 4-binding protein (Ad4BP).

80. A.k.a. liver receptor homologue LRH1 as well as α1-fetoprotein transcription factor, hepatocytic transcription factor, B1-binding factor, and CyP7a promoter-binding factor.

81. A.k.a. dosage-sensitive sex reversal (Dax) and adrenal hypoplasia critical region on chromosome X (AHCH).

82. A.k.a. small heterodimer partner (SHP).

The NR0b2 receptor is expressed in the liver, intestine, heart, pancreas, and adrenal glands [681]. The NR0b1 receptor heterodimerizes with NR5a1 and strongly represses NR5a1 activity.

The NR0B2 gene is a major FXR target in the liver. Protein NR0b2 forms a repressor heterodimer with NR5a2 that silences the CYP7A1 (cholesterol 7α-hydroxylase), CYP8B (sterol 12α-hydroxylase), and NR0B2 genes [681].

6.3.6.11 Cholesterol Homeostasis

Nuclear receptors control sterol homeostasis, as they regulate the transcription of genes that encode enzymes, cofactors, and regulators involved in cholesterol catabolism, storage, absorption, and transport (Table 6.18).

Cholesterol homeostasis relies on a balance among 4 processes: (1) cholesterol synthesis; (2) absorption of dietary cholesterol; and (3) excretion of excess cholesterol via the biliary conduits and intestinal tube.

Many types of molecules are involved: enzymes of cholesterol ana- and catabolism and their eventual cofactors, in particular sterol-esterification enzymes, chaperones, membrane and intracellular carriers, organizers of vesicular transport, buffers, regulators, lipid-transfer proteins, and plasmatic transporters.

Intestinal Cholesterol Absorption

Dietary cholesterol consumption and intestinal cholesterol absorption by enterocytes elevate plasma cholesterol concentration. The small intestine absorbs dietary cholesterol as well as dietary plant sterols and other non-cholesterol sterols [699].

Scavenger receptor ScaRb1 lodges on the apical and basolateral parts of the plasma membrane of enterocytes [699]. Transfer ATPase (ABC transporter) ABCa1 localizes to the basolateral surface of enterocytes.

Niemann-Pick-C1-Like-1 protein (NPC1L1 or NPC3) intervenes in the absorption of cholesterol from the intestinal lumen by enterocytes. The NPC1L1 protein abounds in the small intestine. It lodges in the brush border of the plasma membrane of enterocytes [700]. The NPC1L1 transporter supports the uptake of cholesterol and non-cholesterol sterols into vesicles that then move using a subapical endosomal sorting compartment [700].

Activated PPARδ (NR1c2) reduces cholesterol absorption, as it attenuates the production of NPC1L1 and promotes the transintestinal cholesterol efflux [701].

Endosomes carry cholesterol to the endoplasmic reticulum, where it is esterified by *acylCoA–cholesterol acyltransferase* ACAT2. It is then transferred to chylomicrons for secretion into the blood vessels. Cholesterol synthesized de novo is also esterified by ACAT2 and enters chylomicrons. Plant sterols use a different route to go back to the gut lumen via ABC transporters ABCg5 and ABCg8 [699].

Microsomal triglyceride transfer protein (MTP) binds to *protein disulfide isomerase* (PDI). The resulting heterodimer then transfers neutral lipids into newly formed chylomicrons in the endoplasmic reticulum [699]. The coat protein complex CoP2-associated small guanosine triphosphatase Secretion-associated and Ras-related protein Sar1 is involved in the transfer of chylomicrons from the endoplasmic

Table 6.18. Regulation of cholesterol metabolism by "orphan" nuclear receptors (Source: [556, 681]; +: upregulation; −: downregulation; ABC: ATP-binding cassette transporter; Apo: apolipoprotein; BSEP: bile salt export pump; CETP: cholesterol ester transfer protein; CyP: member of the cytochrome-P450 superfamily; FXR: farnesoid X receptor; HDL: high-density lipoprotein; IBABP: intestinal bile acid-binding protein; LPase: lipoprotein lipase; LRH1: liver receptor homolog-1; LXR: liver X receptor; NTCP: sodium–taurocholate cotransporter polypeptide; PLTP: phospholipid transfer protein; RXR: retinoid X receptor; SF1: steroidogenic factor-1; SHP: small heterodimer partner; SREBP: sterol regulatory element-binding protein). Receptors FXR and LXR operate as bile acid and oxysterol receptors, respectively.

Nuclear receptor	Target gene	Expression regulation	Function
LXR	CyP7a1	+	Bile acid synthesis
(NR1h2/3)	SREBP1c	+	Lipogenesis
	ABCa1	+	Cholesterol efflux
	ABCg5/g8	+	Sitosterol efflux
	ABCg11a	+	Cholesterol transport
	CETP	+	Cholesterol ester transfer
	ApoE	+	Cholesterol efflux (HDL formation)
	LPase	+	In macrophage via ApoC2
FXR	NR0b2	+	Transcriptional silencing
(NR1H4)	BSEP	+	Bile acid efflux
	IBABP	+	Bile acid intracellular transport
	PLTP1b	+	Increase HDL
LRH1	CYP7a1	+	Bile acid synthesis
(NR5a2)	CYP8b1	+	Bile acid synthesis
	NR0b2	+	Transcriptional repression
	CETP	+	Cholesterol ester transfer
SHP	CYP7a1	−	Bile acid synthesis inhibition
(NR0b2)	CYP8b1	−	Bile acid synthesis inhibition
	CETP	−	Cholesterol ester transfer
	NR0b2	−	Relieve inhibition

reticulum. Another small GTPase ADP-ribosylation factor ARF1 contributes to the lipoprotein secretion, as it regulates the vesicular budding from the Golgi body.

Intestinal bile acid-binding protein (IBABP) is a cytosolic bile acid transporter that shuttles bile acids absorbed at the apical membrane to the basolateral membrane of ileocytes. This carrier then is a component of the enterohepatic circulation of bile acids. A small uptake of bile acids occurs in the jejunum. In addition, bile salt micelles are needed to solubilize cholesterol for absorption.

Intestinal Cholesterol Excretion

In addition to hepatobiliary secretion, *transintestinal cholesterol efflux* (TICE) throughout the small intestine allows the excretion of cholesterol in mice [702].[83] Secretion depends on the presence of a cholesterol acceptor. It is strongly stimulated by bile salts and phospholipids. The contribution of this pathway to cholesterol excretion in mice is estimated to be approximately twice that of the biliary pathway.

Cell Uptake of Cholesterol

Cell uptake of cholesterol is mediated by scavenger receptors (Sect. 5.2) that are targeted by high-density lipoproteins. During the selective uptake, only lipids are taken up from lipoproteins, whereas apolipoproteins are not degraded. Cellular uptake of cholesterol can occur at the plasma membrane, possibly after post-endocytosis (retrograde) exocytosis, or in endosomes from internalized HDLs via ScaRb2 [624].

Macrophage uptake of oxidized low-density lipoproteins via scavenger receptors activates PPARγ [632]. Two oxLDL components, 9- and 13-HODE, are ligands of PPARγ. In addition, ligand activation of PPARγ–RXRα heterodimer promotes monocyte differentiation into macrophages as well as uptake of oxLDLs by ScaRb3 [703].

Concept of Reverse Cholesterol Transport

Reverse cholesterol transport (RCT)[84] is a theory that explains the protective effect of high-density lipoproteins. High-density lipoprotein–cholesterol particles (HDL–C) carry cholesterol from the body's cells such as macrophage-derived foam cells in the arterial wall to the liver via blood for ultimate excretion from the body in feces via the biliary and digestive tracts.[85] Reverse cholesterol transport involves many steps and lipoprotein interconversions.

Reverse cholesterol transport relies on a set of molecules: (1) HDL constituents, such as apolipoproteins ApoA1 and ApoE; (2) transporters, such as those of ATP-binding cassette transporter superfamily (ABCa1, ABCg1, and ABCg8; Sect. 4.18), especially those that facilitate egress of phospholipids and cholesterol onto apolipoproteins to initiate the formation of HDLs; (3) scavenger receptors, such as ScaRb1 to ScaRb3; (4) lipid exchangers such as cholesterol ester transfer protein (CETP); and (5) enzymes such as lysolecithin–cholesterol acyltransferase (LCAT).

---

83. Intestinal efflux of cholesterol is such that:

$$\text{proximal} > \text{medial} > \text{distal egress.}$$

84. Because the term "reverse cholesterol transport" is not precise, other expressions have been suggested, based on the fact that the process deals with cholesterol efflux from macrophages followed by transport to the liver and elimination by bile flow and feces.

85. A strong inverse relationship exists between HDL–C and the risk of coronary heart disease.

In atherosclerosis, HDL level is very low and cholesteryl esters accumulate in macrophages in the subendothelial space of arterial walls to give rise to foam cells. Accumulation of cholesterol in foam cells results from uptake of modified lipoproteins with ApoB. Yet reverse cholesterol transport does not seem to be disturbed, resulting from a compensation of defective agents by others.

Therefore, HDL concentration do not control the net cholesterol transport from the cells to liver and bile. Nevertheless, the concentration of apolipoprotein is rate-limiting for cholesterol efflux out of macrophage cytosol into lipoproteins.

Activated PPARδ can affect the reverse cholesterol transport. Activation of PPARδ leads to increased levels of HDL–cholesterol in humans and enhanced apolipoprotein-A1 specific cholesterol efflux from cells of the monocytic lineage.

Macrophage Cholesterol Efflux and its Regulation

Processing of cholesterol by macrophages relies on various molecule types. The ABCa1 transporter and the HDL constituent ApoE promote macrophage cholesterol efflux. Their genes are regulated by the transcription factor LXR (Sect. 6.3.6.3).

Three distinct modes of cellular cholesterol efflux that involve HDL and its apolipoproteins have been described [556]: (1) cholesterol efflux promoted by plasma HDLs after spontaneous desorption of free cholesterol from the plasma membrane, passive aqueous diffusion through the aqueous phase, and incorporation into HDLs by collision; (2) cholesterol efflux through scavenger receptor ScaRb1; and (3) cholesterol and phospholipid efflux through ABCa1 to lipid-poor apolipoproteins such as apoA1 that binds to ABCa1.

The 2 first modes of cholesterol efflux correspond to bidirectional exchange of cholesterol between cells and HDLs. The third mode is a unidirectional transfer. These 3 modes have distinct specificity for cholesterol acceptors. Passive efflux is supported by the phospholipid content of lipoproteins. The ScaRb1 receptor can bind both apolipoproteins and HDLs, but with the greatest affinity for large, spherical HDLs. The ABCa1 transporter interacts weakly with small HDL3 and not with large HDL2 [556]. It then favors the small pool of lipid-poor apolipoproteins secreted by cells or generated by lipid exchange and lipolysis of HDLs. Addition of lipids to ApoA1 by ABCa1 serves as seeds for HDL growth that results from phospholipid transfer by phospholipid transfer protein, cholesterol esterification by LCAT, and acquisition of ApoE molecules.

Plasma Cholesterol

Cholesteryl ester transfer protein (CETP), or plasma lipid transfer protein, secreted mainly from the liver, exchanges triglycerides from very-low- (VLDL) or low-density (LDL) lipoproteins for cholesteryl esters (CE) from high-density lipoproteins (HDL), and conversely. The CETP exchanger thus promotes the transfer of cholesteryl esters from anti-atherogenic HDLs to pro-atherogenic apolipoprotein-B-containing VLDLs and LDLs, as well as VLDL remnants and IDLs. Most plasmatic cholesteryl esters are produced in HDL by lysolecithin–cholesterol acyltransferase (LCAT). Most plasmatic triglycerides originate from the liver in secreted VLDLs or

Table 6.19. Guanylate cyclase receptors (Source: [5]; StaR: heat-stable enterotoxin (Sta) receptor that is predominantly found in intestinal cells). Receptors NP_1 and NP_2 are linked to guanylate cyclases, whereas NP_3 is an unconventional Gi-protein-coupled receptor. The NP_3 receptor binds and removes natriuretic peptides from the circulation; it is hence named clearance receptor. The BNP peptide has auto-, para-, and endocrine effects that are mainly mediated via NP_1 receptor and activation of cGMP messenger.

Type	Other names	Potency order
NP_1, NPR1	NPRa, NP_A, ANPRa, GCa, GuCy2a	ANP $\geq$ BNP $\gg$ CNP
NP_2, NPR2	NPRb, NP_B, ANPRb, GCb, GuCy2b	CNP $\gg$ ANP $\geq$ BNP
NP_3, NPR3	NPRc, NP_C, ANPRc	ANP > CNP > BNP
GuCy2c	GCc, STaR	Uroguanylin > guanylin

intestine in chylomicrons. Hepatic uptake of plasma cholesteryl esters from HDL uses scavenger receptor-B1 (SRb1) and from LDL receptors (LDLR).

6.4 Guanylate Cyclase Receptors

Guanylate cyclases convert guanosine triphosphate (GTP) to cyclic guanosine monophosphate (cGMP), an intracellular second messenger. They include type-1, membrane-bound and type-2, soluble guanylate cyclases. Plasmalemmal guanylate cyclases (mGC) are transmembrane proteins that serve as natriuretic peptide receptors. Soluble, intracellular guanylate cyclase (sGC) is a receptor for nitric oxide.

6.4.1 Plasmalemmal Natriuretic Peptide Receptors

Three natriuretic peptide receptors exist (Tables 6.19 to 6.21): 2 are coupled to guanylate cyclase activity (NP_1 and NP_2, or guanylate cyclase-A [GuCy2a] and -B [GuCy2b]), whereas the third (NP_3) is the natriuretic peptide clearance receptor. The latter internalizes natriuretic peptides for their degradation (Sect. 7.13.41). In many tissues, NP_3 is the most abundant of these 3 natriuretic peptide receptors. It controls the local concentration of natriuretic peptide receptors that are available to bind their ligands and transmit signals. Another family member corresponds to the guanylate cyclase-2C (GuCy2c) receptor that is targeted by guanylin and uroguanylin. Members of the guanylate cyclase receptor family have conserved catalytic and regulatory domains, but divergent ligand-binding domains.

6.4.1.1 Natriuretic Peptides

Natriuretic peptide receptor isoforms are activated by peptide hormones (α)atrial (ANP), brain (type-B; BNP), and type-C natriuretic peptide (CNP; Vol. 2 – Chap. 1. Remote Control Cells – Sect. Endocrine System and Hormones). Type-B natriuretic

Table 6.20. Ligands and functions of natriuretic peptide receptors (Source: [704]; +: activation; −: inhibition; cAMP: cyclic adenosine monophosphate; PLC: phospholipase-C). Receptors NP_1 and NP_2 provoke effects via cGMP upon activation of a guanylate cyclase domain on their cytoplasmic regions (GCa and GCb, respectively, encoded by the GUCY2A [NPR1] and GUCY2B [NPR2] genes). The NP_1 receptor primarily binds to ANP and BNP; NP_2 preferentially to CNP. The NP_3 receptor that connects to ANP, BNP and CNP does not contain a guanylate cyclase domain.

Isoform	Main ligands	Effects
NP_1 (GCa)	ANP, BNP	Vasodilation, diuresis, natriuresis, inhibition of release of renin and aldosterone, inhibition of cell proliferation
NP_2 (GCb)	CNP	Vasodilation, inhibition of cell proliferation
NP_3	ANP, BNP, CNP	Natriuretic peptide degradation, Gi coupling (cAMP −, PLC +)

Table 6.21. Dissociation of complexes formed by natriuretic peptides and their receptors (Source: [705]).

	NP_1	NP_2	NP_3
ANP	Low	Large	Low
BNP	Moderate	Very high	Moderate
CNP	Huge	Moderate	Moderate

peptide was discovered in brain, but is expressed primarily in heart. Urodilatin is an alternatively spliced variant of proANP.[86]

Natriuretic peptides are involved in the regulation of fluid and electrolyte homeostasis, hence blood volume and pressure via their effects on kidneys and vasculature.[87]

Atrial natriuretic peptide undergoes a circadian rhythm, like renin, aldosterone, and cortisol. Acrophase[88] of ANP rhythm (rhythm peak time occurrence at ~4 h) an-

86. In 2005, natriuretic-like peptides have been discovered in venom of Australian inland taipan (Oxyuranus microlepidotus): TNPa to -c.

87. Additional functions include electrophysiology of the heart and central nervous system, muscular contraction in the gastrointestinal tract, and formation and maintenance of bone and musculoskeletal structures.

88. Time of occurrence of the peak of the circadian rhythm. The acrophase value can be the phase angle of the peak of a cosine wave fitted to the raw data obtained during the circadian rhythm. The acrophase of the actual rhythm varies from cycle to cycle.

ticipates those of other selected hormones, but happens almost simultaneously with those of blood pressure and heart rate [706]. [89]

6.4.1.2 Receptors NP_1 and NP_2

Homodimeric, catalytic natriuretic peptide receptors possess a single transmembrane domain and guanylate cyclase activity on their intracellular segment, in addition to the extracellular, glycosylated, ligand-binding region. Their intracellular domain can be further divided into a juxtamembrane region that is similar to protein kinases, hence the name "kinase homology domain" (KHD), hinge, and C-terminal catalytic domain.

In the absence of ligand, NP_1 homodimerizes or tetramerizes [704]. Binding of ANP does not cause further aggregation. The NP_2 receptor is also an oligomer in the absence of ligand. The juxtamembrane regions of natriuretic peptide receptors form intermonomeric disulfide bonds.

Both NP_1 and NP_2 require phosphorylation to be responsive to hormonal signals. Connection of ANP to phosphorylated NP_1 receptors provokes conformational changes and cleavage by peptidases of their extracellular juxtamembrane region.

Moreover, tethered hormone facilitates tethering of adenosine triphosphate to the KHD domain, thereby causing conformational changes to relieve inhibition by KHD on the catalytic domain and connecting the guanylate cyclase domains, which can then form 2 active cyclase sites per dimer [704]. Agent ATP also helps to maintain proper receptor phosphorylation status and may stabilize the oligomeric receptor.

6.4.1.3 NP_3 Receptor

The NP_3 receptor has an extracellular binding domain homologous to that of NP_1 and NP_2, but with a truncated intracellular domain that couples to inhibitory $G\alpha_{i/o}$ subunits of heterotrimeric G proteins, [90] The Gi activator sequence of NP_3 can stimulate $G\alpha_{i1}$ and $G\alpha_{i2}$ proteins, thereby inhibiting adenylate cyclase.

Furthermore, NP_3 operates via $G\beta\gamma$ dimer to activate phospholipase-Cβ, inwardly rectifying potassium channels (K_{IR}) [707].

In addition, CNP binds to NP_3 and inhibits $Ca_V1.2a$ channels in nodal myocytes of the sinoatrial node as well as atrio- and ventriculomyocytes, without influence on either inward rectifier K^+ channels in these cells, or hyperpolarization-activated cyclic nucleotide-gated channels [707]. Hence, adenylate cyclases and/or intracellular cAMP pools are compartmentalized in cardiomyocytes. Hyperpolarization-activated cyclic nucleotide-gated channels are sensitive to acetylcholine, but not C-type natriuretic peptide. Different hormone types activate distinct cAMP pools,

89. In normal subjects, circadian ANP mesors correlate with those of renin and aldosterone. The mesor is a circadian rhythm-adjusted mean based on the parameters of a cosine function fitted to the raw data of the oscillating concentration.

90. Hence its name G protein-coupled receptor. However, NP_3 is not a traditional G-protein-coupled receptor, as it does not possess 7 transmembrane-spanning domains.

possibly due to phosphodiesterases, phosphatases, and protein kinase-A-anchoring proteins.

Cardiac fibroblasts synthesize and secrete collagen-1 and -2, matrix metallopeptidases, as well as various hormones, such as BNP and CNP. They express all the 3 NP receptors. Secreted paracrine natriuretic peptides exert antifibrotic and -proliferative effects on cardiac fibroblasts. In cardiac fibroblasts ($\sim$90% of the non-myocyte population), CNP activates transient receptor potential channels of the TRPC family (Sect. 2.3) via the NP_3–Gi–PLC pathway [707].

Vascular endothelial cells synthesize natriuretic peptides. Their receptors are produced by adjacent vascular smooth muscle cells. The CNP hormone is a potent vasodilator. The pool of CNP stored in endothelial cells causes hyperpolarization, hence relaxation of smooth muscle cells [707]. Once bound to NP_3, CNP stimulates Gi and $G\beta\gamma$ dimer, and, subsequently, GIRK channels. In addition, natriuretic peptides may also prevent vascular smooth muscle cell proliferation and hypertrophy.

6.4.1.4 Regulation of Natriuretic Peptide Receptor Activity

Natriuretic peptide receptors are regulated by phosphorylation and dephosphorylation [704]. Once activated, the rate of cGMP synthesis declines with time (homologous desensitization) due to, at least partly, receptor dephosphorylation by at least 2 protein phosphatases, such as PP2 and PPM1, among others [708]. In addition, heterologous desensitization, also based on receptor dephosphorylation, is entirely independent of natriuretic peptide binding. The phosphatase responsible for both homo- and heterologous NP_1 desensitization is PP3 [710].

Desensitization of natriuretic peptide receptors relies on dephosphorylation. In addition, vasoconstrictors vasopressin, angiotensin-2, and endothelin stimulate phospholipase-Cβ that activates protein kinase-C and causes calcium-dependent increase in phosphodiesterase activity [704]. These events antagonize cGMP-mediated action of natriuretic peptides. Therefore, activities of NP_1 and NP_2 are under the permanent surveillance of the renin–angiotensin orchestrator, among other controllers.

Sphingosine 1-phosphate inhibits NP_2, which is activated by C-type natriuretic peptide [709]. Receptor desensitization mediated by S1P necessitates calcium influx. In addition, gonadotropin-releasing hormone, fibroblast (FGF2) and platelet-derived growth factor, histamine, and lysophosphatidic acid preclude the activity of NP_2 receptor.

6.4.2 Soluble Guanylate Cyclase – Nitric Oxide Receptor

Soluble guanylate cyclase (sGC), or GTP–diphosphate lyase,[91] is a heterodimer that contains α and β chains. The β chain is linked to a iron-bound heme group (hemoprotein), which binds NO for sGC activation [711].

91. GTP–diphosphate lyase (cyclizing, (3′,5′)-cGMP-forming) catalyzes the formation of guanosine (3′,5′)-monophosphate and diphosphate from guanosine triphosphate. This heterodimeric, heme-containing enzyme consists of α and β subunits. Both subunits are required for catalytic activity.

In humans, both chains have 2 subtypes. Heterodimeric hemoprotein sGC thus possesses 4 isoforms. The predominant isoform corresponds to the $\alpha 1\beta 1$ dimer. Soluble guanylate cyclase is targeted by nitric oxide, or nitrogen monoxide (Vol. 4 – Chap. 9. Other Major Signaling Mediators).

Nitric oxide regulates many physiological processes via soluble guanylate cyclase and the resulting cGMP (second messenger). The NO–cGMP pathway triggers vasodilation (Vol. 5 – Chap. 8. Smooth Muscle Cells), inhibits platelet aggregation (Vol. 5 – Chap. 9. Endothelium), impedes inappropriate growth of vascular endothelial and smooth muscle cells, and precludes adhesion of circulating leukocytes to the endothelium in normal conditions. It also intervenes in neurotransmission. This gaseous, signaling molecule belongs to the family of *gasotransmitters*.[92] In addition, nitric oxide is generated by phagocytes (mainly monocytes, macrophages, and neutrophils) as a mediator of the immune response. Nitric oxide reacts with oxygen radicals to create further damaging molecules. Nitric oxide produced by macrophages not only intervenes in killing, but also in phagocytosis. Moreover, reactive oxygen species and nitric oxide contribute to ischemic preconditioning downstream from mitochondrial ATP-dependent potassium channel (K_{ATP}, or $K_{IR}6.2$; Sect. 3.4.4) that is activated by protein kinase-C targeted by reactive oxygen species generated by the mitochondrial electron transport chain [712].[93]

The unbound enzyme produces cGMP at a low level (*basal state*). Nitric oxide binding to the sGC heme partly activates sGC enzyme (*low-activity state*). Additional NO that is not bound to sGC heme is involved in maximal sGC activation (*high-activity state*) [713].[94]

In addition to NO, hydrogen peroxide stimulates soluble guanylate cyclase and relaxes pulmonary arteries [714]. It also causes a dimerization and activation of protein kinase-G that phosphorylates vasodilator-stimulated phosphoprotein, thereby relaxing the artery wall using an additional, cGMP-independent mechanism

6.5 Adenylate Cyclases

6.5.1 Plasmalemmal, G-Protein-Regulated Adenylate Cyclases

Most adenylate cyclases (Vol. 4 – Chap. 10. Signaling Pathways) are transmembrane proteins (mAC) that are regulated by guanine nucleotide-binding proteins (Vol. 4 – Chap. 8. Guanosine Triphosphatases and Their Regulators). G-Protein subunits $G\alpha_s$ and $G\alpha_i$ stimulate and inhibit adenylate cyclases, respectively. G-Protein-regulated adenylate cyclases produce the ubiquitous signaling mediator, cyclic aden-

92. Gasotransmitters include also carbon monoxide (CO) and hydrogen sulfide (H_2S; Vol. 4 – Chap. 9. Other Major Signaling Mediators)). Other hydrophobic diatomic gases, such as N_2, CO, and O_2 do not stimulate sGC enzyme.

93. Ischemic preconditioning protects against ischemia–reperfusion injury in the myocardium. In addition to brief hypoxia and NO, acetylcholine, bradykinin, activators of protein kinase-C, and opioids also favor ischemic preconditioning.

94. A non-heme, NO-binding site is linked to sGC cysteines.

osine monophosphate (cAMP). This second messenger regulates enzymes and ion carriers, such as protein kinase-A and cyclic nucleotide-gated ion channels.

The family of adenylate cyclases encompasses many members. Adenylate cyclases AC1, AC3, and AC8 are stimulated by Ca^{++}–calmodulin; AC9 is inhibited by Ca^{++}–PP3;[95] AC5 and AC6 are inhibited by Ca^{++}. Adenylate cyclases AC2, AC4, and AC7 are stimulated by $G\beta\gamma$ subunits of G proteins; conversely, AC1 and AC8 are inhibited by $G\beta\gamma$ subunits. The activity of AC1, AC5, and AC6 is strongly impeded by Gi subunits of G proteins; that of AC2, AC3, and AC8 to a lesser extent.

6.5.2 Sensor Soluble Adenylate Cyclases

Cellular enzymes are sensitive to pH, hence to respective concentrations of hydrogen (H^+) and bicarbonate (HCO_3^-) ions, i.e., the CO_2–HCO_3^- buffer. Specialized epithelial cells of organs of the hydrogen ion control, such as the kidney with its acid- and base-secreting cells and respiratory epithelia, compensate for any acid–base metabolic and environmental disturbances using suitable sensors and ion carriers.

Soluble adenylate cyclase (sAC) in specialized epithelial cells of acid–base regulatory organs produces cAMP in response to bicarbonate ions and promotes the insertion of proton pumps into the basolateral membrane to absorb H^+ into the blood [715]. Soluble adenylate cyclase forms a regulatory complex with vacuolar proton pump (H^+ vATPase) and carbonic anhydrase in acid–base sensing cells.

6.6 Renin and Prorenin Receptors

Renin is a peptide hormone secreted by granular cells of the juxtaglomerular apparatus (close to the nephron glomerulus and distal convoluted tubule) in response to: (1) reduced blood pressure detected directly by granular cells and arterial baroreceptors; (2) decreased sodium chloride (NaCl) concentration in the distal tubule of the nephron sensed by the macula densa cells of the juxtaglomerular apparatus; and (3) sympathetic nervous stimuli on β1-adrenergic receptors of granular cells. Renin is also an aspartyl peptidase with high substrate specificity, as it has a single known substrate: angiotensinogen. Renin is synthesized as inactive prorenin and secreted as prorenin or mature renin. Receptor-bound prorenin undergoes a conformational change and becomes enzymatically active.

Mannose 6-phosphate receptor that is widely distributed can bind renin and prorenin. In cultured neonatal rat cardiomyocytes and fibroblasts, binding and endocytosis of renin and prorenin and activation of intracellular prorenin[96] is precluded by mannose 6-phosphate [716]. In cultured human umbilical vein endothelial cells, vascular uptake of prorenin and renin as well as activation of internalized prorenin

95. Protein phosphatase-3 is also called calcineurin.
96. Extracellular and surface-bound prorenin remains inactive.

following proteolytic cleavage is also impeded by mannose 6-phosphate, but neither by mannose 1-phosphate nor glucose 6-phosphate [717]. Renin-binding protein (RnBP) [97] blocks renin peptidase activity [718].

Renin binds to a cognate receptor that serves to concentrate the proteolytic activity on the cell surface and to mediate proliferation of arteriolar smooth muscle cells, as renin is neither internalized nor degraded. [98]

Renin receptor (ReR), or prorenin receptor (PRR), binds both renin and prorenin. The PRR receptor is expressed at high levels in heart, brain, placenta, and, at lower levels, in kidneys and liver [719]. It localizes in the mesangium of renal glomeruli and the subendothelial layer of coronary and kidney artery on smooth muscle cells.

The PRR–renin complex increases renin catalytic activity that converts angiotensinogen to angiotensin-1 [99] and triggers activation of P38MAPK and extracellular signal-regulated protein kinases ERK1 and ERK2 [719]. These pathways upregulate profibrotic and cyclooxygenase-2 genes. In addition, renin provokes an increase in plasminogen activator inhibitor PAI1 [720].

Prorenin and renin receptor directly interact with transcription factor zinc finger and BTB domain-containing protein ZBTB16. [100] Once activated by renin, ZBTB16 translocates into the nucleus and represses PRR transcription (negative feedback loop). However, it activates transcription of P85α subunit (PI3KR1) of phosphatidylinositol 3-kinase [721]. In addition, the renin receptor can operate in brain development and cognition.

Full-length, plasmalemmal, transmembrane prorenin receptor (mPRR) intervenes in angiotensin generation and non-proteolytic prorenin activation. It transmits renin and prorenin signals. Soluble prorenin receptor (sPRR) is created in the trans-Golgi network by furin and then shed in extracellular space and plasma, where it is able to bind renin [722].

A PRR fragment, the lysosomal, H^+-transporting adenosine triphosphatase accessory protein ATP6AP2 functions in a renin-independent manner as an adaptor for Wnt receptors low-density lipoprotein receptor-related protein LRP6 and Frizzled Fz8 and vacuolar H^+ ATPase (Sect. 10.3). [101] The PRR extracellular domain, but not its intracellular domain, can then be required for specific, LRP6-mediated Wnt sig-

97. A.k.a. ^{N}acyl ^{D}glucosamine 2-epimerase.

98. Some peptidases such as urokinase and plasmin have a plasmalemmal receptor that focuses their proteolytic activity on the plasma membrane. Urokinase receptor binds not only active enzyme, but also inactive proenzyme. Moreover, peptidase receptors, such as those of urokinase and thrombin, can also activate intracellular signaling pathways.

99. Decapeptide angiotensin-1 is then transformed in the octapeptide angiotensin-2 by soluble or endothelial cell-associated angiotensin-converting enzyme. The rate-limiting step in the renin–angiotensin system corresponds to the generation of angiotensin-1, although angiotensin-2 is the major biologically active peptide of the renin–angiotensin system. Angiotensin-2 is a potent vasoconstrictor. In hearts, the majority of angiotensin-1 is converted by chymase.

100. A.k.a. ZnF145 and promyelocytic leukemia zinc finger protein (PLZF).

101. The vATPase complex localizes to intracellular organelles and plasma membrane. It is involved in embryonic left–right patterning and phagocytosis.

naling [723]. The PRR protein can also intervene in signaling triggered by fibroblast growth factor, Nodal, and bone morphogenetic proteins [723].

6.7 Imidazoline Receptors

Imidazoline receptors (I_1–I_3) contribute to multiple processes, such as cell adhesion and proliferation, neuroprotection, pain, opioid addiction, depression, stress, appetite control, and adipose tissue composition, as well as inflammation, epilepsy, and cancer [727].

6.7.1 Ligands of Imidazoline Receptors

Endogenous ligands of imidazoline receptors include imidazole-acetic acid ribotide, agmatine, as well as harmane, harmalan, and other β-carbolines [727].

6.7.1.1 Arginine and Its Derivative Products

Intracellular arginine (intracellular concentration 0.8–2 mmol) serves as a substrate for synthesis. ^{L}Arginine (^{L}Arg) is one of the 20 common natural amino acids. It is an essential component of proteins. Extracellular arginine targets its receptor to initiate signal transduction. It is able to increase growth hormone secretion.

Arginine is a precursor of signaling molecules, such as nitric oxide (NO), agmatine (decarboxylated arginine), and glutamate, the most abundant excitatory neurotransmitter in the nervous system, as well as creatine involved in the energy metabolism, especially of myocytes and neurons. [102] Nitric oxide is a free radical molecule synthesized in all cell types from ^{L}arginine by nitric oxide synthase (NOS).

Nitric oxide operates as a neurotransmitter, mediator of immune response, [103] and as signaling molecule, especially in vascular endothelia.

Agmatine induces calcium uptake and stimulates nitric oxide synthase in endothelial cells [725]. The extracellular pool of ^{L}arginine must be mobilized and supply NOS3 sites for NO production. This event also depends on calcium ions.

Arginine and its derivative agmatine are able to bind to plasmalemmal receptors, such as non-adrenergic imidazoline and G-protein-coupled α2-adrenergic recep-

102. ^{L}Arginine decarboxylation by arginine decarboxylase forms agmatine. Arginine is also converted to ornithine and urea. Ornithine is transaminated into glutamate 5-semialdehyde. The latter is converted to glutamate.

103. Nitric oxide synthesized by inducible nitric oxide synthase in macrophages contributes to their cytotoxic activity against tumor cells, bacteria, and protozoa.

tors [726].[104] Both imidazoline receptors and $\alpha 2$-adrenoreceptors activate NOS3, using calcium influx via phospholipase-C.

6.7.2 Types of Imidazoline Receptors

Imidazoline I_1 receptor mediates the sympatho-inhibitory actions to lower blood pressure [727]. It interacts with $\alpha 2$a-adrenergic receptors in the rostral ventrolateral medulla.

Imidazoline I_2 receptor localizes to the outer membranes of mitochondria. It yields an allosteric binding site of a subset of monoamine oxidases. Other I_2-binding proteins comprise cerebral creatine kinase and soluble semicarbazide-sensitive amine oxidase [727]. Antagonists of I_2 transiently and reversibly block NMDA receptor-mediated calcium influx [728].

Imidazoline I_3 receptor participates in the regulation of insulin secretion from pancreatic β cells. Insulotropic activity at both normal and elevated glucose levels can be mediated via the closure of K_{ATP} channels [727]. Nonetheless, some agents have a glucose-dependent insulotropic function, as they directly prime insulin exocytosis, without acting on K_{ATP} channels.

6.8 Receptors of the Plasminogen–Plasmin Cascade

Plasminogen is a circulating zymogen that is cleaved (Arg561–Val562) into plasmin by 2 plasminogen activators: urokinase (uPA) and tissue (tPA) plasminogen activators. Tissue and urokinase plasminogen activators use fibrin and urokinase plasminogen activator receptor as a cofactor, respectively.

Plasmin is a serine peptidase that degrades many blood plasma proteins, particularly fibrin (fibrinolysis) and von Willebrand factor, as well as extracellular matrix constituents, such as fibronectin, laminin, and thrombospondin. In addition, plasmin activates matrix metallopeptidases, such as MMP3, MMP9, MMP12, and MMP13.

Tethering of plasminogen to the plasma membrane controls plasmin formation. Both plasminogen and plasmin contribute to inflammatory cell migration via pericellular proteolysis of matrix proteins. Plasminogen binds with moderate and high capacity to proteic and non-proteic receptors.

Conversely, the plasminogen activation system is inhibited by plasminogen activator inhibitors PAI1 (or serine peptidase inhibitor serpin-E1) and PAI2 (or serpin-B2) that prevent activity of tPA, uPA, and $\alpha 2$ antiplasmin (or plasmin inhibitor and serpin-F2; Table 6.22).

104. Endothelial cells express a number of receptors (imidazoline, bradykinin, serotonin, adenosine A_2 receptors, adrenoceptors, purinoceptors, muscarinic, histaminergic receptors, etc.). Bradykinin, histamine, acetylcholine, and adenosine enhance endothelium-dependent NO-mediated vasorelaxation. Synthesis of NO caused by acetylcholine is carried out via Gi, whereas NO production due to bradykinin is not coupled to Gi subunit. A member of C-family of G-protein-coupled receptors GPCRc6a bind arginine, lysine, and ornithine.

Table 6.22. Plasminogen activation and inhibition.

Activators	Inhibitors
Tissue plasminogen activator	Plasminogen activator inhibitors (PAI1–PAI2) (serpin-E1 and -B2)
Urokinase	α2-antiplasmin (serpin-F2)
Clotting factor XIa and XIIa	α2-macroglobulin
Kallikrein	(plasmin and kallikrein inhibitor)

6.8.1 Urokinase-Type Plasminogen Activator Receptor

Urokinase-type plasminogen activator receptor (uPAR) is a plasmalemmal protein that binds with high affinity the serine peptidase uPA, or urokinase. Its precursor is processed by removal of a C-terminal fragment and attachment of a glycosyl phosphatidylinositol anchor to the newly generated C-terminus [729]. The mature protein that is highly glycosylated is composed of 3 homologous domains of similar size (D1–D3).

The uPAR receptor associates with the external surface of the plasma membrane by a GPI anchor. Its concentration rises during inflammation and tissue remodeling such as wound healing. In tissues that experience remodeling, activated monocytes, macrophages, neutrophils, T-lymphocytes, and endothelial cells, as well as keratinocytes in skin are the predominant uPAR-expressing cell types [729].

The uPAR receptor regulates the proteolysis of extracellular matrix components by binding the extracellular serine peptidase urokinase as well as prouPA zymogen. Activated uPA cleaves plasminogen to generate plasmin that reciprocally cleaves and activates pro-uPA (positive feedback loop) [730].

Urokinase can cleave uPAR and create a soluble fragment and another fragment that remains associated with the plasma membrane or is shed away. [105] This cleavage is enhanced by the clustering of uPA–uPAR complexes. In addition, association of uPAR with cation-dependent mannose 6-phosphate (cdM6PR) and cation-independent insulin-like growth factor-2 receptor (ciM6PR or IGF2R) enhances uPA-mediated cleavage of uPAR receptor. [106]

Urokinase-type plasminogen activator receptor not only regulates extracellular proteolysis close to the cell surface, but also mediates internalization of inactive complexes between uPA and serpins PAI1, PAI2, and peptidase nexin-1 (or serpin-E2) in cooperation with receptors of the low-density lipoprotein receptor family [729]. Low-density lipoprotein receptor-related protein (LRP) is the principal clearance receptor for serpins and serpin-peptidase complexes. The uPAR receptor bound to the uPA–PAI1 complex associates with LRP1 (or α2-macroglobulin receptor) that pro-

105. Soluble fragment serves as a chemoattractant for cells thet express G-protein-coupled formyl peptide receptor-like-1 receptors, such as monocytes and basophils [730].

106. Mannose 6-phosphate actually binds 2 transmembrane proteins: cation-dependent and -independent (or IGF2R) mannose 6-phosphate receptors.

vokes clathrin-dependent endocytosis that leads to lysosomal degradation of uPA and PAI1 and subsequent recycling of uPAR and LRP1 to the plasma membrane [730].

The uPAR receptor experiences post-translational modifications [729]: (1) cleavage that causes the release of its domains D1 and D2–D3 fragment by various peptidases, such as urokinase, plasmin, neutrophil elastase, and numerous types of matrix metallopeptidases and (2) shedding that removes the entire protein from the cell surface by hydrolysis of the GPI anchor by phosphatidylinositol-specific phospholipase-D (gpiPLD) or proteolytic cleavage close to the GPI anchor by plasmin, tissue kallikrein-4, and some metallopeptidases. These 2 modifications that occur in isolation or together on a single uPAR molecule explain the existence of 5 uPAR isoforms: GPI-anchored uPAR, GPI-anchored D2–D3, soluble uPAR ($uPAR^S$), soluble D2–D3 fragment, and free D1 fragment [729].

Activity of uPAR can be triggered in the absence of uPA ligand. The uPAR receptor activates many signaling pathways to coordinate intra- and extracellular operations during cell adhesion, migration, proliferation, and survival. Because it lacks transmembrane and intracellular domains, uPAR requires transmembrane coreceptors for signaling. It actually interacts with [729]: (1) plasmalemmal and cortical proteins, such as caveolin, low-density lipoprotein receptor-related proteins (LRP and LRP1b), cation-independent mannose 6-phosphate (or insulin-like growth factor-2 receptor), growth factor (e.g., EGFR and PDGFR), cytokine, and G-protein-coupled chemokine receptors (FPR1–FPR3), as well as integrins; and (2) matrix components, such as a cleaved form of high-molecular-weight kininogen and vitronectin.[107] These interactions enable it to transmit signals to the cytoskeleton. Interaction between uPAR and vitronectin requires uPAR oligomerization [729]. On the other hand, PAI1 antagonizes vitronectin tethering to uPAR receptor.

Signaling effectors include components of the Ras–MAPK module, JaK–STAT, and PI3K–PKB pathways, as well as focal adhesion kinase FAK1, Src kinase, and small GTPase Rac [730]. The signaling cascade downstream from the uPAR–vitronectin complex is mediated by integrins and causes phosphorylation by Src kinase of the substrate domain of CAS (or BCAR1) docker and recruitment of exchange factor dedicator of cytokinesis DOCK1 that activates small GTPase Rac and provokes phosphorylation of ERK1 and -2 kinases.

In vascular smooth muscle cells, uPAR interacts with platelet-derived growth factor receptor-β. It can also link to epidermal growth factor receptor. The uPAR receptor localizes to integrin-containing adhesion complexes. Integrins confer specificity to uPAR-mediated signaling.[108] Interaction between uPAR and $\alpha_M\beta_2$-integrin can contribute to adhesion of macrophages and neutrophils to endothelial cells [730].

107. Vitronectin- and uPA-binding sites are distinct so that uPAR can simultaneously bind both ligands. Furthermore, binding of uPA to uPAR enhances vitronectin linkage to uPAR.

108. Interaction between uPAR and β_1-integrin activates focal adhesion kinase and extracellular signal-regulated protein kinase, whereas the uPAR–β_3Itg linkage stimulates Rac GTPase.

6.8.2 Plasminogen Receptors

Ca^{++}-binding, dimeric protein S100a10 that can reside inside the cell [109] as well as on the extracellular surface of cells binds to tissue plasminogen activator, plasminogen, and plasmin. The binding of tPA and plasmin to S100a10 protects against inhibitors PAI1 and α2-antiplasmin, respectively [731]. The S100a10 receptor also colocalizes plasminogen with the uPA-uPAR complex. Moreover, it complexes with annexin-A2 to form heterotetrameric plasminogen receptors [731]. [110]

Annexin-A2 faces both the cell membrane and the extracellular medium. It contains binding sites for multiple ligands, such as F-actin, fibrin, heparin, and S100a10. Annexin-A2–S100a10 heterotetramer binds tissue-type plasminogen activator (Michaelis constant $K_M \sim 0.68$ μmol), plasminogen ($K_M \sim 0.11$ μmol), and plasmin ($K_M \sim 75$ nmol) [732]. [111]

Receptors for the fibrinolytic molecules plasminogen and urokinase localize at high density to various circulating blood and transformed cell lines. Known plasminogen receptors on cells of the monocytoid lineage include, in addition to annexin-A2 and S100a10, gangliosides, histone-2, and enolase-1. Gangliosides that are components of the outer leaflets of cell membranes interact with plasminogen and urokinase and inhibit binding of both ligands to endothelial cells and granulocytes as well as that of plasminogen to platelets [733]. Histone-2B represents a regulated plasminogen receptor, at least in monocytoid cells and neutrophils [734].

Contribution of specific plasminogen receptors to monocyte recruitment may be organ- and/or stimulus-specific as well as time-dependent, some plasminogen receptors acting earlier and others later [735]. Enolase-1 is primarily an intracellular glycolytic enzyme. Plasmalemmal enolase-1 mediates lipopolysaccharide-induced invasion of monocytes into lungs [735].

6.9 Adipokine Receptors

Brown adipose tissue is mainly found in neonates. White adipose tissue contains adipocytes, pre-adipocytes, macrophages, endothelial cells, fibroblasts, and leukocytes that participate in hormonal regulation and inflammation. Adipokines (or adipocytokines) secreted by adipose tissue include hormones, such as adiponectin, apelin, chemerin, hepcidine, leptin, omentin, resistin, retinol-binding protein RBP4, vaspin, and visfatin, as well as cytokines, such as CC-chemokine ligand CCL2 (or monocyte chemoattractant protein MCP1), interleukin-6, tumor-necrosis factor-α, and plasminogen activator inhibitor PAI1.

109. Protein S100a10 regulates plasmalemmal ion channels and cytosolic phospholipase-A2.

110. Annexin-A2–S100A10 heterotetramer is composed of 2 annexin-A2 (or P36) and 2 S100A10 (or P11) subunits.

111. Phospholipid-associated annexin-A2 subunit binds plasmin ($K_M = 0.78$ μmol), but not tPA and plasminogen. Subunit S100A10 attaches to tPA ($K_M = 0.45$ μmol), plasminogen ($K_M = 1.81$ μmol), and plasmin ($K_M = 0.36$ μmol) [732].

Table 6.23. Receptors of adiponectin and their main transducers (Source: [736]; AMPK: AMP-activated protein kinase; MAPK: mitogen-activated protein kinase). Two adiponectin receptors exist: AdipoR1 and AdipoR2 that predominantly bind globular and full-length adiponectin, respectively. AdipoR1 and AdipoR2 differ in distribution and action. AdipoR1 and AdipoR2 abound in myocytes and hepatocytes, respectively. Both isoforms lodge in adipocytes.

Type	Main transducer	Ligand
AdipoR1	AMPK, MAPK	Globular adiponectin (^{G}Apn)
AdipoR2	AMPK, MAPK	Full-length adiponectin (Apn_{FL})

6.9.1 Adiponectin Receptors

Adiponectin binds to its receptors – adipoR1 and adipoR2 – that are expressed in the brain [112] and peripheral organs (Sect. 7.13.4; Table 6.23). Receptors adipoR1 and -2 abound in skeletal muscle and liver, respectively. They contain 7 transmembrane domains, but differ structurally and functionally from G-protein-coupled receptors.

The extracellular domain of adipoR1 binds to adiponectin (Sect. 7.13.4), whereas its cytoplasmic domain interacts with adaptors containing a pleckstrin homology domain, a phosphotyrosine-binding site, and a leucine zipper motif APPL1 [737]. Receptors adipoR1 and adipoR2 have a high affinity for globular adiponectin and full-length adiponectin, respectively.

Activation of adipoR1 and adipoR2 by adiponectin stimulates the activation of peroxisome-proliferator-activated receptor-α (PPARα, or NR1c1), AMP-activated protein kinase (AMPK) and P38MAPK mitogen-activated protein kinase. Stimulation of AMP-activated protein kinase in the liver and skeletal muscle strongly affects fatty acid oxidation and insulin sensitivity.

6.9.2 Apelin Receptors

Apelin is widespread in the brain, heart, blood, lungs, kidneys, liver, adrenal glands, adipose tissues, and gastrointestinal tract. Apelin binds to Gi/o-protein-coupled receptor APJ (Sect. 7.13.7) and activates the PI3K–PKB and ERK1/2 pathways. The APJ receptor also inhibits adenylate cyclase and stimulates P70 ribosomal S6 kinase S6K1 [738].

Apelin receptor is expressed at high levels in the heart, lung, adipose tissue, kidney, spleen, and in some brain areas (hypothalamic paraventricular and supraoptic nuclei). In the cardiovascular system, apelin and its receptor APJ are synthesized in endothelial cells (heart and vessels). The APJ receptor is also expressed on cardiomyocytes and some types of vascular smooth muscle cells.

112. In the central nervous system, adiponectin receptors are highly synthesized in the paraventricular nucleus of the hypothalamus, the amygdala, and area postrema, and diffusely located in the periventricular areas.

Apelin has a dose-dependent positive inotropic effect and thus belongs to the group of potent stimulators of cardiac contractility [739]. It acts via phospholipase-C and protein kinase-C as well as Na^+–H^+ exchanger isoform NHE1 and Na^+–Ca^{++} exchanger, but not Ca_V1 and K_V channels.

Apelin also causes nitric oxide-dependent arterial vasodilation [740]. Its activity in the blood pressure control is counterbalanced by angiotensin and vasopressin. Conversely, apelin inhibits vasopressin secretion. In addition, angiotensin-1 converting enzyme ACE2 that is predominantly produced by vascular endothelial cells of the heart and kidney targets apelin. In addition, apelin increases blood level of adiponectin, but decreases that of leptin.

Apelin is a 77-amino acid prepropeptide that is cleaved to shorter peptides in various tissues [741]. Several endogenous peptides of various sizes can be generated: $apelin_{36}$ (or $apelin_{(42--77)}$), $apelin_{17}$ (or $apelin_{(61--77)}$), and $apelin_{13}$ (or $apelin_{(65--77)}$). $Apelin_{13}$ and $apelin_{36}$ correspond to predominant apelin peptides. $^{pGlu}Apelin_{13}$ is the highly potent proglutamyl form of $apelin_{13}$. The rate of APJ receptor recycling is faster with $apelin_{13}$ than with $apelin_{36}$ [738]. $Apelin_{13}$ and less potent $apelin_{36}$ have cardioprotective effect via the reperfusion injury salvage kinase, as they delay mitochondrial permeability transition pore opening [742]. $Apelin_{36}$ that is secreted by adipocytes is involved in cardiovascular function and fluid homeostasis, as it inhibits vasopressin secretion, as well as in drink and food intake. $Apelin_{36}$ inhibits glucose-stimulated insulin secretion in pancreatic β cells [743].

6.9.3 Chemerin Receptors

Chemerin is involved in auto- and paracrine signaling as chemerin. Its receptor are expressed in adipocytes and preadipocytes. It acts in adipocyte differentiation [744]. It also stimulates lipolysis and insulin-stimulated glucose uptake by the GluT4 transporter [745]. Chemerin is required for expression of a subset of adipocyte genes, such as the genes ADIPOQ, LEP, PLIN, and GLUT4 that encode adiponectin, leptin, perilipin that coats lipid droplets in adipocytes, and insulin-regulated glucose transporter GluT4, respectively. Chemoattractant chemerin primes chemotaxis of tissue-resident macrophages, plasmacytoid dendritic cells, and natural killer cells to the site of inflammation.

Chemerin is produced in the white adipose tissue, liver, and lung. It is secreted as inactive prochemerin that is activated by inflammatory and coagulation serine peptidases, such as clotting factor XIIa (i.e., serine endopeptidase of the zymogen Hageman factor) and plasmin, neutrophil-derived cathepsin-G (a member of the S1 peptidase family), elastase and peptidase-3, and mastocyte secretory granule-derived serine peptidases, chymase and tryptase [746, 747].

Chemerin is a ligand for the G-protein-coupled, chemokine-like receptor CmklR1 (or ChemR23). The CmklR1 receptor is predominantly expressed in immunocytes and adipocytes. In adipose tissues, its expression is slightly higher in adipocytes than stromal vascular cells. Signaling events triggered by CmklR1 comprise calcium influx, reduction in intracellular cAMP level, and transient ERK1/2 phosphorylation [750]. Gene expression of chemerin and CmklR1 rises in adipose tissue of obese and type-2 diabetes [749]. G-Protein-coupled receptor CCRL2 that resides on mastocyte surface is also targeted by chemerin [748].

6.9.4 Leptin Receptors

Adipocytes constitute the primary source of leptin that regulates food intake and chemical energy use. Leptin production is stimulated by indicators of acute nutritional status such as insulin as well as fat content possibly via auto- and paracrine processes mediated by glucocorticoids and inhibited by counter-regulatory hormones.

Leptin receptors signal via 3 intracellular Tyr residues that are phosphorylated after leptin binding and mediate leptin action. Signaling components link to these phosphorylated residues. Among them, one attracts Janus (pseudo)kinase JaK2 and signal transducers and activators of transduction (STATs). Another one (Tyr985) transduces both positive and negative signals [751]. Connection to suppressor of cytokine signaling protein SOCS3 indeed impedes STAT3 activation. Furthermore, interaction with protein tyrosine phosphatase PTPn11 antagonizes the JAK–STAT signaling.

Leptin acts both as an endo- and paracrine mediator, i.e., targets remote and close cognate receptors. Leptin receptor (LepR) that is encoded by a single LEPR gene and belongs to type-1 cytokine-like receptor class has 6 alternatively spliced isoforms (LepRa–LepRf or ObRa–ObRf [Ob: obese]). Short (LepRa) and long (LepRb) receptor forms contain identical extracellular and transmembrane domains as well as the same segment of the intracellular domain. Leptin receptor LepRb, as are other cytokine receptors, is devoid of intrinsic enzymatic activity and relies upon the activity of constitutively associated Janus kinases to mediate intracellular signaling. Leptin regulates energy homeostasis via the central activation of multiple signaling pathways mediated by the LepRb receptor. Leptin action is indeed relayed by phosphorylation of several tyrosine residues on the LepRb receptor. Mediator $LepRb_{Tyr985^P}$ operates as an age- and diet-dependent regulator that counteracts age-associated or diet-induced obesity [751].

The basomedial hypothalamus, where leptin acts on neurons that regulate the concentration of circulating hormones (e.g., growth, thyroid, and reproductive hormones) and activity of the hypothalamus–pituitary–adrenal axis, has the highest LepRb level. In response to leptin, neurons can change the number and type of their input synapses.

Only the long form (LepRb) associates with Janus kinase JaK2. Moreover, isoform LepRb is the unique type that can signal intracellularly via the JaK–STAT and MAPK pathways. Isoform LepRb initiates signaling via 3 main mechanisms [752]: (1) recruitment of cytosolic protein Tyr phosphatase PTPn11 to regulate extracellular signal-regulated kinase; (2) recruitment of signal transducer and activator of transcription STAT3 to blast off a feedback inhibition mediated by suppressor of cytokine signaling SOCS3 as well as to produce positive effectors of leptin action; and (3) recruitment of signaling mediators by receptor-associated, phosphorylated JaK2. In fact, during leptin signaling, phosphorylated tyrosine residues on the long form of LepRb mediate distinct signals. Phosphorylated Tyr1138 binds STAT3 to mediate its tyrosine phosphorylation and transcriptional activation, whereas phosphorylated Tyr985 associates with PTPn11 to stimulate extracellular signal-regulated protein kinase and inhibit LepRb-mediated STAT3 activation.

The JaK2-dependent activation of the IRS–PI3K pathway [113] regulates the transmembrane potential in LepRb-expressing neurons. The LepRb–STAT3 pathway controls the hypothalamic melanocortin pathway, but does not strongly regulate neuropeptide-Y. As leptin activates JaK2 and STAT3, it increases levels of anorexigenic peptides, such as α-melanocyte-stimulating hormone and cocaine- and amphetamine-regulated transcript and inhibits orexigenic peptides, such as neuropeptide-Y and agouti-related peptide.

Suppressor of cytokine signaling SOCS3 and Tyr phosphatase PTPn1 in neurons (but not in adipocytes, myocytes, and hepatocytes) block leptin signaling [753, 754]. Residue Tyr985 of LepRb not only binds PTPn11 for ERK activation during acute stimulation, but also serves as a node for feedback inhibition of LepRb signaling by binding to LepRb-induced SOCS3. Increased Ser523 phosphorylation of cognate LepRb-associated Janus kinase JaK2 prevents JaK2 activity, hence also contributing to repression of leptin action [755].

6.9.5 Omentin Receptors

Omentin specific to omental fat tissue can be secreted in blood [756]. [114] Omentin is predominantly expressed in visceral, but not in subcutaneous adipose tissue. Adipose tissue stromal cells yield the main source of omentin. Omentin increases phosphorylation of protein kinase-B in the absence and presence of insulin. Omentin can enhance insulin-mediated glucose uptake in adipocytes.

Omentin is identical in amino acid sequence to galactofuranose-binding intelectin-1, a member of the family of X-lectins [758]. [115]

---

113. IRS: insulin receptor substrate; PI3K: phosphatidylinositol 3-kinase.

114. Omental adipose depot originates near stomach and spleen. The omentum is the peritoneum endowed with adipose tissue. The greater omentum is the largest parietal peritoneal fold that is constituted by 4 layers of peritoneum (double sheet of peritoneum). It drapes off the stomach and hangs down underneath the muscles, passing in front of the small intestine, and partly attaches to the transverse colon. The small gastrohepatic omentum is very thin. This double layer of peritoneum extends from the liver to the lesser curvature of the stomach and the entrance segment of the duodenum. Omentin is predominantly found in epicardial and omental human adipose tissue, whereas visfatin is detected to the same extent in epicardial, omental, and subcutaneous fat [757].

115. Lectins mediate carbohydrate recognition. Six major categories of lectins exist: (1) Ca^{++}-dependent lectins (C-type lectins); (2) galectins (S-type lectins); (3) mannose 6-phosphate receptors (P-type lectins); (4) siglec and other immunoglobulin-like sugar-binding lectins (I-type lectins); (5) lectins related in sequence to the leguminous plant lectins (L-type lectins); and (6) lectins that bind carbohydrates only in the presence of Ca^{++} (X-type lectins first studied in oocytes and embryos of Xenopus laevis). They localize to the plasma membrane, cytoplasm, and nucleus. At the cell surface, lectins can act as receptors for intercellular adhesions as well as receptors for elastin, laminin, glycosaminoglycans, and circulating glyco-

Omentin receptor remains to be fully determined. A possible omentin receptor correspond to the EST fragment Loc401397, but the full-length gene product has not been yet cloned.

6.9.6 Resistin Receptors

Resistin is encoded by the gene RETN. It is secreted by adipocytes as well as immune and epithelial cells. [116] Resistin is involved in energy homeostasis and inflammation. It increases expression of several pro-inflammatory cytokines and adhesion molecules.

A non-glycanated decorin isoform (ΔDcn) [117] serves as a resistin receptor on adipose progenitor cells. Interaction between resistin and ΔDcn controls the proliferation, migration, and differentiation of adipose stromal cells that operate as mesenchymal progenitors in white adipose tissue [759].

6.9.7 Visfatin Receptors

Visfatin is mainly an adipocyte hormone. This endo-, auto-, and paracrine peptide binds to *insulin receptor* at a site distinct from that of insulin. It reduces glucose release from hepatocytes and stimulates glucose utilization in adipocytes and myocytes. Visfatin synthesis in the brain, kidney, lung, spleen, testis, and predominantly visceral adipose tissue is heightened by hyperglycemia, hypoxia, and inflammation; it is attenuated by insulin and somatostatin.

6.10 Chemosensors of Olfaction and Taste

In the nervous system, chemosensors convert chemical cues into electrical signals. In mammals, chemosensory cells use metabotropic chemosensory receptors, in addition to ligand-gated ion channels. Ligand-bound receptors indirectly activate ion channels via second messengers, hence initiating signaling that is characterized by a long latency (ranging from a few tens [using proper scaffolds] to several hundred ms), i.e., the duration required to produce second messengers and activate involved effectors. On the other hand, the multicomponent nature of metabotropic signaling

---

proteins. The family of X-lectins is constituted by several mammalian homologs of Xenopus laevis oocyte lectin XL35: galactofuranose-binding intelectin-1 (ItLn1 or HL1), intelectin-2 (ItLn2 or HL2), endothelial lectin, intestinal lactoferrin receptor (LFR on the brush border membrane of the small intestine), and omentin. Lactoferrin is an iron-binding glycoprotein.

116. Resistin is also known as adipocyte-specific secretory factor (ADSF)or CEBPE regulated myeloid-specific secreted cysteine-rich protein precursor XCP1. It has been identified during insulin resistance.

117. Cleavage product of decorin (Dcn) that lacks the glycanation site. Decorin belongs to the small leucine-rich proteoglycan (SLRP) family. It contains a glycosaminoglycan chain, which is either chondroitin (CS) or dermatan (DS) sulfate (Vol. 1 – Chap. 8. Cell Environment).

allow a more sophisticated regulation of neuronal currents in response to chemical signals than that primed by short-latency, ionotropic chemosensory receptors.

In the sensory neuroepithelium, G-protein-coupled olfactory receptors on dendrites of olfactory sensory neurons in the main olfactory or auxiliary vomeronasal sensory epithelia identify odorants. The vomeronasal organ in the nasal septum is the site of pheromone [118] detection in mammals. Sensory neurons in the vomeronasal organ are compartmented structurally and functionally into 2 distinct layers: (1) neurons of the apical layer that express pheromone receptors of the V_1R family and (2) neurons of the basal layer that contain pheromone receptors of the unrelated V_2R family. Various families of 7-transmembrane olfactory sensors encompass [760]: (1) chemosensory receptors that are expressed in olfactory sensory neurons in the main olfactory epithelium, i.e., (1.1) $G\alpha_{olf}$-coupled odorant receptors, [119] and (1.2) trace amine-associated receptors, as well as (2) chemosensory receptors in the vomeronasal organ that comprise (2.1) type-1 and -2 vomeronasal receptors (V_1R and V_2R) [120] and (2.2) vomeronasal formyl peptide receptor-related proteins that are associated with pathogen identification [761].

Most mammalian receptors devoted to taste perception are also G-protein-coupled receptors [760] (Chap. 7). Members of the T_1R family (T_1R1–T_1R3) recognize sweet tastants, and those of the T_2R family (T_2R1–T_2R64 in humans) for bitter taste detection. Many T_2R types can be coexpressed in a given taste receptor cell. Mammalian gustatory G-protein-coupled receptors heteromerize.

118. φερω: to bear, bring, carry, convey; ορμη: attack, effort, impetus, rush, start (ορμηπος: set in motion; ορμησις: rapid motion).

119. Odorant receptors trigger a $G\alpha_{olf}$–AC3–cAMP–CNG–Ca^{++} cascade that elicits the opening of calcium-gated chloride channels for chloride efflux.

120. Type-1 and -2 vomeronasal receptors (V_1R and V_2R) should not be mistaken for type-1a, -1b, and -2 arginine vasopressin receptors (V_{1A}, V_{1B}, and V_2).

7

G-Protein-Coupled Receptors

Receptors coupled to heterotrimeric guanine nucleotide-binding proteins, the so-called G-protein-coupled receptors (GPCR), represent the largest set of plasmalemmal receptors (> 800 members in the human genome) . All GPCRs, also called 7-transmembrane and heptahelical receptors, possess a 7-transmembrane (7TM) scaffold that is adapted to different sets of ligands. The 7 membrane-spanning α-helical segments are separated by alternating intra- and extracellular loops.

7.1 Introduction

Ubiquitous G-protein-coupled receptors transmit the majority of signals from various types of messengers, such as hormones, neurotransmitters, and sensors. They can operate as allosteric signaling mediators. Many GPCRs display various levels of G-protein activation.

G-Protein-coupled receptors are actually implicated in diverse physiological processes, such as hormonal signaling, neurotransmission, cognition, chemosensation, photoreception (vision), taste, olfaction (Sect. 6.10), and pain perception, as well as ensure the cardiac and other physiological functions, metabolism, and cell growth, differentiation, and migration, in addition to inflammation and immunity.

7.1.1 Agonists vs. Antagonists

Ligands bind to transmembrane segments and extracellular domains of their cognate receptors. The basal activity of a GPCR can be activated by *agonists* or inhibited by *antagonists*. Agonist binding at the extracellular side triggers a set of conformational changes of the target GPCR at the intracellular side, where a heterotrimeric G protein and other effectors can then tether to initiate the signaling cascade.

A *neutral antagonist* prevents agonist binding, but do not affect GPCR constitutive (agonist-independent) activity; an *inverse agonist* precludes the constitutive activity of its specific receptors.

Agonist binding does not always cause important conformational changes. Moreover, agonist subtypes can be classified according to their effect on the receptor structure [762]. Some ligands stabilize a given receptor conformation (conformationally selective ligands). Others modify the receptor structure among multiple receptor conformations and shift the dynamical equilibrium.

Human adenosine A_{2A} receptor and β-adrenergic receptors possess binding sites that can bind specifically to agonists rather than antagonists and inverse agonists [763]. Inverse agonists can bind to a given GPCR to prevent the conformational change that primes its activation. An intermediate conformation between the inactive and active states is actually characterized by an occlusion of the G-protein-binding site.

7.1.1.1 Full and Partial Agonists

In addition to neutral antagonists and inverse agonists, full and partial agonists prime various rearrangements of the receptor structure that initiate signaling via G-proteins and/or arrestins (Sect. 7.11). [1] *Full agonists* are capable of maximal receptor stimulation, whereas *partial agonists* are unable to elicit full activity even at high concentration. A given conformation stabilized by a ligand thus determines the efficacy toward a specific pathway.

In summary, a ligand can operate as: (1) a full agonist that is capable of activating the receptor, thereby causing an a priori maximal response; (2) partial agonist that does not activate its cognate receptors thoroughly, hence provoking partial response; (3) inverse agonist that reduces the activity of its receptor by inhibiting their constitutive activity; and (4) antagonist that binds to the receptor, blocks action of agonists, as it prevents access to the receptor by its binding.

7.1.1.2 Dual Agonist and Antagonist Activity

In addition, many GPCRs can stimulate multiple signaling systems, and a given ligand can excite different pathways with different relative efficiency up to opposite activity. For example, agonists of β2-adrenoceptors for the arrestin–MAPK pathway act as antagonists for the Gs–cAMP–PKA axis.

7.1.2 Alternative Splicing of G-Protein-Coupled Receptors

Alternative splicing diversifies the receptorome and augments the number of GPCR types encountered on cell surfaces, hence heterogeneity in GPCR signaling. In humans, airway smooth muscle cells are endowed with multiple types (>350) of G-protein-coupled receptors. More than 190 GPCRs have, on average, 5 different receptor isoforms that can result from various ways of alternative splicing of pre-mRNAs [764] (Vol. 1 – Chap. 5. Protein Synthesis).

1. A huge number ($O[100]$) of GPCR types are coupled to many kinds ($O[10]$) of G proteins and several members ($O[1]$) of the arrestin family.

Table 7.1. Types of subunits of guanine nucleotide-binding (G) proteins and triggered pathways (Source: [765]; ACase: adenylate cyclase; cAMP: cyclic adenosine monophosphate; DAG: diacylglycerol; cGMP: cyclic guanosine monophosphate; GEF: guanine nucleotide-exchange factor; IP_3: inositol (1,4,5)-trisphosphate; PDE: phosphodiesterase; PI3K: phosphatidylinositol 3-kinase; PIP_3: phosphatidylinositol triphosphate; PKC: protein kinase C; PKG: protein kinase G; PLC: phospholipase C; $\ominus\!\longrightarrow$: inhibition).

Type	Targets and signaling axis
	Gα subunit
$G\alpha_{i1/2/3}$	$\ominus\!\longrightarrow$ACase
$G\alpha_{i1/2}$	cGMP–PKG–PDE
$G\alpha_{o1/2}$	$\ominus\!\longrightarrow$ACase
$G\alpha_t$	PDE6–cGMP
$G\alpha_z$	K^+ channel closure, $\ominus\!\longrightarrow$exocytosis
$G\alpha_q$	PLCβ–IP_3–Ca^{++} PLCβ–DAG–PKC
$G\alpha_{11}$	PLCβ
$G\alpha_s$	ACase–cAMP
$G\alpha_{12/13}$	RhoGEF–RhoA
$G\alpha_{14/16}$	PLCβ
	Gβγ dimer
Gβγ	ACase–cAMP PLC–IP_3/DAG Rac-cytoskeleton CDC42-cytoskeleton PI3K–PIP_3 GIRK (depolarization) Ca_V1.2–Ca^{++} influx

7.1.3 GPCR–G-Protein Coupling

Monomeric GPCR embedded in the lipid bilayer[2] is the minimal, but not necessarily optimal, element required to couple heterotrimeric guanine nucleotide-binding (G) proteins that are composed of a Gα subunit and Gβγ dimer (Vol. 4 – Chap. 8. Guanosine Triphosphatases and Their Regulators; Table 7.1). In humans, at least 18 types of Gα, 5 types of Gβ, and 11 types of Gγ subunits exist.

Binding of GPCR to Gα subunit of G proteins especially targets the receptor-binding region composed of the Ras-like domain and possibly its close sequences on the C-terminal region [766].

2. Gαβγ trimers may interact with: (1) the lipid bilayer via the lipid modifications at the N-terminus of Gα and C-terminus of Gγ as well as (2) the cavity formed in the active GPCR via the C-terminus of Gα [766].

Formation and stability of a GPCR–G-protein complex depend on GPCR agonist efficacy and the presence or absence of cytosolic guanine nucleotides. Conversely, signal deactivation is achieved by Gα-mediated GTP hydrolysis (GTPase activity) that is enhanced by the GTPase-accelerating activity of regulators of G-protein signaling (RGS).

The GPCR receptors and G proteins may exist in a precoupled state, via direct, more or less short-lived GPCR–G-protein interactions or scaffold proteins. Constitutive activity of many GPCRs occurs in the absence of agonists. Whereas neutral antagonists inhibit agonist binding, but do not affect GPCR constitutive (agonist-independent) activity, inverse agonists block the constitutive activity. Inverse agonists can prevent the formation of GPCR–G-protein complex and destabilize preformed complexes.

7.1.3.1 Allosteric Activation

Activation of GPCR happens via molecular switches, such as breaking of a hydrogen bond and electrostatic interactions that stabilize the basal state of the receptor, as well as conformational change of amino acid residues. Some switches are interdependent. Different ligands can activate specific sets of switches, as they disrupt various intramolecular interactions, thereby triggering different receptor conformations and distinct effects on signaling mediators. Receptor conformation destabilization and full activation time scales are $\mathcal{O}[10\,\text{ns}]$ and $\mathcal{O}[1\,\text{ms}]$, respectively.

7.1.3.2 GDP- and GTP-Binding Cycle of G-Proteins

Guanine nucleotide-binding heterotrimers regulate signal transduction via a cycle of GDP- and GTP-bound states. Once the ligand is bound, activated GPCRs serve as guanine nucleotide-exchange factors (GEF) that promote GTP binding to heterotrimeric G proteins, hence activating these G proteins (Table 7.2). Interactions between GPCRs and G proteins represent the signaling interface from which signals are transmitted within the cell.

7.2 GPCR Ligands

G-protein-coupled receptors are activated by multiple agonists (Table 7.3). Responses that result from GPCR activation can integrate several intracellular signaling pathways. Ligands are known for multiple GPCRs [767] (Tables 7.4 to 7.11). G-protein-coupled receptors that do not possess any yet identified endogenous ligand are called orphan GPCRs.

7.3 Adhesion G-Protein-Coupled Receptors

Adhesion G-protein-coupled receptors are plasmalemmal molecules that possess a large extracellular domain coupled to a multitransmembrane region. The G-protein-coupled receptor-proteolytic site (GPS) gives rise to an autocatalytic processing of

Table 7.2. Active, GTP-bound G protein and triggered signal (Source: [765]; AMPAR: AMPA-type glutamate receptor; CFTR: cystic fibrosis transmembrane conductance regulator; CNG: cyclic nucleotide-gated channel; FBP(2)P: fructose (2,6)-bisphosphate 2-phosphatase; PDE: phosphodiesterase; PhK: phosphorylase kinase; PLn: phospholamban; RyR: ryanodine-sensitive Ca^{++} channel; SERCA: sarco(endo)plasmic reticulum calcium ATPase).

Cytoskeleton remodeling
G protein^{GTP}–RhoGEF–Rho
Ion channel modulation
Redox signaling via NOx and ROS
MAPK signaling
PI3K signaling
PLC/PLD signaling
cAMP signaling
G protein^{GTP}–ACase–cAMP–PDE G protein^{GTP}–ACase–cAMP–ABCc4–cAMP efflux
G protein^{GTP}–ACase–cAMP–CNG
G protein^{GTP}–ACase–cAMP–RapGEF3/4–Rap1–PLCε
cAMP–PKA axis
Ion fluxes
G protein^{GTP}–ACase–cAMP–PKA–Ca$_V$1.1/1.2
G protein^{GTP}–ACase–cAMP–PKA–RyR
G protein^{GTP}–ACase–cAMP–PKA–Plb–SERCA
G protein^{GTP}–ACase–cAMP–PKA–CFTR
G protein^{GTP}–ACase–cAMP–PKA–AMPAR
G protein^{GTP}–ACase–cAMP–PKA–PDE
Transcription
G protein^{GTP}–ACase–cAMP–PKA–CREB
Metabolism
G protein^{GTP}–ACase–cAMP–PKA–lipase
G protein^{GTP}–ACase–cAMP–PKA–PhK
G protein^{GTP}–ACase–cAMP–PKA–FBP(2)P

the polypeptide into an extracellular α and a membrane-spanning β chain that associate at the plasma membrane.

7.3.1 EGF-TM7 Class Members

Some adhesion G-protein-coupled receptors contain 3 to 5 consecutive epidermal growth factor (EGF) modules linked via a mucin-like spacer to a 7-span transmembrane class-A G-protein-coupled receptor. Receptors of the EGF-TM7 category are expressed predominantly on immunocytes.

First identified members of the EGF-TM7 class comprise CD97 and EGF-like module-containing, mucin-like, hormone receptor-like proteins EMR1 to EMR4

Table 7.3. Various GPCR ligands activate directly and indirectly cytoplasmic and nuclear signaling mediators using G-protein-dependent and -independent pathways (Source: [768]). Some members of the GPCR superclass (e.g., α- and β-adrenergic, muscarinic cholinergic, angiotensin-2, and endothelin-1 receptors) mediate the effect of various hormones and neurotransmitters (e.g., adrenaline, noradrenaline, acetylcholine, angiotensin-2, and endothelin-1) that control the activity of the cardiovascular and ventilatory systems. The GPCR receptors bind not only to their ligands (also called agonists or stimulators), but also associate with heterotrimeric G proteins to exert signaling and inhibitors serine/threonine GPCR kinases and β-arrestins for signaling termination. Yet, β-arrestins also act as signal transducers and adaptors.

Biogenic amines	Adrenaline, noradrenaline, dopamine, serotonin, acetylchonine, histamine
Amino acids	Glutamate, γ-aminobutyric acid
Peptides	Angiotensin, bradykinin, thrombin, bombesin, endorphins, follicle-stimulating hormone, leuteinizing hormone, thyroid-stimulating hormone gastrin-releasing peptide, cholecystokinin, neuromedin-B, neurotensin, vasopressin, galanin
Lipids	Lysophosphatidic acid, sphingosine-1-phosphate, prostaglandins, leukotrienes, platelet-activating factor, anandamine
Ions	Calcium
Nucleosides	Adenosine
Nucleotides	ADP, ATP, UDP, ^{UDP}glucose, UTP
Miscellaneous	Light, odorants, pheronomes

(Vol. 1 – Chap. 7. Plasma Membrane). This class also comprises: (1) brain-specific angiogenesis inhibitor BAI1 to BAI3; (2) cadherin, EGF-like, LAG-like, and seven-pass receptors CELSR1 to CELSR3; (3) latrophilins Lphn1 to Lphn3; and (3) orphan G-protein-coupled receptors GPR56, GPR64, GPR97, GPR110 to GPR116, and GPR123 to GPR128.

Leukocyte-restricted adhesion G-protein-coupled receptors of the EGF-TM7 class include mainly CD97 and EMR1 to EMR4 that undergo alternative splicing. Myeloid-restricted EGF-like module-containing, mucin-like, hormone receptor-like protein EMR2 contains a GPS in the membrane-proximal region. Following the cleavage (Leu517–Ser518), which is independent of the transmembrane domains, the non-covalent association of the resulting extracellular α subunit and transmembrane β subunit can occur [771].[3]

3. The GPS is necessary, but not sufficient for receptor cleavage, which requires the entire extracellular segment. An alternatively spliced EMR2 isoform with a truncated extracellular segment fails to undergo proteolysis.

Table 7.4. Ligands of G-protein-coupled receptors: (**Part 1**) amino acids, dicarboxylic acids, and biogenic amines (Source: [767]; GABA: γ-aminobutyric acid).

Ligand	Receptor	Gα
	Amino acids	
^{L}Arginine, ^{L}lysine	GPRc6a aGq/11	
Glutamate	$mGlu_{1,5}$	Gq/11
	$mGlu_{2,3,4,6,7,8}$	Gi/o
GABA	$GABA_{B1}$ (binding)	Gi/o
	$GABA_{B2}$ (signaling)	
	Dicarboxylic acids	
α-Ketoglutarate	GPR99	Gq/11
Succinate	GPR91	Gq/11, Gi/o
	Biogenic amines	
Acetylcholine	M_1,M_3,M_5	Gq/11
	M_2, M_4	Gi/o
Adrenaline,	α1a,1b,1d	Gq/11
noradrenaline	α2a,2b,2c	Gi/o
	β1,2,3	Gs
Dopamine	D_1,D_5	Gs
	D_2,D_3,D_4	Gi/o
Histamine	H_1	Gq/11
	H_2	Gs
	H_3,H_4	Gi/o
Melatonin	MT_1,MT_2,MT_3	Gi/o
Serotonin	$5HT_{1A/1B/1D/1E/1F}$	Gi/o
	$5HT_{2A/2B/2C}$	Gq/11
	$5HT_4$,$5HT_6$,$5HT_7$	Gs
	$5HT_{5A/5B}$	Gi/o, Gs
Trace amines	TA_1, TA_2	Gs

The vasculature can experience an organ-specific development. In particular, the specialized vasculature of the central nervous system possesses a strong limitation in molecular permeability through the strongly polarized,[4] tightly sealed endothelial cells, extensive pericyte coverage, and reciprocal interactions with neurons and glial cells. Moreover, it acts as a neural stem cell niche. The blood–brain barrier is able to insulate the central nervous system from substances that can be tolerated by peripheral organs. The pro-angiogenic, endothelial, adhesion G-protein-coupled receptor GPR124[5] is involved in the development of endothelial cells and pericytes of the

4. Molecular carriers are heterogeneously distributed between the luminal and abluminal edges of the plasma membrane.

5. A.k.a. tumor endothelial marker-5 (TEM5). During mouse embryogenesis, GPR124 is expressed in both endothelial cells and pericytes, mainly in the neural tube, and, to a lesser extent, the heart, liver, and kidney, as well as the epithelium of embryonic lung and esophagus

Table 7.5. Ligands of G-protein-coupled receptors: (**Part 2**) ions, nucleotides, and nucleosides (Sources: [767, 769, 770]). Proton-sensing G-protein-coupled receptors sense acidic pH. The GPR4 receptor (or G2 accumulation protein [G2A]) has a low affinity for lysophosphatidylcholine (LPC) and high affinity for sphingosylphosphorylcholine (SPC). On the other hand, GPR132 has a high affinity for LPC and low affinity for SPC. Receptors GPR65 and GPR68 are also called T-cell death-associated gene-8 protein (TDAG8) and ovarian cancer GPCR OGR1 as well as SPC receptor SPC_1, respectively. The GPR4 receptor is involved in angiogenesis primed by SPC in endothelial cells via phosphatidylinositol 3-kinase, protein kinase-B, and vascular endothelial growth factor receptor VEGFR2.

Ligand	Receptor	Gα
	Ions	
Ca^{++}	CaSR	Gq/11, Gi/o
H^+	GPR4, GPR65	Gq/11, G12/13
	GPR68, GPR132	Gs
	Nucleotides and nucleosides	
Adenosine	A_1, A_3	Gi/o
	A_{2A} A_{2B}	Gs
ADP	$P2Y_{12}$, $P2Y_{13}$	Gi/o
ADP, ATP	$P2Y_1$	Gq/11
ATP	$P2Y_{11}$	Gq/11, Gs
UDP	$P2Y_6$	Gq/11
^{UDP}glucose	$P2Y_{14}$	Gi/o
UTP, ATP	$P2Y_2$, $P2Y_4$	Gq/11

cerebral vasculature, particularly during CNS-specific angiogenesis that forms the perineural, then periventricular vascular plexi in the forebrain, ventral neural tube, and spinal cord [772, 773].[6] During adulthood, the brain endothelium remains sensitive to GPR124 receptor.

The recruitment of blood-circulating phagocytes to sites of inflammation and infection begins with activation and interactions of cell adhesion receptors on both leukocytes and endothelial cells. Phagocyte activation is regulated partly by cell-

and mesenchyme. In adult mice, it is exclusively produced in the vasculature with more developed pericyte coating, i.e., in the brain and kidney, pancreas, and corpus luteum [772]. The GPR124 receptor is required for invasion and migration of blood vessels into the neuroepithelium, establishment of the blood–brain barrier, and expansion of the cerebral cortex [773].

6. Deletion of GPR124 causes angiogenesis arrest; overexpression hyperproliferative vascular malformations [772]. The GRP124 receptor regulates CDC42-dependent angiogenic migration and sprouting. Mutations of genes of the Wnt–βCtn and VEGF–Nrp1 pathways, of the helix–loop–helix transcriptional repressors, the inhibitors of DNA binding ID1 to ID3 required to maintain the timing of neuronal differentiation in the embryo and invasiveness of the vasculature [774] (ID1 is an inhibitor of angiogenesis via thrombospondin-1 [775]; ID2 is a target of retinoblastoma protein sufficient and necessary for the production of VEGF [776]), and integrin-α_V and -β_8 also impair angiogenesis in the central nervous system, but without CNS tropism.

Table 7.6. Ligands of G-protein-coupled receptors: (**Part 3**) lipids (Source: [767]; 2AG: 2-arachidonyl glycerol; 5oxoETE: 5-oxo (6,8,11,14)-eicosatetraenoic acid; AEA: ^{N}arachidonoyl ethanolamine (anandamide); G_{Tc}; cone-transducin [G_{T2}]; G_{Tr}: rod-transducin [G_{T1}]; LC-, SCFA: long-, short-chain [<6 carbons] fatty acid; OxER: oxoeicosanoid [OXE] receptor; PAF; platelet-activating factor). Agent 11-cis-retinal covalently binds receptors for light-dependent activation. Agent 5oxoETE is produced by oxidation of 5HETE (HETE: hydroxyeicosatrienoic acid) by 5-hydroxyeicosanoid dehydrogenase (5-lipoxygenase pathway).

Ligand	Receptor	Gα
AEA, 2AG	CB_1, CB_2	Gi/o
11-cis-Retinal	Rhodopsin	G_{Tr}
	Opsins	G_{Tc}
	Melanopsin	Gq/11
	Fatty acids (FA)	
SCFA: C2–C5	GPR41, GPR43	Gi/o, Gq/11
LCFA: C12–C20	GPR40	Gq/11
LCFA: C14–C22	GPR120	Gq/11
5OxoETE	OxER1, GPR170	Gi/o
	Leukotrinenes	
LTB4	BLT	Gi/o
LTC4, LTD4	$CysLT_1$, $CysLT_2$	Gq/11
LXA4	FPRL1 (ALXR)	Gi/o
Lysophosphatidic acid	LPA_1–LPA_3	Gi, Gq/11, G12/13
PAF	PAF	Gq/11
	Prostaglandins	
PGI2	IP	Gs
PGD2	DP	Gs
	$CRTH_2$	Gi
PGF2α	FP	Gq/11
PGE2	EP_1	Gq/11
	EP_2, EP_4	Gs
	EP_3	Gs, Gq/11, Gi
TxA2	TP	Gq/11, G12/13
Sphingosine-1-phosphate	$S1P_1$–$S1P_5$	Gi, Gq/11, G12/13
Sphingosylphosphorylcholine	SPC_1, SPC_2	Gi

surface receptors such as adhesion G-protein-coupled receptors. Epidermal growth factor-like module-containing, mucin-like, hormone receptor-like protein EMR2 is synthesized by neutrophils, monocytes, macrophages, and dendritic cells. It binds to chondroitin sulfate on cells and tissue matrix. It regulates the neutrophil response [777]. It potentiates the effects of numerous pro-inflammatory mediators. Upon neutrophil activation, EMR2 is rapidly translocated to membrane ruffles and the leading edge of the migrating cell and its production can rise. Ligation of EMR2 boosts the activation and recruitment of neutrophils, increases neutrophil adhesion

Table 7.7. Ligands of G-protein-coupled receptors: (**Part 4.1**) peptides and proteins (Source: [767]; CG: chorionic gonadotropin [gonadotrophin] CGRP: calcitonin gene-related peptide; CRF: corticitropin-releasing factor; FSH: follicle-stimulating hormone).

Ligand	Receptor	Gα
Adrenocorticotrophin	MC_2	Gs
Adrenomedullin	AM_1, AM_2	Gs
Amylin	AMY_1–AMY_3	Gs
Angiotensin-2	AT_1	Gq/11, G12/13, Gi/o
	AT_2	
Apelin	APJ	Gi/o
Bradykinin	B_1, B_2	Gq/11
Calcitonin	CT	Gs, Gq/11
CGRP	$CGRP_1$	Gs, Gq/11
	Chemokines	
CC	CCR1–CCR10	Gi/o
CXC	CXCR1–CXCR6	Gi/o
CX_3C	XCL1, XCL2, CX_3L1	Gi/o
Cholecystokinin (CCK8)	CCK_1, CCK_2	Gq/11, Gs
Complement C3a, C5a	C3a, C5a	Gi/o
CRF, urocortin	CRF_1, CRF_2	Gs
Endothelin-1, -2	ET_A	Gq/11, G12/13, Gs
Endothelin-1, -2, -3	ET_B	Gq/11, G12/13, Gs
FSH	FSH	Gs
Formyl-Met-Leu-Phe	FPR	Gi/o

and migration under both static and flow conditions. Moreover, liganded EMR2 augments superoxide production and proteolytic enzyme liberation, i.e., leukocyte respiratory burst and degranulation [777].

7.3.2 TRPP1 (Polycystin-1)

Polycystin-1 and -2, or transient receptor potential TRPP1 and TRPP2, are encoded by the genes mutated in autosomal dominant polycystic kidney disease. Mutations in the PKD1 (polycystic kidney disease-1, TRPP1, or polycystin-1) and PKD2 (polycystic kidney disease-2, TRPP2, or polycystin-2) genes account for about 85% and 15% of autosomal dominant polycystic kidney disease, respectively. Polycystin-1 (PC1) operates in renal tubule morphogenesis. Proteins TRPP1 and TRPP2 contribute to calcium flux (Sect. 2.3.4.5), regulation of heterotrimeric G proteins, Wnt and STAT signaling, among other functions.

Protein TRPP1 induces the formation of a complex with tuberin, or tuberous sclerosis complex protein TSC2, and TOR, thereby inhibiting TOR activity, hence unappropriate cell growth and proliferation [778]. The TRPP1–TSC2–TOR complex

Table 7.8. Ligands of G-protein-coupled receptors: (**Part 4.2**) peptides and proteins (Source: [767]; CG: chorionic gonadotropin [gonadotrophin]; GHRH: growth hormone-releasing hormone; GHSR: growth hormone secretagogue receptor; GIP: gastric inhibitory polypeptide; GnRH: gonadotropin-releasing hormone; GRP: gastrin-releasing peptide; LH: luteinizing hormone; LHCGR: luteinizing hormone–choriogonadotropin receptor [a.k.a. lutropin–choriogonadotropin receptor (LCGR) and luteinizing hormone receptor (LHR)]; MCH: melanin-concentrating hormone).

Ligand	Receptor	Gα
Galanin and	GAL_1, GAL_3	Gi/o
galanin-like peptide	GAL_2	Gi/o, Gq/11, G12/13
GIP	GIP	Gs
Gastrin	CCK_2	Gq/11
GRP, bombesin	BB_2	Gq/11
Ghrelin	GHSR	Gq/11
Glucagon	GCGR	Gs
Glucagon-like peptide	GLP_1, GLP_2	Gs
GnRH	GnRH	Gq/11
GHRH	GHRH	Gs
Kisspeptins, metastin	GPR54	Gq/11
LH, CG	LHCGR	Gs, Gi
MCH	MCH_1	Gi/o
	MCH_2	Gq/11
Melanocortins	MC_1, MC_3–MC_5	Gs
Motilin	GPR38	Gq/11

may sense renal insults, possibly by primary cilium-mediated mechanotransduction, and trigger a TOR-initiated repair program.[7]

Proteins TRPP1 and TRPP2 represent an adhesion G-protein-coupled receptor and a mechanosensitive, calcium-permeable, non-selective cation channel of the TRP channel family located at the primary cilium that causes calcium transients at the cilium, respectively. Protein TRPP1 can interact with the C-terminus of TRPP2, thereby regulating the TRPP2 activity.[8]

Protein TRPP1 is a multidomain glycoprotein of 4,303 amino acids. It is made of a large extracellular N-terminal region, a membrane-spanning domain with 11 transmembrane segments, and a cytosolic C-terminal segment (Table 7.12). The last

7. The primary cilium of renal epithelial cells is a non-motile, mechanosensory extensions of the apical (luminal) plasma membrane that bends in response to urine flow and provokes a transient rise in the intracellular calcium concentration. Polycystin-1 links ciliary mechanosensation to changes in gene transcription via flow-regulated proteolytic cleavage of the TRPP1 cytoplasmic tail, its nuclear translocation, and stimulation of STAT6 transcriptional activity [778].

8. The cytoplasmic segment serves in the recruitment to the cell surface and interaction with TRPP2 channel. The TRPP1–TRPP2 complex functions as a cation channel.

Table 7.9. Ligands of G-protein-coupled receptors: (**Part 4.3**) peptides and proteins (Source: [767]; DOR, KOR, MOR: δ-, κ-, μ-opioid receptor; PAR: peptidase-activated receptor; PrRP: prolactin-releasing peptide; PTH: parathyroid hormone; PTHRP: parathyroid hormone-related protein). N-Termini of peptidases released by proteolytic cleavage serves as ligands.

Ligand	Receptor	Gα
Neurokinin-A	NK_2	Gq/11
Neurokinin-B	NK_3	Gq/11
Neuromedin-B, bombesin	BB_1	Gq/11
Neuromedin-U	NMU_1, NMU_2	Gq/11
Neuropeptide-FF, -AF	$NPFF_1$, $NPFF_2$	Gi/o
Neuropeptide-W23, -W30	GRP7, GPR8	Gi/o
Neuropeptide-Y	Y_1, Y_2, Y_4–Y_6	Gi/o
Neurotensin	NTS_1, NTS_2	Gq/11
Opioids	DOR, KOR, MOR, ORL1	Gi/o
Orexin-A/B	OX_1, OX_2	Gs, Gq/11
Oxytocin	OT	Gq/11, Gi/o
PTH, PTHRP	PTH1R	Gs, Gq/11
Prokineticin-1/2	PKR_1, PKR_2	Gq/11
PrRP	PRRP	Gq/11
	Peptidase N-termini	
Thrombin	PAR_1, PAR_3, PAR_4	Gq/11, G12/13, Gi/o
Trypsin	PAR_2	Gq/11

6 transmembrane segments are homologous to TRPP2 and voltage-activated calcium channels.

In renal tubules, bending of the wetted primary cilium by urine flow activates TRPP1 and TRPP2 proteins. It can couple with and activate several heterotrimeric G protein subtypes, such as Gi, Gq, G12/13, and stimulate Jun N-terminal kinase and AP1 transcription factor [779]. Protein TRPP1 regulates the cell cycle, as it upregulates cyclin-dependent kinase inhibitor CKI1a and activates the JaK–STAT pathway [780]. In addition, TRPP1 binds to and stabilizes regulator of G-protein signaling RGS7.

Protein TRPP1, like other adhesion G-protein-coupled receptors, can be cleaved at the *GPCR proteolytic site* (GPS).[9] Protein TRPP1 actually undergoes an autoproteolytic GPS cleavage to form an extracellular N-terminal fragment and a membranous C-terminal fragment that remain non-covalently associated.

Cleavage at GPS is autoproteolytic, but not entirely efficient. Therefore, cleaved and uncleaved adhesion G-protein-coupled receptors coexist. Both full length and cleaved fragment of adhesion G-protein-coupled receptor can be involved in cell signaling.

9. The GPS domain alone is not sufficient to support the cleavage and requires the adjacent receptor for egg jelly domain [781].

Table 7.10. Ligands of G-protein-coupled receptors: (**Part 4.4**) peptides and proteins (Source: [767]; InsL: insulin-like peptide; PACAP: pituitary adenylate cyclase-activating polypeptide; Rln: relaxin; RXFP: relaxin–insulin-like family peptide receptor; TRH: thyrotropin-releasing hormone; TSHR: thyroid-stimulating hormone (thyrotropin) receptor; VIP: vasoactive intestinal polypeptide). Receptors PACAPR1 to PACAPR3 are also named PAC_1 and AdCyAP1R1, $VPAC_1$ and vasoactive intestinal polypeptide receptor VIPR1, and $VPAC_2$ and VIPR2.

Ligand	Receptor	Gα
	Relaxins and insulin-like peptides	
Relaxin	$RXFP_1$	Gs
InsL3	$RXFP_2$	Gs
Relaxin-3	$RXFP_3$	Gi
InsL5	$RXFP_4$	Gi
Secretin	SCTR	Gs
Somatostatin	SST_1–SST_5	Gi/o
Substance-P	NK_1	Gq/11
Thyrotropin	TSHR	Gs, Gq/11, Gi, G12/13
TRH	TRH_1, TRH_2	Gq/11
Urotensin-2	UTS2R	Gq/11
VIP, PACAP	PACAPR1–PACAPR3	Gs
Vasopressin	V_{1A}, V_{1B}	Gq/11
	V_2	Gs

After GPS cleavage, the N-terminal α subunit is anchored to the plasma membrane, as it tethers to transmembrane β subunit or connects to the plasma membrane by itself [782]. In addition, N- and C-termini can be internalized independently. Moreover, α and β subunits of different types of adhesion G-protein-coupled receptors can interact.

However, most of the TRPP1 N-terminal fragment ($TRPP1^{\Delta NT}$) is tethered to the membrane-bound C-terminal fragment ($TRPP1^{\Delta CT}$) in a non-covalent manner (similar to that for long N-terminal class-B GPCR-related 7-transmembrane receptors latrophilin-1 and CD97).

Proteolysis at the GPS that occurs at the endoplasmic reticulum may assist in the transfer to the plasma membrane. However, at least for some adhesion G-protein-coupled receptors such as TRPP1, it is not required for an efficient transport to the plasma membrane [782]. In any case, GPS cleavage is mandatory for a normal receptor functioning. The final, mature, plasmalemmal receptor may be a heterodimer or 2 independent proteins [782]. Membrane-bound N-terminus may either be released in the extracellular medium, or remains connected to the cell surface and bind its cognate ligands and then transmits cues. The C-terminal transmembrane segment subunit may act as a classical GPCR; it can reassociate with the liganded N-terminal subunit [782].

Table 7.11. Ligands of G-protein-coupled receptors: (**Part 5**) sensory receptors (Source: [767]; G_{gust}: gustducin; G_{olf}: olfaction G-protein subunit; G_{Tc} and G_{Tr}: cone- and rod-transducin). Visual signals are sensed by opsins that use a photo-isomerization reaction to translate electromagnetic waves into cellular signals. Opsin signaling is based on the conversion of 11-cis-retinal to all-trans-retinal. Light is an multichromatic electromagnetic wave (visible spectrum wavelength 380–780 nm, i.e. 400–790 THz, with a maximum sensitivity $\sim$ 555 nm). In humans, 4 opsin types exist in addition to rhodopsin. They lodge in different types of cone cells of the retina. They have absorption maxima for yellowish-green, green, and bluish-violet light. Smell sensing is carried out by receptors of the olfactory epithelium that bind odorants (olfactory receptors) and pheromones (vomeronasal receptors). Because of their similar aliases, pheromone receptors V_1R and V_2R should not be mistaken for $V_{1A/1B/2}$ vasopressin receptors.

Ligand	Receptor	Gα
	Light	
Absorption maximum		
$\sim$ 500 nm	Rhodopsin	G_{Tr}
$\sim$ 426 nm	Violet opsin	G_{Tc}
$\sim$ 530 nm	Green opsin	G_{Tc}
$\sim$ 560 nm	Yellow opsin	G_{Tc}
425–480 nm	Melanopsin	Gq/11
	Taste	
Umami	$T_1R1 + T_1R3$	G_{gust}
	mGluR4	Gi/o
Sweet	$T_1R2 + T_1R3$	G_{gust}
Bitter	T_2R ($\sim$25 in humans)	G_{gust}
	Odorants ($\sim$ 350 in humans)	G_{olf}
Pheromones	V_1R group (few in human)	
	V_2R group (none in human)	

Table 7.12. Structure of Polycystin-1 (or TRPP1; Source: Wikipedia). Polycystin-1 contains a large extracellular N-terminal domain with a combination of functional motifs, an odd number of transmembrane segments (TM), and an intracellular C-terminal domain. The extracellular segment is composed of the N-terminus (NT), a cysteine-rich segment (CRS), leucine-rich repeats (LRR), wall cell integrity and stress-response component (WSC), polycystic kidney disease repeats (PKD), C-lectin motif (CLec), low-density lipoprotein-A sequence (LDLa), receptor for egg jelly (REJ), and a G-protein-coupled receptor proteolytic site (GPS). The cytoplasmic segment contains a lipooxygenase homolog-2 (LH2), TMs, a coiled-coil domain (CCD), and the C-terminus (CT).

Extracellular part	NT–CRS–LRR–WSC–PKD–CLec–LDLa–PKD–REJ–GPS
Intracellular and cytosolic part	LH2–TM–CCD–CT– –Polycystin-2 (CT–EF hand–TM–NT)

7.4 Proton-Sensing G-Protein-Coupled Receptors

Proton-sensing G-protein-coupled receptors pertain to the subclass A15. They sense acidic pH; they are indeed activated when the extracellular pH falls below 6.8 (acidosis, i.e., increased hydrogen ion concentration). [10]

They include GPR4, GPR65, [11] GPR68, [12] and GPR132. [13] The latter is an immunoregulatory receptor on macrophages for lysophosphatidylcholine, which is a major phospholipid component of oxidized low-density lipoproteins in atherosclerosis. [783]. The GPR132 receptor resides also on lymphocytes [784]. Activation of GPR132 by lysophosphatidylcholine may increase intracellular calcium concentration, cause receptor internalization, and activate extracellular signal-regulated kinases. The GPR65 receptor on the surface of tumor cells facilitates tumor development by sensing the acidic environment [785].

7.5 GPCR Classification

The superclass of G-protein-coupled receptors contains many hundreds of olfactory receptors to detect a wide variety of olfactory (exogenous) ligands and more than 360 endoGPCRs that are targeted by endogenous (non-olfactory) ligands. The GPCR superclass is constituted by several classes.

On the basis of their sequence and structural similarity, ligand interaction, and phylogeny, GPCRs have been originally grouped into A to F or 1 to 5 classes. [14] Phylogenetic analysis of the human repertoire yields the alternative GRAFS classification with 5 families: glutamate (G), rhodopsin (R), adhesion (A), Frizzled–Taste-2 (F), and secretin (S) [788].

The number of elements in each class varies according to literature data, and especially publication date, as new proteins are discovered (Table 7.13).

The class of rhodopsin GPCRs is the largest set with about 672 members in the human genome (∼388 olfactory receptors). Its members are often encoded by single exons or genes with a small number of introns. Isomerization of 11-cis-retinal

10. Acidosis is usually detected in blood (acidemia; arterial pH <7.35).

11. A.k.a. T-cell death-associated gene-8 (TDAG8) and psychosine receptor. It is overexpressed in various tumor cell types.

12. A.k.a. ovarian cancer G-protein coupled receptor-1 (OGR1).

13. A.k.a. cell cycle phase G2 accumulation protein (G2A).

14. The families comprise Rhodopsin (class A with more than 270 members), Secretin (at least 15 receptors) and Adhesion (33 members; class B), metabotropic glutamate (class C; 15 receptors), fungal mating pheromone receptors (class D), cAMP receptors (class E), and Frizzled–Smoothened–Taste-2 (class F; 24 members) classes [786, 787]. The very large Rhodopsin class A is subdivided into 19 groups (A1–A19). Class-B GPCRs include both Adhesion and Secretin families. Class-B GPCRs are also called class-2 GPCRs. Taste-2 receptors sense bitter substances. Another classification scheme of GPCRs is based on class 1 Rhodopsin-like, class 2 Secretin-like and Adhesion, class 3 Glutamate receptor-like, class 4 Frizzled–Taste-2, and class 5 Miscellaneous (Others, including unclassified GPCRs).

Table 7.13. Estimated GPCR number per class in humans (Source: [789]). Rhodopsin consists of the protein moiety opsin that is reversibly, covalently bound to its cofactor retinal. The latter is produced in the retina from vitamin-A. Photo-isomerization of retinal is the primary vision event that creates photosensory signals via retinal interactions with other proteins.

Class (Group)	Number
Glutamate	22
Rhodopsin-α	101
Rhodopsin-β	43
Rhodopsin-γ	64
Rhodopsin-δ	63
Adhesion	33
Frizzled	11
Taste-2	25
Secretin	15
Pheromone V_1R	3
Olfactory	388
Others	23

Table 7.14. Visible light spectrum.

Color	Wavelength (nm)
Violet	380–450
Blue	450–475
Cyan	476–495
Green	495–570
Yellow	570–590
Orange	590–620
Red	620–750

into all-trans-retinal by light induces a conformational change in opsin that activates the associated G protein and triggers a signaling cascade. In humans, several opsin types exist in addition to rhodopsin. They reside in different types of cone cells of the retina. They have absorption maxima for yellowish-green ($\sim$ 560 nm), green ($\sim$ 530 nm), and bluish-violet ($\sim$ 426 nm) light (Table 7.14).

The class of adhesion GPCRs that comprises 33 members is the second largest GPCR set in humans. Mammalian adhesion GPCRs are encoded by large, complex genomic structures with several introns that give rise to alternatively spliced variants.

The class of secretin GPCRs members have a peptide hormone-binding domain. They likely originates from ancestors to the class of adhesion GPCRs.

Class-1 GPCRs, the largest and most diverse class of GPCRs, encompass more than 400 sensory receptor types involved with the detection of taste, odor, or light, as well as more than 250 non-sensory receptor types for transmitters or chemical messengers, in addition to orphan receptors.

Class-2 GPCRs comprise 20 receptor types that are activated by peptides, in addition to orphan receptors. Their ligand peptides include transmitters, adrenomedullin, calcitonin gene-related peptide, vasoactive intestinal peptide, and urocortins.

Class-3 GPCRs include 11 metabotropic receptors that are activated by glutamate and γ-aminobutyric acid as well as orphan receptors. These receptors localize presynaptically to perivascular nerves, but are generally not expressed in endothelia and smooth muscles.

Class-4 GPCRs incorporate receptors of the Frizzled and Smoothened families that are implicated in vascular remodeling.

Many GPCR classes can be decomposed into GPCR subclasses and families (Tables 7.15 and 7.16). Each GPCR category contains a variable number of receptors.

7.6 Structure and Function

Synthesis of GPCR and their activity are tightly controlled. Availability of GPCRs relies not only on its production and exocytosis, but also on desensitization via endocytosis and resensitization by endosome recycling. Responsiveness of GPCRs depends on associated regulators, such as G proteins, kinases (feedback loops mediated by PKA and PKC) and phosphatases, as well as membrane partners.

Ligand binding on G-protein-coupled receptors can be voltage sensitive, as GPCRs can serve as sensors for both transmembrane potential and external compounds [791]. Voltage-sensitive G-protein-coupled receptors are triggered by the voltage across the plasmalemma. [15] Gating currents act on structural GPCR components.

7.6.1 GPCR Structure

G-protein-coupled receptors share a common structure of 7 membrane-spanning helices (transmembrane domains TM1–TM7) connected by 3 loops on each side of the membrane (intra- [IL1–IL3] and extracellular [EL1–EL3] loops, with an extracellular N- and intracellular C-terminus. Extra- and intracellular loops serve as binding sites for agonists (activators) and G proteins, respectively.

The structure, conformation, and specificity of the G-protein binding site can depends on the ligand type. The ends of the third intracellular loop are involved in G-protein activation and selectivity of GPCR–G-protein interactions, as well as interactions with G-protein-coupled receptor kinases, arrestins, and other signaling molecules.

Ligand binding causes a conformational change in the GPCR cytoplasmic domain. Class-A GPCR activation leads to an outward displacement of transmembrane helix 6 that opens for G-protein binding [792]. The activation rate of α2a-adrenergic

15. The resting potential difference across the 3-nm-thick plasmalemma is equal to about 70 mV. It can affect the conformation of membrane proteins.

Table 7.15. Classes of GPCRs. (**Part 1**) class A (GRH: gonadotropin-releasing hormone; TRH: thyrotropin-releasing hormone; Source: [790]). Platelet-activating factor (PAF) receptor initiates cell response to its phospholipid ligand that has potent platelet aggregating and inflammatory effects and elicits smooth muscle contraction.

Sub-class	Ligand family
Amine	Cathecolamines, dopamine, histamine, acetylcholine (muscarinic), serotonin
Cannabinoid	2-Arachidonoyl glycerol
Peptide	Adrenomedullin, angiogenin-like, angiotensin, bombesin, C5a-anaphylatoxin, galanin-like, interleukin-8, chemokine, bradykinin, cholecystokinin, endothelin, melanocortin, melanin-concentrating hormone, neuromedin-U-like, neuropeptide-Y, neurotensin, orexin and neuropeptides-FF, prokineticin, proteinase-activated-like, somatostatin, tachykinin, vasopressin-like, urotensin-2
GRH	
Hormone	Follicle-stimulating hormone, lutropin-choriogonadotropic hormone, thyrotropin, gonadotropin-1 and -2
Leukotriene-B4	
Lysosphingolipid	Lysophosphatidic acid, sphingosine 1-phosphate
Melatonin	
Nucleoside Nucleotides	Adenosine Purines (ADP, ATP), pyrimidines (UDP, UTP)
(Rhod)opsin	
Olfactory	
PAF	
Prostanoid	Prostaglandin, prostacyclin, thromboxane
TRH	

receptor and parathyroid hormone receptor are 40 ms and 1 s, respectively. The kinetics of the GPCR–G-protein complex formation depends on the cell concentration of G proteins.

Table 7.16. Classes of GPCRs (**Part 2**; Source: [790]).

Class	Types
B	Brain-specific angiogenesis inhibitor Calcitonin Corticotropin-releasing factor Diuretic hormone Gastric inhibitory peptide Growth hormone-releasing hormone Glucagon Latrophilin Methuselah-like proteins Parathyroid hormone Secretin Vasoactive intestinal polypeptide Very large G-protein-coupled receptor
C	Bride of sevenless proteins Calcium-sensing-like $GABA_B$ Metabotropic glutamate Orphan GPCR5 Orphan GPCR6 Putative pheromone receptors Taste receptors
D	Fungal pheromone
E	cAMP receptors
F	Frizzled Smoothened

7.6.2 GPCR Signaling

G-protein-coupled receptors interact with guanine nucleotide–binding (G) proteins (Vol. 4 – Chap. 8. Guanosine Triphosphatases and Their Regulators) to transduce signals transmitted by messengers in the cardiovascular and ventilatory apparatus among others physiological systems. [16] Ligand-activated receptors catalyze the GDP–GTP exchange at a coupled G protein, and thereby promote the activity of effectors (second messenger–producing enzymes and ion channels). G-protein-coupled receptors thus act as guanine nucleotide-exchange factors (GEF) for G protein subunits. Release of GDP is the rate-limiting step in G-protein activation. After receptor activation, G protein binds to the receptor, the subsequent conformational change decreasing the affinity of the G protein for guanosine diphosphate.

16. On the cardiomyocyte sarcolemma, angiotensin-2, endothelin-1, noradrenaline, and prostaglandin-F2α activate receptors coupled to a Gq subunit.

Most receptors are able to activate more than one type of G protein. They thus trigger several signal transduction cascades. However, some receptors interact only with G-protein isoforms of the same class.

Many GPCRs have a complex signaling behavior. In cardiomyocytes, β2-adrenoceptors couple to both stimulatory ($G\alpha_s$) and inhibitory ($G\alpha_i$) subunits that target adenylate cyclase as well as, upon phosphorylation by a G-protein-coupled receptor kinase, arrestins to signal via MAPK pathways (Vol. 4 – Chap. 5. Mitogen-Activated Protein Kinase Modules) in a G-protein-independent manner. Desensitization of GPCRs involves multiple processes, such as receptor phosphorylation and arrestin-mediated endocytosis followed by receptor recycling or lysosomal degradation. Furthermore, GPCR oligomerization and localization to specific membrane compartments influence GPCR activity.

G-protein-coupled receptors signal not only via heterotrimeric G proteins, but also various small monomeric GTPases to regulate the activity of effectors (Tables 7.17 to 7.19). Subunit Gα is composed of a GTPase and a helical domain [792]. The GTPase domain hydrolyzes GTP and provides the binding site for Gβγ dimer, GPCR, and effectors. It contains 3 flexible loops. The helical domain comprises 7 α-helices. Post-translational modifications of Gα regulate membrane location and between-protein interactions. All Gα, except $G\alpha_t$, undergo palmitoylation at their N-termini. Certain $G\alpha_i$ are also myristoylated at their N-termini. Binding of GTP on a Gα subunit induces a structural rearrangement of the GPCR–G-protein complex that destabilizes this complex to interact with effector; some complexes dissociate, but not all.

Subunits Gβ and Gγ form a functional unit. Several Gβγ dimers can interact with the same Gα isoform. Subunit Gβγ binds GPCRs and helps stabilize the GPCR–Gα interface. Subunit Gβ has a 7-bladed β-propeller structure with an α-helical N-terminus that forms a coiled-coil with N-terminus of Gγ [792]. The C-terminus of Gγ binds to blades 5 and 6. All Gγ undergo post-translational isoprenylation of their C-terminus with either a farnesyl (Gγ1, Gγ8, and Gγ11) or geranylgeranyl moiety.

Subunits Gα and Gγ interact sequentially with activated GPCR to dock G protein onto the receptor. Activated Gα and Gβγ proteins positively or negatively regulate various effectors such as phospholipase-A, -C, and -D, [17] adenylate and guanylate cyclases, phosphoinositide 3-kinases, phosphodiesterases, protein kinase-C, and ion channels (Table 7.20, Fig. 7.1). [18]

---

17. Phospholipases PLD1 and PLD2 regulate cytoskeletal organization and endo- and exocytosis. G-protein-coupled receptor signaling directs stimulation of PLD1 by PKC.

18. Transiently activated Gs-coupled β-adrenergic receptors cause a calcium influx through Ca_V1 channels that are phosphorylated by cAMP-dependent protein kinase-A [793]. (Mediator cAMP is produced in response to Gs activation.) On the other hand, sustained activation of β-adrenergic receptors causes their phosphorylation by GPCR kinases for recruitment of β-arrestin-1 and endocytosis. Additional activation of Gi by β2-adrenergic receptors explains the difference in signaling by β1- (the predominant subtype in cardiomyocytes) and β2-adrenergic receptors. Subunit Gi slows down the heart rate and atrioventricular conductance. Isoform Gi2, the major Gi subunit in cardiomyocytes, represses the activity of Ca_V1 channels.

Table 7.17. Subunit Gα of trimeric guanine nucleotide-binding proteins and their distribution and effectors (**Part 1**; Source: [767]; ACase: adenylate cyclase; AC*i*: adenylate cyclase isoform *i*; Ca_V: voltage-gated Ca^{++} channel; GIRK: G$\beta\gamma$-regulated inwardly rectifying K^+ channel [$K_{IR}3$]; G_{Tc}, G_{Tr}: cone and rod transducin; PDE: phosphodiesterase). Gα subunits define the basic properties of a heterotrimeric G protein. They can be classified into 4 families, $G\alpha_s$, $G\alpha_{i/o}$, $G\alpha_{q/11}$, and $G\alpha_{12/13}$. In addition to Gs, 2 transcripts that encode $XL_{\alpha s}$ and neuroendocrine secretory protein NESP55 are generated by promoters upstream of the GNAS (Gs-encoding) promoter of the GNAS complex locus. Extra-large protein Gs_{XL} that has a limited expression pattern (adrenal gland, heart, pancreas, brain, and pituitary pars intermedia) is able to bind to G$\beta\gamma$ dimer and provoke cAMP synthesis. The GNASXL gene also produce ALEX that inhibits Gs_{XL}. In addition, exon 3 of the GNAS gene is alternatively spliced to give rise to long (Gs_L) and short (Gs_S) Gs isoforms.

Type	Distribution	Effectors
	$G\alpha_s$ class	
Gs	Ubiquitous	ACases (+: all types)
Gs_{XL}	Neuroendocrine cells	ACases (+)
$G\alpha_{olf}$	Olfactory epithelium, brain	ACases (+)
	$G\alpha_{i/o}$ class	
Gi1	Wide	AC1/3/5/6/8/9 (−)
Gi2	Ubiquitous	AC1/3/5/6/8/9 (−)
Gi3	Wide	AC1/3/5/6/8/9 (−)
Go	Neurons, neuroendocrine cells	Ca_V (−), GIRK (+) (via G$\beta\gamma$)
Gz	Neurons, platelets	AC5/6 (−), Rap1GAP
$G\alpha_{gust}$	Taste and brush cells	PDE (+)
G_{Tr}	Retinal and taste cells	PDE6 (+)
G_{Tc}	Retina	PDE6 (+)

Subsequently, these effectors activate or inhibit the production of second messengers (cAMP, cGMP, diacylglycerol, inositol trisphosphate, phosphatidylinositol

Go-Coupled muscarinic channels M_2 also inhibit Ca_V1 channels. In enteroendocrine cells, bitter taste receptors (T_2R) and $G\alpha_{gust}$ of the Gi/o group increase Ca^{++} influx via activated Ca_V1 channels [794]. In the hippocampus, activated Gi/o-coupled adenosine receptor A_1 suppresses serotonin release, as it inhibits $Ca_V2.2$ and $Ca_V2.1$ channels (more precisely $Ca_V2.2$–PKC–syntaxin and $Ca_V2.1$–PKA–synaptobrevin axes) [795]. On the other hand, activated Gs-coupled A_2 stimulates serotonin release, as it stimulates $Ca_V2.1$ channels ($Ca_V2.1$–PKA–synaptobrevin axis). The 3 Gi subtypes (Gi1–Gi3) activate K^+ channels $K_{IR}3$ (K_{ACh} current) of atriomyocytes (G$\beta\gamma$ dimer is inactive) [796]. Gi/o-Coupled glutamate receptor mGluR7 is able to inhibit $Ca_V2.1$ channels via G$\beta\gamma$ dimer. Upon Ca^{++} influx, Ca^{++}–calmodulin binds to macrophage myristoylated alanine-rich C-kinase substrate (macMARCKS) that is tethered to mGluR7 [797]. It then displaces macMARCKS and G$\beta\gamma$ from mGluR7 receptor. G$\beta\gamma$ Dimer can then inhibit $Ca_V2.1$ channel. In sympathetic neurons, G$\beta\gamma$ inhibits $Ca_V2.2$ channels upon stimulation by noradrenaline and somatostatin [798].

Table 7.18. Subunit Gα of trimeric guanine nucleotide-binding proteins and their distribution and effectors (**Part 2**; Source: [767]; BTK: Bruton Tyr kinase; HAX1: hematopoietic cell-specific Lyn substrate [HCLS1]-associated protein-X1 [anti-apoptotic]; HSP: heat shock protein [chaperone]; JNK: Jun N-terminal kinase; PP: protein phosphatase; RasA2: Ras P21 protein activator-2 [RasGAP2]). Subunits G15 and G16 are the murine and human ortholog, respectively. Enzymes PLCβ1, -3, and -4 are the major effectors of Gq/11 and PLCβ2 is a minor effector. Subunits G12 and G13 are often activated by Gq/G11-coupled receptors (crosstalk). Cadherins (a portmanteau word for calcium-dependent adhesion) are type-1 transmembrane proteins and constituents of adherens junction. They are connected to nearby actin filaments via α-actinin and cortical actin via catenin and vinculin. Radixin is a cytoskeletal linker of actin to the plasma membrane. Members of the $G\alpha_i$, $G\alpha_q$, and $G\alpha_{12/13}$ families cooperate to regulate Ca^{++} signal and nitric oxide (NO). Subunit $G\alpha_q$ targets phospholipase-C and causes Ca^{++} influx without the assistance of NO. Subunit $G\alpha_{13}$ supports Ca^{++} influx by a mechanism dependent on NO, guanylate cyclase, and cGMP (a process blocked by activated adenylate cyclase, the cAMP–PKA axis preventing Ca^{++} influx and NO production). NO may relieve inhibition of adenylate cyclase caused by Gi subunit via adpribosylation of Gi subunit.

Type	Distribution	Effectors
		$G\alpha_{q/11}$ class
Gq	Ubiquitous	PLCβ1–PLCβ4 (+)
G11	Very wide	PLCβ1–PLCβ4 (+)
G14	Lung, kidney, spleen	PLCβ1–PLCβ4 (+)
G15/16	Hematopoietic cells	PLCβ1–PLCβ4 (+)
		$G\alpha_{12/13}$ class
G12	Ubiquitous	RhoGEF1, RhoGEF11, RhoGEF12, RasA2, BTK, cadherin
G13	Ubiquitous	RhoGEF1, RhoGEF11, RhoGEF12, radixin
G12/13		PLA2, Na^+–H^+ exchanger, JNK, PP5, HAX1, HSP90, AKAP3

(3,4,5)-trisphosphate, and arachidonic and phosphatidic acids), and promote calcium influx, and opening or closure of various ion channels.

A relatively small number of G proteins transduce signals from a large number of GPCRs. Each member of the G-protein family is thus able to interact with many different GPCRs. In addition, many GPCRs can activate multiple G proteins. Suitable signal transduction relies on specific interactions. Each contact site of Gα subunits, as well as other molecular regions, contribute to coupling specificity [792]. In addition, specific isoforms of Gβ interact preferentially with specific GPCRs. [19] The

19. Subunits Gβ4 and Gβ5 have the strongest and the poorest coupling ability with β1-adrenergic and A_{2A} adenosine receptors. Subunits Gβ1 and Gβ4 have a similar coupling capacity with β1-adrenergic receptor, but Gβ1 has 20-fold lower coupling capacity than Gβ4 for adenosine receptor.

Table 7.19. Subunits Gβ and Gγ of trimeric guanine nucleotide-binding proteins and their effectors (Source: [767]). In the absence of signaling, Gα subunit is tethered to its associated Gβγ dimer. This dimer is assembled from a repertoire of 5 Gβ (Gβ1–Gβ5) and 12 Gγ (Gγ1–Gγ5, Gγ7–Gγ8, and Gγ10–Gγ14) subunits. Subunit Gβ5 only binds weakly to Gγ and links to regulators of G-protein signaling. It resides mainly in the brain. Subunits Gβ1 to Gβ4 as well as Gγ2, Gγ4 to Gγ6, and Gγ10 to Gγ12 are widely distributed. Subunits Gγ1, Gγ3, and Gγ13 lodge in the central nervous system. Subunit Gγ8 resides in the olfactory epithelium and Gγ3 also in blood cells. The Gα and Gβγ subunits stimulate various effectors, such as adenylate and guanylate cyclases, phosphodiesterases, phospholipase-A2 and -C (PLC), phosphoinositide 3-kinases (PI3K), besides guanine nucleotide-exchange factor (GEF), thereby activating or inhibiting the production of multiple second messengers such as cAMP, cGMP, diacylglycerol (DAG), inositol (1,4,5)-trisphosphate (IP_3), Ca^{++} ions, phosphatidylinositol (3,4,5)-trisphosphate (PIP_3), arachidonic and phosphatidic acids, and opening or closing of numerous ion channels. Mediators PLC and PI3K via different pathways activate the MAPK module.

Target type	Components
Adenylate cyclases	AC1 (−), AC2/4/7 (+)
Ion channels	Ca_V2.1–Ca_V2.3, Ca_V3.2 (−) K_{IR}3.1–K_{IR}3.4 (GIRK) (+)
Kinases	G-protein-coupled receptor kinases GRK2/3 (+) Phosphatidylinositol 3-kinase-β/γ(+) (Src–SHC–GRB2–SOS–Ras–cRaf–MAPK cascade)
Phospholipases	PLCβ1–PLCβ3 (+) (in increasing order of potency) (DAG–PKC–cRaf–MAPK, IP_3–Ca^{++}–RasGRF–Ras–cRaf–MAPK, cAMP–PKA–RapGEF3/4–Rap–bRaf–MAPK axes)

subtype of Gγ influences the linkage of the Gβγ dimer.[20] Furthermore, homologous Gα binding to GPCRs depends on the GPCR ligand.[21]

Proteic constituents of ubiquitous primary cilia are required for the localization of G-protein-coupled receptors, at least on central neurons [799]. Certain GPCRs such as somatostatin receptor SstR3 and serotonin receptor-6 are specifically located in neuronal cilia. Ciliary localization of somatostatin receptor SstR3 (neurotransmission) and melanin-concentrating hormone receptor MchR1 (feeding regulation) in neurons needs ciliary disorder Bardet-Biedl syndrome BBS2 and BBS4 proteins.

G-protein-coupled receptors also act as scaffolds for the formation and location of signaling complexes in the cell. The response properties are determined by the relationship among the ligand, receptor, G protein, and associated proteins. Recep-

20. Dimers Gβ1γ5 and Gβ1γ7 can mediate binding to M_2 muscarinic receptor, but not Gβ1γ2.

21. Activated peptidase-activated receptor-1 functions according to ligand type. For example, thrombin favors $G\alpha_q$ rather than $G\alpha_{12/13}$.

Table 7.20. Gα effectors (Source: [768]). Trimeric G proteins are made of Gα and Gβγ subunits (IP_3: inositol (1,4,5)-trisphosphate; RhoGEF: Rho-guanine nucleotide exchange factor). Gα Subunits are grouped into 4 main families: Gs, Gi, Gq, and G12. Activated Gα proteins positively or negatively regulate various effectors, such as phospholipases PLA2, PLCβ, PLD, adenylate (ACase) and guanylate (GCase) cyclases, phosphoinositide 3-kinases (PI3K), protein kinase-C (PKC), phosphodiesterases (PDE), Rho GTPases, and ion channels, such as G-protein-regulated inward rectifier potassium channels (GIRK). Both Gα subunit and Gβγ dimer activate or inhibit their effectors. Subunit Gs stimulates adenylate cyclase that primes phosphorylation by protein kinase-A of cAMP response element-binding protein and corresponding transcription. On the other hand, Gi inhibits adenylate cyclase, whereas its associated Gβγ dimer activates mitogen-activated protein kinases via small GTPase Ras. Gi-coupled receptors stimulates phospholipases-C via Gβγ. Family Gq members stimulate phospholipase-C that elevates intracellular calcium concentration and activates PKC to stimulate nuclear factor of activated T cell and MAPK. Subunits G12 and G13 target serum response element-dependent transcription mainly via small GTPase RhoA. Yet, GPCR can signal not only via heterotrimeric G proteins, but also Src kinases, according to ligand level.

G subunit	Effectors
$G\alpha_s$	ACase (+), PKA Na^+ and Cl^- channels (+) $Ca_V2.1$ (+)
$G\alpha_i$	AC1/3/5/6/8/9 (−), phospholipases (+), PDEs (+), K^+ and Cl^- channels (+) Ca_V1 (−), Ca_V2 (−), Ca_V3 (−)
$G\alpha_q$	PLCβ(+), diacylglycerol, PKC, IP_3, Ca^{++}
$G\alpha_{12}$	Rho, RhoGEFs
Gβγ	PLCβ1–PLCβ3 (+) ACase: AC2, AC4, AC7 (+), AC1 (−) GIRK (+), GRK (+), PI3K (+), NRTK (+) Ca_V (−)

tor signaling is regulated by both endogenous and exogenous actions. Activation of G-protein-coupled receptors can be sensitive to the plasmalemma lipid composition. Local cholesterol and sphingolipid concentrations affect the ligand binding and receptor transport.

Although ubiquitous GPCRs classically signal via heterotrimeric G proteins, they can also signal via β-arrestins, even without the participation of G proteins. β-Arrestin adaptors can control receptor signaling, desensitization, and trafficking. Most GPCRs use both β-arrestins and G proteins. However, some GPCR receptors that fail to activate G protein can act as *coreceptors* or *decoy receptors* that eliminate ligands. Agonist binding can prime β-arrestin association with decoy receptor CXCR7, hence β-arrestin-dependent signaling via activation of mitogen-activated protein kinase in the absence of G-protein. Therefore, GPCRs can signal exclu-

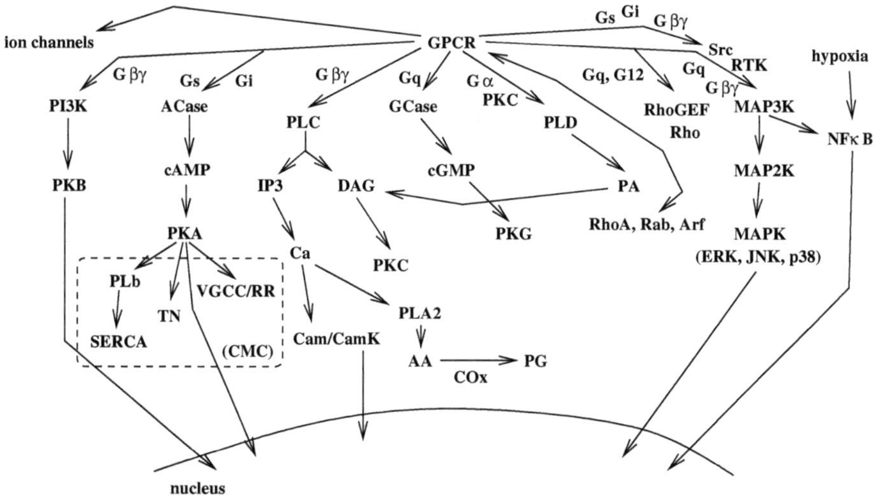

Figure 7.1. Effectors of GPCRs with their associated molecular triggers: Gα (Gs, Gi, Gq, and G12), Gβγ, PKC, Src, and RTK.

sively via β-arrestins (*non-canonical pathway*). Ligand CXCR7 stromal-derived factor SDF1α initiates β-arrestin-2-mediated endocytosis, whereas another ligand, interferon-inducible T-cell α chemoattractant fails to activate Gi and Gq, but promotes vascular smooth muscle cell migration [800].

7.6.3 GPCR Basal Activity

Many G-protein-coupled receptors exhibit a *basal activity* in the absence of their agonists with a variable magnitude according to GPCR type. G proteins can interact with GPCRs before ligand binding (GPCR–G-protein precoupling). The ligand-independent constitutive activity allows the conformation change to an active state upon stimulation, owing to interactions with the plasmalemmal lipids. The plasmalemma tension can then modulate the GPCR activity. The GPCR stimulation then catalyzes G-protein activation.

7.6.4 GPCR Oligomerization

The ligand–receptor complex is supposed to constitute the functional unit of GPCR signaling. However, G-protein-coupled receptors can form homo- or heterodimers. Moreover, some GPCRs form higher-order oligomers upon homo- or heterodimerization during their exocytosis, activation, inactivation, and/or endocytosis. Transient or sustained di- and oligomerization may explain the negative or positive receptor cooperativity. Therefore, G-protein-coupled receptors operate as monomers as well as unstable dimers or oligomers.

Agonist of a GPCR then must choose one protomer of the GPCR complex, its binding influencing the activity of associated protomers, among which at least one functional unit activates signaling effectors. The history of GPCR protomer association and dissociation from synthesis to degradation on cell signaling may play a role.

Heterodimerization can lead to signaling complexes that have a single ligand binding. However, only one GPCR of the dimer can be active in signaling. Two subunits of a receptor dimer coupled to G protein differ in their conformation [801]. A receptor dimer with a single ligand-bound subunit can trigger G-protein activation. The interaction of G protein with the receptor dimer prevents a symmetrical functioning of the dimer and explains the negative cooperativity in ligand binding on GPCR dimers.

Both class-A and class-C GPCRs (Sect. 7.5) can act as oligomers [802]. Metabotropic glutamate receptors assemble into strict dimers. γ-Aminobutyric acid receptor-B spontaneously forms dimers of heterodimers. Homo- and heterodimerization of GPCRs exist for class-C receptors, not only metabotropic γ-aminobutyric acid ($GABA_B$) and glutamate (mGluR) receptors [803],[22] but also taste heterodimeric- (T_1R1–T_1R3) and calcium-sensing receptors, for which only dimers are involved in signal transduction. Whereas, at least, some class-A GPCRs may function as monomers, other class-A GPCRs for glycoproteic hormones, such as luteinizing, follicle-stimulating, and thyroid-stimulating hormone, may transduce their signal in vitro as di- and/or oligomers by transactivation. In vivo, intermolecular cooperation can rescue GPCR defective in binding as well as signaling capacity [804]. Intermolecular cooperation and di- or oligomerization between mutant binding- and signaling-deficient luteinizing hormone receptors, indeed, re-establish luteinizing hormone effect in the absence of functional wild-type receptors.

7.6.5 GPCR Function in the Vasculature

In the cardiovascular system, G-protein-coupled receptors regulate numerous functions, such as heart frequency and contractility as well as vascular tone. More than 100 different GPCRs are involved in the cardiovascular system, such as (muscarinic) acetylcholine, adenosine, adrenergic, angiotensin-2, bradykinin, endothelin, lysophosphatidic acid, serotonin, sphingosine 1-phosphate, and vasopressin receptors.

7.6.5.1 Blood Vessels and Vasomotor Tone

G-protein-coupled receptors abound in the vasculature, particularly those of class-1 (Vol. 5 – Chap. 7. Vessel Wall). In blood vessels, class-1 GPCRs, once they

22. Metabotropic glutamate receptors are constitutive dimers that do not form larger oligomers. Like receptor Tyr kinases and receptor guanylate cyclases, agonist stimulation of mGluR dimer with a protomer that has its extracellular agonist-binding domain uncoupled from its G-protein-activating transmembrane region causes a symmetric activation of twin transmembrane domains with the same efficiency [803]. Agonist binding to one Venus flytrap (VFT) extracellular domain not only cis-activates the transmembrane domain of the same subunit, but also trans-activates that of the other protomer owing to an intersubunit rearrangement.

Table 7.21. Mediators of vasodilation and vasoconstriction (MLCK: myosin light-chain kinase; MLCP: myosin light-chain phosphatase; PKA: cAMP-dependent protein kinase-A; PLC: phospholipase-C; RoCK: Rho kinase). Subunits of the Gq/11 family prime a fast, transient response, whereas those of the G12/13 family causes a slower, but sustained, tonic contraction.

Vasodilation	Vasoconstriction
MLC	MLC^P
Ca^{++} outflux	Ca^{++} influx
MLCP	MLCK
PKA (+; Gs)	PKA (−; Gi)
	RoCK (G12/13–RhoA)
	PLCβ (Gq/11)

are activated by chemical messengers that comprise peptides act directly as vasoconstrictors or vasodilators or indirectly as vasodilators.[23]

Smooth muscle tone is controlled by the phosphorylation state of the regulatory myosin-2 light chain (MLC). Phosphorylation of MLC, hence the vasomotor tone, depends on intracellular calcium concentration. Vasodilation results from: (1) low intracellular calcium concentration by Ca^{++} export (egress from cytosol by reuptake into the sarcoplasmic reticulum and expulsion across the plasma membrane) through Ca^{++} carriers and (2) MLC dephosphorylation by myosin light chain phosphatase (MLCP) under the control of RoCK kinase (Table 7.21).

Conversely, vasoconstriction results from increased Ca^{++} concentration that, in cooperation with calmodulin, activates Ca^{++}–calmodulin-dependent myosin light chain kinase (MLCK). The latter stimulates cross-bridge cycling of actin–myosin stress fibers.

Vasodilation can be primed by: (1) trans-plasmalemmal hyperpolarization (e.g., on elevated activity of ATP-sensitive K^+ channel) that precludes the activity of voltage-gated calcium channels (mechanism used by adenosine) and (2) cAMP and cGMP signaling that stimulates protein kinase-A and -G, respectively, that both phosphorylate MLCK (procedure utilized by prostaglandins [PGD2, PGE2, and PGI2] and nitric oxide, respectively). PKA Kinase also hampers the activity of RhoA GTPase.

Vasoconstriction can be intiated by: (1) trans-plasmalemmal depolarization that promotes the activity of voltage-gated calcium channels (this process can be further enhanced by voltage-gated Na^+ channels and is typically triggered by ATP and tension); (2) decline in PKA activity by Gi subunit; and (3) elevated PLC activity by Gq subunit that causes, via IP_3, Ca^{++} influx, hence MLCK activation, as well as RhoA stimulation (thus RoCK excitation that hinders MLCP action).

23. Blood pressure-lowering drugs target a limited number of class-1 GPCRs. Such drugs include angiotensin-2 receptor and adrenoceptor antagonists.

Vasoconstrictors include adrenaline, angiotensin-2, apelin, adenosine triphosphate, endothelin-1, motilin, neuromedin-U, neuropeptide-Y, sphingosine 1-phosphate, thrombin, thromboxane-A2, urotensin-2, and vasopressin [805].

Vasodilators encompass extracellular adenosine, extracellular ADP and ATP, bradykinin, ghrelin, histamine, natriuretic peptides, nitric oxide, noradrenaline (via β2-adrenergic receptors), platelet-activating factor, some prostaglandins, as well as substance-P ghrelin. Ghrelin targets growth hormone secretagog receptor and nociceptin.

On vascular smooth muscle cells, GPCRs are predominantly coupled to Gs. They thus activate adenylate cyclase and cause direct vasodilation. Among class-2 GPCRs that are activated by peptides, all those with vasoactivity directly operate as vasodilators. They are targeted by adrenomedullin and urocortins.[24]

On vascular endothelial cells, GPCRs can generate and release: (1) nitric oxide; (2) arachidonic metabolites such as prostacyclin; or (3) endothelium-derived hyperpolarizing factor that target adjoining smooth muscle cells, where they excite soluble guanylate cyclase, Gs-coupled prostanoid IP_1 receptors, and induce hyperpolarization, respectively. All these processes create vasodilation.

7.6.5.2 Heart and Its Nervous Command

Cardiac regulation by the sympathetic nervous system is carried out by β-adrenergic receptors that are coupled primarily to Gs. Produced cAMP targets hyperpolarization-activated, cyclic nucleotide-gated channels and activates protein kinase-A. The latter phosphorylates Ca_V channels, phospholamban, and troponin-I [767] (Vol. 5 – Chap. 5. Cardiomyocytes). These events explain positive chronotropic (C+; increased heart rate), dromotropic (D+; elevated conduction velocity through the nodal tissue), lusitropic (L+, enhanced cardiomyocyte relaxation), and inotropic (I+; improved contractility) effects of the sympathetic command.

On the other hand, the parasympathetic nervous system has negative chronotropic (C−) and dromotropic (D−) effects. It operates via muscarinic acetylcholine M_2 Gi/o-coupled receptor. In addition to the inhibition of cAMP production, this receptor activates atrial G-protein-regulated inward rectifier K^+ channels (GIRK; $K_{IR}3.1$ and $K_{IR}3.4$) via released Gβγ dimer [767]. Muscarinic receptors also inhibit voltage-gated Ca_V channels.

7.6.6 Airway Smooth Muscle Tone

Like in vascular smooth muscle cells, the tone of airway smooth muscle cells is mainly regulated by members of the Gq/11 and Gs families of G-protein subunits that cause bronchoconstriction and bronchodilation, respectively.

24. G-protein-coupled receptors on vascular smooth muscle cells that provoke vasoconstriction can be linked to more than one signaling cascade via different families of G-protein subunits (Gi/o, Gq/11, and/or G12/13). Subunits of the Gi/o, Gq/11, and/or G12/13 families activate primarily, but not exclusively, phospholipase-C and small monomeric GTPase Rho and inhibit adenylate cyclase, respectively [805].

Acetylcholine released from postganglionic parasympathetic nerves intervenes mainly via Gq/11-coupled M_3 receptors. Gq/11-coupled receptors in airway smooth muscles also comprise H_1 histamine, B_2 bradykinin, ET_B endothelin, and $CysLT_1$ leukotriene receptor [767]. Subunit Gi impedes the relaxation of airway smooth muscle cells.

Gs-coupled receptors provoke relaxation of contracted airway smooth muscle cells, such as β2-adrenoceptor, EP_2 prostaglandin, and IP prostacyclin receptor [767].

7.6.7 Platelet Activation

Platelet adhesion and activation is initiated by interaction with components of the subendothelial matrix, such as collagen and von Willebrand factor (Vol. 5 – Chap. 9. Endothelium). Collagen is able to cause a firm adhesion of thrombocytes to the subendothelial layer. Additional platelets are recruited by diffusible mediators, such as ADP, ATP, and thromboxane-A2, that are secreted by activated thrombocytes.

G-Protein-coupled receptors operate in platelet aggregation. Twelve most abundant platelet GPCRs include: (1) thrombin receptor, i.e., peptidase-activated receptors PAR_1 and PAR_4;[25] (2) adenosine receptors A_1, A_{2A}, and A_{2B}; (3) nucleotide receptors $P2Y_1$, $P2Y_{12}$, as well as $P2Y_{10}$; (4) GPR183[26] (5) succinate receptor-1 (SucnR1 or GPR91); (6) lysophosphatidic acid receptors LPA_1, LPA_3, LPA_4 (GPR23), and LPA_5 (GPR92); (7) α2a-adrenergic receptor; (8) glutamate $mGlu_3$ and $mGlu_4$; and (9) serotonin receptors $5HT_{1F}$ and $5HT_4$ [836].

Agent ADP activates the G-protein subunits Gq and Gi via $P2Y_1$ and $P2Y_{12}$ receptors, respectively. Thromboxane-A2 targets its cognate Gq-coupled receptor TP. Thrombin binds to peptidase-activated receptors that are coupled to Gq, G12/G13, and in some cases to Gi [767]. Adrenaline connects to Gz-coupled α2-adrenergic receptors to potentiate the effect of other platelet activators.

7.6.8 Leukocyte Migration

Leukocytes move according to a concentration gradient of chemoattractants. They can then enter into lymphoid organs to mature and, then, exit for antigen surveillance. They can also reach sites of infection.

Lymphocyte chemoattractants include chemokines, sphingosine 1-phosphate, and lysophosphatidic acid. Chemokine receptors act mainly via Gi subunits. Lysophospholipid receptors activate Gi, G12/G13, and Gq/G11 subunits. Hematopoietic cells are able to synthesize Gq, G11, and G16 subunits. Members of the G12/13 family target RhoA GTPase.

Neutrophils respond to diverse chemoattractants, such as N-formyl Met-Leu-Phe (fMLP), C5a, platelet-activating factor, or interleukin-8. Complement C5a element

25. Activated receptor PAR_1 is more potent than PAR_4.

26. A.k.a. Epstein-Barr virus-induced gene product-2 EBI_2. Epstein-Barr virus-induced gene product-1 (EBI_1) is the (CCL19 and CCL21) chemokine receptor CCR7. Up- and down-regulation of GPCRs and their ligands result from viral infections.

operates via Gi2, Gi3, and Gβ2 subunits to stimulate PI3Kγ as well as PLCβ2 and PLCβ3. In addition, PIP_3-dependent Rac exchanger-1 (PREx1) that acts as a guanine nucleotide-exchange factor for small GTPases Rac, is boosted by Gβγ dimer.

7.6.9 Mastocyte Activity

Mastocytes also express various GPCRs that intervene during allergy and anaphylaxis (with vasodilation, bronchoconstriction, etc.). Ligands, such as anaphylatoxins C3a and C5a, leukotrienes, sphingosine 1-phosphate, and platelet-activating factor mediate anaphylaxis, acting as transmitters within mastocytes and extracellular medium, including circulating biofluids. Pulmonary mastocytes express a huge number of GPCRs, such as [837]: (1) complement component receptors C3aR1 and C5aR1; (2) cysteinyl-leukotriene receptor $CysLT_1$; (3) sphingosine 1-phosphate receptor $S1P_1$; (4) corticotropin-releasing hormone (CRH) receptor CRF_1; (5) platelet-activating factor receptor (PAFR); (6) cadherin, EGF LAG seven-pass receptor CELSR1;[27] (7) Zn^{++}-activated GPR39; (8) progestin and adiponectin, C1q, and collagen domain (adipoQ)-containing receptor family member PAQR5; (9) lysophosphatidic acid receptor LPA_5; (10) succinate receptor-1 (SucnR1); and (11) GPR12.

Acupuncture is an old medical technique in traditional chinese medicine. In acupuncture, the internal organs are assumed to be interconnected by meridians, i.e., pathways in which the vital energy (Qi) flows throughout the body. Thin needles are manipulated (using lifting, thrusting, and rotation) to stimulate specific points (acupoints) of the body to restore the balance between Yin and Yang energy by removing blocks in the flow of Qi [806].[28] The effects of acupuncture can be explained by interactions between the nervous, circulatory, endocrine, and immune systems. The subcutaneous connective tissue is composed of cells embedded in the extracellular matrix mainly constituted by collagen and elastic fibers in a gel of glycoproteins and proteoglycans. Mastocytes can be presumed to play a major role in acupuncture. They are scattered, but localize predominantly near blood and lymphatic vessels and nerves. Their density is higher near acupoints [809].

A mechanical stress field results from local deformations of the connective tissue imposed by the needle motions. This stress field is sensed by the local population of mastocytes that react, in particular, by degranulation. Released molecules include calcitonin gene-related peptide (CGRP), heparin, histamine, leukotrienes (LTb4, LTc4, LTd4, and LTe4), platelet-activating factor, prostaglandin-E2, serotonin, substance-P, and thromboxane-A2 (Table 7.22). Mastocytes also secrete pep-

27. A.k.a. Flamingo homolog-2.

28. The acupuncture needle is inserted in the subcutaneous connective tissue and manipulated until a special sensation is obtained, which is supposed to prove that the needle has been placed in the proper location. The same acupoints can also be stimulated by other methods, such as moxibuxion, acupressure, and electroacupuncture. Moxibustion relies on heating acupoints by burning a moxa stick made of Artemisia vulgaris (common mugwort) to stimulate the blood circulation and produce a smoother flow of Qi [807]. Acupressure can be referred as needleless acupuncture by applying finger pressure. Electroacupuncture relies on the electrical stimulation of acupoints [808].

Table 7.22. Released molecules by the mastocyte and their effects. Acupuncture can be modeled by an immediate and a late response. Nerves and mastocytes exchange chemical messengers such as substance-P. The latter stimulates histamine and nitric oxide (NO) release. Calcitonin gene-related peptide (CGRP) causes a vasodilation; nitric oxide cooperates with CGRP to increase its positive inotropic effect that raises the local blood flow in dilated vessels. Histamine is quickly catabolized, thereby acting near the site of release. Resulting vasodilation and increased vessel wall permeability support the transfer of chemical mediators into the blood circulation. The NO concentration rises and enhances the vasodilation. Serotonin has a biphasic effect, as it triggers a vasoconstriction and promotes NO release, hence a subsequent vasodilation. Nerve growth factor (NGF), tumor-necrosis factor (TNF) and interleukins (IL) are potent mastocyte chemoattractants. Mastocyte chemotaxis is supported by matrix degradation by secreted peptidases.

Agent	Effects
CGRP	Vasodilation, positive chronotropy, inotropy, and lusitropy, mastocyte degranulation
Heparin	Blood clot prevention
Histamine	Vasodilation (directly and via NO), nerve stimulation
Leukotrienes	Vasodilation, vascular permeability elevation
IL, NGF, TNF	Chemotaxis
Prostaglandin-D2	Nerve stimulation
Prostaglandin-E2	Vasodilation, inhibition of mediator release
Serotonin	Vasoconstriction followed by NO-mediated vasodilation
Thromboxane-A2	Vasoconstriction, platelet aggregation
Tryptase, chymase	Matrix degradation for enhanced cell migration

tidases (e.g., tryptase), growth factors (e.g., FGF, gmCSF, and NGF), and cytokines (e.g., interleukins and tumor-necrosis factor). Nerve endings are stimulated and release substance-P that further activates mastocytes and triggers the production of nitric oxide.

Meridian can be assumed to be a neurovascular signaling tract. A compartmental model can then be designed. Compartment 1 is the acupoint region with its 3 components: (1) mastocytes, (2) blood and lymph vessels, and (3) nerves. Chemoattractants augment the mastocyte population (auto-amplification); autocrine signals bestow a self-sustained response. Nerves and mastocytes exchange cues. Compartment 2 is related to signal transmission to the central nervous system, either very rapidly via nervous impulses, or delayed via messenger convection through the blood circulation. A feedforward loop associated with elevated cardiac function enables an increased blood flow for a relevant material transport. Compartment 3 deals with signal processing with a quick and late responses corresponding to fast and delayed inputs. Compartment 4 represents outputs sent from the central nervous system and the body's response.

7.7 Crosstalk and Transactivations

G-protein-coupled receptors not only act via heterotrimeric G proteins, but can also function in a G-protein-independent manner. β2-Adrenergic receptors signal via either Gα or via tyrosine kinase Src according to the ligand concentration, low or high, respectively [810].

Activated Gi-coupled receptors often contribute to inositol phosphate signal triggered by Gq-coupled receptors (synergistic receptor crosstalk). For example, Gi-coupled adenosine A_1 and α2c-adrenergic receptors collaborate with Gq-coupled bradykinin B_2 and P2Y receptors. Gβ1γ2 Dimer as well as combinations of Gβ (Gβ1–Gβ3) with Gγ (Gγ2–Gγ7) increases potency of bradykinin or UTP messenger without interfering with $G\alpha_q$ signaling [811]. This GPCR crosstalk results from Gβγ exchange between Gi- and Gq-coupled receptors.

Certain GPCR effectors do not depend on G proteins. Crosstalk between GPCRs and small GTPases exists.[29] GPCR signaling through G proteins can activate Ras and Rho GTPases. Activated Ras triggers the mitogen-activated protein kinase cascade. RhoA, Rab, Arf, and ArfGEF can directly associate with GPCRs. GPCRs can also function as guanine nucleotide-exchange factors for small GTPases.

The interactions between GPCRs and small GTPases is required for cellular transport and cell migration. GPCRs are synthesized and modified in the endoplasmic reticulum, then transported to the Golgi body for additional changes, lastly to the plasmalemma. Rab GTPases are implicated in exocytosis. Rab GTPases also regulate GPCR endocytosis into early endosomes, GPCR targeting to lysosomes for degradation, or GPCR recycling from early endosomes. Arf1 and Arf6, with Arf GEFs (adpribosylation factor nucleotide-binding site opener [ARNO]) and β-arrestin, also control GPCR endocytosis [812].

Activated Gi-coupled chemoattractant receptor, the N-formylmethionyl-leucyl-phenylalanine receptor (fMLPR), leads to both Ras-dependent and -independent activation of Ral GTPase. Ral GEF RalGDS interacts with Ras, which activates Ral, hence, actin cytoskeleton reorganization (Ras-dependent mechanism). Plasmalemmal translocation of complexes made by β-arrestin and RalGDS allows binding to fMLPR and uncoupling from G protein, thus activates membrane-bound Ral (Ras-independent mechanism) [812].

Ligand binding to GPCRs activates small GTPases either directly or indirectly.[30] Rho-dependent responses can be activated by Gq and G12 (Fig. 7.2). Furthermore, G12 can bind and activate Rho-specific guanine nucleotide-exchange factors [813]. Mutant Gs-induced cAMP can activate small GTPases Rap1 and B-Raf that, in

29. The small GTPase superfamily includes at least five families: Ras GTPase subfamily (Ras, Rap, and Ral) of regulators of cell signaling, Rho GTPase subfamily (Rho, Rac, and Cdc42) of regulators of actin cytoskeleton, Rab GTPase subfamily of regulators of vesicular transport, Arf GTPase subfamily (Arf, Arl, and Sar) of regulators of vesicular transport, and Ran GTPase subfamily of regulators of microtubule organization and nucleocytoplasmic transport.

30. GPCRs for lysophosphatidic acid and thrombin induce stress fibers, focal adhesions, and cell rounding through Rho-dependent pathways [813].

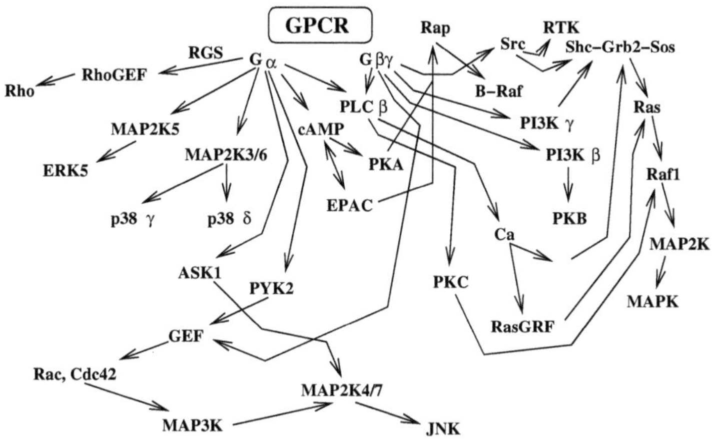

Figure 7.2. Interactions between GPCRs, small GTPAses and MAPKs. G-protein-coupled receptors (GPCR) activate mitogen-activated protein kinase (MAPK) using many pathways. Gβγ can stimulate Ras by the activation of receptor and non-receptor tyrosine kinases, thereby recruiting Sos to the plasmalemma. Activated $G\alpha_q$ can stimulate: (1) Raf1 via phospholipase-Cβ (PLCβ) and protein kinase-C (PKC) and (2) Ras via PLCβ and activation of (2a) Ca^{++}-dependent Ras guanine nucleotide-releasing factor (RasGRF) and (2b) Ca^{++}- and PKC-regulated proline-rich tyrosine kinase-2 (PYK2) of the focal adhesion kinase family. GPCR stimulation also leads to phosphorylation of Src homology and collagen (SHC), which subsequently forms complexes with growth factor receptor-bound-2 (GRB2), and activation of tyrosine kinases acting on Son of sevenless (Sos). PI3Kγ can then act on the SHC–GRB2–Sos–Ras pathway. PI3Kβ stimulated by GPCRs drives activation of Rac and p21-activated kinase to enhance Raf activity. Gi and Gs, via the binding of cAMP to exchange protein activated by cAMP (EPAC) Rap1GEF EPAC and phosphorylation of Rap1 by PKA, can use the Rap1 pathway, which stimulates B-Raf. Gq, via PYK2, with adaptor Crk and paxillin, can activate GEFs for Rac and Cdc42, leading to the activation of MAP2K4/7 and JNK. G12 can stimulate MAP2K4/7 via apoptosis signal-regulating kinase-1 (ASK1). (8) G12/13, with regulators of G-protein signaling (RGS) also targets RhoGEFs for Rho activation. G12/13 and Gq can stimulate ERK5 (or big mitogen-activated kinase BMK1) via MAP2K5. G12, Gq, and Gβγ can activate P38MAPKα in cooperation with Btk and Src kinases. Subunits G12/13 and Gq can also activate P38MAPKγ and P38MAPKδ via MAP2K3/6 (Sources: [768, 812]).

turn, stimulate MAP2K and MAPK. Gβγ activates the MAPK pathway using a Ras-dependent process [814]. Gq-coupled receptors use both PKC-dependent and -independent pathways to stimulate the MAPK pathway.

The transactivation of receptor tyrosine kinases by GPCRs encompasses: (1) the activation of receptor Tyr kinases via non-receptor Tyr kinases; (2) formation of complexes between GPCRs and RTKs; and (3) release of RTK ligands.

Stimulation of GPCRs can stimulate receptor Tyr kinases, such as epidermal growth factor receptors, via: (1) proteolytic cleavage by ADAM metallopeptidase and release of EGF-like ligands, such as heparin-binding EGF-like growth factor, and (2) NRTK activation that can phosphorylate EGFR tyrosine residues.

Activation by GPCRs of the Ras–MAPK pathway uses alternate RTK-dependent and -independent pathways according to cell types. Angiotensin-2 in thoracic aortic smooth muscle cells requires RTK activity to signal to ERK, but not in smooth muscle cells of the renal microvasculature [815].

β2-Adrenergic receptor mediates extracellular signal-regulated kinase activation via assembly of a multireceptor complex with the epidermal growth factor receptor following co-internalization of both receptors into clathrin-coated vesicles [816]. Several NRTKs (CSK, Lyn, BTK, proline-rich tyrosine kinase-2, and focal adhesion kinase; Vol. 4 – Chap. 3. Cytosolic Protein Tyr Kinases) could mediate MAPK activation by Gi and Gq [768]. Src or Src-like kinases can phosphorylate SHC on stimulation by Gβγ and α-adrenoceptors, after recruitment of β-arrestin and GRK2 kinase. Kinase Src can be activated also by interaction with Gi and Gs or β3-adrenoceptors.

Subunit Gq can stimulate components of the MAPK module via: (1) a PKC-dependent, Ras-independent mechanism; (2) PKC- and Ras-dependent process; or (3) PKC-independent, Ras-dependent procedure, depending on the cell type and receptor expression level.

Isoforms of PI3K kinase can be required for GPCR-induced MAPK activation (Fig. 7.2). Subtype PI3Kγ is activated by direct interaction with Gβγ subunit. It can then act on the SHC–GRB2–SOS–Ras pathway. The Gβγ dimer activates PI3Kβ, which then stimulates PKB [817].

Arrestins bind many phosphorylated GPCRs for endocytosis. β2-Adrenoceptors can associate with the Na^+–H^+ exchanger regulatory factor (NHERF); angiotensin receptors AT_{1A} with Janus kinase JaK2; metabotropic glutamate receptors with Homers, which can complex with IP_3 receptors; muscarinic acetylcholine M_3 receptors and AT_{1A} with GTPases Rho and ARF [768].

GPCR ligands, such as lysophosphatidic acid, endothelin-1, platelet-activating factor, and thrombin can induce a rapid transient phosphorylation of epidermal growth factor receptors in fibroblasts and in vascular smooth muscle cells. Catecholamines, angiotensin-2, and endothelin-1 act in cardiomyocytes where they can activate heparin-binding epidermal growth factor with members of a disintegrin and metallopeptidase (ADAM) family and cause GPCR-induced cardiac hypertrophy (Fig. 7.3). Different types of G proteins (Gi, Gq, and G12), different metallopeptidases of the ADAM family (ADAM10, ADAM12, and ADAM17), and different EGF-like ligands (heparin-binding epidermal growth factor, transforming growth factor-α, and amphiregulin) are involved in inter-receptor crosstalk (or transactivation) [818].

7.8 Regulators of G-Protein Signaling

Regulators of G-protein signaling are GTPase-accelerating proteins that activate the GTPase activity of α subunits of heterotrimeric G proteins, thereby inactivating G protein and rapidly switching off the GPCR signaling pathway.

All the RGS proteins contain an RGS domain. Small RGS proteins, such as RGS1 and RGS4, are slightly greater than the RGS domain. Other RGSs contain additional

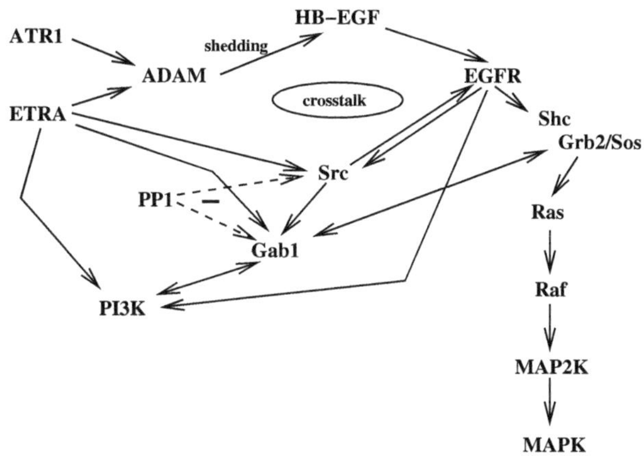

Figure 7.3. Crosstalk between receptors of angiotensin-2 (AT_1) and endothelin-1 (ET_A, or ETRA) and epidermal growth factor receptor (EGFR) in the cardiomyocyte. Transactivation of EGFR requires metallopeptidases of the ADAM family, particularly ADAM12, heparin-binding EGF-like growth factor, docking molecule GRB2-binding protein GAB1, Src Tyr kinase, and phosphoinositide 3-kinase. Transactivation of EGFR leads to MAPK activation.

domains that confer further functionality, such as DEP for membrane targeting, GGL (G-protein-γ subunit-like) for binding Gβ subunits, GoLoco for guanine nucleotide inhibitory activity, PH for guanine nucleotide-exchange stimulating effect, and PDZ for GPCR, PTB for phosphotyrosine, PX for phosphatidylinositol, and RBD for Ras binding.

Regulators of G-protein signaling can also speed up the onset of signaling, in addition to accelerating deactivation without changing amplitude or sensitivity of the signal. This paradoxical activity has been attributed to enzymatic (GAP) or scaffolding (non-GAP) functions. Yet, fast GTPase activity of RGSs may suffice to explain the activation kinetics and agonist sensitivity, i.e., increased onset, deactivation rates, and blunting of sensitivity [819].

7.9 G-Protein-Coupled Receptor Kinases

Agonist–GPCR binding strongly enhances interactions between GPCRs and G proteins, G-protein-coupled receptor kinases (GRK), and β-arrestins. Agonists of GPCRs can simultaneously activate G-protein- and β-arrestin-dependent pathways. Agonist-activated GPCRs are also phosphorylated by G-protein-coupled receptor kinases. Phosphorylation of GPCRs by GRKs decreases GPCR interactions with G proteins and increases GPCR binding of β-arrestins (Sect. 7.11).

β-Arrestins can then further desensitize GPCR-mediated signaling by several mechanisms: (1) receptor hindrance due to β-arrestin attachment; (2) recruitment of second messenger-degrading enzymes; and (3) scaffold proteins that facilitate receptor endocytosis. However, β-arrestin binding can initiate additional signaling programs via recruitment of signaling mediators such as mitogen-activated protein kinases to activated receptors.

Therefore, interaction of GPCRs with β-arrestins prevents GPCR signaling via G proteins, but can simultaneously prime other signaling pathways as well as associate active receptors to clathrin-coated pits to facilitate receptor endocytosis that controls GPCR desensitization and resensitization, in addition to their endocytosis-mediated activity.

Ubiquitous G-protein-coupled receptor kinases constitute a 7-member family (GRK1–GRK7). These kinases that have different distribution patterns among the body's tissues as well as distinct binding preferences for some receptors. They can yield functional specialization. Isoforms of GPCR kinases determine β-arrestin function, as they can lead to functionally distinct pools of β-arrestin, possibly by phosphorylating distinct receptor sites. Enzymes GRK5 and GRK6 are required for β-arrestin-mediated MAPK signaling.

G-protein-coupled receptor kinases, particularly GRK2 isoform, not only target G-protein-coupled receptors, but also numerous other substrates, such as receptor Tyr kinases (PDGFR and EGFR), as well as members of the GIT family of GTPase-activating proteins of adpribosylation factor (ArfGAP), DM2 ubiquitin ligase, and catalytic subunit P110 of PI3K, PKB, MAP2K, and P38MAPK kinases. They also have many interacting proteins. The GRK2 subtype can attenuate PDGF-dependent proliferation of smooth muscle cells, but favors mitogenic signaling induced by EGF in osteoblasts and the Hedgehog receptor Smoothened in fibroblasts [820]. In addition, increased GRK2 concentration potentiates migration of epithelial cells directed by fibronectin and sphingosine 1-phosphate.

G-protein-coupled receptor kinases possess their own regulators. Calcium-binding recoverin is a neuronal calcium sensor that localizes to the retina and serves as a Ca^{++}-dependent regulator of GRK1 kinase. At high Ca^{++} concentration, recoverin sequesters GRK1, whereas at low Ca^{++} level, the complex dissociates and GRK1 becomes active.

In response to GPCR stimulation, GRK2 is rapidly phosphorylated by Src (Tyr13, Tyr86, and Tyr92) and MAPK (Ser670) kinases and then degraded by the proteasome [820]. Cell cycle progression requires changes in the activity or concentration of signaling proteins. The GRK2 kinase is phosphorylated (Ser670) by cyclin-dependent kinase CDK2 during the G2–M transition and subsequently binds to peptidyl-prolyl isomerase Pin1 to be degraded [820].

7.10 G-Protein-Coupled Receptor Phosphatases

G-protein-coupled receptors experience cycles of phosphorylation and dephosphorylation. Membrane-associated G-protein-coupled receptor phosphatase (GRP) aims at dephosphorylating GPCRs that have been phosphorylated by G-protein-coupled receptor kinases to cause agonist-dependent receptor deactivation, hence restoring the ground state of the receptor. [31]

A family of receptor phosphatases contribute to the regulation of G-protein-coupled receptors. G-protein-coupled receptor phosphatase is an unusual form of protein phosphatase-2 with peculiar subcellular distribution and substrate specificity [822].

Desensitization of GPCRs is initiated by phosphorylation of Ser and Thr residues of intracellular loops and C-terminus by second messenger-activated kinases, such as protein kinase-A and GPCR kinases. Once phosphorylated, GPCRs generally bind to arrestin that prevents subsequent coupling between GPCR and G-protein and elicits GPCR integration into clathrin-coated pits for endocytosis. Moreover, when arrestins bind to GPCRs with high affinity, they preclude phosphatase action on receptors.

On the other hand, endocytosis can be a prerequisite for resensitization of G-protein-coupled receptors, such as for β2-adrenoceptors, when dephosphorylation occurs in endosomes, but not at the plasma membrane. Once internalized, low-affinity agonists such as adrenaline dissociate rapidly from β2-adrenoceptors. [32] This event is followed by arrestin dissociation, because the tethering of high-affinity arrestin needs agonist binding, in addition to GRK site phosphorylation. The rates of endocytosis and recycling control the level of β2-adrenoceptors in the plasma membrane.

Class-A GPCRs have higher affinity for arrestin-3 (or β-arrestin-2) than arrestin-2 (or β-arrestin-1). They internalize without arrestin and recycle rapidly to the plasma membrane. On the other hand, class-B GPCRs co-internalize with arrestin-2 or -3 and recycle slowly, often undergoing intracellular degradation. Once liganded, thyrotropin-releasing hormone receptor, neither class-A nor -B GPCR, is rapidly

31. Phosphorylation of GPCRs provokes a slight decrease in receptor activity. Above all, it also enhances GPCR affinity for arrestin. Arrestin binding can inactivate receptors by preventing their coupling to heterotrimeric G proteins triggered by agonist binding. In Drosophila, retinal degeneration-C (RDgC) is an unusual Ser–Thr phosphatase required for rhodopsin dephosphorylation [821]. Unlike vertebrate opsin, most invertebrate photopigments are not bleached after light activation, but can be photoconverted between the rhodopsin form and a thermally stable, active metarhodopsin form.

32. Agonists may be removed quickly from β2-adrenoceptors, especially in synapses, and GPCRs must be rapidly resensitized for appropriate functioning. β2-Adrenergic receptors contain phosphorylation sites for both GRK (Ser355 and Ser356) and PKA (Ser262) [823]. The concentration of agonist and rate constants of phosphorylation and dephosphorylation define the level of PKA- and GRK site phosphorylation. Dephosphorylation of the PKA site does not necessitate endocytosis, as it happens at the plasma membrane. Dephosphorylation of GRK sites occurs with a phase lag with respect to that of PKA site (possibly due to time necessary for phosphatase recruitment) and proceeds slowly. The rapid phase of resensitization (0–5 min) does not depend on dephosphorylation at the plasma membrane, but β-arrestin dissociation upon agonist stimulation of adenylate cyclase [823]. Rapid restoration of adenylate cyclase activity after removal of agonist can explain the fast dissociation of β-arrestin from membrane-bound β2ARs in association with recycling of the receptor to the plasma membrane. Moreover, the pattern of phosphatase inhibitor effects is much more consistent with PP1, rather than PP2 in PKA and GRK site dephosphorylation [823].

phosphorylated at numerous sites by GRKs, but not PKA enzyme. It then internalizes with arrestins-2 and -3. It can be dephosphorylated at the plasma membrane or in endosomes, once its hormonal agonist is removed.

β2-Adrenergic receptor is slowly phosphorylated by both PKA and GRK kinases. The latter recruits arrestin, but internalizes without it. On the other hand, thyrotropin-releasing hormone receptor co-internalizes with arrestin. Both receptors are dephosphorylated following agonist removal, but the latter is dephosphorylated much more rapidly at the plasma membrane.

Resensitization of most GPCRs comprises arrestin dissociation, receptor dephosphorylation, and recycling to the plasma membrane. Endocytosis often enables action of protein kinase-A, whereas GRKs mostly intervene at the plasma membrane. The GPCR cytoplasmic tails support arrestin linkage, but GPCR phosphorylation and dephosphorylation by GRKs and GRPs, respectively, depend on regions outside the cytoplasmic tail, but not their C-termini [824].

7.11 Arrestins

G-protein-coupled receptors stimulate heterotrimeric guanine nucleotide-binding protein and/or β-arrestins to regulate the intracellular concentration of various second messengers. Signaling is terminated by receptor desensitization and degradation of second messengers.

Four arrestin isoforms exist: arrestins-1 and -4 that regulate opsins (GPCRs of photoreceptor cells of the retina) and arrestins-2 and -3 (or β-arrestins-1 and -2, respectively.[33] Ubiquitous β-arrestins control the activity of most G-protein-coupled receptors.[34]

β-Arrestins were originally defined as terminators of GPCR signaling. They actually mediate desensitization and endocytosis of GPCRs and other types of receptors and molecules.[35]

However, β-arrestins also act as signal transducers. They indeed serve as signaling adaptors for GPCRs and other types of receptors. β-Arrestin-biased agonism is related to GPCR signaling after ligand binding via β-arrestins rather than subunits of heterotrimeric G proteins.

33. Some members of the arrestin family of adaptor proteins are called β-arrestins because they were discovered as β2-adrenoceptor-binding proteins.

34. β-Arrestins associate with β2-adrenergic receptor, angiotensin-2 type-1A receptor (AT_{1A}), arginine vasopressin receptor (V_2 [AVPR2]), peptidase-activated receptor-2, and parathyroid hormone receptor, to activate extracellular signal-regulated protein kinases (Vol. 4 – Chap. 5. Mitogen-Activated Protein Kinase Modules). G-Protein-coupled receptor kinases are required, at least, for both AT_{1A} and V_2 receptors.

35. β-Arrestins act on receptor Tyr kinases, such as insulin-like growth factor-1 receptor. In addition, β-arrestins are implicated in endocytosis of VE-cadherin. β-Arrestin-2 also binds to low-density lipoprotein receptor and enhances LDLR endocytosis [825]. β-Arrestin-2 links to type-3 transforming growth factor-β receptor (TβR3; a.k.a. β-glycan) and triggers the receptor phosphorylation by type-2 TGFβ receptor kinase (TβR2), thereby downregulating TGFβ signaling [826].

Mechanical stretch induces β-arrestin-biased signaling from angiotensin-2 receptors AT_1 in the absence of ligand or G-protein activation to promote cell survival [827]. In cardiomyocytes, acute variations in mechanical stress trigger an AT_1-mediated conformational change of β-arrestin similar to that caused by a ligand that primes a β-arrestin-associated process.

7.11.1 Post-Translational Modifications of Arrestins

Phosphorylated β-arrestins pertain to 2 classes: (1) class-A β-arrestin binders that transiently link GPCRs only at the plasma membrane, as they are deubiquitinated, and (2) class-B β-arrestin binders that form stable β-arrestin–GPCR complexes connected to endocytic vesicles.

Ligand-stimulated ubiquitination of β-arrestins governs the stability of receptor-β-arrestin interactions. Ubiquitination of β-arrestins tunes signal strength and localization, as ubiquitination serves to further modify the active conformation of β-arrestin to recruit endocytic and signaling mediators among members of the β-arrestin interactome.

β-Arrestin ubiquitination by DM2 ubiquitin ligase causes rapid endocytosis of β2-adrenergic receptor. Deubiquitinase ubiquitin-specific peptidase USP33 binds β-arrestin-2. Enzymes USP33 and DM2 favor the stability or lability of GPCR–β-arrestin complex, respectively, thereby controlling the longevity and subcellular localization of signalosomes. β2-Adrenoceptors promote a conformational change of β-arrestin for deubiquitinase binding, hence destabilizing β2AR–β-arrestin binding and lowering ERK activation [828]. Conversely, ubiquitination of β-arrestin by DM2 promotes β-arrestin-dependent signaling of activated β2-adrenoceptors. On the other hand, V_2 vasopressin receptor elicits a distinct β-arrestin conformation that favors the dissociation of deubiquitinase, thereby stabilizating β-arrestin–receptor complexes and increasing ERK activation. In summary, DM2 ligase directs endocytosis and signaling, whereas deubiquitinase USP33 causes signal termination.

7.11.2 Receptor Desensitization

β-Arrestins help to restore G-protein-coupled receptors to an inactive state after stimulation, although they can also have positive signaling roles. β-Arrestin-1 and -2 desensitize certain plasmalemmal receptors in conjunction with G-protein-coupled receptor kinases (GRK).[36]

Agonist-promoted desensitization indeed comprises phosphorylation of the receptor by a GRK that promotes the binding of β-arrestin that partially uncouple GPCR from its effectors. This event can lead to GPCR endocytosis, with either a recycling to the cell surface or degradation.

---

36. β-Arrestins bind to activated G-protein-coupled receptors after receptor phosphorylation by G-protein-coupled receptor kinases.

β-Arrestins recruit phosphodiesterases to activated Gs-coupled β2-adrenergic receptors to limit signaling. β-Arrestins also stop Gq-protein-coupled receptor signaling. In particular, β-arrestins regulate signaling primed by Gq-coupled M_1 muscarinic receptors by limiting the production of diacylglycerol and enhancing the degradation rate of this effector [829]. β-Arrestins indeed favor conversion of diacylglycerol into phosphatidic acid and interact with diacylglycerol kinases that degrade diacylglycerol.

7.11.3 Scaffolding of Intracellular Signaling Complexes

β-Arrestins also serve as endocytic and signaling adaptors. β-Arrestins indeed intervene as scaffolds for intracellular assembly of signaling complexes. Dissociation of β-arrestins allows recycling of G-protein-coupled receptors via Rab-dependent mechanisms. β-Arrestins target active kinases to specific locations within the cell and can regulate the lifetime of G-protein-coupled receptors in endosomes [830].

7.11.4 Examples of β-Arrestins–GPCR Linkages

7.11.4.1 β-Arrestins and Angiotensin-2 Receptors

G-protein-coupled receptor kinases and β-arrestins switch off G-protein-dependent signaling, but transduce another kind of signaling from receptors, such as AT_{1A} angiotensin-2 receptor.[37] Such a signaling causes positive inotropy and lusitropy (Vols. 5 – Chap. 6. Heart Wall and 6 – Chap. 3. Cardiovascular Physiology), possibly either by regulating the cytosolic Ca^{++} concentration or Ca^{++} sensitivity of troponin and myosin-binding protein-C [831].

7.11.4.2 β-Arrestins and Adrenoceptors

β2-Adrenergic receptors cooperate with Gi protein and β-arrestin in the cardiomyocyte, in addition to major β1-adrenergic signaling, to stimulate $Ca_V1.2$ channels and increase the intracellular Ca^{++} concentration. On the other hand, β2-adrenergic receptor stimulation via the Gs–PKA signaling inhibits TNFα-induced NFκB activation by stabilizing IκBα via β-arrestin-2. β-Arrestin-2 thus has an opposite role on NFκB signaling.

7.11.4.3 β-Arrestins and Smoothened GPCR

The association of the GPCR Smoothened of the morphogen sonic Hedgehog (Sect. 10.2) with β-arrestin-2 and its phosphorylation by G-protein-coupled receptor kinase-2 promote Smoothened endocytosis via clathrin-coated pits [832].

---

37. This signaling is thus carried out independently from the $G\alpha_q$–PKC pathway. β-Arrestin interacts with the receptor and prevents the coupling between the receptor and G proteins.

7.11.4.4 β-Arrestins and Frizzled GPCR

After binding of Wnt morphogens to Frizzled receptors (Sects. 7.13.22 and 10.3), cytoplasmic Disheveled is recruited to the plasma membrane and phosphorylated. The adaptor β-arrestin-2, also recruited by Frizzled-4, binds to phosphorylated Disheveled and allows Frizzled-4 endocytosis [833].

7.11.4.5 β-Arrestins and Lysophosphatidic Acid Receptors

The bioactive phospholipid lysophosphatidic acid binds to cognate G-protein-coupled receptors (Sect. 7.13.35) to prime the signaling cascades and activate several transcription factors such as nuclear factor-κB (Vol. 4 – Chap. 9. Other Major Signaling Mediators). β-Arrestin-2 (but not β-arrestin-1) links to caspase-recruiting domain and membrane-associated guanylate kinase homolog domain-containing protein CARMA3 that is recruited to LPA receptors for NFκB activation and interleukin-6 production [834]. Scaffold protein CARMA3 associates with BCL10 and MALT1 to activate IκB kinase that target inhibitor of NFκB IκBα. Receptors of LPA trigger the Gq–PKC signaling with different post-translational modification of targets that can then serves as activators or inhibitors.

7.12 Other Partners of G-Protein-Coupled Receptors

Some GPCR-interacting proteins connect to specific GPCRs upon agonist binding to GPCRs to prime particular receptor signaling as mediators, whereas other interactors associate with receptors independently of agonists, thus acting as modulators, but not mediators. Transfer of and signaling from GPCRs depend on the context.

7.12.1 Regulation of GPCR Activity

Activity of G-protein-coupled receptors can be regulated by interactions between GPCRs and not only kinases, phosphatases, arrestins, and other receptor types, but also receptor-selective, cytoplasmic or transmembrane, proteic partners (Tables 7.23 and 7.24).

G-Protein-coupled receptor-interacting proteins, the production of which differs according to the cell type, contribute to fine-tuning of GPCR function. Partners of GPCRs can: (1) regulate GPCR transport, as they can be required for exocytosis of their associated GPCRs down to the plasma membrane or sort endocytosis routes between recycling back to the plasma membrane or lysosomal degradation; (2) anchor GPCRs in suitable subcellular regions; (3) organize receptor signaling via G proteins; (4) act as scaffolds to modulate G-protein signaling, as they enhance signaling by tethering effectors in the vicinity of activated receptors or lower signaling intensity and/or duration either by impeding receptor–G-protein interactions or by

Table 7.23. G-Protein-coupled receptor interactors (modulators of GPCR signaling), excluding G proteins, GPCR kinases, arrestins, and regulators of G-protein signaling (**Part 1**; Source: [835]; AKAP: A-kinase anchor protein [AKAP5 and AKAP12 are also denoted AKAP79 and AKAP250, respectively]; AR: adrenergic receptor; AT_1: angiotensin-2 receptor; D_2: dopamine receptor; DOR, KOR, MOR: δ-, κ-, μ opioid receptor; Fz: Frizzled; GABA: gamma-aminobutyric acid; JaK: Janus kinase; LPAR: lysophosphatidic acid receptor; MAGI: membrane-associated guanylate kinase; mGluR: metabotropic glutamate receptor; MT: melatonin receptor; MuPP: multiple PDZ domain-containing protein; PKA/C: protein kinase-A/C; PAFR: platelet-activating factor receptor; PTH1R: parathyroid hormone-1 receptor; STAT: signal transducers and activators of transcription; V_2R: type-2 vomeronasal receptor).

Interactor	GPCR	Action
AKAP5	β1/2AR	PKA tethering
AKAP12	β1/2AR	PKA tethering
Calmodulin	$5HT_{1A}$	Competition with PKC for GPCR phosphorylation
	$5HT_{2A}$	Impairment of G-protein coupling
	$5HT_{2C}$	Promotion of arrestin-dependent ERK activation
	D_2	Modulation of G-protein signaling
	mGluR7	Regulation of GPCR phosphorylation
	PTH1R	GPCR inhibition
	V_2R	Enhancement of GPCR-induced Ca^{++} signaling
	MOR	Inhibition of G-protein coupling
Homer	mGluR1/5	Regulation of GPCR signaling and location
JaK2	AT_1	Promotion of JaK–STAT signaling
	PAFR	Promotion of JaK–STAT signaling
MAGI3	Fz4, LPAR2	Potentiation of GPCR-mediated ERK activation
MuPP1	$GABA_B$	Enhancement of GPCR-mediated Gi signaling
	MT_1	Enhancement of GPCR-mediated Gi signaling

recruiting inhibitors of G-protein signaling to the receptor; (5) directly mediate receptor signal transmission; and (6) influence binding of ligands, such as hormones, neurotransmitters, or sensory stimuli to GPCRs [835].

Interactors of GPCRs can initiate additional signal transduction cascades. Several GPCRs can initiate signaling via interactions with members of the Janus kinase family (Vol. 4 – Chap. 3. Cytosolic Protein Tyr Kinases) as well as non-receptor protein tyrosine phosphatase such as PTPn11 (or SHP2; Vol. 4 – Chap. 7. Cytosolic Protein Phosphatases) that complement G-protein-mediated signaling. On the other hand, agonist stimulation can dissociate an interacting protein from a GPCR, thereby altering the corresponding intracellular signaling pathway.

Among GPCR-associated proteins that enhance the efficiency of some G-protein-mediated signaling axes, members of the *A-kinase anchoring protein* family (AKAP) connect to β-adrenergic receptors and protein kinase-A to raise PKA-mediated phosphorylation of signaling mediators, such as β2AR and its effectors.

On the other hand, GPCR interactors can reduce G-protein-mediated signaling. In addition to arrestins, spinophilin interacts with some GPCRs, such as dopamine, adrenergic, and muscarinic acetylcholine receptors and several regulators of G-

Table 7.24. G-Protein-coupled receptor interactors, excluding G proteins, GPCR kinases, arrestins, and regulators of G-protein signaling (**Part 2**; Source: [835]; AR: adrenergic receptor; D_2: dopamine receptor; DOR, KOR, MOR: δ-, κ-, μ-opioid receptor; LPAR: lysophosphatidic acid receptor; mAChR: muscarinic acetylcholine receptor; MCHR: melanin-concentrating hormone receptor; mGluR: metabotropic glutamate receptor; MT: melatonin receptor; Ncdn: neurochondrin; NHERF: Na^+–H^+ exchange regulatory factor; P2Y: nucleotide receptor; PKA/C: protein kinase-A/C; PI3KR1/2: PI3K regulatory subunit (P85α/β); Ppl: periplakin [a.k.a. cornified envelope precursor protein]; PPP1R9b: protein phosphatase-1 regulatory (inhibitory) subunit-9B [a.k.a. PPP1R6, PPP1R9, neurabin-2, and spinophilin]; PTH1R: parathyroid hormone-1 receptor; RhoGEF*i*: Rho guanine nucleotide-exchange factor [$i = 11, 12$; a.k.a. ArhGEF*i*]; SSTR: somatostatin receptor).

Interactor	GPCR	Action
Ncdn	MCHR1	Disruption of G-protein signaling
NHERF1	PTH1R	Enhancement of GPCR-mediated Gq signaling
	β2AR	Activation of Na^+–H^+ exchange
	KOR	Activation of Na^+–H^+ exchange
NHERF2	LPAR2	Enhancement of GPCR-mediated Gq signaling
	mGluR5	Prolongation of GPCR-caused Ca^{++} mobilization
	$P2Y_1$	Prolongation of GPCR-caused Ca^{++} signaling
	PTH1R	Enhancement of GPCR-mediated Gq signaling
PI3KR1/2	SSTR2	Mediation of survival
RhoGEF11	LPAR2	Facilitation of GPCR-mediated Rho activation
RhoGEF12	LPAR2	Facilitation of GPCR-mediated Rho activation
Ppl	MCHR1,	
	MOR	Impairment of G-protein signaling
PPP1R9b	D_2	Reduction of G- and arrestin-mediated signaling
	α2AR	Reduction of GPCR-mediated Ca^{++} signaling
	mAChR	Reduction of GPCR-mediated Ca^{++} signaling

protein signaling that act as GαGAP and can also link to some GPCRs to reduce the intensity and duration of GPCR-stimulated G-protein signaling.

Calmodulin associates with Ca^{++} and dopamine, metabotropic glutamate, and serotonin, as well as other receptors to lower G-protein coupling, hence directing a inhibitory feedback to restrain GPCR-initiated G-protein signaling. Nonetheless, G-protein-independent signaling can be potentiated by GPCR–calmodulin interactions. For example, the calmodulin–$5HT_{2C}$ complex promotes arrestin-mediated signal transduction. Periplakin and neurochondrin, the former tethering to μ-type opioid receptor (MOR) and both to melanin-concentrating hormone receptor MCHR1 as well as few other GPCRs, attenuate G-protein-mediated signaling.

7.12.2 Regulation of Intracellular GPCR Transfer and Plasmalemmal Anchoring

Once synthesized, GPCRs take the exocytosis route to localize into the plasma membrane. After agonist stimulation, most GPCRs are internalized into endosomes

Table 7.25. Regulators of GPCR trafficking and/or ligand binding, excluding GPCR kinases, arrestins, and regulators of G protein signaling proteins (**Part 1**; Source: [835]; AR: adrenergic receptor; AT: angiotensin-2 receptor; ATBP50: 50-kDa AT_2 receptor binding protein; CNR: cannabinoid receptor; $D_{1(2)}$: D1 (or D2) dopamine receptor; DOR, KOR, MOR: δ-, κ-, μ opioid receptor; DRIP78: 78-kDa dopamine receptor-interacting protein; EP: prostaglandin receptor; GABA: gamma-aminobutyric acid; GASP: GPCR-associated sorting protein; GEC: glandular epithelial cell protein; M10: M10 family members of the major histocompatibility complex class 1b; MAGI: membrane-associated guanylate kinase, WW and PDZ domain-containing protein; mGluR: metabotropic glutamate receptor; MPP: MAGUK p55 subfamily member; MC2R: melanocortin receptor-2; MRAP: MC2R-accessory protein; MuPP1: multiple PDZ domain protein; NHERF: Na^+–H^+ exchange regulatory factor; NINA: neither inactivation nor afterpotential protein; OdR: odorant response abnormal protein; PICK1: protein interacting with C kinase; PSD: postsynaptic density protein; SSTR: somatostatin receptor; V_2R: type-2 vomeronasal receptor).

Partner	GPCR	Effect
ATBP50	AT_2	Promotes GPCR surface density
DRIP78	AT_1, D_1	Promotes GPCR surface density
GASP1	CNR1, D_2, DOR	Favors GPCR lysosomal degradation
GEC1	EP_3, KOR	Promotes GPCR surface density
Homer	mGluR1/5	Regulates GPCR signaling
M10	V_2R	Promotes GPCR surface density
MAGI2	β1AR	Promotes β1AR endocytosis
MPP3	$5HT_{2C}$	Impedes $5HT_{2C}$ endocytosis
MRAP1/2	MC2R	Promotes GPCR exocytosis
MuPP1	$5HT_{2C}$	Promotes GPCR clustering
	$GABA_B$	Promotes GPCR stability
	SSTR3	Targets GPCR to tight junctions
NHERF1	β2AR, KOR	Promotes GPCR recycling
NINAa	Rhodopsin	Promotes GPCR genesis and trafficking
OdR4	OdR10	Promotes GPCR surface density
PICK1	mGluR7a	Promotes GPCR clustering
PSD95	$5HT_{2A}$	Impedes GPCR endocytosis
	β1AR	Reduces GPCR endocytosis

and take one path among 2 possible routes: lysosomes for degradation or recycling endosomes for reinsertion into the plasma membrane. Interactors of GPCRs can influence exocytosis and post-endocytic sorting of GPCRs [835] (Tables 7.25 and 7.26).

Certain GPCR-interacting proteins regulate GPCR folding after translation before exocytosis as well as transport to and insertion into the plasma membrane.[38]

38. Ran-binding protein RanBP2, dynein light-chain component T-complex testis-specific protein DynLT1 (a.k.a. TCTex1), glandular epithelial cell protein GEC1 (a.k.a. $GABA_A$ receptor-associated protein-like protein GABARAPL1), receptor of activated protein kinase RACK1 (a.k.a. guanine nucleotide-binding-like protein GNB2L1), dopamine receptor-interacting protein DRIP78 (a.k.a. heat shock protein HSP40 [DnaJ] molecular chaperone

Table 7.26. Regulators of GPCR trafficking and/or ligand binding, excluding GPCR kinases, arrestins, and regulators of G protein signaling proteins (**Part 2**; Source: [835]; AR: adrenergic receptor; CalcR: calcitonin receptor; CGRP, calcitonin gene-related peptide; CRLR: calcitonin receptor-like receptor; DOR, KOR, MOR: δ-, κ-, μ opioid receptor; DynLT: dynein light chain Tctex (T-complex testis-specific) protein; mGluR: metabotropic glutamate receptor; OR: olfactory receptor; PAR: proteinase-activated receptor; RACK1: receptor of activated protein kinase-C1; RAMP: receptor activity-modifying protein; RanBP: Ran-binding protein; REEP: receptor expression-enhancing protein; RTP: receptor transporting protein; Shank: SH3- and multiple ankyrin repeat domain-containing protein; SNx: sorting nexin; SSTR: somatostatin receptor; T_2R: taste receptor-2; TXA2R: thromboxane-A2 receptor; USP: ubiquitin-specific-processing peptidase 4).

Partner	GPCR	Effect
RACK1	TXA2R	Promotes GPCR exocytosis
RAMP1	CRLR	Forms functional CGRP receptors
	CalcR	Forms functional amylin receptors
RAMP2	CRLR	Forms functional adrenomedullin receptors
RAMP3	CRLR	Forms functional adrenomedullin receptors
	CalcR	Forms functional amylin receptors
RanBP2	Opsin	Promotes GPCR exocytosis
REEP	OR	Promotes GPCR exocytosis
	T_2R	Promotes GPCR surface density
RTP	OR	Promotes GPCR exocytosis
	T_2R	Promotes GPCR surface density
RTP4	DOR, MOR	Promotes GPCR surface density
Shank	Lphn1	Promotes GPCR clustering
	mGluR1/5	Anchors GPCR in mature dendritic spines
SNx1	PAR1 (F2R)	Facilitates GPCR degradation
Syntrophin	α1dAR	Enhances GPCR stability
DynLT1	Rhodopsin	Promotes apical GPCR delivery
USP4	α2aAR	Promotes GPCR surface density

Other GPCR interactors influence post-endocytic transfer of GPCRs in response to agonist stimulation in a more receptor-selective manner than GRKs and arrestins.[39] Some GPCR interactors, such as Homer proteins and Disc large homologs, control GPCR anchoring to appropriate regions of the plasma membrane.

DnaJC14), AT_2-binding protein ATBP50, and ubiquitin specific-processing peptidase USP4 link to their corresponding GPCRs to enhance their transfer [835].

39. GPCR-associated sorting protein GASP1 promotes endocytosis toward lysosomes of δ-type opioid receptors (DOR), D_2 dopamine receptor, and CB_1 cannabinoid receptor [835]. On the other hand, sodium–hydrogen exchanger-regulatory factor NHERF1 promotes the recycling of β2-adrenoceptors and κ opioid receptors (KOR).

Table 7.27. Metabotropic acetylcholine muscarinic receptors (mAChR; M_1–M_5 or M1R–M5R) and their main targeted G proteins (Source: [736]). Ionotropic nicotinic acetylcholine receptors (nAChR) belong to the superfamily of ligand-gated ion channels (Sect. 2.5.2). Muscarinic M_1, M_3, and M_5 receptors couple preferentially to Gq subunit and activate phospholipase-Cβ. Muscarinic M_2 and M_4 receptors couple preferentially to Gi subunit and inhibit adenylate cyclase.

Type	Main transducer
ACh Muscarinic receptors	
M_1, M_3, M_5	Gq/11
M_2, M_4	Gi/o
ACh Nicotinic receptors	
$\alpha 1$–$\alpha 7$	Ligand-gated ion channels

7.12.3 Regulation of Ligand Binding

Some GPCR interactors influence agonist selectivity of their corresponding GPCRs. Transmembrane receptor activity-modifying proteins RAMP1 to RAMP3 tether to calcitonin receptor-like receptor to construct functional receptor for calcitonin gene-related peptide (RAMP1–CRLR) and adrenomedullin (RAMP2–CRLR and RAMP3–CRLR) as well as calcitonin receptor to form amylin receptors [835].

7.13 Types of G-Protein-Coupled Receptors

G-protein-coupled receptors represent drug targets. Handling of GPCR activation then allows design of agonist or antagonist ligands, in addition to signaling modeling.

7.13.1 Acetylcholine Muscarinic Receptors

Muscarinic acetylcholine receptors (M_1–M_5 or M1R–M5R) pertain to class-A GPCR receptors. Selective enhancers of acetylcholine binding and action exist for various receptor subtypes [5]. Main guanine nucleotide-binding (G) proteins targeted by muscarinic receptors are given in Table 7.27.

Muscarinic receptors M_1 and M_2 couple voltage sensing to acetylcholine binding. Conformational changes are associated with charge motions.[40] Ligand binding of GPCRs can then be modulated by voltage, amplifying or attenuating signals. Effects of muscarinic receptor subtypes are given in Table 7.28.

40. Voltage is sensed not only by voltage-gated ion channels, but also ion transporters, such as sodium–glucose cotransporter, voltage-dependent phosphoinositide phosphatase, and glutamate G-protein-coupled receptors $mGlu_1$ and $mGluR_3$.

Table 7.28. Types of muscarinic acetylcholine receptors and their functions.

Type	Effects
M_1	Cognition and memory (prefrontal cortex), secretion of salivary glands and stomach
M_2	Cardiac frequency reduction, inhibition of β-adrenergic-mediated SMC relaxation
M_3	Smooth muscle contraction (blood vessels and airways), modulation by stimulated NO release from endothelium, endocrine and exocrine gland secretion, learning and memory (hippocampus)
M_4	Locomotion control
M_5	Vasodilation in the cerebral vasculature, behavioral cognition (hippocampus)

7.13.1.1 Muscarinic Receptors in Cardiomyocytes

The activation of muscarinic receptors in cardiomyocytes depends on transsarcolemmal potential. Depolarization reversibly induces changes in affinity state of the receptor [838]. Depolarization raises and reduces the affinity of M_1 and M_2 receptors, respectively [839].

7.13.1.2 Muscarinic Receptors in Nervous Synapses

The actin depolymerizing factor–cofilin complex regulates actin-dependent vesicular transport of acetylcholine receptors to the postsynaptic membrane and contributes to membrane insertion, formation, and maintenance of muscarinic receptor clusters [840].

Scaffold 14-3-3 that sequesters the ADF–cofilin complex attenuates the synaptic insertion of muscarinic receptors and clustering. On the other hand, 2 counteracting nerve-derived factors, agrin and acetylcholine, regulate the redistribution of muscarinic receptors on the sarcolemma at neuromuscular junctions.

Agrin activates muscle-specific Tyr kinase (MuSK) that favors muscarinic receptor clustering on the postsynaptic membrane, whereas acetylcholine disperses extrasynaptic clusters of muscarinic receptors. The agrin–MuSK signaling involves P21-activated kinase, an activator of LIMK kinases that phosphorylates (inactivates) the ADF–cofilin complex. Protein 14-3-3γ colocalizes and interacts with MuSK at neuromuscular junctions.

7.13.1.3 M_1 Receptor

Central Nervous System

The M_1 subtype of muscarinic (acetylcholine) receptors is selectively distributed in the forebrain, where it intervenes in cognition and memory. It can mediate learning and memory via an indirect mechanism, i.e., stimulation of the prefrontal cortex. It also supports the synaptic transmission and excitatory postsynaptic potential in postganglionic nerves of autonomic ganglia.

The M_1 receptor is randomly distributed over the plasma membrane [841]. At any time, about 30% of the receptors are dimers. (Oligomers M_1 are not detected.) Interconversion between M_1 monomers and dimers occurs on a time scale of seconds.

Other Effects

Receptor M_1 transmits signal from the vagus nerves and cause a bronchoconstriction. In the digestive tract, M_1 provokes the secretion of salivary glands and gastric acid in the stomach.

7.13.1.4 M_2 Receptor

Nodal Cells and Cardiomyocytes

In the heart, M_2 receptor: (1) diminishes the speed of depolarization in the sinoatrial node (the natural pacemaker), thereby slowing the cardiac frequency, and (2) reduces the contractile force of atriomyocytes, but have no effect on that of ventriculomyocytes.

Smooth Muscle Cells

Muscarinic M_2 and M_3 receptors are coexpressed in many types of smooth muscle cells, such as those in walls of the gastrointestinal tract, bladder, blood vessels, and airways. The density of M_2 receptors is greater than that of M_3 receptors (relative density $\geq$4:1) [842].

In airway smooth muscle cells, activated Gi-coupled M_2 receptor inhibits β-adrenergic-mediated relaxation; activated Gq-coupled M_3 receptor initiates their contraction.

In smooth muscle cells, M_2 receptor inhibits AC5 and AC6 adenylate cyclases via $G\alpha_{i3}$, whereas M_3 activates AC5 and AC6 via $G\beta\gamma$ [843].

7.13.1.5 M_3 Receptor

Smooth Muscle and Endothelial Cells

Muscarinic M_3 receptor localizes to smooth muscle cells of the wall of blood vessels and respiratory conduits. This Gq-coupled receptor generates an increase in intracellular calcium concentration, thereby priming a contraction of smooth muscle cells (vaso- and bronchoconstriction). [41]

However, in the vasculature, activated M_3 on vascular endothelial cells augment the synthesis of nitric oxide that targets adjacent vascular smooth muscle cells and causes their relaxation. Therefore, nitric oxide modulates the direct action of acetylcholine on smooth muscle cells.

Pancreatic β Cell

Functioning of pancreatic β cell is regulated by many hormones and neurotransmitters. Most of them act on specific G-protein-coupled receptors. Muscarinic receptor M_3 (or M3R) is linked to parasympathetic nerves that innervate the endocrine pancreas.

In pancreatic β cells, M_3 receptor leads to the stimulation of G-protein subunits of the $G\alpha_q$ family and promotes glucose-stimulated insulin secretion. Subunit Gq are controlled by regulators of G-protein signaling. Protein RGS4, the most abundant RGS subtype synthesized in β cells, is a potent inhibitor of M_3 receptors [844].

Similarly to all GPCRs after agonist stimulation, the activity of M_3 is regulated via hyperphosphorylation by diverse protein kinases (GPCR kinases [GRK] and casein kinases CK1α and CK2), hence phosphorylation-dependent receptor endocytosis and β-arrestin recruitment. Additional arrestin-primed phosphorylation of M_3^P by protein kinase-D1 enables the sustained phase of glucose-dependent insulin release, independently of heterotrimeric G proteins [845]. [42]

Central Nervous System

In the hippocampus, M_3 participates in learning and memory [846]. These cognition processes rely on M_3 phosphorylation, receptor endocytosis, and arrestin recruitment.

7.13.1.6 M_4 Receptor

Muscarinic acetylcholine receptors regulate dopaminergic neurotransmission. Receptor M_4 is coexpressed with D_1 dopamine receptor in a specific subset of striatal projection neurons. Activated M_4 receptor in the striatum inhibits D_1-induced locomotor stimulation (in mice) [847].

41. Activation by M_3 of Gq subunit increases phosphoinositide hydrolysis and releases Ca^{++} ions from the sarcoplasmic reticulum.

42. Two phases of insulin release — transient and sustained — result from feeding. The early phase of insulin release depends on activation by Gq of calcium influx and protein kinase-C.

7.13.1.7 M_5 Receptor

Receptor M_5 is the sole mediator of ACh-induced vasodilation in the cerebral vasculature [848]. It is also involved in long-term potentiation at the hippocampal mossy fiber-CA3 synapse and hippocampal-dependent behavioral cognition.

7.13.2 Adenosine Receptors

Adenosine is an ubiquitous, multifunctional purine nucleoside that targets numerous enzymes, such as protein kinases and adenylate cyclases.[43] Intracellular adenosine can be formed by either dephosphorylation of AMP by 5′-nucleotidase or by hydrolysis of ^{S}adenosylhomocysteine. Extracellular adenosine can be produced from released adenine nucleotides by a cascade of ectonucleotidases such as ecto-5′-nucleotidase (NT5E or CD73), which is highly expressed in microglial cells, and ectonucleoside triphosphate diphosphohydrolase ENTPD1 (or CD39) [849].

In the central nervous system, the extracellular concentration of adenosine is controlled by astrocytes. Glial cells and neurons release ATP, a neurotransmitter that activates its P2X and P2Y receptors. Vesicular ATP release from astrocytes (but not from neurons) by exocytosis or through equilibrative nucleoside transporters, possibly pannexin-1 hemichannels, and Maxi anion channels, is a source of extracellular synaptic adenosine via ectonucleotidase action on ATP [849]. Adenosine thus influences synaptic transmission. Endocrine cells also secrete ATP, even without releasing stored hormone.

Adenosine receptors constitute a family of nucleoside receptors (P1 receptors; Table 7.29).[44] Adenosine activates 4 receptor types (A_1, A_{2A}, A_{2B}, and A_3). Gi-Coupled receptors A_1 and A_3 can also be activated by inosine. These adenosine receptors inhibit many intracellular ATP-using enzymes such as adenylate cyclases [5].

Adenosine is a cerebral vasodilator. Smooth muscle cells of cerebral arteries synthesize A_1, A_{2A}, A_{2B}, and A_3 receptors [850]. Adenosine mainly binds to A_{2A} and A_{2B} and generates superoxide using NADPH oxidase and mitochondria.

---

43. Purines adenine and guanine are synthesized in vivo via inosine monophosphate (IMP), which is laso named inosinic acid and inosinate. The latter is also converted into nucleotides, i.e., bases attached to ribose 5-phosphate: (1) adenosine monophosphate (AMP; a.k.a. adenylic acid and adenylate) by adenylsuccinate synthetase and adenylosuccinate lyase and (2) guanosine monophosphate (GMP; a.k.a. guanidylic acid, guanylic acid, and guanylate) by IMP dehydrogenase and GMP synthase. Conversely, AMP can be converted into IMP by myoadenylate deaminase. Adenosine can be transformed into inosine via adenosine deaminase and AMP via adenosine kinase. De novo synthesis of adenosine does not exist.

44. The P1 receptor class is associated with a nucleoside. The term "nucleoside receptor" to define a P1 family member is more appropriate then purinergic receptor. Members of the family of nucleotide receptors (P2) link to both purines and pyrimidines. As both purine and pyrimidine species are nucleotides, the P2 family is better defined as the class of "nucleotide receptor." Nevertheless, the noun "purinergic" should be kept in mind to explain the aliases P1 and P2.

Table 7.29. Classes of the receptor family for nucleosides and nucleotides. Adenosine targets nucleoside receptors of the P1 (purinergic) class. Nucleotides are ubiquitous intercellular messengers that act via specific receptors: ionotropic P2X and metabotropic P2Y nucleotide receptors.

Set	Activator	Molecule type
P1 receptors	Adenosine	G-protein-coupled receptor
P2X receptors	ATP	Ligand-gated ion channel
P2Y receptors	Nucleotides	G-protein-coupled receptor

Adenosine receptors A_1 and A_3 preferentially interact with G-protein subunits of the Gi/o family; A_{2A} and A_{2B} receptors with members of the Gs family [851] (Tables 7.30 and 7.31).

Adenosine receptors A_1 and A_3 stimulate K^+ channels, reduce transient activity of voltage-dependent Ca^{++} channels, and inhibit cAMP formation. Whereas A_{2A} receptor couples to members of the Gs family (e.g., $G\alpha_{olf}$ in striatal neurons), A_{2B} receptor associates with many G-protein subunit types, such as Gs, Gq, and G12 [849]. Adenosine is approximately equipotent on A_1, A_{2A}, and A_3 receptors. The A_{2B} receptor may require higher agonist concentrations.

7.13.2.1 A_1 Receptor

Adenosine A_1 receptor is ubiquitous (Table 7.30). The A_1 receptor attenuates stimulatory actions of catecholamines on β-adrenergic receptors, reduces lipolysis in adipose tissue and urine production, and impedes neuronal activity [852]. Activated A_1 receptor not only decreases cAMP synthesis, but also stimulates phospholipase-C.

Central Nervous System

Adenosine receptor A_1 contributes to sleep regulation, as it is upregulated in cortical and subcortical brain regions, especially in the orbitofrontal cortex after prolonged wakefulness [853]. It hence promotes sleep and inhibits wakefulness via cholinergic neurons, in opposition with xanthine, an adenosine antagonist in coffee and tea. The activity of basal forebrain (substantia innominata) and mesopontine (laterodorsal tegmental nucleus) cholinergic neurons actually bear an inhibitory control by endogenous adenosine in cats [854].

The A_1 receptor in the nucleus of the solitary tract is involved in the baroreflex (Vols. 2 – Chap. 1. Remote Control Cells and 6 – Chap. 3. Cardiovascular Physiology). Its activity inhibits glutamatergic transmission in the nucleus of the solitary tract [855]. This receptor differentially inhibits and resets the baroreflex response of preganglionic adrenal, renal, and lumbar sympathetic nerve.

Table 7.30. Adenosine receptors A_1 and A_3, G mediators, expression loci, and tissue functions (↓: reduction; −: inhibition; +: stimulation; DAG: diacylglycerol; IP_3: inositol trisphosphate; PL(A/C/D): phospholipase-A/C/D; Source: [851]).

A_1	A_3
Gi/o	Gi/o (Gi2/3)
(Gi1/2/3; Go)	Gq/11
PLA2, PLC, PLD	PLC
IP_3/DAG, cAMP −	IP_3/DAG, cAMP −
High expression	
Brain cortex, cerebellum, hippocampus, dorsal horn of spinal cord, eye, adrenal gland, atria	Mastocytes
Intermediate expression	
Other brain regions, skeletal muscle, adipose tissue, liver, kidney, salivary glands, esophagus, colon, antrum, testis	Cerebellum, hippocampus, pineal gland, lung
Low expression	
Lung, pancreas	Brain, adrenal, thyroid, testis, intestine, spleen, liver
Vasoconstriction	Vasodilation
Bradycardia	Preconditioning
Lipolysis −	Mastocyte degranulation
Glomerular filtration ↓	
Ischemic preconditioning	
Sympathetic and parasympathetic activity ↓	

Glial cells that interact with neurons and blood vessels control adenosine that has a neuroprotective role in the central nervous system. Microglial cells, astrocytes, endothelial cells, oligodendrocytes, and neurons produce signals to manage inflammation, hence entry of cells of the innate and adaptive immune system in the brain. These cells release arachidonic acid metabolites, nitric oxide, cytokines, and chemokines as well as ATP and adenosine. The A_1 receptor on microglial cells reduce excessive immune activation of microglial cells [849]. It also stimulates oligodendrocyte migration.

The A_1 receptor on neurons, especially at nerve terminals, contributes to the dampening of neuronal activity mediated by adenosine generated from ATP released from astrocytes [849]. In nerve endings, it may preferentially signal via Go proteins to inhibit calcium channels. In neuron bodies and dendrites, it may preferentially regulate potassium channels via Gi proteins.

Table 7.31. Adenosine receptors A_{2A} and A_{2B}, G mediators, expression loci, and tissue functions (↓: reduction; −: inhibition; +: stimulation; DAG: diacylglycerol; IP_3: inositol trisphosphate; PL(A/C/D): phospholipase-A/C/D; Source: [851]).

A_{2A}	A_{2B}
Gs	Gs
G_{olf}, G15/16	Gq/11
PLA2, PLC, PLD	PLC
cAMP, IP_3/DAG	cAMP, IP_3/DAG
High expression	
Caudate-putamen, nucleus accumbens, tuberculum olfactorium, olfactory bulb spleen, thymus	Colon, bladder
Leukocytes, platelets	
Intermediate expression	
Heart, blood vessels, lungs	Median eminence, blood vessels, lungs, eyes
	Mastocyte
Low expression	
Other brain regions, heart, kidney	Brain, pituitary adrenal, ovary, kidney, liver, adipose tissue
Vasodilation	Vasodilation
Platelet aggregation −	Intestinal SMC relaxation
Neutrophil activation end	Monocyte/macrophage −
Regulation of sensorimotor	Mastocyte degranulation
Integration in basal ganglia	
Sensory nerve activity +	

7.13.2.2 A_2 Receptor

High-affinity A_{2A} and low-affinity A_{2B} receptors are coupled to stimulatory subunit $G\alpha_s$ of heterotrimeric G protein, thereby raising intracellular cAMP concentration (Table 7.31). Activated A_{2A} can protect against tissue injury and act as an anti-inflammatory agent [856].

Human A_{2A} receptor has a binding site that differ in position and orientation with respect to other GPCRs [857]. Extracellular loops in collaboration with the helical core allows ligand recognition.

Central Nervous System

In the brain, A_{2A} receptor is synthesized at high levels in striatal neurons and at low levels in other neurons and glial cells [849]. The A_{2A} production in glial cells rises after brain insult. In astrocytes, activated A_{2A} by extracellular adenosine increases cell proliferation, but impedes NOS2 synthesis. In microglial cells, activated A_{2A} facilitates the release of cytokines and prostaglandin-E2, as it boosts the activity of cyclooxygenase-2, as well as that of nitric oxide synthase and production of nerve growth factor [849].

Activated A_{2A} in the nucleus of the solitary tract in the brainstem increases activity of preganglionic adrenal sympathetic nerves, but decreases that of renal sympathetic nerves [858].

The A_{2B} receptor can couple with Gq and Gs on astrocytes. It is responsible for the adenosine-induced stimulation of interleukin-6 from astrocytes.

Increased neural activity is associated with cerebral vasodilation. Neural activation-associated pial arteriolar dilation involves interactions between A_2 and inward rectifier K_{IR} channels [859]. Large-conductance, Ca^{++}-activated BK channel ($K_{Ca}1$) participates in the process within the glia limitans. The A_2 receptor provokes the synthesis by adenylate cyclase of cAMP that primes the phosphorylation of K^+ channels by protein kinase A.

Skin Microvasculature

In human dermal microvascular endothelial cells, activation of A_{2B}, but not A_{2A}, promotes angiogenesis. On the other hand, activated A_{2A}, but not A_{2B}, promotes angiogenesis in endothelial cells of human umbilical veins and lung microvasculature. Endothelial cells express A_{2A}, A_{2B}, or both.

Airways

In the respiratory tract, A_{2B} receptors regulate chloride channels via cyclic adenosine monophosphate.

7.13.2.3 A_3 Receptor

Like the adenosine receptor A_1, A_3 is coupled to inhibitory subunit $G\alpha_i$ of heterotrimeric G protein, thus inhibiting adenylate cyclase (Table 7.30). It has a protective function during tissue ischemia.

Neutrophil

Adenosine receptor A_3 contributes to inhibition of neutrophil degranulation in tissue injury. In neutrophils, ATP acts on P2Y receptors in synergy with adenosine that binds to A_3 receptors to stimulate migration.

Central Nervous System

Astrocytic A_3 receptor regulates chemokine release [849]. Activation of A_3 by adenosine protects astrocytes from death caused by hypoxia. Adenosine contributes to microglial migration caused by ATP predominantly via $P2Y_{12}$ receptors.

Table 7.32. Kinetics of A_{2A} adenosine receptor activation (Sources: [792, 860]).

Event	Time (ms)
GPCR activation and G-protein binding	~50
Gs activation	~500
Deactivation	~2000

7.13.2.4 Activation and Deactivation Kinetics of Adenosine Receptors

Fluorescence resonance energy transfer-based assays can be used to assess the receptor-specific activation and deactivation kinetics of the different GPCR signaling steps, especially the activation kinetics of initial stages (GPCR–G-protein interaction that depends on the amount of available receptors and determines the activation G-protein and other effectors [860].

Both A_{2A} and β1-adrenergic receptors couple to Gs protein, thereby stimulating adenylate cyclase that produces cAMP messenger. Times of A_{2A} receptor activation, A_{2A}–Gs interaction, and Gs activation following stimulation by A_{2A} or β1-adrenoceptors are smaller than 40, 60, and 550 ms, respectively. Only a fraction of Gs proteins is activated when GPCR–Gs interaction is maximal. The rate of Gs activation limits the signaling initiation via Gs-coupled receptors. Activation time of Gs subunit is also similar to that of Gi stimulation after excitation by β1-adrenergic receptors.

Three main events occur between GPCR–G-protein interaction and G-protein activation: GDP release from Gα, GTP binding, and conformational change. Release of GDP is rate-limiting in G-protein activation. Half-time of cAMP activation is about equal to 35 s, but activation quickly appears (a few ssconds) due to the existence of Gs–ACase complexes.

In addition, although the activation kinetics via A_{2A} and β1-adrenergic receptors are similar, the deactivation kinetics differ. Termination of A_{2A}–Gs interaction and Gs deactivation (following A_{2A} deactivation with a time ~2 s, which is similar to that of α2a-adrenoceptors) are significantly slower than that of β1-adrenoceptor signaling (termination of β1-adrenoceptor–Gi interaction ~13 s). Termination of A_{2A}–Gs interaction is faster than Gs deactivation, which is linked to the activity of accessory proteins (activators and regulators of G-protein signaling) involved in GPCR signaling. Ligand washout is hence not limiting. Kinetics of A_{2A} recptor are given in Table 7.32.

7.13.2.5 Adenosine Receptors in the Cardiovascular Apparatus

Vascular Smooth Muscle Cells

In the vasculature, A_1 receptors as well as A_{2A}, A_{2B}, and A_3 are synthesized by smooth muscle cells [861]. The A_1 receptor causes contraction of vascular smooth muscle via phospholipase-C. It hinders vascular relaxation mediated by other adenosine receptor subtypes.

The A_1 receptor upregulates PKCα, PKCβ1, PKCβ2, PKCγ, PKCε, and PKCζ isoforms (but not PKCδ and PKCμ) in a dose-dependent manner [862]. Activated A_1 tethers to Gi subtypes Gi1 to Gi3 or Go subunit, thereby inhibiting adenylate cyclase and decreasing cAMP concentration.

Gs-Coupled A_{2A} and/or A_{2B} receptors are involved in adenosine-mediated vascular relaxation of coronary and aortic beds in mice [861]. Gi-Coupled A_3 receptor does not cause a vasoconstriction, as it is inhibited by A_{2A} receptor. The A_3 receptor mediates the cardioprotective effect of adenosine.

In the central nervous system, adenosine, a potent vasodilator, intervenes in vasodilation during hypotension within the autoregulatory range of vascular autoregulation via both A_{2A} and A_{2B} receptors [863].

Vascular Endothelial Cells

Hypoxia stimulates hypoxia-inducible HIF1α and HIF2α transcription factors that promote angiogenesis.[45] The HIF2α factor, but not HIF1α, regulates A_{2A} receptor in pulmonary endothelial cells to increase cell proliferation and migration as well as elicit tube formation [856].

Cardiomyocytes

Cardiomyocytes, in particular ventriculoyocytes, express the 4 adenosine receptor subtypes. Receptors A_1 and A_{2A} regulate myocardial oxygen consumption and coronary blood flow.

Adenosine is released from the heart subjected to hypoxia or adrenergic stimulation. The A_1 receptor has a strong anti-adrenergic effect. Therefore, adenosine antagonizes adrenergic stimulation to avoid overstimulation. Stimulation of A_1 activates Gi subunit, releases the G$\beta\gamma$ dimer, and excites phospholipase-C and protein kinase-Cϵ [864]. In addition, activated Gi subunit counteracts effect of Gs subunits stimulated by catecholamines via adrenoceptors.

The A_1 receptor is involved in the protection acquired during ischemic preconditioning against myocardial ischemia–reperfusion injury [865]. Stimulated A_1 re-

45. In almost all cell types, HIF1α targets various genes, such as those that encode vascular endothelial growth factor and its receptors, glucose transporter-1, carbonic anhydrase-9 and -12, hexokinase-2, glucose phosphate isomerase, phosphofructokinase, aldolase-A and -C, glyceraldehyde 3-phosphate dehydrogenase, among others. On the other hand, HIF2α regulates a few genes in specific cell lines, such as those that encode erythropoietin, superoxide dismutase-1 and -2, glutathione peroxidase, catalase, frataxin, Octamer-4, insulin-like growth factor-binding protein IGFBP3, SRY-related HMG-box gene product Sox9, CBP and P300-interacting transactivator with Glu/Asp-rich C-terminal domain CITED2. In most cell types, the gene set regulated by HIF2α overlaps with that of HIF1α [856].

presses the pacemaker cell function and decreases action potential conduction, thus reducing the heart rate [852]. It also lowers the atrial contractility.[46]

Whereas the A_1 receptor reduces β1-adrenergic receptor-induced increase in contractility in rat ventriculomyocytes, the A_{2A} receptor enhances cardiac contractility via Gs and Ca^{++} transient [868]. Cardiac A_1 and A_{2A} receptors use PKCε and the cAMP–PKA pathway, respectively. Cardiac A_{2A} receptor acts like β1-adrenoceptor, as it also primes the cAMP–PKA pathway, but with smaller Ca^{++} transient amplitude.

The A_{2A} receptor, but not the A_{2B} receptor, counteracts the A_1 receptor. It indirectly enhances heart contractility, as it modulates the A_1 anti-adrenergic effect [869]. On the other hand, the A_{2B} receptor exerts a direct contractile effect, but does not alter β-adrenergic or A_1 anti-adrenergic effects.

Activated A_{2A} receptor before reperfusion following coronary artery occlusion reduces infarct size and improves cardiac function, but activation must occur before or less than 1 h after reperfusion [870].

7.13.2.6 Inflammation

The A_2 receptor acts as a sensor of tissue damage. The A_{2A} receptor attenuates inflammation and tissue damage. The A_{2B} receptor abounds in vascular smooth muscle cells. It regulates the activity of the G-protein-coupled CXCR4 receptor that attracts vascular progenitor cells from bone marrow during tissue regeneration, avoiding excessive tissue growth after injury [871]. Chemokine CXCR4 receptor controls migration of inflammatory cells and platelet aggregation. It is thus involved in vascular repair and remodeling after injury.[47] Upregulation of CXCR4 in injured vessels is limited by A_{2B} receptors. In addition, adenosine inhibits smooth muscle cell proliferation via activated A_{2B} receptor.

---

46. Overexpression of A_1 causes supraventricular arrhythmias [866]. A strong A_1 overexpression also induces dilated cardiomyopathy [867].

47. Chemokine CXCR4 receptor is expressed on the surface of macrophages, lymphocytes, hematopoietic stem cells, bone marrow stromal cells, megakaryocytes, and platelets. It is involved in the migration of hematopoietic progenitors and stem cells during cardio- and vasculogenesis and hematopoiesis. It is regulated by TNFα secreted especially by several bone marrow cell types.

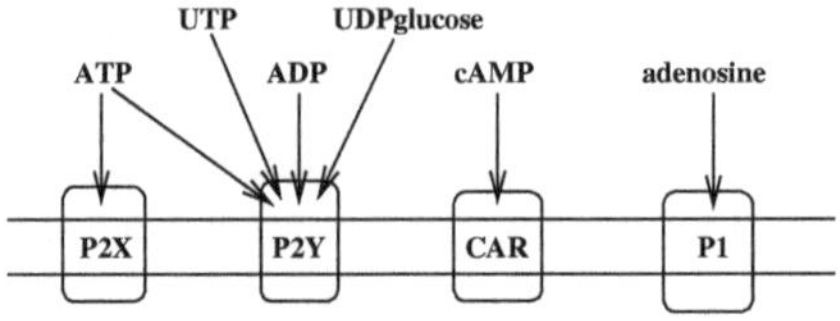

Figure 7.4. Nucleotide receptor types and their ligands.

7.13.3 Nucleotide P2Y Receptors

Nucleotide receptors of the P2Y family bind to both purine and pyrimidine nucleotides.[48] Purines include nucleic acid bases adenine and guanine[49] as well as hypoxanthine, xanthine, theobromine, caffeine, uric acid, and isoguanine.[50]

Ubiquitous nucleoside and nucleotide receptors (or purine and pyrimidine receptors) include P1 receptors (A_1–A_3) and P2 receptors (P2X and P2Y; Table 7.33). P1 Receptors bind adenosine, P2 receptors extracellular adenosine and purine nucleosides and nucleotides [872] (Fig. 7.4).

Ionotropic P2X receptors that encompass 7 known types ($P2X_1$–$P2X_7$) are ion channels (Sect. 2.5.7).[51] Most P2X channels are cation-selective pores that discriminate more or less efficiently different cations, particularly Ca^{++} ions.

Many identified metabotropic, G-protein-coupled P2Y receptors ($P2Y_1$–$P2Y_{14}$) are activated by ATP, ADP, UTP, UDP, and ^{UDP}glucose. The P2Y receptors differ according to their ligands (adenine vs. uracil-derived nucleotides as well as tri- vs. diphosphonucleotides), signaling mode, and cell expression (Tables 7.34 and 7.35). Calcium fluxes generated by P2Y receptors are larger for a given depolarization than equivalent amplitude hyperpolarization [873].

---

48. The term "purinergic receptors" used to designate receptors of the P2Y family is inappropriate. In addition, receptors of the P1 family connect to purine nucleoside adenosine. The term "nucleotide receptors" to denote both members of the P1 and P2Y family of the GPCR superclass should be likewise avoided.

49. Double-stranded DNA relies on base pairing between nucleotide constituents. Purines adenine and guanine couple with pyrimidines thymine and cytosine, respectively. Whereas each DNA strand is composed of 4 bases — adenine, cytosine, guanine, and thymine —, single-stranded RNA contains adenine, cytosine, guanine, and uracil.

50. Purines are components of numerous molecules, such as: (1) purine nucleosides adenosine and guanosine; (2) adenylate (ATP, ADP, AMP, and cAMP) and guanylate (GTP, GDP, GMP, and cGMP) nucleotides; and (3) nucleotides constituents of enzymatic cofactors (coenzyme-A, flavin adenine dinucleotide [FAD], flavin mononucleotide [FMN], and nicotinamide adenine dinucleotide phosphate [NADH]). Pyrimidines include nucleobases cytosine, thymine, and uracil. Uridine is a pyrimidine nucleoside. Pyrimidine nucleotides encompass uridine mono- (or 5′-uridylic acid [UMP]), di- (UDP), and triphosphate (UTP).

51. Nucleotide receptors P2X are fast-acting, ligand-gated channels that are selective for calcium. When ATP binds to plasmalemmal P2X receptors, extracellular Ca^{++} enters and activates signal transduction pathways. These homo- and hetero-oligomeric channels contain 2 transmembrane domains.

Table 7.33. Receptors for purines and pyrimidines (Source: [106]; CNS: central nervous system; NST: nucleus of the solitary tract).

Receptor	Main location
	Receptors of the P1 family
A_1	Brain, spinal cord, autonomic nerve terminals, heart, testis
A_{2A}	Brain, heart, lungs, spleen
A_{2B}	Colon, bladder
A_3	Brain, heart, lung, liver, testis
	Receptors of the P2X family
$P2X_1$	Cerebellum, dorsal horn spinal neurons, smooth muscle cell, platelet
$P2X_2$	CNS, autonomic and sensory ganglia, retina, chromaffin and smooth muscle cells
$P2X_3$	Sensory neurons, NST, some sympathetic neurons
$P2X_4$	CNS, testis, colon
$P2X_5$	Proliferating cells
$P2X_6$	CNS, motor neurons in spinal cord
$P2X_7$	Apoptotic cells
	Receptors of the P2Y family
$P2Y_1$	Epithelial and endothelial cells, platelets, immune cells, osteoclasts
$P2Y_2$	Epithelial (kidney tubule) and endothelial cells, immune cells, osteoblasts
$P2Y_4$	Endothelial cells
$P2Y_6$	Some epithelial cells, T cells
$P2Y_{11}$	Granulocytes
$P2Y_{12}$	Platelets, glial cells
$P2Y_{13}$	Brain, lymph nodes, bone marrow, spleen
$P2Y_{14}$	Brain regions, adipose tissue, digestive tract

Nucleotides can be stored and secreted upon adequate cell excitation to serve as messengers that bind to P2Y receptors. In the chromaffin granules of the adrenal medulla, ATP concentration reaches 100 mmol (cytoplasmic concentration 3–5 mmol) [874]. Dense granules of platelets contain about 500 mmol of ADP and ATP, as well as lower amounts of other nucleotides (adenine dinucleotides, GTP, and UTP). Following secretion and/or degranulation upon appropriate stimulation, [52] concentration of released nucleotides transiently reaches high levels. Nucleotide concentration is much higher near producing cell surface than in bulk extracellular fluid.

52. Change in blood flow pattern induces ATP release from endothelial cells. Release of ATP from human nasal epithelial cells increases upon mechanical stimulation or rising intracellular concentration of calcium ions. Corelease of ATP and nucleotidases from sympathetic nerves terminates the action of a neurotransmitter.

Table 7.34. Adenosine P1 and nucleotide P2Y receptors, their main targeted G proteins, and order of ligand potency (Sources: [736, 874]). The P2X receptors are ATP-gated ion channels for Na^+, K^+, and Ca^{++} ions. Some P2Y receptors couple exclusively to a single $G\alpha$-protein subunit, whereas others are dually coupled to different $G\alpha$-protein subunits (e.g., Gq/11 and Gi). Four subtypes of P1 receptors (A_1, A_{2A}, A_{2B}, and A_3) exist. The P1/A_2 receptors are coupled to Gs subunit and stimulates adenylate cyclase, whereas P1/A_1 and P1/A_3 receptors couple to subunits of the Gi/o family and inhibit adenylate cyclase. The $P2Y_5$ receptor corresponds to LPA_6 receptor that primes the G13–Rho pathway. The $P2Y_9$ receptor (or LPA_4) is most closely related to $P2Y_5$ (both act as lipid and nucleotide receptors).

Type	Main transducer	Potency order
Adenosine receptors (preferentially)		
A_1, A_3	Gi/o	
A_1	Gi1–Gi3	
A_3	Gi2, Gi3, Gq/11	
A_{2A}, A_{2B}	Gs	
Nucleotide receptors		
$P2Y_1$	Gq/11	ADP > ATP
$P2Y_2$	Gi/o, Gq/11	UTP ~ ATP
$P2Y_3$	Gq/11	UDP
$P2Y_4$	Gi/o, Gq/11	UTP > ATP
$P2Y_6$	Gq/11	UDP ≫ UTP > ATP
$P2Y_{11}$	Gs, Gq/11	ATP > UTP
$P2Y_{12}$	Gi/o	ADP ≫ ATP
$P2Y_{13}$	Gs, Gi/o	ADP ≫ ATP
$P2Y_{14}$	Gi/o, Gq/11	^{UDP}glucose (= ^{UDP}galactose)

Table 7.35. Effectors of some nucleotide P2Y receptors (Source: [874]; ACase: adenylate cyclase; Ca_V: voltage-gated Ca^{++} channel; K_{IR}: inwardly rectifying K^+ channel; PLC: phospholipase-C; RoCK: Rho-associated, coiled-coil-containing protein kinase). Voltage-gated K^+ channels of the K_V7 family cause the neuronal M current that regulates action potential firing.

Type	Effectors
$P2Y_1$	PLC, RhoA–RoCK
$P2Y_2$	PLC, K_{IR} channel, $Ca_V2.2$ channel (N-type)
$P2Y_4$	PLC, ACase
$P2Y_6$	PLC, K_V7 channels (M-type)
$P2Y_{11}$	PLC, ACase

Extracellular nucleotides operate as auto- or paracrine mediators (Tables 7.36 and 7.37). Afterward, extracellular nucleotides are rapidly degraded by ubiquitous ectonucleotidases on the cell surface.[53]

53. Two families of integral membrane ectonucleotidases exist: ectonucleotide diphosphohydrolases (ENPDH or CD39) and ecto-phosphodiesterase–nucleotide pyrophosphatase

Table 7.36. Autocrine and paracrine effects of extracellular ATP under various stimulation contexts (Source: [103]; **Part 1**; ↑: increase; IL: interleukin; JNK: Jun N-terminal kinase; NO: nitric oxide; Lφ: lymphoid lineage; MKC: megakaryocytes; RBC: red blood cell; SMC: smooth muscle cell). In some cases, the specific receptor type is unknown (P2).

Cell Type	P2 Type	Effect
		Cardiovascular apparatus
Endothelial	$P2Y_2$	Ca^{++} influx, ↑ NO production
	$P2X_{4/5}$	Ca^{++} influx
	P2	↑ NO production
	$P2X_7$	↑ IL1α release
VSMC	P2	↑ JNK activity
		Blood cells
RBC	$P2Y_1$	↑ Osmolyte permeability
Lφ	$P2X_7$	Proliferation
Macrophage	$P2X_7$	Apoptosis
MKC	$P2Y_1$	Ca^{++} influx
Monocyte	$P2X_7$	Interleukin (IL1β, IL18) secretion
Neutrophil	$P2Y_2$	Cell polarization
Platelet	$P2Y_1$	Ca^{++} influx, inhibition of platelet aggregation
	$P2Y_{12}$	Inhibition of platelet aggregation

Nucleotides exert their effects mainly via binding P2 receptors, although phosphorylation of plasmalemmal proteins by ectoprotein kinases constitutes an additional mechanism by which ATP can modulate cell function. [54]

Nucleotide P2Y receptors can be partly characterized by order of potency of ligand binding. These receptors are activated by a set of nucleotides: adenosine diphosphate (ADP), adenosine triphosphate (ATP), uridine triphosphate (UTP), and uridine diphosphate–glucose (^{UDP}glucose; Table 7.34). [55] The P2Y receptors can possess a main ligand, whereas other nucleotides behave as partial agonists. Nonetheless,

(EPDNP). The latter catalyze the cleavage of ATP into AMP and pyrophosphate, the conversion of cAMP into AMP, as well as several ATP–ADP, ADP–AMP, and AMP–adenosine reactions. Agent CD39 expressed by vascular endothelial cells hydrolyzes both ATP and ADP released from platelets.

54. Various plasmalemmal proteins are phosphorylated in their ectodomains by cell-surface protein kinase, the ectoprotein kinase. Protein CD36 (a.k.a. glycoprotein-3b and -4) is a class-B scavenger receptor that binds many ligands, such as collagen, thrombospondin, lipoproteins, oxidized low-density lipoproteins and phospholipids, as well as long-chain fatty acids. In platelets, cAMP-dependent ectoprotein kinase phosphorylates CD36 and causes platelet reactivity and adhesion to collagen, but reduces thrombospondin binding. Conversely, CD36 dephosphorylation transforms CD36 into a thrombospondin receptor [874].

55. Nucleotide UDP is phosphorylated into UTP by uridine diphosphokinase.

Table 7.37. Autocrine and paracrine effects of extracellular ATP under various stimulation contexts (Source: [103]; **Part 2**; ↑: increase; AA: arachidonic acid; NFAT: nuclear factor of activated T cells; PP: protein phosphatase). In some cases, the specific receptor type is unknown (P2).

Cell Type	P2 Type	Effect
		Respiratory tract
Epithelial	$P2Y_2$	Cl^- channel stimulation, inositol phosphate synthesis, Ca^{++} influx
	$P2Y_1$	Ca^{++} mobilization
	$P2Y_6$	Ca^{++} influx
		Kidney
Epithelial	P2	ENaC activation, Ca^{++} influx, contraction, ↓ AA Release, cAMP production
		Connective tissue
Fibroblast	$P2Y_2$	Ca^{++} influx
	$P2X_7$	Fibronectin ↑, IL6 release, apoptosis
Stromal	P2, $P2Y_1$	Ca^{++} influx, PP3 activation, NFAT translocation, proliferation

some nucleotide P2Y receptors can be activated almost equipotently by 2 nucleotide species. Others can be ligand-selective receptors, such as *purinoceptors* (e.g., $P2Y_1$) and *pyrimidinoceptors* (e.g., $P2Y_6$), for which ADP and ATP are either strong or only active ligands, or very weak stimulators or fully inactive, respectively.

In addition, effect of a given nucleotide on a given nucleotide receptors P2Y can depend on mammalian species, as P2Y can require either full or partial agonist activation. Moreover, several types of P2Y receptors can colocalize in a given cell type.

The P2Y receptors are coupled to distinct G proteins (Gq/11, Gi/o, and Gs; Table 7.34). They are then associated to phospholipase-C (Table 7.35). They can stimulate adenylate cyclase via $G\alpha_s$, although other pathways can be used (activation of some adenylate cyclase isoforms by $G\beta\gamma$ subunits, of other isoforms by Ca^{++}–calmodulin or protein kinase-C, and mediation by prostaglandins). Conversely, they can inhibit adenylate cyclase via $G\alpha_i$.

7.13.3.1 $P2Y_1$ and $P2Y_2$

The $P2Y_1$ receptor is linked to the RhoA-RoCK pathway (Vol. 4 – Chaps. 8. Guanosine Triphosphatases and Their Regulators and 4. Cytosolic Protein Ser/Thr Kinases) in addition to phospholipase-C.

The $P2Y_2$ receptor is linked to PLCβ1 via $G\alpha_{q/11}$ and to PLCβ3 via $G\beta\gamma\alpha_{i3}$. In vascular endothelial and smooth muscle cells, activation of $P2Y_2$ receptors by ATP and UTP induces phosphorylation (activation) of extracellular signal-regulated protein kinases ERK1 and ERK2 downstream from phospholipase-C [874].

In rat cardiac fibroblasts as well as pressure-overloaded mouse cardiomyocytes, ATP stimulates $P2Y_2$ receptor and activates nuclear factors of activated T cells NFAT1 and NFAT3, thereby increasing the concentration of inducible NO synthase (NOS2) [875]. Enzyme NOS2 interacts with the P65 subunit (or RelA) of NFκB and causes its S-nitrosylation.[56] Subunit $P65^{SNO}_{NF\kappa B}$ has a reduced transcriptional activity. Consequently, the density of type-1 angiotensin receptor (AT_1) falls.

7.13.3.2 Effects on Ion Channels

The P2Y receptors can also partner with ion channels. The $P2Y_2$ receptor links to Ca^{++}-dependent Cl^-, inward-rectifier K^+, and $Ca_V2.2$ channels [874].

Receptor $P2Y_6$ can associate with Cl^- channels. Via UDP, it can hamper the activity of $K_V7.2$ and $K_V7.3$ channels.

In neurons and other excitable cells, GPCRs can modulate the activity of voltage-gated ion channels, by mainly closing, or in certain circumstances opening or potentiating, various types of K^+ channels and voltage-gated Ca^{++} and Na^+ channels, via G-protein subunits and protein kinases. Specific couplings between P2Y receptors and certain Ca^{++} and K^+ channels occur in short time scales ($\mathcal{O}$[100 ms]; Table 7.38) [876]. Cells of the superior cervical ganglion (SGC) are endowed with 2 main types of voltage-gated channels: (1) $K_V7.2$ and $K_V7.3$ (M-type K^+ channel) and (2) $Ca_V2.2$ (N-type Ca^{++}) channels. In SCG neurons, Gi/o-coupled $P2Y_{12}$ and $P2Y_{13}$ receptors close $Ca_V2.2$ channel. The $P2Y_1$ and $P2Y_2$ receptors act similarly to Gq/11-linked M_1 receptor. Both UDP and UTP nucleotides are strong agonists of $P2Y_6$ that provoke $Ca_V2.2$ channel closure [876].

7.13.3.3 P2 Receptors in Immunocytes

Immunocytes release the auto- and paracrine controller ATP that activates P2X and P2Y receptors. Stimulated T-cell receptors provoke Ca^{++} influx in the cytosol and then mitochondria, where it promotes ATP synthesis. Messenger ATP, a costimulatory signal to T lymphocytes, released from activated T lymphocytes through pannexin-1 hemichannels excites P2X receptors for a sustained mitogen-activated protein kinase signaling [877]. On the other hand, oxidized ATP, a P2X antagonist, impedes MAPK activation in stimulated T cells.

Nucleotide signaling is also required for neutrophil activation. Activated neutrophils release ATP in response to exogenous stimuli such as formylated bacterial peptides and inflammatory mediators that stimulate Fcγ(Fc receptors for IgG), interleukin-8, C5a complement component, and leukotriene-B4 receptors. Stimulated formyl peptide receptors that colocalize with nucleotide $P2Y_2$ receptors

56. β-Arrestins enable the formation of the quaternary arrestin–P65–IκB–NOS2 complex.

Table 7.38. Interactions between P2Y receptors and ion channels and involved G-protein subunits (Source: [876]; Gβγ (Gi/o, Gq/11): Gβγ dimer associated with Gi/o and Gq/11 family subunits of heterotrimeric G proteins; GIRK: G-protein-activated inwardly rectifying K^+ channel).

Type	$Ca_V2.2$ inactivation	K_V7 inactivation	GIRK activation
$P2Y_1$	Gq/11, Gβγ (Go, Gq)	Gq/11	Gβγ
$P2Y_2$	Gq/11, Gβγ (Go, Gq)	Gq/11	Gβγ
$P2Y_4$	Gβγ (Go)	Gq/11	Unobserved
$P2Y_6$	Gq/11, Gβγ	Gq/11	Unobserved
	Gβγ (Gi/o)	Unobserved	Gβγ

causes ATP release through pannexin-1 hemichannels. Nucleotide signaling that uses pannexin-1 and $P2Y_2$ receptors facilitates neutrophil activation [878].

7.13.3.4 P2 Receptors in the Ventilatory Apparatus

Nucleotides increase the apical permeability to Cl^- in airway epithelial cells and enhance mucin secretion by goblet cells and submucosal gland secretion [874]. They then promote mucociliary clearance. [57] Nucleotides ATP and UTP not only augment formation of inositol phosphates and intracellular calcium concentration in human airway epithelial cells via $P2Y_1$ and $P2Y_2$, but also regulate outwardly rectifying chloride channels via $P2Y_2$ receptors. Like muscarinic receptors, the activity of $P2Y_1$ receptors depends on the plasmalemmal potential.

7.13.3.5 P2 Receptors in the Cardiovascular Apparatus

Nucleotide receptors participate in the regulation of the cardiac function, as P2Y receptors (especially $P2Y_6$, $P2Y_7$, $P2Y_9$, and $P2Y_{14}$) reside in heart (Fig. 7.5). In blood vessel walls, nucleotide signaling controls the vascular tone, smooth muscle cell proliferation, and platelet aggregation. Different types of cells in the cardiovascular system selectively express nucleotide receptor types (Fig. 7.6).

Vascular Smooth Muscle Cells

Contraction of vascular smooth muscle cells results from activated $P2X_1$ and several P2Y subtypes ($P2Y_2$, $P2Y_4$, and $P2Y_6$). [58] The $P2X_1$ receptor causes a rapid

57. In cystic fibrosis, CFTR mutations limit Cl^- secretion and cause formation of underhydrated, viscous mucus that obstructs airways. Aerosolized UTP may be used to treat cystic fibrosis.

58. Contractile type of smooth muscle cells possess larger amounts of $P2Y_4$ and $P2Y_6$, but a lower quantity of $P2Y_2$, than synthetic type of smooth muscle cells.

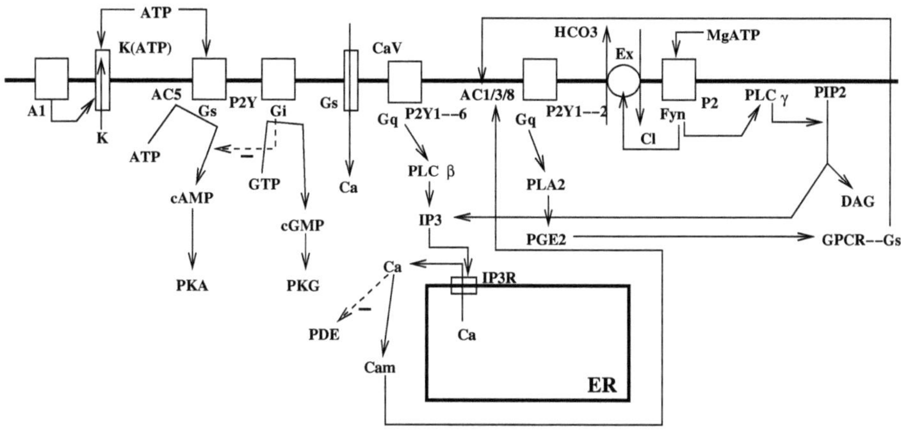

Figure 7.5. Nucleotide receptors in cardiomyocyte, their associated G proteins, and effectors.

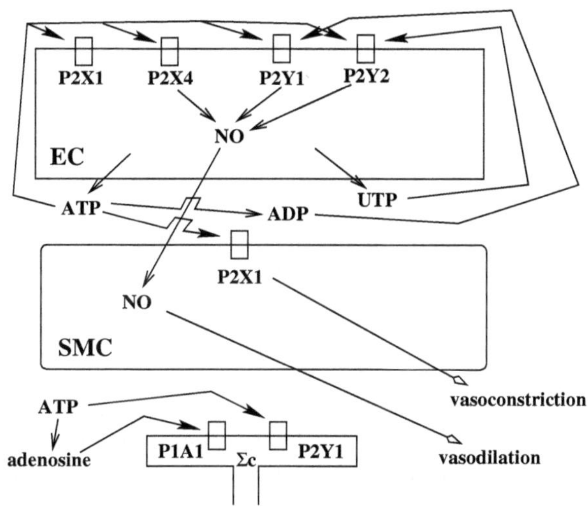

Figure 7.6. Nucleotide receptors in the vascular wall (Source: [879]). Messenger ATP, a cotransmitter of noradrenaline and neuropeptide-Y, released from perivascular sympathetic nerves binds receptors $P2X_1$, $P2X_2$, and $P2X_4$ of smooth muscle cells and induces vasoconstriction. Adenosine produced from ATP degradation binds to receptors A_1 (P1) of sympathetic nerves and inhibits transmitter release. Stressed endothelial cells release: (1) ATP, which binds receptors $P2Y_1$ (with a selective agonist action from ADP) and $P2X_4$, and (2) UTP, a ligand for $P2Y_2$. Both ATP and UTP trigger nitric oxide delivery to smooth muscle cells with subsequent vasodilation.

influx of Ca^{++} to initiate smooth muscle contraction. Gq/11-Coupled P2Y receptors activate phospholipase-C, thereby generating IP_3, which mobilizes Ca^{++} from its intracellular stores.

Prolonged P2Y signaling can cause vasoconstriction, hypertension, vascular smooth muscle hypertrophy, and hyperplasia. However, GPCR-desensitizing GRKs and arrestins prevent prolonged and inappropriate GPCR-initiated signaling.[59] In mesenteric arteries, both GRK2 (but not GRK3, GRK5, or GRK6) and arrestin-2 (but not arrestin-3) regulate UTP-stimulated $P2Y_2$ on smooth muscle cells [880].

Endothelial Cells

Endothelial cells express nucleotide receptors $P2Y_1$, $P2Y_2$, $P2Y_4$, $P2Y_6$, and $P2Y_{11}$,[60] as well as $P2X_1$,[61] to $P2X_7$.[62] The $P2X_4$ receptor is the most abundant P2X receptor subtype in vascular endothelial cells. It contributes to ATP- and flow-induced Ca^{++} influx in endothelial cells. The P2 receptors (particularly $P2X_4$) on endothelial cells bind ATP that triggers secretion of nitric oxide and causes vasodilation. The P2X receptors are implicated in autocrine loops in endothelial cells. Vascular smooth muscle cells abundantly produce $P2X_1$, $P2Y_2$, and $P2Y_6$ receptors. The P2 receptors also localize to the vessel wall adventitia. Perivascular adrenergic and cholinergic nerves have nucleotide receptors.

Endothelial cells that ensure hemostasis release nucleotides, but ectonucleoside triphosphate diphosphohydrolase ENTPD1[63] on the surface of endothelial cells can degrade both anti-aggregatory ATP and pro-aggregatory ADP [874]. Further cleavage by nucleotidase generates adenosine that inhibits platelet aggregation via A_{2A} receptors and cAMP messenger. Moreover, ATP and ADP prime the release of anti-aggregatory mediators prostacyclin and nitric oxide from endothelial cells, as they activate endothelial $P2Y_1$ and $P2Y_2$ receptors.

Erythrocytes and Coronary Endothelial Cells

Erythrocytes release ATP when the blood partial pressure of oxygen decreases. During exercise, when oxygen extraction by the myocardium is high,[64] the coronary blood flow rises. This augmentation is correlated with ATP concentration, as ATP

59. Among 7 GPCR kinases (GRK1–GRK7) and 4 arrestin (arrestin-1–arrestin-4) subtypes, GRK2, GRK3, GRK5, and GRK6, and arrestin-2 and -3 (i.e., β-arrestins) are ubiquitous. The GRK2 kinase is an important subtype for the contraction of vascular smooth muscle cells triggered by AT_1, α1d-adrenergic, and ET_A receptors.

60. The $P2Y_4$ receptor is slightly expressed in endothelial cells, whereas $P2X_4$, $P2Y_1$, $P2Y_2$, and $P2Y_{11}$ are the most abundant among endothelial P2 receptors [872].

61. The $P2X_1$ receptor abounds in endothelial cells of rat mesenteric arteries [881].

62. Receptors $P2X_1$ to $P2X_3$ and $P2X_7$ have similar levels in the endothelium of internal mammary and radial arteries and saphenous veins, whereas levels of $P2X_5$ and $P2X_6$ are lower [882]. Synthesis of $P2X_4$ differs between arteries and veins. Both radial and internal mammary arteries have very low $P2X_4$ levels.

63. A.k.a. ecto-apyrase CD39 and ectoATP diphosphohydrolase EATPDase1.

64. In dogs, the myocardial oxygen consumption increases approximately 3.2 fold, the coronary blood flow about 2.7 fold, and the partial pressure of oxygen in the coronary venous sinus decreases from 2.5 to 1.7 kPa [883].

level increases in blood of the coronary venous sinus (from 31.1 to 51.2 nmol) [883]. Adenine nucleotide ATP is one of the factors that control the coronary blood flow during exercise. It acts on microvascular endothelial cells to produce a vasodilation of upstream arterioles to match the oxygen delivery to myocardial oxygen consumption during exercise.

Adenosine triphosphate released from erythrocytes is catabolized into ADP and AMP in coronary capillaries, where these products target nucleotide receptors such as $P2Y_1$ on the surface of endothelial cells [884]. Activated nucleotide receptors initiate a retrograde vasodilatory signal to upstream arterioles presumably via gap junctions to control coronary blood flow resistance using a feedback loop.

Leukocytes

The $P2Y_{13}$ receptor is synthesized in blood leukocytes. The $P2Y_{11}$ receptor participates in granulocytic differentiation. Agent ATP can trigger hematopoiesis via the $P2Y_{11}$–Gs–ACase–cAMP–PKA pathway [874].

Thrombocytes

Major platelet activators include ADP, an auto- and paracrine regulator secreted by platelets, and thrombin, a peptidase of the coagulation cascade that triggers the release of ADP by platelets. Platelet activators interact with their respective GPCRs to prime and amplify platelet activation and aggregation. Binding of von Willebrand factor, collagen, and thrombin to their corresponding receptors on thrombocyte surface enhances the release of ADP and thromboxane-A2.

Sustained platelet aggregation by ADP requires activation of both $P2Y_1$ and $P2Y_{12}$ receptors. Ligand ADP binds to: (1) Gq-coupled $P2Y_1$ receptor to cause cell shape change for optimal affinity for fibrinogen and von Willebrand factor, calcium mobilization, and initiation of reversible aggregation;[65] and (2) Gi-coupled $P2Y_{12}$ receptor to amplify the platelet aggregation via the ACase–cAMP pathway [885]. Metabotropic receptor $P2Y_1$ also mediates the release of nitric oxide caused by adenine nucleotides.

In platelets, nucleotide (ADP) $P2Y_{12}$ receptor functions downstream from peptidase (thrombin)-activated PAR_1 and PAR_4 receptors to amplify the platelet response to thrombin, but can also operate in collagen-induced platelet activation and thrombin–PAR-independent pathways [886]. Furthermore, $P2Y_{12}$ signaling is more important than that based on PAR receptors in hemostasis, but not in arterial thrombosis for which PAR signaling is predominant.

In addition, ionotropic receptor $P2X_1$ abounds in platelets. Agent ATP, a potent agonist of $P2X_1$, triggers a rapid and quickly desensitized Ca^{++} influx.

---

65. The resulting platelet activation triggers a conformational change in $\alpha_{2B}\beta_3$-integrins (or glycoproteic complex GP2b–GP3a) that increases their affinity for fibrinogen and von Willenbrand factor. These ligands then bind to receptors to bridge adjacent platelets.

Table 7.39. Nucleotide receptors in the nephron (PCT: proximal convoluted tubule, tDL: thin descending limb, TDL: thick descending limb of the Henle loop, tAL: thin ascending limb, TAL: thick ascending limb of the Henle loop, DCT: distal convoluted tubule, CD: collecting duct, Source: [888]).

Type	Location
$P2X_1$	Afferent arteriole, glomerulus,
$P2X_2$	Afferent arteriole
$P2X_4$	PCT, tDL, tAL, TAL, DCT, CD
$P2X_5$	TAL, DCT, CD
$P2X_6$	PCT, TDL, tDL, tAL, TAL, DCT, CD
$P2Y_1$	Afferent and efferent arterioles, glomerulus, TDL
$P2Y_2$	Glomerulus, TAL, CD
$P2Y_4$	PCT, TDL

Table 7.40. Adiponectin receptors (Source: [5]; Apn_{FL}: full-length adiponectin; ^{G}Apn: globular adiponectin).

AdipoR1	AdipoR2
$^{G}Apn > Apn_{FL}$	$^{G}Apn = Apn_{FL}$

Cardiomyocytes, Neurons, and Nephron Epithelial Cells

The $P2X_4$ receptor improves survival of cardiomyocytes in a mouse model of cardiomyopathy [887]. Nucleotides intervene in synaptic transmission. In neurons of the central and peripheral nervous system, ATP colocalizes and is cosecreted with neurotransmitters, such as catecholamines and acetylcholine. The P2 receptors localize in every part of the nephron that regulates water (blood) volume and electrolyte concentration (Table 7.39).

7.13.4 Adiponectin Receptors

Adiponectin receptors (Table 7.40; Sect. 6.9.1) respond to adiponectin synthesized in adipose tissues (Vol. 2 – Chap. 3. Growth Factors). Unlike typical GPCRs, its C- and N-termini are extra- and intracellular, respectively [5]. Receptor AdipoR1 abounds in the skeletal muscle and liver, whereas AdipoR2 is predominantly produced in the liver.

These receptors are produced by endothelial cells and cardiomyocytes. T-cadherin (Vol. 1 – Chap. 7. Plasma Membrane) may act as a coreceptor for mid- (Apn_m) and high-molecular-weight (Apn_h) adiponectin on endothelial and smooth muscle cells, but not low-molecular-weight trimeric (Apn_l) and globular (^{G}Apn) forms [889, 890].

Signaling via adiponectin receptors avoids G proteins and instead stimulates protein phosphorylation via AMP-activated protein kinase (Vol. 4 – Chap. 4. Cytosolic Protein Ser/Thr Kinases) and the mitogen-activated protein kinase module, possibly using Adaptor phosphoTyr interaction, PH and Leu zipper domain-containing protein APPL1. Both Ca^{++}–calmodulin-dependent protein kinase kinase CaMKKβ and protein Ser/Thr kinase STK11 (or LKB1) are required for adiponectin-induced full AMPK activation.

Adiponectin is an antidiabetic and anti-atherogenic adipokine.[66] It induces influx of extracellular Ca^{++} via adiponectin receptor AdipoR1. Subsequent activation of Ca^{++}–calmodulin-dependent protein kinase kinase-β, AMPK, and sirtuin-1. Both AMPK and SIRT1 are required for elevated expression and attenuated acetylation of peroxisome proliferator-activated receptor-γ coactivator PGC1α primed by the Apn–AdipoR1 axis. This pathway raises mitochondrial content and activity in myocytes [891]. It also heightens oxidative type-1 myofibers and oxidative stress-detoxifying enzymes in the skeletal muscle, thereby promoting insulin sensitization and exercise endurance.

7.13.5 Adrenergic Receptors (Adrenoceptors)

Two main groups of ubiquitous adrenergic receptors (or adrenoceptors) include α- and β-adrenergic receptors (AR; Table 7.41). Both groups bind both adrenaline and noradrenaline. These catecholamine messengers regulate diverse physiological systems.

Adrenergic receptors are composed of 3 major types, each of which is further subdivided into 3 subtypes: (1) α1-adrenoceptors with α1a (or α1c), α1b, and α1d subtypes; (2) α2-adrenoceptors with α2a (or α2d), α2b, and α2c subtypes; and (3) β-adrenoceptors with β1, β2, and β3 subtypes. Each adrenoceptor type is characterized by its target pathways, receptor distribution, and effects (Table 7.42).

7.13.5.1 α-Adrenergic Receptors

α1-Adrenergic Receptors

α1-Adrenergic receptors are activated by adrenaline and noradrenaline with about equal potency. Numerous splice variants of α1-adrenoceptors exist. α1d-Adrenoceptors form heterodimers with α1b- or α2-adrenoceptors [5]. Signaling is carried out predominantly via Gq/11, but also Gi/o, Gs, G12/13, and Gh subunit.[67] α1-Adrenoceptor subtypes show differences in regulation and coupling efficiency to Ca^{++} signaling ($\alpha 1a > \alpha 1b > \alpha 1d$) and mitogen-activated protein kinase modules ($\alpha 1d > \alpha 1a > \alpha 1b$) [5].

α1-Adrenergic receptors of vascular smooth muscle cells as well as those of prostate cause contraction and favor cell growth and proliferation [896]. In heart, Gq-coupled α1-adrenoceptor has positive inotropic effect [897]. Activated Gq excites phospholipase-Cβ that hydrolyzes phosphatidylinositol (4,5)-bisphosphate to

66. Plasma adiponectin concentration lowers in obesity, insulin resistance, and type-2 diabetes.

67. Subunit Gh consists of the 74-kDa Gα subunit and a 50-kDa β subunit. Gh-protein-coupled α1-adrenoceptor stimulates membrane-bound phospholipase-C [895].

Table 7.41. Adrenoceptors, their main targeted G proteins, and order of ligand potency (α1c-adrenoceptor is derived from α1a-adrenoceptor; Source: [736]). α1d-Adrenoceptors are able to couple to Gi, at least in human skin fibroblast [892]. All combinations of $G\alpha_{i1}1$, $G\alpha_{i2}$, or $G\alpha_{i3}$ with Gβ1, Gβ2, or Gβ4 are activated by endogenous α2-adrenoceptors [893]. α2-Adrenoceptors couple not only to Gi but also to Gs [894].

Type	Main transducer	Ligand potency order
α1a	Gq/11	Ad ~ NAd
α1b	Gq/11	Ad ~ NAd
α1d	Gq/11	Ad ~ NAd
α1	Gi	
α2a	Gi/o	Ad > NAd
α2b	Gi/o	Ad > NAd
α2c	Gi/o	Ad > NAd
α2	Gs	
β1	Gs	Ad < NAd
β2	Gs, Gi	Ad > NAd
β3	Gs, Gi/o	Ad ~ NAd

inositol (1,4,5)-trisphosphate and diacylglycerol. Increased diacylglycerol level stimulates protein kinase-C, but the latter is not significantly involved in cardiac inotropy. Positive inotropy indeed relies mainly on myosin light-chain kinase and RoCK kinase.

α1a-Adrenoceptors

α1a-Adrenoceptor, or α1c-adrenoceptor,[68] undergoes alternative splicing that generates transcript variants and isoforms with distinct C-termini, but similar ligand-binding properties. α1a-Adrenoceptors do not form heterodimers. α1a-Adrenoceptor mediates G1–S cell cycle arrest that is associated with reduced activities of cyclin-E and cyclin-dependent kinase CDK6 and elevated action of cyclin-dependent kinase inhibitor CKI1b [898].

α1b-Adrenoceptors

α1b-Adrenoceptor (encoded by the ADRA1B gene) contributes to the cell cycle progression of many cell types [898]. It can hence acts as a proto-oncogene product. It interacts with subunit-μ2 (AP2M2: adaptor-related protein complex-2 subunit-μ2 or clathrin-associated adaptor protein ClAPM2; but not μ1, μ3, or μ4 subunits of other adaptor complexes) of the clathrin adaptor complex-2 for clathrin-mediated receptor endocytosis [900]. α1b-Adrenoceptor heterodimerizes.

68. A.k.a. AdRa1a, AdRa1c, and AdRa1L.

Table 7.42. Adrenergic receptors are coupled to: (1) Gs and the ACase–cAMP–PKA pathway; (2) Gi/o and the Gβγ–PI3K–PKB, Gβγ–PLC–PKC, and Gβγ–NOS–PKG pathways; and (3) Gq/11 and the Gβγ–PI3K–PKB pathway (vSMC: vascular smooth muscle cells; bSMC: bronchial smooth muscle cells; CMC: cardiomyocytes; NC: nodal cells; AC: adipocyte; HC: hepatocyte; TC: thrombocyte; PNE: presynaptic nerve ending; A: artery; V: vein; B, C, D, I, and L: bathmo- (nodal excitability [excitation threshold]), chrono- (heart rate), dromo- (nodal conduction velocity), ino- (strength of contraction), and lusitropic (relaxation) effects in cardiac electromechanical coupling; main sources: [736], Wikipedia). In addition to cardiovascular effects, α1-adrenoceptors also cause smooth muscle contraction of ureter, urethral sphincter, vas deferens, uterus at end of pregnancy, arrector pili muscles, and bronchioles. They induce glycogenolysis and gluconeogenesis in adipose tissue and liver, secretion from sweat glands, and sodium reabsorption from kidney. α2-Adrenoceptors inhibit insulin release and provoke glucagon release in pancreas, and induce contraction of sphincters of the gastrointestinal tract. β1-Adrenoceptors elicit renin release from juxtaglomerular cells and lipolysis in adipose tissue. β2-Adrenoceptors relax urinary sphincter, detrusor urinae muscle of bladder wall, and pregnant uterus, but contract sphincters of the gastrointestinal tract, thicken salivary secretions, induce glycogenolysis and gluconeogenesis, inhibit histamine-release from mastocytes, and increase renin secretion from kidney.

Receptor	Pathways	Sites	Effects
α1a	Gq/11	vSMC (A, V)	Vasoconstriction
		CMC	I+
α1b	Gq/11	vSMC	Vasoconstriction
		CMC	I+
		HC	Glycogenolysis and gluconeogenesis
α1d	Gq/11	vSMC	Vasoconstriction
α2a	Gi/o	PNE	Neurotransmitter inhibition (reduction in sympathetic output)
		TC	Platelet aggregation
		AC	Glycogenolysis and gluconeogenesis
		AC	Lipolysis inhibition
α2b	Gi/o	PNE	α2a Counteraction
α2c	Gi/o	vSMC (V)	Vasoconstriction
β1	Gs	NC, CMC	B+, C+, D+, I+, L+
β2	Gs, Gi, Gq	CMC	I+, L+
		vSMC	Vasodilation
		bSMC	Bronchorelaxation
β3	Gs, Gi/o	CMC	I−
		AC	Lipolysis

α1d-Adrenoceptors

α1d-Adrenoceptor (encoded by the ADRA1D gene), like α1a-adrenergic receptor, mediates cell cycle arrest during the G1–S transition [898]. Whereas α1a- and α1b-adrenoceptors localize to the plasma membrane, α1d-adrenergic receptors are intracellular. α1b-Adrenoceptor, but not α1a-adrenoceptor, forms heterodimers with α1d-adrenoceptors to cause their translocation to the cell surface [899].

In the developing fetus, cerebral arteries differ from those in adults. Maturational differences in adrenergic-mediated contractility of cerebral arteries result from distinct expression and/or activities of α1-adrenergic receptor subtypes. Once stimulated, they increase the intracellular inositol (1,4,5)-trisphosphate concentration, hence Ca^{++} influx, as well as activation of protein kinase-C and extracellular signal-regulated kinases ERK1 and ERK2. In sheep fetuses, α1d-adrenoceptors are expressed at a significantly greater level than α1a- and α1b-adrenoceptors [901]. However, α1AR subtype expression is approximately 20% in fetal cerebral arteries compared with that in adults. In fetal cerebral arteries, α1b and α1dAR subtypes contribute to vasocontriction as well as activation of ERK1 and ERK2.

α2-Adrenergic Receptors

α2-Adrenergic receptors are activated by adrenaline and noradrenaline with a potency order (adrenaline > noradrenaline) [5]. α2-Adrenoceptors activate the $G\alpha_{i/o}$ family of guanine nucleotide-binding regulatory (G) proteins that, in turn, regulates several effectors, such as adenylate cyclase (inhibition), Ca^{++} channels (stimulation or inhibition), K^+ channels and H^+–Na^+ antiporter (stimulation) [902].

α2a-Adrenoceptor mediates the central baroreflex control (hypotension and bradycardia) as well as influence platelet aggregation [5]. α2b Subtype is involved in neurotransmission in the spinal cord. α2c-Adrenoceptor regulates the catecholamine release from adrenal chromaffin cells.

α2a-Adrenoceptors

α2a-Adrenoceptor is widespread in the body's tissues (brain, vasculature, submandibular gland, and kidney, as well as platelets). Its desensitization after acute stimulation via that phosphorylation by GRKs depends on tissue and species types. It is constitutively palmitoylated. It can interact with α-subunit of eukaryotic initiation factor-2B and 14-3-3ζ protein, as well as spinophilin [902]. α2a-Adrenoceptor is able to form homo- and heterodimer with α2c subtype as well as β1-adrenoceptor.

Gi-Coupled α2a-adrenergic receptor (encoded by the ADRA2A gene) impedes insulin secretion, as it causes a reduced exocytosis of pancreatic β cells and insulin-granule docking at the plasma membrane [903].

α2b-Adrenoceptors

α2b-Adrenoceptor has a very limited expression in humans. It can promote mitogen-activated protein kinase activation. Desensitization relies on phosphorylation by GRK2 or GRK3 and, then, interaction with β-arrestin. On the other hand, scaffold spinophilin (or neurabin-2) tethers the receptor at the cell surface of dendritic cells. In addition, it can form complexes with α subunit of eukaryotic initiation factor-2B and 14-3-3ζ protein [902].

α2c-Adrenoceptors

α2c-Adrenoceptor is mainly expressed in the brain. It forms homo- or heterodimer with α2a subtype that can impair GRK phosphorylation and β-arrestin recruitment. It can also interact with β2-adrenergic receptors. In addition, it can build complexes with α subunit of eukaryotic initiation factor-2B and 14-3-3ζ protein [902].

7.13.5.2 β-Adrenergic Receptors

Three βAR subtypes (β1AR–β3AR) can bear distinct regulation and triggers different signaling cascades. Specificity of βARs for G-protein coupling is particularly dictated by the third intracellular loop and C-terminal tail. Intracellular domains of βARs are important for binding of βAR regulatory proteins such as G-protein-coupled receptor kinases and β-arrestins. [69]

β-Adrenoceptors possess a set of contacts between helices 3 and 6, the so-called ionic lock that forms a molecular switch for receptor activation. Molecular dynamics simulations of β2-adrenergic receptor show that the ionic lock forms reproducibly [904]. Inactive β2AR conformations vary from formed lock to broken lock, the latter conformation being observed in crystal structures.

β1-Adrenergic receptors activate adenylate cyclase and strongly elevate the cAMP concentration, whereas β2 and β3-adrenergic receptors can couple to Gi. They also stimulate many other signaling pathways, particularly mitogen-activated protein kinases. In rat ventriculomyocytes, β2-adrenoceptors couple to Gs (preferentially) and Gi protein, activating either Gs or both Gs and Gi protein [905]. Adenylate cyclase activation by Gi2-coupled β2-adrenoceptors can be done by preventing Gi2 to interact with the inhibitory $G\alpha_i$ site of adenylate cyclase [906].

The control of cAMP transients via degradation by phosphodiesterases (Vol. 4 – Chap. 10. Signaling Pathways) corresponds to a specific procedure for GPCR–cAMP signaling [907] (Table 7.43). Liganded β2AR recruits a preformed β-arrestin–PDE4d5 complex. Inactivated β1AR forms a signaling complex with another cAMP-specific PDE, PDE4d8, in cardiomyocytes. Ligand-induced dissociation of PDE4d from the β1AR–PDE4d8 complex produces a localized cAMP transient.

β1-Adrenoceptor

Gs-Coupled β-adrenergic receptors activate adenylate cyclase. β1-Adrenoceptor is the predominant subtype in heart. It is also a major mediator of lipolysis in adipocytes and renin release from kidney cells. Lastly, it is the principle subtype in certain brain regions, such as cerebral cortex and hippocampus, where it modulates long-term potentiation and, thereby, synaptic remodeling and memory [908].

---

69. Upon agonist stimulation, GRKs are recruited to the plasma membrane and phosphorylate activated βARs. β-Arrestins then bind to phosphorylated βARs to prime a rapid desensitization and receptor internalization.

Table 7.43. Difference in signaling complex composition (transducer, scaffold, and effector) and production between β-adrenoceptors (Sources: [907]). Different modes of interaction with phosphodiesterases (PDE) generate distinct cAMP signals for β-adrenoceptor signaling specificity.

	β1-Adrenoceptor	β2-Adrenoceptor
Signaling complex	PDE4d8	β-Arrestin–PDE4d5
Agonist binding	Complex dissociation	Complex formation

Human β1AR is glycosylated (Asn15). It heterodimerizes with α2aAR that has a lower potency for agonist stimulation of cAMP production. It interacts with Gs, but, unlike β2AR, not with Gi. Its desensitization results from phosphorylation by protein kinase-A and -C. Its phosphorylation by GRK is much less than that of β2AR; hence arrestin binding and desensitization is weaker [908].

Many class-1 PDZ domain-containing proteins link to β1AR, influencing either β1AR endocytosis or signaling. β1-Adrenoceptor indeed connects to Disc large homologs DLg1 [70] and DLg4, [71] members of the membrane-associated guanylate kinase (MAGuK) family, as well as MAGuK-like proteins with an inverted domain structure (WW and PDZ domain-containing proteins) MAGI2 and MAGI3 [909]. Protein DLg4 inhibits and MAGI2 stimulates β1AR endocytosis, respectively. The association of MAGI3 with β1AR impairs activation of ERK1 and ERK2 without affecting cAMP generation or internalization. Cystic fibrosis transmembrane conductance regulator-associated ligand (CAL) [72] also connects to the β1AR C-terminus. β1-Adrenoceptor also interacts with GAIP-interacting protein C-terminus product GIPC and RapGEF2. The former regulates β1AR-mediated activation of ERK. The latter promotes the activation of Ras by β1AR.

Appropriate β1AR transfer after agonist-promoted endocytosis is needed for resensitization of its signaling pathway. Efficient β1AR recycling requires binding of protein kinase-A-anchoring protein AKAP79 to β1AR C-terminus. Moreover, AKAP79 forms a complex with DLg1 and PKA, the so-called β1AR *receptosome* [910]. This scaffold allows β1AR recycling and PKA-mediated β1AR phosphorylation (Ser312).

In addition, endophilins enhance β1AR internalization [911]. [73] They are involved in clathrin-mediated endocytosis. Endophilins colocalize with endocytic reg-

70. A.k.a. synapse-associated protein SAP97.

71. A.k.a. postsynaptic density protein PSD95.

72. A.k.a. GOPC and FIG.

73. Endophilins are encoded by the genes of the EEN family (EEN, EENB1, and EENB2). Endophilins are also identified as Src homology domain-3-containing proteins (SH3P) and SH3-domain GRB2-like proteins (SH3GL). Endophilin-1 (a.k.a. Een, SH3GL1, and SH3P8), -2 (a.k.a. EenB1, SH3GL2, and SH3P4), and -3 (a.k.a. EenB2, SH3GL3, and SH3P13) bind to β1-adrenoceptors.

ulators, such as GTPase dynamin, phosphatase synaptojanin, and amphiphysin, an adaptor of the cytoplasmic surface of synaptic vesicles in nerve terminals.

β2-Adrenoceptor

β2-Adrenergic receptors are synthesized in most cell types. However, β2-adrenoceptor is the predominant type in the lung, smooth muscle, and liver. In lungs, β2AR resides on airway smooth muscle and epithelial cells. It is also detected in cardiomyocytes and inflammatory cells. It targets MAPK and PI3K kinases. In cardiomyocytes, β2AR-mediated activation of these kinases can counter apoptosis caused by chronic β1AR stimulation [912]. Phosphorylation by GRK followed by β-arrestin binding leads to β2AR endocytosis.

MicroRNA Let7f represses the translation of the ADRB2 mRNA, thereby attenuating the production of β2-adrenergic receptor under stable exposure in a given time range of agonists [913].[74] On the other hand, during agonist activation as well as long-term, chronic agonist exposure, the Let7f level decays (positive feedback loop that relieves the Let7f-mediated repression of ADRB2 translation).

β2-Adrenergic receptor is activated by the hormone and neurotransmitter adrenaline that is strongly involved in cardiovascular and pulmonary physiology. Ligand-binding site is accessible by conformational changes in the extracellular loop [914]. The β2AR basal (activator-independent) activity (and structural instability) involves interactions (hydrogen bonds and charge interactions, the so-called ionic lock) between cytoplasmic ends of transmembrane helices TM3 and TM6 [915].

β2-Adrenergic receptor phosphorylation by G-protein-coupled receptor kinases and protein kinase-A rapidly desensitizes these receptors, hence attenuating airway smooth muscle cell relaxation and bronchodilation. G-protein-coupled receptor kinases GRK2 and GRK5 with subsequent binding of β-arrestins reduce liganded β2AR–Gs coupling. Protein kinase-A also lowers β2AR–Gs coupling, but promotes β2AR–Gi coupling [916]. Protein kinase-A primes the major desensitization pathway in airway smooth muscle cells, especially at low agonist concentrations.

β2-Adrenergic receptor interacts with the Na^+–H^+-exchanger regulatory factors NHERF1 (or SLC9a3r1) and NHERF2 (or SLC9a3r2) to regulate Na^+–H^+ exchanger as well as β2AR postendocytic sorting following agonist-promoted internalization [909]. The recruitment of NHERF1 to β2AR precludes inhibition by NHERF1 on Na^+–H^+ exchanger NHE3 (or SLC9a3) in renal cells. Interactions between NHERF1 with β2AR are regulated by agonists.

Endocytosis of β2-adrenoceptors is regulated by agonist concentration and mediated by pre-existing clathrin-coated pits. β2-Adrenoceptors regulate the duration

74. The so-called "agonist-promoted downregulation" corresponds to the loss of the receptor response under prolonged exposure to an endo- or exogenous agonist. The miR–mRNA interaction prevents mRNA translation into a protein. The Let7f microRNA thus contributes to set the baseline of expression of β2-adrenergic receptors. The 3′ untranslated region of ADRB1 and ADRB2 mRNAs does not contain Let7f-binding sites for all subtypes (βAR-specific mechanism) [913]. Two other microRNAs, miR15 and miR30, have predicted binding sites in the 3′ UTR of ADRB2 mRNAs.

between clathrin coat deposition and membrane scission by a mechanism that involves receptor scaffolding to cortical actin [11]. Activation of protein kinase-A by β2ARs and subsequent receptor phosphorylation regulates Rab4-dependent, rapid recycling, but not slower recycling.

In rodent cardiomyocytes, Gs-coupled β2AR are confined to transverse tubules, whereas β1ARs are distributed across the entire cell surface. Both respond to the same ligands and couple to the same effectors, but they elicit different cellular responses.[75] In cardiomyocytes of a rat model of chronic heart failure, β2ARs redistribute from the transverse tubules (sites of coupling between the plasma membrane depolarization to calcium entry) to the plasma membrane [917]. A diffuse receptor-mediated cAMP signaling then results in contrast to a spatially confined β2AR-induced cAMP signals in normal condition. Loss of proper PKA localization is also observed in human heart failure.

The classical receptor theory postulates that receptors switch between inactivation and activation. Full agonists stimulate all of the signaling pathways of a given receptor to the same extent. The concept of "functional selectivity", or "biased signaling", states that a natural ligand can have distinct performance on different triggered signaling pathways. The endogenous ligands of β2-adrenergic receptor are adrenaline and noradrenaline. β2-Adrenergic receptor has 2 primary modes of activation of adenylate cyclase either via Gs subunit or, upon endocytosis, via β-arrestins and mitogen-activated protein kinases. The type of conformational change differs according to the type of agonists [918]. Both ligand types activate with an almost similar efficiency Gs and adenylate cyclase. However, noradrenaline is less efficient in signaling to β-arrestin. Therefore, noradrenaline is more tightly coupled to the Gs–ACase–cAMP pathway than signaling initiated during receptor endocytosis.

β3-Adrenoceptor

In humans, whereas β1AR and β2AR are ubiquitous, β3AR is essentially produced in the digestive tract and adipose tissue. In humans, β3AR abounds in brown, but not white adipocytes. β3-Adrenergic receptor regulates metabolic and endocrine functions, using the Gs–cAMP–PKA and PKC–P38MAPK pathways [919]. Sustained action of insulin reduces the density of β3AR, the major βAR subtype of adipocyte, without effect on β1AR (β2AR is undetectable) [920]. Activated β3AR stimulates lipolysis and thermogenesis in human adipocytes.

---

75. Upon stimulation, β1ARs generate positive ino- and chronotropic effects via phosphorylation by PKA of regulators of the intracellular concentration of calcium. β1-Adrenoceptors produce cytosolic cAMP that diffuses toward sarcomeres to enhance contractility and frequency. In heart failure, β1AR causes cell apoptosis. β2-Adrenoceptors are also coupled to Gs subunit and adenylate cyclase, but not calcium regulators and myofilaments that control contraction. In addition, β2AR activity opposes β1AR signaling because they switch sequentially from Gs to Gi, hence producing different cAMP pools in the cardiomyocyte and subcellular compartmentation of signals.

Table 7.44. Distinct action of β1- and β2-adrenergic receptors in cardiac physiology and pathology (Sources: [922, 923]; ACase: adenylate cyclase; cAMP: cyclic adenosine monophosphate; PI3K: phosphoinositide 3-kinase; PKA: protein kinase-A). β1- and β2-adrenoceptors are the predominant subtypes expressed in hearts. Both β1- and β2-adrenoceptors trigger the Gs–ACase–cAMP–PKA pathway. β2-Adrenoceptor also activates Gi protein that prime the $G\alpha_i G\beta\gamma$–PI3K–PKB pathway. The latter competes with the former for protein phosphorylation (calcium channels Ca_V1, phospholamban, and troponin) via PI3K during acute stimulation: β2AR–Gi coupling partially inhibits β2AR–Gs complex activity, i.e., positive inotropy (I+) and lusitropy (L+). Persistent β1AR stimulation induces cardiomyocyte hypertrophy and apoptosis via protein kinase-A-independent activation of calmodulin-dependent protein kinase-2 (CamK2). On the other hand, persistent β2AR activation exerts a survival effect in cardiomyocytes via protein kinase-B (PKB).

β1-Adrenoceptor	β2-Adrenoceptor
Gs–ACase–cAMP–PKA pathway	Gs–ACase–cAMP–PKA pathway Gi–Gβγ–PI3K–PKB pathway
I+, L+ Cardiomyocyte apoptosis Cardiomyocyte hypertrophy	Reduced I+, L+ effect Cardiomyocyte survival

In human ventricles, β3AR inhibits cardiac contractility. On the other hand, β3AR activation increases human atrial contractility via the cAMP–PKA axis and Ca^{++} channel Ca_V1 [921].

Cardiomyocyte β-Adrenoceptors

The sympathetic nervous system stimulates blood circulation in response to stress or exercise by activating β1- and β2-adrenoceptors, which activate G-protein subunit stimulatory for adenylate cyclase. Yet, a sustained β-adrenoceptor stimulation can yield pathological cardiac remodeling.

Cardiac β1-adrenoceptors are coupled to Gs and β2-adrenoceptors to both Gs and Gi (Table 7.44). The duration and magnitude of receptor expression determine the cardiomyocyte fate. β2-Adrenoceptors, like most GPCRs, exist at very low concentrations in the sarcolemma.

In heart, α1a- and α1b-adrenoceptors have distinct roles [924]. Overexpression of α1bARs depresses heart contractile response to βAR activation via Gi and predisposes heart to hypertrophy, whereas α1aAR overexpression increases cardiac contractility and improves outcomes after pressure overload or myocardial infarction. Both subtypes hinder the generation of inositol trisphosphate and protect from arrhythmogenesis in early postischemic reperfusion.

In cardiomyocytes, β1- and β2-adrenergic receptors activate mitogen-activated protein kinases P38MAPK and ERK, respectively. α1-Adrenergic receptors can activate the PLC–PKC pathway, stress-activated protein kinases (P38MAPK and Jun N-terminal kinase), Rho and Rac GTPases, and Ca^{++}–calmodulin-dependent protein kinase [925].

A single or several (1–7) ryanodine receptors of a given couplon release Ca^{++} sparks from the sarcoplasmic reticulum in response to its natural trigger, the Ca^{++} sparklet generated by a single Ca_V1 channel. β-Adrenergic receptors regulate the pace and strength of heart contraction, as they cause both prolonged Ca_V1 channel opening duration and accelerated channel coupling. They raise the coordination between Ca_V1 channels and ryanodine receptors via the PKA pathway. The enhanced synchronization of these 2 types of Ca^{++} channels generates higher and quicker Ca^{++} transients to support stronger and faster heart contraction. Channel synchronization relies on: (1) synchronized Ca_V1 channel activity (trigger); (2) synchronized initiation of ryanodine receptor activity (accelerated activation kinetics); and (3) synchronized ryanodine receptor activity (recruitment) [926]. Protein kinase-A phosphorylates Ca_V1 channels, thereby reducing their activation latency and enhancing Ca_V1-mediated Ca^{++} influx that, in turn, by improving connection between the 2 channel types elicits greater Ca^{++} release from the sarcoplasmic reticulum via ryanodine receptors, which are also phosphorylated. Because the Ca^{++} conductance of ryanodine receptors is not sensitive to phosphorylation, β-adrenoceptors increase the kinetics and number of ryanodine receptors that respond nearly simultaneously to the Ca^{++} flux delivered by Ca_V1 channels.

Acute β1-adrenoceptor stimulation triggers the cAMP–PKA pathway and modulates the cardiac excitation–contraction coupling [922]. On the other hand, sustained β1-adrenoceptor stimulation elicits cardiomyocyte hypertrophy and apoptosis via increased intracellular calcium concentration and Ca^{++}–CamK2 activity independently of PKA. However, a persistent β2AR activation can exert a survival effect in cardiomyocytes via protein kinase-B. Activation of ERK1, ERK2, and P38MAPK by the Gi subunit also favors β2AR-mediated anti-apoptotic effect in adult rat cardiomyocytes.

Several transcriptional targets of α1-adrenergic receptors have been identified in cardiomyocytes, particularly α-actin, β-myosin heavy chain, atrial natriuretic factor.[76]

Adipocyte β-adrenoceptors

Four adrenoceptor types are involved in the regulation by catecholamines of lipolysis in adipocytes. The control of adenylate cyclase activity involves stimulatory β1-, β2-, and β3-adrenoceptors. On the other hand, α2-adrenoceptors reduce intracellular cAMP content, thereby reducing lipolysis. Adipocytes, like platelets, possess α2a- rather than α2d-adrenoceptors, at least in rabbits [928]. In humans,

76. α1ARs target transcription factors, such as Jun (Jun stands for avian sarcoma virus-17 oncogene; Japanese: 10 [Ju] + 7 [Nana]), Fos (cellular Finkel Biskis Jinkins murine osteosarcoma virus sarcoma oncogene), and EGR1 (early growth response factor-1), as well as transcriptional corepressor CARP. The zinc finger-only protein Zfp260 of the Krüppel family of transcriptional regulators, highly expressed in the embryonic heart, is downregulated during postnatal development. Protein Zfp260 is a nuclear effector of α1-adrenergic receptors. It is also a transcriptional activator of atrial natriuretic factor and a cofactor for cardiac regulator GATA4, used in MAPK signaling [927].

adrenaline stimulates and inhibits lipolysis in visceral (omental) and subcutaneous adipocytes that have a high and low β- to α2-adrenoceptor ratio, respectively.

In white adipose tissues, types of adrenoceptors on the cell surface depends on mammal species. Adipocytes produce mainly α2- and few β3-adrenoceptors in humans, and conversely in rodent adipocytes [929]. Adrenaline controls lipid mobilization, as it activates antilipolytic α2ARs in human subcutaneous adipocytes during exercise [930]. Estrogens promote and maintain the female type of adipose tissue distribution. They attenuate the lipolytic response by increasing the number of antilipolytic α2a-adrenoceptors only in subcutaneous fat depots, thereby shifting the assimilation of lipids from omental depots to subcutaneous adipose tissues [931]. In addition, increased $[Ca^{++}]_i$ inhibits lipolysis induced by β-adrenoceptor activation via phosphodiesterase inhibition of cAMP signaling. In adipocytes, insulin attenuates β1-adrenoceptor-mediated lipolysis via activation of protein kinase-C. The PKCβ1 isoform interacts with β1AR signaling to decay lipolysis [932].

β3-Adrenoceptors are major mediators of the lipolytic and thermogenic effects of catecholamines. β3-Adrenoceptor activates extracellular signal-regulated kinases ERK1 and ERK2 via epidermal growth factor receptor and Src kinase (Vol. 4 – Chap. 3. Cytosolic Protein Tyr Kinases). This pathway in association with protein kinase-A contributes to maximal β3AR-stimulated lipolysis. Vimentin of intermediate filaments interacts with the β3AR–ERK pathway, whereas actin microfilaments and microtubules do not affect lipolysis and ERK activation [933].

Hepatocyte β-adrenoceptors

Hepatocytes express adrenergic receptors that modulate liver regeneration, hepatocyte proliferation, glycogenolysis, gluconeogenesis, urea synthesis, and fatty acid metabolism. Under basal conditions, the liver supplies glucose only via gluconeogenesis to spare glycogen stores, whereas in situations of enhanced demands primed by glucagon or sympathetic signal, additional glucose release occurs via glycogenolysis.

Glucagon is the main factor that increases cAMP level, which is correlated with the rate of glycogenolysis and exercise intensity [934]. Both α- and β-adrenergic receptors exist in liver. However, α-adrenergic stimulation is much more important than that of β-adrenoceptor to prime formation of cAMP via adenylate cyclase for glycogenolysis, gluconeogenesis, and ureogenesis. Adrenergic control of the hepatic function is mainly due to α1-adrenoceptors. Transactivation from α1AR of epidermal growth factor receptor involves Gq/11, Src kinase, signal transducer and activator of transcription STAT3, and the Src–STAT3 complex (pathway crosstalk) [935]. Normal and regenerating hepatocytes express α1b-, but not α1a-adrenoceptors [936]. Liver regeneration after partial hepatectomy involves β-adrenoceptors in the initial stage and afterward α1-adrenoceptors [937].

G-protein-coupled receptors can signal without G proteins. Stimulated β2-adrenoceptor recruits β-arrestin that forms a complex with phosphodiesterase PDE4 to regulate plasmalemmal PKA activity, which phosphorylates β2-adrenoceptor [938].

Table 7.45. Angiotensin receptors and their main targeted transducers, either G proteins or protein phosphatases (Source: [736]; PSTP: protein Ser/Thr phosphatase; PTP: protein Tyr phosphatase). s Different types of angiotensin receptors exist: AT_{1A}, AT_{1B}, and AT_2 to AT_4. Receptor AT_4 is activated by angiotensin-4.

Type	Main transducer
AT_1	Gq/11, G12/13, Gi/o
AT_2	Gi2/3
	PTP, PSTP

Arrestin-mediated uncoupling of G-protein-mediated signaling hampers Gs activation and switches to Gi stimulation that activates extracellular signal-regulated kinase ERK2.

7.13.6 Angiotensin Receptors

Angiotensins (ATn1–ATn4) act mainly via angiotensin type-1 (AT_1) and type-2 (AT_2) receptors (Table 7.45). Endogenous ligands are mostly angiotensin-2 and -3, whereas angiotensin-1 is weakly active in some systems [5]. In fact, different types of angiotensin receptors exist: AT_{1A}, AT_{1B}, and AT_2 to AT_4. The AT_4 receptor is activated by angiotensin-4, an angiotensin-2 metabolite. This specific receptor of angiotensin-4 resides in the brain and kidney.

Specific receptors for angiotensin-2 in adrenal zona glomerulosa,[77] vascular smooth muscle, kidney, brain, and anterior pituitary gland[78] exhibit generally similar binding properties. Angiotensin receptors of the anterior pituitary are not affected by changes in sodium balance in opposition to adrenal and vascular receptors [939]. In the brain, angiotensin receptors localize to several regions, although they are particularly concentrated in the circumventricular organs.[79] The angiotensin receptors of the renal cortex are localized in glomeruli, whereas in the renal medulla they are distributed diffusely.

Initiated signaling pathways include, in particular, receptor Tyr kinases (e.g., PDGFR and EGFR), IRS1, and cytosolic Tyr kinases, such as Src, Janus, and focal adhesion kinases, as well as mitogen-activated protein kinase.

77. Zona glomerulosa is the superficial layer of the adrenal cortex.

78. Among endocrine cell types - corticotrophic, gonadotrophic, lactotrophic, somatotrophic, and thyrotrophic cells - of anterior pituitary gland, angiotensin-2 receptors are located in corticotrophs and lactotrophs.

79. Circumventricular organs surround the brain ventricular network, primarily the third and fourth ventricles. They correspond to the few cerebral sites that have an incomplete blood–brain barrier, so that neurons can sense various blood compounds, particularly hormones. These organs secrete different neurotransmitters, hormones, and cytokines. Circumventricular organs include pineal gland (circadian rhythm), subfornical organ and organum vasculosum of the lamina terminalis (body fluid regulation), choroid plexi, area postrema (toxin detection), median eminence (regulation of anterior pituitary activity), subcommissural organ (somatostatin secretion), and posterior pituitary (oxytocin and vasopressin sensing).

7.13.6.1 AT_1

Type-1 receptor AT_1 mediates vasoconstriction, secretion of aldosterone and vasopressin, cardiac contractility and hypertrophy, augmented peripheral noradrenergic activity, vascular smooth muscle cells proliferation, decreased renal blood flow, renin inhibition and sodium reuptake in kidneys, modulation of central sympathetic nervous system activity, central osmocontrol, and extracellular matrix formation.

Angiotensin-2 receptor AT_1 activates: (1) $G\alpha_{q/11}$, $G\alpha_{i/o}$, and $G\alpha_{12/13}$ subunits of guanine nucleotide-binding (G) proteins; (2) tyrosine [80] (Vol. 4 – Chap. 3. Cytosolic Protein Tyr Kinases) or serine/threonine kinases (Vol. 4 – Chap. 4. Cytosolic Protein Ser/Thr Kinases); (3) phospholipases PLA2, PLCβ and PLCγ, and PLD (Vol. 4 – Chap. 1. Signaling Lipids); [81] (4) small monomeric GTPases of the Ras superfamily; and (5) ion channels (Chap. 3).

Angiotensin-2 receptor AT_1 predominantly and robustly couples to Gq/11 subunit family. Once bound to AT_1 receptor, angiotensin-2 stimulates vascular cells. In vascular smooth muscle cells, upon AT_1 binding, angiotensin-2 activates phospholipases as well as kinases and various oxidases. G12-Coupled AT_1 can stimulate monomeric GTPase Rho and inhibit PDGF-activated Rac GTPase. The Gβγ subunit associated with G12 and/or Gq/11 family subunits mediates PLC activation by angiotensin-2. Associated with G12 subunit, it also intervenes in angiotensin-2-induced PLD1 activation, possibly via Src- and RhoA-dependent mechanisms. [82] In cultured rat vascular smooth muscle cells, angiotensin-2-mediated activation of phospholipase-D1 via Gβγ provides a major source of sustained generation of second messengers. The G12 subunit contributes to PLD activation, whereas Gi and Gq/11 have no effect [941]. Small GTPase RhoA participates in this signaling pathways. Hydrolysis of phosphatidylcholine by PLD causes a strong production of phosphatidic acid and subsequent generation of diacylglycerol by phosphatidic acid phosphatase. Diacylglycerol contributes to continuous PKC activation. Phosphatidic acid contributes to the activation of NADH/NADPH oxidase that mediates the hypertrophic effects of angiotensin-2.

Angiotensin-2 promotes the association of scaffold proteins (paxillin and talin), leading to focal adhesion. It also stimulates NADPH oxidase via AT_1 receptor. Resulting reactive oxygen species are components of ATn2-mediated signal transduc-

80. Once bound to agonist, AT_1 can recruit a complex of non-receptor protein tyrosine kinase and phosphatase, i.e., Janus kinase JaK2 and PTPn11 that facilitates JaK2 phosphorylation (activation). This additional G-protein-independent AT_1–JaK2 pathway complements signaling primed by G-protein subunit Gq that can further potentiate the JaK–STAT signaling via Ca^{++} import in the cytosol [835].

81. Phospholipase-Cβ1, then PLCγ1, are activated by corresponding stimulated G proteins within 5 s [940]. Diacylglycerol and inositol trisphosphate are thus generated, the production ending within minutes. The transient PLC activation is followed by a sustained activation of phospholipase-D via G proteins. Concomitant stimulation of phospholipase-A2 initiates arachidonic acid signaling.

82. The Gβγ subunit can activate PLD by binding to Rho or ARF GTPases.

tion in vascular smooth muscle cells. In addition, AT_1 and bradykinin B2 receptors can heterodimerize [5].

7.13.6.2 AT_2

The AT_2 receptor is much less abundant than AT_1 in adult tissues. It is upregulated in pathological conditions.[83] Receptor AT_2 counteracts several growth responses initiated by AT_1. Receptors AT_1 and AT_2 stimulates and inhibits ERK1 and ERK2, respectively.

Type-2 receptor AT_2 has a more restricted expression pattern than the AT_1 receptor. It intervenes in cell growth and differentiation and tissue development. It is highly expressed in fetuses, whereas in adults, it is confined to heart, vascular smooth muscle, brain, adrenal cortex, uterus, and ovarian follicles. Activated AT_2 dilates blood vessels, inhibits growth, and causes apoptosis. It provokes a rapid activation of Gi and Go (but not of Gq and Gs) [943].

7.13.6.3 Agonists

Angiotensin-2 causes cell proliferation, hypertrophy, vasoconstriction, and fibrosis [944]. On the other hand, angiotensin$_{(1--7)}$, an active heptapeptide formed from angiotensin-1 and-2, antagonizes angiotensin-2. It provokes vasodilation, as it causes the release of nitric oxide and vasodilatory prostaglandins, and can amplify bradykinin effect [944]. In addition, it has a potent natriuretic and antifibrotic action. It also inhibits the growth of vascular smooth muscle cells and cardiomyocytes.

Angiotensin$_{(1--7)}$ is preferentially formed from angiotensin-2 by Angiotensin-converting enzyme-2 (ACE2). Angiotensin-2 reduces ACE2 synthesis and activity in cardiac myocytes and fibroblasts. Endothelin-1 also attenuates ACE2 synthesis in cardiomyocytes. Decrease in ACE2 production by angiotensin-2 and endothelin-1 requires mitogen-activated protein kinase kinase MAP2K1 and extracellular signal-regulated kinases ERK1 and ERK2. Atrial natriuretic peptide reverses ATn2- and ET1-mediated ACE2 downregulation, when it acts simultaneously with these factors (but not alone).[84]

Angiotensin$_{(1--7)}$ impedes ATn2- and ET1-mediated ACE2 downregulation (but not its synthesis) via a specific $AT_{(1--7)}$ receptor [945].[85] Stimulated $AT_{(1--7)}$ activates the PI3K–PKB signaling cascade and endothelial nitric oxide synthase (NOS3)

83. The AT_1 receptor is involved in the evolution of aortic aneurysm in Marfan syndrome, an autosomal dominant connective tissue disorder characterized by heterozygous mutations in the FBN1 gene that cause a deficiency of fibrillin-1. Transforming growth factor-β signaling augments. Manifestations include aortic aneurysm, emphysema, degeneration of the atrioventricular valves, and myopathy. On the other hand, AT_2 confers protection [942].

84. Atrial natriuretic peptide, a peptide hormone released mainly from the atrium, activates the guanylate cyclase-A receptor (NPR1 or NPRa; Sect. 6.4.1). Activated NPR1 produces cGMP from GTP. Second messenger cGMP causes natriuresis and hypovolemia, and impedes the proliferation of vascular smooth muscle cells.

85. A..k.a. Mas.

in endothelial cells and cardiomyocytes [944]. Moreover, angiotensin$_{(1--7)}$ primes the secretion of atrial natriuretic peptide [944]. In atriomyocytes, angiotensin$_{(1--7)}$ increases ANP release at high atrial pacing via the PI3K–PKB pathway. Conversely, Na^+–H^+ exchanger-1 and Ca^{++}–calmodulin-dependent kinase CamK2 prevent the augmentation of high atrial pacing-induced ANP secretion by ATn$_{(1--7)}$.

Angiotensin$_{(1--12)}$ can trigger a vasoconstriction via AT_1 receptor. It can serve as a substrate for the formation of angiotensin-2 and angiotensin$_{(1--7)}$, according to the available, local type of angiotensin-converting enzyme. In the brain, its concentration is higher than those of angiotensin-1 and -2. It is produced in the nucleus tractus solitarius (baroreceptor reflex center), but do not affect the autonomic control of the cardiac frequency in normal conditions. However, angiotensin$_{(1--12)}$ may be active in hypertension [946]. Exogenous angiotensin$_{(1--12)}$ is converted into angiotensin-2 within the nucleus tractus solitarius. It can cause a transient depressor response [946].

7.13.6.4 Angiotensin Receptors in the Cardiovascular System

The vasculature is characterized by a low density of AT_1 in endothelial cells (ecAT_1) and its relative abundance in vascular smooth muscle cells (smcAT_1). Endothelial cell activation by angiotensin-2 via ecAT_1 reduces the tone of vascular smooth muscle cells, as it elevates expression of endothelial nitric oxide synthase [947]. Therefore, angiotensin-2 stimulation of ecAT_1 reduces smcAT_1-mediated vasoconstriction.

The AT_1 receptor contributes to vascular relaxation in cerebral resistance arteries via transactivation of epidermal growth factor receptor kinase and extracellular signal-regulated protein kinases ERK1 and ERK2 [948].

In hearts, pathways leading to ERK activation differ according to cell types. In cardiac fibroblasts, angiotensin-2 activates ERKs via G$\beta\gamma$ subunit of Gi, Src kinase, adaptors SHC and GRB2, and Ras GTPase.

In cardiomyocytes, mechanical stretch-induced ERK activation that can occur even in the absence of angiotensin-2 involves Gq and protein kinase-C [949]. Cardiomyocyte stretch indeed leads to conformational change and activation of AT_1 independently of angiotensin-2 [950, 951]. The AT_1 receptor is able to detect mechanical stress and transduces it into biochemical signals. Mechanical stretch induces association of AT_1 receptor with Janus kinase-2 and translocation of G proteins into the cytosol.

Angiotensin-2 operates as a potent hypertrophic agent for cardiomyocytes that promotes production of TGFβ1 and atrial natriuretic factor [952].[86] The AT_2 receptor is involved in maladaptive cardiac hypertrophy. On the other hand, cardiac

86. Angiotensin-2 can, hence, indirectly cause cell division, fibrosis, cytokine infiltration, α- to β-myosin heavy chain switch, and stimulation of the renin–angiotensin system. Interactions among cardiomyocytes, fibroblasts, and vascular cells are sources of secreted signals. Locally produced factors, such as growth factors FGF2, IGF1, TNFα, cardiotrophin-1, and endothelin, can also trigger cardiomyocyte hypertrophy. In addition, MAP3K7 can provoke heart failure.

AT_{1A} receptor causes cardiac fibrosis, but not hypertrophy [953]. Transforming growth factor-β1 isoform couples work load and angiotensin-2 to cardiac adaptative hypertrophy

Angiotensin-2 produces reactive oxygen species in vascular smooth muscle cells and cardiomyocytes. In rat neonatal cardiomyocytes, angiotensin-2 binds to $G\alpha_{12/13}$-coupled AT_1 that primes a cascade with the following mediators, which operate successively: (1) small GTPase Rho and RoCK kinase; (2) small GTPase Rac; (3) NADPH oxidase; (4) reactive oxygen species; and (5) JNK and P38MAPK [954].

7.13.6.5 Angiotensin Receptors in the Liver

Angiotensin-2 acts as a pro-inflammatory agent and growth factor in the liver. After injury to the liver, it assists in tissue repair by stimulating: (1) hepatocytes and hepatic stellate cells to synthesize extracellular matrix proteins and secrete cytokines and (2) myofibroblasts to proliferate.

Angiotensin-2 activates nuclear factor NFκB via AT_1 and caspase recruitment domain-containing proteins CaRD10 and B-cell lymphoma/leukemia protein BCL10 as well as paracaspase mucosa-associated lymphoid tissue lymphoma translocation protein MALT1 [955]. [87]

7.13.7 Apelin Receptors

Apelin APJ receptor [88] (Sect. 6.9.2) responds to apelins [741], mainly long $apelin_{36}$ and short $apelin_{13}$ [89] (potency order $apelin_{13} > apelin_{36}$ [5]). Main transducers are subunits of the $G\alpha_{i/o}$ family of guanine nucleotide-binding (G) proteins. The hydrophobic region of APJ is similar to that of the angiotensin receptor [956].

The APJ receptor is particularly produced in the cardiovascular and central nervous systems, and in the latter by neurons, oligodendrocytes, and astrocytes [958]. [90] Apelin is also observed in adipocytes as well as gastric mucosa and Kupffer cells in the liver [961].

87. Caspase recruitment domain-containing proteins (CaRD9 to -11 and -14, as well as BCL10) operate as upstream regulators in NFκB signaling. Proteins CaRDs interact with BCL10 to act as NFκB signaling complexes. Protein BCL10 forms a complex with MALT1 to synergize NFκB activation.

88. The APJ receptor is encoded by chromosome 11.

89. Apelin derives from a precursor, preproapelin (77 amino acids). Apelins include long types (e.g., 36 amino acids; $apelin_{(42--77)}$) expressed in lungs, testes, and uterus, and short types (e.g., 13 amino acids; $apelin_{(65--77)}$) produced in mammary glands, which also synthesize the long apelin type [957]. Long and short forms of apelin differently interact with APJ receptor.

90. Apelin serves a neuropeptide in the hypothalamus, especially in the supraoptic, paraventricular, and arcuate nuclei. Apelin-containing cell bodies of the supraoptic and paraventricular nuclei, such as arginine vasopressin- and oxytocin-containing magnocellular neurons, project toward the internal layer of the median eminence and posterior pituitary (neurohypophysis).

In the central nervous system, the apelin–APJ complex helps maintain the fluid homeostasis and regulates vasopressin release from the hypothalamus [959]. In response to osmotic stimulation, the release of arginine vasopressin, leucine enkephalin (^{Leu}enkephalin), and dynorphin from neurons of the supraoptic and paraventricular nuclei is augmented.[91] Apelin targets vasopressinergic neurons and inhibits their activity, thus impeding arginine vasopressin release. Apelin and arginine vasopressin that coexist in magnocellular neurons have opposite effects (autocrine somatodendritic feedback loop) to maintain body fluid homeostasis.

In the cardiovascular system, apelin is primarily expressed in vascular and endocardial endothelial cells (at least in recesses of the right atrium) [962].[92] Cardiomyocytes can also express apelin. Apelin receptors on vascular smooth muscle and endothelial cells as well as cardiomyocytes are activated by the family of apelin peptides [963]. They operate in both auto- and paracrine signalings.

Apelin, a direct vasoconstrictor,[93] can mediate vasodilation via nitric oxide [961]. In fact, the endothelium-dependent vasodilation results from a prostanoid-dependent mechanism [963]. In mammary artery, the 3 main forms of apelin — N-terminal pyroglutamylate ^{pGlu}apelin$_{13}$, apelin$_{13}$, and apelin$_{36}$ — cause vasodilation, when the endothelium is functional. These regulators act with comparable potency in a concentration-dependent manner.

In human hearts, apelin peptides are among the most potent endogenous positive inotropic agents [963]. The N-terminal pyroglutaminated apelin$_{13}$ is the predominant isoform in hearts from patients with coronary artery disease. Apelin acts on neighboring cardiomyocytes and exerts positive inotropic effect via Na^+–H^+ exchanger NHE1 (or SLC9a1), Na^+–Ca^{++} exchanger (reverse mode), phospholipase-C, and protein kinase-C [739].[94] Autocrine signaling by apelin and APJ receptor in mouse left ventriculomyocytes plays a modest role in basal conditions, but becomes significant under stress, as they both improve sarcomeric shortening and velocity of contraction and relaxation without change in calcium transient [964].

Apelin is also involved in angiogenesis, as it enlarges developing blood vessels through which the flow rate is higher upon stimulation by the Ang1–TIE2 complex [965].[95] Angiopoietin-1 synthesized by mural cells and migrated hematopoietic stem cells activates its receptor TIE2 to generate apelin production in sprout endothelial cells.

---

91. Hypothalamic magnocellular neurons of the paraventricular and supraoptic nuclei synthesize not only vasopressin and oxytocin, but also tyrosine hydroxylase (enzyme of catecholamine synthesis), as well as galanin, dynorphin, and cholecystokinin. High plasma osmolality induces increased mRNA levels for vasopressin, oxytocin, Tyr hydroxylase, galanin, dynorphin, and cholecystokinin [960].

92. Apelin is produced by endothelial cells of, at least, large blood vessels and vessels of heart, kidneys, adrenal glands, and lungs.

93. Apelins are potent vasoconstrictors in endothelium-denuded saphenous veins and mammary arteries.

94. Apelin is downregulated in chronic ventricular pressure overload.

95. Apelin binds to APJ to enlarge blood vessels. Expression of APJ is enhanced by VEGF.

The APJ receptor is coupled to $G\alpha_{i1}$ or $G\alpha_{i2}$, but not $G\alpha_{i3}$ [966]. Apelin inhibits adenylate cyclase and increases the phosphorylation of ERK or PKB. Apelin efficiency in inhibiting adenylate cyclase depends on: (1) activation and deactivation kinetics of the apelin type and (2) cell type. Activation of apelin receptors by short fragments is transient as APJ is quickly internalized. The duration of desensitization depends on the rate of receptor recycling to the plasmalemma ($\sim$1 h). Long subtypes produces a sustained activation.

7.13.8 Bile Acid Receptor

Bile acid receptor GPBAR1 [96] responds to bile acids produced in the liver with the following potency order [5]:

lithocholic acid > deoxycholic acid > chenodeoxycholic acid, cholic acid.

It activates deiodinases that convert prohormone thyroxine (T4) into active hormone tri-iodothyronine (T3). It signals via Gs and cAMP as well as mitogen-activated protein kinase.

7.13.9 Bombesin Receptors

Bombesin receptors (BB_1–BB_3) are activated by bombesin homologs [97] gastrin-releasing peptide (GRP) and neuromedin-B and -C [5]. [98] These receptors couple primarily to Gq/11 subunit family of G proteins. Activated BB_1 and BB_2 receptors stimulate tissue growth, smooth muscle cell contraction, and secretion. The BB_3 receptor may intervene in energy balance and control of body weight [5].

7.13.10 Bradykinin Receptors

Bradykinin receptors (B_1–B_2 or B1R–B2R) are activated by the nonapeptide bradykinin and its derived peptides kinin and kallidin (Table 7.46). [99] Kinins are polypeptides formed locally after tissue damage and inflammation. They provoke smooth muscle contraction as well as vasodilation via the release of nitric oxide and prostaglandins by endothelial cells, increase the vascular permeability, and exert a mitogenic effect using the mitogen-activated protein kinase modules.

Kinins and their derivatives generated from kininogens by kallikreins and other serine peptidases compose the *kallikrein-kinin system*. Kallidin is a bioactive kinin that possesses an additional lysine residue at the N-terminal end, hence its name ^{Lys}bradykinin. This substrate of carboxypeptidase-M and -N signals through the kinin receptor. It is formed in response to injury from kininogen precursors by

96. A.k.a. M-BAR, GPCR19, GPR131, BG37, and TGR5.

97. Bombesin is a tetradecapeptide originally isolated from skins of amphibians.

98. A.k.a. GRP_{18--27}.

99. The B_1 receptor is expressed at a much lower level than B_2 in mice kidneys, but it is induced upon stimulation by endotoxins or cytokines.

Table 7.46. Receptors of the bradykinin/kinin family with their main targeted G proteins (Source: [736]).

Type	Main transducer
B_1, B_2	Gq/11
	Gi ($G\beta\gamma$)

kallikreins. Kallidin can be converted to bradykinin by aminopeptidase. Bradykinin and kallidin are produced by the proteolysis of high- or low-molecular-weight kininogen by plasma and tissue kallikrein, respectively.

Bradykinin binds and stimulates at least the 2 bradykinin receptors: B_1 and B_2. The B_2 receptor is constitutively expressed in various cell types. It relays the majority of the vascular actions of bradykinin and kallidin. On the other hand, the B_1 receptor is scarce in normal tissue and expressed during tissue damage (e.g., inflammation and ischemia), by exposure to pro-inflammatory cytokines, growth factors, or oxidative stress, in various cell types, such as vascular smooth muscle and endothelial cells, cardiomyocytes, fibroblasts, and neurons. Bradykinin receptors are coupled to Gq and/or Gi subunits depending on the cell type.

Carboxypeptidases remove the C-terminal arginine of bradykinin and kallidin to generate desArg9BK (or $BK_{(1-8)}$) and desArg10kallidin (also denoted as ${}^{Lys}BK_{(1-8)}$ or ^{Lys}desArg9BK), respectively. Receptor B_1 is preferentially activated by the metabolites $BK_{(1-8)}$ and ${}^{Lys}BK_{(1-8)}$ peptides.

Angiotensin-2 and endothelin-1 increase the production of B_1 receptor in cardiovascular diseases. These pro-oxidative peptides cooperate via ROS, PI3K, and NFκB on rat vascular smooth muscle cells using AT_1 and ET_A receptors [967]. Newly synthesized B_1 receptors can activate the MAPK module. Once the AT_1 receptor is activated, the effect of angiotensin-2 can be mediated by the subsequent ET1 release and ET_A stimulation.

Regulator of G-protein signaling RGS4 can serve as GTPase-activating proteins for both Gi and Gq (but not for Gs and G12). The RGS4 protein attenuates Gi-mediated inhibition of cAMP synthesis and impedes Gq-mediated activation of phospholipase-Cβ [969]. Both Gq and Gi subunits are able to activate phospholipase-C, the latter via $G\beta\gamma$.

Bradykinin GPCRs can signal without G proteins. In renal cells, the B_2 receptor interacts with protein Tyr phosphatase PTPn11 to prevent cell proliferation [970].

7.13.10.1 Bradykinin Receptors and Angiotensin-Converting Enzyme

Angiotensin-1-converting enzyme (ACE) is a dipeptidyl carboxypeptidase that removes 2 amino acids from the C-terminus of inactive angiotensin-1 to convert it into active angiotensin-2. Moreover, ACE is also a kininase that converts active vasodilatory kinins into inactive metabolites, as it also deletes 2 amino acids from their C-termini.

In humans, the polymorphism of the Ace gene due to the presence or absence of an Alu retrotransposon in intron 16 is associated with up to a 2-fold difference in relative plasma ACE concentration. [100] The kallikrein-kinin system that partly mediates the effects of the polymorphism has a protective role [968]. Kinins indeed bind to Gq-coupled bradykinin B_1 and B_2 receptors to reduce the risk of diabetic complications.

7.13.10.2 Bradykinin Receptors in the Cardiovascular Apparatus

In adult rat cardiomyocytes, bradykinin activates P21 activated kinase PAK1 downstream from monomeric GTPases. Kinase PAK1 stimulates protein phosphatase-2 that dephosphorylates both cardiac troponin-I and phospholamban [971]. Protein phosphatase-2 is able to complex with both PAK1 and phospholamban.

Vasodilation and Vessel Remodeling

Bradykinin is a potent vasodilator of arteries that irrigate muscle, kidney, and heart, among other viscera. Tissue kallikrein does not contribute to uterine artery remodeling during and after pregnancy [972].

Bradykinin synthesized by tissue kallikrein from kininogen stimulates endothelial β2-adrenergic receptors, subsequently releasing nitric oxide [101] and prostacyclin to control the vascular tone.

Both receptors are coupled with Gq proteins. Their stimulation activates phosphatidylinositol-specific phospholipases that augment intracellular Ca^{++} concentration and leads to activation of endothelial nitric oxide synthase (NOS3). Kinins that act via B_2 also support the expression of inducible NOS (NOS2). Nitric oxide reversibly suppresses mitochondrial oxidation, at least partly by inhibition of cytochrome-C oxidase of the electron transport chain. Kinins also facilitate the synthesis of prostanoids, such as prostaglandins PGE2 and PGI2, which heighten intracellular cAMP concentration [968]. Messenger cAMP also decreases mitochondrial respiration, as it activates the NADH-ubiquinone oxidoreductase activity of complex I and inhibits cytochrome-C oxidase. Therefore, nitric oxide and kinins reduce oxidative stress. In particular, bradykinin reduces mitochondrial superoxide generation in human vascular endothelial cells [968].

Sympathoexcitatory Reflex

Bradykinin primes afferent reflex in the heart owing to cardiac sympathetic afferent fibers. Bradykinin causes a sympathoexcitatory reflex [102] characterized by

100. This polymorphism does not significantly affect blood pressure and angiotensin-2 and aldosterone concentration. Nevertheless, it is associated with different risks of diabetic complications.

101. Endothelial nitric oxide synthase expression and, consequently, nitric oxide production, increase during pregnancy.

102. The sympathoexcitation reflex happens when all other cardiovascular reflexes (baroreceptor reflex, cardiopulmonary reflex with vagal afferents) are not efficient enough.

increased arterial blood pressure and renal sympathetic nerve activity via epicardial B_2 receptors and sympathetic cardiac afferent fibers [973]. Intrapericardial bradykinin generates a blood pressure rise, whereas circulating bradykinin acts as a vasodilator, thereby lowering blood pressure.

Cardioprotection

In the heart, bradykinin is released during cardiac ischemia and myocardial infarction. Both bradykinin receptors B_1 and B_2 contribute to the cardioprotective effect of ACE inhibition mediated by bradykinin [974]. The B_2 receptor is the main kinin receptor involved in the cardioprotection yielded by ischemic preconditioning [975]. It forms a complex with angiotensin-converting enzyme that sequesters the latter. This process determines a crosstalk between the renin–angiotensin pathway and the kinin–kallikrein axis.

7.13.11 Calcitonin, Amylin, CGRP, and Adrenomedullin Receptors

Calcitonin (Ct) [103] is a hormone that results from the proteolytic cleavage of a prepropeptide that is encoded by the CALCA gene (or CALC1) primarily in parafollicular cells [104] of the thyroid. Procalcitonin (PCt) belongs to a family of related hormone precursors that include islet amyloid precursor, calcitonin gene-related peptide, and adrenomedullin precursor. It decreases bone resorption.

Amylin (Amy) [105] is a hormone secreted by pancreatic β cells. Amylin reduces nutrient intake and contributes to the control of glycemia, as it optimizes glucose metabolism.

Calcitonin gene related peptide-2 (CGRP2) [106] is a member of the calcitonin family of peptides. It is produced in both peripheral and central neurons. It is a potent vasodilator.

Adrenomedullin-1 (AM1) is ubiquitously synthesized from a precursor (proadrenomedullin) encoded by the AM gene. It abounds in vascular endothelial and smooth muscle cells. It is also a potent vasodilator. Furthermore, it regulates body fluids, as it precludes drinking and sodium intake and causes natri- and kaliuresis [976]. [107] Different excisions form various peptides, such as an inactive adrenomedullin, ePAMP, adrenotensin, and $AM_{95--146}$. In humans, mature adrenomedullin is rapidly catabolized. Circulating adrenomedullin comprises both mature, amidated and inactive, glycated form. The related peptide adrenomedullin-2 (AM2), or intermedin, participates in the regulation of blood circulation. It raises the concentrations

103. A.k.a. calcitonin-related polypeptide-α, CalcA, and Calc1, as well as calcitonin gene-related peptide CGRP1 and α-CGRP (Vol. 2 – Chap. 1. Remote Control Cells – Sect. Endocrine System and Hormones).

104. A.k.a. C cells.

105. A.k.a. (pancreatic) islet amyloid polypeptide (IAPP).

106. A.k.a. calcitonin-related polypeptide-β, β-CGRP, CalcB, and Calc2.

107. Angiotensin-2 is a dipsogenic hormone. Atrial natriuretic peptide also attenuates the drinking rate [977].

Table 7.47. Receptor family of the calcitonin–calcitonin gene-related peptide family of peptide hormones (Source: [5]; AM: adrenomedullin; Amy: amylin; CGRP: calcitonin gene-related peptide; Ct: calcitonin).

Receptor	Composition	Main transducer	Potency order
CT	CT	Gs, Gq	Ct $\geq$ Amy, CGRP $>$ AM1, AM2
Amy_1	CT–RAMP1	Gs	Amy $\geq$ CGRP $>$ AM2 $>$ Ct$>$AM1
Amy_2	CT–RAMP2	Gs	
Amy_3	CT–RAMP3	Gs	Amy $>$ CGRP $>$ AM2 $>$ Ct $>$ AM1
CGRPR	CL–RAMP1	Gs, Gq	CGRP $>$ AM1 $\geq$ AM2 $>$ Amy
AM_1	CL–RAMP2	Gs	AM1 $\gg$ CGRP, AM2 $>$ Amy
AM_2	CL–RAMP3	Gs	AM1 $\geq$ CGRP, AM2 $>$ Amy

of atrial and brain natriuretic peptides and renin in plasma. Adrenomedullin-5 that also derives from a prohormone decreases arterial pressure.

Calcitonin, amylin, calcitonin gene-related peptide, and adrenomedullin receptors are generated by the genes CALCR and CALCRL that encodes the calcitonin receptor (CalcR, CTR, and CTR1) and calcitonin receptor-like receptor (CalcRL and CLR). Calcitonin receptors that activate G-protein subunits Gs and Gq are involved in the maintenance of calcium homeostasis. Their function changes according to their interaction with receptor activity-modifying proteins (RAMP1–RAMP3) that form multimeric amylin receptors Amy_1 (CalcR–RAMP1), Amy_2 (CalcR–RAMP2), and Amy_3 (CalcR–RAMP3; Table 7.47). Splice variants of the CalcR receptor produce variants of Amy receptors.

Calcitonin receptor-like receptors also connect to receptor activity-modifying proteins. Receptor of CGRP2 (CGRPR) is a heteromer composed of the G-protein-coupled receptor CalcRL and receptor activity-modifying protein RAMP1. The CalcRL subunit also produces the adrenomedullin receptor AM_1 and AM_2 with RAMP2 and RAMP3, respectively.

Adrenomedullin-1 targets the CalcRL–RAMP2 or CalcRL–RAMP3 complexes, whereas adrenomedullin-2 interacts non-selectively with all 3 CalcRL–RAMP heteromers [978]. Adrenomedullin-5 tethers to AM_5 receptors (other than CalcRL–RAMP or CalcR–RAMP) to exert its cardiovascular actions in mammals [979].

7.13.12 Calcium-Sensing Receptors

Calcium-sensing receptors (CaR, CaS, or CaSR) are G-protein-coupled receptors (Table 7.48) that sense the concentration of extracellular Ca^{++} ions. Calcium-sensing receptor responds to extracellular concentrations of calcium and magnesium in the millimolar range and of gadolinium and some polycations in the micromolar range [5]. The extracellular Ca^{++}-sensing receptor responds also to many other ligands. In particular, its activity is allosterically regulated by amino acids and H^+ ions.

Table 7.48. Calcium-sensing receptor, targeted G proteins, and order of potency between extracellular concentrations of Ca^{++} and Mg^{++} (Source: [736]).

G-protein transducers	Potency order
Gq/11, Gi/o, G12/13	$[Ca^{++}]_e > [Mg^{++}]_e$

This GPCR resides especially on hematopoietic stem cells. It retains hematopoietic stem cells close to the endosteal surface of the bone marrow. It interacts with extracellular matrix components, particularly collagen-1.

Calcium-sensing receptor also lodges in parathyroids, where it participates in parathyroid hormone response to extracellular Ca^{++} concentration ($[Ca^{++}]_e$). In addition, CaRs are produced by cells of the kidney, liver, and thyroid gland.

In the liver, calcium-sensing receptor is specifically expressed in hepatocytes (not in stellate, endothelial, and Kupffer cells). It mobilizes calcium ions from IP_3-sensitive stores to stimulate bile flow [980].

This major sensor and regulator of $[Ca^{++}]_e$ also localizes to the stomach and intestinal tract, especially on gastrin-secreting G cells in the stomach. Luminal nutrients, particularly Ca^{++} and amino acids are potent stimulators of gastrin and acid secretion [981].

In cardiomyocytes, CaSR predominantly localizes to caveolae. It ensures a cardioprotective role in ischemic preconditioning [982].

7.13.13 Cannabinoid Receptors

Cannabinoid receptors are targeted by endo- and exogenous cannabinoids. Endocannabinoids comprise anandamide, or ^{N}arachidonoyl ethanolamine (AEA), as well as ^{N}homo γ-linolenoyl ethanolamine, ^{N}docosatetra (7,10,13,16)-enoyl ethanolamine, and 2-arachidonyl glycerol (2AG). [108] Anandamide and 2-arachidonoyl glycerol are 2 major endocannabinoids. Endocannabinoids 2-arachidonyl glycerol and anandamide are released from the cell membrane by activated phospholipases.

Cyclooxygenase-2 transforms 2-arachidonyl glycerol and anandamide into prostaglandin glycerol esters (^{PGH2}G) and ethanolamides (^{PGH2}EA). [109] In cells, glycerol esters and ethanolamides of prostaglandins PGE2, PGD2, and PGF2α are endocannabinoid-derived COx2 products (^{PGH2}G and ^{PGH2}EA). The sequential action of COx2 and thromboxane-A synthase (TxaS) on AEA and 2AG yields thromboxane-A2 ethanolamide and glycerol ester, respectively [983]. Similarly, prostaglandin synthases PGdS, PGeS, and PGiS (i.e., prostaglandin-D, -E, -H2, and prosta-

108. 2-Arachidonyl glycerol is also termed 2-arachidonoyl-glycerol, (5,8,11,14)-eicosatetraenoic acid, and 2-hydroxy 1-(hydroxymethyl)ethyl ester.

109. Activated phospholipase releases arachidonic acid from phospholipids. Afterward, cyclooxygenase oxygenates liberated fatty acid to generate the hydroxy-endoperoxide prostaglandin-H2. The latter can be converted into one among various derivatives: prostaglandins, thromboxane, and prostacyclin. Thromboxane-B2 (TxB2) is an inactive TxA2 metabolite. Cyclooxygenase-1 generates thromboxane-A2 in platelets.

cyclin synthases) contribute to the formation of prostaglandin and prostacyclin glycerol esters and ethanolamides.

The endocannabinoid system is involved in signal transduction not only in the central nervous system, but also in other organs. Endocannabinoids are lipid mediators that are released and rapidly degraded by numerous enzymes. The most abundant endocannabinoid in the central nervous system, 2-arachidonoyl glycerol, is synthesized from diacylglycerol by diacylglycerol DAGLα (mainly) and DAGLβ lipases. These enzymes may also produce arachidonic acid. 2-Arachidonoyl glycerol functions as a retrograde signaling molecule that suppresses synaptic transmission in the central nervous system and regulates axonal growth and guidance. [110]

Two known cannabinoid isoreceptors are: (1) type-1 (CB_1) expressed mainly in the brain, but also lung, liver, and kidneys, as well as vascular smooth muscle and endothelial cells, monocytes, and macrophages; and (2) type-2 (CB_2) cannabinoid receptor chiefly produced in hematopoietic cells (B and T lymphocytes as well as macrophages), and also keratinocytes and vascular smooth muscle and endothelial cells. Novel cannabinoid receptors exist in endothelial cells as well as in the central nervous system. Cannabinoid-like receptors encompass GPR18 for ^{N}arachidonoylglycine, GPR55 with a wide spectrum of cannabinoid agonists and antagonists, GPR119 for fatty acid ethanolamides [5]. Anandamide also targets TRPV1 channel.

Cannabinoid receptors participate in the regulation of adult neurogenesis [984]. Receptors CB_1 and CB_2 participate in the genesis of atherosclerosis. Monocytes undergo a significant change in CB_1 and CB_2 expression profile during differentiation into macrophages [985]. The CB_1:CB_2 ratio on monocytes and macrophages, indeed, regulates their inflammatory activity and ability to produce reactive oxygen species [985]. These receptors have opposing influences on ROS production. In macrophages, CB_1, but not CB_2, causes phosphorylation of P38MAPK that causes ROS production and synthesis of pro-inflammatory cytokines such as tumor-necrosis factor-α and CCL2 chemokine. [111] On the other hand, the CB_2 receptor activates small GTPase Rap1 that counteracts CB_1-stimulated ROS production. In endothelial cells, the CB_2 receptor also lowers TNFα-induced proliferation and migration of smooth muscle cells and expression of adhesion molecules and chemokines.

The type of cannabinoid receptor ligand influences Gα-subunit response. A given ligand of cannabinoid receptor CB_1 can act as an agonist for Gi1 and Gi2 and as an antagonist for Gi3. Another type of ligand can behave as an inverse agonist of Gi1 and Gi2 and as an agonist for Gi3.

Cannabinoid receptors target many effectors, such as adenylate cyclase, inwardly rectifying potassium channels (K_{IR}) and other K^+ channels, calcium channels, protein kinase-A and -C, kinases Raf1, ERK, JNK, and P38MAPK, and transcription factors Fos and Jun.

110. Endocannabinoids are released from postsynaptic neurons and cause retrograde suppression of synaptic transmission. In the hippocampus, the postsynaptic release of an endocannabinoid transiently suppresses GABA-mediated transmission at inhibitory synapses [984].

111. A.k.a. monocyte chemoattractant protein-1).

Endocannabinoids released from a neuron that bears endocannabinoid action bind to CB_1 receptor in presynaptic neuron and reduces GABA liberation, hence GABA-mediated neurotransmission. Cannabinoid receptor CB_1 targeted by glucocorticoids in the basolateral nucleus of the amygdala intervenes in the consolidation of emotional memory, although glucocorticoids usually regulate gene transcription by either binding homo- or heterodimers of intracellular glucocorticoid receptors to nuclear DNA or by interacting with transcription factors [986]. Endogenous cannabinoid binding to CB_1 stimulates appetite.

Activated presynaptic CB_1 receptors also repress sympathetic innervation of blood vessels. Anandamide and 2-arachidonyl glycerol produced by macrophages and platelets, respectively, can cause hypotension.

In the liver, activated CB_1 increases fatty acid synthesis via lipogenic transcription factor SREBP1c and acetyl coenzyme-A carboxylase-1 and fatty acid synthase [987].

Cannabinoid CB_2 receptor abounds in mature B lymphocytes. It also exists in myeloid and natural killer cells, and various other cell types. Ligand for CB_2, 2-arachidonoyl glycerol, is generated from arachidonic acid-containing phospholipids. Expression of CB_2 in peripheral B lymphocytes is correlated to the production level of immunoglobulin λ-chain [988].

Production and function of CB_2 are upregulated in immature B lymphocytes. Migration in bone marrow sinusoids of immature B cells depends on Gi-protein-coupled cannabinoid receptor CB_2 [988]. Immunoglobulin-M+ immature B cells can be retained in bone marrow sinusoids owing to integrin-$\alpha_4\beta_1$ and its endothelial ligand VCAM1.

G-protein-coupled receptor GPR119 can also serve as a cannabinoid receptor. It is synthesized predominantly in pancreas and gastrointestinal tract. It regulates the secretion of incretin and insulin. It is activated by endogenous fatty acid ethanolamines, particularly ^{N}oleoylethanolamine and ^{N}palmitoylethanolamine [5]. It is mainly coupled to Gs subunit.

7.13.14 Chemokine Receptors

Chemokine receptors constitute a large class of GPCRs that are activated by at least one chemokine.[112] Chemokines can be subdivided according to their structure into 4 subclasses: 28 known CC,[113] 16 identified CXC,[114] 2 detected C, and 1 CX_3C chemokines. Chemokines can also be classified according to their function into homeostatic and inflammatory groups.

Most chemokine receptors can bind with a high affinity to multiple chemokines (Table 7.49, 7.50, and 7.51). Most chemokines tether to several receptor subtypes. However, chemokines bind almost always to the same structural receptor subclass.

Chemokine receptor CCR7 expressed on mature, naive T lymphocytes intervenes in immune responses that require the coordinated interaction of various cell types

112. I.e., cytokine with chemotactic activity for leukocytes.
113. A.k.a. β-chemokines.
114. A.k.a. α-chemokines.

Table 7.49. Chemokine receptors CCR and their ligands (Source: [5]; ALP: CC chemokine with an ALP (Ala–Leu–Pro) N-terminal sequence; BCA: B-cell attracting chemokine; CTACK (CTAK): cutaneous T-cell-attracting chemokine; ELC: EBV induced gene-1 ligand chemokine; HCC: human C–C motif chemokine; ILC: IL11Rα-locus chemokine; IP: interferon-γ-induced protein; LARC: liver- and activation-regulated chemokine; MCP: monocyte chemoattractant protein; MDC: macrophage-derived chemokine; MEC: mucosa-associated epithelial chemokine; MIP: macrophage inflammatory protein; MPIF: myeloid progenitor inhibitory factor; RANTES: regulated upon activation, normal T-cell expressed, and secreted product; SLC: secondary lymphoid tissue chemokine; STCP: stimulated T-cell chemotactic protein; TARC: thymus and activation-regulated chemokine; TECK: thymus-expressed chemokine).

Type	Main transducer	Agonists
CCR1	Gi/o	CCL3 (MIP1α), CCL5 (RANTES), CCL7 (MCP3), CCL8 (MCP2), CCL13 (MCP4, BCA1), CCL14a (HCC1), CCL15 (HCC2), CCL23 (MIP3, MPIF1)
CCR2	Gi/o	CCL2 (MCP1), CCL7, CCL8, CCL13, CCL16 (HCC4)
CCR3	Gi/o	CCL11 (eotaxin), CCL5, CCL7, CCL8, CCL13, CCL15, CCL24 (eotaxin-2, MPIF2), CCL26 (eotaxin-3), CCL28 (MEC)
CCR4	Gi/o	CCL22 (MDC, STCP1), CCL17 (TARC)
CCR5	Gi/o	CCL3, CCL4 (MIP1β), CCL5, CCL8, CCL11, CCL14a, CCL16
CCR6	Gi/o	CCL20 (LARC)
CCR7	Gi/o	CCL19 (ELC, MIP3β), CCL21 (SLC)
CCR8	Gi/o	CCL1 (I309), CCL4, CCL16, CCL17
CCR9	Gi/o	CCL25 (TECK)
CCR10	Gi/o	CCL27 (eskine, skinkine, ALP, CTACK, ILC), CCL28

within lymphoid tissues. Agonists CCL19 and CCL21 target CCR7 to initiate chemotaxis. Ligand CCL19 leads to CCR7 phosphorylation by both GRK3 and GRK6 and β-arrestin-2 recruitment, whereas CCL21 activates only GRK6. Only CCL19 leads to receptor desensitization, internalization, and degradation, whereas both agonists launch extracellular signal-regulated protein kinase via GRK6 [989].

The Gi-coupled chemokine receptor CXCR4 binds to CXCL12 produced by bone marrow stromal cells. It enables the retention of neutrophils and hematopoietic stem and progenitor cells in the bone marrow.[115] Conversely, disruption of CXCL12-mediated chemoattraction of CXCR4+ cells mobilize neutrophils and hematopoietic stem and progenitor cells into the blood circulation. A pepducin[116] mobilizes bone

115. The CXCR4 receptor resides on hematopoietic stem cells, myeloid progenitors, and immature neutrophils.

116. A pepducin is composed of a peptide derived from the amino acid sequence of one of the intracellular loops of a target GPCR coupled to a lipid. Synthetic pepducins are thus cell-

Table 7.50. Chemokine receptors CXCR and their ligands (Source: [5]; BCA: B-cell-attracting chemokine; BLC: B-lymphocyte chemoattractant; ENA: epithelial-derived neutrophil-activating protein; GCP: granulocyte chemoattractant (chemotactic) protein; GRO: growth-regulated oncogene; IL: interleukin; IP: interferon γ-inducible protein; ITAC: interferon-inducible T-cell alpha chemoattractant; MIG: monokine induced by interferon-γ; NAP: neutrophil-activating peptide; SDF: stroma cell-derived factor; SRPSOx: scavenger receptor for phosphatidylserine and oxidized low-density lipoprotein).

Type	Main transducer	Agonists
CXCR1	Gi/o	CXCL6 (GCP2), CXCL8 (IL8)
CXCR2	Gi/o	CXCL1 (GROα), CXCL2 (GROβ), CXCL3 (GROγ), CXCL5 (ENA78), CXCL6, CXCL7 (NAP2), CXCL8
CXCR3	Gi/o	CXCL9 (MIG), CXCL10 (IP10), CXCL11 (ITAC)
CXCR4	Gi/o	CXCL12α (SDF1α), CXCL12β (SDF1β)
CXCR5	Gi/o	CXCL13 (BLC, BCA1)
CXCR6	Gi/o	CXCL16 (SRPSOx)

Table 7.51. Chemokine receptors CXCR and their ligands (Source: [5]).

Type	Main transducer	Agonists
CX_3CR1	Gi/o	CX_3CL1 (fractalkine)
XCR1	Gi/o	XCL1α and β (lymphotactin-α and -β)

marrow hematopoietic cells [990]. It can be thus used to more easily collect hematopoietic stem and progenitor cells before autologous bone marrow transplantation.

7.13.15 Complement (Anaphylatoxin) and Formyl Peptide Receptors

Anaphylatoxins, or anaphylotoxins, are fragments (C3a, C4a, and C5a) produced once the complement cascade is activated. Anaphylatoxins are able to trigger degranulation of endothelial and mastocytes and phagocytes to produce a local inflammation. Anaphylatoxins indirectly cause smooth muscle cell contraction, increase blood capillary permeability, and provoke chemotaxis of leukocytes. Anaphylatoxin and formyl peptide receptors are activated by anaphylatoxins C3a and C5a (Table 7.52).

penetrating lipopeptides. Any subtype of pepducins inhibits signal transmission from GPCRs to G proteins,

Table 7.52. Receptors of the complement system and anaphylatoxin and formyl peptide receptors, targeted G protein subunits, and order of potency (Source: [5]; fMLP: formyl-Met-Leu-Phe).

Receptor type	G-Subunit transducers	Potency order
C3aR	Gi/o, Gi/z	C3a > C5a
C5aR	Gi/o, Gi/z, Gq/16	C5a > C3a
FPR1	Gi/o, Gz	fMLP > cathepsin-G > annexin-1

Table 7.53. Cholecystokinin receptors, main G-protein subunit transducers, and ligands (Source: [5]).

Type	Transducer	Potency order
CCK_1	Gs, Gq, G11	Cck8 $\gg$ gastrin > Cck4
CCK_2	Gs, Gq	Cck8 $\geq$ gastrin, Cck4

Formyl peptide receptors (FPR) [117] are expressed at high levels on phagocytes. They are involved in chemotaxis, production of reactive oxygen species, and release of peptidases. Activated formyl peptide receptors trigger cytoskeleton rearrangement that, in turn, facilitates cell migration. They lead to activation of: (1) phospholipase-C and the IP_3–Ca^{++} and DAG–PKC cascades; (2) small GTPase Ras and the MAPK module; and (3) CD38 ectoenzyme that converts NAD^+ into cyclic ^{ADP}ribose, which interacts with ryanodine receptors for sustained Ca^{++} influx. In addition, formyl peptide receptor-like proteins of vomeronasal sensory neurons have an olfactory function associated with pathogens.

7.13.16 Cholecystokinin Receptors

Cholecystokinin receptors (CCK_1–CCK_2 or CCKa–CCKb) belong to the class A of G-protein-coupled receptors. They are activated by cholecystokinins Cck4, Cck8, and Cck33, as well as gastrin (Table 7.53).

The CCK_2 receptor is the dominant receptor in the central nervous system [991]. It is highly synthesized in the gastric mucosa, where it mediates gastrin-stimulated gastric acid secretion. It also contributes to gastric epithelial renewal. It resides in exo- and endocrine pancreas, enteric smooth muscle, liver, kidney, and adrenal gland, as well as circulating monocytes. The CCK_2 receptor recognizes each of the mature forms of cholecystokinins and gastrin with similar affinities and responds to them with similar potencies. It binds to progastrin and glycine-extended gastrin with lower affinity. It can couple to Gq G-protein subunit to signal predominantly via phospholipase-C and calcium import. It can also couple to Gs subunit. It

117. Formyl peptide receptors have been originally identified by their ability to bind ^{N}formyl peptides, such as exogenous bacterial product, the formylated tripeptide ^{N}formyl-methionyl-leucyl-phenylalanine (fMLP). The fMLP receptor is now called FPR1.

Table 7.54. Corticotropin-releasing factor receptors, main G-protein subunit transducers, and ligands (Source: [5]). Ligand CRF preferentially activates CRF_1, urocortin-1 can activate both CRF receptors, and urocortin-2 and -3 are selective CRF_2 agonists.

Type	Main transducer	Ligands
CRF_1	Gs	Urocortin-1, CRF (preferentially)
CRF_2	Gs	Urocortin-1–urocortin-3

can stimulate several signaling pathways, such as mitogen-activated protein kinase modules (ERK, JNK, and P38MAPK) as well as the PI3K, JAK2–STAT, and Src pathways [991]. It is phosphorylated by GRKs and protein kinase-C. It can interact with arrestins, regulator of G-protein signaling RGS2, and protein Tyr phosphatase PTPn11.

Besides adipose leptin, gastric leptin can, like cholecystokinin, inhibit splanchnic sympathetic nerve discharge and decreases activity of a subset of presympathetic vasomotor neurons in the rostroventrolateral medulla [992]. Its acute nervous and cardiovascular effects (reduced arterial pressure and heart rate) are exerted via vagal transmission and cholecystokinin receptor CCK_1 activation.

7.13.17 Corticotropin-Releasing Factor Receptors

Corticotropin-releasing factor receptors CRF_1 (or CRFR1) and CRF_2 (or CRFR2) are activated by peptides of the CRF family, i.e., CRF [118] and urocortins (Ucn1–Ucn3; Table 7.54). [119] Urocortin-1 connects to both CRF_1 and CRF_2, whereas Ucn3 is a high-affinity ligand for CRF_2 receptor. Activation of CRF_2 on afferent terminals in the medial nucleus tractus solitarius by Ucn1 and Ucn3 releases glutamate that, in turn, causes a decrease in mean arterial pressure and cardiac frequency via activation of ionotropic glutamate receptors [993] (Sect. 2.5.4).

Corticotropin-releasing factor is a major neuroregulator of the hypothalamus–pituitary–adrenal axis that has many central and peripheral actions. The CRF_1 receptor is expressed in anterior pituitary corticotropes. In response to hypothalamic CRF hormone, CRF_1 triggers the release of adrenocorticotropic hormone. This release causes the secretion of glucocorticoids from the adrenal cortex that, in particular, stimulates liver gluconeogenesis to increase glycemia.

Activity of CRF_1 primed by CRF complements the action of CRF_2 receptor activated by urocortin-3 that is responsible for auto- and paracrine glucose-mediated stimulation of insulin secretion. In pancreatic β cells that produce urocortin-3, activation of CRF_1 promotes insulin secretion [994]. Therefore, the insulinotropic action of pancreatic CRF_1 antagonizes the effect of activated CRF_1 on anterior pituitary corticotropes, as the release of glucocorticoids counteracts the action of insulin.

118. A.k.a. corticotropin-releasing hormone (CRH).

119. Urocortin-2 and -3 are also called stresscopin-related peptide and stresscopin, respectively. Urocortin-1 and -3 are members of the corticotrophin-releasing factor (CRF) peptide family.

Table 7.55. Dopamine receptors, main G-protein subunit transducers, and ligands (Source: [5]).

Type	Main transducer
D_1-like group	
D_1 (D_{1A})	$G\alpha_s$, $G\alpha_{olf}$Golf
D_5 (D_{1B})	$G\alpha_s$
D_2-like group	
D_2	$G\alpha_{i/o}$
D_3	$G\alpha_{i/o}$
D_4	$G\alpha_{i/o}$

The CRF_1 receptor stimulates insulin secretion only when glucose concentration is intermediate to high. The higher the glycemia, the greater the CRF_1-dependent phosphorylation of extracellular signal-regulated protein kinases ERK1 and ERK2 and subsequent cAMP response element-binding phosphorylation. Phosphorylation of ERK1 and ERK2 caused by CRF is similar to glucose-dependent actions of incretins, glucagon-like peptide GLP1 and glucose-dependent insulinotropic peptide (GIP), via incretin receptors GLP1R and GIPR that pertain to the class-B G-protein-coupled receptors.

7.13.18 Dopamine Receptors

Dopamine is the predominant catecholamine neurotransmitter in the central nervous system as well as a neurohormone released by the hypothalamus to inhibit the release of prolactin from the adenohypophysis (anterior lobe of the pituitary gland).[120]

Dopamine receptors include 5 proteins (D_1-D_5) distributed in 2 groups: (1) the D_1-*like* (D_1 and D_5) and (2) D_2-*like* (D_2–D_4) groups (Table 7.55). Two D_1-like receptor subtypes couple to $G\alpha_s$ subunit, thereby activating adenylate cyclase. The other receptor subtypes of the D_2-like group inhibit adenylate cyclase via $G\alpha_{i/o}$ and activate K^+ channels.

The D_1 and D_5 receptor genes are intronless. Pseudogenes of D_5 exist. Alternative splicing yields different species- and tissue-dependent types of D_2 and D_3 receptors. In humans, the D_4 receptor gene exhibits extensive polymorphic variation [995].

In the central nervous system, dopamine receptors are involved in the control of locomotion, cognition, emotion, affect, food intake, and neuroendocrine secretion. In peripheral organs, dopamine receptors are prominent in the vasculature, kidney, and pituitary gland, where they influence mainly sodium homeostasis, vascular tone, renal function, gastrointestinal motility, and hormone secretion [995] (Table 7.56).

120. The adenohypophysis regulates stress response, growth, and reproduction.

Table 7.56. Distribution and function of peripheral dopamine receptors (Source: [995]; ALLH: ascending limb of loop of Henle; CCD: cortical collecting duct; JGA: juxtaglomerular apparatus; PT: proximal tubule).

Tissue	Receptor type	Effect
Blood vessels		
Adventitia	D_2–D_4	Inhibition of noradrenaline release
Media	D_1, D_5	Vasodilation
Intima	D_2–D_4	
Heart	D_4	
Sympathetic ganglia and endings	D_2–D_4	Inhibition of noradrenaline release
Adrenal gland		
Glomerulosa	D_1, D_5	
	D_2–D_4	Inhibition of aldosterone secretion
Medulla	D_1, D_5	Stimulation of adrenaline and noradrenaline release
	D_2–D_4	Inhibition of adrenaline and noradrenaline release
Kidney		
Glomerulus	D_1, D_5	Increase of filtration rate
JGA	D_1, D_5	Stimulation of renin secretion
PT	D_1, D_5	Inhibition of Na^+ reabsorption
ALLH	D_1, D_5	Inhibition of Na^+ reabsorption
CCD	D_1, D_5	Inhibition of Na^+ reabsorption
	D_2–D_4	Inhibition of vasopressin action

The Fos transcription factor, a product of the immediate-early FOS gene required for long-lasting modifications of gene expression in response to acute stimuli, is a final targets in the signaling initiated by dopamine receptors. The D_1 and D_2 receptors operate synergistically on FOS expression [995].

7.13.18.1 D_1 Receptor

Receptor D_1 stimulates phospholipase-C (PLC), thereby leading to the production of inositol trisphosphate that causes Ca^{++} influx by mobilization of calcium from intracellular stores on the one hand, and activating PKC enzyme.

Receptor D_1 can stimulate the activity of Ca_V1 channel via the cAMP–PKA pathway [995]. On the other hand, D_1 reduces Ca^{++} flux through $Ca_V2.1$ and $Ca_V2.2$ channels due to channel dephosphorylation by a phosphatase stimulated by PKA enzyme.

Receptor D_1 precludes the activity of Na^+–H^+ exchanger that regulates intracellular pH and cell volume using both cAMP-dependent and -independent mechanisms [995]. Moreover, D_1 inhibits Na^+–K^+ ATPase that maintains the electrochemical gradient for appropriate excitability of neurons and myocytes and enables the transport of fluid and solutes across epithelial membranes [995]. However, in the kidney, dopamine receptor activates Na^+–K^+ ATPase via phosphorylation by PKA and PKC kinases.

In the kidney, dopamine is produced in renal nerves and epithelial cells of certain nephron segments. Dopamine synthesized in the renal tubular epithelium acts as a para- and autocrine factor that regulates sodium reabsorption in the nephron. In addition, dopaminergic nerve endings lodge near the juxtaglomerular apparatus. Receptor D_1 stimulate the secretion of renin [995]. Dopamine impedes Na^+–K^+ ATPase via both D1-like and D2-like receptors (D_1–D_2 synergism).

Both D_1 and D_2 receptors in the hippocampus mediate the effect of dopamine on learning and memory [995]. Both D_1 and D_2 receptors are involved in mesolimbocortical effect (reward and reinforcement) of dopamine.

Striatal efferent neurons undergo the influence of dopamine. Two major types of neurons are identified by their primary sites of axonal projections and neuropeptide synthesis [995]. A first striatonigral population projects to the entopeduncular nucleus and substantia nigra pars reticulata and synthesizes substance-P and dynorphin, an opioid peptide stored in large (size 80–120 nm), dense-core vesicles. Striatonigral gabaergic neurons preferentially express D_1 receptor that fosters the production of substance-P and dynorphin. The second striatopallidal pool projects to the external segment of the globus pallidus and contains enkephalin. Striatopallidal gabaergic neurons prominently synthesize D_2 receptor that prevents the production of preproenkephalin. A similar organization of dopamine receptors is observed in the nucleus accumbens: D_1 receptor is mostly expressed in substance-P+ neurons, D_2 receptor in enkephalin+, neurotensin+ neurons, and D_3 receptor in substance-P+, neurotensin+ neurons [995]. However, in most of the neuronal populations, D_1 and D_2 are coexpressed rather than segregated.

In the cardiovascular system, more precisely in blood vessels, postjunctional D_1 receptor provokes vasodilation of the renal artery and prejunctional D_2 receptor on postganglionic sympathetic nerve terminals prevents noradrenaline release, thereby indirectly causing vasodilation in the femoral artery [995].

7.13.18.2 D_2 Receptor

Two alternatively spliced D_2 isoforms exist (D_{2L} and D_{2S}) [995]. Both isoforms may prime a phosphatidylinositol-mediated mobilization of intracellular calcium. However, in many other cell types, D_2 receptor do not launch this pathway. It can also inhibit inward calcium currents. In addition, D_2 receptor increase outward potassium currents, thareby provoking a cell hyperpolarization.

Receptor D_2 potentiates the release of arachidonic acid via Gi subunit [995]. Unlike D_1, D_2 receptor activates Na^+–H^+ exchanger in many cells.

The D_2 dimer binds to a single heterotrimeric G protein [996]. The maximal activation of the minimal signaling unit, i.e., a receptor dimer coupled to a single heterotrimeric G protein, is achieved by agonist binding to a single subunit of the dimeric D_2 receptor. The asymmetrically activated dimer can be modulated by the activity state of the second monomer. Ligand-independent constitutive activation of and inverse agonist binding to the second monomer enhances signaling.

In the ventral striatum, activated D_2 autoreceptor reduces dopamine release, hence locomotor activity [995]. On the other hand, activated postsynaptic D_2 receptor slightly increases locomotion.

Receptors of the D_2-like group may promote some aspects of cell differentiation [995].

Prejunctional D_2 receptor on postganglionic sympathetic nerve terminals attenuates the cardiac contractility [995].

Dopamine represses aldosterone secretion by adrenal glomerulosa cells via D_2 receptor [995]. During sodium depletion, dopamine release decays and circulating aldosterone level and plasma aldosterone responsiveness to angiotensin-2 augments.

Dopaminergic neurons of sympathetic ganglia inhibit release of adrenaline and noradrenaline by neuroendocrine chromaffin cells of the medulla of adrenal glands via D_2 receptor [995]. Adrenal chromaffin cells possess also D_1 receptor. The latter causes a rapid catecholamine secretion in response to stress.

7.13.18.3 D_3 Receptor

Receptor D_3 localizes mainly postsynaptically to the nucleus accumbens, where it impedes locomotion, hence antagonizing effect of D_2 receptors [995]. Receptor D_3 have an opposite action w.r.t. that of D_2 on neurotensin gene expression in the nucleus accumbens. [121]

Receptor D_3 stimulates thymidine incorporation in some cell types [995]. It may activate the mitogen-activated protein kinase module.

7.13.18.4 D_4 Receptor

Different forms of the D_4 receptor exist with different number of repeats in the third intracellular loop. The four-repeat form (D_{4_4}) is the predominant in the human population (60%) [995]. [122]

7.13.18.5 D_5 Receptor

Receptor D_5 is highly expressed in the hippocampus, where it may mediate the effects of dopamine on learning and memory [995]. It is coexpressed with D_2 in cholinergic interneurons of the striatum.

---

121. In the ventral shell of the nucleus accumbens, D_3 receptor activates neurotensin synthesis. In the septal pole of the nucleus accumbens, D_2 receptor hinders neurotensin production.
122. The D_{4_7} variant is observed in 14% of the population and the D_{4_2} in 10%.

Table 7.57. Endothelin receptors, their main targeted G proteins, and order of ligand potency (Source: [736]). Endothelin-1 (ET1) is a very strong vasoconstrictor and potent mitogen.

Type	Main transducer	Potency order
ET_A	Gq/11, Gs	ET1 ~ ET2 > ET3
ET_B	Gq/11, Gi/o	ET1 ~ ET2 ~ ET3

7.13.19 Endothelin Receptors

Endothelin is secreted by endothelial cells and targets neighboring smooth muscle cells to regulate the vasomotor tone (Table 7.57; Vol. 5 – Chaps. 8. Smooth Muscle Cells and 9. Endothelium). Among 4 identified isoforms (ET1–ET4), endothelin ET1 is the most commonly found subtype in the cardiovascular system, including blood. Endothelins are synthesized by vascular endothelial cells of both vessel (intima) and heart (endocardium) walls (especially endothelial cells of right atrium and left ventricle), vascular smooth muscle cells, and cardiomyocytes, as well as by extravascular cells of the lung, pancreas, spleen, and nervous system [5].

Vasoconstrictor endothelin-1, the major type in the vasculature, intervenes as an auto- and paracrine regulator. Endothelin-1 indeed either counteracts (paracrine action on vascular smooth muscle cells) or cooperates with vasodilator nitric oxide (via an autocrine loop on endothelial cells). Physical factors such as wall shear stress and chemical agents, such as thrombin, adrenaline, angiotensin-2, growth factors, cytokines, and free radicals trigger ET1 secretion. Several substances, such as nitric oxide, cGMP, atrial natriuretic peptide, and prostacyclin reduce ET1 release from the vascular endothelium.

Two endothelin-1 receptors (ET_A and ET_B) [123] bind ET1 with equal affinity to regulate the gene expression devoted to cell contraction, proliferation, and survival. Receptor subtypes ET_A and ET_B include ET_{A1} and ET_{A2} and and ET_{B1} and ET_{B2}.

The vasoconstrictor activity of endothelins is mediated by Na^+–H^+ and Na^+–Ca^{++} exchangers. An initial activation of Na^+–H^+ exchangers causes an increase in Na^+ concentration inside the smooth muscle cell [997]. Subsequently, Na^+–Ca^{++} exchangers export Na^+ and import Ca^{++} into the smooth muscle cell that provokes contraction. The Na^+–H^+ exchanger also reduces the intracellular H^+ concentration.

Messengers ET1 to ET3 potentiate the mitogenic activity of platelet-derived growth factor via ET_A receptors. Endothelin-1 is a potent mitogen for both vascular endothelial and smooth muscle cells. Ligand ET1 activates extracellular signal-regulated kinase ERK1 and ERK2 by GPCR-induced activation of the receptor Tyr kinase EGFR (Sect. 8.2.5.2). [124]

123. A third receptor subtype specific for ET3 isoform — ET_C — exist in Xenopus laevis, but not in mammals.

124. G-protein-coupled receptors can indeed activate Src kinases with subsequent RTK phosphorylation. Once activated, Src and/or RTK phosphorylate SHC adaptor. Both phosphorylated SHC and activated RTK then bind growth factor receptor-bound protein GRB2 and

7.13.19.1 ET_A Receptor

Type-A endothelin-1 receptor phosphorylates the docking growth factor receptor-bound protein-2 (GRB2)-associated binder-1 (GAB1) and activates ERK1 [998].[125] Phosphorylation of GAB1 induced by ET1 is stimulated by phosphoinositide 3-kinase and Src kinases and inhibited by PP1 phosphatase. Protein Tyr phosphatase PTPn11 potentiates ET1-induced ERK1 activation.

7.13.19.2 ET_B Receptor

Receptor ET_B localizes to the nervous system, heart, vasculature, lung, endocrine system, liver, gastrointestinal tract, pancreas, and placenta [999]. In the cardiovascular apparatus, ET_B is mainly produced in endothelial cells; it is synthesized at low levels in vascular smooth muscle cells and cardiomyocytes. It can lodge in caveolae. igands ET1, ET2, and ET3 activate ET_B with equal affinity and potency.

Among 3 splice variants, variant-1 is the canonical isoform. Variant-1 and -3 have functional characteristics; variant-2 that has similar binding properties, but lacks the functional coupling to phosphoinositide [999].

Both ET_B and ET_A can homo- and heterodimerize. In addition, ET_B can heterodimerize with dopamine D_3 receptor and angiotensin AT_1 receptor in rat renal proximal tubule cells [999].

Most ET_B effects result from coupling to Gq and Gi, but ET_B can also couple to other G-protein subunits, such as Go, Gs and G12/13. Receptor ET_B is thus involved in the inhibition or activation of adenylate cyclase, activation of phospholipase-A2, -C, and -D, guanylate cyclase, protein kinases such as mitogen-activated protein kinases [999]. In vascular smooth muscle cells, liganded ET_B causes a biphasic activation of ERK1 and ERK2 kinases. The delayed phase of ERK activation follows ET_B proteolysis by metallopeptidase and EGFR transactivation.

Two ET_B subtypes may reside on endothelial and vascular smooth muscle cells to cause vasodilation and vasoconstriction, respectively. In any case, activated ET_B counteracts the potent vasoconstrictor effect of ET_A stimulation, as it provokes vasodilation, natriuresis, and clearance of ET1 from the circulation by endothelial cells of the lung, kidney, and liver. However, ET_B is produced at low concentrations in vascular smooth muscle cells, in which it causes contraction [999].

In the lung, ET_B on airway smooth muscle cells launches bronchoconstriction and cell proliferation [999].

Receptor ET_B can be phosphorylated (deactivated) by G-protein coupled receptor kinase GRK2 [999]. The agonist–ET_B complex, like liganded ET_A, dissociates very slowly. Following ligand binding, ET_B receptor is internalized using a dynamin–clathrin-dependent pathway for lysosomal degradation. Endocytosis of the

---

its associated Ras guanine nucleotide-exchange factor Son-of-sevenless to trigger the Ras–MAPK signaling.

125. The docking protein GAB1 can be phosphorylated by receptor Tyr kinases and cytokine receptors to regulate cell survival and proliferation aa well as from activated G-protein-coupled receptors.

Table 7.58. Endothelin receptor distribution in the rat coronary bed (Source: [1000]). Endothelin receptors on endothelial smooth muscle cells mediate vasodilation and -constriction, respectively. Endothelial ETRs thus modulate the activity of smooth muscle ETRs. In other words, endothelin activity depends on a balance between smooth muscle and endothelial endothelin receptors.

ET_A	ET_B
Vascular endothelia	
Capillary	Artery, capillary, vein
Vascular smooth muscles	
Artery	Artery, vein

ET1–ET_B complex and subsequent lysosomal degradation enable clearance of ET1 from plasma.

7.13.19.3 Endothelin Receptor Distribution in the Cardiovascular System

In the coronary bed of rat heart, ET_A is detected in arterial smooth muscle and capillary endothelial cells, whereas ET_B resides in arterial, venous, and capillary endothelial cells as well as arterial and venous smooth muscle cells [1000] (Table 7.58). Receptors ET_A and ET_B control arterial vasoconstriction. Postcapillary vascular resistance is exclusively regulated by ET_B. The presence of ET_A in capillary endothelium attenuates increase in vasomotor tone, especially in cardiac microvascular permeability during ischemia–reperfusion events.

Cultures of human coronary and internal mammary arteries, as well as rat omental, mesenteric, and cerebral arteries during 24 to 48 h modify endothelin receptor expression on endothelial and smooth muscle cells. Both endothelin receptor types are upregulated in vascular smooth muscle cells in vessel culture. Conversely, ET_Bs are downregulated in endothelial cells in vessel culture [1001]. Ligand administration during 5 d in dogs induces a heterogeneous distribution of endothelin receptors in coronary arteries [1002]. Right coronary artery is enriched in ET_B 3 times more than left coronary artery (in vascular smooth muscle cells).

Endothelin-1 increases the myocardial contractility (positive inotropic effect) with a greater potency than endothelin-3. Receptor ET_A is involved in ET-induced increase in myocardial contractility. Activated ET_B receptor has an opposite effect [997]. However, ET_B receptor can lead to positive inotropic effect via coronary vasodilation that increases myocardial contraction and oxygen consumption (*Gregg's phenomenon*).[126]

126. Coronary perfusion pressure affects coronary arterial resistance (autoregulation) as well as myocardial oxygen consumption (Gregg's phenomenon). Gregg's phenomenon results from increased contractility.

Table 7.59. Endothelin receptors ET_A and ET_B have opposite effects.

ET_A	ET_B
Vasoconstriction	Vasodilation
(NO inhibition)	(NOS coupling; NO release)
ROS production	ROS decrease
(vSMC)	(EC)
Extravasation	Extravasation inhibition

7.13.19.4 Endothelin Receptor Desensitization

Signal transduction relies on 2 intertwined processes: (1) signal transmission by stimulating appropriate mediators and (2) signaling termination to avoid prolonged reaction and prepare cells for forthcoming stimulation. G-Protein-coupled receptor kinases (GRK1–GRK7) enable a rapid GPCR desensitization after activation. In vascular smooth muscle cells, the predominant ET1 receptor ET_A is desensitized by GRK2 [1003].

7.13.19.5 Endothelin–NO Interactions

Endothelin receptors ET_A and ET_B have opposite roles (Table 7.59). The ET_B receptor mediates vasodilation [127] as well as endothelial cell survival and proliferation. [128] On the other hand, the ET_A receptor favors vasoconstriction.

Responses to ET_B signaling depends on cell types. Ablation of ET_B exclusively from endothelial cells decreases NO release and increases plasma endothelin-1 [1006]. Endothelial ET_B receptor mediates vasodilation without affecting blood pressure in response to a high-salt diet, whereas non-endothelial ET_B controls blood pressure.

The amount of circulating endothelin-1 and endothelial nitric oxide synthase increases and decreases in persistent pulmonary hypertension, respectively. Endothelin-1 increases ROS production in pulmonary arterial smooth muscle cells in culture. Reactive oxygen species downregulate NOS3 in smooth muscle cells [1007]. Endothelin-1 increases H_2O_2 level in fetal pulmonary arterial smooth muscle cells via ET_A in vitro. On the other hand, ET1 decreases H_2O_2 level in fetal pulmonary arterial endothelial cells in monoculture via ET_B receptor. Furthermore, H_2O_2 at low concentration (12 μmol) increases NOS3 level without affecting NOS3 promoter activity, whereas at higher concentration (100 μmol) it reduces both NOS3 promoter activity and NOS3 level in fetal pulmonary arterial endothelial cells in monoculture.

127. The ET_B receptor is functionally coupled to NOS to produce NO via a Tyr kinase- and calcium–calmodulin-dependent pathway [1004].

128. Endothelin-1 promotes the proliferation and migration of human umbilical vein endothelial cells in a dose-dependent manner, by activating ET_B more effectively than ET_A [1005]. In addition, ET1 stimulates matrix metallopeptidase-2 production. It also enhances VEGF angiogenic effects on endothelial cells in vitro.

In addition, ET1 decreases NOS3 promoter activity in these cells cultured with fetal pulmonary arterial smooth muscle cells.

7.13.19.6 Opposite effects of ET_{B1} and ET_{B2} Receptors

Endothelin is able to exert: (1) a persistent constrictor effect by increasing the intracellular calcium concentration resulting from an influx through activated Na^+–H^+ and Na^+–Ca^{++} exchangers and from cellular stores due to the stimulated PLC–IP_3 pathway in vascular smooth muscle cell and (2) a transient dilator effect via the activation of endothelial nitric oxide synthase mediated by protein kinase-B and receptor ET_{B1} [997]. Receptor subtypes ET_{B1} (endothelial) and ET_{B2} (muscular) mediate opposite effects on vascular tone. Receptors ET_A and ET_{B2} indeed prime vasoconstriction, whereas receptor ET_{B1} triggers transient vasodilation.

Endothelin-1 intervenes during birth when pulmonary vascular resistance falls with the initiation of ventilation and the ductus arteriosus constricts, as well as postnatally. At birth, in both term and preterm, oxygen triggers long-lasting vasoconstriction and closure of ductus arteriosus by activating a specific, cytochrome-P450-mediated reaction that leads to ET1 synthesis; ET1 then operates via predominant constrictor ET_A as well as ET_{B2} [1009]. The number of ET_{B1} dilator receptors is reduced in pulmonary hypertension.

7.13.19.7 Cardiac Endothelin Receptors

Receptor ET_A is able to increase myocardial contractility. This inotropic effect results from an increase in intracellular Ca^{++} concentration induced by Na^+–H^+ and Na^+–Ca^{++} exchangers stimulated by protein kinase-C, which is activated by the PLC–DAG axis initiated by endothelin. Receptors ET_{B1} and ET_{B2} may mediate negative and positive inotropic effects, respectively [1008].

Endothelin-1 is a potent cardiomyocyte survival factor. It causes nuclear translocation of PP3-regulated nuclear factors of activated T cells NFAT1 and production of anti-apoptotic BCL2 protein. Factor P300 potentiates NFAT1 binding to the Bcl2 gene promoter in cardiomyocytes [1010]. Transcription factor NFAT also regulates transcription of several genes, such as those that encode adenylosuccinate synthase AdsS1, heart fatty acid-binding protein hFABP, pyruvate decarboxylase, cytochrome-C oxidase, and succinate dehydrogenase.

α1-Adrenoceptor stimulation of cardiomyocytes causes synthesis of angiotensin-1 that acts as an endothelin-releasing factor on smooth muscle cells in coronary arterioles, as it provokes endothelin release via NADPH oxidase and subsequent vasoconstriction [1011].

7.13.19.8 Diapedesis

The ET_B receptor impedes T-cell adhesion to endothelium, especially in cancers, characterizd by an overexpression of endothelial ET_B receptor.[129] The ET_A receptor

129. Endothelin-1 is produced at high concentrations by ovarian cancer cells. Endothelin-1 inhibits in vitro T-lymphocyte adhesion to endothelial cells, hence extravasation into the

elicits T-cell homing on endothelium of lung vessels during allergy and inflammation. Increased release of NO mediates effects of ET_B receptors with reduced expression of ICAM1 intercellular adhesion molecule.

7.13.19.9 Axonal Growth and Guidance

The sympathetic compartment of the autonomic nervous system is composed of preganglionic sympathetic neurons that have synapses with cell bodies of postganglionic sympathetic neurons in para- and prevertebral sympathetic ganglia. Postganglionic sympathetic neurons innervate smooth muscle layers of blood vessels and intestine, myocardium and nodal tissue, exocrine and endocrine glands and ducts, etc.

During development, sympathetic neurons extend axons along paths, often jointly with artery trajectories, to innervate target tissues. Endothelin-3 released from the neural crest-derived smooth muscle layer of the external carotid arteries acts via ET_A to direct extension of axons of a subset of sympathetic neurons from the superior cervical ganglion (located near the bifurcation of the common carotid artery) along the external carotid artery, which serves with its branches as gateway to appropriate organs [1013]. On the other hand, ET3–ET_A signaling does not participate in projections to the internal carotid artery. Other factors expressed by the mesoderm-derived smooth muscle layer of the internal carotid artery are used for navigating projections from the superior cervical ganglion toward the internal carotid artery and its branches.

7.13.20 Estrogen G-Protein-Coupled Receptor

Estrogens signal via 2 cytoplasmic and 1 plasmalemmal receptor. Nuclear receptors include estrogen receptor-α and -β (NR3a1–NR3a2; Sect. 6.3.5.1). Estrogen receptors ERα and ERβ are able to associate with plasmalemmal components. On the other hand, estrogens bind with high affinity to G-protein-coupled estrogen receptor (GPER or GPR30). Liganded G-protein-coupled receptor GPR30 promotes rapid estrogen signaling.

G-protein-coupled estrogen receptor targets cAMP signaling, as it couples to Gs and Gi/o subunits of guanine nucleotide-binding proteins [5]. Plasmalemmal GPER not only regulates adenylate cyclase, but also primes extracellular release of

---

tumor. The ET_B receptor on the wetted surface of tumor endothelium prevents transendothelial migration from the blood stream of T lymphocytes, which then cannot infiltrate tumors [1012]. Suppression of ET_B activity is required for tumor immunotherapy by increasing homing of tumor-specific T lymphocytes.

proheparin-binding EGF-like growth factor (Sect. 8.2.5.2).[130] In addition, G protein subunits, such as $G\alpha_i$ and $G\beta\gamma$, interact with $ER\alpha$.[131]

The GPER receptor activates phosphatidylinositol 3-kinase and protein kinase-B, as well as members of the mitogen-activated protein kinase modules, such as extracellular signal-regulated protein kinases ERK1 and ERK2. The activation of the PI3K pathway by GPER occurs via GPER-mediated transactivation of the epidermal growth factor receptor (Sect. 7.7).

In arterial smooth muscle cells, activated GPER increases intracellular calcium concentration. In cardiomyocytes, GPER activation ensures cardioprotection during myocardial ischemia–reperfusion events, as it reduces postischemic dysfunction and infarct size via the PI3K–PKB and ERK pathways [1015]. Agent PI3K is required, but not MAPK kinases that phosphorylate (activate) extracellular signal-regulated protein kinases. The GPER receptor inhibits mitochondrial permeability transition pore opening via the ERK pathway, hence protecting the heart against ischemia–reperfusion injury [1016].

7.13.21 Free Fatty Acid Receptors

Fed mammals use glucose as the main metabolic fuel. However, short-chain fatty acids produced by the fermentation of dietary fibers by the intestinal colonic bacterial flora[132] also contribute to a significant proportion of daily energy requirement. Under starvation and ketogenic conditions, ketone bodies produced in the liver from fatty acids are used as the main energy sources. Feeding and fasting regulate chemical energy availability via the sympathetic nervous system. Feeding enables con-

130. Heparin-binding epidermal growth factor-like growth factor (HBEGF) of the EGF family is synthesized as a type-1 transmembrane precursor (proHBEGF). At the cell surface, proHBEGF is a juxtacrine growth factor that signals to neighboring cells via an intercellular contact. The ectodomain of proHBEGF is shed and can serve as an N-terminal soluble ligand of EGFR ($_N$HBEGF or simply HBEGF) that can indirectly transactivate HER2, HER3, and HER4 heterodimerized with EGFR (HER1).

131. Plasmalemmal estrogen receptor-α is a palmitoylated protein. 17β-Estradiol reduces $ER\alpha$ palmitoylation and its interaction with caveolin-1 in a time- and dose-dependent manner [1014]. Palmitoylation of $ER\alpha$ enables rapid E_2 signaling via the ERK and PI3K–PKB pathways. Upon 17β-estradiol binding, $ER\alpha$ undergoes depalmitoylation and dissociates from caveolin-1 (inactivation–activation cycle).

132. Commensal bacterial communities are closely associated with the human skin, oral cavity, gastrointestinal tract, and the female genital tract. Gut flora consists of microorganisms (bacteria, fungi, and protozoa). Most bacteria belong to the species Bacteroides, Clostridium, Fusobacterium, Eubacterium, Ruminococcus, Peptococcus, Peptostreptococcus, and Bifidobacterium, as well as, but to a lesser extent, Escherichia and Lactobacillus. The major species are Bacteroides fragilis, Bifidobacterium adoloscentis, and Eubacterium areofaciens [1017]. The remainder is mainly Escherichia coli, Streptococcus viridans, Streptococcus salivaris, and lactobacilli. Anaerobic species outnumber aerobes (at least 10-fold). The Bacteroides-Prevotella group (Gram− anaerobes) and Clostridium species (Gram+ anaerobes) predominate [1018]. Fungi of the gut flora pertain to the genera Candida, Saccharomyces, Aspergillus, and Penicillium.

Table 7.60. Free fatty acid receptors and their main G-protein subunit transducers (Sources: [5, 1019]; O3FAR: omega-3 fatty acid receptor). Long-chain saturated and unsaturated fatty acids (lcFFA) activate free fatty acid FFA_1 receptor, whereas short-chain fatty acids (scFFA) activate FFA_2 and FFA_3 receptors. The GPR84 receptor is stimulated by medium-chain free fatty acids (mcFFA). The GPR120 receptor mediates the anti-inflammatory and insulin-sensitizing effects of omega-3 fatty acids. Four genes — Ffar1 to Ffar3, and Gpr42 — form a cluster on human chromosome 19q13.

Type	Other alias	Main transducer	Ligand
FFA_1	GPR40	Gq/11	lcFFA
FFA_2	GPR43	Gq/11, Gi/o	scFFA
FFA_3	GPR41	Gi/o	scFFA
GPR42 (gene/pseudogene)	FFAR1L, FFAR3L, GPR41L, GPR42P		No
GPR84	GPCR4, EX33	Gi/o	mcFFA
GPR120	O3FAR1, PGR4 GT01, GPR129	Gq/11	mc/lcFFA

sumption of chemical energy into thermal energy. During fasting, chemical energy is saved by the reduction of the sympathetic activity.

Free fatty acids (FFA) serve not only as nutrients, but also as signaling molecules. Free fatty acid receptors (FFAR) are G-protein-coupled receptors. Four subtypes encoded by different genes exist: FFAR1, or GPR40, FFAR2, or GPR43, FFAR3, or GPR41, and GPR42 (Table 7.60). Receptors GPR84 and GPR120 can also be activated by FFAs. Receptors FFAR1 and GPR120 are activated by medium- and long-chain free fatty acids [1019].[133] The GPR84 receptor is stimulated by medium-, but not long-chain, free fatty acids. Both FFAR2 and FFAR3 are activated by short-chain free fatty acids.

The FFAR1 receptor is mainly expressed in pancreatic β cells, where it fosters insulin secretion [1020]. Receptors FFA_2 and FFA_3 are synthesized in adipocytes; FFA_3 also in enteroendocrine cells. The GPR84 receptor is manufactured in the spleen. The GPR120 receptor abounds in the intestine, where it promotes the secretion of glucagon-like peptide-1.

7.13.21.1 FFAR1 (FFA_1)

Mid- and long-chain saturated and unsaturated fatty acids are ligands for FFAR1 receptor. The latter is coupled to the formation of inositol (1,4,5)-trisphosphate, Ca^{++} influx, and activation of extracellular signal-regulated kinases ERK1 and ERK2. Linolenic acid targets FFA_1 on rat pancreatic β cells and reduces the voltage-gated K^+ current via the cAMP–PKA axis to enhance β-cell excitability and insulin secretion [1019].

133. Eicosatrienoic acid is the most potent agonist of FFAR1 receptor.

In mice, FFAR1 is also expressed in endocrine cells of the gastrointestinal tract, such as those that synthesize the incretin hormones glucagon-like peptide-1 and glucose-dependent insulinotropic polypeptide (GIP) [1019]. It is also observed on splenocytes and human peripheral blood mononuclear cells.

7.13.21.2 FFAR2 and FFAR3 (FFA_2 and FFA_3)

Both FFAR2 and FFAR3 are activated by short-chain fatty acids, such as formate, acetate, propionate, butyrate, and pentanonate [1019]. Whereas FFAR3 is activated equally by propionate, butyrate, and pentanonate, FFAR2 prefers propionate.

The FFAR2 receptor is synthesized predominantly in immunocytes. Its expression is also induced during the differentiation of leukocyte progenitors into monocytes or neutrophils. It can be found in enteroendocrine cells that produce peptide-YY and mucosal mastocytes.

The FFAR2 receptor abounds in adipocytes (but not FFAR3), where it may be implicated in the production of leptin, a potent anorexigenic hormone [1019].

Short-chain fatty acids such as propionate and ketone bodies control the amount of chemical energy, as they regulate the activity of sympathetic ganglia via the Gi/o-coupled receptor FFAR3 that triggers the Gβγ–PLCβ–MAPK pathway [1020]. A major short-chain fatty acid, propionate, supports the GPR41-mediated activation of the sympathetic nervous system. On the other hand, a ketone body — β-hydroxybutyrate — produced during starvation or diabetes antagonizes FFAR3 receptor.

7.13.21.3 GPR84 and GPR120

Among medium-chain FFAs, capric, undecanoic, and lauric acids are the most potent agonists of the GPR84 receptor. In humans, GPR84 is expressed in the brain, heart, muscle, lung, kidney, liver, intestine, colon, thymus, spleen, and placenta, as well as leucocytes. In activated T lymphocytes, GPR84 participates in the regulation of early production of interleukin-4 [1019].

The GPR120 receptor, first identified as an orphan G-protein-coupled receptor, is bound by mid- and long-chain, saturated and unsaturated fatty acids, It is highly expressed in the human intestinal tract. Endocrine cells that produce glucagon-like peptide GLP1 in the large intestine synthesize GPR120. In addition, K cells that release glucose-dependent insulinotropic polypeptide in the duodenal and jejunal epithelia strongly express FFAR1, GPR119, and GPR120 [1019]. In mice, enteroendocrine STC1 cells that secrete GLP1 and cholecystokinin upon stimulation by FFAs are endowed with GPR120.

The GPR120 receptor is also synthesized in adipocytes, type-2 taste cells of rat taste buds and Clara cells in lungs [1019]. In adipose tissue, the expression of GPR120 is higher in adipocytes than in stromal-vascular cells.

Table 7.61. Frizzled receptors (Source: [5]).

Type	Alias and other name
Frizzled-1	Fz1
Frizzled-2	Fz2
Frizzled-3	Fz3
Frizzled-4	Fz4, CD344
Frizzled-5	Fz5
Frizzled-6	Fz6
Frizzled-7	Fz7
Frizzled-8	Fz8
Frizzled-9	Fz9, CD349
Frizzled-10	Fz10, CD350

7.13.22 Frizzled Receptors

Frizzled receptors (Fz; Sect. 10.3) that are activated by Wnt ligands trigger β-catenin-dependent (canonical) and -independent (non-canonical) signaling. The Wnt–Fz axis can raise intracellular calcium concentration, signal via Disheveled phosphoproteins to monomeric small GTPase Rac1 and Jun N-terminal kinase, as well as Rho and RoCK kinases, activate cGMP-specific phosphodiesterase PDE6, and heighten cAMP via heterotrimeric G proteins. Like other GPCRs, members of the Frizzled family (Table 7.61) depend on β-arrestin scaffold protein for internalization and signaling. The Wnt–Fz signaling is controlled by: (1) additional ligands that can enhance (extracellular Norrin or R-spondins-1 to -4) or inhibit Fz (secreted Frizzled-related protein-1 to -5, extracellular Wnt inhibitory factor WIF1, and Dickkopf-1) as well as (2) modulatory proteins with positive (RYK and ROR1 and ROR2 kinases) and negative (extracellular Kremen-1 and -2) regulatory effects [5].

7.13.23 γ-Aminobutyric Acid Receptor

γ-Aminobutyric acid (GABA) is the main inhibitory neurotransmitter in the central nervous system. Ionotropic γ-aminobutyric acid receptors ($GABA_A$; Sect. 2.5.3) are pentameric, ligand-gated, Cl^- channels.

Metabotropic γ-aminobutyric acid receptors ($GABA_B$: Table 7.64) are expressed in almost all neurons of the brain. They regulate synaptic transmission and signal propagation. They indeed control the activity of voltage-gated calcium (Ca_V) and inward rectifier potassium channels (K_{IR}).

The $GABA_B$ receptor can indeed associate with $Ca_V2.1$ and $Ca_V2.2$ and $K_{IR}3$ channels [5]. Its main transducer is a subunit of the Gi/o family. Selective agonists of $GABA_B$ include 3-aminopropyl pmethylphosphinic acid (3APMPA) and 3-aminopropylphosphinic acid (3APPA).

Metabotropic $GABA_B$ receptor is a heterodimer formed by 2 similar subunits $GABA_{B1}$ and $GABA_{B2}$. The $GABA_{B1}$ subunit alone binds cognate ligands (antagonists and agonists), but with much lower affinity (10–100 fold less) [5]. Calcium

Table 7.62. Galanin receptors, main G-protein subunit transducers, and ligands (Source: [5]; GaLP: galanin-like peptide).

Type	Main transducer	Potency order
Gal_1	Gi/o	Galanin > GaLP
Gal_2	Gi/o, Gq/11	GaLP ≥ galanin
Gal_3	Gi/o	GaLP > galanin

ion binds to $GABA_{B1}$ to act as a positive modulator. Coexpression of $GABA_{B1}$ and $GABA_{B2}$ enables the transport of $GABA_{B1}$ to the plasma membrane and generates a functional receptor. [134] The $GABA_{B2}$ subunit mediates G-protein-coupled signaling.

The $GABA_{B2}$ subunit increases $GABA_{B1}$ affinity for agonists [5]. Reciprocally, $GABA_{B1}$ facilitates coupling of $GABA_{B2}$ to G proteins. Therefore, the $GABA_{B2}$ subunit does not signal by itself in response to γ-aminobutyric acid, but is essential for cell surface expression of the binding $GABA_{B1}$ subunit and for signaling of the heterodimer, as $GABA_{B1}$ transactivates $GABA_{B2}$.

Several $GABA_{B1}$ isoforms exist. The $GABA_{B1a}$ and $GABA_{B1b}$ isoforms are predominant subtypes in neonatal and adult central nervous system, respectively [5]. [135]

In the central nervous system, the $GABA_B$ receptor is constituted by not only $GABA_{B1}$ and $GABA_{B2}$, but also members of a subfamily of auxiliary K^+ channel tetramerization domain-containing proteins (KCTD) [1021]. These KCTD proteins increase the potency of agonists and accelerate receptor–effector coupling, hence G-protein signaling.

7.13.24 Galanin Receptors

Galanin receptors (Gal_1–Gal_3 or GalR1–GalR3) are activated by galanin and galanin-like peptide (Table 7.62). They localize to the peripheral and central nervous systems and endocrine system.

The galanin family includes galanin and galanin message-associated peptide (GMAP) that are generated from the same peptide precursor on the one hand, and galanin-like peptide (GalP) encoded by a different gene and alarin, a GALP splice variant, on the other. Several neuropeptides are involved in the pathophysiology of depression, in addition to noradrenaline and serotonin. Galanin mainly inhibits both the noradrenergic and serotonergic networks via Gal_1 and Gal_3 receptors. It also impedes neurotransmission via Gal_2 and Gi/o subunits, but can stimulate neurotransmitter release via Gal_2 and Gq/11 subunits. Galanin hampers neurotransmission in memory acquisition, modulates food intake and sexual behavior, and participates in the control of pain. Galanin is also a neuroendocrine peptide of the gastrointestinal tract. Galanin precludes insulin release.

134. Subunit $GABA_{B1}$ alone is not transported to the cell surface and remains non-functional.
135. The $GABA_{B1a}$ heterodimers on distal axons impede glutamate release in CA1–CA3 nerve terminals and GABA release onto the layer 5 pyramidal neurons. The $GABA_{B1b}$ heterodimers on dendritic spines mediate slow postsynaptic inhibition.

Table 7.63. Main G-protein subunit transducers and ligands of glucagon receptors (Source: [5]; GHRH: growth hormone-releasing hormone; GIP: glucose-dependent insulinotropic polypeptide; GLP: glucagon-like peptide).

Selective ligands	Main transducer
Glucagon	Gs
GLP1, exendins-3/4	Gs
GLP2	Gs
GIP	Gs
GHRH	Gs
Secretin	Gs

7.13.25 Ghrelin Receptor

Ghrelin is cleaved from its precursor preproghrelin. Alternative splicing generates a second peptide, des-Gln14-ghrelin with equipotent activity [5]. Both peptides undergo a post-translational octanoylation (Ser3) for full activity on binding to Gq/11-coupled ghrelin receptor (a.k.a. growth hormone secretagog receptor type-1 and growth hormone-releasing peptide receptor) in the hypothalamus and pituitary gland and release of growth hormone from the pituitary gland.

7.13.26 Glucagon Receptors

Glucagon receptors are activated by glucagon, glucagon-like peptides GLP1 and GLP2, [136] glucose-dependent insulinotropic polypeptide, [137] growth hormone-releasing hormone (GHRH), and secretin.

Growth hormone-releasing peptide stimulates growth hormone release via the GH secretagog-receptor GHSR1a, the ghrelin G-protein-coupled receptor. In addition, growth hormone-releasing peptide protects the cardiovascular apparatus independently of growth hormone. Growth hormone releasing-peptide is a ligand for leukocyte differentiation antigen cluster of differentiation CD36, or ScaRb3 scavenger receptor. [138] Transmembrane protein CD36 binds to collagen, thrombospondin, anionic phospholipids, long chain fatty acids, and oxidized LDLs. It is involved in cell adhesion and transport. Growth hormone-releasing peptides enhance the synthesis of ATP-binding cassette transporters ABCa1 and ABCg1 that are responsible for cholesterol efflux from macrophages, hence having a potent anti-atherosclerotic activity (Sect. 6.3.6.4).

136. Glucagon as well as GLP1 and GLP2 have a common precursor.

137. A.k.a. gastric inhibitory polypeptide.

138. A.k.a. collagen-1 and thrombospondin receptor, coronary heart disease susceptibility protein CHDS7, fatty acid translocase (FAT), periodic acid-Schiff staining platelet-agglutinating substance PAS4, and glycoproteins GP3b, GP4, and GP88.

Table 7.64. Metabotropic glutamate and GABA receptors, and their main transducers (Source: [736]). Glutamate ionotropic receptors include members of ^{N}methyl ^{D}aspartate (NMDA; GluN1, GluN2a–GluN2d, and GluN3a–GluN3b) and α-amino 3-hydroxy 5-methyl 4-isoxasole propionic acid (AMPA; GluR1–GluR4), and kainate (GluK1–GluK5) receptor families. The $GABA_A$ receptors are ligand-gated ion channels. The $GABA_B$ receptors can couple to $Ca_V2.1$ and $Ca_V2.2$ as well as $K_{IR}3$ channels.

Type	Main transducer
Glutamate metabotropic receptors	
$mGlu_1/_5$	Gq/11
$mGlu_{2-4}$, $mGlu_{6-8}$	Gi/o
γ-Aminobutyric acid receptors	
$GABA_B$	Gi/o

7.13.27 Glutamate Receptors

Ionotropic glutamate receptors include members of NMDA- and AMPA-type glutamate receptor channel family [139] that have a high relative permeability to calcium ions and are blocked by magnesium at rest potential. Hydrogen and zinc ions also inhibit NMDA receptor channel.

Metabotropic glutamate receptors ($mGlu_1$–$mGlu_8$ or mGluR1–mGluR8) are (Table 7.64) activated by ^{L}glutamate as well as ^{L}aspartate, ^{L}serine^{O_P}, [140] ^{N}acetylaspartylglutamate, and ^{L}cysteine sulfonic acid.

Three groups of receptors have been defined according to their sequence, G-protein coupling, and ligands: group 1 ($mGlu_1$ and $mGlu_5$), 2 ($mGlu_2$ and $mGlu_3$), and 3 ($mGlu_4$ and $mGlu_6$ to $mGlu_8$). Positive (potentiators) and negative allosteric modulators operate in the presence of agonists.

^{L}glutamate is the main neurotransmitter in the central nervous system that mediates fast excitatory neurotransmission by activating ionotropic glutamate receptors. Metabotropic glutamate receptors that are not located in the synaptic cleft prime slower effects. Type-1 mGluRs that operate in synaptic transmission are associated with slow synaptic excitatory currents and long-term remodeling. Kinetics of mGluR activation and deactivation determine features of synaptic transmission. Metabotropic glutamate receptors can be characterized by fast activation and slow deactivation. Activation time course can actually reach a minimum of approximately 10 ms, whereas deactivation time course nearly equals to 50 ms [1022]. Sensitization can develop over about 400 ms.

Neuronal stimulation of astrocytic metabotropic glutamate receptors induces an astrocytic calcium wave that propagates to astrocytic endfeet close to brain arterioles.

139. NMDA: ^{N}methyl ^{D}aspartate; AMPA: α-amino 3-hydroxy 5-methyl 4-isoxasole propionic acid.

140. This ^{L}serine precursor (^{L}SOP) in the serine synthesis pathway is produced by phosphoserine (Ser^P) aminotransferase (PSAT) and metabolized to ^{L}serine by phosphoserine phosphatase (PSP). It is an agonist at group-3 metabotropic glutamate receptors.

Table 7.65. Glycoprotein hormone receptors and their main G-protein subunit transducers (Source: [5]; FSHR: follicle-stimulating hormone receptor; LHR: luteinizing hormone receptor; TSHR: thyroid-stimulating hormone receptor).

Type	Main transducer
FSHR	Gs
LHR	Gs, Gi, Gq/11
TSHR	Gs, Gi, Gq/11, G12/13

Active neurons then rapidly (latency < 2 s) trigger dilation of intracerebral arterioles. Astrocytic calcium signaling activates astrocytic large-conductance, Ca^{++}-sensitive K^+ channels (BK) for local release of potassium into the perivascular space. A modest rise in extracellular potassium activates inward rectifier K^+ channels ($K_{IR}2.1$) of brain arteriolar smooth muscle cells, inducing membrane potential hyperpolarization and relaxation [1023].[141] Augmented neuronal activity thus increases the local cerebral blood flow to supply sufficient amounts of glucose and oxygen (functional hyperemia).

Homer proteins associate with metabotropic glutamate receptors mGluR1 and mGluR5 as well as inositol trisphosphate receptors to increase mGluR-stimulated Ca^{++} signaling. In addition, Homer proteins facilitate anchoring and clustering of mGluR1 and mGluR5 in postsynaptic dendritic spines. Scaffolds of the Shank family of SH3- and multiple ankyrin repeat domain-containing proteins interact with both Homers and mGluRs to strengthen mGluR anchoring [835].

7.13.28 Glycoprotein Hormone Receptors

Glycoprotein hormone receptors are activated by heterodimeric glycoproteins made up of a common αand a β chain that confers the specificity of these glycoproteic heterodimers. These glycoprotein hormones include: (1) follicle-stimulating hormone (FSH, or follitropin); (2) luteinizing hormone (LH, a.k.a. lutropin and, in males, interstitial cell-stimulating hormone [ICSH]); (3) choriogonadotropin (or chorionic gonadotropin); and (4) thyroid-stimulating hormone (TSH, or thyrotropin; Table 7.65).

Follicle-stimulating hormone receptor (FSHR) in gonads is targeted by pituitary FSH glycoprotein for gonadal development, maturation at puberty, and sustained gamete production during the reproductive phase of life. Follicle-stimulating and luteinizing hormones that are both synthesized and secreted by the anterior pituitary gland act synergistically in reproduction. At the moment of menstruation, FSH initiates follicular growth. Stimulated FSHR not only prevents atresia of early antral follicles, but also promotes granulosa cell proliferation, estrogen synthesis, and LHR expression [1024]. An acute release of LH (LH surge) triggers ovulation and corpus

141. Elevation of the concentration of external K^+ that depolarizes vascular smooth muscle cells induces vasoconstriction of cerebral arteries and arterioles.

Table 7.66. Gonadotropin-releasing hormone receptors, main G-protein subunit transducers, and ligands (Source: [5]).

Type	Main transducer	Potency order
$GnRH_1$	Gq/11	GnRH1 > GnRH2
$GnRH_2$	Gq/11	GnRH2 > GnRH1

luteum development. In the ovary, the number of FSHRs increases during follicular maturation under low levels of FSH and decays after ovulatory LH surge.

In both males and females, FSH stimulates the maturation of germ cells. In males, it stimulates Leydig cell production of testosterone, enables persistance of Sertoli cells during gonadal development, and causes inhibin secretion in these cells. Unlike LHR, gonadal FSHR expression is specific to cell type [1024]. In the ovary and testis, FSHRs are expressed on granulosa cells of developing follicles and Sertoli cells, respectively.

Follicle stimulating hormone receptor interacts with arrestins and G-protein-coupled receptor kinases [1024]. Immature FSHR also associates with chaperone proteins calnexin, calreticulin, and disulfide isomerase as well as 14-3-3 adaptor protein, APPL (adaptor protein, phosphoTyr interaction, pleckstrin homology domain- and leucine zipper-containing protein), and FOXO1a.

Signaling mediated by the thyroid-stimulating hormone receptor and the G-protein subunit Gs can be enhanced by receptor endocytosis [11].

7.13.29 Gonadotropin-Releasing Hormone Receptors

Gonadotropin-releasing hormone (GnRH) is a hypothalamic decapeptide (Vol. 2 – Chap. 1. Remote Control Cells – Sect. Endocrine System and Hormones). Isoforms GnRH1 and GnRH2 as well as their cognate receptors $GnRH_1$ and $GnRH_2$ exist in mammals (Table 7.66). The $GnRH_1$ receptor is produced primarily in pituitary gonadotrophs and contribute to the central control of reproduction. Receptors $GnRH_1$ and $GnRH_2$ couple primarily to Gq/11, but also to Gs and Gi subunits. The $GnRH_2$ receptor can also target protein kinases [5].

7.13.30 Histamine Receptors

Histamine [142] is a neurotransmitter of the nervous system [143] and a signaling molecule in the immune system, skin, and gut. [144] Endothelial cells do not synthesize histamine and histamine receptors, but they can take up histamine.

Histamine binds to 4 known metabotropic histamine receptors (H_1–H_4 or H1R–H4R). These receptor subtypes associate with different types of G proteins to prime various effects (Table 7.67). Receptor H_3 abounds in the brain; H_4 receptor lodges mainly in peripheral tissues. Receptors H_1 and H_2 are mostly excitatory; H_3 homo- and heterodimers are inhibitory [1025].

Bordetella pertussis that can cause encephalomyelitis and disrupts the blood–brain barrier. On the other hand, H_1 on endothelial cells counteracts Bordetella pertussis effect. It actually reduces the permeability of the blood–brain barrier [1029].

Histamine mediates the stress-induced surges of adrenocorticotropic hormone, β-endorphin, and vasopressin from the pituitary gland (hypophysis) [1025]. It controls stress-related activity of aminergic networks (serotonin-, noradrenaline-, dopamine-, and acetylcholine-containing neurons).

The preferential site of histamine-induced suppression of food intake is Histamine stimulates neurons in the supraoptic nucleus that release the antidiuretic hor-

---

142. I.e., amine in biological tissues. This organic nitrogen compound s also called imidazolethylamine or imidazolethanamine. Histamine is synthesized from histidine taken up into the cerebrospinal fluid and neurons by amino acid transporter via oxidative decarboxylation by histidine decarboxylase. It is stored and transported into a vesicle using the vesicular monoamine transporter VMAT2, and, upon appropriate cues, released and methylated (inactivated) by neuronal histamine ^{N}methyltransferase using ^{S}adenosyl methionine. The latter enzyme also resides in blood vessel walls and processes circulating and mastocyte-released histamine. The main histamine-degrading enzyme in peripheral organs is diamine oxidase that converts histamine into imidazoleacetic acid [1025].

143. In the mammalian brain, histaminergic neurons localize exclusively in the tuberomamillary nucleus of the posterior hypothalamus and send their axons to the entire central nervous system, like other amine networks [1025]. All amine networks rely on autoreceptors that yield a negative feedback on excitability and amine neurotransmitter release and synthesis. Histaminergic neurons are active only during waking; they maintain wakefulness and attention. Mutual interactions with other transmitter networks are involved in higher brain functions (sleep–wake regulation, circadian and feeding rhythms, immunity, learning, and memory. Although histamine is the main neurotransmitter in the tuberomamillary nucleus, cotransmitters include GABA, galanin, enkephalins, thyrotropin-releasing hormone, and substance-P [1025]. Glutamatergic fibers from the cortex and hypothalamus secrete glutamate that excites neurons of the tuberomamillary nucleus. On the other hand, glycine inhibits a subpopulation of histaminergic neurons. In addition, gabaergic cues from mostly hypothalamic origin suppress the firing of histaminergic neurons. Aminergic (noradrenergic via gabaregic and serotoninregic) and cholinergic neurons also send projections to histaminergic neurons. Nucleotides excite histaminergic neurons using ionotropic and metabotropic receptors.

144. In neuroepithelial and hematopoietic cells, histamine participates in contraction of airway and intestinal smooth muscles, but vasodilation, as well as gastric acid secretion, mastocyte-based innate and acquired immunity, allergy, and inflammation, T-cell-associated immunity modulation, and control of epi- and endothelial barriers [1025].

Table 7.67. Histamine receptors and their effects (Sources: [1026–1028]; CNS: central nervous system; PN: peripheral nerves; AM: adrenal medulla; SMC: smooth muscle cell [v: vascular]; EC: endothelial cell; TL: T lymphocyte; GC: gastric cells; EcC: enterochromaffin cell; NKC: natural killer cell; MC: monocyte; DC: dendritic cells; Nϕ: neutrophil; C: chronotropy; I: inotropy [−: negative; + positive]; VGCaC: voltage-dependent Ca^{++} current).

Type	Gα	Site	Effects
H_1	Gq/11	CNS AM, heart, SMC, EC	Vasodilation, I−, bronchoconstriction, ileum contraction, modulation of the circadian cycle,
H_2	Gs	CNS, heart, vSMC GC TL	Smooth muscle cell relaxation, C+, I+, stimulation of gastric acid secretion, inhibition of T-cell function, facilitation of signal transduction in CNS
H_3	Gi/o	CNS, PN, EC, EcC	Inhibition of neurotransmitter release, increase VGCaC in smooth muscle cells, firing inhibition in tuberomammilary (histaminergic) neurons
H_4	Gi/o	NKL MC, DC, Nϕ	Innate immune cell chemotaxis, neutrophil release from bone marrow mastocyte chemotaxis

mone [1025]. Histamine and vasopressin-containing neurons exert reciprocal interactions.

likely the satiety center of the ventromedial hypothalamus. Histamine effects on food intake are associated with other facors, such as neuropeptide-Y, peptide-YY, and bombesin. In addition, orexigenic action of orexins and anorexigenic effects of leptin and glucagon-like peptide-1, which depend on corticotropin-releasing hormone released by neurons of the paraventricular nucleus rely on H_1 receptor [1025].

The brain histamine network controls thermogenesis, mostly via feeding and motor activity [1025]. The body's autonomic response that regulates heat conservation and production relies on the paraventricular nucleus and dorsomedial hypothalamus, and the nucleus raphe pallidus, respectively. Most of the structures implicated in thermoregulation are targets of histaminergic innervation.

Histaminergic activity exhibits a circadian rhythm with high levels during the day and low levels during the night [1025]. Histamine can entrain molecular clockworks outside the suprachiasmatic nucleus.

7.13.30.1 H_1

The Gq/11-coupled H_1 receptor activates phospholipase-C, thereby supporting IP_3-dependent release of Ca^{++} from its intracellular stores and diacylglycerol-primed activation of protein kinase-C. The latter forsters Ca^{++} influx through voltage-dependent calcium channels, transient receptor potential channels of the TRPC family, and Na^+–Ca^{++} exchanger. Other effects include production of arachidonic acid, nitric oxide, and cGMP using Gi/o-mediated activation of phospholipase-A2, Ca^{++}-dependent NOS, and NO-dependent guanylate cyclases, respectively. Moreover, H_1 activate AMPK kinase and nuclear factor-κB [1025].

Glia cells produce H_1 and H_2 receptors. Histamine promotes release of neurotrophins and cytokines from astrocytesas well as ATP in the hypothalamus [1025]. Histamine causes the opening of the blood–brain barrier, mainly via H_2 receptor.

7.13.30.2 H_2

Gs-coupled H_2 receptor stimulates adenylate cyclase, thereby elevating intracellular cAMP concentration, then activating protein kinase-A and CREB transcription factor [1025]. In addition, cAMP can interact with hyperpolarization-activated cation HCN2 channel. Upon PKA-mediated phosphorylation, H_2 represses Ca^{++}-activated, small conductance K^+ channel (SK), which enables the long-lasting (seconds) afterhyperpolarization after action potentials in pyramidal cells, as well as $K_V3.2$ channel in interneurons involved in fast spiking [1025].

7.13.30.3 H_3

Gi/o-coupled H_3 receptor links to $Ca_V2.1$ and $Ca_V2.2$ channels [1025]. Due to crosstalk with other GPCRs, H_3 can initiate Gq/11 signaling and activate phospholipase-A2, the PKB–GSK3 axis, and MAPK module to control axonal and synaptic remodeling.

7.13.30.4 H_4

Gi/o-coupled H_4 receptor is expressed mainly in peripheral cells of blood, lung, liver, spleen, and gut [1025]. It may also reside in some regions of the brain.

7.13.31 Kiss1, NPff, PRP, and QRFP Receptors

Endogenous peptides with an arginine-phenylalanine-amide motif: (1) chemotaxis-inhibitory kisspeptin-54 (KP54 or metastin) cleaved from its precursor, a Kiss1 gene product; (2) pain-modulatory neuropeptide-FF (NPff); (3) prolactin-releasing peptide (PRP); and (4) pyroglutamylated arginine-phenylalanine (QRF)-amide peptide (QRFP) bind to their cognate receptors [5] (Table 7.68).

Table 7.68. Receptors for kisspeptin (KP), neuropeptide-AF and -FF (NPaf and NPff), prolactin-releasing peptide (PRP), and pyroglutamylated arginine-phenylalanine-amide peptide (QRFP), main G-protein subunit transducers, and ligands (Source: [5]).

Type	Other names	Main transducer	Potency order
GPR10	PrlHR, PrRPR	Gq/11	PRP
GPR54	Kiss1R, AXOR12	Gq/11	KP
GPR74	NPffR2	Gq/11	NPaf, NPff > PRP > QRFP
GPR103	QRFPR, AQ27	Gq/11, Gi/o	QRFP
GPR147	NPffR1	Gi/o	NPff > NPaf > QRFP, PRP

7.13.32 Latrophilin Receptors

Latrophilin receptors (Lphn1–Lphn3; a.k.a. calcium-independent receptors for latrotoxin [$CIRL_1$–$CIRL_3$]) constitute a group of GPCRs of the class-B secretin family [835]. They operate in both cell adhesion and signal transduction. They are located in pre- and postsynaptical membranes in the central nervous system. They connect to postsynaptic proteins, such as SH3- and multiple ankyrin repeat domain-containing proteins Shank1 to Shank3, a.k.a. somatostatin receptor-interacting protein (SstRIP) and proline-rich synapse-associated proteins ProSAP1 and ProSAP2, respectively [835].

7.13.33 Leukotriene Receptors

Leukotrienes comprise 5-lipoxygenase-derived family of short-lived eicosanoid lipid mediators that modulate vascular function and inflammatory cell reactivity. Leukotrienes as well as lipoxins are active metabolites derived from arachidonic acid. They are mainly involved in auto- and paracrine signaling.

5-Lipoxygenase (5LOx) in conjunction with arachidonate 5-lipoxygenase-activang protein (ALOx5AP), [145] an integral membrane protein of the nucleus membrane, converts arachidonic acid into 5-hydroperoxy eicosatetraenoic acid (5HPETE) that spontaneously reduces to 5-hydroxy eicosatetraenoic acid (5HETE). 5-Lipoxygenase converts 5HETE into leukotriene-A4, an unstable epoxide. In cells equipped with LTa4 hydrolase, such as neutrophils and monocytes, LTa4 is converted to leukotriene-B4. In cells that possess LTc4 synthase, such as mastocytes and eosinophils, LTa4 gives rise to leukotriene-C4. Hydrolysis of LTc4 produces leukotriene-D4 and -E4.

Leukotrienes include *cysteinyl-leukotrienes* (cysLTs: LTc4, LTd4, and LTe4) and *dihydroxy-leukotriene* (LTb4). Cysteinyl leukotrienes secreted by basophils and mastocytes together constitute the slow-reacting substance of anaphylaxis, as they trigger a prolonged, slow contraction of smooth muscle cells, especially vasoconstriction and bronchoconstriction.

Leukotriene receptors are activated by leukotrienes LTb4, LTc4, LTd4, LTe4, 12^SHETE, and 12^RHETE (Table 7.69). In addition to their receptors and ALX

145. A.k.a. five-LOx-activation protein (FLAP).

Table 7.69. Leukotriene receptors, their main targeted G proteins, and order of ligand potency (Sources: [5, 736]; EPA: eicosapentaenoic acid; fMLP: formyl methionyl-leucyl-phenylalanine; HETE: hydroxyeicosatetraenoic acid; HPETE: hydroperoxyeicosatetraenoic acid; 5oxoETE: 5-oxo-eicosatetraenoic acid; RARRes2CTP [ChCTP]: retinoic acid receptor [RAR] responder-2 C-terminal peptide [chemerin C-terminal peptide; chemoattractant chemerin is also called tazarotene-induced gene-2 protein (TIG2)]; RvE1: resolvin-E1 [EPA-derived (in the presence of aspirin), non-classic eicosanoid; inflammation reducer]). Receptors $CysLT_1$ and $CysLT_2$ are coexpressed by most myeloid cells. The $CysLT_2$ receptor suppresses $CysLT_1$ expression. The RS nomenclature (Latin rectus [R]: right behavior, straight, in a straight line, upright, direct path [used in the sense of dexter: right, on the right side]; sinister [S]: left, at the left side) for an enantiomer (Greek εναντιος: opposite) defines the chirality (Greek χειρ: strong-handed) of a subtype of amino acid or carbohydrate (X) among 2 stereoisomers (^{R}X and ^{S}X).

Type	Main transducer	Potency order
BLT_1	Gq/11, Gi/o (G16, Gi2)	LTb4 > 12^{R}HETE
BLT_2	Gq/11, Gi/o (Gz-like)	LTb4 > 12^{S}HETE ~ 12^{S}HPETE > 12^{R}HETE
$CysLT_1$	Gq/11	LTd4 > LTc4 > LTe4
$CysLT_2$	Gq/11	LTc4 ~ LTd4 ≫ LTe4
$CysLT_E$		LTe4 > LTc4, LTd4
ALX	Gi, Gq/16	LXA4 > LTc4 ~ LTd4 ≫ fMLP
OxER1	Gi/o	5oxoETE ≫ 5^{S}HPETE ≫ 5^{S}HETE
RvE1R		RvE1 > RARRes2CTP > 18^{R}HEPE > EPA

lipoxin receptor, leukotrienes connect to enzymes of their metabolism (glutathione S-transferase-2, or LTc4 synthase, γ-glutamyl transpeptidase, and several aminopeptidases) and peroxisome proliferator-activated PPARα (nuclear receptor NR1c1).

Lipoxins are short-lived, endogenous, non-classical eicosanoids generated by lipoxygenases that act as anti-inflammatory mediators. Leukotriene-B4 and related hydroxyacids target BLT receptors BLT_1 and BLT_2. Cysteinyl-leukotrienes activate CysLT receptors $CysLT_1$ and $CysLT_2$, as well as LTe4-specific $CysLT_E$ receptor. These leukotriene GPCRs are classified as either chemoattractants (BLT_1 and BLT_2) or nucleotide receptors ($CysLT_1$ and $CysLT_2$). Both BLT and CysLT receptors via Gq and Gi elicit a large increase in intracellular calcium concentration and a decrease in cAMP level, as well as kinase activation (i.e., extracellular signal-regulated protein kinases ERK1 and ERK2) to promote cell differentiation and chemotaxis. Main effects of leukotrienes and lipoxins are given in Table 7.70.

Lipoxins LXA4 and LXB4 [146] activate lipoxin receptor ALX, a.k.a. formyl peptide receptor FPR2, formyl peptide receptor-like protein FPRL1, and formyl peptide

146. Lipoxins are eicosanoids that are generated within the vascular lumen by leukocytes. Lipoxins-A4 and -B4, as well as some peptides, are high-affinity ligands for LXA4 receptor (LXA4R), also called formyl peptide receptor-like receptor FPRL1. Similarly to leukotrienes, LXA4 can form cysteinyl-lipoxins LXC4, LXD4, and LXE4. At subnanomolar concentra-

Table 7.70. Main effects of leukotrienes and lipoxins (Source: [1032]).

LTb4	Leukocyte activation, cytokine secretion, IgE synthesis
LTc4, LTd4, LTe4	Vasoconstriction (via LTc4/D4), bronchospasm, vasodilation (pulmonary artery and vein via LTd4) cardiodepression, eosinophil recruitment, smooth muscle cell proliferation, plasma exudation, mucus secretion
Lipoxin-A4	Anti-angiogenesis, inhibition of cell proliferation, resolution of pulmonary edema, chemokine and cytokine expression, enhancement of macrophage phagocytosis of leukocytes

receptor homolog FPRH1, that is also included in the chemoattractant receptor class with formyl peptide receptors. The ALX receptor is synthesized mainly by leukocytes (granulocytes, monocytes, macrophages, and dendritic cells). It is also detected in microglia cells and astrocytes, enterocytes, and synovial fibroblasts [1030]. Most peptide ligands activate signaling pathways similar to those primed by FPR1 receptor. All ALX peptide ligands, such as formyl peptides and host-derived (annexin-A1 and its related N-terminal peptides, serum amyloid-A, soluble cleaved form of urokinase-type plasminogen activator, humanin, pituitary adenylate cyclase activating polypeptide-27 [a vasoactive intestinal peptide homolog], vasoactive intestinal peptide, amyloid $A\beta_{(1--42)}$, neutrophil and epithelial cell-derived cathelicidin LL37, and truncated splice variant of chemokine CCL23) [147] and pathogen-derived non-formyl peptides, mobilize Ca^{++} in neutrophils [1030]. They then stimulate neutrophil functions, hence causing chemotaxis, degranulation, and superoxide production. On the other hand, the only lipid ligand LXa4 of ALX receptor binds ALX at nanomolar concentrations to exert anti-inflammatory function [1030]. The ALX receptor thus has both pro- and anti-inhibitory effects. In human neutrophils,

tions, LXA4 and LXB4 rapidly inhibit leukotriene-stimulated interactions of human neutrophils and endothelial cells [1031].

147. Annexin-A1 and its related N-terminal peptides target ALX to prime their anti-inflammatory effects, in particular, neutrophil transmigration [1030]. Serum amyloid-A binds to ALX receptors of monocytes. Several uPA fragments connect to ALX to cause chemotaxis of monocytes and basophils [1030]. Neuroprotective peptide non-formylated humanin (as well as much more potent N-formyl humanin) tethers ALX and prevents Aβ(1-42) activity [1030]. Peptide PACAP27 that also activates G-protein-coupled vasoactive intestinal peptide receptor $VIPR_1$ ($VPAC_1$) stimulates α_M-integrin (CD11b) in human neutrophils [1030]. Vasoactive intestinal peptide also links to Gs-coupled vasoactive intestinal peptide receptors $VIPR_1$ and $VIPR_2$ ($VPAC_2$) to provoke smooth muscle relaxation, exocrine and endocrine secretion, and water and ion flux in lung and intestinal epithelia. It stimulates ALX to activate phosphatidylinositol 3-kinase and extracellular signal-regulated kinase and cause α_M-integrin synthesis [1030].

Table 7.71. Leukotriene-producing leukocytes (Source: [1032]). Leukotriene-C4 is metabolized to LTd4 and LTe4 by LTc4-synthesizing cells.

LTb4	Eosinophils, neutrophils, monocytes, macrophages
LTc4	Eosinophils, neutrophils, monocytes, macrophages
LTd4	Eosinophils, neutrophils, monocytes, macrophages
LTe4	Eosinophils, neutrophils, monocytes, macrophages

Table 7.72. Major cells that express leukotriene receptors (Source: [1032]). Signaling from LTb4 in granulocytes is mediated by Gi as granulocytes abundantly express Gi, mainly Gi2, whereas nervous cells mainly produce Gi1 and Go. LTb4-induced calcium mobilization results from Gq activation. BLT1-mediated phospholipase-C activation is mediated by Gi6 and $G\beta\gamma$. BLT2 is less efficient for calcium mobilization than BLT1. BLT2 induces chemotaxis via Gi and the Rac–ERK–ROS cascade. LTd4 activates CysLT1 to produce second messengers diacylglycerol, inositol 3-phosphate, and Ca^{++}. Subsequently, protein kinase-C and phospholipase-A2 are activated, the latter generating arachidonic acid. LTd4 primes constriction of smooth muscle cells in bronchi and intestine via protein kinase-Cϵ.

Type	Location
BLT_1	Thymus, spleen
	Leukocytes
BLT_2	Ovary, liver
	Leukocytes (ubiquitous)
$CysLT_1$	Spleen
	Smooth muscle cells (lung, intestine), blood leukocytes
$CysLT_2$	Heart, spleen, adrenal medulla, brain
	Blood leukocytes
ALX	Lung, spleen
	Blood leukocytes

ALX is involved in serum amyloid-A-stimulated interleukin-8 production and nuclear factor-κB activation. Although ALX activity regulation is similar to that of FPR1, FPR2/ALX differs from FPR1 receptor. Gelsolin-derived peptide PBP10 selectively blocks ALX-mediated mobilization of granules and production of oxygen radicals [1030].

Oxoeicosanoid receptor OxER1 (or GPR170) is activated by endogenous chemotactic eicosanoids oxidized at the C5 position [5]. 5-oxo-Eicosatetraenoic acid (5oxoETE) is its most potent agonist. *Resolvin receptors* are activated by anti-inflammatory resolvin-E1 (RvE1) that derives from ω-3 fatty acids eicosapentaenoic acid (EPA) and docosahexaenoic acid (DHA) by aspirin-modified cyclooxygenase and lipoxygenase [5].

Leukotrienes are synthesized not only by multiple cell types, but also via transcellular metabolism involving different cells, such as neutrophils, platelets, and vascular cells (Table 7.71). Primary expression of leukotriene receptors is given in Table 7.72.

Leukotriene-B4, a chemotactic and immune-modulating agent implicated in allergic and inflammatory reactions, acts via G-protein-coupled receptors BLT_1 and BLT_2. Receptors BLT_1 and BLT_2 signal via 3 classes of G protein subunits, Gi, Gq-like, and Gz. [148] Leukotriene-B4 [149] is a chemoattractant produced by neutrophils, macrophages, and mastocytes for polymorphonuclear leukocytes and T lymphocytes. Leukotrienes are involved in arteriosclerosis.

In mature endothelial cells, LTb4–BLT signaling enables angiogenesis. Endothelial cell migration depends on CysLT receptors. Moreover, 12LOx-dependent pathway via 12HETE and BLT_2 is involved in VEGFa-induced angiogenesis. Embryoid bodies derived from human embryonic pluripotent cells differentiate into endothelial cells as well as different hematopoietic lineages (monocytes and macrophages, T lymphocytes, NK cells, and dendritic cells). Leukotrienes LTb4 and, to a lesser extent, LTd4 concomitantly with their cognate receptors BLT and CysLT are expressed in embryonic stem cells during embryogenesis, particularly endothelial plexus during the later phase of vasculogenesis and angiogenesis, at least in rodents. [150] Myeloid-like cells that synthesize 5LOx and its cofactor FLAP produce LTb4 that operates as a paracrine agent to stimulate endothelial progenitors via BLT receptors to initiate vasculogenesis [1034]. Leukotriene action in differentiating embryonic stem cells is mediated via extracellular signal-regulated kinase ERK1 and ERK2.

7.13.34 Lysophospholipid Receptors

Lysophospholipids are built on a glycerol backbone and are defined by the nature of the phosphate groups linked to the carbon atom at position 3 (stereospecific nomenclature sn3). They contain a single fatty acid side chain. On the other hand, phospholipids typically contain saturated and unsaturated fatty acids at the sn1 and sn2 positions, respectively.

Lysophospholipids can be generated from phospholipids by phospholipases that cleave the sn1 or sn2 acyl fatty acids. The sn2 acyl side chain of lysophosphatidic acid (LPA) is unstable and moves to the sn1 position, generating the predominant form of LPA: 1-acyl 2-hydroxy sn-glycero 3-phosphate (Table 7.73). The sequential actions of phospholipases PLD and PLA2 can generate lysophosphatidic acid from phosphatidic acid (PA). Structural and functional diversity arises from differences in fatty acid moieties esterified to sn1 that may influence LPA binding to its receptors. The synthesis of LPA involves either the sequential action of secretory phospholipase-A and lysophospholipase-D, or phospholipase-D and cytosolic

148. The expression of BLT receptors can be enhanced in endothelial cells by lipopolysacharides, cytokines, such as tumor-necrosis factor-α and interleukin-1β, and LTb4 [1033].

149. LTb4 Leukotriene is synthesized from arachidonic acid by the concerted action of LTa4 hydrolase and 5-lipoxygenase, which is assisted by 5-lipoxygenase-activating protein.

150. Myeloid-like, CD11b+ (α_M-integrin) embryonic stem cells are also 5LOx+ and FLAP+, whereas endothelial-like, VEGFR2+, CD105+ (endoglin of the TGFβ receptor complex) embryonic stem cells are non-synthesizing leukotriene cells.

Table 7.73. Main axes of lysophosphatidic acid (LPA) synthesis, either from phosphatidic acid (PA) using phospholipases (PL), or from lysophosphatidylcholine (lysoPC) using autotaxin, or lysophospholipase-D (lysoPLD). Autotaxin is produced in high endothelial venules, where it can support transendothelial migration and lymphocyte entry into lymph nodes and secondary lymphoid organs, as well as lymphocyte migration into inflammatory sites.

Axis	Enzymes
PC–LysoPC–LPA	sPLA2 and LysoPLD, or PLD and cPLA2
PA–LPA	PLD and PLA2

Table 7.74. Endothelial differentiation gene (EDG) family of GPCRs.

Gene	Name	Ligand
EDG1	$S1P_1$	S1P
EDG2	LPA_1	LPA
EDG3	$S1P_3$	S1P
EDG4	LPA_2	LPA
EDG5	$S1P_2$	S1P
EDG6	$S1P_4$	S1P
EDG7	LPA_3	LPA
EDG8	$S1P_5$	S1P

PLA. The second major axis of LPA synthesis uses lysophospholipase-D (lysoPLD), or autotaxin, that cleaves the choline group from lysophosphatidylcholine. In addition, membrane-bound phosphatidic acid-selective phospholipase-1α (MPAPLA1), or lipase-H (LipH), is a 2-acylLPA-producer. The generation of LPA can also be achieved by phosphorylation of monoacylglycerol (MAG) by MAG kinases.

The most prominent source of LPA in the blood circulation (at micromolar concentrations) arises from autotaxin. Lysophosphatidic acid circulates bound to serum proteins, especially albumin and gelsolin, a member of the extracellular actin scavenger pool that depolymerises and removes from the blood actin released in the extracellular medium to avoid thrombus creation. In response to inflammation, LPA diffuses into the extravascular space, where it supports cell chemotaxis and proliferation, cytokine and chemokine secretion, platelet aggregation, and smooth muscle cell contraction (vasoconstriction).

Glycerol- and sphingosine-based phospholipids are abundant components of cellular membranes. They are metabolized into eicosanoids and lysophospholipids, such as lysophosphatidic acid, lysophosphatidylcholine, sphingosylphosphoryl choline, and sphingosine 1-phosphate. G-protein-coupled receptors for lysophospholipids include 12 identified members. The lysophospholipid GPCR family, i.e., the endothelial differentiation gene (EDG) family of GPCRs, encompasses receptors for sphingosine 1-phosphate and lysophosphatidic acid, as well as lysophosphatidylcholine and sphingosylphosphoryl choline psychosine groups [1035] (Table 7.74).

Upon receptor activation, plasmalemmal phospholipids are metabolized into lysophospholipid mediators, such as sphingosine 1-phosphate and lysophosphatidic acid. These messengers regulate angiogenesis, cardiac development, neuronal survival, and immunity [1035].

7.13.35 Lysophosphatidic Acid Receptors

Lysophosphatidic acid (LPA), the simplest natural phospholipid,[151] is produced by: (1) hydrolysis of phosphatidic acid that localizes mainly at the inner leaflet of the plasma membrane by phospholipase-A1 or -A2 and (2) conversion of lysophospholipids, such as lysophosphatidylcholine (LPC),[152] lysophosphatidylethanolamine (LPE), and lysophosphatidylserine (LPS), which are also generated from membrane phospholipids by phospholipase-A1 or -A2, to LPA by lysophospholipase-D, or autotaxin.[153]

This bioactive lipid mediator involved in stress fiber formation using the Rho–RoCK axis[154] and cell proliferation, differentiation, survival, and motility. Lysophosphatidic acid is dephosphorylated (inactived) by lipid phosphate phosphohydrolases (LPP) into monoacylglycerol. The LPP phosphatase has its catalytic site oriented toward the external face of the plama membrane.

Lysophosphatidic acid can be synthesized outside the cell by membrane or secreted enzymes, such as secreted phospholipase-A2 (sPLA2) and lysophospholipase-D (sLysoPLD), especially after membrane lipid reorganization during inflammation.

Extracellular LPA that serves as lipid mediator is water soluble. It resides in many biological fluids, where it is maintained in solution owing to its connection to

---

151. Lysophosphatidic acid is composed of a single fatty acyl chain, a glycerol backbone, and a phosphate group. In fact, many LPA species exist. These LPA species possess an alkyl or alkenyl ether-linked fatty acyl chain of various lengths and unsaturation degrees. A saturated or unsaturated fatty acid chain is esterified either at the sn1 (1-acylLPA; e.g., 1-palmitoylLPA) or sn2 (2-acylLPA; e.g., 2-arachidonoylLPA) position (sn: stereospecific nomenclature) of the glycerol backbone of LPA phospholipid. Ether-linked LPAs carry an alkyl- (alkylLPA; e.g., hexadecylLPA) or an alkenyl-linkage (alkenylLPA) at the sn1 position [1036]. Activity of LPA depends on: (1) the length and the unsaturation level of the carbon chain attached to the glycerol backbone and (2) the type of linkage and its position. Lysophosphatidic acid is indeed synthesized via numerous routes. In particular, autotaxin, a lysophospholipase-D, removes the choline group from lysophosphatidylcholine.

152. Lysophosphatidylcholine abounds in biological fluids linked to albumin or lipoproteins. It is synthesized from phosphatidylcholine by PLA2 or lecithin cholesterol acyltransferase (LCAT).

153. Autotaxin (Atx; a.k.a. lysophospholipase-D and type-2 ecto-nucleotide pyrophosphatase phosphodiesterase) is a transmembrane ecto-enzyme. It is upregulated by certain peptide growth factors. Soluble autotaxin derives from a membrane-bound form by proteolytic cleavage.

154. In 1960, Vogt observed that LPA causes the contraction of isolated rabbit duodenum. Afterward, the contraction of smooth muscle cells generated by several LPA species was demonstrated on vascular cell types.

Table 7.75. Lysophosphatidic acid (LPA) receptors, main G-protein subunit transducers, and ligands (Sources: [5, 1036]; EDG: endothelial differentiation gene product). Lysophosphatidic acid also binds PPARγ receptors. In addition to the 3 G-protein-coupled receptors of the endothelial differentiation gene (EDG) family (LPA_1–LPA_3), lysophosphatidic acid operates via 3 other homologous receptors (LPA_4–LPA_6) that are structurally distinct from the EDG family LPA receptors. The latter are involved, in particular, in angiogenesis and platelet activation. The LPA receptors of the EDG family that preferntially link to acylLPA than alkylLPA provoke intracellular Ca^{++} influx, inositol phosphate production, adenylate cyclase inhibition, and mitogen-activated protein kinase activation. Receptors LPA_6, or $P2Y_5$, and LPA_8, or $P2Y_{10}$, are activated by LPA and LPA and S1P, respectively.

Type	Other names	Main transducers
LPA_1	EDG2	Gi/o, Gq/11, G12/13
LPA_2	EDG4	Gi/o, Gq/11, G12/13
LPA_3	EDG7	Gs, Gi/o, Gq/11
LPA_4	GPR23, $P2Y_9$	Gs, Gi/o, Gq/11, G12/13
LPA_5	GPR92, GPR93	Gq, G12/13
LPA_6	GPR87, $P2Y_5$ LPA_7	G12/13, Gs, Gi/o
LPA_8	$P2Y_{10}$	Gq/11

its carrier albumin. In addition to albumin, other proteins, such as fatty acid-binding proteins and gelsolin, link to LPA.

Almost all cells, except in the liver, coexpress several types of LPA receptors (LPA_1–LPA_6; Table 7.75). Receptors LPA_4 and LPA_6 correspond to $P2Y_9$ and $P2Y_5$, respectively [1037, 1038]. The $P2Y_9$ receptor is most closely related to $P2Y_5$ that is coupled to the G13–Rho pathway. It is relatively close to nucleotide receptors $P2Y_1$, $P2Y_4$, and $P2Y_6$ as well as lipid receptors of the group of *proton-sensing GPCRs* [155] and other lipid receptors, such as platelet-activating factor [156] receptor and leukotriene receptors $CysLT_1$ and $CysLT_2$.

Lysophosphatidic acid receptors are coupled to G-protein subunits Gs, Gq, Gi, and/or G12/13 (Tables 7.75 and 7.76). Activated Gq stimulates phospholipase-C that activates protein kinase-C for calcium influx. Activated Gi targets 3 pathways, as it: (1) inhibits adenylate cyclase; (2) stimulates Ras–mitogen-activated protein ki-

155. Lipid ligands modulate action of protons that activate proton-sensing G-protein-coupled receptors. Proton-sensing GPCRs include: (1) Gs-coupled GPR4 that is targeted by sphingosylphosphorylcholine and lysophosphatidylcholine; (2) psychosine receptor GPR65 (or T-cell death-associated gene product TDAG8) that activates adenylate cyclase; (3) GPR68 (or ovarian cancer GPCR OGR1); and (4) GPR132 (or G2A), a high-affinity receptor for lysophosphatidylcholine. Sphingosyl-phosphorylcholine, lysophosphatidylcholine and psychosine are high- and low-affinity ligands for GPR4 and GPR65, respectively. Lipids favor effects of protons as activators of GPR4. Psychosine and its related lysosphingolipids behave as antagonists of protein-sensing receptors [1039].

156. Platelet-activating factor is a phospholipid (acetyl-glyceryl-ether phosphorylcholine) produced by neutrophils, basophils, platelets, and endothelial cells.

nase cascade; and (3) activates phosphatidylinositol 3-kinase, guanosine nucleotide-exchange factor TIAM1 (RacGEF), hence small GTPase Rac, as well as protein kinase-B. Hence, LPA has mitogenic and anti-apoptotic effects.

Activated LPA_1 and LPA_2 coupled to G12/13 stimulates small GTPase RhoA that causes cytoskeletal contraction and favors cell motility and adhesion.

Lysophosphatidic acid receptor LPA_2 interacts with scaffold membrane-associated guanylate kinase, WW and PDZ domain-containing protein MAGI3 to enhance receptor-mediated activation of extracellular signal-regulated kinase. On the other hand, LPA_2 connects to 2 other PDZ domain-containing scaffolds, RhoGEF11 and RhoGEF12 to potentiate LPAR2-induced stimulation of Rho signaling on cytoskeleton dynamics [835]. Therefore, LPA_2 can activate PLCβ, ERK, or Rho GTPase according to the synthesis level of PDZ domain-containing scaffold, hence cell type.

Lysophosphatidic acid receptor LPA_2 (or LPAR2), like purinergic receptor $P2Y_1$, parathyroid hormone receptor PTH1R (or PTH_1), and metabotropic glutamate receptor mGluR5 can link to Na^+–H^+ exchanger regulatory factors NHERF1 and NHERF2 independently of agonists to enhance Gq-protein- and PLCβ-mediated signaling.[157] Protein NHERF2 can also bridge LPA_2 to cystic fibrosis transmembrane regulator Cl^- channel. On the other hand, β2AR and KOR opioid receptors also connect to NHERF proteins, but in an agonist-dependent manner [835].

Due to its unique C-terminal binding domain, LPA_2 bind to several PDZ motif-containing proteins (NHERFs, RhoGEF11, RhoGEF12, and membrane-associated guanylate kinase with an inverted domain structure protein MAGI3) [1040]. These scaffolds modulate LPA-induced activation of ERK and/or RhoA GTPase. Both NHERF2 and MAGI3 can recruit phosphatase and tensin homolog to repress the PI3K–PKB signaling. In addition, NHERF2 can also serve as a scaffold for 3-phosphoinositide-dependent protein kinase PDK1 that activates AGC family kinases such as PKB. Therefore, NHERF2 restricts or promotes the PI3K–PKB pathway according to the relative synthesis rate of PTen and PDK1 enzymes.

In addition to NHERF2 scaffold, LPA_2 interacts with thyroid hormone receptor interactor TRIP6 and apoptosis-inducing factor Siva-1. The LPA_2–TRIP6–Siva-1 complex and NHERF2 enable LPA_2-mediated protection against apoptosis [1040].

Whereas LPA_1 and LPA_2 can be coupled to Gi, thus inhibiting adenylate cyclase, LPA_3 can be tethered to Gs, thereby activating adenylate cyclase.

Liganded LPA_4 coupled to Gs and Gq subunits increases intracellular concentrations of Ca^{++} and cAMP messengers. Receptor LPA_4 coupled to G12/13 proteins activates Rho GTPase, thereby causing cytoskeleton remodeling. Therefore, LPA_4 can counteract the effect of LPA_1, as Rho and Rac exert a mutual inhibition.

Receptor LPA_5 can activate adenylate cyclase independently of $G\alpha_s$ [1036]. It also mobilizes Ca^{++} ions. In mice, LPA_5 is highly produced in the small intestine and stomach as well as dorsal root ganglion. In mouse small intestine, LPA_5 is synthesized not only in epithelial cells, but also in intraepithelial lymphocytes.

---

157. Many NHERF-binding partners, such as TRPC channels, various isoforms of PLCβ, protein kinases PKC and PKD, are components of Gq–PLCβ pathways.

Plasmalemmal protein LPA_5 is the most prevalent LPA receptor in human mastocytes [1042].[158] Cell proliferation caused by LPA stimulation of mastocytes results from the activity of LPA_1 and LPA_3, whereas cytokine generation is induced by LPA_2 receptor. Receptor LPA_5 is responsible for release of CCL4 chemoattractant launched by LPA [1042].

Lysophosphatidic acid is a potent pro-aggregating factor for platelets. In humans, platelets synthesize LPA_1 to LPA_6 transcript; the most abundant are LPA_4 and LPA_5 transcripts [1041]. At the protein level, among LPA receptors, platelets produce mainly LPA_5 receptor. In humans, platelets aggregate more potently in response to alkylLPA than to corresponding acylLPA [1036].

Lysophosphatidic acid induces not only smooth muscle contraction, platelet aggregation, and increase in endothelial permeability [1043],[159] but also cell survival, proliferation, and migration, as well as attachment of myeloid cells to endothelium (Table 7.76). In addition it causes gap-junction closure and tight-junction opening, favors wound healing, and elicits the production of endothelin, angiogenic factors (vascular endothelial growth factor and interleukin-6 and -8), urokinase-type plasminogen activator,[160] metallopeptidases such as MMP2 and ADAM17 [1044]. Lysophosphatidic acid is a ligand for the transcription factor perixosome proliferating activating receptor-γ. Five G12/13-coupled LPA receptors can contribute to angio- and lymphangiogenesis.

Transcription of LPA is activated by growth factors. Lysophosphatidic acid can indirectly regulate cell functions by stimulating G-protein-coupled receptor-regulated transmembrane metallopeptidase that cleaves precursor heparin-binding EGF-like growth factor (HBEGF) at the plasma membrane and activating EGF receptor (Sect. 8.2.5.2).

7.13.36 Mas1-Related G-Protein-Coupled Receptors

Signaling molecules expressed in nociceptive sensory neurons include vanilloid receptor VR1, purinergic receptors such as $P2X_3$, and tetrodotoxin-insensitive voltage-gated sodium channels that convey most of inward flux during upstroke and shoulder during the falling phase of nociceptor action potentials, with significant contribution from tetrodotoxin-sensitive sodium channels and voltage-dependent calcium channels, respectively, and regulate duration of action potentials. Several GPCRs are expressed in nociceptive sensory neurons, such as bradykinin,

158. Mastocytes lodge particularly around the microvasculature and in the skin and respiratory and gastrointestinal tracts, ready to respond to damages and infections.

159. LPA causes a rapid, reversible, dose-dependent decrease in paracellular flux resistance in brain endothelial cells. LPA modulates tight junction permeability without relocalization of adherens junction- or tight junction-associated proteins.

160. In some types of migrating cells, urokinase-type plasminogen activator receptor (uPAR) associates urokinase-type plasminogen activator (uPA) to the leading edges and controls the activation of small GTPase Rac. As uPAR is not a transmembrane receptor, it interacts with plasmalemmal and cortical proteins such as integrins as well as matrix components such as vitronectin to transmit signals to the cytoskeleton.

Table 7.76. Lysophosphatidic acid activity (Source: [1044]). Upon LPA receptor stimulation, G12/13 proteins activate Rho GTPase and RoCK and LIMK1 kinases that phosphorylate cytoskeletal proteins, such as myosin light chain, moesin, and cofilin. (Calcium influx and Rac activation are not involved in LPA-induced cell shape change.)

Action	Pathway
Cell survival	Gi–PI3K–PKB
Cell growth	Gi–Ras–ERK1/2
Cell migration	Gi–PI3K–TIAM1–Rac, RhoA, Cdc42
Cell shape (rounding)	G12/13–RhoGEF–RhoA
Cell differentiation	G12/13–RhoGEF–RhoA
(inhibition or reversal)	ERK, P38MAPK
Endothelial permeability	G12/13–RhoGEF–RhoA
Gap junction communication	Gq–PLC
SMC contraction	G12/13–RhoGEF–RhoA
Platelet aggregation	G12/13–RhoGEF–RhoA

Table 7.77. Melanin-concentrating hormone receptors and their main G-protein subunit transducers (Source: [5]).

Type	Other names	Main transducer
MCH_1	MCHR1, GPR24, SLC1	Gi/o, Gq/11
MCH_2	MCHR2, GPR145, SLT	Gq/11

neuropeptide-Y, opioid, peptidase-activated, and prostaglandin receptors, as well as Mas1-related G-protein-coupled receptors (MRGPRd–MRGPRg; MRGPRx1–MRGPRx4) [1045].

7.13.37 Melanin-Concentrating Hormone Receptors

Melanin-concentrating hormone receptors (Table 7.77) bind to melanin-concentrating hormone (MCH) that is a nonadecameric cyclic orexinogenic hypothalamic peptide. The latter regulates the feeding behavior, hence energy balance. Melanin-concentrating hormone is generated from a precursor, proMCH, that also produces neuropeptide-EI and -GE. It interacts with neuropeptide-Y [5]. [161]

7.13.38 Melanocortin Receptors

Melanocortin has a protective effect in the cardiovascular system. It attenuates inflammation during myocardial infarction and stroke. The melanocortin receptor fam-

161. Melanin-concentrating hormone and α-melanocyte-stimulating hormone (αMSH) have mostly antagonistic, but also agonistic effects. Neuropeptide-EI is a MCH-antagonist and MSH-agonist. Neuropeptide-GE derived from proMCH can mimic MSH signaling via MCHR1 receptor.

Table 7.78. Melanocortin receptors, main G-protein subunit transducers, and ligands (Source: [5]).

Type	Main transducer	Potency order
MC_1	Gs	αMSH > βMSH ≥ ACTH, γMSH
MC_2	Gs	ACTH
MC_3	Gs	γMSH, βMSH ≥ ACTH, αMSH
MC_4	Gs	βMSH ≥ αMSH, ACTH > γMSH
MC_5	Gs	αMSH ≥ βMSH ≥ ACTH > γMSH

ily includes 5 GPCRs (MC_1–MC_5 or MC1R–MC5R; Table 7.78). The MC_2 receptor is a component of the hypothalamic–pituitary–adrenal axis, whereas MC_3 and MC_4 intervene in energy homeostasis. They are activated by members of the melanocortin family of peptide hormones that include different forms of melanocyte-stimulating hormone (MSHα–MSHγ) and adrenocorticotrophin (ACTH).

Endogenous antagonists include Agouti and agouti-related protein. Melanocortin receptor accessory proteins MRAP1 and MRAP2 heighten the plasmalemmal density of MC_2 as well as their responsiveness to melanocyte-stimulating hormone, but lower those of MC_1 and MC_3 to MC_5 [1046]. Melanocortin MC_2 receptor on the plasma membrane requires the presence of MRAP1 for its activation. On the other hand, MRAP2 is an endogenous inhibitor, as it competes with MRAP1 for MC_2 binding and decreases the action potency of adrenocorticotropic hormone [1047]. The ACTH hormone indeed binds with high affinity to MC_2 in the presence of MRAP1, but not MRAP2. The influence of MRAP1 and MRAP2 on ligand-binding affinity is specific to MC_2. The balance between stimulatory and inhibitory accessory proteins controls MC_2 sensitivity to its agonist. These proteins actually have little effect on the binding of α-melanocyte-stimulating hormone to the MC_4 receptor.

Thyroid hormones regulate metabolism and appetite. Active triiodothyronine (T_3) stimulates the basal metabolic rate.[162] Increased T_3 level prevents transcription of the thyrotropin-releasing hormone (Trh) gene in TRH+ neurons in the paraventricular nucleus of the hypothalamus. These neurons produce both nuclear thyroid hormone receptors[163] and MC_4 melanocortin receptor. Triiodothyronine exerts a negative feedback on MC_4 in hypothalamic paraventricular thyrotropin-releasing hormone neurons [1048].

The MC_4 receptor intervenes in signaling by leptin, a satiety hormone, and links the central energy control to the status of peripheral metabolic reserve. The leptin–melanocortin pathway operates in the hypothalamo–pituitary–thyroid axis to adapt energy expenditure to metabolic reserves. Leptin stimulates TRH production via MC_4 receptor. Activation of MC_4 reduces food intake and increases energy expenditure. Both thyroid hormone receptors ThRα (NR1a1) and ThRβ (NR1a2) con-

162. Hyper- and hypothyroidism causes weight loss and gain, respectively.

163. Functional thyroid hormone nuclear receptors (ThRα1, or NR1a1-1, as well as ThRβ1 and ThRβ2, or NR1a2-1 and NR1a2-2) colocalize in many hypothalamic nuclei. Each receptor isoform exerts specific transcriptional activities on the Trh promoter.

Table 7.79. Melatonin receptors and their main G-protein subunit transducers (Source: [5]). Melatonin-related receptor MT_3, or Mel_{1C}, is also called GPR50 in humans [608].

Type	Other names	Main transducer
MT_1	Mel_{1A}	Gi/o
MT_2	Mel_{1B}	Gi/o
MT_3	Mel_{1C}, GPR50	

tribute to MC_4 regulation. The T_3 hormone represses Mc4r gene transcription together with that of Trh gene. It prevents Mc4 expression not only via hypothalamic neurons, but also those of other energy-related brain regions, such as Agouti-related protein-containing neurons [1048]. The T_3 hormone dampens cerebral responsiveness to melanocortin anorectic signaling, but stimulates orexigenic pathways via neuropeptide-Y.

7.13.39 Melatonin Receptors

Melatonin receptors (MT_1–MT_3) are activated by the endogenous ligands melatonin secreted by the pineal gland to set the circadian rhythm and ^{N}acetylserotonin [5] (Table 7.79). They lodge in the brain and some other organs.

7.13.40 Motilin Receptors

Gq/11-coupled motilin receptor (MlnR, MtlR1, or GPR38) is activated by the polypeptide motilin that derives from a precursor that also generates motilin-associated peptide. Motilin is secreted by endocrine M cells in crypts of the small intestine to stimulate gastric activity.

7.13.41 G-Protein-Coupled Natriuretic Peptide Receptor

Natriuretic peptide receptors include 2 main sets of receptors. Receptors NPR1 and NPR2 are linked to guanylate cyclases (Sect. 6.4.1), whereas NPR3 is a G-protein-coupled receptor. The NPR3 receptor — the *clearance receptor* — binds to and removes natriuretic peptides from the blood circulation. It has an extracellular binding domain homologous to that of NPR1 and NPR2, but possesses a truncated intracellular domain that couples via Gi/o proteins to phospholipase-C and inwardly rectifying potassium channels, besides its inhibition of adenylate cyclase.

7.13.42 Receptors of Neuromedin-U and Neuromedin-S

Neuromedin-U receptors (NMU_1–NMU_2; [164] Table 7.80) are activated by neuromedin-U (NmU). Neuromedin-U is produced at its highest levels in the central ner-

164. The NMU_1 receptor was designated as the orphan GPR66 or FM3 and NMU_2 as FM4 or TGR1.

Table 7.80. Neuromedin-U receptors and their main G-protein subunit transducers (Source: [5]).

Type	Other names	Main transducer
NMU_1	NMUR1, FM3, GPR66, SNORF62	Gq/11
NMU_2	NMUR2, FM4, TGR1, SNORF72	Gq/11

vous system, bone marrow, upper gastrointestinal tract, and fetal liver. The structurally related peptide neuromedin-S is also an endogenous agonist of NMU receptors with equivalent potency to that of neuromedin-U [5].

Receptors NMU_1 and NMU_2 couple mainly to Gq/11, but also Gi/o. Receptor NMU_1 is predominantly expressed in peripheral tissues, particularly the gastrointestinal tract, whereas NMU_2 abounds within the central nervous system (brain and spinal cord) [1049]. Neuromedin-U causes vasoconstriction predominantly via NMU_1 and nociception and bone remodeling via NMU_2. It also prevents obesity. Neuromedin-S is more potent when it binds to NMU_2 receptors in vivo. Like neuromedin-U, it suppresses feeding and contributes to the regulation of the circadian rhythm [1049].

7.13.43 Receptors of Neuropeptide-B and Neuropeptide-W

Neuropeptide-B/W receptors ($NPBW_1$–$NPBW_2$; Table 7.81) are activated not only by neuropeptide-B (23 amino acid peptides NPB23) and -W (NPW23), but also their C-terminally extended forms (NPB29–NPW30) [5]. These receptors are predominately produced in the central nervous system.

7.13.44 Neuropeptide-S Receptor

Neuropeptide-S (NPS) is synthesized in the central nervous system. It suppresses anxiety. It targets neuropeptide-S receptor (NPSR; Table 7.81).

7.13.45 Neuropeptide-Y Receptors

Neuropeptide-Y (NPY) receptors (Table 7.81) are activated by neuropeptide-Y (NPY) and its fragment NPY_{3--36}, peptide-YY (PYY) and its fragment PYY_{3--36}, and pancreatic polypeptide (PP) [5]. Neuropeptide-Y is a sympathetic neurotransmitter involved in the behavior (control of the circadian rhythm and appetite, integration of emotional behavior, and cerebrocortical excitability). Under strong stress, NPY is released with noradrenaline from the peripheral sympathetic nerve terminals to possibly enhance permeability of vascular walls to large molecules, proliferation of vascular smooth muscles, and formation of vascular sprouts in ischemic tissues.

Neuropeptide-Y receptors are connected to Gi subunit of heterotrimeric G protein. Five cloned mammalian neuropeptide-Y receptors exist (Y_1–Y_2 and Y_4–Y_6;

Table 7.81. Receptors for neuropeptide-B (NPB), -W (NPW), -S (NPS) and -Y (NPY), their main G-protein subunit transducers, and ligands (Source: [5]; GPRA: G protein-coupled receptor for asthma susceptibility; PP: pancreatic polypeptide; PYY: peptide-YY; VRR: vasopressin receptor-related receptor). The Y_3 receptor is the CXCR4 chemokine receptor.

Type	Other names	Main transducer	Potency order
$NPBW_1$	NPBWR1, GPR7	Gi/o	NPB29>NPB23>NPW23>NPW30
$NPBW_2$	NPBWR2, GPR8	Gi/o	NPW23>NPW30>NPB29>NPB23
NPSR1	GPR154, GPRA, VRR1, PGR14, ASRT2	Gs, Gq/11	
Y_1		Gi/o	$NPY \geq PYY \gg PP$
Y_2		Gi/o	NPY $\geq$ PYY » PP
Y_4		Gi/o	$PP > NPY \sim PYY$
Y_5		Gi/o	$NPY \geq PYY \geq PP$
Y_6		Gi/o	$NPY \sim PYY > PP$

Table 7.81).[165] Neuropeptide-Y heightens the permeability of endothelial monolayers to large molecules via Gi-coupled Y_3 receptor [1050]. Noradrenaline enhances neuropeptide-Y effect via Y_3 receptor subtype on lung vascular permeability [1051]. The PLC–PKC and PI3K pathways may be activated for both permeability and cell proliferation primed by NPY under hypoxia [1050].

7.13.46 Neurotensin Receptors

Neurotensin receptors (NTS_1–NTS_2; Table 7.82) are activated by tridecapeptide neurotensin as well as neuromedin-N. Both ligands of neurotensin receptors derive from a common precursor. Receptors NTS_1 and NTS_2 are high- and low-affinity neurotensin receptors, respectively.

7.13.47 Nicotinic Acid Receptors

Receptors of the nicotinic acid family (GPR81 and niacin receptors NiacR1 [or GPR109a] and NiacR2 [or GPR109b]) are targeted by organic acids, such as

165. Receptors Y_1, Y_2, Y_4, and Y_5 are also designated as NPY1R, NPY2R, PPYR1, and NPY5R, respectively. The Y_3 receptor has not been cloned. However, Y_3 receptor is characterized biologically by its inability to be activated by peptide-YY and an ability to be inhibited by NPY_{3-36} [1050].

Table 7.82. Neurotensin receptors, main G-protein subunit transducers, and ligands (Source: [5]).

Type	Other names	Main transducer	Potency order
NTS_1	NTRH, NTR1	Gq/11	Neurotensin > neuromedin-N
NTS_2	NTRL, NTR2	Gq/11	Neurotensin ∼ neuromedin-N

Table 7.83. Nicotinic acid receptors and their main G-protein subunit transducers (Source: [5]; PUMaG: protein upregulated in macrophages by interferon-γ). Receptors GPR109a and GPR109b have a high and low affinity for nicotinic acid. The organic compound and essential nutrients — nicotinic acid — is also called niacin (hence the alias NiacR), vitamin-B3, and vitamin–PP.

Type	Other names	Main transducer
GPR81	GPR104	Gi/o
GPR109a	NiacR1, HM74a, HM74b, PUMaG	Gi/o
GPR109b	NiacR2, HM74, PUMaG	Gi/o

lipid-lowering nicotinic acid (niacin or vitamin-B3). Niacin is converted to nicotinamide and then nicotinamide adenine dinucleotide (NAD) and dinucleotide phosphate (NADP). Therefore, niacin is a precursor to NAD^+ involved in redox reactions, NADH used as a reducing agent (electron donor), and coenzyme $NADP^+$ and its reduced form NADPH. [166] Receptors GPR109a and GPR109b are activated by submicromolar and millimolar concentrations of nicotinic acid, respectively, whereas GPR81 does not respond to nicotinic acid [5] (Table 7.83).

7.13.48 Opioid and Opioid-like Receptors

Opioid and opioid-like receptors are activated by various peptides, such as endogenous opioid peptide neurotransmitters ^{Met}enkaphalin, ^{Leu}enkephalin [167] and β-endorphin, [168] as well as active peptides released from prodynorphin cleavage by pro-

166. Coenzyme NAD^+ is reduced to NADH in glycolysis and tricarboxylic acid cycle. $NADP^+$ Enzyme cofactor serves as a reducing agent in fatty acid and nucleic acid synthesis.

167. Enkephalins are pentapeptides that contribute to the regulation of nociception. Three different preprohormones contain the enkephalin sequence: preproopiomelanocortin (prePOMC or PPOMC) and preproenkephalin-A and -B (PPEnkA and PPEnkB).

168. β-Endorphin results from the cleavage of proopiomelanocortin that is also the precursor for adrenocorticotrophic hormone. Endorphins (endogenous morphine) are opioid peptides produced by the pituitary gland and the hypothalamus that function as neurotransmitters. The anterior pituitary gland produces the prohormone proopiomelanocortin (POMC). The latter is cleaved into adrenocorticotropin (ACTH) and β-lipotropin. Lipotropin-β is a POMC C-terminal fragment. Lipotropin-β is also involved in lipolysis and steroidogenesis. In particular, hypothalamic histaminergic neurons mediate the ACTH and β-endorphin response to lipopolysaccharides [1052].

protein convertase-2 (α-neodynorphin and dynorphin-A and -B) and prodynorphin partial processing, i.e., big dynorphin, [169] in addition to nociceptin, and tetrapeptides endomorphin-1 and -2 [5]. [170] Endogenous opioids act on plasmalemmal receptors to modulate synaptic transmission.

Precursors of endogenous opioid peptides encompass preprodynorphin (PPDyn, PDyn, or PPEnkB), preproenkephalin (PPEnk or PPEnkA; also PEnk or PEnkA), prepronociceptin (PPNoc or PNoc), and preproopiomelanocortin (PPOMC). They are encoded by different genes (PPDYN/PDYN, PPENK/PENK, PPNOC/PNOC, and PPOMC). [171] Prodynorphin (proDyn or PDyn; a.k.a. neoendorphin–dynorphin and preprodynorphin, and proenkephalin-B) is cleaved by proprotein convertase-2 to generate dynorphin-A and -B and α (β)-neoendorphin [1054]. Proenkephalin (proEnk or PEnk), or proenkephalin-A is cleaved into non-opioid synenkephalin (or proenkephalin$_{(1--70)}$), ^{Met}enkaphalin, ^{Leu}enkaphalin, ^{Met}-enkephalin$^{Arg--Gly--Leu}$, ^{Met}-enkephalin$^{Arg--Phe}$, proenkephalin$_{(114--133)}$, proenkephalin$_{(143--183)}$, and proenkephalin$_{(237--258)}$. The proenkephalin gene is expressed in neurons, hematopoietic (granulocytes, or polymorphonuclear leukocytes, lymphocytes, mastocytes, monocytes, and macrophages) and reproductive cells. Prepronociceptin gives rise to nociceptin (Noc), [172] as well as nocistatin, prepronociceptin$_{(154--181)}$, prepronociceptin$_{(169--176)}$, Noc2, [173] a nociceptin antagonist in pain transmission, and biologically inactive Noc3 protein. [174] Proopiomelanocortin (POMC) [175] is processed into adrenocorticotropin (ACTH) and β-lipotropin. [176] Lipotropin-β can also be cleaved into smaller peptides, such as γ-lipotropin, β-melanocyte-stimulating hormones (MSH), α- to γ-endorphin, and ^{Met}enkephalin (Table 7.84).

Four major types of opioid receptors exist: δ- (or Op_1), κ- (or Op_2), μ- (or Op_3), and nociceptin receptors (or Op_4; Table 7.85). δ-Opioid receptors (δ1–δ2 or DOR1–DOR2) are bound by enkephalins. Dynorphin is the primary endogenous ligand

169. Dynorphin inhibits neurotransmitter release at presynaptic terminals.

170. ενδον: within [ενδογενης: endogenous, born in the house]; Μορφευς: god of sleep and dream. Endomorphins are 2 endogenous opioid peptides of the nucleus of the solitary tract and the periventricular and dorsomedial hypothalamus, especially in hypothalamic histaminergic neurons.

171. Increase of preproopiomelanocortin mRNA in arcuate nucleus and decrease of preprodynorphin mRNA in dentate gyrus of spontaneously hypertensive rats may be associated with the genesis of spontaneous hypertension [1053].

172. Nociceptin is a heptadecapeptide (content 17 amino acid; επτα: 7; δεκα: 10; επτακαιδεκα: 17) that modulates ion channel activity (hence synaptic transmission) and analgesia.

173. The other heptadecapeptide Noc2 corresponds to the immediately downstream segment from nociceptin in the PNoc precursor.

174. Protein Noc3 corresponds to Noc2 with an added motif of 3 arginine residues (Noc2–Arg–Arg–Arg).

175. Proopiomelanocortin (POMC) is a precursor polypeptide that contains 241 amino acid residues. It is synthesized from a precursor, preproopiomelanocortin (prePOMC), that possesses 285-amino acid residues.

176. Hence its other name corticotropin–lipotropin.

Table 7.84. Opioid precursors and products (ACTH: adrenocorticotropic hormone; CLIP: corticotropin-like intermediate peptide; LPH: lipotropin hormones; MSH: melanocyte-stimulating hormones [a.k.a. melanotropins and intermedins]). Endogenous opioid peptides include dynorphins, endomorphins, endorphins, and enkephalins. They share the common N-terminal sequence (opioid motif Tyr–Gly–Gly–Phe–[Met or Leu]). Dynorphins are produced in the hypothalamus, hippocampus, midbrain, medulla, pons, and spinal cord. Endomorphin-1 is widely distributed in the brain and upper brainstem, whereas endomorphin-2 is more prevalent in the spinal cord and lower brainstem. Endorphins (αEnd–γEnd and σEnd) are produced by the hypophysis and hypothalamus. The set of melanocortins includes ACTH, αMSH, βMSH, and γMSH. Proopiomelanocortin, β-lipotropin, corticotropin (ACTH), and corticotropin-like intermediate peptide are secreted by corticotrope cells of the adenohypophysis, or anterior pituitary gland.

Precursor	Opioid peptides
Prodynorphin	Dynorphin-A, dynorphin-B, neoendorphin
Proenkephalin ^{Leu}enkaphalin, ^{Met}-enkephalin${}^{Arg--Phe}$ proenkephalin$_{(143--183)}$,	^{Met}enkaphalin, ^{Met}-enkephalin${}^{Arg--Gly--Leu}$, proenkephalin$_{(114--133)}$, proenkephalin$_{(237--258)}$, synenkephalin
Pronociceptin	Nociceptin, nocistatin, Noc2, Noc3, prepronociceptin$_{(154--181)}$, prepronociceptin$_{(169--176)}$
Proopiomelanocortin Corticotropin β-Lipotropin	γMSH, ACTH, βLPH, αMSH, CLIP γLPH, βMSH, β-endorphin, ^{Met}enkaphalin
	Endomorphin-1, endomorphin-2

of κ-opioid receptors (κ1–κ3 or KOR1–KOR3). κ-Opioid receptors can associate with $Ca_V2.2$ channels [1055] and G-protein-linked inward rectifier potassium channel GIRK1 ($K_{IR}3.1$) [1056]. Pre- and postsynaptic μ-opioid receptors (μ1–μ3 or MOR1–MOR3) typically inhibit neurotransmitter release.

β-Endorphin has its highest affinity for μ1-, slightly lower affinity for μ2- and δ-, and low affinity for κ1-opioid receptors. Endomorphin-1 and -2 have the highest known affinity for μ-opioid receptors. [177]

Nociceptin, or orphanin-FQ (OFQ), is an opioid-related peptide that derives from prepronociceptin (PPNoc). Nociceptin binds to nociceptin receptor (a.k.a. κ3-related

177. In the mouse pons and medulla, G-protein activation by endomorphin-1 and -2 is mediated by both MOR1 and MOR2 receptors and that by β-endorphin via MOR2, but not by MOR1 receptors [1058].

Table 7.85. Opioid and opioid-like receptors, main G-protein subunit transducers, and ligands (Sources: [5, 1057]; DOR, KOR, MOR: δ-, κ-, μ-opioid receptor; NOP: nociceptin receptors; TyrMIF: Tyr-Pro-Leu-Gly-NH_2; TyrWMIF: Tyr-Pro-Trp-Gly-NH_2). Endomorphin-1 and -2 are endogenous opioid peptides highly selective for Op_3 receptor, whereas β-endorphin, another endogenous opioid peptide, targets Op_1 and Op_3 receptors. In general, endogenous opioid receptor ligands modulate effects of various neurotransmitters (acetylcholine, dopamine, noradrenaline, and serotonin) and neurohormones (oxytocin and vasopressin). The majority of opioid peptides undergo a rapid extracellular degradation mainly by integral membrane exo- and endopeptidases. Enkephalin is a pentapeptide derived from preproopiomelanocortin and preproenkephalin- and -B (^{Met}enkaphalin: Tyr-Gly-Gly-Phe-Met [Met in position 5]; ^{Leu}Met-enkephalin: Tyr-Gly-Gly-Phe-Leu [Leu in position 5]).

Type	Other Aliases	Main Transducer	Ligands (endogenous opioids)
Op_1	OpRδ, DOR	Gi/o	[Met^5]/[Leu^5]enkephalins, Deltorphin-1/2
Op_2	OpRκ, KOR	Gi/o	Dynorphin-A/B, dynorphin-$A_{(1-8)}$,
Op_3	OpRμ, MOR	Gi/o	Endomorphin-1/2, β-endorphin, β-Casomorphins-5/7, hemorphins-4/7 morphiceptin, Tyr-MIF-1, Tyr-W-MIF-1
Op_4	NOP	Gi/o	Nociceptin (orphanin-FQ)

opioid receptor KORL3, ORL1, OpRL1, OOR, NOCIR, and NOP). This receptor is considered as opioid-related rather than opioid receptor, because, despite its structural homology with conventional opioid receptors, it exhibits a distinct pharmacology. Nociceptin and its receptor NOP lodge in the central and peripheral nervous system, where they modulate nociception [1057]. Nociceptin is, indeed, a potent anti-analgesic.

Exogenous nociceptin decreases blood pressure and heart rate via OP_4 receptor in rodents. In the cardiovascular system, OP_4 receptor localizes to sensory afferent fibres, the nucleus tractus solitarius, and rostral ventrolateral medulla, on preganglionic and/or postganglionic sympathetic and parasympathetic nerve fibers innervating blood vessels and heart, as well as in these target organs [1059].

Opioid receptor-δ that belongs to the B-group of rhodopsin subfamily of GPCRs is widely distributed throughout the central nervous system. Opioid Op_1 receptor has also been detected in peripheral tissues (cochlea, cornea, eyelid, and lip, as well as skin fibroblast-like cells, monocytes and T and B lymphocytes [1060]. Endorphins and enkephalins have the highest affinity for Op_1 receptor. The Op_1 receptor can form homodimers as well as heterodimers with other opioid receptors and GPCRs (α2a- and β2-adrenergic receptors, CCR5, CXCR2, and CXCR4 chemokine receptors, and sensory neuron-specific receptor SNSR4) [1060]. Agonist-induced monomerization process may be required for receptor endocytosis. It is not only coupled to

Table 7.86. Orexin receptors, main G-protein subunit transducers, and ligands (Source: [5]).

Type	Main transducer	Potency order
OX_1	Gq/11	Orexin-A > orexin-B
OX_2	Gq/11	Orexin-A ~ orexin-B

Table 7.87. Parathyroid hormone receptors, main G-protein subunit transducers, and ligands (Source: [5]).

Type	Main transducer	Potency order
PTH_1	Gs, Gq/11	PTH ~ PTHRP
PTH_2	Gs, Gq/11	PTH ≫ PTHRP

Gi, but also Gz and G16 subunits [1060]. Coupled to Gi, it inhibits adenylate cyclase and Ca^{++} channels. Coupled to Gz, it also inhibit adenylate cyclase, but stimulates phospholipase-C and activates the extracellular signal-regulated kinases ERK1 and ERK2. Coupled to G16, it also stimulates phospholipase-C. In addition, it can regulates P38MAPKs, phospholipase-D2, voltage-gated sodium $Na_V1.7$ channel, and potassium channels. Furthermore, it intervenes in signaling cascades stimulated by kinases PKB, Src, Raf1 (MAP3K), and receptor Tyr kinases as well as small GTPases CDC42, Rac, and Ras [1060]. The Op_1 receptor can be palmitoylated, glycosylated, phosphorylated (by GRKs and protein kinase-C), and ubiquitinated.

Opioid receptor-κcan regulate the Na^+–H^+ exchange via agonist-induced interactions with Na^+–H^+ exchanger regulatory factor NHERF1 that thus confers cell type-specific signaling to this GPCR. Stimulation of KOR, indeed, activates Na^+–H^+ exchanger NHE3 in cell lines that express high levels of NHERF1, but not other cell types [835].

7.13.49 Orexin Receptors

Orexin receptors are activated by orexin-A and -B [178] that derive from a common precursor, the preproorexin, by proteolytic cleavage [5] (Table 7.86).

7.13.50 Parathyroid Hormone Receptors

Receptors PTH_1 (or PTH1R) and PTH_2 (or PTH2R; Table 7.87) are activated by parathyroid hormone (PTH) and parathyroid hormone-related peptide (PTHRP) as well as related peptides ($PTHRP_{(1--36)}$, $PTHRP_{(38--94)}$, and osteostatin) [5]. Parathyroid hormone regulates calcium homeostasis and bone metabolism via type-1 PTH receptor.

Binding of PTH to PTH_1 (or PTH1R) activates protein kinase-A and -C. Signaling initiated by PTH terminates by PTH_1 endocytosis triggered by β-arrestins. The

178. A.k.a. hypocretin-1 and -2.

PTH_1 receptor interacts with Na^+–H^+ exchanger regulatory factor NHERF2 in cells that contain high NHERF levels, such as endothelial cells. In the latter, PTH_1 signals mainly via Gαq and PLCβ, whereas in osteoblasts that do not express detectable levels of NHERFs, PTH_1 transduces signals mainly via Gαs subunit and adenylate cyclase [835]. Scaffold NHERF not only binds to PTH_1, but also phospholipase-Cβ, protein kinase-C, and transient receptor potential channels that can then reside in close proximity to PTH_1 receptor.

The PTH_1 receptor also associates with type-2 transforming growth factor-β receptor TβR2 and forms an endocytic complex in response to PTH [1061]. The TβR2 kinase phosphorylates the PTH_1 cytoplasmic domain to promote PTH-induced endocytosis of the PTH_1–TβR2–Arrβ2 complex.

7.13.51 Platelet-Activating Factor Receptor

Platelet-activating factor (PAF, alkyl acetyl-glycerophosphocholine) is a phospholipid mediator that binds to a single G-protein-coupled receptor — PAFR — to activate multiple signaling pathways via Gq/11 and Gi/o family subunits of guanine nucleotide-binding proteins (Table 7.92). The PAFR receptor may also be activated by lysophosphatidylcholine, oxidized phosphatidylcholine, as well as bacterial lipopolysaccharide [5].

Platelet-activating factor receptor can associate with members of the Janus kinase family of cytosolic protein Tyr kinases such as the TyK2–JaK2 complex composed of tyrosine kinase-2 and Janus kinase-2.

7.13.52 Prokineticin Receptors

Two prokineticin receptors (PK_1–PK_2 or PKR1–PKR2) are activated by prokineticin-1 (PK1) [179] and -2 (PK2) [180] (Table 7.88). [181] Prokineticins are widespread. In the cardiovascular system, PK_1 is observed in cardiomyocytes, endothelial cells, and epicardial-derived progenitor cells, [182] but not in vascular smooth muscle cells.

Prokineticins potently contract gastrointestinal smooth muscle cells and favor angiogenesis, hence cardiomyocyte survival [1062], among other functions (Table 7.89). In mouse hearts, PK_1 receptor upregulates prokineticins that, in turn, acts

179. Prokineticin-1 was originally named endocrine gland-derived vascular endothelial growth factor (egVEGF). It is also termed mambakine.

180. A.k.a. Bv8.

181. Initially, the peptide hormones prokineticins (PK1–PK2) were identified in the gastrointestinal tract as potent constrictors. They were investigated mostly in the context of angiogenesis in the digestive and reproductive tracts. Subtype PK2 exerts an auto- and paracrine control via PK_1 in the heart and kidney.

182. The actin regulator thymosin-β4 that assists wound repair is able to activate epicardial-derived progenitor cells in adults that then differentiate into endothelial cells, smooth muscle cells, and fibroblasts. Prokineticin-2 is also capable of stimulating these progenitors that then differentiate into endothelial and smooth muscle cells, but not fibroblasts.

Table 7.88. Prokineticin receptors, main G-protein subunit transducers, and ligands (Source: [5]; ProK: prokineticin). Receptor PK_2 predominates in the adult brain, whereas PK_1 is widely distributed in peripheral organs.

Type	Other names	Main transducer	Potency order
PK_1	PKR1, PROKR1, GPR73a, ZAQ	Gq/11	ProK2 $\geq$ ProK1
PK_2	PKR2, PROKR2, GPR73b, GPR73L1, GPRg2, KAL3	Gq/11	ProK2 $\geq$ ProK1

Table 7.89. Examples of effects of prokineticins (PK1–PK2; Source: [1063]). Signaling from PK_1 receptor is required for cardiomyocyte survival and angiogenesis. This receptor is involved in postnatal cardiac and renal vascularization, as it activates organ-specific progenitor cells.

Role	Target cell types or framework
Cell survival	Cardiomyocytes, endothelial cells, hematopoietic cells, neurons
Cell motility	Hematopoiesis, vasculo-, angio-, neurogenesis
Cell excitability	Circadian rhythm
Behavior	Feeding, drinking

as paracrine factors to promote the differentiation of transcription factor TCF21+ [183] progenitor cells into a vasculogenic cell type [1063]. In isolated renal TCF21+ progenitors, PK2 binds to PK_1 and provokes the differentiation of these progenitors into endothelial and smooth muscle cells.

In the reproductive tract, PK1 production rises upon stimulation by estrogen, progesterone, and human chorionic gonadotrophin, as well as hypoxia-inducible factor-1α [1063]. In neurons of the olfactory bulb, PK2 synthesis is primed by basic helix–loop–helix (bHLH) transcription factors neurogenin-1 (NeuroG1 or bHLHa6) and achaete–scute complex-like factor ASCL1 (or bHLHa46) [184] and repressed by transcriptional factors distal-less homeobox DLx1 and DLx2.

Prokineticins can cause differentiation of human bone marrow cells into the monocyte–macrophage lineage, monocyte proliferation and differentiation, and macrophage migration. In monocytes and neutrophils, PK2 is upregulated in response to granulocyte colony-stimulating factor (CSF3).

183. Transcription factor TCF21 is also called epicardin, capsulin, class-A basic helix–loop–helix protein bHLHa23, and podocyte-expressed protein Pod1.

184. A.k.a. mammalian achaete–scute homolog mASH1. The achaete–scute complex is a set of 4 genes (achaete, scute, lethal of scute, and asense) in Drosophila melanogaster that encode basic helix–loop–helix transcription factors.

Prokineticin receptors activate many signaling pathways and mobilize calcium messenger. Activated PKRs stimulate PKB, MAPK modules, and the PP3–NFAT axis, as well as phosphoinositol turnover [1063]. In addition, they reduce the production of zonula occludens-1 of tight junctions and can activate protein kinase-C and subsequently transient receptor potential vanilloid TRPV1 channel.

Hepatic sinusoidal endothelial cells express only PK_2 receptor. In coronary endothelial cells, PK_1 predominates over PK_2 [1064]. Activated PK_1 that colocalizes with $G\alpha_{11}$ primes mitogen-activated protein kinases and protein kinase-B signaling to stimulate cell proliferation and migration for angiogenesis. Once stimulated by prokineticin-2, PK_1 undergoes endocytosis. On the other hand, prokineticins also cause fenestrations in endothelia [1064]. When PK_2 predominates over PK_1 in coronary endothelial cells, these cells contain a large number of multivesicular bodies and caveolar clusters. Prokineticin-2 activates PK_2 associated with $G\alpha_{12}$ subunit. The latter binds to tight junction protein zonula occludens-1 and trigger its degradation.

7.13.53 Prostanoid Receptors

Prostanoid receptors are activated by various types of prostaglandins, i.e., PGD2, PGE2, PGF2α, PGH2, PGI2 (prostacyclin), and thromboxane-A2 [185] (TxA2; Table 7.90). Inflammatory cytokines stimulate the production from arachidonic acid of lipid mediators, such as prostanoids. Rate-limiting enzymes in prostanoid production include cyclooxygenases, [186] in particular constitutively expressed COx1 and inducible COx2. Cyclooxygenase COx2 induced in macrophages and endothelial cells by inflammation stimuli, such as oxidized low-density lipoprotein and interleukin-1, produces eicosanoids that enhance vascular permeability and promote cell chemotaxis.

The effects of prostacyclin and prostaglandin-E2 are transduced by the I prostanoid receptor (IP) and E prostanoid receptors (EP), respectively (Table 7.90; Vol. 5 – Chap. 7. Vessel Wall). Receptors EP_2 and EP_4 are linked to Gs that activates adenylate cyclase, whereas EP_1 and EP_3 are coupled to Gq and/or Gi subunits. Receptors EP_1 and EP_3 mediate PGE2-induced vasoconstriction. Prostaglandin PGE2 can activate platelets by EP_3. Receptor EP_4 mediates anti-inflammatory effects and activation of MMP9 metallopeptidase.

Deletion of EP_2 and IP causes salt-sensitive hypertension. In cardiomyocytes, dominant COx2-derived products are prostacyclin and prostaglandin-E2 that target I prostanoid and E prostanoid EP_2 receptor, respectively. Both PGI2 and PGE2 serves as mediators of cardioprotection. In mice, COx2-deficient cardiomyocytes have impaired ventricular function subsequent to pressure overload [1066].

Although prostanoid receptors are expressed in many of the body's tissues, the distribution of prostanoid receptor types and cell identities that produce each receptor

185. Thromboxanes of the eicosanoid superfamily include thromboxane-A2 and -B2. Thromboxane-A synthase that in platelets converts prostaglandin-H2 to thromboxane. Thromboxane-B2 is an inactive metabolite of thromboxane-A2.

186. There are 3 isoforms of cyclooxygenases, COx1, COx2, and COx3.

Table 7.90. Prostanoid receptors, their main targeted G proteins, and order of ligand potency (Sources: [736, 1065]).

Type	Main transducer	Potency order
DP_1	Gs	PGD2 ≫ PGE2 > PGF2α > PGI2, TxA2
DP_2	Gi/o	PGD2 ≫ PGE2, PGF2α > PGI2, TxA2
EP_1	Gq/11	PGE2 > PGF2α, PGI2 > PGD2, TxA2
EP_2	Gs	PGE2 > PGF2α, PGI2 > PGD2, TxA2
EP_3	Gi/o	PGE2 > PGF2α, PGI2 > PGD2, TxA2
EP_{3A}	Gi	
EP_{3B}	Gs	
EP_{3C}	Gs	
EP_{3D}	Gi, Gs, Gq	
EP_4	Gs	PGE2 > PGF2α, PGI2 > PGD2, TxA2
FP	Gq/11	PGF2α > PGD2 > PGE2 > PGI2, TxA2
IP	Gs, Gq	PGI2 ≫ PGD2, PGE2, PGF2α > TxA2
TP	Gq/11	TxA2, PGH2 ≫ PGD2, PGE2, PGF2α, PGI2
TPα	Gq, Gi	
TPβ	Gq, Gs	

Table 7.91. Distribution of prostanoid receptors (Source: [1065]).

Type	Distribution
DP	Low levels in humans (least abundant among prostanoid receptors)
	Small intestine and leptomeninges
EP_1	Kidney, lung, stomach, thalamus
EP_2	Least abundant among EP receptors
EP_3	Widely distributed throughout the body
EP_4	Widely distributed throughout the body
FP	Corpus luteum, kidney, heart, lung, and stomach
IP	Neurons of dorsal root ganglion, megakaryocytes,
	Vascular smooth muscle cells, mature thymocytes, splenic lymphocyes
TP	Abundant in lung, kidney, and heart,
	Lymphoid organs (thymus and spleen)
	Platelets

type is difficult to determine, as prostanoid receptors are expressed at a relatively low level (Table 7.91). A given cell can synthesize several types of prostanoid receptors.

Prostaglandin-D2 is produced by cyclooxygenase-2 that is inducible by mitogens, cytokines, and tumor promoters in activated mast cells, macrophages, and TH2 cells. Prostaglandin-D2 targets receptor DP that is highly expressed by tumor endothelial cells to suppress vascular hyperpermeability via an increase in intracellular cAMP production in tumors [1067]. On the other hand, PGE2 promotes tumor growth and angiogenesis.

Prostaglandins PGI2 and TxA2 impede and cause platelet activation and aggregation, respectively. The balance between PGI2 and TxA2 prevents thrombosis and vasospasm, but maintains the body ready to perform efficient hemostasis.

7.13.53.1 Prostaglandins and the Vasomotor Tone

Prostaglandins PGI2 and TxA2 are a potent vasodilator and vasoconstrictor, respectively. They abound in vascular endothelial cells and platelets, respectively. However, PGI2 and its main *IP receptor* are less potent regulators of blood circulation than endothelium-derived vasorelaxant nitric oxide.

Prostaglandin-E2 produced by cyclooxygenase-2 can promote relaxation of vascular smooth muscle cells. Prostaglandin PGE2 is synthesized in response to a high-salt diet and works via relaxant EP_2 receptor. Its dysfunction elicits salt-sensitive hypertension. Prostaglandin-E2 EP_1 receptor impairs Na^+–Ca^{++} exchanges and leads to neurotoxicity [1068]. Membrane-associated proteins involved in eicosanoid and glutathione metabolism (MAPEG) affect the expression of endogenous PGE2 and PGI2. Microsomal prostaglandin-E synthase-1 (mPGES1), a member of the MAPEG family, is a major source of PGE2 formation. Deletion of mPGES1 does not increase blood pressure and retards atherogenesis, whereas mPGES1-derived PGE2 accelerates atherogenesis [1069]. On the other hand, prostacyclin protects the cardiovascular function.

In pulmonary arteries, chronic hypoxia increases cyclooxygenase COx2 expression and PGF2α release that activates thromboxane *TP receptors* [1070]. In pulmonary vascular smooth muscle cells, hypoxia upregulates COx2 expression, thereby increasing production of vasorelaxant (PGI2, PGE2, and PGD2) and vasoconstrictors (PGF2α). The COx1 enzyme, the major TxA2 source, is produced in all wall layers of pulmonary arteries. Its expression remains unaltered in chronic hypoxia. TxA2 synthase that is mainly located in the media of pulmonary arteries is markedly downregulated in chronic hypoxia. In addition, COx enzymes are potential sources of reactive oxygen species in pulmonary arteries, especially in pulmonary arterial hypertension associated with hypoxia. Prostaglandin-F2α activates TP receptors to generate vasocontriction of pulmonary arteries. Whereas cyclooxygenase-catalyzed metabolism of arachidonic acid leads to an endothelium- and COx1-dependent relaxation of pulmonary arteries in normoxia, it causes an endothelium-independent, COx2- and TP receptor-dependent vasocontriction in hypoxia [1070].

7.13.53.2 Prostaglandins in Heart

In cardiomyocytes, prostaglandins target prostanoid F-receptor (FP) that increases myofilament sensitivity to Ca^{++} by enhancing myosin light chain-2 phosphorylation. Prostanoid F-receptor is coupled to Gq/11 protein that leads to production of inositol trisphosphate and diacylglycerol. This second messenger activates PKC, but its effect extends beyond this classical pathway, possiby via G12/13 protein [1071]. Positive inotropic effect of Gq/11-protein-coupled receptors (e.g., α1-adrenoceptor, $5HT_{2A}$, and endothelin-1 receptors) results, at least partly, from myofilament Ca^{++} sensitization by increased MLC2 phosphorylation. The level of

Table 7.92. Peptidase-activated and platelet-activating factor receptors and their main targeted G proteins (Source: [736]).

Type	Main transducer
Peptidase-activated receptors	
PAR_1	Gq/11, Gi/o, G12/13
$PAR_{2/3/4}$	Gq/11, Gi/o
Platelet-activating factor receptor	
PAFR	Gq/11, Gi/o

MLC2 phosphorylation is regulated by both Ca^{++}–calmodulin-dependent activation of myosin light chain kinase and Rho kinase-mediated inhibition of myosin light chain phosphatase. On the other hand, slow positive inotropic effect of receptor FP results from increased phosphorylation of myosin phosphatase targeting subunit MyPT2 and myosin light chain MLC2 by stimulating the RhoA–RoCK pathway, hence decreasing MLCP activity, and activating MLCK, respectively.

7.13.54 Tissue Factor and Peptidase-Activated Receptors

Peptidase-activated receptors (PAR_1–PAR_4; Table 7.92) are highly expressed in platelets, but also in endothelial cells, myocytes, and neurons. Peptidase-activated receptors are activated by proteolytic cleavage of their N-terminal exodomains by serine peptidases.

Among serine peptidases, *thrombin* and *trypsin* can be agonist peptidases for certain peptidase-activated receptors in vivo.[187] Alternative endogenous peptidases or ligands to thrombin for PAR_1, PAR_3, and PAR_4 exist. Several peptidases, such as *cathepsin-G* and *chymotrypsin*, inhibit PAR_1 receptor, as they prevent activation by thrombin [5]. Activated *protein-C*,[188] an anticoagulant peptidase that elicits an anti-inflammatory response and protects against endothelial barrier dysfunction caused by thrombin, also activates PAR_1 receptor. Therefore, PAR_1 is equipped for peptidase-selective signaling. *Tissue factor* can also bind peptidase-activated receptor (Table 7.92). Peptidase-activated receptor PAR_1 is activated by both thrombin and PAR_1-*activating peptide* (PAR_1-AP). Like PAR_1, PAR_2 on smooth muscle and respiratory epithelial cells can be activated by tryptase, trypsin, and PAR_2-*activating peptide* (PAR_2-AP).

187. Thrombin is produced from prothrombin by factor Xa to mediate the formation of fibrin from fibrinogen that forms the fibrous matrix of blood clots. Thrombin is also a strong activator of platelet aggregation. Moreover, thrombin acts on various cell types, such as monocytes, smooth muscle cells, endothelial cells, fibroblasts, and lymphocytes, among others [1072].

188. Protein-C on the endothelial cell surface is activated by the thrombin–thrombomodulin complex. Activated protein-C binds to endothelial protein-C receptor (EPCR) and cleaves (inactivates) clotting factors Va and VIIa, thereby reducing thrombin generation. Moreover, it provokes cellular responses via PAR_1 receptor.

Table 7.93. Human peptidase-activated receptors and their activators and loci (Source: [1073]).

	Primary activator peptidase	Localization
PAR_1	Thrombin	Platelets, endothelial and epithelial cells, fibroblasts, myocytes, neurons, astrocytes
PAR_2	Trypsin Tryptase	Endothelial and epithelial cells, Fibroblasts, myocytes, neurons, astrocytes
PAR_3	Thrombin	Endothelial cells, myocytes, astrocytes
PAR_4	Thrombin, Trypsin	Platelets, endothelial cells, Myocytes, astrocytes

Four known peptidase-activated receptors are expressed in the entire body (Table 7.93). Thrombin targets mainly PAR_1, PAR_3, and PAR_4, and trypsin chiefly activates PAR_2 (Tables 7.94 and 7.95). Most PARs act via Gi, G12/13 (Ras–Raf pathway), and Gq (PLC–DAG–PKC and PLC–IP_3–Ca^{++} axes) signaling (Table 7.96). Peptidase-activated receptors participate in the regulation of the vasomotor tone and operate in inflammation, muscle growth, and bone cell differentiation and proliferation. Sorting nexin-1 links to peptidase-activated receptor PAR_1 to favor lysosomal path among endocytosis routes.

7.13.54.1 Thrombin and PAR Receptors

Thrombin facilitates production and release of growth factors, such as PDGF TGFβ, and ET1, as well as formation of VEGF receptors. Activated PAR_1 participates in synthesis of extracellular factors that contribute to wound healing. Thrombin stimulates procollagen synthesis by smooth muscle cells and pulmonary fibroblasts. It also regulates the release of matrix metallopeptidases, such as MMP1, MMP2, MMP3, and progelatinase-A.

PAR_1 Receptor and prothrombin are expressed in numerous regions of the central nervous system, such as thalamus, hypothalamus, cortex, and cerebellum. Thrombin causes increased synthesis of nerve growth factor and endothelin-1. However, in some circumstances, thrombin induces cell death.

Thrombin provokes endothelial-dependent relaxation of smooth muscle cells of aortic and coronary arteries via prostacyclin and by nitric oxide. However, in certain territories, it generates vasoconstriction via an increase in intracellular calcium concentration [1072]. Relative expression and function of PAR_1 in endothelial and smooth muscle cells in different vessels can explain the difference in effect on the vasomotor tone.

7.13.54.2 PAR Receptors and the Vasomotor Tone

Peptidase-activated receptors PAR_1 and PAR_2 modulate the endothelium-dependent regulation of the vasomotor tone in arteries and veins, such as internal mam-

Table 7.94. Cellular distribution and effects of PAR_1 (Source: [1072]; GEC: glomerular epithelial cells; GMC: glomerular mesangial cells; SMC: smooth muscle cell; IL: interleukin; TNFα: tumor-necrosis factor-α; gmCSF: granulocyte-macrophage colony-stimulating factor; IgCAM: immunoglobulin-like cell adhesion molecule).

Cell	Effects
Platelets	Degranulation; ↑ aggregation; ↑ TxA2
Vascular SMC	Vasorelaxation or contraction; mitogenesis
Endothelial cell	↑ Release of von Willebrand factor, NO, ↑ expression of selectins and IgCAMs, proliferation
Airway wall cells	↑ Release of procollagen, PDGF, prostanoids, relaxation or contraction
Osteoblasts	Proliferation
Synovial fibroblasts	↑ proliferation; ↑ IL6 and gmCSF
Keratinocytes	↑ proliferation; inhibition of differentiation
Monocytes, T cells	↑ IL1, IL6, TNFα, LProliferation
Mastocytes	Degranulation
Intestinal SMC	Relaxation or contraction
Stomachal SMC	Relaxation or contraction
Skeletal myocytes	↑ $[Ca^{++}]_i$
GEC, GMC	↑ Production of clusterin and TGFβ
neuron, glial cell	↑ Proliferation; neuronal apoptosis
astrocyte	Proliferation

mary artery and greater saphenous vein that are used in coronary artery bypass grafting. Amounts of PAR_1 and PAR_2 are similar in both vessel types [1074]. Although selective PAR_2-activating peptide (PAR_2-AP) fails to induce vasorelaxation, PAR_1-activating peptide (PAR_1-AP) generates vasodilation. The endothelium-dependent relaxation is greater in internal mammary artery than in greater saphenous vein. In addition, inflammatory stimuli such as TNFα enhance endothelium-dependent relaxation selectively via PAR_2-AP in internal mammary artery.

Trypsin and peptidase-activated receptor-2-activating peptides, via Ca^{++} mobilization, endothelin receptor ET_{B1}, and subsequent activation of endothelial NO synthase (NOS3), as well as via prostacyclin, are able to cause in vitro an endothelium-dependent (nitric oxide-mediated) relaxation of vascular smooth muscle cells. On the other hand, trypsin and high concentrations of PAR_2-activating peptides can also initiate endothelium-dependent contraction in both rat pulmonary artery and human umbilical vein [1072]. Whereas endothelium-dependent vasodilation initiated by PAR_2-APs is due to PAR2 activation, endothelium-independent contraction primed by PAR_2-APs is done via a receptor different than PAR_2 [1075].

Table 7.95. Cellular distribution and effects of PAR_2 (Source: [1072]; GEC: glomerular epithelial cells; GMC: glomerular mesangial cells; SMC: smooth muscle cell; IL: interleukin; TNFα: tumor-necrosis factor-α; gmCSF: granulocyte-macrophage colony-stimulating factor; IgCAM: immunoglobulin-like cell adhesion molecule).

Cell	Effects
Vascular SMC	Vasorelaxation or contraction; mitogenesis
Endothelial cell	↑ Release of von Willebrand factor, NO
Airway wall cells	↑ PGE2 release, relaxation or contraction, proliferation
Osteoblasts	↑ $[Ca^{++}]_i$
Keratinocytes	Inhibition of differentiation and proliferation, ↑ IL6 and gmCSF
Leukocytes, mastocytes	↑ Adhesion, rolling, and migration
Salivary gland cells	↑ Saliva production, ↑ amylase and mucin secretion
Gallbladder SMC	Contraction, prostanoid release
Intestinal SMC	Contraction, PGE2 release
Stomachal SMC	Contraction
Skeletal myocytes	↑ $[Ca^{++}]_i$
Kidney cortex cells	↑ Cl^- secretion
Ureter sphincters	↓ Beating
Neurons, glial cells, astrocytes	↑ $[Ca^{++}]_i$

7.13.54.3 PAR_1

PAR_1 Receptor and Heterotrimeric G Protein

The PAR_1 receptor interacts with Gi/o, Gq/11, and G12/13. The Gi subunit inhibits cAMP and triggers various effects via its activated partner Gβγ dimer. The Gq/11 subunit stimulates phospholipase-C that hydrolyzes phosphoinositides into IP_3 and DAG that increases the cytosolic calcium content and activates protein kinase-C, respectively. Subunits of the G12/13 family activate the RhoGEF–Rho–RoCK pathway for cell remodeling and migration, as well as JNK using small Ras or Rac GTPases.

Thrombin also stimulates hydrolysis of other phospholipids via PLA2, PLC, and PLD. The PAR_1 receptor activates Ras via Gi and Tyr kinases after recruitment of adaptors SHC and GRB2, SOS, and Raf1 to the cell cortex then extracellular-regulated kinase. The Gβγ subunit activates phosphatidylinositol 3-kinase.

PAR_1 Receptor and Peptidase-Selective Signaling

Both activated protein-C and thrombin target PAR_1, but have antagonistic effects, especially on blood coagulation, endothelial barrier permeability, and inflammation.

Table 7.96. Signaling pathways activated by PAR_1 in platelets (Source: [1073]; DAG: diacylglycerol; GEF: guanine nucleotide-exchange factor; GRK: G-protein-coupled receptor kinase; IP_3: inositol trisphosphate; NRTK: non-receptor tyrosine kinase; PI3K: phosphatidylinositol 3-kinase; PKC: prtoein kinase-C; PLC: phospholipase-C; RoCK: Rho-associated, coiled-coil-containing protein kinase).

G Protein subunit	Pathways (effects)
Gαi/z	RhoGEF–Rho–RoCK
Gαq	PLCβ–IP_3–Ca^{++}
	PLCβ–DAG–PKC
	(granule secretion)
	(activation of Ca^{++}-regulated kinases and phosphatases)
	(activation of RTKs and RasGEFs)
	(growth factor shedding)
Gα12/13	ACase Inhibition
Gβγ	PI3K
	PLCβ
	K^+ channels
	GRK
	NRTK
	(Recruitment to plasma membrane of kinases, GEFs, and scaffold proteins)

Thrombin excites small GTPase RhoA-associated signaling without Rac1 activation, whereas activated protein-C stimulates small GTPase Rac1, but not RhoA GTPase. Peptidase-selective signaling via PAR_1 relies on caveolin-1, hence PAR_1 compartmentation, as activated protein-C requires caveolin-1, but not thrombin [1076]. Activated protein-C cofactor endothelial protein-C receptor, PAR_1, Gq and Gi subunits are partitioned into caveolin-1-containing membrane rafts (caveolae). Signal transduction mediated by Gi/o family subunits protects endothelial barrier. In addition, activated protein-C causes PAR_1 phosphorylation. It also desensitizes endothelial cells to thrombin. Moreover, it limits receptor proteolysis, its endocytosis, and degradation, even after prolonged exposure to activated protein-C.

7.13.54.4 PAR_2

The PAR_2 receptor on both endothelial cells and leukocytes intervenes in inflammation. Trypsin and PAR2-activating peptides stimulate activation of T lymphocytes and neutrophils, and promote leukocyte recruitment (rolling, adhesion, and extravasation) by a mechanism dependent on platelet-activating factor release [1077].

7.13.54.5 PAR_3 and PAR_4

The PAR_3 receptor resides in heart, small intestine, and bone marrow, among other organs, especially on airway smooth muscle cells, vascular endothelial cells, and astrocytes. In mouse platelets, PAR_3 expression is necessary for full activation by thrombin. Tissue distribution of PAR_4 differs from that of other PARs, with the highest levels in the lung, small intestine, pancreas, thyroid, and testis. The PAR_4 receptor operates as a low-affinity thrombin receptor.

7.13.54.6 Tissue Factor in Coagulation, Inflammation, and Angiogenesis

Tissue factor can have 2 distinct structural and functional modes, as it is an initiator of blood clotting,[189] or a cofactor for cell signaling that is unable to promote coagulation.[190] In other words, plasmalemmal tissue factor binds to: (1) serine peptidase factor VIIa to activate coagulation or (2) peptidase-activated receptor-2 to trigger inflammation and angiogenesis, as tissue factor belongs to the cytokine receptor family.[191] Tissue factor–PAR_2 complex formation can be inhibited without preventing the coagulation activity of tissue factor.

Protein disulfide isomerase (PDI)[192] stabilizes a distinct tissue factor–factor VIIa complex that does not bind factor X [1078]. PDI inhibits the coagulation activity of the tissue factor, and switches tissue factor to cell signaling.[193]

Inflammation and coagulation initiated by tissue factor are coupled by a crosstalk between peptidase-activated receptor-1 and sphingosine 1-phosphate receptor-3, at least in dendritic cells [1079]. Signaling from PAR_1 is also coupled to S1P in endothelial cells. Dendritic cells in the lymphatics can cause exacerbated inflammation and coagulation in deregulated innate immune responses such as disseminated intravascular coagulation. Block of PAR_1 is then sufficient to interrupt systemic dissemination of inflammation and coagulation.

---

189. Tissue factor, an integral membrane protein, is normally excluded from the endothelial wetted surface. Tissue factor also circulates and becomes active only with a growing thrombus. Coagulation (Vol. 5 – Chap. 9. Endothelium) is triggered when tissue factor is exposed to zymogen coagulation peptidases in plasma after vessel wall damage. It binds and activates factor VII. The complex made of coagulant tissue factor and factor VIIa binds and activates factor X. Factor Xa acts with its cellular cofactor Va for thrombin production. Thrombin cleaves fibrinogen to generate fibrin and causes platelet aggregation.

190. Tissue factor is also activated in inflammation, vascular development, and cancer.

191. The non-coagulant form of tissue factor bound to factor VIIa (binary signaling complex) activates receptor PAR_2. Factor Xa signals via peptidase-activated receptors PAR_1 and PAR_2 either as a monomer or as a ternary complex with tissue factor and factor VIIa. Factor Xa is inhibited by antithrombin-3 and tissue factor pathway inhibitor. Factor VIIa is also inhibited by tissue factor pathway inhibitor bound to factor Xa. Thrombin is a potent activator of peptidase-activated receptors, except PAR_2 receptor. It activates PAR_1 and PAR_4 for platelet aggregation. Like Factor Xa, thrombin is inhibited by antithrombin-3. Thrombomodulin hampers the binding of thrombin with PAR_1 on the endothelial cell surface.

192. Protein disulfide isomerase cleaves disulfide bonds in the extracellular domains of certain receptors to regulate protein activity. Protein disulfide isomerase breaks a disulfide bond that is required to activate the coagulation. Nitric oxide regulates PDI activity, hence, suppresses the coagulant activity of tissue factor.

193. Disulfide/thiol exchange is required for the formation of Tissue factor–PAR_2 complex.

Table 7.97. Receptors of the relaxin family peptides, main G-protein subunit transducers, and ligands (Source: [5]; InsL*i*R: insulin-like peptide type-*i* receptor; LGR: leucine-rich repeat-containing G-protein-coupled receptor; Rln3R: relaxin-3 receptor; SALPR: somatostatin and angiotensin-like peptide receptor). The $G\alpha_{oB}$ subunit is a member of the $G\alpha_{i/o}$ family (with $G\alpha_{oA}$, $G\alpha_{i1}$, α_{i2}, α_{i3}, $G\alpha_{Tr}$, and $G\alpha_{z}$).

Type	Other names	Main transducer	Ligands Potency order
$RXFP_1$	RXFPR1, RX_1, LGR7	Gs, GoB, Gi3 (ACase, PI3K, PKA, PKC, ERK1/2)	Relaxins-1/2/3 Relaxin-2 > relaxin-3 ≫ InsL3
$RXFP_2$	RXFPR2, RX_2, LGR8, GREAT, InsL3R, GPR106	Gs, GoB (ACase)	Relaxins-1/2/3, InsL3 InsL3 > relaxin-2 ≫ relaxin-3
$RXFP_3$	RXFPR3, RX_3, Rln3R1, GPCR135, SALPR	Gi/o (ERK1/2)	Relaxin-3
$RXFP_4$	RXFPR4, RX_4, Rln3R2, InsL5R, GPR100, GPCR142	Gi/o	Relaxin-3, InsL3/5 InsL5 ∼ relaxin-3

7.13.55 Receptors of the Relaxin Family Peptides

The set of relaxin family peptide receptors are divided into 2 categories: group 1 ($RXFP_1$–$RXFP_2$) and 2 ($RXFP_3$–$RXFP_4$; Table 7.97). They are targeted by heterodimeric peptide hormones relaxin-1 to -3 and insulin-like peptide InsL3 and InsL5 [5]. Relaxin receptors reside in central and autonomous nervous system, heart, smooth muscle, and connective tissue. At least 2 binding sites, with high and low affinity, exist on $RXFP_1$ and $RXFP_2$ receptors [5].

The $RXFP_1$ receptor activates adenylate cyclase, protein kinase-A and -C, phosphatidylinositol 3-kinase, and extracellular signaling regulated kinases ERK1 and ERK2 [1080]. It also intervenes in nitric oxide signaling. Relaxin targets $RXFP_1$ for connective tissue remodeling.

Gs-Coupled $RXFP_2$ activates and Gi-coupled $RXFP_3$ and $RXFP_4$ and inhibit adenylate cyclase, respectively. The $RXFP_3$ receptor also activates ERK1 and ERK2 kinases.

7.13.56 Serotonin (5-Hydroxytryptamine) Receptors

Except ionotropic receptors of the $5HT_3$ class that are ligand-gated Na^+ and K^+ cation channels, all other serotonin (or 5-hydroxytryptamine) receptors ($5HT_1$–$5HT_2$ and $5HT_4$–$5HT_7$; Table 7.98) are G-protein-coupled receptors. The diversity of metabotropic 5HT receptors is augmented by alternatively spliced variants. Subtypes $5HT_{2A}$, $5HT_{2C}$, and $5HT_6$ are non-functional [5]. Furthermore, RNA editing produces $5HT_{2C}$ isoform that differ in efficiency and specificity of Gq/11 coupling.

Serotonin metabotropic receptors reside in the central and peripheral nervous system, where they activate an intracellular cascade to produce an excitatory or inhibitory response (Table 7.99), as well as other tissues, particularly vasculature. The bioamine serotonin is synthesized from tryptophan. Once released, serotonin increases the smooth muscle tone.

The central serotoninergic system innervates diverse brain regions. In the central nervous system, cell bodies that contain serotonin localize to nuclei raphé, near the midline of the brainstem, and project to the brainstem and spinal cord as well as forebrain [1081]. The activity of the raphé network is modulated by activators and inhibitors of the release of serotonin.

Various serotonin receptors modulate the synaptic transmission and postsynaptic excitability. Serotonin receptors indeed influence the release of many neurotransmitters, such as acetylcholine, adrenaline, dopamine, GABA, glutamate, and noradrenaline, as well as many hormones, such as corticosterone, corticotropin, cortisol, oxytocin, prolactin, substance-P, vasopressin, etc.

Disc large homolog DLg4 [194] connects to $5HT_{2A}$ and $5HT_{2C}$ to facilitate their anchoring and clustering to postsynaptic dendrites and agonist-dependent internalization, respectively [835].

Breathing is controlled by the nervous system. The serotoninergic apparatus belongs to the arousal system that controls the body's ventilation. Adjustment of excitability of the ventilatory command is also regulated by acetylcholine, adenosine, catecholamines, opioids, and other neuropeptides [1081].

Apneusis consists of a pause at inspiration and prolonged breath holding. Central apnea results from the loss of inspiratory motions. [195] Serotonin receptors can counteract both apneusis and opioid-induced apnea [1081]. In mammals, the respiratory rhythm is generated by the *pre-Bötzinger complex* (PBC), a cluster of interneurons in the ventrolateral medulla of the brainstem. Many neurons of the pre-Bötzinger complex produce adenosine (A_1 and A_{2A}), GABA, glutamate, neurokinin-1 (NK_1), and serotonin receptors, such as $5HT_{1A}$, $5HT_{2A}$, $5HT_{2B}$, $5HT_4$, and $5HT_7$. Serotonin receptors colocalize with opioid Op_3 receptor in respiratory neurons of the pre-Bötzinger complex to promote activity of neurotrophins. Gq-Coupled $5HT_{2A}$ and $5HT_{2B}$ receptor isoforms are coexpressed in these neurons. They does not excite breathing. Only $5HT_{1A}$ and $5HT_4$ isotypes are able to counteract opioid-induced respiratory depression [1081].

194. A.k.a. PSD95.

195. These respiratory disorders can be caused by brainstem tumors or degeneration such as olivopontocerebellar hypoplasia, as well as opioid or barbiturate administration.

Table 7.98. Subtypes of serotonin receptors and their direct functions in the circulatory and ventilatory systems as well as via the central (CNS) and peripheral (PNS) nervous system (Source: Wikipedia). Among the 7 categories of 5HT receptors ($5HT_1$–$5HT_7$), 1 ionotropic serotonin receptor ($5HT_3$ transmitter-gated Na^+–K^+ ion channel), and 15 different metabotropic isoforms have been identified.

Subtype	Distribution	Function
$5HT_{1A}$	Blood vessel CNS	Vasoconstriction, blood pressure, inhibitory nervous signal, ventilation depression
$5HT_{1B}$	Blood vessel CNS	Vasoconstriction
$5HT_{1C}$	see $5HT_{2C}$	
$5HT_{1D}$	Blood vessel CNS	Vasoconstriction
$5HT_{1E}$	Blood vessel CNS	
$5HT_{1F}$	Blood vessel CNS	Vasoconstriction
$5HT_{2A}$	Blood vessel CNS, PNS Platelet, SMC	Vasoconstriction
$5HT_{2B}$	Blood vessel CNS, PNS Platelet, SMC	Vasoconstriction, excitatory neural input
$5HT_{2C}$	Blood vessel CNS, PNS Platelet, SMC	Vasoconstriction
$5HT_3$	CNS, PNS	
$5HT_4$	CNS, PNS	Ventilation stimulation
$5HT_{5A}$	CNS	
$5HT_{5B}$	absent in humans	
$5HT_6$	CNS	
$5HT_7$	Blood vessel CNS	Vasoconstriction Ventilation stimulation

7.13.56.1 $5HT_{1A}$

In the central nervous system, Gi-coupled $5HT_{1A}$ is predominantly expressed in the hippocampus, lateral septum, and brain cortex, as well as in brainstem regions, such as nuclei raphé and hypoglossal nuclei. On presynaptic terminals, it precludes presynaptic transmitter release. At postsynaptic membranes, it activates via $G\beta\gamma$ subunit inwardly rectifying K^+ channels and inhibits Ca^{++} channels.

Table 7.99. Subfamilies of serotonin receptors in the nervous system (Sources: Wikipedia and [5]).

Type	Gαsubunit (preferential)	Mediator	Response
$5HT_1$	Gi/o	cAMP	Inhibition
$5HT_2$	Gq/11	IP_3, DAG	Excitation
$5HT_4$	Gs	cAMP	Excitation
$5HT_5$	Gi/o	cAMP	Inhibition
$5HT_6$	Gs	cAMP	Excitation
$5HT_7$	Gs	cAMP	Excitation

In the central respiratory network, serotonin lowers excitability of respiratory neurons [1081]. Activated $5HT_{1A}$ receptor enhances opioid-induced bradycardia, but partially compensates the decrease in vascular resistance and sympathetic activity, in addition to breathing restoration [1082].

7.13.56.2 $5HT_{1B}$

The Gi/o-coupled $5HT_{1B}$ receptor controls several signaling axes [1083]: (1) inhibition of adenylate cyclase; (2) inhibition of Rap1 guanine nucleotide-exchange factor (Rap1GAP); (3) activation of phospholipase-C and stimulation of Ca^{++} release; (4) phosphorylation of extracellular signal regulated kinase, phosphatidylinositol 3-kinase, P70 ribosomal S6 kinase (S6K), and protein kinase-B; (5) activation of Ca^{++}-dependent K^+ channels; and (6) stimulation of nitric oxide production.

It is primarily expressed in the central nervous system, both pre- and postsynaptically (on serotoninergic axon terminals), where it regulates the release of various neurotransmitters (glutamate, dopamine, GABA, acetylcholine, and ^{N}acetyl aspartyl glutamate), including serotonin itself (e.g., terminal autoreceptors of serotoninergic neurons) [1083]. It also lodges in sympathetic nerves of the gastrointestinal tract.

The $5HT_{1B}$ receptor is also synthesized in smooth muscle cells of cerebral arteries and other blood vessels. It impedes noradrenaline release in the vena cava as well as trigeminal ganglion-stimulated plasma extravasation in rodents [1083]. It mediates the vasoconstriction of rat caudal arteries.

The $5HT_{1B}$ receptor can homo- and heterodimerize with $5HT_{1A}$ and $5HT_{1D}$. It can interact with GSK3β, glutathione S-transferase-κ1, and non-metastatic cell subunit NME2 of nucleoside diphosphate kinase. It is internalized and interacts with various regulators.

7.13.56.3 $5HT_{2A}$

Most activated GPCRs associate with arrestin. A single GPCR can then trigger multiple signaling events according to the presence or absence of β-arrestin and the type of binding agonist. Different patterns of signal transduction and functional selectivity of $5HT_{2A}$ receptor [196] is determined by β-arrestin. The absence of β-arrestin-2 abrogates many effects of serotonin from $5HT_{2A}$ receptor, but has no or little effect on the same signaling pathways with other kinds of agonists.

7.13.56.4 $5HT_{2B}$

In the central nervous system, $5HT_{2B}$ receptor [197] participates in embryo- and fetogenesis. It is also involved in anxiety, migraine, schizophrenia, autism, and depression [1081].

7.13.56.5 $5HT_{2C}$

The $5HT_{2C}$ receptor (previously called $5HT_{1C}$) is coupled to $G\alpha_q$, hence to phospholipase-Cβ. It is exclusively expressed in the central nervous system. It is the single GPCR that undergoes adenosine-to-inosine RNA editing at 5 positions, hence generating multiple functional variants with different G-protein-coupling properties and transfer modes [1084]. [198]

The $5HT_{2C}$ receptor intervenes in various behavioral and physiological processes, such as regulation of mood, nociception, motor behavior, endocrine secretion, thermoregulation, modulation of appetite, cerebrospinal fluid production by choroid plexi, and control of exchanges between the central nervous system and cerebrospinal fluid [1084]. It contributes to the control of mesocorticolimbic and nigrostriatal dopaminergic systems, as it inhibits dopamine release.

The $5HT_{2C}$ receptor has both constitutive and inducible activity [1084]. Desensitization of its agonist-dependent activation results from receptor phosphorylation by G-protein-coupled receptor kinase GRK2 that is followed by recruitment of β-arrestins, which uncouples $5HT_{2C}$ from G protein and promotes its internalization into endosomes. Endocytosis not only ensures receptor desensitization, but also allows receptor dephosphorylation and recycling to the plasma membrane for resensitization.

Owing to transactivation of the small GTPase RhoA, $5HT_{2C}$ receptor can activate phospholipase-D. It can also stimulate phospholipase-A2 and extracellular signal-regulated kinases ERK1 and ERK2. It also interacts with phosphatase and tensin homolog.

---

196. In the central nervous system, $5HT_{2A}$, preferentially coupled to Gq localizes to the neocortex, entorhinal and pyriform cortex, claustrum, caudate nucleus, nucleus accumbens, olfactory tubercle, hippocampus, and cerebellum [1081].

197. The $5HT_{2B}$ receptor resides in the neocortex, cerebellum, dorsal hypothalamus, and medial amygdala [1081].

198. Pre-mRNA editing generates multiple functional variants with decayed constitutive activity, agonist affinity, and PLC activation potency, as well as additional G13-coupling capacity, but modified mediator selectivity and trafficking.

7.13.56.6 $5HT_4$

In the central nervous system, Gs-coupled $5HT_4$ receptor [199] contributes to the control of transmitter secretion and cognition. It localizes also to the myocardium, adrenal glands, digestive tract (from esophagus to colon), and bladder.

Splice variants have distinct C-terminus ($5HT_{4A}$–$5HT_{4H}$) [1081]. Types $5HT_{4A}$ and $5HT_{4B}$ couple also to G13 and Gi/o family subunit, respectively. Unlike $5HT_{1A}$, $5HT_{4A}$ receptor further reduces the decrease in opioid-induced vascular resistance, heart rate, and sympathetic activity, but partially rescues breathing [1082].

7.13.56.7 $5HT_7$

In the central nervous system, $5HT_7$ receptor [200] intervenes in the circadian rhythm. It also resides in smooth muscle cells of blood vessels and digestive tract.

Alternate splicing of intron 2 generate 4 isoforms of the $5HT_7$ receptor ($5HT_{7A}$–$5HT_{7D}$) that differ in their C-termini, but neither in their tissue distribution, nor pharmacological and signaling features [1081]. Gs-Coupled $5HT_7$ receptor activates adenylate cyclase AC5 as well as Ca^{++}–calmodulin-regulated AC1 and AC8 isoforms. In addition, it can trigger the mitogen-activated protein kinase cascade to activate extracellular signal-regulated protein kinases ERK1 and ERK2 [1081].

7.13.57 Somatostatin Receptors

Somatostatin [201] regulates the endocrine system and influences neurotransmission and cell proliferation. Two active forms of somatostatin derive from alternative cleavage of a single preproprotein: $somatostatin_{14}$ [202] and $somatostatin_{28}$ [5]. [203] It acts on 5 subtypes of the somatostatin receptor (Sst_1–Sst_5) to inhibit the secretion of many hormones (Table 7.100).

Agonist stimulation can dissociate P85 regulatory subunit of phosphoinositide 3-kinase (PI3K) and somatostatin receptor Sst_2, thereby reducing the cell survival PI3K–PKB pathway [835]. In the absence of agonist, constitutive association of P85 subunit with Sst_2 enhances PI3K activity.

199. The HT_4 receptor lodges in basal ganglia, hippocampus, olfactory tubercle, and limbic structures [1081].
200. The $5HT_7$ receptor is detected in the thalamus, hypothalamus, hippocampus, and cerebral cortex [1081].
201. A.k.a. somatotropin release-inhibiting factor (SRIF) and growth hormone-inhibiting hormone (GHIH). It is produced by neuroendocrine neurons of the periventricular nucleus of the hypothalamus.
202. A.k.a. SRIF14 (with 14 amino acids).
203. A.k.a. SRIF28 (with 28 amino acids).

Table 7.100. Somatostatin receptors and their main G-protein subunit transducers (Source: [5]).

Type	Other name	Main transducer
Sst_1	SstR1	Gi
Sst_2	SstR2	Gi
Sst_3	SstR3	Gi
Sst_4	SstR4	Gi
Sst_5	SstR5	Gi

Table 7.101. Sphingosine 1-phosphate receptors and their main targeted G proteins (Source: [736]; SPC: sphingosylphosphorylcholine). Receptors $S1P_4$ and $S1P_5$ are observed in hematopoietic cells and neurons, whereas the others are ubiquitous.

Type	Main transducer	Potency order
$S1P_1$	Gi/o	S1P > SPC
$S1P_2$	Gs, Gq, G12/13	S1P > SPC
$S1P_3$	Gs, Gi/o, Gq	S1P > SPC
$S1P_4$	Gs, Gi/o, G12/13	S1P > SPC
$S1P_5$	Gi/o, G12/13	S1P > SPC

7.13.58 Sphingosine 1-Phosphate Receptors

Sphingosine 1-phosphate is a lipid growth factor (Vol. 2 – Chap. 3. Growth Factors) that acts via its specific G-protein-coupled receptors (Table 7.101). Sphingosylphosphorylcholine (SPC) also activates this receptor with equal, smaller, or greater potency than that of sphingosine 1-phosphate according to the receptor type. The S1P receptor family comprises 5 receptors ($S1P_1$–$S1P_5$) that regulate cell survival, proliferation, and migration.

Sphingosine 1-phosphate receptors are members of the *endothelial differentiation gene* (EDG) family with lysophosphatidic acid receptors. Receptors $S1P_1$ to $S1P_3$ are widespread, whereas $S1P_4$ and $S1P_5$ reside in cells of the nervous and immune systems. Endocytosis of S1P GPCRs can trigger Gi-mediated signaling similarly to that initiated from the plasma membrane [11].

7.13.58.1 $S1P_1$ Receptor

The major $S1P_1$ receptor is observed on vascular endothelial and smooth muscle cells. It couples to Gi protein. It regulates endothelium-dependent vasorelaxation, smooth muscle contraction, and blood vessel maturation during embryo- and fetogenesis [1085]. It also controls the transfer of immunocytes.

Lipidic S1P ligand is predominantly stored and released by erythrocytes. It is present at high concentrations (100 nmol–1 μmol) in the blood and lymph. A large S1P fraction is bound to plasma lipoproteins and albumin.

Table 7.102. Signaling from $S1P_2$ receptor (Source: [1085]; ACase: adenylate cyclase; ERK: extracellular signal-regulated protein kinase; JNK: Jun N-terminal kinase; MAPK: mitogen-activated protein kinase; PI3K: phosphatidylinositol 3-kinase; PKB: protein kinase-B; PLC(D): phospholipase-C(D); PTen: phosphatase and tensin homolog deleted on chromosome 10; RacGAp: Rac GTPase-activating protein).

Gα type	Pathways
Gi	ERK PI3K–PKB–Rac
Gq	PLC–Ca^{++}
G12/13	Rho–RacGAP–Rac (−), Rho–PKB (−) ACase–cAMP PTen
	PLD; JNK; P38MAPK

Upon S1P binding, $S1P_1$ receptor primes egress of lymphocytes from secondary lymphoid organs toward the blood (low S1P concentration) or lymph (high S1P concentration) [1086].

In endothelial cells, transient exposure of exogenous $S1P_1$ ligand[204] triggers ligand-bound $S1P_1$ endocytosis and long-lasting signaling with inhibition of adenylate cyclase and increased ERK phosphorylation that causes augmented cell migration, whereas calcium response is abrogated [1086]. On the other hand, S1P agonist does not provoke persistent signaling from internalized $S1P_1$ receptors.

7.13.58.2 $S1P_2$ Receptor

Widespread $S1P_2$ couples to Gi, Gq, and G12/13 types of Gα subunit of heterotrimeric G-proteins to regulate several mediators, such as Rho GTPase, PTen phosphatase, ERK kinase, and VE-cadherin (Table 7.102). However, it preferentially activates G12/13 family subunits.

Activated $S1P_2$ on endothelial cells provokes disruption of VE-cadherin-based adherens junctions, hence increasing endothelial paracellular permeability [1085]. Because $S1P_2$ activates RhoA, it impedes Rac activity. The $S1P_2$ receptor actually inhibits small GTPase Rac, as it stimulates a GTPase-activating protein for Rac GTPase, thereby impeding cell migration. In addition, $S1P_2$ stimulates cAMP synthesis via G13 (but not Gs). Once coupled to G12/13 family subunits, it can also activate monomeric Rho GTPase.

The $S1P_2$ receptor is able to activate the Gi–Ras–MAPK axis to regulate cell proliferation and survival. Phosphorylation of ERK kinases, in turn, stimulates Jun and Fos transcription factors. In addition, $S1P_2$ activates other members of the MAPK

204. E.g., immunomodulator drug fingolimod.

Table 7.103. Expression of S1P receptor types in various cell types of the cardiovascular apparatus (Source: [1087]). Five known receptors bind to S1P ligand. Receptors $S1P_1$ to $S1P_3$ are widespread; $S1P_4$ and $S1P_5$ reside in blood cells. In cells of the cardiovascular apparatus, $S1P_1$ couples exclusively to Gi protein; $S1P_2$ and $S1P_3$ to subunits of the Gi, Gq, and G12/13 families; and $S1P_4$ and $S1P_5$ to those of the Gi and G12/13 families. The Gi subunit inhibits adenylate cyclase, whereas $G\beta\gamma$ can activate ion channels and kinases; Gq subunit targets phospholipase-C; and G12/13 subunit connects to RhoA guanine nucleotide-exchange factor. Because Gi protein prevents cAMP formation, it precludes PKA-mediated activation of Ca_V1 channel. The $G\beta\gamma$ dimer associated with Gi acts on Gi-regulated inward rectifier $K_{IR}3$ channel (K_{ACh} current) that contributes to resting membrane potential in human atriomyocytes.

Tissue	S1P Receptor expression
Cardiomyocytes	$S1P_1 \gg S1P_3 > S1P_2$
Cardiac fibroblasts	$S1P_3 \gg S1P_1 > S1P_2$
Aortic smooth muscle cells	$S1P_2 > S1P_3 \gg S1P_1$
Vascular endothelial cells	$S1P_1 > S1P_3 \gg S1P_2$

family — JNK and P38MAPK — that operate in response to cell stress, in particular for cytokine production. It also activates the Gq–PLC–Ca^{++} pathway as well as phospholipase-D.

7.13.58.3 S1P Receptors in the Cardiovascular System

Vascular endothelial and smooth muscle cells as well as cardiomyocytes and cardiac fibroblasts express $S1P_1$ to $S1P_3$ receptors, but with different expression patterns (Table 7.103). Because $S1P_1$ is the most abundant type in endothelial cells and cardiomyocytes, most S1P-primed responses occur via Gi-coupled $S1P_1$ receptor alone or in combination with Gi-, Gq-, or G12/13-coupled $S1P_3$ receptor. In vascular smooth muscle cells, $S1P_2$ has the highest density. Stimulated $S1P_1$ or $S1P_3$ activates Rac GTPase, whereas stimulated $S1P_2$ inhibits Rac GTPase. Cardiac fibroblasts express predominantly $S1P_3$ receptor. In addition, sphingosine kinase activity is higher in cardiac fibroblasts than in cardiomyocytes [1087].

Cardiomyocytes

In cardiomyocytes, $S1P_1$ receptor inhibits cAMP formation and antagonizes adrenergic receptor-mediated inotropy [1087]. The $S1P_1$ receptor localizes to caveolae in ventriculomyocytes. The $S1P_3$ receptor induces bradycardia. Both $S1P_2$ and $S1P_3$ receptors favor cardioprotection in ischemia–reperfusion injury. Receptors of S1P also participate in remodeling, proliferation, and differentiation of cardiac fibroblasts.

Endothelial Cells

In endothelial cells, caveolar domains of the plasma membrane are enriched in NOS3 and contain $S1P_1$ receptors. Lipid S1P activates NOS3 via: (1) Ca^{++}–calmodulin that relieves caveolin inhibition and (2) protein kinase-B that phosphorylates NOS3.

Sphingosine 1-phosphate enhances the endothelial barrier. In particular, it prevents the permeability increase caused by platelet-activating factor or bradykinin. Among $S1P_1$ to $S1P_3$ located on endothelial cells, only $S1P_1$ protects the endothelial barrier against the effect of platelet-activating factor and bradykinin in rat venules [1088].

Sphingosine 1-phosphate stabilizes newly formed vessels and antagonizes thrombin that disrupts the endothelial barrier via $S1P_1$ and $S1P_3$ receptors [1087]. It interacts with $S1P_1$ during angiogenesis and vasculature maturation [1089]. Gi-Coupled $S1P_1$ receptor regulates cell survival, proliferation, migration, and morphogenesis in response to S1P messenger. Activated $S1P_1$ also stimulates the assembly of cadherin complexes in endothelial cells. In endothelial cells, $S1P_1$ activates $\alpha_V\beta_3$- and β_1-integrins via GTPase Rho [1090]. Liganded $S1P_1$ primes the Gi–PI3K–PKB, PI3K–Rac, PLC–DAG–PKC and –IP_3–Ca^{++}, and Ras–ERK pathways. Protein kinase-B then binds to and phosphorylates $S1P_1$, hence regulating GTPase Rac and cortical actin assembly, lamellopodium formation, and chemotaxis [1091]. Activated $S1P_1$ favors cell survival and proliferation of vascular endothelial cells and stimulates nitric oxide synthase in endothelial cells via Gi and PI3K [1092]. Whereas $S1P_2$ inhibits Rac and growth factor-induced chemotaxis [1093], $S1P_3$ activates small GTPase Rho.

Smooth Muscle Cells

In vascular smooth muscle cells, S1P activates Rho GTPase and promotes myosin light-chain phosphorylation that provokes vasoconstriction mainly via $S1P_3$ receptor. On the other hand, S1P can promote the formation of nitric oxide in endothelial cells that, once released, relaxes smooth muscle cells.

7.13.58.4 S1P Receptors in the Immune System

Sphingosine 1-phosphate and lysophosphatidic acid are produced by mastocytes, platelets, and macrophages. Ligands S1P and LPA affect immunocyte survival, differentiation, proliferation, migration, receptor expression, and protein synthesis and secretion (Table 7.104).

The emigration of thymocytes from the thymus, transfer of lymphocytes between blood and secondary lymphoid organs (thus avoiding sequestration in lymph nodes as well as persistence in blood), and immigration of B lymphocytes into splenic follicles (white pulp of spleen) is assisted by S1P and its receptors [1094]. Sphingosine 1-phosphate in dendritic and T cells enhances the IgE production.

Table 7.104. $S1P_1/2$ receptors and their effects on immunocytes (Source: [1094]).

Receptor	Cells	Effects
$S1P_1$	Mast, B, and T cells, macrophage	Cell survival, chemotaxis (at low level)
	Dendritic and NK cells, eosinophil	Chemotaxis
$S1P_2$	Mastocyte	Migration inhibition
	Macrophage	Migration stimulation
	Dendritic cell, eosinophil	
$S1P_3$	Eosinophil, some dendritic and B cells	
$S1P_4$	B and T cells	Cell survival
	Dendritic and NK cells, macrophage	
$S1P_5$	Monocyte, NK cell	

In an autocrine manner, via interactions with different GPCRs, S1P enhances mastocyte migration and release of pro-inflammatory mediators in allergy. Whereas LPA promotes mastocyte development, S1P-bound $S1P_1$ increases mastocyte chemotaxis. In contrast, chemotaxis of mastocytes is prevented by IgE-upregulated S1P-bound $S1P_2$ receptor. In addition, $S1P_2$ augments IgE-mediated release of allergic mediators.

7.13.59 Tachykinin Receptors

Tachykinin receptors are activated by members of the tachykinin family of neuropeptide neurotransmitters: substance-P, neurokinin-A [205] and -B, [206] and neuropeptide-K, [207] and -γ. [208]

Three distinct tachykinin receptors have been identified (NK_1–NK_3, or NK1R–NK3R). Substance-P preferentially activates NK_1, neurokinin-A NK_2, and neurokinin-B NK_3 (Table 7.105). In the respiratory tract, both NK_1 and NK_2 receptors are

205. A.k.a. substance-K, neurokinin-α, and neuromedin-L.

206. A.k.a. neurokinin-β and neuromedin-K.

207. A.k.a. neurokinin-K.

208. The human tachykinin genes TAC1, TAC3 (mouse TAC2 gene), and TAC4 produce tachykinins, as they encode tachykinin precursors, the *preprotachykinins* that undergo proteolytic cleavage to form smaller peptides, and multiple splice variants that leads to different sets of peptides. Each preprotachykinin type is the product of one gene. The TAC1 gene encodes preprotachykinin-A (PPTKa; a.k.a. protachykinin-1 and tachykinin precursor-1), leads to formation by alternative splicing to neurokinins-A and -K and neuropeptide-γ. All TAC1 splice variants (α, β, and γ splice variants) manufacture substance-P, but only β and γ splice variants produce neurokinin-A, neuropeptide-K, and neuropeptide-γ. Neuropeptide-K and -γ are versions of neurokinin-A with a longer N-terminus. The genes TAC3 and TAC4 encode preprotachykinins-B (PPTKb) and -C (PPTKc) that produce neurokinin-B and hemokinin-1, respectively.

Table 7.105. Tachykinin receptors, main G-protein subunit transducers, and ligands (Source: [5]; NKa(b): neurokinin-A(B); SP: substance P).

Type	Main transducer	Potency order
NK_1	Gq/11	SP > NKa > NKb
NK_2	Gq/11	NKa > NKb ≫ SP
NK_3	Gq/11	NKb > NKa > SP

Table 7.106. Trace amine-associated receptors, main G-protein subunit transducers, and ligands (Source: [5]; PEA: phenylethylamine).

Type	Other names	Main Transducer	Potency order
TA_1	TAA_1, TAR1, TAAR1	Gs	Tyramine ≤ PEA > octopamine ∼ dopamine
TA_2	TAA_2, TAR2, TAAR2 TAAR3, GPR57, GPR58	Gs	PEA > tryptamine

detected in bronchial glands and vessels, as well as bronchial smooth muscle, but not in the respiratory epithelium [1095].[209]

7.13.60 Trace Amine Receptors

Trace amine receptors, or trace amine-associated receptors, (TA_1–TA_2) are bound by trace amines (Table 7.106). Trace amines are endogenous compounds that are structurally related to classical biogenic amines. They are produced in the nervous system. They colocalize in neurons in which these biogenic amines serves as neurohormones, neuromodulators, and/or neurotransmitters, such as monoamines that encompass catecholamines, which derive from phenylalanine and tyrosine[210] (adrenaline, dopamine, and noradrenaline), serotonin, histamine, and melatonin.

Trace amines include: (1) tyramine,[211] a monoamine derived from tyrosine;[212] (2) phenylethylamine (PEA), a natural monoamine alkaloid that is synthesized from

209. Tachykinin receptors can also be found in nerves (NK_1) and inflammatory cells (NK_2), such as T lymphocytes, macrophages, and mastocytes [1095].

210. Tyrosine is either generated from phenylalanine by phenylalanine hydroxylase or processed from ingested proteins.

211. A.k.a. 4-hydroxyphenethylamine and para-tyramine.

212. Tyramine releases catecholamines (adrenaline, dopamine, and noradrenaline). It is metabolized by monoamine oxidase. A large dietary intake of tyramine can cause the tyramine pressor response, with an increase in systolic blood pressure associated with vasoconstriction, but repeated exposure to tyramine reduces tyramine pressor response.

Table 7.107. Thyrotropin-releasing hormone receptors and their main G-protein subunit transducers (Source: [5]).

Type	Main transducer
TRH_1	Gq
TRH_2	Gq

phenylalanine; (3) tryptamine, another monoamine alkaloid;[213] (4) octopamine;[214] (5) 3-iodothyronamine that has negative inotropic and chronotropic effects [1097]; and (6) psychostimulant amphetamines[215] (α-methylphenethylamines).

Endogenous amphetamines, like all trace amines, bind to trace amine receptors, or trace amine-associated receptors. The TA_1 receptor localizes to the central nervous system as well as peripheral organs such as the kidney. It may be targeted by thyronamines, endogenous derivatives of thyroid hormones (decarboxylated and deiodinated metabolites).

7.13.61 Thyrotropin-Releasing Hormone Receptors

Thyrotropin-releasing hormone receptors (Table 7.107) are activated by the endogenous tripeptide thyrotropin-releasing hormone (TRH) that is produced by medial neurons of the hypothalamic paraventricular nucleus to stimulate the release of thyroid-stimulating hormone and prolactin by the adenohypophysis (endocrine anterior lobe of the pituitary gland).

7.13.62 Urotensin-2 Receptor

Urotensin-2 receptor[216] is a Gq/11-coupled receptor. In humans, its highest levels are detected in skeletal muscle and cerebral cortex and lower levels in kidney cortex and left ventricle [1098]. Urotensin-2 is synthesized from preprourotensin-2 isoforms that contain 124 and 139 amino acids. Urotensin-2 causes vasoconstriction in human arteries (e.g., coronary, mammary, and radial arteries) and veins (e.g.,

213. Tryptamine serves as a backbone for members of the tryptamine group. Tryptamine derivatives comprise neurotransmitter serotonin (5-hydroxytryptamine) and sleep–wake cycle regulator neurohormone melatonin (^{N}acetyl 5-methoxytryptamine). Serotonin is synthesized from tryptophan (α-carboxyltryptamine) successively by tryptophan hydroxylase and amino acid decarboxylase. Serotonin, in turn, can be converted to melatonin by ^{N}acetyltransferase and 5-hydroxyindole ^{O}methyltransferase.

214. Tyramine is hydroxylized to octopamine by dopamine β-hydroxylase to be subsequently packaged in synaptic vesicles with noradrenaline. In mammals, octopamine may release lipids from adipocytes. Whereas serotonin increases peristaltic movements in the digestive tract, octopamine suppresses them [1096].

215. Endogenous amphetamines are manufactured in the central and peripheral nervous system. They modulate the level of excitement and alertness.

216. A.k.a. hypocretin receptor; UTR, UTR2, UTs2R, or GPR14.

Table 7.108. Arginine vasopressin (AVP) and oxytocin (OxT) receptors, main G-protein subunit transducers, and order of potency of ligands (Source: [5]; CNS: central nervous system, CVS: cardiovascular system).

Type	Other names	Main transducer	Potency order	Location
V_{1A}	AVPR1a	Gq/11	AVP > OxT	CNS, CVS, kidney, liver
V_{1B}	AVPR1b	Gq/11	AVP > OxT	CNS
V_2	AVPR2	Gs	AVP > OxT	Kidney
OT	OXTR	Gq/11, Gi/o	AVP ≤ OxT	CNS, uterus (at birth), mammary gland

Table 7.109. Vasopressin and oxytocin receptors, mediators, and effects (Source: Wikipedia; ACTH: adrenocorticotropic hormone; vWF: von Willebrand factor). Cullin-5 has been originally named vasopressin-activated, calcium-mobilizing receptor VACM1. It reaches its highest levels in cardiac and skeletal tissues. It is specifically expressed in vascular endothelium and renal collecting tubules.

Type	Mediators	Effects
V_{1A}	Calcium	Vasoconstriction, gluconeogenesis, platelet aggregation, release of factor VIII and vWF,
V_{1B}	Calcium	ACTH secretion in response to stress
V_2	ACase–cAMP	Water reabsorption, release of vWF from endothelial cells
Cullin-5	Calcium	Calcium signaling

saphenous and umbilical veins) [1098]. On the other hand, urotensin-2 can trigger a potent vasodilation in human small muscular pulmonary and abdominal resistance arteries [1099].

7.13.63 Vasopressin and Oxytocin Receptors

Magnocellular neurosecretory cells of the paraventricular and supraoptic nuclei of the hypothalamus synthesize one of the neurohypophysial hormones – nonapeptides arginine vasopressin (AVP), [217] or antidiuretic hormone (ADH), and oxytocin (OT) – along with different neuropeptides or neuromodulators. Vasopressin and oxytocin receptors are activated by vasopressin and oxytocin. Vasopressin signals via 3 cognate receptors (Tables 7.108 and 7.109). Oxytocin receptor requires Mg^{++} ion and cholesterol.

Hormones oxytocin and vasopressin not only act on peripheral organs via the blood circulation, but are also released in the brain, where they influence social be-

217. Lysine vasopressin (LVP), or lypressin, plays the same role in pigs.

havior.[218] Vasopressin contributes to water reabsorption in the renal collecting ducts via aquaporin-2. Vasopressin also causes a moderate vasoconstriction. Neuropeptide oxytocin acts as a hormone with several peripheral actions and a neurotransmitter in the central nervous system released from centrally projecting oxytocin neurons. Oxytocin intervenes in parturition initiation and milk ejection during lactation, in addition to psychosocial behavior.

Oxytocin and vasopressin are stored in vesicles in and released by the neurohypophysis into the blood circulation.[219] Oxytocin is synthesized as an inactive precursor that includes its carrier neurophysin-1.[220] It is stored with neurophysin-1 in Herring bodies at axon terminals in the neurohypophysis. Vasopressin and its carrier neurophysin-2 also derives from the same precursor.[221]

Regulation of water transport across the epithelium of the collecting duct of the nephron enables a precise control of renal water excretion, thereby regulating the osmolality of body fluids. Water transport control is mainly achieved by vasopressin that binds to V_2 receptor in the basolateral plasma membrane of principal cells of collecting duct to cause an increase in intracellular cAMP and Ca^{++} concentrations. Vasopressin, indeed, favors via V_2 the transfer of aquaporin-2 to the apical plasma membrane, as it triggers phosphorylation by myosin light-chain kinase of the myosin regulatory light chain, as well as aquaporin-2 phosphorylation (Ser256, and Ser264, and S269) and dephosphorylation (Ser261) [1101]. Vasopressin primes activation of several kinases of the AGC family, such as cAMP-dependent protein kinase-A and calmodulin-dependent kinase-2 (via autophosphorylation at Thr286), but lowers phosphorylation of cyclin-dependent kinases and members of the mitogen-activated protein kinase family, such as JNK1 and JNK2 (Thr183 and Tyr185) and ERK1 and ERK2 (Thr183 and Tyr185; Thr203 and Tyr205). In normal circumstances, vasopressin attenuates activities of ERK1 and -2 and JNK1 and -2, but not P38MAPK. Aquaporin-2 is strongly phosphorylated (Ser261) by JNK, P38MAPK, and CDK5 and -9, as well as weakly by ERK kinases. It is phosphorylated (Ser256) by protein kinase-A.

---

218. In rats, vasopressin signals in the olfactory system for proper social recognition [1102]. A population of vasopressin neurons resides in the olfactory bulb. In humans, oxytocin participates in the processing of social stimuli [1100]. Oxytocin reduces activation in lateral and dorsal regions of the anterior amygdala in the presence of negative social stimuli, but enhances activity for positive social stimuli. Furthermore, oxytocin increases activity in the posterior amygdala and promotes the coupling of this subregion to the superior colliculi [1100]. Oxytocin also operates in behavioral and endocrine stress responses. It contributes to increase trust and reduce betrayal aversion.

219. The pituitary gland consists of adeno- (anterior pituitary) and neurohypophysis (posterior pituitary). The pituitary secretes hormones that regulate water homeostasis, blood pressure, development, and reproduction.

220. Neurophysin-1 and oxytocin result from the cleavage of a common precursor: preprooxyphysin, or preprooxytocin-neurophysin-1.

221. Neurophysins also act as neurohypophysial hormone. Neurophysin-2 stimulates prolactin secretion. It is generated from the same precursor as vasopressin: preproArg vasopressin–neurophysin-2.

Table 7.110. Vasoactive intestinal peptide (VIP) and pituitary adenylate cyclase-activating peptide (PACAP) receptors, main G-protein subunit transducers, and ligands (Source: [5]; GRF: growth hormone-releasing factor; PHI: peptide histidine isoleucineamide). Two – short and long – PACAP isoforms (PACAP27 and PACAP38) derive from proteolytic cleavage of a precursor (proPACAP(131–157) and proPACAP(131–168), respectively).

Type	Other names	Main transducer	Potency order
$VPAC_1$	VIP_1, VIPR1, PACAPR2	Gs	VIP, PACAP27 $\sim$ PACAP38 $>$ GRF $\gg$ $\gg$ PHI $\gg$ secretin
$VPAC_2$	VIP_2, VIPR2, PACAPR3	Gs	VIP, PACAP38 $>$ PACAP27 $>$ PHI $\gg$ GRF, secretin
PAC_1	PACAPR1, AdCyAP1R1	Gs	PACAP27, PACAP38 $\gg$ VIP $>$ PHI

7.13.64 Receptors for VIP and PACAP Peptides

Vasoactive intestinal peptide and pituitary adenylate cyclase-activating peptide receptors are activated by endogenous peptides vasoactive intestinal peptide (VIP), pituitary adenylate cyclase-activating peptides PACAP27 and PACAP38, peptide histidine isoleucineamide (PHI), peptide histidine methionine amide (PHM), peptide histidine valine (PHV), and growth hormone-releasing factor (GRF) [5]. Vasoactive intestinal peptide causes smooth muscle relaxation, exo- and endocrine secretion, and water and ion flux in respiratory and intestinal epithelia. Pituitary adenylate cyclase-activating polypeptides are neuropeptides that belong to the VIP–PACAP–glucagon–secretin family. They can operate as neurotransmitters and neuromodulators. In addition, they contribute to para- and autocrine regulation of some cell types.

The VIP–PACAP receptors constitute a family ($VPAC_1$, $VPAC_2$, and PAC_1; Table 7.110). The PAC_1 receptor is expressed in brain, adrenal medulla, pancreatic acini, uterus, and myenteric plexus [1106].

The $VPAC_1$ receptor is widely distributed, but predominantly detected in the lung, small intestine, thymus, and, within brain, in the cerebral cortex and hippocampus [1103]. (It is also found in the heart, kidney, liver, spleen, colon, prostate, testis, and placenta, as well as T lymphocytes [1104]). The $VPAC_1$ receptor is also activated by secretin.

The $VPAC_2$ receptor is also broadly distributed (central nervous system, skeletal and smooth muscles, heart, kidney, adipose tissue, pancreas,[222] stomach, and testis). In the central nervous system, $VPAC_2$ receptor is present in regions associated with neuroendocrine function, such as several hypothalamic nuclei. It is also detected in the pituitary gland and pancreatic islets [1103]. In the human respiratory tract, $VPAC_2$ is observed in tracheal and bronchial ciliated epithelial cells as well

222. It has been isolated from a mouse insulin-secreting β-cell line.

as mucous and serous cells of submucosal glands, bronchiolar epithelial cells, and alveolar macrophages, but not airway and vascular smooth muscle and endothelial cells [1105].

8

Receptor Protein Kinases

Catalytic receptors at the cell surface encompass: (1) particulate guanylate cyclases (Sect. 6.4); (2) intrinsic receptor tyrosine kinases that initiate a signaling cascade by autophosphorylation; (3) extrinsic receptor tyrosine kinases that cooperate with a required, distinct signaling mediator; (4) intrinsic heterodimeric receptor serine/threonine kinase; and (5) receptor phosphotyrosine phosphatases.

Functional versions of plasmalemmal catalytic receptors are usually dimers that contain several regions: (1) an extracellular ligand-binding segment; (2) a single transmembrane-spanning sequence; and (3) an intracellular functional domain with enzymatic activity.

Binding of peptidic ligands on these receptors can cause receptor dimerization. Once a catalytically silent member of the heterodimer is activated, the second constituent that lacks ligand-binding capacity becomes able to initiate a signaling via Tyr phosphorylation.

8.1 Receptor Tyrosine Pseudokinases

Pseudokinases are molecules that have a kinase-like domain that lacks at least one of 3 motifs in the catalytic domain required for catalytic activity. Therefore, they are catalytically inactive. Nevertheless, they contribute to the regulation of cellular processes, as they possess accessory sequences, such as catalytic, lipid-binding, GTPase-binding, RNA-binding, interproteic interaction, and intracellular sensor domains.

Among 518 protein kinases encoded by the human genome (kinome), 48 are pseudokinases [1107]. They are scattered among the distinct protein kinase families. Human pseudokinases are distributed in 7 groups (A–G) according to motifs lacking in the pseudokinase domain (Table 8.1).

Yet, certain kinases, the catalytic domain of which lacks canonical, conserved (among species), invariant catalytic residues, can be active, as the missing residue (e.g., Lys in VAIK motif) is substituted by the same residue type in another subdomain of the catalytic domain or a lacking motif of the catalytic domain is substituted

Table 8.1. Classification of pseudokinases (Source: [1107]; SgK: Sugen kinase). The 3 possible lacking motifs of the catalytic domain are: (1) DFG (Asp-Phe-Gly) motif, in which aspartic acid binds Mg^{++} that coordinates phosphates of ATP in the ATP-binding cleft; (2) HRD (His-Arg-Asp) motif, in which aspartic acid is the catalytic residue that functions as a base acceptor to achieve proton transfer; and (3) VAIK (Val-Ala-Ile-Lys) motif, in which the lysine residue interacts with phosphates of ATP to anchor and orient ATP. Several pseudokinases (isoforms of StRAd, Trb, and NRBP; SgK495, Slob, VRK3, and SuRTK106) have a simple structure that consists essentially of a single pseudokinase domain. Others (isoforms of ANP; CASK, PTK7, EPHa10, EPHb6, and TRRAP) are parts of large multidomain proteins that likely operate independently of their pseudokinase domain. The 4 Janus kinases (JAK1, JAK2, JAK3 and Tyk2) and GCN2 have a pseudokinase domain and a functional kinase domain within the same polypeptide.

Group	A	B	C	D	E	F	G
Missing Motif(s)	DFG	HRD	DFG HRD	VAIK	DFG, VAIK	HRD, VAIK	DFG, HRD, VAIK
Types	CASK	ANPRa	ILK	KSR1	ULK4	GCN2	EPHa10
	PTK7	ANPRb	IRAK2	KSR2	RSKL1		EPHb6
	SgK223	CyGd	MLKL		RSKL2		NRBP1
	SgK269	CyGf	SgK307				NRBP2
	SgK495	HER3	StRAdβ				SCYL1
	SuRTK106	HSER					SCYL2
	Trb1	JaK1					SCYL3
	Trb2	JaK2					SgK196
	Trb3	JaK3					SgK424
		PSKH2					Slob
		SgK071					StRAdα
		SgK396					TBCK
		TyK2					TrrAP
		VACAMKL					VRK3

by another catalytic sequence (e.g., WNK1 among the 4 isoforms of the WNK kinase, PRPK, haspin, in which lacking Mg^{++}-binding DFG motif is substituted by a DYTLS [Asp-Tyr-Thr-Leu-Ser] motif, titin, and RYK, a Wnt signaling associated receptor protein Tyr kinase).

Among receptor protein Tyr kinases, some are pseudokinases [1107]: The member HER3 of the epidermal growth factor receptor (EGF) family, [1] but HER3 retains a

1. Upon binding to its ligand such as neuregulin, HER3 heterodimerizes with the active receptor protein Tyr kinase HER2 that stimulates autophosphorylation (activation) of HER2 to initiate a signaling cascade.

weak phosphoryl transferase activity (Sect. 8.2.5.2), receptors EPHa10 and EPHb6, [2] and orphan PTK7 receptor.

8.2 Receptor Protein Tyrosine Kinases

Receptor protein tyrosine kinases (RTK) serve as information relays between extracellular messengers and intracellular cascades of chemical reactions that involve a set of signaling mediators, thereby integrating and coordinating cellular decisions. These transmembrane receptors contain a regulated protein Tyr kinase activity within their cytoplasmic domains. They sense environmental stimuli using their extracellular ligand-binding domain that, upon ligand binding, triggers receptor dimerization and kinase activation, i.e., autophosphorylation (activation) of their cytoplasmic domains. Their activated cytoplasmic kinase domains then recruit and phosphorylate (activate) signaling effectors that control diverse cellular processes from cell differentiation, proliferation, and migration, to death.

8.2.1 Classification

The superclass of plasmalemmal protein Tyr kinases includes prototypical growth factor receptors, such as vascular endothelial growth factor and angiopoietin receptors (Vols. 2 – Chap. 3. Growth Factors and 5 – Chap. 10. Vasculature Growth). Although growth factor receptors are typical examples of receptor Tyr kinases, other RTK family members couple with other ligand types. Hormone insulin also targets a receptor Tyr kinase.

The family of receptor Tyr kinases includes numerous types distributed in many classes according to the set of related ligands (Tables 8.2 and 8.3).

Class-1 RTKs are epidermal growth factor receptor (EGFR) [3] as well as other types of human epidermal growth factor receptors HER2 to HER4.

Class-2 RTKs comprise: (1) insulin receptor (InsR); (2) insulin receptor-related receptor (InsRR); (3) insulin-like growth factor-1 receptor (IGF1R); [4] and (4) receptor-like Ros1 kinase.

Class-3 RTKs are characterized by 5 to 7 immunoglobulin-like domains in their extracellular segment. The class 3 includes: (1) platelet-derived growth factor receptor, which are homo- or heterodimers of α and β chains; (2) colony-stimulating factor-1 receptor (CSF1R), or macrophage colony-stimulating factor

2. Receptor for erythropoietin-producing hepatocyte (EPH) receptor interactors (ephrins) EPHb6 can activate protein Tyr kinase ZAP70 upon ephrin binding by recruiting other signaling mediator.

3. A.k.a. erythroblastoma viral gene products-B ErbB1 and human epidermal growth factor receptor HER1.

4. Insulin-like growth factor-2 receptor (IGF2R) is the mannose 6-phosphate receptor. Mannose 6-phosphate indeed binds to 2 transmembrane proteins: cation-dependent (cdM6PR) and -independent (ciM6PR or IGF2R) mannose 6-phosphate receptors.

Table 8.2. Classes of receptor protein tyrosine kinases. Alias ROR is also used to designate ambiguously RAR-related orphan nuclear receptors (NR1f1–NR1f3), when the subtype is not specified (RORα to RORγ); the proposed $ROR_{(RTK)}$ alias for receptor Tyr kinase (RTK)-like orphan receptor is unequivoqual as well as ROR1 and ROR2 for its isoforms. The obsolete designation of ROR as neurotrophic Tyr receptor kinase-related protein (NTRKR) seems better. Because $ROR_{(RTK)}$ protein is a Wnt receptor, Its designation as orphan receptor should be updated. In the present text, alias WNRRTK is proposed (Wnt and neurotrophin receptor-related receptor Tyr kinase).

RTK class	Prototypical member
1	Epidermal growth factor receptor
2	Insulin receptor
3	Platelet-derived growth factor receptor
4	Fibroblast growth factor receptor
5	Vascular endothelial growth factor receptor
6	Hepatocyte growth factor receptor
7	Neurotrophin receptor kinase
8	Ephrin receptor
9	Adhesion-related kinase receptor (Axl)
10	Leukocyte Tyr kinase receptor (LTK)
11	Angiopoietin receptor (TIE)
12	RTK orphan receptor ($ROR_{(RTK)}$ [WNRRTK])
13	Discoidin domain receptor (DDR)
14	Rearranged during transfection receptor (Ret)
15	Receptor protein Tyr kinase PTK7 receptor (KLG)
16	Receptor-like Tyr kinase (RYK)
17	Muscle-specific kinase receptor (MuSK)

(mCSF);[5] (3) stem cell factor receptor (SCFR);[6] and (4) fetal liver kinase FLK2 on the surface of hematopoietic progenitor cells (especially multipotent progenitors and common lymphoid progenitors.[7]

Four types of fibroblast growth factor receptors (FGFR1–FGFR4) constitute the RTK class 4.[8] The 23 mammalian members of the FGF family interact with heparan sulfate proteoglycans to activate FGFR kinases. In addition to the 4 FGFRs, 2 additional receptors, FGFR5, or FGFR-like protein-1, and FGFR6 exist.

5. A.k.a. CD115 and McDonough feline sarcoma viral [v-fms] oncogene product homolog (Fms).

6. A.k.a. (cellular) kinase in tyrosine (c)KIT and CD117).

7. A.k.a. cytokine receptor cluster of differentiation CD135, Fms-like tyrosine kinase receptor FLT3, and stem cell tyrosine kinase STK1. It is important for lymphocyte development, but not for that of oher blood cells.

8. The Fgfr4 transcript, unlike other FGFR transcripts, is alternatively spliced to produce a single isoform.

Table 8.3. Receptor tyrosine kinases and their cognate ligands. Orphan receptors such as V-ros UR2 sarcoma avian (chick retrovirus) virus proto-oncogene homolog-1 (Ros1) that have a Tyr kinase activity do not bind known ligands (receptor-like Tyr kinases). Pseudokinases have kinase-like domain that lacks at least one of the conserved catalytic residues, hence they cannot phosphorylate substrates, but can regulate cellular processes.

RTK	Ligands
AATyK1–3	STK39
ALK	Pleiotrophin
Axl	GAS6
CSF1R	Macrophage colony-stimulating factor
CD135	FLT3 ligand (FLT3L)
DDR1/2	Collagen
EGFR	Epidermal growth factor, TGFα
EPH	Ephrins
FGFR	Fibroblast growth factor
HGFR	Hepatocyte growth factor
IGF1R	Insulin-like growth factor
IR	Insulin
LKT	Absence of known ligand (receptor-like Tyr kinase)
Mer	GAS6
MuSK	Agrin
PDGFR	Platelet-derived growth factor
PTK7	Absence of substrates (pseudokinase)
Ret	Glial-cell-line-derived neurotrophic factor, Neurturin, artemin, persephin
ROR [WNRRTK]	Wnt
Ros1	Absence of known ligand (receptor-like Tyr kinase)
RYK	Wnt
SCFR	Stem cell factor
Sky (Tyro3)	Gas, protein-S
TIE	Angiopoietins-1 and 4 (agonists) Angiopoietins-2 and 3 (antagonists)
VEGFR	Vascular endothelial growth factor

Class-5 RTKs correspond to: (1) vascular endothelial growth factor receptors VEGFR1,[9] VEGFR2,[10] and VEGFR3.[11]

Class-6 RTKs comprise (1) hepatocyte growth factor receptor (HGFR)[12] and (2) macrophage-stimulating-1 receptor (MSt1R), or Met-related Tyr kinase RON.

9. A.k.a. FLT1.
10. A.k.a. FLK1 and KDR.
11. A.k.a. FLT4.
12. A.k.a. Met.

Neurotrophin Tyr receptor kinases (NTRK1–NTRK3) constitute the class 7.[13] They regulate synaptic strength and adaptivity (plasticity) in the nervous system.

Class-8 RTKs are EPH receptors (EPHa1–EPHa8, EPHa10, and EPHb1–EPHb6), which constitute the largest class of the RTK superfamily. They are involved in the body's development.

Class-9 RTKs correspond to receptor Tyr kinases Axl, MerTK (or Mer), and TyrO3. The MerTK kinase is a proto-oncogene product. The Axl kinase binds to vitamin-K-dependent growth arrest-specific gene product GAS6 that is involved in cell proliferation. The TyrO3 kinase also possesses an oncogenic activity. In fact, GAS6 can stimulate the phosphorylation of Axl, MerTK, and TyrO3 kinases.

Class-10 RTKs are: (1) leukocyte Tyr kinase (LTK) with different isoforms that result from 2 alternatively spliced transcript variants and (2) anaplastic lymphoma kinase (ALK)[14] that is particularly involved in neurogenesis.

Class-11 RTKs are Ig- and EGF-like domain-containing TIE1 and endothelial-specific TIE2 receptors that are activated by angiopoietins. The latter promote angiogenesis. The TIE1 receptor is synthesized exclusively in endothelial cells, in which it heightens the production of cell adhesion molecules (E-selectin, ICAM1, and VCAM1) via P38MAPK enzyme.

Class-12 RTKs are receptor Tyr kinase-like orphan receptors ROR1 (a.k.a. neurotrophic Tyr receptor kinase-related protein NTRKR1 and, in the present text, Wnt and neurotrophin receptor-related receptor Tyr kinase isoform WNRRTK1) and ROR2 (NTRKR2 or WNRRTK2). The ROR1 receptor is expressed on the surface of B lymphocytes in acute lymphoblastic leukemia, but not in normal B lymphocytes and other blood cell. Signaling launched by Wnt can be mediated by the plasmalemmal ROR2 kinase.

Class-13 RTKs are discoidin domain-containing DDR1[15] and DDR2[16] receptors. Protein DDR1 is restricted to epithelial cells (particularly in the lung, kidney, gastrointestinal tract, and brain).

Class-14 RTKs are Rearranged during transfection (Ret) with naturally variants generated from alternatively spliced transcript leading to 3 different isoforms Ret9, Ret43, and Ret51.

Class-15 RTK is the protein Tyr kinase-7 (PTK7)[17] that represents a KLG-related transmembrane glycoprotein, and its 4 isoforms that result from transcript variants.

Class-16 RTK corresponds to receptor-like Tyr kinase (RYK) with its 2 isoforms generated by alternative splicing.

---

13. A.k.a. tropomyosin-receptor-kinases TrkA–TrkC, respectively. Most Slit and NTRK-like integral membrane proteins (SLITRK), but not SLITRK1, have a C-terminus homolog to neurotrophin receptors.

14. This alias followed by a integer denotes the type of activin receptor-like kinase, a mediator of the transforming growth factor-β pathway.

15. A.k.a. NEP and CD167a.

16. A.k.a. TyrO10 or NTRKR3.

17. A.k.a. colon carcinoma kinase CCK4.

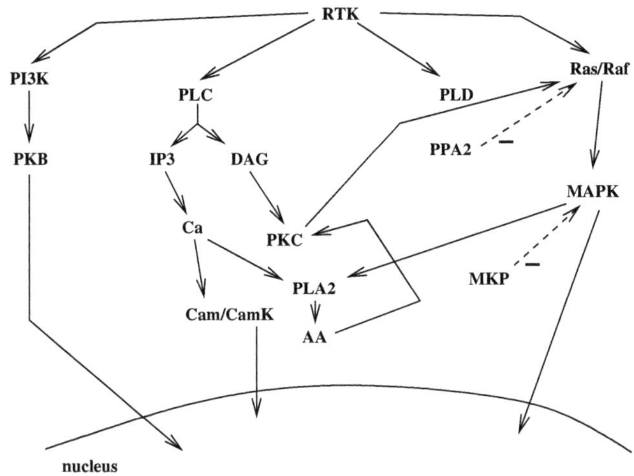

Figure 8.1. Receptor Tyr kinases intervene in cell growth, differentiation, and apoptosis, and metabolism. Main RTK effectors include phosphatidylinositol 3-kinase (PI3K), phospholipase-A2 (PLA2), -C (PLC), and -D (PLD), calmodulin-dependent kinases (CamK), protein kinase-C (PKC), protein phosphatase-2 (PP2), and Ras–Raf (MAP3K; MKP: mitogen-activated protein kinase phosphatase).

Class-17 RTK is skeletal muscle-specific receptor Tyr kinase that is required for the formation of the neuromuscular junction.

8.2.2 Functions

Functions of receptor Tyr kinases during embryo- and fetogenesis include embryonic polarity, formation of the germ layers, specification of cell types, and regulation of cell migration. Morphogenesis requires intercellular communications, during which receptor Tyr kinases, such as NTRKs and EPHs, intervene.[18]

Many cell types, such as cardiac precursors, melanocytes, germ cells, neurons, and neuronal growth cones migrate over long distances during development to reach their target sites. The interactions of RTKs with their soluble ligands (growth factors or membrane-associated affectors expressed on neighboring cells) elicit intracellular signaling cascades.

Proximal regulators of receptor Tyr kinase signaling are assigned to different pathways (Fig. 8.1). They can specifically act according to the receptor type or cell type. They include Raf kinase of the MAPK module (Vol. 4 – Chap. 5. Mitogen-Activated Protein Kinase Modules) and protein Ser/Thr phosphatases (Vol. 4 – Chap. 7. Cytosolic Protein Phosphatases).

18. These receptor Tyr kinases relay signals emitted from the family of nerve growth factors (nerve growth factor, brain-derived neurotrophic factor, and neurotrophin-3 and -4) and ephrins, respectively.

8.2.3 Structure

All receptor Tyr kinases share structural and functional features. Any RTK is composed of 3 regions: (1) a single transmembrane domain; (2) an extracellular sequence with a ligand-binding domain that varies in structure according to the RTK type; and (3) a cytoplasmic part made of a conserved bilobular Tyr kinase region that nests ATP in a cleft between 2 lobes.

Both extra- and intracellular domains are maintained in an auto-inhibited conformation that changes when a ligand binds. The extracellular N-terminus, which is responsible for the signaling specificity of ligand binding, is highly variable and displays a modular architecture based on combinations of protein domains, such as immunoglobulin, fibronectin-3, cadherin, discoidin, kringle, EGF, WIF, and plexin domains. Ligand binding induces dimerization or, in the case of pre-existing tethered dimer, detaches intramolecular tether to stabilize the active dimer. The active conformation activates the kinase domain.

8.2.4 Signaling

Most RTKs use the same type of signaling pathways, certain RTKs having some specificity. Protein kinases and phosphatases, ubiquitin ligases, and adaptor proteins primarily regulate RTK signaling that mainly relies on the Ras–MAPK, PI3K–PKB, PI3K–Rac, PLCγ–IP_3, and STAT pathways. Feedback loops are delayed controllers of RTK signaling (Table 8.4; Fig. 8.2).

8.2.4.1 Direct Signaling Effectors

Signaling proteins that bind to the intracellular domain of receptor Tyr kinases include RasGAP, phosphatidylinositol 3-kinase, phospholipase-Cγ, Tyr^P phosphatase PTPn6, [19] and adaptors. [20]

8.2.4.2 Activation of Immediate- and Delayed-Early Gene Products

Immediate-early genes are activated rapidly and transiently without protein synthesis. Many IEG products are transcription factors or other types of DNA-binding proteins. (Other products include cytoskeletal components, receptor subunits, and secreted proteins.) On the other hand, late response genes are activated upon the synthesis of early response gene products. The signaling time window is determined by the kind of negative feedback loops.

19. The PTPn6 phosphatase inhibits several messengers implicated in nitric oxide generation (Vol. 4 – Chap. 9. Other Major Signaling Mediators), such as JaK2, ERK1, and ERK2 kinases.

20. The recruitment of transmembrane proteins to the clathrin coat involves Adaptor protein complexes that interact directly with both clathrin and membrane receptors (Vol. 1 – Chap. 9. Intracellular Transport).

Table 8.4. Immediate-early (IEG) and delayed-early (DEG) genes targeted by receptor tyrosine kinases (Sources: [1108]; AP1: Activator protein-1; CREB: cAMP-responsive element-binding protein; MAPK: mitogen-activated protein kinase; NFE2: nuclear factor, erythroid-derived, type 2; NFκB: nuclear factor κ light-chain-enhancer of activated B cells). Immediate-early gene products stimulate a second type of transcription factor that targets delayed-early genes.

Gene	Time of peak mRNA expression (mn)	Transcription complex, signaling component
	Immediate-early genes	
FOSa	10–20	AP1
JUNA	10–20	AP1
EGR1	30	EGR
	Delayed-early genes	
KLF2	40–60	AP1, NFκB
KLF6	40–60	AP1, NFκB
JUNB	40–60	AP1
FOSL1	60–120	AP1
FOSL2	60	AP1
MAFF	120	AP1, NFE2
NAB2	120	EGR
EGR3	40–60	EGR
DUSP	30–120	MAPK
ATF3	40–60	AP1, NFκB
CREM	60–120	CREB
NFKBIA (IκBα)	30	NFκB
NFKBIE (IκBϵ)	90–120	NFκB

Sequential activation of a kinase cascade can lead to MAPK activation via double phosphorylation. Translocation of MAPKs to the nucleus enables phosphorylation of a first type of transcription factors that activate IEG transcription. Subsequently, IEG products stimulate a second type of transcription factors that targets DEG genes. The latter encode numerous proteins such as inhibitors that act at various pathway storeys (i.e., close pathway nodes or farther upstream): (1) transcriptional repressors (e.g., ATF3, KLF2, and NAB2) that impede the second type of transcription factors (short-distance inhibition for brief signaling window); (2) RNA-binding proteins (e.g., zinc finger protein homolog ZFP36)[21] that act at the IEG level (mid-distance inhibition); and (3) MAPK phosphatases (i.e., dual-specificity phosphatases DUSP) that inhibit MAPK (long-distance inhibition for longer signaling duration).

21. G0–G1 switch regulatory protein-24 (G0S24), growth factor-inducible nuclear protein NuP475, NUP475, and tristetraprolin.

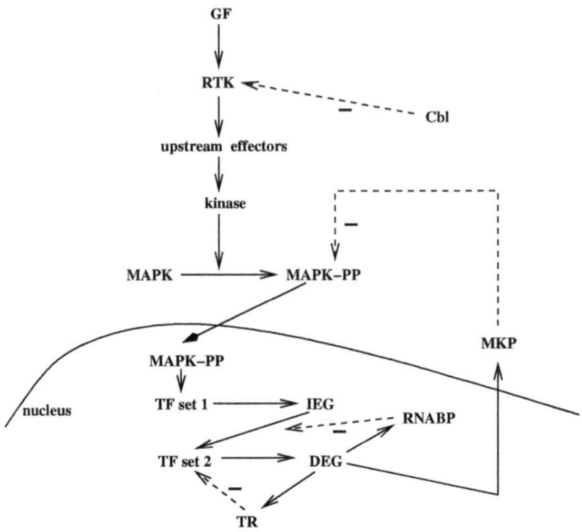

Figure 8.2. The type of feedback loop determines the duration of the activity of receptor tyrosine kinases (Sources: [1108]). Feedback circuits of kinase cascades use either pre-existing components, such as phosphatases and ubiquitin ligase CBL (that promotes receptor endocytosis and lysosomal degradation) or newly synthesized components (e.g., transcriptional negative feedback). Newly synthesized regulators can be produced by the mitogen-activated protein kinase (MAPK) module. Activated receptor Tyr kinase (RTK) by growth factor (GF) leads to the stimulation of the MAPK module via double phosphorylation ($MAPK^{PP}$; MAPK-PP in this figure). Translocation of $MAPK^{PP}$ to the nucleus phosphorylates set 1 of transcription factors (TF set 1) that activates transcription of immediate-early genes (IEG). Products of IEGs regulate a second set of transcription factors (TF set 2) that activate transcription of delayed-early genes (DEG). The latter encode inhibitors: (1) transcription repressors (TR) for short-duration signaling; (2) RNA-binding proteins (RNABP) for mid-duration signaling; and (3) mitogen-activated protein kinase phosphatases (MKP) for long-duration signaling.

8.2.4.3 Receptor Dimerization

Some non-autonomous RTKs cannot undergo direct activation by a ligand. Epidermal growth factor receptor subtypes HER2 and HER3 must heterodimerize with each other or with other RTKs to generate signaling [1108].

Ligand binding to 2 adjacent receptors forms an active dimer that causes a cross-phosphorylation. Activating autophosphorylation [22] leads to a phosphorylation cascade within the cytosol, using the activated cytosolic Tyr kinases.

22. Kinases often need to undergo autophosphorylation to trigger their phosphorylating activity. Dual-specificity Tyr phosphorylation-regulated kinases (DYRK) phosphorylate themselves on tyrosine (alias Tyr and Y), and subsequently phosphorylate their exogenous substrates (on Ser and Thr) [1109].

8.2.4.4 Phosphatase-Mediated Inhibition

Protein Tyr kinase signaling is counteracted by protein Tyr phosphatases, (Chap. 9 and Vol. 4 – Chap. 7. Cytosolic Protein Phosphatases) such as protein Tyr phosphatase receptor PTPRb. [23]

8.2.4.5 Ganglioside Interactions

Activated receptor Tyr kinases, such as receptors for epidermal, fibroblast, and platelet-derived growth factor, are inhibited by glycosphingolipids, such as ganglioside GM3 [1110]. [24] Ganglioside GM3 binds to and inhibits epidermal growth factor receptor. It also interacts with fibroblast growth factor receptor and prevents receptor endocytosis.

8.2.5 Growth Factor Receptors

Signaling cascades start with the binding of the signaling molecule with its plasmalemmal receptor. Growth factors (Vol. 2 – Chap. 3. Growth Factors) are major ligands of receptor Tyr kinases [1111]. Growth factor binding initiates interactions with GRB2 and SHC adaptors as well as guanine nucleotide-exchange SOS factor.

Signaling components are often pre-assembled into complexes. Scaffold proteins help to maintain these complexes and can contribute to specificity in various signaling systems.

8.2.5.1 Inhibitors of Growth Factor Receptor Signaling

Sproutys (Spry1–Spry4) are palmitoylated phosphoproteins ($Spry_P^P$) that translocate to the plasma membrane upon growth factor stimulation. They operate as negative feedback regulators of multiple receptor Tyr kinases, such as receptors of epidermal, fibroblast, and hepatocyte growth factor. [25] All Sprouty isoforms impede the ERK pathway. They can also homo- and heterodimerize to enhance inhibition. All 4 members of the Sprouty family associate with caveolin-1 [1112]. Hence, they preclude the activation by growth factors of mitogen-activated protein kinases.

Sprouty-1 and -2 not only hinder the activation of ERK1 and ERK2 by FGF and EGF, but also the signaling by hepatocyte, insulin, nerve, neurotrophic, platelet-derived, and vascular endothelial growth factor, as well as that initiated from stem cell growth factor receptor [1112].

On the other hand, Sprouty-2 can also promote EGFR signaling, as it sequesters the ubiquitin ligase Casitas B-lineage lymphoma protein (CBL). Human Spry2 localizes to microtubules in unstimulated cells. Fibroblast growth factor FGF10 that

23. A.k.a. vascular endothelial Tyr^P phosphatase (VEPTP).

24. Plasmalemmal glycosphingolipids are implicated in nanodomains with signal transducers, tetraspanins, growth factor receptors, and integrins.

25. Sproutys were first known as antagonists of FGF signaling. Protein FGF provokes production of sproutys. Activation of ERK1 and ERK2 by FGF is prevented by sproutys.

promotes bronchus morphogenesis stimulates Spry2 synthesis and phosphorylation [1113]. Sprouty-2 then interacts with some components of the FGFR signalosome, such as fibroblast growth factor receptor substrates FRS2 and FRS3 [26] and Raf kinase (MAP3K). On the other hand, it links to protein Tyr phosphatase-2 and GAP1 GTPase-activating protein.

Sprouty-3 phosphorylation may cause its binding to SH2 domain-containing proteins. Sprouty-4 impedes the formation of active Ras^{GTP} that triggers the MAPK pathway.

In addition, PTen phosphatase lowers EGFR activity, as it promotes ubiquitination by CBL Ub ligase and degradation of activated receptors [1114].

8.2.5.2 Human Epidermal Growth Factor Receptor Family

The family of human epidermal growth factor receptors (HER) [27] is activated by members of the epidermal growth factor superfamily, i.e., EGF, amphiregulin, β cellulin, epigen, epiregulin, heparin-binding EGF-like growth factor, neuregulins (NRg1–NRg4), and transforming growth factor-α (Vol. 2 – Chap. 3. Growth Factors).

These receptors have a tyrosine kinase activity that stimulates multiple signal transduction pathways. Intracellular signal transduction transmitted by the epidermal growth factor is governed by phosphorylation of protein kinases and their substrates. Signaling actually begins with activation of its receptor that associates with its cluster of intracellular signaling initiators. Signaling then extends via a cascade of activated kinases that phosphorylate a large number of substrates. The set of intracellular kinases includes those that prevent a sustained EGF signaling (negative feedback), as some phosphorylated sites of EGFR and components of mitogen-activated protein kinase modules (e.g., Raf and MAP2K2) impede with a given delay the activity of the EGF pathway. Among the HER family members, HER2 acts as a partner for the other members of the family, but it is not activated by a cognate ligand.

Activation duration depends on the type of ligands as well as the HER subtype. In mammary cells, stimulation by HER3- and HER4-specific ligand neuregulin-1 causes a prolonged ERK activation responsible for cell differentiation, whereas EGF provokes a transient ERK activation that supports cell proliferation.

Neuregulin-1 operates via HER2 to HER4 receptor. In ventriculomyocytes, the Nrg1–HER signaling is a paracrine regulator during cardiogenesis [28] as well as in the adult heart, [29] where it intervenes in sarcomere structure, [30] myocardial perfor-

26. A.k.a. Suc1-associated neurotrophic factor target SNT2.

27. A.k.a. erythroblastoma viral gene products-B (ErbB).

28. In the fetal heart, Nrg1 is produced in the endocardial endothelium; HER2 and HER4 receptors in cardiomyocytes; HER3 receptor in mesenchymal cells of the endocardial cushions that form cardiac valves.

29. In the adult heart, endocardial and myocardial microvascular endothelium express Nrg1; cardiomyocytes HER2 and HER4.

30. The ERK1/2 pathway is involved in the regulation of sarcomere formation, organization, and stability.

Table 8.5. Potential EGFR phosphorylation sites (Source: [1116]). Many sites (number N) of the intracellular domain can undergo phosphorylation–dephosphorylation cycles. Each phosphorylated site can attach at least one partner, hence possessing 3 possible states (free, phosphorylated, effector-bound). The EGFR receptors must form dimers to become active. Therefore, the number of potential EGFR states that results from combinations of all target site states is theoretically equal to $3^{2\times N}$, i.e., $O[10^6]$ for $N = 7$ (K_d: EGFR–DOK2 dissociation constant; Abl: Abelson kinase; DOK2: downstream of Tyr kinase-related docking protein; GRB: growth factor receptor-bound protein; PTPn: protein Tyr phosphatase non-receptor; SHC: Src homology-2 domain-containing-transforming adaptor; Src: sarcoma-associated kinase).

Phosphorylation sites	Partners (no mapping, except for DOK2)
Tyr845	Adaptors CRK, CBL, NCK, GRB2, SHC
Tyr992	Kinases Abl, Src
Tyr1045	Phosphatases PTPn6, PTPn11
Tyr1068	Phospholipase-Cγ
Tyr1086	DOK2 ($K_d = 132$ nmol)
Tyr1148	DOK2 ($K_d = 117$ nmol)
Tyr1173	PI3K
	RasGAP
	STAT3/5
	EGFR

mance, and cardiomyocyte survival, growth, and proliferation. Neuregulin-1 signals via ERK1 and ERK2, the PI3K–PKB axis, FAK, and NOS3. Agent Nrg1β1 rapidly (in mn) enhances NO production using the PI3K–PKB axis that phosphorylates NOS3 [1115]. Synthesized NO triggers the cGMP–PKG pathway that phosphorylates phospholamban without targeting other PKG substrates (e.g., Ca_V1) due to signaling compartmentation.

Epidermal Growth Factor Receptor

Epidermal growth factor receptor is implicated in hyperproliferative diseases as well as proliferation, migration, and metastasis of tumor cells.

Ligand binding to the extracellular domain of the receptor causes its dimerization [31] and subsequent autophosphorylation of 2 receptors in several Tyr residues. Phosphorylated tyrosines provide docking sites for effectors and regulators (Table 8.5), such as SHC, GRB2, and CBL, to activate intracellular signaling cascades and ubiquitination (Table 8.6).

Therefore, signal propagation via EGFR (as well as other RTKs) results not only on kinase activity, but also its scaffolding and docking function. Phosphorylation of

31. Four distinct structural states can be identified: (1) an autoinhibited, tethered monomer; (2) an extended monomer; (3) an inactive, symmetric dimer; and (4) an active, asymmetric dimer. Ligand binding to the tethered monomer releases the inhibitory link and allows the formation of an extended monomer, necessary to generate a stable dimer.

Table 8.6. Steps of EGFR signaling. The cell must differentiate between Son of sevenless (SOS) recruited to the membrane via growth factor receptor-bound protein (GRB2) bound to Src homology-2 domain-containing transforming protein (SHC) attached to EGFR and SOS recruited via GRB2 bound to EGFR directly.

Step 1	EGF–EGFR Binding
Step 2	EGFR Dimerization
Step 3	EGFR Phosphorylation
	Signaling cascade example 1
Step 4	EGFR–GRB2 Binding
Step 5	GRB2–SOS Binding
	Signaling cascade example 2
Step 4	EGFR–SHC Binding
Step 5	SHC Phosphorylation
Step 6	SHCP–GRB2 Binding
Step 7	GRB2–SOS Binding
	Signaling cascade example 3
Step 4	EGFR–DOK2 Binding
Step 5	DOK Phosphorylation (either Tyr276 and Tyr304 or Tyr351)
Step 6	Association with RasGAP or NCK

Tyr residues of the receptor cytoplasmic tail allows the recruitment of some adaptors. Subsequently, these adaptors recruit signaling effectors, such as guanine nucleotide-exchange SOS factor (via GRB2) and PI3K enzyme (via GAB1). The latter phosphorylates membrane phospholipids, thereby creating docking sites for PDK1 and PKB kinases.

In addition, ubiquitin ligase CBL, a major GRB2 partner, can then target EGFR. The EGFR receptor is then several times mono- or polyubiquitinated for lysosomal degradation. Proteins, such as hepatocyte growth factor-regulated Tyr kinase substrate (HRS) or signal-transducing adaptor molecules STAM1 and STAM2 recruit ubiquitinated EGFR to the ESCRT complex. On the other hand, cytoplasmic histone deacetylase deacetylates α-tubulin, thereby delaying the EGFR transfer, hence degradation.

Structural and functional diversity of the surrounding membrane modulates the receptor function. Lipids can actually operate as allosteric regulators of EGFR function. The lipid composition of the adjoining environment does not influence EGF binding to EGFR, but regulates the activation of the kinase domain [1117]. Ganglioside GM3 [32] prevents the autophosphorylation of the EGFR kinase domain in response to ligand binding. [33]

32. Gangliosides are glycosphingolipids with a head group that contains a neuraminic acid. Specific gangliosides influence the activity of receptor Tyr kinases, such as EGF, FGF, NTRK1, PDGF, and insulin receptors [1117].

33. In vitro, a 3-component lipid mixture with unsaturated phosphatidylcholine, sphingomyelin, and cholesterol in molar ratios that phase separate into coexisting liquid-disordered

EGFR-Regulated Proteins

Proteins regulated by EGFR, i.e., initiators, activators, and terminators of the EGF signaling, can be grouped into various functional protein classes [33]: (1) kinases (from EGFR to kinases of the MAPK module, e.g., ERK1, ERK2, and P38MAPK; Vol. 4 – Chap. 5. Mitogen-Activated Protein Kinase Modules) and phosphatases; (2) actin-binding proteins and small GTPases; (3) transcription factors and coregulators (Table 8.7); [34] and (4) enzymes of ubiquitination.

Phosphorylation of Tyr residues of EGFR enables signal transduction; $EGFR^{Tyr^P}$ engages various signaling pathways, using Tyr^P-binding adaptors (e.g., growth factor receptor-bound protein GRB2 and SH2 domain-containing protein SHC) or enzymes.

EGFR Activation and Deactivation

Widespread EGFR is activated by epidermal growth factor, heparin-binding epidermal growth factor-like growth factor (HBEGF), [35] transforming growth factor-α, amphiregulin (Areg), β-cellulin (BtC), epiregulin (Ereg or Epr), epigen (Epgn), and cryptic (Table 8.8). [36]

Indirect EGFR phosphorylation (activation) occurs during hyperosmolarity and oxidative and mechanical stresses, due to possible inactivation of phosphatases that antagonize the intrinsic receptor kinase activity. The EGFR receptor can also be stimulated by chemokines, cell adhesion molecules, and G-protein-coupled receptors.

---

and -ordered domains prevents EGFR autophosphorylation in the absence of EGF, but, upon EGF binding, allows receptor dimerization and activation. A cholesterol-poor uniform liquid-disordered phase enables EGFR autophosphorylation, whereas cholesterol-rich coexisting liquid-disordered and liquid-ordered medium promotes EGFR auto-inhibition [1117]. When GM3 is added to this artificial lipidic environment, EGFR autophosphorylation is impeded, but not ligand binding. Removal of neuraminic acid from GM3 by neuraminidase and mutation of a single membrane proximal lysine residue rescue EGFR autophosphorylation [1117].

34. Signaling from EGF involves members of the Activator protein-1 complex, such as JunD, FRA2, ATF2, and ATF7. They are activated by phosphorylations (e.g., Ser308 and Ser320 of FRA2 and Thr69 and Thr71 for AFT2).

35. Heparin-binding EGF-like growth factor is synthesized as a type-1 transmembrane precursor (proHBEGF). At the cell surface, proHBEGF is a juxtacrine growth factor that signals to neighboring cells via an intercellular contact. The ectodomain of proHBEGF ($_N$HBEGF or simply HBEGF) is cleaved at the juxtamembrane domain by an adamlysin and shed into the extracellular space, where it can serve as an N-terminal soluble ligand of EGFR that can indirectly transactivate HER2, HER3, and HER4 heterodimerized with EGFR (HER1). On the other hand, the plasma membrane-associated remnant, i.e., the C-terminal fragment ($_C$HBEGF), which consists of the transmembrane and cytoplasmic domains, can translocate from the plasma membrane to the nucleus. It can operate as a signaling mediator and transcriptional regulator. Lysophosphatidic acid and vasoactive peptides that bind to G-protein-coupled receptors stimulate metallopeptidases that cleave the transmembrane precursor of HBEGF (proHBEGF) into a soluble EGFR ligand, thereby transactivating EGFR.

36. Except Cryptic, these EGFR ligands are synthesized as membrane precursors that undergo proteolysis to become active.

Table 8.7. Transcriptional regulators activated by EGF signaling pathways (Source: [33]; BCL2: apoptosis regulator B-cell leukemia–lymphoma-2; CASP: cut alternatively spliced; CDP: CCAAT displacement protein; WBRS9: Williams-Beuren syndrome chromosomal region 9 protein). Members of the Jun family, components of the early-response transcription factor Activating protein-1, are activated via a double phosphorylation by Jun N-terminal kinases. Activator protein-1 is a dimer composed of combinations of members of the Jun, Fos, and ATF families of transcription factors. The dimer composition of AP1 determines the specificity of the DNA-binding site, hence the expression of a distinct subset of genes. CCAAT-box-binding transcription factors of the nuclear factor-1 family (NF1a–NF1c and NF1x, encoded by 4 different genes) tether their recognition sequence as homo- or heterodimers. Alternatively spliced variants of NF1b exist (NF1b1–NF1b3).

Type	Alias
Activating transcription factors	ATF2, ATF7
AT Hook-containing transcription factor-1	AHCTF1 (TMBS62)
BCL2-associated transcription factor	BclAF1
Bromodomain adjacent to zinc finger domain-1B	BAZ1b
Cut-like homeobox-1	Cux1 (CutL1, CASP, CDP)
ETS domain-containing repressor factor	ERF
Fos-related antigen-2	FRA2
Juunana	JunD
Leucine zipper protein-1	LuZP1
Myocardin-related transcription factors	MRTFa, MRTFb
Mitochondrial tumor suppressor gene product	MtSG1
Myc Proto-oncogene product	Myc
Nuclear factor-1B/1C	NF1b, NF1c
PHD Finger protein-2	PHF2
RB1-Inducible coiled-coil protein-1	RB1CC1
Ring finger protein-4	RnF4
SMAD4-Interacting transcriptional coactivator	SMIF
Sperm-associated antigen-9	SpAG9
Signal transducer and activator of transcription (nuclear and cytoplasm)	STAT5
General transcription factor-3C1	GTF3c1 (TF3cα)

Ligand binding to the epidermal growth factor receptor and its subsequent dimerization induces receptor autophosphorylation. In turn, EGFR autophosphorylation on multiple Tyr residues initiates the transmission of EGF signal inside the cell.[37] Stimulation by EGF can lead to: (1) rapid, transient activation of extracellular signal-regulated kinases ERK1 and ERK2 as well as (2) delayed, prolonged activation of P38MAPK.

Growth factor receptor-bound protein GRB2 recruited by phosphorylated receptors recruits guanine nucleotide-exchange SOS factor (Son of sevenless) to activate

37. Target Tyr residues comprise Tyr1069, Tyr1092, Tyr1110, Tyr1138, Tyr1172, and Tyr1197 [33].

Ras GTPase and, then, the Raf–MAP2K–ERK module as well as the PI3K–PDK1–PKB axis.

In addition, GRB2 attracts CBL ubiquitin ligase [38] that enables EGFR sorting to endosomes. Ubiquitinated EGFR is recognized by ubiquitin- and NEDD8-binding proteins of the clathrin coat. It thus recruits EGFR pathway substrate EPS15, members of the EPS15-interacting protein family (epsin1–epsin3), and hepatocyte growth factor-regulated Tyr kinase substrate (HRS). Ubiquitinated EGFR is then targeted to lysosomes for degradation. In addition, recruitment of adaptors SH3 domain-containing kinase-binding protein SH3KBP1 [39] and endophilins leads to EGFR endocytosis. Endocytosis and proteolytic degradation of EGFR after ligand binding regulate EGF signaling (signaling termination by negative feedback). In addition, P38MAPK is required for destruction of ligand-stimulated CBL-bound EGFR [1119]. On the other hand, deubiquitinating enzymes can prevent lysosomal degradation and support EGFR recycling to the plasma membrane.

In endosomes, ubiquitin ligases Itch and NEDD4 ubiquitinate ubiquitin-binding proteins such as HRS to promote their dissociation from EGFR.

Therefore, once activated, EGFR stimulates numerous signaling cascades and is internalized. Yet, EGFR can be recycled to the cell surface. The endocytosis–exocytosis cycle is controlled by the regulators SNARE proteins and clathrin-associated AP2 Adaptor protein complex (Vol. 1 – Chap. 9. Intracellular Transport).

Spatiotemporal Control

The HER set consists of 4 receptors and 11 ligands that bind to more than one receptor, but the ligandless receptor HER2 operates as an amplifier and the pseudokinase HER3 cannot signal alone. Numerous proteic interactions that constitute the HER network ensure signal specificity and outcome diversity. The HER network actually receives diverse inputs and reliably integrates them to generate a specific output. In addition, redundant modules achieve output reproducibility.

Temporal Dynamics of EGF Signaling and Feedback Loops

Once the mitogen activated protein kinase module is activated, some microRNAs such as miR101 that prevents erroneous, signal-independent protein synthesis, are destroyed to initiate gene expression.

On the other hand, Ser and Thr phosphorylation (Ser991, Thr693, and Thr993) often precludes signaling (large negative feedback loop). Moreover, EGFR phosphorylation is regulated by Tyr phosphatases that terminate signaling, such as PTPRj.

38. Ubiquitin ligase CBL is phosphorylated by EGFR and ubiquitinates this receptor. Ubiquitin ligase CBL also modifies EGFR with ubiquitin-like molecule NEDD8 (neural precursor cell expressed developmentally downregulated protein-8). Neddylation enhances subsequent ubiquitination [1118]. A set of clathrin coat-associated binders of ubiquitin also bind NEDD8, but they undergo ubiquitination, not neddylation.

39. A.k.a. 85-kDa CBL-interacting protein CIN85 and human SRC family kinase-binding protein HSB1.

The retrocontrol not only targets EGFR, but also its effectors. The canonical Raf–MAP2K–ERK cascade undergoes phosphorylation (e.g., Ser186 on aRaf and Ser23 on MAP2K2 [33]) by ERK1 and ERK2 kinases (short negative feedback loop).

Spatial Regulation

Phosphorylation allows translocation of signaling mediators, such as kinases and transcription regulators, into the nucleus. For example, signal transducer and activator of transcription STAT5 is phosphorylated (Tyr694), then dimerizes and localizes to the nucleus.

Early and Late Regulation

Regulation of signaling from EGFR relies on feedforward and negative and positive feedback loops. Feedforward and feedback are 2 distinct control strategies. Feedforward monitors the system's environment and applies appropriate compensatory signals. It actually anticipates the relation between a physiological apparatus and its environment to determine a course of action. The feedforward controller achieves a correction of a disturbance before the controlled variable is affected in a pre-defined way without adjustment of the load reaction. [40] The associated mathematical model is characterized by limits, rhythms, mechanisms, and patterns. A feedforward controller is often combined with a feedback controller to optimize the performance. A feedback controller enables an adjustment of the output in response to the load reaction. Feedback monitors, indeed, the output to adjust, i.e., to carry out a compensatory change in the input without instability, but with an efficient correction of noise.

A *coherent feedforward* is characterized by the cooperation between 2 mediators, such as the binding of both RB and Myc to the E2F promoter, the latter acting directly (weak response) or via RB (strong response). This mechanism attenuates noise. An *incoherent feedforward* results from an activation of a mediator that is later inhibited by another effector. For example, EGFR rapidly activates Fos, but later represses Fos by activating the RNA-binding ZFP36 protein. This process enables a transient response.

A *positive feedback* transforms a transient signal into a self-sustained loop. A *negative feedback* enables signal termination. A reciprocal negative regulation causes *bistability*. A delayed negative feedback resulting from the combination of a slower and faster component provokes oscillations that can stabilize a composite loop.

Controllers organize the HER network into immediate and late controllers, which determine the duration, amplitude, and frequency of signals, thereby enabling a specific response to an external stimulus.

40. E.g., anticipatory regulation of the cardiac output in advance of a scheduled physical exercise and hematopoiesis with its irreversible cell commitments.

Early Regulation The early signaling phase (seconds to minutes) relies on pre-existing signaling components, using protein translocations and post-translational modifications, but not on time-consuming synthesis of new proteins. In addition, the ligand–receptor complex is internalized with 2 possible routes: recycling or degradation in lysosomes. The latter represents a negative feedback loop. Yet, in both routes, endocytosed receptors remain active and continue to signal.

Ubiquitination enables receptor internalization and degradation, hence determining the signaling duration. Conversely, deubiquitination that can happen in multivesicular endosomes by ubiquitin isopeptidases STAM-binding protein (STAMBP)[41] and USP8 [1120].[42]

Immediate regulation of receptor signaling also depends on phosphorylations and other protein modifications as well as a very rapid clearance of microRNAs [1120].

Protein Tyr phosphatases dephosphorylates EGFR and its effectors. Cytosolic PTPn1 that localizes to the endoplasmic reticulum may operate on EGFR or ESCRT0 components such as signal transducing adaptor molecule STAM2 in multivesicular endosomes close to the endoplasmic reticulum [1120]. Ubiquitous cytosolic phosphatase PTPn2 and receptor phosphatases, ubiquitous PTPRj and PTPRk in keratinocytes, may also dephosphorylate EGFR. In early endosomes, KIF16b of the kinesin-3 family tethers to PTPn21 and the resulting complex fosters microtubule-dependent motility and EGFR recycling [1120].

Transphosphorylation of EGFR modulates its activity and transfer. Receptor EGFR can be phosphorylated on Tyr (activation) or Ser and Thr (mostly inactivation) residues by kinase Src and P38MAPK, respectively.

Secondary phosphorylation (or backward phosphorylation) corresponds to phosphorylation of upstream effectors of a signaling cascade. In particular, protein kinase-B phosphorylates cRaf (Ser259) to recruit 14-3-3 protein [1120].[43] Enzyme cRAF is also phosphorylated by Src, P21-activated kinase (activation), protein kinase-A and -C, and extracellular signal-regulated kinases (negative feedback [desensitization]).

Rapid turnover due to degradation of some miRs, the so-called immediately downregulated miRs (IDmiR) can result from EGFR signaling, especially via ERK enzymes, to relieve the inhibition of these miRs. The latter prevent the activity of transcription factors encoded by immediate early genes, such as Fos and early growth response protein EGR1. This derepression is an important aspect of both the transcription-independent, early and late regulation of EGF signaling.

Late Regulation Late regulation (time scale of hour) comprises newly synthesized adaptors, transcriptional and translational repressors, phosphatases of mitogen-activated protein kinase modules [1120].

41. A.k.a. associated molecule with the SH3 domain of STAM (AMSH).

42. A.k.a. ubiquitin isopeptidase-Y (UBPY).

43. Upon EGFR activation, the cRaf inhibitory site Ser259, which is phosphorylated in resting conditions, is dephosphorylated by 2 phosphatases, PP1 and PP2. On the other hand, phosphorylation of the cRaf stimulatory site Ser338 by P21-activated kinase enhances cRaf activity and affinity for MAP2K, and promotes the translocation of cRAF to mitochondria [1120].

Once active, immediate-early gene products foster the transcription of delayed-early genes, which exert a negative feedback on immediate early genes (via inhibitor of DNA binding ID2 and the repressor cAMP-dependent transcription factor ATF3). In addition, IEG as well as DEG products stimulate secondary response genes (SRG) that encode actin-binding proteins, cell adhesion molecules (cadherins and integrins) and transcription factors [1120].

The transcriptional response to EGF can be decomposed into 3 phases [1120]. Phase 1 is an initial wave with a peak at approximately 45 mn from stimulation that is associated with immediate-early gene products, removal of negative elongation factor (NELF) from promoters, and nucleosome destabilization that enables access to promoters of transcriptional activators and recruitment of RNA polymerase-2. Phase 2 is a second wave from 45 to 120 minutes after EGF-mediated excitation linked to delayed-early gene products with positive and negative regulators. Phase 3 is a third wave of secondary response gene products that starts about 1 hour after process initialization and maximal activity in the 4–5 h range.

Autocrine Regulation

Autocrine regulation yields an extracellular positive feedback that converts a transient stimulus into a sustained signal. Autocrine regulators can stimulate the migration of mammary epithelial cells more efficiently than paracrine growth factors [1120].[44]

HER Receptor Family

The family of epidermal growth factor receptor-related protein Tyr kinases consists of 4 members that are called erythroblastoma viral gene products-B (ErbB1–ErbB4) or, more conveniently, human epidermal growth factor receptor (HER1–HER4; Table 8.8).

Many extracellular ligands bind to epidermal growth factor receptors. Whereas neuregulin NRg1 is produced in the endocardium, HER2 and HER4 are expressed in the myocardium [1122]. The HER receptors are hierarchically configured into redundant modules to generate an output despite disturbances. Heterodimerization and phosphorylation of HER family members trigger signaling.

HER1

The order of magnitude of the number of epidermal growth factor HER1 receptors produced by a cell ranges from $O[10^4]$ to $O[10^6]$ receptors (i.e., from 50×10^3 to either less than 500×10^3 or up to 1000×10^3 according to the explored cell) [1123]. Inactive HER1 is monomeric at a physiological density. Yet, they can form transient,

44. Colony-stimulating factor CSF1 produced by mammary epithelial cells promotes the synthesis of EGF by macrophages; the latter, in turn, fosters the expression of CSF1 (positive feedback).

Table 8.8. Family of human epidermal growth factor receptor-related protein tyrosine kinases and their ligands (Source: [1121]; AReg: amphiregulin; BtC: β-cellulin; Epgn: epigen; EReg: epiregulin; HBEGF: heparin-binding epidermal growth factor; NRg: neuroregulin: TGF: transforming growth factor). Except Cryptic and EGF, EGFR ligands are synthesized as membrane precursors that undergo proteolysis to become active. In particular, transmembrane precursor and juxtacrine growth factor proHBEGF is cleaved at the juxtamembrane domain to generate the soluble HBEGF ligand.

Type	Ligands
HER1	EGF, HBEGF, AReg, BtC, EReg, TGFα
HER2	No
HER3	NRg1/2
HER4	NRg1–NRg4, HBEGF, BtC, EReg

unstable aggregates at large concentrations. Hence, inactive dimers differs from the stable, activated dimers formed after ligand stimulation.

In cells that possess 50,000 to 200,000 receptors, HER1 exists only as a monomer in the absence of ligand stimulation. Liganded HER1 receptor subunits then dimerize to fulfill their role. On the other hand, when the receptor number is greater than 500,000, up to 30% of HER1 are preformed dimers [1123]. Therefore, although inactive monomeric receptors undergo ligand-induced dimerization and activation: (1) inactive receptors are not necessarily monomeric; (2) HER1 is monomeric at relatively low concentration at the cell surface; and (3) HER1 forms ligand-independent dimers at higher density.

Polymerization of HER1 depends on its density and cell type. Membrane rafts and gangliosides may help HER1 to form preformed inactive dimers. However, gangliosides are neither necessary, nor sufficient for HER1 dimerization.

Ligand binding provokes a transition from a closed conformation of HER1 to an extended configuration that is able to dimerize via intermolecular interactions. Dimerization of HER1 can then activate the kinase domain. Messenger EGF causes the formation not only of HER1 dimers, but also oligomers (up to pentamers) that colocalize with clathrin-coated pits [1123].

Phosphorylated HER1 recruits signal transducers, such as growth factor receptor-bound adaptor GRB2 and SH2 domain-containing adaptor (SHC) that attract small Ras GTPase and activate the mitogen-activated protein kinase cascades. The HER1 receptor also recruits STAT5 factor.

The presenilin–γ-secretase complex controls the expression of the EGFR gene via the intracellular domain of the transmembrane β-amyloid precursor protein. The latter can bind to the EGFR promoter and prevent transcription of the EGFR gene [1124]. As it alters the EGFR pathway, the presenilin–γ-secretase complex can be considered as a tumor suppressor with respect to EGFR-mediated tumorigenesis.

HER2

The distribution of unliganded HER2 in cells that contain 3×10^5 to 10^6 receptors differs from that of HER1 [1123]. The HER2 receptor constitutes preformed homo-oligomers of 5 to 10 HER2 molecules in the plasma membrane. These constitutive HER2 clusters colocalize with caveolae. These homo-oligomers depolymerize upon EGF stimulation.

The HER2 receptor does not bind any known ligands of the EGF superfamily (EGFSF), but is a partner of the 3 other HER receptors, favoring their heterodimerization. It constitutively adopts an extended conformation that enables the formation of heterodimers. Upon transactivation, HER2 possesses the most potent kinase activity of the HER family, thereby improving the efficiency of signaling mediated by HER2-containing heterodimers.

Ligandless HER2 also serves as a coreceptor for GPCRs that can form RTK–GPCR complexes in cardiomyocytes, where HER2 can activate extracellular signal-regulated kinases ERK1 and -2 [1125].[45] Receptors HER2 and HER3 can be activated only via heterodimerization.

Estrogens bind to estrogen receptor and repress HER2 receptor. Expression of the Her2 gene is controlled by a competitive binding between Amplified in breast cancer AIB1 and Paired box-2 gene product Pax2 to Her2 regulatory element [1126]. Transcription repressor PAX2 hinders Her2 expression and inhibits cell proliferation, whereas estrogen receptor coactivator AIB1 favors Her2 expression.

HER3

The HER3 receptor binds neuregulins and heterodimerizes with HER2 receptor.[46] The HER3 heterodimer activates phosphatidylinositol 3-kinase. Intracellular kinase domain of HER3 is supposed to be a pseudokinase, as it lacks several catalytically important residues. However, HER3 retains phosphoryl-transferase kinase activity to trans-autophosphorylate its intracellular region, although it is weaker than EGFR [1127]. HER3 Kinase domain binds ATP with an affinity similar to that of known active kinases (coefficient $\sim$1.1 μmol).

HER4

The HER4 Receptor shares recognition and signaling features with HER1 receptor. It is activated by neuregulin-2 and -3, heparin-binding EGF-like growth factor,

45. Kinases ERK1 and ERK2 acts in recruitment of β-arrestin, activation of cytosolic Tyr kinase Src, and transactivation of receptor Tyr kinases.

46. The HER3 receptor is a type-1 receptor Tyr kinase similar in sequence to the epidermal growth factor receptor. The ERBB3 gene (v-erb-b2 erythroblastic leukemia viral oncogene homolog-3) encodes HER3 of the EGFR family. The HER3 receptor does not have a kinase domain, but does form heterodimers with other EGFR family members to get a kinase activity and activate signaling pathways, especially via PI3K. It mediates EGF responses in cells expressing both HER3 and EGFR.

and β-cellulin. Ligand binding causes cell division and differentiation. Proteolysis by ADAM17 sheddase[47] and presenilin-dependent γ-secretase releases cytoplasmic ($HER4^{ICD}$) and extracellular ($HER4^{ECD}$) fragments. Alternatively spliced variants have been detected.

The HER4 receptor is the single member of the HER family that encodes both nuclear localization and export signals. It intervenes during embryogenesis. It is also involved during breast development and lactation. Its activity in the breast is partly mediated by STAT5a signal transducer and activator of transcription [1128]. It potentiates the expression of STAT5a target genes.

In the nervous system, HER4 interacts with the 3 related PDZ domain-containing member of the membrane-associated guanylate kinase (MAGUK) group, the Disc large homologs DLg2,[48] DLg3,[49] and DLg4,[50] as well as β2-syntrophin [1129]. Its indirect binding partners include neuroligin, NMDA- and AMPA-type glutamate receptors, and potassium channels.

Several genes that encode neurotransmitter receptors and voltage-gated channels in postsynaptic cells are regulated by NRg1 via HER3 and HER4 during muscle and neural development [1129]. In muscles, NRg1 enhances transcription of nicotinic acetylcholine receptor δ and ϵ nAChR subunits and voltage-gated sodium channels. In neurons, NRg1 heightens expression of the ^{N}methyl ^{D}aspartate (NMDA)-type glutamate receptor-2C subunit in cerebellar granule cells and α7 nAChR subunit in the superior cervical ganglia.

EGFR Interactome

Even in the absence of stimulation by epidermal growth factor, EGFR can associate with diverse proteins to build an EGFR interactome. Binding of epidermal growth factor to its receptor leads not only to receptor dimerization, but also to assembly of proteic complexes associated with liganded EGFR receptor.

In the EGFR-associated interactome, cytoplasmic HDAC6 lysine deacetylase participates in the control of EGFR endocytosis and degradation. Histone deacetylase HDAC6 impedes EGFR endocytosis and degradation, as it controls the acetylation level of α-tubulin and, subsequently, receptor transfer along microtubules [1130]. After EGF stimulation, EGFR phosphorylates (inactivates) HDAC6 (Tyr570), thereby increasing acetylation of α-tubulin that enhances interactions between microtubules and nanomotors and accelerates endocytic transport along microtubules. A feedback loop relies on phosphorylation by EGFR of HDAC6.

Cytoprotection ensured by EGF is mediated via the Ras–ERK pathway and nestin that prevents caspase activation due to BCL2 phosphorylation by CDK5 kinase [1131]. Nestin is a type-6 intermediate filament protein that assembles other filament constituents, such as vimentin and α-internextin. It is mainly expressed in

47. A.k.a. TNFα-converting enzyme (TACE).
48. A.k.a. postsynaptic density adaptor PSD93 and chapsyn110.
49. A.k.a. synapse-associated protein SAP102.
50. A.k.a. PSD95 or SAP90.

muscle and neural progenitors. In mature cells, such as skeletal, cardiac, and vascular smooth muscle cells, nestin expression resumes during injuries.

EGFR Endocytosis: Signaling, Recycling, or Degradation

Signaling from the epidermal growth factor receptor that determines cell fate is governed by EGFR endocytosis and accessibility to signaling proteins. Epidermal growth factor receptor undergoes a rapid endocytosis via clathrin-coated pits by directly interacting with clathrin coat Adaptor protein complex AP2 or indirectly via other adaptor proteins such as EGFR pathway substrate EPS15 and ubiquitination.[51]

Endocytosis captures both transmembrane receptor and its extracellular ligand. Clathrin-coated pits give rise to cytoplasmic clathrin-coated vesicles, owing to GTPase dynamin, that then merge with early Rab5+ endosomes. Receptors then can rapidly recycle back to the plasma membrane owing to small GTPase Rab4 via Rab11a+ recycling endosomes or remain in endosomes that mature due to Rab7 into multivesicular bodies and late endosomes that fuse with lysosomes. The latter contains proteolytic enzymes that degrade vesicular content.[52]

The EGFR receptor remains bound to its ligand, phosphorylated, and active in endosomes until late endosomes. All the components of the extracellular signal-regulated kinase activation cascade (GRB2, SHC, SOS, Ras, Raf, MAP2K1, and MAP2K2) are detected in endosomes [11].

Resensitization of EGFR occurs when EGFR is activated by ligands that dissociate from EGFR in endosomes. Agonist TGFα is released from the receptor in acidic endosomes. Subsequently, recycled EGFR can be repetitively reactivated by TGFα [11].

The EGFR receptor is internalized using the clathrin-mediated route and recycled back to the plasma membrane more efficiently, hence degraded less potently, than receptors endocytosed by clathrin-independent, caveolin-mediated route [11].

Distinct features of endocytic sorting influence EGFR activity in signaling directed by protein kinase-B and extracellular signal-regulated protein kinases. Activation of EGFR can raise the pool of clathrin at the plasma membrane. This event depends on Src kinases, as these RTK effectors heighten the rate of EGFR–GRB2 cargo-specific clathrin-coated pit formation by phosphorylating clathrin heavy chain. Yet, activated EGFRs can also be recruited into pre-existing clathrin-coated pits.

Signaling can also regulate later stages in endocytosis. Activated EGFR slows endosome maturation and increases the number of intraluminal vesicles in multivesicular bodies [11].

Signaling by effector kinases affects endocytosis. Stress-induced P38MAPK stimulates the formation of complexes of small Rab5 GTPase and GDP-dissociation inhibitor to accelerate endocytosis. Activated P38MAPK also promotes internalization of unliganded EGFR [11]. Kinases ERK1 and ERK2 may regulate endosomal

51. Growth factor receptor-bound protein GRB2 is an adaptor that recruits ubiquitin ligase Casitas B lineage lymphoma (CBL) to EGFR receptor.

52. Ubiquitin-directed sorting into multivesicular bodies is mediated by endosomal sorting complexes required for transport (ESCRT0–ESCRT3).

maturation and cargo degradation. The aRaf isoform localizes to recycling endosomes, where it is required for the activity of adpribosylation ARF6 factor.

On the other hand, ubiquitin signals direct sorting into multivesicular endosomes (or bodies) destined for lysosomal compartment and degradation within the lysosome. Receptor sorting into intraluminal vesicles in endosomes that create multivesicular bodies, indeed, primes their delivery into the lumen of lysosomes, and hence terminates signaling. Entry into the degradative route is regulated. Ubiquitin acts as a cis-acting signal (i.e., acts from the ubiquinated molecule itself without requiring any other molecule) in phosphatidylinositol 3-phosphate-enriched endosomes, whereas endosomal sorting complex required for transport (ESCRT) operates as trans-acting factors for multivesicular endosome sorting.

Ubiquitination of EGFR results from a balance between ubiquitinating and deubiquitinating enzymes. Endosome-associated deubiquitinase Associated molecule with SH3 domain (AMSH)[53] reduces EGFR ubiquitination and subsequent degradation. However, ubiquitin-specific peptidase USP8 enhances EGFR degradation [11]. Ubiquitinases have different effects in endosomal sorting according to their specific activity.

Once internalized, EGFR is able to interact with protein Tyr phosphatase PTPn1 that localizes to the cytoplasmic face of the endoplasmic reticulum. Interaction between EGFR and PTP1b results from direct membrane contacts between membranes of multivesicular bodies and endoplasmic reticulum [1132]. Insulin receptor and HGFR also interact with PTPn1 after endocytosis.

Vesicular SNARE VAMP7[54] localizes to endosomes and lysosomes, as well as the trans-Golgi network. It is required for exocytosis from the Golgi body to the plasma membrane of substances such as Tspan27. Tetraspanin-27 (or CD82) controls EGFR localization in plasmalemmal nanodomains. Both VAMP7 and Tspan27 limits endocytosis of activated EGFRs [1133].[55]

EGFR-Mediated Signaling Pathways

The PI3K–PKB pathway is predominantly primed on phosphorylation of epidermal growth factor receptors, particularly HER3 receptor. Once phosphorylated by EGFR (HER1), HER3 associates with phosphatidylinositol 3-kinase and couples growth factor receptor Tyr kinases to phosphatidylinositol 3-kinase [1134].

Similarly, HER3 can undergo a cross-phosphorylation by HER2 receptor. The HER2 receptor is activated by EGFR family members, but does not bind any known ligand. It efficiently interacts with HER1 and HER3, whereas HER1 and HER3 do not form stable complexes [1135]. Effectors mitogen-activated protein kinase and

53. A.k.a. STAMBP.

54. A.k.a. synaptobrevin-like gene-1 protein and tetanus neurotoxin-insensitive vesicle-associated membrane protein (TIVAMP).

55. Depletion of VAMP7 reduces Tspan27 plasmalemmal density and augments the recruitment of Adaptor protein AP2 complex. The latter favors EGFR endocytosis, thereby reducing signaling by MAPK kinases.

Jun N-terminal kinase are durably inhibited when both HER1 and HER2 activities are prevented.

Several components of the EGFR pathways are substrates for NEDD4-like ubiquitin ligases, such as PTen phosphatase, CBL adaptor and Ub ligase, and Jun transcription factor. The endosomal membrane proteins — NEDD4 family-interacting proteins NDFIP1 and NDFIP2 — act as recruiters, activators, and adaptors of ubiquitin ligases of the NEDD4 family, as they bind to the 9 human members of this family [1136].[56] The NDFIP proteins link to EGFR and PTen [1136]. They then control ubiquitination and abundance of PTen, CBL, and Src family kinases. Agent NDFIP2, but not NDFIP1, is phosphorylated by Src and Lyn kinases. It then can act as a scaffold for Src kinase that phosphorylates NDFIP1 protein. Protein NDFIP1 is required for PKB activation. On the other hand, NDFIP1 hampers JNK signaling.

Epidermal growth factor generates a transient hRas activity at the plasma membrane as well as a sustained activity at the Golgi body [1137]. The acylation cycle transmits Ras activity between subcellular compartments.

Disregulation of EGFR signaling promotes cell proliferation and survival and, hence, cancer. A set of kinases modulates the generation, amplitude, propagation, and duration (desensitization or termination) of EGFR signals. Extracellular signal-regulated kinases ERK1 and ERK2 as well as signal transducer and activator of transcription STAT3 are activated in response to EGFR stimulation. Maximal activation of ERK1 and ERK2 can result from a relatively small number of EGFR receptors; STAT3 activation depends more on the number of EGFR phosphorylation events [1138]. Many (52) EGFR signal modulators have reproducible effects on ERK1, ERK2, and/or STAT3 activation [1138]. In particular, AMPK-related kinase hormonally upregulated HER2 (Neu)-associated kinase (HUNK) assists liganded EGFR activation.[57] Pyrimidine kinase UCK1 and deoxyguanosine kinase (DGK or DGuoK) may sustain aberrant EGFR signaling in vivo [1138].[58] On the other hand, MerTK receptor Tyr kinase reduces the accumulation of EGFR in response to EGF.

56. The NEDD4 family includes (NEDD4-1 and NEDD4-2, NEDD4-like ligase NEDL2 [or HECW2], Itch (or atrophin-1-interacting protein AIP4), WW domain-containing Ub protein ligases WWP1 and WWP2, SMAD ubiquitination regulatory factors SMURF1 and SMURF2, and Ub ligase-binding protein BUL1 homolog. These ligases are involved in cell signaling (NEDD4-1, Itch, SMURF1, and SMURF2) and protein transfer (NEDD4-1 and NEDD4-2) as well as immune response (Itch and NEDD4-1), blood pressure (NEDD4-2), and bone homeostasis (SMURF1). Adaptor NDFIP1 regulates Itch ligase in T lymphocytes to dampen signaling from T-cell receptors, thereby avoiding excessive inflammation.

57. Receptor protein Tyr kinase HER2 is also called Neu, as well as CD340, metastatic lymph node gene product MLN19, in addition to ErbB2 (for v-Erb-b2 erythroblastic leukemia viral oncogene homolog).

58. Deoxyguanosine kinase and thymidine kinase TK2 are the 2 rate-limiting enzymes involved in the first step of the mitochondrial salvage pathway of the purine and pyrimidine deoxyribonucleosides, respectively.

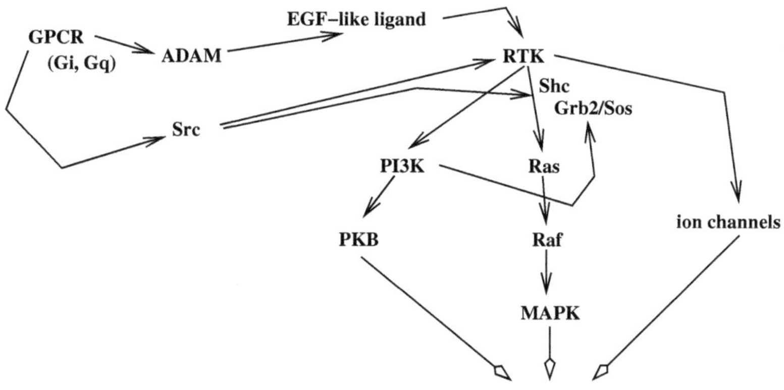

Figure 8.3. Inter-receptor crosstalk. Receptor tyrosine kinase (RTK) such as epidermal growth factor receptor can be a downstream signaling partner of G-protein-coupled receptors, especially for the activation of mitogen-activated protein kinases (MAPK) and phosphoinositide 3-kinase (PI3K) that lead to cell survival, proliferation, and migration. The crosstalk involves a disintegrin and metallopeptidase (ADAM) and EGF-like ligand. Such pathways are implicated in hyperproliferative diseases.

GPCR-Induced EGFR Transactivation

G-protein-coupled receptors can use EGFR as a downstream signaling partner to generate mitogenic signals [818].[59] Inter-receptor crosstalk involve various GPCR types. Subsequently, GPCR-stimulated EGFR activates effectors, such as mitogen-activated protein kinases, phosphoinositide 3-kinase[60] (and PKB), and modulates ion channel activity, for cell survival, proliferation, and migration, especially in cardiac hypertrophy, and cancer cells (Fig. 8.3).

Transactivation of EGFR induced by GPCRs depends on the cell type and signaling context. Crosstalk can involve Src kinases. Protein kinase-C and calcium ions can be required in cardiomyocytes, cardiac fibroblasts, and vascular smooth muscle cells.

---

59. Transduction of a mitogenic signal from the plasma membrane to the nucleus either involves receptor Tyr kinases, adaptors SHC and GRB2, and the Ras–MAPK pathway or Ras–MAPK activation by G-protein-coupled receptors and docking protein GAB1 [1139]. Receptor Tyr kinases can act as downstream mediators in GPCR mitogenic signaling owing to an intracellular crosstalk. Epidermal growth factor receptor is quickly phosphorylated upon stimulation of fibroblasts with GPCR ligands, such as endothelin-1, lysophosphatic acid, and thrombin.

60. Inhibition of PI3K does not affect GPCR-induced EGFR phosphorylation, but prevents MAPK stimulation in the presence of both GPCR ligands and low doses of EGF [1140]. Enzyme PI3K plays an important role upstream from Ras at GRB2 level in GPCR-mediated MAPK stimulation via both Gq and Gi proteins.

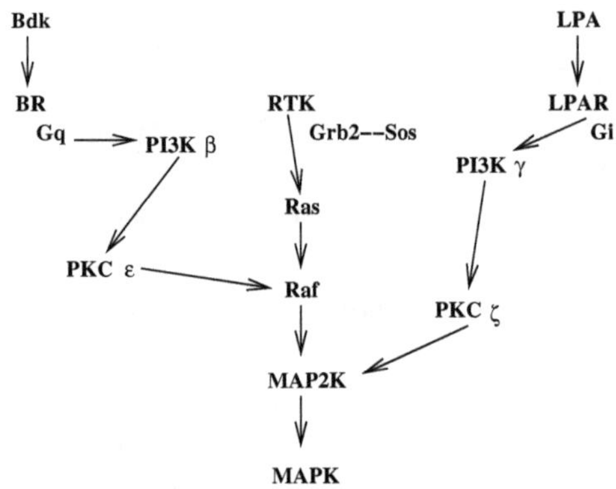

Figure 8.4. MAPK activation by bradykinin and lysophosphatidic acid via their GPCRs coupled to Gq (BR) and Gi (LPAR), respectively. Protein kinase-C isoforms (PKCϵ and PKCζ) are stimulated by corresponding phosphatidylinositol 3-kinase isoforms (PI3Kϵ and PI3Kγ) to activate MAPK pathway at Raf (MAP3K) or MAP2K level.

Activation of MAPKs by Gq-coupled bradykinin receptor depends on phosphoinositide 3-kinase-β and protein kinase-Cϵ [1141] (Fig. 8.4).[61] Activation of MAPKs by Gi-coupled lysophosphatidic acid receptor is regulated by PI3Kγ and PKCζ [1142] (Fig. 8.4).[62]

Transactivation of various types of RTKs can be induced by GPCRs, in addition to EGFR receptor. Angiotensin-2–induced phosphorylation of platelet-derived growth factor receptor-β and thrombin-stimulated phosphorylation of insulin-like growth factor-1 receptor is observed in smooth muscle cells. Angiotensin-2–induced activation of mitogen-activated protein kinase mainly results from calcium-dependent activation of receptor Tyr kinase via Gq-coupled ATR1 in vascular smooth muscle cells. Angiotensin-2 rapidly induces the phosphorylation of the epidermal growth factor receptor and its association with SHC and GRB2 [1143]. Angiotensin-2 induces Ca^{++}-dependent transactivation of EGFR which serves as a scaffold for pre-activated Src and adaptors, leading to MAPK activation in vascular smooth muscle cells.

Multisubstrate dockers of the family of IRS1-like proteins — GRB2-associated binder-1 (GAB1) and insulin receptor substrates IRS1 and IRS2 — regulate RTK-induced PI3K activity.[63]

61. Gq (Gα11) coupled to the bradykinin receptor increases the production of phosphatidylinositol 3-kinase-β and activates PI3Kβ, which then associates with protein kinase-Cϵ and induces the translocation of PKCϵ from the cytosol to the plasma membrane to transmit the mitogenic signal to the MAPK pathway.

62. Lysophosphatidic acid receptor coupled to the inhibitory G-protein subunit (Gi) activates MAPK using a Ras-dependent or -independent cascade associated with phosphatidylinositol 3-kinase-γ and protein kinase-Cζ. Lysophosphatidic acid also induces the activation of MAP2K, but not of Raf1 kinase.

63. Dockers GAB1, IRS1, also targeted by growth factors, and IRS2 are used by numerous signaling pathways. These substances form the family of IRS1-like multisubstrate docking

Zinc-dependent metallopeptidases of a disintegrin and metallopeptidase family (ADAM) are involved in EGFR transactivation. In cardiomyocytes, G-protein-coupled receptor ligands, such as angiotensin-2 or endothelin-1, activate adamlysin ADAM12, thereby causing the shedding of heparin-binding EGF-like growth factor (HBEGF) and subsequent transactivation of the epidermal growth factor receptor to provoke maldaptive cardiac hypertrophy resulting from pressure overload [1146].

Epidermal Growth Factor-like Domain-Containing Protein

Secreted epidermal growth factor-like domain-containing protein EGFL7 contributes to cell migration and angiogenesis. It is secreted by endothelial cells. The EGFL7 protein binds to transmembrane receptors of the Notch family (Notch1–Notch4) and acts as an antagonist of Notch signaling. It indeed competes with Jagged-1 and -2 [1147]. It thus reduces the proliferation and self-renewal of neural stem cells.

EGFR-Mediated Cell Proliferation

Angiotensin-2 and endothelin-1 increase the production of oxidants by NADPH oxidase, thereby activating the cytosolic Tyr kinase Src that phosphorylates EGFR. Activated EGFR triggers MPAK signaling, in particular extracellular signal-regulated kinases ERK1 and ERK2 to contribute to the proliferation of vascular smooth muscle cells in spontaneously hypertensive rats [1148].

Carcinogenic mineral fibers such as asbestos also elicits oxidative stress and cell proliferation. Asbestos causes activation of ERK4 (or ERK5) via an EGFR-independent, Src-dependent pathway and activation of ERK1 and ERK2 using the Src-mediated transactivation of EGFR to provoke the proliferation of lung epithelial cells [1149].

EGFR-Mediated Chemotaxis

Membrane-bound chemokine CX_3CL1 (a.k.a. fractalkine and neurotactin), the single member of the CX_3C family of chemoattractant cytokines, is produced by activated endothelial cells, smooth muscle cells, and macrophages, especially upon inflammatory stimuli, such as TNFα and Ifnγ. It can be cleaved from the cell surface in response to inflammatory stimuli by adamlysin ADAM17.[64] It can then act as a classical chemoattractant for leukocytes and smooth muscle cells that express

proteins. GAB1 is phosphorylated following activation of RTKs and cytokine receptor. GAB1 is a substrate for EGF and insulin receptors and a docking protein for PI3K, among others. Overexpression of GAB1 enhances cell growth [1144]. Stimulation of the hepatocyte growth factor receptor induces, via GAB1, mitosis, motility, and branching tubulogenesis of epithelial and endothelial cell lines in culture. Following stimulation of epithelial cells with HGF, GAB1 associates with phosphatidylinositol 3-kinase and Tyr phosphatase PTPn11 (SHP2) [1145].

64. A.k.a. TNFα-converting enzyme (TACE).

the G-protein-coupled receptor CX_3CR1 to provoke chemotaxis primed by CX_3CL1 chemokine.

In coronary artery smooth muscle cells, CX_3CL1 has anti-apoptotic and mitogenic effects [1150]. It causes phosphorylation of extracellular signal-regulated kinases and protein kinase-B via epidermal growth factor receptor upon the release of the soluble EGFR ligand epiregulin.

Migration–Proliferation Dichotomy

Growth factor receptors, such as EGFR, can trigger 2 mutually exclusive processes: cell migration and division. [65] These processes rely on different set of signaling mediators (phospholipase-Cγ1 and phosphatidylinositol 3-kinase for cell migration; the MAPK module and Src–STAT5b pathway for cell division).

Subunit $G\alpha_i$ and its cytosolic guanine nucleotide-exchange factor, girder of actin filament (girdin), [66] program EGFR signaling and orchestrate this dichotomy [1151]. Girdin interacts directly with EGFR (when its GEF function is intact). Subunit $G\alpha_i$ tethers to the girdin–EGFR complex via girdin to form the ternary $G\alpha_i$–girdin–EGFR complex, enhance EGFR autophosphorylation, and prolong EGFR localization to the plasma membrane. [67] Girdin promotes migratory signals such as phosphorylation of the EGFR sites implicated in the activation of PKB and PLCγ1 as well as phosphorylation (activation) of these effectors.

On the other hand, the alternatively spliced variant of girdin, a truncated protein that lacks the C-terminal GEF domain (girdin$^{\Delta CT}$), is a mitogenic factor [1151]. Agent girdin$^{\Delta CT}$ fosters mitogenic signals, such as phosphorylation of the EGFR sites implicated in the activation of Src kinase as well as phosphorylation (activation) of extracellular signal-regulated kinases ERK1 and ERK2 and STAT5b transcription factor. [68]

8.2.5.3 Fibroblast Growth Factor Receptors

Fibroblast growth factors that comprise 22 detected secreted proteins that participate in metabolic regulations, cell proliferation and differentiation, the body's development, and angiogenesis, as well as in the maintenance of the cardiovascular and nervous systems and wound healing in adults.

65. Receptors EGFR, VEGFR, and PDGFR can initiate both cell motility and division. The type and concentration of their ligands and the amount and distribution of receptor influence the type of response.

66. A.k.a. coiled-coil domain-containing CCDC88a, GIV, Gα-interacting vesicle-associated protein, and Akt (PKB) phosphorylation enhancer (APE).

67. The $G\alpha_i$–girdin complex regulates EGFR signaling both in time and space [1151].

68. On the other hand, the $G\alpha_i$–girdin–EGFR complex does not appear; EGFR autophosphorylation is reduced, and EGFR residence in endosomes is prolonged.

Table 8.9. Family of fibroblast growth factor receptors and their ligands (Source: [1121]).

Type	Ligands
FGFR1	FGF1–6/8/10/19/21–23
FGFR2	FGF1–7/9/10/16/19/20–23
FGFR3	FGF1/2/4/8/9/15–21/23
FGFR4	FGF1/2/4/6/8/17/19/21/23

Fibroblast growth factors signal via their cognate protein Tyr kinase receptors. The latter constitute a family of 4 known members (FGFR1–FGFR4) [69] with multiple isoforms (Table 8.9) [70]

Fibroblast growth factor receptors consist of: (1) an extracellular ligand-binding domain, a heparin-binding region for interactions with extracellular matrix components, a cell adhesion molecule homology motif, and 3 immunoglobulin-like domains; (2) a single transmembrane domain; and (3) an intracellular domain with tyrosine kinase activity and a short C-terminal tail.

FGFR1 and its Splice Variants

Alternative splicing (Vol. 1 – Chap. 5. Protein Synthesis) of Fgfr1 transcripts (pre-mRNAs) generates diverse isoforms that are differentially expressed according to the cell type [1153]. Isoforms of FGFR1 can also reside in the nucleus complexed with various FGF types, where they can regulate transcription of target genes. The FGFR1 receptor can be truncated in the extra- or intracellular region, lack Ig-like domains, or derive from different exons encoding variants of particular Ig-like domains. Isoforms with a variable Ig-like domain 3 are termed FGFR1-3a to FGFR1-3c with different ligand-binding specificities. Isoforms of FGFR1 with either 2 or 3 Ig-like domains are called FGFR1β and FGFR1α, respectively. Subtype FGFR1β has a 10-fold higher affinity for FGF1 and FGF2 than FGFR1α isoforms. Yet, only FGFR1α translocates to the nucleus.

FGFR2

Members of the FGF family stimulate angiogenesis. Overexpressed FGFR2 causes endothelial dysfunction and has pro-inflammatory effect that may destabilize atheromatous plaques [1154].

69. The Fgfr5 gene encodes a FGFR5 protein that lacks a cytoplasmic Tyr kinase domain.
70. Alternative mRNA splicing generates different versions of FGFR Ig-like domain-3 (3a–3c) that determine ligand-binding specificity. Isoform FGFR1-3b binds to FGF1 with a higher affinity than FGF2 [1152]. Receptor FGFR2-3b links to FGF7 and has a reduced affinity for FGF2 with respect to FGFR2-3c. The FGFR3-3c splice variant is preferentially activated by FGF1 and FGF4 as well as, to a lesser extent, by FGF2, whereas FGFR3-3b tethers exclusively to FGF1. Isotype FGF9 is able to activate FGFR-3c and FGFR-3b.

FGFR2 Splice Variants

During alternative splicing of the human FGFR2 pre-mRNA, inclusion of exon-3b or -3c is modulated by levels of trimethylation of histone-3 ($H3K_4me^3$ and $H3K_{36}me^3$). Trimethylation of histone-3 Lys36 promotes the binding of chromatin protein MRG15 [71] that, in turn, recruits the splicing regulator polypyrimidine tract-binding protein. The latter represses alternative exon inclusion [1155].

FGFR4

At the cell surface, FGFR4 complexes with and stimulates or suppresses the activity of membrane type-1 matrix metallopeptidase (mt1MMP) in tumor cells, thereby supporting or precluding invasion and metastasis [1156].

FGFR Coreceptors and Cofactors

Fibroblast growth factor FGF19, in particular, has coreceptors and cofactors, such as heparin and β-Klotho [1157].

Klotho

Fibroblast growth factors have coreceptors and cofactors for receptor activation. Single-pass transmembrane proteins α- or β-Klotho stabilize FGF–FGFR interactions, thereby enhancing signaling. In particular, β-Klotho is required for FGF21 signaling. [72]

β-Klotho is required for FGF19 interaction with FGFR1c, -2c, and -3c. On the other hand, FGF19 interacts with FGFR4 in the absence of β-Klotho, as well as with heparin. Heparin potentiates the effects of both α- and β-Klotho in FGF19 signaling. Both α- and β-Klotho provoke phosphorylation of ERK1 and ERK2 in response to FGF19–FGFR4 interactions [1159].

Agent FGF23, which regulates the sodium–phosphate cotransporter and enzymes of the vitamin-D metabolism in kidneys, specifically binds to FGFR1 owing to plasmalemmal Klotho [1160]. Interaction of Klotho and FGFR1-3c forms a specific FGF23 receptor.

---

71. The transcription factor-like protein Mortality factor on chromosome 4 (MORF4)-related gene (MRG) product MRG15, a histone acetylase complex subunit, is also called mortality factor 4-like protein MorF4L1. The predominant members of the human MRG family are MRG15 and MRGX.

72. In adipocytes, the potent metabolic regulator FGF21 activates protein Ser/Thr kinase STK11 (or LKB1) that phosphorylates (stimulates) AMP-activated protein kinase and increasing the cellular NAD^+ concentration. Consequently, FGF21 stimulates the activity of NAD^+-dependent type-3 deacetylase sirtuin-1 and enhances mitochondrial oxidative phosphorylation. Activated sirtuin-1 deacetylate peroxisome proliferator-activated receptor-γ coactivator PGC1α and histone-3 [1158].

Heparan Sulfate Proteoglycans

Fibroblast growth factors interact with plasmalemmal heparan sulfate proteoglycans (Vol. 1 – Chap. 8. Cell Environment) for signal transduction. [73] Heparan sulfate proteoglycans facilitate FGF binding to FGFR receptors. They act as coreceptors and may also determine the specificity of GFG binding to their receptors.

Among members of the FGF19 subfamily, FGF21 does not interact with heparin, whereas FGF19 and FGF23 bind to heparin and heparan sulfate weakly compared to other FGF factors.

FGFR Pathway

Phosphorylated FGFR1 (Tyr766) has a high-affinity binding site for phospholipase-Cγ. Activated PLCγ hydrolyzes phosphatidylinositol bisphosphate to inositol trisphosphate and diacylglycerol that causes Ca^{++} release from intracellular stores and PKC activation, respectively.

The Ras–MAPK cascade is another important signaling pathway that uses the FGFR substrate FRS2α adaptor. Growth factors FGF1 and FGF2 also connects to FGFR1 to regulate the Delta–Notch pathway [1153].

Signals are transmitted to the nucleus not only via the Ras–MAPK pathway, but also nuclear translocation of FGF–FGFR1 [1153]. The FGFR1 receptor can interact with nuclear import protein importin-β. Nuclear FGFR1 has kinase activity. It can regulate the cell cycle, as it stimulates Jun that induces cyclin-D1 expression.

Liganded receptor Tyr kinases homo- or heterodimerize. In addition, receptor activation can recruit and activate other Tyr kinases. For example, activated FGFR recruits Src kinase via FRS2 adaptor. Factor FGF2 induces the corecruitment to and accumulation at the plasma membrane of FGFR and a small amount of Src [1162]. Kinase Src regulates colocalized FGFR1, coactivated by FGF2, driving FGFR to activity loci for signal amplification. Kinase Src controls actin-dependent FGFR1 transport in RhoB-dependent endosomes, [74] and signaling dynamics by phosphorylation of attenuator Sprouty. Kinase Src may control the signaling dynamics by regulating the spatial location of activated signaling complexes of the ERK and PI3K–PKB pathways. Factor FGF2 induces a quick activation of protein kinase-B. Activity of

73. Heparan sulfate proteoglycans are sulfated glycoproteins that bind many growth factors. They are required for proper signaling in TGFβ, Hedgehog, Wnt, and FGF pathways. They indeed regulate diffusion of signaling agents (TGFβ, Hedgehog, and Wnt), thereby determining morphogen gradients. In addition, heparan sulfate proteoglycans collaborate in the formation of the FGF–FGFR complex. Fibroblast growth factors have a low affinity for FGFRs, but a high affinity for heparan sulfate proteoglycans. Heparan sulfate proteoglycans stabilize FGF–FGFR interaction for organo- and morphogenesis. Monomeric FGF9 binds to heparin with a lower affinity than homodimeric FGF9, thereby having an increased diffusion and spreading FGF9 signaling [1161].

74. Small GTPase RhoB is involved in Src transport and activation. Agents Src and RhoB coordinate to control the actin assembly required for endosome transport, Src–RhoB-dependent endosomes being associated with actin-regulatory proteins.

PKB is attenuated in the abscence of Src, whereas ERK activation level does not change, but is delayed in its initial phase and fails to decay.

Regulation Overview

Several distinct mechanisms repress FGF signaling: (1) digestion of the sugar backbone of heparan sulfate proteoglycans by heparanases or other glycosaminoglycan-degrading enzymes; (2) sequestration of extracellularly immobilized FGFs to secreted FGF-binding proteins; and (3) other feedback attenuators of the Ras–MAPK pathway, such as dual-specificity phosphatase MKP3, Sonic hedgehog, and Sprouty inhibitors.

Secreted FGF-binding protein can release FGFs from their local extracellular matrix storage, chaperone them to their cognate receptors, and thus modulate FGF signaling. On the other hand, Similar expression to FGFs (Sef), operate in coordination with feedback attenuators of the Ras–MAPK pathway.

Fibroblast growth factor-binding protein FGFBP1 interacts at least with FGF1, -2, -7, -10, and -22. Heparan sulfateproteoglycans and other heparinoids compete with FGFBP1 binding to FGF2. It interacts with perlecan in the basement membrane. Moreover, FGFBP1 can supplement HSPG assistance in FGF-bound FGFR oligomerization and signaling [1163]. During embryo- and fetogenesis, different FGFs contribute to the fine tuning of the amplitude of FGF activity.

8.2.5.4 Hepatocyte Growth Factor Receptor

Hepatocyte growth factor receptor (HGFR)[75] is secreted as a single-chain, inert precursor and post-translationally cleaved by extracellular serine proteases, such as plasmin and matriptase to produce the α and β subunits. These subunits are disulfide linked to form a functional 2-chain heterodimer [1164].

Hepatocyte Growth Factor Agonist

Similarly to the formation of its mature receptor, the ligand hepatocyte growth factor (HGF; Vol. 2 – Chap. 3. Growth Factors)[76] is synthesized as a full-length single-chain precursor proHGF that is cleaved to form an active heterodimer made of an α (heavy) chain linked by a disulfide bridge to a β (light) chain.[77] The Hgf gene not only encodes full-length proHGF precursor, but also 2 truncated forms,

75. A.k.a. Tyr kinase receptor mesenchymal–epithelial transition factor (MET or cMET), scatter factor receptor, and renal cell carcinoma papillary protein RCCP2. It is encoded by the Hgfr gene.

76. A.k.a. hepapoietin-A, lung fibroblast-derived epithelial cell mitogen, and scatter factor.

77. The HGF light chain has the structure of a serine peptidase with amino acid substitutions that renders HGF devoid of proteolytic activity. In vitro, several serine peptidases can cleave (activate) HGF, such as HGF activator (HGFA), matriptase, hepsin (Hpn), urokinase (uPA), tissue plasminogen activator (tPA), plasma kallikrein, and activated coagulation factors XIa and XIIa [1165]. Hepsin is a cell-surface serine protease. Matriptase (a.k.a. Suppression of tu-

HGF^{NK1} and HGF^{NK2}, that consist of the N terminus linked in tandem with the first kringle domain (K1) or two kringle domains (K1 and K2), respectively.[78] In fact, 5 splice variants of the human Hgf gene have been identified. Transcript variant-1 encodes the longest isoform; transcript variant-2 HGF^{NK2}; transcript variant-3, -4, and -5 HGF isoform-3, -4, and -5, the latter being HGF^{NK1}NK1 [1165]. Receptor HGFR is widely distributed from early embryogenesis to adulthood.

Plasma HGF concentration depends on HGF clearance from the blood, uptake by organs, and excretion into bile. Blood HGF distributes primarily to the liver and kidneys.

Hepatocyte growth factor receptor is sequestered, mainly in its inactive form, by proteoglycans in the extracellular matrix (Vol. 1 – Chap. 8. Cell Environment). Similarly, HGF interacts with heparan sulfate proteoglycans. In particular, circulating HGF is rapidly sequestered by heparan sulfates on the wetted (luminal) vascular wall. Heparan sulfate is a component of proteoglycans on most cell type surfaces. The closely related glycosaminoglycan heparin, which is produced by mastocytes also binds to hepatocyte growth factor. Moreover, HGF connects to syndecan-1, -2, and -4 [1165]. Unlike, other heparan sulfate-binding growth factors such as members of the FGF family, HGF can tether to dermatan sulfate, which are chains of small interstitial proteoglycans decorin (DSPG1), a small cellular or pericellular matrix proteoglycan, and biglycan (DSPG2).[79]

Hepatocyte growth factor is a potent mitogen, motogen, and morphogen for epithelial cells. Whereas it remains inactive as a single-chain form in the normal state, it is converted to an active heterodimer in response to tissue injury, in particular, by the serine peptidase hepatocyte growth factor activator (HGFA). The latter is also synthesized as an inactive precursor in the liver. It circulates in blood as an inactive zymogen. This plasma protein is activated by thrombin.

The activated, cell-surface bound HGFA acquires a heparin-binding ability, hence becoming sedentary [1166]. Proteolysis of full-length proHGF before its conversion into the active heterodimer is controlled by Kunitz-type serine peptidase in-

---

morigenicity ST14, membrane-type serine protease MTSP1, prostamin, and tumor-associated differentially-expressed gene product TADG15) is a type-2 transmembrane serine protease of most epithelia, where it is coexpressed with its cognate transmembrane serine peptidase inhibitor SPInt1. Activation of matriptase zymogen requires sequential N-terminal cleavage, activation site autocleavage, and transient link to SPInt1. Matriptase also cleaves (activates) urokinase plasminogen activator. It is activated by sphingosine 1-phosphate.

78. The heavy chain of HGF dimer derives from the N terminus of the precursor and contains 4 kringle domains. Kringle domains have a characteristic folding due to 3 internal disulfide bonds. Like HGF, HGF^{NK1}NK1 is a mitogen, motogen, and morphogen, but with lower potency than that of HGF [1165]. On the other hand, HGF^{NK2} activates HGFR and can competitively antagonize the mitogen effects of HGF and HGF^{NK1}NK1, but HGF^{NK2} retains motogenic activity.

79. Dermatan sulfate abounds in matrix of many organs. Hence, the HGF retention by this glycosaminoglycan must be overcome to deliver HGF to target epithelial and endothelial cells. In basement membranes. heparan sulfate predominates over dermatan sulfate. Therefore, dermatan sulfate and, then, heparan sulfate control HGF transfer to target cells.

hibitors SPInt1, its splice variant SPInt1b, and SPint2.[80] Protein SPint1 that transiently binds to and sequesters HGFA promotes the localized activation of proHGF on target cell surface for concentrated release under appropriate circumstances (tissue injury or local inflammation).

Hepatocyte Growth Factor Signaling

Activated HGFR intervenes in cell survival, proliferation, and motility,[81] epithelial–mesenchymal transition, thereby participating in morphogenesis during embryo- and fetogenesis as well as postnatal development, especially angiogenesis, wound healing,[82] and tissue remodeling, as well as tumor invasiveness and metastasis.

In the lung and kidney, it contributes to repair and regeneration after an injury, avoiding fibrosis [1165]. It antagonizes the profibrotic action of transforming growth factor-β.

In the vasculature, HGF production is supported by prostaglandins and HGF itself, but repressed by angiotensin-2, TGFβ, glucose, and hypoxia [1165]. Hepatocyte growth factor can also assist angiogenesis.

Mechanism of Action

Signaling Effectors

Once liganded, HGFR dimerizes and is phosphorylated[83] Receptor HGFR then cooperates with adaptors that serves as signal amplifiers. In particular, it can recruit the adaptor GRB2-associated-binding protein GAB1 [1164] (Table 8.10).

The adaptor GRB2-associated-binding protein GAB2 binds to HGFR directly or indirectly via growth factor receptor-bound protein GRB2. Adaptor GAB1 enables the recruitment of other SH2 domain-containing molecules, such as adaptors CRK and SHC, phosphoinositide 3-kinase, Src kinase, phosphatase PTPn11, phospholipase Cγ1, and Ras GTPase-activating protein RasA1 [1164].

Receptor HGFR activates diverse signaling cascades, such as those that involve protein kinase-B, extracellular signal-regulated kinase, GTPase Rac, and signal transducer and activator of transcription STAT3.

---

80. A.k.a. hepatocyte growth factor activator inhibitor HAI1, HAI1b, and HAI2 (or placental bikunin), respectively. Both SPInt1 and SPInt1b are potent inhibitors of HGFA, matriptase, and hepsin. Protein SPint2 inhibits various serine proteases, such as plasma kallikrein and clotting factor XIa [1165].

81. The HGF–HGFR complex is involved in survival and proliferation of epithelial cells and migration of muscle progenitors [1164]. It contributes to morphogenesis, as it operates in shape change and extracellular matrix turnover. Mesenchymal cells are major sources of HGF that acts as a paracrine regulator on neighboring, HGFR+ epithelial cells.

82. During tissue repair, the cytokines interleukin-1 and -6, tumor-necrosis factor-α, and transforming growth factor-β, heighten the production of HGF in fibroblasts and resident macrophages and HGFR in epithelial cells [1164].

83. Activated HGFR is first transphosphorylated in its kinase activation sequence (Tyr1234 and Tyr1235) and secondarily phosphorylated in its docking site (Tyr1349 and Tyr1356).

Table 8.10. Effects of HGFR (Source: [1164]; ERK: extracellular signal-regulated kinase; GAB: GRB2-associated-binding protein; GRB: growth factor receptor-bound protein; JNK: Jun N-terminal kinase; MAPK: mitogen-activated protein kinase; NFκB: nuclear factor-κB; PI3K: phosphoinositide 3-kinase; SHC: Src homology-2 domain-containing transforming protein; STAT: signal transducer and activator of transcription).

Transducer	Interaction with HGFR	Effect
	Adaptors	
CRK	Via GAB1	Cell growth and migration
GAB1	Direct or via GRB2	Tubulogenesis
GRB2	Direct or via SHC	Cell proliferation and migration
SHC	Direct	Cell proliferation and migration
	Kinases	
ERK1/2	Effector	Tubulogenesis, cell survivall
JNK	Effector	Cell growth and apoptosis
P38MAPK	Effector	Cell proliferation and apoptosis
PI3K	Direct or via GAB1	Cell survival, proliferation, migration
PKB	Effector	Cell survival
Src	Direct	Cell growth and migration
	Phosphatases	
PTPn11	Direct or via GAB1	Tubulogenesis
	Transcription factors	
NFκB	Effector	Tubulogenesis, cell survival and proliferation
STAT3	Direct	Tubulogenesis, cell growth

Moreover, it can undergo an intracellular cleavage by γ-secretase and caspases. Once cleaved by caspases, the soluble intracellular fragment ($HGFR^{ICD}$) can signal. After cleavage by γ-secretase, the intracellular fragment is destroyed by proteasome. Signaling by HGFR is also interrupted by extracellular shedding and degradation of its extracellular domain or ubiquitination. Shedding by an adamlysin generates a soluble extracellular fragment ($HGFR^{ECD}$) that not only sequesters the HGF ligand, but also tethers to full-length HGFR ($HGFR_{FL}$) and, then, hinders its dimerization and transactivation.[84]

In addition, HGFR is a substrate for protein Tyr phosphatases, such as receptor phosphatases PTPRf and PTPRj and cytosolic PTPn1 and PTPn2.[85]

Receptor HGFR also interacts with $\alpha_6\beta_4$-integrin[86] and epican [1164].[87] The coreceptor $\alpha_6\beta_4$-integrin enhances HGF signaling. Epican is a linker of the extracel-

84. The plasma membrane-anchored cytoplasmic remnant is cleaved by γ-secretase into a labile intracellular fragment that is rapidly cleared by a proteasomal degradation.

85. These phosphatases dephosphorylate either the catalytic (PTPn1 and PTPn2) or docking tyrosines (PTPRj).

86. Activated HGFR phosphorylates the cytoplasmic domain of the β_4 subunit. The latter can then recruit SHC, PI3K, and PTPn11.

87. A.k.a. heparan sulfate proteoglycan HCAM and CD44.

lular matrix to the actin cytoskeleton that possesses many alternatively spliced variants (epican$_{V1}$–epican$_{V10}$). Isoform epican$_{V6}$ connects HGFR to the cytoskeleton.[88] The ternary HGF–HGFR–epican complex primes the assembly of a juxtamembrane aggregate made of HGFR, GRB2, ^{F}actin, ezrin, radixin and moesin proteins, and the guanine nucleotide-exchange factor Son of sevenless. This complex boosts the activation of Ras. In endothelial cells, the epican$_{V10}$ contributes to the association of HGFR with caveolin, the Rac1 GEF TIAM1, cortactin, and dynamin-2 [1164].

Furthermore, class-B plexins associate with HGFR and transactivate it in response to semaphorins, even in the absence of HGF [1164]. Therefore, cells in which HGFR and class-B plexins coexist react to either HGF or semaphorins.

Receptor HGFR regulates the activity of several transcription factors that operate in the synthesis of cytoskeletal and intercellular junction components, cell cycle regulators, and anti-apoptotic mediators [1164]. The HGF–HGFR complex upregulates numerous matrix metallopeptidases as well as several proteases that convert MMP precursors into active enzymes.

Last, but not least, the HGF–HGFR complex is a potent antifibrotic agent that antagonizes fibrosis generated by interstitial myofibroblasts under the control of transforming growth factor-β.

Signaling Pathways

Receptor HGFR targets the MAPK modules. It activates Ras via the GRB2–SOS complex or phosphatase PTPn11 that dephosphorylates Ras GTPase-activating protein RasA1 to foster Ras activation. Stimulated Ras activates Raf, MAP2K1 and MAP2K2, and then extracellular signal-regulated kinases (Table 8.11). This axis can be counteracted by Notch and Sprouty (negative feedback) [1164]. Terminal effectors also include Jun N-terminal kinases and P38MAPKs that are activated via Rac GTPase.

Phosphoinositide 3-kinase generates phosphatidylinositol (3,4,5)-trisphosphate that creates a docking site for protein kinase-B. The latter then phosphorylates several substrates involved in cell survival, growth, and proliferation. In particular, PKB phosphorylates BAD that then dissociates from BCL2, thereby freeing BCL2 that can inhibit pro-apoptotic BAX and BAK proteins.

Upon HGFR stimulation, activated PI3K and Src kinases leads to the activation of IKK that phosphorylates the inhibitor IκB to stimulate nuclear factor-κB, which translocates to the nucleus to initiate the transcription of genes implicated in cell survival and division.

Transphosphorylation of HGFR offers a docking site for STAT3 that is then phosphorylated. Afterward, STAT3 dissociates from HGFR, homodimerizes, and travels to the nucleus, where this transcription factor regulates the transcription of several genes involved in cell proliferation or differentiation.

88. Its intracellular domain links HGFR to actin filaments via GRB2 and ezrin, radixin, and moesin proteins.

Table 8.11. Signaling pathways tiggered by activated HGFR (Source: [1164]; BAD: BCL2 antagonist of cell death; GAB: GRB2-associated-binding protein; GRB: growth factor receptor-bound protein; GSK: glycogen synthase kinase; PI3K: phosphoinositide 3-kinase; PKB: protein kinase-B; SOS: Son of sevenless; STAT: signal transducer and activator of transduction; TOR: target of rapamycin).

MAPK modules	HGFR–GRB2–SOS–Ras–Raf–MAP2K1/2–ERK1/2
	HGFR–Ras–PI3K–Rac–MAP3K–MAP2K4/7–JNK
	HGFR–Ras–PI3K–Rac–MAP3K–MAP2K3/6–P38MAPK
PKB	HGFR–PI3K and HGFR–Ras–PI3K
	PI3K–PKB–TOR
	PI3K–PKB–GSK3β
	PI3K–PKB–BAD
NFκB	PI3K–PKB–IKK
	Src–IKK
STAT3	HGFR–STAT3

Plasmalemmal vs. Endosomal Signaling

These different pathways are specifically regulated and coordinated. Furthermore, signaling from HGFR emanates not only from the plasma membrane, but also from endosomes. Following ligand binding, HGFR is rapidly internalized by a clathrin-mediated process. In peripheral early endosomes, HGFR can also signal. [89] Therefore, the functional outcome of HGFR activation depends on signaling mediator transport from the plasma membrane to the nucleus, i.e., diffusion or endosomal transport.

Protein kinase-Cϵ allows HGFR recruitment to early endosomes, hence the delivery of activated extracellular signal-regulated kinase to focal adhesions to support cell migration. Protein kinase-Cα enables HGFR sorting to non-degradative perinuclear endosomes, i.e., in a subcellular compartment where STAT3 can be efficiently phosphorylated, as the activity of Tyr^P phosphatases is low [1164].

The magnitude of the signaling response is controlled not only by the type of signaling effectors, but also intracellular transport of receptors. However, signaling and transport features using similar transport mechanism can differ according to the type of signaling mediators. Receptor HGFR can cause strong or weak STAT3 signaling whether phosphorylated (activated) STAT3 diffuses from the cytosol into the nucleus or STAT3 colocalizes with HGFR in endosomes that are conveyed via microtubules to the perinuclear region, where STAT3 is phosphorylated before reaching the nucleus [1167]. [90] In the latter case, HGFR ensures that weak STAT3 signals are

89. Endocytosis of HGFR enables an efficient signal transmission using extracellular signal-regulated kinases and signal transducer and activator of transcription STAT3 [1164].

90. Signaling endosomes such as APPL-positive endosomes operate in signal specificity. Factor STAT3 can bind to microtubule-binding protein stathmin.

protected from inactivation by cytosolic phosphatases. On the other hand, HGF elicits a potent activation of extracellular signal-regulated kinases ERK1 and ERK2 via internalized HGFR into endosomes without needing intermediate localization to the perinuclear compartment. However, strong ERK1 and ERK2 signals can also proceed via cytosolic diffusion.

Endocytosis is also involved in other types of HGF-mediated signaling, such as HGF-induced clathrin- and Rab5-mediated endocytosis of GTPase Rac during cell migration [1168]. Activation of Rac occurs on early endosomes, where RacGEF TIAM1 is also recruited. Small GTPase Rac is subsequently recycled to the plasma membrane to ensure localized signaling that leads to the formation of actin-based migratory protrusions.

On the other hand, ubiquitin ligase CBL primes HGFR endocytosis in multivesicular bodies for lysosomal degradation.

Cell Apoptosis vs. Survival

After a sequential cleavage of its cytoplasmic domain by caspases,[91] the resulting intracellular soluble fragment that contains the kinase domain can trigger cell death.

In cell stress conditions associated with an anti-apoptotic HGF stimulation, the activation of caspases and, then, the generation of the intracellular pro-apoptotic fragment is repressed [1164].

The HGFR receptor interacts with TNFRSF6 and precludes TNFRSF6–TNFSF6 linkage. However, an elevated HGF concentration can sensitize at least some cell types to apoptosis via TNFSF6 [1164].[92]

8.2.5.5 Insulin and Insulin-like Growth Factor Receptors

Cells receive, interpret, and respond to insulin (Vol. 2 – Chap. 1. Remote Control Cells – Sect. Endocrine System and Hormones) that attaches to insulin receptor Tyr kinase to trigger upon rapid autophosphorylation intracellular signaling cascades that are mediated via multiple intracellular adaptor proteins. Insulin suppresses hepatic glucose production by gluconeogenesis and glycogenolysis on the one hand and increase glucose uptake in myocytes and adipocytes via translocation of sequestered glucose transporter GluT4 (Sect. 4.17.2) to the plasma membrane on the other.

The functional insulin receptor derives from a single gene product that is cleaved post-translationally into 2 peptides. These peptides then form a heterotetramer. Similarly, their endogenous ligands are post-translationally processed to produce heterodimers.

---

91. A first caspase-dependent cleavage at an aspartic acid residue of the C-terminus (Asp1374) is followed by a second one at an aspartic acid in the juxtamembrane portion (Asp1000) [1164].

92. Hepatocyte growth factor, also named tumor cytotoxic factor, can prime death of tumor cells.

Table 8.12. Receptors of insulin and insulin-like growth factor isotypes IGF1 and IGF2 (Source: [1169, 1170]).

Type	Affinity
Insulin receptor	Insulin $\gg$ IGF2 $>$ IGF1
IGF1 receptor	IGF1 $>$ IGF2 $\gg$ insulin
IGF2 receptor	IGF2 $>$ IGF1 (no insulin binding)

Isoform IGF1 is much more efficient in stimulating cell growth than insulin. Whereas insulin and IGF1 binds to a specific, glycosylated, tetrameric receptor Tyr kinase, IGF2 preferentially targets a distinct receptor (Table 8.12). Receptor IGF2 is the mannose 6-phosphate receptor does not have protein kinase activity.

Insulin Receptors

Insulin receptor (IR or InsR) is composed of 2 extracellular insulin-binding α and 2 transmembrane β subunits. Subunit β has a cytoplasmic ATP-binding and Tyr kinase domain. Domains L2 and/or fibronectin-1 of insulin receptor create a high-affinity insulin-binding site of the dimeric receptor.

Different insulin receptor isoforms — IR_A (or IRa) and IR_B (or IRb) — result from alternative splicing of exon 11 of the Ir transcript.[93] The IR_A isoform is expressed predominantly in the developing fetus as well as a few adult cells such as pancreatic β cells. During development, IR_A is controlled mainly by insulin-like growth factor IGF2.[94] The IR_B isoform is produced predominantly in adult tissues. It mediates the metabolic effects of insulin in muscle, adipose, and liver [536].[95]

Upon insulin binding, the kinase domain of the insulin receptor autophosphorylates, recruits, and, then, phosphorylates effectors, such as insulin receptor substrates (IRS1–IRS3) and SHC adaptors. In turn, IRS adaptors recruit and promote the activation of the PI3K–PKB and Ras-MAPK pathways to elicit metabolic responses, such as glucose uptake and glycogen synthesis.

Transient signal transduction relies on proper initiation of signaling following signal reception, suitable spatial and temporal localization of mediators for signal transmission, and appropriate termination of signaling. Adaptors GRB10 and GRB14 shut off signal transduction [1171].[96]

93. Human IR isoforms IR_A and IR_B differ by the inclusion or exclusion for IR_B and IR_A, respectively, of 12 amino acids at the α-subunit C-terminus. In adult tissues, insulin is the major ligand for IR_B receptor.

94. Insulin-like growth factor IGF2 is mainly a growth-promoting hormone during gestation.

95. In pancreatic β cells, insulin activates both IR_A and IR_B.

96. The GRB adaptors differ in the relative efficiency to bind their partners. Adaptors GRB10 and GRB14 are potent binders of phosphoinositol phosphates and activated Ras, re-

Insulin-like Growth Factor-1 Receptor

Like insulin receptor, insulin-like growth factor-1 receptor (IGF1R) is usually a homodimer. Both IR and IGFR homodimers bind to insulin, but the latter with a lower affinity than that of the former. These receptors can also form heterodimers (IR–IGFR hybrid receptors) that associate with insulin with an affinity that is intermediate between that of IR and IGFR [1172]. Both IR_A–IGFR and IR_B–IGFR hybrids link to IGF1 and IGF2 with high affinity, similarly to IGFR homodimer.

Insulin-like growth factor-1 receptor transmits signals via the PI3K and MAPK pathways. It promotes the modification of IGF1R by small ubiquitin-like modifier protein SUMo1 and its translocation to the nucleus [1173]. Nuclear IGF1R then associates with enhancer-like elements to increase transcription.

Insulin Receptor-Related Protein

Insulin receptor-related protein (InsRR) is a member of the RTK class-2 with insulin and insulin-like growth factor receptors. However, ligands that activate IR or IGF1R, such as proinsulin, insulin, IGF1, IGF2, and relaxin, are not able to bind and activate InsRR protein. Nevertheless, insulin receptor substrates IRS1 and IRS2 are InsRR substrates. The InsRR protein influences IGF signaling, as it can colocalize with IGF1R. It also dimerizes with the insulin receptor.

Insulin receptor-related protein has a restricted cellular distribution, in particular in some cells of neuronal origin, on which it is closely associated with IR, IGF1R, and, in sympathetic neurons, NTRK1 nerve growth factor receptor [1174]. It is also produced in thecal-interstitial cells of large antral follicles in the ovary, where, like NTRK1, it is transiently activated during the preovulatory surge of gonadotropins [1175].

Substrates of Insulin Receptor

Insulin receptor phosphorylates several scaffold proteins that, in turn, recruit various effectors [536]. Scaffold proteins comprise the 4 members of the insulin receptor substrate family (IRS1–IRS4) [97] as well as adaptors CBL, GAB1, and SH2 domain-containing B-lymphocyte-expressed adaptor SH2B2, [98] in addition to signal-

---

spectively. The GRB adaptors contain a Ras-association (RA), a pleckstrin homology (PH), and a Tyr^P-binding Src homology-2 (SH2) domain, as well as, between the 2 last-mentioned motifs, a between PH and SH2 sequence (BPS) [1171]. The latter sequence can inhibit the catalytic activity of the insulin receptor by binding to the kinase active site. The BPS and SH2 domains of GRB proteins tether the insulin receptor kinase domain at 2 distinct sites. The RA and PH domains of GRBs that bind to small Ras GTPase and phosphoinositides, respectively, are also involved in the inhibition of insulin receptor signaling. All these interactions block further phosphorylation and may disrupt association of IRS with the insulin receptor.

97. The IRS3 isoform is much shorter than the 3 others (IRS1, IRS2, and IRS4).

98. A.k.a. adaptor with PH and SH2 domains (APS).

regulatory proteins [99] that correspond to inhibitory and stimulatory receptors in neurons and hematopoietic cells. [100]

Insulin-primed Tyr phosphorylation of insulin receptor substrates generates docking sites for effectors, such as PI3K1a, [101] protein Tyr phosphatase PTPn11, SRC family member kinase Fyn, and adaptors GRB2 and NCK [536]. Insulin-stimulated production of $PI(3,4,5)P_3$ that is counteracted by inositol 5-phosphatase SHIP2 and 3-specific $PI(3,4,5)P_3$ phosphatase PTen allows GluT4 translocation.

Phosphatidylinositol 3-kinase that tethers to and is activated by phosphorylated IRSs to be recruited to the plasma membrane to target $PI(4,5)P_2$ regulates cell growth, protein and glycogen synthesis, and glucose transport. Produced $PI(3,4,5)P_3$ serves as a lipidic platform that recruits and anchors signaling mediators. It particularly activates phosphoinositide-dependent kinase PDK1 that phosphorylates and activates both atypical protein kinases-C and protein kinase-B (Thr308). [102] Enzyme PKBβ participates in persistent residence of GluT4 in the plasma membrane [536]. Kinases PKCζ and PKCι favor GluT4 translocation to the plasma membrane.

Caveolin-enriched membrane rafts contain structural flotillin, glycosyl-phosphatidylinositol-anchored proteins, palmitoylated signaling mediators, glycolipids, sphingolipids, and cholesterol (Vol. 1 – Chaps. 7. Plasma Membrane and 9. Intracellular Transport). Insulin-bound insulin receptors phosphorylate adaptor and Ub ligase CBL via adaptor SH2B2. The recruitment of CBL to activated receptor Tyr kinases leads to receptor ubiquitination. However, in the insulin pathway, CBL associates with actin-binding adenylate cyclase-associated protein (CAP). The CBL–CAP complex is recruited by IRSs to membrane rafts via SH2B2 adaptor. Once CBL is phosphorylated by insulin receptor, the CAP–CBL heterodimer dissociates from insulin receptors and accumulate in rafts, as CAP interacts with integral membrane, caveola-linked flotillin to favor plasmalemmal insertion of GluT4. Adaptor CBL then recruits the CRK-RapGEF1 complex to regulate small GTPase RhoQ, a resident of membrane rafts, that also contributes to GluT4 translocation and glucose uptake. Furthermore, GRB2 adaptor links insulin to Ras activation via guanine-nucleotide exchange factor Son of sevenless and the Raf–ERK pathway.

Several protein Ser/Thr kinases desensitize insulin receptor, such as PI3K, PKB, GSK3, TOR, ERK, JNK, and IKKβ. Protein Tyr phosphatases can also attenuate signaling from insulin receptor such as PTPn1 that dephosphorylates both insulin re-

---

99. A.k.a. SH2 domain-containing phosphatase substrates (SHPSs), macrophage fusion receptor (MFR), and brain immunoglobulin-like molecule with a Tyr-based activation motif (BIT).

100. Transmembrane glycoprotein SIRPα1 not only inhibits signaling from receptor Tyr kinases, but also favors the MAPK pathway in response to insulin and potentiate integrin-induced MAPK activation. Protein SIRPα1 is Tyr phosphorylated following cell stimulation with insulin, epidermal growth factor, or platelet-derived growth factor.

101. Class-1A PI3Ks are heterodimers composed of regulatory (P50, P55, or P85 encoded by Pik3r1, -2, and -3 genes) and catalytic P110 subunits (Vol. 4 – Chap. 1. Signaling Lipids). Kinase PI3K1a phosphorylates preferentially phosphatidylinositol (4,5)-bisphosphate.

102. Protein kinase-B can also be phosphorylated (Ser473) by integrin-linked kinase, P21-activated kinase PAK2, and MAPKAPK2.

Table 8.13. Family of platelet-derived growth factor receptors and their ligands (Sources: [1121], Wikipedia).

Type	Ligands
PDGFRαα	PDGFaa, PDGFab, PDGFbb, PDGFcc
PDGFRαβ	PDGFbb, PDGFcc, PDGFdd
PDGFRββ	PDGFaa, PDGFab, PDGFbb

ceptors and insulin receptor substrates [536]. In addition, insulin receptor substrates are ubiquitinated for 26S proteasomal degradation. Suppressors of cytokine signaling promote ubiquitination of IRS1 and IRS2 proteins.

8.2.5.6 Macrophage-Stimulating-1 Receptor

Macrophage-stimulating-1 receptor (MSt1R) is a mesenchymal–epithelial transition factor-related Tyr kinase. [103] The MST1R transcript is transformed into a glycosylated single chain precursor that is cleaved into α and β chains before exposure at the cell surface to form a heterodimer [1176]. The β chain has Tyr kinase activity. Autophosphorylation of MSt1R results from stimulation by macrophage-stimulating protein that is a chemoattractant.

The MSt1R receptor resides at the apical surface of ciliated epithelia in airways and oviduct. Activation of MSt1R by macrophage-stimulating protein (MSP) [104] increases ciliary beat frequency of human nasal cilia [1177]. It signals via the GRB2–SOS–Ras pathway [1178]. Receptor MSt1R also interacts with SHC1 and PLCγ1 enzyme.

8.2.5.7 Platelet-Derived Growth Factors Receptors

Platelet-derived growth factor receptors constitute a family of 4 elementary inactive monomers (PDGFa–PDGFd) that form homo- and heterodimers (PDGFaa, -ab, -bb, -cc, and -dd). These dimers bind to protein Tyr kinase PDGFRα and PDGFRβ. Once bound to PDGF dimers, these 2 receptor isoforms dimerize to generate 3 possible receptor complexes: PDGFRαα, PDGFRββ, and PDGFRαβ (Table 8.13).

Platelet-derived growth factor receptors recruit and activate several proteins, such as PI3K, PLCγ, PTPn11 phosphatase, and Src kinases. PDGFRs also bind adaptors

103. A.k.a. PTK8, stem cell-derived Tyr kinase (STK; and not Ser/Thr-specific protein kinase), and RON.

104. A.k.a. macrophage stimulatory protein and hepatocyte growth factor-like protein (HGFL).

NHERF1 [105] and NHERF2. [106] Protein PDGF activates Rac GTPases in lamellipodia of migrating cells.

Platelet-derived growth factor-BB stimulates vasculogenesis (Vol. 5 – Chap. 10. Vasculature Growth) and maintains vascular stability by attracting mural cells that express PDGFRβ. Platelet-derived growth factor-BB and its receptor PDGFR-β are mainly expressed in developing vasculature, as PDGFbb and PDGFR-β are produced by endothelial and mural cells, respectively. Growth factor PDGFbb activates ERK1, ERK2, and JNK, but neither P38MAPK nor PI3K kinase.

The PDGFR-β receptor is expressed by hemangioprecursors. Reactive oxygen species operate in differentiating embryonic stem cells involved in endothelial cell commitment to promote vasculogenesis via the VEGF–HIF1α–MAPK pathway. In embryonic stem cells, PDGFbb augments intracellular Ca^{++} level that activates several NADPH oxidase isoforms. Subsequent production of reactive oxygen species stimulate extracellular-regulated kinase-1 and -2 and initiate smooth muscle cell proliferation and migration [1180]. Jun N-terminal kinases are weakly activated.

105. Two types of sodium–hydrogen exchanger isoform-3 regulatory factors exist: NHERF1 (a.k.a. ezrin–radixin–moesin-binding phosphoprotein-50 [EBP50]), and NHERF2 (a.k.a. Na^+–H^+ exchanger type 3 kinase-A regulatory protein [E3KARP]). Both NHERF1 and NHERF2 contain 2 tandem PDZ domains for the assembly of proteic complexes. They thus interacts with various receptors, such as β2-adrenergic and P2Y receptors. The regulatory cofactor of the Na^+–H^+ exchanger-3 (NHE3) mediates the regulation of the activity of ion transporters, in particular NHE3 by protein kinase-A. Sodium–hydrogen exchanger isoform-3 regulatory factor-1 was indeed first found to be a cofactor in cAMP-associated inhibition of the renal brush border NHE3 exchanger. Proteins NHERF1, ezrin, and PKA form a complex that links the actin cytoskeleton and downregulates NHE3 exchanger. Scaffold NHERF1 recruits various receptors, ion carriers, and other proteins to the plasma membrane. In particular cytosolic adaptor NHERF1 binds to actin-associated proteins, such as ezrin, radixin, moesin, and merlin (moesin–ezrin–radixin-like protein merlin is the product of the neurofibromatosis-2 gene). Membrane-associated ezrin, radixin, and moesin regulate the structure and function of certain domains of the cell cortex. The ERM proteins associate with either directly the cytoplasmic region of membrane proteins or indirectly other membrane proteins via NHERFs. They also connect to filamentous actin. They link to signaling molecules of the Rho pathway such as Rho guanine dinucleotide-dissociation inhibitor. Therefore, ERMs participate in microfilament–membrane attachment and Rho signaling. Binding sites for NHERFs, ^{F}actin, and RhoGDI are masked in inactive monomers. The binding regulation of ERMs and merlin is carried out by phosphorylation in combination with acidic phospholipids such as phosphatidylinositol (4,5)bisphosphate [1179]. The ERM proteins then serve as a regulated linkage between plasmalemmal proteins and the cortical cytoskeleton. Attachment of plasmalemmal proteins to ^{F}actin determines cell shape and is required in cell adhesion, motility, and transport associated with signaling pathways.

106. Protein NHERF2 is strongly expressed in the vasculature. It interacts with PDGFR, favoring its dimerization and subsequent activation of Tyr kinase activity. It leads to the formation of a cytoskeleton-associated signaling complex able to trigger the MAPK pathway.

Table 8.14. Family of vascular endothelial growth factor receptors and their ligands (Sources: [1121] and Wikipedia). Vascular endothelial growth factors exist in multiple isoforms generated by alternative splicing and post-translational processing. They have distinct receptor specificities. Signaling initiated by VEGFa via VEGFR2 is the major pathway that regulates endothelial cell sprouting, proliferation, migration, and survival. All VEGFa isoforms bind to VEGFR1 and VEGFR2, whereas VEGFb and placental growth factor (PlGF) are specific for VEGFR1. Agent VEGFc stimulates lymphangiogenesis and contributes to angiogenesis via VEGFR3 receptor. Once their C-termini are cleaved, VEGFc and VEGFd are high-affinity ligands for VEGFR3 receptor. Once propeptides are processed, these growth factors acquire binding affinity for VEGFR2 to promote angiogenesis. Pox viruses encode VEGF variants that are collectively called VEGFe, which specifically bind to VEGFR2 receptor.

Type	Ligands
VEGFR1	VEGFa, VEGFb, PlGF
VEGFR2	VEGFa, VEGFc, VEGFd, VEGFe
VEGFR3	VEGFa, VEGFc, VEGFd

8.2.5.8 Vascular Endothelial Growth Factor Receptors

All endogenous members of the family of vascular endothelial growth factor (VEGFa–VEGFd and placental growth factor [PlGF]) regulate angiogenesis and lymphangiogenesis.[107] They bind with high affinity to and specifically activate their cognate receptor Tyr kinases: the VEGF receptors (VEGFR; Table 8.14).

Activated VEGFRs dimerize and transphosphorylate. Three VEGF receptor Tyr kinases exist (VEGFR1–VEGFR3). The VEGFa isoform binds to VEGFR1[108] and VEGFR2[109] The VEGFR2 receptor is the predominant transducer of signals for angiogenesis, as it mediates almost all known VEGF responses. On the other hand, VEGFR1 modulates VEGFR2 signaling and can sequester VEGF out of reach from VEGFR2 receptor. Receptors VEGFR2 and VEGFR3 are involved in the functioning of the endothelium of blood and lymph vessels, respectively. The VEGFR3 receptor is mainly targeted by VEGFc and VEGFd factors. The VEGFc isotype is able to induce the formation of VEGFR3, but not VEGFR2 homodimers.[110] In addition, VEGFc supports the formation of VEGFR2–VEGFR3 heterodimers.

Signaling primed by VEGF is modulated by interactions with distinct heparan sulfate proteoglycans and neuropilins that act as coreceptors.[111] Among VEGFR coreceptors, neuropilin-1 is expressed in arteries; neuropilin-2 in veins and lymphatics. Once stimulated by VEGFc, neuropilin-2 associates with VEGFR3.

Vascular endothelial growth factor receptors are not exclusively expressed on vascular cells. Numerous cell types synthesize both VEGFR2 and VEGFR3. Yet,

107. I.e., blood and lymph vessel formation and growth from pre-existing conduits.
108. A.k.a. Fms-like Tyr kinase-1 (FLT1).
109. A.k.a. kinase insert domain-containing receptor (KDR) and fetal liver kinase-1 (FLK1).
110. Agent VEGFc has its highest affinity for VEGFR3, but can bind to VEGFR2.
111. Neuropilins are VEGF-binding molecules without VEGF-induced catalytic function.

VEGFR1 is mainly produced in hematopoietic and endothelial cells and VEGFR2 and VEGFR3 in vascular and lymphatic endothelial cells.

VEGFR1

The function of VEGFR receptors depends on their types [1181]. The VEGFR1 receptor is required for the recruitment of hematopoietic precursors and migration of monocytes. It also exists in a secreted, inactive form ($VEGFR1^S$). It has higher affinity for VEGF, but weak kinase activity. It is regulated by hypoxia-inducible factors. It can regulate VEGFR2 signaling.

In endothelial cells, VEGFb heightens the abundance of fatty acid transport proteins (FATP) via VEGFR1 and neuropilin-1 [1184]. Overexpression of FATP3 or FATP4 raises the uptake of long-chain fatty acids. Isotype VEGFb abounds in tissues enriched in mitochondria that use fatty acids as an energy source (heart, skeletal muscle, and brown adipose tissue).

VEGFR2

The VEGFR2 receptor promotes endothelial cell differentiation and proliferation, as well as vascular sprouting.[112] It is inhibited by Tyr^P phosphatases PTPn6 and PTPn11. It activates PLCγ and PI3K, therefore the ERK cascade and protein kinase-B. It forms mechanosensory complexes with platelet–endothelial cell adhesion molecule-1 and vascular endothelial cadherin.

Cerebral cavernous malformation protein CCM3[113] is required for signaling from VEGFR2 in embryos, especially in endothelial cells [1182]. In response to VEGF stimulation, CCM3 is recruited to and stabilizes VEGFR2 receptor.

Vascular endothelial growth factors VEGFa and VEGFc are crucial regulators of vascular development on dimerization and activation of their cognate receptors VEGFR2 and VEGFR3. Both VEGFa and VEGFc provoke the formation of VEGFR2–VEGFR3 heterodimers in both developing blood vessels and immature lymphatic structures in embryoid bodies [1183]. In addition, these heterodimers frequently localize to tip cell filopodia. Angiogenic sprouting caused by VEGFa requires VEGFR3 that is observed throughout the length of the sprout. Sprouts induced by VEGFc have higher expression of VEGFR3 and higher extent of VEGFR2–VEGFR3 heterodimerization in the tip. Owing to VEGFc, VEGFR2–VEGFR3 heterodimers abound in the leading tip cells with respect to trailing stalk cells of growing sprouts.

---

112. Sprouting requires flipping of cell polarity (i.e., cell polarity needs to be reversed when new sprouts emerge from the basal side of the endothelium, as the apical edge does not remain the wetted interface), induction of motility, and local matrix degradation. Tip endothelial cells selected for sprouting drive the growing sprout. Growing endothelial sprout is guided by attractive or repulsive signals, especially VEGFa and PDGFb gradients.

113. Cerebral cavernous malformations (CCM) are caused by mutations in 3 genes (Ccm1–Ccm3). Protein CCM3 is also called TF1 cell apoptosis-related protein TFAR15 and programmed cell death PdCD10.

VEGFR3

Lymphangiogenesis occurs later than angiogenesis. It is initiated by the budding of endothelial cells from the cardinal vein after a subpopulation of vascular endothelial cells differentiated into Sox18+, Prox1+, lymphatic endothelial cells. The VEGFR3 receptor is widely expressed in blood vessels with high expression in blood capillaries and veins, but low in arteries during embryogenesis, at least in mice.

Lymphangiogenesis, but not angiogenesis, relies on VEGFR3. [114] Nonetheless, VEGFR3 may intervene in angiogenesis, as VEGFR3 activated by its cognate ligands VEGFc and VEGFd modulates VEGFR2 signaling, independently of its ligand-binding capacity or kinase activity. The VEGFR3 receptor can indeed form heterodimers with VEGFR2. Upon VEGFa binding, these VEGFR2–VEGFR3 dimers reduce the level of $VEGFR2^P$ as well as those of effectors $ERK1^P$ and $ERK2^P$ in endothelial cells [1185]. This process is an additional regulatory mechanism of VEGFR2 signaling independently of ligand binding or kinase activity, as decoy receptor VEGFR1 that has a higher affinity, but lower kinase activity than VEGFR2, precludes VEGFR2 signaling by sequestering VEGF ligand.

8.2.5.9 Stem Cell Factor Receptor

The stem cell factor (SCF) binds to the cytokine type-3 receptor Tyr kinase, the so-called stem cell factor receptor (SCFR) [115] involved in hematopoiesis [1186] (Vol. 5 – Chap. 2. Hematopoiesis). When SCFR is bound to stem cell factor, it dimerizes and activates cell signaling. Signaling mediated by SCFR supports cell survival, proliferation, and differentiation. The SCF–SCFR complex is particularly involved in mastocyte development and melanocyte migration, as well as in spermatogenesis and hematopoiesis.

The SCFR receptor is constituted of an extracellular ligand-binding domain with 5 immunoglobulin-like motifs, a single transmembrane domain, and intracellular juxtamembrane and kinase domains. Multiple transcript variants generate different isoforms. [116]

Regulation of SCFR Activity

Binding of SCF, either plasma membrane-tethered homodimer ($_m$SCF) or soluble form (SCF^S), to SCFR causes receptor dimerization and autophosphoryla-

114. When VEGFR3 ligand-binding domain is defective, lymph sac are observed, but lymphangiogenic sprouting lacks [1185]. Both impaired VEGFR3 ligand-binding and Tyr kinase domains do not affect blood vessel development, but impede lymphatic growth.

115. A.k.a. kinase in tyrosine KIT, cellular KIT (cKIT), and cluster of differentiation CD117. It identifies some types of hematopoietic progenitors in the bone marrow, such as hematopoietic stem cells, multipotent progenitors, common myeloid and lymphoid (at low level) progenitors, and early thymocyte progenitors (DN1 and DN2 thymocytes).

116. Alternatively spliced variants Ser715+ and Ser715− are expressed in the same cell types in humans.

tion [1187].[117] In addition, SCF promotes the recruitment of SCFR to membrane rafts for signal transmission, in particular to PI3K enzyme.

Protein kinase-C phosphorylates (represses) SCFR [1187].[118] Other inhibitors include suppressors of cytokine signaling SOCS1 and SOCS6, megakaryocyte-associated Tyr kinase (MATK; or CSK homologous kinase), and PTPn6 and PTPn11 phosphatases.

Proteolytic cleavage of SCFR receptor generates a soluble fragment, which can compete with plasmalemmal receptors for SCF binding, thereby attenuating SCF signaling.

Like other hematopoietic growth factor receptors of the plasma membrane, SCFR is quickly ($\sim$3 mn) internalized after ligand binding using Src kinase. Activated SCFR is degraded in proteasomes or lysosomes, once ubiquitinated by CBL ligase.

Interactors

Partners of SCFR, via its docking Tyr^P sites, include adaptors CBL, downstream of Tyr kinase docking protein DOK1, GRB2, GRB7, GRB2-related adaptor protein (GRAP), SH2B2, and SH2B3, multiple PDZ domain protein (MPDZ), regulatory subunit PI3KR1, SRC family kinases, MATK, phosphatases PTPn6 and PTPn11, and SOCS6 [1187]. The SCFR receptor associates with JaK2 to recruit STAT1α, STAT3, STAT5a, and STAT5b. It also interacts with PLCγ, as well as erythropoietin receptor and α subunit of granulocyte–macrophage colony-stimulating factor receptor (CSF2R), in addition to tetraspanin-28 to -30.

Production Cells

The SCFR receptor participates in growth and differentiation of early multipotent hematopoietic stem cells [1187]. During maturation of bone marrow progenitor cells into their respective lineages, SCFR expression is downregulated, except in the mastocyte lineage.[119] The SCFR receptor is indeed required for mastocyte growth and differentiation [1187]. In addition, SCFR may raise the response to IgE-mediated activation of mastocytes.

The SCFR receptor participates in the development and activation of various immunocytes. On dendritic cells, it promotes T_{H2} and T_{H17} responses via interleukin-6 and Jagged-2 [1187]. It also contributes to early T-cell development and lineage commitment mediated by Notch.

In the central nervous system, it is produced in glial cells and some neurons. In the cardiovascular system, it serves as a marker of progenitor cells that can differentiate into endothelial and cardiac and smooth myocytes [1187].

During embryogenesis, SCFR contributes to melanocyte migration from neural crest to skin [1187]. The SCFR receptor is also synthesized by interstitial cells of

117. At juxtamembrane Tyr568, Tyr570, Tyr703, Tyr721, Tyr730, Tyr823, Tyr900 and Tyr936 in humans, which yield docking sites.

118. At Ser741 and Ser746 residues.

119. Hence, its name mastocyte–stem cell growth factor receptor.

Cajal in the gastrointestinal tract, where it supports their survival, proliferation, and network formation.

The SCFR receptor is expressed in germ cells (spermatogonia and primordial oocytes) [1187]. It is implicated in gametocyte development; it enables germ cell precursors to migrate to genital ridges during embryogenesis.

8.2.5.10 Colony-Stimulating Factor-1 Receptor

Colony-stimulating factor-1 receptor (CSF1R), or macrophage colony-stimulating factor receptor (mCFSR), [120] is a cytokine class-3 receptor Tyr kinase. [121] Plasmalemmal CSF1R kinase is expressed in cells of the macrophage and dendritic lineages. It has alternative splice forms [1284]. A soluble isoform that lacks the transmembrane domain is secreted, whereas the others are anchored to the plasma membrane. The latter are able to inhibit macrophage proliferation.

The CSF1R receptor controls the proliferation, differentiation, and survival of macrophages. Liganded (activated) CSF1R undergoes oligomerization and transphosphorylation. The CSF1R receptor interacts with Fyn kinase, suppressor of cytokine signaling SOCS1, and GRB2 and CBL adaptors. Inhibitor SOCS1 is phosphorylated by activated CSF1R receptor [1188].

The CSF1R receptor belongs to the *prolactin family* of homodimeric receptor Tyr kinases that couple primarily to the JaK–STAT signaling cascade. [122] The CSF1R receptor is targeted by prolactin, chorionic somatomammotropin hormones CSH1 and CSH2, chorionic somatomammotropin hormone-like-1 (CSHL), erythropoietin, thrombopoietin, growth hormones GH1 and GH2, granulocyte–macrophage colony-stimulating factor (CSF2), and granulocyte colony-stimulating factor (CSF3; Table 8.15).

Prolactin and growth hormone receptors are, in fact, non-kinase receptors that activate signaling pathways via receptor-associated kinases, such as Janus or Src kinases. They are members of the type-1 cytokine receptor class (Sect. 11.1.1) of the cytokine receptor superclass that are activated by ligand-induced homodimerization.

8.2.6 Fetal Liver Kinase-2 (CD135)

Fetal liver kinase-2 (FLK2), or cytokine receptor CD135, [123] This class-3 cytokine receptor Tyr kinase lodges on the surface of hematopoietic progenitors, especially multipotent and common lymphoid progenitors. Ligand FLT3L is particularly

120. A.k.a. Fms (McDonough feline sarcoma viral oncogene homolog), and cluster of differentiation CD115.

121. Colony-stimulating factor-2 receptor CSF2Rα is a low-affinity granulocyte–macrophage colony-stimulating factor receptor (gmCSFRα) that is also termed CSF2R, gmCSFR, GMR, as well as cluster of differentiation CD116.

122. Janus kinases phosphorylate the receptor, thereby facilitating the recruitment of signal transducers and activators of transcription.

123. A.k.a. Fms-like Tyr kinase receptor-3 (FLT3) and stem cell protein Tyr kinase receptor STK1.

Table 8.15. The prolactin family of receptor tyrosine kinases (Source: [5]; CSF: colony-stimulating factor; CSH: chorionic somatomammotropin hormone [CSH1 is also called lactogen and chorionommammotropin]; CSHL: chorionic somatomammotropin hormone-like; Epo: erythropoietin; GH: growth hormone [GH2 is also named placenta-specific growth hormone and growth hormone variant]; gCSF: granulocyte colony-stimulating factor [a.k.a. CSF3, pluripoietin, filgrastim, and lenograstim]; gmCSF, granulocyte–macrophage colony-stimulating factor [a.k.a. sargramostim and molgramostin]; Tpo: thrombopoietin [a.k.a. megakaryocyte colony-stimulating factor, myeloproliferative leukemia (MPL) virus oncogene ligand, and megakaryocyte growth and development factor (MGDF)]). Prolactin and growth hormone receptors are, in fact, members of the class of type-1 cytokine receptors.

Type	Ligands
CSF1R	gCSF
CSF2R	gmCSF
PRLR	Prolactin, CSH1
EPOR	Epo
GHR	GH1, GH2
TPOR	Tpo

involved in the development of conventional and plasmacytoid dendritic cells. When this receptor is bound to its ligand, it forms a homodimer that activates signaling for lymphocyte survival, proliferation, and differentiation (but not of other blood cells).

8.2.7 Apoptosis-Associated Tyrosine Kinases

Apoptosis-associated Tyr kinases (AATK or AATyK) [124] causes growth arrest and/or apoptosis of myeloid precursor cells. Three identified members (AATK1–AATK3) possess highly similar kinase and non-kinase domains. The kinase domain is located in the N-terminus. The C-terminus is constituted by a proline-rich domain. The AATK kinase is expressed in different regions of the brain, where it promotes neuronal differentiation [1190].

The AATK kinase interacts with protein Ser/Thr kinase STK39 that operates in cell stress response [1189]. The latter is indeed activated in response to hypotonic stress. It phosphorylates several cation–chloride cotransporters such as Na^+–K^+–$2Cl^-$ cotransporter and activates the P38MAPK pathway.

Both AATK splice variants, long (AATK or $AATK_L$) and short ($AATK_S$), can be phosphorylated by CDK5, the single active cyclin-dependent kinase in postmitotic neurons [1191]. Like other CDKs, CDK5 requires binding to an CDK5 activator, such as neuronal-specific CDK5R1 or CDK5R2 regulatory subunits. [125] The CDC5–CDK5R1-regulated kinase (CPRK) that is expressed in various tissues, but is en-

124. A.k.a. lemur Tyr kinase-1 (Lmr1 and LmTK1).

125. The CDK5 kinase associates with β-catenin to regulate neuron migration during the body's development. It also participates in synaptic vesicle release and endocytosis, as well as dopamine signaling and neuregulin-induced expression of acetylcholine receptor at the neuromuscular junction.

riched in brain and muscle, is phosphorylated (inactivated) by the CDK5–CDK5R1 complex [1192]. [126]

8.2.8 Axl–Mer–TyrO3 (Sky) Class

Adhesion-related kinase receptor (Axl) class of RTKs is based on protein Axl encoded by the AXL gene that transduces signals from the extracellular matrix by binding vitamin-K-dependent growth arrest-specific gene product GAS6. Two different alternatively spliced transcript variants give rise to 2 isoforms Axl_{1409} and Axl_{1419} that differ only in their N-termini.

Protein growth arrest-specific gene product GAS6 is structurally homologous to anticoagulant protein-S. It binds members Axl, Mer, and Sky (or Tyr03) of the AXL–SKY class of receptor Tyr kinases [1193]. Protein GAS6 is a γ-carboxyglutamic acid domain-containing protein that contributes to cell proliferation.

The AXL–MER–SKY class is constituted by receptor Tyr kinases Axl and its homologs, Sky and Mer. Signaling launched by Axl is particularly implicated in platelet functions. Interaction between Axl and $\alpha_{2B}\beta_3$-integrin participates in platelet activation and thrombus stabilization via phosphatidylinositol 3-kinase and protein kinase-B [1194]. Receptor Tyr kinase Axl is activated by the vitamin-K-dependent GAS6 [1195]. [127]

8.2.9 Discoidin Domain-Containing Receptors

Discoidin domain-containing receptors are specific collagen-binding receptor Tyr kinases. [128] Two members of this family are encoded by the human genome: DDR1 [129] and DDR2. [130]

---

126. The CDK5–CDK5R1 and CDK5–CDK5R2 complexes latter also phosphorylate neurofilaments, protein phosphatase inhibitor-1, P53, presenilin-1, ERK1, JNK3, and the microtubule-associated Tau protein.

127. Vitamin-K–dependent plasma proteins (prothrombin, coagulation factor-VII, -IX, and -X, protein-C, -Z, and -S, and GAS6) bind to negatively charged phospholipid membranes via a γ-carboxyglutamic acid-containing motif. γ-Carboxyglutamic acid domain-containing protein GAS6, a platelet-response amplifier, is structurally similar to anticoagulant protein-S, a cofactor for activated protein-C. However, GAS6 lacks the anticoagulant activity of protein-S. Inactivation of GAS6 prevents venous and arterial thrombosis in mice with normal bleeding.

128. The discoidin domain has been discovered in discoidin-1 of Dictyostelium discoideum. Coagulation factor-V and -VIII as well as neuropilin (a.k.a. vascular endothelial cell growth $factor_{165}$ receptor) possess a discoidin domain.

129. Discoidin domain-containing receptor-1 is also called cell adhesion kinase (CAK), epithelial discoidin domain receptors EDDR1 and EDDR2, neuroepithelial Tyr kinase (NEP), tropomyosin receptor kinase TRKe, neurotrophic Tyr kinase receptor NTRK4, protein Tyr kinase PTK3, protein Tyr kinase RTK6, mammalian carcinoma kinase MCK10, human homolog of the chicken and mouse FAK gene product HGK2, and CD167a.

130. Discoidin domain-containing receptor-2 is also termed receptor protein-Tyr kinase TKT, Tyr-protein kinase Tyro10, neurotrophic Tyr kinase receptor-related NTRKR3, and CD167b.

Five DDR1 isoforms are generated by alternative splicing. The DDR1c isoform is the longest DDR1 (919 amino acids). Isoforms DDR1a and DDR1b are the shorter (882 and 913 amino acids, respectively) [1196]. Isoforms DDR1d and DDR1e are truncated variants that lack either the entire kinase region or parts of the juxtamembrane region and the ATP-binding site.

In several cell types, DDR1 is processed into a membrane-anchored β subunit and soluble extracellular domain-containing α subunit, especially upon DDR1 activation [1196].

8.2.9.1 DDRs and Other Collagen Receptors

Unlike most receptor Tyr kinases, DDRs are not activated by soluble growth factors, but by collagens. The DDR1 receptor autophosphorylates, i.e., is activated by collagen-1 to -6 and -8, whatever their types (fibrils or monomers) [1197]. The DDR2 receptor is only activated by fibrillar collagens (not monomeric collagens), in particular collagen-1 to -3, and -5 (but not collagen-4). The DDR receptors are activated only when collagen is in its native, triple-helical form [1196]. Maximal activation of DDRs occurs several hours after the initial stimulation by collagen.

Fibrillar and non-fibrillar collagens can anchor cell-adhesion receptors. Five different types of collagen receptors have been identified: (1) integrins, particularly 4 integrin heterodimers that contain β_1 subunit associated with α_1, α_2, α_{10}, or α_{11} subunit; (2) glycoprotein-6 (platelet glycoprotein receptor for collagen) expressed on thrombocytes; (3) leukocyte-associated Ig-like receptor LAIR1 (or CD305) on B, NK, and T cells; (4) members of the mannose receptor family of the C-type lectin subclass, in particular on dendritic cells; and (5) discoidin domain-containing receptors DDR1 and DDR2.

Both integrin and discoidin domain-containing receptors monitor the integrity of collagens in the extracellular matrix and can trigger matrix degradation and renewal [1198]. Binding of collagen to integrins causes Tyr phosphorylations, independently of DDRs, by integrin-associated kinases of the SRC and FAK family.

8.2.9.2 DDR Expression Pattern

Discoidin domain-containing receptors are more or less widespread (Table 8.16). The DDR1 isoform is synthesized in epithelial cells of the brain, lung, kidney, mammary gland, and intestine. In human bronchial epithelial cells, DDR1 predominantly localizes to the basolateral surface, where it colocalizes and interacts with its ligand collagen-4 of the basement membrane [1199]. The DDR2 isoform is expressed in the heart, skeletal muscle, skin, liver, and renal connective tissue.

8.2.9.3 DDR Effects

Collagen receptors DDR1 and DDR2 regulate extracellular matrix remodeling, as they elevate both the synthesis and activity of matrix metallopeptidases [1200]. The

Table 8.16. Distribution of DDR1 and DDR2 (Source: [1196]).

Type	Type of cell, tissue, or organ
DDR1	Bronchial epithelium, vascular smooth muscle cells, cerebellum, renal mesangial cells, collecting tubules, colon epithelium, mammary gland, cornea, keratinocytes
DDR2	Collecting tubules, heart and skin fibroblasts, hepatic lipocytes (pericytes or stellate cells)

Table 8.17. Collagen DDR receptors, their partners, and signaling network (Sources: [59, 1196]; NCK: non-catalytic region of tyrosine kinase adaptor; PTPn: protein Tyr phosphatase non-receptor; RGS: regulator of G-protein signaling; SHC1: SH2 domain-containing-transforming protein [or SHCa; adaptor]; snapin: SNARE-associated protein [or SNAPAP]; TM4SF: transmembrane-4 superfamily member [tetraspanin]; snRNP40: 40-kDa U5 small nuclear ribonucleoprotein). The DDR receptors are involved in cell differentiation, proliferation, and migration, as well as matrix remodeling. Full activation of DDR2 by fibrillar collagen requires SHC1 adaptor and Src kinases.

Type	Partners
DDR1	NCK2, SHC1, RGS2, snapin, TM4SF1, snRNP40 PI3K, PTPn11 (SHP2), STAT5 (indirect interaction) SHC1–TRAF6–P38MAPK pathway SHC1–TRAF6–NKκB pathway
DDR2	SHC1, Src (indirect interaction)

DDR1 isoform upregulates the secretion of MMP2 and MMP9 enzymes. In response to collagen-1, phosphorylated DDR2 also mediates MMP2 release. In human smooth muscle cells, DDR1 or DDR2 enhances MMP1 synthesis, but only DDR2 heightens production of MMP2 that degrades collagen and elastin [1201].

Moreover, DDRs contribute to the control of cell adhesion and migration, as well as proliferation and survival [1202]. Collagen-1-dependent upregulation of DDR2 enhances cell proliferation and migration (positive feedback loop) [1200].

8.2.9.4 DDR Signaling

Stimulation by collagen-1 causes aggregation of discoidin domain-containing receptor DDR1 and, then, incorporation of DDR1 dimers into early endosomes [1203]. Isotype DDR1 suppresses cell spreading.

In macrophages, DDR1 phosphorylates SHC1 adaptor that then activates TRAF6 Ub ligase. The latter can triggers the P38MAPK and NFκB pathways (Table 8.17). Several other molecules interact directly with DDRs [1196].

In humans, both DDR1 and DDR2 upregulate P-selectin glycoprotein ligand (or CD162) that resides on the surface of leukocytes (myeloid cells and stimulated T lymphocytes) and endothelial cells. It tethers to P-selectin of activated platelets or endothelial cells. Conversely, activated DDRs repress gene expression of agrin, syndecan-1, and α_3-integrin [1204].

Both collagen receptors $\alpha_2\beta_1$-integrin and DDR1 participate in N-cadherin upregulation that promotes tumor growth via Jun N-terminal kinase [1205]. Each receptor signals via separate pathways, but involves BCAR1 scaffold [131] Integrin-$\alpha_2\beta_1$ and DDR1 target focal adhesion kinases FAK1 and FAK2, respectively. [132] Small GTPase Rap1, but not Rho, is required in the response to collagen-1.

E-cadherin complexes with both DDR1 isoforms DDR1a and DDR1b. E-cadherin impedes DDR1 activity by sequestering DDR1 to cell junctions to prevent its contact with collagen ligand [1207].

8.2.9.5 DDRs in Diseases

In atherosclerosis, both DDRs can be highly expressed by smooth muscle cells within the fibrous cap [1196]. Overexpression of DDR1 or DDR2 in human smooth muscle cells provokes MMP1 production. Overexpression of DDR2 can also cause MMP2 synthesis.

During the progression of lung fibrosis, DDR1b is selectively produced. The expression of CCL2 chemokine, interleukins, and MMP9 thus raises [1196].

8.2.10 Leukocyte Receptor Tyrosine Kinase

The ALK–LTK (anaplastic lymphoma and leukocyte Tyr kinase) family of receptor tyrosine kinases represents class-10 RTKs. [133]

Extracellular domain of ALK is highly similar to, but much larger than that of LTK receptor. The receptor kinase ALK has a restricted expression pattern. It is produced in the central nervous system, i.e., scattered neurons and glial and endothelial cells, especially in the developing brain [1210]. It activates extracellular signal-regulated protein kinase [1209]. *Pleiotrophin* [134] of the neurite growth-promoting

131. Breast cancer anti-estrogen resistance docking protein is also called P130CAS and CAS (Crk-associated substrate).

132. Kinase FAK2 that particularly abounds in pulmonary vascular endothelial cells and lung is required for the synthesis of FAK1 and BCAR1 proteins [1206].

133. The Alk gene was first described as the typical chromosomal translocation (nucleophosmin–anaplastic lymphoma kinase gene fusion) in non-Hodgkin's lymphoma such as anaplastic CD30+, large-cell lymphoma [1208]. Molecule TNFRSF8 (or CD30), is a plasmalemmal protein of the tumor-necrosis factor receptor family that serves as tumor marker. Rearrangements of the Alk gene is involved in many genetic translocation events such as inflammatory myofibroblastic tumors.

134. A;k.a. heparin-binding growth factor HBGF8, heparin affinity regulatory peptide (HARP), heparin-binding growth-associated molecule (HBGAM), heparin-binding brain mitogen (HBBM), neurite growth-promoting factor NEGF1, and osteoblast-specific factor OSF1. It is encoded by the PTN gene.

factor (NEGF) family causes proliferation of epithelial, endothelial, and mesenchymal cells. It is a ligand for ALK, but not for LTK [1210]. [135]

8.2.11 Muscle-Specific Kinase

Muscle-specific kinase (MuSK) is involved in the neuromuscular junction, where it enables acetylcholinesterase anchoring [1211]. At the neuromuscular junction, acetylcholinesterase links to acetylcholinesterase-associated collagen-Q (ColQ) [136] of asymmetrical acetylcholinesterase [137] and, then, accumulates at the synaptic basement membrane to control the temporal and spatial cholinergic neurotransmission from motoneuron to skeletal myofiber. [138] The ColQ–perlecan–MuSK complex is required for acetylcholinesterase clustering [1211].

Muscle-specific kinase is activated by agrin, [139] a nerve-derived proteoglycan. The MuSK receptor recruits casein kinase-2 for clustering [1212]. It also binds Disheveled of the Wnt pathway that also operates in MuSK-mediated clustering of acetylcholine receptors [1213]. Acetylcholine receptor aggregation in fact relies on a MuSK platform formed by scaffolds rapsyn, a cytoplasmic membrane-associated protein, and DOK7 adaptor that is required for synaptogenesis.

8.2.12 Neurotrophic Tyrosine Receptor Kinases

Neurotrophic Tyr receptor kinases (NTRK1–NTRK3; Table 8.18) [140] and low-affinity nerve growth factor receptor (LNGFR) [141] are targeted by members of the neurotrophin family, i.e., neurotrophins NT3 and NT4, as well as nerve growth factor (NGF) and brain-derived neurotrophic factor (BDNF).

Both NTRK1 and NTRK2 exist as monomers or dimers. Low-affinity nerve growth factor receptor, i.e., the common pan-neurotrophin receptor, has some similarities with the tumor-necrosis factor receptor, [142] lacks a Tyr kinase domain, but can signal via released ceramide as well as nuclear factor-κB activation [5].

Neurotrophic receptors have different binding affinity for diverse types of neurotrophins. The NTRK1 receptor has the highest affinity for nerve growth factor;

135. Protein Tyr phosphatase receptor PTPRb and syndecan-3 are other receptors for pleiotrophin.

136. Collagen-like tail subunit that homotrimerizes.

137. I.e., collagen-tailed tetramer of catalytic subunits (A form), which links to a single (A4), 2 (A8, i.e., 8 catalytic subunits per collagen tail), or mostly 3 strands (A12) of ColQ.

138. This accumulation results from the interaction between ColQ and perlecan bound to dystroglycan [1211].

139. Agrin is involved in the aggregation of acetylcholine receptors during synaptogenesis. Agrin binds MuSK as well as several other proteins, such as dystroglycan and laminin, to stabilize the neuromuscular junction. Low density lipoprotein receptor LRP4 operates as a coreceptor for agrin.

140. A.k.a. tropomyosin receptor kinases (TRKa–TRKc).

141. A.k.a. P75 neurotrophin receptor (P75NTR).

142. Hence its other name tumor-necrosis factor receptor superfamily member TNFRSF16.

Table 8.18. Family of neurotrophic Tyr receptor kinases and their ligands (Source: [1121]). Receptor NTRK1 (TRKa) has the highest binding affinity for nerve growth factor (NGF), NTRK2 (TRKb) for brain-derived neurotrophic factor (BDNF) and neurotrophin NT4, and NTRK3 (TRKc) to neurotrophin-3. Low-affinity nerve growth factor receptor (LNGFR), or P75 neurotrophin receptor (P75NTR), is coexpressed with NTRK receptors. It binds to all members of the neurotrophin family. It can interact with other receptors to promote signaling. It can also produce its own signaling that require its proteolysis by the γ-secretase peptidase complex. The LNGFR receptor raises the binding affinity and specificity for NTRKs, lowers ligand-induced receptor ubiquitination, and delays receptor endocytosis and degradation.

Type	Ligands
NTRK1	NGF, (NT3)
NTRK2	BDNF, NT4, (NT3)
NTRK3	NT3
LNGFR	All neurotrophin family members Immature neurotrophins *gg* mature neurotrophins

NTRK2 for brain-derived neurotrophic factor and NT4; and NTRK3 for NT3 neurotrophin. Both NTRK1 and NTRK2 also bind NT3, but to a lesser extent.

Neurotrophic Tyr receptor kinases form a family of regulators of synaptic activity in the nervous system as well as neuronal differentiation, growth, and survival, among other functions.[143]

8.2.12.1 Low-Affinity Nerve Growth Factor Receptor

Neurotrophins are synthesized in immature forms and, then, undergo a cleavage by peptidases. Immature neurotrophins connect to LNGFR receptor. Mature neurotrophins have higher affinity for their corresponding NTRK receptors. They can still bind to LNGFR, but with a much lower affinity. In addition, LNGFR promotes high-affinity binding of NGF to NTRK1 receptor. LNGFR Receptor can also reduce receptor ubiquitination and delay receptor internalization and degradation [1215].

In sympathetic neurons, LNGFR can form a high-affinity receptor complex with NTRK1 and facilitates the growth and survival effects of nerve growth factor [1216]. The LNGFR–NTRK1 interaction enhances selectivity for NGF over NT3 ligand.[144] On the other hand, BDNF produced by neurons and/or their innervated cells act via through LNGFR to antagonize NGF effects. Agent BDNF promotes apoptosis via LNGFR alone to establish the suitable balance between the density of the neuronal population and that of cells of innervated tissue as well as between sympathetic and

143. Neurotrophin signaling regulates axon and dendrite growth and patterning, hence synaptic strength and remodeling, cell survival and proliferation, as well as neural precursor fate, in addition to control of expression and activity of ion channels and neurotransmitter receptors [1214].

144. Developing sympathetic neurons use NT3 during their growth toward their targets and switch to NGF near their destination.

Table 8.19. Positional situation of tyrosine residues targeted by NTRK1 transphosphorylation (Source: [1218]). After neurotrophin binding, NTRK receptors dimerization triggers the activation of the kinase domain that transphosphorylates tyrosine residues on the complementary receptor. Phosphotyrosines then provide docking sites for various adaptors, such as fibroblast growth factor receptor substrate FRS2, receptor-associated protein of the synapse (RAPS), SH2-containing collagen-related protein SHC, SH2 domain-containing adaptor SH2b, and signaling mediators, such as CSK homologous kinase (CHK) and phospholipase-Cγ1 (PLCγ1), to activate mitogen-activated protein kinase (MAPK), phosphatidylinositol 3-kinase (PI3K), protein kinase-C (PKC), and Ca^{++} signaling pathways.

Phosphorylation sites	Partners (mapping)
Tyr490	FRS2, SHC
Tyr670	RAPS, SH2b
Tyr674	RAPS, SH2b
Tyr675	RAPS, SH2b
Tyr785	CHK, PLCγ1

parasympathetic neuronal populations. In addition, LNGFR can interact with plasmalemmal reticulon-4 receptor (Rtn4R; a.k.a. Nogo receptor) [145] to transmit growth-repulsive signals activated by a family of myelin proteins Cholinergic neurons of the intrinsic cardiac system also express LNGFR [1216].

Besides its interactions with other receptors, LNGFR can also create its own signals. Its extracellular domain can be cleaved by a α-secretase. Afterward, a transmembrane presenilin-associated γ-secretase peptidase complex provokes intramembrane cleavage to release an intracellular domain ($LNGFR^{ICD}$). Its intracellular domain fragment can then moves to the nucleus to regulate gene expression such as the CCNE1 gene that encodes cyclin-E1 [1217]. The $LNGFR^{ICD}$ fragment can also interact with multiple intracellular proteins to influence numerous signalings.

8.2.12.2 Signaling

Neurotrophic receptors dimerize in response to ligand binding. These dimers then bear transphosphorylation that enhances their catalytic activity (Table 8.19). They activate the PLC–DAG–PKC and PLC–IP_3–Ca^{++}, Ras–MAPK, and PI3K–PDK1–PKB pathways [1214] (Table 8.20). They activates several small GTPases, such as Ras, Rap1, and members of the RHO family (CDC42, Rac, and Rho; Vol. 4 – Chap. 8. Guanosine Triphosphatases and Their Regulators).

145. Reticulon-4 receptor is targeted by reticulon-4, neurite outgrowth inhibitor (Nogo), oligodendrocytemyelin, and myelin-associated glycoprotein(MAG). This receptor inhibits axonal growth via Rho kinase (ROCK). It is involved in neuronal adaptivity and regeneration.

Table 8.20. Stages and pathways of neurotrophin-induced NTRK signaling (Source: [1218]; CHK: CSK homologous kinase; CSK: C-terminal Src kinase; CRK: chicken tumor virus regulator of kinase; DAG: diacylglycerol; ERK: extracellular signal-regulated kinase; FRS: fibroblast growth factor receptor substrate; GAB: Grb2-associated binder; GRB: growth factor receptor-binding protein; IP_3: inositol (1,4,5)-trisphosphate; MAP2K: mitogen-activated protein kinase kinase; PDK: phosphoinositide-dependent protein kinase; PI3K: phosphatidylinositol 3-kinase; PKB: protein kinase-B; PKC: protein kinase-C; PLC: phospholipase-C; SH2b: SH2 domain-containing adaptor; SHC: SH2-containing collagen-related proteins; SOS: Son of sevenless).

NTRK Dimerization
Transphosphorylation (Tyr)
FRS2 Binding FRS2–CRK–RapGEF1–Rap1–bRaf–MAP2K1/2–ERK1/2 (sustained stimulation)
SHC Binding SHC–GRB2–SOS–Ras–cRaf–MAP2K1/2–ERK1/2 (transient stimulation) SHC–GRB2–SOS–Ras–PI3K–PDK1–PKB SHC–GRB2–GAB1/2–PI3K–PDK1–PKB
SH2B Binding SH2B–GRB2–GAB1/2–PI3K–PDK1–PKB SH2B–GRB2–SOS–Ras–Raf–ERK1/2 SH2B–GRB2–SOS–Ras–PI3K–PDK1–PKB
RAPS Binding RAPS–GRB2–GAB1/2–PI3K–PDK1–PKB RAPS–GRB2–SOS–Ras–Raf–ERK1/2 RAPS–GRB2–SOS–Ras–PI3K–PDK1–PKB
CHK Binding
PLCγBinding PLCγ–DAG–PKCδ–MAP2K1/2 PLCγ–IP_3–Ca^{++}

8.2.12.3 Neurotrophin Receptors in the Cardiovascular System

In the cardiovascular system, members of the neurotrophin family participate in various processes directly on vasculature cells or via nervous control. They are actually implicated in the heart development, control of blood vessel, as well as smooth muscle cell growth. Nerve growth factor is produced in the heart and vasculature. It promotes not only sympathetic neuron survival and axonal growth and branching, but also cardiomyocyte survival.

The sympathetic innervation of the heart develops owing to chemoattractants such as nerve growth factor and chemorepulsants such as semaphorin-3A. Nerve growth factor stimulates sympathetic axon extension into the heart via NTRK1 receptor [1219]. It also acts via LNGFR to modulate the activity of multiple coreceptors that can stimulate or inhibit axon outgrowth [1219]. Different neurotrophins may have distinct effects. Agent NT3 may counter maladaptive hypertrophy.

Both sympathetic and parasympathetic networks of the autonomic nervous system control the state of the cardiovascular system via their coordinated activity transmitted from the central nervous system to corresponding autonomic ganglia (Vol. 6 – Chap. 3. Cardiovascular Physiology). [146] Sympathetic nerves stimulate cardiac function via noradrenaline and β1-adrenoceptors. A transient imbalance can occur between sympathetic and parasympathetic inputs. Abnormal cardiac parasympathetic activity concomitant with sympathetic hyperactivation can persist, hence causing the so-called *autonomic neural remodeling* with high risk of life-threatening events.

Plasmalemmal LNGFR contributes to the regulation of the sympathetic innervation of the heart. Depletion of LNGFR causes a loss in nerve input to the subendocardium of the left ventricle, but not subepicardium and right ventricle [1219].

As sympathetic neurons innervate the heart, LNGFR blunts the repulsive effects of Sema3a that is expressed in the subendocardium, thereby allowing axonal arborization. In the absence of LNGFR, Sema3a potently repels the growing axons away from the subendocardium.

8.2.13 Protein Tyrosine Kinase-7

Protein tyrosine kinase PTK7 lacks detectable catalytic activity. Nevertheless, it operates in signal transduction. Four alternatively splice transcript variants encode 4 isoforms [1284].

The PTK7 kinase participates in cell migration. It cooperates with Wnt receptor Frizzled (alternative signaling) to recruit Disheveled, a regulator of both the Wnt canonical and alternative signalings, to the plasma membrane [1220]. Signaling initiated by semaphorin-6 and plexin-A via PTK7 influences Wnt alternative signaling via VanGl protein [1221].

The PTK7 kinase contributes to capillary-like tube formation from human vascular endothelial cells [1222]. It promotes the activity of vascular endothelial growth factor. It actually favors VEGF-induced phosphorylation of focal adhesion kinase and paxillin, relocalization of paxillin to focal adhesions, and formation of stress fibers.

8.2.14 Ret Receptor Family – GDNF Family Receptors

Receptor Tyr kinase Rearranged during transfection (ReT) is targeted by members of the glial cell line-derived neurotrophic factor (GDNF) family. The GDNF protein promotes survival and differentiation of neurons.

---

146. Activated cardiac sympathetic efferents increase heart rate and contractility. In the vasculature, the sympathetic nervous system reduces venous capacitance, constricts resistance vessels, hence heightening blood pressure. Conversely, the parasympathetic nervous system decreases heart rate. It counteracts β-adrenergic action and inhibits directly Ca_V1 channels.

Table 8.21. GDNF Family coreceptors (Source: [5]).

Type	Ligands (potency order)
GFRα1	GDNF > neurturin > artemin
GFRα2	Neurturin > GDNF
GFRα3	Artemin
GFRα4	Persephin

The GDNF family of dimeric ligands (GFL) consists of 4 neurotrophic factors that function as homodimers: GDNF, [147] artemin, neurturin, and persephin. These GFLs are synthesized as inactive precursors (preproGDNF family ligands) that are processed successively into proGDNF family ligands and active GDNF family ligands.

The natural alternative splicing of the RET gene produces 3 isoforms: ReT_{51}, ReT_{43}, and ReT_9 (that contain 51, 43 and 9 amino acids in their C-termini, respectively).

The ReT receptor interacts with adaptors GRB2, GRB7, and GRB10, SHC1, DOK1 and DOK5, as well as transcription factor STAT3 and GDNF family receptor GFRα1.

At the cell surface of target cells, GDNF family ligands bind to complexes composed of ReT receptor and a coreceptor of the GFRα family (Table 8.21). [148] In order to activate ReT, GFLs first complex with coreceptor GFRα. Coreceptors GFRα1 to GFRα4 are bound by their respective primary ligands GDNF, neurturin, artemin, and persephin [1223]. Once the GFL–GFRα complex is formed, it couples 2 ReT receptors that then transautophosphorylate. Phosphorylated ReT initiates intracellular signaling.

8.2.15 Receptor-like Tyrosine Kinase

Receptor-like Tyr (Y) kinase (RYK) that is encoded by the Ryk gene differs from other RTKs by its activation and nucleotide-binding domains. It launches its activity by recruiting an auxiliary protein. Two alternative splice variants generate to distinct isoforms.

Receptor RYK participates in alternative modes of Wnt signaling (Sect. 10.3), as it operates as a Wnt-binding receptor [1224]. In mammals, RYK binds both Wnt1 and Wnt3a. It is required for canonical Wnt signaling that is based on β-catenin stabilization and its association with T-cell factor to activate target genes. It forms a

147. In addition to the transcript encoding by gene Gdnf, 2 additional alternatively spliced transcripts lead to distinct proteins that are referred to as astrocyte-derived trophic factors.

148. The GFRα1 member of the GDNF family receptor encoded by the GFRA1 gene is a glycosyl phosphatidylinositol-linked plasmalemmal receptor for both glial cell line-derived neurotrophic factor and neurturin. It is also called retinoic acid early transcript ReT1L, or ReTL1, as well as GDNFRαA and TrnR1. It activates ReT receptor .

ternary complex with Frizzled and Wnt, hence acting as a Frizzled coreceptor that binds to Disheveled.

8.2.16 Receptor Tyrosine Kinase-like Orphan Receptor Family (ROR / WNRRTK)

Receptor Tyr kinase-like orphan receptors ($ROR_{(RTK)}$) [149] that are related to the NTRK family members possess extracellular Frizzled-like cystein-rich, kringle, [150] and N-terminal immunoglobulin-like domains. Two genes encode WNRRTK1 (or ROR1) and WNRRTK2 (or ROR2). [151] The group of WNRRTK ($ROR_{(RTK)}$) kinases is included in the NTRK superfamily of receptor Tyr kinases, with the MuSK, neuro-specific receptor kinase (NRK), discoidin domain-containing receptors (DDR), and neurotrophic Tyr kinase (NTRK) families.

Splice variants of the Wnrrtk1 transcript encode truncated proteins lacking either the extracellular domains (i.e., truncated WNRRTK1 [$_{t}$WNRRTK1] that can be artifactual) or transmembrane and intracellular domains [1225].

Receptor WNRRTK kinases are expressed in many tissue types during development, particularly skeletal development. They actually function in cell migration and polarity. In mice, during the organism development, WNRRTK1 and WNRRTK2 are produced according to a spatiotemporal regulation [1226]. They contribute to the heart development, in particular cardiac septal formation [1227]. Like many other RTKs, they form homodimers.

Despite their original name due to unknown ligand, $ROR_{(RTK)}$ proteins are Wnt receptors, thereby renaming WNRRTK enzymes. Depending on the cellular context, single-pass WNRRTK receptors can activate or repress transcription of Wnt target genes. Moreover, they can modulate Wnt signaling by sequestering Wnt ligands. Tissue polarity is determined by the integration of multiple, opposite Wnt pathways, as signaling is processed from Wnt receptor Frizzled (a GPCR) and receptor RYK and WNRRTK kinases [1228].

The Wnt5a ligand inhibits Wnt3a-induced canonical Wnt signaling in a dose-dependent manner without influencing β-catenin level. Receptor kinase WNRRTK2 acts as a Wnt receptor or coreceptor to: (1) activate canonical β-catenin signaling via Frizzled-4 and LRP5 [1229]; (2) hamper βCtn–TCF cascade triggered by other Wnt isoforms; and (3) activate the non-canonical Wnt–JNK axis [1230].

---

149. Alias ROR is also used to designate RAR-related orphan nuclear receptor (NR1f1–NR1f3), hence alias $ROR_{(RTK)}$ used in the present text, but its subtypes are simply abbreviated as ROR1 and ROR2, which differ from aliases RORα to RORγ used for nuclear receptor isoforms. Alias NTRKR that stands for neurotrophic Tyr receptor kinase-related protein, despite its abandoned usage, avoids ambiguity. Because $ROR_{(RTK)}$ protein is a Wnt receptor, its designation as an orphan receptor should be updated. In the present text, alias WNRRTK is proposed (Wnt and neurotrophin receptor-related receptor Tyr kinase).

150. Sponge Frizzled Kringle protein could be representative of an ancestral family of proteins present in early metazoa, before $ROR_{(RTK)}$, MuSK, and NTRK split.

151. Paralogs ROR1 and ROR2 were formerly called neurotrophic Tyr kinase receptors NTRKR1 and NTRKR2, respectively.

The Wnt5a ligand and WNRRTK2 cooperate during cell migration. The former provokes phosphorylation of WNRRTK2 by glycogen synthase kinase GSK3 [1225]. Phosphorylated WNRRTK2 can then operates in cell migration. However, WNRRTK2 can influence the cytoskeleton independently of Wnt5a, especially in filopodia.

Scaffold 14-3-3β protein is a WNRRTK2-binding partner that is phosphorylated by WNRRTK2 [1225]. Kinase Src is activated by the Wnt5a–WNRRTK2 axis. Agonist Wnt5a causes WNRRTK2 phosphorylation followed by rapid internalization. It also binds to WNRRTK1 and can then activate nuclear factor-κB.

Kinase WNRRTK can function independently from Wnt ligands. In mice, the melanoma-associated antigen (MAGE) family member MAGeD1 [152] binds to WNRRTK2, but not to WNRRTK1 subtype. Isoform WNRRTK2 recruits MAGeD1 from the cytoplasm to the plasma membrane. In the absence of WNRRTK2, MAGeD1 localizes to the nucleus. Kinase WNRRTK2 thus indirectly affects the transcriptional activity of MAGeD1 interactor Msx2 [1225]. Casein kinase-1ϵ is also a WNRRTK2-binding partner that phosphorylates WNRRTK2 and provokes its autophosphorylation. Kinase WNRRTK2 phosphorylates GRK2 (GPCR) kinase. Growth differentiation factor GDF5 and bone morphogenetic protein receptor BMPR1b are also WNRRTK-interacting proteins.

8.2.17 Ros1 Receptor Tyrosine Receptors

Ros1 receptor-like Tyr kinase encoded by human ortholog of V-ros UR2 sarcoma avian (chick retrovirus) virus proto-oncogene ROS1 is an orphan receptor that belongs to class-2 RTKs (insulin receptor). It resides exclusively in specific epithelia. Transient, restricted Ros1 expression is observed during development in the lung, kidney, and intestine [1231].

Phosphorylated Ros (Tyr2267) strongly and directly tethers to protein Tyr phosphatase PTPn6 (or SHP1) that is expressed in hematopoietic and, to a lesser extent, epithelial cells, to prevent ROS1 signaling [1232, 1233]. Glioblastoma-associated, ligand-independent rearrangement product of Ros activates phosphatase SHP2 and the phosphatidylinositol 3-kinase–protein kinase-B–target of rapamycin axis [1234].

8.2.18 Ephrin Receptors

Erythropoetin-producing hepatocyte (or hepatoma; EPH) receptor interactors — the so-called *ephrins* — are transmembrane proteic ligands that bind to EPH receptor protein Tyr kinases on adjacent cells to transmit juxtacrine bidirectional signaling via intercellular, between-plasma membrane contacts, i.e., with signal propagation within both cells, especially neurons and epithelial cells. Information is then processed by the 2 interacting cells.

Ephrins and EPH receptors participate in organ development, in particular angiogenesis, and repair, as well as insulin secretion and immunity. Ephrins and EPH

152. A.k.a. Dlxin-1 and neurotrophin receptor-interacting MAGe homolog (NRAGe).

Table 8.22. Family of EPH receptors and their ligands (Source: [1121]). Ephrins and EPHs are mainly involved in morphogenesis (cell attraction and repulsion, survival, differentiation, sorting, and motility). In mammals, 14 known ephrin receptors (EPHa1–EPHa8, EPHa10, and EPHb1–EPHb6) are targeted by 8 identified ephrin ligands (ephrin-A1–ephrin-A5 and ephrin-B1–ephrin-B3). In humans, 9 type-A receptors (EPHa) bind to 5 glycosyl-phosphatidylinositol-anchored ephrin-A ligands; 5 type-B EPH receptors (EPHb) connect to 3 transmembrane ephrin-B ligands that have a transmembrane domain with a short cytoplasmic region. Ligand–receptor specificity is low within the A and B categories; some inter-category associations (ephrin-A–EPHb and ephrin-B–EPHa) exist. Liganded EPHs trigger a bidirectional mode of signaling after tetramerization.

Type	Ephrin-A ligands	Ephrin-B ligands
EPHa1	Ephrin-A1	
EPHa2	Ephrin-A1, ephrin-A3–ephrin-A5	
EPHa3	Ephrin-A1–ephrin-A3, ephrin-A5	
EPHa4	Ephrin-A1–ephrin-A3, ephrin-A5	Ephrin-B1–ephrin-B3
EPHa5	Ephrin-A1–ephrin-A5	
EPHa6	Ephrin-A1–ephrin-A5	
EPHa7	Ephrin-A1–ephrin-A3	
EPHa8	Ephrin-A2, ephrin-A3, ephrin-A5	
EPHb1	Ephrin-A3	Ephrin-B1–ephrin-B2
EPHb2	Ephrin-A5	Ephrin-B1–ephrin-B3
EPHb3		Ephrin-B2–ephrin-B2
EPHb4		
EPHb5		
EPHb6		

receptors modulate the reorganization of the actin cytoskeleton. They control cell shape, adhesion, separation, and migration (attraction and repulsion). Signals generated by the engagement of ephrins to their EPH receptors generally cause repulsive responses, such as retraction of the cell periphery and cell rounding.

The EPH receptors constitute the largest group of receptor Tyr kinases (about a quarter of all known human RTKs; Table 8.22). Members of the ephrin family are classified according to their composition and cell-surface binding. Two types of EPH receptors and ephrins exist: type-A (EPHa1–EPHa8; ephrin-A1–ephrin-A5), type-A ephrins being connected to the plasma membrane by a glycosyl-phosphatidylinositol anchor, and type-B (EPHb1–EPHb6; ephrin-B1–ephrin-B3), type-B ephrins being tethered to the plasma membrane by a single transmembrane segment. Type-A and -B EPH receptors bind to most or all type-A and -B ephrins, respectively. However, EPHa4 can bind to both type-A and most type-B ephrins. In addition, 3 EPHa10 isoforms, one soluble and 2 transmembrane isoforms, are mainly expressed in the testis [1235]. Subtypes EPHa10 and EPHb6 are pseudokinases.

The EPH receptors possess: (1) an extracellular region with an N-terminal ligand-binding domain, a cysteine-rich motif, and 2 fibronectin-3 sequences; (2) transmembrane domain; and (3) intracellular region with a Tyr kinase domain, a sterile-α motif, and a C-terminal PDZ-binding sequence [1236]. Their ephrin ligands contain an N-terminal extracellular receptor-binding domain and C-terminus linked to the plasma membrane by a lipid anchor (class-A ephrins) or transmembrane helix (class-B ephrins).

Receptors EPHs may associate in homotypic complexes via their extracellular or/and cytoplasmic domains. When the density of EPH receptor remains low, preclustered ephrins initialize EPH clustering. [153] Above a concentration threshold, free EPHa receptors can aggregate via interactions of their ectodomains, independently of ephrin binding [639].

8.2.18.1 Signal Processing by Interacting Cells

Upon contact between receptors and ligands from apposed cells, preclustered ephrins form homo-oligomers that bind EPHs (1:1 stoichiometry). Once this initial contact is achieved, the EPH–ephrin complex dimerizes and, then, tetramerizes. This aggregation causes conformational changes in both the receptor and ligand. Signaling occurs only after tetramerization of the EPH–ephrin complex [639]. The EPH Tyr kinase domains can transphosphorylate each other to launch a forward signaling. Moreover, tetramers can further assemble to regulate the mode and strength of signaling.

Small amounts of ephrins can support assembling of EPHs that can then recruit more receptors in the region of initial intercellular contact and strengthen intercellular communications. Cooperative homo- and heteromeric binding enable amplification of weak signals. However, different multimeric states of the EPH–ephrin complex (di-, tetra-, and higher-order multimers) provoke different cell responses [639].

Interactions between cells that express transmembrane EPH receptor Tyr kinases and those that produce membrane-bound ephrin ligands initiate cell-specific bidirectional signaling. Cells thet express EPH receptors and ephrin ligands use different types of protein Tyr kinases and subtrates such as adaptors to process signals induced by intercellular contacts [1237]. For example, recruitment of kinases of the SRC family to ephrins and phosphorylation of ephrin-B initiate reverse signaling.

Contact between cells that express either EPHb receptors or their transmembrane ephrin ligands initiates bidirectional, commonly asymmetrical signals that regulate cell positioning. Adjoining cells with intercellular contact use different cytosolic protein Tyr kinases to process ephrin-B1–EPHb2 signaling. In addition, phosphorylation of adaptors preferentially occurs in EPHb2-expressing cells than ephrin-B1-producing cells.

153. Only clustered or membrane-bound (and not free, soluble) ephrin-B1 can cause phosphorylation of EPHb1 receptor [639]. In addition, although ephrin-B1 dimer can provoke EPHb1 phosphorylation, receptor multimers are necessary for the recruitment of effectors.

8.2.18.2 Ephrin–EPH Signaling

A constitutively active form of EPHa4 that is phosphorylated albeit unliganded achieves its developmental activity only when it becomes clustered by ephrin-B, which enables accumulation of signaling effectors [639].

Receptor EPH kinases, like other RTKs, initiate signal transduction associated with intercellular communication via autophosphorylation after ephrin binding (*forward signaling*). Upon activation, they can form signaling clusters. Moreover, ephrins are plasmalemmal compounds with signaling capability (*reverse signaling*). Ephrins hence act in bidirectional communications, to and from a given cell (Fig. 8.5).

Binding affinities vary among types of EPH receptors and ephrin ligands, but, in general, binding within classes is predominant. Type-A ephrins bind to EPHa receptors on the contacting cell membrane. They are cleaved by adamlysins. Release of type-A ephrins from cells may then cause long-range action.

For example, once bound to its ligands such as ephrin-A1, EPHa2 undergoes dimerization and transphosphorylation of its cytoplasmic domains, then recruits a molecular complex with SHC and GRB2 adaptor proteins. Activated EPHa2 stimulates the mitogen-activated protein kinase and the phosphoinositide 3-kinase pathways and recruits CBL adaptor and adamlysin ADAM10. Both CBL and ADAM10 regulate receptor degradation.

Type-B ephrins link to homo- and heterodimeric EPHb receptors that possess a cytoplasmic tail with a kinase signaling activity and an extracellular part. In reverse signaling, ephrin-B cytoplasmic tail undergoes modifications, such as tyrosine and serine phosphorylation, and recruits signaling effectors.

The EPH receptors are inhibited by protein Tyr phosphatases. Protein Tyr phosphatase receptor PTPRo specifically dephosphorylates both EPHa and EPHb receptors for termination of signaling [1238].

Once activated by ephrins, phosphorylation of Tyr residues of EPH receptors in the cytoplasmic domain enables docking of many effectors. The EPH receptors bind to several effectors, such as small GTPases Ras, Src kinases, and Jun N-terminal kinases, hence regulating many pathways.

Once EPHa2, a target of the mitogen-activated protein kinase pathway, is transported to the plasma membrane, it binds to ephrin-A1 on the apposed cell surface to be activated, and downregulates MAPK signaling (negative feedback loop).

Size and pattern of EPHa2 clusters do not influence the phosphorylation status of the receptor, but impinge on the intracellular distribution of effectors, such as ^{F}actin and ADAM2 [639].

Stimulation of EPHa3 by ephrin-A5 inhibits migration, reduces neurite outgrowth, and causes growth cone collapse in hippocampal neurons [1239]. [154]

---

154. Mutations that target either Tyr602 or Tyr779 partially decreases EPHa-mediated responses, but those of both Tyr602 or Tyr779 completely abrogate these responses. Several SH2 domain-containing proteins bind to phosphorylated Tyr residues of the juxtamembrane domain of EPH receptors, such as Src, RasGAP, Vav, and CRK and NCK adaptors.

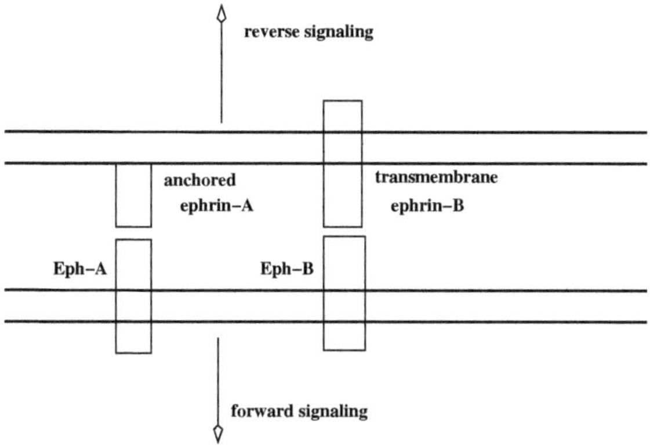

Figure 8.5. Bidirectional signaling by the ephrin–EPH complex. Type-A ephrins are attached to the plasma membrane by a glycosylphosphatidylinositol anchor, whereas type-B ephrins have a transmembrane domain with a short cytoplasmic region. Type-A and -B EPH receptors bind to most or all type-A and -B ephrins, respectively (except EPHa4 that can bind to type-A and most type-B ephrins). The Ephrins–EPH complexes initiate signaling in both directions from EPH receptors (forward signaling) and ephrin ligands (reverse signaling).

Signaling pathways can involve not only ligands and receptors but also coreceptors and matrix components. The ephrin–EPH complexes signal via additional regulators. Ephrin-A3 binds to heparan sulfate proteoglycans for full activity [1240]. Heparan sulfate proteoglycans serve to capture released ephrin-A3 on the cell surface and thus to increase the local concentration. [155]

The EPHb1 receptor modulate integrin-mediated cell–matrix adhesion according to the density of ephrin-B1 [639]. Although EPH receptors are phosphorylated once exposed to a low density of ephrin-B1, they only support cell attachment via $\alpha_V\beta_3$-integrins when ephrin-B1 density is greater than a given threshold. Below this threshold, EPH receptors reduce cell adhesion.

The EPHb2 receptor and its ligands are involved in the growth of the vascular smooth muscle cells [1241]. Furthermore, EPHb2 receptor and its ligands regulate interactions between endothelial and mesenchymal cells in developing arteries or smooth muscle cells in adult arteries.

Tyrosine phosphorylation of ephrin-B1 leads to recruitment of adaptor growth factor-receptor-bound protein GRB4. The ephrin-B1^P–GRB4 complex modifies the distribution of FAK-binding paxillin and increases the activity of focal adhesion kinase (FAK) for reorganization of the actin cytoskeleton. Using the 3 SH3 domains of GRB4, ephrin-B1 can be linked to numerous signaling molecules. Moreover, ephrin-B1 also contains PDZ-binding motif for additional association with proteins, such as phosphatases and GTPase-activating proteins. The extracellular region of ephrin-B1 also binds to the extracellular domains of adjacent claudin-1 or -4 in tight junc-

155. However, among ephrin-A1, -A2, -A5, -B1, and -B2, only ephrin-A3 binds to heparan sulfate proteoglycans.

tions [1242] (Vol. 1 – Chap. 7. Plasma Membrane). Tyrosine phosphorylation of the cytoplasmic domain of ephrin-B1 also depends on claudins. Phosphorylated ephrin-B1 increases the intercellular permeability, as interactions of claudins with ephrin-B1 hinder the formation of the tight adhesion between adjacent cells. Ephrin-B1 reverse signaling also activates the transcriptional activity of signal transducer and activator of transcription STAT3 after STAT3 recruitment to ephrin-B1 and phosphorylation by Janus kinase JaK2 [1243].

Signaling influences endocytosis. Activation of EPHb by ephrin-B causes phosphorylation, hence activity attenuation of phosphoinositol 5-phosphatase synaptojanin-1 that is involved in uncoating of clathrin-coated vesicles after internalization [11].

8.2.18.3 The Ephrin–EPH Complexes in Morphogenesis and Remodeling

Vascular endothelial growth factor, angiopoietin, ephrin, and Notch, among others, cooperate for the development of a functional vasculature. Agents DLL4, Notch-4, and ephrin-B2 are gradually upregulated during hepatocarcinoma evolution and expressed on tumor sinusoidal endothelial cells [1244]. In venous endothelial cells, VEGF stimulates the production of DLL4 and presenilin and assists in Notch-4 activation, thereby upregulating ephrin-B2, but downregulating EPHb4. The activation of Notch-4 enables VEGF-induced upregulation of ephrin-B2 and the differentiation of human venous endothelial cells in vitro.

Vasculature

Pericytes and smooth muscle cells apposed to endothelial cells express ephrin-B2 and EPHb2. Receptors EPHs and ephrins mediate a cell-repulsive response in migrating cells, destabilizing intercellular contacts.

Ephrin-B1 and -B2 are synthesized by endothelial cells in most arteries, ephrin-B1 as well as EPHb3 and EPHb4 in veins [1245, 1248, 1249].[156] Ephrin-B2 and EPHb4 are also detected in the lymphatic endothelium.

Ephrin-B2 and EPHb4, one of the ephrin-B2 cognate receptors predominantly expressed in the venous endothelium, regulate vasculature morphogenesis and blood vessel permeability during vasculo- and angiogenesis, including postnatal angiogenesis.

Ephrin-B2, a marker of arterial endothelial cells, promotes sprouting behavior and motility in the angiogenic endothelium. In fact, ephrin-B2 controls VEGF-induced angio- and lymphangiogenesis, as it regulates internalization of and signaling from VEGFR2 and VEGFR3, the receptor for VEGFc [1246].

156. Whereas ephrin-B1 is expressed by arterial and venous endothelial cells, ephrin-B2 is expressed on arteries but not veins. Ephrin-B2 interacts with multiple EPHb receptor classes, such as EPHb4 preferentially expressed on veins.

Nervous System

Ephrin ligands and their cognate EPH receptors guide axons during neural development and regulate synapse formation and neuron functional flexibility (synaptic remodeling) [157] in adults [1247].

The ephrin–EPH complexes operate in bidirectional neural–neural and neural–glial communications. Receptors EPHs expressed on dendrites are activated by ephrins on contacting axons or astrocytes. The ephrin–EPH complexes not only control spine [158] and synapse formation, [159] but also participate in synaptic strength, [160]

8.2.18.4 Mechanotransduction and Ephrin–EPH Complexes

Many receptors in the plasma membrane assemble into dimers, trimers, or higher-order oligomers, i.e., form clusters to trigger signaling. Receptor-ligand binding, clustering, and subsequent lateral transport is indeed observed in the junctional region between apposed cells.

Applied mechanical forces can influence lateral displacement of receptors within the membrane and change the degree of oligomerization of these receptors, i.e., the size of receptor clusters. As mechanical forces contribute to the spatial organization of clusters of plasmalemmal receptors, they regulate cell fate. Intra- and intercellular tensions actually participate in proper cell differentiation, growth, proliferation, and survival.

In particular, signaling between EPHa2 receptor Tyr kinases on a given cell membrane and their ephrin-A1 ligands on apposed membranes of adjoining cells is sensitive to mechanical forces [1250]. In normal cells, once bound to their ligands, EPH

157. Synaptic adaptivity (plasticity) means a change in strength of a synapse between 2 neurons due to a change in: (1) synaptic proteins; (2) number of synaptic ion channels; (3) amount of released neurotransmitters; and/or (4) density of postsynaptic membrane receptors.

158. Spines are small, actin-enriched dendritic protrusions that can be sites of excitatory input. Spine formation and retraction participate in synaptic remodeling.

159. In dendrites and spines, type-B ephrins colocalize with synaptic markers. Ephrin-B3 interacts with glutamate receptor-interacting protein GRIP1 and promotes both synapse and spine formation (reverse signaling). Ephrin-B1 forms a complex with adaptor GRB4 and G-protein-coupled receptor kinase-interacting protein GIT1 that is involved in spine maturation. All EPHb receptors (EPHb1–EPHb3) are involved in spine morphogenesis (forward signaling) [1247]. Receptor EPHb activates P21-activated kinase that regulates actin dynamics needed for motility of dendritic filopodia aimed at enhancing short-range filopodial exploration for synaptic partners. Receptor EPHb regulates spine morphology by modulating the activity of small Rho GTPases in cooperation with focal adhesion kinase, Src kinase, and paxillin. Activated EPHa4 interacts with spine-associated RapGAP that inactivates Rap1 GTPase. In addition, ephrin-A-stimulated EPHa activates phospholipase-Cγ1 to maintain mature spine morphology.

160. Ephrins of axon terminals enhance neurotransmitter release [1247]. Receptor EPHb2 interacts with and promotes phosphorylation of NMDA receptor by Src kinases (forward signaling). Ephrin-B2 engaged with EPHs is phosphorylated by Src kinases to serve in synaptic remodeling (reverse signaling).

receptors oligomerize into dimers or small oligomers and activate signaling to control cell growth, survival, and motion.

On contact with EPHa2-expressing cells, ephrin-A1 also forms clusters that rely on actomyosin-dependent radial and lateral transport in a fluid membrane. EPHa2–ephrin-A1 clusters localize at sites of EPHa2 phosphorylation. Once stimulated by its membrane-bound ligand, large ligand–receptor clusters are transported inside the cell by contractile actomyosin filaments. Mechanical forces result from actomyosin contractility aimed at moving EPHa2–ephrin-A1 complexes. The EPHa2 clusters can then serve as mechanical stress sensing.

Mechanical forces influence receptor spatial organization. Added mechanical forces also affect actin reorganization, hence displacement of ligand-receptor complexes along cytoskeletal elements, as well as recruitment of mediators such as the adamlysin ADAM10, thereby signaling primed by ligand-activated EPH receptor [1250].

In tumor cells, ligand-bound EPHa2 receptors move much more efficiently and assemble into very large clusters (size several μm) [1250]. Normal mechanical constraints on EPHa2 movement in the membrane may lack in cancer cells.

8.2.19 Angiopoietin Receptors TIE

Angiopoietin receptor Tyr kinases with immunoglobulin and epidermal growth factor homology domains TIE1 and TIE2 are specific to endothelial cells. [161] They are activated by growth factors angiopoietins (Ang1–Ang4) [162] that intervene in angiogenesis, blood vessel maturation and stability, as well as in early development of the heart. The TIE receptors mutually interacts.

Angiopoietin-1 primarily mediates reciprocal interactions between the endothelium and surrounding matrix (mesenchymal connective tissue in the case of vasculogenesis during embryogenesis). [163] Angiopoietin-2 participates in vascular remodeling. Angiopoietin-3 [164] is widespread in highly vascularized tissues and does not seem to act as an endothelial cell mitogen. Angiopoietin-4 contributes to interactions of endothelial cell with supporting perivascular cells.

Whereas TIE1 mainly links to angiopoietin-4, TIE2 associates with angiopoietin-1, -2, and -4 ligands. The latter also interacts with DOK2 docking protein.

Once activated by angiopoietin-1, TIE2 localizes to cell–matrix adhesions in mobile cells and intercellular junctions of contacting cells. Adhesive structures are

161. The TIE1 receptor is also identified as TIE and JTK14; TIE2 receptor is also labeled Tyr endothelial kinase TEK and venous malformations, multiple cutaneous and mucosal kinase VMCM1.

162. Angiopoietin-1 and -2 are encoded by the ANGPT1 and ANGPT2 genes. The ANGPT4 gene is also designated as ANG3. The ANG4 or ANGPTL1 gene was formerly termed ANGPT3 or ANG3.

163. Cells of the mesenchyme are able to develop into cells of the lymphatic and circulatory systems, in addition to connective tissue, bone, and cartilage.

164. A.k.a. angiopoietin-like protein-1 (AngL1).

close, but distinct from focal adhesions. The angiopoietin–TIE2 complex has different functions during angiogenesis and vessel quiescence. Angiopoietin-1 stimulates TIE2 to remodel vessels and ensure vessel integrity. In migrating endothelial cells, matrix-bound angiopoietin-1 acts both in cell adhesion and motility; cell–matrix adhesions are created successively in the leading and trailing cell parts. In addition, TIE2 promotes endothelial cell polarization, as it colocalizes with adaptor DOK2 [165] in the rear of migrating cells, like caveolin-1, a marker of cell rear polarization during migration.

Angiopoietin-1 generates *TIE2 complexes* in mobile and confluent endothelial cells [1251]. In contacting cells, Ang1 induces TIE2 translocation to intercellular contacts and formation of homotypic TIE2 complexes with mandatory participation of TIE2 receptors from both cells. The TIE2 complex includes protein Tyr phosphatase receptor PTPRb, [166] which inhibits paracellular permeability. Endothelial nitric oxide synthase NOS3 colocalizes with activated TIE2 to intercellular adhesions.

Angiopoietin-2 behaves as a context-dependent agonist or antagonist of TIE2, as it can counteract the activity of angiopoietin-1 on endothelial cell adhesion and vascular integrity.

Cell—matrix and intercellular contacts contain distinct constituents. The PTPRb phosphatase interacts with TIE2 only in contacting cells. Angiopoietin-2 can compete with angiopoietin-1 and reduce TIE2 phosphorylation (activation) in intercellular junctions. Angiopoietin-1 and TIE2 form distinct complexes at cell–matrix and –cell adhesions to activate the ERK and PKB cascades, respectively. Stimulated Protein kinase-B phosphorylates FoxO1 transcription factor and NOS3 enzyme that both operate in endothelial cell survival and stability. Matrix-bound angiopoietin-1 promotes lamellipodium formation and focal adhesions. Activation of ERK by TIE2 partially depends on focal adhesion kinase [1252].

Angiopoietin-2 activates TIE2 during lymphangiogenesis. It either acts as a TIE2 antagonist or TIE2 activator [1253]. [167] Angiopoietin-2 is also expressed by stressed endothelial cells and then acts as an autocrine regulator of TIE2 and the PKB pathway to protect the cell.

---

165. A.k.a. DOK-related protein (DOKR).

166. A.k.a. vascular endothelial phosphoTyr phosphatase (VEPTP).

167. Rather than antagonizing angiopoietin-1 activity, angiopoietin-2 could instead compensate for Ang1 signaling defect. When protein kinase-B activity is high, after strong Ang1 or Ang2 signaling, Ang2 expression could be stopped to prevent overstimulation of the PKB pathway. Angiopoietin-2 expression by endothelial cells that is strong during angiogenesis is induced by the transcription factor FoxO1 after the inhibition of the PI3K–PKB pathway. The FoxO1 factor may also regulate the production of proteins that modulate Ang2 responsiveness.

8.3 Receptor Serine/Threonine Kinases: TGF Superfamily Receptors

A set of transmembrane receptors is characterized by intracellular serine/threonine kinase domains. Receptor Ser/Thr kinases (RSTK) belong to the family of transforming growth factor-β receptors (TβR; Tables 8.23 to 8.25).

Table 8.23. Mediators of TGFβ signaling at the cell surface. (**Part 1**) Extracellular ligands and plasmalemmal receptors (BMP: bone morphogenetic protein; GDF: growth differentiation factor; TGF: transforming growth factor; TβR: TGFβ receptor). Ligand accessibility to their cognate receptors is modulated by antagonists. In addition, receptor activation is controlled by coreceptors and localization to membrane nanodomains, possible endocytosis, and receptor-associated proteins. Plasmalemmal β-glycan, a component of the TβR3 receptor, and homodimeric glycoprotein endoglin present ligands to receptors. Endoglin, a type-1 membrane glycoprotein, is a constituent of the TβR complex (detected on endothelial and smooth muscle cells, fibroblasts, and activated macrophages). β-Glycan binds to various members of the TGFβ ligand superfamily and serves as a reservoir of ligands for TGFβ receptors. Cryptic, a member of the EGF-CFC family of proteins that serve as membrane-bound extracellular cofactor in Nodal signaling, is required for mesoderm and endoderm induction and the body's laterality formation. Cryptic acts as a required coreceptor for Nodal as well as growth and differentiation factors GDF1 and GDF3. Cryptic binds to TGFβ and reduces its linkage with TβR1.

Extracellular ligands (TGFβ superfamily)	
TGFβ family	TGFβ1–TGFβ3
BMP family	BMP2–BMP7, BMP8a/b, BMP10, BMP15
GDF family	GDF1–GDF3, GDF5–GDF11, GDF15
Miscellaneous	Activins, inhibins, anti-Müllerian hormone, Nodal
Plasmalemmal receptors	
Type-1	TβR1, AcvR1a–AcvR1c, AcvRL1, BMPR1 (BMPR1a–BMPR1b)
Type-2	TβR2, AcvR2a–AcvR2b, AMHR2, BMPR2
Type-3	TβR3 (β-glycan), endoglin, Cryptic

Members of the TGFβ receptor superfamily are plasmalemmal kinases that participate in the regulation of cell proliferation and differentiation, especially during angiogenesis and wound healing. They signal via 2 specific transmembrane, catalytic receptors: type-1 and -2 receptors.

In mammals, 28 ligands (Table 8.23; Vol. 2 – Chap. 3. Growth Factors), 7 type-1 (TβR1, BMPR1a–BMPR1b, AcvR1a–AcvR1c, and AcvRL1), i.e., activin receptor-like kinases (ALK1–ALK7; Sect. 8.3.9)) and 5 type-2 receptors (TβR2, BMPR2, AcvR2a–AcvR2b, and AMHR2) have been identified. After ligand binding, TGFβ receptors remain active at least 3 to 4 h [1254].

Table 8.24. Mediators of TGFβ signaling at the cell surface. (**Part 2**) Enhancers, coreceptors and intracellular transducers (SARA: SMAD [small mothers against decapentaplegia] anchor for receptor activation [a.k.a. ZFYVE9]; ZFYVE: zinc finger FYVE domain-containing protein). BMP and activin membrane-bound inhibitor homolog (BAMBI) is a pseudoreceptor that inhibit TGFβ signaling. Many TGFβ inhibitors and neutralizing partners intervene during embryogenesis. Ubiquitous follistatin focuses mainly on activin. Gremlin-1 antagonizes BMP agents. Proteins Lefty regulate the body's left–right asymmetry in opposition to Nodal agonist. Latent TGFβ-binding proteins (LTBP) sequester TGFβ precursors in the latency-associated proteins (LAP). Thrombospondin-1 antagonizes TGFβ, especially in endothelial cells. Noggin prevents TGFβ ligands from binding to their receptors. Receptor-regulated SMADs (rSMAD) are anchored at the plasma membrane by various cytosolic proteins. Inhibitory SMADs (iSMAD) antagonize signaling transmitted by receptor-regulated SMADs (rSMAD) that are assisted by common mediator SMAD (coSMAD).

Binding enhancers and plasmalemmal facilitators and potentiators	
Glycoproteins	Endoglin
Proteoglycans	β-Glycan
Anchoring proteins	ZFYVE9 (SARA), ZFYVE16 (endofin)
Coreceptors (pseudo or accessory receptors)	
Inhibitors	BAMBI, Cryptic
Neutralizing partners	
Cerberus, chordin, decorin, follistatin, gremlin, Lefty-1/2, LTBP1–LTBP4, noggin, thrombospondin-1	
Intracellular transducers	
rSMADs	SMAD1–SMAD3, SMAD5, SMAD9
iSMADs	SMAD6, SMAD7
coSMAD	SMAD4

8.3.1 TGFβ Receptor- and SMAD Activation

In the absence of ligand, type-1 and -2 transmembrane receptor Ser/Thr kinases (Sects. 8.3.6 and 8.3.7) exist as homodimers in the plasma membrane. At the cell surface, the ligand binds a complex of type-1 and -2 receptors and induces transphosphorylation of type-1 receptor by type-2 receptor (Table 8.26). The TGFβ ligands cause receptor endocytosis for efficient signaling via SMAD proteins. [168]

Agonists TGFβ1, TGFβ3, and activins bind their type-2 receptors without needing a type-1 receptor [1254]. On the other hand, TGFβ2 ligand interacts only with TβR1 (TβR1–TβR2 complex).

168. SMAD: small (son of/similar to) mothers against decapentaplegia (DPP) homolog. This word results from the concatenation (Latin cum: with, together with, in the company of, in connection with, along with; and catena: chain, fetter, shackle) of Sma (small [a.k.a. dwarfin]) protein in Caenorhabditis elegans and its related molecule MAD (mothers against decapentaplegia) in Drosophila melanogaster.

Table 8.25. Family of transforming growth factor-β receptors and their ligands (Source: [1121]; AcvR: activin receptor; AMHR: anti-Müllerian hormone receptor; BMPR: bone morphogenetic protein receptor; GDF: growth differentiation factor). Activins and inhibins are dimers made of 2 activin β chains and 1 inhibin α chain and 1 activin β chain, respectively. Two types of inhibins exist: inhibin-A (α–β_A dimer) and -B (α–β_B dimer). Seven types of activins can be detected: activin-A (β_A–β_A dimer), -AB (β_A–β_B dimer), -AC (β_A–β_C dimer), -B (β_B–β_B dimer), -BC (β_B–β_C dimer), -C (β_C–β_C dimer), and -E (β_E–β_E dimer). Activin receptor AcvR1b, or activin receptor-like kinase ALK4, transduces signals from activin and activin-like ligands (e.g., inhibins), as well as nodal and growth differentiation factor GDF1, both of the TGFβ superfamily. Transducers SMADs (small mothers against decapentaplegia homologs) include receptor-regulated rSMADs (SMAD1–SMAD3, SMAD5, and SMAD9), inhibitory iSMADs (SMAD6 and SMAD7), and common mediator coSMAD (SMAD4). All rSMADs cooperate with SMAD4 mediator. Inhibitory iSMADs interact with type-1 receptors activated by type-2 receptors.

Type-2 receptor	Ligands	Type-1 receptor	rSMADs
TβR2	TGFβ1–3	TβR1 (ALK5)	SMAD2/3
		AcvRL1 (ALK1)	SMAD1/5
		AcvR1a (ALK2)	
AcvR2a	Activin	AcvR1b (ALK4)	SMAD2/3
		AcvRL1 (ALK1)	
	Inhibin-β_A		
AcvR2b	Activin,	ALK4,	SMAD2/3
	nodal	AcvR1c (ALK7)	SMAD2/3
BMPR2	BMP6/7	AcvR1a (ALK2)	SMAD1/5
	BMP2/4/7	BMPR1a (ALK3)	SMAD1/5/9
	BMP2/4/7, GDF5	BMPR1b (ALK6)	SMAD1/5/9
AMHR2	AMH	ALK2, ALK3, ALK6	SMAD1/5

Table 8.26. Stages of TGFβ receptor-mediated signaling via SMAD activation.

Step	Event
1	Ligand binding to type-1–type-2 receptor complex
2	Phosphorylation of rSMAD (activation)
3	Formation of the $rSMAD^P$–SMAD4 complex
4	Nuclear translocation of the $rSMAD^P$–SMAD4 complex
5	Gene transcription
6	SMURF-mediated degradation

Agonists BMP2, BMP4, and BMP7 bind primarily to their type-1 receptors BMPR1a or BMPR1b, although heteromeric BMPR complexes yield higher-affinity ligand binding.

A given receptor combination often binds different ligands. Type-2 receptors AcvR2a and AcvR2b can combine with type-1 receptor AcvR1b and mediate activin signaling [1254]. On the other hand, AcvR2a and AcvR2b connect to BMPR1a or BMPR1b and bind BMP proteins. A given ligand can induce different signals according to the composition of the receptor complex. Type-2 BMP receptor can combine with BMPR1a, BMPR1b, and AcvR1 to bind several BMP proteins. The TβR2 receptor interacts not only with TβR1 to activate SMAD2 and SMAD3, but also with AcvRL1 to activate SMAD1 and SMAD5 [1254].

Accessory proteins influence ligand-binding specificity. Nodal acts via the activin AcvR2b–AcvR1b receptor complex to stimulate SMAD2. Yet, efficient ligand binding requires association of fucosylated TGFα-like transmembrane protein Cryptic (or Crypto homolog) with AcvR1b receptor [1254]. In addition, β-glycan (or TβR3) and endoglin yield high-affinity TGFβ presentation to the TGFβ receptor complex.

Endoglin (or CD105) is an integral membrane glycoprotein and coreceptor for members of the TGFβ superfamily. It regulates TGFβ-dependent angiogenesis and vascular remodeling [1255].[169] In addition to the predominant, long endoglin isoform (endoglin_L), 2 other forms of endoglin operate in vascular pathology and homeostasis [1255]: (1) the alternatively spliced, short endoglin isoform (endoglin_S) and (2) the soluble endoglin form (endoglin^S), which is proteolytically cleaved from membrane-bound endoglin.

Receptor-interacting proteins, such as ZFYVE9 (SARA), sorting component of the ubiquitin-binding complex hepatocyte growth factor-regulated Tyr kinase substrate (HGS, HRS, or ZFYVE8),[170] Disabled homolog Dab2, RasGAP-binding protein DOK1, TβR-associated protein TGFβRAP1 (encoded by the TGFBRAP1 gene),[171] axin, and activin receptor-interacting protein ARIP1, interact with type-1 or -2 receptors and rSMADs [1254]. Proteins ZFYVE9 or HGS and Dab2 that reside in endosomes and clathrin-coated vesicles bind to both receptors and rSMADs, stabi-

169. The endoglin gene is mutated in type-1 hereditary hemorrhagic telangiectasia characterized by vascular malformations. Endoglin is overexpressed in angiogenic vessels. Endoglin regulates the production and activity of nitric oxide synthase NOS3 involved in angiogenesis [1255].

170. Hepatocyte growth factor-regulated Tyr kinase substrate participates in both exo- and endocytosis. It interacts with ubiquitinated cargos via its ubiquitin-interacting motif (UIM) domain. Ubiquitination of plasmalemmal proteins serves as a signal for endosomal sorting into lysosomes. Yet, ubiquitin-independent binding of HGS mediates endosomal sorting of some receptors such as interleukin-2 receptor [1256]. Protein HGS recruits clathrin onto early endosomes for trafficking from early to late endosomes [1257]. In addition, both FYVE domain-containing proteins HGS (ZFYVE8) and SARA (ZFYVE9) are necessary for receptor-mediated activation of SMAD2 [1258].

171. A.k.a. TRAP1. However, this alias is ambiguous, as TRAP1 also designates a chaperone with an ATPase activity, the 75-kDa heat shock protein, mitochondrial precursor (HSP95), or tumor-necrosis factor receptor (TNFR)-associated protein-1. Yet, the latter is a functional partner of type-1 and type-2 TGFβ receptors as well as TGFβ receptor-associated protein-1. Its other interactors comprise small mothers against decapentaplegia homolog SMAD4, tumor-necrosis factor receptor superfamily member TNFRSF1a, retinoblastoma-associated protein RB1, mitochondrial protein Ser/Thr kinase PINK1, endoplasmic reticulum-resident type-2 transmembrane glycosyltransferases exostosin-1 and -2 involved in the chain elongation of heparan sulfates, and Toll-like receptor TLR6.

lize interactions of SMAD2 and SMAD3 with TβR1, and operate with the endocytic machinery. The TβR receptors can also associate with caveolin.

Cytoskeletal proteins contributes to the localization and function of SMAD proteins. Unphosphorylated SMAD2 and SMAD3 bind to microtubules. Activator TGFβ dissociates SMADS from microtubules [1254]. The SMAD regulators also interact with filamin, an actin cross-linking and scaffold protein for intracellular signaling mediators [1254]. In addition, TGFβ provokes phosphorylation of non-erythrocytic spectrin β chain Sptbn1,[172] hence its association with SMAD3 and SMAD4 [1254].

Activated type-1 receptor subsequently phosphorylates (activates) SMAD proteins. Once their C-termini is phosphorylated, monomeric SMADs oligomerize with a neighboring rSMAD and/or SMAD4, hence generating heterodimers or -trimers with 2 rSMADs and 1 SMAD4 protein. Activated SMAD complexes translocate into the nucleus to regulate transcription of target genes. Nuclear import of rSMADs owing to nuclear localization sequence (Vol. 1 – Chap. 9. Intracellular Transport) does not require SMAD4 factor. However, SMAD4 cotranslocates with rSMAD factors. In the nucleus, SMAD heterodimers and -trimers bind to and cooperate with transcription factors and CBP or P300 coactivators. The SMAD2–SMAD4 complex with forkhead box transcription factor FoxH1[173] contains 2 SMAD2s and 1 SMAD4, whereas the SMAD3–SMAD4 complex at the Jun promoter may be a heterodimer.

In the nucleus, rSMADs are dephosphorylated. Consequently, they dissociate from SMAD complexes. Inactive SMADs are then exported to the cytoplasm [1254]. Protein SMAD4 permanently shuttles between the nucleus and cytoplasm owing to its nuclear localization sequence and export signal via exportin-1.

Activation of rSMADs by type-1 receptor is inhibited by SMAD6 or SMAD7 protein. In some cells, SMAD6 and SMAD7 localize to the nucleus, where they can interact with transcriptional coactivators CBP and E1A-binding protein P300, which are also histone acetyltransferases.[174] Ligands TGFβ and BMPs cause SMAD7 and SMAD6 nuclear export, respectively.

The ubiquitin ligases SMAD ubiquitination regulatory factors SMURF1 and SMURF2 ubiquitinate rSMADs for degradation (Table 8.26). They also interact with inhibitory SMAD6 and SMAD7 to ubiquitinate type-1 receptors. In addition, nuclear phosphorylated SMAD3 can interact with a Ring-box protein RBx1 (or Roc1) to promote its association with SCF ubiquitin ligase complex (that contains RBx1, Skp1, Cullin1, and β-transducin repeat-containing protein βTRCP1) for nuclear export and proteasomal degradation in the cytoplasm [1254]. Yet, only a small fraction of SMAD2 and SMAD3 is ubiquitinated and the major part of nuclear SMAD2 or SMAD3 is dephosphorylated and relocates to the cytoplasm. Besides, SMAD4 sumoylation enhances its stability.

172. A.k.a. embryonic liver β-fodrin.

173. A.k.a. forkhead activin signal transducer FAST1 or FAST2.

174. Transcriptional coactivators CBP and P300 relax the chromatin structure at gene promoters, recruit the transcriptional machinery such as RNA polymerase-2 to promoters, and act as adaptors.

8.3.2 TGFβ Signaling in Endosomes

Signaling mediated by SMADs can be initiated from the plasma membrane, but in some cases it can be triggered during endocytosis [11]. Specific signaling complexes can be assembled, once components have been recruited to early endosomes by early endosomal resident Rab5 GTPase and Rab5 effectors, the adaptors phosphotyrosine interaction, PH domain and leucine zipper-containing proteins APPL1 and APPL2 that are anchored to the endosomal membrane. The subpopulation of APPL1/2+ early endosomes contains small GTPase Rab5, but lack EEA1 early endosomal antigen. Recruitment by early endosomes of vacuolar protein sorting-associated protein VPS34, or class-3 PIK3 catalytic subunit ($PIK3C3_c$), converts PI into PI(3)P. Subsequently, early endosomal antigen EEA1 that binds to lipids via its FYVE domain accumulates in maturing early endosomes, whereas APPL1 and APPL2 concomitantly dissociate. Proteins APPL1 and APPL2 are required for signaling from early endosomes via protein kinase-B and glycogen synthase kinase GSK3β as well as endosomal activation of extracellular signal-regulated protein kinases ERK1 and ERK2 [11].

Endosomes recruit proteins containing PI3P-binding domains for transforming growth factor-β signaling from internalized heterotetrameric receptors constituted by TβR1 dimers and TβR2 dimers [11]. Endosomal recruitment of SMAD2 and SMAD4 effectors by FYVE domain-containing adaptors ZFYVE9 (or SMAD anchor for receptor activation [SARA]) and ZFYVE16 (or endofin), respectively, allows efficient phosphorylation of SMAD2 by internalized TβR receptors and the formation of active SMAD2–SMAD4 complexes. In early endosomes, TβR1 indeed connects to ZFYVE9 anchor that also associates with SMAD2. Effector SMAD2 then dissociates from the TβR1–ZFYVE9–SMAD2 complex to interact with SMAD4. In parallel, ZFYVE16 interacts with TβR1 and SMAD4, therefore potentiating TGFβ signaling, as it facilitates the formation of a SMAD2–SMAD4 complex in endosomes. The SMAD2–SMAD4 complex translocates to the nucleus, where it regulates gene transcription.

8.3.3 TGFβ Factors

Transforming growth factor-β participates mainly in the regulation of cell proliferation and differentiation. It regulates the cell cycle via MyC transcription factor. The expression of multiple type-1 receptors within different cell types can drive cell-specific responses to TGFβ factor.

Three secreted isoforms exist (TGFβ1–TGFβ3). TGFβ1 has been first detected in human platelets. It intervenes in wound healing. It also controls the immune reponse, as it is secreted by most leukocytes. TGFβ2 Cytokine possesses 2 alternatively spliced variants. Cytokine TGFβ3 [175] regulate cell adhesion and extracellular matrix formation.

Transforming growth factor-β is synthesized as a homodimeric proprotein. The latter is cleaved intracellularly into *small latent TGFβ complex* (SLTC) that includes

175. A.k.a. arrhythmogenic right ventricular dysplasia-1 (ARVD1).

Table 8.27. Family of bone morphogenetic proteins (BMP) and growth and differentiation factors (GDF) and common members of these 2 subfamilies.

BMP Member	Corresponding GDF member
BMP1	GDF1
BMP2	GDF2
BMP3	GDF3
BMP4	GDF5
BMP5	GDF6
BMP6	GDF7
BMP7	GDF8
BMP8a	GDF8 (myostatin [Mstn])
BMP8b	GDF9
BMP10	GDF10
BMP15	GDF11
	GDF15
Common members	
BMP3b	GDF10
BMP9	GDF2
BMP11	GDF11
BMP12	GDF7
BMP13	GDF6
BMP14	GDF5

the future TGFβ and *latency-associated protein* (LAP). The SLTC dimer is then secreted. It cannot bind to TGFβ receptors, but commonly tethers to latent TGFβ-binding proteins (LTBP1–LTBP4) to form a *large latent TGFβ complex* (LLTC, i.e., TGFβ propeptide [LAP]–TGFβ–LTBP) [1259].

8.3.4 TGFβ Superfamily

The TGFβ superfamily can be subdivided into 4 major families: (1) the decapentaplegic-Vg-related (DVR; Table 8.27) family that includes the 2 subfamilies of *bone morphogenetic proteins* (BMP2–BMP7, BMP8a/b, BMP10, and BMP15) and *growth and differentiation factors* (GDF1–GDF3, GDF5–GDF11, and GDF15); (2) the family of *activins* and *inhibins*, which are antagonists in hypothalamic, pituitary, and gonadal hormone secretion, and act in growth and differentiation of various cell types; (3) the TGFβ family (TGFβ1–TGFβ3); and (4) a family that encompasses various other members: left–right determination factors *Lefty*-1 and -2, *anti-Müllerian hormone* (AMH), [176] *myostatin* (Mstn), [177] *neurturin*, *artemin*, and *persephin*.

176. A.k.a. Müllerian inhibiting factor (MIF), hormone (MIH), and substance (MIS).
177. Formerly called growth differentiation factor GDF8.

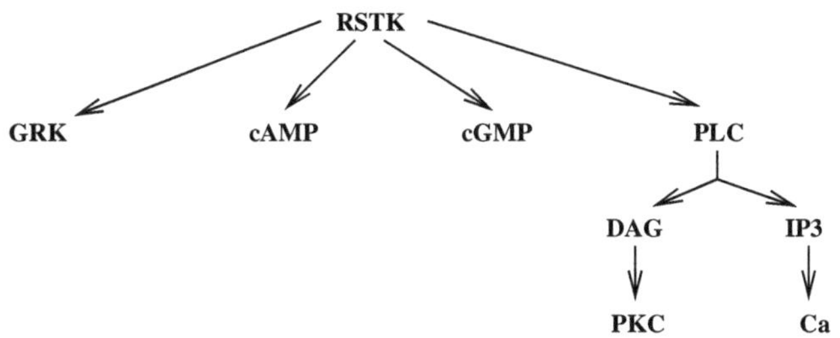

Figure 8.6. Receptor Ser/Thr kinases (RSTK) comprise transforming growth factor-β family receptors. RSTKs are implicated in cell growth inhibition. Main effectors are cAMP, cGMP, phospholipase-C (PLC), and GPCR kinases (GRK).

Activins and *Inhibins* are formed by dimerization of inhibin-α and activin-β (AcvβA, -B, -C, and -E) subunits. Activin-A [178] synthesized in hematopoietic cells is an AcvβA homodimer; activin-B an AcvβB homodimer; activin-AB an AcvβA–AcvβB heterodimer; activin-C expressed in liver and prostate an AcvβC homodimer; activin-AC an AcvβA–AcvβC heterodimer; activin-BC an AcvβB–AcvβC heterodimer; and activin-E produced in liver an AcvβE–AcvβE homodimer.

Inhibins-A and -B that are synthesized by granulosa cells of the ovarian follicles and Sertoli cells in the seminiferous tubules of testes to inhibit FSH and GnRH secretion from hypophysis and hypothalamus (feedback), respectively, are Inhα–AcvβA and Inhα–AcvβB heterodimers, respectively.

Members of the TGFβ superfamily are not only implicated in cell proliferation and differentiation (Fig. 8.6), but also regulate migration and adhesion of various cell types. As TGFβ is a multifunctional cytokine that is strongly involved in embryo- and fetogenesis and tissue homeostasis, any aberration in its functioning directs cancer.

8.3.5 TGFβ Receptor Types and Their Regulators

Receptor serine/threonine kinases can be divided into subsets. TGFβ acts via both type-1 (TβR1 or ALK5) and -2 (TβR2) receptors. In the absence of ligand, TGFβ receptors are homodimers. Ligands first bind to type 2 receptors that then phosphorylate (activate) type 1 receptors, thus initiating the signaling cascade. The affinity between TβR1 and TβR2 rises 300-fold due to initial recruitment of a pair of constitutively active TβR2s by dimeric TGFβ. Receptors TβR1 and TβR2 localize at the basolateral surfaces of polarized endothelial cells. These receptors concentrate predominantly at lateral sites of intercellular contacts adjacent to gap junctions.

Type-3 TGFβ receptor (TβR3), or β-*glycan*, and endoglin present ligands to the receptors (Table 8.28). β-Glycan is a multifunctional transmembrane proteoglycan that binds TGFβ isoforms and enhances their ability to bind and assemble signaling

178. A.k.a. erythroid differentiation factor [EDF], follicle-stimulating hormone releasing protein [FRP], restrictin-P, and WEHI-mesoderm inducing factor.

Table 8.28. Coreceptors of the TGFβ superfamily and their ligands (Sources: [59, 600, 608]; BAMBI: BMP and activin membrane-bound inhibitor homolog; FRL: fibroblast growth factor (FGF)-related ligand, or FGF receptor ligand]; HHT: hereditary hemorrhagic telangiectasia; IGSF1: immunoglobulin superfamily member-1, a.k.a. inhibin-binding protein [InhBP], immunoglobulin-like domain-containing protein IGDC1 [and IGCD1], and pituitary gland-specific factor PGSF2; NMA: non-metastatic gene A protein; ORW: Osler-Rendu-Weber syndrome). The epidermal growth factor–Cripto, FRL1, and Cryptic (EGF–CFC) family includes human Cryptic (CFC1) and Cripto homolog, frog FRL1, and zebrafish one-eyed pinhead protein (OEP). These proteins are characterized by 2 cysteine-rich motifs: an EGF-like and a CFC domain. Most EGF–CFC members are extracellular membrane-anchored proteins via a glycosyl-phosphatidylinositol link. Human Cryptic shares 25% amino acid idenitity with human Cripto. Cryptic, like Cripto, is a Nodal coreceptor. It is involved in left–right asymmetry. Other TGFβ superfamily members involved in left–right patterning such as GDF1 also require EGF–CFC family cofactors. Endoglin resides on endothelial cells, erythroid cell precursors, and activated macrophages, as well as, transiently, on syncytiotrophoblasts and cytotrophoblasts. Agent IGSF1 potentiates the antagonizing effects of inhibin-B on gene expression stimulated by activin-A. Two IGSF1 isoforms exist ($IGSF1_L$ and $IGSF1_S$).

Type	Aliases	Ligands
TβR3	β-glycan, BGcan	TGFβ1–TGFβ3, inhibin-A, TβR2
BAMBI	NMA	BMP, activin-A, Sox30
Cryptic	FRL1	Nodal, GDF1
Endoglin	CD105, EnG, End, HHT1, ORW1	TGFβ1–TGFβ3, activin-A, BMP2/7
IGSF1	InhBP, IGCD1, IGDC1, IgSF1, PGSF2	Inhibin-A/B, IGF1

receptors. β-Glycan potentiates TGFβ binding to type-2 receptor. It also tethers to inhibins to facilitate binding of these inhibitory ligands to activin type-2 receptors, thereby precluding signaling by activins and BMP proteins.

Endoglin connects to multiple ligands of the TGFβ superfamily. In particular, it determines whether TGFβ signals via TβR1 or distinct type-1 receptor ALK1 (AcvRL1; Sect. 8.3.9). In fact, endoglin is a transmembrane glycoprotein that belongs to the TGFβ receptor complex. It contributes to vascular development and remodeling. It is expressed in proliferating endothelial cells as well as other cell types. It binds to TGFβ1 and TGFβ3, activin-A, and BMP2 and BMP7 [1260]. Endoglin enhances TGFβ binding to ALK1 in endothelial cells.

Transmembrane glycoprotein BMP and activin membrane-bound inhibitor homolog (BAMBI) is a pseudoreceptor without intracellular kinase domain related to the type-1 TGFβ receptors. It interacts with various type-1 and -2 receptors to inhibit signaling. It inhibits BMP, activin, and TGFβ signaling, as it forms inactive complexes with these ligands and their respective signaling receptors. Expression of BAMBI is induced by BMPs and TGFβ [1260]. Pseudoreceptor BAMBI acts as a decoy receptor that competes with type-1 receptor [1261]. Glycosylphosphatidylinositol-anchored membrane protein *Cryptic*, acts as a coreceptor that increases the binding of some TGFβ family ligands such as growth and differentiation factor GDF1 to activin receptors, but blocks activin signaling [1261].

Multiple other agents modulate TGFβ and BMP signaling. Access of TGFβ to receptors is inhibited by soluble latency-associated peptide, decorin, and α2-macroglobulin that sequester TGFβ [1261]. Soluble BMP antagonists include noggin, chordin, chordin-like, Cerberus, and sclerostin.

TGFβ receptor ubiquitination (Sect. 1.2.5 and Vol. 1 – Chap. 5. Protein Synthesis) by ubiquitin ligases such as SMURF proteins recruited by inhibitory SMAD6 or SMAD7 regulates the amplitude and duration of TGFβ signaling using receptor degradation by proteasome. The TGFβ1 receptor is sumoylated after TGFβ stimulation to enhance binding of SMAD coactivators and TGFβ signaling [1262].

The TβR1 kinase activates the *canonical SMAD pathway*, whereas ubiquitin ligase TRAF6 that interacts with TβR1 activates MAP3K7 [179] that is an effector of TGFβ-induced P38MAPK activation (Vol. 4 – Chap. 5. Mitogen-Activated Protein Kinase Modules). (SMAD2 Activation does not depend on TRAF6.) The TβR1–TRAF6 complex leads to TRAF6 autoubiquitination and subsequent activation of the MAP3K7–P38MAPK and MAP3K7–JNK pathways that can induce apoptosis [1263]. Kinase MAP3K7 hastens P38MAPK and JNK via MAP2K3 and MAP2K6, as well as NFκB using the TNFα and IL1 pathways. Kinase MAP3K7 can bind to MAP3K7-interacting proteins (MAP3K7IP1–MAP3K7IP3) [180] to form complex, the activity of which depends on TRAF6 and interleukin-1 and its IL1R receptor.

8.3.6 Type-1 TGFβ Receptor (TβR1 or ALK5)

Ubiquitous type-1 transforming growth factor-β receptor consists of an extracellular, a single membrane spanning, and an intracellular domain. In humans, its highest expression is detected in the lung, kidney, skeletal muscle, and placenta, whereas its lowest levels are observed in the brain and heart [1264]. Two alternative spliced variants of TβRI exist: $ALK5_L$ and $ALK5_S$.

The extracellular domain facilitates ligand binding to the receptor complex formed with TβR2, as TβR1 is unable to directly bind TGFβ ligands, to then initiate the canonical TGFβ cascade. Upon assembly of the TβR1–TβR2 complex, the intracellular domain of TβR1 is phosphorylated by TβR2, thus activating the kinase activity of TβR1. Afterward, TβR1 phosphorylates SMAD2 and -3 transcription factors.

On the other hand, in addition to the inhibitory SMAD7, TβR1-binding proteins FKBP12, TGFβRAP1, STRAP1, SMURF1, and SMURF2 inhibit TβR1 [1264]. The

179. A.k.a. TGFβ-associated kinase TAK1.

180. A.k.a. TAK1-binding proteins (TAB1–TAB3).

WD domain-containing protein Ser/Thr kinase receptor-associated protein STRAP1 can link to both type-1 and -2 TGFβ receptors as well as SMAD2, SMAD3, and SMAD7 [1254]. It stabilizes the interaction of SMAD7 with these receptors, thereby cooperating with SMAD7 to inhibit TGFβ-dependent transcription.

Receptor TβR1 also forms proteic complex with TβR3 and endoglin. Cytoplasmic domain of TβR1 directly interacts with several proteins, such as SMAD anchor for receptor activation (SARA), [181] apolipoprotein-J, phosphatases PP1c, PP2 regulatory Bα subunit, [182] caveolin-1, clathrin-associated adaptor complex AP2β2–adaptin subunit, Disabled-2, [183] early endosome antigen EE1A, [184] vacuolar protein sorting VPS39 (or TLP), [185] 14-3-3ε, and TGFβ regulator cytoplasmic promyelocytic leukemia protein (PML) [1264].

Type-1 transforming growth factor-β receptor (TβR1) directly phosphorylates eukaryotic elongation factor eEF1a1 (Ser300) to prevent translation and, hence, protein synthesis and cell proliferation [1265]. [186]

8.3.7 Type-2 TGFβ Receptor

Homodimers TGFβ1 and TGFβ3 directly bind to TβR2, or requires accessory TβR3 to bind to TGFβ2 (Table 8.29). In endothelial cells, upon ligand binding, TβR2 can also form a complex with and activate ALK1, which then phosphorylates SMAD1, SMAD5, and SMAD8 [1266].

Type-2 TGFβ receptor is ubiquitous in adults. A TβR2 splice variant, TβR2b, is also an ubiquitous, functional TGFβ receptor [1266]. Autophosphorylation (Ser213) is required for its kinase activity. Autophosphorylation (Ser409) is enhanced by receptor dimerization and promotes kinase activity, whereas autophosphorylation (Ser416) precludes receptor function. Autophosphorylation (Tyr259, Tyr336, and Tyr424) also favors kinase activity [1266].

The TβR2 receptor interacts not only with TβR1, but also ALK1 via accessory receptor endoglin [1266]. The enroller *SMAD anchor for receptor activation* (SARA or ZFYVE9) binds to and recruits SMAD2 to the cytoplasmic domains of the TβR1–TβR2 complex. *Disabled protein* Dab2 is needed for SMAD2 phosphorylation by TβR1 receptor. Inhibitory SMAD7 recruits ubiquitin ligase SMURF2 for degradation and enhanced turnover of the TβR1–TβR2–SMAD7 complex.

TGFβ Receptor-interacting protein TRIP1, [187] is phosphorylated by TβR2 receptor. Protein TRIP1 impedes TGFβ to prime transcription from the plasminogen

---

181. A.k.a. zinc finger FYVE domain-containing protein ZFYVE9.

182. This agent enhances antiproliferative effect of TGFβ.

183. This adaptor associates both TβR1 and TβR2 to SMADs.

184. This protein resides in vesicles. It is responsible for TGFβ1 endocytosis. It promotes TGFβ1 signaling.

185. This molecule couples TGFβ receptor to the SMAD pathway.

186. Residue Ser300 pertains to the region of eEF1a1 that interacts with amino acid-loaded transfer RNA.

187. A.k.a. eukaryotic translation initiation factor-3 subunit 2 (eIF3S2 or eIF3β. Protein TRIP1 participates in the translation–initiation factor complex eIF3.

Table 8.29. Type-2 receptors of the TGFβ superfamily and their known ligands (Sources: [59, 600, 608]; CTBP: C-terminal-binding protein; Dab: disabled homolog; LIMK: Lin1, Isl1, and Mec3 motif-containing kinase; PI3KR: phosphatidylinositol 3-kinase, regulatory subunit; PKC: protein kinase-C; STRAP: Ser/Thr kinase receptor-associated protein; TRPC: canonical transient receptor potential channel). Aortic aneurysm, familial thoracic lesions (AAT) are dominantly inherited disorders. Types AAT3 and AAT5 result from mutations in the TGFBR2 and TGFBR1 genes that produce TβR2 and TβR1 (or ALK5), respectively. Patients may have aneurysms of the aorta and other arteries. Mutations in the TGFBR2 gene are responsible for about 5% of familial thoracic aortic aneurysms and dissections (FAA and TAAD). (Type AAT4 (thoracic aortic aneurysm and/or dissection with ductus arteriosus) is caused by mutations in the myosin heavy chain MYH11 gene; type AAT6 is caused by mutations in the smooth muscle α actin ACTA2 gene; type AAT7 is caused by mutations in the myosin light chain kinase MYLK gene.) Many symptoms of the Loeys-Dietz syndrome (LDS) are common with the Marfan syndrome (MFS) such as ascending aortic aneurysm and aortic dissection. Two forms of Loeys-Dietz syndrome are further subdivided into 2 subtypes (1A and 2A associated with the TGFBR1 gene (locus 9q22), 1B and 2B with the TGFBR2 gene (locus 3p22). Receptor kinase BMPR2 (a.k.a. BMP receptor kinase BRK3 and type-two activin receptor-like kinase tALK) binds to bone morphogenetic protein BMP7, BMP2, and, less efficiently, BMP4. (Binding is weak but enhanced by BMPR1.) Mutations in the BMPR2 gene cause primary pulmonary hypertension (PPH). Anti-Müllerian hormone (AMH), also called Müllerian-inhibiting substance (MIS), links to its type-2 receptor (AMHR or MISR2).

Type	Other aliases	Ligands
TβR2	AAT3, FAA3, LDS1b/2b, MFS2, RIIC, TAAD2	TGFβ1–TGFβ3, TβR1/3, ALK2, endoglin, SMAD7, Dab2, STRAP, PIK3R1/2, SMURF1/2, UbC,
AcvR2a	AcvR2, ActR2a	Activin-A, inhibin-A/B, BMP2/3/6/7/10/15, GDF5/6/7/8/11, ALK4, endoglin,
AcvR2b	ActR2b BMP2/6/7, GDF1/5/6/8/9/11,	Activin-A, inhibin-A/B, Nodal, ALK4, endoglin
BMPR2	BRK3, PPH1, tALK	Inhibin-A, BMP2/4/6/7/15, GDF5/6/9, ALK2/3–6, endoglin, CTBP1, FoxL1, TRPC1, LIMK1, PKCβ
AMHR2	AMHR, MISR2	AMH

activator inhibitor-1 promoter via SMADs, but not TGFβ inhibition of cyclin-A transcription [1267]. It can also inhibit activated TβR1 receptors. Protein STRAP1 interacts with TβR1 and TβR2 and prevents TGFβ signaling. Protein STRAP1 cooperates with SMAD7, but not SMAD6, to hinder TGFβ-induced transcription [1268].

Table 8.30. Subsets of bone morphogenetic proteins and growth and differentiation factors.

Group	Members
1	BMP2/4
2	BMP5–BMP8
3	BMP9–BMP10
4	GDF5–GDF7

TNF Receptor-associated protein TRAP1 connects to the TβR1–TβR2 complex and SMAD4, and promotes association of SMAD2 with SMAD4. The *TRAP1-like protein* (TLP) [188] interacts with TβR1 and TβR2 and recruits SMAD4 [1266].

Dynein light chain is required for SMAD2-dependent TGFβ signaling. Mammalian homolog of the LC7 family of dynein light chains is a TβR-interacting protein that is phosphorylated by TβR2 after ligand-receptor engagement in the TβR2-activated JNK pathway, which leads to phosphorylation of Jun transcription factor and primes apoptosis [1269]. The TGFβ factor causes the recruitment of mLC7-1, or dynein light-chain roadblock DynLRB1, to dynein intermediate chain. [189]

Death-associated protein DAP6 (or Daxx) interacts with TβR2 and regulates the TGFβ response via Jun N-terminal kinase. *Apolipoprotein*-J also directly interacts with TβR2. *Sorting nexin* SNx6 binds to TβR2 and inactive TβR1.

8.3.8 Bone Morphogenetic Proteins and Their Receptors

Bone morphogenetic proteins, originally identified as factors that induce the formation of bone and cartilage, reside in various tissues (bone, cartilage, tooth, and muscle, as well as heart, blood vessels, neurons, lung, kidney, liver, skin, hair, etc.). They are members of the TGFβ family that bind to type-1 and type-2 Ser/Thr kinase receptors and transduce signals via SMAD (e.g., BMP type-1 receptors and SMAD1, -5, and -8) and non-canonical (non-SMAD) signaling pathways (Sect. 8.3.11).

Bone morphogenetic proteins can be classified into several groups [1260] (Table 8.30). [190] Certain BMP or GDF factors have been misnamed, because they do not activate BMP type-1 receptors and SMAD1/5/8, such as BMP3 and myostatin (or GDF8).

The BMPR1 isoforms (BMPR1a and BMPR1b) and BMPR2 constitute homo- (BMPR1a–BMPR1a, BMPR1b–BMPR1b, and BMPR2–BMPR2), as well as heteromeric (BMPR1a–BMPR1b, BMPR2–BMPR1a, and BMPR2–BMPR1b) complexes [1270]. Before ligand binding, BMPRs reside at the cell surface as monomers, homodimers, and pre-assembled heteromeric BMPR1–BMPR2 complexes.

The BMP ligands bind to type-1–type-2 receptor heterotetramers. In the absence of ligand stimulation, only a small fraction of type-1 and -2 receptor homo- and

188. A.k.a. vacuolar protein sorting VPS39.
189. Roadblock dynein light chain can bind dynein intermediate chains.
190. Group 2 corresponds to osteogenic protein-1 group.

heterodimers exist on the cell surface. Ligand-binding increases receptor hetero- and homo-oligomerization and causes conformational changes. [191] The BMP agonists can bind to type-1 receptors in the absence of constitutively active type-2 receptors, but, when both types of receptors are linked, the binding affinity increases strongly. Constitutively active type-1 receptor kinases work in the absence of ligands or type-2 receptors [1260]. However, type-1 receptors act downstream from type-2 receptors and determine the specificity of intracellular signals. Type-2 receptor kinase transphosphorylates type-1 receptor that transmits specific intracellular signals. Upon phosphorylation by type-1 receptors, receptor-regulated R-Smads form complexes with common partner Smad4, translocate into the nucleus, and regulate transcription of target genes.

Three type-2 receptors are used by BMPs: BMPR2 and AcvR2a and AcvR2b. The BMPR2 receptor is specific for BMPs, whereas AcvR2a and AcvR2b are also shared by activins and myostatin. Proteins BMP2 and BMP4 bind to ALK3 (or BMPR1a) and ALK6 (or BMPR1b; Table 8.25). [192] Proteins BMP6 and BMP7 connect strongly to ALK2 (or AcvR1a), and weakly to ALK3 and ALK6 receptors. Proteins BMP9 and BMP10 tether to ALK1 and ALK2 receptors. Protein GDF5 preferentially links to ALK6 [1260].

Ligand binding to preformed complexes primes phosphorylation of BMPR1 (ALK3 and ALK6) by BMPR2 to propagate signal by phosphorylation (activation) of receptor-specific SMADs, i.e., SMAD1, SMAD5, and SMAD8, to trigger the canonical pathway. Release of SMADs from receptors requires clathrin-mediated endocytosis of the receptors. Then, rSMADs associate with coSMAD (SMAD4) and subsequently translocate to the nucleus. Nucleocytoplasmic transfer of SMADs is regulated by phosphorylations (Sect. 8.3.10). The non-canonical signaling is initiated by BMP2 binding to high-affinity receptor BMPR1 that consequently recruits BMPR2 to activate the MAPK modules (Sect. 8.3.11).

Type-1 and -2 receptors of the TGFβ superfamily are regulated by dephosphorylation, ubiquitination, sumoylation, and proteolysis. Signaling transduction from BMPs via types-1 and -2 receptors is regulated by coreceptors. Glycosylphosphatidylinositol-anchored proteins of the *repulsive guidance molecule* (RGM) family, such as RGMa, RGMb, [193] and RGMc, [194] are coreceptors for BMP2 and BMP4 (but not BMP7) that enhance BMP signaling. Coreceptor RGMa facilitates the use of AcvR2a by BMP2 and BMP4 [1260]. Coreceptor RGMb associates directly with type-1 (ALK2, ALK3, and ALK6) and type-2 (AcvR2a and AcvR2b) BMP receptors [1261]. In hepatocytes, BMP6 binds to RGMc (or hemojuvelin) and induces the expression of hepcidin that lowers iron absorption by intestine and iron release from macrophages [1260].

191. The BMPR2 homo-oligomers bind ligand poorly in the absence of BMPR1 receptor.

192. Dimer BMP2 possesses high- and low-affinity binding sites for BMPR1a and BMPR2, respectively.

193. A.k.a. Dragon.

194. A.k.a. hemojuvelin.

In addition, some transmembrane Tyr and cytoplasmic Ser/Thr kinases interact with BMP receptors [1260]. Receptor Tyr kinase NTRK3 binds to BMPR2 and inhibits signaling. Another receptor Tyr kinase WNRRTK2 (or ROR2) forms a heteromer with BMPR1b independently of ligands. The WNRRTK2 receptor inhibits SMAD1 and SMAD5 signaling downstream from the GDF5–BMPR1b pathway. Protein kinase-G1 phosphorylates BMPR2 to potentiate BMPR signaling [1271]. [195]

In developmental pathways, bistability that relies on signaling thresholds refers to the generation of binary outputs from graded or noisy inputs, such as the specification of the left–right axis in embryogenesis. The left–right laterality involves the bistable expression of transforming growth factor-β signaling member Nodal. The BMP–SMAD1 signaling sets a repressive threshold for Nodal activation in the left lateral plate mesoderm that limits availability of SMAD4 [1272].

Bone morphogenic proteins signal by communicating with other signaling pathways, such as FGF–MAPK and Wnt–GSK3 axes, as well as SMADs and retinoic acid and their receptors. Retinoic acid limits BMP signal duration, as it supports ubiquitination and proteasomal degradation of phosphorylated SMAD1 via GADD45 and MAPK [1273].

8.3.9 Activin Receptor-like Kinases

Activin receptor-like kinases (ALK1–ALK7) belong to the group of receptor protein Ser/Thr kinases. [196] Activin-like receptor corresponds to TβR1 that pairs with specific type-2 TGFβ receptors (TβR2, BMPR2, AcvR2a, AcvR2b, and AMHR2).

Activin dimers operate as growth and differentiation factors that belong to the transforming growth factor-β superfamily. Activins signal via a complex that includes at least 2 type-1 and 2 type-2 receptors that serve for signaling and ligand binding, respectively. Type-1 and -2 receptors form a stable complex after ligand binding and phosphorylation of type-1 receptors by type-2 receptors (Table 8.31).

Type-2 activin-A receptor-like protein AcvRL1, [197] type-1 activin-A receptor AcvR1, [198] type-1a bone morphogenetic protein receptor, [199] type-1b activin receptor

195. Soluble cytosolic Ser/Thr kinase PKG1 that is highly expressed in vascular smooth muscle cells also intervenes in nitric oxide signaling that controls the vasomotor tone. As receptor Tyr kinase mast and stem cell growth factor receptor (SCFR or KIT), PKG1 is a BMPR2-associated kinase that phosphorylates the receptor at its tail. Kinase PKG1 dissociates from BMPR2 on BMP2 stimulation.

196. Alias ALK denotes a receptor Tyr kinase, the anaplastic lymphoma kinase, which forms a chimeric NPM-ALK protein in anaplastic TNFRSF8+ large cell lymphoma. This disease results from a chromosome translocation with fusion of the nucleophosmin gene on chromosome 5q35 to a portion of the Alk gene on chromosome 2p23.

197. A.k.a. type-2 activin-A receptor-like kinase AcvRLK1.

198. A.k.a. AcvR1a and AcvRLK2.

199. A.k.a. AcvRLK3.

Table 8.31. Activin receptor-like kinases and their main ligands (Source: [59]; AAT: aortic aneurysm, familial thoracic protein [AAT5 and AAT3 are linked to mutations in the TGFBR1 and TGFBR2 genes, respectively] AcvRL; AcvRLK: activin receptor-like kinase; AMH: anti-Müllerian hormone; BMP: bone morphogenetic protein; BMPR: BMP receptor; BRK: BMP receptor kinase [BRK3 is BMPR2]; GDF: growth differentiation factor; FOP: fibrodysplasia ossificans progressiva protein; HHT: hereditary hemorrhagic telangiectasia protein; LDS: Loeys-Dietz aortic aneurysm syndrome; ORW: Osler-Rendu-Weber syndrome protein [ORW1: endoglin]; SKR: protein Ser/Thr kinase receptor; TGF: transforming growth factor; TSR: TGFβ superfamily receptor). Ligand types of ALK receptors and affinity vary according to ALK subtypes. For example, ALK2 connects to both activin and BMP7, but not to BMP4; ALK3 binds to BMP2 and BMP4 more efficiently than to BMP7; ALK6 links to both BMP4 and BMP7 efficiently.

Type	Alternative names	Ligands
ALK1	AcvRL1, AcvRLK1, HHT1/2, ORW2, SKR3, TSR1	TGFβ1, activin-A
ALK2	AcvRL2, AcvRLK2, AcvR1a, FOP, SKR1, TSR1	AcvR1b, TβR2, BMPR2, SMAD1/5, activin-A, AMH, BMP6/7, GDF5
ALK3	BMPR1a, AcvLK3, CD292 BRK1, SKR5	BMPR1a/b, BMPR2, SMAD7, AMH, BMP2/4/6/7, GDF5/6/7/8
ALK4	AcvR1b, AcvRLK4, SKR2	AcvR1a, AcvR2a/b, SMAD2/3, activin-A/AB/B, BMP3/11, GDF1/8, Nodal
ALK5	TβR1 AcvRLK5, AAT5, LDS1a/2a, SKR4	TβR2, BMPR2, SMAD2/4/7/8, TGFβ1, GDF8/9
ALK6	BMPR1b, AcvRLK6, BRK2, CDw293	BMPR1a/b, BMPR2, SMAD6/7, AMH, BMP2/4/6/7/10/15, GDF5/6/7
ALK7	AcvR1c, AcvRLK7	Nodal, activin-B

AcvR1b,[200] TβR1,[201] type-1b bone morphogenetic protein receptor,[202] and type-1c activin receptor AcvR1c[203] correspond to ALK1 to ALK7, respectively.

200. A.k.a. AcvRLK4.
201. A.k.a. AcvRLK5.
202. A.k.a. AcvRLK6.
203. A.k.a. AcvRLK7.

Activin-A receptor-like receptors are classified into 3 groups based on structural and functional similarities: (1) BMPR1 group with BMPR1a and BMPR1b (or ALK3 and ALK6, respectively); (2) ALK1 group with ALK1 and ALK2 (or AcvRL1 and AcvR1a, respectively); and (3) TβR1 group with ALK4, ALK5, and ALK7 (or AcvR1b, TβR1, and AcvR1c, respectively). Receptors of the ALK1 and BMPR1 groups activate SMAD1, SMAD5, and SMAD8, whereas receptors of the TβR1 group activate SMAD2 and SMAD3. Receptors ALK2 and ALK3 are widespread, whereas ALK6 has a more restricted expression pattern and ALK1 production is limited to endothelial cells and some other cell types [1260].

ALK1

The ALK1 receptor is a component of the TGFβ pathway in endothelial cells and, to a lesser extent, vascular smooth muscle cells. It consists of an extracellular, single membrane spanning, and intracellular domain. Ligands of its extracellular domain include TGFβ1 and -3 as well as activin-A [1274]. However, TGFβ1 and activin-A binding to ALK1 only occurs in the presence of TβR2 and activin receptor-2 or -2B, respectively. Upon assembly of the TβR2–ALK1 complex, TβR2 phosphorylates (activates) ALK1 intracellular domain. Then, ALK1 phosphorylates SMAD1, -5, and -8. ALK1 inhibits ALK5 signaling. According to concentration and type of endothelial cells, TGFβ signaling via ALK1 either favors, inhibits, or has no effect on endothelial cell proliferation and migration.

The ALK1 receptor interacts with some members of the TGFβ receptor superfamily: TβR2, activin receptor ActR2, ALK5, and endoglin. It can associate with TβR2, ActR2, and endoglin in a ligand-independent manner [1274]. However, its linkage to TβR2 and ALK5 is enhanced by TGFβ1. The ALK1 receptor raises the affinity of TGFβ1 for TβR2 (3-fold), but does not change that of activin-A for ActR2.

Inhibitors of ALK1 comprise FKBP12 and liver X LXRβ receptor. Conversely, ALK5 and endoglin can enhance signaling via the TGFβ1–ALK1 pathway [1274]. However, endoglin can also attenuate ALK1 signaling.

8.3.10 SMAD Mediators – The Canonical Pathway

Eight SMAD mediators exist (SMAD1–SMAD7 and SMAD9 [or SMAD8]). Activated TβR1 phosphorylates receptor-associated SMAD mediators. [204] Afterward, these mediators, especially SMAD2 and SMAD3 in the TGFβ pathway, translocate to the nucleus, where they participitate with SMAD4 and other transcriptional regulators in gene activity, as SMADs form complexes to serve as transcription factors.

There are 3 classes of SMAD modulators of TGFβ1 activity: (1) receptor-regulated SMADs (rSMAD) that include SMAD1 to -3, -5, and -9; (2) common-mediator SMAD (coSMAD) that comprises only SMAD4, which interacts with rS-

204. SMAD proteins are homologs of both Drosophila protein "mothers against decapentaplegia" (MAD) and protein Sma of the nematode Caenorhabditis elegans.

Table 8.32. Structural domains of SMADs and their roles (Source: [1254]). The MH1 Domain serves to bind DNA (except SMAD2) and transcription factors; MH2 for receptor–SMAD interaction, SMAD oligomerization, transcription activation, CREB-binding protein or P300 binding, and transcription factor tethering (CamK: calmodulin-dependent kinase; MAPK: mitogen-activated protein kinase; NES: nuclear export signal; NLS: nuclear localization signal; PKC: protein kinase-C; SMURF: Smad ubiquitination regulatory factor).

rSMAD (SMAD1–SMAD3, SMAD5, SMAD9)	coSMAD (SMAD4)	iSMAD (SMAD6–SMAD7)
MH1	MH1	
MH2	MH2	MH2
Receptor binding		
MAPK-binding	MAPK binding	MAPK binding
PKC binding		
CamK2 binding		
SMURF binding		SMURF binding
NLS	NLS	
	NES	

MADs; and (3) inhibitory SMADs (iSMAD), i.e., SMAD6 and SMAD7, that prevent activation of rSMADs and coSMADs.

Regulators rSMADs and SMAD4 are expressed in most, if not all, cell types. Production of inhibitory SMAD6 or SMAD7 is highly regulated by extracellular signals. Expression of SMAD6 and SMAD7 by TGFβ and BMPs constitutes an autoinhibitory feedback. Activation of epidermal growth factor receptor via mitogen-activated protein kinase and possibly other receptor Tyr kinases, interferon-γ signaling via signal transducer and activator of transcription STAT1, and activation of transcription factor NFκB by tumor-necrosis factor-α also induce SMAD7 expression, thereby inhibiting TGFβ signaling [1254].

Regulators of SMAD proteins possess various binding sites inside or outside their MH domains(Table 8.32). Inhibitory SMADs lack a MH1 domain, but have an MH2 domain. Activated SMADs translocate into the nucleus owing to nuclear localization sequence. The SMAD4 protein permanently shuttles between the nucleus and cytoplasm owing to its nuclear localization sequence and export signal via exportin-1.

Proteins SMAD2 and SMAD3 are phosphorylated (activated) by ALK5 (TβR1) and ALK4 (AcvR1b), whereas SMAD1, SMAD5, and SMAD9 are activated by ALK1 (AcvRL1), ALK2 (AcvR1a), ALK3 (BMPR1a), and ALK6 (BMPR1b) in response to BMP1 to BMP4 or other ligands [1254]. Some TGFβ responses can occur in the absence of SMAD4 protein. However, the absence of SMAD4 limits TGFβ responsiveness.

Receptor-regulated SMADs are anchored at the plasma membrane by various cytosolic proteins. Anchoring protein SARA, or ZFYVE9, binds to rSMADs SMAD2 and SMAD3 as well as type-1 TGFβ receptors. It then presents SMAD2 and SMAD3 to type-1 receptors and facilitates their activation [1260]. The ZFYVE9 protein does not tether to BMP-specific rSMADs. On the other hand, *endofin*, or ZFYVE16, interacts with SMAD1 to enhance BMP signaling [1260]. It is also an endosomal protein that contributes to the regulation of membrane trafficking. In addition, plasmalemmal glycoprotein CD44, a receptor for hyaluronan that also interacts with osteopontin, collagens, and matrix metallopeptidases, connects to SMAD1 in chondrocytes.

Nuclear BMP signaling depends on SMAD interactions with proteins of the nuclear envelope and recruitment of and cooperation with specific transcription factors. In the nucleus, SMADs regulate transcription of target genes, as they bind to DNA or interact with other DNA-binding proteins (Runx1–Runx3, Schnurri-1–Schnurri-3, P300 and CREB-binding protein, MyoD, etc. [1260]) and recruit transcriptional coactivators and/or corepressors.[205] Besides their role in gene transcription, rSMADs can promote the processing of primary transcripts of microRNA-21 into miR21 precursor [1260].

Common SMAD (SMAD4) can form complexes with rSMADs that serve as transcription factors and bind to target genes. However, rSMAD transducers are able to regulate transcription of certain target genes without complexing with coSMAD. Inhibitory iSMADs antagonize rSMADs and coSMAD mediators. Expression of iSMADs is strongly induced by TGFβ, activins, and BMPs (negative feedback loop). The iSMAD inhibitors interact with type-1 receptors, thus competing with rSMADs for receptor binding. Effect of iSMADs occurs rapidly (within 2 h) and transiently on BMP stimulation, followed by possible sustained activity. Unlike rSMADs, iSMADs are not released from type-1 receptors, thereby preventing rSMAD activation. Mediator SMAD7 inhibits both TGFβ and BMP signals, whereas SMAD6 preferentially represses BMP, but not TGFβ cues. SMAD7 interacts with receptors of the TβR1, BMPR1, and ALK1 groups, whereas SMAD6 inhibits signals from the BMPR1 group and only weakly those from the ALK1 group [1260]. Transducer SMAD6 can form a complex with SMAD1, hence competing with SMAD4 for oligomer formation. Besides, iSMADs also recruit ubiquitin ligases SMURF1 and SMURF2 for receptor degradation. Furthermore, iSMADs interfere with TGFβ family signaling in the nucleus. Inhibitor SMAD7 binds to the SMAD-responsive gene element, hence impeding the formation of the SMAD–DNA complex.

Phosphorylation of the SMAD C-terminus by type-1 receptor triggers SMAD activation. However, other kinases phosphorylate SMAD proteins. (1) Modulator SMAD2 is phosphorylated in response to epidermal and hepatocyte growth factors that bind to their cognate receptor Tyr kinases. Moreover, extracellular signal-regulated protein kinase stimulated by activated receptor Tyr kinases and/or small Ras GTPase phosphorylates rSMADs. Activated receptor Tyr kinases and Ras GTPase can impede nuclear translocation of activated SMADs. Activated mitogen-activated protein kinase kinase MAP2K1 can also phosphorylates SMAD proteins.

205. Runx transcription factors regulate hematopoiesis and osteogenesis. Schnurri-2 causes nuclear translocation of chloride intracellular channel CLIC4 and protect phosphoSMADs2 and -3 from dephosphorylation by nuclear phosphatases [1260]. Histone acetyltransferases P300 and CBP as well as GCN5 enhance SMAD-dependent transcription of target genes. Transcriptional corepressors (Ski, SnoN, and Evi1) recruit histone deacetylases to SMAD complexes.

It enhances SMAD2 heteromerization with SMAD4, nuclear translocation, and transcriptional activity [1254]. Jun N-terminal kinase phosphorylates SMAD3 to improve its migration toward the nucleus. (2) Activated Ca^{++}–calmodulin-dependent protein kinase CamK2 also phosphorylates SMAD2 to SMAD4 proteins [1254]. It precludes SMAD2 nuclear import. (3) Protein kinase-C phosphorylates SMAD2 and SMAD3.

Phosphorylation of SMAD1 by members of the mitogen-activated protein kinase family ERKs, P38MAPKs, and JNKs and then glycogen synthase kinase GSK3 leads to proteasomal degradation [1260]. Therefore, SMAD1 undergoes a triple phosphorylation. Signaling by BMP, indeed, induces a first phosphorylation of SMAD1 and its nuclear translocation. SMAD1 Phosphorylation in the linker region counteracts SMAD1 function. Phosphorylation of rSMAD in the linker region by MAPKs inhibits rSMAD nuclear translocation. Phosphorylation in the linker region by GSK3 leads to SMAD1 proteasomal degradation.

Conversely, phosphatases dephosphorylate rSMADs. Among metal ion-dependent protein phosphatases, magnesium-dependent PPM1a [206] interacts with triply phosphorylated SMAD1 to SMAD3 [1260]. Other phosphatases, such as pyruvate dehydrogenase phosphatase and nuclear small C-terminal domain-containing phosphatases (SCP), [207] repress BMP and enhance TGFβ signaling by dephosphorylating $SMAD1^P$. [208] In addition, PP2 dephosphorylates SMAD1 to increase its nuclear translocation, thereby amplifying BMP signaling.

Upon BMP stimulation, STAM-binding protein (STAMBP) [209] interacts with SMAD6, thereby hampering SMAD6 association with BMPR1 and SMAD1 [1260]. Transducer of ErbB2 (TOB) inhibits BMP signal transduction in osteoblasts, as it connects to BMP-specific rSMADs. It also influences BMPR2 activity.

---

206. A.k.a. PP2cα.

207. Among protein Ser/Thr phosphatases, that also include the PPM and PPP families, the FCP–SCP family of phosphatases with RNA polymerase-2 C-terminal domain phosphatase domain homology comprises 3 known members (SCP1–SCP3). Small C-terminal domain phosphatases localize to the nucleus, where they inhibit RNA polymerase-2 by dephosphorylating its C-terminus. C-Terminal domain, RNA polymerase-2 subunit-A phosphatase CTDP1 is also termed FCP1 (as CTD phosphatase in Saccharomyces cerevisiae). All members SCP1 to SCP3 are related to the catalytic subunit of FCP1, which is a highly conserved enzyme that dephosphorylates the CTD of RNA polymerase-2.

208. The SMAD proteins transduce signals from bone morphogenetic proteins and transforming growth factor-β upon phosphorylation (activation) of their C-terminus (activating site) by receptor kinases. On the other hand, CDK2, CDK4, MAPK, P38MAPK, JNK, among others, phosphorylate (repress) SMADs at the linker region (inhibitory site) between the N-terminal DNA-binding domain and C-terminal transcriptional sequence. Small C-terminal domain phosphatases SCP1 to SCP3 dephosphorylate SMAD1 C-terminus and linker, thereby resetting SMAD1 to the basal unphosphorylated state [1276]. These phosphatases dephosphorylate SMAD2 and SMAD3 only at the linker, thus enhancing TGFβ signaling.

209. A.k.a. STAM-associated, SH3-containing adaptor AMSH (Associated molecule with the SH3 domain of signal transducing adaptor molecule (STAM)).

Protein kinase-G1 dissociates from BMPR receptors upon ligand stimulation, interacts with activated rSMADs, and undergoes nuclear translocation [1271].[210] In the nucleus,[211] Protein kinase-G1 binds to SMAD1 and general transcription factor GTF2i[212] at promoters of BMP target genes to potentiate gene transcription. Therefore, PKG1 has a dual function in BMP signaling [1271]: (1) modulation of BMPR-induced SMAD activity at the plasma membrane and (2) regulation after relocalization to the nucleus of transcription as a nuclear SMAD cofactor.

In response to extracellular ligand, SMAD mediators activate transcription, as they form nuclear proteic complex with other transcription factors to target SMAD-binding element (SBE [CAGACA sequence]), activin-response element (ARE),[213] TPA-response element (TRE),[214] an AP1-binding site), and X-factor-binding element (XBE) [1275].,[215] The SMAD2 regulator interacts with forkhead box FoxH1 transcription factor for activin- and SMAD2–SMAD4-mediated transcription [1275]. The SMAD3 protein interacts with Fos and Jun transcription factors (e.g., transcription from the collagenase-1 promoter). Effectors SMAD3 and SMAD4 cooperate with basic helix–loop–helix TFE3 protein.

Agent SMADs cooperate with transcription factors as well as coactivators CBP and P300 (Table 8.33). Besides CBP or P300 that are coactivators of rSMADs, SMAD4 also serves as rSMAD coactivator, as it stabilizes rSMAD interaction with CBP or P300 coactivator [1254]. Other SMAD coactivators, such as ubiquitous mRNA-decapping enzyme Dcp1a,[216] CBP/P300-interacting transactivator with glutamic (E) and aspartic acid (D)-rich C-terminus-containing protein CITED1,[217] and mediator of RNA polymerase-2 transcription subunit MED15,[218] determine the level of SMAD-mediated transcription activation [1254]. The SMAD-interacting corepressors, Sloan-Kettering Institute proto-oncoprotein SKI and SKI-like protein SKIL,[219] cell cycle regulator MyC, dual corepressor and coactivator myelodysplasia syndrome MDS1 and ecotropic viral integration site protein EVI1 complex locus (MECOM)[220] TGFβ-induced factor (TGIF), survival of motor neuron protein-interacting protein SIP1 that downregulates E-cadherin expression, and

---

210. Consequences of mutations in BMPR2 in pulmonary arterial hypertension patients can be compensated via PKG1.

211. In the nucleus, the SMAD1–SMAD4 complex interacts with other transcription factors (TF1 , TF2, and TF3).

212. A.k.a. TFII-I, Bruton Tyr kinase-associated protein BAP135 and BTKAP1, SRF-Phox1-interacting protein (SPIN), and Williams-Beuren syndrome chromosomal region 6 protein WBS and WBSCR6.

213. I.e., TGT(G/T)(T/G)ATT sequence.

214. I.e., TGACTCA sequence.

215. I.e., 5′-ATGGAGAATTGCCTCCC-3′ forward sequence.

216. A.k.a. SMAD4-interacting factor (SMIF).

217. A.k.a. melanocyte-specific gene product MSG1.

218. A.k.a. ARC105, PCQAP, and TIG1.

219. A.k.a. SnoA and SnoN.

220. A.k.a. EVI1. This alias is also used for Runt-related transcription factor RUNX1.

Table 8.33. The rSMAD–SMAD4 complex cooperate with transcription factors (coactivators and corepressors) that bind with high affinity to a X-transcription factor-binding promoter element (XBE) and lower affinity to a SMAD-binding DNA element (SBE; Source: [1254]; CBP: cAMP-responsive element-binding protein (CREB)-binding protein; CITED: CBP/P300-interacting transactivator with glutamic and aspartic acid-rich C-terminus-containing protein; Dcp1a: mRNA-decapping enzyme homolog (1)-A [a.k.a. SMAD4-interacting factor (SMIF)]; EVI: ecotropic viral integration site protein; MED: mediator of RNA polymerase-2; SIP: survival of motor neuron protein-interacting protein; SKI: Sloan-Kettering Institute proto-oncoprotein; SNIP: SMAD nuclear-interacting protein; TGIF: TGFβ-induced factor).

Coactivators	Corepressors
CBP	EVI1
CITED1	MyC
Dcp1a	SIP1
MED15	SKI, SKIL
P300	SNIP1
	TGIF

nuclear SMAD-interacting protein SNIP1, prevent SMAD-mediated transactivation [1254].[221]

The rSMAD transducers are degraded by the ubiquitin–proteasome pathway. Type-HECT ligases of the NEDD4 family (SMURF1, SMURF2, and NEDD4-2) connect to rSMAD proteins. The SMURF ligases also interact with iSMADs and provoke their nuclear export. Activity of SMAD4 is controlled by monoubiquitination by the type-RING Ub ligase tripartite motif family member TRIM33[222] that prevents its association with rSMAD proteins. Ubiquitin-specific peptidase-9, X chromosome (USP9X)[223] relieves SMAD4 sequestration.

Transforming growth factor-β has a dual role in cancer, as it acts as a tumor suppressor in early tumor development and promotes epithelial–mesenchymal transition and tumor cell invasion in later stages. Its effectors SMAD3 and SMAD4 complex with transcription factor Snai1 that impedes the production of cell junction components, such as E-cadherin, claudins, occludin, and desmoplakin. The Snai1–SMAD3–SMAD4 complex is a repressor for CAR, occludin, claudin-3, and E-cadherin promoters [1277].

221. Factor TGFβ inhibits cell cycle progression, as it hinders activity of MyC and members of the family of inhibitors of DNA-binding and differentiation protein (ID). Factor SMAD3 represses Myc transcription in association with E2F4 and E2F5 transcriptional suppressors and corepressor retinoblastoma-like protein RBL1 (P107) [1254]. Once activated by TGFβ, SMAD3 provokes expression of cAMP-dependent activating transcription factor ATF3 that then complexes with SMAD3 to repress the promoter of inhibitor of DNA-binding ID1 [1254]. Factor SMAD3 represses transcription by Runx2 in osteoblastic differentiation, MyoD and other myogenic bHLH transcription factors in myoblasts, and CCAAT/enhancer-binding proteins in adipocyte differentiation [1254].

222. A.k.a. transcriptional intermediary factor-1γ (TIF1γ).

223. A.k.a. fat facets-like deubiquitinase (FaF).

Table 8.34. Non-canonical TGFβ pathways (Source: [1254]).

MAP3K1–MAP2K4–JNK
MAP3K1–MAP2K3/6–P38MAPK
MAP3K7–MAP2K4–JNK
MAP3K7–MAP2K3/6–P38MAPK
Ras–cRaf–MAP2K1/2–ERK
RhoA–RoCK1
PP2–S6K

SMAD5

Receptor-regulated SMAD5 is a transcription factor that contributes to developmental pattern formation and homeostasis upon stimulation from anti-Müllerian hormone, bone morphogenetic proteins, and growth and differentiation factors [1278]. The SMAD5 factor possesses N-terminal MH1 and C-terminal MH2 domains involved in DNA binding and proteic interaction, respectively.

8.3.11 Non-Canonical Pathways

Interactions between TGFβ receptors and SMAD complexes as well as receptor- and SMAD-interacting proteins, in addition to cooperation with transcription factors generate diversification of responses induced by members of the TGFβ family. Moreover, TGFβ receptors activate SMAD-independent signaling cascades that not only regulate SMAD signaling, but also allow SMAD-independent TGFβ responses [1254].

SMAD-independent TGFβ receptor signaling relies on proteins that can tether to TβR1 and TβR2 receptors, such as immunophilin FKBP1a. [224] The latter lowers TGFβ signaling, as it hampers type-1 receptor phosphorylation by type-2 receptor [1254].

Activated receptor complexes are able to activate non-SMAD signaling cascades, such as many MAPK modules, the PP2–S6K axis, as well as small GTPases of the Ras superfamily [1254] (Table 8.34). According to the cell line, TGFβ can activate RhoA GTPase (via RhoA-specific guanine nucleotide-exchange factor RhoGEF8, or NET1), Rac, and CDC42 (Vol. 4 – Chap. 8. Guanosine Triphosphatases and Their Regulators), as well as enhance RhoB expression. Small RhoB and Rac1 GTPases counteracts and contributes to TGFβ-induced gene expression [1254]. Activated RhoA as well as Rac1 and CDC42 that target P38MAPK are required for cell migration in response to TGFβfactor. Activation of the RhoA–RoCK1 axis as well as CDC24, P38MAPK, and SMADs contribute to stress-fiber formation.

Non-canonical pathways of signal transduction from members of the TGFβ superfamily use components of mitogen-activated protein kinase modules, such as

224. A.k.a. rotamase, or peptidyl-prolyl cis–trans isomerase, FKBP12.

P38MAPK and ERK kinases. Kinase MAP3K7 [225] is activated by BMP2 and BMP4 proteins. It stimulates MAP2K3 and MAP2K6, hence P38MAPK [1260].

Protein phosphatase-1 can interact with ZFYVE9 (SARA) protein. Upon TGFβ stimulation, trimeric protein phosphatase PP2 [226] links to S6 kinase to trigger cell cycle arrest [1254].

Signaling initiated by TGFβ is also transmitted by several molecules that interact with BMPR2 C-terminus, [227] such as LIM kinase LIMK1, Tribbles-like protein Trb3, dynein light chain Tctex-type-1 (Tctex1), [228] and Src kinase. Kinase LIMK1 phosphorylates (inactivates) actin-depolymerizing cofilin, hence regulating actin dynamics. The BMPR2 receptor inhibits LIMK1 action on cofilin [1279]. On the other hand, in tips of neurites (i.e., axons and dendrites), LIMK1 binds to BMPR2 and the LIMK1–BMPR2 complex cooperates with CDC42 GTPase to activate LIMK1 [1280]. Tribbles-like protein Trb3 that associates with BMPR2 C-terminus is released from BMPR2 on ligand binding. It then induces SMURF1 degradation to stabilize BMP-specific rSMAD factor [1260].

225. A.k.a. TGFβ-activated kinase-1 (TAK1).

226. Its catalytic C and regulatory A subunits interact upon stimulation with regulatory B subunit.

227. Mutations of the Bmpr2 gene that remove BMPR2 C-terminus generate primary pulmonary hypertension.

228. A.k.a. T-complex-associated testis-expressed-1-like TCTEL1 and DynLT1.

9

Receptor Tyrosine Phosphatases

According to the primary structure of their catalytic domains, the subclass of protein Tyr phosphatases (PTP) can be decomposed into 4 distinct infraclasses. Infraclass-1 Cys-based PTPs are subdivided into: (1) 22 classical, high-molecular-weight, transmembrane, receptor protein Tyr phosphatases that constitute the RPTP superfamily; (2) 19 classical, low-molecular-weight, cytosolic protein Tyr phosphatases that form the PTPn (n: non-receptor) superfamily (Vol. 4 – Chap. 7. Cytosolic Protein Phosphatases); as well as (3) the hyperfamily of dual-specific phosphatases that encompasses 7 superfamilies, i.e., (3.1) 10 MAPK phosphatases and 1 MAPK pseudophosphatase; (3.2) 3 Slingshots; (3.3) 3 protein Tyr phosphatases-4A (or phosphatases of regenerating liver [PRL]); (3.4) 4 cell division cycle-14 homolog phosphatases (CDC14a–CDC14b, cyclin-dependent kinase inhibitor CDKn3 [1] and PTPDC1 [or PTP9Q22]); (3.5) 20 atypical dual-specific phosphatases and pseudophosphatases; (3.6) 5 phosphatase and tensin homolog (PTen)-like phosphatases; and (3.7) 16 myotubularins. The infraclass 2 is represented by a single member, the low-molecular-weight phosphotyrosine phosphatase (LMWPTP). Infraclass-3 PTPs correspond to CDC25 phosphatases (CDC25a–CDC25c). The infraclass 4 contains 4 Eyes absent homolog phosphatases (EyA1–EyA4) and haloacid dehydrogenases.

Receptor-like protein Tyr phosphatases are also called protein Tyr phosphatase receptors (PTPR). The cytosolic domain of these transmembrane enzymes contains the phosphatase domain. Receptor Tyr phosphatases form symmetrical dimers, each monomer blocking the active site of its partner. Ligand binding to receptor Tyr phosphatases leads to dephosphorylation of target proteins.

Receptor-like Tyr phosphatases contribute to the dynamics of cell adhesion that yields a structural support and communication path. Cell adhesion molecules are specialized plasmalemmal receptors that form clusters for intercellular contacts via their ectodomains. These dynamical molecular clusters act in signaling via phosphorylation–dephosphorylation cycles. Intracellular catalytic regions of RPTPs

1. A.k.a. CDK2-associated dual-specificity phosphatase, cyclin-dependent kinase interactor CDI1, cyclin-dependent kinase-interacting protein CIP2, and kinase-associated phosphatase (KAP).

dephosphorylate cadherin–catenin complexes, thereby stabilizing the intercellular contacts [1281].

Heparan sulfate proteoglycans regulate receptor-like Tyr phosphatases. Reactive oxygen species such as hydrogen peroxide as well as post-translational reversible oxidation affect RPTP functioning. Many RPTPs can indeed be transiently oxidized during signaling. [2]

Among PTPRs, 12 proteins possess 2 phosphatase domains, whereas the rest contain a single phosphatase domain [1283]. The catalytic domain (D1) adjacent to the membrane provides the main phosphatase activity, while the second catalytic domain (D2) may be involved in the regulation of enzyme activation, between-protein interaction, substrate specificity and presentation of substrates to the active catalytic domain.

Therefore, PTPRs can be categorized according to the presence, number, and position of each identified domain. They can also be classified based on structural differences in the extracellular domain.

The PTPR superfamily can be decomposed into 8 families according to the structural features (Table 9.1). Family-1 PTRPs contain heavily glycosylated and cysteine-rich extracellular regions that include fibronectin-3-like domains. Family-2 PTRPs possess multiple extracellular immunoglobulin-G-like domains and fibronectin-3-like domains. Isoforms PTPRd, PTPRf, and PTPRs constitute the subfamily 2A, whereas PTPRk, PTPRl (PTPRu), PTPRm, and PTPRt that are involved in homophilic cell adhesion form the subfamily 2B characterized by the presence of an N-terminal meprin-A5 antigen-PTPμ(MAM) domain. Family-3 PTRPs have a single cytosolic protein Tyr phosphate domain and multiple extracellular fibronectin-3-like domains. Family-4 PTRPs are constituted by short, highly glycosylated, extracellular domains and 2 active cytosolic PTP domains. Family-5 PTRPs are composed of a single extracellular fibronectin-3-like domain along with a carbonic anhydrase (CA)-like motif, in addition to 2 catalytic domains. Family-6 PTRPs comprise heavily glycosylated and cysteine-rich extracellular regions with fibronectin-3-like domains. Family-7 PTRPs consist of a short extracellular domain, a cytosolic kinase-interaction domain, and single PTP domain. Family-8 PTRPs are made up of an extracellular RDGS-adhesion recognition motif and a single cytosolic PTP domain.

Subtypes of PTPRs can have a tissue- or cell-specific expression. In mice, PTPRd is synthesized in the hypothalamus; PTPRg in the brain, lung, white adipose tissue, and eye; PTPRo in the brain; PTPRs in the thalamus; PTPRk in blastocysts; and PTPRe and PTPRo in the monocyte lineage [1284].

The LAR group of protein Tyr phosphatases corresponds to the PTPR subfamily 2A. It includes leukocyte common-antigen-related receptor (LAR or PTPRf) as well as PTPRd and PTPRs. With family-3 PTPR member PTPRo, they participate in the regulation of axon growth and guidance as well as mammary gland development.

2. Reactive oxygen species produced by NADPH oxidase are controlled, since NADPH oxidase subunits must assemble into a complex with scaffold and regulators and bind to inositol phospholipids and GTPase Rac, the latter acting as a molecular switches [1282].

Table 9.1. Families of the subclass of receptor-like protein Tyr phosphatases (RPTP) or protein Tyr phosphatase receptors (PTPR; Source: National Library of Medicine - Medical Subject Headings).

Family	Members
1	PTPRc
2	PTPRd, PTPRf, PTPRk, PTPRm, PTPRs, PTPRt, PTPRu
3	PTPRb, PTPRh, PTPRj, PTPRo, PTPRq, PTPRv
4	PTPRa, PTPRe
5	PTPRg, PTPRz
6	PTPRn2
7	PTPRr
8	PTPRn

Liprin (LAR-interacting protein) can regulate the disassembly of focal adhesions and recruit subfamily-2A PTPRs at specific sites on the plasma membrane to regulate their interaction with the extracellular matrix and their association with substrates. Liprin-α and -β form homo- and heterodimers that interact with the second phosphatase domain of PTPRd, PTPRf, and PTPRs (Tables 9.2 to 9.5). Alternatively spliced transcript variants encode distinct isoforms of ubiquitous cytoplasmic liprins.

Protein Tyr phosphatase receptors can homo- and heterodimerize. However, monomeric PTPRs can be constitutively active and dimerization leads to inactivation. Certain PTPRs, such as PTPRf and PTPRm, contain a wedge-shaped helix–loop–helix located near the catalytic domain. The helix–loop–helix domain favors homophilic binding and inhibits functioning.

Phosphatases of the RPTP subfamily 2A (LAR) as well as others possess 2 cytoplasmic catalytic – active D1 and inactive D2 – domains. The wedge sequence in D1 mediates D1–D1 homodimerization in PTPRa. The PTPRs phosphatase can associate with the PTPRd phosphatase. Heterodimerization yields a negative regulatory function [1285]. Binding between $PTPRs^{D1}$ and $PTPRd^{D2}$ domains happens, but is not reciprocal, as $PTPRd^{D1}$ does not bind $PTPRs^{D2}$.

9.1 Protein Tyrosine Phosphatase Receptor-A

Protein Tyr phosphatase receptor-A (PTPRa)[3] is a transmembrane PTP with a relatively short, glycosylated, extracellular domain and a tandem repeat of 2 cytoplasmic domains. Its N- and C-termini possess a strong and weak phosphatase activity, respectively. It is constitutively phosphorylated (Tyr789) on a binding site. Three alternatively spliced variants of the Ptpra mRNA generate 2 distinct isoforms.

The PTPRa phosphatase dephosphorylates (activates) protein Tyr kinases of the SRC family. It thus contributes to the regulation of integrin signaling, hence cell

3. A.k.a. receptor protein Tyr phosphatase-α (RPTPα). It is an example of type-4 PTPRs.

Table 9.2. Partners and substrates of protein Tyr phosphatase receptors (**Part 1**; Sources: Wikipedia, BioGRID; CD4: T-cell surface antigen T4; CD8a: T-lymphocyte differentiation antigen T8 [a.k.a. T-cell surface glycoprotein CD8α chain]; CD10: neprilysin [a.k.a. membrane metallo-endopeptidase (MME), neutral endopeptidase (NEP), common acute lymphocytic leukemia antigen (CALLA), enkephalinase, and GP100]; CD22: B-lymphocyte cell adhesion molecule [BLCAM]; CD44: homing-associated cell adhesion molecule [HCAM; a.k.a. 85-kDa hyaluronic acid-binding protein (HABP85), extracellular matrix receptor-3 (ECMR3); Indian blood group antigen; and phagocytic glycoprotein-1 (PGP1)]; CD79a and -b: Igα and -β [parts of B-cell antigen receptor on B lymphocytes and plasmocytes]; CD247: T-cell receptor T3ζ chain; DLg: Disc large homolog; LGALS: lectin, galactoside-binding, soluble protein; MAGI: membrane-associated guanylate kinase, WW and PDZ domain-containing protein; NrCAM: neuronal cell adhesion molecule; PTPRcAP: PTPRc-associated protein; PrKcSh: protein kinase-C substrate 80K-H; Sema: semaphorin (Sema-, Ig-, transmembrane- and short cytoplasmic domain) SKAP: Src kinase-associated phosphoprotein; SLAMF: signaling lymphocytic activation molecule family member). PTPRb is a receptor-type phosphoTyr phosphatase that is restricted to endothelial cells. Molecule CD10 that is expressed on various cell types (B- and T cells and their precursors, neutrophils, monocytes, bone marrow stromal cells, keratinocytes, and endothelial cells) is a membrane-bound metallopeptidase that cleaves many inflammatory and vasoactive peptides (fMLP, bradykinin, endothelin-1, angiotensin-1, substance-P, bombesin, enkephalins, and atrial natriuretic peptide).

Type	Enzymes	Adaptors, dockers, scaffolds GEFs, GAPs	Ion channels, conventional receptors, adhesion and cytoskeletal molecules, transcription factors, cell markers
PTPRa	Src, Fyn	GRB2	$K_V1.2$
PTPRb	TIE2 MAGI3	DLg4	Na_V, NrCAM, Cadherin-5, contactin-1, tenascin-C
PTPRc	Src, LcK, Lyn JaK PLK	SKAP1, SLAMF1 PTPRcAP PrKcSh	Glucosidase-α, LGALS1 CD4, CD8a, CD10, CD22, CD44, CD79a/b, CD247, Sema4D

adhesion and proliferation. It binds to adaptor GRB2 [1286]. Binding of GRB2 to RasGEF Son of sevenless and PTPRa are mutually exclusive.

The PTPRa phosphatase also interacts with $K_V1.2$ channel [1287]. Activated muscarinic acetylcholine M_1 receptor leads to phosphorylation (inactivation) of $K_V1.2$ as well as PTPRa to enhance the magnitude of K^+ channel phosphorylation and repression.

Phosphatase PTPRa can form inactive homo- and heterodimers, which may repress the phosphatase activity.

Table 9.3. Partners and substrates of protein Tyr phosphatase receptors (**Part 2**; Sources: Wikipedia, BioGRID; BCAR: Breast cancer anti-estrogen resistance docking protein; DLg: Disc large homolog; GHR: growth hormone receptor; LAT: linker of activated T lymphocytes; PPfIα: PTPRf polypeptide-interacting protein-α1 [liprin or liprin-α]; SdcBP: syndecan binding protein [syntenin]).

Type	Enzymes	Adaptors, dockers, scaffolds GEFs, GAPs	Ion channels, conventional receptors, adhesion and cytoskeletal molecules, transcription factors, cell markers
PTPRd	PTPRs	PPfIα1	
PTPRe		GRB2	K_V1.5/2.1
PTPRf	Ret	PPfIα1–3 Trio, GIT1	Catenins-βand -γ, Nidogen-1, caveolin-1
PTPRg		PPfIα1–3 DLg4	
PTPRh		BCAR1	GHR, paxillin
PTPRj	PLCγ1 PDGFβ	LAT SdcBP	Catenins-β1 and -δ1

9.2 Protein Tyrosine Phosphatase Receptor-B

Protein Tyr phosphatase receptor-B (PTPRb) [4] a chondroitin sulfate proteoglycan receptor-like protein Tyr phosphatase. This type-3 RPTP is synthesized in glial cells and neurons of the brain, as well as some cell types of germinal centers of lymphoid tissues, blood-brain barrier, adrenal medulla, gastric mucosa, and male genitalia.

It contains an extracellular domain with a carbonic anhydrase-like (CAH) and a fibronectin-3-like repeats, a single transmembrane segment, and a single intracytoplasmic catalytic domain. Its extracellular domain functions as a ligand for neuronal receptor complexes, growth hormones, and interleukins.

The extracellular domain interacts with cell adhesion molecules, such as VE-cadherin in endothelial adherens junctions and immunoglobulin superclass member contactin of the neuronal membrane, as well as anti-adhesive glycoproteins tenascins-C and -R of the extracellular matrix. This domain also binds neural cell adhesion molecules, such as nrCAM, ngCAM, nCAM, and axonal contactin-2 [5] (Vol. 1 – Chap. 7. Plasma Membrane). Moreover, it tethers growth factors, such as fibroblast growth factor-2, midkine, [6] and pleiotrophin [1288]. [7] Last, it anchors to members of

4. A.k.a. receptor protein Tyr phosphatase-β (RPTPβ), PTPζ, and vascular endothelial protein Tyr phosphatase (VE-PTP).

5. A.k.a. TAG1 and axonin-1.

6. A.k.a. neurite growth-promoting factor.

7. A.k.a. heparin-binding growth-associated molecule or heparin-binding growth factor-8.

Table 9.4. Partners and substrates of protein Tyr phosphatase receptors (**Part 3**; Sources: Wikipedia, BioGRID; Cdh: cadherin; Ctn: catenin [cadherin-associated protein]; Gβ2L1: G protein-β polypeptide 2-like 1; RACK: receptor for activated C-kinase; SCFR: stem cell factor receptor; SCFR: stem cell factor receptor; TK: thymidine kinase; Unc119: uncoordinated-119 homolog or retinal protein-4 [RG4; expressed in photoreceptors; promotes IL4 production and eosinophilic inflammation of airways; activator of LcK and Fyn kinases for T-cell stimuation]).

Type	Enzymes	Adaptors, dockers, scaffolds	Ion channels, receptors, adhesion and cytoskeletal molecules, transcription factors
PTPRk	TK1		Ctnβ1/γ Proteasome 26S non-ATPase-11
PTPRm		RACK1 Gβ2L1	Ctnδ1, Cdh5 (VE-cadherin)
PTPRn			Spectrin-β4, syntrophin-β2 Calpain-2
PTPRn2			Spectrin non-erythroid-β3
PTPRo	SCFR		
PTPRq			Ctnα1/β1δ1/γ, Cdh1/2/5, α-actinin, desmoglein-1

the DLg (Disc large homolog) family via its PDZ domain.[8] Both PTPRb and DLg family proteins are located in postsynaptic densities that are specialized elements of the cytoskeleton at neuronal synapses of pyramidal neurons of the hippocampus and neocortex.

The PTPRb protein hence participates in intercellular and cell–matrix interactions. It reverses effect of Tyr phosphorylation of VE-cadherin by vascular endothelial growth factor receptor VEGFR2, hence protecting VE-cadherin-mediated barrier integrity, but independently of its enzymatic activity [1289].

The PTPRb phosphatase interacts with membrane-associated guanylate kinase, WW and PDZ domain-containing protein MAGI3 that is concentrated in specific sites at the plasma membrane and in the nucleus [1290]. In particular, MAGI3 localizes with zonula occludens-1 and cingulin at tight junctions in epithelial cells and with E-cadherin in cellular contacts of astrocytes. Protein MAGI3 can serve as a scaffold that links PTPRb with its substrates at the plasma membrane.

8. The DLg family, or membrane-associated guanylate kinase (MAGuK)-family, includes DLg1 (or SAP97), DLg2 (a.k.a. PSD93 and chapsyn110, DLg3 (or SAP102), and DLg4 (a.k.a. PSD95 and SAP90 that are concentrated in synapses. These proteins bind to membrane proteins, such as NR2 subunits of ^{NMDA}Glu receptor and voltage-gated K^+ channels. They form large synaptic complexes that organize and regulate synapses.

Table 9.5. Partners and substrates of protein Tyr phosphatase receptors (**Part 4**; Sources: Wikipedia, BioGRID; CHD: chromodomain helicase DNA-binding protein; EEF: eukaryotic translation elongation factor; EIF: eukaryotic translation initiation factor; FEZ: fasciculation and elongation protein-ζ(zygin); PPflα: PTPRf polypeptide-interacting protein-α1 (liprin or liprin-α); SetDB: ERG-associated protein with a SET domain, bifurcated; UTP14a: UTP14, U3-small nucleolar ribonucleoprotein, homolog A). High-mobility group protein HMGB1 is a cytokine involved in inflammation. Pleiotrophin, or neurite growth-promoting factor NEGF1, and midkine, or NEGF2, are growth factors that constitute a family of secreted heparin-binding proteins. Midkine is a pleiotropic protein that binds to the receptor complex formed by receptor-type Tyr phosphatase PTPRz1, low-density lipoprotein receptor-related protein LRP1, ALK kinase, and syndecans. Contactin-1 is a glycosylphosphatidylinositol (GPI)-anchored neuronal cell adhesion molecule. Tenascin-R is an extracellular matrix protein.

Type	Enzymes	Adaptors, dockers, scaffolds	Ion channels, receptors, adhesion and cytoskeletal molecules, cytokines, transcription factors
PTPRr	MAPK1/3/7/14, ERK1/2/4 P38MAPK		
PTPRs	PTPRd	PPflα1–3	Pleiotrophin, tubulin-β2A Profilin-2, SetDB1, Unc119, CHD3, EEF1γ, EIF4A2, FEZ1, UTP14a, Ub Ligase UbR1
PTPRt			Paxillin
PTPRu			SCFR Catenin-β1
PTPRv		Pseudogene product	
PTPRz1			$P2X_7$, HMGB1, APAP1 (ArfGAP), catenin-β1, contactin-1, tenascin-R, midkine, pleiotrophin

The PTPRb phosphatase also links to voltage-gated sodium channels [1291]. Dephosphorylation slows sodium channel inactivation and shifts its voltage dependence.

In endothelial cells, PTPRb is a specific phosphatase for angiopoietin TIE2 receptor, but not VEGFR2, which is coexpressed with TIE2 during embryogenesis [1292]. The PTPRb phosphatase attenuates TIE2 activity.

The PTPRb phosphatase and the angiopoietin–TIE2 pathway are antagonists in tumoral angiogenesis. Whereas TIE2 receptor Tyr kinase phosphorylates its subtrates to govern the blood vessel size, PTPRb dephosphorylates TIE2 [1293].

9.3 Protein Tyrosine Phosphatase Receptor-C

The prototypical Tyr phosphatase (type-1 or family-1 RPTP) — the plasmalemmal PTPRc — [9] is observed exclusively in leukocytes, [10] other nucleated blood cells, and their precursors. It is hence absent in erythrocytes as well as plasmocytes.

Four alternatively spliced transcripts variants generate distinct isoforms. The Ptprc gene actually contains 34 exons; 3 exons of the primary transcript can be alternatively spliced to generate 8 different mature mRNAs and, after translation, 8 different proteins (CD45ra–CD45rc, CD45rab, CD45rac, CD45rbc, CD45ro, CD45rabc). These splice variants allow the classification of different populations of a given cell types. For example, human thymus, lymph nodes, spleen, and blood contain 2 populations of regulatory CD25+, CD4+ T lymphocytes: (1) $CD25^{bright}$, CD45ro+ and (2) $CD25^{low}$, CD45ra+ cell population [1294].

The PTPRc phosphatase is heavily glycosylated. It contains an extracellular region with a cysteine-rich sequence and a fibronectin-3-like motif, a single membrane-spanning segment, and 2 tandem intracytoplasmic catalytic domains. It possesses one of the largest cytoplasmic domains that has an intrinsic phosphatase activity. It removes inhibitory phosphate groups on Tyr kinases, especially members of the SRC family, such as leukocyte-specific Tyr kinase and spleen Tyr kinase (Vol. 4 – Chap. 3. Cytosolic Protein Tyrosine Kinases) in T and B lymphocytes, respectively, to activate these kinases. It also dephosphorylates other types of Tyr kinases.

The PTPRc phosphatase is involved in differentiation of hematopoietic cell lineages and immunity. It indeed contributes to the regulation of T- and B-cell antigen receptor signaling. It directly interacts with components of antigen receptor complexes or activates various SRC family kinases that are required for the antigen receptor signaling. It positively or negatively regulates Src kinases according to the cell type and context [1295]. In addition, costimulator PTPRc suppresses JaK activity and thus precludes the cytokine receptor signaling, thereby impeding hematopoietic cell differentiation and proliferation, as well as antiviral immunity launched of these cells [1296]. [11] Dimerized PTPRc is inactive.

9.4 Protein Tyrosine Phosphatase Receptor-D

Protein Tyr phosphatase receptor-D (PTPRd) [12] contains an extracellular region, a single transmembrane segment, and 2 tandem intracytoplasmic catalytic domains. Multiple tissue-specific alternatively spliced transcript variants generate PTPRd isoforms.

---

9. A.k.a. cluster determinant protein, or cluster of differentiation, CD45.

10. Hence, its label leukocyte common antigen.

11. Once cytokine receptors are stimulated, JaK1 to JaK3, and Tyk2 are phosphorylated. Activated JaKs can then phosphorylate transcription activators STATs for translocation to the nucleus.

12. A.k.a. receptor protein Tyr phosphatase-δ (RPTPδ).

Phosphatase PTPRd interacts with PTPRs [1285] as well as PTPRf-interacting protein liprin-α1. Liprins interact with members of the LAR family, such as PTPRd and PTPRs, to localize these phosphatases to cell adhesion sites. This interaction can regulate focal adhesion disassembly.

9.5 Protein Tyrosine Phosphatase Receptor-E

Protein Tyr phosphatase receptor-E (PTPRe) [13] contains a short extracellular domain, a single transmembrane region, and 2 tandem intracytoplasmic catalytic domains. Unlike ubiquitous PTPRa, PTPn6, and PTPn11, PTPRe has a narrow tissue distribution and can regulate the proliferation of specific cell types. It is actually expressed preferentially in endothelial cells and also strongly in neurons. Major sites of PTPRe synthesis include the brain, lung, and testis [1297].

The Ptpre gene has 2 promoters that generate a receptor type (PTPRe) and a non-receptor type, the cytoplasmic PTPRe (^{C}PTPRe). Alternative splicing produces an additional PTPRe isoform (^{P67}PTPRe). In addition, PTPRe, ^{C}PTPRe, and ^{P67}PTPRe can be cleaved by calpain to yield cytosolic, catalytically active ^{P65}PTPRe. The transmembrane and 3 cytoplasmic isoforms (^{C}PTPRe, ^{P65}PTPRe, and ^{P67}PTPRe) differ only in their extreme N-termini. The different isoforms can exert different functions and have distinct substrate selectivity owing to their subcellular location and expression in distinct cell types [1298]. The PTPRe phosphatase can homo- and heterodimerize with ^{C}PTPRe and with related PTPRa.

Activity of RPTPe can be inhibited by dimerization (e.g., in response to EGFR signaling) and by oxidation of the cysteine residue at its catalytic domain [1297]. Cytosolic PTPe isoforms ($_{c}$RPTPe) are less active than membrane-bound type (RPTPe or $_{m}$RPTPe) toward membrane-associated substrates. Plasmalemmal localization of PTPRe participates in inhibition of cell rounding and detachment induced by hormones such as insulin [1298]. Phosphorylation of $_{c}$RPTPe and RPTPe causes association with adaptor growth factor receptor-bound protein GRB2. In addition, PTPRe dephosphorylates insulin receptor (Sect. 8.2.5.5). Furthermore, PTPRe stabilizes the FAK–Src complex that can associate with several molecules involved in cell adhesion (GRB2, GRB7, BCAR1 [Cas], paxillin, and tensin) and in insulin signaling (insulin receptor substrate-1, phosphatidylinositol 3-kinase, and phospholipase-Cγ). However, $_{c}$RPTPe is less efficient in opposing insulin-induced cell rounding and detachment.

The PTPRe phosphatase can interact with voltage-gated delayed rectifiers $K_V1.5$ and $K_V2.1$ to reduce their activity [1299]. Among other substrates, Src kinase is dephosphorylated (activated) by PTPRe phosphatase.

The PTPRe phosphatase attenuates proliferation of endothelial cells, but not that of smooth muscle cells and fibroblasts [1300]. However, activity of antimitogenic PTPRe can rise in response to proliferative stimuli. Similarly, antimitogenic dual-specificity phosphatases MKP1 and MKP2 can be stimulated by growth factors to modulate cell proliferation.

---

13. A.k.a. receptor protein Tyr phosphatase-ϵ (RPTPϵ).

9.6 Protein Tyrosine Phosphatase Receptor-F

Protein Tyr phosphatase receptor-F (PTPRf)[14] possesses an extracellular region with 3 immunoglobulin-like and 8 fibronectin-3-like sequences, a single transmembrane region, and 2 tandem intracytoplasmic catalytic domains ($PTPR^{D1}$ and $PTPR^{D2}$). Two alternatively spliced variants exist.

Phosphatase PTPRf operates in the regulation of between-epithelial cell contacts at adherens junctions. It interacts with liprins and β-catenin. β-Catenins are indeed able to tether to both HER2 and PTPRf receptors [1301]. A LAR isoform can serve as a receptor for the laminin–nidogen complex and influence the actin cytoskeleton structure.

9.7 Protein Tyrosine Phosphatase Receptor-G

Protein Tyr phosphatase receptor-G (PTPRg)[15] contains an extracellular region with a carbonic anhydrase-like domain and a fibronectin-like-3 motif, a single transmembrane region, and 2 tandem intracytoplasmic catalytic domains. It is produced in many tissues. The expression of the Ptprg gene is suppressed by estrogens.

Type-5 receptor protein Tyr phosphatases-γ (PTPRg) and -ζ (PTPRz) are expressed primarily in the nervous system. Among members of the contactin family of neural recognition molecules, PTPRz binds only to contactin-1, whereas PTPRg interacts with contactins-3 to -5 and -6 [1302].

9.8 Protein Tyrosine Phosphatase Receptor-H

Protein Tyr phosphatase receptor-H (PTPRh)[16] is made of an extracellular region with 8 fibronectin-3-like repeats and multiple N-glycosylation sites, a single transmembrane region, and a single intracytoplasmic catalytic domain. It is expressed primarily in brain and liver and, at a lower level, in heart and stomach.

Phosphatase PTPRh causes apoptosis, as it inhibits cell survival signaling mediated by the PI3K–PKB axis and ILK (integrin-linked kinase) and activates a caspase-dependent pro-apoptotic pathway [1303].

9.9 Protein Tyrosine Phosphatase Receptor-J

Protein Tyr phosphatase receptor-J (PTPRj)[17] possesses an extracellular region that contains 5 fibronectin-3 repeats, a single transmembrane region, and a single

14. A.k.a. leukocyte common antigen-related receptor (LAR).

15. A.k.a. receptor protein Tyr phosphatase-γ (RPTPγ).

16. A.k.a. stomach cancer-associated protein-Tyr phosphatase-1 (SAP1) and brain-enriched membrane-associated phosphatase-2 (PTP-BEM2).

17. A.k.a. receptor protein Tyr phosphatase-η (RPTPη), susceptibility to colon cancer SCC1, density enhanced phosphatase DEP1, and CD148.

intracytoplasmic catalytic domain. It is produced by hematopoietic cells. It impedes T-cell receptor signaling.

9.10 Protein Tyrosine Phosphatase Receptor-K

Protein Tyr phosphatase receptor-K (PTPRk) [18] contains an extracellular region with a MAM domain, an Ig-like sequence, and 4 fibronectin-3-like repeats, a single transmembrane region, and 2 tandem catalytic domains. It interacts with β- and γ-catenin [19] to mediate homophilic intercellular interaction.

9.11 Protein Tyrosine Phosphatase Receptor-M

Protein Tyr phosphatase receptor-M (PTPRm) [20] contains an extracellular region with a meprin–A5–PTPμ(MAM) domain, an Ig-like sequence, and fibronectin-3-like repeats, a single transmembrane region, and 2 tandem catalytic domains.

It can interact with RACK1 scaffold or GNB2L1 (guanine nucleotide-binding protein-β polypeptide-2-like-1). It is highly expressed in the lung, where it is almost exclusively restricted to the endothelium.

The PTPRm phosphatase may function as cell-adhesion receptors and dephosphorylates its substrates in response to cell contact. It interacts with cadherins in the nervous system, vascular endothelia, and intercalated discs of cardiomyocytes. It favors endothelial barrier integrity, as it directly binds to vascular endothelial cadherins [1304].

9.12 Protein Tyrosine Phosphatases Receptor-N and -N2

Protein Tyr phosphatase receptor-N (PTPRn) [21] contains an extracellular region, a single transmembrane region, and a single catalytic domain. It is produced in the brain and pancreas.

At the plasma membrane, PTPRn cytoplasmic domain can be cleaved and relocalize to the nucleus, where it enhances the transcription of the insulin gene. It can homo- and heterodimerize. It interacts with spectrin-β4.

Protein Tyr phosphatase receptor-N2 (PTPRn2), or PTPRp, [22] and related PTPRn are both major autoantigens in insulin-dependent diabetes mellitus.

---

18. A.k.a. receptor protein Tyr phosphatase-κ (RPTPκ).
19. A.k.a. plakoglobin.
20. A.k.a. receptor protein Tyr phosphatase-μ (RPTPμ). It is an example of type-2 PTPRs.
21. A.k.a. islet cell antigen ICA512 and insulinoma-associated protein-2 ([PTP] IA2).
22. A.k.a. RPTPx, pancreatic islet cell antigen IA2β, IAR, ICAAR, and nervous system and pancreatic PTP-NP.

9.13 Protein Tyrosine Phosphatase Receptor-O

Owing to its identification in several different cell types, protein Tyr phosphatase receptor-O (PTPRo) has many names.[23] It contains an extracellular segment containing a MAM domain, an immunoglobulin-like sequence, 4 fibronectin-3 repeats, a transmembrane segment, and 2 tandem intracellular PTP domains. Several tissue-specific alternatively spliced transcript variants exist, some of which encode different isoforms of the protein. In fetal tissues, PTPRo is expressed in the brain and lung, and, at a lower level, in the kidney. In adult tissues, PTPRo is less restricted, as it is detected in the lung, heart, skeletal muscle, prostate, testis, and in various areas of the brain [1305].

9.14 Protein Tyrosine Phosphatase Receptor-Q

Protein Tyr phosphatase receptor-Q (PTPRq) regulates the concentration of phosphatidylinositol (4,5)-bisphosphate, as it can dephosphorylate many phosphatidylinositol phosphates. It is synthesized in the lung and kidney.[24] The PTPRq phosphatase is downregulated during the early phase of adipogenesis [1306].

This receptor-like inositol lipid phosphatase is a chondroitin sulfate proteoglycan (like PTPRb, another type-3 PTPR) [1307]. Multiple PTPRq isoforms may exist in most sensory hair cells of the inner ear.

9.15 Protein Tyrosine Phosphatase Receptor-R

Protein Tyr phosphatase receptor-R (PTPRr)[25] is expressed predominantly in the brain. It possesses an extracellular region, a single transmembrane segment, and a single intracellular catalytic domain. Three isoforms exist: PTPRrα in the cell membrane and PTPRrδ and PTPRrγ in the perinuclear region.

It interacts with MAPK7 and complexes with ERK1 and ERK2 to suppress MAPK signaling [1308]. It sequesters mitogen-activated protein kinases, such as MAPK1, MAPK3, and MAPK14, in the cytoplasm in an inactive form. Upon phosphorylation of the kinase-interacting motif by protein kinase-A, MAPK is released for activation and translocation into the nucleus.

---

23. It is indeed also named receptor protein Tyr phosphatase-o(RTPRo), PTP-U2, PTPRl, PTPπ (or RTPπ), PTPψ (or RTPψ), PTP-J, osteoclastic transmembrane protein-tyrosine phosphatase (PTP-oc), pancreatic carcinoma phosphatase-2 (PCP2; not Purkinje cell protein-2), and glomerular epithelial protein-1 (GlEpP1).

24. Hence, its alias PTP-GMC for glomerular mesangial cell PTP.

25. A.k.a. PC12-derived PTP (PCPTP1 or PC12; PC12 is a cell line derived from a pheochromocytoma of adrenal medulla), PTPBR7, EC-PTP, and PTP-SL.

9.16 Protein Tyrosine Phosphatase Receptor-S

Protein Tyr phosphatase receptor-S (PTPRs)[26] contains an extracellular region with multiple Ig-like and fibronectin-3-like motifs, a single transmembrane segment, and 2 tandem intracytoplasmic catalytic domains. It is involved in intercellular interaction, axonogenesis, and axon guidance during embryogenesis, as well as adult nerve repair. Four alternatively spliced transcript variants exist. The PTPRs phosphatase interacts with PTPRd and liprin.

9.17 Protein Tyrosine Phosphatase Receptor-T

Protein Tyr phosphatase receptor-T (PTPRt)[27] possesses an extracellular region with a MAM domain, an Ig-like sequence, and fibronectin-3-like repeats, a single transmembrane segment, and 2 tandem intracellular catalytic domains. It intervenes in homophilic intercellular adhesion in the central nervous system. It is the most frequently mutated Tyr phosphatase receptor in human cancers.

Paxillin, an adaptor involved in cell adhesion, migration, proliferation, and apoptosis, is a direct substrate of PTPRt (Tyr88) [1309]. Paxillin directly interacts with numerous proteins, such as focal adhesion kinase, GAB1 adaptor, RasA1 (or P120RasGAP), among others.

9.18 Protein Tyrosine Phosphatase Receptor-U

Protein Tyr phosphatase receptor-U (PTPRu)[28] influences concentrations of glucose, insulin, triglycerides, fatty acids, cholesterol, and glycosylated form of hemoglobin (HbA1c).

Phosphatase PTPRu possesses an extracellular region with a MAM domain, an Ig-like sequence, and fibronectin-3-like repeats, a single transmembrane region, and 2 tandem intracellular catalytic domains.

It is a negative modulator of FGF8-induced Wnt1 expression at the mid-hindbrain boundary and an inhibitor of canonical Wnt signaling, possibly by interacting with and sequestering β-catenin [1310].

9.19 Protein Tyrosine Phosphatase Receptor-V

Protein Tyr phosphatase receptor-V (PTPRv)[29] is markedly produced in cells that undergo P53-dependent cell cycle arrest, but not in cells that bear P53-mediated

26. A.k.a. receptor protein Tyr phosphatase-σ (RPTPσ).
27. A.k.a. receptor protein Tyr phosphatase-ρ (RPTPρ).
28. A.k.a. PTPRl. Other aliases include RPTPλ (receptor protein Tyr phosphatase-λ), PTPπ, PTPψ, PTP-U2, GlEpP1, PTP, FMI, PCP2, and PTP-J.
29. A.k.a. osteotesticular protein Tyr phosphatase (OST-PTP).

apoptosis [1311]. It is thus stimulated by P53 in certain circumstances. In fact, PTPRv is a pseudogene [608].

9.20 Protein Tyrosine Phosphatase Receptor-Z1

Protein Tyr phosphatase receptor-Z1 (PTPRz1)[30] encoded by the PtprZ1 gene that is expressed in remyelinating oligodendrocytes of multiple sclerosis lesions, human embryonic kidney cells under hypoxia, and gastric cancer cells.

This type-5 PTPR contains a single-pass type-1 membrane segment with 2 cytoplasmic protein Tyr phosphatase domains, an α-carbonic anhydrase sequence and a fibronectin-3-like motif.

Brain-specific PTPRz2 phosphatase[31] may be generated by another gene [608] (Ptpr1 on chromosome 7 and PtprZ2 on chromosome 1). However, the PtprZ2 gene may correspond to the PtprZ1 gene.

Three splice variants can be detected: a full-length ($PTPRz1_A$ or $PTPRz1_{FL}$); a short ($PTPRz1_B$ or $PTPRz1_S$); and a soluble ($PTPRz1^S$) form, or phosphacan.[32]

The PTPRz1 phosphatase interacts with high-mobility group box HMGB1 (amphoterin) that itself interacts with P53, midkine,[33] pleiotrophin,[34] tenascin-R,[35] catenin-β1, APAP1 [59].[36]

Like type-3 PTPRs, PTPRz1 functions as a ligand for neuronal receptor complexes. Phosphatase PTPRz possesses several other binding partners, such as cell adhesion molecule of the Ig superfamily, such as L1CAM (or CD171), neural cell adhesion molecule (NCAM, or CD56), and neuron–glia cell adhesion molecule-related protein (nrCAM), and glycosyl-phosphatidylinositol-anchored neural recognition molecule contactin-1 [1302]. The associations between PTPRz on glial cells and contactin-1 on neurons promotes the outgrowth of neurites and induces bidirectional signaling between glial cells and neurons.

9.21 Transmembrane RPTPs and RPTKs in Vasculo- and Angiogenesis

Vasculo- and angiogenesis requires several plasmalemmal receptors as well as cell adhesion molecules. Several receptor protein Tyr kinases and phosphatases are implicated in the formation and remodeling of nascent blood vessels and primitive vascular plexus during embryogenesis.

---

30. A.k.a. PTPRz, PTPRb, RPTPβ, PTPζ, hPTPζ, PTP18, and phosphacan.
31. A.k.a. protein Tyr phosphatase receptor-Z polypeptide-2 (hTPZP2) and PTPRz.
32. A.k.a. 6B4 proteoglycan.
33. A.k.a. neurite growth-promoting factor-2.
34. A.k.a. heparin-binding growth factor-8 and neurite growth-promoting factor-1.
35. A.k.a. restrictin and janusin.
36. A.k.a. G-protein-coupled receptor kinase interactor GIT1.

Among RPTKs, VEGFR1 and VEGFR2, PDGFRβ, TIE2, and EPH receptors are activated by their respective ligands, VEGF, PDGFb, angiopoietins, and ephrins. The VEGFR2 receptor has a prominent role among these RPTKs, as it can prime endothelial cell migration to construct the primitive vascular network. However, excessive VEGFR2 activation provokes vascular dysmorphogenesis, as action of RPTKs must be balanced by RPTPs. The EPHb receptors participate in establishing the primary vessel architecture and maturation. The PDGFRβ receptor is entailed in the recruitment of pericytes to newly formed vessels. The TIE2 receptor contributes to blood vessel sprouting, remodeling, and integrity. The absence of TIE2 causes only mild vascular defects.

Among involved RPTPs, PTPRb (or vePTP) associates with TIE2 and VE-cadherin to enhance adhesive function of the latter, particularly in arterial endothelia. The TIE2 kinase is a partner and substrate of PTPRb. The PTPRm phosphatase participates in flow-induced dilation of resistance arteries, but it is not vital for their development. Both PTPRb and PTPRj play a major role, as their deficiency leads to enlarged primitive blood vessels that fail to remodel appropriately as well as discontinuities of endocardium. These discontinuities result from a weakening in adhesion between endothelial cells and between endothelial and mesothelial or endodermal cells due to an impaired phosphorylation level of VE-cadherin [1312]. Although PTPRb is dispensable for the initial formation of blood vessels, it is essential for their maintenance and remodeling.

10

Morphogen Receptors

Tissue specification and patterning are regulated by controlled concentration gradients of morphogens. Morphogens are signaling molecules that can induce distinct cellular responses according to their concentrations. Free Brownian diffusion of morphogens from their sources through the extracellular space and sinks represented by receptor-mediated endocytosis in target cells and morphogen degradation can set up these morphogen concentration gradients [1313].

Developmental proteins Hedgehog, Notch, and Wnt regulate cell differentiation, proliferation, and migration in several tissue types, thereby tissue morphogenesis. Their related signaling pathways are frequently activated in cancers.

10.1 Notch Receptors

Notch signaling relies on direct intercellular contact. The resulting intercellular communication participates in tissue morphogenesis. In particular, Notch is involved in the formation and growth of the vasculature.

Notch receptors form a family of large, single-pass transmembrane glycoproteins that transmit juxtacrine signals initiated by single-pass transmembrane Delta-like and Jagged ligands of the Delta–Serrate–LAG2 (DSL) family. [1]

Transmembrane signaling protein Delta can be secreted. Afterward, before binding to Notch, it is internalized in the sending cell and glycosylated.

Notch receptors interact with plasmalemmal ligands expressed on adjacent cells, hence linking the fate of one cell to that of its apposed neighbor (short-range interaction; Fig. 10.1). Notch proteins act both as plasmalemmal receptors and as regulators of gene transcription via their intracellular domains. Ligand binding to Notch receptors indeed triggers Notch proteolysis that releases the messenger intracellular Notch fragments ($Notch^{ICD}$), i.e., the Notch transcription factor that migrate to the nucleus.

1. I.e., Delta and Serrate in Drosophila melanogaster (fruit fly) and Delta-like and Jagged in mammals.

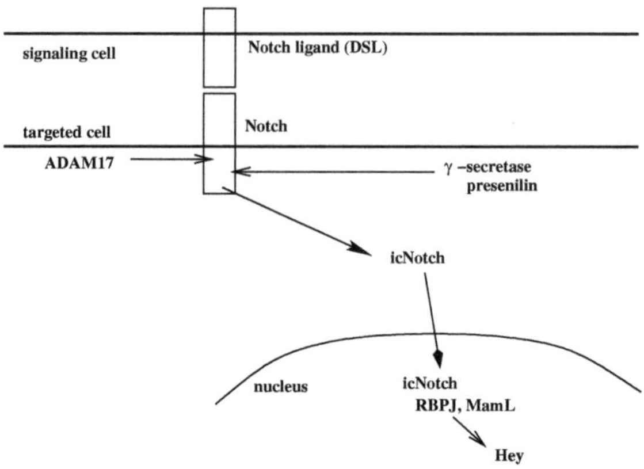

Figure 10.1. Notch signaling pathway for communication between 2 apposed cells. Notch agonists include Jagged (Jag1–Jag2) and Delta-like ligands (DLL1, DLL3, and DLL4). The binding of Delta–Serrate–LAG2 (DSL) domain-containing ligand to Notch receptor primes Notch cleavage by a proteic complex involving presenilin. Released Notch intracellular fragment (icNotch or $\text{Notch}^{\text{ICD}}$), i.e. Notch transcription factor, then migrates to the nucleus and modulates expression of target genes. Notch targets include Hairy enhancer of Split (HES) and HES-related transcriptional regulators (HRT). In the absence of icNotch, sequence-specific DNA-binding proteins called recombination signal-binding protein for immunoglobulin-Jκ region proteins (RBPJ) bind to specific DNA sequences in the regulatory elements of target genes and repress transcription by recruiting histone deacetylases and other components to form a corepressor complex. In the nucleus, Notch intracellular domain displaces histone deacetylase–corepressor complex from RBPJ proteins. Intranuclear icNotch forms a transcriptional complex with RBPJ, histone acetyltransferases, and coactivators such as mastermind-like (MamL). Usual Notch target genes include those that encode basic helix–loop–helix transcriptional repressors of the Hes and Hrt families.

10.1.1 Notch and DSL Family Members

Four Notch genes that encode receptors are identified (Notch-1–Notch-4; Table 10.1). They have at least 5 different mammalian Notch ligands that are subdivided into 2 classes: (1) Delta-like ligands (DLL1, DLL3, and DLL4) encoded by the Delta-like genes (Dll1 and Dll3–Dll4) and (2) Jagged ligands (Jagged-1–Jagged-2) [2] encoded by the Jagged genes (JAG1 and JAG2). [3] Agonists Delta-like and Jagged act redundantly to activate Notch.

10.1.2 Notch Signaling

Notch is maintained in a resting auto-inhibited conformation in the absence of ligand. Because most Notch ligands are transmembrane proteins, Notch signaling is

2. A.k.a. Serrate-1 and Serrate-2.

3. Notch ligands are characterized by an N-terminal Delta, Serrate, and LAG2 (DSL) domain for interactions with Notch receptor. The location of both ligands and receptors on the cell influences Notch signaling.

Table 10.1. Main mediators of the Notch pathway (CBF1: C promoter-binding factor-1; CSL: CBF1, suppressor of Hairless [SuH], and LAG1 [a.k.a. recombining binding protein suppressor of hairless (RBPSuH)]; HES: Hairy and enhancer of Split; HRT: HES-related transcription factor; RBPJ: recombination signal sequence-binding protein for immunoglobulin-κ J region protein). The transcriptional repressor RBPJ (a.k.a. RBPJκ, CSL and CBF1), a classical target of Notch signaling, recognizes the specific sequence GTGGGAA. γ-secretase complex (γSC) consists of at least 4 components: presenilin, nicastrin, anterior pharynx-defective homolog APH1, and presenilin enhancer PEn2.

Type	Aliases, subtypes	Role, effects
Notch	Notch1–Notch4	Receptors
	Ligands	
Delta-like	DLL1/3/4	Auto- and juxtacrine messengers
Jagged ligands	Jag1/2	(cis/trans-signaling), trans-activation, cis-inhibition
	Peptidases	
adamlysin	ADAM17	Cleavage and shedding of the extracellular domain ($Notch^{ECD}$)
γ-secretase	γSC	Cleavage of the extracellular domain ($Notch^{ICD}$)
	Ligases	
Deltex homologs	Dtx1–Dtx4	Notch ubiquitination and endocytosis
	Transcription factors	
HES	HES	Gene transcription
HRTs	HRT1–HRT3, HEY	Gene transcription
Mastermind-like	MamL	Notch transcription complex
RBPJκ	CBF1, CSL	Transcription repression

restricted to neighboring cells and exerted via between-cell contacts. A transmembrane Notch (DSL) ligands on a signal-emitting cell interact with a Notch receptor on a signal-receiving cell.

10.1.2.1 Notch Activation

Notch exists either as a heterodimer after cleavage by furin in the trans-Golgi network or as an intact protein. Notch activation requires the association between Notch ligands and receptors of homo- or heterotypic cells, both in contact. However, ligand–receptor interactions do not necessary cause signaling, because some ligands may impede Notch signaling.

Notch integrates signals from Delta-like molecules on the same cell (*cis-signaling*) with those from an apposed cell (*trans-signaling*). The Notch–Delta pathway is devoted to communication between neighboring cells during the body's develop-

Table 10.2. Notch signaling associated with blood circulation.

Cardio- and angiogenesis
Neurogenesis
T-cell lineage commitment from common lymphoid precursor
Osteogenesis with expansion of the hematopoietic stem cell compartment

ment. It generates distinct cell fates in populations of initially equivalent adjacent cells and delineates adjoining regions in developing tissues (Table 10.2).

Ligands Delta-like activate Notch in neighboring target cells (*Notch trans-activation*) and inhibit Notch in source cells (*Notch cis-inhibition*). The response to trans-Delta is graded, whereas that to cis-Delta is sharp and occurs at a fixed threshold, independent of trans-Delta [1314]. Mutual inactivation between Notch and Delta in the same cell directs cells into predominantly sending (high Delta and low Notch density) or receiving (high Notch- and low Delta level) states. Cell signaling status switches between these 2 mutually exclusive signaling (sending or receiving) states.

10.1.2.2 Notch Cleavage

Ligand–receptor binding causes successive Notch proteolytic cleavages. The cleavage of the ectodomain of ligand-bound Notch receptor by adamlysins is followed by a cutting catalyzed by γ-secretase to release $\text{Notch}^{\text{ICD}}$ and initiate Notch signaling.

The first extracellular cleavage requires ADAM17 [4] The cleaved and shed extracellular domain undergoes endocytosis by the Notch ligand-expressing neighboring cell. Signaling primed by the interaction between Notch-3 receptor and Jagged-1 ligand is enhanced by extracellular thrombospondin-2 that connects to numerous receptors such as low-density lipoprotein receptor-related protein LRP1 and adhesion molecules. Receptor LRP1 potentiates the trans-endocytosis of Notch-3 stimulated by Tsp2 into the signal-sending cell [1315]. [5] Extracellular thrombospondin-2 and LRP1 on the signal-sending cell bind to the ectodomains of Notch-3 and Jagged-1. The transient membrane-tethered Notch species is subsequently cleaved.

The second cleavage is executed by a proteic complex, the γ-secretase complex (presenilin, nicastrin, anterior pharynx defective phenotype homolog APH1, and presenilin enhancer PEN2). The cleaved cytoplasmic domain of the Notch receptor is released and translocates into the nucleus using the endosomal transport (Vol. 1 – Chap. 9. Intracellular Transport). [6]

4. A.k.a. tumor-necrosis factor-α–converting enzyme (TACE).
5. Endocytosis of Notch-3 occurs, even in the absence of LRP1.
6. Four RING ubiquitin ligases, mind bomb homologs Mib1 (a.k.a. DAPK-interacting protein DIP1 and zinc finger Z type with ankyrin repeat domain-containing protein ZZAnk2) and Mib2 (a.k.a. novel zinc finger protein, skeletrophin, and ZZAnk1) as well as plasma membrane-associated neuralized homolog NeurL1 (a.k.a. NeurL1a and RING finger protein RNF67) and NeurL2, in cooperation with epsin and Rab11a, are implicated in endocytosis

The transcription of target genes relies on the formation of transcriptional activator complexes. The DNA-bound *Notch transcription complex* include Notch^{ICD}, Mastermind-like coactivator (MamL), and transcription factor CSL (C promoter-binding factor CBF1, suppressor of Hairless [SuH], and LAG1) bound as monomer to DNA [1316]. These complexes can oligomerize on clusters of CSL binding sites. Phosphorylation, ubiquitination, and proteosomal degradation of intracellular Notch terminate Notch signaling.

10.1.2.3 Notch Signaling in Endosomes

Notch can also be activated in a ligand-independent fashion by Notch stimulators ubiquitin ligases of the Deltex family. Deltex homologs Dtx1 and Dtx2 causes Notch ubiquitination and endocytosis. Subsequently, full-length Notch accumulates in late endosomes and multivesicular bodies, but still can signal [11]. Notch ectodomain can be cleaved by proteolytic enzymes in the lumen of late endosomes and multivesicular bodies. This event is followed by the release into the cytoplasm of Notch intracellular fragment by γ-secretase of the membrane of late endosomes and multivesicular bodies. Simultaneously, members of the Deltex family prevent Notch incorporation inside multivesicular bodies, which effectively insulates receptors from cytoplasmic effectors and can lead to the degradation pathway.

10.1.2.4 Notch Regulators

The spatiotemporal expression pattern of Delta-like and Notch with feedback loops in the Delta–Notch pathway as well as colocalization of Delta-like and Notch in a single cell control intercellular interactions between adjacent cells.

In the *canonical* Delta–Notch signaling, Delta-like activates Notch in neighboring cells that downregulates Delta-like in these cells. This *lateral inhibition* is mediated by ankyrin repeat and SOCS box-containing protein ASB11, a member of the family of ASB proteins that interact with Cullin-5 and RING-box protein Rbx2 to form an Ub-ligase complex [1317]. The latter ubiquinates Delta-like for degradation [1318]. However, lateral inhibition is avoided when neighboring cells must coordinate their activity. Notch signaling then keeps neighbors in synchrony.

Cytosolic, ubiquitous β-arrestins [7] participate in signaling from plasmalemmal receptors (classical GPCRs, non-classical GPCRs, such as Smoothened and Frizzled, receptor Tyr kinases, as well as Notch and TGFβ receptors) as well as their internalization followed by recycling or degradation. They can also serve as scaffolds for kinases of the MAPK module and the PI3K–PKB axis. The Drosophila

---

and recycling [11]. On the other hand, endocytic adaptors membrane-bound NUMB and complex AP2 as well as Notch polyubiquitination by Ub ligase ITCH promote the endocytosis of non-liganded Notch and its degradation in lysosome of signal-sending cell [11].

7. The arrestin family can be divided in 2 classes: (1) the visual arrestins — arrestin-1 and-4 — located almost exclusively in photoreceptor cells and (2) ubiquitous (non-visual) β-arrestin-1 and -2 (arrestin-2 and -3). In Drosophila melanogaster, the single ubiquitous β-arrestin, kurtz operates in development, survival, and nervous function.

homolog of mammalian, ubiquitous β-arrestins, kurtz, forms a trimer with Notch receptor, leading to Notch ubiquitination and Notch signaling attenuation. [8] The basic Notch pathway is the same in most Notch-dependent processes, but the pathway regulations are different [1319].

Ubiquitin ligases that interact with Notch ligands are required for Notch activation, thereby controlling the activity of the Notch pathway.

Notch signaling is modulated not only by post-translational modifications and cleavage of Notch receptors, but also exo- and endocytosis of Notch receptors, hence on intracellular transfer machinery as well as actin polymerization regulators. Transmembrane Delta-like ligand is imported from and then re-exported to the plasma membrane of signal-sender cells to then bind and activate Notch receptor on neighboring recipient cells. In addition to myosin nanomotors, Wiskott-Aldrich syndrome proteins and actin-related proteins ARP2 and ARP3 that participate in the dynamical organization of the actin cytoskeleton and formation of branched actin filaments, contribute to Notch ligand transfer [1320].

Lethal giant disc (LGD) [9] regulates Notch endosomal transport and prevents Notch delivery to a compartment where it can be aberrantly activated. Moreover, Notch regulators are themselves controlled for efficient interactions with other signaling pathways.

In endothelial cells, NAD^+-dependent deacetylase sirtuin-1 impedes Notch signaling [1321]. Acetylation of $Notch^{ICD}1$ stabilizes this messenger and, thus, controls the amplitude and duration of the Notch response. On the other hand, sirtuin-1 deacetylase promotes $Notch^{ICD}$ turnover. In response to DLL4 stimulation, activated sirtuin-1 supports adequate endothelial growth and sprout elongation, hence, proper vascular branching and density. Therefore, reversible acetylation of $Notch^{ICD}$ enables adjustment of the dynamics of Notch signaling. In addition, the regulation by SIRT1 of the Notch pathway includes other family members than Notch-1 (e.g., Notch-4) [1321].

10.1.2.5 Transcription Factors of Notch Signaling

Together with cofactors, kinases, and ubiquitin ligases, Notch activates gene expression. Notch receptors regulate cell growth and death in various cell types, as well as differentiation in embryos and tissue development in adults.

HES and HRT Factors

Notch activation enables the transcription of the specific target genes that encode Hairy enhancer of Split-3 (HES3) and HES-related transcription factors (HRT), [10]

---

8. The formation of a trimeric Notch–Deltex–Kurtz complex mediates the ubiquitination and degradation of the Notch receptor.
9. A.k.a. coiled-coil and C2 domain-containing protein 1-like protein.
10. A.k.a. HES-related with YRPW motif (HEY).

Table 10.3. Hairy-related transcription factor (HRT) family (CHF: cardiovascular basic helix–loop–helix factor; HERP: HES-related repressor protein; HES: Hairy and enhancer of Split; HESR: Hairy and enhancer of Split-related protein; HEY: Hairy–enhancer of Split related with YRPW motif protein). Proteins HRTs hinder Notch-dependent activation of their genes.

Type	Aliases
HRT1	HEY1, HESR1, HERP2, CHF2, OAF1
HRT2	HEY2, HESR2, HERP1, CHF1, GRL, Gridlock
HRT3	HEY3, HEYL

as well as sonic Hedgehog via protein kinase-B and signal transducer and activator of transduction STAT3, [11] and promotes regenerative responses following hypoxia [1322].

The Hairy-related transcription factor family of basic helix–loop–helix (bHLH) proteins includes 3 members (HRT1–HRT3; Table 10.3). The transcription of Hrt genes is activated by Notch in cooperation with Notch partner RBPJκ transcription factor. The HRT proteins interact with GATA factors and repress the transcriptional activity of GATA-dependent genes.

The transcriptional repressor HRT2 regulates cardiomyocyte identity. It is expressed in ventriculomyocytes, as well as in vascular endothelial and smooth muscle cells, but not in atriomyocytes. It prevents activation of atrial genes in ventricular myocytes [1323].

NRF2

Nuclear factor erythroid-derived-2 (NFE2)-related factor NRF2 is a prosurvival transcription factor that recognizes the anti-oxidant response element (ARE) in the promoter of Notch-1 [1324]. Crosstalk between the KEAP1–NRF2 [12] and Notch-1 pathways operates in liver regeneration after partial hepatectomy.

RBPJ Repressor and Activator

In the nucleus, Notch^{ICD} interacts with the transcription factor recombination signal sequence-binding protein for immunoglobulin-κ J region protein (RBPJκor RBPJ). [13] The intracellular fragment degrades the *RBPJ transcriptional repressor*

11. STAT3 is activated at the plasma membrane.

12. KEAP1: kelch-like ECH-associated protein-1. A.k.a. cytosolic inhibitor of NRF2 (INRF2).

13. A.k.a. recombining binding protein suppressor of hairless (RBPSuH); immunoglobulin-κJ region recombination signal-binding protein (IgκRB); C promoter-binding factor CBF1; CBF1, suppressor of Hairless, and Lag1 (CSL); and κB factor KBF2 (that binds to the NFκB [nuclear factor of κ light polypeptide gene enhancer in B-cells-1 (KBF1)] site of major histocompatibility complex class-1 genes. In the absence of Notch^{ICD} in the nucleus, RBPJ inhibits the transcription of Notch target genes, as it recruits the RBPJ transcriptional repres-

complex and, with RBPJκ, recruits coactivator Mastermind-like (MamL) and histone acetyltransferase to form a *RBPJ transcriptional activator complex*, among others. Mastermind-like is required to activate the transcription. However, Notch not only uses the RBPJ-dependent (*canonical*), but also the RBPJ-independent (*non-canonical*) pathway.

RBPJ-interacting and tubulin-associated protein (RITA) binds to tubulin in the cytoplasm and shuttles rapidly between cytoplasm and nucleus. It exports RBPJ from the nucleus [1325]. [14] It can repress Notch-mediated transcription.

10.1.3 Notch Effects

10.1.3.1 Anti-Apoptotic Notch

In response to apoptotic stimuli, BCL2-associated X protein (BAX) nucleates proteic complexes on mitochondria, thereby committing cells to irrevocable damage. On the other hand, survival of differentiated cells is regulated by Notch, as it has anti-apoptotic activity. Notch intracellular fragment signals via protein kinase-B to prevent the loss of mitochondrial function and subsequent nuclear damage via mitochondrial remodeling proteins mitofusin-1 and -2 independently of the canonical pathway [1326].

10.1.3.2 Notch in Stem and Progenitor Cells

Notch operates in self-renewal of adult cells, particularly in hematopoiesis. Notch either initiates cell differentiation or maintains the undifferentiated state of progenitor cells. Notch signaling is required for cardiovascular system development, maturation, and functioning before birth as well as during postnatal life. Notch receptors are expressed in cardiac, vascular endothelial, and smooth muscle cells. Notch roles in the cardiovascular system include regulation of artery–vein differentiation in endothelial and smooth muscle cells, regulation of blood vessel sprouting and branching, and activity of vascular smooth muscle cells. Notch is also involved in the cardiac valve and trabecula formation. The development of the anastomotic circle of Willis requires the Notch pathway in vascular smooth muscle cells [1327].

Neural stem cell can divide asymmetrically to generate another neural stem cell and a committed cell or symmetrically to either expand or reduce the population by

---

sor complexes composed of MINT (Msx2-interacting protein [a.k.a. SpEn (split ends) homolog and SMART/HDAC1-associated repressor protein (SHARP)]), Runx1T1 (cyclin-D- and Runt-related transcription factor-1 translocated to protein-1, or eight twenty one protein [ETO]), CTBP (C-terminal-binding protein), HDAC, and $H3K_4D$ ($H3K_4D$ demethylase; i.e., the RBPJ–MINT–Runx1T1–CTBP–HDAC–$H3K_4D$ complex). When $Notch^{ICD}$ resides in the nucleus, the RBPJ corepressor complexes disassemble; RBPJ and $Notch^{ICD}$ then constitute a RBPJ transcriptional activator complex.

14. Protein RITA possesses a functional nuclear localization signal (NLS), nuclear export signal (NES), and RBPJ-interaction domain [1325].

giving rise to 2 stem cells or 2 committed cells, respectively. In adults, periventricular neural stem cells localize in vascular niches, where they are close neighbors to blood vessels, thereby being subjected to vascular signals. Stem cell environment with its vasculature contributes to the regulation of the balance between self-renewal and commitment, particularly via controlled segregation of Notch activities. Notch and vascular niche-derived serpin-F1 [15] cooperate to regulate self-renewal of neural stem cells [1328]. Serpin-F1 prevents choroidal neovascularization, thereby avoiding vision loss. It also influences neuronal differentiation. It stimulates Notch transcriptional effector HES1 to regulate self-renewal of neural stem cells. It induces a non-canonical activation of the NFκB pathway [16] that provokes the displacement of the nuclear receptor corepressor (NCoR) from specific Notch-responsive promoters, i.e., the promoters of Hes1 and Egfr genes, as well as nuclear export of NCoR corepressor. PEDF Factor favors division of neural stem cells into self-renewing cells. Both daughter cells produce high levels of epidermal growth factor receptor, thereby raising responsiveness to mitogenic EGF factor.

10.1.3.3 Notch in Embryonesis

Notch signaling intervenes in the development and homeostasis of the 3 germ layers and their derivatives. Notch is required in somitogenesis. [17] During the development of the nervous system, neural progenitor cells first express Notch effector HES1 and then proneural genes and Notch ligands. [18] Oscillations generated by

15. A.k.a. pigment epithelium-derived factor (PEDF). Serpin-F1 is a member of the serine peptidase inhibitor (portmanteau serpin) secreted by pigment epithelial cells of human retina. It has neurotrophic and anti-angiogenic activity.

16. Serpin-F1 excites Notch signaling not via cleavage of Notch to its transcriptionally active, intracellular fragment, although activities of both $Notch^{ICD}$ and Notch partner RBPJκ transcription factor are required, but via RBPJκ-dependent activation of transcription of HES1 transcription factor and epidermal growth factor receptor.

17. Somites are clusters of mesodermal cells that give birth to segmental body structures, such as blood vessels during embryonesis.

18. Upon activation by Notch ligands from neighboring cells, Notch releases $Notch^{ICD}$ that translocates into the nucleus. The $Notch^{ICD}$–RBPJ complex then induces expression of HES1 and HES5 that repress expression of proneural genes and hence Notch ligand genes to maintain neural progenitors in an undifferentiated state and, thus, avoid premature differentiation of all neighboring cells (lateral inhibition). Neural progenitors in fact initially undergo proliferation, and then subsets of progenitors bear neuronal differentiation, whereas others remain progenitors. Some cells express higher DLL1 levels and more efficiently activate Notch signaling within neighboring cells (classical model by amplification of stochastic difference). The latter then express higher levels of HES1 and HES5 that repress proneural and Notch ligand genes, whereas the former that are less affected by Notch further upregulate expression of proneural and Notch ligand genes (amplification). However, HES1 expression in neural progenitors oscillates with a period of about 2 to 3 h. Proneural genes neurogenin-2 and Mash1 induce expression of Notch ligand genes such as Dll1 during early developmental stages by directly binding to Dll1 enhancer. Expression of HES1 is regulated by a negative feedback, as HES1 represses Hes1 gene transcription. Under control of HES1 oscillations, Mash1, Ngn2,

HES1 [19] in proneural and Notch ligand gene expression maintain cells in an undifferentiated state by mutual activation of Notch signaling. Therefore, mutual activation of Notch signaling in neighboring neural progenitors with opposite phase challenges lateral inhibition concept [1329].

Notch in Cardiogenesis

Notch signaling pathway is required in cardiovascular development. Mutations in Notch signaling components are involved in congenital cardiovascular diseases. Defects of the cardiac outflow tract result from abnormal Jagged-1 and Notch-2; aortic valve disease from defective Notch-1; cardiomyopathy from dysfunctional presenilin-1 and -2, [20] and cerebral arteriopathy from altered Notch-3 [1330].

During early stages of cardiogenesis, Notch inhibits the differentiation of cardiomyocytes from mesoderm in embryos and embryonic stem cells. Later in cardiogenesis, Notch positively acts in ventricular maturation. Moreover, Notch participates in crosstalk with other signaling pathways involved in cardiogenesis, such as TGFβ–BMP, Wnt, and Ras pathways.

Notch targets genes of the family of Hairy enhancer of Split-related transcription factors. HES-related transcription factors regulate heart chamber construction, atrioventricular canal development, and valve formation. T-box transcription factor TBX2 expressed by atrioventricular canal myocardium and regulated by bone morphogenetic protein BMP2 is an inhibitor of chamber-specific gene expression. HES-related transcription factors HRT1 and HRT2 expressed in atria and ventricles, respectively, can inhibit BMP2 and TBX2.

Epithelial-to-mesenchymal transition is very important for cardiac valve genesis. Notch promotes epithelial–mesenchymal transition in the developing endocardial cushions in the atrioventricular canal and outflow tract, which are precursors of heart valves. Notch activates Snail and Slug that downregulate VE-cadherin gene. Endocardial cells that undergo an epithelial–mesenchymal transition invade a thick layer of extracellular matrix between the endocardium and myocardium, the *cardiac jelly*, which exists before the onset of cushion formation. Notch1 activation also upregulates the expression of mesenchymal markers such as α-smooth muscle actin.

---

and DLL1 are expressed cyclically in neural progenitors, but show an inverse correlation with HES1 expression. Agonist DLL1 then activates Notch signaling in adjacent neural progenitors.

19. Oscillations in expression of signal transducer and activator of transcription STAT3 and suppressor of cytokine signaling SOCS3 are coupled with HES1 oscillations. Janus kinase phosphorylates (activates) STAT3 that then forms a dimer and enters the nucleus to activate expression of target genes such as SOCS3. The latter in turn inhibits phosphorylation of STAT3. This negative feedback loop induces oscillations in SOCS3 expression and STAT3 formation. The STAT3–SOCS3 signaling regulates HES1 oscillatory expression. On the other hand, the helix–loop–helix factor, inhibitor of DNA binding (ID), leads to sustained HES1 expression.

20. Presenilin acts as the catalytic subunit of γ-secretase complex for Notch cleavage and release of Notch intracellular domain. Once cleaved, $Notch^{ICD}$ translocates to the nucleus where it forms a transcriptional complex with DNA-binding protein RBPJ and coactivators such as mastermind-like.

Trabecular formation depends on ephrin-B2 and its receptor EPHb4, neuregulin Nrg1 and its receptors HER2 and HER4, and bone morphogenetic protein BMP10. Notch activity in the endocardium promotes ventricular trabeculation by stimulating ephrin-B2 and BMP10, which elicit proliferation and differentiation of adjacent cardiomyocytes. Notch activated by DLL1 or DLL4 is restricted to endocardial cells at the trabecular base. Ephrin-B2 regulates Nrg1 that yields adjacent cell differentiation into trabecular myocytes. Endocardial Notch also activates BMP10 expressed in the adjacent myocardium.

The cardiac outflow tract includes the cardiac arterial pole, from which the right and left ventricles expel blood into the pulmonary artery and aorta, respectively. The heart outflow tract initially is a single vessel, which circulates blood into 3 major pairs of aortic arch arteries to join 2 paired dorsal aortae that distribute the blood throughout the embryo. Notch-2, Jagged-1, and presenilin-1 are implicated in cardiac outflow tract genesis. Development of the cardiac outflow tract requires precursors from the secondary heart field, cardiac neural crest cells, [21] and endothelial cells. Cardiac neural crest cells migrate toward and invest nascent vessels (endothelial tubes of the aortic arch) andvdifferentiate into vascular smooth muscle cells. Subsequently, the aortic arch arteries remodel to form the mature aortic arch. Notch in cardiac neural crest cells is required to form the smooth muscle layer of vessel walls. Notch activates smooth muscle genes, such as smooth muscle α-actin and myosin heavy chain.

Notch in Vasculo- and Angiogenesis

The Notch pathway is involved in a feedback loop with vascular endothelial growth factor, where Notch lies downstream from VEGF. Activated Notch can downregulate the expression of VEGFR2. Notch signaling not only regulates angiogenesis, particularly controlling the fate of endothelial tip cells, but also guides arteriovenous specification upstream from ephrin signaling. Notch receptors and ligands in endothelial and vascular smooth muscle cells are given in Table 10.4.

The vascular system develops from hemangioblasts that form the primitive vascular plexus, which remodels into a vascular system. Vascular morphogenesis comprises 4 overlapping stages: (1) endothelial cell commitment and differentiation; (2) vasculature morphogenesis and expansion; (3) remodeling; and (4) maturation. Notch enables the transition from the primitive vascular plexus to the hierarchical development of a branched network of arteries, capillaries, and veins. This transition includes: (1) sprouting angiogenesis; (2) regression or maintenance of vascular segments; (3) incorporation of smooth muscle cells or pericytes; and (4) progressive differentiation in arterial, venous, or capillary cell types.

Notch-1 and RBPJκ intervene in hematopoiesis and establishment of arterial identity in mouse embryos. Jagged-1-activated Notch-1 also regulates hematopoiesis via its Notch-1 transcriptional target GATA2, a hematopoietic transcription factor, but Jagged-1 is not required for arteriogenesis [1332].

21. The neural crest contains precursors derived from the dorsal neural tube that give rise to melanocytes, nerve tissue, muscle, cartilage and bone, among other cell types.

Table 10.4. Notch receptors and ligands in the vasculature (Sources: [1331]). Notch-2 is involved in the heart and lymphatics, as well as in the vascular morphogenesis of specific vascular beds (hyaloid vasculature of the eye and glomerular capillaries). Agonist DLL4 is mainly expressed in the endothelium, although low levels are detected in smooth muscle cells. Ligand DLL1 is detected in arterial endothelium, but not in veins and capillaries, although it is expressed in the endothelium of both arteries and veins of midgestational embryos. The distribution of Jag1 is complementary rather than overlapping with that of DLL4.

Endothelial cell	Vascular smooth muscle cell
Notch-1	Notch-1
Notch-4	Notch-3
DLL1	DLL1
DLL4	(DLL4)
Jag1	Jag1
Jag2	Jag2

Vascular Sprouting

Vascular expansions sprout owing to migrating tip cells [22] guided along gradients in matrix-bound VEGF. Neuropilin-1 is a transmembrane receptor with a repulsive function in axon guidance that binds to VEGFa and stimulates angiogenesis as it enables endothelial tip cell filopodia to protrude in a new direction at a specific location in the developing brain according to the substrate [1333].

Notch is used iteratively in a cell-autonomous manner to regulate vessel caliber, to limit branching, and to establish a mature, quiescent pattern [1334]. The DLL4–Notch-1 signaling coordinates cell specification in vascular sprouts. The DLL4–Notch signaling downstream from VEGFa reduces tip cell number and filopodia in tip cells, hence prevents branching at the tip of vascular sprouts. [23] Notch-1 is frequently absent in tip cells, but is prominently expressed in stalk cells. Yet, Notch signaling is not required for re-establishment of endothelial cell polarity or for lumen formation.

Arterioveinous Differentiation

Notch signaling drives the specification of somitic cells into endothelial precursors and their migration toward and assembly into the dorsal aorta, the first embryonic vessel in avian embryos [1335]. Ephrin-B2 that acts downstream from Notch to mediate arterial differentiation is also required for migration of Notch-activated cells.

22. Tip cells are specialized endothelial cells that display morphological and functional features distinct from stalk cells. They migrate in response to VEGFa gradients, whereas stalk cells proliferate and rearrange to form intercellular lumens also in response to VEGFa.

23. Suppression of Notch signaling leads to tip cell division, with both daughter cells being specified as tip cells, therefore causing vessel bifurcation.

Notch-1 and -4 interact with Jag1 and DLL4 to specify arterial fate in endothelial cells [1336]. Genetic prepatterning, largely mediated by the Notch pathway, prior to the initiation of blood flow, regulates arteriovenous differentiation. Hemangioblasts that differentiate from mesenchymal progenitors are bipotential precursors of a subset of hematopoietic and endothelial cells. They give rise to endothelium of both arteries and veins.

Numerous types of factors harmonize angiogenesis and blood vessel types. The Notch pathway interacts with the pathway primed by vascular endothelial growth factor-A that collaborates with sonic Hedgehog-A. Reduced VEGFa activity causes a loss of arterial marker expression and ectopic arterial expression of vein markers. Notch and sonic Hedgehog-A work downstream and upstream from the VEGFa pathway, respectively. The VEGFa factor via VEGFR1 and VEGFR2 and the PI3K–PKB pathway induces Notch-1 and DLL4 expression in arterial endothelial cells, but not in venous endothelial cells [1337]. In addition, both VEGFa induces ephrin-B2 expression in endothelial cells of mouse cardiac capillaries, hence favoring arterial networks in the heart, but reduces the number of EPHb4 receptors (venular marker; Table 10.5) [1338]. Notch induces arterial-specific markers, such as ephrin-B2, connexin-40, and transcriptional repressor Hairy-related transcription factor HRT2 [1339]. [24] Stimulation by DLL4 (but not Jagged-1) induces ephrin-B2 expression via HRT2-independent mechanism. Transcription factors Forkhead box FoxC1 and FoxC2 directly activate DLL4 transcription [1341]. Both FoxC1 and FoxC2 are required for arterial specification and lymphatic sprouting from veins in association with VEGFc during angiogenesis. The VEGF factor also collaborates with angiopoietins to regulate angiogenesis. Angiopoietin-1 that may build in cooperation with VEGF mature (non-leaky) vessels can offset VEGF-induced angiogenesis in vivo at least in the heart. In contrast, angiopoietin-2 and VEGF act synergistically in angiogenesis [1338]. Furthermore, angiopoietins and VEGF control the ratio of arteries to veins during angiogenesis.

Notch signaling regulates arterial specification of vascular smooth muscle cells. [25] Arterial identity of endothelial cells and neighboring vascular smooth muscle cells is specified independently [1336].

10.1.3.4 Notch and Vascular Homeostasis

Notch1 is active in endothelial and vascular smooth muscle cells beyond vascular morphogenesis. Notch-1 and Jag1 can be upregulated after vascular injury. Notch is activated in the vasculature via 4 mechanisms according to involved cell

---

24. Hairy-related transcription factor-2 (HRT2) is also named HES-related repressor protein HERP1, HEY2, and HESR2. This transcriptional repressor abounds in the vasculature, where it is targeted by Notch. In particular, HRT2 inhibits the activity of serum response factor with its coactivator myocardin that is required for vascular smooth muscle cell differentiation [1340]. Expression of HRT2 is associated with SMC proliferation and dedifferentiation during vascular injury and atherosclerosis.

25. Vascular smooth muscle cells that surround arteries in Notch-3−/− mice acquire a venous fate.

Table 10.5. Arterial and venous markers. The EPHb4 receptor and corresponding ligand ephrin-B2 are implicated in arteriovenous specification. Arterial fate is acquired by the combined effect of forkhead transcription factors FoxC1 and FoxC2 and VEGF. Notch-1 signaling pertains to arterial specification. HRT1 and HRT2 produce arterial markers. The expression of nuclear receptor NR2F1 (or COUP-TF2) initiates the venous program, as it represses Notch. Smoothelin is a constituent of the cytoskeleton specific to smooth muscle cells [1342]. SM22α or transgelin is a calponin that is expressed exclusively in differentiated smooth muscle cells and represses MMP9 expression [1343]. Promoter of the Sm22 gene is active in the smooth muscle cells of arteries but not veins. Nuclear receptor NR2F1 suppresses neuropilin-1 expression, thereby suppressing VEGFa signal reception and Notch activation.

Artery	Vein
Endothelial cell	
Ephrin-B2	EPHb4
FoxC	NR2F1 (COUP-TF2)
Notch-1	
Neuropilin-1	
Connexin-40	
HES1	
HRT1, HRT2	
Vascular smooth muscle cell	
Notch-3	
Smoothelin	
Transgelin	

types [1331]: (1) homotypic cis-activation or inhibition by endothelial cell itself (autocrine loop), as endothelial cells express both receptors and ligands; (2) homotypic trans-activation by neighboring endothelial cells; (3) heterotypic activation by adjoining smooth muscle cells; and (4) interaction with the microfibrillar proteins (MAGP1 and MAGP2) that are often present on microfibrils of elastic tissue such as the inner elastic lamina and can provoke Notch1 dissociation. Both MAGP proteins that can be stimulated by stretch can interact with Notch (possible mechanotransduction).

Notch reduces proliferation and prevents apoptosis of endothelial cells. In addition, Notch maintains the arterial phenotype. The expression profile of arterial genes enables vascular cells to sustain the time-varying high blood pressure. Venous endothelial cells are more adapted to leukocyte extravasation. Notch may regulate plasticity in differentiated endothelium, when vein arterialization occurs caused by high blood pressure.

10.1.3.5 Notch and Vascular Smooth Muscle Cells

The Jag1–Notch pathway promotes smooth muscle cell differentiation from mesenchymal cells via RBPJκ and myocardin–SRF–CArG complex [1344]. Notch targets the genes that encode smooth muscle myosin heavy chain and α-actin [1344, 1345]. Notch regulates the expression of smooth muscle α-actin in vascular endothelial and smooth muscle cells via activation of RBPJκ. Vascular smooth muscle cells exposed to cyclic mechanical strain exhibit reduced proliferation, augmented apoptosis, diminished expression of Notch-1 and Notch-3 receptors.

In adult tissues, Notch-3 expression is restricted to vascular smooth muscle cells. Notch-3 signaling is associated with ischemic stroke via regulation of the vascular SMC function [1346]. [26] The vasoconstrictive response is lacking during ischemia in the abscence of Notch-3 in smooth muscle cells of cerebral arteries. [27]

10.1.3.6 Notch in Immune System

In the immune system, B and T lymphocytes develop from common bone marrow lymphoid-restricted progenitors. Notch is involved in the T-cell development, T lymphocytes being formed once progenitors have left the bone marrow and reached the thymus. The thymus produces multiple Notch activators. On the other hand, Notch repression leads progenitors toward B-cell development, as B lymphocytes are essentially produced in the bone marrow.

Notch-1 upregulates expression of interferon-γ in peripheral T lymphocytes via activation of nuclear factor-κB (NFκB). The intracellular domain of Notch-1 directly interacts with NFκB and competes with IκBα, leading to NFκB retention in the nucleus and activation of IFNγ promoter [1347]. On the other hand, IκBα can enter into the nucleus and bind to NFκB, exporting the latter to the cytosol.

10.1.3.7 Notch and Pathologies

Aberrant Notch signaling is involved in many cancers. Notch-1 facilitates melanoma development, as it stimulates protein kinase-B [1348]. Conversely, protein kinase-B increases Notch-1 expression via nuclear factor-κB during hypoxia, i.e., in the presence of transcription factor hypoxia-inducible factor-1α. Hypoxia-inducible factor-1 causes a decrease in mitochondrial activity that favors tumor growth [1349].

In the cardiovascular system, defects in Notch signaling lead to abnormal vascular development and hereditary disorders, such as cadasil (cerebral autosomal dominant arteriopathy with subcortical infarcts and leukoencephalopathy) [28] and Alagille

---

26. The contractile activity in isolated aortas does not reveal differences between the absence and the presence of Notch-3. Notch-3 expression in vascular smooth muscle cells is associated with the expression of genes involved in vascular tone in cerebral arteries but not aortas.

27. Mutations of the NOTCH-3 gene cause cerebral arteriopathy with subcortical infarcts and leukoencephalopathy.

28. Cadasil due to mutations in NOTCH-3 is characterized by degeneration of smooth muscle layers surrounding small arteries and arterioles of the brain and skin. Wall media is substituted by connective tissue, leading to fibrosis and narrowing of the lumen. Cadasil manifests by strokes and progressive dementia.

syndrome. [29] The Notch pathway is also involved in many cancers, particularly human lung cancers.

10.2 Hedgehog Receptors

The Hedgehog (Hh) family of secreted proteic morphogens is involved in various developmental processes, especially during embryo- and fetogenesis. Nonetheless, Hedgehog signaling remains active in adult life. Hedgehog regulates adult stem cells for the maintenance and regeneration of adult tissues. Hedgehog is particularly implicated in maintenance and self-renewal of neural progenitors in stem cell niches of the postnatal brain. Defects in Hedgehog signaling lead to organ defects and the initiation and growth of certain cancers (Gorlin's syndrome).

The family of Hedgehog morphogens comprises 3 types of secreted proteins: sonic (SHh), indian (IHh), and desert Hedgehog (DHh). Hedgehog can move far from its site of synthesis and control outcomes of remote cells.

Moreover, different ligand concentrations drive different cell fates. Several kinases have both positive and negative effects on Hedgehog signaling. Specific phosphorylations of Hedgehog effectors influence their subcellular location, stability, and activity.

Signals from sonic Hedgehog stimulates 7 genes that encode thrombomodulin, TSC22 domain family TSC22D3 (or glucocorticoid-induced leucine zipper protein GILZ), [30] FoxG1, [31] early-response protein NR4a1, [32] insulin-like growth factor IGF2, peripheral myelin protein PMP22, and LIM and SH3 domain-containing protein LASP1. On the other hand, these signals inhibit 4 genes that encode secreted Frizzled-related proteins sFRP1 and sFRP2, macrophage inflammatory protein MIP1γ (chemokine CCL9), and anti-Müllerian hormone (AMH) [1350].

10.2.1 Hedgehog Synthesis and Release

Mature Hedgehog is synthesized from a precursor that bears a set of post-translational modifications, i.e., attachment of lipid adducts (cholesterol and palmitic acid) that confer a high affinity for the plasma membrane. [33]

29. Alagille syndrome is a disorder associated with mutations in Jag1 (most often) or Notch-2. It is exhibited by abnormal blood vessels, arterial stenosis, and heart disease, in addition to hepatic lesions and skeletal defects.

30. Transcription factor TSC22 (TGFβ-stimulated clone-22) belongs to the family of early-response genes.

31. A.k.a. brain factor BF1 or BF2.

32. A.k.a. nuclear receptor NuR77, testicular receptor TR3, regenerating liver nuclear receptor RNR1, Nakai factor NaK1, hormone receptor (HmR), growth factor-inducible nuclear protein N10, nerve growth factor-like nuclear receptor-1B (NGF1b), and growth factor response protein GFRP1.

33. The mature protein has a palmitic acid covalently attached to the N-terminus and a cholesterol moiety anchored to the C-terminus.

Hedgehog precursor is cleaved into 2 proteins. The first Hedgehog end is responsible for its activity. It can be palmitoylated. The other acts as a cholesterol transferase, as the cholesterol modification of Hedgehog is involved for its transport.

Sonic Hedgehog precursor undergoes a cholesterol-mediated cleavage. Lipid-modified Hedgehog can form multimers. Furthermore, lipid-modified Hedgehog can be carried by lipoproteins. In addition, Hh signaling is linked to lipid metabolism.

Hedgehog reaches both apical and basolateral segments of the plasma membrane of polarized epithelial cells and can be secreted by both segment types. The lipid modification of Hedgehog ensure its secretion, but secreted lipid-linked Hedgehog protein can travel only over short distances w.r.t. lipid-free Hedgehog [1351]. Nevertheless, Hedgehog is able to signal over both short and long ranges. For a long-range transfer, attached lipids must be removed. Hedgehog then is able to signal to remote cells. Lipid detachment can be carried out during transcytosis. Apically released Hedgehog can be subsequently internalized in the producing cells and transferred to the basolateral surface, where Hedgehog is released for long-range signaling [1351].

Endocytosis followed by apical-to-basolateral plasma membrane recycling may be required for acquisition by Hedgehog of its full long-range signaling potency. The transcytosis in the sending cells involves dynamin (for the initial internalization of extracellular Hedgehog), Rab5 (for the detachment of endocytic vesicles from the plasma membrane), and Rab4 and Rab8 GTPases (for the fusion of Hedgehog-containing vesicles to early endosomes before the release of Hedgehog to the basolateral surface) [1351].

10.2.2 Hedgehog Signal Reception

Hedgehog receptor is composed of 2 identified proteins: (1) 12-transmembrane receptor Patched (Ptc) that is encoded by a tumor suppressor gene, and (2) 7-transmembrane Smoothened (Smo), encoded by a proto-oncogene.

The Patched receptor is a lipoprotein receptor [1352]. Smoothened is related to the Frizzled family of G-protein-coupled receptors. Although Smoothened possesses a structural similarity to GPCRs, it signals in a distinct manner.

Once Hedgehog is secreted, it targets its receptor Patched. The expression of Patched is upregulated by Hedgehog signaling (positive feedback). After binding to its cognate receptor, signaling, and internalization, Hedgehog is degraded in the receiving cell. Dispatched assists Patched in Hedgehog reception and endocytosis.

Patched is an inhibitor of Smoothened activation. Binding of Hedgehog (e.g., SHh) to Patched relieves Patched inhibition of Smoothened. The Patched–Hedgehog complex then undergoes endocytosis.

Two Patched receptors exist (PtcH1 and PtcH2). The 3 mammalian Hedgehogs bind both receptors with similar affinity. PtcH2 is mainly expressed in the testis, where it binds desert Hedgehog.

Patched and Dispatched belong to the *resistance nodulation division* (RND) family of homotrimeric *proton-driven pumps* that transport small molecules across membranes. Patched could serve as a carrier to modify the concentration or location of a small second messenger that regulates Smo [1353].

Hedgehog-interacting protein (HhIP) is a cell-surface receptor antagonist that is equipotent against the 3 mammalian Hh homologs. It actually binds and inhibits Hedgehog proteins. Cations Ca^{++} and Zn^{++} modulate interactions between Hedgehog proteins and Hedgehog-interacting proteins [1354].

10.2.3 Hedgehog Signaling

Sonic Hedgehog is involved in the growth of various tissues (lung, skeleton, muscles, gastrointestinal tract, and prostate). It induces the proliferation of neuronal precursor cells. It is implicated in angiogenesis not only during embryo- and fetogenesis, but also in adults. Hedgehog upregulates Notch target genes.

Sonic Hedgehog can serve as an autocrine signal. Paracrine Hedgehog signaling requires Dispatched protein. Once sonic Hedgehog is bound to Patched, Smoothened cytoplasmic tail is extensively phosphorylated by protein kinase-A and casein kinase-1 (Fig. 10.2). Smoothened phosphorylation leads to Smo plasmalemmal accumulation and favors dimerization (if constitutive Smo dimer does not exist), which is required for Hedgehog signal transduction.

Binding of Hedgehog to Patched abrogates Smoothened inhibition and activates Smo via dimerization and a conformational switch [1355].[34] The Smoothened autoinhibitory domain (SAID) translates graded Hedgehog signals into distinct intracellular responses: the higher the number of phosphorylated sites, the larger the Smo plasmalemmal level and activity.[35]

10.2.3.1 Hedgehog and the Primary Cilium

The primary cilium (height $<5\mu m$, width $<1\mu m$), a solitary, tiny microtubule-based protrusion on the periphery of most cell types serves as a major organelle for organization of Hedgehog signal transduction. It can detect and interpret numerous extracellular signals.

When cells are stimulated with Hedgehog ligands, mediators of the Hh pathway enter into and exit out the primary cilium. This spatial organization of the Hh axis ensures the specificity of signaling. The dynamic motion of proteins into and out of the primary cilium regulates the activity of signaling complexes.

Multiple components of the Hh pathway localize to the primary cilium or its basal body, such as sonic Hedgehog, Patched-1, Smoothened, Suppressor of Fused, and Gli transcription factors. Stimulation by extracellular Hedgehog changes ciliary

34. The Smoothened auto-inhibitory domain contains multiple regulatory modules made of arginine clusters that restrict plasmalemmal concentration of Smoothened and impedes Smo activity. Hedgehog Binding to Patched activates Smo by dimerization. Increased proximity of Smo cytoplasmic tails is accompanied by a conformational change. Both dimerization and conformational change are regulated by phosphorylation. Clusters of Smo cytoplasmic tails promote close proximity by conformational changes and phosphorylation, then trigger Hedgehog pathway.

35. Phosphorylation of Smoothened neutralizes Arg function.

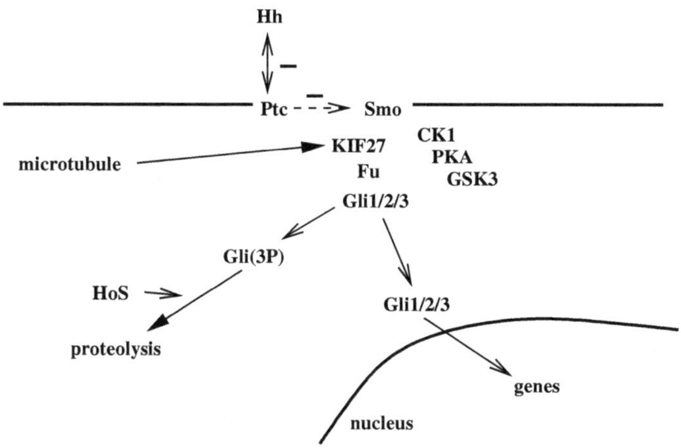

Figure 10.2. Hedgehog signaling (Source: [1356]). Hedgehog activation occurs by autocatalytic cleavage of a precursor into a cholesterol-modified N-terminal signaling domain (HhN) that is subsequently palmitoylated by Hedgehog acyltransferase (HHAT), or skinny Hedgehog, then released from the plasma membrane by Dispatched. Initiation of the pathway entails Hh binding to Patched. Inhibition of Hh-bound Patched in turn derepresses transmembrane Smoothened (Smo). Unphosphorylated Smoothened forms and acts via a proteic complex that contains kinesin-like protein KIF27, Ser/Thr kinase Fused (Fu), and transcription factors glioma-associated oncogene homologs (Gli1–Gli3). Protein KIF27 sequesters protein kinase-A (PKA), casein kinase-1 (CK1), and glycogen synthase kinase-3 (GSK3). Phosphorylation of Gli by PKA, CK1, and GSK3 is required for the efficient processing into transcriptional repressor form. In addition, hyperphosphorylated Gli is targeted by ubiquitin ligase homolog of Slimb (HoS or F box–WD repeat-containing protein FBXW11) for proteolytic cleavage. Protein KIF27 also inhibits Smo phosphorylation and accumulation at the plasma membrane, possibly by binding Smoothened. Hedgehog signaling leads to phosphorylation of Smo, Fu, KIF27, and Suppressor of Fused (Sufu). Derepressed Smo activates Fu that phosphorylates KIF27 that then dissociates from the kinase complex and releases Gli. The latter then translocates to the nucleus to regulate Hh target genes. Suppressor of Fused binds to Gli.

localization of Hh receptor components Smoothened and Patched as well as gene expression via transfer of mediators of Hh-dependent transcriptional activation.

Transmembrane receptor Smoothened can localize to the primary cilium on: (1) rapid delivery by (1.1) vesicles from the Golgi body along microtubules to the base of the cilium, as the Golgi body is close to the basal body, and (1.2) recycling endosomes, as well as (2) transport from the plasma membrane to the cilium membrane (the so-called *lateral transport*), when the Hedgehog pathway is activated [1357].

Protein kinase-A is a component of Hedgehog signaling that is involved in protein transfer (Vol. 4 – Chap. 4. Cytosolic Protein Ser/Thr Kinases). Protein kinase-A localizes near the primary cilium, at a location suitable to control the recruitment of other proteins in the Hedgehog pathway [1358].

10.2.3.2 Gli Family of Transcription Factors

The pathway downstream from Smoothened relies on transcription factors of the GLI family [36] that act as activators or repressors. The 3 detected members of the GLI family (Gli1–Gli3) are encoded by 3 genes (GLI1–GLI3).

When Hedgehog signaling is absent, transcription factor Gli [37] is cleaved to become a transcriptional repressor (Gli^R); when it is present, it acts as an activator (Gli^A) [1359]. Therefore, the Hedgehog gradient is associated with antagonist gradients of Gli activator and repressor that compete for the same DNA-binding sites; gradients of signaling effectors control spatial patterns of gene expression.

The transcriptional response to Hh gradients results from: (1) the competition between repressor and activator forms for binding sites; (2) the affinity of the transcription factor binding sites; and (3) cooperative binding of repressor forms. [38]

Transcriptional regulator Gli1 recruits the histone deacetylase complexes via Suppressor of Fused homolog (SuFu), which can impede gene transcription. Components of the primary cilia, such as intraflagellar transport proteins (Vol. 1 – Chap. 9. Intracellular Transport), participate not only in Gli activation, but also in Gli3 processing [1360].

Member Gli2 of the GLI–Krüppel family has 4 isoforms (Gli2α–Gli2δ). It is the main mediator of Hh-dependent transcriptional activation. Agent Gli2 and, to a lesser extent, Gli1, but not Gli3, upregulates anti-apoptotic BCL2 protein.

Factors Gli2 and Gli3 possess a C-terminal activator and a N-terminal repressor region. Factor Gli3 exists in 2 forms: (1) a full-length (Gli_{FL}) transcriptional activator $Gli3_A$ and (2) a repressor fragment $Gli3_R$ [1353].

Factors Gli2 and Gli3 contribute to the formation of the respiratory tract. In stratified epidermis, Gli2 operates as an activator of keratinocyte proliferation and a repressor of epidermal differentiation [1361].

10.2.3.3 Hedgehog Pathway

Mediators of the Hedgehog pathway are given in Table 10.6. In the absence of SHh, the primary cilium is enriched in Ptc1 receptors. Both Smoothened receptor and Gli2 transcription factor enter into and exit out of the primary cilium. Mediator Gli2, but not Smoothened, requires cytoplasmic microtubules for ciliary entry [1362]. Both

---

36. Effectors Gli of Hedgehog signaling were originally isolated in human glioblastoma.

37. Cubitus interruptus in Drosophila.

38. Low-affinity binding sites for Gli in the enhancer of the Hh target gene decapentaplegia (Dpp) in Drosophila are required for proper spatial expression of the corresponding protein in regions of low Hh signal. High-affinity binding sites are present in the enhancer of another Hh target gene, Patched, adapted for a more restricted, high-magnitude signal. Repressor cooperativity corresponds to easier binding of additional Gli^R, when one Gli^R is already bound. However, repressor cooperativity must maintain possible activation by Gli^A in regions of low Hh signaling. In the repressor cooperativity model, repressors outcompete activators for binding at high-affinity enhancers, but not at low-affinity enhancers.

Table 10.6. Main mediators of the Hedgehog pathway (DHh, IHh, SHh; desert, indian, and sonic Hedgehog; BOC: brother of CDO; CDO: cell adhesion molecule-related–downregulated by oncogenes; CK1: casein kinase-1; Gli: glioma-associated oncogene homologs; GSK3: glycogen synthase kinase-3; HhIP: Hedgehog-interacting protein; PKA: protein kinase-A; SuFu: Suppressor of Fused homolog). All 3 mammalian Hedgehog types bind both PtcH1 and PtcH2 receptors with similar affinity.

Type and subtypes	Role effects
DHh, IHh, SHh	Messengers
	Receptors and coreceptors
BOC homolog	Modulation of Hh signaling
CDO homolog	Modulation of Hh signaling
Dispatched	Proton-driven pump (transport small molecules)
HhIP	Sequestration of Hedgehog ligands
Patched homologs	Repressors of Smoothened,
PtcH1, PtcH2	proton-driven pump
Smoothened	GPCR
	Kinases
Fused	Ser/Thr kinase; stabilization of Gli
SuFu	Inhibition of Gli nuclear translocation
CK1, GSK3, PKA	Phosphorylation of Gli (transcriptional repressor)

Gli2 and Smo need cytoplasmic nanomotor dynein-2 for ciliary exit. Changes in ciliary and nuclear transfer of Gli2 characterized by a shift from primarily cytoplasmic localization to accumulation at the distal tip of the cilium and ultimately in the nucleus are triggered by Hh-dependent accumulation of activated Smo in the primary cilium [1362].

Upon Hh excitation, the primary cilium becomes enriched in Hh pathway stimulators such as Smoothened and 20α-hydroxycholesterol [1363]. Hedgehog stimulation modifies the ciliary transport of both Smo and Gli2 such that these mediators accumulate within the cilium and primarily at the ciliary tip, respectively [1362].

Translocation to the primary cilium of Smo from an intracellular source happens via an intraflagellar transport [1363]. Assembly and maintenance of the primary cilium involve intraflagellar transport using intraflagellar transport particles, which are made of 2 proteic complexes – IFTa and IFTb – as well as nanomotors kinesin-2 and dynein for anterograde and retrograde transport, respectively.

When sonic Hedgehog binds to Patched-1, Ptc1 leaves the primary cilium, and relieves inhibition of Smoothened. The latter translocates to the primary cilium [1353]. Smoothened inhibits the repressor Suppressor of Fused that prevents translocation to the nucleus of Gli transcription factor.

Hedgehog signaling mediators Ptc, Smo, SuFu kinase, and Gli are located mainly, but not exclusively, in the primary cilium [1353]. Patched in the primary cilium controls Smo entry into the primary cilium. Inactive Smo is sequestered in

Table 10.7. Positive and negative regulators of the Hedgehog signaling pathway. The Hedgehog family of secreted signaling proteins regulates in a dose-dependent manner cell fate during embryogenesis and tissue homeostasis after birth. Hedgehog intervenes in the maintenance and regulation of stem cells in adults. Hedgehog is implicated in certain cancers. Several proteins regulate the cellular response to secreted Hedgehog: Transmembrane proteins Patched (Ptc) and plasmalemmal proteins cell adhesion molecule-related protein-downregulated by oncogenes (CDO; interference Hedgehog [IHog] in Drosophila) and brother of CDO (BOC) affect Hedgehog signaling. Both CDO and BOC act either directly as coreceptors or indirectly downstream from Patched.

Activators	Repressors
Hedgehog	Patched
Smoothened	Suppressor of Fused
Gli_{FL}	Gli fragment (Gli^{R})
	PKA, CK1, GSK3

vesicles and excluded from the primary cilium, whereas active Smo is conveyed into the primary cilium.

Hedgehog triggers translocation of Ptc to vesicles and Smo to the plasma membrane. Without Hh, Ptc travels between the plasma membrane and intracellular vesicles. Patched reduces Smo levels and prevents Smo motions from vesicles to the plasma membrane.

10.2.4 Regulators of the Hedgehog Pathway

Various proteins regulate the cellular response to secreted Hedgehog (Table 10.7). Many molecules influence Hedgehog signaling in the extracellular space, at the plasma membrane, or in the cytoplasm. Certain receptors enhance Hh binding to Ptc, such as members of the cell adhesion molecule-related–down-regulated by oncogenes family [1360].

10.2.4.1 Growth Factors

The epidermal and insulin-like growth factor pathways modulate Hh signaling, as they act on Gli via Ras GTPase and MAPK and PKB kinases in some cell types.

10.2.4.2 Kinases

Protein kinase-A, glycogen synthase kinase GSK3β, and casein kinase CK1 phosphorylate Gli and prevent activation of SHh pathway. On the other hand, other protein kinases, such as casein kinase CK2 and cyclin-dependent kinases CDK9 and CDK11, and protein phosphatase PP2, promote Hh signaling.

10.2.4.3 Proteoglycans

Proteoglycans act as crucial anchors and modulators of sonic Hedgehog. Interactions between SHh and proteoglycans promote proliferation in mitogenic niches rather than cell differentiation.[39] Cyclin-D1 and -D2 and the transcriptional repressor Polycomb complex protein Bmi1 are major effectors of SHh–proteoglycan interactions.

10.2.4.4 Hedgehog-Interacting Protein-1

Hedgehog-interacting protein HhIP1 is a feedback antagonist of Hh signaling. It sequesters Hedgehog, but unlike Patched, it has no effect on the activity of Smoothened. Patched homolog-1 and HhIP1 thus share a feedback control of Hh morphogens [1365].

10.2.4.5 Inhibitor Numb

Developmental protein homolog and cell fate regulator Numb, a localized intracellular receptor, precludes the Hedgehog pathway that is downregulated in cell progenitors and cancer cells.[40] Inhibitor Numb not only hinders the Notch-1 pathway by ubiquitination, but also suppresses Hedgehog signaling by Itch-dependent ubiquitination of Gli1 [1366].

10.2.4.6 CDO and BOC Homologs

The Hh signaling is modulated in space and time by Cell adhesion molecule-related downregulated by oncogene homolog (CDO) and brother of CDO homolog (BOC). These molecules act either directly as coreceptors or indirectly downstream of Patched.

10.2.4.7 β-Arrestins and Kinesins

Activated Smoothened has some common behavior with canonical G-protein-coupled receptors, as it is capable of undergoing phosphorylation by GPCR kinase and to recruit β-arrestin-2 for endocytosis.

---

39. In particular, proteoglycans establish and maintain a favorable microenvironment for proliferation of neural precursors. Proteoglycans recruit SHh to mitogenic niches in developing brain to promote the proliferation of neural stem and precursor cells. Morphogen SHh interacts with heparan sulfate proteoglycans on membrane of cerebellar granule neuron precursors to alter the nature and timing of Gli2-dependent transcription [1364].

40. The cytoplasmic membrane-associated precursor cell specifier Numb is involved in the proper differentiation of a type of sensory neurons, the multiple dendritic (md) neurons. It is asymmetrically distributed in neuronal and mesodermal (precursors of striated myocytes) progenitors. It enables the determination and segregation of alternative daughter cell fates during during the terminal asymmetrical division in cell differentiation.

In cultured cells, β-arrestin-1 and -2 form a complex with Smo and kinesin family KIF3a member. Smoothened-dependent interaction of β-arrestins with KIF3a is required for Smo endocytosis and localization to the primary cilium, as well as activation of Gli transcription factors [1367].

10.2.4.8 Oxysterol Activators and Vitamin-D3 Repressor

Small lipophilic molecules, such as oxysterols [41] and vitamin-D3, regulate signaling mediated by Smoothened, as they act as activators and inhibitors, respectively.

10.2.4.9 Glucocorticoids

Some glucocorticoids can operate as Smoothened agonists [1368]. 3β-Hydroxysteroid dehydrogenase (3BHSD) produces all the 3 groups of adrenal steroids: mineralocorticoids, glucocorticoids, and sex steroids. 3β-Hydroxysteroids can regulate Smoothened.

10.2.4.10 Lipids

Endogenous sterol derivatives modulate the Hedgehog pathway [1369]. Some lipid types can indeed influence Hh transport and signaling. Hedgehog C-terminal domain catalyzes its own removal. It is then replaced by cholesterol, as the C-terminus of the resulting N-terminal Hedgehog fragment links to cholesterol. This fragment is then palmitoylated by the endoplasmic reticulum transmembrane Hedgehog acyltransferase (HHAT), or skinny Hedgehog. [42] Cholesterol-modified Hedgehog requires heparan sulfate proteoglycans that interact with lipoproteins to act.

Patched-mediated repression of Smoothened signaling in the absence of Hedgehog can operate indirectly. Agents Patched and Dispatched and the cholesterol transporter Niemann-Pick protein NPC1, which participates in cholesterol efflux from late endosomes, are members of the resistance-nodulation division family that have sterol-sensing domain. Patched may have transporter activity for cholesterol and/or its derivatives.

10.2.4.11 Mechanical Stress

Hedgehog signals respond to mechanical loading. In particular, Hedgehog signaling is affected by equibiaxial cyclic strain (10% stretch, 1 Hz, 24 h duration) and pulsatile flow (either 0.3 ml/mn, 0.2 Hz, pressure amplitude of ~0.8 kPa, or 25 ml/mn, 2 Hz, pressure amplitude of ~6.65 kPa) in cultured adult rat vascular smooth muscle cells [1370]. Cyclic strains reduce the expression of sonic Hedgehog and Patched-1, reducing Gli2 activity and SMC proliferation, and increasing SMC apoptosis.

---

41. Oxysterols are synthesized from cholesterol by hydroxylation.
42. Skinny Hedgehog is also called Sightless, Rasp, or Central missing.

Although Notch signaling also decays when smooth muscle cells bear cyclic strains, Notch3^{ICD}, which acts on Hh signaling, is overexpressed in loaded cells. Messenger Notch3^{ICD} can thus compensate the strain-induced inhibition of Hedgehog signaling. Pulsatile flow also decreases Hh signaling, with subsequent reduction in cell proliferation and increase in apoptosis. Hedgehog expression decays in cells subjected to chronic exposure of time-dependent wall stresses.

10.3 Wnt Morphogens

The Hedgehog and Wnt pathways target common genes [1371]. In addition, the multitask kinase STK11 (or LKB1) ensures the maintenance of the primary cilium, thus promoting Hh signaling; it also mediates the Wnt pathways.

The Wnt morphogens are secreted, palmitoylated glycoproteins.[43] Ligands of the family of Wnt specify cell fate and intervene in the body's development[44] and regeneration. Moreover, Wnt ligands maintain the undifferentiated state of various types of stem cells in their niche and promotes self-renewal, particularly of hematopoietic stem cells. They actually prevent stem-cell differentiation in opposition to other growth factors, such as FGF and EGF factors. However, dysfunctional Wnt pathways (e.g., disruption of adenomatous polyposis coli) trigger tumorigenesis.

10.3.1 Wnt Family and Their Receptors

Nineteen human Wnt ligands have been identified: Wnt1, Wnt2, Wnt2b (or Wnt13), Wnt3, Wnt3a, Wnt4, Wnt5a, Wnt5b, Wnt6, Wnt7a, Wnt7b, Wnt8a, Wnt8b,

---

43. Secretion of Wnt proteins from producing cells relies on the Wnt-secretion factor GPR177, a cargo for the retromer complex (a.k.a. Wntless homolog [Wls], evenness interrupted homolog [EvI], and abnormal cell migration Mig14 in Caenorhabditis elegans) that binds Wnt in the Golgi network and transports it to the cell surface for release. The myotubularin lipid phosphatase MTM6–MTM9 complex that dephosphorylates phosphatidylinositol-3-phosphate is required for the retromer-dependent recycling of Wls from the plasma membrane to the trans-Golgi network for efficient Wnt secretion [1372].

44. The Wnt pathway has been discovered in Drosophila melanogaster, in which it directs the spatial organization of the body plan. The name Wnt corresponds to a combination of segment polarity gene wingless (Wg) and cellular proto-oncogene product Int1. The Wnt pathway diverges into a canonical axis that relies on transcriptional activator β-catenin, non-canonical route, and the planar cell polarity pathway for synchronous polarity of sheets of cells. These axes share some components with dual or distinct roles. In Drosophila melanogaster wings, Frizzled, Disheveled, and Diego accumulate at the distal edge, whereas VanGogh and Prickle localize to the proximal side of each cell (Wnt–Frizzled axis for planar cell polarity). The distal edge can be bridged to the proximal side of the adjacent cell by between-cell Flamingo–Flamingo, Flamingo–VanGogh, Flamingo–Frizzled, and Frizzled–VanGogh interactions. In addition, mutually repressive interactions exist in the cell between distal and proximal complexes. Binding of Dachsous to Fat also regulates the planar cell polarity. Fat Mediator recruits the transcriptional corepressor Atrophin (or Grunge) that suppresses transcription of Four-jointed.

Wnt9a (or Wnt14), Wnt9b (or Wnt15), Wnt10a, Wnt10b or (Wnt12), Wnt11, and Wnt16. Secreted Wnt lipoglycoproteins [45] are hydrophobic and thus mostly associated with cell membranes and extracellular matrix components.

Ligands of the Wnt family bind to their cognate 7-pass transmembrane receptors (GPCRs) of the Frizzled family. [46] This family is constituted by 10 members (Fz1–Fz10). Frizzled receptors homo- and heteromerize. They connect to their single-pass transmembrane coreceptors low-density lipoprotein receptor-related proteins LRP5 and LRP6. [47] Signals from Wnt agonists are transmitted by specific Frizzled or Fz–LRP5/6 complexes. [48]

10.3.2 Wnt Signaling

Binding of a Wnt ligand to Frizzled–LRP5/6 complexes inhibits phosphorylation and degradation of its transcriptional coactivators β-catenins (Table 10.8). The resulting signal is then propagated via Disheveled proteins (Dvl) that direct *canonical* (β-catenin-dependent) or *non-canonical* (β-catenin-independent) signaling (Table 10.9). The canonical pathway controls the concentration of the pool of non-cadherin-associated β-catenins.

Disheveled mediates both the canonical and non-canonical Wnt signaling pathways. Three homologs of the DVL family exist in mammals (Dvl1–Dvl3). Disheveled may function as a heterotrimer. All Dvl proteins contain 3 regions: an N-terminal DIX domain used for β-catenin activation; a central PDZ motif for βCtn and JNK activation; and a C-terminal DEP sequence for JNK activation. Disheveled has multiple partners (~30 interacting proteins), including numerous signaling effectors that are not components of Wnt signaling.

Some free G-protein subunits, such as $G\alpha_o$, $G\alpha_q$, $G\alpha_{i2}$, as well as $G\beta\gamma$, act cooperatively to inhibit β-catenin degradation [1374]. Subunit $G\beta 1\gamma 2$ promotes the phosphorylation and activation of Wnt coreceptor LRP6, as it recruits glycogen synthase kinase-3 to the plasma membrane and enhances its kinase activity. Furthermore, these components of the canonical Wnt pathway form the $G\beta 1\gamma 2$–LRP6–GSK3–axin–Dvl complex.

Coreceptor LRP6 that is required for signal transduction in the canonical pathway is sufficient to activate Wnt signaling when it is overexpressed. Heterotrimeric

45. Palmitate attachment (palmitoylation) is required for Wnt signaling. Glycosylation might increase Wnt interactions with heparin sulfate proteoglycans on the surface of target cells [1373].

46. Frizzled receptors share features with prototypical GPCRs, such as glycosylation and phosphorylation by protein kinase-A and -C as well as casein kinase-2. However, they lack other hallmarks of GPCRs.

47. The low-density lipoprotein receptor (LDLR) family includes the low-density lipoprotein receptor-related proteins (LRP), apoliprotein-E receptor-2, and very-low-density lipoprotein receptors.

48. When a Wnt ligand simultaneously binds to its receptors Frizzled and coreceptors low-density receptor-related protein, Fz-associated, intracellular components phosphorylate (activate) LRPs to trigger the canonical Wnt–β-catenin signaling.

Table 10.8. Main mediators of the Wnt pathway (CELSR: cadherin, EGF LAG seven-pass G-type receptor; DACT: dapper, antagonist of β-catenin, homolog; Dkk: Dickkopf homolog; Dvl: Disheveled homolog; Krm: Kremen (Kringle-containing transmembrane protein); LEF: lymphoid enhancer-binding factor; Nkd: naked cuticle homolog; PCP: planar cell polarity; RSpo: R-spondin homolog; TCF: T-cell factor; VangL: VanGogh-like protein; WNRRTK: Wnt and neurotrophin receptor-related receptor Tyr kinase [$ROR_{(RTK)}$]). Disheveled can activate small GTPases CDC42, Rac, and Rho and JNK kinase, among others to regulate the cell fate. In vertebrates, the planar cell polarity requires the activation of different Wnt ligands (Wnt4, Wnt5a, Wnt7a, and Wnt11).

Type and subtypes	Role, effects
Wnt	Messenger
Wnt1/2/2b/3/3a/4/5a/5b/6/7a/7b/8a/8b/9a/9b/10a/10b/11/16	
Axin-1–axin-2	Inhibitors
Ca^{++}	Signaling (Wnt–Ca^{++} axis)
β-catenin	Gene transcription (canonical pathway)
CELSR1–CELSR3	Planar cell polarity
CK1	Phosphorylation of LRP5 and LRP6
DACT1–DACT3	Lysosomal degradation of Dvl
Dkk1–4	Endocytosis of LRP5 and LRP6
Diego (Dgo)	Establishment and maintenance of asymmetrical Fz localization (Drosophila), compartmentation of Dvl for PCP
Dvl1–Dvl3	Canonical and non-canonical axes (adaptors)
Frizzled	GPCR (canonical and non-canonical axes)
GPR177	Exocytosis of Wnt proteins
GSK3	Phosphorylation of LRP5 and LRP6
Inversin	Tubule formation
Klotho	Antagonist of secreted Wnt
Krm1–Krm2	Dickkopf receptors
LRP5–LRP6	Phosphorylated LRP5 and LRP6 interact with axin
Nkd1–Nkd2	Antagonists
PKCδ	Plasmalemmal accumulation of Dvl
Prickle-1–Prickle-4	Planar cell polarity
RSpo1–RSpo4	Inhibition of Krm–Dkk-dependent LRP6 endocytosis
TCF, LEF	β-catenin interactor; gene transcription
VangL1–VangL2	Planar cell polarity
WNRRTK1–WNRRTK2	Receptor kinases

G protein subunits, such as $G\alpha_i$ and $G\alpha_o$ (Vol. 4 – Chap. 8. Guanosine Triphosphatases and Their Regulators) can be required in the Wnt–βCtn axis (Table 10.10).

Frizzled receptors can respond to Wnt proteins in the absence of LRP5 or LRP6 to activate non-canonical pathways. β-Catenin-independent signaling can also depend on G proteins. Activation by Wnt5a of Ca^{++}–calmodulin-dependent protein

Table 10.9. Canonical (β-catenin-dependent) and non-canonical (β-catenin-independent) Wnt signaling. Disheveled (Dvl) mediates both the canonical and non-canonical Wnt pathways. In the canonical pathway, binding of a Wnt ligand to Frizzled receptor (Fz) and its coreceptor lipoprotein receptor-related protein LRP5 or LRP6 causes phosphorylation of these coreceptors that then recruit axin-1, hence diverting axin-1 from β-catenin destruction. β-Catenins can then accumulate and move to the nucleus, where they connect to members of the T-cell factor (TCF) and lymphoid enhancer-binding factor LEF1 family of proteins. They thus relieve the inhibition by transcriptional repressors such as transducin-like enhancer proteins (TLE). This cascade of events promotes transcriptional activation of target genes, such as Myc and axin-2. Axin-2, in turn, inhibits β-catenins via the assembly of destruction complexes (negative feedback loop). A crosstalk exists between non- and canonical Wnt pathways. Non-canonical Wnt signaling can be initiated by Wnt binding to other coreceptors, such as class-12 receptor Tyr kinases WNRRTK1 and WNRRTK2 (ROR1 and ROR2) and receptor-like Tyr (Y) kinase (RYK).

Canonical pathway	Wnt–Fz–LRP5/6–Dvl–βCtn–TCF/LEF1
Planar cell polarity	Wnt–Fz–Dvl–VangL–Prickle-1–CELSR Wnt–Fat-1–Dachsous-1
Non-canonical pathway	Wnt–PLC–Ca^{++}–CamK2, Wnt–PLC–Ca^{++}–PKC, Wnt–Rho–RoCK, Wnt–Rac1–JNK, Wnt–CK1–Rap1, Wnt–RYK/WNRRTK–Rac–JNK, Wnt–RYK/WNRRTK–CDC42

kinase CamK2, protein kinase-C, and nuclear factor of activated T cells actually depends on G proteins [1375].

Signaling by Wnt ligands is characterized by several main distinct intracellular signaling cascades: (1) β-catenin-dependent canonical pathway that leads to gene expression; (2) β-catenin-independent non-canonical pathway that is mainly related to the cell cytoskeleton; (3) a special non-canonical pathway, the "planar cell polarity" pathway, with its cognate mediators that control various polarized cell behaviors;[49] and (4) Wnt calcium and protein kinase-C-dependent pathway.

10.3.2.1 Non-Canonical Wnt Pathways

Pathways initiated by different Wnt ligands depend on the types and densities of receptors and coreceptors on a given target cell as well as on the ligand type. The non-canonical Wnt signaling intervenes in actin cytoskeletal organization, cell adhesion, migration, lymphopoiesis, and inflammation.

Non-canonical pathways function with effectors Disheveled, Rho and Rac GTPases (Vol. 4 – Chap. 8. Guanosine Triphosphatases and Their Regulators), and Jun N-terminal kinases (Vol. 4 – Chap. 5. Mitogen-Activated Protein Kinase Modules).

49. β-Catenin-independent signaling process controls planar cell patterning of cochlear hair cells, convergent extension movements, tissue boundary formation, and dorsoventral patterning during gastrulation in vertebrates [1375].

Table 10.10. Frizzled receptors and corresponding pathways (Source: [1375]). Disheveled and/or heterotrimeric G proteins that are composed of α, β, and γ subunits can prime several signaling axes. Two proteic complexes formed by Disheveled and either Disheveled-associated activator of morphogenesis DAAM1 and small GTPase Rho or Rac acts via protein kinase RoCK1 and Jun N-terminal kinase, respectively, for actin remodeling. Another axis is constituted by phospholipase-C that provokes Ca^{++} influx and then excites Ca^{++}–calmodulin-dependent protein kinase CamK2, protein kinase-C (PKC), and nuclear factor of activated T cells (NFAT). Receptor Tyr kinase-like orphan receptor WNRRTK2 stimulates a non-canonical pathway based on Jun N-terminal kinase (JNK) stimulation of activating transcription factor ATF2 and Jun-dependent transcriptional regulation of the paraxial protocadherin gene. Receptor Tyr kinase RYK controls axon guidance via cytosolic kinase Src. Morphogen Wnt can also promote the nuclear localization of the intracellular domain of RYK after its cleavage to control neuronal differentiation. In addition RYK that can form a complex with Frizzled and Disheveled could also intervene in β-catenin-dependent signaling.

Mediators	Effect
Disheveled-associated Wnt signaling	
GTPases Rho and Rac	Actin cytoskeleton regulation
Phospholipase-C	Ca^{++} influx, Activation of CamK2, PKC, and NFAT
Receptor Tyr kinase-associated Wnt signaling	
WNRRTK2	Inhibition of Wnt–β-catenin-dependent signaling Activation of JNK and ATF2
RYK	Neuron differentiation, axon guidance

In particular, non-canonical Wnt pathways target actin via small RhoA GTPase. According to the mediator type, non-canonical Wnt cascades include the Wnt–Rho–RoCK, Wnt–Rac1–JNK, Wnt–Ca^{++}–PKC, Wnt–Ca^{++}–CamK2, Wnt–WNRRTK2–CDC42, and Wnt–CK1–Rap1 pathways.[50] G Proteins may be involved in the transduction of some Wnt signals.

The β-catenin-independent Wnt signaling antagonizes the Wnt–β-catenin pathway in tissue regeneration, limb development, and survival of progenitor thymocytes [1375]. Furthermore, Wnt signaling interacts with other pathways, such as the Notch and fibroblast growth factor axes.

Both Wnt3a or Wnt5a activate the common Fz–Dvl pathway that is linked to their specific coreceptors. Whereas Wnt3a triggers the canonical axis, Wnt5a utilizes the non-canonical signaling cascade, i.e., Wnt3a causes the phosphorylation of LRP5 and LRP6, and Wnt5a that of WNRRTK1 and WNRRTK2 [1376].[51] Binding

50. Alias ROR also designates RAR-related orphan nuclear receptor (NR1f1–NR1f3), hence alias ROR(RTK) used in the present text.

51. Coreceptor WNRRTK2 can bind various Wnt ligands, such as canonical Wnt1, Wnt3a, and Wnt8 when Wnt ligands and WNRRTK2 are overexpressed. In vitro, WNRRTK2 specifically interacts only with non-canonical Wnt5a.

to Frizzled and recruitment of shared intracellular effectors Dvl, axin, and GSK3 are required for both activation of LRPs and WNRRTKs by Wnt3a and Wnt5a, respectively. Glycogen synthase kinase-3 phosphorylates LRP6 in canonical and WNRRTK2 (Ser864) in non-canonical Wnt signaling. Furthermore, a mutual antagonism exists between Wnt3a and Wnt5a, as Wnt3a prevents Wnt5a-induced phosphorylation of WNRRTK2, and Wnt5a precludes Wnt3a-induced phosphorylation of LRP6.

10.3.3 Canonical Wnt Pathways

Activated β-catenins in the canonical Wnt signaling transmits signals from the extracellular environment and the Wnt receptor to the nucleus. β-Catenin resides in the cytoplasm, in which it is destroyed or activated. In the latter case, it can relay the Wnt signal into the nucleus.

In the absence of Wnt stimulation, cytosolic β-catenins associate with the *destruction complex* composed of various β-catenin-binding proteins: scaffold proteins axin-1 and axin-2, adenomatous polyposis coli protein (APC), casein kinase CK1α, and glycogen-synthase kinase GSK3β. This destruction complex phosphorylates β-catenin for ubiquitin-mediated degradation. β-Catenin is sequentially phosphorylated by casein kinase CK1α (Ser45) and glycogen synthase kinase GSK3β (Ser33, Ser37, and Thr41; negative role in Wnt signaling).

Phosphorylated β-catenin is then polyubiquitinated by ubiquitin ligase Skp1–cullin–F-box β-transducin repeat-containing protein $SCF^{\beta TRCP}$, which operates in phosphorylation-dependent ubiquitination, and destroyed by the proteasome.

Consequently, cytosolic β-catenin level remains low and Wnt-responsive genes are repressed. On the other hand, both polyadpribosylating enzymes tankyrase-1 and -2 promote axin ubiquitination for proteasomal degradation [1377].

When a cell receives a Wnt signal, stimulated Fz and LRP5/6 coreceptors form a Fz–LRP5/6 complex. This Wnt signal is relayed from the ternary Wnt–Fz–LRP complex in the cytoplasm by scaffold protein Disheveled. The phosphoprotein Disheveled is required for the Frizzled-LRP6 aggregation and LRP6 phosphorylation.

Signaling through the canonical Wnt–β-catenin pathway relies on the inhibition of β-catenin phosphorylation by glycogen synthase kinase GSK3 that enables β-catenin to escape proteasomal destruction and activates the transcription of target genes. Enzyme GSK3 that phosphorylates (activates) LRP6 (positive role in Wnt signaling) is recruited to the receptor complex (*LRP6 signalosome*), as well as casein kinase CK1γ that also phosphorylates LRP5 and/or LRP6. Sequential phosphorylation of LRP5 and/or LRP6 by GSK3β and CK1γ within the LRP6 signalosome[52] allows axin recruitment to the receptor complex [1378–1380]. Due to the recruitment of components of the destruction complex by Disheveled, β-catenin avoids axin-mediated degradation (signaling initiation). Axin binds to LRP6 further impeding β-catenin degradation by sequestering axin in regions of high LRP concentration

52. Phosphorylated LRP6, Dvl, Fz, axin, GSK3β and CK1γ colocalize to LRP6 signalosomes. Sequestration of axin by LRP5 or LRP6 inhibits the formation of the β-catenin degradation complex and leads to β-catenin stabilization.

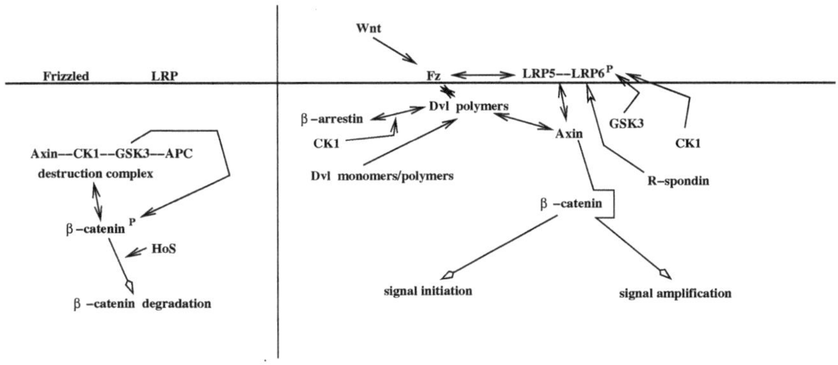

Figure 10.3. The Wnt–β-catenin pathway (Wnt signaling leads to β-catenin-independent and -dependent responses; Source: [1381]). (**Left**) In the absence of Wnt, β-catenin is phosphorylated for degradation by a destruction complex made of axin, casein kinase CK1γ, glycogen synthase kinase GSK3β, adenomatous polyposis coli protein (APC), and ubiquitin ligase Homolog of Slimb (HoS) for proteolytic cleavage. (**Right**) Protein Wnt binds to and stimulates its Frizzled (Fz) receptors that recruit and bind Disheveled (Dvl; Disheveled is the branching molecule that couples Frizzled receptors to β-catenin-dependent or -independent pathways). Disheveled transduces signals from Frizzled, leading to the cytosolic accumulation and stabilization of β-catenins, and subsequent activation of transcription factors. Disheveled forms polymers to transduce Wnt signals. The sequence of events initiated by Wnt proteins involves the formation of Frizzled–LRP5/6 signalosomes in about 15 mn that recruit Dvl and scaffold axin to activate cytoplasmic mediators. Disheveled polymers bound to Frizzled interact with and inactivate axin, hence preventing the destruction complex activity and initiating Wnt signaling. β-Arrestin binds Dvl and axin to form a trimer for Dvl activation. Coreceptors low-density lipoprotein receptor-related proteins LRP5–LRP6 associated with Frizzled receptors impedes β-catenin degradation by inactivating the destruction complex. Kinases GSK3β and CK1γ phosphorylate LRP6, which subsequently docks axin, thereby inhibiting the destruction complex. Once the degradation complex is inactivated, β-catenins accumulate in the cytosol and translocate into the nucleus to regulate the cell fate. R-spondins inhibit Kremen–DKK1-dependent internalization of LRP6 in the presence of Wnt proteins.

(amplification). Unphosphorylated β-catenin translocates to the nucleus where it activates transcription.

Two mechanisms can eliminate axin-based degradation of β-catenin. Monomeric and polymeric Disheveled can be detected in the cytosol. Stimulated plasmalemmal Frizzled receptors recruit the cytoplasmic adaptor Disheveled. Disheveled transduces both canonical and non-canonical Wnt signals and forms polymers required for signal transduction [1381]. Disheveled polymers serve as scaffolds for axin recruitment and inactivation (Fig. 10.3). Frizzled receptors can also function with LRP5–LRP6 coreceptors. Phosphorylated LRP5–LRP6 by CK1 and GSK3 can interact with axin, thus potentiating axin inactivation at the plasma membrane.

In addition, coreceptor LRP6 directly inhibits phosphorylation of β-catenin by GSK3, thereby preventing β-catenin degradation without requiring inhibition of axin-associated destruction complex [1382].[53] Both LRP-mediated inhibition of β-catenin phosphorylation (for degradation) and stimulation of axin degradation contribute to Wnt–β-catenin signaling.

10.3.3.1 Wnt Signaling Specificity

Specific responses to extracellular messengers of the Wnt family bound to their cognate receptors is partly yielded by interaction with distinct receptors that can act in combinatorial manner. Activated Wnt receptors of the Frizzled family can trigger distinct cellular responses.

Frizzled regulates cell fate decisions via the canonical Wnt–Fz–β-catenin axis and non-canonical pathway. On apical plasma membrane of epithelial cells, Fz receptors foster non-canonical signaling; on basolateral membrane, canonical output. In addition, local heterogeneities in plasmalemmal nanodomains due to different intracellular pH and/or lipid composition modulate signaling from Fz activation.

Disheveled from which canonical and non-canonical pathways diverge contributes to response specificity. The choice between canonical and non-canonical signaling involves a pH-dependent interaction between Disheveled and negatively charged plasmalemmal phospholipids [1385]. Response specificity arises from Na^+–H^+ exchanger NHE3 that is needed for Fz-dependent Dvl recruitment at the plasma membrane.

Agonist Wnt5a competes with Wnt3a for binding to Fz2 receptor. Ligand Wnt3a excites the β-catenin-dependent axis via the Fz2–LRP6 complex. Ligand Wnt5a prevents Wnt3a-dependent LRP6 phosphorylation. It activates the β-catenin-independent axis and inhibits the Fz2-triggered canonical pathway. It uses receptor Tyr kinase-like orphan receptors WNRRTK1 and WNRRTK2 to stimulate Rac GTPase [1386].

In addition, Fz2 endocytosis in response to Wnt3a uses mainly caveolin rather than clathrin. On the other hand, Wnt5a-dependent activation of small GTPase Rac relies on clathrin-mediated endocytosis of Fz2 receptor. Abundance of a given Wnt type influences Fz signaling and receptor endocytosis.

Agonist Wnt5a also triggers endocytosis of Fz4 and Fz5 receptors via recruitment of clathrin adaptor proteic complex AP2, Dvl2, and β-arrestin-2 to the plasma membrane [1386].

53. Like β-catenin, the stability of numerous proteins is regulated by proteasomal degradation following GSK3-dependent phosphorylation. Among these substrates, transcription factors Snai and SMAD1 are controlled by the Wnt pathway. In the presence of Wnt ligands, mitogen-activated protein kinase phosphorylates (activates) SMAD1, when bone morphogenetic proteins are active, thereby enhancing BMP signaling. Myosin regulatory light chain and transmembrane protein TMEM4 that can interact are regulated by axin-associated destruction complex [1383]. RNA-binding proteins Fused in sarcoma (FuS), FuS-interacting protein FuSIP1, and KH domain-containing, RNA-binding, signal transduction-associated protein KHDRBS1 (a.k.a. 68-kDa Src-associated in mitosis protein [SAM68]) that regulate mRNA splicing are also degraded by axin-associated destruction complex.

Wnt signaling also relies on endocytosis of the Wnt–Fz–LRP complex that clusters polymerized Disheveled to form endocytic signalosomes.[54] Internalization of the Wnt–Fz–LRP complex begins with the recruitment of prorenin receptor adaptor and vacuolar ATPase.[55] Receptor clustering triggers recruitment of CK1γ and GSK3 kinases and the destruction complex (Sect. 10.3.3. The subsequent formation of multivesicular endosomes sequesters GSK3, Disheveled, axin, and β-catenin.[56] Consequently, numerous cytoplasmic proteins escape from GSK3-labeled degradation, especially β-catenins, either newly synthesized or released from plasmalemmal stores, that can run to the nucleus [1388]. Hence, the sustained inhibition of GSK3 results from its sequestration with Rab7 GTPase, vacuolar protein sorting VPS4 AAA ATPase, and hepatocyte receptor Tyr kinase substrate (a components of the endosomal-sorting complexes required for transport [ESCRT]) in multivesicular endosomes caused by the canonical Wnt signaling [1388]. The fast inhibition of GSK3 during Wnt signaling results from the binding to $LRP6^P$, and its sustained inhibition from sequestration in multivesicular endosomes.

10.3.3.2 Wnt Signaling-Associated Molecules

Wnt Synthesis, Exocytosis, and Secretion

The release of Wnt ligands is regulated by post-translational modifications, interaction partners, and subcellular locations. The Wnt morphogens are synthesized in the endoplasmic reticulum. They then undergo various modifications (e.g., glycosylation and palmitoylation) for transport and functioning. Porcupine, an endoplasmic reticulum-resident acyltransferase, may be responsible for lipid modifications of Wnts that lead to Wnt hydrophobicity and membrane association [1389]. Hydrophobic morphogens form multimers for long-range signaling, whereas acylated Wnts can distribute in the vicinity of secreting cells to induce short-range cue transduction. Therefore, separate secretion pathways may regulate morphogen release for short- and long-range signaling.

Transcriptional coactivators β-catenins recruit many cofactors. The Wnt ligands can signal via catenin-δ1 to relieve the Kaiso-mediated repression.

---

54. At least in some cell types, multivesicular endosomes can generate specialized, secretory granules, such as Weibel-Palade bodies in endothelial cells, α-granules in platelets, and mastocyte granules [1387]. Multivesicular endosomes can give rise to exosomes (secreted vesicles), or cargosomes, that travel from one cell to another.

55. Vacuolar ATPase enables vesicle acidification that is indispensable for endocytosis and vesicular transfer. Endocytosis is also required for non-canonical Wnt signaling.

56. β-Catenin is required for the formation of multivesicular endosomes (non-transcriptional role).

Many factors are involved in Wnt protein secretion: lipoproteins, the retromer complex,[57] and G-protein-coupled receptor GPR177, or Wntless (Wls) [1390].[58] The GPR177 receptor specifically produced in Wnt-secreting cells promotes the secretion of Wnt ligands.

Lipoproteins Argosomes are involved in the transport of lipid-modified proteins, such as Wnts and Hedgehogs [1373]. Lipoproteins that link to Wnts also enables the morphogen displacement over long distances before their transmission to receiving cells via heparan sulfate proteoglycans and low-density lipoprotein receptors.

Disheveled

Disheveled not only mediates Wnt signaling from the plasma membrane to the intracellular medium, but also operates in the nucleus. It can either stabilize the βCtn–TCF transcriptional complex with the formation of a quaternary Dvl–Jun–βCtn–TCF complex or act as a repressor [1391].

Among its numerous partners, Dvl binds to P65 of the class-2 NFκB proteins (or RelA) in the nucleus and prevents P65-mediated and TNFα-stimulated activation of the NFκB-dependent reporter. This effect of Dvl depends neither on Wnt nor β-catenin [1391]. Therefore, Dvl can sensitize cells to TNFα-mediated apoptosis.

Biglycan

Among extracellular controllers of Wnt signaling,[59] biglycan, a proteoglycan, enhances the canonical Wnt pathway [1392]. Biglycan interacts with Wnt3a ligand and LRP6 coreceptor and possibly forms a trimer. Biglycan may serve as a Wnt reservoir in the extracellular medium that modulates Wnt availability. Several other proteoglycans and/or their glycosaminoglycan chain components can also stimulate the Wnt–βCtn pathway, such as free heparan sulfate, heparin chains, and heparan sulfate proteoglycans (glypican-3 and syndecan-1).

Chibby

Nuclear import and export of β-catenin is regulated by a complex between β-catenin and 14-3-3 proteins and Chibby. Chibby that can localize to both the nucleus

57. The retromer complex can act within Wnt-producing cells to generate Wnt molecules suitable for exocytosis. The retromer complex especially operates in basal-to-apical transcytosis. It is composed of members of the vacuolar protein sorting family.

58. Protein Wntless works for Wnt ligands like multipass transmembrane protein Dispatched in the Hedgehog pathway for Hh release from the producing cell. It is also called Evenness interrupted (EvI) or sprinter (Srt). It is located in the Golgi body and/or plasma membrane. It controls the transport and secretion of Wnt proteins in secretory vesicles.

59. The extracellular control of Wnt signaling depends on the release and extracellular storage of Wnt proteins as well as the secretion of Wnt antagonists, such as secreted frizzled-related proteins, Wnt inhibitory factor-1, sclerostin, and Dickkopf proteins.

and cytoplasm interacts with β-catenin. It sequesters β-catenin in the cytoplasm, thereby hindering its binding to T-cell factor–lymphoid enhancer factor [1393].

Chibby-binding partners include 14-3-3ε and 14-3-3ζ. Protein kinase-B phosphorylates the 14-3-3-binding motif of Chibby to allow 14-3-3–Chibby interaction that sequesters Chibby into the cytoplasm. Chibby actually can form a ternary complex with 14-3-3 and β-catenin. Therefore, membrane-tethered protein kinase-B stimulates β-catenin signaling, whereas nuclear PKB impedes binding of β-catenin to transcription factors TCF–LEF and causes nuclear export of β-catenin bound to the phosphorylated 14-3-3–Chibby–β-catenin ternary complex.

MicroRNAs

MicroRNA-8 inhibits Wnt signaling from ligand secretion to transcription of target genes, as it downregulates transcription factor TCF among others [1394]. This inhibitor promotes adipogenesis in bone marrow from stromal cells. On the other hand, miR315 activates the Wnt pathway, as it targets axin inhibitor.

Phosphoinositides

Glycoprotein Wnt3a binds to Frizzled recruits Disheveled scaffold. The latter then binds to and activates membrane-associated phosphatidylinositol 4-phosphate 5-kinase PIP(5)K1 that produces phosphatidylinositol (4,5)-bisphosphate [1395]. This product (PI(4,5)P_2) participates in aggregation of LRP6 and subsequently activated PIP5K1 phosphorylates LRP6. Phosphoinositide PI(4,5)P_2 also contributes to axin translocation.

Ubiquitin

Several ubiquitin ligases control mammalian Disheveled localization and turnover [1396]. Ubiquitin ligase HECW1 [60] ubiquitinates Disheveled-1. The KlhL12–Cul3 ubiquitin ligase [61] targets Dvl3 for degradation. Prickle homolog-1 binds to Dvl3 to facilitate Dvl3 ubiquitination and breakdown. Dapper, antagonist of β-catenin, homolog-1 (DACt1) targets Dvl for lysosomal degradation. Inversin (Invs: Inversion of embryo turning homolog) that switches between canonical and non-canonical Wnt signaling connects to cytoplasmic Dvl for ubiquitin-mediated destruction via the anaphase-promoting complex–cyclosome complex [1396].

The complex formed by Naked cuticle homolog (Nkd) and protein phosphatase-2 regulatory subunit-Bα (PPP2R3α) [62] regulates the Dvl concentation via ubiquitin ligases. Naked cuticle homolog is a Wnt-induced antagonist of the Wnt signaling cascade that restricts the classical Wnt axis, but activates another Wnt pathway

60. HECT, C2, and WW domain-containing ubiquitin–protein ligase HECW1 is also called NEDD4-like ubiquitin–protein ligase NEDL1.

61. Kelch-like protein KlhL12 is also named cullin-3 (Cul3)-interacting protein C3IP1.

62. A.k.a. PR72 and PR130.

to control planar cell polarity during gastrulation movements in vertebrates. Subunit PPP2R3α interacts with Naked cuticle homolog to regulate cell morphogenetic movements during body axis formation [1397].

The density of plasmalemmal Frizzled is controlled by a cycle of monoubiquitination and deubiquitination [1398]. The latter results from the action of deubiquitinating enzyme ubiquitin-specific peptidase USP8. [63] The amount of Frizzled on the plasma membrane is a major rate-limiting factor in Wnt responsiveness. Ubiquitinated Fz4 is rapidly internalized and transferred to lysosome for degradation via Rab7+ late endosome and multivesicular body. However, its deubiquitination allows Fz recycling back to the plasma membrane via Rab11+ endosomes.

Secreted (Soluble) Frizzled-Related Proteic Inhibitors

The Wnt inhibitory factor (WIF) and secreted Frizzled-related protein sFRP1 [64] are soluble molecules that bind selectively to and sequester Wnt ligands. They thus prevent Wnt signaling by precluding Wnt interaction with Frizzled receptor and LRP5 and LRP6 coreceptors.

Secreted (soluble) Frizzled-related proteins constitute the largest family of Wnt inhibitors, with 5 known distinct members in humans (sFRP1–sFRP5). They form 2 groups of closely related proteins: group 1 with sFRP1, sFRP2, and sFRP5, and group-2 with sFRP3 and sFRP4 proteins [1399]. [65] Secreted Frizzled-related and Frizzled proteins are able to homo- and heteromerize.

Soluble Fz-related proteins can operate as secreted apoptosis-related proteins. They are not only Wnt–Fz-binding proteins, but also behave as reciprocal antagonists [1400]. [66]

Soluble Fz-related proteins can interact with other molecules, especially matrix constituents. In particular, sFRPs bind to proteins distinct from Wnt ligands. Therefore, secreted Fz-related proteins have activities that are unrelated to Wnt signaling. Isoform sFRP1 operates in axon guidance by interacting with Frizzled-2 in a Wnt-independent manner [1401].

Proteins sFRPs can interfere with the fibronectin–$\alpha_5\beta_1$-integrin complex independently of Wnt activity [1399]. They intervene as peptidase inhibitors in bone morphogenetic protein signaling [1402]. They bind to TNFSF11 [1399]. [67] Conversely,

63. A.k.a. ubiquitin isopeptidase UbPy.

64. A.k.a. FrzA and secreted apoptosis-related protein SARP2.

65. Group-1 sFRPs are encoded by 3 exons on chromosome 8p12-p11.1, 4q31.3 and 10q24.1, respectively. Group-2 sFRPs are encoded by 6 exons on chromosome 2q31-q33 and 7p14-p13, respectively.

66. Isotypes sFRP1 and sFRP2 compete locally to regulate Wnt signaling during renal organogenesis. Isoform sFRP1 binds Wnt4 with strong affinity and inhibits DNA-binding activity of Wnt effector TCF, whereas sFRP2 does not intervene in TCF activation. Antagonistic effect of sFRPs prevents inappropriate development and maintains a population of cortical blastemal cells to facilitate further renal expansion. On the other hand, they can also cooperate, as sFRP2 can promote tubule formation owing to Wnt4 signaling in the presence of sFRP1.

67. A.k.a. receptor activator for nuclear factor-κB (RANKL).

components of the morphogen Sonic hedgehog pathway can regulate transcription of the Sfrp1 and Sfrp2 genes [1350].

Soluble Fz-related proteins can undergo post-translational modifications, such as Tyr sulfation and glycosylation. Glycosaminoglycans modulate activity of sFRP–Wnt complexes, as they bind and regulate Wnt diffusion [1403] and facilitate sFRP secretion and accumulation [1404]. Heparin stabilizes unsulfated sFRP1, thereby causing sFRP1 accumulation.

Heparan sulfate proteoglycans affect the transport of extracellular Wnt proteins. Plasmalemmal syndecan-4 that binds to fibronectin regulates Wnt signaling. It activates protein kinase-Cα and RhoA GTPase. With integrins, it stabilizes focal adhesions.

Antagonist sFRP1 hinders proliferation of vascular endothelial and smooth muscle cells, hence hampering angiogenesis in hindlimb ischemia [1405]. It decreases concentrations of cyclins and cyclin-dependent kinases, but increases levels of cytosolic β-catenin^P, a marker of β-catenin degradation.

Adipocytes and macrophages of adipose tissue secrete adipo-, chemo-, and cytokines that modulate inflammation and insulin sensitivity. Secreted Frizzled-related protein sFRP5 released by adipocytes can suppress activation of inflammatory cells such as macrophages in adipose tissue, as it inhibits Jun N-terminal kinase [1406]. In adipocytes, the Frizzled-stimulated non-canonical pathway triggered by Wnt5a stimulates RhoA and Rac1 and leads to activation of JNK1 enzyme. The latter impedes the activity of insulin receptor substrate IRS1 and decreases insulin signaling. Protein sFRP5 acts as an anti-inflammatory adipokine and decoy receptor that sequesters Wnt5a. Therefore, the balance between Wnt5a and sFRP5 controls the non-canonical Wnt signaling in macrophages and adipocytes of adipose tissue and modulates inflammation and the state of insulin sensitivity.

Dickkopf, Kremen Receptors, and R-Spondins

Concentration of LRP5 and LRP6 in the plasma membrane is regulated by Dickkopf (Dkk). Transmembrane Dkk1 receptors Kremen-1 and -2 preclude Wnt signaling. Inhibitor Dkk1 prevents Wnt signaling by binding to LRP5 and LRP6, thereby impeding their interaction with Wnt and Frizzled. They also prime prime LRP5/6 internalization via formation of complexes with Kremen receptors.

R-Spondins that constitute a set of 4 secreted proteins (RSpo1–RSpo4) activate the Wnt–βCtn pathway, as they modulate the plasmalemmal concentration of coreceptor LRP6 [1407]. Modulator RSpo1 activate neither Fz nor LRP6, but inhibits Kremen–Dkk-dependent internalization of LRP6. Activity of RSpo1 depends on the available pool of Wnt proteins.

R-Spondins bind to leucine-rich repeat-containing G-protein-coupled receptors LGR4 (or GPR48) and LGR5 (a.k.a. GPR49 and GPR67) [1408]. Receptor LGR5 is specifically expressed in stem cells of the intestinal crypt and hair follicle.[68] It enhances Wnt-induced LRP6 phosphorylation. The action potency of RSpo1 to RSpo4

68. Receptor LGR5 serves as a marker of crypt basal columnar stem cells in the gastrointestinal tract and bulge stem cells in hair follicles.

is determined by LGR4 and LGR5 concentrations; the maximum β-catenin activity is directed by Wnt3a and LRP6 concentrations. When LGR4 and LGR5 are stimulated by the R-spondins, they couple neither to heterotrimeric G proteins nor to β-arrestins. Receptors LGR4 and LGR5, bound by R-spondins or not, are internalized. The RSpo-LGR complex may assist in the endocytosis of the Fz–Wnt–LRP6 signalosome and improve the activity of cofactors of Wnt–βCtn–TCF axis [1408]. In addition, LGR5 accelerates the degradation of $LRP6^P$ (desensitization).

Small GTPase Rac1

Small GTPase Rac1 is a crucial component of the canonical Wnt signaling pathway. In response to Wnt3a, Rac1 activates Jun N-terminal kinase JNK2 that phosphorylates β-catenin and promotes its nuclear translocation [1409]. Small GTPase Rac1 interacts with β-catenin and Dickkopf-1 that antagonizes the canonical Wnt pathway to control limb outgrowth.

Silent Information Regulator-2 (Sirtuin)

Sirtuin-1 is a class-3 NAD^+-dependent histone deacetylase that deacetylates histone as well as non-histone proteins to regulate gene transcription and protein activity. It mediates the epigenetic silencing of Wnt antagonists. Sirtuin-1 also complexes with Disheveled [1410]. Sirtuin-1-mediated regulation of Disheveled may explain the diverse Sirt1-dependent responses in different cellular contexts.

Sirtuin-1 is a regulator of transient and constitutive Wnt signaling, especially during Wnt-stimulated cell migration. It ensures adequate concentration of all 3 Disheveled proteins [1410]. Yet, the magnitude of the effect on specific Disheveled types depends on the cell type.

β-Arrestins and Casein Kinases

β-Arrestins regulate Wnt–β-catenin signaling as well as non-canonical Disheveled-mediated Wnt pathways such as the Wnt–Rac1 cascade. Small RhoA and Rac1 GTPases, but not CDC42, are target mediators of β-arrestin and Disheveled.

Disheveled recruits many different intracellular proteins. In particular, Disheveled interacts with scaffold protein β-arrestin. The latter does not affect LRP6 phosphorylation, but is required for Dvl activation. Furthermore, Disheveled is phosphorylated by different kinases, such as casein kinases CK1δ and CK1ε.

β-Arrestin binds Dvl and axin to form a trimer [1411]. β-Arrestin associates with a Dvl region that contains phosphorylation site targeted by casein kinases. Inhibition of CK1 reduced the binding of β-arrestin to Dvl. On the other hand, β-arrestin binding to Dvl is potentiated by CK2 phosphorylation. Casein kinases thus regulate β-arrestin–Dvl interaction.

β-Arrestin and casein kinase-1 and -2 also contribute to distinct branches of non-canonical Wnt signaling to small GTPase Rac1 [1412]. The balance between β-arrestin and CK1 and CK2 determines the type of activated non-canonical Wnt pathways. β-Arrestin is necessary and sufficient for the Wnt–Rac1 pathway, whereas casein kinases prevent Rac1 activation and promote other non-canonical Wnt transductions via Disheveled phosphorylation. Activation of Rac1 requires CK1 and CK2 inhibition. Casein kinases CK1 and CK2 act as switches to select a branch of β-catenin-independent Wnt signaling.

Nucleoredoxin

Nucleoredoxin interacts with Disheveled, and inhibits the Wnt–β-catenin pathway [1413]. The redox-dependent regulation of Wnt–β-catenin signaling contributes to cell proliferation induced by oxidative stress.

Receptor Tyr Kinases

Receptor Tyr kinase RYK (related to receptor Tyr kinase) is required for Wnt3a-mediated canonical Wnt signaling. Single-pass receptor Tyr kinase WNRRTK2 (or ROR2) is involved in Wnt5a signaling that inhibits the activity of transcription factors of the T-cell factor and lymphoid enhancer factor family [1373]. The TCF–LEF family includes members LEF1 (or TCF1α) and TCF1 to TCF4.

Transmembrane receptor Tyr kinases WNRRTK2 and RYK, as well as Frizzled GPCRs that act independently of low-density lipoprotein receptor-related proteins LRP5 or LRP6 serve as receptors for Wnt and activate β-catenin-independent pathways [1375].

At least in mice and in vitro, Wnt5a interacts with WNRRTK2 to activate Jun N-terminal kinase. On the other hand, RYK interacts with Wnt1 and Wnt3a as well as Disheveled. Agonist Wnt3a promotes the nuclear translocation of the intracellular domain of RYK (RYK^{ICD}), especially during neurogenesis.

Non-canonical Wnt and planar cell polarity axes are distinct signaling pathways with overlapping. They control cell motility. Non-canonical Wnt signals are regulated by interaction of Wnt, Wnt antagonists, Frizzled, and WNRRTK2, and use mediators PKC, RoCK, MAP3K7, JNK, and NFAT [1221].

VanGogh-like, Prickle, and CELSR Proteins

Planar cell polarity axis is controlled by the interaction of the VangL–Prickle-1 complex,[69] CELSR,[70] and Fz–Dvl complex, and operates via PTK7, RoCK, and

69. The VangL alias stands for Vangogh-like (Strabismus-related) proteins. Proteins Vangl is a membrane molecule involved in planar cell polarity. Human ortholog Prickle-1 is another protein in the planar cell polarity signaling pathway. Prickle-1 is recruited to the plasma membrane by Vangl protein.

70. Cadherin, EGF LAG seven-pass G-type receptor paralogs (CELSR1–CELSR3) belong to the cadherin superfamily. They are encoded by the Celsr1–Celsr3 genes, respectively. They have a distinct distribution in some tissues, but overlapping expression in others. Isoform CELSR3 in renal mesenchymal cells is involved together with members of the Wnt cascade in the polarization of the forming nephron [1414].

JNK kinases. Receptor Tyr kinase PTK7, a pseudokinase that is regulated by Sema6 and plexin-A, operates via Vangl protein.

Primary cilium controls the balance between β-catenin-dependent and -independent Wnt signaling, as *Vangogh-like protein* VangL2, a β-catenin-independent signaling mediator as well as *inversin* [71] and *Bardet-Biedl syndrome proteins* that interact with effectors of the β-catenin-independent axis are located in primary cilia or basal bodies.

Prorenin and Renin Receptor

Prorenin and renin receptor (PRR; Vol. 3 – Chap. 6. Receptors) can serve as components of the Wnt receptor complex. (Pro)renin receptor actually binds to Wnt receptor proteins Fz8 and LRP6 as well as proton pump vacuolar H^+ adenosine triphosphatase (vATPase) that contributes to phosphorylation of LRP6 and Wnt by acidification of the vicinity of activated LRP6, similarly to its influence on endocytosis by vesicle acidification [723].

Adenomatosis Polyposis Coli Downregulated Inhibitor

The membrane-bound glycoprotein Adenomatosis polyposis coli downregulated protein APCDd1 abounds in human hair follicles. It prevents Wnt signaling, especially in human hair growth. It can interact with Wnt3a and LRP5 to preclude Wnt signaling upstream from β-catenin [1415]. Moreover, APCDd1 represses activation of Wnt reporters and target genes. [72]

Histone Methylase

Histone methylases contribute to the transcriptional regulation. In addition to $H3K_4$ and $H3K_9$ methylation, $H4K_{20}$ monomethylation ($H4K_{20}me^1$) at the T-cell factor (TCF)-binding element using the SET domain-containing histone methylase SET8 [73] is primed by Wnt3a [1416]. Enzyme SET8 connects directly to 2 β-catenin cofactors, LEF1 and TCF4 (or TCF7L2), under the control of Wnt3a. [74]

---

71. In kidneys, inversin (Invs) encoded by gene INVS may control the balance between canonical and non-canonical signaling during renal tubule morphogenesis. β-Catenin signaling regulates cellular proliferation, whereas non-canonical signaling modulates the correct orientation and elongation of cells during tubule formation.

72. Protein APCDd1 impedes Wnt effects during neuron differentiation from progenitors in the developing chicken nervous system as well as axis specification in Xenopus laevis embryos.

73. A.k.a. SET07, $H4K_{20}$HMTase (histone lysine methyltransferase), lysine (K) ^{N}methyltransferase KMT5A.

74. Lymphoid enhancer factor LEF1 and T-cell factors (TCF1, TCF3, and TCF4), members of the high-mobility group (HMG) family of transcription factors, bind to Wnt response elements to yield docking sites for β-catenin.

Transcription Factors of the SOX Family

In the spinal cord, proteins of the SOXB1 group (Sox1–Sox3) preserve cells in an undifferentiated state.[75] On the other hand, factors of the SOXB2 group promote the initiation of the differentiation. In spinal neural progenitors, transcription factor Sox5 of the SOXD group precludes the WNT–β-catenin activity on cell cycle progression via cyclin-D1 and -D2 as well as MyC, hence terminal differentiation, as it promotes cell cycle exit at the G1–S transition during spinal cord genesis [1417].

Transcription Factor-2D

Signaling by Wnts targets the TF2d complex that is devoted to promoter recognition and coactivation [1418]. The TF2d transcription factor is constituted by TATA box-binding protein (TBP) and TBP-associated factor (TAF). Activator-mediated recruitment of TF2d corresponds to an early stage for transcription initiation by RNA polymerase-2.

T-Cell Factor–Lymphoid Enhancer Factor

In the canonical Wnt pathway, transcription of Wnt target gene is regulated by nuclear β-catenin and various cofactors. Translocated β-catenins interact with DNA-binding proteins of the T-cell factor–lymphoid enhancer factor family to form transcription factors that can bind to the *Wnt response element*. In resting condition, a low concentration of β-catenin exists that is able to activate the transcription factors T-cell factor and lymphoid enchancer factor, which stimulate target genes.

Genes targeted by the Wnt signaling are inhibited by members of the Groucho family of transcriptional repressors that bind to TCF–LEF proteins. Many Wnt ligands promote the nuclear accumulation of β-catenins that then bind to TCF–LEF family members, thus relieving TCF–LEF repression and activating target genes. In the nucleus, histone acetyltransferase cAMP-responsive element-binding protein-binding protein (CBP) and P300 promotes Wnt signaling.

Lymphocyte enhancer factor LEF1[76] chaperones numerous controllers of gene transcription. In the canonical Wnt signaling cascade, complexes of LEF1 and transcriptional repressors such as Groucho are replaced by LEF1–β-catenin-containing complexes to activate transcription of Wnt-target genes [1419].

Upon Wnt stimulation, β-catenins replace T-cell factor-bound corepressors and (or) co-import additional T-cell factors. Once bound to Wnt response elements via T-cell factors, β-catenins act as scaffolds that coordinate recruitment of coactivators, RNA polymerase-2 regulators, and chromatin recognition cofactors and remodeling factors [1420].

---

75. The SOX genes are classified into 9 categories according to their structures: SOXA (sex-determining region Y (SRY)-box gene); SOXB1 (SOX1–SOX3); SOXB2 (SOX14 and SOX21); SOXC (SOX4, SOX11, and SOX12); SOXD (SOX5, SOX6, and SOX13); SOXE (SOX8–SOX10); SOXF (SOX7, SOX17, and SOX18); SOXG (SOX15); and SOXH (SOX30).

76. A.k.a. T-cell-specific transcription factor TCF1α.

Many chromatin-modifying factors are recruited by scaffold β-catenins: histone acetyltransferases,[77] histone methyltransferases,[78] ubiquitin ligase, SWI/SNF ATPases,[79] the scaffold Mediator complex,[80] and parafibromin.[81]

10.3.4 Wnt Signaling in Heart and Blood Vessels

10.3.4.1 Wnt in Cardiogenesis

Wnt Proteins as well as bone morphogenetic proteins and fibroblast growth factors are required for cardiomyocyte differentiation. Embryonic stem cells implicated in cardiogenesis, both in cell differentiation and spatial organization, are subjected to a biphasic pattern of Wnt signaling, being first activated then inhibited, with a possible cross-regulation between the canonical and non-canonical Wnt pathways [1421].[82]

The canonical Wnt–β-catenin signaling is required in early events that direct the fate of embryonic stem cells and mesoderm generation, driving the evolution of mesoderm cells of the cardiac compartment into the cardiac lineage. Conversely, during later stages of cardiogenesis, i.e., maturation of settled cardiac cells in suitable regions of the embryo, Wnt signaling is inhibited, thus limiting the size of the heart-forming field.

Cells of the first and the second heart fields contribute to the left ventricular myocardium and the outflow tract, right ventricular myocardium, endocardium, and smooth muscle of the great vessels, respectively. The pluripotent progenitors of the second heart field, which are initially located outside the heart, migrate into the heart and give birth to cardiomyocyte, endothelial, and smooth muscle lineages.

The transcription factor Islet1 intervenes in the fate and migration of second heart field progenitors. Factor Islet1 expression in cardiac progenitors requires β-catenin via the canonical Wnt signaling [1422].

---

77. Post-translational modifications such as acetylation of histone tails that protrude from nucleosomes to interact with adjacent histones or other nuclear proteins and form an inaccessible environment for transcription change interaction surface and then allow gene transcription. The cAMP-responsive element-binding protein-binding protein and P300 serve as histone acetyltransferases. Histone acetyltransferase TIP60 is involved in chromatin regulation with transcription factors such as E2F and P53.

78. Mono-, di-, or tri-methylation of histone tail residues activate or repress gene transcription.

79. SWI/SNF ATPases can shuffle histones and disassemble histone octamers. Nucleosome position can then be rearranged by SWI/SNF complexes to expose regulatory DNA sequences.

80. The Mediator complex is recruited by various transcription factors, such as P53 or nuclear receptors.

81. Parafibromin is a positive nuclear Wnt signaling component that interacts with RNA polymerase-2. It also serves as a scaffold for histone ubiquitination.

82. Wnt–β-Catenin signaling before gastrulation promotes cardiac differentiation, whereas signaling during gastrulation inhibits heart formation in both mouse and zebrafish embryos. Ligand Wnt3a stimulates mesoderm formation and activates a negative feedback that represses Wnt pathway and increases cardiogenesis. However, the kinetics of the expression and the concentrations of effectors of Wnt signaling depend on the animal species.

Signaling by Wnt promotes right ventricular and interventricular myocardial expansion [1423]. On the other hand, the canonical Wnt signaling hinders the first heart field specification, the cells of which are characterized by Hand1 and Tbx5 transcriptional regulators.

Insulin-like growth factor-binding protein IGFBP4 (Vol. 2 – Chap. 3. Growth Factors) operates as a cardiogenic growth factor. It interacts with Frizzled-8 and LRP6 to inhibit the canonical Wnt pathway. It actually prevents binding of Wnt3a ligand to the Fz8–LRP6 complex [1424]. Although independent of insulin-like growth factors, the cardiogenic effect of IGFBP4 is attenuated by IGFs via IGFBP4 sequestration.

10.3.4.2 Wnt and Cardiac Regeneration

Bone marrow multipotent mesenchymal stem cells can be used for heart regeneration, as they reduce the extent of necrotic myocardium and promote the regeneration of a contractile myocardium. The Wnt–β-catenin signaling is involved in the differentiation of mesenchymal cells, whereas Wnt inhibitors such as Dkk1 favor mesenchymal stem cell self-renewal. The other Wnt inhibitor secreted Frizzled-related protein sFRP2, by its intra-, auto-, and paracrine actions, enhances mesenchymal stem cell proliferation and engraftment, and improves organization and vascularization of granulation tissue as well as myocardial repair [1425].

In addition to Wnt signaling inhibition, sFRP2 enhances procollagen cleavage by Tolloid-like metallopeptidases, as sFRP2 binds both procollagen and Tolloid-like peptidase BMP1 [1431].[83] Overexpression of sFRP2 and BMP1 in infarcted heart can lead to excessive collagen fiber formation that causes cardiac fibrosis.

10.3.4.3 Wnt in Cardiac Electrophysiology

A non-canonical pathway launched by Wnt11 ensures the functional heterogeneity in electrochemical features that enables an efficient coupling between developing cardiomyocytes in the plane of the developing cardiac tissue of embryonic zebrafish hearts [1426].[84] The patterning of the electrochemical gradient requires Wnt11-mediated attenuation of the conductance of Ca_V1 channels.

---

83. Four mammalian Tolloid-like peptidases exist: BMP1, mTLD, mTLL1, and mTLL2. Sizzled of the sFRP family that is absent in mammals acts as an inhibitor of bone morphogenetic protein. Sizzled suppresses cleavage of BMP antagonist Chordin by Tolloid-like metallopeptidase, as it binds to Tolloid-like metallopeptidase, thereby blocking BMP signaling [1432]. Only BMP1 and mTLL1 cleave Chordin, but all Tolloid-like metallopeptidases act as procollagen-C peptidase. They convert procollagen into fibrillar components of the extracellular matrix. Peptidase BMP1 has the highest procollagen-C peptidase activity. Mammalian Tolloid-like peptidases also process other precursors into mature functional proteins involved in the formation of the extracellular matrix and activate TGFβ family members (GDF8, GDF11, and TGFβ).

84. Transmural and apex-to-base gradients appear during cardiogenesis in higher vertebrates. Electrochemical gradients intervene not only in the functioning of excitable cells, but

10.3.4.4 Wnt in Angiogenesis

Secreted Frizzled-related proteins sFRP1 and -4 overexpression in endothelial cells increases angiogenesis by regulating Wnt signaling [1427]. Protein sFRP1 interacts with Frizzled-4 and -7 on endothelial cells to stimulate GTPase Rac1 activity in cooperation with glycogen synthase kinase-3β.

In addition, bone marrow-derived mesenchymal stem cells (MSC) quickly organize into primitive tubes connected to host circulation and recruit endothelial (EC) and smooth muscle cells (SMC) to form new mature vessels [1428]. Protein sFRP1 upregulates platelet-derived growth factor-BB in mesenchymal stem cells and elicits β-catenin-dependent MSC–MSC, MSC–EC, and MSC–SMC adhesions. The pathway that promotes MSC-based angiogenesis uses GSK3β kinase. On the other hand, sFRP1 activation neither modifies VEGF, FGF2, and TGFβ expression in MSCs nor induces MSC transdifferentiation into endothelial or smooth muscle cells.

Angiogenesis in the Central Nervous System

Agonists Wnt7a and Wnt7b produced by the neuroepithelium of the developing central nervous system promote both the specific assembly and differentiation of cerebral vasculature that is characterized by its tightly sealed endothelium (blood–brain barrier) by targeting endothelial cells owing to the canonical Wnt pathway [1429].

The Wnt–β-catenin pathway is required for angiogenesis in the central nervous system and formation of the blood–brain barrier [1430]. Different Wnt ligands are indeed expressed by neural progenitor cells, such as Wnt7a and Wnt7b in ventral regions and Wnt1, Wnt3, Wnt3a, and Wnt4 in dorsal regions. However, Wnt–β-catenin signaling is not involved in angiogenesis in other body regions during development.

10.3.5 Wnt Signaling in the Nervous System

In midbrain, Wnts regulate dopaminergic neuron precursor differentiation as well as neuron morphogenesis, specification, proliferation, and survival [1433]. Agonists Wnt1 and Wnt5a that signal via the Wnt–β-catenin axis and planar cell polarity pathway, respectively, control midbrain dopaminergic neuron development.

---

also left–right patterning and organogenesis, as well as wound healing.

Zebrafish cardiomyocytes begin to depolarize spontaneously in the bilateral cardiac primordia before they assemble into the linear heart tube and about 24 h after fertilization synchronous contractions start, i.e., action potentials propagate slowly and uniformly throughout the linear heart tube [1426]. Over the next 24 h, the cardiac atrium and ventricle form and the heart loops. From 48 h after fertilization, the propagation speed of action potentials rises in both chambers, except in regions of slow conduction at sinoatrial, atrioventricular, and ventriculoaortic boundaries. The conduction is heterogeneous, as the myocardium at the outer curvature (the future ventricular apex) conducts action potentials 3 times faster than that at the inner curvature (future ventricular base) [1426].

Signaling primed by Wnt participates in synaptic differentiation, maintenance, and function. In the central nervous system, Wnt ligands modulate the synaptic vesicle cycle at presynaptic sites as well as the transfer of neurotransmitter receptors and their interaction with scaffold proteins in postsynaptic regions [1433].

In presynaptic sites, several Wnt ligands cause clustering of several presynaptic proteins and regulate transfer of certain receptors. For example, Wnt7a increases the clustering of synapsin-1 in granule cells and stimulates the recycling of presynaptic vesicles in neurons [1433].

In postsynaptic sites, non-canonical Wnt signaling triggered by Wnt5a provokes rapid changes in clustering of Disc large homolog DLg4 (or postsynaptic density protein PSD95) as well as glutamate receptors, such as NMDAR [85] and AMPAR [86] via Ca^{++}–calmodulin-dependent protein kinase CamK2 and Jun N-terminal kinase.

10.3.6 Wnt-Mediated Tissue Repair

Messenger Wnt7b is produced by macrophages that invade an injured tissue and re-establish tissue development to stimulate repair and regeneration [1434]. Macrophages also synthesize numerous other repair factors, such as growth factors FGF2, IGF, and HGF, as well as interleukin-10.

10.3.7 Wnt Signaling and Cell Fate

The Wnt–βCtn signaling regulates cell fate during embryogenesis as well as regeneration in adults. Nevertheless, inappropriate activation of the Wnt pathway can lead to cancer. Agonist Wnt controls both cell polarity and movement via Wnt-mediated receptor–actin–myosin polarity (WRAMP) complex that accumulates asymmetrically at the cell periphery in response to Wnt5a protein.

10.3.7.1 Wnt Signaling in Normoxic and Hypoxic Conditions

Hypoxia activates the heterodimeric transcription factor hypoxia-inducible factor-1 (Vol. 4 – Chap. 9. Other Major Signaling Mediators). Factor HIF1 is composed of 2 subunits: oxygen-sensitive HIF1α and constitutive HIF1β. It activates genes associated with glycolysis, angiogenesis, and pH regulation. It particularly targets glucose transporters, vascular endothelial growth factor, carbonic anhydrase-9, and cyclooxygenase-2. Hypoxia-regulated genes also include those that encode octamer-binding transcription factor Oct4, the growth factors erythropoietin and insulin-like growth factor IGF2, inhibitor of DNA binding ID2, and targets of Notch signaling such as Hairy enhancer of Split (HES) and HES-related transcriptional regulators.

In normoxia, β-catenin that acts as both a transducer of the Wnt pathway and a coactivator of T-cell factor and leukocyte enhancer factor activity, promotes cell proliferation (Fig. 10.4). In hypoxia, β-catenin potentiates HIF1 activity and favors

85. NMDA: ^{N}methyl ^{D}aspartic acid.
86. AMPA: α-amino 3-hydroxy 5-methyl 4-isoxazole propionic acid.

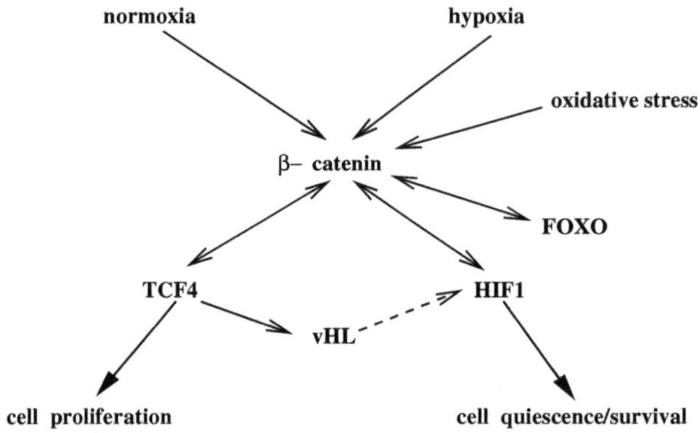

Figure 10.4. Shift in β-catenin signaling according to oxygen availability (Source: [1436]). During normoxia, β-catenin–T-cell factor-4 (TCF4) complex stimulates von Hippel-Lindau ligase (vHL) that degrades HIF1, thereby, avoiding cell cycle arrest. On the other hand, during hypoxia, β-catenin potentiates HIF1 activity leading to cell cycle arrest.

adaptation to hypoxia and cell survival [1435]. β-Catenin binds HIF1 factor. The complex then binds to DNA via HIF1.

Hypoxia inhibits the formation of complexes between β-catenin and T-cell factor-4, hence inducing cell-cycle arrest, because HIF1α competes with T-cell factor-4 for binding to β-catenin. In normoxia, the complex formed by β-catenin and T-cell factor-4 stimulates von Hippel-Lindau ligase for degradation of hydroxylated HIF1α [1436].

The von Hippel-Lindau protein not only promotes degradation of hypoxia-inducible factor-α, but also stabilizes Jade-1 ubiquitin ligase and component of histone acetylation complexes.[87] Enzyme Jade-1 hampers Wnt signaling by ubiquitinating both phosphorylated and non-phosphorylated β-catenin [1439].[88] However, Jade-1–β-catenin interaction is greater in the absence of phosphorylated β-catenin than in the presence of non-phosphorylated β-catenin, as Jade-1 and β-catenin colocalize in the cytosol in the absence of Wnt stimulation, whereas they have different subnuclear localizations in the presence of Wnt stimulation.

Although some stem cells lodge in perivascular regions, other stem cell populations reside in hypoxic niches during development as well as postnatal life [1437].[89] In addition to the Notch pathway, hypoxia-inducible transcription factor regulates the Wnt–βCtn pathway to maintain both adult and embryonic stem cells. Both hypoxia and Wnt agonists (canonical signaling with production of the effectors LEF1 and TCF1) support stem cell self-renewal. Both HIF-mediated and Wnt–βCtn sig-

87. A.k.a. PHD finger protein PHF17.

88. Cytosolic ubiquitin ligase SCF also binds β-catenin, but only ubiquitinates phosphorylated β-catenin.

89. Hematopoietic stem cells reside in 2 different niches in the bone marrow: predominant, hypoxic, endosteal and vascular niche (Vol. 5 – Chap. 2. Hematopoiesis). Lin−, Sca1+, SCFR+, CD148−, CD150+ stem cells move to a vascular niche when mobilization is needed.

nalings synergistically contribute to the maintenance of stem cell niches for both embryonic stem cells and adult neural stem cells [1438]. This regulation is not used in differentiated cells.

10.3.7.2 Cell Polarity, Orientation, and Migration

The Wnt pathways coordinate organization of organelles and cytoskeletal proteins to confer cell polarity and movement [1384]. Ligand Wnt5a signals via small guanosine triphosphatases Rab4 and RhoB [90] to redistribute molecules that control cell polarity, orientation, and migration. Messenger Wnt5a polarizes the location of cytoskeleton elements, such as actin, myosin-2B, and melanoma cell adhesion molecule (MCAM). Cell movement is associated with intracellular MCAM translocation using endosome transport. The Wnt-mediated receptor–actin–myosin polarity (W-RAMP) complexes (made of actin, myosin-2B, Frizzled-3, and MCAM) accumulate asymmetrically at the cell periphery, where they trigger membrane contractility and nuclear movement in the direction of membrane retraction.

10.3.7.3 Wnt Signaling during Cell Division

Signaling triggered by Wnt stimulates the expression of various genes, such as MyC and cyclin-D1. Therefore, the Wnt–βCtn pathway promotes G1–S progression. Components of the Wnt–βCtn axis, such as β-catenin, axin, glycogen synthase kinase GSK3β, and adenomatous polyposis coli localize to the mitotic spindle or centrosomes. In addition, LRP6 is phosphorylated by a mitotic cyclin-Y-dependent protein kinase CDK14 and β-catenin-dependent transcription rises during mitosis.

Faithful segregation of chromosomes in mitosis is ensured by the spindle assembly checkpoint. [91] The regulation of the dynamics of microtubule plus-ends during mitosis, i.e., the proper spindle orientation and stable microtubule–kinetochore attachment, relies on Disheveled-2 localization to the spindle and spindle poles during mitosis and its phosphorylation by Polo-like kinase PLK1 (Thr206) [1440]. Furthermore, Dvl2 is required for the activation of a spindle assembly checkpoint kinase TTK and the recruitment of the spindle assembly checkpoint components, budding uninhibited by benzimidazoles homologs BUB1α and BUB1β, to the kinetochores, independently of PLK1 [1440]. [92]

The cell centrosome consists of 2 centrioles surrounded by a dense pericentriolar material. During the cell division cycle, more precisely during the progression from G1 to S phase, the centriole duplication begins to create 2 centrosomes that

90. Small GTPase Rab4 controls the displacement of endosomes and RhoB regulates the formation of multivesicular bodies.

91. The spindle checkpoint impedes the separation of sister chromatids until all pairs of opposing sister kinetochores attach to microtubules that emanate from the 2 spindle poles (Vol. 2 – Chap. 2. Cell Growth and Proliferation).

92. The mitotic checkpoint complex contains MAD2, BUB1β, BUB3, and CDC20. It binds to the anaphase-promoting complex/cyclosome (APC/C) and prevents its ubiquitin-ligase activity on securin and cyclin-B1.

each contains 2 closely juxtaposed centrioles by the end of the S–G2 transition. The 2 centrosomes are held together by a linker composed of centrosomal Never in mitosis gene-A (NIMA)-related kinase-2 (NeK2)-associated protein CNAP1 and rootletin. They function as one microtubule nucleation center. Before the onset of mitosis, the centrosome linker is degraded and the centrosomes separate to form the spindle poles. Phosphorylation of rootletin and CNAP1 by NeK2 leads to separation of the centrosome pair. After mitosis, each daughter cell inherits a single centrosome with 2 centrioles. Axin-2 (conductin), an inhibitor of β-catenin, binds to CNAP1 and localizes to centrosomes and mitotic spindle. Axin-2 controls centrosome cohesion. Axin-2 indeed prevents premature centrosome splitting, as it promotes phosphorylation of centrosome-associated β-catenin (Ser33, Ser37, and Thr41) [1441]. Messenger Wnt and inhibitors of glycogen synthase kinase GSK3 impede β-catenin phosphorylation and cause centrosomal splitting.

10.3.7.4 Wnt Signaling and Aging

The Wnt proteins are not only implicated during development, but also operate in cell aging, especially that of stem and progenitor cells. The Wnt signaling is more active in aged cells [1442]. Moreover, the activation of the canonical Wnt pathway in aged myogenic progenitors leads to the conversion from myogenic to fibrogenic lineages.

Klotho antagonizes secreted Wnt. The Klotho gene encodes transmembrane protein Klotho with a large extracellular domain that interact with various Wnt family members, particularly morphogen Wnt3, to block Wnt activity, especially in stem cells [1443] and tissue fibrosis [1442]. Chronic exposure to Wnt3a accelerates cellular senescence (in mice). During early stages of cell senescence (in humans), repression of Wnt2 signaling allows phosphorylation by GSK3β of the histone chaperone histone cell cycle regulation-defective homolog-A (HiRa) that participates in the formation of senescence-associated heterochromatin foci [1444].

10.4 Transmembrane Glycoprotein EpCAM

EpCAM[93] is a transmembrane glycoprotein supposed to act as a homophilic cell adhesion molecule. This signal transducer operates in cell proliferation, as it upregulates cyclin-A and -E. Glycoprotein EpCAM is overexpressed in human stem and progenitor cells. Agent EpCAM transmits signals after *regulated intramembrane proteolysis* that releases its intracellular domain ($EpCAM^{ICD}$) and $EpCAM^{ICD}$ nuclear translocation [1445]. Transducer EpCAM is sequentially cleaved by sheddase ADAM17 and presenilin-2. Released $EpCAM^{ICD}$ associates with FHL2, β-catenin, and Lef1 to form a nuclear complex that targets Lef1 consensus sites to induce transcription of cell proliferation-promoting gene such as the MYC gene. Product $EpCAM^{ICD}$ can then associate with components of the Wnt pathway.

93. Glycoprotein EpCAM is also called CD326.

Shedding of released EpCAM ectodomain (EpCAM^{ECD}) produces a soluble ligand that can induce auto- or paracrine regulation. Homotypic aggregation of EpCAM extracellular domain in either membrane-bound or soluble form indeed yields an activation signal. In addition, EpCAM^{ECD} shedding is a prerequisite for subsequent intramembrane cleavage by presenilin and release of EpCAM^{ICD} similarly to that of Notch receptors, L1CAM, epican (or CD44), amyloid-β precursor protein (APP), sortilin-related receptor, LDLR class-A repeats-containing protein SorLA (or SorL1), and Deleted in colorectal cancer (DCC). Agent EpCAM^{ICD} can be sequestered in the cytosol or quickly degraded like short-lived Notch^{ICD} and APP^{ICD}.

10.5 Semaphorins and Plexins

Semaphorins (or collapsins) are one of the largest families of messengers that signal to the actin cytoskeleton for cell navigation (Vol. 2 – Chap. 6. Cell Motility). Certain transmembrane semaphorins can function as both ligand and receptor (reverse or bidirectional signaling). Semaphorins are subdivided into 8 subfamilies of membrane-attached and secreted glycoproteins that form homodimers.

The main receptors for semaphorins are plexins, which regulate Rho GTPases. The binding of soluble or membrane-anchored semaphorins to the plexin ectodomain activates intrinsic GTPase-activating activity of plexins to modulate cell adhesion.

The family of plexins can be split into 4 subfamilies (plexin-A–plexin-D).[94] Plexins are large type-1 single transmembrane-spanning cell surface receptors. Plexins possess a resting, auto-inhibited configuration. Semaphorin binding activates the GAP domain and/or recruits Rho GTPases.

Plexins are receptors for semaphorins, either alone or in combination with *neuropilin*-1 or -2. Members of the plexin family bestow specificity on neuropilin receptors, as they form complexes with them and act as coreceptors for semaphorins, in particular semaphorin-3A. Ligand binding activates signal transduction pathways that control axon guidance in the nervous system.

Subfamily-4 semaphorins signal via members of subfamily-B plexins. The prototypical interaction of subfamily-B plexins is the Sema4d–PlxnB1 linkage implicated in migration and proliferation of neuronal, endothelial, and tumor cells, especially during angiogenesis and axonal guidance [1446]. Subfamily-6 semaphorins (Sema6a–Sema6d) typically interact with subfamily-A plexins.[95]

Effective dimerization of plexins results from semaphorin binding, before plexin-generated signaling [1446]. However, before binding, the Sema6a ectodomain is in the homodimer arrangement, similar to that adopted by Sema3a and Sema4d [1447]. Protein PlxnA2 is also in an homodimer arrangement. The structure of the Sema6a–PlxnA2 signaling complex revealed a 2:2 heterotetramer in which the 2 PlxnA2

---

94. Human plexins are also referred to as the SEX family with NOV (plexin-1), OCT (plexin-2), and SEX (plexin-4).

95. The Sema6a–PlxnA2 signaling controls axon guidance in the hippocampus and granule cell migration in the cerebellum [1446].

monomers dissociate from one another, but remains docked onto the Sema6a homodimer [1447]. This transition provokes a GAP domain rearrangement and activation.

The plexin-A subfamily is constituted by plexins that transduce signals from semaphorin-1A and -1B as well as -3A and -3C. Plexin-A1 (or plexin-1) encoded by the PLXNA1 gene interacts with protein kinase-B1. Plexin-1 can form a complex with neuropilin-1 that has a higher affinity for semaphorin-3A than neuropilin-1 alone [1448]. Plexin-A2 (or plexin-2) is encoded by the PLXNA2 gene. Plexin-A3 (or plexin-4) that is encoded by the PLXNA3 gene is expressed predominantly in the brain. Plexin-A4 that is encoded by the PLXNA4A gene forms a receptor complex with neuropilin-1 and can propagate semaphorin-3A inhibitory signals. In humans, plexin-A4 precursor has 4 isoforms. It bears post-translational modification such as phosphorylatation upon DNA damage.

The plexin-B subfamily contains plexins that regulates Rho GTPase via guanine nucleotide-exchange factors RhoGEF11 [96] and RhoGEF12 [1449]. [97] Plexin-B1 [98] stimulated by semaphorin-4D interacts with RhoGEF11 and -12 and Rnd1 GTPase. It is expressed in endothelial cells and can elicit chemotaxis and tubulogenesis. Plexin-B2 [99] and -B3 [100] also interact with RhoGEF11.

Plexin-C1 [101] is a receptor for semaphorin-7A. Plexin-D1, [102] together with semaphorin-3 and neuropilins, is required in endothelial cells for the cardiovascular development [1450].

Plexin-A and semaphorin-3 are expressed by endothelial cells. Cardiomyocytes synthesize plexin-A1 and -D1. Plexins interact with semaphorins both on adjacent cells and in the extracellular medium. Plexins can function as both signaling receptors and ligand-binding receptors in coreceptor complexes for both plexins and vascular endothelial growth factor receptors in particular. Phosphorylated plexins stimulate kinases. Plexin-A regulates Fes and Fyn kinases. Plexin-B1 binds to and stimulates HGHR and HER2 receptor Tyr kinases. Semaphorin signaling activates extracellular signal-regulated kinase.

Semaphorin-6 yields reverse signaling in cardiogenesis. Semaphorin-3 reduces endothelial cell motility, inhibiting angiogenesis. On th other hand, semaphorin-4 promotes endothelial cell migration and angiogenesis, its receptor plexin-B1 being expressed in endothelial cells. [103] Semaphorin-4D may act both remotely and locally like vascular endothelial growth factor VEGFa.

Microtubule-associated monoxygenase, calponin, and LIM domain-containing proteins of the Mical family of redox enzymes connect to plexins to mediate axon

96. A.k.a. PDZ-RhoGEF or ArhGEF11.
97. A.k.a. leukemia-associated RhoGEF (LARG).
98. A.k.a. plexin-5 or SEP. It is encoded by the PLXNB1 gene.
99. A.k.a. MM1. It is encoded by the PLXNB2 gene.
100. A.k.a. plexin-6. It is encoded by the PLXNB3 gene.
101. A.k.a. virus-encoded semaphorin protein receptor (VESPR) and CD232. It is encoded by the PLXNC1 gene.
102. It is encoded by the PLXND1 gene.
103. Semaphorin-4D produced by malignant cells, in breast and lung neoplasms among others, stimulates endothelial cell migration for angiogenesis [1451].

Table 10.11. Slit–Robo signaling complex (Source: [1453]).

Ligand	Receptor
Slit1	Robo1–Robo3
Slit2	Robo1–Robo4
Slit3	Robo1–Robo3

guidance, synaptogenesis, and dendritic pruning. The Rab1-interacting protein Mical1 binds to actin filament and links repellent semaphorins to their plexin receptors to control ^{F}actin reorganization (semaphorin–plexin–Mical complex) [1452]. It disassembles both individual and bundled actin filaments to ensure changes in cell morphology for axon navigation in response to semaphorins that yield repulsive guidance cues.

10.6 Roundabout Receptors

Roundabout receptors (Robo1–Robo4) belong to the immunoglobulin superfamily of transmembrane signaling molecules (Table 10.11). Roundabout proteins are guidance receptors that operate in the nervous and cardiovascular systems. Robo receptors bind Slit ligands (Slit1–Slit3). Slit–Robo signaling induces repulsion. Slits are involved in heart morphogenesis, angiogenesis, and tumor metastasis.

Subtype Robo4 is specific to the vascular endothelium of both embryos and adults. It is not synthesized by vascular smooth muscle cells and pericytes. During angiogenesis (Vol. 5 – Chap. 10. Vasculature Growth), Robo4 is highly expressed in the endothelial cells of the stem of developing blood vessels that form lumens of interconnected vessel networks,[104] but is absent from many of the tip cells at the sprouting front of the vascular plexus that use filopodia to sense and respond to extracellular signals such as vascular endothelial growth factor. Activation of Robo4 by Slit2 inhibits $VEGF_{165}$-induced migration of endothelial cells and vascular tube formation and permeability in vitro and $VEGF_{165}$-stimulated vascular leak in vivo by blocking Src family kinase activation [1454].[105] Members of the SRC family of non-receptor Tyr kinases stimulated by VEGF activate Rac1 GTPase. In mouse models of retinal and choroidal vascular disease, the Slit2–Robo4 signaling inhibits angiogenesis and vascular leakage.

104. Endothelial stalk cells are similar to differentiated, stabilized type of endothelial cells in mature vascular tube and blood vessels.

105. The monolayer of endothelial cells prevents the free motion of water, proteins, and circulating cells from the blood vessel lumen into the surrounding tissue (Vol. 5 – Chap. 9. Endothelium).

11

Receptors of the Immune System

The immune system, which ensures the body's protection, encompasses 2 main components: (1) *humoral immunity* that involves antibodies and complement in the body's fluid and (2) *cellular immunity* supported by activated cells, such as macrophages, natural killer cells, and antigen-specific cytotoxic T lymphocytes, that release *cytokines*, i.e., signaling molecules that serve as immunomodulating agents.

Immune cells, or immunocytes, permanently travel in blood circulation, some scouting into tissues of perfused organs, to search for invading foreign elements. Leukocytes encompass various specialized cell types that operate in *innate* and *adaptive* (acquired) immunity in response to antigens.

The innate immune system is the first line of defense against invading pathogens. Adaptive immunity recognizes and remembers specific pathogens, thereby preparing future challenges. Immunological memory [1] results from gene restructuring (accelerated somatic mutations and irreversible genetic recombination of antigen receptor gene segments) that enables the generation of a huge number of antigen receptor types carried on the lymphocyte surface from a small number of genes. Irreversible gene rearrangement is transmissible to the progeny of memory B and T lymphocytes. Upon interaction with a previously encountered antigen, the appropriate memory cells are activated.

Cells of the innate immunity comprise monocytes and macrophages, neutrophils, eosinophils, and basophils, as well as mast, dendritic, and natural killer cells (Vols. 1 – Chap. 2. Cells of the Blood Circulation and 5 – Chap. 3. Blood Cells).

Cells of the adaptive immunity include B and T lymphocytes, such as CD8+ cytotoxic and CD4+ helper T lymphocytes. Blood contains 20 to 50% of circulating lymphocytes; the remaining lymphocytes reside in the lymphatic system. Peripheral lymphoid organs contain a mixture of B and T lymphocytes, which are: (1) naive cells, which nonetheless have already matured and migrated to the lymphatic sys-

1. Active long-term memory is acquired by infections (as well as vaccines) that activate B and T lymphocyte (immunization). On the other hand, passive short-term (a few days to several months) memory is yielded by maternal antibodies from blood and breast milk to the fetus and newborn.

tem, but do not yet encounter antigens; (2) effector cells, which were activated by cognate antigen to eliminate a pathogen; and (3) memory cells, which transmit their experience of past infections.

Foreign microorganisms are detected by the immune system via several families of *pattern-recognition receptors* and a set of signaling molecules to produce factors of immune response. Antigen receptors of B lymphocytes – the so-called *B-cell receptors* – correspond to immunoglobulins. Antigens are also recognized by specific *T-cell receptors*. Dendritic cells express *Toll-like receptors*. Secretion by cytotoxic T and natural killer cells is triggered by *antigen* and *NK receptors*, respectively.

The differentiation and activation of monocytes and macrophages, granulocytes, microglial cells, dendritic cells, osteoclasts, and platelets are regulated by signals that activate plasmalemmal and intracellular receptors. These receptors yield specific tasks, such as growth factor and Toll-like receptors, or modulate the response magnitude (setting thresholds for cell responses to specific stimuli), without initiating any activation of signaling pathways, such as macrophage interferon-γ receptors.

Many activated receptors augment the intracellular calcium concentration. The ubiquitous PP3–NFAT (calcineurin–nuclear factor of activated T cells) pathway, discovered in lymphocytes, operates in many aspects of organogenesis (hemato- and lymphopoiesis, axonal guidance, and cardio- and vasculogenesis). Calcium–calmodulin stimulates PP3 phosphatase. Calcineurin regulatory subunit PP3R1 is not required for the development of the myeloid lineage (erythrocytes, granulocytes, megakaryocytes, monocytes, and myeloid-derived dendritic cells), whereas it is involved in that of the lymphoid lineage (T and B lymphocytes, natural killer cells, and lymphoid-derived dendritic cells) [1455].[2]

11.1 Cytokine Receptors

Cytokines are peptides, proteins, or glycoproteins that serve as intercellular communication molecules. These auto- or paracrine regulators that are released by multiple cell types bind to their cognate receptors.

Various categories of cytokines inhibit feeding, such as interleukins IL1α, IL1β, IL6, and IL8, as well as leptin, tumor-necrosis factor-α, and interferon-α [1457].

11.1.1 Type-1 Cytokine Receptors

Type-1 cytokine transmembrane receptors share a common motif in their extracellular domain that recognize cytokines. Type-1 cytokine receptors rely on JaK kinases (Vol. 4 – Chap. 3. Cytosolic Protein Tyrosine Kinases) to phosphorylate (activate) effectors of their signaling cascades.

2. Although CD8α+ and CD8α− dendritic cells may be considered as lymphoid- and myeloid-derived dendritic cells, respectively, both CD8α+ and CD8α− dendritic cells are generated from CD4^{low} lymphoid-committed precursors [1456].

Type-1 cytokine receptors include: (1) many *interleukin receptors* (IL2R–IL7R, IL9R, IL11R–IL13R, IL15R, IL21R, IL23R, and IL27R; Sect. 11.1.3); (2) *colony-stimulating factor receptors*, which are targeted by erythropoietin (EpoR), granulocyte colony-stimulating factor (gCSFR, CSF3R, or CD114), and low-affinity granulocyte–macrophage colony-stimulating factor (gmCSFR, CSF2Rα, or CD116); and (3) other cytokine receptors, which bind to growth hormone (GHR), leukemia inhibitory factor (LIFR or CD118), oncostatin-M (OSMR), and prolactin (PrlR).

Type-1 cytokine receptors possess different chains that are involved in cytokine linkage or signal transduction. Signal transduction chains are often shared between different receptors. Common β chain (CD131) is shared by receptors of gmCSF, IL3, and IL5 and common γ chain (CD132) by receptors of interleukins IL2, IL4, IL7, IL9, IL13, and IL15.

11.1.2 Type-2 Cytokine Receptors

Type-2 cytokine receptors possess an intracellular domain that is typically associated with a cytoplasmic tyrosine kinase of the Janus kinase family. They include heterodimeric type-2 interleukin receptors (IL10R, IL20R, IL22R, and IL28R) and interferon receptors (Ifnα/βR and IfnγR; Vol. 2 – Chap. 3. Growth Factors). Interferon-α/β receptor (IFNAR) binds type-1 interferons, i.e., interferon-α and -β. This heteromeric receptor is composed of 2 subunits IfnαR1 and IfnαR2. Interferon-γ receptor binds to type-1 interferon, i.e., interferon-γ. This other heterodimeric receptor is made of 2 chains IfnγR1 and IfnγR2.

11.1.3 Families of Interleukins and Their Receptors

Among cytokines, interleukins constitute a superfamily that can be decomposed into several families. Interleukins are cytokines that are especially produced by leukocytes and act on leukocytes. Yet interleukins are produced by multiple cell types (Tables 11.1 and 11.2). They promote the development and differentiation of B and T lymphocytes and hematopoietic cells via their receptors (Table 11.3).

11.1.3.1 Interleukin-1 Family

The interleukin-1 family includes 11 members: IL1α, IL1β, IL1 receptor antagonist (IL1RA), IL18 (a.k.a. IL1F4), IL33 (a.k.a. IL1F11), and IL1F5 to IL1F10 (Vol. 2 – Chap. 3. Growth Factors). Members of the IL1 family signal via a group of interleukin receptors of the immunoglobulin superfamily [1458].

Interleukin-1α and -β– IL1 Receptors

Both IL1α and IL1β target type-1 IL1 receptor (IL1R1) and recruit a receptor subunit such as IL1R accessory protein (IL1RAP) for the IL1 receptor. The receptor heterodimer then recruits myeloid differentiation primary response protein

Table 11.1. Interleukins and their cellular sources and targets (**Part 1**).

IL	Sources	Targets
IL1	Monocyte, macrophage, B and dendritic cell	Macrophage, endothelial cell, B, NK, T_H cell
IL2	T_{H1} cell	Macrophage, oligodendrocyte, B, NK, T cell
IL3	Eosinophil, endothelial, mast, NK, T_H cell	Hematopoietic stem cell, mastocyte
IL4	Macrophage, mastocyte, T_{H2}, memory CD4+ cell	Endothelial cell, B, T cell
IL5	Eosinophil, mastocyte, T_{H2} cell	Eosinophil, basophil B cell
IL6	Macrophage, astrocyte, endothelial, B, T_{H2} cell, osteoblast, myocyte, VSMC	Hematopoietic stem cell, plasma, B, and T cell
IL7	Bone marrow and thymus stromal cells	PreproB and T cell, NK cell
IL8	Epithelial, endothelial cell, macrophage, lymphocyte	Basophils, neutrophils, lymphocytes
IL9	T_{H2} cell	B and T cell
IL10	Monocyte, macrophage, B, T_{H2}, CD4+ and CD8+ T cells, mastocyte	Macrophage, mastocyte, B, T_{H1}, T_{H2} cell
IL11	Bone marrow stromal cell	Bone marrow stromal cell
IL12	Macrophage, dendritic cell, B, T cell	NK, T_{H1}, T cell
IL13	Mastocyte, NK, T_{H2} cell	Monocyte, macrophage, B, T_{H2} cell
IL14	T cell	B cell
IL15	Mononuclear phagocytes	B, NK, T cell
IL16	Epithelial cell, eosinophil, fibroblast, neurons, lymphocyte, mast, dendritic cell	Neuron, monocyte, eosinophil, T cell, proB cell
IL17	T_{H17} cell	Epithelial, endothelial cell
IL18	Macrophage	NK, T_{H1} cell

MyD88, IL1R-associated kinase IRAK4 (Vol. 4 – Chap. 4. Cytosolic Protein Ser/Thr Kinases), TNFR-associated factor TRAF6, among others, to activate the nuclear factor-κB and mitogen-activated protein kinase pathways.

Type-2 IL1 receptor (IL1R2) that also binds to IL1 possesses an extracellular region and a short cytoplasmic domain that cannot signal; it thus operates as a *decoy receptor*. The IL1R2 receptor is mainly synthesized by cells that produce IL1 (e.g., monocytes, macrophages, B lymphocytes, and many subpopulations of dendritic cells). It prevents excessive autocrine activation of IL1 signaling.

Table 11.2. Interleukins and their cellular sources and targets (**Part 2**).

IL	Sources	Targets
IL19	Monocyte, B cell	
IL20	Monocyte, keratinocyte	
IL21	CD4+ T, NKT, T_{H2}, T_{H17}, T follicular cells	NK, cytotoxic T cell
IL22	Dendritic, T cell	
IL23		Memory CD4+ T cell
IL24	Monocyte, macrophage, T_{H2} cell	
IL25		
IL26		
IL27		Monocyte, mastocyte, CD4+ T, NK, T_{H1} cell
IL28		
IL29		
IL31	T_{H2} cell	Monocyte, epithelial cell
IL32	Lymphocytes, NK, epithelial cell	Monocyte, macrophage
IL33		
IL34		Monocyte, osteoclast progenitor
IL35	T_{Reg} cell	Monocyte

Like IL1R1, once bound to its ligand, IL1R2 can interact with the accessory protein IL1RAP [1458]. It thus sequesters both the ligand and the accessory protein. In addition, IL1R2 binds to IL1 receptor antagonist (IL1RA) with low affinity, so that IL1R2 does not impede antagonist action of the IL1RA cytokine.

The IL1R2 receptor can also be shed from the cell surface as a soluble protein ($IL1R2^S$) that retains a ligand-inhibiting activity. This inhibition can be enhanced by a soluble form of IL1RAP ($IL1RAP^S$) that can interact with $IL1R2^S$ [1458].

Interleukin-18 – IL18 Receptor

Interleukin-18 binds to IL18R and this complex then recruits the accessory protein IL18RAP to initiate signaling. It connects to soluble IL18-binding protein (IL18BP). The latter then impedes IL18 interaction with its receptor [1458]. Regulatory T cells can produce IL1R, IL18R, and IL1RL1 receptors.

Interleukin-33 – IL1RL1 Receptor

Interleukin-33 tethers to interleukin-1 receptor-like IL1RL1 receptor. [3] The IL33–IL1RL1 complex recruits the receptor subunit IL1RAP to initiate signal transduction. Soluble form of IL1RL1 ($IL1RL1^S$), which is generated from alternative splicing, binds tightly to IL33, thereby inhibiting its activity. Soluble IL1RAP reinforces this inhibition [1458].

3. A.k.a. IL1 receptor-related protein IL1RRP1, ST2, T1, Fit1, and IL33Rα.

Table 11.3. Interleukins and their receptor complexes. Interleukins IL2, IL4, IL7, IL9, IL15, and IL21 share the common cytokine receptor γ chain; IL12 and IL23 IL12Rβ1; IL4 and IL13 the type 2 IL4 receptor, and IL7 and thymic stromal lymphopoietin (TSLP) the receptor IL7Rα. Interleukin-4 receptor chain α can bind IL4 and IL13 to regulate IgE production by B cells. Receptor subunits ILiRα and ILiRβ are also denoted ILiR1 and ILiR2.

Type	Receptor
IL1	IL1R1, IL1R2, IL1R1–IL1RAP, IL1R2–IL1RAP
IL1F5/6/8/9	IL1RL2
IL1F7	IL18R
IL1F10	IL1R1
IL2	IL2Rα–IL2Rβ–IL2Rγ
IL3	IL3Rα–IL3Rβ
IL4	IL4Rα–IL2Rγ
IL5	IL5Rα–IL3Rβ
IL6	IL6Rα–GP130–GP130
IL7	IL7Rα–IL2Rγ
IL8	IL8Rα, IL8Rβ(CXCR1 and CXCR2)
IL9	IL9Rα–IL2Rγ
IL10	IL10Rα–IL10Rβ
IL11	IL11Rα–GP130–GP130
IL12	IL12Rβ1–IL12Rβ2
IL13	IL13Rα1–IL13Rα2, IL4Rα
IL14	IL14R
IL15	IL15Rα–IL2Rβ–IL2Rγ
IL16	CD4
IL17	IL17Rβ1–IL17Rβ2
IL18	IL18R, IL18R–IL18RAP
IL19	IL20R (IL20R1–IL20R2)
IL20	IL20Rα–IL20Rβ (IL20R)
IL21	IL21Rα–IL2Rγ
IL22	IL22R1–IL10R2
IL23	IL23Rα–IL12Rβ1
IL24	IL20R (IL20R1–IL20R2); IL22R1–IL20R2
IL25	
IL26	IL20R1–IL10R2
IL27	IL27Rα–IL30R (GP130)
IL28	IL28Rα–IL10Rβ
IL31	IL31Rα–OSMR
IL33	IL1RL1
IL34	CSF1R
IL35	CD169, CD274

IL1 Family Members – IL1R1 and IL1RL2 Receptor

Members of the IL1 family — IL1F5, IL1F6, IL1F8, and IL1F9, interact with IL1RL2 (or IL1RRP2). The IL1RL2 receptor mainly resides on macrophages and dendritic cells.

Cytokine IL1F5 acts as an antagonist. Once bound to IL1RL2, it precludes agonist binding and IL1RAP recruitment. Cytokine IL1F7 can link to both IL18R and IL18BP; IL1F10 to IL1R1 [1458].

IL1R Family Members

In addition to interleukin-1 receptors IL1R1 and IL1R2 and related receptors IL1RL1 and IL1RL2, other IL1R family members have been identified, but only the Single immunoglobulin IL1R-related molecule (SIGIRR) [4] has a known action. The SIGIRR receptor actually inhibits IL1 signaling [1458]. Therefore, SIGIRR is an additional regulator that limits the magnitude of the response to IL1 family cytokines.

11.1.3.2 Interleukin-2 Family

The interleukin-2 family comprises IL2, IL3, IL4, IL7, IL9, IL13, IL15, and IL21 that belong to type-1 cytokine receptor ligands.

Interleukin-2 – IL2R Receptor

Interleukin-2 binds to IL2 receptors expressed by lymphocytes. It acts as a lymphocytotrophic hormone. Heterotrimeric IL2 receptor consists of: (1) type-1 transmembrane IL2-specific receptor α chain (CD25); (2) β chain (CD122); and (3) common γ chain (IL2Rγ or CD132), which is shared by interleukins-2, -4, -7, -9, -15, and -21 and carries out signal transduction. Binding of IL2 to its receptor (more precisely to the $\alpha\beta$ dimer) activates the Ras–MAPK, JaK–STAT, and PI3K–PKB pathways.

Interleukin-3 – IL3R Receptor

Interleukin-3 [5] stimulates the differentiation of multipotent hematopoietic stem cells into myeloid progenitor cells as well as proliferation of all the cell types of the myeloid lineage (erythrocytes, megakaryocytes, granulocytes, monocytes, and dendritic cells). It is secreted by activated T cells.

Interleukin-3 dimeric receptor is composed of: (1) an IL3-specific α chain (CD123) and (2) signal-transducing β subunit, which is also shared by the receptors of colony-stimulating factor-2 (gmCSF) and interleukin-5 [1459].

Interleukin-4 – IL4R Receptor

Interleukin-4 causes the differentiation of naive T helper cells (T_{H0}) to T_{H2} cells and stimulates proliferation of activated B and T cells. Eosinophils, effector cells in

4. A.k.a. TIR8.
5. A.k.a. multiple colony-stimulating factor (multiCSF).

allergy, synthesize IL4 and IL5. Basophils that operate as antigen-presenting cells are required for allergen-induced activation of T_{H2} cells [1460]. Interleukin-4 targets the interleukin-4 receptor. The latter comprises a α (IL4Rα) and the common γ chain (IL2Rγ).

Interleukin-7 – IL7R Receptor

Interleukin-7 is a hematopoietic cytokine secreted by stromal cells of the bone marrow and thymus. It is also produced by dendritic cells, neurons, epi- and endothelial cells, keratinocytes, and multipotent human adipose-derived stem cells [1461], but not lymphocytes. It stimulates the differentiation of hematopoietic stem cells into lymphoid progenitor cells as well as proliferation of all the cells of the lymphoid lineage (B, T, and NK cells).

Interleukin-7 and a β chain variant of hepatocyte growth factor can form the IL7–HGFβ heterodimer [6] that stimulates the growth of preproB lymphocytes [1462]. [7]

Interleukin-7 binds its cognate receptor IL7R that is constituted by a α (IL7Rα; a.k.a. CD127) and the common γ chain. The IL7–HGFβ complex signals via both IL7R and HGFR to provoke the proliferation of spleen colony-forming units (CFUs), common lymphoid progenitors, and preproB cells [1463].

Interleukin-9 – IL9R Receptor

Interleukin-9 is produced by CD4+ helper T cells, preferentially T_{H2} lymphocytes. It acts on various cell types, such as T and B-lymphocytes, mastocytes, and hematoopoietic progenitors. It tethers to interleukin-9 receptor that is composed of α (IL9Rα or CD129) and the common γ chain (IL2Rγ) [1464].

Ligand binding to IL9R activates various types of JaK kinases and STAT proteins. Interleukin-9 also causes the rapid production of cytokine-inducible SH2-containing protein (CIS) and suppressor of cytokine signaling SOCS2 and SOCS3. However, only overexpression of SOCS3 inhibits STAT activation, gene transcription, and anti-apoptotic activity of IL9 [1464].

Interleukin-13 – IL13R Receptor

Interleukin-13 is secreted by many cell types, especially T_{H2} cells. It acts on B cells and monocytes, in which it promotes IgE class switching and inhibits inflammatory cytokine production. In the lung, IL13 is the main mediator of allergic asthma [1465].

Interleukin-13 connects to its cognate receptor that includes IL4Rα chain and at least one of 2 IL13-binding chains. It activates a single transcription factor, signal transducer and activator of transcription STAT6.

6. A.k.a. preproB-cell growth-stimulating factor (PPBSF).

7. Pleiotropic hepatocyte growth factor is synthesized in numerous cell types, particularly bone marrow stromal cells and some hematopoietic cell lines, but not epithelial cells. PreproB cells differentiate into proB cells.

Interleukin-15 – IL15R Receptor

Interleukin-15 is secreted by mononuclear phagocytes and some other cell types. It provokes proliferation of natural killer cells. It helps in the maintenance of memory CD8+ T-cell population [1466].

It signals via a receptor made of IL15Rα, IL2Rβ(CD122), and the common γ chain (CD132). It activates JaK kinases that phosphorylate transcription activators STAT3, STAT5, and STAT6.

Interleukin-21 – IL21R Receptor

Interleukin-21 is expressed in activated human CD4+ T, NKT, T_{H2} and T_{H17}, and follicular T cells [1467]. It regulates natural killer and cytotoxic T cells.

Interleukin-21 links to IL21 receptor that is expressed on the surface of B, NK, and T cells. It signals via JaK1 and JaK3, then STAT1 and STAT3, to activate its target genes.

11.1.3.3 Interleukin-6 Family

The interleukin-6 family encompasses IL6, IL11, IL27, IL30, and IL31, as well as cardiotrophin-1 (CT1) and cardiotrophin-like cytokine (CLC), ciliary neurotrophic factor (CNTF), leukemia inhibitory factor (LIF), and oncostatin-M (OSM).

In addition to their functions in inflammation and immunity, members of the interleukin-6 family intervene in hematopoiesis, hepatic and neuronal regeneration, embryo- and fetogenesis, and fertility. They activate target genes involved in differentiation, survival, proliferation, and apoptosis.

The IL6 subfamily is also associated with type-1 cytokine receptors. Receptors that recognize cytokines of the IL6 family comprise a non-signaling α subunit (IL6Rα, IL11Rα, and CNTFRα) and a signaling protomer (Table 11.4). Cytokines of the IL6 family exert their action via glycoprotein GP130 (or IL30R) and LIF and OSM receptors that prime activation of the JaK–STAT axis, the PI3K pathway, and mitogen-activated protein kinase modules (ERK1, ERK2, JNK, and P38MAPK) cascades [1468].

The signal-transducing chains GP130, LIFR, and OSMR bind to JaK1 and JaK2 as well as TYK2 pseudokinase of the JAK family. The interaction between GP130 and JaK1 is long-lasting [1468]. Other kinases, such as Src and Tec family kinases, associate with IL6-family receptors. Hematopoietic cell kinase (HCK) of the SRC family binds to GP130 and provokes activation of extracellular-regulated kinase. Cyclin-dependent kinase CDK9 and protein kinase-Cδ also connects to GP130 upon IL6 stimulation. All IL6 family cytokines potently activate STAT3, and, to a lesser extent, STAT1 factor. In addition, OSMR is the most potent activator of STAT5 factor.

In addition to phosphorylation (on Tyr or Ser) of STAT1, STAT3, STAT5a, and STAT5b, methylation by protein arginine methyltransferase PRMT1 influences STAT function.

Table 11.4. Receptor complexes of IL6-family cytokines (Source: [1468]). Cytokines IL6 and IL11 use GP130–GP130 homodimers for signaling. On the other hand, cardiotrophin CT1, cardiotrophin-like cytokine (CLC), ciliary neurotrophic factor (CNTF), leukemia-inhibitory factor (LIF), and oncostatin-M (OSM) utilize GP130–LIFR heterodimers. Alternatively, OSM can signal via GP130–OSMR heterodimer.

Cytokine	Receptor
IL6	GP130–IL6Rα–GP130
IL11	GP130–IL11Rα–GP130
CLC	LIFR–CNTFR–GP130
CNTF	LIFR–CNTFR–GP130
CT1	LIFR–GP130
LIF	LIFR–GP130
OSM	LIFR–GP130
OSM	OSMR–GP130

Table 11.5. Half-lives of IL6-type cytokine signaling components (Source: [1468]). Isoform STAT3β is a splice variant of STAT3α subtype.

Mediator	Half-life (h)
SOCS2	1
SOCS1	1.5
SOCS3	1.6
JAK2	1.9
TYK2	2
JAK1	3.2
IL6Rα	2–3
GP130	2.5
STAT3β	4.5
STAT3α	8.5
STAT1	16
PTPn11	18–20

Signaling Termination

Various mediators of IL6 signaling differ in half-lives [1468] (Table 11.5). Feedback inhibitors SOCS1 to SOCS3 have very short lives. On the other hand, STAT1, STAT3α, and PTPn11 have slow turnover rates.

Signal termination is carried out by plasmalemmal protein Tyr phosphatase PTPRc as well as cytosolic protein Tyr phosphatase cytosolic PTPn1, PTPn2, PTPn6, and PTPn11 [1468]. The cytoplasmic variant of PTPRe ($_c$PTPRe) that is expressed in hematopoietic cells selectively inhibits IL6-activated GP130, JaK, and STAT factor. Moreover, the feedback inhibitor SOCS3 also contributes to Tyr759^P-mediated inhibition.

The family of protein inhibitor of activated STAT (PIAS) comprises 5 members (PIAS1, PIAS3, PIASxα, PIASxβ, and PIASy) that have small ubiquitin-related modifier (SUMO) ligase activity. Inhibitors PIAS1 and PIAS3 preclude STAT1 and STAT3 signaling, respectively [1468]. Inhibitor PIAS1 preferentially associates with unmethylated STAT1.

The 8 members of the SOCS [8] family (i.e., cytokine-inducible SH2 domain-containing protein [CIS] and SOCS1–SOCS7) can be rapidly upregulated by IL6 (CIS and SOCS1–SOCS3), IL11 (SOCS3), LIF (CIS and SOCS1–SOCS3), and OSM (CIS, SOCS1, and SOCS3) [1468].

Interleukin-6 – IL6R Receptor

Interleukin-6 secreted by T lymphocytes and macrophages acts as both a pro- and anti-inflammatory cytokine. It can cross the blood–brain barrier.

Interleukin-6 stimulates the appearance of other anti-inflammatory cytokines, such as IL1 receptor antagonist (IL1RA) and IL10 and inhibits the production of pro-inflammatory TNFα [1469]. [9] The IL1R antagonist inhibits signaling from IL1 receptor. Interleukin-10 impedes the production of IL1α and IL1β, TNFα, and chemokines, such as IL8 (CXCL8) and CCL3. [10]

Interleukin-6 is released by T lymphocytes and macrophages. Osteoblasts secrete IL6 to stimulate osteoclast formation. In addition, IL6 is produced in myocytes (hence, the name myokine). Its concentration rises in response to muscle contraction, during which it mobilizes extracellular substrates and/or augments substrate delivery [1469]. Vascular smooth muscle cells also produce IL6.

Interleukin-6 signals via its cognate receptor that consists of: (1) the ligand-binding IL6Rα chain (or CD126) and (2) signal-transducing component GP130 (or CD130) that is the common subunit for several cytokine receptors, such as those of IL11, cardiotrophin-1, ciliary neurotropic factor, leukemia inhibitory factor, and oncostatin-M. Subunit CD130 is expressed in most tissues, but IL6Rα chain is restricted to some cell types.

When IL6 links to its receptor, GP130 and IL6Rα chain complexes and initiates a signaling cascade via Janus kinases and signal transducers and activators of transcription. Interleukin-6 enhances lipid turnover, as it stimulates lipolysis and fat oxidation [1469]. Moreover, it contributes to glucose homeostasis during exercise.

---

8. Suppressor of cytokine signaling proteins are also named STAT-induced STAT inhibitors (SSI).

9. The local inflammatory response is associated with a systemic response, the so-called *acute-phase response*. This event includes the production of numerous hepatocyte-derived acute phase proteins such as C-reactive protein. Initial cytokines in the cytokine cascade successively encompass TNFα, IL1β, IL6, IL1RA, and soluble TNFα receptors. Cytokine TNFα stimulates IL6 synthesis in adipose tissue and blood mononuclear cells. Interleukin-6 enhances systemic levels of IL1RA, IL10, soluble TNF receptor, and C-reactive protein [1469].

10. Chemokine CCL3 is also termed macrophage inflammatory protein MIPα. Both CCL3 and CXCL8 activate granulocytes, monocytes, macrophages, and natural killer, T, and B cells and recruit them to inflammation sites.

Interleukin-11 – IL11R Receptor

Interleukin-11 [11] was first isolated from bone marrow-derived stromal cells. It participates in the regulation of hematopoiesis, especially megakaryocyte maturation [1470]. It targets its cognate receptor that is composed of the common coreceptor GP130 and IL11-specific IL11Rα subunit.

Interleukin-27 – IL27R Receptor

Interleukin-27 is a member of the `IL12` group that includes IL12, IL23, and IL27 proteins. Innate and T_{H1}-based adaptive immune responses are necessary to eliminate bacterial infections. The T_{H1}-stimulating cytokines include IL12, IL18, IL23, and IL27 agents.

Interleukin-27 is a heterodimer composed of: (1) an IL27β subunit, a soluble type-1 cytokine receptor homologous to the P40 subunit of interleukin-12, the Epstein-Barr virus (EBV)-induced gene-3 (EBI3) product [12] and (2) IL30 subunit [1471]. [13] Subunit IL27β may intervene, independently from its association to IL30, in the regulation of antiviral and -tumoral immunity. Interleukin-27 induces proliferation of naive CD4+ T and NK cells. Human mastocytes also respond to IL27 as well as monocytes.

Interleukin-27 receptor is constituted of IL27Rα [14] and GP130 [1472]. It is expressed in lymphocytes, particularly naive CD4+ and CD8+ T cells. Unlike the receptors for IL12 and IL23, IL27 receptor subunits are coexpressed on many cell types, such as monocytes, mastocytes, B and T lymphocytes, and dendritic and NK cells.

Interleukin-27 provokes STAT1 and STAT3 phosphorylation. It induces the expression of IL18 that, in combination with IL12, supports Ifnγ production in activated CD4+ T lymphocytes.

Interleukin-30

Interleukin-30 corresponds to a subunit of the interleukin-27 heterodimer.

Interleukin-31 – IL31R Receptor

Interleukin-31 is preferentially produced by T_{H2} cells. It signals via a receptor complex composed of IL31Rα and oncostatin-M receptor (OSMR) subunits [1473]. This receptor is expressed on activated monocytes and unstimulated epithelial cells.

---

11. A.k.a. adipogenesis inhibitory factor (AGIF) and oprelvekin.
12. In humans, 2 viruses, γ-herpesvirus Epstein-Barr virus (EBV) and type-1 T-cell leukemia virus (HTLV1), can cause lymphomas.
13. A.k.a. IL12 P35-related subunit IL27P28.
14. A.k.a. T-cell cytokine receptor (TCCR), cytokine receptor-like-1 (CRL1), and WSX1.

Ciliary Neurotrophic Factor

Ciliary neurotrophic factor (CNTF) serves as a hormone and nerve growth factor that promotes neurotransmitter synthesis and neurite outgrowth. Moreover, it is a potent survival factor for neurons and oligodendrocytes. It targets the LIFR–CNTFR–GP130 heterotrimer.

Leukemia Inhibitory Factor

Leukemia inhibitory factor (LIF) elicits terminal differentiation of myeloid leukemic cells. This cytokine influences cell differentiation, survival, and proliferation in embryos as well as adults. It targets a high-affinity receptor complex composed of low-affinity LIF receptor and high-affinity converter subunit GP130.[15]

Oncostatin-M

Oncostatin-M (OSM) produced by activated macrophages and T lymphocytes contributes to the development of the central nervous system and liver, as well as hematopoeisis and inflammation. It particularly acts as an autocrine growth stimulator and a mitogen for vascular smooth muscle cells. It links to the GP130–LIFR and GP130–OSMR heterodimers.

Cardiotrophin-1

Cardiotrophin-1 (CT1) is a cytokine and a cardiac hypertrophic factor. It connects to the LIFR–GP130 heterodimer.

11.1.3.4 Interleukin-10 Family

The interleukin-10 family encompasses anti-inflammatory cytokine IL10, as well as IL19, IL20, IL22, IL24, and IL26. It is connected to type-2 cytokine receptors. CD4+ and CD8+ T lymphocytes synthesize large and small amounts of IL10, respectively. Activated T lymphocytes produce IL19, IL22, and IL26, whereas IL24 is produced not only by activated T lymphocytes, but also by activated monocytes. The IL19 group is a part of the IL10 family that includes IL19, IL20, IL22, and IL24. Interleukin-28 and -29 are distantly related to the interleukin-10 family.

---

15. In addition to leukemia inhibitory factor, other cytokines that signal via GP130-containing receptors comprise interleukins IL11 and IL27, ciliary neurotrophic factor, cardiotrophin-1, cardiotrophin-like cytokine, oncostatin-M, and Kaposi's sarcoma-associated herpes virus IL6-like protein.

Interleukin-10

Interleukin-10 [16] is an anti-inflammatory dimeric cytokine. It is synthesized primarily in monocytes and, to a lesser extent, lymphocytes (e.g., type-2 helper, cytotoxic, and CD4+, CD25+, FoxP3+, regulatory T cells) and mastocytes. It lowers the production of cytokines and costimulatory molecules in macrophages. It enhances B-cell survival and proliferation, and antibody production.

Interleukin-10 targets IL10 receptor that is composed of a high-affinity α subunit (IL10Rα or IL10R1) and accessory, low-affinity IL10Rβ chain. [17]

Interleukin-19

Interleukin-19 is expressed in resting monocytes and B cells. It signals via the same receptor IL20R used by IL20 and IL24 cytokines.

Interleukin-20

Interleukin-20 is produced by monocytes and activated keratinocytes. Interleukin-20 receptor is a type-2 cytokine receptor. This heterodimer is made up of IL20Rα and IL20Rβ subunits (IL20R1–IL20R2 dimer).

Interleukin-22

Interleukin-22 is a potent mediator of inflammation. It is produced by activated dendritic and T cells. Activity of IL22 is initiated by binding to IL22 receptor composed of IL22R1 and IL10R2 chains [1474]. It activates STAT1, STAT3, and STAT5. It interacts with soluble IL22BP partner.

Interleukin-24

Interleukin-24 [18] is chiefly released by activated monocytes, macrophages, and T_{H2} cells. It signals via 2 heterodimeric receptors: IL20R1–IL20R2 and IL22R1–IL20R2 [1475].

Interleukin-26

Interleukin-26 binds to a dimeric receptor composed of IL20R1 and IL10R2 subunits [1476]. This receptor primes the phosphorylation of STAT1 and STAT3 transcription factors.

---

16. A.k.a. cytokine synthesis inhibitory factor (CSIF).

17. A.k.a. IL10R2, CDw210b, and cytokine receptor family 2 member 4 (CRF2-4 or CRFb4).

18. A.k.a. melanoma differentiation-associated MDA7 and suppression of tumorigenicity ST16.

Interleukins-28 and -29

Interleukins IL28a, IL28b, and IL29 constitute the family of type-III interferons and belong to the class-2 cytokines with the interleukin-10 family. In addition to type-I and -II interferons, they have an antiviral activity.

Interleukin-28 and -29 interact with a heterodimeric class-2 cytokine receptor that consists of IL10Rβ and IL28Rα (or IL28R1–IL10R2) [1477].

11.1.3.5 Interleukin-12 Family

The interleukin-12 family includes IL12, IL23, and IL27. Activated monocytes produce these 3 types of interleukins.

Interleukin-12

Interleukin-12 is a pro-inflammatory cytokine that is involved in the differentiation of naive T cells into T_{H0} cells, which further develop into T_{H1} or T_{H2} cells. It activates and maintains T_{H1} cells, and stimulates natural killer cells. It is naturally produced by dendritic cells and macrophages. Interleukin-12 has an anti-angiogenic activity via interferon-γ and CXCL10 chemokine. [19]

Interleukin-12 is composed of IL12α [20] and IL12β subunit. [21] Homodimer IL12α can act as an IL12 antagonist, as it competes for its receptor [1478].

Interleukin-12 binds to the IL12 heterodimeric receptor formed by IL12Rβ1 and IL12Rβ2. Upon IL12 binding, IL12Rβ2 undergoes phosphorylation and provides binding sites for TyK2 and JaK2 kinases [1479].

Interleukin-23

Interleukin-23 is made up of IL12α-related protein IL12α (or P19) and IL12β (or P40) subunit [1478]. Heterodimer IL23 promotes the production of the matrix metallopeptidase MMP9 and angiogenesis, and reduces CD8+ T-cell infiltration.

The IL23 receptor is formed by IL12Rβ1 and IL23-specific IL23R subunit. Like IL12, IL23 can activate the transcription activator STAT4 and stimulate the production of interferon-γ. However, whereas IL12 acts mainly on naive CD4+ T cells, IL23 preferentially operates on memory CD4+ T cells [1478].

Interleukin-27

Interleukin-27 is a IL27β–IL30 heterodimer (Sect. 11.1.3.3) that is involved in early T_{H1}-cell initiation [1478].

---

19. A.k.a. 10-kDa interferon-γ-induced protein (IP10).
20. A.k.a. P35, cytotoxic lymphocyte maturation factor CLMF1, and natural killer cell stimulatory factor NKSF1.
21. A.k.a. P40, CLMF2, and NKSF2.

Table 11.6. Interleukin IL17 family and their receptors (Source: [1481]; LTI: lymphoid tissue inducer cell; NK: natural killer cell; NKT: natural killer T cell; T_H: T helper.). The IL17R family encompasses 5 single transmembrane domain-containing receptor subunits (IL17Ra–IL17Re). Interleukins IL17a and IL17f form homodimers as well as IL17a–IL17f heterodimers. Receptors IL17Ra and IL17Rd colocalize. Receptor IL17Ra is expressed ubiquitously, but mainly in hematopoietic cells. Epi- and endothelial cells and fibroblasts, indeed, respond to IL17a. Interleukins IL15 and IL21 upregulate the expression of IL17Ra by CD8+ T lymphocytes. Interleukin IL17a stimulates the production of granulocyte colony-stimulating factor and interleukins IL6 and IL8. Interleukin IL17e is a T_{H2}-promoting cytokine, but limits T_{H17} cell development, as it causes IL13 production by dendritic cells and inhibits that of IL23 by macrophages. Effects of IL17b, IL17c, and IL17d remain poorly defined.

Interleukins	Receptors	Producing cells and tissues
IL17a	IL17Ra–IL17Rc	T_{H17}, CD8+ and $\gamma\delta$T cells, NK, NKT, LTI
IL17b	IL17Rb	Gastrointestinal tract, pancreas, neurons
IL17c	IL17Re	Prostate, fetal kidney
IL17d	Unknown	Muscles, heart, lung, pancreas, adipose tissue, central nervous system
IL17e	IL17Ra–IL17Rb	Intra-epithelial lymphocyte, lung epithelial cells, alveolar macrophage, eosinophils, basophils, NKT, T_{H2}, mastocytes, gastrointestinal tract, uterus
IL17f	IL17Ra–IL17Rc	T_{H17}, CD8+ and $\gamma\delta$T cells, NK, NKT, LTI

Interleukin-35

Interleukin-35 is produced by regulatory T cells [1480]. This anti-inflammatory dimer is composed of IL12α and IL27β chain. It causes proliferation of T_{Reg} cells, but reduces activity of T_{H17} cells. Interleukin-35 acts via CD274 [22] and siglec-1 [23] of the immunoglobulin superfamily. [24]

11.1.3.6 Interleukin-17 Family

The interleukin-17 family is constituted by IL17a to IL17d, IL17e (or IL25), and IL17f with its isoforms IL17f1 and IL17f2 (or ML1) that target cytokine receptors IL17Ra to IL17Re (Table 11.6). They induce inflammatory and allergic responses.

Interleukin-17, but not IL25 (IL17e), induces the production of many other cytokines (IL6, gCSF, gmCSF, IL1β, TGFβ, and TNFα), chemokines (IL8, CCL2,

22. A.k.a. programmed death ligand PDL1 and B7H1.
23. A.k.a. sialoadhesin and CD169.
24. Siglec-1 binds predominantly to neutrophils, but also monocytes, natural killer, B, and some cytotoxic T cells.

and CXCL1),[25] and prostaglandins (e.g., PGE2) from many cell types (fibroblasts, macrophages, endo- and epithelial cells, and keratinocytes). They promote the proliferation of cytotoxic T lymphocytes. Interleukin-17a is the hallmark cytokine of CD4+ helper-17 T cells.[26]

Ubiquitous receptor IL17Ra that is targeted by both IL17a and IL17f functions as a coreceptor with at least 2 other members of the IL17R family. It is actually produced as a pre-associated multimeric receptor that is regulated in some cell types.

The IL17Rb receptor associates with IL17Ra to prime IL17e signaling [1481]. It is expressed in various endocrine cells as well as kidney and liver cells and T_{H2} cells. It binds to both IL17b and IL17e. It forms a functional receptor complex with IL17Ra for IL17e.

The IL17Rc receptor functions as a coreceptor with IL17Ra to trigger IL17a and IL17f signaling. It binds with higher affinity to IL17f than IL17a. It is highly synthesized in non-immune cells of liver, kidney, thyroid, prostate, and joints.

The IL17Rd receptor can tether to fibroblast growth factor receptors FGFR1 and FGFR2 and impedes FGF signaling. The IL17Re receptor is targeted by interleukin-17c.

Whereas cytokines involved in T_{H1} and T_{H2} responses trigger the JaK–STAT pathway, IL17Ra engages ACT1 adaptor[27] to recruit TNFR-associated factor TRAF6 and activate nuclear factor-κB as well as mitogen-activated protein kinases, especially extracellular signal-regulated ERK1 and ERK2 kinases. The latter activate activator protein AP1 and transcription factors CCAAT–enhancer-binding proteins C/EBPβ and C/EBPδ.

11.1.3.7 Others Interleukins

Interleukin-5

Interleukin-5 is produced by eosinophils, T_{H2} cells, and mastocytes. It stimulates B-cell activity and growth. It also mediates eosinophil activation. This homodimer targets IL5 receptor of the type-1 cytokine receptor family.

---

25. Interleukin-8, or CXCL8, is a chemokine produced by macrophages and epithelial and endothelial cells. It binds to many receptors, especially G-protein-coupled receptors CXCR1 and CXCR2.

26. Helper T_{H17} cell operates prominently at mucosal surfaces and triggers pro-inflammatory danger signals that elicit neutrophil mobilization. These cells produce IL17a, IL17f, IL21, IL22, and IL26, as well as chemokines such as CC-chemokine ligand CCL20 (or MIP3α). Various T_{H17}-like cells that produce a similar range of cytokines include γδ T cells, natural killer (NK), natural killer T (NKT), and lymphoid tissue inducer (LTi) cells [1481]. Differentiation of T_{H17} cell is caused by transforming growth factor-β, in addition to IL1 and IL6 [1481]. This differentiation is regulated by transcription factors, such as signal transducer and activator of transcription STAT3, retinoic acid receptor-related orphan receptor RORγ2, and aryl hydrocarbon receptor. Interleukin-23 is required to expand and stabilize the cell population.

27. A.k.a. CIKS.

The IL5R heterodimer is composed of IL5-binding IL5Rα and signal-transducing IL5Rβ subunits. The (common) IL5Rβ subunit is also used by granulocyte–macrophage colony-stimulating factor and interleukin-3 [1482]. Alternative splicing of IL5Rα yields a membrane-bound ($_m$IL5Rα) and soluble form (IL5Rα^S). The latter does not support signal transduction, but antagonizes IL5 signaling.

Interleukin-8

Interleukin-8 is the chemokine CXCL8 synthesized by macrophages and epithelial and endothelial cells. It binds to its cognate CXCR1 and CXCR2 G-protein-coupled receptors.

Interleukin-14

Interleukin-14 [28] controls the growth and proliferation of B cells. This syntaxin-binding protein binds to syntaxin-1A, -3A, and -4A, as well as regulator of Gz (of the Gi subset)-selective protein signaling RGS20 [59]. It is produced mainly by T lymphocytes. It connects to IL14R receptor on lymphocyte surface.

Interleukin-16

Prointerleukin-16 is a pleiotropic cytokine that acts as a chemoattractant and a modulator of T-cell activation. It undergoes a proteolytic cleavage to yield 2 functional proteins. Interleukin-16 is synthesized by various types of immunocytes, such as lymphocytes, eosinophils, and mast and dendritic cells, and non-immune cells, such as neurons, fibroblasts, and some epithelial cells.

Interleukin-16 [29] targets cluster of differentiation CD4, a glycoproteic coreceptor of T-cell receptor that is expressed on the surface of helper and regulatory T cells, eosinophils, monocytes, macrophages, and dendritic cells, as well as proB cells and neurons [1483].

Interleukin-16 increases intracellular concentrations of 2 second messengers: Ca^{++} and inositol (1,4,5)-trisphosphate. It also provokes autophosphorylation of leukocyte-specific protein Tyr kinase (LcK). Tetraspanin-29 (or CD9) acts as an alternate interleukin-16 receptor.

Interleukin-32

Interleukin-32 [30] stimulates monocytes and macrophages to secrete tumor-necrosis factor-α and chemokines, such as CXCL2 and CXCL8 [1484]. It is produced in

28. A.k.a. high-molecular-weight B-cell growth factor (HMW-BCGF) and taxilin TxlnA (not Txnl1 (thioredoxin-like-1).

29. A.k.a. lymphocyte chemoattractant factor (LCF).

30. A.k.a. natural killer-cell protein NK4 and tumor-necrosis factor-α-inducing factor (TAIFa–TAIFd).

peripheral lymphocytes, epithelial cells, and NK cells. This pro-inflammatory mediator activates typical cytokine signal pathways, such as nuclear factor-κB and P38MAPK, as well as ERK1, ERK2, and PKB kinases. Four human IL32 splice variants exist.

Interleukin-34

Interleukin-34 promotes growth or survival of monocytes [1485]. This homodimer binds to the colony-stimulating factor-1 receptor (CSF1R) [31] It stimulates osteoclastogenesis, as it promotes the adhesion and proliferation of osteoclast progenitors [1486]. It primes phosphorylation of ERK1, ERK2, and PKB kinases.

11.1.4 Cytokine Receptors of the Immunoglobulin Superclass

A large group of plasmalemmal proteins that are involved in material recognition, molecular binding, or cellular adhesion belongs to the immunoglobulin superclass. Ubiquitous cytokine receptors of the immunoglobulin superclass are represented by: (1) certain *interleukin receptors*, such as IL1R chains IL1Rα (type-1 IL1R) that transmits inflammatory effects of IL1 and IL1Rβ (type-2 IL1R) [32] that suppresses IL1 activity, as well as IL18R; (2) macrophage colony-stimulating factor receptor (mCFSR), or colony-stimulating factor-1 receptor (CSF1R; [33] Sect. 8.2.5.10); and (3) stem cell factor receptor (SCFR; [34] Sect. 8.2.5.9).

Binding of IL1 to its receptor IL1R triggers dimerization of IL1R with transmembrane IL1R accessory protein (IL1RAP) and requires TRAF6 adaptor to activate JNKs and NFκB. The IL1R–IL1RAP dimer recruits death domain-containing adaptor MyD88 that tethers IL1R-associated IRAK1 and IRAK2 kinases. The latter can link to TRAF6 ligase. The MyD88–IRAK1–TRAF6 complex stimulate JNKs and NFκB factor.

11.1.5 Tumor-Necrosis Factor Receptor Superfamily

Members of the superfamily of tumor-necrosis factor receptors (TNFRSF), or tumor-necrosis factor and nerve growth factor receptors, mediate signaling from tumor-necrosis factor-related cytokines. They are activated by at least 18 different human TNF homologs of the TNF superfamily (TNFSF) that induce either cell survival, proliferation, and differentiation, or apoptosis on the one hand and contribute to inflammation and immune response on the other. Many of TNFRSF receptors and TNFSF ligands operate as multimers.

These transmembrane glycoproteins possess an extracellular ligand-binding domain, a single membrane-spanning region, and a cytoplasmic tail. Their cytoplasmic

31. A.k.a. macrophage colony-stimulating factor receptor.
32. A.k.a. CD121b
33. A.k.a. CD115.
34. A.k.a. KIT or CD117.

Table 11.7. Members of the family of TNFRSF receptors and their aliases (**Part 1**; Act35: lymphoid activation antigen; Apo: surface apoptosis antigen; DcR: death decoy receptor; DR: death receptor; FDCR: follicular dendritic cell-derived receptor; LTαR: lymphotoxin-α receptor; LTβR: lymphotoxin-β receptor; OCIF: osteoclastogenesis inhibitory factor; ODFR: osteoclast differentiation factor receptor; Opg: osteoprotegerin; RANK: receptor activator of nuclear factor-κB; TR: TNFR-related molecule; TRAILR: TNF-related apoptosis-inducing ligand receptor; TRAnCeR: TNF-related activation-induced cytokine receptor; TRID: TRAILR without an intracellular domain; TrunDD: TRAILR containing a truncated death domain; TxGP1R: Tax-transcriptionally activated glycoprotein-1 receptor).

Type	Aliases
TNFRSF1a	TNFR1, CD120a, TNFαR, P55TNFR, TNFR60
TNFRSF1b	TNFR2, CD120b, TNFbR, P75TNFR, P80TNFR, LTαR
TNFRSF3	TNFR3, CD18, TNFcR, TNFRRP, TNFR2RP, LTβR
TNFRSF4	CD134, OX40, TxGP1R, Act35
TNFRSF5	CD40, BP50
TNFRSF6a	CD95, FasR, Apo1, Apt1
TNFRSF6b	DcR3, TR6
TNFRSF7	CD27, S152, Tp55, T14
TNFRSF8	CD30, Ki1
TNFRSF9	CD137, ILA, and 4-1BB
TNFRSF10a	CD261, Apo2, DR4, TRAILR1
TNFRSF10b	CD262, DR5, KILLER, TRAILR2, and TRICK2
TNFRSF10c	CD263, DcR1, TRID, TRAILR3
TNFRSF10d	CD264, DcR2, TrunDD, TRAILR4
TNFRSF11a	CD265, RANK, ODFR, TRAnCeR
TNFRSF11b	Opg, OCIF, FDCR1, TR1

regions differ much more than extracellular domains. Most of TNFRSF members recruits adaptors of the TNFR-associated factor (TRAF) family to promote cell survival, as they lead to activation transcription factors NFκB and AP1 Activator protein.

The family of TNFRSF receptors includes many members (Tables 11.7 to 11.9).

The TNFRSF receptors can then be classified into 3 classes: (1) *death domain-containing receptors* that have cytoplasmic death domains for recruitment of death domain and death effector domain-containing proteins to prime cell apoptosis; (2) signaling *receptors without a death domain*; and (3) soluble or membrane-anchored *decoy receptors* that lack transmembrane and cytoplasmic domains and do not trigger signaling pathways.

Apoptosis-inducing TNFRSF members, such as TNFRSF1a, -6a, -10a and -10b, -16, -21, -25, and -27 possess a death domain that is associated with activation of apoptosis. Death receptors (DR3–DR6, i.e., TNFRSF10a and -10b, -21, and -25) transmit an apoptotic signal, whereas death decoy receptors (DcR1–DcR3 and osteoprotegerin, i.e., TNFRSF6a, -10c and -10d, and -11b) inhibit it (Table 11.10).

Table 11.8. Members of the family of TNFRSF receptors and their aliases (**Part 2**; AIR: apoptosis-inducing receptor; AITR: activation-inducible TNFR family member; Apo: surface apoptosis antigen; ATAR: another TRAF-associated receptor; BAFFR: B-cell-activating factor receptor; BCMA: B-cell maturation antigen; DDR: death domain receptor; DR: death receptor; EDa2R: ectodysplasin A2 receptor; GITR: glucocorticoid-induced TNFR family-related protein; HVEM: herpes virus entry mediator; LARD: lymphocyte-associated receptor of death; LIGHTR: lymphotoxin homolog inducible competitor of herpes simplex virus glycoprotein-D for HVEM receptor expressed by T-lymphocyte receptor; NGFR: low-affinity nerve growth factor receptor; RELT: receptor expressed in lymphoid tissue; TAJ: toxicity and JNK inducer; TR: TNFR-related molecule; TRAMP: TNFR-related apoptosis-mediating receptor; TRoy: TNFRSF expressed in the mouse embryo; TWeakR: TNF-like weak inducer of apoptosis receptor; XEDaR: X-linked EDaR).

Type	Aliases
TNFRSF12a	CD266, TWeakR
TNFRSF13b	TNFRSF14b, CD267
TNFRSF13c	CD268, BAFFR, BR3
TNFRSF14	CD258, TR2, ATAR, HVEM, LIGHTR
TNFRSF16	CD271, NGFR, GP80LNGFR, $P75^{NTR}$, NTR
TNFRSF17	TNFRSF13, TNFRSF13a, CD269, BCMA
TNFRSF18	AITR, GITR
TNFRSF19	TAJ, TRADE, TRoy
TNFRSF19L	RELT
TNFRSF21	DR6
TNFRSF22	SOBa; TNFRh2, mDCTRAILR2, TNFRSF1aL2
TNFRSF23	mSOB, TNFRh1, mDCTRAILR1
TNFRSF25	TNFRSF12, DR3, DDR3, Apo3, LARD, AIR, TR3, TRAMP, WSL1, WSL2, and WSL-LR
TNFRSF27	EDa2R, XEDaR, TR14

Death domain-containing receptors are upregulated during T-cell activation. They stimulate nuclear factor-κB. Death domains mediate homo- and heterotypic proteic interactions that allow formation of receptor–effector complexes to implement signaling programs.

Death domain-containing proteins (adaptors and kinases), such as FADD (FasR [TNFRSF6a]-associated death domain)-containing protein, receptor (TNFRSF)-interacting protein kinases (RIPK; Vol. 4 – Chap. 4. Cytosolic Protein Ser/Thr Kinases),[35] and TNFR-associated death domain-containing protein (TRADD), interact with TNFRSF1a or TNFRSF6a following ligand-mediated receptor aggregation (Table 11.11). Protein FADD then recruits caspase-8 to the receptor complex, thereby

35. Receptor-interacting protein kinase can interact with death domain-containing receptors such as TNFRSF6a receptor. It is involved in TNF-mediated activation of NFκB factor. It possesses a C-terminal death domain and N-terminal Ser/Thr kinase motif. Like TRAF2, it is an activator of both JNK and P38MAPK kinases. However, RIPK deletion does not disturb JNK activation by TNF. Affectors MAP4K2, MAP4K5, MAP3K1, and MAP3K5 actually bind to and are activated by TRAFs independently of RIPK protein.

Table 11.9. Members of the tumor-necrosis factor receptor superfamily (TNFRSF) and their adaptors and ligands (Source: [5]; BDNF: brain-derived neurotrophic factor; BTLA: B and T-lymphocyte attenuator; FADD: Fas-associated death domain; NGF: nerve growth factor; NT: neurotrophin; TRADD: TNF receptor-associated death domain; TRAF: TNF receptor-associated factor TNFSF1: TNFβ or lymphotoxin-α; TNFSF2: TNF or TNFα; TNFSF3: lymphotoxin-β; TNFSF4: CD252; TNFSF5: CD40 ligand or CD154; TNFSF6: Fas ligand, CD95L, or CD178; TNFSF7: CD70 or CD27 ligand; TNFSF8: CD30 ligand or CD153; TNFSF9: CD137L; TNFSF10: P53-regulated pro-apoptotic TRAIL or CD253; TNFSF11: receptor activator of nuclear factor-κB ligand [RANKL]; TNFSF12: CD255; TNFSF13: CD256; TNFSF13b: CD257; TNFSF14: CD258; TNFSF15: TNF ligand-related molecule-1; and TNFSF18: activation-inducible TNF-related ligand). Shed TNF form (TNF^S) preferentially activates TNFRSF1a, whereas membrane-bound TNF form ($_m$TNF) stimulates TNFRSF1a and TNFRSF1b equally.

Type	Adaptors	Ligands
TNFRSF1a	TRADD	TNFSF1, TNFSF2
TNFRSF1b	TRAF1–2/5	TNFSF1, TNFSF2
TNFRSF3	TRAF3–5	TNFSF3, TNFSF14
TNFRSF4	TRAF1–3/5	TNFSF4
TNFRSF5	TRAF1–3/5–6	TNFSF5
TNFRSF6a	FADD	TNFSF6
TNFRSF6b		TNFSF6, TNFSF14, TNFSF15
TNFRSF7	TRAF2, SIVA	TNFSF7
TNFRSF8	TRAF1–3/5	TNFSF8
TNFRSF9	TRAF1–3	TNFSF9
TNFRSF10a	FADD	TNFSF10
TNFRSF10b	FADD	TNFSF10
TNFRSF10c		TNFSF10
TNFRSF10d		TNFSF10
TNFRSF11a	TRAF1–3/5–6	TNFSF11
TNFRSF11b		TNFSF11
TNFRSF12a	TRAF1–3/5	TNFSF12
TNFRSF13b	TRAF2/5–6	TNFSF13b
TNFRSF13c	TRAF3	TNFSF13b
TNFRSF14	TRAF1–3/5	TNFSF1, TNFSF14, BTLA
TNFRSF16	TRAF2/4/6	NGF, BDNF, NT3, NT4
TNFRSF17(13a)	TRAF1–3/5–6	TNFSF13, TNFSF13b
TNFRSF18	TRAF1–3	TNFSF18
TNFRSF19	TRAF1–3/5	
TNFRSF19L	TRAF1	
TNFRSF21	TRADD	
TNFRSF22		
TNFRSF23		
TNFRSF25(12)	TRADD	TNFSF12, TNFSF15
TNFRSF27		

Table 11.10. Death and death decoy receptors of the TNFRSF superfamily.

Type	TNFRSF alias
Death receptors	
DR3	TNFRSF25
DR4	TNFRSF10a
DR5	TNFRSF10b
DR6	TNFRSF21
Death decoy receptors	
DCR1	TNFRSF10c
DcR2	TNFRSF10d
DcR3	TNFRSF6b

activating the apoptotic peptidase cascade. However, death domain-containing members of the TNFR superfamily are also able to stimulate alternative signaling pathways, hence preventing rather than triggering apoptosis. Besides, TRADD contains a TRAF-binding domain. It can thus recruit TNFR-associated factor TRAF2 and RIPK that are affectors of the Jun N-terminal kinases and P38MAPK pathways (Vol. 4 – Chap. 5. Mitogen-Activated Protein Kinase Modules). However, TRADD overexpression does not activate JNK kinases, but nuclear factor-κB.

Multimerization of TNFR superfamily members that results from ligand binding activates these receptors, hence triggering signal transduction by revealing binding domains for kinases or adaptors, such as TRAF, TRADD, and FADD proteins. Tumor-necrosis factor receptor-associated factors directly interact with some TNFRSF members, such as TNFRSF1b, TNFRSF3, TNFRSF5, and TNFRSF8. Members of the TNFRSF superfamily without a death domain, such as TNFRSF9 and TNFRSF10, interact with cytoplasmic TRAF adaptors either directly or indirectly. These TNFRSF members stimulate NFκB and AP1 transcription factors.

Signaling from TNFRSF via tumor-necrosis factor receptor-associated factors can be influenced by multiple factors [1487]: (1) TRAF abundance and location and (2) affinity of TRAFs for TRAF-binding proteins, such as TRAF family member-associated NFκB activator (TANK), TRAF-interacting protein (TRAIP or TRIP), and inhibitors of apoptosis, as well as other TRAF proteins. Protein TANK binds to and sequesters inactive TRAFs in the cytoplasm. The TRAF-interacting protein also interacts with TRAFs and prevents TRAF-mediated NFκB activation, hence protecting cells against apoptosis.

A subgroup of TNFRSF receptors that comprises TNFRSF1b, TNFRSF4, TNFRSF7, TNFRSF8, TNFRSF9, TNFRSF14, and TNFRSF18 is expressed on T lymphocytes. These receptors mediate distinct effects of costimulation in specific T-cell subsets. They can contribute to T-cell priming and polarization into T-helper cell subtypes (T_{H1}, T_{H2}, or T_{H17}) or local T-cell accumulation and cytokine production.

Table 11.11. Death domain-containing adaptors (Sources: [59, 608]; AnkK, AnkRD: ankyrin repeat domain-containing kinase; CaRD: caspase recruitment domain-containing protein; CaRDIAK: CaRD-containing interleukin-1β-convertase (ICE)-associated kinase; CFLAR: caspase-8 and FADD-like apoptosis regulator; DAPK: death-associated protein kinase; DSTYK: dual protein Ser/Thr and Tyr kinase; FADD: FasR (TNFRSF6)-associated via death domain-containing protein; GIG3: growth-inhibiting gene-3 protein; MORT: mediator of receptor-induced toxicity; PKK: PKCδ-interacting protein kinase; RIPK: receptor (TNFRSF)-interacting Ser/Thr kinase; TICAM: Toll-interleukin-1 receptor domain-containing adaptor; TNFαIP: TNFα-induced protein; TRADD: TNFR-associated death domain-containing protein; TRAF: TNFR-associated protein: TRPC4AP: transient receptor potential cation channel-C4-associated protein [or TNFR-associated ubiquitous scaffolding and signaling protein (TRUSS)]; UbC: polyubiquitin-C).

Type	Aliases	Interactors
FADD	GIG3, MORT1	TNFSF6, TNFRSF1a/6a/10a/10b, TRADD, RhoA, RalGEF, RhoGDI, DAPK1, RIPK1, MAPK8, PKCζ, Casp8/10, CFLAR, SUMo1, UbC
RIPK1	RIP1	TNFSF2, TNFRSF1a/6a/10b, TNFαIP3, TRAF1–3/5, FADD, TRADD, TRPC4AP, EGFR, RIPK3, MAP3K7/14, MAP3K7IP2, IKKγ, Casp8/10, Itch, UbC, USP21
RIPK2	RIP2, GIG30, CaRD3, CaRDIAK	TNFRSF1a TRAF1/2/5/6, CaRD6, IKKγ, Casp1, CFLAR, UbC9, UbC,
RIPK3	RIP3	TNFRSF1a, RIPK1, TRAF2, TICAM1
RIPK4	RIP4, PKK, AnkK2, AnkRD3	PKCβ/δ, RhoA
RIPK5	RIP5, DSTYK, DuSTYPK	IKKϵ
TRADD		TNFSF2, TNFRSF1a/6a/10a/10b/16/21/25, TRAF2, FADD, TRPC4AP, BCL10, RIPK1, HIPK2, STAT1, Cav1, Casp8

11.1.5.1 Receptors TNFRSF1a and -b

The TNFRSF1a receptor can recruit all of the known signaling pathways activated by TNFα. Upon ligand binding, TNFRSF1a binds adaptor TRADD and recruit both pro-apoptotic adaptor FADD and pro-inflammatory adaptor TRAF2. In addition, RIPK that does not interact directly with TNFRSF1a binds to TRADD to then form a TRADD–RIPK–TRAF2 trimer. Once complexed, TNFRSF1b that does not possess a death domain, binds TRAF adaptor.

In mammals, tumor-necrosis factor causes the translocation of Embryonic ectodermal development protein (EED) from the nucleus to the plasma membrane, where it forms the TNFRSF1a–EED–nSMase2–FAN–RACK1 complex [1488]. The EED protein of the Polycomb group couples TNFRSF1a to neutral sphingomyelinase (nSMase; Vol. 4 – Chap. 1. Signaling Lipids) that generates ceramide. The latter intervenes in tissue development and inflammation as well as cell differentiation, growth, and death. Neutral sphingomyelinase-2 is activated exclusively by TNFRSF1a receptor that can bind to Factor associated with nSMase activation (FAN). The latter recruits scaffold Receptor for activated C-kinase RACK1. [36]

Tumor-necrosis factor is a prominent signal for the production of reactive oxygen species by NADPH oxidase. Reactive oxygen species serve as signals in immunity. The TNFRSF1a-binding riboflavin kinase (RFK) [37] of the class of ATP:riboflavin 5′-phosphotransferase couples TNFRSF1a to NADPH oxidase [1489]. [38] Bridging between TNFRSF1a and NADPH oxidase mediated by RFK is required for ROS production primed by TNF, but not Toll-like receptors.

The immune response is controlled by positive and negative signals delivered by antigen-independent, costimulatory signals from receptors of the TNFRSF superfamily to maintain and terminate the immune response, once CD4+ or CD8+ T lymphocytes have received antigen-specific signals from T-cell receptors. [39] The TNFRSF1a receptor is a T-cell costimulatory molecule that operates in T-cell activation mediated by T-cell receptors for the timing of cytokine responses [1490].

11.1.5.2 Receptors TNFRSF11a and -b

The TNFRSF11a Receptor interacts with the adaptors TNFR-associated factors (TRAF1–TRAF3, TRAF5, and TRAF6). [40] Binding to TRAF contributes to

---

36. A.k.a. G-protein subunit-β2-like protein-1 (Gβ2L1).

37. Formerly termed flavokinase.

38. RFK Kinase binds to both TNFRSF1a and 22-kD polypeptide P22PhOx (Vol. 4 – Chap. 9. Other Major Signaling Mediators), the common subunit (light α chain) of NADPH oxidase isoforms.

39. Resting T lymphocytes express both TNFRSF1a and -1b. Their synthesis augments after T-cell activation. In addition to costimulatory receptors of the CD28 family, several members of the TNFRSF superfamily (TNFSF1, -2, -4, -7 to -9, -14, and -18) exert costimulatory effects on T lymphocytes. Costimulatory signals from TNFRSF7 (CD27) and CD28 (receptor for CD80 and CD86 that prime T cells against antigens presented by antigen-presenting cells) contribute to the initiation of T-cell activation, as they enhance T-cell proliferation and interleukin-2 production. Besides, TNF enhances TCR- or mitogen-induced T-cell proliferation and IL2Rα production. Members of the TNFRSF superfamily participate in the control of effector T-cell differentiation and survival [1490].

40. The TNFRSF11a receptor contains 2 independent TRAF-binding sites: a binding domain for multiple TRAFs (TRAF1–TRAF3, TRAF5, and TRAF6) and a binding domain only for TRAF6.

TNFRSF11a-dependent activation of transcription factor NFκB and Jun N-terminal kinases. Interaction with TRAF6 is required in the TNFRSF11a-mediated NFκB signaling, whereas it partially contributes to JNK activation [1487].

The receptor TNFRSF11a, specific to TNFSF11, is particularly expressed in osteoclast progenitors, mature osteoclasts, activated T lymphocytes, and dendritic cells. Yet, TNFRSF11a can be widely distributed (skeletal muscle, thymus, liver, colon, adrenal gland, and small intestine).

Once activated, TNFRSF11a initiates signaling cascades that not only involve NFκB, JNKs, and Jun, but also PKB kinase. The complex made of TNFSF11 and TNFRSF11a promotes dendritic cell survival, clustering, and immunostimulatory capacity. The TNFSF11–TNFRSF11a complex also fosters the survival of T lymphocytes that are subjected to TGFβ. It also contributes to the regulation of interactions between T lymphocytes and dendritic cells.

Protein TNFSF11 self-assembles as a trimer. It is detected in thymus, lymph node, spleen, bone marrow stroma, and trabecular bone. Specific lymphoid cells that express TNFSF11 include both CD4+ and CD8+ T lymphocytes and B-lymphocyte progenitors. It is expressed in osteoblastic lineage cells as a cell-bound form that can give rise to a truncated ectodomain variant after cleavage by adamlysin ADAM17. It is also observed in activated T lymphocytes in a soluble form ($TNFSF11^S$). Its production is modulated by various cytokines (IL1, IL6, and IL11 and TNFα) and hormones, such as glucocorticoids and parathyroid hormone. It promotes osteoclast differentiation, activation, and survival.

Protein TNFRSF11b[41] impedes the TNFSF11 activity. Secreted glycoprotein TNFRSF11b (sTNFRSF11b) exists as monomer and homodimer. It blocks osteoclast differentiation from precursor cells and prevents osteoporosis. It acts as a soluble decoy receptor in osteoclast-mediated bone remodeling as well as immunocyte interactions, dendritic cell survival, and lymph node genesis. It indeed competes with TNFRSF11a receptor. Its C-terminus contains 2 death domains.

Agent TNFRSF11b is produced in various tissues, such as the heart, arteries, veins, lung, kidney, intestine, and bone, in addition to hematopoietic cells. Cytokines (TNFα, IL1α and IL18, TGFβ, and BMPs) and steroid hormones (e.g., 17β-estradiol [E_2]) upregulate TNFRSF11b expression, whereas glucocorticoids, parathyroid hormone, prostaglandin-E2, and FGF2 suppress TNFRSF11b production.

Protein TNFRSF11b not only prevents the interaction between TNFSF11 and TNFRSF11a, but also inhibits TNFSF10, a potent activator of apoptosis. Conversely, TNFSF10 can block TNFRSF11b activity.

Last, but not least, TNFRSF11b is involved in arterial calcification [1493]. It can be synthesized by vascular smooth muscle and endothelial cells. In endothelial cells, TNFRSF11b acts as an autocrine survival factor.

41. A.k.a. osteoprotegerin.

11.1.5.3 Other TNFRSF Types

The TNFRSF3 receptor is prominently produced in epithelial cells, but is absent in B and T lymphocytes.

The TNFRSF5 receptor binds TRAF6 independently of other TRAF connections. It then can prime NFκB signaling in the absence of TRAF2 or TRAF5 binding.

The TNFRSF6a receptor is a major effector of cytotoxic T lymphocytes and natural killer cells. It is also involved in the establishment of immunological tolerance.

The TNFRSF6b receptor is a soluble receptor without death domain and cytoplasmic domain that is secreted by a wide set of human cell types. It binds to TNFSF6, TNFSF14, and TNFSF15 [1491].[42] As a decoy receptor, it blocks TNFSF14 interaction with its receptor. It exerts apoptosis inhibition, as it tethers cytotoxic ligands. It then modulates the immune response.

The TNFRSF14 receptor is expressed mainly in the lung, spleen, thymus, bone marrow, and small intestine. It binds to 2 secreted ligands: TNFSF2 and TNFSF14 proteins. The latter is produced by activated T lymphocytes and macrophages. It binds not only TNFRSF14, but also TNFRSF3 receptor. The TNFRSF14 protein interacts with both TRAF2 and TRAF5 adaptors and ligases. Like most other TRAF-binding TNFRSF members, TNFRSF14 is able to activate NFκB, when it interacts with TRAF5, but not TRAF2 [1494].

Death domain-containing receptor TNFRSF25 is preferentially expressed in immunocytes. The TNFRSF25 receptor abounds on activated lymphocytes and thymocytes. Its cytoplasmic region causes NFκB activation and cell death. It is indeed able to recruit Fas-associating protein with death domain (FADD) and TNFR1-associated death domain protein (TRADD) that can connect TNFRSFs to the apoptosis-related cysteine peptidase caspase-1[43] via adaptor FADD and procaspase-8.[44]

11.1.5.4 Tumor-Necrosis Factor Signaling in Endosomes

Components of the death-inducing signaling complex (DISC) — Fas-associated death domain (FADD), TNFR-associated death-domain (TRADD), and receptor-interacting protein RIPK1 — are recruited to ligand-bound TNF receptors at the plasma membrane. However, caspase-8 is recruited to DISC in endosomes and then activated by autoproteolytic cleavage, and not at the plasma membrane, to promote apoptosis [11]. In endosomes, caspase-8 binds to DISC, whereas RIPK1 is removed

42. Protein TNFSF15 (or endothelial cell-derived TNF-like factor TL1a) is a ligand for both death receptor DR3 (TNFRSF25) and decoy receptor DcR3 (TNFRSF6b). Its expression is inducible by TNF and IL1α cytokines. Ligand TNFSF15 causes NFκB activation and apoptosis in TNFRSF25-expressing cells. This costimulator also increases IL2 responsiveness and pro-inflammatory cytokine secretion. Effects of TNFSF15 are antagonized by TNFRSF6b [1492].

43. A.k.a. interleukin-1β-converting enzyme (IL1βCE), or interleukin 1β convertase (IL1βC and ICE).

44. A.k.a. FADD-homologous ICE-like peptidase (FLICE) and mediator of receptor-induced toxicity (MORT1)MORT1-associated CeD3 (cell death) homolog (MACH).

from the complex. Protein RIPK1 is polyubiquitinated by Ub ligase CARP2[45] in endosomes and then degraded by the proteosome.

11.1.6 Chemokine Receptors

Chemokines constitute a family of chemotactic cytokines that provoke chemotaxis of nearby responsive cells (Vol. 2 – Chap. 6. Cell Motility). They signal once bound to their cognate G-protein-coupled chemokine receptors (Sect. 7.13.14). Some chemokines control immunocytes during immune surveillance. Others operate in tissue development, remodeling, and repair, such as angiogenesis and wound healing. Inflammatory chemokines guide cells of both innate and adaptive immune system to sites of infection or tissue damage.

Members of the chemokine family are classified into 4 groups according to the spacing of 2 N-terminal cysteine residues: (1) CC chemokines (CCL1–CCL28), or β-chemokines, that have 2 adjacent cysteines (C); (2) CXC chemokines (CXCL1–CXCL17), or α-chemokines, with the 2 N-terminal cysteines separated by one amino acid (X); (3) C chemokines (XCL1–XCL2 or lymphotactinsα and -β), or γ-chemokines; and (4) a single CX_3C chemokine, CX_3CL1 or fractalkine.

Therefore, chemokine receptors are subdivided into 4 distinct classes that correspond to the 4 chemokine groups: (1) CXCR receptors (CXCR1–CXCR7), (2) CCR receptors (CCR1–CCR11); (3) a single XCR1 receptor (CCXCR1 or GPR5), and (4) CX_3CR1 receptor.

11.1.7 Other Cytokine Receptors

Other cytokine receptors include the receptors of interleukin-8, transforming growth factor-β, and thrombopoietin. *Interleukin-8 receptor* is a dimer composed of 2 chains: IL8Rα and IL8Rβ. It acts preferentially on neutrophils.

Transforming growth factor-β receptors (TβR1–TβR3) are Ser/Thr kinase receptors (Sect. 8.3). The TβR isoforms can homo- or heterodimerize. Receptors TβR1 and TβR2 have a high and low affinity for TGFβ1 and TGFβ2, respectively, whereas TβR3 has a high affinity for both TGFβ1 and TGFβ2.

Thrombopoietin receptor (TpoR or CD110) is involved in megakaryocyte and platelet development. Upon binding of thrombopoietin, it is phosphorylated and forms a dimer that stimulates the JaK–STAT and MAPK pathways.

11.2 Other Receptors of the Immune System

Pathogen recognition by the innate immune system can be specific. *Pattern recognition receptors* (PRR) are actually expressed by cells of the innate immune

45. Caspase-8 and -10-associated RING finger protein is also called RING finger and FYVE-like domain-containing protein RFFL1.

system to identify components of microbes, the so-called *pathogen-associated molecular patterns* (PAMP) that are associated with pathogens or cellular stress. Once they have sensed pathogen invasion, PRRs trigger innate immune response and antigen-specific adaptive immunity.

Pathogen-associated molecular pattern (PAMP) receptors include: (1) *Toll-like* (TLR), (2) *NOD-like* (NLR), and (3) *C-type lectin* (CLR) receptors, and (4) *RIG-I-like receptors* (RLR), or cytosolic RIG-I-like RNA helicases (RLH). Toll-like receptors are also involved in the recognition of self molecules to avoid autoimmune diseases.

According to functional mode, PRRs are subdivided into *endocytic* and *signaling* pattern recognition receptors. Signaling pattern recognition receptors include plasmalemmal Toll-like and cytoplasmic NOD-like receptors. Endocytic pattern recognition receptors promote attachment, engulfment, and destruction of microorganisms by phagocytes, without relaying an intracellular signal. They comprise *mannose receptors* of macrophages and dendritic cells, *glucan receptors* on phagocytes, and *scavenger receptors* for charged ligands on phagocytes. Plasmalemmal pattern recognition receptors comprise receptor kinases, Toll-like receptors, and mannose receptor.

11.2.1 B-Cell Receptors

Immunoglobulins (IgA, IgD, IgE, IgG, and IgM), or antibodies, are γ-globulins that are produced by B lymphocytes to identify and neutralize foreign objects. Soluble forms are detected in the body's fluids, whereas membrane-bound forms remain attached to B-lymphocyte plasma membranes, where they constitute a part of B-cell receptors (BCR). Developing B lymphocytes use a preB-cell receptor (preBCR) that monitors successful adaptative rearrangement at the gene that encode the immunoglobulin heavy chains.

The B-cell receptors enable B lymphocytes to detect antigen and trigger B-cell activation (Tables 11.12 and 11.13). They are composed of surface-bound IgD or IgM and associated Igα–Igβ heterodimers. Upon antigen binding, they cluster in large patches isolated from most other cell signaling receptors.

Receptors BCRs and TCRs are encoded by genes that are assembled from dispersed variable (V), diversity (D), and joining (J) genic elements in lymphocyte precursors. Precursor preB- and preT-cell receptors mediates lymphocyte proliferation and development progression. Afterward, lymphocytes stop dividing and recombine the genes that encode receptor chains (IgLC and TCRα and -γ). V(D)J Recombination at specific recombination signal sequences is catalyzed by RAG1 and RAG2 recombination-activating gene products. Inclusion of D elements generates higher diversity. Each antigen receptor heterodimer is encoded by a first receptor-chain locus that includes V, D, and J elements, which encodes IgHC, TCRβ, and TCRδ, and a second receptor-chain locus that comprises only V and J elements, which is usually rearranged and encodes IgLC and TCRα and -γ [1495]. For example, TCRA locus that encodes TCRα contains about 50 Vα and 60 Jα elements. Selection ultimately

Table 11.12. Signaling from B- (BCR) and T-cell (TCR) receptors. The depicted TCR pathway corresponds to the MHC class-2 pathway. Only major effectors and transcription factors are given. Activation of these receptors leads to gene transcription to synthesize products for cell differentiation and proliferation, as well as secretion of multiple cytokines and synthesis of their cognate receptors (**Part 1**; CD: cluster of differentiation; DCSIGN: dendritic cell-specific intercellular adhesion molecule-3-grabbing non-integrin [a.k.a. CLec4l and CD209]; ITAM: immunoreceptor tyrosine-based activation motif-containing protein). Unlike dendritic cells and macrophages, B lymphocytes have specific plasmalemmal immunoglobulins that efficiently trap antigens for endocytosis, degradation, and presentation of derived peptides to MHC class-2 molecules on B lymphocytes. The latter then present processed antigen particules to T helper cells. The MHC class-1 pathway is similar, but all the nucleated cells express MHC class-1 molecules (not only antigen-presenting cells, i.e., B lymphocytes, dendritic cells, and macrophages). Moreover, the MHC class-1 pathway mainly aims at producing T cytotoxic cells, instead of T helper cells. Lymphocyte function-associated antigens (LFA1–LFA3) are cell adhesion molecules on leukocytes. Molecule LFA1 (CD11a–CD18 dimer) is the $\alpha_L\beta_2$-integrin. It localizes to plasma membranes of all T and B lymphocytes, macrophages, and neutrophils. It binds to ICAM1 on antigen-presenting cells. Molecule LFA3 (CD58) is expressed on antigen-presenting cells, especially macrophages. It binds to LFA2 (CD2), a member of the immunoglobulin superfamily, on T lymphocytes and natural killer cells. The tetraspanin family includes Tspan24 (CD151), Tspan25 (CD53), Tspan26 (CD37), Tspan27 (CD82), Tspan28 (CD81), Tspan29 (CD9), and Tspan30 (CD63). Tetraspanin-28 interacts directly with immunoglobulin superfamily member IgSF8 (CD316) and CD36. It forms a signaling complex with CD19, CD21 (complement receptor CR2), and CD225 (interferon-induced transmembrane protein IFITM1) on the surface of B lymphocytes. On T lymphocytes, Tspan28 associates with CD4 and CD8 and provides a costimulatory signal with CD3 antigen.

	BCR	TCR
Inducer cell	Helper T cell	Antigen-presenting cell
Ligand	Antigen	MHC class-2 molecule (MHC class-1 molecule)
Interaction site	Lymph node, spleen	Thymus, lymph node, spleen
Costimulators	B cell: CD19/21/40/81 T_H cell: TNFSF5	T cell: CD2/4/28; LFA1; ICAM3 Antigen-presenting cell: MHC2, CD80/86, LFA3, ICAM1/2, DCSIGN
ITAM	Igα/β	CD3

stops gene rearrangements and, in T lymphocytes, promotes the loss of either CD4 or CD8 marker.

11.2.2 Fc Receptors

The Fc receptor on the surface of natural killer cells, neutrophils, macrophages, and mastocytes binds the crystallizable fragment (Fc) of antibodies attached to in-

Table 11.13. Signaling from B- (BCR) and T-cell (TCR) receptors. Only major effectors and transcription factors are given. Activation of these receptors leads to gene transcription to synthesize products for cell differentiation and proliferation, as well as secretion of multiple cytokines and synthesis of their cognate receptors (**Part 2**; BLnk: B-cell linker; CD: cluster of differentiation; GAB: growth factor receptor-bound protein GRB2-associated binder; ITAM: immunoreceptor tyrosine-based activation motif-containing protein; LAT: linker for activated T cells; LCP2: lymphocyte cytosolic protein-2 [also called SLP76]; PIK3AP1: phosphoinositide 3-kinase adaptor protein-1 [also called BCAP]; PTPRc: protein tyrosine phosphatase receptor-C [also called CD45]; RIBP: resting lymphocyte kinase (RLK–TXK) and interleukin-2-inducible T-cell kinase (ITK)-binding protein [a.k.a. leukocyte-specific non-receptor Tyr kinase (LcK)-associated adapter (LAd)]; TRAT1: T-cell receptor associated transmembrane adaptor-1 [a.k.a. TRIM]). The depicted TCR pathway corresponds to the MHC class-2 pathway.

	BCR	TCR
Initiating Src kinases	BLK, Fyn, Lyn	LcK
Initiating phosphatase	PTPRc	PTPRc
ITAM-Binding SYK	SYK	ZAP70
Step-1 mediators (enzymes and adaptors)	BTK, Tec PI3K CBL, CD19, PIK3AP1 GAB1	ITK, TxK PI3K RIBP, LAT, LCP2, TRAT1
Effectors	PLCγ, IP_3, DAG, Ca^{++}, PKC Ras, MAPK, PP3 BLnk	 LAT, LCP2, GAds
Transcription factors	AP1, NFAT, NKκB, SRE	
Outcomes	Proliferation Differentiation into memory B cells and plasmocytes Ig class switching	IL2 and IL2R synthesis Proliferation Differentiation into T_H cells

fected cells or pathogens. Activited Fc receptor stimulates phagocytic or cytotoxic cells to destroy microbes or infected cells. Several types of Fc receptors exist according to the type of recognized antibody. Fcα- (FcαR), Fcγ- (FcγR), and Fcϵ receptors (FcϵR) bind to IgA, IgG, and IgE, respectively.

11.2.2.1 Fcα Receptors

A single Fc receptor pertains to the FcαR subgroup, the FcαR1 (CD89). It resides on eosinophils, neutrophils, monocytes, some macrophages, and some dendritic cells. This member of the immunoglobulin superfamily also belongs to the multichain immune recognition receptor family (MIRR). It connects to 2 FcγR to signal. Although it has a higher affinity for IgM antibodies, the Fcα–μ receptor (Fcα–μR), a member of the immunoglobulin superfamily, binds also to IgA, hence its name. It is observed on B lymphocytes, macrophages, and mesangial cells in kidneys.

11.2.2.2 Fcγ Receptors

All Fcγ receptors belong to the immunoglobulin superfamily. They prime phagocytosis of opsonized (coated) microbes. The FcγR subgroup includes several members with different structure: FcγR1 (CD64), FcγR2a (CD32a), FcγR2b (CD32b), FcγR3a (CD16a), FcγR3b (CD16b).

The FcγR1 receptor is detected on eosinophils, neutrophils, dendritic cells, and macrophages; FcγR2a on eosinophils, neutrophils, macrophages, platelets, and Langerhans cells; FcγR2b1 on B lymphocytes and mastocytes; FcγR2b2 on eosinophils, neutrophils, and macrophages; FcγR3a on natural killer cells and macrophages; FcγR3b on eosinophils, neutrophils, macrophages, mastocytes, and follicular dendritic cells. The FcγR2a receptor has a low affinity for IgG antibodies.

The neonatal Fc receptor (FcRn) is expressed on multiple cell types. It binds also to IgG. In particular, it contributes to the transfer of IgG from a mother to her fetus via the placenta or to her infant via breast milk. It localizes to monocytes, macrophages, dendritic cells, epithelial and endothelial cells, and hepatocytes.

11.2.2.3 Fcϵ Receptors

Two types of Fcϵ receptors (FcϵR) exist: (1) high-affinity FcϵR1 receptor of the immunoglobulin superfamily and (2) low-affinity FcϵR2 receptor (CD23), a C-type lectin.

The FcϵR1 receptor is especially the high-affinity heterotetrameric receptor for immunoglobulin-E. It localizes on antigen-presenting cells, basophils, eosinophils, mastocytes, and epidermal Langerhans cells. It intervenes in allergy. The FcϵR1 receptor is connected to Lyn of the SRC family of protein Tyr kinases. It also activates cytosolic SYK protein Tyr kinase.

The low-affinity FcϵR2 receptor (CD23) lodges on B lymphocytes, eosinophils, and Langerhans cells. It operates as a membrane-bound ($_m$FcϵR2) or soluble (FcϵR2^S) receptor. It controls B-cell growth and differentiation and prevents IgE binding of basophils, eosinophils, and monocytes.

11.2.2.4 Allergy Inhibitory Receptor-1

The immunoglobulin-like receptor Allergy inhibitory receptor-1 (allergin-1) is preferentially expressed on mastocytes. It recruits phosphatases PTPn6 and PTPn11 as well as SHIP phosphatase. Coligation of allergin-1 and FcεR1 suppresses IgE-mediated degranulation of mastocytes to avoid anaphylaxis [1496].[46]

11.2.3 T-Cell Receptors

T-cell receptors are members of the immunoglobulin superfamily on the surface of T lymphocytes. These heterodimers are composed of α and β chains in 95% of T cells, whereas other TCRs consist of γ and δ chains. T-Cell receptor complex formed by TCR, CD3, and ζ chain signal by simultaneous binding of MHC molecules to a specific coreceptor: CD4 on T helper cell that binds class-2 MHC molecules and CD8 on cytotoxic T cells that associates to class-1 MHC molecules (Tables 11.12 and 11.13).

T-cell receptor (TCR), targeted by a combination of antigen fragments with a glycoprotein of the major histocompatibility complex, leads to the activation of protein phosphatase-3 that dephosphorylates nuclear factor of activated T cells. Dephosphorylated NFAT enters the nucleus and, with accessory transcription factors, binds to gene promoters.

The TCR$\alpha\beta$ dimer of T-cell receptor[47] on the surface of T lymphocytes requires antigen recognition to detect antigens and function in adaptive immunity, but not its predecessor, the preT-cell receptor (preTCRα–TCRβ dimer).[48] The preT-cell antigen receptor is involved in early T-cell development (TCRβ selection, survival and proliferation of CD4−, CD8−, double-negative thymocytes, and subsequent $\alpha\beta$ T-cell lineage differentiation) [1498]. PreTCRs can form functional dimers independently of ligand stimulation. Invariant preTCRα chain couples with any TCRβ chain after successful rearrangement of the TCRB gene [1498].

11.2.3.1 Phosphorylations in Macrophages Primed by TLR Stimulation

Recognition of microbial danger signals by Toll-like receptors causes reprogramming of macrophages that are tissue-resident sensors of invading pathogens via pattern recognition receptors.[49] Once TLR4 bound to its ligand lipopolysaccharide, cascades of kinase activation are triggered in macrophages [1499] (Table 11.14).

---

46. Degranulation of mastocytes and basophils releases stored histamine, tryptase, carboxypeptidase-A, and proteoglycans. Pro-inflammatory lipid mediators, such as prostaglandins and leukotrienes, as well as platelet-activating factor are synthesized and released in the early phase (5–30 mn) after exposure to antigen, and cytokines and chemokines in the late phase (2–6 h).

47. Each chain (α and β) contains a variable (V) domain and a constant (C) immunoglobulin-like domain. Upon antigen recognition, TCR $\alpha\beta$ dimers on mature T lymphocytes can dimerize.

48. Subunit pTCRα is an invariant transmembrane glycoprotein that does not undergo rearrangements. Immature T lymphocytes (thymocytes) constitutively produce preTCRα. Subunit preTCRα contains a C-type immunoglobulin domain. This domain interacts with the C domain of the associated TCRβ chain.

49. To avoid excessive inflammation, macrophage activation is controlled by immunosuppressive cytokine interleukin-10.

Table 11.14. Kinases of TLR4 signaling triggered by lipopolysaccharide (Source: [1499]). Many signaling pathway are stimulated, such as the adipocytokine, insulin, MAPK, PI3K–PKB, PKB–BAD, Rho GTPase, TGFβ, TNFα, and TOR signaling. The PI3K–PKB axis is implicated in the production of interferon type-1 and interleukin-10, as well as STAT3-dependent control of caspase-1. Enzymes ATMK and ATRK phosphorylate the cell cycle checkpoint kinases ChK2 and ChK1, respectively. Macrophage motility and phagocytosis depend on cytoskeletal remodeling, hence Rho GTPases. Several transcription factors are activated, such as CCAAT-enhancer-binding proteins sucha as C/EBPδ, cAMP-responsive element-binding protein (CREB), heat shock transcription factors, interferon regulatory factors (IRF), myocyte enhancer factor MEF2, nuclear factor of activated T cells (NFAT), and NFκB.

Time	Kinase types
15 mn	ATMK, ATRK, Aurora, CamK2, CHK1, CK1, ERK1/2, GSK3, NEK6, PKA, PKB, PLK
4 h	PKD, CHK1, ERK1/2

11.2.3.2 Signaling Adaptors

T Lymphocytes are activated by engagement of T-cell receptor with processed antigen fragments presented by antigen-presenting cells (e.g., macrophages, dendritic cells, and B lymphocytes). Engagement of T-cell receptor initiates activation of several tyrosine kinases, hence phosphorylation of many intracellular proteins that primes Ca^{++} influx and multiple pathways such as the Ras–MAPK cascade (Vol. 4 – Chap. 5. Mitogen-Activated Protein Kinase Modules).

Proteic adaptors integrate these pathways by their ability to simultaneously bind multiple signaling components. Furthermore, adaptors contribute to the spatiotemporal regulation of signaling cascades that control lymphocyte differentiation and activation (Tables 11.12 and 11.13). Among them, many plasma membrane-associated adaptors exert an important role in T-cell activation by coupling engaged TCR to signaling mediators [1500].

Transmembrane adaptor Linker for activation of T cells (LAT), once phosphorylated, associates with adaptors GRB2 and GAds, [50] PLCγ1, among other signaling molecules, hence leading to aggregation of signaling complexes. Protein GAds participates in leukocyte-specific protein Tyr kinase signaling.

Adaptor GRB2 attracts various signaling proteins to phosphorylated LAT, such as guanine nucleotide-exchange factor Son of sevenless (SOS) and the adaptor Casitas B-lineage lymphoma ligase (CBL).

Adaptor GAds interacts with adaptors GAB1 and Lymphocyte cytosolic protein LCP2 [51] to recruit its partners Vav guanine nucleotide-exchange factor, IL2-inducible T-cell kinase ITK, and adaptors Non-catalytic region of tyrosine kinase NCK1 and Adhesion and degranulation-promoting adaptor protein (ADAP) to LAT adaptor. Cytokine production and integrin clustering in T lymphocytes depend on

50. A.k.a. GRB2-related adaptor protein GRAP2.
51. A.k.a. SLP76.

lipid-interacting adaptor ADAP that binds to acidic lipids such as phosphatidylinositides [1501].

Upon TCR commitment, cytosolic leukocyte-specific Tyr kinase (LcK) becomes activated and, in turn, phosphorylates the ITAM motif of the CD3 coreceptor. [52] Phosphorylated CD3ζ subunit binds to ZAP70 kinase [53] and enables ZAP70 to phosphorylate LAT adaptor. Afterward, phosphorylated LAT docks numerous signaling proteins.

The adaptors T-cell receptor-associated transmembrane protein TRAT1 [54] and SHP2 (PTPn11)-interacting transmembrane protein (SIT), [55] which constitutively associate with several plasmalemmal molecules, bind, after T-cell activation, to PI3K kinase and PTPn11 phosphatase, respectively.

Adaptor in lymphocytes of unknown function-X (ALX) [56] and RLK- and ITK-binding protein (RIBP) [57] are related by structure and coexpressed in T lymphocytes. Nonetheless, ALX and RIBP do not exhibit any functional redundancy or synergy during lymphocyte development, selection, activation, or survival [1502].

T-cell adaptor RIBP is also expressed in B lymphocytes. It localizes at the immunological synapse (T-cell contact with antigen-loaded presenting cells). It recruits mitogen-activated protein kinase kinase kinase MAP3K2, but not MAP3K3, during T-cell activation [1503]. Therefore, both upstream kinases of the MAPK module are primed by different signaling axes to activate their common target extracellular signal-regulated protein kinase ERK5 (last node of the MAPK module). [58]

In addition, RIBP binds to G-protein β subunit. Consequently, RIBP intervene in T-cell migration via G-protein-coupled chemokine receptors [1504]. In response to chemokine stimulation, RIBP associates with Tyr kinases of the Src and SYK families, leukocyte-specific protein Tyr kinase (LcK), and ζ-chain-associated Zap70 kinase. The latter is expressed in T and natural killer cells. Moreover, RIBP is required for chemokine-dependent tyrosine phosphorylation of focal adhesion molecules, such as focal adhesion kinase FAK2 and paxillin.

Adaptor Abl Src homology-3 domain-binding protein SH3BP2 (or 3BP2) regulate the transcriptional activity of T lymphocytes. SH3BP2 regulates the cytotoxicity of natural killer cells, as it mobilizes relevant mediators [1505]. Its binding partners include guanine nucleotide-exchange factor Vav1 and phospholipase-Cγ.

In turn, Vav1 is able to interact with many adaptors, such as: (1) docking protein DOK2; (2) growth factor receptor-bound proteins Ras pathway-associated GRB2

---

52. The transmembrane CD3 complex, a TCR coreceptor, is composed of several distinct chains. Subtypes include: CD3δ(CD3d), CD3ϵ(CD3e), CD3η(CD3h or CD3ζ [CD3z]), and CD3γ(CD3g). It is expressed on thymocytes and T lymphocytes. Therefore, the CD3–TCR complex contains 1 CD3γ, 1 CD3δ, 2 CD3ϵ, and 1 CD3ζ-chain. These chains associate with the T-cell receptor to generate an activation signal in T lymphocytes.
53. Hence, its alias CD3ζ.
54. A.k.a. T-cell receptor-interacting molecule (TRIM).
55. A.k.a. signaling threshold regulating transmembrane adaptor-1 (SIT1).
56. A.k.a. HSH2.
57. A.k.a. LcK-interacting adaptor (LAd), TSAd, and VRAP.
58. A.k.a. ERK4, MAPK6, and MAPK7.

and its isoform GRB3 (that is associated with the MAPK module and apoptosis); (3) B-cell linker (BLnk); (4) Src homology-2 domain-containing adaptor SHB, (5) Src-like adaptor SLA2 (or SLAP2); (6) cell cycle progression protein CCPg1 (a scaffold regulator of RhoGEF Dbs [1506]); (7) adaptor with a PH and SH2 domain (APS; that is involved in a PI3K-independent pathway initiated by its binding to insulin receptor [1507]); (8) mastocyte immunoreceptor signal transducer MIST (that contributes to IgE receptor-mediated mastocyte degranulation [1508]); (9) noncatalytic region of tyrosine kinase adaptor NCK1; (10) Krüppel-like zinc finger protein Znf655 (a.k.a. VIK that can link to cyclin-dependent kinase-4 of the cell cycle machinery [1509]); (11) CBL-related protein (CBLb; that forms a complex distinct from GRB2–CBL and GRB2–SOS1 complexes and impedes Vav1 activity [1510]). In addition, Vav1 links to phospholipase-Cγ1 and ubiquitin ligase Seven in absentia homolog SIAH2, without leading to its degradation. Protein Vav1 thus participates in: (1) BCR-, TCR-, and FcεR1-initiated signaling pathway; (2) Rac1-assisted cell motility; and (3) Ras-independent cell-mediated cytotoxicity.

11.2.4 Toll-like Receptors

In humans, 10 plasmalemmal Toll-like receptors (TLR; Table 11.15)[59] participate in the innate immune response to microbes (bacteria, viruses, parasites, and fungi), as they recognize pathogen-associated molecular patterns (lipids, lipoproteins, proteins, and nucleic acids). Pattern-recognition receptors distinguish microbial molecules from hosts. The PAMP recognition by TLRs occurs in various cellular compartments, such as the plasma membrane, endosomes, lysosomes, and endolysosomes [1511]. Each TLR has a distinct function in PAMP recognition and immune response. They not only detect pathogen components, but also host-derived proteins, such as heat shock proteins and fibronectin.

These single type-1 transmembrane proteins possess an intracellular domain for signal transduction and multiple leucine-rich motifs in their ectodomain. Stimulation of Toll-like receptors leads to activation of intracellular protein kinases and regulation of gene transcription.

Expression of isoforms of these immune system sentinels depends on cell type. Isoforms TLR2 to TLR4 and TLR6 are synthesized by cardiomyocytes, but not TLR1 and TLR5 [1497]. Therefore, cell type-specific signaling pathways are associated with TLR immunological properties. On the other hand, inappropriate TLR responses contribute to acute and chronic inflammation as well as to systemic autoimmune diseases.

11.2.4.1 Toll-Like Receptor Classes

Toll-like receptors can be classified into several groups according to ligand type. These innate immune receptors actually recognize bacterial and fungal cell wall

59. Toll receptor was discovered in Drosophila melanogaster as it is involved in embryonic development.

Table 11.15. Toll-like receptors and their selective agonists (Sources: [5, 1511]). Functional TLR1 to TLR9 have been identified in humans and mice. Mouse TLR10 is not functional. Types TLR11, TLR12, and TLR13 do not exist in the human genome. Each TLR recruits its selected adaptors to yield a specific immunological response suitable to infecting microbes. Toll-like receptors, members of pattern-recognition receptors, operate in the early detection of products from infectious agents, such as viruses, bacteria, fungi, and parasites, to activate both innate and adaptive immune response (ssRNA: single-stranded RNA; dsRNA: double-stranded RNA; CpG: unmethylated 2′-deoxyribocytidine-phosphate-guanosine). Toll-like receptors thus must distinguish between self and foreign constituents.

Type	Other names	Selective agonists
TLR1		Bacterial lipopolysaccharides
TLR2	CD282	Bacterial lipopeptides, peptidoglycan, lipoteichoic acid, lipoarabinomannan, viral hemagglutinin, trypanosomal tGPI-mucin, fungal zymosan
TLR3	CD283	Polyinosine–polycytosine, dsRNA
TLR4	CD284	Lipopolysaccharides, taxol
TLR5		Flagellin
TLR6		
TLR7		Imidazoquinoline derivatives (imiquimod, resiquimod), guanine analogs (loxoribine), ssRNA
TLR8		Viral ssRNA, imiquimod, resiquimod
TLR9	CD289	DNA enriched in cytosine–guanosine pairs, CpG DNA motifs

components, bacterial lipoproteins, highly conserved microbial proteins, and bacterial and viral nucleic acids. Lipoproteins and glycolipids are recognized by TLR2, double-stranded RNA by TLR3, lipopolysaccharides by TLR4, and microbial nucleic acids by TLR7 to TLR9 receptors.

Toll-like receptors are expressed mainly on antigen-presenting cells, such as dendritic cells and macrophages, as well as on B lymphocytes. Many TLR agonists trigger antibody responses as well as helper T_{H1} and T_{H17} responses. Toll-like receptors thus support adaptive immunity. They trigger B and T-cell responses and facilitate adaptive immune responses primed by non-antigen-presenting cells and mediated by dendritic cells.

Two TLR groups can be defined according to cellular location and PAMP ligands [1511]. Class-1 TLRs is composed of TLR1, TLR2, and TLR4 to TLR6. They are expressed on cell surface and recognize mainly microbial membrane constituents, such as lipids, lipoproteins, and proteins.

The TLR2 receptor generally forms heterodimers with TLR1 or TLR6. Moreover, TLR2 is able to act together with other plasmalemmal coreceptors, such as the collagen-1 and thrombospondin receptor CD36 and C-type lectin dectin-1 that binds fungus β-glycan [1511].

The TLR5 receptor recognizes flagellin of bacterial flagella. It is, in particular, synthesized by CD11b+ (α_M), CD11c+ (α_X) dendritic cells of the small intestine lamina propria. These cells promotes the differentiation of helper T_{H1} and T_{H17} cells, as well as the differentiation of naive B cells into IgA-producing plasmocytes [1511].

Class-2 TLRs encompasses TLR3 and TLR7 to TLR9 that are produced exclusively in intracellular vesicles, such as endosomes, lysosomes, and endolysosomes, where they recognize microbial nucleic acids.

The TLR3 receptor triggers antiviral immune responses via type-1 interferon and inflammatory cytokines. It is restricted to dendritic cells. It recognizes double-stranded viral RNA. It then causes the activation of NFκB to increase the production of type-I interferons.

The TLR7 receptor is highly synthesized in plasmacytoid dendritic cells. These cells are able to produce large amounts of type-1 interferon upon virus infection [1511]. In conventional dendritic cells, TLR7 senses RNA species from bacteria.

The TLR8 receptor is widespread, with its highest level in monocytes. It is also particularly strongly produced in the lung as well as blood leukocytes.

The TLR9 receptor is strongly produced in plasmacytoid dendritic cells. It detects viral DNA and insoluble crystal hemozoin generated as a by-product of the detoxification process after digestion of host hemoglobin by Plasmodium falciparum [1511].

11.2.4.2 TLR Adaptors and Signaling

Five known adaptors modulate TLR signaling: (1) myeloid differentiation primary response protein MyD88; (2) Toll–IL1 receptor (TIR) domain-containing adaptor protein (TIRAP or MAL); (3) TIR domain-containing inducing Ifnβ factor (TRIF);[60] (4) TRIF-related adaptor molecule (TRAM);[61] and (5) Sterile-α and Armadillo motif-containing protein (SARM). Adaptor SARM is a TIR domain-containing protein that suppresses TLR4 signaling (Fig. 11.1). All TLRs use MyD88 adaptor, except the TLR3 receptor. The TLR4 receptor is the single TLR family member that activates both MyD88- and TRIF-dependent pathways. Moreover, TLR4 is the only TLR that uses all the 4 adaptors (MyD88, TIRAP, TRIF, and TRAM).

Toll-like receptors trigger different signal transduction axes, as they cooperate with multiple TIR domain-containing adaptors. Adaptors MyD88 and TIRAP contribute to activation of mitogen-activated protein kinase and nuclear factor-κB that generate a pro-inflammatory response. On the other hand, TRAM and TRIF activate TRAF family member-associated NFκB activator (TANK)-binding kinase TBK1 and IκB kinase-ϵ that cause type-1 Ifn production [1513]. Receptors TLR3 and TLR4

60. A.k.a. TIR domain-containing adaptor molecule TICAM1.
61. A.k.a. TICAM2 and TIR domain-containing adaptor protein (TIRP).

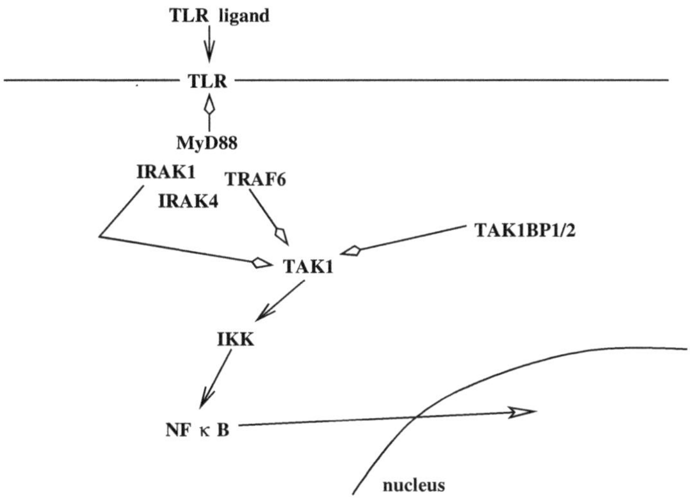

Figure 11.1. Toll-like receptors (TLR) regulate both innate and adaptive immune responses. Cytoplasmic regions of TLRs and related interleukin-1 receptors (IL1R) are homologous. In addition to this structural similarity, TLRs and IL1Rs also share adaptors such as myeloid differentiation primary response protein MyD88. Ligand-bound Toll-like receptors recruit MyD88, as well as IL1R-associated protein kinases IRAK1 and IRAK4 and tumor-necrosis factor receptor-associated factor TRAF6. IRAK1 and TRAF6 then dissociate from this complex and associate with TGFβ-activated kinase TAK1 (MAP3K7) and TAK1-binding proteins-1 and -2. MAP3K7 Kinase is activated and, in turn, activates inhibitor of NFκB kinase complex (IKK). Therefore, the pathway leads to IκB phosphorylation and degradation and nuclear translocation of NFκB (Source: [1512]).

prime both a pro-inflammatory and type-1 interferon response, whereas TLR2 and TLR5 initiate only a pro-inflammatory response.

Signaling by TLRs initiates acute inflammation, as it stimulates antimicrobial genes and triggers the production of inflammatory cytokines and chemokines in various cell types, especially cells of the myeloid lineage. Subsequent recruitment of neutrophils and activation of macrophages lead to destruction of invasive microorganisms. The TLR receptors also contribute to the activation of adaptive immune responses. Signaling by TLRs provokes maturation of lymphocytes and dendritic cells that become efficient antigen-presenting cells via costimulatory molecules, upregulation of major histocompatibility complex molecules, and secretion of cytokines and chemokines.

11.2.4.3 MyD88-Mediated Signaling

The MyD88-dependent pathway that activates NFκB is used by all TLRs except TLR3 (Table 11.16). Adaptor MyD88 recruits the IL1R-associated kinases IRAK1 to IRAK4. Protein IRAK4 is activated initially. The IRAK1 kinase and IRAK2 pseudokinase are activated sequentially to strenghten NFκB and MAPK ac-

Table 11.16. Signaling initiated by TLRs (Source: [1511]).

Macrophages and conventional dendritic cells
TLR2–TIRAP–MyD88–IRAK–TRAF6–IRF5
TLR3–TRIF–TRADD–Pellino-1–RIPK1–MAP3K7–MAPK/NFκB
TLR3–TRIF–TRAF6–MAP3K7–MAPK/NFκB
TLR3–TRIF–TRAF3–TBK1/IKKε–IRF3
TLR3–TRIF–inflammasome
TLR4–TIRAP–MyD88–IRAK–TRAF6–IRF5
TLR4–TRAM–TRIF–TRADD–Pellino-1–RIPK1–MAP3K7–MAPK/NFκB
TLR4–TRAM–TRIF–TRAF6–MAP3K7–MAPK/NFκB
TLR4–TRAM–TRIF–TRAF3–TBK1/IKKε–IRF3
TLR4–TRAM–TRIF–inflammasome
TLR5–MyD88–IRAK–TRAF6–IRF5
Plasmacytoid dendritic cells
TLR7–MyD88–IRAK–TRAF6–IRF5
TLR7–MyD88–IRAK–TRAF6–NFκB
TLR7–MyD88–IRAK–TRAF6–TRAF3–IRF7
TLR7–MyD88–IRAK–TRAF6–OPNI–IRF7
TLR9–MyD88–IRAK–TRAF6–IRF5
TLR9–MyD88–IRAK–TRAF6–NFκB
TLR9–MyD88–IRAK–TRAF6–TRAF3–IRF7
TLR9–MyD88–IRAK–TRAF6–OPNI–IRF7

tivation [1511]. Activation of IRAK proteins causes interaction with TRAF6 Ub ligase. Resulting ubiquitination elicits binding to MAP3K7IP2 and MAP3K7IP3 as well as IKK complex for MAP3K7 and NFκB activation. Kinase MAP3K7 activates ERK1 and ERK2 as well as P38MAPK and JNK. The latter, in turn, activate various transcription factors, such as AP1 factor.

11.2.4.4 TRIF-Mediated Signaling

The TRIF-dependent pathway is characterized by the activation of the MAP3K7–NFκB (Table 11.16) as well as interferon-regulatory protein IRF3 for interferon-β transcription. It is used by TLR3 and TLR4 receptors. TRIF recruits and forms a signaling complex with [1511]: (1) Ub ligases TRAF6 and Pellino homolog-1 (Peli1); (2) adaptor receptor-interacting protein RIPK1; (3) TRADD adaptor to activate MAP3K7 that, in turn, activates the NFκB and MAPK module. Ubiquitination of RIPK1 is required for NFκB activation. In addition, TRIF forms another signaling complex with TRAF3, TANK-binding kinase TBK1, and IKKε that phosphorylates IRF3 and induce its nuclear translocation.

11.2.4.5 TLR Compartmentalization

In addition to initiation of distinct signaling pathways, TLRs reside in different compartments within cells that influence ligand accessibility. These subcellular loci are stable or not. Receptors TLR1, TLR2, and TLR4 to TLR6 localize to the cell surface [1513]. However, plasmalemmal TLRs can undergo endocytosis once they have been activated. These receptors are, indeed, rapidly recruited to phagosomes or endosomes that contain microbial cargos.

Activation of innate immune Toll-like receptors and associated signal transduction are regulated by subcellular compartmentalization of receptors and their signaling effectors [1513]. Bacterium-sensoring TLRs (e.g., TLR2 and TLR4) may be transported from the Golgi body to the plasma membrane, whereas nucleic acid-detector TLRs (e.g., TLR3 and TLR7 to TLR9) are delivered to *endolysosomes* using intrinsic membrane chaperone Unc93B1.[62] The latter matures by cleavage by lysosomal cathepsins. Endolysosomes trigger TLR-dependent interferon response. Hence, all Ifn-inducing TLRs that use spatially restricted ubiquitin ligase tumor-necrosis factor receptor-associated factor TRAF3 signal from endolysosomes [1513]. Bacterium-sensoring TLRs uses sorting adaptors for their subcellular localization as functional receptors. Sorting adaptors TIRAP and TRAM recruit signaling mediators to the plasma membrane and endosomes, respectively.

11.2.4.6 TLR Signaling in Endosomes

Toll-like receptor TLR4 uses depletion in $PI(4,5)P_2$ in endosomes to run from the plasma membrane to endosomes. At the plasma membrane, activated TLR4 interacts with an adaptor complex constituted by TIR domain-containing adaptor protein (TIRAP) and myeloid differentiation primary response protein MyD88 to stimulate P38MAPK and inhibitor of NFκB kinase-β. After its endocytosis, the TIRAP–MyD88 complex dissociates from TLR4, whereas adaptor complex constituted by TRIF-related adaptor molecule (TRAM) and TIR-domain-containing adaptor protein inducing Ifnβ (TRIF) binds to internalized TLR4. Endosome-specific activation of Ifn-regulatory factors and interferon Ifnβ can then occur [11]. Several other members of the TLR family (TLR3, TLR7, and TLR9) can also reside in endosomes, where they detect viral nucleic acids released after viral degradation.

11.2.4.7 Inhibitors of TLR Signaling

Inhibitors of TLR signaling include some ubiquitin ligases, deubiquitinases, transcriptional regulators, and microRNAs to avoid persistent inflammation [1511].

62. In mouse macrophages and dendritic cells, nascent TLR7 and TLR9 transit through the Golgi body for glycosyl modification before being retained in endolysosomes, where they are cleaved to create a functional receptor that binds to adaptor MyD88 [1514]. Membrane protein Unc93B1 that delivers nucleic-acid-sensing TLRs to endolysosomes favors TLR9 cleavage. Endolysosomal sequestration of functional TLR9 avoid plasmalemmal localization and meeting of self genetic material.

TRAF-associated NFκB activator (TANK) precludes TRAF6 ubiquitination in both macrophages and B lymphocytes. Zinc finger and RNase domain-containing regulator ZC3h12a that enters into action on TLR stimulation impedes autoimmunity. Protein TNFαIP3 (or A20) that is also stimulated during TLR stimulation operates as both an Ub ligase and a deubiquitinase. It restricts NFκB activation, as it modulates RIPK1 and TRAF6 agents. Phosphatase PTPn6 hinders the MyD88-dependent pathway, as it suppresses the activity of both IRAK1 and IRAK2 kinases.

11.2.4.8 TLR Endogenous Ligands

In addition to PAMPs, TLRs respond to endogenous molecules, such as products of cell injury, chromatin–DNA and ribonucleoprotein complexes liberated from dying cells, substances released by tumor cells, as well as degradation products of the extracellular matrix, to trigger inflammation.

In particular, heat shock proteins and high-mobility group box-1 (HMGB1) proteins, as well as amyloid-β in Alzheimer's disease and oxidized low-density lipoproteins in atherosclerosis stimulate TLRs [1511]. In fact, cytoplasmic HMGB1, HMGB2, and HMGB3 operate as sentinels for nucleic acids that activate cytosolic TLRs and RLRs. After its export, nuclear, non-histone, HMGB1 can link to TLR2, TLR4, and TLR9 [1511]. In addition, HMGB1 can bind to pathogens and self DNAs. The HMGB1-DNA complexes can activate plasmacytoid dendritic cells and B lymphocytes via TLR9 [1511]. Heat shock proteins, such as HSP22, HSP60, and HSP70, activate macrophages and dendritic cells via TLR2 and TLR4 [1511].

Once cleaved by cellular peptidases, components of the extracellular matrix, such as biglycan, hyaluronic acid, versican, fibronectin, and surfactant proteins can activate TLR2 and/or TLR4 [1511].

11.2.5 NOD-like Receptors

Nucleotide-binding oligomerization domain (NOD)-containing protein-like receptors, the so-called NOD-like receptors (NLR), intervene in innate immune responses. Whereas plasmalemmal Toll-like receptors are sensors aimed at detecting extracellular microbes, cytosolic pattern recognition receptors, such as RIG-like helicases and NOD-like receptors are intracellular surveillance molecules. The NLR proteins can operate outside of the innate immune system, as they regulate cell death and the major histocompatibility complex to influence adaptive immunity.

11.2.5.1 NLR Protein Structure

The NLR proteins typically contain: (1) a central nucleotide-binding domain (NBD; more precisely NACHT domain);[63] (2) N-terminus with an effector domain,

63. Alias NACHT stands for NAIP (NLR family, apoptosis inhibitory protein), CIITA (Class-2 [II] major histocompatibility complex transactivator), HETe protein of the fungus Podospora anserina, and TEP1 (telomerase-associated protein) [1515]. The NACHT domain consists of 7 motifs, including the nucleoside triphosphatase (ATPase–GTPase) region and Mg^{++}-binding site.

the Pyrin (PyD), caspase recruitment (CaRD; involved in inflammation and apoptosis), or baculovirus inhibitior of apoptosis repeat (BIR) domain, that binds signaling mediators; and (3) C-terminus with leucine-rich repeat (LRR) domain. Oligomerization of NACHT domains is used to form an active signaling platform, the so-called *inflammasome* or *nodosome*, that links to adaptors and signaling effectors.

11.2.5.2 NLR Sensors

NOD-like receptors recognize not only pathogen-associated molecular patterns (PAMP; i.e., microbial products), but also intracellular danger signals, i.e., danger-associated molecular patterns (DAMP) released by cellular damage or stress in response to either invading pathogens or sterile inflammation. They initiate a host defense program by activating the NFκB pathway and caspases.

11.2.5.3 NLR Signaling

Several NLRs can activate an inflammasome platform, i.e., a signaling complex composed of an NLR protein, adaptor apoptotic speck-containing protein with a CaRD sequence (ASC), and procaspase-1, which cleaves pro-inflammatory cytokines, interleukin-1β and -18.

Inflammasomes also stimulate proteolytic processing of inflammatory cytokines. Furthermore, NOD-like receptors regulate inflammasome-independent functions in the immune system, such as [1516]: (1) activation of canonical and non-canonical NFκB axes and mitogen-activated protein kinase modules; (2) production of type-1 interferon, cytokines, chemokines, as well as antimicrobial reactive oxygen species; and (3) ribonuclease-L activity.

11.2.5.4 NLR Proteins and Mitochondrial Antiviral Signaling Protein

The NLR proteins can associate with mitochondrial antiviral signaling protein (MAVS),[64] an adaptor that promotes type-1 interferon production [1516]. The NLR–MAVS complex also includes helicases that directly bind to viral single-stranded RNA, double-stranded RNA, RNase-L-cleaved RNA, and other regulators to form the mitosignalosome. The MAVS protein activates the transcription factors interferon response factor IRF3 and NFκB. The latter increases production of type-1 interferons and pro-inflammatory cytokines such as interleukin-6. Agent MAVS also regulates inflammation via retinoic acid-inducible protein RIG1-like family of pathogen recognition receptors, such as RIG1 and interferon-induced with helicase-C domain-containing protein IfIH1 [1516].[65] The IfIH1 protein recognizes long dsRNA, whereas RIG1 detects short dsRNA and 5′-triphosphate-bearing ssRNA.

---

64. A.k.a. virus-induced signaling adapter (VISA), interferon-β promoter stimulator IPS1, and CaRD adaptor inducing interferon-β (Cardif).

65. A.k.a. melanoma differentiation-associated gene product MDA5.

11.2.5.5 NLR Family

NOD-like receptors constitute a large family of pattern recognition receptors. The human NLR family contains 22 members: (1) 14 NLR family, pyrin domain-containing proteins (NLRP); (2) 5 members of the NLRC (NLR family CaRD domain-containing protein) subfamily; as well as (3) NLR family, apoptosis inhibitory protein (NAIP), NLR family member-X1 (NLRX), and class-2 (II) major histocompatibility complex transactivator (CIITA), a major activator of MHC class-2 gene transcription. Several respond to various PAMPs, non-PAMP particles, and cellular stresses to trigger pro-inflammatory responses, such as IL1β secretion.

Agent NLRP12[66] is restricted to the myeloid-monocytic lineage [1516]. It can activate the canonical NFκB pathway as well as caspase-1. On the other hand, it impedes the activation of the slow-kinetics, non-canonical NFκB pathway downstream from the tumor-necrosis factor receptor. The non-canonical NFκB pathway depends on NFκB-inducing kinase. Enzyme NIK associates with NFκB2 (P100) and causes its cleavage into its active form (P52) that provokes the expression of a subset of pro-inflammatory genes. Therefore, NLRP12 downregulates NIK, RIPK1, TRAF2, and TRAF6 [1516].

The CaRD domain-containing NLRs, such as NLRC1[67] and NLRC2,[68] interact with CaRD-containing receptor-interacting protein kinase RIPK2 to activate the CaRD9 and nuclear factor-κB pathways.

Several NALP[69] proteins form inflammasomes (Vol. 5 – Chap. 11. Tissue Growth, Repair, and Remodeling), that binds adaptor apoptosis-associated speck-like protein (ASC) for caspase activation.

The NLRC1 protein (or NOD1) is an intracellular pattern recognition receptor. The NLRC2 protein (or NOD2) activates NFκB via RIPK2, to which it connects, once translocated to the plasma membrane [1516]. The NLRC2–RIPK2 complex activates IκB kinase via Lys63-linked polyubiquitination of its γ subunit (NEMO), as it recruits ubiquitin ligases cIAP1 and cIAP2. The NLRC2–RIPK2 complex also stimulates P38MAPK and JNK kinases. In addition, NLRC2 can link to NLRP1 to cooperate in caspase-1 activation. It also associates with dual oxidase DuOx2 of the reduced nicotinamide adenine dinucleotide phosphate (NADPH) oxidase family to promote reactive oxygen species production [1516].

The NLRC3 protein (or NOD3) inhibits the activity of T cells. The NLRC4 protein[70] is expressed in myeloid cells, where it regulates caspase-1 activation as well as interleukin IL1β and IL18 processing [1519]. Like NLRP3, NLRC1, and NLRC2,

66. A.k.a. Regulated by nitric oxide [RNO], Pyrin-containing APAF1-like protein PYPAF7, and Monarch-1.

67. A.k.a. nucleotide-binding oligomerization domain-containing protein NOD1 and caspase recruitment domain-containing protein CaRD4.

68. A.k.a. NOD2, CaRD15, and inflammatory bowel disease protein IBD1.

69. NALP stands for NACHT (NB-ARC [Apaf-1/R gene/CED4] nucleotide-binding domain), LRR (leucine-rich repeat), and pyrin domain-containing proteins.

70. A.k.a. caspase recruitment domain-containing protein CaRD12.

it interacts with heat shock protein-90 and ubiquitin ligase-associated protein SGT1, an interactor of Skp1 of the SCF ubiquitin ligase complex.

The NLRC5 protein (or NOD4) is a pattern recognition receptor implicated in innate immunity against viruses, as it is involved in a JaK–STAT-mediated autocrine loop with interferon-γ [1517].

Mitochondrial NLRX1 inhibitor interacts with mitochondrial antiviral signaling protein to prevent its binding to RIG1, hence precluding the activation of NFκB and IRF3 factors [1516].

11.2.5.6 NLR Partners of Inflammasomes

Three types of inflammasomes have been identified: (1) NALP1 inflammasome composed of ASC, caspase-1, and caspase-5, in addition to NALP1;[71] (2) NALP2 and NALP3 inflammasomes, which contain, in addition to NALP2 or NALP3, Cardinal,[72] ASC, and caspase-1; and (3) NLRC4 inflammasome made of caspase-1 in addition to NLRC4[73] Members of the NLR family — NLRP1, NLRP3, and NLRC4 — can form caspase-1-activating inflammasomes.

The NLRP3 inflammasome promotes the maturation and secretion of interleukin-1β. It can integrate signals from interleukin-1β, reactive oxygen species, and thioredoxin-interacting protein,[74] thereby acting as a sensor for metabolic stress, such as in insulin resistance and type-2 diabetes mellitus [1518].

The NLRP3 inflammasome contains the adaptor apoptosis-associated speck-like protein (ASC) and pro-inflammatory caspase-1, in addition to NLRP3. The production of active IL1β requires at least 2 independent signals for synthesis of an inactive form and maturation. Prointerleukin-1β production is stimulated by Toll-like receptors or cytokines, such as tumor-necrosis factor or IL1β. ProIL1β is processed by caspase-1 for maturation and secretion.

Reactive oxygen species generated by PAMPs or DAMPs operate in NLRP3 activation. Activators of NLRP3 elicit ROS production such as adenosine triphosphate that stimulates rapid K^+ efflux from ATP-gated $P2X_7$ receptor channel, and triggers gradual recruitment and pore formation by the pannexin-1 hemichannel. Moreover, ROS favor binding between NLRP3 with thioredoxin-interacting protein.

---

71. Protein NALP1 is also called NLR family, pyrin domain-containing protein NLRP1, NB and CaRD domain-containing protein NAC, caspase recruitment domain-containing protein CaRD7, or Death effector filament-forming Ced-4-like apoptosis protein (DEFCAP)

72. Cardinal is also known as caspase recruitment domain-containing CaRD8 or tumor-upregulated CaRD-containing antagonist of caspase-9 (TUCAn).

73. The CaRD domain-containing protein of the NLR family NLRC4 is also called interleukin-1β convertase (ICE) peptidase-activating factor (IPAF), CaRD-, LRR-, and NACHT-containing protein (CLAN), and caspase-recruitment domain-containing protein CaRD12.

74. A.k.a. vitamin-D3-upregulated protein VDUP1.

Table 11.17. Signaling initiated by NOD-like receptors (NLR; Source: [1516]; IRF: interferon-regulatory protein; MAPK: mitogen-activated protein kinase module; NFκB: nuclear factor κ light chain-enhancer of activated B cells; NIK: NFκB-inducing kinase). A given NLR can have multiple functions. Different NLRs can have similar functions.

Signalosome (Function)	Type(s)
CIITA transcriptosome (MHC expression)	CIITA
Inflammasomes (caspase-1 activation)	NLPR1/3; NLRC2/4; NAIP
Nodosome (NFκB–MAPK activation)	NLRC1/2
Mitosignalosome (NFκB–IRF activation)	NLRC2; NLRX1
Non-canonical NFκB axis (NIK activation)	NLPR12; NLRC2

11.2.5.7 NLR Functions

The NLR receptors signal via different signalosomes, among which: (1) the caspase-1-activating *inflammasome*; (2) the interferon–cytokine-inducing *mitosignalosome*; (3) the NFκB–MAPK-activating *nodosome* (NOD1–NOD2 complex); (4) the CIITA *transcriptosome* (or C2TA for class-2 [II] major histocompatibility complex transactivator); and (5) NIK pathway. A given NLR can have multiple functions; different NLRs can have similar functions [1516] (Table 11.17).

11.2.6 C-Type Lectin Receptors

Lectins [75] are innate immune defense proteins that recognize bacterial cell wall components. [76] C-Type (Ca^{++}-dependent) lectin receptors (CLR) are transmembrane and soluble proteins that are expressed by dendritic cells (Vol. 1 – Chap. 7. Plasma Membrane).

C-Type lectin receptors support the immune response to most pathogen classes from recognition of mannose, fucose, and glucan carbohydrates [1520]. Mannose actually enables the recognition of viruses, fungi, and mycobacteria. Fucose is more

75. Latin lectio: choice, recruitment, picking out, selecting.

76. The name C-type lectin has been introduced to distinguish between Ca^{++}-dependent and -independent carbohydrate-binding lectins. However, some members of the CLR family do not bind carbohydrates, although they have at least one domain that is homologous to carbohydrate recognition domains.

specifically expressed by certain bacteria and helminths. Glucans reside on mycobacteria and fungi. Recognition by C-type lectin receptors causes pathogen endocytosis, degradation, and subsequent antigen presentation.

Following pathogen binding, C-type lectin receptors trigger diverse signaling pathways that prime the production of specific cytokines to determine T-lymphocyte fate. Some C-type lectin receptors have a specific expression pattern, whereas others are synthesized by several subsets of dendritic cells (Table 11.18).

The CLR signaling involves: (1) calcium mobilization; (2) many mediators, such as (2.1) immunoreceptor tyrosine-based activation motif-containing adaptors (Fc receptor-γ and DAP12); (2.2) protein kinases and phosphatases, such as spleen tyrosine kinase,[77] mitogen-activated protein kinase module components cRaf and ERK kinases, and protein phosphatase PP3 and protein Tyr phosphatases PTPn6 or PTPn11; (2.3) small GTPases and their partners, such as RhoA and guanine nucleotide-exchange factor RhoGEF12; (2.4) transcription factors such as nuclear factor-κB; and (2.5) histone acetyltransferases cAMP-responsive element-binding protein (CREB)-binding protein (CBP) and P300; as well as (3) signaling crosstalk with Toll-like receptors. Some C-type lectin receptors signal via nuclear factor-κB, whereas others use Toll-like receptors.

C-Type lectin receptors orchestrate signal transduction pathways that regulate adaptive immune responses, as they: (1) trigger their own signaling pathways; (2) initiate carbohydrate-specific signaling; and (3) intervene in crosstalk with other pattern recognition receptors.

Crosstalk between pattern recognition receptors on dendritic cells primes the differentiation of T helper cells via adequate cytokines to specifically target invading pathogens. C-Type lectin receptors contribute to T helper cell differentiation into: (1) T_{H1} cells that secrete interferon-γ; (2) T_{H2} cells that produce interleukins IL4, IL5, and IL13; and (3) T_{H17} cells that release IL17 [1520].

11.2.7 Triggering Receptors Expressed on Myeloid Cells

The family of plasmalemmal triggering receptors expressed on myeloid cells (TREM) are members of the immunoglobulin superfamily [1521]. The TREM family is encoded by the TREM cluster on chromosome 6p21 that includes Trem1 and Trem2 genes, as well as the *TREM-like* subset with Treml1 and Treml2 genes. The TREM-like genes encode TREM-like transcripts TLT1 and TLT2.

77. Spleen tyrosine kinase (SYK) forms a signaling complex with caspase recruitment domain-containing protein CaRD9, B-cell lymphoma BCL10, and mucosa-associated lymphoid tissue lymphoma translocation gene product MALT1 that activates IκB kinase, which phosphorylates inhibitor of nuclear factor-κB, hence priming NFκB nuclear translocation [1520]. In addition, activated SYK stimulates a complex that consists of B-cell linker, Bruton's Tyr kinase, and phospholipase-Cγ2. This complex impedes production of interferon-α and -β, tumor-necrosis factor, and interleukin-6 from excited Toll-like receptor TLR9 by plasmacytoid dendritic cells. The SYK pathway also synergistically activates NFκB subunit P65 (or RelA) via Raf1 kinase that phosphorylates P65 and finely tunes NFκB-induced cytokine response via a cross-regulation of canonical and non-canonical NFκB signaling.

Table 11.18. C-type lectin receptors and their expression (Source: [1520]; BDCA: blood dendritic cell antigen; DCIR: dendritic cell immunoreceptor; CLec: C-type lectin domain family member; DCSIGN: dendritic cell-specific ICAM3-grabbing non-integrin; dectin: dendritic cell-associated C-type lectin; MICL: myeloid inhibitory C-type lectin-like receptor [a.k.a. CTL]; mincle: macrophage-inducible C-type lectin; R: receptor).

CLR	Alias	Expression
	Group-1 CLRs (mannose receptor class)	
Mannose R	CD206	Myeloid dendritic cells, macrophages
Dec205	CD205, LY75	Myeloid dendritic cells,
	Group-2 CLRs (asialoglycoprotein receptor class)	
DCSIGN	CD209	Myeloid dendritic cells
CLec4k	CD207, Langerin	Dermal dendritic cells, Langerhans cells
CLec5a	MDL1	Monocytes, macrophages
CLec10a	CD301, MGL	Myeloid dendritic cells, macrophages
	Group-2 CLRs (dectin-1 class)	
CLec2	CLec1b	Platelets
CLec7a	Dectin-1	Myeloid dendritic cells, B lymphocytes, monocytes, macrophages
CLec9a	DNGR1	BDCA3+ dendritic cells, monocytes, B lymphocytes
CLec12a	MICL, DCAL2	Myeloid dendritic cells, monocytes, macrophages, neutrophils
CLec12b		Macrophages
	Group-2 CLRs (DCIR class)	
CLec4a	DCIR	Myeloid and plasmacytoid dendritic cells, monocytes, macrophages, B cells, neutrophils
CLec4c	CD303, BDCA2	Plasmacytoid dendritic cells, monocytes, macrophages, neutrophils
CLec4e	Mincle	Myeloid dendritic cells, monocytes, macrophages
CLec6a	Dectin-2	Myeloid and plasmacytoid dendritic cells, monocytes, macrophages, B cells, neutrophils

TREM substances participate in particular to inflammation and coagulation, among other functions. Proteins TREM1 and TREM2 form a complex with DAP12 adaptor. Ligand binding to the receptor activates Src kinase that phosphorylates DAP12. The latter then binds to SYK kinase. Downstream effectors, such as protein kinases PKB and PKCθ, mitogen-activated protein kinases, are then activated. Protein TREM1 amplifies inflammatory signaling, acting synergistically with Toll- and NOD-like receptors. Protein TREM2 is involved in osteoclastogenesis. It reduces the expression of inflammatory cytokines by microglial cells. It is also implicated in clearing apoptotic neurons. It inhibits TLR-induced cytokine production by macrophages. TREM2 binds plexin-A1, particularly after plexin activation by semaphorin-6D (Sect. 10.5).

TREM-like transcript-1 is expressed by platelets. In the absence of cell stimulation, TLT1 is sequestered in α-granules of circulating platelets. After activation, TLT1 translocates to the plasma membrane and contributes to thrombin-mediated platelet aggregation. Protein TLT1 recruits Src homology protein Tyr phosphatases PTPn6 and PTPn11 (Vol. 4 – Chap. 7. Cytosolic Protein Phosphatases). Expression of TLT2 in B lymphocytes, granulocytes, and tissue-resident macrophages is upregulated by inflammation.

11.2.8 Tyro3, Axl, and Mer (TAM) Receptors

The TAM receptor tyrosine kinases include 3 types: TyrO3, Axl, and Mer (Sect. 8.2.8).[78] They are expressed on dendritic cells, macrophages, and immature natural killer cells, as well as Sertoli cells of the testis and retinal pigment epithelial cells, among other cell types.

Receptor tyrosine kinases of the TAM group are implicated in innate immunity as well as homeostasis of the nervous, reproductive, and vascular systems [1522].[79] The TAM kinases regulate cell survival, adhesion, proliferation, and migration, as well as blood clot stabilization and regulation of cytokine.

Signaling primed by TAM kinases is particularly involved in: (1) inhibition of the inflammatory response to pathogens by dendritic cells and macrophages; (2) phagocytosis of apoptotic cells by these cells; and (3) maturation and killing of natural killer cells. Both TAM-mediated inhibition of inflammation and stimulation of NK-cell differentiation require interaction between TAM and cytokine receptors.

The TAM receptors are closely related to the class-6 RTKs: macrophage-stimulating RTK RON[80] and hepatocyte growth factor receptor.

The TAM ligands comprise Growth arrest-specific GAS6 and protein-S. Ligand GAS6 binds and activates all the 3 TAM receptors, albeit with different affinities (Axl $\geq$ Tyro3 $\gg$ Mer). In addition to its blood anticoagulant activity,[81] protein-S targets Tyro3 and Mer kinases.

Both TAM ligands and TAM receptors heterodimerize. TAM Receptors indeed signal as dimers. The TAM kinases mainly target the phosphoinositide 3-kinase–protein kinase-B pathway. In addition, they prime the Janus kinase–signal transducer and activator of transcription pathway.

The TAM receptors act via suppressor of cytokine signaling SOCS1 and SOCS3 proteins. The SOCS mediators are stimulated by cytokine receptors to inhibit the JaK–STAT pathway. Factor STAT1, but neither STAT2 nor STAT3, is stimulated by

78. Kinase Tyro3 is also called BRT, DTK, RSE, SKY, and TIF. Kinase Axl is also termed ARK, Tyro7 and UFO. Kinase Mer is also designated as EYK, NYM, and Tyro12.

79. Signaling primed by TAMs regulates the activity of vascular smooth muscle cells and platelets.

80. The RON receptor is also called CD136 or MSt1R.

81. Protein-S is a cofactor for activated peptidase protein-C^a that degrades factors V^a and-VIII^aa.

activated TAM receptors. Inhibition of inflammation mediated by TAM receptors activated by Toll-like receptors requires STAT1 factor. The TAM receptors also interact with interleukin-15 receptor.

Activated interferon-α (β/ω) receptor IfnAR interacts with TAM kinases. The TAM receptors bind to and antagonize the IfnAR–STAT1 complex, thereby switching inflammation stimulation to inhibition. Consequently, TAMs inhibit both Toll-like receptors and TLR-induced cytokine-receptor cascades to restrain TLR signaling, hence avoiding chronic inflammation [1523].

Effect of TAMs on phagocytosis depends on phosphatidylserine. Kinase Mer and its ligand GAS6 are upregulated by interleukin-10 that stimulates the phagocytosis of apoptotic cells by monocytes and macrophages. Natural killer cells that recognize and destroy infected cells target cells that express inhibitory (e.g., Ly49 and CD94 family members) and stimulatory (NK1.1, DX5, and CD69) receptors. The expression of these receptors require TAM signaling. In the bone marrow, Tyro3, Axl, and Mer expressed by immature NK cells are activated by GAS6 and protein-S produced by stromal cells.

11.2.9 Signaling Lymphocytic Activation Molecules and SLAM-Associated Proteins

Signaling lymphocytic activation molecules (SLAM) constitute a family of immunomodulatory receptors that regulate lymphocyte interactions and adhesion. The SLAM Receptors are thus required for normal development and functioning of the immune system.

The SLAM receptor family includes 9 known members (SLAMF1–SLAMF9).[82] Except SLAMF4 in NK cells and CD8+ cytotoxic T cells that interacts with CD48, most SLAM family members act as self-ligands [1524].

The adaptor SLAM-associated protein (SAP) intervenes in the signaling pathway initiated by activated SLAM receptors. SAP Adaptor is expressed by T cells, natural killer cells, natural killer T cells, eosinophils, platelets, and some B-cell types.[83]

Adaptor SAP operates in: (1) development of natural killer T cells and other innate T-cell lineages that are selected in the thymus,[84] i.e., innate T-cell populations

82. Agent SLAMF1 is also designated as SLAM or CD150; SLAMF2 as CD48; SLAMF3 as LY9 or CD229; SLAMF4 as CD244 or 2B4; SLAMF5 as CD84; SLAMF6 as natural killer–T-cell–B-cell antigen (NTBA); SLAMF7 as CD2-like receptor-activating cytotoxic cells (CRACC), CS1, or CD319; SLAMF8 as B-lymphocyte activator macrophage expressed protein (BLAME); and SLAMF9 as CD2F0.

83. Mutations of the SRC homology-2 domain protein-1A gene (sh2d1A) that encodes SLAM-associated protein cause X-linked lymphoproliferative syndrome (fulminant infections, lymphoproliferative disorders such as B-cell lymphomas, and dysgammaglobulinemia that can progress to hypogammaglobulinemia. IgG+ memory B-cell number decays, thereby altering long-term humoral immunity.

84. Although most T lymphocytes (conventional T lymphocytes) are selected by cortical thymic epithelial cells, a small fraction of T lymphocytes (innate-like T lymphocytes) such as NKT cells that express a restricted TCR repertoire specific for MHC class-1b molecules is

selected by double-positive thymocytes; and (2) regulation of interactions between B and T cells to form germinal centers [85] and long-term humoral immunity [1524].

Adaptor SAP selectively recruits and activates protein Tyr kinase Fyn that phosphorylates SLAM. The latter can then bind to its signaling effectors, such as inositol 5-phosphatase SHIP, docking proteins DOK1 (a.k.a. $P62^{dok}$) and DOK2 (a.k.a. DOK-R, FRIP, or $P56^{dok}$), and Ras GTPase-activating protein [1524].

Engagement of SLAM also cooperates with T-cell receptors for sustained recruitment of protein kinase-Cθ (Vol. 4 – Chap. 4. Cytosolic Protein Ser/Thr Kinases) and activation of nuclear factor-κB to participate in T helper-2 cytokine production.

Adaptor SAP can compete with protein Tyr phosphatases PTPn6 and PTPn11, as they bind to the same SLAM motif. It can also interact with RhoGEF6 or RhoGEF7, [86] a guanine nucleotide-exchange factor for CDC42 and Rac1 GTPases, upon SLAMF4 commitment to activate transcription factor nuclear factor of activated T cells.

Linkage between 2 SLAMF6 or SLAMF2 and SLAMF4 of 2 adjacent cells causes receptor phosphorylation (activation), SAP binding, and Fyn kinase recruitment. Activated receptors SLAMF4 and SLAMF6 induce phosphorylation of guanine nucleotide-exchange factor Vav1 and adaptor Casitas B-lineage lymphoma protein (CBL).

11.2.10 Intracellular RNA Helicases – RIG-like Receptors

Intracellular viral double- and single-stranded RNAs are recognized by a group of RNA helicases that in turn recruit factors to activate an antiviral gene program. In particular, RIG-like receptors detect viral RNA species and signal via IPS1 adaptor to trigger antiviral response.

---

selected by thymic hematopoietic cells. Selection of conventional T cells occurs when a T-cell receptor on double-positive thymocyte interacts with a antigen–MHC class-1a complex on thymic epithelial cells. Selection of many types of innate-like T lymphocytes happens when a T-cell receptor and a SLAM receptor on double-positive thymocyte interacts with a lipid- or peptide antigen-bound MHC class-1b molecule and a SLAM receptor on thymic hematopoietic cell, respectively. Adaptor SAP and kinases Fyn and protein kinase-Cθ are necessary for development of innate-like T lymphocytes, but not for conventional T-cell maturation [1524].

85. Germinal centers are sites of antibody affinity maturation and memory B-cell generation for long-term humoral immunity. Helper CD4+ T cells are needed for development of a germinal-center response, as they promote proliferation and differentiation of activated B lymphocytes. Activation and differentiation of helper CD4+ T cells depend on antigen-specific interactions between T lymphocytes and dendritic cells. Subsequently, activated B lymphocytes specifically interact with T lymphocytes in the interfollicular region and at the border between the follicle and T-cell zone. Thereafter, T lymphocytes relocate from the T-cell zone to the follicle and reside inside germinal centers. CD4+ helper T lymphocytes located within germinal centers represent a separate effector CXCR5+, TNFSF5+, ICOS+, IL21+ T-cell lineage (follicular helper T cells [TFH]; ICOS: inducible T-cell costimulator; CXCR: CXC-chemokine receptor) that induces antibody production by B lymphocytes [1524].

86. A.k.a. PAK-interacting exchange factors PIXα and PIXβ.

Three RIG1-like receptors (RLR), or RNA helicases, include RIG1 and MDA5 that detect 5′-triphosphate RNA and dsRNA, respectively, and LGP2 that acts as a dominant-negative inhibitor.

Concluding Remarks

Volume 3 of the set of textbooks devoted to Circulatory and Ventilatory Systems in the framework of Biomathematical and Biomechanical Modeling gives the basic knowledge on the first phases of signal transduction triggered by extracellular messengers. Signaling determines cell behavior.

Numerous process can be modeled to quickly assess effects of parameters, all other agents remaining constant, once the mathematical model has been validated. Advantage of mathematical models is to provide the complete quantity fields, whereas measurements are made in some points or corresponds to averages of exploration windows of the field of the investigated variable.

Mathematical description of large, complex biochemical reaction networks, in which molecules are nodes, and modeling of the dynamics of interactions (scaffolding, reaction, transcription, etc.) relies on computational simulations. A biochemical reaction network is defined by: (1) a set of variables — the state variables — that define the state of the system and (2) rules of temporal changes and possible transport of involved variables. The network behavior can be analyzed in a single cells and extended to multicellular systems, especially in tumor models.

Mathematical models of cascades of chemical reactions are aimed at describing evolution of molecular concentrations [1] using the mass action law or a reaction–diffusion equation set. A given reaction occurs with a certain probability. Stochastic models rely on a stochastic update of system variables. On the other hand, deterministic models are carried out in systems with a large number of molecules, the time and variable states uniquely defining the state at the next time step. In a deterministic continuous formulation based on the mass action formalism, [2] molecular reactions

1. Typical molecular concentrations, i.e., continuous variables related to the number of molecules per unit volume, range from 0.1 nmol to 1 μmol.

2. The law of mass action states that the rate of a given chemical reaction is proportional to the product of the concentrations of reactants with eventual exponents that are stoichiometric coefficients for the reaction, in the absence of one-to-one stoichiometry. It is based on several assumptions: (1) well mixed media; and (2) low molecular concentrations; and (3) a probability of diffusing molecules to be available for the reaction proportional to the concentration.

are described by differential equations defining the rate of change of molecular concentrations. A rate constant (or affinity parameter) describes the occurrence rate of a reaction when reactants are close each other. In general, a mass action reaction can be written as:

$$\sum_{i=1}^{N_R} s_i R_i \xrightarrow{k_f} \sum_{j=1}^{N_P} s_j P_j, \tag{11.1}$$

where N_R reactants R_i generate N_P products P_j with a given stoichiometry s_k (relative quantities of reactants and products in the chemical reaction) with a forward rate parameter k_f (and a backward rate parameter k_b when the reaction is reversible.[3]

The modeling procedure for a biochemical network incorporates several steps: (1) definition of the molecular players and their interactions; (2) description of the major molecular interactions by a mathematical model; (3) estimation of values of the model parameters (e.g., diffusion coefficients and reaction rates); and (4) analysis of the dynamical behavior.

The dynamics of a continuous system can be described by a set of ordinary differential equations (ODE). In general, this ODE set cannot be solved analytically and, thus require numerical integrations. On the other hand, reaction–diffusion models are based on partial differential equations (PDE). For example, an activator (act)–inhibitor (inh) reaction-diffusion model (Gierer-Meinhardt model) in morphogenesis is given by:

$$\begin{aligned} \frac{\partial c_{act}}{\partial t} &= \mathsf{p}\frac{c_{act}^2}{c_{inh}} + \mathsf{p}_{act} - \mathsf{d}_{act} c_{act} + \mathcal{D}_{act} \nabla^2 c_{act}, \\ \frac{\partial c_{inh}}{\partial t} &= \mathsf{p}a^2 + \mathsf{p}_{inh} - \mathsf{d}_{inh} c_{inh} + \mathcal{D}_{inh} \nabla^2 c_{inh}, \end{aligned} \tag{11.2}$$

where c_{act} is the concentration of a short-range, autocatalytic activator, c_{inh} that of a long-range inhibitor, p the production rate, p_{act} the production rate of the activator required to initiate the activator autocatalysis, p_{inh} the low baseline production rate of the inhibitor, d the decay rate, and $\mathcal{D}$ the diffusion coefficient.

Cells react to various types of external stimuli, in particular, mechanical stresses. In physiological systems associated with flows, the magnitude and direction of mechanical stresses applied by the flowing fluid on the wetted surface of conduit wall (i.e., vascular endothelium with its glycocalyx or respiratory epithelium with the mucus layer and periciliary fluid) as well as within the vessel wall varies during the cardiac and breathing cycles. The heart generates an unsteady flow with a given frequency spectrum in a network of blood vessels characterized by complicated architecture and variable geometry both in space and time. Vessel geometry varies over short distances. The vascular network of curved blood vessels is composed of successive geometrical singularities, mainly branchings. The thoracic muscular cage cyclically inflates and deflates, thereby lowering and heightening the intrathoracic pressure, and hence dilating and collapsing lung alveoli and airways to inhale and exhale air. The respiratory tract is characterized by a large wetted surface inside a

3. The reaction constant of a reversible reaction is the ratio k_f/k_b.

small volume, especially in the nose and thorax. In the nose, turbinates allows heat and water exchange, but renders air currents less simple. In addition, the laryngeal constriction, the aperture of which varies during the ventilatory cycle, provokes air jet. The bronchial tree is a network of successive branchings at inspiration, or junctions at expiration, between short, more or less curved pipes of corrugated walls in large bronchi due to the presence of partial or complete cartilaginous rings.

Therefore, blood and air streams correspond to time-dependent, three-dimensional, developing, as they are conveyed in conduit entrance length, where boundary layer develop (Vol. 7). Moreover, blood vessels and airways are deformable. Changes in transmural pressure (the pressure difference between the pressure at the wetted surface of the lumen applied by the moving fluid on the deformable conduit wall and the pressure at the external wall side that depends on the activity on the neighbor organs) can also influence the shape of vessel cross-section, especially when it becomes negative. In addition, in the arterial compartment, the change in cross-section shape can result from taper. More generally, possible prints of adjacent organs with more or less progressive constriction and enlargment, and adaptation to branching (transition zone) also give rise to three-dimensional flows. These flows are commonly displayed by virtual transverse currents, even if the vessel is considered straight. Furthermore, geometrical singularities influence flow pattern both upstream and downstream from it.

Local changes in the direction of stress components can also be caused by flow separation and flow reversal during the cardiac and respiratory cycles. Flow separation is set by an adverse pressure gradient when inertia forces and fluid vorticity are high enough, especially in branching segments. Due to its time-dependent feature, flow separation regions spread over a variable length during the flow cycle and can move. The location and variable size of the flow separation region depends on the flow distribution between branches that can vary during the flow cycle. Flow reversal occurs during the diastole of the left ventricle in elastic arteries, such as the aorta, and most of the muscular arteries, such as brachial and femoral arteries (but not in the carotid arteries), as well as during alternates from inspiratory decelerating flow phase and expiratory accelerating flow phase and conversely. In arteries, flow reversal can be observed either in a region near the wall, more or less wide with respect to the position of the local center of vessel curvature, or in the entire lumen.

Consequently, the stress field experienced by the wall tissues are strongly variable both in time and space. Cellular sensors then process mechanical signals by ensemble averaging not only to raise the signal-to-noise ratio, but also to adequately adapt the local size of the conduit lumen, i.e., the local flow resistance to maintain either flow rate or pressure, only in the case of sustained, abnormal stress field.

References

Introduction

1. Lucretius (1997) De rerum natura (L I-685) [On the Nature of Things]. Garnier Flammarion, Paris

Chap. 1. Signal Transduction

2. Huang R, Martinez-Ferrando I, Cole PA (2010) Enhanced interrogation: emerging strategies for cell signaling inhibition. Nature Structural and Molecular Biology 17:646–649
3. Lim WA (2010) Designing customized cell signalling circuits. Nature Reviews – Molecular Cell Biology 11:393–403
4. Manz BN, Groves JT (2010) Spatial organization and signal transduction at intercellular junctions. Nature Reviews – Molecular Cell Biology 11:342–352
5. Alexander SPH, Mathie A, Peters JA (2009) Guide to Receptors and Channels (GRAC), 4th ed., British Journal of Pharmacology 158:S1–S254 (www3.interscience.wiley.com/journal/122684220/issue)
6. Liu YY, Slotine JJ, Barabási AL (2011) Controllability of complex networks. Nature 473:167–173
7. Rall TW, Sutherland EW (1958) Formation of a cyclic adenine ribonucleotide by tissue particles. Journal of Biological Chemistry 232:1065–1076
8. Sturm OE, Orton R, Grindlay J, Birtwistle M, Vyshemirsky V, Gilbert1 D, Calder1 M, Pitt A, Kholodenko B, W Kolch (2010) The mammalian MAPK/ERK pathway exhibits properties of a negative feedback amplifier. Science Signaling 3:ra90
9. Shimizu TS, Tu Y, Berg HC (2010) A modular gradient-sensing network for chemotaxis in Escherichia coli revealed by responses to time-varying stimuli. Molecular Systems Biology 6:382
10. Lage K, Møllgård K, Greenway S, Wakimoto H, Gorham JM, Workman CT, Bendsen E, Hansen NT, Rigina O, Roque FS, Wiese C, Christoffels VM, Roberts AE, Smoot LB, Pu WT, Donahoe PK, Tommerup N, Brunak S, Seidman CE, Seidman JG, Larsen LA

(2010) Dissecting spatio-temporal protein networks driving human heart development and related disorders. Molecular Systems Biology 6:381

11. Sorkin A, von Zastrow M (2009) Endocytosis and signalling: intertwining molecular networks. Nature Reviews – Molecular Cell Biology 10:609–622
12. Murphy JE, Padilla BE, Hasdemir B, Cottrell GS, Bunnett NW (2009) Endosomes: a legitimate platform for the signaling train. Proceedings of the National Academy of Sciences of the United States of America 106:17615–17622
13. Janes KA, Reinhardt HC, Yaffe MB (2008) Cytokine-induced signaling networks prioritize dynamic range over signal strength. Cell 135:343–354
14. Hersen P, McClean MN, Mahadevan L, Ramanathan S (2008) Signal processing by the HOG MAP kinase pathway. Proceedings of the National Academy of Sciences of the United States of America 105:7165–7170
15. Spencer SL, Gaudet S, Albeck JG, Burke JM, Sorger PK (2009) Non-genetic origins of cell-to-cell variability in TRAIL-induced apoptosis. Nature 459:428–432
16. Goentoro L, Shoval O, Kirschner MW, Alon U (2009) The incoherent feedforward loop can provide fold-change detection in gene regulation. Molecular Cell 36:894–899
17. Goentoro L, Kirschner MW (2009) Evidence that fold-change, and not absolute level, of β-catenin dictates Wnt signaling. Molecular Cell 36:872–884
18. Cohen-Saidon C, Cohen AA, Sigal A, Liron Y, Alon U (2009) Dynamics and variability of ERK2 response to EGF in individual living cells. Molecular Cell 36:885–893
19. Yang Q, Pando BF, Dong G, Golden SS, van Oudenaarden A (2010) Circadian gating of the cell cycle revealed in single cyanobacterial cells. Science 327:1522–1526
20. Perkins TJ, Swain PS (2009) Strategies for cellular decision-making. Molecular Systems Biology 5:326
21. Paszek P, Ryan S, Ashall L, Sillitoe K, Harper CV, Spiller DG, Rand DA, White MR (2010) Population robustness arising from cellular heterogeneity. Proceedings of the National Academy of Sciences of the United States of America 107:11644–11649
22. Bandyopadhyay S, Mehta M, Kuo D, Sung MK, Chuang R, Jaehnig EJ, Bodenmiller B, Licon K, Copeland W, Shales M, Fiedler D, Dutkowski J, Guénolé A, van Attikum H, Shokat KM, Kolodner RD, Huh WK, Aebersold R, Keogh MC, Krogan NJ, Ideker T (2010) Rewiring of genetic networks in response to DNA damage. Science 330:1385–1389
23. Lee S, Mandic J, Van Vliet KJ (2007) Chemomechanical mapping of ligand–receptor binding kinetics on cells. Proceedings of the National Academy of Sciences of the United States of America 104:9609–9614
24. Pawson CT, Scott JD (2010) Signal integration through blending, bolstering and bifurcating of intracellular information. Nature Structural and Molecular Biology 17:653–658
25. Mody A, Weiner J, Ramanathan S (2009) Modularity of MAP kinases allows deformation of their signalling pathways. Nature – Cell Biology 11:484–491
26. Matsuzawa A, Tseng PH, Vallabhapurapu S, Luo JL, Zhang W, Wang H, Vignali DAA, Gallagher E, Karin M (2008) Essential cytoplasmic translocation of a cytokine receptor-assembled signaling complex. Science 321:663–668
27. Miaczynska M, Pelkmans L, Zerial M (2004) Not just a sink: endosomes in control of signal transduction. Current Opinion in Cell Biology 16:400–406

28. Di Guglielmo GM, Baass PC, Ou WJ, Posner BI, Bergeron JJ (1994) Compartmentalization of SHC, GRB2 and mSOS, and hyperphosphorylation of Raf-1 by EGF but not insulin in liver parenchyma. EMBO Journal 13:4269–4277
29. Groves JT, Kuriyan J (2010) Molecular mechanisms in signal transduction at the membrane. Nature Structural and Molecular Biology 17:659–665
30. Zotenko E, Mestre J, O'Leary DP, Przytycka TM (2008) Why do hubs in the yeast protein interaction network tend to be essential: reexamining the connection between the network topology and essentiality. PLoS Computational Biology 4:e1000140
31. Zhao C, Du G, Skowronek K, Frohman MA, Bar-Sagi D (2007) Phospholipase D2-generated phosphatidic acid couples EGFR stimulation to Ras activation by Sos. Nature Cell Biology 9:707–712
32. Pike ACW, Rellos P, Niesen FH, Turnbull A, Oliver AW, Parker SA, Turk BE, Pearl LH, Knapp S (2008) Activation segment dimerization: a mechanism for kinase autophosphorylation of non-consensus sites. EMBO Journal 27:704–714
33. Olsen JV, Blagoev B, Gnad F, Macek B, Kumar C, Mortensen P, Mannet M (2006) Global, in vivo, and site-specific phosphorylation dynamics in signaling networks. Cell 127:635–648
34. Grueneberg DA, Degot S, Pearlberg J, Li W, Davies JE, Baldwin A, Endege W, Doench J, Sawyer J, Hu Y, Boyce F, Xian J, Munger K, Harlow E (2008) Kinase requirements in human cells: I. Comparing kinase requirements across various cell types. Proceedings of the National Academy of Sciences of the United States of America 105:16472–16477
35. Yoshida-Moriguchi T, Yu L, Stalnaker SH, Davis S, Kunz S, Madson M, Oldstone MBA, Schachter H, Wells L, Campbell KP (2010) O-Mannosyl phosphorylation of α-dystroglycan is required for laminin binding. Science 327:88–92
36. Pincet F (2007) Membrane recruitment of scaffold proteins drives specific signaling. PLoS One 2:e977
37. Deribe YL, Pawson T, Dikic I (2010) Post-translational modifications in signal integration. Nature Structural and Molecular Biology 17:666–672
38. Zhao S, Xu W, Jiang W, Yu W, Lin Y, Zhang T, Yao J, Zhou L, Zeng Y, Li H, Li Y, Shi J, An W, Hancock SM, He F, Qin L, Chin J, Yang P, Chen X, Lei Q, Xiong Y, Guan KL (2010) Regulation of cellular metabolism by protein lysine acetylation. Science 327:1000–1004
39. Hwang CS, Shemorry A, Varshavsky A (2010) N-terminal acetylation of cellular proteins creates specific degradation signals. Science 327:973–977
40. Kinoshita T, Fujita M, Maeda Y (2008) Biosynthesis, remodelling and functions of mammalian GPI-anchored proteins: recent progress. Journal of Biochemistry 144:287–294
41. Johnson D, Bennett ES (2006) Isoform-specific effects of the β2 subunit on voltage-gated sodium channel gating. Journal of Biological Chemistry 281:25875–25881
42. Montpetit ML, Stocker PJ, Schwetz TA, Harper JM, Norring SA, Schaffer L, North SJ, Jang-Lee J, Gilmartin T, Head SR, Haslam SM, Dell A, Marth JD, Bennett ES (2009) Regulated and aberrant glycosylation modulate cardiac electrical signaling. Proceedings of the National Academy of Sciences of the United States of America 106:16517–16522

43. Miyagi T, Wada T, Yamaguchi K, Hata K, Shiozaki K (2008) Plasma membrane-associated sialidase as a crucial regulator of transmembrane signalling. Journal of Biochemistry 144:279–285
44. Laczy B, Hill BG, Wang K, Paterson AJ, White CR, Xing D, Chen YF, Darley-Usmar V, Oparil S, Chatham JC (2009) Protein O-GlcNAcylation: a new signaling paradigm for the cardiovascular system. American Journal of Physiology – Heart and Circulatory Physiology 296:H13–H28
45. Golks A, Guerini D (2008) The O-linked N-acetylglucosamine modification in cellular signalling and the immune system. EMBO Reports 9:748–753
46. Wang Z, Udeshi ND, Slawson C, Compton PD, Sakabe K, Cheung WD, Shabanowitz J, Hunt DF, Hart GW (2010) Extensive crosstalk between O-GlcNAcylation and phosphorylation regulates cytokinesis. Science Signaling 3:ra2
47. Wang Z, Gucek M, Hart GW (2008) Cross-talk between GlcNAcylation and phosphorylation: Site-specific phosphorylation dynamics in response to globally elevated O-GlcNAc. Proceedings of the National Academy of Sciences of the United States of America 105:13793–13798
48. Hakmé A, Wong HK, Dantzer F, Schreiber V (2008) The expanding field of poly(ADP-ribosyl)ation reactions. EMBO Reports 9:1094–1100
49. McBride AE, Silver PA (2001) State of the Arg: protein methylation at arginine comes of age. Cell 106:5–8
50. Hoffmann R, Valencia A (2004) A gene network for navigating the literature. Nature – Genetics 36:664 (Information Hyperlinked over Proteins (www.ihop-net.org/)
51. Huang J, Berger SL (2008) The emerging field of dynamic lysine methylation of non-histone proteins. Current Opinion in Genetics and Development 18:152–158
52. Jansson M, Durant ST, Cho EC, Sheahan S, Edelmann M, Kessler B, La Thangue NB (2008) Arginine methylation regulates the p53 response. Nature Cell Biology 10:1431–1439
53. Lu T, Jackson MW, Wang B, Yang M, Chance MR, Miyagi M, Gudkov AV, Stark GR (2010) Regulation of NF-κB by NSD1/FBXL11-dependent reversible lysine methylation of p65. Proceedings of the National Academy of Sciences of the United States of America 107:46–51
54. Ea CK, Baltimore D (2009) Regulation of NF-κB activity through lysine monomethylation of p65. Proceedings of the National Academy of Sciences of the United States of America 106:18972–18977
55. Huang B, Chen SC, Wang DL (2009) Shear flow increases S-nitrosylation of proteins in endothelial cells. Cardiovascular Research 83:536–546
56. Benhar M, Forrester MT, Stamler JS (2009) Protein denitrosylation: enzymatic mechanisms and cellular functions. Nature Reviews – Molecular Cell Biology 10:721–732
57. Guo Z, Wu YW, Das D, Delon C, Cramer J, Yu S, Thuns S, Lupilova N, Waldmann H, Brunsveld L, Goody RS, Alexandrov K, Blankenfeldt W (2008) Structures of RabGGTase-substrate/product complexes provide insights into the evolution of protein prenylation. EMBO Journal 27:2444–2456
58. Moore KL (2009) Protein tyrosine sulfation: a critical posttranslation modification in plants and animals. Proceedings of the National Academy of Sciences of the United States of America 106:14741–14742

59. BioGRID: General Repository for Interaction Datasets; database of physical and genetic interactions for model organisms (www.thebiogrid.org)
60. Husnjak K, Elsasser S, Zhang N, Chen X, Randles L, Shi Y, Hofmann K, Walters KJ, Finley D, Dikic I (2008) Proteasome subunit Rpn13 is a novel ubiquitin receptor. Nature 453:481–488
61. Mukhopadhyay D, Riezman H (2007) Proteasome-independent functions of ubiquitin in endocytosis and signaling. Science 315:201–205
62. Xia ZP, Sun L, Chen X, Pineda G, Jiang X, Adhikari A, Zeng W, Chen ZJ (2009) Direct activation of protein kinases by unanchored polyubiquitin chains. Nature 461:114–119
63. Ulrich HD, Walden H (2010) Ubiquitin signalling in DNA replication and repair. Nature Reviews – Molecular Cell Biology 11:479–489
64. Singh M, Roginskaya M, Dalal S, Menon B, Kaverina E, Boluyt MO, Singh K (2010) Extracellular ubiquitin inhibits β-AR-stimulated apoptosis in cardiac myocytes: role of GSK-3β and mitochondrial pathways. Cardiovascular Research 86:20–28
65. Daino H, Matsumura I, Takada K, Odajima J, Tanaka H, Ueda S, Shibayama H, Ikeda H, Hibi M, Machii T, Hirano T, Kanakura Y (2000) Induction of apoptosis by extracellular ubiquitin in human hematopoietic cells: possible involvement of STAT3 degradation by proteasome pathway in interleukin 6-dependent hematopoietic cells. Blood 95:2577–2585
66. Komander D, Clague MJ, Urbé S (2009) Breaking the chains: structure and function of the deubiquitinases. Nature Reviews – Molecular Cell Biology 10:550–563
67. Samara NL, Datta AB, Berndsen CE, Zhang X, Yao T, Cohen RE, Wolberger C (2010) Structural insights into the assembly and function of the SAGA deubiquitinating module. Science 328:1025–1029
68. Meulmeester E, Melchior F (2008) Cell biology: SUMO. Nature 452:709–711
69. Geiss-Friedlander R, Melchior F (2007) Concepts in sumoylation: a decade on. Nature Reviews – Molecular Cell Biology 8:947–956
70. Van der Veen AG, Schorpp K, Schlieker C, Buti L, Damon JR, Spooner E, Ploegh HL, Jentsch S (2011) Role of the ubiquitin-like protein Urm1 as a noncanonical lysine-directed protein modifier. Proceedings of the National Academy of Sciences of the United States of America 108:1763–1770
71. Page-McCaw A, Ewald AJ, Werb Z (2007) Matrix metalloproteinases and the regulation of tissue remodelling. Nature Reviews – Molecular Cell Biology 8:221–233
72. Kemble DJ, Sun G (2009) Direct and specific inactivation of protein tyrosine kinases in the Src and FGFR families by reversible cysteine oxidation. Proceedings of the National Academy of Sciences of the United States of America 106:5070–5075
73. Tonks NK (2005) Redox redux: revisiting PTPs and the control of cell signaling. Cell 121:667–670
74. Singh DK, Kumar D, Siddiqui Z, Basu SK, Kumar V, Rao KV (2005) The strength of receptor signaling is centrally controlled through a cooperative loop between Ca^{2+} and an oxidant signal. Cell 121:281–293
75. Keene JD (2007) RNA regulons: coordination of post-transcriptional events. Nature Reviews – Genetics 8,:533–543
76. Alberts B, Bray D, Lewis J, Raff M, Roberts K, Watson JD (2002) Molecular biology of the cell, 4th ed. Garland Science, New York

77. Pokholok DK, Zeitlinger J, Hannett NM, Reynolds DB, Young RA (2006) Activated signal transduction kinases frequently occupy target genes. Science 313:533–536
78. Williams RL, Urbé S (2007) The emerging shape of the ESCRT machinery. Nature Reviews – Molecular Cell Biology 8:355–368
79. Simons K, Toomre D (2000) Lipid rafts and signal transduction. Nature Reviews – Molecular Cell Biology 1:31–39
80. Harmer NJ, Chirgadze D, Kim KH, Pellegrini L, Blundell TL (2003) The structural biology of growth factor receptor activation Biophysical Chemistry 100:545–553
81. Helikar T, Konvalina J, Heidel J, Rogers JA (2008) Emergent decision-making in biological signal transduction networks. Proceedings of the National Academy of Sciences of the United States of America 105:1913–1918
82. Gore J, van Oudenaarden A (2009) Synthetic biology: The yin and yang of nature. Nature 457:271-272
83. Brandman O, Meyer T (2008) Feedback loops shape cellular signals in space and time. Science 322:390–395
84. Stricker J, Cookson S, Bennett MR, Mather WH, Tsimring LS, Hasty J (2008) A fast, robust and tunable synthetic gene oscillator. Nature 456:516–519
85. Tsai TYC, Choi YS, Ma W, Pomerening JR, Tang C, Ferrell JE (2008) Robust, tunable biological oscillations from interlinked positive and negative feedback loops. Science 321:126–129
86. Tigges M, Marquez-Lago TT, Stelling J, Fussenegger M (2009) A tunable synthetic mammalian oscillator. Nature 457:309–312
87. Miller-Jensen K, Janes KA, Brugge JS, Lauffenburger DA (2007) Common effector processing mediates cell-specific responses to stimuli. Nature 448:604–608
88. Sacks DB (2006) The role of scaffold proteins in MEK/ERK signalling. Biochemical Society Transactions 34:833–836
89. Vandecasteele G, Rochais F, Abi-Gerges A, Fischmeister R (2006) Functional localization of cAMP signalling in cardiac myocytes. Biochemical Society Transactions 34:484–488
90. Behar M, Dohlman HG, Elston TC (2007) Kinetic insulation as an effective mechanism for achieving pathway specificity in intracellular signaling networks. Proceedings of the National Academy of Sciences of the United States of America 104:16146–16151
91. Kholodenko BN (2006) Cell-signalling dynamics in time and space. Nature Reviews – Molecular Cell Biology 7:165–176
92. Kholodenko BN, Demin OV, Moehren G, Hoek JB (1999) Quantification of short term signaling by the epidermal growth factor receptor. Journal of Biological Chemistry 274:30169–30181
93. Bhalla US, Ram PT, Iyengar R (2002) MAP kinase phosphatase as a locus of flexibility in a mitogen-activated protein kinase signaling network. Science 297:1018–1023
94. Natarajan M, Lin KM, Hsueh RC, Sternweis PC, Ranganathan R (2006) A global analysis of cross-talk in a mammalian cellular signalling network. Nature Cell Biology 8:571–580
95. Whitehurst A, Cobb MH, White MA (2004) Stimulus-coupled spatial restriction of extracellular signal-regulated kinase 1/2 activity contributes to the specificity of signal-response pathways. Molecular and Cellular Biology 24:10145–10150

96. Harding A, Tian T, Westbury E, Frische E, Hancock JF (2005) Subcellular localization determines MAP kinase signal output. Current Biology 15:869–873
97. Santos SD, Verveer PJ, Bastiaens PI (2007) Growth factor-induced MAPK network topology shapes Erk response determining PC-12 cell fate. Nature Cell Biology 9:324–330
98. Karlebach G, Shamir R (2008) Modelling and analysis of gene regulatory networks. Nature Reviews – Molecular Cell Biology 9:770–780
99. Berezhkovskii AM, Coppey M, Shvartsman SY (2009) Signaling gradients in cascades of two-state reaction-diffusion systems. Proceedings of the National Academy of Sciences of the United States of America 106:1087–1092
100. Inui M, Martello G, Piccolo S (2010) MicroRNA control of signal transduction. Nature Reviews – Molecular Cell Biology 11:252–263
101. Fitz JG (2007) Regulation of cellular ATP release. Transactions of the American Clinical and Climatological Association 118:199–208
102. Burnstock G (2008) Unresolved issues and controversies in purinergic signalling. Journal of Physiology 586:3307–3312
103. Corriden R, Insel PA (2010) Basal release of ATP: an autocrine-paracrine mechanism for cell regulation. Science Signaling 3:re1
104. Okada Y, Sato K, Numata T (2009) Pathophysiology and puzzles of the volume-sensitive outwardly rectifying anion channel. Journal of Physiology 587:2141-2149
105. Sabirov RZ, Okada Y (2004) Wide nanoscopic pore of maxi-anion channel suits its function as an ATP-conductive pathway. Biophysical Journal 87:1672-1685
106. Burnstock G (2006) Historical review: ATP as a neurotransmitter. Trends in Pharmacological Sciences 27:166–176
107. Fields RD, Ni Y (2010) Nonsynaptic communication through ATP release from volume-activated anion channels in axons. Science Signaling 3:ra73
108. Birk AV, Broekman MJ, Gladek EM, Robertson HD, Drosopoulos JH, Marcus AJ, Szeto HH (2002) Role of extracellular ATP metabolism in regulation of platelet reactivity. Journal of Laboratory and Clinical Medicine 140:166–175
109. Krötz F, Sohn HY, Keller M, Gloe T, Bolz SS, Becker BF, Pohl U (2002) Depolarization of endothelial cells enhances platelet aggregation through oxidative inactivation of endothelial NTPDase. Arteriosclerosis, Thrombosis, and Vascular Biology 22:2003–2009
110. Eltzschig HK, Macmanus CF, Colgan SP (2008) Neutrophils as sources of extracellular nucleotides: functional consequences at the vascular interface. Trends in Cardiovascular Medicine 18:103–107
111. Sprague RS, Ellsworth ML, Stephenson AH, Lonigro AJ (1996) ATP: the red blood cell link to NO and local control of the pulmonary circulation. American Journal of Physiology – Heart and Circulatory Physiology 271:H2717–2722
112. Wan J, Ristenpart WD, Stone HA (2008) Dynamics of shear-induced ATP release from red blood cells. Proceedings of the National Academy of Sciences of the United States of America 105:16432–16437
113. Lazarowski ER, Homolya L, Boucher RC, Harden TK (1997) Identification of an ecto-nucleoside diphosphokinase and its contribution to interconversion of P2 receptor agonists. Journal of Biological Chemistry 272:20402–20407

114. Lazarowski ER, Paradiso AM, Watt WC, Harden TK, Boucher RC (1997) UDP activates a mucosal-restricted receptor on human nasal epithelial cells that is distinct from the P2Y2 receptor. Proceedings of the National Academy of Sciences of the United States of America 94:2599–2603

Chap. 2. Membrane Ion Carriers

115. Scemes E, Suadicani SO, Dahl G, Spray DC (2007) Connexin and pannexin mediated cell–cell communication. Neuron Glia Biology 3:199–208

116. Palacios-Prado N, Briggs SW, Skeberdis VA, Pranevicius M, Bennett MV, Bukauskas FF (2010) pH-Dependent modulation of voltage gating in connexin45 homotypic and connexin45/connexin43 heterotypic gap junctions. Proceedings of the National Academy of Sciences of the United States of America 107:9897–9902

117. Jiang H, Zhu AG, Mamczur M, Falck JR, Lerea KM, McGiff JC (2009) Stimulation of rat erythrocyte P2X7 receptor induces the release of epoxyeicosatrienoic acids. British Journal of Pharmacology 151:1033–1040

118. Pelegrin P, Surprenant A (2009) The $P2X_7$ receptor-pannexin connection to dye uptake and IL-1β release. Purinergic Signal 5:129–137

119. Sridharan M, Adderley SP, Bowles EA, Egan TM, Stephenson AH, Ellsworth ML, Sprague RS (2010) Pannexin 1 is the conduit for low oxygen tension-induced ATP release from human erythrocytes. American Journal of Physiology – Heart and Circulatory Physiology 299:H1146–H1152

120. Krishnamurthy H, Piscitelli CL, Gouaux E (2009) Unlocking the molecular secrets of sodium-coupled transporters. Nature 459:347–355

121. Saier MH, Tran CV, Barabote RD (2006) TCDB: the Transporter Classification Database for membrane transport protein analyses and information. Nucleic Acids Research 34:D181–D186

122. Gadsby DC (2009) Ion channels versus ion pumps: the principal difference, in principle. Nature Reviews – Molecular Cell Biology 10:344–352

123. Zagotta WN (2006) Membrane biology: permutations of permeability. Nature 440:427–429

124. Cymes GD, Grosman C (2008) Pore-opening mechanism of the nicotinic acetylcholine receptor evinced by proton transfer. Nature Structural and Molecular Biology 15:389–396

125. Okamura Y (2007) Biodiversity of voltage sensor domain proteins. Pflügers Archiv – European Journal of Physiology 454:361–371

126. Tao X, Lee A, Limapichat W, Dougherty DA, MacKinnon R (2010) A gating charge transfer center in voltage sensors. Science 328:67–73

127. Jensen MØ, Borhani DW, Lindorff-Larsen K, Maragakis P, Jogini V, Eastwood MP, Dror RO, Shaw DE (2010) Principles of conduction and hydrophobic gating in K^+ channels. Proceedings of the National Academy of Sciences of the United States of America 107:5833–5838

128. Sasaki M, Takagi M, Okamura Y (2006) A voltage sensor-domain protein is a voltage-gated proton channel. Science 312:589–592

129. Berkefeld H, Sailer CA, Bildl W, Rohde V, Thumfart JO, Eble S, Klugbauer N, Reisinger E, Bischofberger J, Oliver D, Knaus HG, Schulte U, Fakler B (2006) BK_{Ca}–Ca_V channel complexes mediate rapid and localized Ca^{2+}-activated K^+ signaling. Science 314:615–620

130. Lape R, Colquhoun D, Sivilotti LG (2008) On the nature of partial agonism in the nicotinic receptor superfamily. Nature 454:722–727

131. Coste B, Mathur J, Schmidt M, Earley TJ, Ranade S, Petrus MJ, Dubin AE, Patapoutian A (2010) Piezo1 and Piezo2 are essential components of distinct mechanically activated cation channels. Science 330:55-60

132. Gottlieb PA, Suchyna TM, Ostrow LW, Sachs F (2004) Mechanosensitive ion channels as drug targets. Current Drug Targets. CNS and Neurological Disorders 3:287–295

133. Kohl P, Bollensdorff C, Garny A (2006) Effects of mechanosensitive ion channels on ventricular electrophysiology: experimental and theoretical models. Experimental Physiology 91:307–321

134. Christensen AP, Corey DP (2007) TRP channels in mechanosensation: direct or indirect activation? Nature Reviews – Neuroscience 8:510–521

135. Gudi SRP, Clark CB, Frangos JA (1996) Fluid flow rapidly activates G proteins in human endothelial cells. Involvement of G proteins in mechanochemical signal transduction. Circulation Research 79:834-839

136. Kung C (2005) A possible unifying principle for mechanosensation. Nature 436:647–654.

137. Vanin EF (1985) Processed pseudogenes: characteristics and evolution. Annual Review of Genetics 19:253-272

138. Putney JW (2007) Multiple mechanisms of TRPC activation (Chap. 1). In Liedtke WB, Heller S (Eds.) TRP ion channel function in sensory transduction and cellular signaling cascades, CRC Press (Taylor and Francis Group), Boca Raton, FL, USA

139. Spassova MA, Hewavitharana T, Xu W, Soboloff J, Gill DL (2006) A common mechanism underlies stretch activation and receptor activation of TRPC6 channels. Proceedings of the National Academy of Sciences of the United States of America 103:16586–16591

140. Yuan JP, Zeng W, Huang GN, Worley PF, Muallem S (2007) STIM1 heteromultimerizes TRPC channels to determine their function as store-operated channels. Nature Cell Biology 9:636–645

141. Minke B, Cook B (2002) TRP channel proteins and signal transduction. Physiological Review 82:429–472

142. Wegierski T, Hill K, Schaefer M, Walz G (2006) The HECT ubiquitin ligase AIP4 regulates the cell surface expression of select TRP channels. EMBO Journal 25:5659–5669

143. Tominaga M (2007) The role of TRP channels in thermosensation (Chap. 20). In Liedtke WB, Heller S (Eds.) TRP ion channel function in sensory transduction and cellular signaling cascades, CRC Press (Taylor and Francis Group), Boca Raton, FL, USA

144. Latorre R, Vargas G, Orta G, Brauchi S (2007) Voltage and temperature gating of thermoTRP channels (Chap. 21). In Liedtke WB, Heller S (Eds.) TRP ion channel function in sensory transduction and cellular signaling cascades, CRC Press (Taylor and Francis Group), Boca Raton, FL, USA

145. Patel S, Docampo R (2009) In with the TRP channels: intracellular functions for TRPM1 and TRPM2. Science Signaling 2:pe69

146. Wegierski T, Steffl D, Kopp C, Tauber R, Buchholz B, Nitschke R, Kuehn EW, Walz G, Köttgen M (2009) TRPP2 channels regulate apoptosis through the Ca^{2+} concentration in the endoplasmic reticulum. EMBO Journal 28:490–499

147. Yu Y, Ulbrich MH, Li MH, Buraei Z, Chen XZ, Ong AC, Tong L, Isacoff EY, Yang J (2009) Structural and molecular basis of the assembly of the TRPP2/PKD1 complex. Proceedings of the National Academy of Sciences of the United States of America 106:11558–11563

148. Nilius B, Owsianik G, Voets T (2008) Transient receptor potential channels meet phosphoinositides. EMBO Journal 27:2809–2816

149. Earley S, Reading S, Brayden JE (2007) Functional significance of transient receptor potential channels in vascular function (Chap. 26). In Liedtke WB, Heller S (Eds.) TRP ion channel function in sensory transduction and cellular signaling cascades, CRC Press (Taylor and Francis Group), Boca Raton, FL, USA

150. Wu X, Eder P, Chang B, Molkentin JD (2010) TRPC channels are necessary mediators of pathologic cardiac hypertrophy. Proceedings of the National Academy of Sciences of the United States of America 107:7000–7005

151. Pozsgai G, Bodkin JV, Graepel R, Bevan S, Andersson DA, Brain SD (2010) Evidence for the pathophysiological relevance of TRPA1 receptors in the cardiovascular system in vivo. Cardiovascular Research 87:760–768

152. Guimaraes MZP, Jordt SE (2007) TRPA1: a sensory channel of many talents (Chap. 11). In Liedtke WB, Heller S (Eds.) TRP ion channel function in sensory transduction and cellular signaling cascades, CRC Press (Taylor and Francis Group), Boca Raton, FL, USA

153. Planells-Cases R, Ferrer-Montiel A (2007) TRP Channel trafficking (Chap. 23). In Liedtke WB, Heller S (Eds.) TRP ion channel function in sensory transduction and cellular signaling cascades, CRC Press (Taylor and Francis Group), Boca Raton, FL, USA

154. Pani B, Ong HL, Brazer SC, Liu X, Rauser K, Singh BB, Ambudkar IS (2009) Activation of TRPC1 by STIM1 in ER-PM microdomains involves release of the channel from its scaffold caveolin-1. Proceedings of the National Academy of Sciences of the United States of America 106:20087–20092

155. Eder P, Schindl R, Romanin C, Groschner K (2007) Protein–protein interactions in TRPC channel complexes (Chap. 24). In Liedtke WB, Heller S (Eds.) TRP ion channel function in sensory transduction and cellular signaling cascades, CRC Press (Taylor and Francis Group), Boca Raton, FL, USA

156. Rychkov G, Barritt GJ (2007) TRPC1 Ca^{2+}-Permeable Channels in Animal Cells. In Flockerzi V, Nilius B (eds) Transient Receptor Potential (TRP) Channels, Handbook of Experimental Pharmacology, Vol. 179, Part I, 23–52, Springer, Berlin

157. Poteser M, Schleifer H, Lichtenegger M, Schernthaner M, Stockner T, Kappe CO, Glasnov TN, Romanin C, Groschner K (2011) PKC-dependent coupling of calcium permeation through transient receptor potential canonical 3 (TRPC3) to calcineurin signaling in HL-1 myocytes. Proceedings of the National Academy of Sciences of the United States of America 108:10556–10561

158. Park HW, Kim JY, Choi SK, Lee YH, Zeng W, Kim KH, Muallem S, Lee MG (2011) Serine–threonine kinase with-no-lysine 4 (WNK4) controls blood pressure via transient receptor potential canonical 3 (TRPC3) in the vasculature. Proceedings of the National Academy of Sciences of the United States of America 108:10750–10755

159. Cavali A (2007) TRPC4. In Flockerzi V, Nilius B (eds) Transient Receptor Potential (TRP) Channels, Handbook of Experimental Pharmacology, Vol. 179, Part I, 93–108, Springer, Berlin

160. Xu SZ, Sukumar P, Zeng F, Li J, Jairaman A, English A, Naylor J, Ciurtin C, Majeed Y, Milligan CJ, Bahnasi YM, Al-Shawaf E, Porter KE, Jiang LH, Emery P, Sivaprasadarao A, Beech DJ (2008) TRPC channel activation by extracellular thioredoxin. Nature 451:69–72

161. Weissmann N, Dietrich A, Fuchs B, Kalwa H, Ay M, Dumitrascu R, Olschewski A, Storch U, Mederos y Schnitzler M, Ghofrani HA, Schermuly RT, Pinkenburg O, Seeger W, Grimminger F, Gudermann T (2006) Classical transient receptor potential channel 6 (TRPC6) is essential for hypoxic pulmonary vasoconstriction and alveolar gas exchange. Proceedings of the National Academy of Sciences of the United States of America 103:19093-19098

162. Numaga T, Wakamori M, Mori Y (2007) TRPC7. In Flockerzi V, Nilius B (eds) Transient Receptor Potential (TRP) Channels, Handbook of Experimental Pharmacology, Vol. 179, Part I, 143–151, Springer, Berlin

163. Du J, Xie J, Yue L (2009) Intracellular calcium activates TRPM2 and its alternative spliced isoforms. Proceedings of the National Academy of Sciences of the United States of America 106:7239–7244

164. Harteneck C, Schultz G (2007) TRPV4 and TRPM3 as volume-regulated cation channel (Chap. 10). In Liedtke WB, Heller S (Eds.) TRP ion channel function in sensory transduction and cellular signaling cascades, CRC Press (Taylor and Francis Group), Boca Raton, FL, USA

165. Liman ER (2007) The Ca^{2+}-activated TRP channels: TRPM4 and TRPM5 (Chap. 15). In Liedtke WB, Heller S (Eds.) TRP ion channel function in sensory transduction and cellular signaling cascades, CRC Press (Taylor and Francis Group), Boca Raton, FL, USA

166. Inoue K, Xiong ZG (2009) Silencing TRPM7 promotes growth/proliferation and nitric oxide production of vascular endothelial cells via the ERK pathway. Cardiovascular Research 83:547–557

167. Johnson CD, Melanaphy D, Purse A, Stokesberry SA, Dickson P, Zholos AV (2009) Transient receptor potential melastatin 8 channel involvement in the regulation of vascular tone. American Journal of Physiology – Heart and Circulatory Physiology 296:H1868–H1877

168. McKemy DD (2007) TRPM8: the cold and menthol receptor (Chap. 13). In Liedtke WB, Heller S (Eds.) TRP ion channel function in sensory transduction and cellular signaling cascades, CRC Press (Taylor and Francis Group), Boca Raton, FL, USA

169. Giamarchi A, Delmas P (2007) Activation mechanisms and functional roles of TRPP2 cation channels (Chap. 14). In Liedtke WB, Heller S (Eds.) TRP ion channel function in sensory transduction and cellular signaling cascades, CRC Press (Taylor and Francis Group), Boca Raton, FL, USA

170. Rosenbaum T, Simon SA (2007) TRPV1 Receptors and signal transduction (Chap. 5). In Liedtke WB, Heller S (Eds.) TRP ion channel function in sensory transduction and cellular signaling cascades, CRC Press (Taylor and Francis Group), Boca Raton, FL, USA

171. Liedtke WB (2007) TRPV Channel function in osmo- and mechanotransduction (Chap. 22). In Liedtke WB, Heller S (Eds.) TRP ion channel function in sensory transduction and cellular signaling cascades, CRC Press (Taylor and Francis Group), Boca Raton, FL, USA

172. Kojima I, Nagasawa M (2007) TRPV2: a calcium-permeable cation channel regulated by insulin-like growth factors (Chap. 7). In Liedtke WB, Heller S (Eds.) TRP ion channel function in sensory transduction and cellular signaling cascades, CRC Press (Taylor and Francis Group), Boca Raton, FL, USA

173. Shukla AK, Kim J, Ahn S, Xiao K, Shenoy SK, Liedtke W, Lefkowitz RJ (2010) Arresting a transient receptor potential (TRP) channel: β-arrestin 1 mediates ubiquitination and functional down-regulation of TRPV4. Journal of Biological Chemistry 285:30115–30125

174. Plant TD, Strotmann R (2007) TRPV4: a multifunctional nonselective cation channel with complex regulation (Chap. 9). In Liedtke WB, Heller S (Eds.) TRP ion channel function in sensory transduction and cellular signaling cascades, CRC Press (Taylor and Francis Group), Boca Raton, FL, USA

175. Köhler R, Hoyer J (2007) Role of TRPV4 in the mechanotransduction of shear stress in endothelial cells (Chap. 27). In Liedtke WB, Heller S (Eds.) TRP ion channel function in sensory transduction and cellular signaling cascades, CRC Press (Taylor and Francis Group), Boca Raton, FL, USA

176. Lorenzo IM, Liedtke W, Sanderson MJ, Valverde MA (2008) TRPV4 channel participates in receptor-operated calcium entry and ciliary beat frequency regulation in mouse airway epithelial cells. Proceedings of the National Academy of Sciences of the United States of America 105:12611–12616

177. Andrade YN, Fernandes J, Lorenzo IM, Arniges M, Valverde MA (2007) The TRPV4 channel in ciliated epithelia (Chap. 30). In Liedtke WB, Heller S (Eds.) TRP ion channel function in sensory transduction and cellular signaling cascades, CRC Press (Taylor and Francis Group), Boca Raton, FL, USA

178. Inoue R, Jensen LJ, Shi J, Morita H, Nishida M, Honda A, Ito Y (2006) Transient receptor potential channels in cardiovascular function and disease. Circulation Research 99:119–131

179. Herrmann S, Stieber J, Stöckl G, Hofmann F, Ludwig A (2007) HCN4 provides a "depolarization reserve" and is not required for heart rate acceleration in mice. EMBO Journal 26:4423–4432

180. Alig J, Marger L, Mesirca P, Ehmke H, Mangoni ME, Isbrandt D (2009) Control of heart rate by cAMP sensitivity of HCN channels. Proceedings of the National Academy of Sciences of the United States of America 106:12189–12194

181. Monteggia LM, Eisch AJ, Tang MD, Kaczmarek LK, Nestler EJ (2000) Cloning and localization of the hyperpolarization-activated cyclic nucleotide-gated channel family in rat brain. Brain Research – Molecular Brain Research 81:129–139

182. Ludwig A, Budde T, Stieber J, Moosmang S, Wahl C, Holthoff K, Langebartels A, Wotjak C, Munsch T, Zong X, Feil S, Feil R, Lancel M, Chien KR, Konnerth A, Pape HC,

Biel M, Hofmann F (2003) Absence epilepsy and sinus dysrhythmia in mice lacking the pacemaker channel HCN2. EMBO Journal 22:216–224

183. Hibbs RE, Gouaux E (2011) Principles of activation and permeation in an anion-selective Cys-loop receptor. Nature 474:54–60

184. Lo WY, Lagrange AH, Hernandez CC, Harrison R, Dell A, Haslam SM, Sheehan JH, Macdonald RL (2010) Glycosylation of β2 subunits regulates GABAA receptor biogenesis and channel gating. Journal of Biological Chemistry 285:31348–31361

185. Changeux JP (2010) Nicotine addiction and nicotinic receptors: lessons from genetically modified mice. Nature Reviews – Neuroscience 11:389–401

186. Braithwaite SP, Xia H, Malenka RC (2002) Differential roles for NSF and GRIP/ABP in AMPA receptor cycling. Proceedings of the National Academy of Sciences of the United States of America 99:7096–7101

187. Farrant M, Cull-Candy SG (2010) AMPA receptors – another twist? Science 327:1463–1465

188. von Engelhardt J, Mack V, Sprengel R, Kavenstock N, Li KW, Stern-Bach Y, Smit AB, Seeburg PH, Monyer H (2010) CKAMP44: a brain-specific protein attenuating short-term synaptic plasticity in the dentate gyrus. Science 327:1518–1522

189. Banke TG, Bowie D, Lee H, Huganir RL, Schousboe A, Traynelis SF (2000) Control of GluR1 AMPA receptor function by cAMP-dependent protein kinase. Journal of Neuroscience 20:89–102

190. Boehm J, Kang MG, Johnson RC, Esteban J, Huganir RL, Malinow R (2006) Synaptic incorporation of AMPA receptors during LTP is controlled by a PKC phosphorylation site on GluR1. Neuron 51:213–225

191. Derkach V, Barria A, Soderling TR (1999) Ca^{2+}/calmodulin-kinase II enhances channel conductance of α-amino 3-hydroxy 5-methyl 4-isoxazolepropionate type glutamate receptors. Proceedings of the National Academy of Sciences of the United States of America 96:3269–3274

192. Hayashi Y, Shi SH, Esteban JA, Piccini A, Poncer JC, Malinow R (2000) Driving AMPA receptors into synapses by LTP and CaMKII: requirement for GluR1 and PDZ domain interaction. Science 287:2262–2267

193. Sommer B, Keinänen K, Verdoorn TA, Wisden W, Burnashev N, Herb A, Köhler M, Takagi T, Sakmann B, Seeburg PH (1990) Flip and flop: a cell-specific functional switch in glutamate-operated channels of the CNS. Science 249:1580–1585

194. Joiner MLA, Lisé MF, Yuen EY, Kam AY, Zhang M, Hall DD, Malik ZA, Qian H, Chen Y, Ulrich JD, Burette AC, Weinberg RJ, Law PY, El-Husseini A, Yan Z, Hell JW (2010) Assembly of a β2-adrenergic receptor-GluR1 signalling complex for localized cAMP signalling. EMBO Journal 29:482–495

195. Krugers HJ, Hoogenraad CC, Groc L (2010) Stress hormones and AMPA receptor trafficking in synaptic plasticity and memory. Nature Reviews – Neuroscience 11:675–681

196. Piña-Crespo JC, Talantova M, Micu I, States B, Chen HS, Tu S, Nakanishi N, Tong G, Zhang D, Heinemann SF, Zamponi GW, Stys PK, Lipton SA (2010) Excitatory glycine responses of CNS myelin mediated by NR1/NR3 "NMDA" receptor subunits. Journal of Neuroscience 30:11501–11505

197. Yu XM, Askalan R, Keil GJ, Salter MW (1997) NMDA channel regulation by channel-associated protein tyrosine kinase Src. Science 275:674–678

198. Hardingham GE, Bading H (2010) Synaptic versus extrasynaptic NMDA receptor signalling: implications for neurodegenerative disorders. Nature Reviews – Neuroscience 11:682–696

199. Matsuda K, Miura E, Miyazaki T, Kakegawa W, Emi K, Narumi S, Fukazawa Y, Ito-Ishida A, Kondo T, Shigemoto R, Watanabe M, Yuzaki M (2010) Cbln1 is a ligand for an orphan glutamate receptor $\delta 2$, a bidirectional synapse organizer. Science 328:363–368

200. Kishore U, Gaboriaud C, Waters P, Shrive AK, Greenhough TJ, Reid KB, Sim RB, Arlaud GJ (2004) C1q and tumor necrosis factor superfamily: modularity and versatility. Trends in Immunology 25:551–561

201. Lummis SC (2010) 5-Hydroxytryptamine receptor 3A. UCSD-Nature Molecule Pages, UCSD-Nature Signaling Gateway (www.signaling-gateway.org)

202. Vassort G (2001) Adenosine $5'$-triphosphate: a P2-purinergic agonist in the myocardium. Physiological Reviews 81:767–806

203. Jacques-Silva MC, Correa-Medina M, Cabrera O, Rodriguez-Diaz R, Makeeva N, Fachado A, Diez J, Berman DM, Kenyon NS, Ricordi C, Pileggi A, Molano RD, Berggren PO, Caicedo A (2010) ATP-gated $P2X_3$ receptors constitute a positive autocrine signal for insulin release in the human pancreatic β cell. Proceedings of the National Academy of Sciences of the United States of America 107:6465–6470

204. Schenk U, Frascoli M, Proietti M, Geffers R, Traggiai E, Buer J, Ricordi C, Westendorf AM, Grassi F (2011) ATP inhibits the generation and function of regulatory T cells through the activation of purinergic P2X receptors. Science Signaling 4:ra12

205. Chizh BA, Illes P (2001) P2X receptors and nociception. Pharmacological Reviews 53:553–568

206. Clark K, Langeslag M, van Leeuwen B, Ran L, Ryazanov AG, Figdor CG, Moolenaar WH, Jalink K, van Leeuwen FN (2006) TRPM7, a novel regulator of actomyosin contractility and cell adhesion. EMBO Journal 25:290–301

207. Casey JR, Grinstein S, Orlowski J (2010) Sensors and regulators of intracellular pH. Nature Reviews – Molecular Cell Biology 11:50–61

Chap. 3. Main Classes of Ion Channels and Pumps

208. Gadsby DC (2009) Ion channels versus ion pumps: the principal difference, in principle. Nature Reviews – Molecular Cell Biology 10:344–352

209. Lai HC, Jan LY (2006) The distribution and targeting of neuronal voltage-gated ion channels. Nature Reviews – Neuroscience 7:548–562

210. Orrenius S, Zhivotovsky B, Nicotera P (2003) Regulation of cell death: the calcium-apoptosis link. Nature Reviews – Molecular Cell Biology 4:552–565

211. Hénaff M, Antoine S, Mercadier JJ, Coulombe A, Hatem SN (2002) The voltage-independent B-type Ca^{2+} channel modulates apoptosis of cardiac myocytes. FASEB Journal 16:99–101

212. Cartwright EJ, Schuh K, Neyses L (2005) Calcium transport in cardiovascular health and disease–the sarcolemmal calcium pump enters the stage. Journal of Molecular and Cellular Cardiology 39:403–406

213. Bano D, Young KW, Guerin CJ, Lefeuvre R, Rothwell NJ, Naldini L, Rizzuto R, Carafoli E, Nicotera P (2005) Cleavage of the plasma membrane Na^{+}/Ca^{2+} exchanger in excitotoxicity. Cell 120:275–285

214. Boscia F, Gala R, Pignataro G, de Bartolomeis A, Cicale M, Ambesi-Impiombato A, Di Renzo G, Annunziato L (2006) Permanent focal brain ischemia induces isoform-dependent changes in the pattern of Na^{+}/Ca^{2+} exchanger gene expression in the ischemic core, periinfarct area, and intact brain regions. Journal of Cerebral Blood Flow and Metabolism 26:502–517

215. Calcraft PJ, Ruas M, Pan Z, Cheng X, Arredouani A, Hao X, Tang J, Rietdorf K, Teboul L, Chuang KT, Lin P, Xiao R, Wang C, Zhu Y, Lin Y, Wyatt CN, Parrington J, Ma J, Evans AM, Galione A, Zhu MX (2009) NAADP mobilizes calcium from acidic organelles through two-pore channels. Nature 459:596–600

216. Dammermann W, Zhang B, Nebel M, Cordiglieri C, Odoardi F, Kirchberger T, Kawakami N, Dowden J, Schmid F, Dornmair K, Hohenegger M, Flügel A, Guse AH, Potter BV (2009) NAADP-mediated Ca^{2+} signaling via type 1 ryanodine receptor in T cells revealed by a synthetic NAADP antagonist. Proceedings of the National Academy of Sciences of the United States of America 106:10678–10683

217. Alix JJP, Dolphin AC, Fern R (2008) Vesicular apparatus, including functional calcium channels, are present in developing rodent optic nerve axons and are required for normal node of Ranvier formation. Journal of Physiology 586:4069–4089

218. Perocchi F, Gohil VM, Girgis HS, Bao XR, McCombs JE, Palmer AE, Mootha VK (2010) MICU1 encodes a mitochondrial EF hand protein required for Ca^{2+} uptake. Nature 467:291–296

219. Feske S, Gwack Y, Prakriya M, Srikanth S, Puppel SH, Tanasa B, Hogan PG, Lewis RS, Daly M, Rao A (2006) A mutation in Orai1 causes immune deficiency by abrogating CRAC channel function. Nature 441:179–185

220. Zhou Y, Meraner P, Kwon HT, Machnes D, Oh-hora M, Zimmer J, Huang Y, Stura A, Rao A, Hogan PG (2009) STIM1 gates the store-operated calcium channel ORAI1 in vitro. Nature – Structural and Molecular Biology 17:112–116

221. Penna A, Demuro A, Yeromin AV, Zhang SL, Safrina O, Parker I, Cahalan MD (2008) The CRAC channel consists of a tetramer formed by Stim-induced dimerization of Orai dimers. Nature 456:116–120

222. Dellis O, Dedos SG, Tovey SC, Rahman TU, Dubel SL, Taylor CW (2006) Ca^{2+} entry through plasma membrane IP3 receptors. Science 313:229–233

223. Peinelt C, Vig M, Koomoa DL, Beck A, Nadler MJS, Koblan-Huberson M, Lis A, Fleig A, Penner R, Kinet JP (2006) Amplification of CRAC current by STIM1 and CRACM1 (Orai1). Nature Cell Biology 8:771–773

224. Cahalan MD (2009) STIMulating store-operated Ca^{2+} entry. Nature – Cell Biology 11:669–677

225. Park CY, Shcheglovitov A, Dolmetsch R (2010) The CRAC channel activator STIM1 binds and inhibits L-type voltage-gated calcium channels. Science 330:101–105

226. Wang Y, Deng X, Mancarella S, Hendron E, Eguchi S, Soboloff J, Tang XD, Gill DL (2010) The calcium store sensor, STIM1, reciprocally controls Orai and $Ca_V1.2$ channels. Science 330:105–109

227. Schindl R, Frischauf I, Bergsmann J, Muik M, Derler I, Lackner B, Groschner K, Romanin C (2009) Plasticity in Ca^{2+} selectivity of Orai1/Orai3 heteromeric channel. Proceedings of the National Academy of Sciences of the United States of America 106:19623–19628

228. Lee KP, Yuan JP, Zeng W, So I, Worley PF, Muallem S (2009) Molecular determinants of fast Ca^{2+}-dependent inactivation and gating of the Orai channels. Proceedings of the National Academy of Sciences of the United States of America 106:14687–14692

229. Takahashi Y, Watanabe H, Murakami M, Ono K, Munehisa Y, Koyama T (2007) Functional role of stromal interaction molecule 1 (STIM1) in vascular smooth muscle cells. Biochemical and Biophysical Research Communications 361:934–940

230. Li J, Sukumar P, Milligan CJ, Kumar B, Ma ZY, Munsch CM, Interactions, functions, and independence of plasma membrane STIM1 and TRPC1 in vascular smooth muscle cells. Circulation Research 103:e97–e104

231. Ishii K, Hirose K, Iino M (2006) Ca^{2+} shuttling between endoplasmic reticulum and mitochondria underlying Ca^{2+} oscillations. EMBO Reports 7:390–396

232. Yazawa M, Ferrante C, Feng J, Mio K, Ogura T, Zhang M, Lin PH, Pan Z, Komazaki S, Kato K, Nishi M, Zhao X, Weisleder N, Sato C, Ma J, Takeshima H (2007) TRIC channels are essential for Ca^{2+} handling in intracellular stores. Nature 448:78–82

233. Mullins FM, Park CY, Dolmetsch RE, Lewis RS (2009) STIM1 and calmodulin interact with Orai1 to induce Ca^{2+}-dependent inactivation of CRAC channels. Proceedings of the National Academy of Sciences of the United States of America 106:15495-15500

234. Srikanth S, Jung HJ, Kim KD, Souda P, Whitelegge J, Gwack Y (2010) A novel EF-hand protein, CRACR2A, is a cytosolic Ca^{2+} sensor that stabilizes CRAC channels in T cells. Nature Cell Biology 12:436–446

235. Calderón-Sánchez, Fernández-Tenorio M, Ordóñez A, López-Barneo J, Ureña J (2009) Hypoxia inhibits vasoconstriction induced by metabotropic Ca^{2+} channel-induced Ca^{2+} release in mammalian coronary arteries. Cardiovascular Research 82:115–124

236. Davies A, Kadurin I, Alvarez-Laviada A, Douglas L, Nieto-Rostro M, Bauer CS, Pratt WS, Dolphin AC (2010) The $\alpha 2\delta$ subunits of voltage-gated calcium channels form GPI-anchored proteins, a posttranslational modification essential for function. Proceedings of the National Academy of Sciences of the United States of America 107:1654–1659

237. Schredelseker J, Shrivastav M, Dayal A, Grabner M (2010) Non-Ca^{2+}-conducting Ca^{2+} channels in fish skeletal muscle excitation-contraction coupling. Proceedings of the National Academy of Sciences of the United States of America 107:5658–5663

238. Dick IE, Tadross MR, Liang H, Tay LH, Yang W, Yue DT (2008) A modular switch for spatial Ca^{2+} selectivity in the calmodulin regulation of Ca_V channels. Nature 451:830–834

239. Kim EY, Rumpf CH, Van Petegem F, Arant RJ, Findeisen F, Cooley ES, Isacoff EY, Minor DL (2010) Multiple C-terminal tail Ca^{2+}/CaMs regulate CaV1.2 function but do not mediate channel dimerization. EMBO Journal 29:3924–3938

240. Fuller MD, Emrick MA, Sadilek M, Scheuer T, Catterall WA (2010) Molecular mechanism of calcium channel regulation in the fight-or-flight response. Science Signaling 3:ra70

241. Cheng X, Pachuau J, Blaskova E, Asuncion-Chin M, Liu J, Dopico AM, Jaggar JH (2009) Alternative splicing of $Ca_V1.2$ channel exons in smooth muscle cells of resistance-size arteries generates currents with unique electrophysiological properties. American Journal of Physiology – Heart and Circulatory Physiology 297:H680–H688

242. Blaich A, Welling A, Fischer S, Wegener JW, Köstner K, Hofmann F, Moosmang S (2010) Facilitation of murine cardiac L-type $Ca_V1.2$ channel is modulated by Calmodulin kinase II-dependent phosphorylation of S1512 and S1570. Proceedings of the National Academy of Sciences of the United States of America 107:10285–10289

243. Zeng Q, Han Y, Bao Y, Li W, Li X, Shen X, Wang X, Yao F, O'Rourke ST, Sun C (2010) 20-HETE increases NADPH oxidase-derived ROS production and stimulates the L-type Ca^{2+} channel via a PKC-dependent mechanism in cardiomyocytes. American Journal of Physiology – Heart and Circulatory Physiology 299:H1109–H1117

244. Keef KD, Hume JR, Zhong J (2001) Regulation of cardiac and smooth muscle Ca^{2+} channels (CaV1.2a,b) by protein kinases. American Journal of Physiology – Cell Physiology 281:C1743–C1756

245. Smith IF, Wiltgen SM, Shuai J, Parker I (2009) Ca^{2+} Puffs originate from preestablished stable clusters of inositol trisphosphate receptors. Science Signaling 2:ra77

246. Foskett JK, White C, Cheung KH, Mak DO (2007) Inositol trisphosphate receptor Ca^{2+} release channels. Physiological Reviews 87:593–658

247. Nakagawa T, Okano H, Furuichi T, Aruga J, Mikoshiba K (1991) The subtypes of the mouse inositol 1,4,5-trisphosphate receptor are expressed in a tissue-specific and developmentally specific manner. Proceedings of the National Academy of Sciences of the United States of America 88:6244–6248

248. Wojcikiewicz RJ (1995) Type I, II, and III inositol 1,4,5-trisphosphate receptors are unequally susceptible to down-regulation and are expressed in markedly different proportions in different cell types. Journal of Biological Chemistry 270:11678–11683

249. Lipp P, Laine M, Tovey SC, Burrell KM, Berridge MJ, Li W, Bootman MD (2000) Functional InsP3 receptors that may modulate excitation–contraction coupling in the heart. Current Biology 10:939–942

250. Li X, Zima AV, Sheikh F, Blatter LA, Chen J (2005) Endothelin-1-induced arrhythmogenic Ca^{2+} signaling is abolished in atrial myocytes of inositol-1,4,5-trisphosphate (IP3)-receptor type 2-deficient mice. Circulation Research 96:1274–1281

251. Domeier TL, Zima AV, Maxwell JT, Huke S, Mignery GA, Blatter LA (2008) IP3 receptor-dependent Ca^{2+} release modulates excitation-contraction coupling in rabbit ventricular myocytes. American Journal of Physiology – Heart and Circulatory Physiology 294:H596–H604

252. Nakayama H, Bodi I, Maillet M, DeSantiago J, Domeier TL, Mikoshiba K, Lorenz JN, Blatter LA, Bers DM, Molkentin JD (2010) The IP3 receptor regulates cardiac hypertrophy in response to select stimuli. Circulation Research 107:659–666

253. Woodcock EA, Kistler PM, Ju YK (2009) Phosphoinositide signalling and cardiac arrhythmias. Cardiovascular Research 82:286–295

254. Janowski E, Berríos M, Cleemann L, Morad M (2010) Developmental aspects of cardiac Ca^{2+} signaling: interplay between RyR- and IP3R-gated Ca^{2+} stores. American Journal of Physiology – Heart and Circulatory Physiology 298:H1939–H1950

255. Vermassen E, Parys JB, Mauger JP (2004) Subcellular distribution of the inositol 1,4,5-trisphosphate receptors: functional relevance and molecular determinants. Biology of the Cell 96:3–17

256. Grayson TH, Haddock RE, Murray TP, Wojcikiewicz RJ, Hill CE (2004) Inositol 1,4,5-trisphosphate receptor subtypes are differentially distributed between smooth muscle and endothelial layers of rat arteries. Cell Calcium 36:447–458

257. Zhao G, Adebiyi A, Blaskova E, Xi Q, Jaggar JH (2008) Type 1 inositol 1,4,5-trisphosphate receptors mediate UTP-induced cation currents, Ca^{2+} signals, and vasoconstriction in cerebral arteries. American Journal of Physiology – Cell Physiology 295:C1376–C1384

258. Sundivakkam PC, Kwiatek AM, Sharma TT, Minshall RD, Malik AB, Tiruppathi C (2008) Caveolin-1 scaffold domain interacts with TRPC1 and IP3R3 to regulate Ca^{2+} store release-induced Ca^{2+} entry in endothelial cells. American Journal of Physiology – Cell Physiology 296:C403–C413

259. Bergner A, Sanderson MJ (2002) ATP stimulates Ca^{2+} oscillations and contraction in airway smooth muscle cells of mouse lung slices. American Journal of Physiology – Lung Cellular and Molecular Physiology 283:L1271–L1279

260. Antigny F, Norez C, Cantereau A, Becq F, Vandebrouck C (2008) Abnormal spatial diffusion of Ca^{2+} in F508del-CFTR airway epithelial cells. Respiratory Research 9:70

261. Healy JA, Nilsson KR,. Hohmeier HE, Berglund J, Davis J, Hoffman J, Kohler M, Li LS, Berggren PO, Newgard CB, Bennett V (2010) Cholinergic augmentation of insulin release requires ankyrin-B. Science Signaling 3:ra19

262. Antl M, von Brühl ML, Eiglsperger C, Werner M, Konrad I, Kocher T, Wilm M, Hofmann F, Massberg S, Schlossmann J (2007) IRAG mediates NO/cGMP-dependent inhibition of platelet aggregation and thrombus formation. Blood 109:552–559

263. Casteel DE, Boss GR, Pilz RB (2005) Identification of the interface between cGMP-dependent protein kinase Iβ and its interaction partners TFII-I and IRAG reveals a common interaction motif. Journal of Biological Chemistry 280:38211–38218

264. Schlossmann J, Ammendola A, Ashman K, Zong X, Huber A, Neubauer G, Wang GX, Allescher HD, Korth M, Wilm M, Hofmann F, Ruth P (2000) Regulation of intracellular calcium by a signalling complex of IRAG, IP3 receptor and cGMP kinase Iβ. Nature 404:197–201

265. van Rossum DB, Patterson RL, Cheung KH, Barrow RK, Syrovatkina V, Gessell GS, Burkholder SG, Watkins DN, Foskett JK, Snyder SH (2006) DANGER, a novel regulatory protein of inositol 1,4,5-trisphosphate-receptor activity. Journal of Biological Chemistry 281:37111–37116

266. Taufiq-Ur-Rahman, Skupin A, Falcke M, Taylor CW (2009) Clustering of InsP3 receptors by InsP3 retunes their regulation by InsP3 and Ca^{2+}. Nature 458:655–659

267. Wellman GC, Nelson MT (2003) Signaling between SR and plasmalemma in smooth muscle: sparks and the activation of Ca^{2+}-sensitive ion channels. Cell Calcium 34:211–229

268. M Fill, JA Copello (2002) Ryanodine receptor calcium release channels. Physiological Reviews 82:893-922

269. Ledbetter MW, Preiner JK, Louis CF, Mickelson JR (1994) Tissue distribution of ryanodine receptor isoforms and alleles determined by reverse transcription polymerase chain reaction. Journal of Biological Chemistry 269:31544–31551

270. Bose DD, Pessah IN (2011) Ryanodine receptor type I UCSD-Nature Molecule Pages, UCSD-Nature Signaling Gateway (www.signaling-gateway.org)

271. Siekierka JJ, Wiederrecht G, Greulich H, Boulton D, Hung SH, Cryan J, Hodges PJ, Sigal NH (1990) The cytosolic-binding protein for the immunosuppressant FK-506 is both a ubiquitous and highly conserved peptidyl-prolyl cis-trans isomerase. Journal of Biological Chemistry 265:21011–21015

272. Qi Y, Ogunbunmi EM, Freund EA, Timerman AP, Fleischer S (1998) FK-binding protein is associated with the ryanodine receptor of skeletal muscle in vertebrate animals. Journal of Biological Chemistry 273:34813–34819

273. Chen SR, Li X, Ebisawa K, Zhang L (1997) Functional characterization of the recombinant type 3 Ca^{2+} release channel (ryanodine receptor) expressed in HEK293 cells. Journal of Biological Chemistry 272:24234–24246

274. Vanterpool CK, Vanterpool EA, Pearce WJ, Buchholz JN (2006) Advancing age alters the expression of the ryanodine receptor 3 isoform in adult rat superior cervical ganglia. Journal of Applied Physiology 101:392–400

275. Hart JD, Dulhunty AF (2000) Nitric oxide activates or inhibits skeletal muscle ryanodine receptors depending on its concentration, membrane potential and ligand binding. Journal of Membrane Biology 173:227–236

276. Sonnleitner A, Fleischer S, Schindler H (1997) Gating of the skeletal calcium release channel by ATP is inhibited by protein phosphatase 1 but not by Mg^{2+}. Cell Calcium 21:283–290

277. Benkusky NA, Weber CS, Scherman JA, Farrell EF, Hacker TA, John MC, Powers PA, Valdivia HH (2007) Intact β-adrenergic response and unmodified progression toward heart failure in mice with genetic ablation of a major protein kinase A phosphorylation site in the cardiac ryanodine receptor. Circulation Research 101:819–829

278. Noguchi N, Yoshikawa T, Ikeda T, Takahashi I, Shervani NJ, Uruno A, Yamauchi A, Nata K, Takasawa S, Okamoto H, Sugawara A (2008) FKBP12.6 disruption impairs glucose-induced insulin secretion. Biochemical and Biophysical Research Communications 371:735–740

279. Zhang F, Jin S, Yi F, Xia M, Dewey WL, Li PL (2008) Local production of $O2^{-}$ by NAD(P)H oxidase in the sarcoplasmic reticulum of coronary arterial myocytes: cADPR-mediated Ca^{2+} regulation. Cellular Signalling 20:637–644

280. Arendshorst WJ, Thai TL (2009) Regulation of the renal microcirculation by ryanodine receptors and calcium-induced calcium release. Current Opinion in Nephrology and Hypertension 18:40–49

281. Yamasaki-Mann M, Demuro A, Parker I (2009) cADPR stimulates SERCA activity in Xenopus oocytes. Cell Calcium 45:293–299

282. Jiang D, Xiao B, Li X, Chen SR (2003) Smooth muscle tissues express a major dominant negative splice variant of the type 3 Ca^{2+} release channel (ryanodine receptor). Journal of Biological Chemistry 278:4763–4769

283. Fabiato A (1983) Calcium-induced release of calcium from the cardiac sarcoplasmic reticulum. American Journal of Physiology – Cell Physiology 245:C1-14

284. Fabiato A (1985) Time and calcium dependence of activation and inactivation of calcium-induced release of calcium from the sarcoplasmic reticulum of a skinned canine cardiac Purkinje cell. Journal of General Physiology 85:247–289

285. Nelson MT, Cheng H, Rubart M, Santana LF, Bonev AD, Knot HJ, Lederer WJ (1995) Relaxation of arterial smooth muscle by calcium sparks. Science 270:633–637

286. Launikonis BS, Zhou J, Royer L, Shannon TR, Brum G, Ríos E (2006) Depletion "skraps" and dynamic buffering inside the cellular calcium store. Proceedings of the National Academy of Sciences of the United States of America 103:2982–2987

287. Lamb GD, Stephenson DG (1996) Effects of FK506 and rapamycin on excitation-contraction coupling in skeletal muscle fibres of the rat. Journal of Physiology 494:569–576

288. Tripathy A, Xu L, Mann G, Meissner G (1995) Calmodulin activation and inhibition of skeletal muscle Ca^{2+} release channel (ryanodine receptor). Biophysical Journal 69:106–119

289. Ono M, Yano M, Hino A, Suetomi T, Xu X, Susa T, Uchinoumi H, Tateishi H, Oda T, Okuda S, Doi M, Kobayashi S, Yamamoto T, Koseki N, Kyushiki H, Ikemoto N, Matsuzaki M (2010) Dissociation of calmodulin from cardiac ryanodine receptor causes aberrant Ca^{2+} release in heart failure. Cardiovascular Research 87:609–617

290. Fruen BR, Black DJ, Bloomquist RA, Bardy JM, Johnson JD, Louis CF, Balog EM (2003) Regulation of the RYR1 and RYR2 Ca^{2+} release channel isoforms by Ca^{2+}-insensitive mutants of calmodulin. Biochemistry 42:2740–2747

291. Fruen BR, Bardy JM, Byrem TM, Strasburg GM, Louis CF (2000) Differential Ca(2+) sensitivity of skeletal and cardiac muscle ryanodine receptors in the presence of calmodulin. American Journal of Physiology – Cell Physiology 279:C724–C733

292. Smith JS, Rousseau E, Meissner G (1989) Calmodulin modulation of single sarcoplasmic reticulum Ca^{2+}-release channels from cardiac and skeletal muscle. Circulation Research 64:352–359

293. Wright NT, Prosser BL, Varney KM, Zimmer DB, Schneider MF, Weber DJ (2008) S100A1 and calmodulin compete for the same binding site on ryanodine receptor. Journal of Biological Chemistry 283:26676–26683

294. Kushnir A, Shan J, Betzenhauser MJ, Reiken S, Marks AR (2010) Role of CaMKIIδ phosphorylation of the cardiac ryanodine receptor in the force frequency relationship and heart failure. Proceedings of the National Academy of Sciences of the United States of America 107:10274–10279

295. Beard NA, Wei L, Dulhunty AF (2009) Ca^{2+} signaling in striated muscle: the elusive roles of triadin, junctin, and calsequestrin. European Biophysics Journal 39:27–36

296. Knudson CM, Stang KK, Jorgensen AO, Campbell KP (1993) Biochemical characterization of ultrastructural localization of a major junctional sarcoplasmic reticulum glycoprotein (triadin). Journal of Biological Chemistry 268:12637–12645

297. Knudson CM, Stang KK, Moomaw CR, Slaughter CA, Campbell KP (1993) Primary structure and topological analysis of a skeletal muscle-specific junctional sarcoplasmic reticulum glycoprotein (triadin). Journal of Biological Chemistry 268:12646–12654

298. Lim KY, Hong CS, Kim DH (2000) cDNA cloning and characterization of human cardiac junctin. Gene 255:35-42

299. Park H, Park IY, Kim E, Youn B, Fields K, Dunker AK, Kang C (2004) Comparing skeletal and cardiac calsequestrin structures and their calcium binding: a proposed mechanism for coupled calcium binding and protein polymerization. Journal of Biological Chemistry 279:18026–18033

300. Knollmann BC (2009) New roles of calsequestrin and triadin in cardiac muscle. Journal of Physiology 587:3081–3087

301. Wei L, Gallant EM, Dulhunty AF, Beard NA (2009) Junctin and triadin each activate skeletal ryanodine receptors but junctin alone mediates functional interactions with calsequestrin. International Journal of Biochemistry and Cell Biology 41:2214–2224

302. Ziviani E, Lippi G, Bano D, Munarriz E, Guiducci S, Zoli M, Young KW, Nicotera P (2011) Ryanodine receptor-2 upregulation and nicotine-mediated plasticity. EMBO Journal 30:194–204

303. Olesen C, Picard M, Lund Winther AM, Gyrup C, Morth JP, Oxvig C, Møller JV, Nissen P (2007) The structural basis of calcium transport by the calcium pump. Nature 450:1036–1042

304. Park SW, Zhou Y, Lee J, Lee J, Ozcan U (2010) Sarco(endo)plasmic reticulum Ca^{2+}-ATPase 2b is a major regulator of endoplasmic reticulum stress and glucose homeostasis in obesity. Proceedings of the National Academy of Sciences of the United States of America 107:19320–19325

305. Oceandy D, Buch MH, Cartwright EJ, Neyses L (2006) The emergence of plasma membrane calcium pump as a novel therapeutic target for heart disease. Mini Reviews in Medicinal Chemistry 6:583–588

306. Strehler EE, Zacharias DA (2001) Role of alternative splicing in generating isoform diversity among plasma membrane calcium pumps. Physiological Reviews 81:21-50

307. Abramowitz J, Aydemir-Koksoy A, Helgason T, Jemelka S, Odebunmi T, Seidel CL, Allen JC (2000) Expression of plasma membrane calcium ATPases in phenotypically distinct canine vascular smooth muscle cells. Journal of Molecular and Cellular Cardiology 32:777–789

308. Gros R, Afroze T, You XM, Kabir G, Van Wert R, Kalair W, Hoque AE, Mungrue IN, Husain M (2003) Plasma membrane calcium ATPase overexpression in arterial smooth muscle increases vasomotor responsiveness and blood pressure. Circulation Research 93:614–621

309. Palty R, Ohana E, Hershfinkel M, Volokita M, Elgazar V, Beharier O, Silverman WF, Argaman M, Sekler I (2004) Lithium-calcium exchange is mediated by a distinct potassium-independent sodium-calcium exchanger. Journal of Biological Chemistry 279:25234–25240

310. Henderson SA, Goldhaber JI, So JM, Han T, Motter C, Ngo A, Chantawansri C, Ritter MR, Friedlander M, Nicoll DA, Frank JS, Jordan MC, Roos KP, Ross RS, Philipson KD (2004) Functional adult myocardium in the absence of Na^+–Ca^{2+} exchange. Circulation Research 95:604–611

311. Ottolia M, Philipson KD, John S (2004) Conformational changes of the Ca^{2+} regulatory site of the Na^+–Ca^{2+} exchanger detected by FRET. Biophysical Journal 87:899–906

312. Altimimi HF, Schnetkamp PP (2007) Na^+/Ca^{++}–K^+ exchangers (NCKX): functional properties and physiological roles. Channels 1:62–69

313. Palty R, Silverman WF, Hershfinkel M, Caporale T, Sensi SL, Parnis J, Nolte C, Fishman D, Shoshan-Barmatz V, Herrmann S, Khananshvili D, Sekler I (2010) NCLX is an essential component of mitochondrial Na^+/Ca^{2+} exchange. Proceedings of the National Academy of Sciences of the United States of America 107:436–441

314. Lipskaia L, Lompré AM (2004) Alteration in temporal kinetics of Ca^{2+} signaling and control of growth and proliferation. Biology of the Cell 96:55–68

315. House SJ, Potier M, Bisaillon J, Singer HA, Trebak M (2008) The non-excitable smooth muscle: calcium signaling and phenotypic switching during vascular disease. Pflügers Archiv (European Journal of Physiology) 456:769–785

316. Afroze T, Sadi AM, Momen MA, Gu S, Heximer S, Husain M (2007) c-Myb-dependent inositol 1,4,5-trisphosphate receptor type-1 expression in vascular smooth muscle cells. Arteriosclerosis, Thrombosis, and Vascular Biology 27:1305–1311

317. Kellenberger S, Schild L (2002) Epithelial sodium channel/degenerin family of ion channels: a variety of functions for a shared structure. Physiological Reviews 82:735–767

318. Drummond HA, Price MP, Welsh MJ, Abboud FM (1998) A molecular component of the arterial baroreceptor mechanotransducer. Neuron 21:1435–1441

319. Lu Y, Ma X, Sabharwal R, Snitsarev V, Morgan D, Rahmouni K, Drummond HA, Whiteis CA, Costa V, Price M, Benson C, Welsh MJ, Chapleau MW, Abboud FM (2009) The ion channel ASIC2 is required for baroreceptor and autonomic control of the circulation. Neuron 64:885–897

320. Bingle CD, LeClair EE, Havard S, Bingle L, Gillingham P, Craven CJ (2004) Phylogenetic and evolutionary analysis of the PLUNC gene family. Protein Science 13:422–430

321. Wang S, Publicover S, Gu Y (2009) An oxygen-sensitive mechanism in regulation of epithelial sodium channel. Proceedings of the National Academy of Sciences of the United States of America 106:2957–2962

322. Goldfarb SB, Kashlan OB, Watkins JN, Suaud L, Yan W, Kleyman TR, Rubenstein RC (2006) Differential effects of Hsc70 and Hsp70 on the intracellular trafficking and functional expression of epithelial sodium channels. Proceedings of the National Academy of Sciences of the United States of America 103:5817–5822

323. Soundararajan R, Melters D, Shih IC, Wang J, Pearce D (2009) Epithelial sodium channel regulated by differential composition of a signaling complex. Proceedings of the National Academy of Sciences of the United States of America 106:7804–7809

324. Garcia-Caballero A, Rasmussen JE, Gaillard E, Watson MJ, Olsen JC, Donaldson SH, Stutts MJ, Tarran R (2009) SPLUNC1 regulates airway surface liquid volume by protecting ENaC from proteolytic cleavage. Proceedings of the National Academy of Sciences of the United States of America 106::11412–11417

325. Ziemann AE, Allen JE, Dahdaleh NS, Drebot II, Coryell MW, Wunsch AM, Lynch CM, Faraci FM, Howard MA 3rd, Welsh MJ, Wemmie JA (2009) The amygdala is a chemosensor that detects carbon dioxide and acidosis to elicit fear behavior. Cell 139:1012–1021

326. Darmellah A, Rücker-Martin C, Feuvray D (2009) ERM proteins mediate the effects of Na^+/H^+ exchanger (NHE1) activation in cardiac myocytes. Cardiovascular Research 81:294–300

327. Li X, Karki P, Lei L, Wang H, Fliegel L (2009) Na^+/H^+ exchanger isoform 1 facilitates cardiomyocyte embryonic stem cell differentiation. American Journal of Physiology – Heart and Circulatory Physiology 296:H159–H170

328. Boedtkjer E, Aalkjaer C (2009) Insulin inhibits Na^+/H^+ exchange in vascular smooth muscle and endothelial cells in situ: involvement of H_2O_2 and tyrosine phosphatase SHP-2. American Journal of Physiology – Heart and Circulatory Physiology 296:H247–H255

329. Orlowski J, Grinstein S (2004) Diversity of the mammalian sodium/proton exchanger SLC9 gene family. Pflügers Archiv – European Journal of Physiology 47:549-65

330. Lee LJ, Zachos NC, Donowitz M (2011) NHE3. UCSD-Nature Molecule Pages, UCSD-Nature Signaling Gateway (www.signaling-gateway.org)

331. Catterall WA, Goldin AL, Waxman SG (2005) International Union of Pharmacology. XLVII. Nomenclature and structure-function relationships of voltage-gated sodium channels. Pharmacological Reviews 57:397–409

332. Bosmans F, Martin-Eauclaire MF, Swartz KJ (2008) Deconstructing voltage sensor function and pharmacology in sodium channels. Nature 456:202–208

333. Yu FH, Catterall WA (2003) Overview of the voltage-gated sodium channel family. Genome Biology 4:207

334. Herzog RI, Liu C, Waxman SG, Cummins TR (2003) Calmodulin binds to the C terminus of sodium channels NaV1.4 and NaV1.6 and differentially modulates their functional properties. Journal of Neuroscience 23:8261–8270

335. Casini S, Tan HL, Demirayak I, Remme CA, Amin AS, Scicluna BP, Chatyan H, Ruijter JM, Bezzina CR, van Ginneken AC, Veldkamp MW (2010) Tubulin polymerization modifies cardiac sodium channel expression and gating. Cardiovascular Research 85:691–700

336. Lorincz A, Nusser Z (2010) Molecular identity of dendritic voltage-gated sodium channels. Science 328:906–909

337. Meguro K, Iida H, Takano H, Morita T, Sata M, Nagai R, Nakajima (2009) T Function and role of voltage-gated sodium channel NaV1.7 expressed in aortic smooth muscle cells. American Journal of Physiology – Heart and Circulatory Physiology 296:H211–H219

338. Weiss J, Pyrski M, Jacobi E, Bufe B, Willnecker V, Schick B, Zizzari P, Gossage SJ, Greer CA, Leinders-Zufall T, Woods CG, Wood JN, Zufall F (2011) Loss-of-function mutations in sodium channel $Na_V1.7$ cause anosmia. Nature 472:186–190

339. Morth JP, Pedersen BP, Toustrup-Jensen MS, Sørensen TLM, Petersen J, Andersen JP, Vilsen B, Nissen P (2007) Crystal structure of the sodium–potassium pump. Nature 450:1043–1049

340. Ratheal IM, Virgin GK, Yu H, Roux B, Gatto C, Artigas P (2010) Selectivity of externally facing ion-binding sites in the Na/K pump to alkali metals and organic cations. Proceedings of the National Academy of Sciences of the United States of America 107:18718–18723

341. Morth JP, Pedersen BP, Buch-Pedersen MJ, Andersen JP, Vilsen B, Palmgren MG, Nissen P (2011) A structural overview of the plasma membrane Na^+,K^+-ATPase and H^+-ATPase ion pumps. Nature Reviews – Molecular Cell Biology 12:60–70

342. Song H, Lee MY, Kinsey SP, Weber DJ, Blaustein MP (2006) An N-terminal sequence targets and tethers Na+ pump $\alpha 2$ subunits to specialized plasma membrane microdomains. Journal of Biological Chemistry 281:12929-12940

343. James PF, Grupp IL, Grupp G, Woo AL, Askew GR, Croyle ML, Walsh RA, Lingrel JB (1999) Identification of a specific role for the Na,K-ATPase $\alpha 2$ isoform as a regulator of calcium in the heart. Molecular Cell 3:555–563

344. Takeuchi A, Reyes N, Artigas P, Gadsby DC (2008) The ion pathway through the opened Na^+,K^+-ATPase pump. Nature 456:413–416

345. Pritchard TJ, Bowman PS, Jefferson A, Tosun M, Lynch RM, Paul RJ (2010) Na^+-K^+-ATPase and Ca^{2+} clearance proteins in smooth muscle: a functional unit. American Journal of Physiology – Heart and Circulatory Physiology 299:H548–H556

346. Scheiner-Bobis G (2011) The Na^+,K^+-ATPase: more than just a sodium pump. Cardiovascular Research 89:6–8

347. Radzyukevich TL, Lingrel JB, Heiny JA (2009) The cardiac glycoside binding site on the Na,K-ATPase-$\alpha 2$ isoform plays a role in the dynamic regulation of active transport in skeletal muscle. Proceedings of the National Academy of Sciences of the United States of America 106:2565–2570

348. Dostanic-Larson I, Van Huysse JW, Lorenz JN, Lingrel JB (2005) The highly conserved cardiac glycoside binding site of Na,K-ATPase plays a role in blood pressure regulation. Proceedings of the National Academy of Sciences of the United States of America 102:15845–15850

349. Zheng J, Koh X, Hua F, Li G, Larrick JW, Bian JS (2011) Cardioprotection induced by Na^+,/K^+-ATPase activation involves extracellular signal-regulated kinase 1/2 and phosphoinositide 3-kinase/Akt pathway. Cardiovascular Research 89:51–59

350. Cordero-Morales JF, Cuello LG, Perozo E (2006) Voltage-dependent gating at the KcsA selectivity filter. Nature Structural and Molecular Biology 13:319–322

351. Ocorr K, Reeves NL, Wessells RJ, Fink M, Chen HSV, Akasaka T, Yasuda S, Metzger JM, Giles W, Posakony JW, Bodmer R (2007) KCNQ potassium channel mutations cause cardiac arrhythmias in Drosophila that mimic the effects of aging. Proceedings of the National Academy of Sciences of the United States of America 104:3943–3948

352. Chakrapani S, Cordero-Morales JF, Jogini V, Pan AC, Cortes DM, Roux B, Perozo E (2011) On the structural basis of modal gating behavior in K^+ channels. Nature – Structural and Molecular Biology 18:67–74

353. Tao X, Avalos JL, Chen J, MacKinnon R (2009) Crystal structure of the eukaryotic strong inward-rectifier K^+ channel $K_{ir}2.2$ at 3.1 resolution. Science 326:1668–1674

354. Vikstrom KL, Vaidyanathan R, Levinsohn S, O'Connell RP, Qian Y, Crye M, Mills JH, Anumonwo JM (2009) SAP97 regulates $K_{ir}2.3$ channels by multiple mechanisms. American Journal of Physiology – Heart and Circulatory Physiology 297:H1387–H1397

355. Rosenhouse-Dantsker A, Sui JL, Zhao Q, Rusinova R, Rodríguez-Menchaca AA, Zhang Z, Logothetis DE (2008) A sodium-mediated structural switch that controls the sensitivity of K_{ir} channels to PtdIns(4,5)P_2. Nature Chemical Biology 4:624–631

356. Luján R, Maylie J, Adelman JP (2009) New sites of action for GIRK and SK channels. Nature Reviews – Neuroscience 10:475–480

357. Lüscher C, Slesinger PA (2010) Emerging roles for G protein-gated inwardly rectifying potassium (GIRK) channels in health and disease. Nature Reviews – Neuroscience 11:301–315

358. Yue P, Lin DH, Pan CY, Leng Q, Giebisch G, Lifton RP, Wang WH (2009) Src family protein tyrosine kinase (PTK) modulates the effect of SGK1 and WNK4 on ROMK channels. Proceedings of the National Academy of Sciences of the United States of America 106:15061–15066

359. Alekseev AE, Reyes S, Yamada S, Hodgson-Zingman DM, Sattiraju S, Zhu Z, Sierra A, Gerbin M, Coetzee WA, Goldhamer DJ, Terzic A, Zingman LV (2010) Sarcolemmal ATP-sensitive K^+ channels control energy expenditure determining body weight. Cell Metabolism 11:58–69

360. Nichols CG (2006) K_{ATP} channels as molecular sensors of cellular metabolism. Nature 440:470–476

361. Hund TJ, Mohler PJ (2011) Differential roles for SUR subunits in K_{ATP} channel membrane targeting and regulation. American Journal of Physiology – Heart and Circulatory Physiology 300:H33–H35

362. Kline CF, Kurata HT, Hund TJ, Cunha SR, Koval OM, Wright PJ, Christensen M, Anderson ME, Nichols CG, Mohler PJ (2009) Dual role of K_{ATP} channel C-terminal motif in membrane targeting and metabolic regulation. Proceedings of the National Academy of Sciences of the United States of America 106:16669–16674

363. Enkvetchakul D, Loussouarn G, Makhina E, Shyng SL, Nichols CG (2000) The kinetic and physical basis of K_{ATP} channel gating: toward a unified molecular understanding. Biophysical Journal 78:2334–2348

364. Enkvetchakul D, Loussouarn G, Makhina E, Nichols CG (2001) ATP interaction with the open state of the K_{ATP} channel. Biophysical Journal 80:719–728

365. Garg V, Jiao J, Hu K (2009) Regulation of ATP-sensitive K^+ channels by caveolin-enriched microdomains in cardiac myocytes. Cardiovascular Research 82:51–58

366. Bao L, Hadjiolova K, Coetzee WA, Rindler MJ (2011) Endosomal K_{ATP} channels as a reservoir after myocardial ischemia: a role for SUR2 subunits. American Journal of Physiology – Heart and Circulatory Physiology 300:H262–H270

367. Tang G, Wu L, Liang W, Wang R (2005) Direct stimulation of K_{ATP} channels by exogenous and endogenous hydrogen sulfide in vascular smooth muscle cells. Molecular Pharmacology 68:1757–1764

368. Yang W, Yang G, Jia X, Wu L, Wang R (2005) Activation of K_{ATP} channels by H_2S in rat insulin-secreting cells and the underlying mechanisms. Journal of Physiology 569:519–531

369. Papapetropoulos A, Pyriochou A, Altaany Z, Yang G, Marazioti A, Zhou Z, Jeschke MG, Branski LK, Herndon DN, Wang R, Szabó C (2009) Hydrogen sulfide is an endogenous stimulator of angiogenesis. Proceedings of the National Academy of Sciences of the United States of America 106:21972–21977

370. Gutman GA, Chandy KG, Grissmer S, Lazdunski M, McKinnon D, Pardo LA, Robertson GA, Rudy B, Sanguinetti MC, Stühmer W, Wang X (2005) International Union of Pharmacology. LIII. Nomenclature and molecular relationships of voltage-gated potassium channels. Pharmacological Reviews 57:473–508

371. Salinas M, Duprat F, Heurteaux C, Hugnot JP, Lazdunski M (1997) New modulatory α subunits for mammalian Shab K^+ channels. Journal of Biological Chemistry 272:24371–24379

372. Burgoyne RD (2007) Neuronal calcium sensor proteins: generating diversity in neuronal Ca^{2+} signalling. Nature Reviews – Neuroscience 8:182–193

373. Amberg GC, Koh SD, Imaizumi Y, Ohya S, Sanders KM (2003) A-type potassium currents in smooth muscle. American Journal of Physiology – Cell Physiology 284:C583–C595

374. Gubitosi-Klug RA, Mancuso DJ, Gross RW (2005) The human $K_V1.1$ channel is palmitoylated, modulating voltage sensing: identification of a palmitoylation consensus sequence. Proceedings of the National Academy of Sciences of the United States of America 102:5964–5968

375. Jindal HK, Folco EJ, Liu GX, Koren G (2008) Posttranslational modification of voltage-dependent potassium channel $K_V1.5$: COOH-terminal palmitoylation modulates its biological properties. American Journal of Physiology – Heart and Circulatory Physiology 294:H2012-H2021

376. Takimoto K, Yang EK, Conforti L (2002) Palmitoylation of KChIP splicing variants is required for efficient cell surface expression of $K_V4.3$ channels. Journal of Biological Chemistry 277:26904–26911

377. Cuello LG, Jogini V, Cortes DM, Pan AC, Gagnon DG, Dalmas O, Cordero-Morales JF, Chakrapani S, Roux B, Perozo E (2010) Structural basis for the coupling between activation and inactivation gates in K^+ channels. Nature 466:272–275

378. Cuello LG, Jogini V, Cortes DM, Perozo E (2010) Structural mechanism of C-type inactivation in K^+ channels. Nature 466:203–208

379. Oliver D, Lien CC, Soom M, Baukrowitz T, Jonas P, Fakler B (2004) Functional conversion between A-type and delayed rectifier K^+ channels by membrane lipids. Science 304:265–270

380. Loussouarn G, Park KH, Bellocq C, Baró I, Charpentier F, Escande D (2003) Phosphatidylinositol-4,5-bisphosphate, PIP_2, controls KCNQ1/KCNE1 voltage-gated potassium channels: a functional homology between voltage-gated and inward rectifier K^+ channels. EMBO Journal 22:5412–5421

381. Ramu Y, Xu Y, Lu Z (2006) Enzymatic activation of voltage-gated potassium channels. Nature 442:696–699

382. Xu Y, Ramu Y, Lu Z (2008) Removal of phospho-head groups of membrane lipids immobilizes voltage sensors of K^+ channels. Nature 451:826–829

383. Schmidt D, Jiang QX, MacKinnon R (2006) Phospholipids and the origin of cationic gating charges in voltage sensors. Nature 444:775–779

384. Xu X, Kanda VA, Choi E, Panaghie G, Roepke TK, Gaeta SA, Christini DJ, Lerner DJ, Abbott GW (2009) MinK-dependent internalization of the IKs potassium channel. Cardiovascular Research 82:430–438

385. Decher N, Streit AK, Rapedius M, Netter MF, Marzian S, Ehling P, Schlichthörl G, Craan T, Renigunta V, Köhler A, Dodel RC, Navarro-Polanco RA, Preisig-Müller R, Klebe G, Budde T, Baukrowitz T, Daut J (2010) RNA editing modulates the binding of drugs and highly unsaturated fatty acids to the open pore of K_V potassium channels. EMBO Journal 29:2101–2113

386. Williams MR, Markey JC, Doczi MA, Morielli AD (2007) An essential role for cortactin in the modulation of the potassium channel Kv1.2. Proceedings of the National Academy of Sciences of the United States of America 104:17412–17417

387. Benson MD, Li QJ, Kieckhafer K, Dudek D, Whorton MR, Sunahara RK, Iñiguez-Lluhí JA, Martens JR (2007) SUMO modification regulates inactivation of the voltage-gated potassium channel Kv1.5. Proceedings of the National Academy of Sciences of the United States of America 104:1805–1810

388. Balse E, El-Haou S, Dillanian G, Dauphin A, Eldstrom J, Fedida D, Coulombe A, Hatem SN (2009) Cholesterol modulates the recruitment of $K_V1.5$ channels from Rab11-associated recycling endosome in native atrial myocytes. Proceedings of the National Academy of Sciences of the United States of America 106:14681–14686

389. Miguel-Velado E, Pérez-Carretero FD, Colinas O, Cidad P, Heras M, López-López JR, Pérez-García MT (2010) Cell cycle-dependent expression of $K_V3.4$ channels modulates proliferation of human uterine artery smooth muscle cells. Cardiovascular Research 86:383–391

390. He W, Jia Y, Takimoto K (2009) Interaction between transcription factors Iroquois proteins 4 and 5 controls cardiac potassium channel Kv4.2 gene transcription. Cardiovascular Research 81:64–71

391. Anderson D, Mehaffey WH, Iftinca M, Rehak R, Engbers JD, Hameed S, Zamponi GW, Turner RW (2010) Regulation of neuronal activity by Ca_V3-K_V4 channel signaling complexes. Nature – Neuroscience 13:333–337

392. Keskanokwong T, Lim HJ, Zhang P, Cheng J, Xu L, Lai D, Wang Y (2011) Dynamic $K_V4.3$-CaMKII unit in heart: an intrinsic negative regulator for CaMKII activation. European Heart Journal 32:305–315

393. Chen J, Sroubek J, Krishnan Y, Li Y, Bian JS, McDonald TV (2009) PKA phosphorylation of HERG protein regulates the rate of channel synthesis. American Journal of Physiology – Heart and Circulatory Physiology 296:H1244–H1254

394. Lin EC, Holzem KM, Anson BD, Moungey BM, Balijepalli SY, Tester DJ, Ackerman MJ, Delisle BP, Balijepalli RC, January CT (2010) Properties of WT and mutant hERG K^+ channels expressed in neonatal mouse cardiomyocytes. American Journal of Physiology – Heart and Circulatory Physiology 298:H1842–H1849

395. Rainbow RD, Norman RI, Everitt DE, Brignell JL, Davies NW, Standen NB (2009) Endothelin-I and angiotensin-II inhibit arterial voltage-gated K^+ channels through different protein kinase C isoenzymes. Cardiovascular Research 83:493–500

396. Feher A, Rutkai I, Beleznai T, Ungvari Z, Csiszar A, Edes I, Bagi Z (2010) Caveolin-1 limits the contribution of BK(Ca) channel to EDHF-mediated arteriolar dilation: implications in diet-induced obesity. Cardiovascular Research 87:732–739

397. Pantazis A, Gudzenko V, Savalli N, Sigg D, Olcese R (2010) Operation of the voltage sensor of a human voltage- and Ca^{2+}-activated K^+ channel. Proceedings of the National Academy of Sciences of the United States of America 107:4459–4464

398. Yuan P, Leonetti MD, Pico AR, Hsiung Y, Mackinnon R (2010) Structure of the human BK channel Ca^{2+}-activation apparatus at 3.0 Åresolution. Science 329:182–186

399. Wu Y, Yang Y, Ye S, Jiang Y (2010) Structure of the gating ring from the human large-conductance Ca^{2+}-gated K^+ channel. Nature 466:393–397

400. Tian L, Jeffries O, McClafferty H, Molyvdas A, Rowe IC, Saleem F, Chen L, Greaves J, Chamberlain LH, Knaus HG, Ruth P, Shipston MJ (2008) Palmitoylation gates phosphorylation-dependent regulation of BK potassium channels. Proceedings of the National Academy of Sciences of the United States of America 105:21006–21011

401. Zhou XB, Wulfsen I, Utku E, Sausbier U, Sausbier M, Wieland T, Ruth P, Korth M (2010) Dual role of protein kinase C on BK channel regulation. Proceedings of the National Academy of Sciences of the United States of America 107:8005–8010

402. Alioua A, Li M, Wu Y, Stefani E, Toro L (2011) Unconventional myristoylation of large-conductance Ca^{2+}-activated K^+ channel (Slo1) via serine/threonine residues regulates channel surface expression. Proceedings of the National Academy of Sciences of the United States of America 108:10744–10749

403. Hou S, Xu R, Heinemann SH, Hoshi T (2008) Reciprocal regulation of the Ca^{2+} and H^+ sensitivity in the SLO1 BK channel conferred by the RCK1 domain. Nature Structural and Molecular Biology 15:403–410

404. Williams SE, Wootton P, Mason HS, Bould J, Iles DE, Riccardi D, Peers C, Kemp PJ (2004) Hemoxygenase-2 is an oxygen sensor for a calcium-sensitive potassium channel. Science 306:2093–2097

405. Leffler CW, Parfenova H, Jaggar JH (2011) Carbon monoxide as an endogenous vascular modulator. American Journal of Physiology – Heart and Circulatory Physiology 301:H1–H11

406. Wilkinson WJ, Kemp PJ (2011) Carbon monoxide: an emerging regulator of ion channels. Journal of Physiology (under press)

407. Kwan HY, Shen B, Ma X, Kwok YC, Huang Y, Man YB, Yu S, Yao X (2009) TRPC1 associates with BK_{Ca} channel to form a signal complex in vascular smooth muscle cells. Circulation Research 104:670–678

408. Ivanov A, Gerzanich V, Ivanova S, Denhaese R, Tsymbalyuk O, Simard JM (2006) Adenylate cyclase 5 and $K_{Ca}1.1$ channel are required for EGFR up-regulation of PCNA in native contractile rat basilar artery smooth muscle. Journal of Physiology 570:73–84

409. Cheong A, Bingham AJ, Li J, Kumar B, Sukumar P, Munsch C, Buckley NJ, Neylon CB, Porter KE, Beech DJ, Wood IC (2005) Downregulated REST transcription factor is a switch enabling critical potassium channel expression and cell proliferation. Molecular Cell 20:45–52

410. Li M, Tanaka Y, Alioua A, Wu Y, Lu R, Kundu P, Sanchez-Pastor E, Marijic J, Stefani E, Toro L (2010) Thromboxane A2 receptor and MaxiK-channel intimate interaction supports channel trans-inhibition independent of G-protein activation. Proceedings of the National Academy of Sciences of the United States of America 107:19096–19101

411. Gardos G (1958) The function of calcium in the potassium permeability of human erythrocytes. Biochimica et Biophysica Acta 30:653

412. Tharp DL, Bowles DK (2009) The intermediate-conductance Ca^{2+}-activated K^+ channel ($K_{Ca}3.1$) in vascular disease. Cardiovascular and Hematological Agents in Medicinal Chemistry 7:1–11

413. Li W, Aldrich RW (2009) Activation of the SK potassium channel-calmodulin complex by nanomolar concentrations of terbium. Proceedings of the National Academy of Sciences of the United States of America 106:1075–1080

414. Lu L, Timofeyev V, Li N, Rafizadeh S, Singapuri A, Harris TR, Chiamvimonvat N (2009) α-Actinin-2 cytoskeletal protein is required for the functional membrane localization of a Ca^{2+}-activated K^+ channel (SK2 channel). Proceedings of the National Academy of Sciences of the United States of America 106:18402–18407

415. Faber ES, Delaney AJ, Power JM, Sedlak PL, Crane JW, Sah P (2008) Modulation of SK channel trafficking by βadrenoceptors enhances excitatory synaptic transmission and plasticity in the amygdala. Journal of Neuroscience 28:10803–10813

416. Yang B, Desai R, Kaczmarek LK (2007) Slack and Slick K_{Na} channels regulate the accuracy of timing of auditory neurons. Journal of Neuroscience 27:2617–2627

417. Gao SB, Wu Y, Lv CX, ZH Guo ZH, Li CH, Ding JP (2008) Slack and Slick K_{Na} channels are required for the depolarizing afterpotential of acutely isolated, medium diameter rat dorsal root ganglion neurons. Acta Pharmacologica Sinica 29:899–905

418. Santi CM, Ferreira G, Yang B, Gazula VR, Butler A, Wei A, Kaczmarek LK, Salkoff L (2006) Opposite regulation of Slick and Slack K^+ channels by neuromodulators. Journal of Neuroscience 26:5059–5068

419. Brown MR, Kronengold J, Gazula VR, Chen Y, Strumbos JG, Sigworth FJ, Navaratnam D, Kaczmarek LK (2010) Fragile X mental retardation protein controls gating of the sodium-activated potassium channel Slack. Nature – Neuroscience 13:819–821

420. Galindo BE, de la Vega-Beltrán JL, Labarca P, Vacquier VD, Darszon A (2007) Sp-tetraKCNG: a novel cyclic nucleotide gated K^+ channel. Biochemical and Biophysical Research Communications 354:668–675

421. Honoré E (2007) The neuronal background K2P channels: focus on TREK1. Nature Reviews – Neuroscience 8:251–261

422. Duprat F, Lesage F, Fink M, Reyes R, Heurteaux C, Lazdunski M (1997) TASK, a human background K^+ channel to sense external pH variations near physiological pH. EMBO Journal 16:5464–5471

423. Plant LD, Dementieva IS, Kollewe A, Olikara S, Marks JD, Goldstein SA (2010) One SUMO is sufficient to silence the dimeric potassium channel K2P1. Proceedings of the National Academy of Sciences of the United States of America 107:10743–10748

424. Gestreau C, Heitzmann D, Thomas J, Dubreuil V, Bandulik S, Reichold M, Bendahhou S, Pierson P, Sterner C, Peyronnet-Roux J, Benfriha C, Tegtmeier I, Ehnes H, Georgieff M, Lesage F, Brunet JF, Goridis C, Warth R, Barhanin J (2010) Task2 potassium channels set central respiratory CO_2 and O_2 sensitivity. Proceedings of the National Academy of Sciences of the United States of America 107:2325–2330

425. Warth R, Barrière H, Meneton P, Bloch M, Thomas J, Tauc M, Heitzmann D, Romeo E, Verrey F, Mengual R, Guy N, Bendahhou S, Lesage F, Poujeol P, Barhanin J (2004) Proximal renal tubular acidosis in TASK2 K^+ channel-deficient mice reveals a mechanism for stabilizing bicarbonate transport. Proceedings of the National Academy of Sciences of the United States of America 101:8215–8220

426. Sandoz G, Thümmler S, Duprat F, Feliciangeli S, Vinh J, Escoubas P, Guy N, Lazdunski M, Lesage F (2006) AKAP150, a switch to convert mechano-, pH- and arachidonic acid-sensitive TREK K^+ channels into open leak channels. EMBO Journal 25:5864–5872

427. Garry A, Fromy B, Blondeau N, Henrion D, Brau F, Gounon P, Guy N, Heurteaux C, Lazdunski M, Saumet JL (2007) Altered acetylcholine, bradykinin and cutaneous

pressure-induced vasodilation in mice lacking the TREK1 potassium channel: the endothelial link. EMBO Reports 8:354–359

428. Feng L, Campbell EB, Hsiung Y, MacKinnon R (2010) Structure of a eukaryotic CLC transporter defines an intermediate state in the transport cycle. Science 330:635–641
429. Lísal JI, Maduke M (2008) The ClC-0 chloride channel is a "broken" Cl^-/H^+ antiporter. Nature – Structural and Molecular Biology 15:805–810
430. Estévez R, Boettger T, Stein V, Birkenhäger R, Otto E, Hildebrandt F, Jentsch TJ (2001) Barttin is a Cl^- channel β-subunit crucial for renal Cl^- reabsorption and inner ear K^+ secretion. Nature 414:558–561
431. Scholl U, Hebeisen S, Janssen AG, Müller-Newen G, Alekov A, Fahlke C (2006) Barttin modulates trafficking and function of ClC-K channels. Proceedings of the National Academy of Sciences of the United States of America 103:11411–11416
432. Gentzsch M, Cui L, Mengos A, Chang XB, Chen JH, Riordan JR (2003) The PDZ-binding chloride channel ClC-3B localizes to the Golgi and associates with cystic fibrosis transmembrane conductance regulator-interacting PDZ proteins. Journal of Biological Chemistry 278:6440–6449
433. Günther W, Lüchow A, Cluzeaud F, Vandewalle A, Jentsch TJ (1998) ClC-5, the chloride channel mutated in Dent's disease, colocalizes with the proton pump in endocytotically active kidney cells. Proceedings of the National Academy of Sciences of the United States of America 95:8075–8080
434. Novarino G, Weinert S, Rickheit G, Jentsch TJ (2010) Endosomal chloride–proton exchange rather than chloride conductance is crucial for renal endocytosis. Science 328:1398–1401
435. Eggermont J, Buyse G, Voets T, Tytgat J, De Smedt H, Droogmans G, Nilius B (1997) Alternative splicing of ClC-6 (a member of the CIC chloride-channel family) transcripts generates three truncated isoforms one of which, ClC-6c, is kidney-specific. Biochemical Journal 325:269–276
436. Meadows NA, Sharma SM, Faulkner GJ, Ostrowski MC, Hume DA, Cassady AI (2007) The expression of Clcn7 and Ostm1 in osteoclasts is coregulated by microphthalmia transcription factor. Journal of Biological Chemistry 282:1891–1904
437. Weinert S, Jabs S, Supanchart C, Schweizer M, Gimber N, Richter M, Rademann J, Stauber T, Kornak U, Jentsch TJ (2010) Lysosomal pathology and osteopetrosis upon loss of H^+-driven lysosomal Cl^- accumulation. Science 328:1401–1403
438. Caputo A, Caci E, Ferrera L, Pedemonte N, Barsanti C, Sondo E, Pfeffer U, Ravazzolo R, Zegarra-Moran O, Galietta LJV (2008) TMEM16A, a membrane protein associated with calcium-dependent chloride channel activity. Science 322:590–594
439. Marmorstein LY, McLaughlin PJ, Stanton JB, Yan L, Crabb JW, Marmorstein AD (2002) Bestrophin interacts physically and functionally with protein phosphatase 2A. Journal of Biological Chemistry 277:30591–30597
440. Matchkov VV, Larsen P, Bouzinova EV, Rojek A, Briggs Boedtkjer DM, Golubinskaya V, Pedersen FS, Aalkjaer C, Nilsson H (2008) Bestrophin-3 (vitelliform macular dystrophy 2-like 3 protein) is essential for the cGMP-dependent calcium-activated chloride conductance in vascular smooth muscle cells. Circulation Research 103:864–872
441. Tsunenari T, Sun H, Williams J, Cahill H, Smallwood P, Yau KW, Nathans J (2003) Structure-function analysis of the bestrophin family of anion channels. Journal of Biological Chemistry 278:41114–41125

442. Liu HT, Toychiev AH, Takahashi N, Sabirov RZ, Okada Y (2008) Maxi-anion channel as a candidate pathway for osmosensitive ATP release from mouse astrocytes in primary culture. Cell Research 18:558–565

443. Dutta AK, Korchev YE, Shevchuk AI, Hayashi S, Okada Y, Sabirov RZ (2008) Spatial distribution of maxi-anion channel on cardiomyocytes detected by smart-patch technique. Biophysical Journal 94:1646–1655

444. Toychiev AH, Sabirov RZ, Takahashi N, Ando-Akatsuka Y, Liu H, Shintani T, Noda M, Okada Y (2009) Activation of maxi-anion channel by protein tyrosine dephosphorylation. American Journal of Physiology – Cell Physiology 297:C990–C1000

445. He Y, Ramsay AJ, Hunt ML, Whitbread AK, Myers SA, Hooper JD (2008) N-glycosylation analysis of the human Tweety family of putative chloride ion channels supports a penta-spanning membrane arrangement. Impact of N-glycosylation on cellular processing of Tweety homologue 2 (TTYH2). Biochemical Journal 412:45–55

446. He Y, Hryciw DH, Carroll ML, Myers SA, Whitbread AK, Kumar S, Poronnik P, Hooper JD (2008) The ubiquitin-protein ligase Nedd4-2 differentially interacts with and regulates members of the Tweety family of chloride ion channels. Journal of Biological Chemistry 283:24000–24010

447. Raucci FJ, Wijesinghe DS, Chalfant CE, Baumgarten CM (2010) Exogenous and endogenous ceramides elicit volume-sensitive chloride current in ventricular myocytes. Cardiovascular Research 86:55–62

448. Okada SF, O'Neal WK, Huang P, Nicholas RA, Ostrowski LE, Craigen WJ, Lazarowski ER, Boucher RC (2004) Voltage-dependent anion channel-1 (VDAC-1) contributes to ATP release and cell volume regulation in murine cells. Journal of General Physiology 124:513–526

449. Lee SH, Park JH, Jung HH, Lee SH, Oh JW, Lee HM, Jun HS, Cho WJ, Lee JY (2005) Expression and distribution of ion transport mRNAs in human nasal mucosa and nasal polyps. Acta Oto-Laryngologica 125:745–752

450. Gruber AD, Elble RC, Ji HL, Schreur KD, Fuller CM, Pauli BU (1998) Genomic cloning, molecular characterization, and functional analysis of human CLCA1, the first human member of the family of Ca^{2+}-activated Cl^- channels. Genomics 54:200–214

451. Gruber AD, Schreur KD, Ji HL, Fuller CM, Pauli BU (1999) Molecular cloning and transmembrane structure of hCLCA2 from human lung, trachea, and mammary gland. American Journal of Physiology – Cell Physiology 276:C1261–C1270

452. Abdel-Ghany M, Cheng HC, Elble RC, Pauli BU (2001) The breast cancer β4 integrin and endothelial human CLCA2 mediate lung metastasis. Journal of Biological Chemistry 276:25438–25446

453. Connon CJ, Kawasaki S, Yamasaki K, Quantock AJ, Kinoshita S (2005) The quantification of hCLCA2 and colocalisation with integrin β4 in stratified human epithelia. Acta Histochemica 106:421–425

454. Gruber AD, Pauli BU (1999) Molecular cloning and biochemical characterization of a truncated, secreted member of the human family of Ca^{2+}-activated Cl^- channels. Biochimica et Biophysica Acta 1444:418–423

455. Huang F, Rock JR, Harfe BD, Cheng T, Huang X, Jan YN, Jan LY (2009) Studies on expression and function of the TMEM16A calcium-activated chloride channel. Proceedings of the National Academy of Sciences of the United States of America 106:21413–21418

456. Singh H, Cousin MA, Ashley RH (2007) Functional reconstitution of mammalian "chloride intracellular channels" CLIC1, CLIC4 and CLIC5 reveals differential regulation by cytoskeletal actin. FEBS Journal 274:6306–6316

457. Board PG, Coggan M, Watson S, Gage PW, Dulhunty AF (2004) CLIC-2 modulates cardiac ryanodine receptor Ca^{2+} release channels. International Journal of Biochemistry and Cell Biology 36:1599–1612

458. Qian Z, Okuhara D, Abe MK, Rosner MR (1999) Molecular cloning and characterization of a mitogen-activated protein kinase-associated intracellular chloride channel. Journal of Biological Chemistry 274:1621–1627

459. Shukla A, Malik M, Cataisson C, Ho Y, Friesen T, Suh KS, Yuspa SH (2009) TGFβ signalling is regulated by Schnurri-2-dependent nuclear translocation of CLIC4 and consequent stabilization of phospho-Smad2 and 3. Nature Cell Biology 11:777–784

460. Suginta W, Karoulias N, Aitken A, Ashley RH (2001) Chloride intracellular channel protein CLIC4 (p64H1) binds directly to brain dynamin I in a complex containing actin, tubulin and 14-3-3 isoforms. Biochemical Journal 359:55–64

461. Shanks RA, Larocca MC, Berryman M, Edwards JC, Urushidani T, Navarre J, Goldenring JR (2002) AKAP350 at the Golgi apparatus. II. Association of AKAP350 with a novel chloride intracellular channel (CLIC) family member. Journal of Biological Chemistry 277:40973–40980

462. Berryman MA, Goldenring JR (2003) CLIC4 is enriched at cell-cell junctions and colocalizes with AKAP350 at the centrosome and midbody of cultured mammalian cells. Cell Motility and the Cytoskeleton 56:159–172

463. Berryman M, Bruno J, Price J, Edwards JC (2004) CLIC-5A functions as a chloride channel in vitro and associates with the cortical actin cytoskeleton in vitro and in vivo. Journal of Biological Chemistry 279:34794–34801

464. Griffon N, Jeanneteau F, Prieur F, Diaz J, Sokoloff P (2003) CLIC6, a member of the intracellular chloride channel family, interacts with dopamine D(2)-like receptors. Brain Research – Molecular Brain Research 117:47–57

465. Larkin D, Murphy D, Reilly DF, Cahill M, Sattler E, Harriott P, Cahill DJ, Moran N (2004) ICln, a novel integrin $\alpha_{IIb}\beta_3$-associated protein, functionally regulates platelet activation. Journal of Biological Chemistry 279:27286–27293

466. Zhang WK, Wang D, Duan Y, Loy MMT, Chan HC, Huang P (2010) Mechanosensitive gating of CFTR. Nature – Cell Biology 12:507–512

467. Decoursey TE (2003) Voltage-gated proton channels and other proton transfer pathways. Physiological Reviews 83:475–579

468. Koch HP, Kurokawa T, Okochi Y, Sasaki M, Okamura Y, Larsson HP (2008) Multimeric nature of voltage-gated proton channels. Proceedings of the National Academy of Sciences of the United States of America 105:9111–9116

469. Ramsey IS, Mokrab Y, Carvacho I, Sands ZA, Sansom MS, Clapham DE (2010) An aqueous H^+ permeation pathway in the voltage-gated proton channel H_V1. Nature – Structural and Molecular Biology 17:869–875

470. Morgan D, Capasso M, Musset B, Cherny VV, Ríos E, Dyer MJ, DeCoursey TE (2009) Voltage-gated proton channels maintain pH in human neutrophils during phagocytosis. Proceedings of the National Academy of Sciences of the United States of America 106:18022–18027

Chap. 4. Membrane Compound Carriers

471. Thevelein JM, Voordeckers K (2009) Functioning and evolutionary significance of nutrient transceptors. Molecular Biology and Evolution 26:2407–2414

472. Klaassen CD, Aleksunes LM (2010) Xenobiotic, bile acid, and cholesterol transporters: function and regulation. Pharmacological Reviews 62:1–96

473. Kanai Y, Hediger MA (2004) The glutamate/neutral amino acid transporter family SLC1: molecular, physiological and pharmacological aspects. Pflügers Archiv (European Journal of Physiology) 447:469–479

474. Palacín M, Kanai Y (2004) The ancillary proteins of HATs: SLC3 family of amino acid transporters. Pflügers Archiv (European Journal of Physiology) 447:490–494

475. Verrey F, Closs EI, Wagner CA, Palacin M, Endou H, Kanai Y (2004) CATs and HATs: the SLC7 family of amino acid transporters. Pflügers Archiv (European Journal of Physiology) 447:532–542

476. Daniel H, Kottra G (2004) The proton oligopeptide cotransporter family SLC15 in physiology and pharmacology. Pflügers Archiv (European Journal of Physiology) 447:610-618

477. Reimer RJ, Edwards RH (2004) Organic anion transport is the primary function of the SLC17/type I phosphate transporter family. Pflügers Archiv (European Journal of Physiology) 447:629–635

478. Eiden LE, Schäfer MK, Weihe E, Schütz B (2004) The vesicular amine transporter family (SLC18): amine/proton antiporters required for vesicular accumulation and regulated exocytotic secretion of monoamines and acetylcholine. Pflügers Archiv (European Journal of Physiology) 447:636–640

479. Mount DB, Romero MF (2004) The SLC26 gene family of multifunctional anion exchangers. Pflügers Archiv (European Journal of Physiology) 447:710–721

480. Gasnier B (2004) The SLC32 transporter, a key protein for the synaptic release of inhibitory amino acids. Pflügers Archiv (European Journal of Physiology) 447:756–759

481. Boll M, Daniel H, Gasnier B (2004) The SLC36 family: proton-coupled transporters for the absorption of selected amino acids from extracellular and intracellular proteolysis. Pflügers Archiv (European Journal of Physiology) 447:776–779

482. Mackenzie B, Erickson JD (2004) Sodium-coupled neutral amino acid (System N/A) transporters of the SLC38 gene family. Pflügers Archiv (European Journal of Physiology) 447784–795

483. Edinger AL, Thompson CB (2002) Antigen-presenting cells control T cell proliferation by regulating amino acid availability. Proceedings of the National Academy of Sciences of the United States of America 99:1107–1109

484. Weyand S, Shimamura T, Yajima S, Suzuki S, Mirza O, Krusong K, Carpenter EP, Rutherford NG, Hadden JM, O'Reilly J, Ma P, Saidijam M, Patching SG, Hope RJ, Norbertczak HT, Roach PCJ, Iwata S, Henderson PJF, Cameron AD (2008) Structure and molecular mechanism of a nucleobase–cation-symport-1 family transporter. Science 322:709–713

485. Dawson PA, Rao A (2007) Ntcp. UCSD-Nature Molecule Pages, UCSD-Nature Signaling Gateway (www.signaling-gateway.org)

486. Poole RC, Halestrap AP (1993) Transport of lactate and other monocarboxylates across mammalian plasma membranes. American Journal of Physiology – Cell Physiology 264:C761–C782

487. Bröer S, Schneider HP, Bröer A, Rahman B, Hamprecht B, Deitmer JW (1998) Characterization of the monocarboxylate transporter 1 expressed in Xenopus laevis oocytes by changes in cytosolic pH. Biochemical Journal 333:167–174

488. Halestrap AP (2009) Monocarboxylate transporter 1. UCSD-Nature Molecule Pages, UCSD-Nature Signaling Gateway (www.signaling-gateway.org)

489. Becker HM, Bröer S, Deitmer JW (2004) Facilitated lactate transport by MCT1 when coexpressed with the sodium bicarbonate cotransporter (NBC) in Xenopus oocytes. Biophysical Journal 86:235–247

490. Robinson NJ, Winge DR (2010) Copper metallochaperones. Annual Review of Biochemistry 79:7.1–7.26

491. Kaplan JH, Lutsenko S (2009) Copper transport in mammalian cells: special care for a metal with special needs. Journal of Biological Chemistry 284:25461–25465

492. Culotta VC, Yang M, O'Halloran TV (2006) Activation of superoxide dismutases: putting the metal to the pedal. Biochimica et Biophysica Acta 1763:747–758

493. Kim BE, Turski ML, Nose Y, Casad M, Rockman HA, Thiele DJ (2010) Cardiac copper deficiency activates a systemic signaling mechanism that communicates with the copper acquisition and storage organs. Cell Metabolism 11:353–363

494. Banci L, Bertini I, Ciofi-Baffoni S, Kozyreva T, Zovo K, Palumaa P (2010) Affinity gradients drive copper to cellular destinations. Nature 465:645–648

495. Chen W, Paradkar PN, Li L, Pierce EL, Langer NB, Takahashi-Makise N, Hyde BB, Shirihai OS, Ward DM, Kaplan J, Paw BH (2009) Abcb10 physically interacts with mitoferrin-1 (Slc25a37) to enhance its stability and function in the erythroid mitochondria. Proceedings of the National Academy of Sciences of the United States of America 106:16263–16268

496. Zhou H, Clapham DE (2009) Mammalian MagT1 and TUSC3 are required for cellular magnesium uptake and vertebrate embryonic development. Proceedings of the National Academy of Sciences of the United States of America 106:15750–15755

497. Shen MR, Lin AC, Hsu YM, Chang TJ, Tang MJ, Alper SL, Ellory JC, Chou CY (2004) Insulin-like growth factor 1 stimulates KCl cotransport, which is necessary for invasion and proliferation of cervical cancer and ovarian cancer cells. Journal of Biological Chemistry 279:40017–40025

498. Gamba G (2009) The thiazide-sensitive Na^+–Cl^- cotransporter: molecular biology, functional properties, and regulation by WNKs. American Journal of Physiology – Renal Physiology 297:F838–F848

499. Russell JM (2000) Sodium–potassium–chloride cotransport. Physiological Reviews 80:211–276

500. Haas M (1994). The Na-K-Cl cotransporters. American Journal of Physiology – Cell Physiology 267:C869-C885

501. Hediger MA, Kanai Y, You G, Nussberger S (1995) Mammalian ion-coupled solute transporters. Journal of Physiology 482:7S–17S

502. Weihe E, Tao-Cheng JH, Schäfer MK, Erickson JD, Eiden LE (1996) Visualization of the vesicular acetylcholine transporter in cholinergic nerve terminals and its targeting to a specific population of small synaptic vesicles. Proceedings of the National Academy of Sciences of the United States of America 93:3547–3552

503. Carneiro AM, Blakely RD (2009) SERT. UCSD-Nature Molecule Pages, UCSD-Nature Signaling Gateway (www.signaling-gateway.org)

504. Sitte HH, Freissmuth M (2009) Gat1. UCSD-Nature Molecule Pages, UCSD-Nature Signaling Gateway (www.signaling-gateway.org)

505. Kanai Y, Nussberger S, Romero MF, Boron WF, Hebert SC, Hediger MA (1995) Electrogenic properties of the epithelial and neuronal high affinity glutamate transporter. Journal of Biological Chemistry 270:16561–16568

506. Watabe M, Aoyama K, Nakaki T (2007) Regulation of glutathione synthesis via interaction between glutamate transport-associated protein 3-18 (GTRAP3-18) and excitatory amino acid carrier-1 (EAAC1) at plasma membrane. Molecular Pharmacology 72:1103–1110

507. Chen H, Bai J, Ye J, Liu Z, Chen R, Mao W, Li A, Zhou J (2007) JWA as a functional molecule to regulate cancer cells migration via MAPK cascades and F-actin cytoskeleton. Cell Signaling 19:1315–1327

508. Wersinger E, Schwab Y, Sahel JA, Rendon A, Pow DV, Picaud S, Roux MJ (2006) The glutamate transporter EAAT5 works as a presynaptic receptor in mouse rod bipolar cells. Journal of Physiology 577:221–234

509. Weiss MD, Derazi S, Kilberg MS, Anderson KJ (2001) Ontogeny and localization of the neutral amino acid transporter ASCT1 in rat brain. Developmental Brain Research 130:183–190

510. Yamamoto T, Nishizaki I, Nukada T, Kamegaya E, Furuya S, Hirabayashi Y, Ikeda K, Hata H, Kobayashi H, Sora I, Yamamoto H (2004) Functional identification of ASCT1 neutral amino acid transporter as the predominant system for the uptake of L-serine in rat neurons in primary culture. Neuroscience Research 49:101–111

511. Zerangue N, Kavanaugh MP (1996) ASCT-1 is a neutral amino acid exchanger with chloride channel activity. Journal of Biological Chemistry 271:27991–27994

512. Bröer A, Brookes N, Ganapathy V, Dimmer KS, Wagner CA, Lang F, Bröer S (1999) The astroglial ASCT2 amino acid transporter as a mediator of glutamine efflux. Journal of Neurochemistry 73:2184–2194

513. MacLean MR, Herve P, Eddahibi S, Adnot S (2000) 5-hydroxytryptamine and the pulmonary circulation: receptors, transporters and relevance to pulmonary arterial hypertension. British Journal of Pharmacology 131:161–168

514. Erickson JD, Varoqui H (2000) Molecular analysis of vesicular amine transporter function and targeting to secretory organelles. FASEB Journal 14:2450–2458

515. Deniaud A, Le Bras M, Lecellier G, Brenner C, Kroemer G (2005) Ant1. UCSD-Nature Molecule Pages, UCSD-Nature Signaling Gateway (www.signaling-gateway.org)

516. Le Bras M, Deniaud A, Lecellier G, Kroemer G, Brenner C (2005) Ant2. UCSD-Nature Molecule Pages, UCSD-Nature Signaling Gateway (www.signaling-gateway.org)

517. Pastor-Anglada M, Errasti-Murugarren E, Aymerich I, Casado FJ (2007) Concentrative nucleoside transporters (CNTs) in epithelia: from absorption to cell signaling. Journal of Physiology and Biochemistry 63:97–110

518. Young JD, Yao SY, Sun L, Cass CE, Baldwin SA (2008) Human equilibrative nucleoside transporter (ENT) family of nucleoside and nucleobase transporter proteins. Xenobiotica 38:995–1021

519. Baldwin SA, Beal PR, Yao SY, King AE, Cass CE, Young JD (2004) The equilibrative nucleoside transporter family, SLC29. Pflügers Archiv (European Journal of Physiology) 447:735–743

520. Rose JB, Naydenova Z, Bang A, Eguchi M, Sweeney G, Choi DS, Hammond JR, Coe IR (2010) Equilibrative nucleoside transporter 1 plays an essential role in cardioprotection. American Journal of Physiology – Heart and Circulatory Physiology 298:H771–H777

521. Gournas C, Papageorgiou I, Diallinas G (2008) The nucleobase–ascorbate transporter (NAT) family: genomics, evolution, structure–function relationships and physiological role. Molecular BioSystems 4:404–416

522. Noy N (2000) Retinoid-binding proteins: mediators of retinoid action. Biochemical Journal 348:481–495

523. Maestro A, Terdoslavich M, Vanzo A, Kuku A, Tramer F, Nicolin V, Micali F, Decorti G, Passamonti S (2010) Expression of bilitranslocase in the vascular endothelium and its function as a flavonoid transporter. Cardiovascular Research 85:175–183

524. Mazurek MP, Prasad PD, Gopal E, Fraser SP, Bolt L, Rizaner N, Palmer CP, Foster CS, Palmieri F, Ganapathy V, Stühmer W, Djamgoz MB, Mycielska ME (2010) Molecular origin of plasma membrane citrate transporter in human prostate epithelial cells. EMBO reports 11:431–437

525. Wang W, Hart PS, Piesco NP, Lu X, Gorry MC, Hart TC (2003) Aquaporin expression in developing human teeth and selected orofacial tissues. Calcified Tissue International 72:222–227

526. Nielsen S, Frøkiaer J, Marples D, Kwon TH, Agre P, Knepper MA (2002) Aquaporins in the kidney: from molecules to medicine. Physiological Reviews 82:205–244

527. Masyuk AI, LaRusso NF (2006) Aquaporins in the hepatobiliary system. Hepatology 43:S75–S81

528. Miller EW, Dickinson BC, Chang CJ (2010) Aquaporin-3 mediates hydrogen peroxide uptake to regulate downstream intracellular signaling. Proceedings of the National Academy of Sciences of the United States of America 107:15681–15686

529. Li Y, Konings IB, Zhao J, Price LS, de Heer E, Deen PM (2008) Renal expression of exchange protein directly activated by cAMP (Epac) 1 and 2. American Journal of Physiology – Renal Physiology 295:F525–F533

530. Musa-Aziz R, Chen LM, Pelletier MF, Boron WF (2009) Relative CO_2/NH_3 selectivities of AQP1, AQP4, AQP5, AmtB, and RhAG. Proceedings of the National Academy of Sciences of the United States of America 106:5406–5411

531. Sidhaye VK, Schweitzer KS, Caterina MJ, Shimoda L, King LS (2008) Shear stress regulates aquaporin-5 and airway epithelial barrier function. Proceedings of the National Academy of Sciences of the United States of America 105:3345–3350

532. Li SZ, McDill BW, Kovach PA, Ding L, Go WY, Ho SN, Chen F (2007) Calcineurin-NFATc signaling pathway regulates AQP2 expression in response to calcium signals and osmotic stress. American Journal of Physiology – Cell Physiology 292:C1606–C1616

533. Moeller HB, Praetorius J, Rützler MB, Fenton RA (2010) Phosphorylation of aquaporin-2 regulates its endocytosis and protein–protein interactions. Proceedings of the National Academy of Sciences of the United States of America 107:424–429

534. Hibuse T, Maeda N, Nakatsuji H, Tochino Y, Fujita K, Kihara S, Funahashi T, Shimomura I (2009) The heart requires glycerol as an energy substrate through aquaporin 7, a glycerol facilitator. Cardiovascular Research 83:34–41

535. Kawedia JD, Nieman ML, Boivin GP, Melvin JE, Kikuchi KI, Hand AR, Lorenz JN, Menon AG (2007) Interaction between transcellular and paracellular water transport pathways through aquaporin 5 and the tight junction complex. Proceedings of the National Academy of Sciences of the United States of America 104:3621–3626

536. Watson RT, Kanzaki M, Pessin J (2004) Regulated membrane trafficking of the insulin-responsive glucose transporter 4 in adipocytes. Endocrine Reviews 25:177–204

537. Stöckli J, James DE (2009) GLUT4. UCSD-Nature Molecule Pages, UCSD-Nature Signaling Gateway (www.signaling-gateway.org)

538. Yang Q, Graham TE, Mody N, Preitner F, Peroni OD, Zabolotny JM, Kotani K, Quadro L, Kahn BB (2005) Serum retinol binding protein 4 contributes to insulin resistance in obesity and type 2 diabetes. Nature 436:337–338

539. Lalioti V, Muruais G, Dinarina A, van Damme J, Vandekerckhove J, Sandoval IV (2009) The atypical kinase Cdk5 is activated by insulin, regulates the association between GLUT4 and E-Syt1, and modulates glucose transport in 3T3-L1 adipocytes. Proceedings of the National Academy of Sciences of the United States of America 106:4249–4253

540. Schurmann A, Joost HG (2006) GLUT8. UCSD-Nature Molecule Pages, UCSD-Nature Signaling Gateway (www.signaling-gateway.org)

541. Schurmann A, Augustin R, Joost HG (2006) GLUT9. UCSD-Nature Molecule Pages, UCSD-Nature Signaling Gateway (www.signaling-gateway.org)

542. Preitner F, Bonny O, Laverrière A, Rotman S, Firsov D, Da Costa A, Metref S, Thorens B (2009) Glut9 is a major regulator of urate homeostasis and its genetic inactivation induces hyperuricosuria and urate nephropathy. Proceedings of the National Academy of Sciences of the United States of America 106:15501–15506

543. Dawson PA, Mychaleckyj JC, Fossey SC, Mihic SJ, Craddock AL, Bowden DW (2001) Sequence and functional analysis of GLUT10: a glucose transporter in the Type 2 diabetes-linked region of chromosome 20q12-13.1. Molecular Genetics and Metabolism 74:186–199

544. Lee YC, Huang HY, Chang CJ, Cheng CH, Chen YT (2010) Mitochondrial GLUT10 facilitates dehydroascorbic acid import and protects cells against oxidative stress: mechanistic insight into arterial tortuosity syndrome. Human Molecular Genetics 19:3721–3733

545. Rees DC, Johnson E, Lewinson O (2009) ABC transporters: the power to change. Nature Reviews – Molecular Cell Biology 10:218–227

546. Klein I, Sarkadi B, Varadi A (1999) An inventory of the human ABC proteins. Biochimica et Biophysica Acta 1461:237–262

547. Vazquez de Aldana CR, Marton MJ, Hinnebusch AG (1995) GCN20, a novel ATP binding cassette protein, and GCN1 reside in a complex that mediates activation of the eIF-2 alpha kinase GCN2 in amino acid-starved cells. EMBO Journal 14:3184–3199

548. Oldham ML, Chen J (2011) Crystal structure of the maltose transporter in a pretranslocation intermediate state. Science 332:1202–1205

549. Hollenstein K, Frei DC, Locher KP (2007) Structure of an ABC transporter in complex with its binding protein. Nature 446:213–216

550. Borycz J, Borycz JA, Kubów A, Lloyd V, Meinertzhagen IA (2008) Drosophila ABC transporter mutants white, brown and scarlet have altered contents and distribution of biogenic amines in the brain. Journal of Experimental Biology 211:3454–3466

551. Ricardo S, Lehmann R (2009) An ABC transporter controls export of a Drosophila germ cell attractant. Science 323:943–946

552. Bréchot JM, Hurbain I, Fajac A, Daty N, Bernaudin JF (1998) Different pattern of MRP localization in ciliated and basal cells from human bronchial epithelium. Journal of Histochemistry and Cytochemistry 46:513–517

553. Kobayashi N, Kobayashi N, Yamaguchi A, Nishi T (2009) Characterization of the ATP-dependent sphingosine 1-phosphate transporter in rat erythrocytes. Journal of Biological Chemistry 284:21192–21200

554. Kawahara A, Nishi T, Hisano Y, Fukui H, Yamaguchi A, Mochizuki N (2009) The sphingolipid transporter Spns2 functions in migration of zebrafish myocardial precursors. Science 323:524–527

555. Brown MS, Ye J, Goldstein JL (2010) HDL MiR-ed down by SREBP introns. Science 328:1495–1496

556. Tall AR, Costet P, Wang N (2002) Regulation and mechanisms of macrophage cholesterol efflux. Journal of Clinical Investigation 110:899–904

557. Najafi-Shoushtari SH, Kristo F, Li Y, Shioda T, Cohen DE, Gerszten RE, Näär AM (2010) MicroRNA-33 and the SREBP host genes cooperate to control cholesterol homeostasis. Science 328:1566–1569

558. Rayner KJ, Suárez Y, Dávalos A, Parathath S, Fitzgerald ML, Tamehiro N, Fisher EA, Moore KJ, Fernández-Hernando C (2010) MiR-33 contributes to the regulation of cholesterol homeostasis. Science 328:1570–1573

559. Kaminski WE, Piehler A, Püllmann K, Porsch-Ozcürümez M, Duong C, Bared GM, Büchler C, Schmitz G (2001) Complete coding sequence, promoter region, and genomic structure of the human ABCA2 gene and evidence for sterol-dependent regulation in macrophages. Biochemical and Biophysical Research Communications 281:249–258

560. Matsumura Y, Ban N, Inagaki N (2009) Abca3. UCSD-Nature Molecule Pages, UCSD-Nature Signaling Gateway (www.signaling-gateway.org)

561. Beharry S, Zhong M, Molday RS (2004) N-retinylidene-phosphatidylethanolamine is the preferred retinoid substrate for the photoreceptor-specific ABC transporter ABCA4 (ABCR). Journal of Biological Chemistry 279:53972–53979

562. Wang N, Lan D, Gerbod-Giannone M, Linsel-Nitschke P, Jehle AW, Chen W, Martinez LO, Tall AR (2003) ATP-binding cassette transporter A7 (ABCA7) binds apolipoprotein A-I and mediates cellular phospholipid but not cholesterol efflux. Journal of Biological Chemistry 278:42906–42912

563. Piehler A, Kaminski WE, Wenzel JJ, Langmann T, Schmitz G (2002) Molecular structure of a novel cholesterol-responsive A subclass ABC transporter, ABCA9. Biochemical and Biophysical Research Communications 295:408-416

564. Wenzel JJ, Kaminski WE, Piehler A, Heimerl S, Langmann T, Schmitz G (2003) ABCA10, a novel cholesterol-regulated ABCA6-like ABC transporter. Biochemical and Biophysical Research Communications 306:1089–1098

565. Jiang YJ, Lu B, Kim P, Paragh G, Schmitz G, Elias PM, Feingold KR (2008) PPAR and LXR activators regulate ABCA12 expression in human keratinocytes. Journal of Investigative Dermatology 128:104–109

566. Oancea G, O'Mara ML, Bennett WF, Tieleman DP, Abele R, Tampé R (2009) Structural arrangement of the transmission interface in the antigen ABC transport complex TAP. Proceedings of the National Academy of Sciences of the United States of America 106:5551–5556

567. Rigor RR, Hawkins BT, Miller DS (2010) Activation of PKC isoform β I at the blood-brain barrier rapidly decreases P-glycoprotein activity and enhances drug delivery to the brain. Journal of Cerebral Blood Flow and Metabolism 30:1373–1383

568. Kunzelmann K, Schreiber R (1999) CFTR, a regulator of channels. Journal of Membrane Biology 168:1–8

569. Stutts MJ, Canessa CM, Olsen JC, Hamrick M, Cohn JA, Rossier BC, Boucher RC (1995) CFTR as cAMP-dependent regulator of sodium channels. Science 269: 847–850

570. Schwiebert EM, Egan ME, Hwang TH, Fulmer SB, Allen SS, Cutting GR, Guggino WB (1995) CFTR regulates outwardly rectifying chloride channels through an autocrine mechanism involving ATP. Cell 81:1063–1073

571. Player MR, Torrence PF (1998) The 2-5A system: modulation of viral and cellular processes through acceleration of RNA degradation. Pharmacology and Therapeutics 78:55–113

572. Tyzack JK, Wang X, Belsham GJ, Proud CG (2000) ABC50 interacts with eukaryotic initiation factor 2 and associates with the ribosome in an ATP-dependent manner. Journal of Biological Chemistry 275:34131–34139

573. Schmitz G, Langmann T, Heimerl S (2001) Role of ABCG1 and other ABCG family members in lipid metabolism. Journal of Lipid Research 42:1513–1520

574. Yvan-Charvet L, Pagler T, Gautier EL, Avagyan S, Siry RL, Han S, Welch CL, Wang N, Randolph GJ, Snoeck HW, Tall AR (2010) ATP-Binding cassette transporters and HDL suppress hematopoietic stem cell proliferation. Science 328:1689–1693

575. Graf GA, Yu L, Li WP, Gerard R, Tuma PL, Cohen JC, Hobbs HH (2003) ABCG5 and ABCG8 are obligate heterodimers for protein trafficking and biliary cholesterol excretion. Journal of Biological Chemistry 278:48275–48282

Chap. 5. Receptors of Cellular Trafficking

576. May P, Herz J, Bock HH (2005) Molecular mechanisms of lipoprotein receptor signalling. Cellular and Molecular Life Sciences 62:2325–2338

577. Strickland DK, Kounnas MZ (1997) Mechanisms of cellular uptake of thrombin-antithrombin II complexes. Role of the low-density lipoprotein receptor-related protein as a serpin-enzyme complex receptor. Trends in Cardiovascular Medicine 7:9–16

578. Sakaguchi H, Takeya M, Suzuki H, Hakamata H, Kodama T, Horiuchi S, Gordon S, van der Laan LJ, Kraal G, Ishibashi S, Kitamura N, Takahashi K (1998) Role of macrophage scavenger receptors in diet-induced atherosclerosis in mice. Laboratory Investigation 78:423–434

579. Mehta KD, Radominska-Pandya A, Kapoor GS, Dave B, Atkins BA (2002) Critical role of diacylglycerol- and phospholipid-regulated protein kinase Cϵ in induction of low-density lipoprotein receptor transcription in response to depletion of cholesterol. Molecular and Cellular Biology 22:3783–3793

580. Maurer ME, Cooper JA (2006) The adaptor protein Dab2 sorts LDL receptors into coated pits independently of AP-2 and ARH. Journal of Cell Science 119:4235–4246

581. Bräuer AU, Nitsch R (2008) Plasticity-related genes (PRGs/LRPs): a brain-specific class of lysophospholipid-modifying proteins. Biochimica et Biophysica Acta – Molecular and Cell Biology of Lipids 1781:595-600

582. Huang SS, Ling TY, Tseng WF, Huang YH, Tang FM, Leal SM, Huang JS (2003) Cellular growth inhibition by IGFBP-3 and TGF-β1 requires LRP-1. FASEB Journal 17:2068–2081

583. Leucht C, Simoneau S, Rey C, Vana K, Rieger R, Lasmézas CI, Weiss S (2003) The 37 kDa/67 kDa laminin receptor is required for PrP^{Sc} propagation in scrapie-infected neuronal cells. EMBO Reports 4:290–295

584. Taylor DR, Hooper NM (2007) The low-density lipoprotein receptor-related protein 1 (LRP1) mediates the endocytosis of the cellular prion protein. Biochemical Journal 402:17–23

585. Huo Y, Zhao L, Hyman MC, Shashkin P, Harry BL, Burcin T, Forlow SB, Stark MA, Smith DF, Clarke S, Srinivasan S, Hedrick CC, Praticò D, Witztum JL, Nadler JL, Funk CD, Ley K (2004) Critical role of macrophage 12/15-lipoxygenase for atherosclerosis in apolipoprotein E-deficient mice. Circulation 110:2024–2031

586. Sears DD, Miles PD, Chapman J, Ofrecio JM, Almazan F, Thapar D, Miller YI (2009) 12/15-Lipoxygenase is required for the early onset of high fat diet-induced adipose tissue inflammation and insulin resistance in mice. PLoS One 4:e7250

587. Orlando RA, Farquhar MG (1994) Functional domains of the receptor-associated protein (RAP). Proceedings of the National Academy of Sciences of the United States of America 91:3161–3165

588. Tanaga K, Bujo H, Zhu Y, Kanaki T, Hirayama S, Takahashi K, Inoue M, Mikami K, Schneider WJ, Saito Y (2004) LRP1B attenuates the migration of smooth muscle cells by reducing membrane localization of urokinase and PDGF receptors. Arteriosclerosis, Thrombosis, and Vascular Biology 24:1422–1428

589. Li Y, Knisely JM, Lu W, McCormick LM, Wang J, Henkin J, Schwartz AL, Bu G (2002) Low density lipoprotein (LDL) receptor-related protein 1B impairs urokinase receptor regeneration on the cell surface and inhibits cell migration. Journal of Biological Chemistry 277:42366–42371

590. Stefansson S, Chappell DA, Argraves KM, Strickland DK, Argraves WS (1995) Glycoprotein 330/low density lipoprotein receptor-related protein-2 mediates endocytosis of low density lipoproteins via interaction with apolipoprotein B100. Journal of Biological Chemistry 270:19417–19421

591. Fisher C. New Perspectives in Shh Signalling? NCBI – Bookshelf – Madame Curie Bioscience Database – Development (www.ncbi.nlm.nih.gov/bookshelf). Landes Bioscience and Springer Science+Business Media

592. Fisher CE, Howie SE (2006) The role of megalin (LRP-2/Gp330) during development. Developmental Biology 296:279–297

593. Nykjaer A, Fyfe JC, Kozyraki R, Leheste JR, Jacobsen C, Nielsen MS, Verroust PJ, Aminoff M, de la Chapelle A, Moestrup SK, Ray R, Gliemann J, Willnow TE, Christensen EI (2001) Cubilin dysfunction causes abnormal metabolism of the steroid hormone 25(OH) vitamin D_3. Proceedings of the National Academy of Sciences of the United States of America 98:13895–13900

594. Gekle M, Knaus P, Nielsen R, Mildenberger S, Freudinger R, Wohlfarth V, Sauvant C, Christensen EI (2003) Transforming growth factor-beta1 reduces megalin- and cubilin-mediated endocytosis of albumin in proximal-tubule-derived opossum kidney cells. Journal of Physiology 552:471–481

595. Li J, Ji C, Zheng H, Fei X, Zheng M, Dai J, Gu S, Xie Y, Mao Y (2005) Molecular cloning and characterization of a novel human gene containing 4 ankyrin repeat domains. Cellular and Molecular Biology Letters 10:185–193

596. Ishii H, Kim DH, Fujita T, Endo Y, Saeki S, Yamamoto TT (1998) cDNA cloning of a new low-density lipoprotein receptor-related protein and mapping of its gene (LRP3) to chromosome bands 19q12-q13. Genomics 51:132-135

597. Tian QB, Suzuki T, Yamauchi T, Sakagami H, Yoshimura Y, Miyazawa S, Nakayama K, Saitoh F, Zhang JP, Lu Y, Kondo H, Endo S (2006) Interaction of LDL receptor-related protein 4 (LRP4 [?]) with postsynaptic scaffold proteins via its C-terminal PDZ domain-binding motif, and its regulation by Ca/calmodulin-dependent protein kinase II. European Journal of Neuroscience 23:2864–2876

598. Choi HY, Dieckmann M, Herz J, Niemeier A (2009) Lrp4, a novel receptor for Dickkopf 1 and Sclerostin, is expressed by osteoblasts and regulates bone growth and turnover in vivo. PLoS One 4:e7930

599. Ohazama A, Johnson EB, Ota MS, Choi HY, Porntaveetus T, Oommen S, Itoh N, Eto K, Gritli-Linde A, Herz J, Sharpe PT (2008) Lrp4 modulates extracellular integration of cell signaling pathways in development. PLoS One 3:e4092

600. Ibelgaufts H (2010) Cytokines and Cells Online Pathfinder Encyclopaedia (www.copewithcytokines.de/cope.cgi)

601. Schmidt V, Sporbert A, Rohe M, Reimer T, Rehm A, Andersen OM, Willnow TE (2007) SorLA/LR11 regulates processing of amyloid precursor protein via interaction with adaptors GGA and PACS-1. Journal of Biological Chemistry 282:32956–32964

602. Taira K, Bujo H, Hirayama S, Yamazaki H, Kanaki T, Takahashi K, Ishii I, Miida T, Schneider WJ, Saito Y (2001) LR11, a mosaic LDL receptor family member, mediates the uptake of ApoE-rich lipoproteins in vitro. Arteriosclerosis, Thrombosis, and Vascular Biology 21:1501–1506

603. Lintzel J, Franke I, Riedel IB, Schaller HC, Hampe W (2002) Characterization of the VPS10 domain of SorLA/LR11 as binding site for the neuropeptide HA. Biological Chemistry 383:1727–1733

604. Boels K, Glassmeier G, Herrmann D, Riedel IB, Hampe W, Kojima I, Schwarz JR, Schaller HC (2001) The neuropeptide head activator induces activation and translocation of the growth-factor-regulated Ca^{2+}-permeable channel GRC. Journal of Cell Science 114:3599–3606

605. Kojima I, Nagasawa M. TRPV2: A calcium-permeable cation channel regulated by insulin-like growth factors. NCBI Bookshelf – Frontiers in Neuroscience – TRP

Ion Channel Function in Sensory Transduction and Cellular Signaling Cascades (www.ncbi.nlm.nih.gov/bookshelf)

606. Hisanaga E, Nagasawa M, Ueki K, Kulkarni RN, Mori M, Kojima I (2009) Regulation of calcium-permeable TRPV2 channel by insulin in pancreatic β-cells. Diabetes 58:174–184

607. Brodeur J, Larkin H, Boucher R, Thériault C, Chayer St-Louis S, Gagnon H, Lavoie C (2009) Calnuc binds to LRP9 and affects its endosomal sorting. Traffic 10:1098–1114

608. Hoffmann R, Valencia A (2004) A gene network for navigating the literature. Nature – Genetics 36:664 (Information Hyperlinked over Proteins www.ihop-net.org/)

609. Marks N, Berg MJ (2008) Neurosecretases provide strategies to treat sporadic and familial Alzheimer disorders. Neurochemistry International 52:184–215

610. Vincent JB, Herbrick JA, Gurling HMD, Bolton PF, Roberts W, Scherer SW (2000) Identification of a novel gene on chromosome 7q31 that is interrupted by a translocation breakpoint in an autistic individual. American Journal of Human Genetics 67:510–514

611. Battle MA, Maher VM, McCormick JJ (2003) ST7 is a novel low-density lipoprotein receptor-related protein (LRP) with a cytoplasmic tail that interacts with proteins related to signal transduction pathways. Biochemistry 42:7270–7282

612. Gray JP, Davis JW, Gopinathan L, Leas TL, Nugent CA, Vanden Heuvel JP (2006) The ribosomal protein rpL11 associates with and inhibits the transcriptional activity of peroxisome proliferator-activated receptor-α. Toxicological Sciences 89:535–546

613. Moore KJ, Freeman MW (2006) Scavenger receptors in atherosclerosis: beyond lipid uptake. Arteriosclerosis, Thrombosis, and Vascular Biology 26:1702–1711

614. Limmon GV, Arredouani M, McCann KL, Corn Minor RA, Kobzik L, Imani F (2008) Scavenger receptor class-A is a novel cell surface receptor for double-stranded RNA. FASEB Journal 22:159–167

615. Fong LG, Le D (1999) The processing of ligands by the class A scavenger receptor is dependent on signal information located in the cytoplasmic domain. Journal of Biological Chemistry 274:36808–36816

616. Friedman G, Ben-Yehuda A, Dabach Y, Hollander G, Babaey S, Ben-Naim M, Stein O, Stein Y (2000) Macrophage cholesterol metabolism, apolipoprotein E, and scavenger receptor AI/II mRNA in atherosclerosis-susceptible and -resistant mice. Arteriosclerosis, Thrombosis, and Vascular Biology 20:2459–2464

617. Ouchi N, Kihara S, Arita Y, Nishida M, Matsuyama A, Okamoto Y, Ishigami M, Kuriyama H, Kishida K, Nishizawa H, Hotta K, Muraguchi M, Ohmoto Y, Yamashita S, Funahashi T, Matsuzawa Y (2001) Adipocyte-derived plasma protein, adiponectin, suppresses lipid accumulation and class A scavenger receptor expression in human monocyte-derived macrophages. Circulation 103:1057–1063

618. Iwaki M, Matsuda M, Maeda N, Funahashi T, Matsuzawa Y, Makishima M, Shimomura I (2003) Induction of adiponectin, a fat-derived antidiabetic and antiatherogenic factor, by nuclear receptors. Diabetes 52:1655–1663

619. Kanno S, Furuyama A, Hirano S (2007) A murine scavenger receptor MARCO recognizes polystyrene nanoparticles. Toxicological Sciences 97:398–406

620. Han HJ, Tokino T, Nakamura Y (1998) CSR, a scavenger receptor-like protein with a protective role against cellular damage caused by UV irradiation and oxidative stress. Human Molecular Genetics 7:1039–1046

621. Ohtani K, Suzuki Y, Eda S, Kawai T, Kase T, Keshi H, Sakai Y, Fukuoh A, Sakamoto T, Itabe H, Suzutani T, Ogasawara M, Yoshida I, Wakamiya N (2001) The membrane-type collectin CL-P1 is a scavenger receptor on vascular endothelial cells. Journal of Biological Chemistry 276:44222–44228

622. Connelly MA, Williams DL (2004) Scavenger receptor BI: a scavenger receptor with a mission to transport high density lipoprotein lipids. Current Opinion in Lipidology 15:287–295

623. Silver DL (2002) A carboxyl-terminal PDZ-interacting domain of scavenger receptor B, type I is essential for cell surface expression in liver. Journal of Biological Chemistry 277:34042–34047

624. Eckhardt ER, Cai L, Shetty S, Zhao Z, Szanto A, Webb NR, Van der Westhuyzen DR (2006) High density lipoprotein endocytosis by scavenger receptor SR-BII is clathrin-dependent and requires a carboxyl-terminal dileucine motif. Journal of Biological Chemistry 281:4348–4353

625. Mulcahy JV, Riddell DR, Owen JS (2004) Human scavenger receptor class B type II (SR-BII) and cellular cholesterol efflux. Biochemical Journal 377:741–747

626. Kuronita T, Eskelinen EL, Fujita H, Saftig P, Himeno M, Tanaka Y (2002) A role for the lysosomal membrane protein LGP85 in the biogenesis and maintenance of endosomal and lysosomal morphology. Journal of Cell Science 115:4117–4131

627. Zhang X, Fitzsimmons RL, Cleland LG, Ey PL, Zannettino AC, Farmer EA, Sincock P, Mayrhofer G (2003) CD36/fatty acid translocase in rats: distribution, isolation from hepatocytes, and comparison with the scavenger receptor SR-B1. Laboratory Investigation 83: 317–332

628. Silverstein RL, Asch AS, Nachman RL (1989) Glycoprotein IV mediates thrombospondin-dependent platelet-monocyte and platelet-U937 cell adhesion. Journal of Clinical Investigation 84:546–552

629. Oquendo P, Hundt E, Lawler J, Seed B (1989) CD36 directly mediates cytoadherence of Plasmodium falciparum parasitized erythrocytes. Cell 58:95–101

630. Endemann G, Stanton LW, Madden KS, Bryant CM, White RT, Protter AA (1993) CD36 is a receptor for oxidized low density lipoprotein. Journal of Biological Chemistry 268:11811–11816

631. Calvo D, Gómez-Coronado D, Suárez Y, Lasunción MA, Vega MA (1998) Human CD36 is a high affinity receptor for the native lipoproteins HDL, LDL, and VLDL. Journal of Lipid Research 39:777–788

632. Nagy L, Tontonoz P, Alvarez JG, Chen H, Evans RM (1998) Oxidized LDL regulates macrophage gene expression through ligand activation of PPARγ. Cell 93:229–240

633. Amoruso A, Bardelli C, Fresu LG, Palma A, Vidali M, Ferrero V, Ribichini F, Vassanelli C, Brunelleschi S (2009) Enhanced peroxisome proliferator-activated receptor-γ expression in monocyte/macrophages from coronary artery disease patients and possible gender differences. Journal of Pharmacology and Experimental Therapeutics 331:531–538

634. Xue JH, Yuan Z, Wu Y, Liu Y, Zhao Y, Zhang WP, Tian YL, Liu WM, Liu Y, Kishimoto C (2010) High glucose promotes intracellular lipid accumulation in vascular smooth muscle cells by impairing cholesterol influx and efflux balance. Cardiovascular Research 86:141–150

635. Tchoukalova Y, Koutsari C, Jensen M (2007) Committed subcutaneous preadipocytes are reduced in human obesity. Diabetologia 50:151–157

636. Sawamura T, Kume N, Aoyama T, Moriwaki H, Hoshikawa H, Aiba Y, Tanaka T, Miwa S, Katsura Y, Kita T, Masaki T (1997) An endothelial receptor for oxidized low-density lipoprotein. Nature 386:73–77

637. Murphy JE, Vohra RS, Dunn S, Holloway ZG, Monaco AP, Homer-Vanniasinkam S, Walker JH, Ponnambalam S (2008) Oxidised LDL internalisation by the LOX-1 scavenger receptor is dependent on a novel cytoplasmic motif and is regulated by dynamin-2. Journal of Cell Science 121:2136–2147

Chap. 6. Receptors

638. Gong H, Shen B, Flevaris P, Chow C, Lam SCT, Voyno-Yasenetskaya TA, Kozasa T, Du X (2010) G Protein subunit Gα13 binds to integrin $\alpha_{IIb}\beta_3$ and mediates integrin "outside-in" signaling. Science 327:340–343

639. Bethani I, Skånland SS, Dikic I, Acker-Palmer A (2010) Spatial organization of transmembrane receptor signalling. EMBO Journal 29:2677–2688

640. Krasteva G, Canning BJ, Hartmann P, Veres TZ, Papadakis T, Mühlfeld C, Schliecker K, Tallini YN, Braun A, Hackstein H, Baal N, Weihe E, Schütz B, Kotlikoff M, Ibanez-Tallon I, Kummer W (2011) Cholinergic chemosensory cells in the trachea regulate breathing. Proceedings of the National Academy of Sciences of the United States of America 108:9478–9483

641. Shah AS, Ben-Shahar Y, Moninger TO, Kline JN, Welsh MJ (2009) Motile cilia of human airway epithelia are chemosensory. Science 325:1131–1134

642. Cain WS, Jalowayski AA, Kleinman M, Lee NS, Lee BR, Ahn BH, Magruder K, Schmidt R, Hillen BK, Warren CB, Culver BD (2004) Sensory and associated reactions to mineral dusts: sodium borate, calcium oxide, and calcium sulfate. Journal of Occupational and Environmental Hygiene 1:222–236

643. Waldmann M, Thompson GW, Kember GC, Ardell JL, Armour JA (2006) Stochastic behavior of atrial and ventricular intrinsic cardiac neurons. Journal of Applied Physiology 101:413–419

644. Armour JA (2008) Potential clinical relevance of the 'little brain' on the mammalian heart. Experimental Physiology 93:165–176

645. Schultz HD (2001) Cardiac vagal chemosensory afferents. Function in pathophysiological states. Annals of the New York Academy of Sciences 940:59–73

646. Digby GJ, Sethi PR, Lambert NA (2008) Differential dissociation of G protein heterotrimers. Journal of Physiology 586:3325–3335

647. Lu B, Su Y, Das S, Wang H, Wang Y, Liu J, Ren D (2009) Peptide neurotransmitters activate a cation channel complex of NALCN and UNC-80. Nature 457:741–744

648. Bezbradica JS, Medzhitov R (2009) Integration of cytokine and heterologous receptor signaling pathways. Nature – Immunology 10:333–339

649. Ivashkiv LB (2009) Cross-regulation of signaling by ITAM-associated receptors. Nature Immunology 10:340–347

650. Wilson NS, Dixit V, Ashkenazi A (2009) Death receptor signal transducers: nodes of coordination in immune signaling networks. Nature Immunology 10:348–355

651. Tremblay AM, Giguère V (2007) The NR3B subgroup: an ovERRview. Nuclear Receptor Signaling 5:e009XS

652. Olefsky JM (2001) Nuclear receptor – Minireview series. Journal of Biological Chemistry 276:36863–36864

653. Rosenfeld MG, Glass CK (2001) Coregulator codes of transcriptional regulation by nuclear receptors. Journal of Biological Chemistry 276:36865–36868

654. Bin LH, Nielson LD, Liu X, Mason RJ, Shu HB (2003) Identification of uteroglobin-related protein 1 and macrophage scavenger receptor with collagenous structure as a lung-specific ligand-receptor pair. Journal of Immunology 171:924–930

655. Feng Q, Yi P, Wong J, O'Malley BW (2006) Signaling within a coactivator complex: methylation of SRC-3/AIB1 is a molecular switch for complex disassembly. Molecular and Cellular Biology 26:7846–7857

656. Yoshida H, Liu J, Samuel S, Cheng W, Rosen D, Naora H (2005) Steroid receptor coactivator-3, a homolog of Taiman that controls cell migration in the Drosophila ovary, regulates migration of human ovarian cancer cells. Molecular and Cellular Endocrinology 245:77–85

657. Long W, Yi P, Amazit L, LaMarca HL, Ashcroft F, Kumar R, Mancini MA, Tsai SY, Tsai MJ, O'Malley BW (2010) SRC-3Δ4 mediates the interaction of EGFR with FAK to promote cell migration. Molecular Cell 37:321

658. Stevens JL, Cantin GT, Wang G, Shevchenko A, Shevchenko A, Berk AJ (2002) Transcription control by E1A and MAP kinase pathway via Sur2 mediator subunit. Science 296:755–758

659. Misra P, Qi C, Yu S, Shah SH, Cao WQ, Rao MS, Thimmapaya B, Zhu Y, Reddy JK (2002) Interaction of PIMT with transcriptional coactivators CBP, p300, and PBP differential role in transcriptional regulation. Journal of Biological Chemistry 277:20011–20019

660. Calderon MR, Laperrière D, Mader S, JH White (2010) LCoR. UCSD-Nature Molecule Pages, UCSD-Nature Signaling Gateway (www.signaling-gateway.org)

661. Bagattin A, Hugendubler L, Mueller E (2010) Transcriptional coactivator PGC-1α promotes peroxisomal remodeling and biogenesis. Proceedings of the National Academy of Sciences of the United States of America 107:20376–20381

662. Teyssier C, Ma H, Emter R, Kralli A, Stallcup MR (2005) Activation of nuclear receptor coactivator PGC-1α by arginine methylation. Genes and Development 19:1466–1473

663. Strahl BD, Briggs SD, Brame CJ, Caldwell JA, Koh SS, Ma H, Cook RG, Shabanowitz J, Hunt DF, Stallcup MR, Allis CD (2001) Methylation of histone H4 at arginine 3 occurs in vivo and is mediated by the nuclear receptor coactivator PRMT1. Current Biology 11:996-1000

664. Barrero MJ, Malik S (2006) Two functional modes of a nuclear receptor-recruited arginine methyltransferase in transcriptional activation. Molecular Cell 24:233–243

665. Lei NZ, Zhang XY, Chen HZ, Wang Y, Zhan YY, Zheng ZH, Shen YM, Wu Q (2009) A feedback regulatory loop between methyltransferase PRMT1 and orphan receptor TR3. Nucleic Acids Research 37:832–848

666. Herrmann F, Fackelmayer FO (2009) Nucleo-cytoplasmic shuttling of protein arginine methyltransferase 1 (PRMT1) requires enzymatic activity. Genes to Cells 14:309–317

667. Qi C, Chang J, Zhu Y, Yeldandi AV, Rao SM, Zhu YJ (2002) Identification of protein arginine methyltransferase 2 as a coactivator for estrogen receptor α. Journal of Biological Chemistry 277:28624–28630

668. Jung DJ, Na SY, Na DS, Lee JW (2002) Molecular cloning and characterization of CAPER, a novel coactivator of activating protein-1 and estrogen receptors. Journal of Biological Chemistry 277:1229–1234.

669. Kolodkin AN, Bruggeman FJ, Plant N, Moné MJ, Bakker BM, Campbell MJ, van Leeuwen JPTM, Carlberg C, Snoep JL, Westerhoff HV (2010) Design principles of nuclear receptor signaling: how complex networking improves signal transduction. Molecular Systems Biology 6:446

670. Mendelsohn ME, Karas RH (2005) Molecular and cellular basis of cardiovascular gender differences. Science 308:1583–1587

671. Zhang Y, Zhang H, Liang J, Yu W, Shang Y (2007) SIP, a novel ankyrin repeat containing protein, sequesters steroid receptor coactivators in the cytoplasm. EMBO Journal 26:2645–2657

672. Hall JM, Couse JF, Korach KS (2001) The multifaceted mechanisms of estradiol and estrogen receptor signaling. Journal of Biological Chemistry 276:36869–36872

673. Wong WP, Tiano JP, Liu S, Hewitt SC, Le May C, Dalle S, Katzenellenbogen JA, Katzenellenbogen BS, Korach KS, Mauvais-Jarvis F (2010) Extranuclear estrogen receptor-α stimulates NeuroD1 binding to the insulin promoter and favors insulin synthesis. Proceedings of the National Academy of Sciences of the United States of America 107:13057–13062

674. Carascossa S, Dudek P, Cenni B, Briand PA, Picard D (2010) CARM1 mediates the ligand-independent and tamoxifen-resistant activation of the estrogen receptor α by cAMP. Genes and Development 24:708–719

675. Chalopin M, Tesse A, Martinez MC, Rognan D, Arnal JF, Andriantsitohaina R (2010) Estrogen receptor α as a key target of red wine polyphenols action on the endothelium. PLoS One 5(1):e8554

676. Egan KM, Lawson JA, Fries S, Koller B, Rader DJ, Smyth EM, FitzGerald GA (2004) COX-2-derived prostacyclin confers atheroprotection on female mice. Science 306:1954–1957

677. Billon-Galés A, Fontaine C, Filipe C, Douin-Echinard V, Fouque MJ, Flouriot G, Gourdy P, Lenfant F, Laurell H, Krust A, Chambon P, Arnal JF (2009) The transactivating function 1 of estrogen receptor α is dispensable for the vasculoprotective actions of 17β-estradiol. Proceedings of the National Academy of Sciences of the United States of America 106:2053–2058

678. Li HJ, Haque Z, Lu Q, Li L, Karas R, Mendelsohn M (2007) Steroid receptor coactivator 3 is a coactivator for myocardin, the regulator of smooth muscle transcription and differentiation. Proceedings of the National Academy of Sciences of the United States of America 104:4065–4070

679. Bougarne N, Paumelle R, Caron S, Hennuyer N, Mansouri R, Gervois P, Staels B, Haegeman G, De Bosscher K (2009) PPARα blocks glucocorticoid receptor α-mediated transactivation but cooperates with the activated glucocorticoid receptor α

for transrepression on NF-κB. Proceedings of the National Academy of Sciences of the United States of America 106:7397–7402

680. Kino T, Hurt DE, Ichijo T, Nader N, Chrousos GP (2010) Noncoding RNA Gas5 is a growth arrest- and starvation-associated repressor of the glucocorticoid receptor. Science Signaling 3:ra8
681. Lu TT, Repa JJ, Mangelsdorf DJ (2001) Orphan nuclear receptors as eLiXiRs and FiXeRs of sterol metabolism. Journal of Biological Chemistry 276:37735–37738
682. Fang C, Dean J, Smith JW (2007) A novel variant of ileal bile acid binding protein is up-regulated through nuclear factor-κB activation in colorectal adenocarcinoma. Cancer Research 67:9039–9046
683. Zelcer N, Hong C, Boyadjian R, Tontonoz P (2009) LXR regulates cholesterol uptake through Idol-dependent ubiquitination of the LDL receptor. Science 325:100–104
684. Marquart TJ, Allen RM, Ory DS, Baldán A (2010) miR-33 links SREBP-2 induction to repression of sterol transporters. Proceedings of the National Academy of Sciences of the United States of America 107:12228-12232
685. Rosen ED, Spiegelman BM (2001) PPARγ: a nuclear regulator of metabolism, differentiation, and cell growth. Journal of Biological Chemistry 276:37731–37734
686. Zhou J, Wang KC, Wu W, Subramaniam S, Shyy JY, Chiu JJ, Li JY, Chien S (2011) MicroRNA-21 targets peroxisome proliferators-activated receptor-α in an autoregulatory loop to modulate flow-induced endothelial inflammation. Proceedings of the National Academy of Sciences of the United States of America 108:10355–10360
687. Wen X, Li Y, Liu Y (2010) Opposite action of peroxisome proliferator-activated receptor-γ in regulating renal inflammation: functional switch by its ligand. Journal of Biological Chemistry 285:29981–29988
688. Wang D, Wang H, Guo Y, Ning W, Katkuri S, Wahli W, Desvergne B, Dey SK, DuBois RN (2006) Crosstalk between peroxisome proliferator-activated receptorδand VEGF stimulates cancer progression. Proceedings of the National Academy of Sciences of the United States of America 103:19069–19074
689. Ali F, Ali NS, Bauer A, Boyle JJ, Hamdulay SS, Haskard DO, Randi AM, Mason JC (2010) PPARδ and PGC1α act cooperatively to induce haem oxygenase-1 and enhance vascular endothelial cell resistance to stress. Cardiovascular Research 85:701–710
690. Bujold K, Rhainds D, Jossart C, Febbraio M, Marleau S, Ong H (2009) CD36-mediated cholesterol efflux is associated with PPARγ activation via a MAPK-dependent COX-2 pathway in macrophages. Cardiovascular Research 83:457–464
691. Waku T, Shiraki T, Oyama T, Maebara K, Nakamori R, Morikawa K (2010) The nuclear receptor PPARγ individually responds to serotonin- and fatty acid-metabolites. EMBO Journal 29:3395–3407
692. Germain P, Gaudon C, Pogenberg V, Sanglier S, Van Dorsselaer A, Royer CA, Lazar MA, Bourguet W, Gronemeyer H (2009) Differential action on coregulator interaction defines inverse retinoid agonists and neutral antagonists. Chemistry and Biology 16:479–489
693. le Maire A, Teyssier C, Erb C, Grimaldi M, Alvarez S, de Lera AR, Balaguer P, Gronemeyer H, Royer CA, Germain P, Bourguet W (2010) A unique secondary-structure switch controls constitutive gene repression by retinoic acid receptor. Nature Structural and Molecular Biology 17:801–807

694. Naltner A, Ghaffari M, Whitsett JA, Yan C (2000) Retinoic acid stimulation of the human surfactant protein B promoter is thyroid transcription factor 1 site-dependent. Journal of Biological Chemistry 275:56–62

695. Journiac N, Jolly S, Jarvis C, Gautheron V, Rogard M, Trembleau A, Blondeau JP, Mariani J, Vernet-der Garabedian B (2009) The nuclear receptor RORα exerts a bi-directional regulation of IL-6 in resting and reactive astrocytes. Proceedings of the National Academy of Sciences of the United States of America 106:21365–21370

696. Phelan CA, Gampe RT Jr, Lambert MH, Parks DJ, Montana V, Bynum J, Broderick TM, Hu X, Williams SP, Nolte RT, Lazar MA (2010) Structure of Rev-erbα bound to N-CoR reveals a unique mechanism of nuclear receptor–co-repressor interaction. Nature – Structural and Molecular Biology 17:808–814

697. Yin L, Joshi S, Wu N, Tong X, Lazar MA (2010) E3 ligases Arf-bp1 and Pam mediate lithium-stimulated degradation of the circadian heme receptor Rev-erb α. Proceedings of the National Academy of Sciences of the United States of America 107:11614–11619

698. Mueller M, Cima I, Noti M, Fuhrer A, Jakob S, Dubuquoy L, Schoonjans K, Brunner T (2006) The nuclear receptor LRH-1 critically regulates extra-adrenal glucocorticoid synthesis in the intestine. Journal of Experimental Medicine 203:2057–2062

699. Klett EL, Patel SB (2004) Will the real cholesterol transporter please stand up? Science 303:1149–1150

700. Altmann SW, Davis HR, Zhu LJ, Yao X, Hoos LM, Tetzloff G, Iyer SP, Maguire M, Golovko A, Zeng M, Wang L, Murgolo N, Graziano MP (2004) Niemann-Pick C1 Like 1 protein is critical for intestinal cholesterol absorption. Science 303:1201–1204

701. Vrins CL, Van der Velde AE, Van den Oever K, Levels JH, Huet S, Oude Elferink RP, Kuipers F, Groen AK (2009) PPARδ Activation leads to increased transintestinal cholesterol efflux. Journal of Lipid Research 50:2046–2054

702. van der Velde AE, Vrins CL, van den Oever K, Kunne C, Oude Elferink RP, Kuipers F, Groen AK (2007) Direct intestinal cholesterol secretion contributes significantly to total fecal neutral sterol excretion in mice. Gastroenterology 133:967–975

703. Tontonoz P, Nagy L, Alvarez JG, Thomazy VA, Evans RM (1998) PPARγ promotes monocyte/macrophage differentiation and uptake of oxidized LDL. Cell 93:241–252

704. Potter LR, Hunter T (2001) Guanylyl cyclase-linked natriuretic peptide receptors: structure and regulation. Journal of Biological Chemistry 276:6057–6060

705. Koller KJ, Goeddel DV (1992) Molecular biology of the natriuretic peptides and their receptors. Circulation 86:1081–1088

706. Portaluppi F, Bagni B, degli Uberti E, Montanari L, Cavallini R, Trasforini G, Margutti A, Ferlini M, Zanella M, Parti M (1990) Circadian rhythms of atrial natriuretic peptide, renin, aldosterone, cortisol, blood pressure and heart rate in normal and hypertensive subjects. Journal of Hypertension 8:85–95

707. Rose RA, Giles WR (2008) Natriuretic peptide C receptor signalling in the heart and vasculature. Journal of Physiology 586:353–366

708. Bryan PM, Potter LR (2002) The atrial natriuretic peptide receptor (NPR-A/GC-A) is dephosphorylated by distinct microcystin-sensitive and magnesium-dependent protein phosphatases. Journal of Biological Chemistry 277:16041–16047

709. Abbey-Hosch SE, Cody AN, Potter LR (2004) Sphingosine 1-phosphate inhibits C-type natriuretic peptide activation of guanylyl cyclase B (GC-B/NPR-B). Hypertension 43:1103–1109

710. Henesy MB, Rich TC (2009) Calcineurin mediates natriuretic peptide receptor-A desensitization in MA-10 cells. FASEB Journal 23:888.2

711. Hardman JG, Sutherland EW (1969) Guanyl cyclase, an enzyme catalyzing the formation of guanosine 3′,5′-monophosphate from guanosine trihosphate. Journal of Biological Chemistry 244:6363–6370

712. Lebuffe G, Schumacker PT, Shao ZH, Anderson T, Iwase H, Van den Hoek TL (2003) ROS and NO trigger early preconditioning: relationship to mitochondrial KATP channel. American Journal of Physiology – Heart and Circulatory Physiology 284:H299–H308

713. Fernhoff NB, Derbyshire ER, Marletta MA (2009) A nitric oxide/cysteine interaction mediates the activation of soluble guanylate cyclase. Proceedings of the National Academy of Sciences of the United States of America 106:21602–21607

714. Neo BH, Kandhi S, Wolin MS (2010) Roles for soluble guanylate cyclase and a thiol oxidation-elicited subunit dimerization of protein kinase G in pulmonary artery relaxation to hydrogen peroxide. American Journal of Physiology – Heart and Circulatory Physiology 299:H1235–H1241

715. Tresguerres M, Parks SK, Salazar E, Levin LR, Goss GG, Buck J (2010) Bicarbonate-sensing soluble adenylyl cyclase is an essential sensor for acid/base homeostasis. Proceedings of the National Academy of Sciences of the United States of America 107:442–447

716. van Kesteren CA, Danser AH, Derkx FH, Dekkers DH, Lamers JM, Saxena PR, Schalekamp MA (1997) Mannose 6-phosphate receptor-mediated internalization and activation of prorenin by cardiac cells. Hypertension 30:1389–1396

717. Admiraal PJ, van Kesteren CA, Danser AH, Derkx FH, Sluiter W, Schalekamp MA (1999) Uptake and proteolytic activation of prorenin by cultured human endothelial cells. Journal of Hypertension 17:621–629

718. Maru I, Ohta Y, Murata K, Tsukada Y (1996) Molecular cloning and identification of N-acyl-D-glucosamine 2-epimerase from porcine kidney as a renin-binding protein. Journal of Biological Chemistry 271:16294–16299

719. Nguyen G, Delarue F, Burcklé C, Bouzhir L, Giller T, Sraer JD (2002) Pivotal role of the renin/prorenin receptor in angiotensin II production and cellular responses to renin. Journal of Clinical Investigation 109:1417–1427

720. Nguyen G, Delarue F, Berrou J, Rondeau E, Sraer JD (1996) Specific receptor binding of renin on human mesangial cells in culture increases plasminogen activator inhibitor-1 antigen. Kidney International 50:1897–1903

721. Schefe JH, Menk M, Reinemund J, Effertz K, Hobbs RM, Pandolfi PP, Ruiz P, Unger T, Funke-Kaiser H (2006) A novel signal transduction cascade involving direct physical interaction of the renin/prorenin receptor with the transcription factor promyelocytic zinc finger protein. Circulation Research 99:1355–1366

722. Cousin C, Bracquart D, Contrepas A, Corvol P, Muller L, Nguyen G (2009) Soluble form of the (pro)renin receptor generated by intracellular cleavage by furin is secreted in plasma. Hypertension 53:1077–1082

723. Cruciat CM, Ohkawara B, Acebron SP, Karaulanov E, Reinhard C, Ingelfinger D, Boutros M, Niehrs C (2010) Requirement of prorenin receptor and vacuolar H^+-ATPase-mediated acidification for Wnt signaling. Science 327:459–463

724. Head GA, Mayorov DN (2006) Imidazoline receptors, novel agents and therapeutic potential. Cardiovascular and Hematological Agents in Medicinal Chemistry 4:17–32

725. Morrissey JJ, Klahr S (1997) Agmatine activation of nitric oxide synthase in endothelial cells. Proceedings of the Association of American Physicians 109:51–57

726. Joshi MS, Ferguson TB, Johnson FK, Johnson RA, Parthasarathy S, Lancaster JR (2007) Receptor-mediated activation of nitric oxide synthesis by arginine in endothelial cells. Proceedings of the National Academy of Sciences of the United States of America 104:9982–9987

727. Head GA, Mayorov DN (2006) Imidazoline receptors, novel agents and therapeutic potential. Cardiovascular and Hematological Agents in Medicinal Chemistry 4:17–32

728. Jiang SX, Zheng RY, Zeng JQ, Li XL, Han Z, Hou ST (2010) Reversible inhibition of intracellular calcium influx through NMDA receptors by imidazoline I_2 receptor antagonists. European Journal of Pharmacology 629:12–19

729. Sidenius N, Blasi F (2009) Urokinase-type plasminogen activator receptor. UCSD-Nature Molecule Pages, UCSD-Nature Signaling Gateway (www.signaling-gateway.org)

730. Smith HW, Marshall CJ (2010) Regulation of cell signalling by uPAR. Nature Reviews – Molecular Cell Biology 11:23–36

731. Kwon M, MacLeod TJ, Zhang Y, Waisman DM (2005) S100A10, annexin A2, and annexin a2 heterotetramer as candidate plasminogen receptors. Frontiers in Bioscience 10:300–325

732. MacLeod TJ, Kwon M, Filipenko NR, Waisman DM (2003) Phospholipid-associated annexin A2-S100A10 heterotetramer and its subunits: characterization of the interaction with tissue plasminogen activator, plasminogen, and plasmin. Journal of Biological Chemistry 278:25577–25584

733. Miles LA, Dahlberg CM, Levin EG, Plow EF (1989) Gangliosides interact directly with plasminogen and urokinase and may mediate binding of these fibrinolytic components to cells. Biochemistry 28:9337–9343

734. Herren T, Burke TA, Das R, Plow EF (2006) Identification of histone H2B as a regulated plasminogen receptor. Biochemistry 45:9463–9474

735. Wygrecka M, Marsh LM, Morty RE, Henneke I, Guenther A, Lohmeyer J, Markart P, Preissner KT (2009) Enolase-1 promotes plasminogen-mediated recruitment of monocytes to the acutely inflamed lung. Blood 113:5588–5598

736. Alexander SPH, Mathie A, Peters JA (2007) Guide to receptors and channels. British Journal of Pharmacology 150:S1–S168

737. Goldstein BJ, Scalia RG, Ma XL (2008) Protective vascular and myocardial effects of adiponectin. Nature – Clinical Practice – Cardiovascular Medicine 6:27–35

738. Audigier Y (2006) Apelin receptor. UCSD-Nature Molecule Pages, UCSD-Nature Signaling Gateway (www.signaling-gateway.org)

739. Szokodi I, Tavi P, Földes G, Voutilainen-Myllylä S, Ilves M, Tokola H, Pikkarainen S, Piuhola J, Rysä J, Tóth M, Ruskoaho H (2002) Apelin, the novel endogenous ligand of the orphan receptor APJ, regulates cardiac contractility. Circulation Research 91:434–440

740. Kagiyama S, Fukuhara M, Matsumura K, Lin Y, Fujii K, Iida M (2005) Central and peripheral cardiovascular actions of apelin in conscious rats. Regulatory Peptides 125:55–59

741. Tatemoto K, Hosoya M, Habata Y, Fujii R, Kakegawa T, Zou MX, Kawamata Y, Fukusumi S, Hinuma S, Kitada C, Kurokawa T, Onda H, Fujino M (1998) Isolation and characterization of a novel endogenous peptide ligand for the human APJ receptor. Biochemical and Biophysical Research Communications 251:471–476

742. Simpkin JC, Yellon DM, Davidson SM, Lim SY, Wynne AM, Smith CCT (2007) Apelin-13 and apelin-36 exhibit direct cardioprotective activity against ischemia–reperfusion injury. Basic Research in Cardiology 102:518–528

743. Sörhede Winzell M, Magnusson C, Ahrén B (2005) The apj receptor is expressed in pancreatic islets and its ligand, apelin, inhibits insulin secretion in mice. Regulatory Peptides 131:12–17

744. Goralski KB, Sinal CJ (2009) Elucidation of chemerin and chemokine-like receptor-1 function in adipocytes by adenoviral-mediated shRNA knockdown of gene expression. Methods in Enzymology 460:289–312

745. Takahashi M, Takahashi Y, Takahashi K, Zolotaryov FN, Hong KS, Kitazawa R, Iida K, Okimura Y, Kaji H, Kitazawa S, Kasuga M, Chihara K (2008) Chemerin enhances insulin signaling and potentiates insulin-stimulated glucose uptake in 3T3-L1 adipocytes. FEBS Letters 582:573–578

746. Zabel BA, Allen SJ, Kulig P, Allen JA, Cichy J, Handel TM, Butcher EC (2005) Chemerin activation by serine proteases of the coagulation, fibrinolytic, and inflammatory cascades. Journal of Biological Chemistry 280:34661–34666

747. Guillabert A, Wittamer V, Bondue B, Godot V, Imbault V, Parmentier M, Communi D (2008) Role of neutrophil proteinase 3 and mast cell chymase in chemerin proteolytic regulation. Journal of Leukocyte Biology 84:1530–1538

748. Zabel BA, Nakae S, Zúñiga L, Kim JY, Ohyama T, Alt C, Pan J, Suto H, Soler D, Allen SJ, Handel TM, Song CH, Galli SJ, Butcher EC (2008) Mast cell-expressed orphan receptor CCRL2 binds chemerin and is required for optimal induction of IgE-mediated passive cutaneous anaphylaxis. Journal of Experimental Medicine 205:2207–2220

749. Bozaoglu K, Bolton K, McMillan J, Zimmet P, Jowett J, Collier G, Walder K, Segal D (2007) Chemerin is a novel adipokine associated with obesity and metabolic syndrome. Endocrinology 148:4687–4694

750. Wittamer V, Franssen JD, Vulcano M, Mirjolet JF, Le Poul E, Migeotte I, Brézillon S, Tyldesley R, Blanpain C, Detheux M, Mantovani A, Sozzani S, Vassart G, Parmentier M, Communi D (2003) Specific recruitment of antigen-presenting cells by chemerin, a novel processed ligand from human inflammatory fluids. Journal of Experimental Medicine 198:977–985

751. You J, Yu Y, Jiang L, Li W, Yu X, Gonzalez L, Yang G, Ke Z, Li W, Li C, Liu Y (2010) Signaling through Tyr985 of leptin receptor as an age/diet-dependent switch in the regulation of energy balance. Molecular and Cellular Biology 30:1650–1659

752. Myers MG (2004) Leptin receptor signaling and the regulation of mammalian physiology. Recent Progress in Hormone Research 59:287–304

753. Bjorbak C, Lavery HJ, Bates SH, Olson RK, Davis SM, Flier JS, Myers MG (2000) SOCS3 mediates feedback inhibition of the leptin receptor via Tyr985. Journal of Biological Chemistry 275:40649–40657

754. Bence KK, Delibegovic M, Xue B, Gorgun CZ, Hotamisligil GS, Neel BG, Kahn BB (2006) Neuronal PTP1B regulates body weight, adiposity and leptin action. Nature – Medicine 12:917–924

755. Ishida-Takahashi R, Rosario F, Gong Y, Kopp K, Stancheva Z, Chen X, Feener EP, Myers MG (2006) Phosphorylation of Jak2 on Ser(523) inhibits Jak2-dependent leptin receptor signaling. Molecular and Cellular Biology 26:4063–4073

756. Yang RZ, Lee MJ, Hu H, Pray J, Wu HB, Hansen BC, Shuldiner AR, Fried SK, McLenithan JC, Gong DW (2006) Identification of omentin as a novel depot-specific adipokine in human adipose tissue: possible role in modulating insulin action. American Journal of Physiology – Endocrinology and Metabolism 290:E1253–E1261

757. Fain JN, Sacks HS, Buehrer B, Bahouth SW, Garrett E, Wolf RY, Carter RA, Tichansky DS, Madan AK (2008) Identification of omentin mRNA in human epicardial adipose tissue: comparison to omentin in subcutaneous, internal mammary artery periadventitial and visceral abdominal depots. International Journal of Obesity (London) 32:810–815

758. Lee JK, Pierce M (2009) X-Lectins: a new family with homology to the Xenopus laevis oocyte lectin XL35. In Vasta GR, Ahmed H (eds) Animal Lectins. A Functional View. CRC Press, Boca Raton, FL, USA

759. Daquinag AC, Zhang Y, Amaya-Manzanares F,Daquinag Simmons PJ, Kolonin MG (2011) An isoform of decorin is a resistin receptor on the surface of adipose progenitor cells. Cell Stem Cell 8 (in press)

760. Silbering AF, Benton R (2010) Ionotropic and metabotropic mechanisms in chemoreception: chance or design? EMBO Reports 11:173–179

761. Rivière S, Challet L, Fluegge D, Spehr M, Rodriguez I (2009) Formyl peptide receptor-like proteins are a novel family of vomeronasal chemosensors. Nature 459:574–577

Chap. 7. G-Protein-Coupled Receptors

762. Xu F, Wu H, Katritch V, Han GW, Jacobson KA, Gao ZG, Cherezov V, Stevens RC (2011) Structure of an agonist-bound human A_{2A} adenosine receptor. Science 332:322–327

763. Lebon G, Warne T, Edwards PC, Bennett K, Langmead CJ, Leslie AGW, Tate CG (2011) Agonist-bound adenosine A_{2A} receptor structures reveal common features of GPCR activation. Nature 474:521–525

764. Einstein R, Jordan H, Zhou W, Brenner M, Moses EG, Liggett SB (2008) Alternative splicing of the G protein-coupled receptor superfamily in human airway smooth muscle diversifies the complement of receptors. Proceedings of the National Academy of Sciences of the United States of America 105:5230–5235

765. Berridge MJ (2009) Module 2: Cell Signalling Pathways. Cell Signalling Biology. Biochemical Journal's Signal Knowledge Environment Portland Press Ltd., London, UK (www.biochemj.org/csb/004/csb002.pdf)

766. Tesmer JJG (2010) The quest to understand heterotrimeric G protein signaling. Nature Structural and Molecular Biology 17:650–652

767. Wettschureck N, Offermanns S (2005) Mammalian G proteins and their cell type specific functions. Physiological Reviews 85:1159–1204

768. Marinissen MJ, Gutkind JS (2001) G-protein-coupled receptors and signaling networks: emerging paradigms. Trends in Pharmacological Sciences 22:368–376

769. Zhu K, Baudhuin LM, Hong G, Williams FS, Cristina KL, Kabarowski JH, Witte ON, Xu Y (2001) Sphingosylphosphorylcholine and lysophosphatidylcholine are ligands for the G protein-coupled receptor GPR4. Journal of Biological Chemistry 276:41325–41335

770. Kim KS, Ren J, Jiang Y, Ebrahem Q, Tipps R, Cristina K, Xiao YJ, Qiao J, Taylor KL, Lum H, Anand-Apte B, Xu Y (2005) GPR4 plays a critical role in endothelial cell function and mediates the effects of sphingosylphosphorylcholine. FASEB Journal 19:819–821

771. Chang GW, Stacey M, Kwakkenbos MJ, Hamann J, Gordon S, Lin HH (2003) Proteolytic cleavage of the EMR2 receptor requires both the extracellular stalk and the GPS motif. FEBS Letters 547:145–150

772. Kuhnert F, Mancuso MR, Shamloo A, Wang HT, Choksi V, Florek M, Su H, Fruttiger M, Young WL, Heilshorn SC, Kuo CJ (2010) Essential regulation of CNS angiogenesis by the orphan G protein-coupled receptor GPR124. Science 330:985–989

773. Cullen M, Elzarrad MK, Seaman S, Zudaire E, Stevens J, Yang MY, Li X, Chaudhary A, Xu L, Hilton MB, Logsdon D, Hsiao E, Stein EV, Cuttitta F, Haines DC, Nagashima K, Tessarollo L, St Croix B (2011) GPR124, an orphan G protein-coupled receptor, is required for CNS-specific vascularization and establishment of the blood-brain barrier. Proceedings of the National Academy of Sciences of the United States of America 108:5759–5764

774. Lyden D, Young AZ, Zagzag D, Yan W, Gerald W, O'Reilly R, Bader BL, Hynes RO, Zhuang Y, Manova K, Benezra R (1999) Id1 and Id3 are required for neurogenesis, angiogenesis and vascularization of tumour xenografts. Nature 401:670–677

775. O Volpert OV, Pili R, Sikder HA, Nelius T, Zaichuk T, Morris C, Shiflett CB, Devlin MK, Conant K, Alani RM (2002) Id1 regulates angiogenesis through transcriptional repression of thrombospondin-1. Cancer Cell 2:473–483

776. Lasorella A, Rothschild G, Yokota Y, Russell RG, Iavarone A (2005) Id2 mediates tumor initiation, proliferation, and angiogenesis in Rb mutant mice. Molecular and Cellular Biology 25:3563–3574

777. Yona S, Lin HH, Dri P, Davies JQ, Hayhoe RP, Lewis SM, Heinsbroek SE, Brown KA, Perretti M, Hamann J, Treacher DF, Gordon S, Stacey M (2008) Ligation of the adhesion-GPCR EMR2 regulates human neutrophil function. FASEB Journal 22:741–751

778. Weimbs T (2006) Regulation of mTOR by polycystin-1: is polycystic kidney disease a case of futile repair? Cell Cycle 5:2425–2429

779. Parnell SC, Magenheimer BS, Maser RL, Zien CA, Frischauf AM, Calvet JP (2002) Polycystin-1 activation of c-Jun N-terminal kinase and AP-1 is mediated by heterotrimeric G proteins. Journal of Biological Chemistry 277:19566–19572

780. Bhunia AK, Piontek K, Boletta A, Liu L, Qian F, Xu PN, Germino FJ, Germino GG (2002) PKD1 induces $p21^{waf1}$ and regulation of the cell cycle via direct activation of the JAK-STAT signaling pathway in a process requiring PKD2. Cell 109:157–168

781. Qian F, Boletta A, Bhunia AK, Xu H, Liu L, Ahrabi AK, Watnick TJ, Zhou F, Germino GG (2002) Cleavage of polycystin-1 requires the receptor for egg jelly domain

and is disrupted by human autosomal-dominant polycystic kidney disease 1-associated mutations. Proceedings of the National Academy of Sciences of the United States of America 99:16981–16986

782. Lin HH, Stacey M, Yona S, Chang GW (2010) GPS Proteolytic Cleavage of Adhesion-GPCRs. In Yona S, Stacey M (Eds.) Adhesion-GPCRs: Structure to Function, Landes Bioscience, Austin, TX, and Springer, New York

783. Rikitake Y, Hirata K, Yamashita T, Iwai K, Kobayashi S, Itoh H, Ozaki M, Ejiri J, Shiomi M, Inoue N, Kawashima S, Yokoyama M (2002) Expression of G2A, a receptor for lysophosphatidylcholine, by macrophages in murine, rabbit, and human atherosclerotic plaques. Arteriosclerosis, Thrombosis, and Vascular Biology 22:2049–2053

784. Kabarowski JH, Zhu K, Le LQ, Witte ON, Xu Y (2001) Lysophosphatidylcholine as a ligand for the immunoregulatory receptor G2A. Science 293:702–705

785. Ihara Y, Kihara Y, Hamano F, Yanagida K, Morishita Y, Kunita A, Yamori T, Fukayama M, Aburatani H, Shimizu T, Ishii S (2010) The G protein-coupled receptor T-cell death-associated gene 8 (TDAG8) facilitates tumor development by serving as an extracellular pH sensor. Proceedings of the National Academy of Sciences of the United States of America 107:17309–17314

786. Fredriksson R, Lagerstrom MC, Lundin LG, Schioth HB (2003). The G-protein-coupled receptors in the human genome form five main families. Phylogenetic analysis, paralogon groups, and fingerprints. Molecular Pharmacology 63:1256–1272

787. Kroeze WK, Sheffler DJ, Roth BL (2003) G-protein-coupled receptors at a glance. Journal of Cell Science 116:4867–4869

788. Nordström KJV, Lagerström MC, Wallér LMJ, Fredriksson R, Schiöth HB (2009) The secretin GPCRs descended from the family of adhesion GPCRs. Molecular Biology and Evolution 26:71–84

789. Bjarnadóttir TK, Gloriam DE, Hellstrand SH, Kristiansson H, Fredriksson R, Schiöth HB (2006) Comprehensive repertoire and phylogenetic analysis of the G protein-coupled receptors in human and mouse. Genomics 88:263–273

790. G-protein-coupled receptor data base of EC BIOTECH FP4 program. http://www.gpcr.org

791. Ben-Chaim Y, Chanda B, Dascal N, Bezanilla F, Parnas I, Parnas H (2006) Movement of "gating charge" is coupled to ligand binding in a G-protein-coupled receptor. Nature 444:106–109

792. Oldham WM, Hamm HE (2008) Heterotrimeric G protein activation by G-protein-coupled receptors. Nature Reviews – Molecular Cell Biology 9:60–71

793. Lipsky R, Potts EM, Tarzami ST, Puckerin AA, Stocks J, Schecter AD, Sobie EA, Akar FG, Diversé-Pierluissi MA (2008) β-Adrenergic receptor activation induces internalization of cardiac $Ca_V1.2$ channel complexes through a β-arrestin-1-mediated pathway. Journal of Biological Chemistry 283:17221–17226

794. Chen MC, Wu SV, Reeve JR, Rozengurt E (2006) Bitter stimuli induce Ca^{2+} signaling and CCK release in enteroendocrine STC-1 cells: role of L-type voltage-sensitive Ca^{2+} channels. American Journal of Physiology – Cell Physiology 291:C726–C739

795. Okada M, Zhu G, Yoshida S, Iwasa H, Kaneko S (2002) [Mechanisms of interaction between adenosine receptor subtypes on hippocampal serotonin release]. Nihon Shinkei Seishin Yakurigaku Zasshi (Japanese Journal of Psychopharmacology 22:61–69

796. Yatani A, Mattera R, Codina J, Graf R, Okabe K, Padrell E, Iyengar R, Brown AM, Birnbaumer L (1988) The G protein-gated atrial K^+ channel is stimulated by three distinct Gi alpha-subunits. Nature 336:680–682

797. Bertaso F, Lill Y, Airas JM, Espeut J, Blahos J, Bockaert J, Fagni L, Betz H, El-Far O (2006) MacMARCKS interacts with the metabotropic glutamate receptor type 7 and modulates G protein-mediated constitutive inhibition of calcium channels. Journal of Neurochemistry 99:288–298

798. Delmas P, Brown DA, Dayrell M, Abogadie FC, Caulfield MP, Buckley NJ (1998) On the role of endogenous G-protein $\beta\gamma$ subunits in N-type Ca^{2+} current inhibition by neurotransmitters in rat sympathetic neurones. Journal of Physiology 506:319–329

799. Berbari NF, Lewis JS, Bishop GA, Askwith CC, Mykytyn K (2008) Bardet-Biedl syndrome proteins are required for the localization of G protein-coupled receptors to primary cilia. Proceedings of the National Academy of Sciences of the United States of America 105:4242–4246

800. Rajagopal S, Kim J, Ahn S, Craig S, Lam CM, Gerard NP, Gerard C, Lefkowitz RJ (2010) β-Arrestin- but not G protein-mediated signaling by the "decoy" receptor CXCR7. Proceedings of the National Academy of Sciences of the United States of America 107:628–632

801. Damian M, Martin A, Mesnier D, Pin JP, Banères JL (2006) Asymmetric conformational changes in a GPCR dimer controlled by G-proteins. EMBO Journal 25:5693–5702

802. Maurel D, Comps-Agrar L, Brock C, Rives ML, Bourrier E, Ayoub MA, Bazin H, Tinel N, Durroux T, Prézeau L, Trinquet E, Pin JP (2008) Cell-surface protein-protein interaction analysis with time-resolved FRET and snap-tag technologies: application to GPCR oligomerization. Nature Methods 5:561–567

803. Brock C, Oueslati N, Soler S, Boudier L, Rondard P, Pin JP (2007) Activation of a dimeric metabotropic glutamate receptor by intersubunit rearrangement. Journal of Biological Chemistry 282:33000–33008

804. Rivero-Müller A, Chou YY, Ji I, Lajic S, Hanyaloglu AC, Jonas K, Rahman N, Ji TH, Huhtaniemi I (2010) Rescue of defective G protein-coupled receptor function in vivo by intermolecular cooperation. Proceedings of the National Academy of Sciences of the United States of America 107:2319–2324

805. Maguire JJ, Davenport AP (2005) Regulation of vascular reactivity by established and emerging GPCRs. Trends in Pharmacological Sciences 26:448–454

806. Cheng X (1987) Chinese Acupuncture and Moxibustion. Foreign Language Press, Beijing

807. Yi SH (2009) Thermal properties of direct and indirect moxibustion. Journal of Acupuncture and Meridian Studies 2:273–279

808. Ulett GA, Han S, Han JS (1998) Electroacupuncture: mechanisms and clinical application. Biological Psychiatry 44:129–138

809. Zhang D, Ding G, Shen X, Yao W, Zhang Z, Zhang Y, Lin J, Gu Q (2008) Role of mast cells in acupuncture effect: a pilot study. Journal of Science and Healing 4:170–177

810. Sun Y, Huang J, Xiang Y, Bastepe M, Jüppner H, Kobilka BK, Zhang JJ, Huang XY (2007) Dosage-dependent switch from G protein-coupled to G protein-independent signaling by a GPCR. EMBO Journal 26:53–64

811. Quitterer U, Lohse MJ (1999) Crosstalk between Gα_i- and Gα_q-coupled receptors is mediated by G$\beta\gamma$ exchange. Proceedings of the National Academy of Sciences of the United States of America 96:10626–10631

812. Bhattacharya M, Babwah AV, Ferguson SSG (2004) Small GTP-binding protein-coupled receptors. Biochemical Society Transactions 32:1040–1044

813. Sah VP, Seasholtz TM, Sagi SA, Brown JH (2000) The role of Rho in G protein-coupled receptor signal transduction. Annual Review of Pharmacology and Toxicology 40:459–489

814. Crespo P, Xu N, Simonds WF, Gutkind S (1994) Ras-dependent activation of MAP kinase pathway mediated by G-protein $\beta\gamma$ subunits. Nature 369:418–420.

815. Escano CS, Keever LB, Gutweiler AA, Andresen BT (2008) Angiotensin II activates extracellular signal-regulated kinase independently of receptor tyrosine kinases in renal smooth muscle cells: implications for blood pressure regulation. Journal of Pharmacology and Experimental Therapeutics 324:34-42

816. Maudsley S, Pierce KL, Zamah AM, Miller WE, Ahn S, Daaka Y, Lefkowitz RJ, Luttrell LM (2000) The β2-adrenergic receptor mediates extracellular signal-regulated kinase activation via assembly of a multi-receptor complex with the epidermal growth factor receptor. Journal of Biological Chemistry 275:9572–9580

817. Murga C, Fukuhara S, Gutkind JS (2000) A novel role for phosphatidylinositol 3-kinase β in signaling from G protein-coupled receptors to Akt. Journal of Biological Chemistry 275:12069–12073

818. Fischer OM, Hart S, Gschwind A, Ullrich A (2003) EGFR signal transactivation in cancer cells. Biochemical Society Transactions 31:1203–1208

819. Lambert NA, Johnston CA, Cappell SD, Kuravi S, Kimple AJ, Willard FS, Siderovski DP (2010) Regulators of G-protein signaling accelerate GPCR signaling kinetics and govern sensitivity solely by accelerating GTPase activity. Proceedings of the National Academy of Sciences of the United States of America 107:7066–7071

820. Penela P, Rivas V, Salcedo A, Mayor F (2010) G protein-coupled receptor kinase 2 (GRK2) modulation and cell cycle progression. Proceedings of the National Academy of Sciences of the United States of America 107:1118–1123

821. Vinós J, Jalink K, Hardy RW, Britt SG, Zuker CS (1997) A G protein-coupled receptor phosphatase required for rhodopsin function. Science 277:687–690

822. Pitcher JA, Payne ES, Csortos C, DePaoli-Roach AA, Lefkowitz RJ (1995) The G-protein-coupled receptor phosphatase: a protein phosphatase type 2A with a distinct subcellular distribution and substrate specificity. Proceedings of the National Academy of Sciences of the United States of America 92:8343–8347

823. Tran TM, Friedman J, Baameur F, Knoll BJ, Moore RH, Clark RB (2007) Characterization of β2-adrenergic receptor dephosphorylation: Comparison with the rate of resensitization. Molecular Pharmacology 71:47–60

824. Gehret AU, Hinkle PM (2010) Importance of regions outside the cytoplasmic tail of G-protein-coupled receptors for phosphorylation and dephosphorylation. Biochemical Journal 428:235-245

825. Wu JH, Peppel K, Nelson CD, Lin FT, Kohout TA, Miller WE, Exum ST, Freedman NJ (2003) The adaptor protein β-arrestin-2 enhances endocytosis of the low density lipoprotein receptor. Journal of Biological Chemistry 278:44238–44245

826. Chen W, Kirkbride KC, How T, Nelson CD, Mo J, Frederick JP, Wang XF, Lefkowitz RJ, Blobe GC (2003) β-Arrestin 2 mediates endocytosis of type III TGF-β receptor and down-regulation of its signaling. Science 301:1394–1397
827. Rakesh K, Yoo B, Kim IM, Salazar N, Kim KS, Rockman HA (2010) β-Arrestin-biased agonism of the angiotensin receptor induced by mechanical stress. Science Signaling 3:ra46
828. Shenoy SK, Modi AS, Shukla AK, Xiao K, Berthouze M, Ahn S, Wilkinson KD, Miller WE, Lefkowitz RJ (2009) β-Arrestin-dependent signaling and trafficking of 7-transmembrane receptors is reciprocally regulated by the deubiquitinase USP33 and the E3 ligase Mdm2. Proceedings of the National Academy of Sciences of the United States of America 106:6650–6655
829. Nelson MT, Perry SJ, Regier DS, Prescott SM, Topham MK, Lefkowitz RJ (2007) Targeting of diacylglycerol degradation to M1 muscarinic receptors by β-arrestins. Science 315:663–666
830. Luttrell LM, Lefkowitz RJ (2002) The role of β-arrestins in the termination and transduction of G-protein-coupled receptor signals. Journal of Cell Science 115:455–465
831. Rajagopal K, Whalen EJ, Violin JD, Stiber JA, Rosenberg PB, Premont RT, Coffman TM, Rockman HA, Lefkowitz RJ (2006) β-Arrestin2-mediated inotropic effects of the angiotensin II type 1A receptor in isolated cardiac myocytes. Proceedings of the National Academy of Sciences of the United States of America 103:16284–16289
832. Chen W, Ren XR, Nelson CD, Barak LS, Chen JK, Beachy PA, de Sauvage F, Lefkowitz RJ (2004) Activity-dependent internalization of smoothened mediated by β-arrestin 2 and GRK2. Science 306:2257–2260
833. Chen W, ten Berge D, Brown J, Ahn S, Hu LA, Miller WE, Caron MG, Barak LS, Nusse R, Lefkowitz RJ (2003) Dishevelled 2 recruits β-arrestin 2 to mediate Wnt5A-stimulated endocytosis of Frizzled 4. Science 301:1391–1394
834. Sun J, Lin X (2008) β-Arrestin 2 is required for lysophosphatidic acid-induced NF-κB activation. Proceedings of the National Academy of Sciences of the United States of America 105:17085–17090
835. Ritter SL, Hall RA (2009) Fine-tuning of GPCR activity by receptor-interacting proteins. Nature Reviews – Molecular Cell Biology 10:819-830
836. Amisten S, Braun OO, Bengtsson A, Erlinge D (2008) Gene expression profiling for the identification of G-protein-coupled receptors in human platelets. Thrombosis Research 122:47-57
837. Kajiwara N, Sasaki T, Bradding P, Cruse G, Sagara H, Ohmori K, Saito H, Ra C, Okayama Y (2010) Activation of human mast cells through the platelet-activating factor receptor. Journal of Allergy and Clinical Immunology 125:1137–1145
838. Cohen-Armon M, Sokolovsky M (1991) Depolarization-induced changes in the muscarinic receptor in rat brain and heart are mediated by pertussis-toxin-sensitive G-proteins. Journal of Biological Chemistry 266:2595–2605
839. Ben-Chaim Y, Tour O, Dascal N, Parnas I, Parnas H (2003) The M2 muscarinic G-protein-coupled receptor is voltage-sensitive. Journal of Biological Chemistry 278:22482–22491
840. Lee CW, Han J, Bamburg JR, Han L, Lynn R, Zheng JQ (2009) Regulation of acetylcholine receptor clustering by ADF/cofilin-directed vesicular trafficking. Nature – Neuroscience 12:848–856

841. Hern JA, Baig AH, Mashanov GI, Birdsall B, Corrie JE, Lazareno S, Molloy JE, Birdsall NJ (2010) Formation and dissociation of M_1 muscarinic receptor dimers seen by total internal reflection fluorescence imaging of single molecules. Proceedings of the National Academy of Sciences of the United States of America 107:2693–2698

842. Eglen RM, Reddy H, Watson N, Challiss RA (1994) Muscarinic acetylcholine receptor subtypes in smooth muscle. Trends in Pharmacological Sciences 15:114–119

843. Murthy KS, Makhlouf GM (1997) Differential coupling of muscarinic m2 and m3 receptors to adenylyl cyclases V/VI in smooth muscle. Concurrent M2-mediated inhibition via $G\alpha i3$ and m3-mediated stimulation via $G\beta\gamma q$. Journal of Biological Chemistry 272:21317–21324

844. Ruiz de Azua I, Scarselli M, Rosemond E, Gautam D, Jou W, Gavrilova O, Ebert PJ, Levitt P, Wess J (2010) RGS4 is a negative regulator of insulin release from pancreatic β-cells in vitro and in vivo. Proceedings of the National Academy of Sciences of the United States of America 107:7999-8004

845. Kong KC, Butcher AJ, McWilliams P, Jones D, Wess J, Hamdan FF, Werry T, Rosethorne EM, Charlton SJ, Munson SE, Cragg HA, Smart AD, Tobin AB (2010) M_3-muscarinic receptor promotes insulin release via receptor phosphorylation/arrestin-dependent activation of protein kinase D1. Proceedings of the National Academy of Sciences of the United States of America 107:21181–21186

846. Poulin B, Butcher A, McWilliams P, Bourgognon JM, Pawlak R, Kong KC, Bottrill A, Mistry S, Wess J, Rosethorne EM, Charlton SJ, Tobin AB (2010) The M_3-muscarinic receptor regulates learning and memory in a receptor phosphorylation/arrestin-dependent manner. Proceedings of the National Academy of Sciences of the United States of America 107:9440–9445

847. Jeon J, Dencker D, Wörtwein G, Woldbye DP, Cui Y, Davis AA, Levey AI, Schütz G, Sager TN, Mørk A, Li C, Deng CX, Fink-Jensen A, Wess J (2010) A subpopulation of neuronal M_4 muscarinic acetylcholine receptors plays a critical role in modulating dopamine-dependent behaviors. Journal of Neuroscience 30:2396–2405

848. Bridges TM, Marlo JE, Niswender CM, Jones CK, Jadhav SB, Gentry PR, Plumley HC, Weaver CD, Conn PJ, Lindsley CW (2009) Discovery of the first highly M_5-preferring muscarinic acetylcholine receptor ligand, an M_5 positive allosteric modulator derived from a series of 5-trifluoromethoxy N-benzyl isatins. Journal of Medicinal Chemistry 52:3445–3448

849. Boison D, Chen JF, Fredholm BB (2010) Adenosine signaling and function in glial cells. Cell Death and Differentiation 17:1071–1082

850. Gebremedhin D, Weinberger B, Lourim D, Harder DR (2010) Adenosine can mediate its actions through generation of reactive oxygen species. Journal of Cerebral Blood Flow and Metabolism 30:1777–1790

851. Fredholm BB, Ijzerman AP, Jacobson KA, Klotz KN, Linden J (2001) International Union of Pharmacology. XXV. Nomenclature and classification of adenosine receptors. Pharmacological Reviews 53:527–552

852. Dhalla AK, Shryock JC, Shreeniwas R, Belardinelli L (2003) Pharmacology and therapeutic applications of A_1 adenosine receptor ligands. Current Topics in Medicinal Chemistry 3:369–385

853. Elmenhorst D, Meyer PT, Winz OH, Matusch A, Ermert J, Coenen HH, Basheer R, Haas HL, Zilles K, Bauer A (2007) Sleep deprivation increases A_1 adenosine receptor

binding in the human brain: a positron emission tomography study. Journal of Neuroscience 27:2410–2415

854. Portas CM, Thakkar M, Rainnie DG, Greene RW, McCarley RW (1997) Role of adenosine in behavioral state modulation: a microdialysis study in the freely moving cat. Neuroscience 79:225–235

855. Scislo TJ, Ichinose TK, O'Leary DS (2008) Stimulation of NTS A_1 adenosine receptors differentially resets baroreflex control of regional sympathetic outputs. American Journal of Physiology – Heart and Circulatory Physiology 294:H172–H182

856. Ahmad A, Ahmad S, Glover L, Miller SM, Shannon JM, Guo X, Franklin WA, Bridges JP, Schaack JB, Colgan SP, White CW (2009) Adenosine A_{2A} receptor is a unique angiogenic target of HIF-2α in pulmonary endothelial cells. Proceedings of the National Academy of Sciences of the United States of America 106:10684–10689

857. Jaakola VP, Griffith MT, Hanson MA, Cherezov V, Chien EYT, Lane JR, Ijzerman AP, Stevens RC (2008) The 2.6 angstrom crystal structure of a human A_{2A} adenosine receptor bound to an antagonist. Science 322:1211–1217

858. Ichinose TK, O'Leary DS, Scislo TJ (2009) Activation of NTS A_{2a} adenosine receptors differentially resets baroreflex control of renal vs. adrenal sympathetic nerve activity. American Journal of Physiology – Heart and Circulatory Physiology 296:H1058–H1068

859. Paisansathan C, Xu H, Vetri F, Hernandez M, Pelligrino DA (2010) Interactions between adenosine and K^+ channel-related pathways in the coupling of somatosensory activation and pial arteriolar dilation. American Journal of Physiology – Heart and Circulatory Physiology 299:H2009–H2017

860. Hein P, Rochais F, Hoffmann C, Dorsch S, Nikolaev VO, Engelhardt S, Berlot CH, Lohse MJ, Bünemann M (2006) GS Activation is time-limiting in initiating receptor-mediated signaling. Journal of Biological Chemistry 281:33345–33351

861. Tawfik HE, Schnermann J, Oldenburg PJ, Mustafa SJ (2005) Role of A_1 adenosine receptors in regulation of vascular tone. American Journal of Physiology – Heart and Circulatory Physiology 288:H1411–H1416

862. Nayeem MA, Mustafa SJ (2002) Protein kinase C isoforms and A_1 adenosine receptors in porcine coronary smooth muscle cells. Vascular Pharmacology 39:47–54

863. Kusano Y, Echeverry G, Miekisiak G, Kulik TB, Aronhime SN, Chen JF, Winn HR (2010) Role of adenosine A_2 receptors in regulation of cerebral blood flow during induced hypotension. Journal of Cerebral Blood Flow and Metabolism 30:808–815

864. Fenton RA, Shea LG, Doddi C, Dobson JG (2010) Myocardial adenosine A_1-receptor-mediated adenoprotection involves phospholipase C, PKC-varepsilon, and p38 MAPK, but not HSP27. American Journal of Physiology – Heart and Circulatory Physiology 298:H1671–H1678

865. Lankford AR, Yang JN, Rose'Meyer R, French BA, Matherne GP, Fredholm BB, Yang Z (2006) Effect of modulating cardiac A_1 adenosine receptor expression on protection with ischemic preconditioning. American Journal of Physiology – Heart and Circulatory Physiology 290:H1469–H1473

866. Kirchhof P, Fabritz L, Fortmuller L, Matherne GP, Lankford A, Baba HA, Schmitz W, Breithardt G, Neumann J, Boknik P (2003) Altered sinus nodal and atrioventricular nodal function in freely moving mice overexpressing the A_1 adenosine receptor. American Journal of Physiology – Heart and Circulatory Physiology 285:H145–H153

867. Funakoshi H, Chan TO, Good JC, Libonati JR, Piuhola J, Chen X, MacDonnell SM, Lee LL, Herrmann DE, Zhang J, Martini J, Palmer TM, Sanbe A, Robbins J, Houser SR, Koch WJ, Feldman AM (2006) Regulated overexpression of the A_1-adenosine receptor in mice results in adverse but reversible changes in cardiac morphology and function. Circulation 114:2240–2250

868. Dobson JG, Shea LG, Fenton RA (2008) Adenosine A_{2A} and β-adrenergic calcium transient and contractile responses in rat ventricular myocytes. American Journal of Physiology – Heart and Circulatory Physiology 295: H2364–2372

869. Chandrasekera PC, McIntosh VJ, Cao FX, Lasley RD (2010) Differential effects of adenosine A_{2a} and A_{2b} receptors on cardiac contractility. American Journal of Physiology – Heart and Circulatory Physiology 299:H2082–H2089

870. Yang Z, Linden J, Berr SS, Kron IL, Beller GA, French BA (2008) Timing of adenosine 2A receptor stimulation relative to reperfusion has differential effects on infarct size and cardiac function as assessed in mice by MRI. American Journal of Physiology – Heart and Circulatory Physiology 295:H2328–H2335

871. Yang D, Koupenova M, McCrann DJ, Kopeikina KJ, Kagan HM, Schreiber BM, Ravid K (2008) The A_{2b} adenosine receptor protects against vascular injury. Proceedings of the National Academy of Sciences of the United States of America 105:792–796

872. Wang L, Karlsson L, Moses S, Hultgardh-Nilsson A, Andersson M, Borna C, Gudbjartsson T, Jern S, Erlinge D (2002) P2 receptor expression profiles in human vascular smooth muscle and endothelial cells. Journal of Cardiovascular Pharmacology 40:841–853

873. Martinez-Pinna J, Tolhurst G, Gurung IS, Vandenberg JI, Mahaut-Smith MP (2004) Sensitivity limits for voltage control of P2Y receptor-evoked Ca^{2+} mobilization in the rat megakaryocyte. Journal of Physiology 555:61–70

874. Communi D, Janssens R, Suarez-Huerta N, Robaye B, Boeynaems JM (2000) Advances in signalling by extracellular nucleotides. The role and transduction mechanisms of P2Y receptors. Cellular Signalling 12:351–360

875. Nishida M, Ogushi M, Suda R, Toyotaka M, Saiki S, Kitajima N, Nakaya M, Kim KM, Ide T, Sato Y, Inoue K, Kurose H (2011) Heterologous down-regulation of angiotensin type 1 receptors by purinergic $P2Y_2$ receptor stimulation through S-nitrosylation of NF-κB. Proceedings of the National Academy of Sciences of the United States of America 108:6662–6667

876. Abbracchio MP, Burnstock G, Boeynaems JM, Barnard EA, Boyer JL, Kennedy C, Knight GE, Fumagalli M, Gachet C, Jacobson KA, Weisman GA (2006) International Union of Pharmacology LVIII: update on the P2Y G protein-coupled nucleotide receptors: from molecular mechanisms and pathophysiology to therapy. Pharmacological Reviews 58:281–341

877. Schenk U, Westendorf AM, Radaelli E, Casati A, Ferro M, Fumagalli M, Verderio C, Buer J, Scanziani E, Grassi F (2008) Purinergic control of T cell activation by ATP released through pannexin-1 hemichannels. Science Signaling 1:ra6

878. Chen Y, Yao Y, Sumi Y, Li A, To UK, Elkhal A, Inoue Y, Woehrle T, Zhang Q, Hauser C, Junger WG (2010) Purinergic signaling: a fundamental mechanism in neutrophil activation. Science Signaling 3:ra45

879. Burnstock G (2006) Vessel tone and remodeling. Nature – Medicine 12:16–17

880. Morris GE, Nelson CP, Everitt D, Brighton PJ, Standen NB, Challiss RA, Willets JM (2011) G protein-coupled receptor kinase 2 and arrestin2 regulate arterial smooth muscle P2Y-purinoceptor signalling. Cardiovascular Research 89:193–203

881. Harrington LS, Mitchell JA (2004) Novel role for P2X receptor activation in endothelium-dependent vasodilation. British Journal of Pharmacology 143:611–617

882. Ray FR, Huang W, Slater M, Barden JA (2002) Purinergic receptor distribution in endothelial cells in blood vessels: a basis for selection of coronary artery grafts. Atherosclerosis 162:55–61

883. Farias M, Gorman MW, Savage MV, Feigl EO (2005) Plasma ATP during exercise: possible role in regulation of coronary blood flow. American Journal of Physiology – Heart and Circulatory Physiology 288:H1586–H1590

884. Gorman MW, Rooke GA, Savage MV, Jayasekara MP, Jacobson KA, Feigl EO (2010) Adenine nucleotide control of coronary blood flow during exercise. American Journal of Physiology – Heart and Circulatory Physiology 299:H1981–H1989

885. Jin J, Kunapuli SP (1998) Coactivation of two different G protein-coupled receptors is essential for ADP-induced platelet aggregation. Proceedings of the National Academy of Sciences of the United States of America 95:8070–8074

886. Cornelissen I, Palmer D, David T, Wilsbacher L, Concengco C, Conley P, Pandey A, Coughlin SR (2010) Roles and interactions among protease-activated receptors and P2ry12 in hemostasis and thrombosis. Proceedings of the National Academy of Sciences of the United States of America 107:18605–18610

887. Shutt RH, Shen JB, Pappano AJ, Liang BT (2009) Stimulation of P2X purinergic receptors increases calcium spark frequency, but does not normalize calcium transient synchronization, in mouse cardiomyocytes from the calsequestrin model of cardiomyopathy (CSQ). Biophysical Journal 96:516a

888. Burnstock G (2006) Purinergic signalling British Journal of Pharmacology 147:S172–S187

889. Rosen ED, Spiegelman BM (2006) Adipocytes as regulators of energy balance and glucose homeostasis. Nature 444:847–853

890. Tilg H, Moschen AR (2006) Adipocytokines: mediators linking adipose tissue, inflammation and immunity. Nature Reviews – Immunology 6:772–783

891. Iwabu M, Yamauchi T, Okada-Iwabu M, Sato K, Nakagawa T, Funata M, Yamaguchi M, Namiki S, Nakayama R, Tabata M, Ogata H, Kubota N, Takamoto I, Hayashi YK, Yamauchi N, Waki H, Fukayama M, Nishino I, Tokuyama K, Ueki K, Oike Y, Ishii S, Hirose K, Shimizu T, Touhara K, Kadowaki T (2010) Adiponectin and AdipoR1 regulate PGC-1α and mitochondria by Ca^{2+} and AMPK/SIRT1. Nature 464:1313–1319

892. Sterin-Borda L, Furlan C, Orman B, Borda E (2007) Differential regulation on human skin fibroblast by α1 adrenergic receptor subtypes. Biochemical Pharmacology 74:1401–1412

893. Gibson SK, Gilman AG (2006) Giα and Gβ subunits both define selectivity of G protein activation by α2-adrenergic receptors. Proceedings of the National Academy of Sciences of the United States of America 103:212–217

894. Eason MG, Liggett SB (1995) Identification of a Gs coupling domain in the amino terminus of the third intracellular loop of the α2A-adrenergic receptor. Evidence for

distinct structural determinants that confer Gs versus Gi coupling. Journal of Biological Chemistry 270:24753–24760

895. Baek KJ, Das T, Gray C, Antar S, Murugesan G, Im MJ (1993) Evidence that the Gh protein is a signal mediator from α1-adrenoceptor to a phospholipase C. I. Identification of α1-adrenoceptor-coupled Gh family and purification of Gh7 from bovine heart. Journal of Biological Chemistry 268:27390–27397
896. Hoffman BB, Hu ZW (2000) α1-Adrenoceptors (α1-AR) and vascular smooth muscle cell growth. Prostate Supplement 9:29–33
897. Andersen GØ, Qvigstad E, Schiander I, Aass H, Osnes JB, Skomedal T (2002) α1-AR-induced positive inotropic response in heart is dependent on myosin light chain phosphorylation. American Journal of Physiology – Heart and Circulatory Physiology 283:H1471–H1480
898. Gonzalez-Cabrera PJ, Shi T, Yun J, McCune DF, Rorabaugh BR, Perez DM (2004) Differential regulation of the cell cycle by α1-adrenergic receptor subtypes. Endocrinology 145:5157–5167
899. Hague C, Uberti MA, Chen Z, Hall RA, Minneman KP (2004) Cell surface expression of α1D-adrenergic receptors is controlled by heterodimerization with α1B-adrenergic receptors. Journal of Biological Chemistry 279:15541–15549
900. Diviani D, Lattion AL, Abuin L, Staub O, Cotecchia S (2003) The adaptor complex 2 directly interacts with the α1b-adrenergic receptor and plays a role in receptor endocytosis. Journal of Biological Chemistry 278:19331–19340
901. Goyal R, Mittal A, Chu N, Zhang L, Longo LD (2010) α1-Adrenergic receptor subtype function in fetal and adult cerebral arteries. American Journal of Physiology – Heart and Circulatory Physiology 298:H1797–H1806
902. Bylund DB (2009) Adrenergic receptor α2a; adrenergic receptor α2b; adrenergic receptor α2c. UCSD-Nature Molecule Pages, UCSD-Nature Signaling Gateway (www.signaling-gateway.org)
903. Hansson O, Li DQ, Nagaraj V, Reinbothe TM, Tuncel J, Eliasson L, Groop L, Rorsman P, Salehi A, Lyssenko V, Luthman H, Renström E (2010) Overexpression of α2A-adrenergic receptors contributes to type 2 diabetes. Science 327:217–220
904. Dror RO, Arlow DH, Borhani DW, Jensen MØ, Piana S, Shaw DE (2009) Identification of two distinct inactive conformations of the β2-adrenergic receptor reconciles structural and biochemical observations. Proceedings of the National Academy of Sciences of the United States of America 106:4689–4694
905. Ponicke K, Groner F, Heinroth-Hoffmann I, Brodde OE (2006) Agonist-specific activation of the β2-adrenoceptor/Gs-protein and β2-adrenoceptor/Gi-protein pathway in adult rat ventricular cardiomyocytes. British Journal of Pharmacology 147:714–719
906. Seifert R, Wenzel-Seifert K, Arthur JM, Jose PO, Kobilka BK (2002) Efficient adenylyl cyclase activation by a β2-adrenoceptor-$G_i\alpha 2$ fusion protein. Biochemical and Biophysical Research Communications 298:824–828
907. Richter W, Day P, Agrawal R, Bruss MD, Granier S, Wang YL, Rasmussen SGF, Horner K, Wang P, Lei T, Patterson AJ, Kobilka B, Conti M (2008) Signaling from β1- and β2-adrenergic receptors is defined by differential interactions with PDE4. EMBO Journal 27:384–393
908. Fishman PH (2005) Adrenergic receptor β1. UCSD-Nature Molecule Pages, UCSD-Nature Signaling Gateway (www.signaling-gateway.org)

909. He J, Bellini M, Inuzuka H, Xu J, Xiong Y, Yang X, Castleberry AM, Hall RA (2006) Proteomic analysis of β1-adrenergic receptor interactions with PDZ scaffold proteins. Journal of Biological Chemistry 281:2820–2827

910. Gardner LA, Naren AP, Bahouth SW (2007) Assembly of an SAP97-AKAP79-cAMP-dependent protein kinase scaffold at the type 1 PSD-95/DLG/ZO1 motif of the human β1-adrenergic receptor generates a receptosome involved in receptor recycling and networking. Journal of Biological Chemistry 282:5085–5099

911. Tang Y, Hu LA, Miller WE, Ringstad N, Hall RA, Pitcher JA, DeCamilli P, Lefkowitz RJ (1999) Identification of the endophilins (SH3p4/p8/p13) as novel binding partners for the β1-adrenergic receptor. Proceedings of the National Academy of Sciences of the United States of America 96:12559–12564

912. Kobilka B (2009) Adrenergic receptor β2. UCSD-Nature Molecule Pages, UCSD-Nature Signaling Gateway (www.signaling-gateway.org)

913. Wang WC, Juan AH, Panebra A, Liggett SB (2011) MicroRNA let-7 establishes expression of β2-adrenergic receptors and dynamically down-regulates agonist-promoted down-regulation. Proceedings of the National Academy of Sciences of the United States of America 108:6246–6251

914. Cherezov V, Rosenbaum DM, Hanson MA, Rasmussen SGF, Thian FS, Kobilka TS, Choi HJ, Kuhn P, Weis WI, Kobilka BK, Stevens RC (2007) High-resolution crystal structure of an engineered human β2-adrenergic G protein-coupled receptor. Science 318:1258–1265

915. Rasmussen SGF, Choi HJ, Rosenbaum DM, Kobilka TS, Thian FS, Edwards PC, Burghammer M, Ratnala VRP, Sanishvili R, Fischetti RF, Schertler GFX, Weis WI, Kobilka BK (2007) Crystal structure of the human β2 adrenergic G-protein-coupled receptor. Nature 450:383–387

916. Wang WC, Mihlbachler KA, Brunnett AC, Liggett SB (2009) Targeted transgenesis reveals discrete attenuator functions of GRK and PKA in airway β2-adrenergic receptor physiologic signaling. Proceedings of the National Academy of Sciences of the United States of America 106:15007–15012

917. Nikolaev VO, Moshkov A, Lyon AR, Miragoli M, Novak P, Paur H, Lohse MJ, Korchev YE, Harding SE, Gorelik J (2010) β2-Adrenergic receptor redistribution in heart failure changes cAMP compartmentation. Science 327:1653–1657

918. Reiner S, Ambrosio M, Hoffmann C, Lohse MJ (2010) Differential signaling of the endogenous agonists at the β2-adrenergic receptor. Journal of Biological Chemistry 285:36188–36198

919. Tchivileva IE, Tan KS, Gambarian M, Nackley AG, Medvedev AV, Romanov S, Flood PM, Maixner W, Makarov SS, Diatchenko L (2009) Signaling pathways mediating β3-adrenergic receptor-induced production of interleukin-6 in adipocytes. Molecular Immunology 46:2256–2266

920. Fève B, Elhadri K, Quignard-Boulangé A, Pairault J (2004) Transcriptional down-regulation by insulin of the β3-adrenergic receptor expression in 3T3-F442A adipocytes: a mechanism for repressing the cAMP signaling pathway. Proceedings of the National Academy of Sciences of the United States of America 91:5677–5681

921. Skeberdis VA, Gendviliene V, Zablockaite D, Treinys R, Macianskiene R, Bogdelis A, Jurevicius J, Fischmeister R (2008) β3-Adrenergic receptor activation increases human atrial tissue contractility and stimulates the L-type Ca^{2+} current. Journal of Clinical Investigation 118:3219–3227

922. Xiao RP, Zhu W, Zheng M, Chakir K, Bond R, Lakatta EG, Cheng H (2004) Subtype-specific β-adrenoceptor signaling pathways in the heart and their potential clinical implications. Trends in Pharmacological Sciences 25:358–365

923. Zheng M, Zhu W, Han Q, Xiao RP (2005) β-adrenergic receptor subtype signaling. Pharmacology and Therapeutics 108:257–268

924. Woodcock EA (2007) Roles of α1A- and α1B-adrenoceptors in heart: insights from studies of genetically modified mice. Clinical and Experimental Pharmacology and Physiology 34:884–888

925. Sabri A, Pak E, Alcott SA, Wilson BA, Steinberg SF (2000) Coupling function of endogenous α1- and β-adrenergic receptors in mouse cardiomyocytes. Circulation Research 86:1047–1053

926. Zhou P, Zhao YT, Guo YB, Xu SM, Bai SH, Lakatta EG, Cheng H, Hao XM, Wang SQ (2009) β-Adrenergic signaling accelerates and synchronizes cardiac ryanodine receptor response to a single L-type Ca^{2+} channel. Proceedings of the National Academy of Sciences of the United States of America 106:18028–18033

927. Debrus S, Rahbani L, Marttila M, Delorme B, Paradis P, Nemer M (2005) The zinc finger-only protein Zfp260 is a novel cardiac regulator and a nuclear effector of α1-adrenergic signaling. Molecular and Cellular Biology 25:8669-8682

928. Naselsky DP, Ashton D, Ruffolo RR Jr, Hieble JP (2001) Rabbit α2-adrenoceptors: both platelets and adipocytes have α2A-pharmacology. Journal of Pharmacology and Experimental Therapeutics 298:219–225

929. Valet P, Saulnier-Blache JS (1999) Rôle métabolique et trophique des catécholamines sur le développement du tissu adipeux blanc [Metabolic and trophic role of catecholamines in the development of white adipose tissue]. Annales d'Endocrinologie (Paris) 60:167–174

930. Stich V, de Glisezinski I, Crampes F, Suljkovicova H, Galitzky J, Riviere D, Hejnova J, Lafontan M, Berlan M (1999) Activation of antilipolytic α2-adrenergic receptors by epinephrine during exercise in human adipose tissue. American Journal of Physiology 277:R1076–R1083

931. Pedersen SB, Kristensen K, Hermann PA, Katzenellenbogen JA, Richelsen B (2004) Estrogen controls lipolysis by up-regulating α2A-adrenergic receptors directly in human adipose tissue through the estrogen receptor α. Implications for the female fat distribution. Journal of Clinical Endocrinology and Metabolism 89:1869–1878

932. Nakamura J (2008) Protein kinase Cβ I interacts with the β1-adrenergic signaling pathway to attenuate lipolysis in rat adipocytes. Biochimica et Biophysica Acta 1781:277–281

933. Kumar N, Robidoux J, Daniel KW, Guzman G, Floering LM, Collins S (2007) Requirement of vimentin filament assembly for β3-adrenergic receptor activation of ERK MAP kinase and lipolysis. Journal of Biological Chemistry 282:9244–9250

934. Winder WW (1988) Role of cyclic AMP in regulation of hepatic glucose production during exercise. Medicine and Science in Sports and Exercise 20:551–559

935. Han C, Bowen WC, Michalopoulos GK, Wu T (2008) α-1 adrenergic receptor transactivates signal transducer and activator of transcription-3 (Stat3) through activation of Src and epidermal growth factor receptor (EGFR) in hepatocytes. Journal of Cellular Physiology 216:486–497

936. Kost DP, DeFrances MC, Lee CR, Michalopoulos GK (1992) Patterns of α-1-adrenergic receptor expression in regenerating and neoplastic hepatic tissue. Pathobiology 60:303–308

937. Huerta-Bahena J, Villalobos-Molina R, García-Saínz JA (1983) Roles of α1- and β-adrenergic receptors in adrenergic responsiveness of liver cells formed after partial hepatectomy. Biochimica et Biophysica Acta 763:112–119

938. Houslay MD, Baillie GS (2003) The role of ERK2 docking and phosphorylation of PDE4 cAMP phosphodiesterase isoforms in mediating cross-talk between the cAMP and ERK signalling pathways. Biochemical Society Transactions 31:1186–1190

939. Catt KJ, Mendelsohn FA, Millan MA, Aguilera G (1984) The role of angiotensin II receptors in vascular regulation. Journal of Cardiovascular Pharmacology 6:S575–S586

940. Lassegue B, Clempus RE (2003) Vascular NAD(P)H oxidases: specific features, expression, and regulation. American Journal of Physiology – Regulatory, Integrative and Comparative Physiology 285:R277–R297

941. Ushio-Fukai M, Alexander RW, Akers M, Lyons PR, Lassègue B, Griendling KK (1999) Angiotensin II receptor coupling to phospholipase D is mediated by the $\beta\gamma$ subunits of heterotrimeric G proteins in vascular smooth muscle cells. Molecular Pharmacology 55:142–149

942. Habashi JP, Doyle JJ, Holm TM, Aziz H, Schoenhoff F, Bedja D, Chen Y, Modiri AN, Judge DP, Dietz HC (2011) Angiotensin II type 2 receptor signaling attenuates aortic aneurysm in mice through ERK antagonism. Science 332:361–365

943. Hansen JL, Servant G, Baranski TJ, Fujita T, Iiri T, Sheikh SP (2000) Functional reconstitution of the angiotensin II type 2 receptor and G(i) activation. Circulation Research 87:753–759

944. Shah A, Gul R, Yuan K, Gao S, Oh YB, Kim UH, Kim SH (2010) Angiotensin-(1-7) stimulates high atrial pacing-induced ANP secretion via Mas/PI3-kinase/Akt axis and Na^+/H^+ exchanger. American Journal of Physiology – Heart and Circulatory Physiology 298:H1365–H1374

945. Gallagher PE, Ferrario CM, Tallant EA (2008) Regulation of ACE2 in cardiac myocytes and fibroblasts. American Journal of Physiology – Heart and Circulatory Physiology 295:H2373-H2379

946. Arnold AC, Isa K, Shaltout HA, Nautiyal M, Ferrario CM, Chappell MC, Diz DI (2010) Angiotensin-(1-12) requires angiotensin converting enzyme and AT1 receptors for cardiovascular actions within the solitary tract nucleus. American Journal of Physiology – Heart and Circulatory Physiology 299:H763–H771

947. Ramchandran R, Takezako T, Saad Y, Stull L, Fink B, Yamada H, Dikalov S, Harrison DG, Moravec C, Karnik SS (2006) Angiotensinergic stimulation of vascular endothelium in mice causes hypotension, bradycardia, and attenuated angiotensin response. Proceedings of the National Academy of Sciences of the United States of America 103:19087–19092

948. McEwen ST, Balus SF, Durand MJ, Lombard JH (2009) Angiotensin II maintains cerebral vascular relaxation via EGF receptor transactivation and ERK1/2. American Journal of Physiology – Heart and Circulatory Physiology 297:H1296–H1303

949. Yamazaki T, Komuro I, Yazaki Y (1999) Role of the renin-angiotensin system in cardiac hypertrophy. American Journal of Cardiology 83:53H–57H

950. Zou Y, Akazawa H, Qin Y, Sano M, Takano H, Minamino T, Makita N, Iwanaga K, Zhu W, Kudoh S, Toko H, Tamura K, Kihara M, Nagai T, Fukamizu A, Umemura S, Iiri T, Fujita T, Komuro I (2004) Mechanical stress activates angiotensin II type 1 receptor without the involvement of angiotensin II. Nature – Cell Biology 6:499-506

951. Yasuda N, Miura SI, Akazawa H, Tanaka T, Qin Y, Kiya Y, Imaizumi S, Fujino M, Ito K, Zou Y, Fukuhara S, Kunimoto S, Fukuzaki K, Sato T, Ge J, Mochizuki N, Nakaya H, Saku K, Komuro I (2008) Conformational switch of angiotensin II type 1 receptor underlying mechanical stress-induced activation. EMBO Reports 9:179–186

952. Schultz Jel J, Witt SA, Glascock BJ, Nieman ML, Reiser PJ, Nix SL, Kimball TR, Doetschman T (2002) TGF-β 1 mediates the hypertrophic cardiomyocyte growth induced by angiotensin II. Journal of Clinical Investigation 109:787–796

953. Billet S, Bardin S, Verp S, Baudrie V, Michaud A, Conchon S, Muffat-Joly M, Escoubet B, Souil E, Hamard G, Bernstein KE, Gasc JM, Elghozi JL, Corvol P, Clauser E (2007) Gain-of-function mutant of angiotensin II receptor, type 1A, causes hypertension and cardiovascular fibrosis in mice. Journal of Clinical Investigation 117:1914–1925

954. Nishida M, Tanabe S, Maruyama Y, Mangmool S, Urayama K, Nagamatsu Y, Takagahara S, Turner JH, Kozasa T, Kobayashi H, Sato Y, Kawanishi T, Inoue R, Nagao T, Kurose H (2005) Gα12/13- and reactive oxygen species-dependent activation of c-Jun NH2-terminal kinase and p38 mitogen-activated protein kinase by angiotensin receptor stimulation in rat neonatal cardiomyocytes. Journal of Biological Chemistry 280:18434–18441

955. McAllister-Lucas LM, Ruland J, Siu K, Jin X, Gu S, Kim DSL, Kuffa P, Kohrt D, Mak TW, Nuñez G, Lucas PC (2007) CARMA3/Bcl10/MALT1-dependent NF-kappaB activation mediates angiotensin II-responsive inflammatory signaling in nonimmune cells. Proceedings of the National Academy of Sciences of the United States of America 104:139–144

956. O'Dowd BF, Heiber M, Chan A, Heng HH, Tsui LC, Kennedy JL, Shi X, Petronis A, George SR, Nguyen T (1993) A human gene that shows identity with the gene encoding the angiotensin receptor is located on chromosome 11. Gene 136:355–360

957. Kawamata Y, Habata Y, Fukusumi S, Hosoya M, Fujii R, Hinuma S, Nishizawa N, Kitada C, Onda H, Nishimura O, Fujino M (2001) Molecular properties of apelin: tissue distribution and receptor binding. Biochimica et Biophysica Acta 1538:162–171

958. Croitoru-Lamoury J, Guillemin GJ, Boussin FD, Mognetti B, Gigout LI, Chéret A, Vaslin B, Le Grand R, Brew BJ, Dormont D (2003) Expression of chemokines and their receptors in human and simian astrocytes: evidence for a central role of TNFα and IFNγ in CXCR4 and CCR5 modulation. Glia 41:354–370

959. De Mota N, Reaux-Le Goazigo A, El Messari S, Chartrel N, Roesch D, Dujardin C, Kordon C, Vaudry H, Moos F, Llorens-Cortes C (2004) Apelin, a potent diuretic neuropeptide counteracting vasopressin actions through inhibition of vasopressin neuron activity and vasopressin release. Proceedings of the National Academy of Sciences of the United States of America 101:10464–10469

960. Meister B, Cortés R, Villar MJ, Schalling M, Hökfelt T (1990) Peptides and transmitter enzymes in hypothalamic magnocellular neurons after administration of hyperosmotic stimuli: comparison between messenger RNA and peptide/protein levels. Cell and Tissue Research 260:279–297

961. Tatemoto K, Takayama K, Zou MX, Kumaki I, Zhang W, Kumano K, Fujimiya M (2001) The novel peptide apelin lowers blood pressure via a nitric oxide-dependent mechanism. Regulatory Peptides 99:87–92

962. Kleinz MJ, Davenport AP (2004) Immunocytochemical localization of the endogenous vasoactive peptide apelin to human vascular and endocardial endothelial cells. Regulatory Peptides 118: 119–125

963. Maguire JJ, Kleinz MJ, Pitkin SL, Davenport AP (2009) [Pyr^1]apelin-13 identified as the predominant apelin isoform in the human heart: vasoactive mechanisms and inotropic action in disease. Hypertension 54:598–604

964. Charo DN, Ho M, Fajardo G, Kawana M, Kundu RK, Sheikh AY, Finsterbach TP, Leeper NJ, Ernst KV, Chen MM, Ho YD, Chun HJ, Bernstein D, Ashley EA, Quertermous T (2009) Endogenous regulation of cardiovascular function by apelin-APJ. American Journal of Physiology – Heart and Circulatory Physiology 297:H1904–H1913

965. Kidoya H, Ueno M, Yamada Y, Mochizuki N, Nakata M, Yano T, Fujii R, Takakura N (2008) Spatial and temporal role of the apelin/APJ system in the caliber size regulation of blood vessels during angiogenesis. EMBO Journal 27:522–534

966. Masri B, Morin N, Pedebernade L, Knibiehler B, Audigier Y (2006) The apelin receptor is coupled to Gi1 or Gi2 protein and is differentially desensitized by apelin fragments. Journal of Biological Chemistry 281:18317–18326

967. Morand-Contant M, Anand-Srivastava MB, Couture R (2010) Kinin B1 receptor upregulation by angiotensin II and endothelin-1 in rat vascular smooth muscle cells: receptors and mechanisms. American Journal of Physiology – Heart and Circulatory Physiology 299:H1625–H1632

968. Kakoki M, Sullivan KA, Backus C, Hayes JM, Oh SS, Hua K, Gasim AM, Tomita H, Grant R, Nossov SB, Kim HS, Jennette JC, Feldman EL, Smithies O (2010) Lack of both bradykinin B_1 and B_2 receptors enhances nephropathy, neuropathy, and bone mineral loss in Akita diabetic mice. Proceedings of the National Academy of Sciences of the United States of America 107:10190–10195

969. Huang C, Hepler JR, Gilman AG, Mumby SM (1997) Attenuation of Gi- and Gq-mediated signaling by expression of RGS4 or GAIP in mammalian cells. Proceedings of the National Academy of Sciences of the United States of America 94:6159–163

970. Duchene J, Schanstra JP, Pecher C, Pizard A, Susini C, Esteve JP, Bascands JL, Girolami JP (2002) A novel protein-protein interaction between a G protein-coupled receptor and the phosphatase SHP-2 is involved in bradykinin-induced inhibition of cell proliferation. Journal of Biological Chemistry 277:40375-40383

971. Ke Y, Sheehan KA, Egom EE, Lei M, Solaro RJ (2010) Novel bradykinin signaling in adult rat cardiac myocytes through activation of p21 activated kinase. American Journal of Physiology – Heart and Circulatory Physiology 298:H1283–H1289

972. Hilgers RHP, Bergaya S, Schiffers PMH, Meneton P, Boulanger CM, Henrion D, Lévy BI, De Mey JGR (2003) Uterine artery structural and functional changes during pregnancy in tissue kallikrein-deficient mice. Arteriosclerosis, Thrombosis, and Vascular Biology 23:1826–1832

973. Veelken R, Glabasnia A, Stetter A, Hilgers KF, Mann JF, Schmieder RE (1996) Epicardial bradykinin B2 receptors elicit a sympathoexcitatory reflex in rats. Hypertension 28:615–621

974. Duka A, Kintsurashvili E, Duka I, Ona D, Hopkins TA, Bader M, Gavras I, Gavras H (2008) Angiotensin-converting enzyme inhibition after experimental myocardial infarct: role of the kinin B1 and B2 receptors. Hypertension 51:1352–1357

975. Messadi-Laribi E, Griol-Charhbili V, Gaies E, Vincent MP, Heudes D, Meneton P, Alhenc-Gelas F, Richer C (2008) Cardioprotection and kallikrein-kinin system in acute myocardial ischaemia in mice. Clinical and Experimental Pharmacology and Physiology 35:489–493

976. Israel A, Diaz E (2000) Diuretic and natriuretic action of adrenomedullin administered intracerebroventricularly in conscious rats. Regulatory Peptides 89:13–18

977. Ogoshi M, Nobata S, Takei Y (2008) Potent osmoregulatory actions of homologous adrenomedullins administered peripherally and centrally in eels. American Journal of Physiology – Regulatory, Integrative and Comparative Physiology 295:R2075–R2083

978. Roh J, Chang CL, Bhalla A, Klein C, Hsu SY (2004) Intermedin is a calcitonin/calcitonin gene-related peptide family peptide acting through the calcitonin receptor-like receptor/receptor activity-modifying protein receptor complexes. Journal of Biological Chemistry 279:7264–7274

979. Takei Y, Hashimoto H, Inoue K, Osaki T, Yoshizawa-Kumagaye K, Tsunemi M, Watanabe TX, Ogoshi M, Minamino N, Ueta Y (2008) Central and peripheral cardiovascular actions of adrenomedullin 5, a novel member of the calcitonin gene-related peptide family, in mammals. Journal of Endocrinology 197:391–400

980. Canaff L, Petit JL, Kisiel M, Watson PH, Gascon-Barré M, Hendy GN (2001) Extracellular calcium-sensing receptor is expressed in rat hepatocytes. Coupling to intracellular calcium mobilization and stimulation of bile flow. Journal of Biological Chemistry 276:4070–4079

981. Feng J, Petersen CD, Coy DH, Jiang JK, Thomas CJ, Pollak MR, Wank SA (2010) Calcium-sensing receptor is a physiologic multimodal chemosensor regulating gastric G-cell growth and gastrin secretion. Proceedings of the National Academy of Sciences of the United States of America 107:17791–17796

982. Sun J, Murphy E (2010) Calcium-sensing receptor: a sensor and mediator of ischemic preconditioning in the heart. American Journal of Physiology – Heart and Circulatory Physiology 299:H1309–H1317

983. Kozak KR, Crews BC, Morrow JD, Wang LH, Ma YH, Weinander R, Jakobsson PJ, Marnett LJ (2002) Metabolism of the endocannabinoids, 2-arachidonylglycerol and anandamide, into prostaglandin, thromboxane, and prostacyclin glycerol esters and ethanolamides. Journal of Biological Chemistry 277:44877–44885

984. Gao Y, Vasilyev DV, Goncalves MB, Howell FV, Hobbs C, Reisenberg M, Shen R, Zhang MY, Strassle BW, Lu P, Mark L, Piesla MJ, Deng K, Kouranova EV, Ring RH, Whiteside GT, Bates B, Walsh FS, Williams G, Pangalos MN, Samad TA, Doherty P (2010) Loss of retrograde endocannabinoid signaling and reduced adult neurogenesis in diacylglycerol lipase knock-out mice. Journal of Neuroscience 30:2017–2024

985. Han KH, Lim S, Ryu J, Lee CW, Kim Y, Kang JH, Kang SS, Ahn YK, Park CS, Kim JJ (2009) CB_1 and CB_2 cannabinoid receptors differentially regulate the production of reactive oxygen species by macrophages. Cardiovascular Research 84:378–386

986. Hill MN, McEwen BS (2009) Endocannabinoids: the silent partner of glucocorticoids in the synapse. Proceedings of the National Academy of Sciences of the United States of America 106:4579–4580

987. Osei-Hyiaman D, DePetrillo M, Pacher P, Liu J, Radaeva S, Bátkai S, Harvey-White J, Mackie K, Offertáler L, Wang L, Kunos G (2005) Endocannabinoid activation at hepatic CB1 receptors stimulates fatty acid synthesis and contributes to diet-induced obesity. Journal of Clinical Investigation 115:1298–1305

988. Pereira JP, An J, Xu Y, Huang Y, Cyster JG (2009) Cannabinoid receptor 2 mediates the retention of immature B cells in bone marrow sinusoids. Nature – Immunology 10:403–411

989. Zidar DA, Violin JD, Whalen EJ, Lefkowitz RJ (2009) Selective engagement of G protein coupled receptor kinases (GRKs) encodes distinct functions of biased ligands. Proceedings of the National Academy of Sciences of the United States of America 106:9649–9654

990. Tchernychev B, Ren Y, Sachdev P, Janz JM, Haggis L, O'Shea A, McBride E, Looby R, Deng Q, McMurry T, Kazmi MA, Sakmar TP, Hunt S 3rd, Carlson KE (2010) Discovery of a CXCR4 agonist pepducin that mobilizes bone marrow hematopoietic cells. Proceedings of the National Academy of Sciences of the United States of America 107:22255–22259

991. Miller CJ (2010) Cholecystokinin type B receptor. UCSD-Nature Molecule Pages, UCSD-Nature Signaling Gateway (www.signaling-gateway.org)

992. Sartor DM, Verberne AJM (2010) Gastric leptin: a novel role in cardiovascular regulation. American Journal of Physiology – Heart and Circulatory Physiology 298:H406–H414

993. Nakamura T, Sapru HN (2009) Cardiovascular responses to microinjections of urocortins into the NTS: role of inotropic glutamate receptors. American Journal of Physiology – Heart and Circulatory Physiology 296:H2022–H2029

994. Huising MO, van der Meulen T, Vaughan JM, Matsumoto M, Donaldson CJ, Park H, Billestrup N, Vale WW (2010) CRFR1 is expressed on pancreatic β cells, promotes β cell proliferation, and potentiates insulin secretion in a glucose-dependent manner. Proceedings of the National Academy of Sciences of the United States of America 107:912–917

995. Missale C, Nash SR, Robinson SW, Jaber M, Caron MG (1998) Dopamine receptors: from structure to function. Physiological Reviews 78:189–225

996. Han Y, Moreira IS, Urizar E, Weinstein H, Javitch JA (2009) Allosteric communication between protomers of dopamine class A GPCR dimers modulates activation. Nature – Chemical Biology 5:688–695

997. Penna C, Rastaldo R, Mancardi D, Cappello S, Pagliaro P, Westerhof N, Losano G (2006) Effect of endothelins on the cardiovascular system. Journal of Cardiovascular Medicine 7:645–652

998. Bisotto S, Fixman ED (2001) Src-family tyrosine kinases, phosphoinositide 3-kinase and Gab1 regulate extracellular signal-regulated kinase 1 activation induced by the type A endothelin-1 G-protein-coupled receptor. Biochemical Journal 360:77–85

999. Hilal-Dandan R, Brunton LL (2011) Endothelin type B receptor. UCSD-Nature Molecule Pages, UCSD-Nature Signaling Gateway (www.signaling-gateway.org)

1000. Wendel-Wellner M, Noll T, König P, Schmeck J, Koch T, Kummer W (2002) Cellular localization of the endothelin receptor subtypes ET A and ET B in the rat heart and their differential expression in coronary arteries, veins, and capillaries. Histochemistry and Cell Biology 118:361–369

1001. Nilsson D, Wackenfors A, Gustafsson L, Ugander M, Paulsson P, Ingemansson R, Edvinsson L, Malmsjö M (2008) Endothelin receptor-mediated vasodilatation: effects of organ culture. European Journal of Pharmacology 579:233–240

1002. Louden CS, Nambi P, Pullen MA, Thomas RA, Tierney LA, Solleveld HA, Schwartz LW (2000) Endothelin receptor subtype distribution predisposes coronary arteries to damage. American Journal of Pathology 157:123–134

1003. Morris GE, Nelson CP, Standen NB, Challiss RA, Willets JM (2010) Endothelin signalling in arterial smooth muscle is tightly regulated by G protein-coupled receptor kinase 2. Cardiovascular Research 85:424–433

1004. Tsukahara H, Ende H, Magazine HI, Bahou WF, Goligorsky MS (1994) Molecular and functional characterization of the non-isopeptide-selective ETB receptor in endothelial cells. Receptor coupling to nitric oxide synthase. Journal of Biological Chemistry 269:21778–21785

1005. Salani D, Taraboletti G, Rosanò L, Di Castro V, Borsotti P, Giavazzi R, Bagnato A (2000) Endothelin-1 induces an angiogenic phenotype in cultured endothelial cells and stimulates neovascularization in vivo. American Journal of Pathology 157:1703–1711

1006. Bagnall AJ, Kelland NF, Gulliver-Sloan F, Davenport AP, Gray GA, Yanagisawa M, Webb DJ, Kotelevtsev YV (2006) Deletion of endothelial cell endothelin B receptors does not affect blood pressure or sensitivity to salt. Hypertension 48:286–293

1007. Wedgwood S, Black SM (2005) Endothelin-1 decreases endothelial NOS expression and activity through ETA receptor-mediated generation of hydrogen peroxide. American Journal of Physiology – Lung Cellular and Molecular Physiology 288:L480-L487

1008. Leite-Moreira AF, Brás-Silva C (2004) Inotropic effects of ETB receptor stimulation and their modulation by endocardial endothelium, NO, and prostaglandins. American Journal of Physiology – Heart and Circulatory Physiology 287:H1194–H1199

1009. Perreault T, Coceani F (2003) Endothelin in the perinatal circulation. Canadian Journal of Physiology and Pharmacology 81:644–653

1010. Kawamura T, Ono K, Morimoto T, Akao M, Iwai-Kanai E, Wada H, Sowa N, Kita T, Hasegawa K (2004) Endothelin-1-dependent nuclear factor of activated T lymphocyte signaling associates with transcriptional coactivator p300 in the activation of the B cell leukemia-2 promoter in cardiac myocytes. Circulation Research 94:1492–1499

1011. Yamaguchi O, Kaneshiro T, Saitoh S, Ishibashi T, Maruyama Y, Takeishi Y (2009) Regulation of coronary vascular tone via redox modulation in the α1-adrenergic-angiotensin-endothelin axis of the myocardium. American Journal of Physiology – Heart and Circulatory Physiology 296:H226–H232

1012. Buckanovich RJ, Facciabene A, Kim S, Benencia F, Sasaroli D, Balint K, Katsaros D, O'Brien-Jenkins A, Gimotty PA, Coukos G (2008) Endothelin B receptor mediates the endothelial barrier to T cell homing to tumors and disables immune therapy. Nature Medicine 14:28–36

1013. Makita T, Sucov HM, Gariepy CE, Yanagisawa M, Ginty DD (2008) Endothelins are vascular-derived axonal guidance cues for developing sympathetic neurons. Nature 452:759–763

1014. Acconcia F, Ascenzi P, Bocedi A, Spisni E, Tomasi V, Trentalance A, Visca P, Marino M (2005) Palmitoylation-dependent estrogen receptor α membrane localization: regulation by 17β-estradiol. Molecular Biology of the Cell 16:231–237

1015. Deschamps AM, Murphy E (2009) Activation of a novel estrogen receptor, GPER, is cardioprotective in male and female rats. American Journal of Physiology – Heart and Circulatory Physiology 297:H1806–H1813

1016. Bopassa JC, Eghbali M, Toro L, Stefani E (2010) A novel estrogen receptor GPER inhibits mitochondria permeability transition pore opening and protects the heart against ischemia-reperfusion injury. American Journal of Physiology – Heart and Circulatory Physiology 298:H16–H23

1017. Hill MJ, Drasar BS (1975) The normal colonic bacterial flora. Gut 16:318–323

1018. Mai V, Morris JG (2004) Colonic bacterial flora: changing understandings in the molecular age. Journal of Nutrition 134:459–464

1019. Ichimura A, Hirasawa A, Hara T, Tsujimoto G (2009) Free fatty acid receptors act as nutrient sensors to regulate energy homeostasis. Prostaglandins and Other Lipid Mediators 89:82–88

1020. Kimura I, Inoue D, Maeda T, Hara T, Ichimura A, Miyauchi S, Kobayashi M, Hirasawa A, Tsujimoto G (2011) Short-chain fatty acids and ketones directly regulate sympathetic nervous system via G protein-coupled receptor 41 (GPR41). Proceedings of the National Academy of Sciences of the United States of America 108:8030–8035

1021. Schwenk J, Metz M, Zolles G, Turecek R, Fritzius T, Bildl W, Tarusawa E, Kulik A, Unger A, Ivankova K, Seddik R, Tiao JY, Rajalu M, Trojanova J, Rohde V, Gassmann M, Schulte U, Fakler B, Bettler B (2010) Native $GABA_B$ receptors are heteromultimers with a family of auxiliary subunits. Nature 465:231–235

1022. Marcaggi P, Mutoh H, Dimitrov D, Beato M, Knöpfel T (2009) Optical measurement of mGluR1 conformational changes reveals fast activation, slow deactivation, and sensitization. Proceedings of the National Academy of Sciences of the United States of America 106:11388–11393

1023. Filosa JA, Bonev AD, Straub SV, Meredit AL, Wilkerson MK, Aldrich RW, Nelson MT (2006) Local potassium signaling couples neuronal activity to vasodilation in the brain. Nature Neuroscience 9:1397–1403

1024. Buech TR, Gudermann T (2010) Follicle stimulating hormone receptor. UCSD-Nature Molecule Pages, UCSD-Nature Signaling Gateway (www.signaling-gateway.org)

1025. Haas HL, Sergeeva OA, Selbach O (2008) Histamine in the nervous system. Physiological Reviews 88:1183–1241

1026. Hill SJ, Ganellin CR, Timmerman H, Schwartz JC, Shankley NP, Young JM, Schunack W, Levi R, Haas HL (1997) International Union of Pharmacology. XIII. Classification of histamine receptors. Pharmacological Reviews 49:253–278

1027. Damaj BB, Becerra CB, Esber HJ, Wen Y, Maghazachi AA (2007) Functional expression of H4 histamine receptor in human natural killer cells, monocytes, and dendritic cells. Journal of Immunology 179:7907–7915

1028. Takeshita K, Bacon KB, Gantner F (2004) Critical role of L-selectin and histamine H4 receptor in zymosan-induced neutrophil recruitment from the bone marrow: comparison with carrageenan. Journal of Pharmacology and Experimental Therapeutics 310:272–280

1029. Lu C, Diehl SA, Noubade R, Ledoux J, Nelson MT, Spach K, Zachary JF, Blankenhorn EP, Teuscher C (2010) Endothelial histamine H_1 receptor signaling reduces blood-brain barrier permeability and susceptibility to autoimmune encephalomyelitis. Proceedings of the National Academy of Sciences of the United States of America 107:18967–18972

1030. Ye RD (2009) FPR2/ALX. UCSD-Nature Molecule Pages, UCSD-Nature Signaling Gateway (www.signaling-gateway.org)

1031. Papayianni A, Serhan CN, Brady HR (1996) Lipoxins A4 and B4 inhibit leukotriene-stimulated interactions of human neutrophils and endothelial cells. Journal of Immunology 156:2264–22672

1032. Brink C, Dahlén SE, Drazen J, Evans JF, Hay DW, Nicosia S, Serhan CN, Shimizu T, Yokomizo T (2003) International Union of Pharmacology XXXVII. Nomenclature for leukotriene and lipoxin receptors. Pharmacological Reviews 55:195–227

1033. Qiu H, Johansson AS, Sjöström M, Wan M, Schröder O, Palmblad J, Haeggström JZ (2006) Differential induction of BLT receptor expression on human endothelial cells by lipopolysacharide, cytokines, and leukotriene B4. Proceedings of the National Academy of Sciences of the United States of America 103:6913–6918

1034. Finkensieper A, Kieser S, Bekhite MM, Richter M, Mueller JP, Graebner R, Figulla HR, Sauer H, Wartenberg M (2010) The 5-lipoxygenase pathway regulates vasculogenesis in differentiating mouse embryonic stem cells. Cardiovascular Research 86:37–44

1035. Hla T, Lee MJ, Ancellin N, Paik JH, Kluk MJ (2001) Lysophospholipids–receptor revelations. Science 294:1875–1878

1036. Yanagida K, Ishii S (2011) Non-Edg family LPA receptors: the cutting edge of LPA research. Journal of Biochemistry 150:223–232

1037. Noguchi K, Ishii S, Shimizu T (2003) Identification of p2y9/GPR23 as a novel G protein-coupled receptor for lysophosphatidic acid, structurally distant from the Edg family. Journal of Biological Chemistry 278:25600–25606

1038. Yanagida K, Masago K, Nakanishi H, Kihara Y, Hamano F, Tajima Y, Taguchi R, Shimizu T, Ishii S (2009) Identification and characterization of a novel lysophosphatidic acid receptor, p2y5/LPA6. Journal of Biological Chemistry 284:17731–17741

1039. Wang JQ, Kon J, Mogi C, Tobo M, Damirin A, Sato K, Komachi M, Malchinkhuu E, Murata N, Kimura T, Kuwabara A, Wakamatsu K, Koizumi H, Uede T, Tsujimoto G, Kurose H, Sato T, Harada A, Misawa N, Tomura H, Okajima F (2004) TDAG8 is a proton-sensing and psychosine-sensitive G-protein-coupled receptor. Journal of Biological Chemistry 279:45626–45633

1040. Shuyu E, Lai YJ, Tsukahara R, Chen CS, Fujiwara Y, Yue J, Yu JH, Guo H, Kihara A, Tigyi G, Lin FT (2009) Lysophosphatidic acid 2 receptor-mediated supramolecular complex formation regulates its antiapoptotic effect. Journal of Biological Chemistry 284:14558–14571

1041. Khandoga AL, Pandey D, Welsch U, Brandl R, Siess W (2011) GPR92/LPA_5 lysophosphatidate receptor mediates megakaryocytic cell shape change induced by human atherosclerotic plaques. Cardiovascular Research 90:157–164

1042. Lundequist A, Boyce JA (2011) LPA5 is abundantly expressed by human mast cells and important for lysophosphatidic acid induced MIP-1β release. PLoS One 6:e18192

1043. Schulze C, Smales C, Rubin LL, Staddon JM (1997) Lysophosphatidic acid increases tight junction permeability in cultured brain endothelial cells. Journal of Neurochemistry 68:991–1000

1044. Mills GB, Moolenaar WH (2003) The emerging role of lysophosphatidic acid in cancer. Nature Reviews – Cancer 3:582–591

1045. Dong X, Han S, Zylka MJ, Simon MI, Anderson DJ (2001) A diverse family of GPCRs expressed in specific subsets of nociceptive sensory neurons. Cell 106:619–632

1046. Chan LF, Webb TR, Chung TT, Meimaridou E, Cooray SN, Guasti L, Chapple JP, Egertová M, Elphick MR, Cheetham ME, Metherell LA, Clark AJ (2009) MRAP and MRAP2 are bidirectional regulators of the melanocortin receptor family. Proceedings of the National Academy of Sciences of the United States of America 106:6146–6151

1047. Sebag JA, Hinkle PM (2010) Regulation of G-protein-coupled receptor signaling: specific dominant-negative effects of melanocortin 2 receptor accessory protein 2. Science Signaling 3:ra28

1048. Decherf S, Seugnet I, Kouidhi S, Lopez-Juarez A, Clerget-Froidevaux MS, Demeneix BA (2010) Thyroid hormone exerts negative feedback on hypothalamic type 4 melanocortin receptor expression. Proceedings of the National Academy of Sciences of the United States of America 107:4471–4476

1049. Mitchell JD, Maguire JJ, Davenport AP (2009) Emerging pharmacology and physiology of neuromedin U and the structurally related peptide neuromedin S. British Journal of Pharmacology 158:87–103

1050. Chen ZY, Feng GG, Nishiwaki K, Shimada Y, Fujiwara Y, Komatsu T, Ishikawa N (2007) Possible roles of neuropeptide Y Y_3-receptor subtype in rat aortic endothelial cell proliferation under hypoxia, and its specific signal transduction. American Journal of Physiology – Heart and Circulatory Physiology 293:H959–H967

1051. Hamdy O, Nishiwaki K, Yajima M, Murakami HO, Maekawa H, Moy RT, Shimada Y, Hotta Y, Ishikawa N (2000) Presence and quantification of neuropeptide Y in pulmonary edema fluids in rats. Experimental Lung Research 26:137–147

1052. Knigge U, Kjaer A, Jørgensen H, Garbarg M, Ross C, Rouleau A, Warberg J (1994) Role of hypothalamic histaminergic neurons in mediation of ACTH and beta-endorphin responses to LPS endotoxin in vivo. Neuroendocrinology 60:243–251

1053. Yin X, Zhu YH, Xu SF (1997) Expression of preproopiomelanocortin mRNA and preprodynorphin mRNA in brain of spontaneously hypertensive rats. Zhongguo Yao Li Xue Bao – Acta Pharmacologica Sinica 18:391–394

1054. Day R, Lazure C, Basak A, Boudreault A, Limperis P, Dong W, Lindberg I (1998) Prodynorphin processing by proprotein convertase 2. Cleavage at single basic residues and enhanced processing in the presence of carboxypeptidase activity. Journal of Biological Chemistry 273:829–836

1055. Tallent M, Dichter MA, Bell GI, Reisine T (1994) The cloned κopioid receptor couples to an N-type calcium current in undifferentiated PC-12 cells. Neuroscience 63:1033–1040

1056. Henry DJ, Grandy DK, Lester HA, Davidson N, Chavkin C (1995) κ-Opioid receptors couple to inwardly rectifying potassium channels when coexpressed by Xenopus oocytes. Molecular Pharmacology 47:551–557

1057. Fichna J, Janecka A, Costentin J, Do Rego JC (2007) The endomorphin system and its evolving neurophysiological role. Pharmacological Reviews 59:88–123

1058. Mizoguchi H, Narita M, Wu H, Narita M, Suzuki T, Nagase H, Tseng LF (2000) Differential involvement of mu_1-opioid receptors in endomorphin- and β-endorphin-induced G-protein activation in the mouse pons/medulla. Neuroscience 100:835–839

1059. Malinowska B, Godlewski G, Schlicker E (2002) Function of nociceptin and opioid OP4 receptors in the regulation of the cardiovascular system. Journal of Physiology and Pharmacology 53:301–324

1060. Qiu Y, Loh HH, Law PY (2009) Opioid receptor δ. UCSD-Nature Molecule Pages, UCSD-Nature Signaling Gateway (www.signaling-gateway.org)

1061. Qiu T, Wu X, Zhang F, Clemens TL, Wan M, Cao X (2010) TGF-β type II receptor phosphorylates PTH receptor to integrate bone remodelling signalling. Nature – Cell Biology 12:224–234

1062. Urayama K, Guilini C, Messaddeq N, Hu K, Steenman M, Kurose H, Ert G, Nebigil CG (2007) The prokineticin receptor-1 (GPR73) promotes cardiomyocyte survival and angiogenesis. FASEB Journal 21:2980–2993

1063. Boulberdaa M, Urayama K, Nebigil CG (2011) Prokineticin receptor 1 (PKR1) signalling in cardiovascular and kidney functions. Cardiovascular Research 92:191–198

1064. Guilini C, Urayama K, Turkeri G, Dedeoglu DB, Kurose H, Messaddeq N, Nebigil CG (2010) Divergent roles of prokineticin receptors in the endothelial cells: angiogenesis and fenestration. American Journal of Physiology – Heart and Circulatory Physiology 298(3):H844–H852

1065. Narumiya S, Sugimoto Y, Ushikubi F (1999) Prostanoid receptors: structures, properties, and functions. Physiological Reviews 79:1193–1226

1066. Wang D, Patel VV, Ricciotti E, Zhou R, Levin MD, Gao E, Yu Z, Ferrari VA, Lu MM, Xu J, Zhang H, Hui Y, Cheng Y, Petrenko N, Yu Y, Fitzgerald GA (2009) Cardiomyocyte cyclooxygenase-2 influences cardiac rhythm and function. Proceedings of the National Academy of Sciences of the United States of America 106:7548–7552

1067. Murata T, Lin MI, Aritake K, Matsumoto S, Narumiya S, Ozaki H, Urade Y, Hori M, Sessa WC (2008) Role of prostaglandin D2 receptor DP as a suppressor of tumor hyperpermeability and angiogenesis in vivo. Proceedings of the National Academy of Sciences of the United States of America 105:20009–20014

1068. Kawano T, Anrather J, Zhou P, Park L, Wang G, Frys KA, Kunz A, Cho S, Orio M, Iadecola C (2006) Prostaglandin E2 EP1 receptors: downstream effectors of COX-2 neurotoxicity. Nature Medicine 12, 225–229

1069. Wang M, Zukas AM, Hui Y, Ricciotti E, Puré E, FitzGerald GA (2006) Deletion of microsomal prostaglandin E synthase-1 augments prostacyclin and retards atherogenesis. Proceedings of the National Academy of Sciences of the United States of America 103:14507–14512

1070. Delannoy E, Courtois A, Freund-Michel V, Leblais V, Marthan R, Muller B (2010) Hypoxia-induced hyperreactivity of pulmonary arteries: role of cyclooxygenase-2, isoprostanes, and thromboxane receptors. Cardiovascular Research 85:582–592

1071. Riise J, Nguyen CHT, Qvigstad E, Sandnes DL, Osnes JB, Skomedal T, Levy FO, Krobert KA (2008) Prostanoid F receptors elicit an inotropic effect in rat left ventricle by enhancing myosin light chain phosphorylation. Cardiovascular Research 80:407–415

1072. Macfarlane SR, Seatter MJ, Kanke T, Hunter GD, Plevin R (2001) Proteinase-activated receptors. Pharmacological Reviews 53:245–282

1073. Angiolillo DJ, Capodanno D, Goto S (2010) Platelet thrombin receptor antagonism and atherothrombosis. European Heart Journal 31:17–28

1074. Ballerio R, Brambilla M, Colnago D, Parolari A, Agrifoglio M, Camera M, Tremoli E, Mussoni L (2007) Distinct roles for PAR1- and PAR2-mediated vasomotor modulation in human arterial and venous conduits. Journal of Thrombosis and Haemostasis 5:174–180

1075. McGuire JJ, Dai J, Andrade-Gordon P, Triggle CR, Hollenberg MD (2002) Proteinase-activated receptor-2 (PAR2): vascular effects of a PAR2-derived activating peptide via a receptor different than PAR2. Journal of Pharmacology and Experimental Therapeutics 303:985–392

1076. Russo A, Soh UJ, Paing MM, Arora P, Trejo J (2009) Caveolae are required for protease-selective signaling by protease-activated receptor-1. Proceedings of the National Academy of Sciences of the United States of America 106:6393–6397

1077. Vergnolle N (1999) Proteinase-activated receptor-2-activating peptides induce leukocyte rolling, adhesion, and extravasation in vivo. Journal of Immunology 163:5064–5069

1078. Ahamed J, Versteeg HH, Kerver M, Chen VM, Mueller BM, Hogg PJ, Ruf W (2006) Disulfide isomerization switches tissue factor from coagulation to cell signaling. Proceedings of the National Academy of Sciences of the United States of America 103:13932–13937

1079. Niessen F, Schaffner F, Furlan-Freguia C, Pawlinski R, Bhattacharjee G, Chun J, Derian CK, Andrade-Gordon P, Rosen H, Ruf W (2008) Dendritic cell PAR1–S1P3 signalling couples coagulation and inflammation. Nature 452:654–658

1080. Bathgate RA, Ivell R, Sanborn BM, Sherwood OD, Summers RJ (2006) International Union of Pharmacology LVII: recommendations for the nomenclature of receptors for relaxin family peptides. Pharmacological Reviews 58:7–31

1081. Manzke T (2005) Expression and function of serotonin receptor isoforms in the respiratory system. PhD Thesis, Göttingen

1082. Dutschmann M, Waki H, Manzke T, Simms AE, Pickering AE, Richter DW, Paton JF (2009) The potency of different serotonergic agonists in counteracting opioid evoked cardiorespiratory disturbances. Philosophical Transactions of the Royal Society – London – B Biological Sciences 364:2611–2623

1083. Dong S, Roth B (2010) 5-Hydroxytryptamine receptor 1B. UCSD-Nature Molecule Pages, UCSD-Nature Signaling Gateway (www.signaling-gateway.org)

1084. Becamel C (2008) 5-Hydroxytryptamine receptor 2C. UCSD-Nature Molecule Pages, UCSD-Nature Signaling Gateway (www.signaling-gateway.org)

1085. Skoura A, Hla T (2009) Regulation of vascular physiology and pathology by the S1P2 receptor subtype. Cardiovascular Research 82:221–228

1086. Mullershausen F, Zecri F, Cetin C, Billich A, Guerini D, Seuwen K (2009) Persistent signaling induced by FTY720-phosphate is mediated by internalized $S1P_1$ receptors. Nature – Chemical Biology 5:428–434

1087. Means CK, Brown JH (2009) Sphingosine-1-phosphate receptor signalling in the heart. Cardiovascular Research 82:193–200

1088. Zhang G, Xu S, Qian Y, He P (2010) Sphingosine-1-phosphate prevents permeability increases via activation of endothelial sphingosine-1-phosphate receptor 1 in rat venules. American Journal of Physiology – Heart and Circulatory Physiology 299:H1494–H1504

1089. Lee MJ, Thangada S, Claffey KP, Ancellin N, Liu CH, Kluk M, Volpi M, Sha'afi RI, Hla T (1999) Vascular endothelial cell adherens junction assembly and morphogenesis induced by sphingosine-1-phosphate. Cell 99:301–312

1090. Paik JH, Chae SS , Lee MJ, Thangada S, Hla T (2001) Sphingosine 1-phosphate-induced endothelial cell migration requires the expression of EDG-1 and EDG-3 receptors and Rho-dependent activation of $\alpha_v\beta_3$- and β_1-containing integrins. Journal of Biological Chemistry 276:11830–11837

1091. Lee MJ, Thangada S, Paik JH, Sapkota GP, Ancellin N, Chae SS, Wu M, Morales-Ruiz M, Sessa WC, Alessi DR, Hla T (2001) Akt-mediated phosphorylation of the G protein-coupled receptor EDG-1 is required for endothelial cell chemotaxis. Molecular Cell 8:693–704

1092. Morales-Ruiz M, Lee MJ, Zöllner S, Gratton JP, Scotland R, Shiojima I, Walsh K, Hla T, Sessa WC (2001) Sphingosine 1-phosphate activates Akt, nitric oxide production, and chemotaxis through a Gi protein/phosphoinositide 3-kinase pathway in endothelial cells. Journal of Biological Chemistry 276:19672–19677

1093. Okamoto H, Takuwa N, Yokomizo T, Sugimoto N, Sakurada S, Shigematsu H, Takuwa Y (2000) Inhibitory regulation of Rac activation, membrane ruffling, and cell migration by the G protein-coupled sphingosine-1-phosphate receptor EDG5 but not EDG1 or EDG3. Molecular and Cellular Biology 20:9247–9261

1094. Rosen H, Goetz EJ (2005) Sphingosine 1-phosphate and its receptors: an autocrine and paracrine network. Nature Reviews – Immunology 5:560–570

1095. Mapp CE, Miotto D, Braccioni F, Saetta M, Turato G, Maestrelli P, Krause JE, Karpitskiy V, Boyd N, Geppetti P, Fabbri LM (2000) The distribution of neurokinin-1 and neurokinin-2 receptors in human central airways. American Journal of Respiratory and Critical Care Medicine 161:207–215

1096. Cheng JT, Hsieh-Chen SC (1988) Octopamine relaxes rabbit jejunal smooth muscle by selective activation of dopamine D1 receptors. Naunyn Schmiedebergs Archives of Pharmacology 338:373–378

1097. Chiellini G, Frascarelli S, Ghelardoni S, Carnicelli V, Tobias SC, DeBarber A, Brogioni S, Ronca-Testoni S, Cerbai E, Grandy DK, Scanlan TS, Zucchi R (2007) Cardiac effects of 3-iodothyronamine: a new aminergic system modulating cardiac function. FASEB Journal 21:1597–1608

1098. Maguire JJ, Kuc RE, Davenport AP (2000) Orphan-receptor ligand human urotensin II: receptor localization in human tissues and comparison of vasoconstrictor responses with endothelin-1. British Journal of Pharmacology 131:441–446

1099. Stirrat A, Gallagher M, Douglas SA, Ohlstein EH, Berry C, Kirk A, Richardson M, MacLean MR (2001) Potent vasodilator responses to human urotensin-II in human pulmonary and abdominal resistance arteries. American Journal of Physiology – Heart and Circulatory Physiology 280:H925–H928

1100. Gamer M, Zurowski B, Büchel C (2010) Different amygdala subregions mediate valence-related and attentional effects of oxytocin in humans. Proceedings of the National Academy of Sciences of the United States of America 107:9400–9405

1101. Rinschen MM, Yu MJ, Wang G, Boja ES, Hoffert JD, Pisitkun T, Knepper MA (2010) Quantitative phosphoproteomic analysis reveals vasopressin V2-receptor-dependent signaling pathways in renal collecting duct cells. Proceedings of the National Academy of Sciences of the United States of America 107:3882–3887

1102. Tobin VA, Hashimoto H, Wacker DW, Takayanagi Y, Langnaese K, Caquineau C, Noack J, Landgraf R, Onaka T, Leng G, Meddle SL, Engelmann M, Ludwig M (2010) An intrinsic vasopressin system in the olfactory bulb is involved in social recognition. Nature 464:413–417

1103. Usdin TB, Bonner TI, Mezey E (1994) Two receptors for vasoactive intestinal polypeptide with similar specificity and complementary distributions. Endocrinology 135:2662–2680

1104. Sreedharan SP, Huang JX, Cheung MC, Goetzl EJ (1995) Structure, expression, and chromosomal localization of the type I human vasoactive intestinal peptide receptor gene. Proceedings of the National Academy of Sciences of the United States of America 92:2939–2943

1105. Groneberg DA, Hartmann P, Dinh QT, Fischer A (2001) Expression and distribution of vasoactive intestinal polypeptide receptor $VPAC_2$ mRNA in human airways. Laboratory Investigation 81:749–755

1106. Reubi JC, Läderach U, Waser B, Gebbers JO, Robberecht P, Laissue JA (2000) Vasoactive intestinal peptide/pituitary adenylate cyclase-activating peptide receptor subtypes in human tumors and their tissues of origin. Cancer Research 60:3105–3112

Chap. 8. Receptor Protein Kinases

1107. Boudeau J, Miranda-Saavedra D, Barton GJ, Alessi DR (2006) Emerging roles of pseudokinases. Trends in Cell Biology 16:443–452

1108. Amit I, Wides R, Yarden Y (2007) Evolvable signaling networks of receptor tyrosine kinases: relevance of robustness to malignancy and to cancer therapy. Molecular Systems Biology 3:151

1109. Lochhead PA, Sibbet G, Morrice N, Cleghon V (2005) Activation-loop autophosphorylation is mediated by a novel transitional intermediate form of DYRKs. Cell 121:925–936

1110. Yoon SJ, NakayamavKI, Hikita T, Handa K, Hakomori SI (2006) Epidermal growth factor receptor tyrosine kinase is modulated by GM3 interaction with N-linked GlcNAc termini of the receptor. Proceedings of the National Academy of Sciences of the United States of America 103:18987–18991

1111. Corey SJ, Anderson SM (1999) Src-related protein tyrosine kinases in hematopoiesis. Blood 93:1–14

1112. Cabrita MA, Jäggi F, Widjaja SP, Christofori G (2006) A functional interaction between sprouty proteins and caveolin-1. Journal of Biological Chemistry 281:29201–29212

1113. Tefft D, Lee M, Smith S, Crowe DL, Bellusci S, Warburton D (2002) mSprouty2 inhibits FGF10-activated MAP kinase by differentially binding to upstream target proteins. American Journal of Physiology – Lung Cellular and Molecular Physiology 283:L700–L706

1114. Vivanco I, Rohle D, Versele M, Iwanami A, Kuga D, Oldrini B, Tanaka K, Dang J, Kubek S, Palaskas N, Hsueh T, Evans M, Mulholland D, Wolle D, Rajasekaran S, Rajasekaran A, Liau LM, Cloughesy TF, Dikic I, Brennan C, Wu H, Mischel PS, Perera T, Mellinghoff IK (2010) The phosphatase and tensin homolog regulates epidermal growth factor receptor (EGFR) inhibitor response by targeting EGFR for degradation. Proceedings of the National Academy of Sciences of the United States of America 107:6459–6464

1115. Brero A, Ramella R, Fitou A, Dati C, Alloatti G, Gallo MP, Levi R (2010) Neuregulin-1β1 rapidly modulates nitric oxide synthesis and calcium handling in rat cardiomyocytes. Cardiovascular Research 88:443–452

1116. Jones N, Dumont DJ (1999) Recruitment of Dok-R to the EGF receptor through its PTB domain is required for attenuation of Erk MAP kinase activation. Current Biology 9:1057–1060

1117. Coskun U, Grzybek M, Drechsel D, Simons K (2011) Regulation of human EGF receptor by lipids. Proceedings of the National Academy of Sciences of the United States of America 108:9044–9048

1118. Oved S, Mosesson Y, Zwang Y, Santonico E, Shtiegman K, Marmor MD, Kochupurakkal BS, Katz M, Lavi S, Cesareni G, Yarden Y (2006) Conjugation to Nedd8 instigates ubiquitylation and down-regulation of activated receptor tyrosine kinases. Journal of Biological Chemistry 281:21640–21651

1119. Frey MR, Dise RS, Edelblum KL, Polk DB (2006) p38 kinase regulates epidermal growth factor receptor downregulation and cellular migration. EMBO Journal 25:5683–5692

1120. Avraham R, Yarden Y (2011) Feedback regulation of EGFR signalling: decision making by early and delayed loops. Nature Reviews – Molecular Cell Biology 12:104–117

1121. Offermanns S, Rosenthal W (Eds.) (2008) Encyclopedia of Molecular Pharmacology, (2nd ed.) Springer, Berlin

1122. Citri A, Yarden Y (2006) EGF–ERBB signalling: towards the systems level. Nature Reviews – Molecular Cell Biology 7:505–516

1123. Nagy P, Claus J, Jovin TM, Arndt-Jovin DJ (2010) Distribution of resting and ligand-bound ErbB1 and ErbB2 receptor tyrosine kinases in living cells using number and brightness analysis. Proceedings of the National Academy of Sciences of the United States of America 107:16524–16529

1124. Zhang YW, Wang R, Liu Q, Zhang H, Liao FF, Xu H (2007) Presenilin/γ-secretase-dependent processing of β-amyloid precursor protein regulates EGF receptor expression. Proceedings of the National Academy of Sciences of the United States of America 104:10613–10618

1125. Negro A, Brar BK, Gu Y, Peterson KL, Vale W, Lee KF (2006) erbB2 is required for G protein-coupled receptor signaling in the heart. Proceedings of the National Academy of Sciences of the United States of America 103:15889–15893

1126. Hurtado A, Holmes KA, Geistlinger TR, Hutcheson IR, Nicholson RI, Brown M, Jiang J, Howat WJ, Ali S, Carroll JS (2008) Regulation of ERBB2 by oestrogen receptor PAX2 determines response to tamoxifen. Nature 456:663–666

1127. Shi F, Telesco SE, Liu Y, Radhakrishnan R, Lemmon MA (2010) ErbB3/HER3 intracellular domain is competent to bind ATP and catalyze autophosphorylation. Proceedings of the National Academy of Sciences of the United States of America 107:7692–7697

1128. Williams CC, Allison JG, Vidal GA, Burow ME, Beckman BS, Marrero L, Jones FE (2004) The ERBB4/HER4 receptor tyrosine kinase regulates gene expression by functioning as a STAT5A nuclear chaperone. Journal of Cell Biology 167:469–478

1129. Garcia RA, Vasudevan K, Buonanno A (2000) The neuregulin receptor ErbB-4 interacts with PDZ-containing proteins at neuronal synapses. Proceedings of the National Academy of Sciences of the United States of America 97:3596–3601

1130. Deribe YL, Wild P, Chandrashaker A, Curak J, Schmidt MH, Kalaidzidis Y, Milutinovic N, Kratchmarova I, Buerkle L, Fetchko MJ, Schmidt P, Kittanakom S, Brown KR, Jurisica I, Blagoev B, Zerial M, Stagljar I, Dikic I (2009) Regulation of epidermal growth factor receptor trafficking by lysine deacetylase HDAC6. Science Signaling 2:ra84

1131. Huang YL, Wu CM, Shi GY, Wu GC, Lee H, Jiang MJ, Wu HL, Yang HY (2009) Nestin serves as a prosurvival determinant that is linked to the cytoprotective effect of epidermal growth factor in rat vascular smooth muscle cells. Journal of Biochemistry 146:307–315

1132. Eden ER, White IJ, Tsapara A, Futter CE (2010) Membrane contacts between endosomes and ER provide sites for PTP1B–epidermal growth factor receptor interaction. Nature – Cell Biology 12:267–272

1133. Danglot L, Chaineau M, Dahan M, Gendron MC, Boggetto N, Perez F, Galli T (2010) Role of TI-VAMP and CD82 in EGFR cell-surface dynamics and signaling. Journal of Cell Science 123:723–735

1134. Kim HH, Sierke SL, Koland JG (1994) Epidermal growth factor-dependent association of phosphatidylinositol 3-kinase with the erbB3 gene product. Journal of Biological Chemistry 269:24747–24755

1135. Wehrman TS, Raab WJ, Casipit CL, Doyonnas R, Pomerantz JH, Blau HM (2006) A system for quantifying dynamic protein interactions defines a role for Herceptin in modulating ErbB2 interactions. Proceedings of the National Academy of Sciences of the United States of America 103:19063–19068

1136. Mund T, Pelham HR (2010) Regulation of PTEN/Akt and MAP kinase signaling pathways by the ubiquitin ligase activators Ndfip1 and Ndfip2. Proceedings of the National Academy of Sciences of the United States of America 107:11429–11434

1137. Lorentzen A, Kinkhabwala A, Rocks O, Vartak N, Bastiaens PIH (2010) Regulation of Ras localization by acylation enables a mode of intracellular signal propagation. Science Signaling 3:ra68

1138. Komurov K, Padron D, Cheng T, Roth M, Rosenblatt KP, White MA (2010) Comprehensive mapping of the human kinome to epidermal growth factor receptor signaling. Journal of Biological Chemistry 285:21134–21142

1139. Daub H, Weiss FU, Wallasch C, Ullrich A (1996) Role of transactivation of the EGF receptor in signalling by G-protein-coupled receptors. Nature 379:557–560

1140. Daub H, Wallasch C, Lankenau A, Herrlich A, Ullrich A (1997) Signal characteristics of G protein-transactivated EGF receptor. EMBO Journal 16:7032–7044

1141. Graness A, Adomeit A, Heinze R, Wetzker R, Liebmann C (1998) A novel mitogenic signaling pathway of bradykinin in the human colon carcinoma cell line SW-480 involves sequential activation of a Gq/11 protein, phosphatidylinositol 3-kinase β, and protein kinase Cepsilon. Journal of Biological Chemistry 273:32016–32022

1142. Takeda H, Matozaki T, Takada T, Noguchi T, Yamao T, Tsuda M, Ochi F, Fukunaga K, Inagaki K, Kasuga M (1999) PI 3-kinase γ and protein kinase C-ζ mediate RAS-independent activation of MAP kinase by a Gi protein-coupled receptor. EMBO Journal 18:386–395

1143. Eguchi S, Numaguchi K, Iwasaki H, Matsumoto T, Yamakawa T, Utsunomiya H, Motley ED, Kawakatsu H, Owada KM, Hirata Y, Marumo F, Inagami T (1998) Calcium-dependent epidermal growth factor receptor transactivation mediates the angiotensin II-induced mitogen-activated protein kinase activation in vascular smooth muscle cells. Journal of Biological Chemistry 273:8890–8896

1144. Holgado-Madruga M, Emlet DR, Moscatello DK, Godwin AK, Wong AJ (1996) A Grb2-associated docking protein in EGF- and insulin-receptor signalling. Nature 379:560–564

1145. Maroun CR, Holgado-Madruga M, Royal I, Naujokas MA, Fournier TM, Wong AJ, Park M (1999) The Gab1 PH domain is required for localization of Gab1 at sites of cell-cell contact and epithelial morphogenesis downstream from the met receptor tyrosine kinase. Molecular and Cellular Biology 19:1784–1799

1146. Asakura M, Kitakaze M, Takashima S, Liao Y, Ishikura F, Yoshinaka T, Ohmoto H, Node K, Yoshino K, Ishiguro H, Asanuma H, Sanada S, Matsumura Y, Takeda H, Beppu S, Tada M, Hori M, Higashiyama S (2002) Cardiac hypertrophy is inhibited by antagonism of ADAM12 processing of HB-EGF: metalloproteinase inhibitors as a new therapy. Nature Medicine 8:35–40

1147. Schmidt MH, Bicker F, Nikolic I, Meister J, Babuke T, Picuric S, Müller-Esterl W, Plate KH, Dikic I (2009) Epidermal growth factor-like domain 7 (EGFL7) modulates Notch signalling and affects neural stem cell renewal. Nature – Cell Biology 11:873–880

1148. Li Y, Lévesque LO, Anand-Srivastava MB (2010) Epidermal growth factor receptor transactivation by endogenous vasoactive peptides contributes to hyperproliferation of vascular smooth muscle cells of SHR. American Journal of Physiology – Heart and Circulatory Physiology 299:H1959–H1967

1149. Scapoli L, Ramos-Nino ME, Martinelli M, Mossman BT (2004) Src-dependent ERK5 and Src/EGFR-dependent ERK1/2 activation is required for cell proliferation by asbestos. Oncogene 23:805–813

1150. White GE, Tan TC, John AE, Whatling C, McPheat WL, Greaves DR (2010) Fractalkine has anti-apoptotic and proliferative effects on human vascular smooth muscle cells via epidermal growth factor receptor signalling. Cardiovascular Research 85:825–835

1151. Ghosh P, Beas AO, Bornheimer SJ, Garcia-Marcos M, Forry EP, Johannson C, Ear J, Jung BH, Cabrera B, Carethers JM, Farquhar MG (2010) A Gαi-GIV molecular complex binds epidermal growth factor receptor and determines whether cells migrate or proliferate. Molecular Biology of the Cell 21:2338–2354

1152. Chellaiah AT, McEwen DG, Werner S, Xu J, Ornitz DM (1994) Fibroblast growth factor receptor (FGFR) 3. Alternative splicing in immunoglobulin-like domain III creates a receptor highly specific for acidic FGF/FGF-1. Journal of Biological Chemistry 269:11620–11627

1153. M Groth C, Lardelli M (2002) The structure and function of vertebrate fibroblast growth factor receptor 1. International Journal of Developmental Biology 46:393–400

1154. Che J, Okigaki M, Takahashi T, Katsume A, Adachi Y, Yamaguchi S, Matsunaga S, Takeda M, Matsui A, Kishita E, Ikeda K, Yamada H, Matsubara H (2011) Endothelial FGF receptor signaling accelerates atherosclerosis. American Journal of Physiology – Heart and Circulatory Physiology 300:H154–H161

1155. Luco RF, Pan Q, Tominaga K, Blencowe BJ, Pereira-Smith OM, Misteli T (2010) Regulation of alternative splicing by histone modifications. Science 327:996–1000

1156. Sugiyama N, Varjosalo M, Meller P, Lohi J, Chan KM, Zhou Z, Alitalo K, Taipale J, Keski-Oja J, Lehti K (2010) FGF receptor-4 (FGFR4) polymorphism acts as an activity switch of a membrane type 1 matrix metalloproteinase-FGFR4 complex. Proceedings of the National Academy of Sciences of the United States of America 107:15786–15791

1157. Wu X, Ge H, Lemon B, Weiszmann J, Gupte J, Hawkins N, Li X, Tang J, Lindberg R, Li Y (2009) Selective activation of FGFR4 by an FGF19 variant does not improve glucose metabolism in ob/ob mice. Proceedings of the National Academy of Sciences of the United States of America 106:14379–14384

1158. Chau MD, Gao J, Yang Q, Wu Z, Gromada J (2010) Fibroblast growth factor 21 regulates energy metabolism by activating the AMPK-SIRT1-PGC-1α pathway. Proceedings of the National Academy of Sciences of the United States of America 107:12553–12558

1159. Wu X, Ge H, Gupte J, Weiszmann J, Shimamoto G, Stevens J, Hawkins N, Lemon B, Shen W, Xu J, Veniant MM, Li YS, Lindberg R, Chen JL, Tian H, Li Y (2007) Co-receptor requirements for fibroblast growth factor-19 signaling. Journal of Biological Chemistry 282:29069–29072

1160. Urakawa I, Yamazaki Y, Shimada T, Iijima K, Hasegawa H, Okawa K, Fujita T, Fukumoto S, Yamashita T (2006) Klotho converts canonical FGF receptor into a specific receptor for FGF23. Nature 444:770–774

1161. Harada M, Murakami H, Okawa A, Okimoto N, Hiraoka S, Nakahara T, Akasaka R, Shiraishi YI, Futatsugi N, Mizutani-Koseki Y, Kuroiwa A, Shirouzu M, Yokoyama S, Taiji M, Iseki S, Ornitz DM, Koseki H (2009) FGF9 monomer–dimer equilibrium regulates extracellular matrix affinity and tissue diffusion. Nature – Genetics 41:289–298

1162. Sandilands E, Akbarzadeh S, Vecchione A, McEwan DG, Frame MC, Heath (2007) Src kinase modulates the activation, transport and signalling dynamics of fibroblast growth factor receptors. EMBO Reports 8:1162–1169

1163. Gibby KA, McDonnell K, Schmidt MO, Wellstein A (2009) A distinct role for secreted fibroblast growth factor-binding proteins in development. Proceedings of the National Academy of Sciences of the United States of America 106:8585–8590

1164. Trusolino L, Bertotti A, Comoglio PM (2010) MET signalling: principles and functions in development, organ regeneration and cancer. Nature Reviews – Molecular Cell Biology 11:834–848

1165. Rubin JS, Bottaro DP (2011) HGF UCSD-Nature Molecule Pages, UCSD-Nature Signaling Gateway (www.signaling-gateway.org)

1166. Miyazawa K, Shimomura T, Kitamura N (1996) Activation of hepatocyte growth factor in the injured tissues is mediated by hepatocyte growth factor activator. Journal of Biological Chemistry 271:3615–3618

1167. Kermorgant S, Parker PJ (2008) Receptor trafficking controls weak signal delivery: a strategy used by c-Met for STAT3 nuclear accumulation. Journal of Cell Biology 182:855–863

1168. Palamidessi A, Frittoli E, Garré M, Faretta M, Mione M, Testa I, Diaspro A, Lanzetti L, Scita G, Di Fiore PP (2008) Endocytic trafficking of Rac is required for the spatial restriction of signaling in cell migration. Cell 134:135–147

1169. Humbel RE (1990) Insulin-like growth factors I and II. European Journal of Biochemistry 190:445–462

1170. Stiles AD, D'Ercole AJ (199O) The insulin-like growth factors and the lung. American Journal of Physiology – Lung Cellular and Molecular Physiology 3:93–100

1171. Depetris RS, Wu J, Hubbard SR (2009) Structural and functional studies of the Ras-associating and pleckstrin-homology domains of GRB10 and GRB14. Nature – Structural and Molecular Biology 16:833–839

1172. Benyoucef S, Surinya KH, Hadaschik D, Siddle K (2007) Characterization of insulin/IGF hybrid receptors: contributions of the insulin receptor L2 and Fn1 domains and the alternatively spliced exon 11 sequence to ligand binding and receptor activation. Biochemical Journal 403:603–613

1173. Sehat B, Tofigh A, Lin Y, Trocmé E, Liljedahl U, Lagergren J, Larsson O (2010) Sumoylation mediates the nuclear translocation and signaling of the IGF-1 receptor. Science Signaling 3:ra10

1174. Weber A, Huesken C, Bergmann E, Kiess W, Christiansen NM, Christiansen H (2003) Coexpression of insulin receptor-related receptor and insulin-like growth factor 1 receptor correlates with enhanced apoptosis and dedifferentiation in human neuroblastomas. Clinical Cancer Research 9:5683–5692

1175. Dissen GA, Garcia-Rudaz C, Tapia V, Parada LF, Hsu SY, Ojeda SR (2006) Expression of the insulin receptor-related receptor is induced by the preovulatory surge of luteinizing hormone in thecal-interstitial cells of the rat ovary. Endocrinology 147:155–165

1176. Gaudino G, Follenzi A, Naldini L, Collesi C, Santoro M, Gallo KA, Godowski PJ, Comoglio PM (1994) RON is a heterodimeric tyrosine kinase receptor activated by the HGF homologue MSP. EMBO Journal 13:3524–3532

1177. Sakamoto O, Iwama A, Amitani R, Takehara T, Yamaguchi N, Yamamoto T, Masuyama K, Yamanaka T, Ando M, Suda T (1997) Role of macrophage-stimulating protein and its receptor, RON tyrosine kinase, in ciliary motility. Journal of Clinical Investigation 99:701–709

1178. Li BQ, Wang MH, Kung HF, Ronsin C, Breathnach R, Leonard EJ, Kamata T (1995) Macrophage-stimulating protein activates Ras by both activation and translocation of SOS nucleotide exchange factor. Biochemical and Biophysical Research Communications 216:110–118

1179. Bretscher A, Edwards KE, Fehon RG (2002) ERM proteins and merlin: integrators at the cell cortex. Nature Reviews – Molecular Cell Biology 3:586–599

1180. Lange S, Heger J, Euler G, Wartenberg M, Piper HM, Sauer H (2009) Platelet-derived growth factor BB stimulates vasculogenesis of embryonic stem cell-derived endothelial cells by calcium-mediated generation of reactive oxygen species. Cardiovascular Research 81:159–168

1181. Olsson AK, Dimberg A, Kreuger J, Claesson-Welsh L (2006) VEGF receptor signalling in control of vascular function Nature Reviews – Molecular Cell Biology 7:359–371

1182. He Y, Zhang H, Yu L, Gunel M, Boggon TJ, Chen H, Min W, (2010) Stabilization of VEGFR2 signaling by cerebral cavernous malformation 3 is critical for vascular development. Science Signaling 3:ra26

1183. Nilsson I, Bahram F, Li X, Gualandi L, Koch S, Jarvius M, Söderberg O, Anisimov A, Kholová I, Pytowski B, Baldwin M, Ylä-Herttuala S, Alitalo K, Kreuger J, Claesson-Welsh L (2010) VEGF receptor 2/-3 heterodimers detected in situ by proximity ligation on angiogenic sprouts. EMBO Journal 29:1377–1388

1184. Hagberg CE, Falkevall A, Wang X, Larsson E, Huusko J, Nilsson I, van Meeteren LA, Samen E, Lu L, Vanwildemeersch M, Klar J, Genove G, Pietras K, Stone-Elander S, Claesson-Welsh L, Ylä-Herttuala S, Lindahl P, Eriksson U (2010) Vascular endothelial growth factor B controls endothelial fatty acid uptake. Nature 464:917–921

1185. Zhang L, Zhou F, Han W, Shen B, Luo J, Shibuya M, He Y (2010) VEGFR-3 ligand-binding and kinase activity are required for lymphangiogenesis but not for angiogenesis. Cell Research 20:1319–1331

1186. Broudy VC, Lin NL, Liles WC, Corey SJ, O'Laughlin B, Mou S, Linnekin D (1999) Signaling via Src family kinases is required for normal internalization of the receptor c-Kit. Blood 94:1979–1986

1187. Akin C (2011) Kit. UCSD-Nature Molecule Pages, UCSD-Nature Signaling Gateway (www.signaling-gateway.org)

1188. Bourette RP, De Sepulveda P, Arnaud S, Dubreuil P, Rottapel R, Mouchiroud G (2001) Suppressor of cytokine signaling 1 interacts with the macrophage colony-stimulating factor receptor and negatively regulates its proliferation signal. Journal of Biological Chemistry 276:22133–22139

1189. Piechotta K, Garbarini N, England R, Delpire E (2003) Characterization of the interaction of the stress kinase SPAK with the Na^+–K^+–$2Cl^-$ cotransporter in the nervous system: evidence for a scaffolding role of the kinase. Journal of Biological Chemistry 278:52848–52856

1190. Raghunath M, Patti R, Bannerman P, Lee CM, Baker S, Sutton LN, Phillips PC, Damodar Reddy C (2000) A novel kinase, AATYK induces and promotes neuronal differentiation in a human neuroblastoma (SH-SY5Y) cell line. Brain Research – Molecular Brain Research 77:151–162

1191. Honma N, Asada A, Takeshita S, Enomoto M, Yamakawa E, Tsutsumi K, Saito T, Satoh T, Itoh H, Kaziro Y, Kishimoto T, Hisanaga S (2003) Apoptosis-associated tyrosine kinase is a Cdk5 activator p35 binding protein. Biochemical and Biophysical Research Communications 310:398–404

1192. Kesavapany S, Lau KF, Ackerley S, Banner SJ, Shemilt SJ, Cooper JD, Leigh PN, Shaw CE, McLoughlin DM, Miller CC (2003) Identification of a novel, membrane-associated neuronal kinase, cyclin-dependent kinase 5/p35-regulated kinase. Journal of Neuroscience 23:4975–4983

1193. Nagata K, Ohashi K, Nakano T, Arita H, Zong C, Hanafusa H, Mizuno K (1996) Identification of the product of growth arrest-specific gene 6 as a common ligand for Axl, Sky, and Mer receptor tyrosine kinases. Journal of Biological Chemistry 271:30022–30027

1194. Angelillo-Scherrer A, de Frutos P, Aparicio C, Melis E, Savi P, Lupu F, Arnout J, Dewerchin M, Hoylaerts M, Herbert J, Collen D, Dahlback B, Carmeliet P (2001) Deficiency or inhibition of Gas6 causes platelet dysfunction and protects mice against thrombosis. Nature – Medicine 7:215–221.

1195. Sasaki T, Knyazev PG, Clout NJ, Cheburkin Y, Gohring W, Ullrich A, Timpl R, Hohenester E (2006) Structural basis for Gas6–Axl signalling. EMBO Journal 25:80–87

1196. Vogel WF, Abdulhussein R, Ford CE (2006) Sensing extracellular matrix: an update on discoidin domain receptor function. Cellular Signalling 18:1108–1116

1197. Abdulhussein R, McFadden C, Fuentes-Prior P, Vogel WF (2004) Exploring the collagen-binding site of the DDR1 tyrosine kinase receptor. Journal of Biological Chemistry 279:31462–31470

1198. Vogel WF (2001) Collagen-receptor signaling in health and disease. European Journal of Dermatology 11:506–514

1199. Sakamoto O, Suga M, Suda T, Ando M (2001) Expression of discoidin domain receptor 1 tyrosine kinase on the human bronchial epithelium. European Respiratory Journal 17:969–974

1200. Olaso E, Ikeda K, Eng FJ, Xu L, Wang LH, Lin HC, Friedman SL (2001) DDR2 receptor promotes MMP-2-mediated proliferation and invasion by hepatic stellate cells. Journal of Clinical Investigation 108:1369–1378

1201. Ferri N, Carragher NO, Raines EW (2004) Role of discoidin domain receptors 1 and 2 in human smooth muscle cell-mediated collagen remodeling: potential implications in atherosclerosis and lymphangioleiomyomatosis. American Journal of Pathology 164:1575–1585

1202. Labrador JP, Azcoitia V, Tuckermann J, Lin C, Olaso E, Mañes S, Brückner K, Goergen JL, Lemke G, Yancopoulos G, Angel P, Martínez C, Klein R (2001) The collagen receptor DDR2 regulates proliferation and its elimination leads to dwarfism. EMBO Reports 2:446–452

1203. Mihai C, Chotani M, Elton TS, Agarwal G (2009) Mapping of DDR1 distribution and oligomerization on the cell surface by FRET microscopy. Journal of Molecular Biology 385:432–445

1204. Faraci E, Eck M, Gerstmayer B, Bosio A, Vogel WF (2003) An extracellular matrix-specific microarray allowed the identification of target genes downstream of discoidin domain receptors. Matrix Biology 22:373–381

1205. Shintani Y, Fukumoto Y, Chaika N, Svoboda R, Wheelock MJ, Johnson KR (2008) Collagen I-mediated up-regulation of N-cadherin requires cooperative signals from integrins and discoidin domain receptor 1. Journal of Cell Biology 80:1277–1289

1206. Tang H, Hao Q, Fitzgerald T, Sasaki T, Landon EJ, Inagami T (2002) Pyk2/CAKβ tyrosine kinase activity-mediated angiogenesis of pulmonary vascular endothelial cells. Journal of Biological Chemistry 277:5441–5447

1207. Wang CZ, Yeh YC, Tang MJ (2009) DDR1/E-cadherin complex regulates the activation of DDR1 and cell spreading. American Journal of Physiology – Cell Physiology 97:C419–C429

1208. Duyster J, Bai RY, Morris SW (2001) Translocations involving anaplastic lymphoma kinase (ALK). Oncogene 20:5623–5637

1209. Lorén CE, Scully A, Grabbe C, Edeen PT, Thomas J, McKeown M, Hunter T, Palmer RH (2001) Identification and characterization of DAlk: a novel Drosophila melanogaster RTK which drives ERK activation in vivo. Genes to Cells 6:531–544

1210. Pulford K, Lamant L, Morris SW, Butler LH, Wood KM, Stroud D, Delsol G, Mason DY (1997) Detection of anaplastic lymphoma kinase (ALK) and nucleolar protein nucleophosmin (NPM)-ALK proteins in normal and neoplastic cells with the monoclonal antibody ALK1. Blood 89:1394–1404

1211. Cartaud A, Strochlic L, Guerra M, Blanchard B, Lambergeon M, Krejci E, Cartaud J, Legay C (2006) MuSK is required for anchoring acetylcholinesterase at the neuromuscular junction. Journal of Cell Biology 165:505–515.

1212. Cheusova T, Khan MA, Schubert SW, Gavin AC, Buchou T, Jacob G, Sticht H, Allende J, Boldyreff B, Brenner HR, Hashemolhosseini S (2006) Casein kinase 2-dependent serine phosphorylation of MuSK regulates acetylcholine receptor aggregation at the neuromuscular junction. Genes and Development 20:1800–1816

1213. Luo ZG, Wang Q, Zhou JZ, Wang J, Luo Z, Liu M, He X, Wynshaw-Boris A, Xiong WC, Lu B, Mei L (2002) Regulation of AChR clustering by Dishevelled interacting with MuSK and PAK1. Neuron 35:489–505

1214. Huang EJ, Reichardt LF (2003) Trk receptors: roles in neuronal signal transduction. Annual Review of Biochemistry 72:609–642

1215. Makkerh JP, Ceni C, Auld DS, Vaillancourt F, Dorval G, Barker PA (2005) p75 neurotrophin receptor reduces ligand-induced Trk receptor ubiquitination and delays Trk receptor internalization and degradation. EMBO Reports 6:936–941

1216. Carter BD, Feng N, Paolocci N (2010) The p75 neurotrophin receptor, semaphorins, and sympathetic traffic in the heart. American Journal of Physiology – Heart and Circulatory Physiology 298:H1633–H1636

1217. Parkhurst CN, Zampieri N, Chao MV (2010) Nuclear localization of the p75 neurotrophin receptor intracellular domain. Journal of Biological Chemistry 285:5361–5368

1218. Ng YP, Lo KY, Cheung ZH, Ip NY (2006) Signaling through the neurotrophin receptors (Chap. 2; p 11–41.) In Lajtha A (Ed.) Handbook of Neurochemistry and Molecular Neurobiology. Lim R (Volume Ed.) Neuroactive Proteins and Peptides. Springer, New York, USA

1219. Lorentz CU, Alston EN, Belcik T, Lindner JR, Giraud GD, Habecker BA (2010) Heterogeneous ventricular sympathetic innervation, altered β-adrenergic receptor expression, and rhythm instability in mice lacking the p75 neurotrophin receptor. American Journal of Physiology – Heart and Circulatory Physiology 298:H1652–H1660

1220. Shnitsar I, Borchers A (2008) PTK7 recruits dsh to regulate neural crest migration. Development 135:4015–4024

1221. Katoh M, Katoh M (2007) Comparative integromics on non-canonical WNT or planar cell polarity signaling molecules: transcriptional mechanism of PTK7 in colorectal cancer and that of SEMA6A in undifferentiated ES cells. International Journal of Molecular Medicine 20:405–409

1222. Shin WS, Maeng YS, Jung JW, Min JK, Kwon YG, Lee ST (2008) Soluble PTK7 inhibits tube formation, migration, and invasion of endothelial cells and angiogenesis. Biochemical and Biophysical Research Communications 371:793–798

1223. Arighi E, Borrello MG, Sariola H (2005) RET tyrosine kinase signaling in development and cancer. Cytokine and Growth Factor Reviews 16:441–467

1224. Lu W, Yamamoto V, Ortega B, Baltimore D (2004) Mammalian Ryk is a Wnt coreceptor required for stimulation of neurite outgrowth. Cell 119:97-108

1225. Green JL, Kuntz SG, Sternberg PW (2008) Ror receptor tyrosine kinases: orphans no more. Trends in Cell Biology 18:536–544

1226. Oishi I, Takeuchi S, Hashimoto R, Nagabukuro A, Ueda T, Liu ZJ, Hatta T, Akira S, Matsuda Y, Yamamura H, Otani H, Minami Y (1999) Spatio-temporally regulated expression of receptor tyrosine kinases, mRorl, mRor2, during mouse development: implications in development and function of the nervous system. Genes to Cells 4:41–56

1227. Takeuchi S, Takeda K, Oishi I, Nomi M, Ikeya M, Itoh K, Tamura S, Ueda T, Hatta T, Otani H, Terashima T, Takada S, Yamamura H, Akira S, Minami Y (2000) Mouse Ror2 receptor tyrosine kinase is required for the heart development and limb formation. Genes to Cells 5:71–78

1228. Green JL, Inoue T, Sternberg PW (2008) Opposing Wnt pathways orient cell polarity during organogenesis. Cell 134:646–656

1229. Mikels AJ, Nusse R (2006) Purified Wnt5a protein activates or inhibits β-catenin-TCF signaling depending on receptor context. PLoS Biology 4:e11

1230. Oishi I, Suzuki H, Onishi N, Takada R, Kani S, Ohkawara B, Koshida I, Suzuki K, Yamada G, Schwabe GC, Mundlos S, Shibuya H, Takada S, Minami Y (2003) The receptor tyrosine kinase Ror2 is involved in non-canonical Wnt5a/JNK signalling pathway. Genes to Cells 8:645–654

1231. Sonnenberg E, Gödecke A, Walter B, Bladt F, Birchmeier C (1991) Transient and locally restricted expression of the ros1 protooncogene during mouse development. EMBO Journal 10:3693–3702

1232. Keilhack H, Müller M, Böhmer SA, Frank C, Weidner KM, Birchmeier W, Ligensa T, Berndt A, Kosmehl H, Günther B, Müller T, Birchmeier C, Böhmer FD (2001) Negative regulation of Ros receptor tyrosine kinase signaling. An epithelial function of the SH2 domain protein tyrosine phosphatase SHP-1. Journal of Cell Biology 152:325–334

1233. Biskup C, Böhmer A, Pusch R, Kelbauskas L, Gorshokov A, Majoul I, Lindenau J, Benndorf K, Böhmer FD (2004) Visualization of SHP-1-target interaction. Journal of Cell Science 117:5165–5178

1234. Charest A, Wilker EW, McLaughlin ME, Lane K, Gowda R, Coven S, McMahon K, Kovach S, Feng Y, Yaffe MB, Jacks T, Housman D (2006) ROS fusion tyrosine kinase activates a SH2 domain-containing phosphatase-2/phosphatidylinositol 3-kinase/mammalian target of rapamycin signaling axis to form glioblastoma in mice. Cancer Research 66:7473–7481

1235. Aasheim HC, Patzke S, Hjorthaug HS, Finne EF (2005) Characterization of a novel Eph receptor tyrosine kinase, EphA10, expressed in testis. Biochimica et Biophysica Acta 1723:1–7

1236. Seiradake E, Harlos K, Sutton G, Aricescu AR, Jones EY (2010) An extracellular steric seeding mechanism for Eph-ephrin signaling platform assembly. Nature – Structural and Molecular Biology 17:398–402

1237. Jørgensen C, Sherman A, Chen GI, Pasculescu A, Poliakov A, Hsiung M, Larsen B, Wilkinson DG, Linding R, Pawson T (2009) Cell-specific information processing in segregating populations of Eph receptor ephrin-expressing cells. Science 326:1502–1509

1238. Shintani T, Ihara M, Sakuta H, Takahashi H, Watakabe I, Noda M (2006) Eph receptors are negatively controlled by protein tyrosine phosphatase receptor type O. Nature Neuroscience 9:761–769

1239. Shi G, Yue G, Zhou R (2010) EphA3 functions are regulated by collaborating phosphotyrosine residues. Cell Research 20:1263–1275

1240. Irie F, Okuno M, Matsumoto K, Pasquale EB, Yamaguchi Y (2008) Heparan sulfate regulates ephrin-A3/EphA receptor signaling. Proceedings of the National Academy of Sciences of the United States of America 105:12307-12312

1241. Woods TC, Blystone CR, Yoo J, Edelman ER (2002) Activation of EphB2 and its ligands promotes vascular smooth muscle cell proliferation. Journal of Biological Chemistry 277:1924–1927

1242. Tanaka M, Kamata R, Sakai R (2005) Phosphorylation of ephrin-B1 via the interaction with claudin following cell-cell contact formation. EMBO Journal 24:3700–3711

1243. Bong YS, Lee HS, Carim-Todd L, Mood K, Nishanian TG, Tessarollo L, Daar IO (2007) EphrinB1 signals from the cell surface to the nucleus by recruitment of STAT3. Proceedings of the National Academy of Sciences of the United States of America 104:17305–17310

1244. Hainaud P, Contrerès JO, Villemain A, Liu LX, Plouët J, Tobelem G, Dupuy E (2006) The role of the vascular endothelial growth factor–delta-like 4 ligand/Notch4–Ephrin B2 cascade in tumor vessel. Cancer Research 66:8501–8510

1245. Adams RH, Wilkinson GA, Weiss C, Diella F, Gale NW, Deutsch U, Risau W, Klein R (1999) Roles of ephrinB ligands and EphB receptors in cardiovascular development: demarcation of arterial/venous domains, vascular morphogenesis, and sprouting angiogenesis. Genes and Development 13:295–306

1246. Wang Y, Nakayama M, Pitulescu ME, Schmidt TS, Bochenek ML, Sakakibara A, Adams S, Davy A, Deutsch U, Lüthi U, Barberis A, Benjamin LE, Mäkinen T, Nobes CD, Adams RH (2010) Ephrin-B2 controls VEGF-induced angiogenesis and lymphangiogenesis. Nature 465:483–486

1247. Klein R (2009) Bidirectional modulation of synaptic functions by Eph/ephrin signaling. Nature – Neuroscience 12:15–20

1248. Gerety SS, Wang HU, Chen ZF, Anderson DJ (1999) Symmetrical mutant phenotypes of the receptor EphB4 and its specific transmembrane ligand ephrin-B2 in cardiovascular development. Molecular Cell 4:403–414

1249. Wang HU, Chen ZF, Anderson DJ (1998) Molecular distinction and angiogenic interaction between embryonic arteries and veins revealed by ephrin-B2 and its receptor Eph-B4. Cell 93:741–753

1250. Salaita K, Nair PM, Petit RS, Neve RM, Das D, Gray JW, Groves JT (2010) Restriction of receptor movement alters cellular response: physical force sensing by EphA2. Science 327:1380–1385

1251. Saharinen P, Eklund L, Miettinen J, Wirkkala R, Anisimov A, Winderlich M, Nottebaum A, Vestweber D, Deutsch U, Koh GY, Olsen BR, Alitalo K (2008) Angiopoietins assemble distinct Tie2 signalling complexes in endothelial cell–cell and cell–matrix contacts. Nature Cell Biology 10:527–537

1252. Fukuhara S, Sako K, Minami T, Noda K, Kim HZ, Kodama T, Shibuya M, Takakura N, Koh GY, Mochizuki N (2008) Differential function of Tie2 at cell–cell contacts and cell–substratum contacts regulated by angiopoietin-1. Nature Cell Biology 10:513–526

1253. Daly C, Pasnikowski E, Burova E, Wong V, Aldrich TH, Griffiths J, Ioffe E, Daly TJ, Fandl JP, Papadopoulos N, McDonald DM, Thurston G, Yancopoulos GD, Rudge JS (2006) Angiopoietin-2 functions as an autocrine protective factor in stressed endothelial cells. Proceedings of the National Academy of Sciences of the United States of America 103:15491–15496

1254. Derynck R, Zhang YE (2003) Smad-dependent and Smad-independent pathways in TGF-β family signalling. Nature 425:577–584

1255. López-Novoa JM, Bernabeu C (2010) The physiological role of endoglin in the cardiovascular system. American Journal of Physiology – Heart and Circulatory Physiology 299:H959-H974

1256. Yamashita Y, Kojima K, Tsukahara T, Agawa H, Yamada K, Amano Y, Kurotori N, Tanaka N, Sugamura K, Takeshita T (2008) Ubiquitin-independent binding of Hrs mediates endosomal sorting of the interleukin-2 receptor β-chain. Journal of Cell Science 121:1727–1738

1257. Raiborg C, Bache KG, Mehlum A, Stang E, Stenmark H (2001) Hrs recruits clathrin to early endosomes. EMBO Journal 20:5008–5021

1258. Miura S, Takeshita T, Asao H, Kimura Y, Murata K, Sasaki Y, Hanai JI, Beppu H, Tsukazaki T, Wrana JL, Miyazono K, Sugamura K (2000) Hgs (Hrs), a FYVE domain protein, is involved in Smad signaling through cooperation with SARA. Molecular and Cellular Biology 20:9346–9355

1259. Rifkin DB (2005) Latent transforming growth factor-β (TGF-β) binding proteins: orchestrators of TGF-β availability. Journal of Biological Chemistry 280:7409–7412

1260. Miyazono K, Kamiya Y, Morikawa M (2010) Bone morphogenetic protein receptors and signal transduction. Journal of Biochemistry 147:35–51

1261. Samad TA, Rebbapragada A, Bell E, Zhang Y, Sidis Y, Jeong SJ, Campagna JA, Perusini S, Fabrizio DA, Schneyer AL, Lin HY, Brivanlou AH, Attisano L, Woolf CJ (2005) DRAGON, a bone morphogenetic protein co-receptor. Journal of Biological Chemistry 280:14122–14129

1262. Kang JS, Saunier EF, Akhurst RJ, Derynck R (2008) The type I TGF-β receptor is covalently modified and regulated by sumoylation. Nature Cell Biology 10:654–664

1263. Sorrentino A, Thakur N, Grimsby S, Marcusson A, von Bulow V, Schuster N, Zhang S, Heldin CH, Landström M (2008) The type I TGF-β receptor engages TRAF6 to activate TAK1 in a receptor kinase-independent manner. Nature Cell Biology 10:1199–1207

1264. Karthikeyan M, Blobe GC (2007) TGF-β type I receptor. UCSD-Nature Molecule Pages, UCSD-Nature Signaling Gateway (www.signaling-gateway.org)

1265. Lin KW, Yakymovych I, Jia M, Yakymovych M, Souchelnytskyi S (2010) Phosphorylation of eEF1A1 at Ser300 by TβR-I results in inhibition of mRNA translation. Current Biology 20:1615–1625

1266. Hempel N, Blobe GC (2006) TGF-β type II receptor. UCSD-Nature Molecule Pages, UCSD-Nature Signaling Gateway (www.signaling-gateway.org)

1267. Choy L, Derynck R (1998) The type II transforming growth factor (TGF)-β receptor-interacting protein TRIP-1 acts as a modulator of the TGF-β response. Journal of Biological Chemistry 273:31455–31462

1268. Datta PK, Moses HL (2000) STRAP and Smad7 synergize in the inhibition of transforming growth factor β signaling. Molecular and Cellular Biology 20:3157–3167

1269. Tang Q, Staub CM, Gao G, Jin Q, Wang Z, Ding W, Aurigemma RE, Mulder KM (2002) A novel transforming growth factor-β receptor-interacting protein that is also a light chain of the motor protein dynein. Molecular Biology of the Cell 13:4484–4496

1270. Gilboa L, Nohe A, Geissendörfer T, Sebald W, Henis YI, Knaus P (2000) Bone morphogenetic protein receptor complexes on the surface of live cells: a new oligomerization mode for serine/threonine kinase receptors. Molecular Biology of the Cell 11:1023–1035

1271. Schwappacher R, Weiske J, Heining E, Ezerski V, Marom B, Henis YI, Huber O, Knaus P (2009) Novel crosstalk to BMP signalling: cGMP-dependent kinase I modulates BMP receptor and SMAD activity. EMBO Journal 28:1537–1550

1272. Furtado MB, Solloway MJ, Jones VJ, Costa MW, Biben C, Wolstein O, Preis JI, Sparrow DB, Saga Y, Dunwoodie SL, Robertson EJ, Tam PP, Harvey RP (2008) BMP/SMAD1 signaling sets a threshold for the left/right pathway in lateral plate mesoderm and limits availability of SMAD4. Genes and Development 22:3037–3049

1273. Sheng N, Xie Z, Wang C, Bai G, Zhang K, Zhu Q, Song J, Guillemot F, Chen YG, Lin A, Jing N (2010) Retinoic acid regulates bone morphogenic protein signal duration by promoting the degradation of phosphorylated Smad1. Proceedings of the National Academy of Sciences of the United States of America 107:18886–18891

1274. Favaro JP, Wiley K, Blobe GC (2005) Alk1. UCSD-Nature Molecule Pages, UCSD-Nature Signaling Gateway (www.signaling-gateway.org)

1275. Derynck R, Zhang Y, Feng XH (1998) Transcriptional activators of TGF-β responses: Smads. Cell 95:737–740

1276. Sapkota G, Knockaert M, Alarcón C, Montalvo E, Brivanlou AH, Massagué J (2006) Dephosphorylation of the linker regions of Smad1 and Smad2/3 by small C-terminal domain phosphatases has distinct outcomes for bone morphogenetic protein and transforming growth factor-β pathways. Journal of Biological Chemistry 281:40412–40419

1277. Vincent T, Neve EPA, Johnson JR, Kukalev A, Rojo F, Albanell J, Pietras K, Virtanen I, Philipson L, Leopold PL, Crystal RG, Garcia de Herreros A, Moustakas A, Pettersson RF, Fuxe J (2009) A SNAIL1–SMAD3/4 transcriptional repressor complex promotes TGFβ mediated epithelial–mesenchymal transition. Nature – Cell Biology 11:943–950

1278. Stefancsik R (2006) Smad5 UCSD-Nature Molecule Pages, UCSD-Nature Signaling Gateway (www.signaling-gateway.org)

1279. Foletta VC, Lim MA, Soosairajah J, Kelly AP, Stanley EG, Shannon M, He W, Das S, Massagué J, Bernard O (2003) Direct signaling by the BMP type II receptor via the cytoskeletal regulator LIMK1. Journal of Cell Biology 162:1089–1098

1280. Lee-Hoeflich ST, Causing CG, Podkowa M, Zhao X, Wrana JL, Attisano L (2004) Activation of LIMK1 by binding to the BMP receptor, BMPRII, regulates BMP-dependent dendritogenesis. EMBO Journal 23:4792–4801

Chap. 9. Receptor Tyrosine Phosphatases

1281. Aricescu AR, Hon WC, Siebold C, Lu W, van der Merwe PA, Jones EY (2006) Molecular analysis of receptor protein tyrosine phosphatase mu-mediated cell adhesion. EMBO Journal 25:701–712

1282. Bokoch GM, Zhao T (2006) Regulation of the phagocyte NADPH oxidase by Rac GTPase. Antioxidants and Redox Signaling 8:1533–1548

1283. Gandhi TK, Chandran S, Peri S, Saravana R, Amanchy R, Prasad TS, Pandey A (2005) A bioinformatics analysis of protein tyrosine phosphatases in humans. DNA Research 12:79–89

1284. Forrest AR, Taylor DF, Crowe ML, Chalk AM, Waddell NJ, Kolle G, Faulkner GJ, Kodzius R, Katayama S, Wells C, Kai C, Kawai J, Carninci P, Hayashizaki Y, Grimmond SM (2006) Genome-wide review of transcriptional complexity in mouse protein kinases and phosphatases. Genome Biology 7:R5

1285. Wallace MJ, Fladd C, Batt J, Rotin D (1998) The second catalytic domain of protein tyrosine phosphatase δ (PTP δ) binds to and inhibits the first catalytic domain of PTP σ. Molecular and Cellular Biology 18:2608–2616

1286. den Hertog J, Hunter T (1996) Tight association of GRB2 with receptor protein-tyrosine phosphatase α is mediated by the SH2 and C-terminal SH3 domains. EMBO Journal 15:3016–3027

1287. Tsai W, Morielli AD, Cachero TG, Peralta EG (1999) Receptor protein tyrosine phosphatase α participates in the m1 muscarinic acetylcholine receptor-dependent regulation of K_V1.2 channel activity. EMBO Journal 18:109–118

1288. Kawachi H, Tamura H, Watakabe I, Shintani T, Maeda N, Noda M (1999) Protein tyrosine phosphatase ζ/RPTPβ interacts with PSD-95/SAP90 family. Brain Research – Molecular Brain Research 72:47–54

1289. Nawroth R, Poell G, Ranft A, Kloep S, Samulowitz U, Fachinger G, Golding M, Shima DT, Deutsch U, Vestweber D (2002) VE-PTP and VE-cadherin ectodomains interact to facilitate regulation of phosphorylation and cell contacts. EMBO Journal 21:4885–4895

1290. Adamsky K, Arnold K, Sabanay H, Peles E (2003) Junctional protein MAGI-3 interacts with receptor tyrosine phosphatase β (RPTPβ) and tyrosine-phosphorylated proteins. Journal of Cell Science 116:1279–1289

1291. Ratcliffe CF, Qu Y, McCormick KA, Tibbs VC, Dixon JE, Scheuer T, Catterall WA (2000) A sodium channel signaling complex: modulation by associated receptor protein tyrosine phosphatase β. Nature – Neuroscience 3:437–444

1292. Fachinger, Deutsch U, Risau W (1999) Functional interaction of vascular endothelial-protein-tyrosine phosphatase with the Angiopoietin receptor Tie-2. Oncogene 18:5948–5953

1293. Li Z, Huang H, Boland P, Dominguez MG, Burfeind P, Lai KM, Lin HC, Gale NW, Daly C, Auerbach W, Valenzuela D, Yancopoulos GD, Thurston G (2009) Embryonic stem cell tumor model reveals role of vascular endothelial receptor tyrosine phosphatase in regulating Tie2 pathway in tumor angiogenesis. Proceedings of the National Academy of Sciences of the United States of America 106:22399–22404

1294. Seddiki N, Santner-Nanan B, Tangye SG, Alexander SI, Solomon M, Lee S, Nanan R, Fazekas de Saint Groth B (2006) Persistence of naive CD45RA+ regulatory T cells in adult life. Blood 107:2830–2838

1295. Irie-Sasaki J, Sasaki T, Penninger JM (2003) CD45 regulated signaling pathways. Current Topics in Medicinal Chemistry 3:783–796

1296. Penninger JM, Irie-Sasaki J, Sasaki T, Oliveira-dos-Santos AJ (2001) CD45: new jobs for an old acquaintance. Nature – Immunology 2:389–396

1297. Elson A (2006) RPTPe. UCSD-Nature Molecule Pages, UCSD-Nature Signaling Gateway (www.signaling-gateway.org)

1298. Andersen JN, Elson A, Lammers R, Rømer J, Clausen JT, Møller KB, Møller NP (2001) Comparative study of protein tyrosine phosphatase-ϵ isoforms: membrane localization confers specificity in cellular signalling. Biochemical Journal 354:581–590

1299. Peretz A, Gil-Henn H, Sobko A, Shinder V, Attali B, Elson A (2000) Hypomyelination and increased activity of voltage-gated K^+ channels in mice lacking protein tyrosine phosphatase ϵ. EMBO Journal 19:4036–4045

1300. Thompson LJ, Jiang J, Madamanchi N, Runge MS, Patterson C (2001) PTP-ϵ, a tyrosine phosphatase expressed in endothelium, negatively regulates endothelial cell proliferation. American Journal of Physiology – Heart and Circulatory Physiology 281:H396–H403

1301. Bonvini P, An WG, Rosolen A, Nguyen P, Trepel J, Garcia de Herreros A, Dunach M, Neckers LM (2001) Geldanamycin abrogates ErbB2 association with proteasome-resistant β-catenin in melanoma cells, increases β-catenin-E-cadherin association, and decreases β-catenin-sensitive transcription. Cancer Research 61:1671–1677

1302. Bouyain S, Watkins DJ (2010) The protein tyrosine phosphatases PTPRZ and PTPRG bind to distinct members of the contactin family of neural recognition molecules. Proceedings of the National Academy of Sciences of the United States of America 107:2443–2448

1303. Takada T, Noguchi T, Inagaki K, Hosooka T, Fukunaga K, Yamao T, Ogawa W, Matozaki T, Kasuga M (2002) Induction of apoptosis by stomach cancer-associated protein-tyrosine phosphatase-1. Journal of Biological Chemistry 277:34359–34366

1304. Sui XF, Kiser TD, Hyun SW, Angelini DJ, Del Vecchio RL, Young BA, Hasday JD, Romer LH, Passaniti A, Tonks NK, Goldblum SE (2005) Receptor protein tyrosine phosphatase micro regulates the paracellular pathway in human lung microvascular endothelia. American Journal of Pathology 166:1247–1258

1305. Avraham S, London R, Tulloch GA, Ellis M, Fu Y, Jiang S, White RA, Painter C, Steinberger AA, Avraham H (1997) Characterization and chromosomal localization of PTPRO, a novel receptor protein tyrosine phosphatase, expressed in hematopoietic stem cells. Gene 204:5–16

1306. Jung H, Kim WK, Kim do H, Cho YS, Kim SJ, Park SG, Park BC, Lim HM, Bae KH, Lee SC (2009) Involvement of PTP-RQ in differentiation during adipogenesis of human mesenchymal stem cells. Biochemical and Biophysical Research Communications 383:252–257

1307. Nayak G, Goodyear RJ, Legan PK, Noda M, Richardson GP (2011) Evidence for multiple, developmentally regulated isoforms of Ptprq on hair cells of the inner ear. Developmental Neurobiology 71:129–141

1308. Ogata M, Oh-Hora M, Kosugi A, Hamaoka T (1999) Inactivation of mitogen-activated protein kinases by a mammalian tyrosine-specific phosphatase, PTPBR7. Biochemical and Biophysical Research Communications 25:52–56

1309. Zhao Y, Zhang X, Guda K, Lawrence E, Sun Q, Watanabe T, Iwakura Y, Asano M, Wei L, Yang Z, Zheng W, Dawson D, Willis J, Markowitz SD, Satake M, Wang Z (2010) Identification and functional characterization of paxillin as a target of protein tyrosine phosphatase receptor T. Proceedings of the National Academy of Sciences of the United States of America 107:2592–2597

1310. Badde A, Schulte D (2008) A role for receptor protein tyrosine phosphatase λin midbrain development. Journal of Neuroscience 28:6152–6164

1311. Doumont G, Martoriati A, Marine JC (2005) PTPRV is a key mediator of p53-induced cell cycle exit. Cell Cycle 4:1703–1705

1312. Bäumer S, Keller L, Holtmann A, Funke R, August B, Gamp A, Wolburg H, Wolburg-Buchholz K, Deutsch U, Vestweber D (2006) Vascular endothelial cell-specific phosphotyrosine phosphatase (VE-PTP) activity is required for blood vessel development. Blood 107:4754–4762

Chap. 10. Morphogen Receptors

1313. Yu SR, Burkhardt M, Nowak M, Ries J, Petrás Z, Scholpp S, Schwille P, Brand M (2009) FGF8 morphogen gradient forms by a source-sink mechanism with freely diffusing molecules. Nature 461:533–536

1314. Sprinzak D, Lakhanpal A, Lebon L, Santat LA, Fontes ME, Anderson GA, Garcia-Ojalvo J, Elowitz MB (2010) Cis-interactions between Notch and Delta generate mutually exclusive signalling states. Nature 465:86–90

1315. Meng H, Zhang X, Lee SJ, Strickland DK, Lawrence DA, Wang MM (2010) Low density lipoprotein receptor-related protein-1 (LRP1) regulates thrombospondin-2 (TSP2) enhancement of Notch3 signaling. Journal of Biological Chemistry 285:23047–23055

1316. Arnett KL, Hass M, McArthur DG, Ilagan MXG, Aster JC, Kopan R, Blacklow SC (2010) Structural and mechanistic insights into cooperative assembly of dimeric Notch transcription complexes. Nature – Structural and Molecular Biology 17:1312–1317

1317. Kohroki J, Nishiyama T, Nakamura T, Masuho Y (2005) ASB proteins interact with Cullin5 and Rbx2 to form E3 ubiquitin ligase complexes. FEBS Letters 579:6796–6802

1318. Diks SH, Sartori da Silva MA, Hillebrands JL, Bink RJ, Versteeg HH, van Rooijen C, Brouwers A, Chitnis AB, Peppelenbosch MP, Zivkovic D (2008) d-Asb11 is an essential mediator of canonical Delta–Notch signalling. Nature Cell Biology 10:1190–1198

1319. Bray SJ (2006) Notch signalling: a simple pathway becomes complex. Nature Reviews – Molecular Cell Biology 7:678–689

1320. Rajan A, Tien AT, Haueter CM, Schulze KL, Bellen HJ (2009) The Arp2/3 complex and WASp are required for apical trafficking of Delta into microvilli during cell fate specification of sensory organ precursors. Nature – Cell Biology 11:815–824

1321. Guarani V, Deflorian G, Franco CA, Krüger M, Phng LK, Bentley K, Toussaint L,Dequiedt F, Mostoslavsky R, Schmidt MHH, Zimmermann B, Brandes RP, Mione M, Westphal CH, Braun T, Zeiher AM, Gerhardt H, Dimmeler S, Potente M (2011) Acetylation-dependent regulation of endothelial Notch signalling by the SIRT1 deacetylase. Nature 473:234–238

1322. Androutsellis-Theotokis A, Leker RR, Soldner F, Hoeppner DJ, Ravin R, Poser SW, Rueger MA, Bae SK, Kittappa R, McKay RDG (2006) Notch signalling regulates stem cell numbers in vitro and in vivo. Nature 442:823–826

1323. Xin M, Small EM, van Rooij E, Qi X, Richardson JA, Srivastava D, Nakagawa O, Olson EN (2007) Essential roles of the bHLH transcription factor Hrt2 in repression of atrial gene expression and maintenance of postnatal cardiac function. Proceedings of the National Academy of Sciences of the United States of America 104:7975–7980

1324. Wakabayashi N, Shin S, Slocum SL, Agoston ES, Wakabayashi J, Kwak MK, Misra V, Biswal S, Yamamoto M, Kensler TW (2010) Regulation of notch1 signaling by nrf2: implications for tissue regeneration. Science Signaling 3:ra52

1325. Wacker SA, Alvarado C, von Wichert G, Knippschild U, Wiedenmann J, Clauß K, Nienhaus GU, Hameister H, Baumann B, Borggrefe T, Knöchel W, Oswald F (2011) RITA, a novel modulator of Notch signalling, acts via nuclear export of RBP-J. EMBO Journal 30:43–56

1326. Perumalsamy LR, Nagala M, Sarin A (2010) Notch-activated signaling cascade interacts with mitochondrial remodeling proteins to regulate cell survival. Proceedings of the National Academy of Sciences of the United States of America 107:6882–6887

1327. Proweller A, Wright AC, Horng D, Cheng L, Lu MM, Lepore JJ, Pear WS, Parmacek MS (2007) Notch signaling in vascular smooth muscle cells is required to pattern the cerebral vasculature. Proceedings of the National Academy of Sciences of the United States of America 104:16275–16280

1328. Andreu-Agulló C, Morante-Redolat JM, Delgado AC, Fariñas I (2009) Vascular niche factor PEDF modulates Notch-dependent stemness in the adult subependymal zone. Nature – Neuroscience 12:1514–1523

1329. Kageyama R, Ohtsuka T, Shimojo H, Imayoshi I (2008) Dynamic Notch signaling in neural progenitor cells and a revised view of lateral inhibition. Nature Neuroscience 11:1247–1251

1330. High FA, Epstein JA (2008) The multifaceted role of Notch in cardiac development and disease. Nature Reviews – Genetics 9:49–61

1331. Hofmann JJ, Iruela-Arispe ML (2007) Notch signaling in blood vessels. Who is talking to whom about what? Circulation Research 100:1556

1332. Robert-Moreno A, Guiu J, Ruiz-Herguido C, López MA, Inglés-Esteve J, Riera L, Tipping A, Enver T, Dzierzak E, Gridley T, Espinosa L, Bigas A (2008) Impaired embryonic haematopoiesis yet normal arterial development in the absence of the Notch ligand Jagged. EMBO Journal 27:1886–1895

1333. Gerhardt H, Ruhrberg C, Abramsson A, Fujisawa H, Shima D, Betsholtz C (2004) Neuropilin-1 is required for endothelial tip cell guidance in the developing central nervous system. Developmental Dynamics 231:503–509

1334. Sainson RC, Aoto J, Nakatsu MN, Holderfield M, Conn E, Koller E, Hughes CC (2005) Cell-autonomous notch signaling regulates endothelial cell branching and proliferation during vascular tubulogenesis. FASEB Journal 19:1027–1029

1335. Sato Y, Watanabe T, Saito D, Takahashi T, Yoshida S, Kohyama J, Ohata E, Okano H, Takahashi Y (2008) Notch mediates the segmental specification of angioblasts in somites and their directed migration toward the dorsal aorta in avian embryos. Developmental Cell 14:890–901

1336. Gridley T (2007) Notch signaling in vascular development and physiology. Development 134:2709–2718

1337. Liu ZJ, Shirakawa T, Li Y, Soma A, Oka M, Dotto GP, Fairman RM, Velazquez OC, Herlyn M (2003) Regulation of Notch1 and Dll4 by vascular endothelial growth factor in arterial endothelial cells: implications for modulating arteriogenesis and angiogenesis. Molecular and Cellular Biology 23:14–25

1338. Visconti RP, Richardson CD, Sato TN (2002) Orchestration of angiogenesis and arteriovenous contribution by angiopoietins and vascular endothelial growth factor (VEGF). Proceedings of the National Academy of Sciences of the United States of America 99:8219–8224

1339. Iso T, Maeno T, Oike Y, Yamazaki M, Doi H, Arai M, Kurabayashi M (2006) Dll4-selective Notch signaling induces ephrinB2 gene expression in endothelial cells. Biochemical and Biophysical Research Communications 341:708–714

1340. Doi H, Iso T, Yamazaki M, Akiyama H, Kanai H, Sato H, Kawai-Kowase K, Tanaka T, Maeno T, Okamoto EI, Arai M, Kedes L, Kurabayashi M (2005) HERP1 inhibits myocardin-induced vascular smooth muscle cell differentiation by interfering with SRF binding to CArG box. Arteriosclerosis, Thrombosis, and Vascular Biology 25:2328–2334

1341. Seo S, Fujita H, Nakano A, Kang M, Duarte A, Kume T (2006) The forkhead transcription factors, Foxc1 and Foxc2, are required for arterial specification and lymphatic sprouting during vascular development. Developmental Biology 294:458–470

1342. van der Loop FT, Gabbiani G, Kohnen G, Ramaekers FC, van Eys GJ (1997) Differentiation of smooth muscle cells in human blood vessels as defined by smoothelin, a novel marker for the contractile phenotype. Arteriosclerosis, Thrombosis, and Vascular Biology 17:665–671

1343. Moessler H, Mericskay M, Li Z, Nagl S, Paulin D, Small JV (1996) The SM 22 promoter directs tissue-specific expression in arterial but not in venous or visceral smooth muscle cells in transgenic mice. Development 122:2415–2425

1344. Doi H, Iso T, Sato H, Yamazaki M, Matsui H, Tanaka T, Manabe I, Arai M, Nagai R, Kurabayashi M (2006) Jagged1-selective notch signaling induces smooth muscle differentiation via a RBP-Jkappa-dependent pathway. Journal of Biological Chemistry 281:28555–28564

1345. Noseda M, Fu Y, Niessen K, Wong F, Chang L, McLean G, Karsan A (2006) Smooth Muscle α-actin is a direct target of Notch/CSL. Circulation Research 98:1468–1470

1346. Arboleda-Velasquez JF, Zhou Z, Shin HK, Louvi A, Kim HH, Savitz SI, Liao JK, Salomone S, Ayata C, Moskowitz MA, Artavanis-Tsakonas S (2008) Linking Notch signaling to ischemic stroke. Proceedings of the National Academy of Sciences of the United States of America 105:4856–4861

1347. Shin HM, Minter LM, Cho OH, Gottipati S, Fauq AH, Golde TE, Sonenshein GE, Osborne BA (2006) Notch1 augments NF-κB activity by facilitating its nuclear retention. EMBO Journal 25:129–138

1348. Bedogni B, Warneke JA, Nickoloff BJ, Giaccia AJ, Powell MB (2008) Notch1 is an effector of Akt and hypoxia in melanoma development. Journal of Clinical Investigation 118:3660–3670

1349. Denko NC (2008) Hypoxia, HIF1 and glucose metabolism in the solid tumour. Nature Reviews – Cancer 8:705–713

1350. Ingram WJ, Wicking CA, Grimmond SM, Forrest AR, Wainwright BJ (2002) Novel genes regulated by Sonic Hedgehog in pluripotent mesenchymal cells. Oncogene 21:8196–8205

1351. Callejo A, Bilioni A, Mollica E, Gorfinkiel N, Andrés G, Ibáñez C, Torroja C, Doglio L, Sierra J, Guerrero I (2011) Dispatched mediates Hedgehog basolateral release to form the long-range morphogenetic gradient in the Drosophila wing disk epithelium. Proceedings of the National Academy of Sciences of the United States of America 108:12591–12598

1352. Callejo A, Culi J, Guerrero I (2008) Patched, the receptor of Hedgehog, is a lipoprotein receptor. Proceedings of the National Academy of Sciences of the United States of America 105:912–917

1353. Rohatgi R, Scott MP (2007) Patching the gaps in Hedgehog signalling. Nature Cell Biology 9:1005–1009

1354. Bishop B, Aricescu AR, Harlos K, O'Callaghan CA, Jones EY, Siebold C (2009) Structural insights into hedgehog ligand sequestration by the human hedgehog-interacting protein HHIP. Nature – Structural and Molecular Biology 16:698-703

1355. Zhao Y, Tong C, Jiang J (2007) Hedgehog regulates smoothened activity by inducing a conformational switch. Nature 450:252–258

1356. Aikin RA, Ayers KL, Thérond PP (2008) The role of kinases in the Hedgehog signalling pathway. EMBO Reports 9:330–336

1357. Milenkovic L, Scott MP, Rohatgi R (2009) Lateral transport of Smoothened from the plasma membrane to the membrane of the cilium. Journal of Cell Biology 187:365–374

1358. Milenkovic L, Scott MP (2010) Not lost in space: trafficking in the Hedgehog signaling pathway. Science Signaling 3:pe14

1359. Parker DS, White MA, Ramos AI, Cohen BA, Barolo S (2011) The cis-regulatory logic of hedgehog gradient responses: key roles for Gli binding affinity, competition, and cooperativity. Science Signaling 4:ra38

1360. Jacob L, Lum L (2007) Deconstructing the Hedgehog pathway in development and disease. Science 318:66–68

1361. Regl G, Kasper M, Schnidar H, Eichberger T, Neill GW, Ikram MS, Quinn AG, Philpott MP, Frischauf AM, Aberger F (2004) The zinc-finger transcription factor GLI2 antagonizes contact inhibition and differentiation of human epidermal cells. Oncogene 23:1263–1274

1362. Kim J, Kato M, Beachy PA (2009) Gli2 trafficking links Hedgehog-dependent activation of Smoothened in the primary cilium to transcriptional activation in the nucleus. Proceedings of the National Academy of Sciences of the United States of America 106:21666–21671

1363. Wang Y, Zhou Z, Walsh CT, McMahon AP (2009) Selective translocation of intracellular Smoothened to the primary cilium in response to Hedgehog pathway modulation. Proceedings of the National Academy of Sciences of the United States of America 106:2623–2628

1364. Chan JA, Balasubramanian S, Witt RM, Nazemi KJ, Choi Y, Pazyra-Murphy MF, Walsh CO, Thompson M, Segal RA (2009) Proteoglycan interactions with Sonic Hedgehog specify mitogenic responses. Nature – Neuroscience 12:409–417

1365. Jeong J, McMahon AP (2005) Growth and pattern of the mammalian neural tube are governed by partially overlapping feedback activities of the hedgehog antagonists patched 1 and Hhip1. Development 132:143–154

1366. Di Marcotullio L, Ferretti E, Greco A, De Smaele E, Po A, Sico MA, Alimandi M, Giannini G, Maroder M, Screpanti I, Gulino A (2006) Numb is a suppressor of Hedgehog signalling and targets Gli1 for Itch-dependent ubiquitination. Nature Cell Biology 8:1415–1423

1367. Kovacs JJ, Whalen EJ, Liu R, Xiao K, Kim J, Chen M, Wang J, Chen W, Lefkowitz RJ (2008) β-Arrestin-mediated localization of Smoothened to the primary cilium. Science 320:1777–1781

1368. Wang J, Lu J, Bond MC, Chen M, Ren XR, Lyerly HK, Barak LS, Chen W (2010) Identification of select glucocorticoids as Smoothened agonists: potential utility for regenerative medicine. Proceedings of the National Academy of Sciences of the United States of America 107:9323–9328

1369. Eaton S (2008) Multiple roles for lipids in the Hedgehog signalling pathway. Nature Reviews – Molecular Cell Biology 9:437–445

1370. Morrow D, Sweeney C, Birney YA, Guha S, Collins N, Cummins PM, Murphy R, Walls D, Redmond EM, Cahill PA (2007) Biomechanical regulation of hedgehog signaling in vascular smooth muscle cells in vitro and in vivo. American Journal of Physiology – Cell Physiology 292:C488–C496

1371. Jacob LS, Wu X, Dodge ME, Fan CW, Kulak O, Chen B, Tang W, Wang B, Amatruda JF, Lum L (2011) Genome-wide RNAi screen reveals disease-associated genes that are common to Hedgehog and Wnt signaling. Science Signaling 4:ra4

1372. Silhankova M, Port F, Harterink M, Basler K, Korswagen HC (2010) Wnt signalling requires MTM-6 and MTM-9 myotubularin lipid-phosphatase function in Wnt-producing cells. EMBO Journal 29:4094–4105

1373. Nusse R (2008) Wnt signaling and stem cell control. Cell Research 18:523–527

1374. Jernigan KK, Cselenyi CS, Thorne CA, Hanson AJ, Tahinci E, Hajicek N, Oldham WM, Lee LA, Hamm HE, Hepler JR, Kozasa T, Linder ME, Lee E (2010) Gβγ activates GSK3 to promote LRP6-mediated beta-catenin transcriptional activity. Science Signaling 3:ra37

1375. Angers S, Moon RT (2009) Proximal events in Wnt signal transduction. Nature Reviews – Molecular Cell Biology 10:468–477

1376. Grumolato L, Liu G, Mong P, Mudbhary R, Biswas R, Arroyave R, Vijayakumar S, Economides AN, Aaronson SA (2010) Canonical and noncanonical Wnts use a common mechanism to activate completely unrelated coreceptors. Genes and Development 24:2517–2530

1377. Huang SMA, Mishina YM, Liu S, Cheung A, Stegmeier F, Michaud GA, Charlat O, Wiellette E, Zhang Y, Wiessner S, Hild M, Shi X, Wilson CJ, Mickanin C, Myer V, Fazal A, Tomlinson R, Serluca F, Shao W, Cheng H, Shultz M, Rau C, Schirle M, Schlegl J, Ghidelli S, Fawell S, Lu C, Curtis D, Kirschner MW, Lengauer C, Finan PM, Tallarico JA, Bouwmeester T, Porter JA, Bauer A, Cong F (2009) Tankyrase inhibition stabilizes axin and antagonizes Wnt signalling. Nature 461:614–620

1378. Davidson G, Wu W, Shen J, Bilic J, Fenger U, Stannek P, Glinka A, Niehrs C (2005) Casein kinase 1γ couples Wnt receptor activation to cytoplasmic signal transduction. Nature 438:867–872

1379. Zeng X, Tamai K, Doble B, Li S, Huang H, Habas R, Okamura H, Woodgett J, He X (2005) A dual-kinase mechanism for Wnt co-receptor phosphorylation and activation. Nature 438:873–877

1380. Bilic J, Huang YL, Davidson G, Zimmermann T, Cruciat CM, Bienz M, Niehrs C (2007) Wnt induces LRP6 signalosomes and promotes Dishevelled-dependent LRP6 phosphorylation. Science 316:1619–1622

1381. Schwarz-Romond T, Fiedler M, Shibata N, Butler JG, Kikuchi A, Higuchi Y, Bienz M (2007) The DIX domain of Dishevelled confers Wnt signaling by dynamic polymerization. Nature Structural and Molecular Biology 14:484–492

1382. Cselenyi CS, Jernigan KK, Tahinci E, Thorne CA, Lee LA, Lee E (2008) LRP6 transduces a canonical Wnt signal independently of Axin degradation by inhibiting GSK3's phosphorylation of β-catenin. Proceedings of the National Academy of Sciences of the United States of America 105:8032–8037

1383. Kim NG, Xu C, Gumbiner BM (2009) Identification of targets of the Wnt pathway destruction complex in addition to β-catenin. Proceedings of the National Academy of Sciences of the United States of America 106:5165–5170

1384. Witze ES, Litman ES, Argast GM, Moon RT, Ahn NG (2008) Wnt5a control of cell polarity and directional movement by polarized redistribution of adhesion receptors. Science 320:327–328

1385. Simons M, Gault WJ, Gotthardt D, Rohatgi R, Klein TJ, Shao Y, Lee HJ, Wu AL, Fang Y, Satlin LM, Dow JT, Chen J, Zheng J, Boutros M, Mlodzik M (2009) Electrochemical cues regulate assembly of the Frizzled/Dishevelled complex at the plasma membrane during planar epithelial polarization. Nature – Cell Biology 11:286–294

1386. Sato A, Yamamoto H, Sakane H, Koyama H, Kikuchi A (2010) Wnt5a regulates distinct signalling pathways by binding to Frizzled 2. EMBO Journal 29:41–54

1387. Niehrs C, Acebron SP (2010) Wnt signaling: multivesicular bodies hold GSK3 captive. Cell 143:1044–1046

1388. Taelman VF, Dobrowolski R, Plouhinec JL, Fuentealba LC, Vorwald PP, Gumper I, Sabatini DD, De Robertis EM (2010) Wnt signaling requires sequestration of glycogen synthase kinase 3 inside multivesicular endosomes. Cell 143:1136–1148

1389. Bartscherer K, Boutros M (2008) Regulation of Wnt protein secretion and its role in gradient formation. EMBO Reports 9:977–982

1390. Hausmann G, Banziger C, Basler K (2007) Helping Wingless take flight: how WNT proteins are secreted. Nature Reviews – Molecular Cell Biology 8:331–336

1391. Deng N, Ye Y, Wang W, Li L (2010) Dishevelled interacts with p65 and acts as a repressor of NF-Ŏ3baB-mediated transcription. Cell Research 20:1117–1127

1392. Berendsen AD, Fisher LW, Kilts TM, Owens RT, Robey PG, Gutkind JS, Young MF (2011) Modulation of canonical Wnt signaling by the extracellular matrix component biglycan. Proceedings of the National Academy of Sciences of the United States of America 108:17022–17027

1393. Li FQ, Mofunanya A, Harris K, Takemaru K (2008) Chibby cooperates with 14-3-3 to regulate β-catenin subcellular distribution and signaling activity. Journal of Cell Biology 181:1141–1154

1394. Kennell JA, Gerin I, MacDougald OA, Cadigan KM (2008) The microRNA miR-8 is a conserved negative regulator of Wnt signaling. Proceedings of the National Academy of Sciences of the United States of America 105:15417–15422

1395. Pan W, Choi SC, Wang H, Qin Y, Volpicelli-Daley L, Swan L, Lucast L, Khoo C, Zhang X, Li L, Abrams CS, Sokol SY, Wu D (2008) Wnt3a-mediated formation of phosphatidylinositol 4,5-bisphosphate regulates LRP6 phosphorylation. Science 321:1350–1353

1396. Ganner A, Lienkamp S, Schäfer T, Romaker D, Wegierski T, Park TJ, Spreitzer S, Simons M, Gloy J, Kim E, Wallingford JB, Walz G (2009) Regulation of ciliary polarity by the APC/C. Proceedings of the National Academy of Sciences of the United States of America 106:17799–17804

1397. Creyghton MP, Roël G, Eichhorn PJ, Hijmans EM, Maurer I, Destrée O, Bernards R (2005) PR72, a novel regulator of Wnt signaling required for Naked cuticle function. Genes and Development 19(3):376–386

1398. Mukai A, Yamamoto-Hino M, Awano W, Watanabe W, Komada M, Goto S (2010) Balanced ubiquitylation and deubiquitylation of Frizzled regulate cellular responsiveness to Wg/Wnt. EMBO Journal 29:2114–2125

1399. Bovolenta P, Esteve P, Ruiz JM, Cisneros E, Lopez-Rios J (2008) Beyond Wnt inhibition: new functions of secreted Frizzled-related proteins in development and disease. Journal of Cell Science 121:737–746

1400. Yoshino K, Rubin JS, Higinbotham KG, Uren A, Anest V, Plisov SY, Perantoni AO (2001) Secreted Frizzled-related proteins can regulate metanephric development. Mechanisms of Development 102:45–55

1401. Rodriguez J, Esteve P, Weinl C, Ruiz JM, Fermin Y, Trousse F, Dwivedy A, Holt C, Bovolenta P (2005) SFRP1 regulates the growth of retinal ganglion cell axons through the Fz2 receptor. Nature Reviews – Neurosciences 8:1301–1309

1402. Lee HX, Ambrosio AL, Reversade B, De Robertis EM (2006) Embryonic dorsal-ventral signaling: secreted frizzled-related proteins as inhibitors of tolloid proteinases. Cell 124:147–159

1403. Logan CY, Nusse R (2004) The Wnt signaling pathway in development and disease. Annual Review of Cell and Developmental Biology 20:781–810

1404. Zhong X, Desilva T, Lin L, Bodine P, Bhat RA, Presman E, Pocas J, Stahl M, Kriz R (2007) Regulation of secreted Frizzled-related protein-1 by heparin. Journal of Biological Chemistry 282:20523–20533

1405. Ezan J, Leroux L, Barandon L, Dufourcq P, Jaspard B, Moreau C, Allières C, Daret D, Couffinhal T, Duplàa C (2004) FrzA/sFRP-1, a secreted antagonist of the Wnt-Frizzled pathway, controls vascular cell proliferation in vitro and in vivo. Cardiovascular Research 63:731–738

1406. Ouchi N, Higuchi A, Ohashi K, Oshima Y, Gokce N, Shibata R, Akasaki Y, Shimono A, Walsh K (2010) Sfrp5 is an anti-inflammatory adipokine that modulates metabolic dysfunction in obesity. Science 329:454–457

1407. Binnerts ME, Kim KA, Bright JM, Patel SM, Tran K, Zhou M, Leung JM, Liu Y, Lomas WE, Dixon M, Hazell SA, Wagle M, Nie WS, Tomasevic N, Williams J, Zhan X, Levy MD, Funk WD, Abo A (2007) R-Spondin1 regulates Wnt signaling by inhibiting internalization of LRP6. Proceedings of the National Academy of Sciences of the United States of America 104:14700–14705

1408. Carmon KS, Gong X, Lin Q, Thomas A, Liu Q (2011) R-spondins function as ligands of the orphan receptors LGR4 and LGR5 to regulate Wnt/beta-catenin signaling. Proceedings of the National Academy of Sciences of the United States of America 108:11452–11457

1409. Wu X, Tu X, Joeng KS, Hilton MJ, Williams DA, Long F (2008) Rac1 activation controls nuclear localization of β-catenin during canonical Wnt signaling. Cell 133:340–353

1410. Holloway KR, Calhoun TN, Saxena M, Metoyer CF, Kandler EF, Rivera CA, Pruitt K (2010) SIRT1 regulates Dishevelled proteins and promotes transient and constitutive Wnt signaling. Proceedings of the National Academy of Sciences of the United States of America 107:9216–9221

1411. Bryja V, Gradl D, Schambony A, Arenas E, Schulte G (2007) β-Arrestin is a necessary component of Wnt/β-catenin signaling in vitro and in vivo. Proceedings of the National Academy of Sciences of the United States of America 104:6690–6695

1412. Bryja V, Schambony A, Scaronajánek S, Dominguez I, Arenas E, Schulte G (2008) β-Arrestin and casein kinase 1/2 define distinct branches of non-canonical WNT signalling pathways. EMBO Reports 9:1244–1250

1413. Funato Y, Michiue T, Asashima M, Miki H (2006) The thioredoxin-related redox-regulating protein nucleoredoxin inhibits Wnt–β-catenin signalling through Dishevelled. Nature Cell Biology 8:501–508

1414. Stuart RO, Bush KT, Nigam SK (2003) Changes in gene expression patterns in the ureteric bud and metanephric mesenchyme in models of kidney development. Kidney International 64:1997–2008

1415. Shimomura Y, Agalliu D, Vonica A, Luria V, Wajid M, Baumer A, Belli S, Petukhova L, Schinzel A, Brivanlou AH, Barres BA, Christiano AM (2010) APCDD1 is a novel Wnt inhibitor mutated in hereditary hypotrichosis simplex. Nature 464:1043–1047

1416. Li Z, Nie F, Wang S, Li L (2011) Histone H4 Lys 20 monomethylation by histone methylase SET8 mediates Wnt target gene activation. Proceedings of the National Academy of Sciences of the United States of America 108:3116–3123

1417. Martinez-Morales PL, Quiroga AC, Barbas JA, Morales AV (2010) SOX5 controls cell cycle progression in neural progenitors by interfering with the WNT-beta-catenin pathway. EMBO Reports 11:466–472

1418. Wright KJ, Tjian R (2009) Wnt signaling targets ETO coactivation domain of TAF4/TFIID in vivo. Proceedings of the National Academy of Sciences of the United States of America 106:55-60

1419. Behrens J, Von Kries JP, Kühl M, Bruhn L, Wedlich D, Grosschedl R, Birchmeier W (1996) Functional interaction of β-catenin with the transcription factor LEF-1. Nature 382:638–642

1420. Mosimann C, Hausmann G, Basler K (2009) β-Catenin hits chromatin: regulation of Wnt target gene activation. Nature Reviews – Molecular Cell Biology 10:276–286

1421. Ueno S, Weidinger G, Osugi T, Kohn AD, Golob JL, Pabon L, Reinecke H, Moon RT, Murry CE (2007) Biphasic role for Wnt/β-catenin signaling in cardiac specification in zebrafish and embryonic stem cells. Proceedings of the National Academy of Sciences of the United States of America 104:9685–9690

1422. Lin L, Cui L, Zhou W, Dufort D, Zhang X, Cai CL, Bu L, Yang L, Martin J, Kemler R, Rosenfeld MG, Chen J, Evans SM (2007) β-Catenin directly regulates Islet1 expression in cardiovascular progenitors and is required for multiple aspects of cardiogenesis. Proceedings of the National Academy of Sciences of the United States of America 104:9313–9318

1423. Ai D, Fu X, Wang J, Lu MF, Chen L, Baldini A, Klein WH, Martin JF (2007) Canonical Wnt signaling functions in second heart field to promote right ventricular growth. Proceedings of the National Academy of Sciences of the United States of America 104:9319–9324

1424. Zhu W, Shiojima I, Ito Y, Li Z, Ikeda H, Yoshida M, Naito AT, Nishi JI, Ueno H, Umezawa A, Minamino T, Nagai T, Kikuchi A, Asashima M, Komuro I (2008) IGFBP-4 is an inhibitor of canonical Wnt signalling required for cardiogenesis. Nature 454:345–349

1425. Alfaro MP, Pagni M, Vincent A, Atkinson J, Hill MF, Cates J, Davidson JM, Rottman J, Lee E, Young PP (2008) The Wnt modulator sFRP2 enhances mesenchymal stem cell engraftment, granulation tissue formation and myocardial repair. Proceedings of the National Academy of Sciences of the United States of America 105:18366-18371

1426. Panáková D, Werdich AA, Macrae CA (2010) Wnt11 patterns a myocardial electrical gradient through regulation of the L-type Ca^{2+} channel. Nature 466:874–878

1427. Dufourcq P, Leroux L, Ezan J, Descamps B, Lamazière JM, Costet P, Basoni C, Moreau C, Deutsch U, Couffinhal T, Duplàa C (2008) Regulation of endothelial cell cytoskeletal reorganization by a secreted frizzled-related protein-1 and frizzled 4- and frizzled 7-dependent pathway: role in neovessel formation. American Journal of Pathology 172:37–49

1428. Dufourcq P, Descamps B, Tojais NF, Leroux L, Oses P, Daret D, Moreau C, Lamazière JM, Couffinhal T, Duplàa C (2008) Secreted frizzled-related protein-1 enhances mesenchymal stem cell function in angiogenesis and contributes to neovessel maturation. Stem Cells 26:2991–3001

1429. Stenman JM, Rajagopal J, Carroll TJ, Ishibashi M, McMahon J, McMahon AP (2008) Canonical Wnt signaling regulates organ-specific assembly and differentiation of CNS vasculature. Science 322:1247–1250

1430. Daneman R, Agalliu D, Zhou L, Kuhnert F, Kuo CJ, Barres BA (2009) Wnt/β-catenin signaling is required for CNS, but not non-CNS, angiogenesis. Proceedings of the National Academy of Sciences of the United States of America 106:641–646

1431. Kobayashi K, Luo M, Zhang Y, Wilkes DC, Ge G, Grieskamp T, Yamada C, Liu TC, Huang G, Basson CT, Kispert A, Greenspan DS, Sato TN (2009) Secreted Frizzled-related protein 2 is a procollagen C proteinase enhancer with a role in fibrosis associated with myocardial infarction. Nature – Cell Biology 11:46–55

1432. Nathan E, Tzahor E (2009) sFRPs: a declaration of (Wnt) independence. Nature – Cell Biology 11:13

1433. Inestrosa NC, Arenas E (2010) Emerging roles of Wnts in the adult nervous system. Nature Reviews – Neuroscience 11:77–86

1434. Lin SL, Li B, Rao S, Yeo EJ, Hudson TE, Nowlin BT, Pei H, Chen L, Zheng JJ, Carroll TJ, Pollard JW, McMahon AP, Lang RA, Duffield JS (2010) Macrophage Wnt7b is critical for kidney repair and regeneration. Proceedings of the National Academy of Sciences of the United States of America 107:4194–4199

1435. Kaidi A, Williams AC, Paraskeva C (2007) Interaction between β-catenin and HIF-1 promotes cellular adaptation to hypoxia. Nature Cell Biology 9:210–217

1436. Giles RH, Lolkema MP, Snijckers CM, Belderbos M, van der Groep P, Mans DA, van Beest M, van Noort M, Goldschmeding R, van Diest PJ, Clevers H, Voest EE (2006) Interplay between VHL/HIF1α and Wnt/β-catenin pathways during colorectal tumorigenesis. Oncogene 25:3065–3070

1437. Kaufman DS (2010) HIF hits Wnt in the stem cell niche. Nature – Cell Biology 12:926–927

1438. Mazumdar J, O'Brien WT, Johnson RS, LaManna JC, Chavez JC, Klein PS, MC Simon (2010) O_2 regulates stem cells through Wnt/β-catenin signalling. Nature – Cell Biology 12:1007–1013

1439. Chitalia VC, Foy RL, Bachschmid MM, Zeng L, Panchenko MV, Zhou MI, Bharti A, Seldin DC, Lecker SH, Dominguez I, Cohen HT (2008) Jade-1 inhibits Wnt signalling by ubiquitylating β-catenin and mediates Wnt pathway inhibition by pVHL. Nature Cell Biology 10:1208–1216

1440. Kikuchi K, Niikura Y, Kitagawa K, Kikuchi A (2010) Dishevelled, a Wnt signalling component, is involved in mitotic progression in cooperation with Plk1. EMBO Journal 29:3470–3483

1441. Hadjihannas MV, Brückner M, Behrens J (2010) Conductin/axin2 and Wnt signalling regulates centrosome cohesion. EMBO Reports 11:317–324

1442. Brack AS, Conboy MJ, Roy S, Lee M, Kuo CJ, Keller C, Rando TA (2007) Increased Wnt signaling during aging alters muscle stem cell fate and increases fibrosis. Science 317:807–810

1443. Liu H, Fergusson MM, Castilho RM, Liu J, Cao L, Chen J, Malide D, Rovira II, Schimel D, Kuo CJ, Gutkind JS, Hwang PM, Finkel T (2007) Augmented Wnt signaling in a mammalian model of accelerated aging. Science 317:803–806

1444. Ye X, Zerlanko B, Kennedy A, Banumathy G, Zhang R, Adams PD (2007) Downregulation of Wnt signaling is a trigger for formation of facultative heterochromatin and onset of cell senescence in primary human cells. Molecular Cell 27:183–196

1445. Maetzel D, Denzel S, Mack B, Canis M, Went P, Benk M, Kieu C, Papior P, Baeuerle PA, Munz M, Gires O (2009) Nuclear signalling by tumour-associated antigen EpCAM. Nature Cell Biology 11:162–171

1446. Janssen BJ, Robinson RA, Pérez-Branguli F, Bell CH, Mitchell KJ, Siebold C, Jones EY (2010) Structural basis of semaphorin–plexin signalling. Nature 467:1118–1122

1447. Nogi T, Yasui N, Mihara E, Matsunaga Y, Noda M, Yamashita N, Toyofuku T, Uchiyama S, Goshima Y, Kumanogoh A, Takagi J (2010) Structural basis for semaphorin signalling through the plexin receptor. Nature 467:1123–1127

1448. Takahashi T, Fournier A, Nakamura F, Wang LH, Murakami Y, Kalb RG, Fujisawa H, Strittmatter SM (1999) Plexin-neuropilin-1 complexes form functional semaphorin-3A receptors. Cell 99:59–69

1449. Perrot V, Vazquez-Prado J, Gutkind JS (2002) Plexin B regulates Rho through the guanine nucleotide exchange factors leukemia-associated Rho GEF (LARG) and PDZ-RhoGEF. Journal of Biological Chemistry 277:43115–43120

1450. Gitler AD, Lu MM, Epstein JA (2004) PlexinD1 and semaphorin signaling are required in endothelial cells for cardiovascular development. Developmental Cell 7:107–116

1451. Basile JR, Castilho RM, Williams VP, Gutkind JS (2006) Semaphorin 4D provides a link between axon guidance processes and tumor-induced angiogenesis. Proceedings of the National Academy of Sciences of the United States of America 103:9017–9022

1452. Hung RJ, Yazdani U, Yoon J, Wu H, Yang T, Gupta N, Huang Z, van Berkel WJH, Terman JR (2010) Mical links semaphorins to F-actin disassembly. Nature 463:823–827

1453. Morlot C, Thielens NM, Ravelli RBG, Hemrika W, Romijn RA, Gros P, Cusack S, McCarthy AA (2007) Structural insights into the Slit-Robo complex. Proceedings of the National Academy of Sciences of the United States of America 104:14923–14928

1454. Jones CA, London NR, Chen H, Park KW, Sauvaget D, Stockton RA, Wythe JD, Suh W, Larrieu-Lahargue F, Mukouyama YS, Lindblom P, Seth P, Frias A, Nishiya N, Ginsberg MH, Gerhardt H, Zhang K, Li DY (2008) Robo4 stabilizes the vascular network by inhibiting pathologic angiogenesis and endothelial hyperpermeability. Nature Medicine 14:448–453

Chap. 11. Receptors of the Immune System

1455. Gallo EM, Ho L, Winslow MM, Staton TL, Crabtree GR (2008) Selective role of calcineurin in haematopoiesis and lymphopoiesis. EMBO Reports 9:1141–1148

1456. Martín P, del Hoyo GM, Anjuère F, Ruiz SR, Arias CF, Marín AR, Ardavín C (2000) Concept of lymphoid versus myeloid dendritic cell lineages revisited: both CD8$\alpha-$ and CD8$\alpha+$ dendritic cells are generated from CD4^{low} lymphoid-committed precursors. Blood 96:2511–2519

1457. Plata-Salamán CR (1998) Cytokines and feeding. News in Physiological Sciences 13:298–304

1458. Sims JE, Smith DE (2010) The IL-1 family: regulators of immunity. Nature Reviews – Immunology 10:89–102

1459. Stomski FC, Sun Q, Bagley CJ, Woodcock J, Goodall G, Andrews RK, Berndt MC, Lopez AF (1996) Human interleukin-3 (IL-3) induces disulfide-linked IL-3 receptor α- and β-chain heterodimerization, which is required for receptor activation but not high-affinity binding. Molecular and Cellular Biology 16:3035–3046

1460. Sokol CL, Chu NQ, Yu S, Nish SA, Laufer TM, Medzhitov R (2009) Basophils function as antigen-presenting cells for an allergen-induced T helper type 2 response. Nature – Immunology 10:713–720

1461. Kilroy GE, Foster SJ, Wu X, Ruiz J, Sherwood S, Heifetz A, Ludlow JW, Stricker DM, Potiny S, Green P, Halvorsen YD, Cheatham B, Storms RW, Gimble JM (2007) Cytokine profile of human adipose-derived stem cells: expression of angiogenic, hematopoietic, and pro-inflammatory factors. Journal of Cellular Physiology 212:702–709

1462. Lai L, Goldschneider I (2001) Cutting edge: identification of a hybrid cytokine consisting of IL-7 and the β-chain of the hepatocyte growth factor/scatter factor. Journal of Immunology 167:3550–3554

1463. Lai L, Zeff RA, Goldschneider I (2006) A recombinant single-chain IL-7/HGFβ hybrid cytokine induces juxtacrine interactions of the IL-7 and HGF (c-Met) receptors and stimulates the proliferation of CFU-S_{12}, CLPs, and pre-pro-B cells. Blood 107:1776–1784

1464. Lejeune D, Demoulin JB, Renauld JC (2001) Interleukin 9 induces expression of three cytokine signal inhibitors: cytokine-inducible SH2-containing protein, suppressor of cytokine signalling (SOCS)-2 and SOCS-3, but only SOCS-3 overexpression suppresses interleukin 9 signalling. Biochemical Journal 353:109–116

1465. Wynn TA (2003) IL-13 effector functions. Annual Review of Immunology 21:425–456

1466. Klebanoff CA, Finkelstein SE, Surman DR, Lichtman MK, Gattinoni L, Theoret MR, Grewal N, Spiess PJ, Antony PA, Palmer DC, Tagaya Y, Rosenberg SA, Waldmann TA, Restifo NP (2004) IL-15 enhances the in vivo antitumor activity of tumor-reactive CD8+ T cells. Proceedings of the National Academy of Sciences of the United States of America 101:1969–1974

1467. Søndergaard H, Skak K (2009) IL-21: roles in immunopathology and cancer therapy. Tissue Antigens 74:467–479

1468. Heinrich PC, Behrmann I, Haan S, Hermanns HM, Müller-Newen G, Schaper F (2003) Principles of interleukin (IL)-6-type cytokine signalling and its regulation. Biochemical Journal 374:1–20

1469. Petersen AM, Pedersen BK (2005) The anti-inflammatory effect of exercise. Journal of Applied Physiology 98:1154–1162

1470. Paul SR, Bennett F, Calvetti JA, Kelleher K, Wood CR, O'Hara RM Jr, Leary AC, Sibley B, Clark SC, Williams DA, Yang YC (1990) Molecular cloning of a cDNA encoding interleukin 11, a stromal cell-derived lymphopoietic and hematopoietic cytokine. Proceedings of the National Academy of Sciences of the United States of America 87:7512–7516

1471. Larousserie F, Bardel E, Pflanz S, Arnulf B, Lome-Maldonado C, Hermine O, Brégeaud L, Perennec M, Brousse N, Kastelein R, Devergne O (2005) Analysis of interleukin-27 (EBI3/p28) expression in Epstein-Barr virus- and human T-cell leukemia virus type 1-associated lymphomas: heterogeneous expression of EBI3 subunit by tumoral cells. American Journal of Pathology 166:1217–1228

1472. Pflanz S, Hibbert L, Mattson J, Rosales R, Vaisberg E, Bazan JF, Phillips JH, McClanahan TK, de Waal Malefyt R, Kastelein RA (2004) WSX-1 and glycoprotein 130 constitute a signal-transducing receptor for IL-27. Journal of Immunology 172:2225–2231

1473. Dillon SR, Sprecher C, Hammond A, Bilsborough J, Rosenfeld-Franklin M, Presnell SR, Haugen HS, Maurer M, Harder B, Johnston J, Bort S, Mudri S, Kuijper JL, Bukowski T, Shea P, Dong DL, Dasovich M, Grant FJ, Lockwood L, Levin SD, LeCiel C, Waggie K, Day H, Topouzis S, Kramer J, Kuestner R, Chen Z, Foster D, Parrish-Novak J, Gross JA (2004) Interleukin 31, a cytokine produced by activated T cells, induces dermatitis in mice. Nature – Immunology 5:752–760

1474. Xie MH, Aggarwal S, Ho WH, Foster J, Zhang Z, Stinson J, Wood WI, Goddard AD, Gurney AL (2000) Interleukin (IL)-22, a novel human cytokine that signals through the interferon receptor-related proteins CRF2-4 and IL-22R. Journal of Biological Chemistry 275:31335–31339

1475. Wang M, Liang P (2005) Interleukin-24 and its receptors. Immunology 114:166–170

1476. Sheikh F, Baurin VV, Lewis-Antes A, Shah NK, Smirnov SV, Anantha S, Dickensheets H, Dumoutier L, Renauld JC, Zdanov A, Donnelly RP, Kotenko SV (2004) Cutting edge: IL-26 signals through a novel receptor complex composed of IL-20 receptor 1 and IL-10 receptor 2. Journal of Immunology 172:2006–2010

1477. Sheppard P, Kindsvogel W, Xu W, Henderson K, Schlutsmeyer S, Whitmore TE, Kuestner R, Garrigues U, Birks C, Roraback J, Ostrander C, Dong D, Shin J, Presnell S, Fox B, Haldeman B, Cooper E, Taft D, Gilbert T, Grant FJ, Tackett M, Krivan W, McKnight G, Clegg C, Foster D, Klucher KM (2003) IL-28, IL-29 and their class II cytokine receptor IL-28R. Nature – Immunology 4:63–68

1478. Brombacher F, Kastelein RA, Alber G (2003) Novel IL-12 family members shed light on the orchestration of Th1 responses. Trends in Immunology 24:207–212

1479. Bacon CM, McVicar DW, Ortaldo JR, Rees RC, O'Shea JJ, Johnston JA (1995) Interleukin 12 (IL-12) induces tyrosine phosphorylation of JAK2 and TYK2: differential use of Janus family tyrosine kinases by IL-2 and IL-12. Journal of Experimental Medicine 181:399-404

1480. Collison LW, Workman CJ, Kuo TT, Boyd K, Wang Y, Vignali KM, Cross R, Sehy D, Blumberg RS, Vignali DA (2007) The inhibitory cytokine IL-35 contributes to regulatory T-cell function. Nature 450:566–569

1481. Gaffen SL (2009) Structure and signalling in the IL-17 receptor family. Nature Reviews – Immunology 9:556–567

1482. Woodcock JM, Zacharakis B, Plaetinck G, Bagley CJ, Qiyu S, Hercus TR, Tavernier J, Lopez AF (1994) Three residues in the common β chain of the human GM-CSF, IL-3 and IL-5 receptors are essential for GM-CSF and IL-5 but not IL-3 high affinity binding and interact with Glu21 of GM-CSF. EMBO Journal 13:5176–5185

1483. Cruikshank WW, Kornfeld H, Center DM (2000) Interleukin-16. Journal of Leukocyte Biology 67:757–766

1484. Kim SH, Han SY, Azam T, Yoon DY, Dinarello CA (2005) Interleukin-32: a cytokine and inducer of TNFα. Immunity 22:131–142

1485. Lin H, Lee E, Hestir K, Leo C, Huang M, Bosch E, Halenbeck R, Wu G, Zhou A, Behrens D, Hollenbaugh D, Linnemann T, Qin M, Wong J, Chu K, Doberstein SK, Williams LT (2008) Discovery of a cytokine and its receptor by functional screening of the extracellular proteome. Science 320:807–811

1486. Baud'huin M, Renault R, Charrier C, Riet A, Moreau A, Brion R, Gouin F, Duplomb L, Heymann D (2010) Interleukin-34 is expressed by giant cell tumours of bone and plays a key role in RANKL-induced osteoclastogenesis. Journal of Pathology 221:77-86

1487. Galibert L, Tometsko ME, Anderson DM, Cosman D, Dougall WC (1998) The involvement of multiple tumor necrosis factor receptor (TNFR)-associated factors in the signaling mechanisms of receptor activator of NF-κB, a member of the TNFR superfamily. Journal of Biological Chemistry 273:34120–34127

1488. Philipp S, Puchert M, Adam-Klages S, Tchikov V, Winoto-Morbach S, Mathieu S, Deerberg A, Kolker L, Marchesini N, Kabelitz D, Hannun YA, Schütze S, Adam D (2010) The Polycomb group protein EED couples TNF receptor 1 to neutral sphingomyelinase. Proceedings of the National Academy of Sciences of the United States of America 107:1112–1117

1489. Yazdanpanah B, Wiegmann K, Tchikov V, Krut O, Pongratz C, Schramm M, Kleinridders A, Wunderlich T, Kashkar H, Utermöhlen O, Brüning JC, Schütze S, Krönke M (2009) Riboflavin kinase couples TNF receptor 1 to NADPH oxidase. Nature 460:1159–1163

1490. Evangelidou M, Tseveleki V, Vamvakas SS, Probert L (2010) TNFRI is a positive T-cell costimulatory molecule important for the timing of cytokine responses. Immunology and Cell Biology 88:586–595

1491. Yu KY, Kwon B, Ni J, Zhai Y, Ebner R, Kwon BS (1999) A newly identified member of tumor necrosis factor receptor superfamily (TR6) suppresses LIGHT-mediated apoptosis. Journal of Biological Chemistry 274:13733–13736

1492. Migone TS, Zhang J, Luo X, Zhuang L, Chen C, Hu B, Hong JS, Perry JW, Chen SF, Zhou JX, Cho YH, Ullrich S, Kanakaraj P, Carrell J, Boyd E, Olsen HS, Hu G, Pukac L, Liu D, Ni J, Kim S, Gentz R, Feng P, Moore PA, Ruben SM, Wei P (2002) TL1A is a TNF-like ligand for DR3 and TR6/DcR3 and functions as a T cell costimulator. Immunity 16:479–492

1493. Schoppet M, Preissner KT, Hofbauer LC (2002) RANK ligand and osteoprotegerin: paracrine regulators of bone metabolism and vascular function. Arteriosclerosis, Thrombosis, and Vascular Biology 22:549–553

1494. Hsu H, Solovyev I, Colombero A, Elliott R, Kelley M, Boyle WJ (1997) ATAR, a novel tumor necrosis factor receptor family member, signals through TRAF2 and TRAF5. Journal of Biological Chemistry 272:13471–13474

1495. Nemazee D (2006) Receptor editing in lymphocyte development and central tolerance. Nature Reviews – Immunology 6:728–740

1496. Hitomi K, Tahara-Hanaoka S, Someya S, Fujiki A, Tada H, Sugiyama T, Shibayama S, Shibuya K, Shibuya A (2010) An immunoglobulin-like receptor, Allergin-1, inhibits immunoglobulin E-mediated immediate hypersensitivity reactions. Nature – Immunology 11:601–607

1497. Frantz S, Bauersachs J, Ertl G (2009) Post-infarct remodelling: contribution of wound healing and inflammation. Cardiovascular Research 81:474–481

1498. Pang SS, Berry R, Chen Z, Kjer-Nielsen L, Perugini MA, King GF, Wang C, Chew SH, La Gruta NL, Williams NK, Beddoe T, Tiganis T, Cowieson NP, Godfrey DI, Purcell AW, Wilce MC, McCluskey J, Rossjohn J (2010) The structural basis for autonomous dimerization of the pre-T-cell antigen receptor. Nature 467:844–848

1499. Weintz G, Olsen JV, Frühauf K, Niedzielska M, Amit I, Jantsch J, Mages J, Frech C, Dölken L, Mann M, Lang R (2010) The phosphoproteome of toll-like receptor-activated macrophages. Molecular Systems Biology 6:371

1500. Zhang W, Samelson LE (2000) The role of membrane-associated adaptors in T cell receptor signalling. Seminars in Immunology 12:35–41

1501. Heuer K, Arbuzova A, Strauss H, Kofler M, Freund C (2005) The helically extended SH3 domain of the T cell adaptor protein ADAP is a novel lipid interaction domain. Journal of Molecular Biology 348:1025–1035

1502. Perchonock CE, Pajerowski AG, Nguyen C, Shapiro MJ, Shapiro VS (2007) The related adaptors, adaptor in lymphocytes of unknown function X and Rlk/Itk-binding protein, have nonredundant functions in lymphocytes. Journal of Immunology 179:1768–1775

1503. Sun W, Kesavan K, Schaefer BC, Garrington TP, Ware M, Johnson NL, Gelfand EW, Johnson GL (2001) MEKK2 associates with the adapter protein Lad/RIBP and regulates the MEK5-BMK1/ERK5 pathway. Journal of Biological Chemistry 276:5093–5100

1504. Park D, Park I, Lee D, Choi YB, Lee H, Yun Y (2007) The adaptor protein Lad associates with the G protein β subunit and mediates chemokine-dependent T-cell migration. Blood 109:5122–5128

1505. Jevremovic D, Billadeau DD, Schoon RA, Dick CJ, Leibson PJ (2001) Regulation of NK cell-mediated cytotoxicity by the adaptor protein 3BP2. Journal of Immunology 166:7219–7228

1506. Kostenko EV, Olabisi OO, Sahay S, Rodriguez PL, Whitehead IP (2006) Ccpg1, a novel scaffold protein that regulates the activity of the Rho guanine nucleotide exchange factor Dbs. Molecular and Cellular Biology 26:8964–8975

1507. Wilcox A, Katsanakis KD, Bheda F, Pillay TS (2004) Asb6, an adipocyte-specific ankyrin and SOCS box protein, interacts with APS to enable recruitment of elongins B and C to the insulin receptor signaling complex. Journal of Biological Chemistry 279:38881-38888

1508. Goitsuka R, Kanazashi H, Sasanuma H, Fujimura Y, Hidaka Y, Tatsuno A, Ra C, Hayashi K, Kitamura D (2000) A BASH/SLP-76-related adaptor protein MIST/Clnk involved in IgE receptor-mediated mast cell degranulation. International Immunology 12:573–580

1509. Houlard M, Romero-Portillo F, Germani A, Depaux A, Regnier-Ricard F, Gisselbrecht S, Varin-Blank N (2005) Characterization of VIK-1: a new Vav-interacting Kruppel-like protein. Oncogene 24:28–38

1510. Lavagna-Sévenier C, Marchetto S, Birnbaum D, Rosnet O (1998) The CBL-related protein CBLB participates in FLT3 and interleukin-7 receptor signal transduction in pro-B cells. Journal of Biological Chemistry 273:14962–14967

1511. Kawai T, Akira S (2010) The role of pattern-recognition receptors in innate immunity: update on Toll-like receptors. Nature – Immunology 11:373–384

1512. Cook DN, Pisetsky DS, Schwartz DA (2004) Toll-like receptors in the pathogenesis of human disease. Nature Immunology 5:975–979

1513. Barton GM, Kagan JC (2009) A cell biological view of Toll-like receptor function: regulation through compartmentalization. Nature Reviews – Immunology 9:535–542

1514. Ewald SE, Lee BL, Lau L, Wickliffe KE, Shi GP, Chapman HA, Barton GM (2008) The ectodomain of Toll-like receptor 9 is cleaved to generate a functional receptor. Nature 456:658–662

1515. Koonin EV, Aravind L (2000) The NACHT family – a new group of predicted NTPases implicated in apoptosis and MHC transcription activation. Trends in Biochemical Sciences 25:223–224

1516. Ting JPY, Duncan JA, Lei Y (2010) How the noninflammasome NLRs function in the innate immune system. Science 327:286–290

1517. Kuenzel S, Till A, Winkler M, Häsler R, Lipinski S, Jung S, Grötzinger J, Fickenscher H, Schreiber S, Rosenstiel P (2010) The nucleotide-binding oligomerization domain-like receptor NLRC5 is involved in IFN-dependent antiviral immune responses. Journal of Immunology 184:1990–2000

1518. Schroder K, Zhou R, Tschopp J (2010) The NLRP3 inflammasome: a sensor for metabolic danger? Science 327:296–300

1519. Sutterwala FS, Flavell RA (2009) NLRC4/IPAF: a CARD carrying member of the NLR family. Clinical Immunology 130:2–6

1520. Geijtenbeek TBH, Gringhuis SI (2009) Signalling through C-type lectin receptors: shaping immune responses. Nature Reviews – Immunology 9:465–479

1521. Klesney-Tait JK, Turnbull IR, Colonna M (2006) The TREM receptor family and signal integration. Nature Immunology 7:1266–1273

1522. Lemke G, Rothlin CV (2008) Immunobiology of the TAM receptors. Nature Reviews – Immunology 8:327–336

1523. Rothlin CV, Ghosh S, Zuniga EI, Oldstone MB, Lemke G (2007) TAM receptors are pleiotropic inhibitors of the innate immune response. Cell 131:1124–1136

1524. Schwartzberg PL, Mueller KL, Qi H, Cannons JL (2009) SLAM receptors and SAP influence lymphocyte interactions, development and function. Nature Reviews – Immunology 9: 39–46

A

Notation Rules: Aliases and Symbols

Aliases that designate different types of molecules as well as those that do not carry an obvious meaning are not used in the present text. For example, P35 is an alias for annexin-A1, brain syntaxin-1A, ficolin-2, interleukin-12A, the cyclin-H assembly factor ménage á trois homolog-1, the regulatory subunit-1 of cyclin-dependent kinase CDK5, and uroplakin-3B, among others. It is substituted by AnxA1, Stx1a, Fcn2, IL12a, MAT1, CDK5R1, and UPK3B aliases, respectively. The P39 protein corresponds to the subunit D1 of the lysosomal V-type H^+ ATPase (ATP6v0d1), Jun transcription factor, a component of the Activator protein AP1, and the regulatory subunit-2 of cyclin-dependent kinase CDK5 (CDK5R2). Extracellular signal-regulated protein kinases ERK1 and ERK2, members of the mitogen-activated protein kinase (MAPK) module (last tier), are also abbreviated P44 and P42 (also P40 and P41). However, both P42 and P44 correspond to the 26S protease regulatory AAA ATPase subunit (PSMC6). The alias P42 is also utilized for cyclin-dependent kinase CDK20, cyclin-dependent kinase-like protein CDKL1, and 43-kDa nucleoporin NuP43. The alias P44 can also refer to interferon-induced protein IFI44 (or microtubule-associated protein MTAP44) and androgen receptor cofactor P44 (a.k.a. methylosome protein MeP50 and WD repeat-containing protein WDR77). In the present text, P38 members (P38α–P38δ) of the mitogen-activated protein kinase modules (i.e., MAPK11–MAPK14) [1] are designated as P38MAPKs to avoid confusion with other molecules, the alias of which is also P38. [2]

---

1. Protein P38α is also known as MAPK14, cytokine suppressive anti-inflammatory drug (CSAID)-binding protein CSBP, CSBP1, or CSBP2, and stress-activated protein kinase SAPK2a; P38β as MAPK11 and SAPK2b; P38γ as MAPK12, ERK6, and SAPK3; P38δ as MAPK13 and SAPK4.

2. Alias P38 is used for: (1) mitogen-activated protein kinase MAPK1, extracellular signal-regulated kinase ERK2, as well as P40, P41, and P42; (2) adaptor CRK (chicken tumor virus regulator of kinase, or v-crk sarcoma virus CT10 oncogene homolog); (3) growth factor receptor-binding protein GRB2-related adaptor protein GRAP2 (a.k.a. GRID, GADS, GRB2L, GRF40, GRPL, and Mona); (4) ubiquitin ligase ring finger protein RNF19a, or dorfin; (5) 38-kDa DNA polymerase-δ-interacting protein PolδIP2 (a.k.a. polymerase [DNA-directed] PDIP38 and PolD4); (6) activator of 90-kDa heat shock protein ATPase homo-

A.1 Aliases for Molecules

Aliases include all written variants, i.e., any abbreviation [3] such as acronyms. An *acronym* corresponds to a word made from the initial letters or syllables of nouns that are pronounceable as a word. Acronyms are generally written with all letters in upper case. Yet, some acronyms are treated as words and written in lower case (e.g., laser [originally LASER] is an acronym for light amplification by stimulated emission of radiation, sonar [originally SONAR] for sound navigation and ranging). A substance name can derive from its chemical name (e.g., amphetamine: α-methylphenethylamine).

Acronyms can give rise to molecule names by adding a scientific suffix such as "-in", a common ending of molecule nouns (e.g., sirtuin, a portmanteau, that comes from the alias SIRT, which stands for silent information regulator-2 [two]). Other scientific prefixes and suffixes can be frequently detected throughout the present text. Their meaning is given in appendix Notations – Prefixes and Suffixes, particularly for readers from Asia. Many prefixes are used to specify position, configuration and behavior, quantity, direction and motion, structure, timing, frequency, and speed.

A *portmanteau* is a word that combines initials and some inner letters of at least 2 words (e.g., calmodulin stands for calcium modulated protein; caspase, a cysteine-dependent aspartate-specific protease; chanzyme, an ion channel and enzyme; chemokine, a chemoattractant cytokine; [4] emilin, an elastin microfibril interfacer; endorphins and endomorphins, endogenous morphines; porin, a pore-forming protein; restin, a Reed-Steinberg cell-expressed intermediate filament-associated protein, an alias for cytoplasmic linker protein CLiP1 (or CLiP170); serpin, a serine protease inhibitor; siglec, a sialic acid-binding Ig-like lectin; transceptor, a transporter-related receptor; and Prompt, a promoter upstream transcript). [5]

Initialisms are abbreviations that are formed from initial letters of a single long noun or several nouns and, instead of being pronounced like an ordinary word, are read letter-by-letter (e.g., DNA that stands deoxyribonucleic acid).

Some abbreviations can give rise to alphabetisms that are written as new words (e.g., Rho-associated, coiled-coil-containing protein kinase [RoCK] that is also called Rho kinase). In biochemistry, multiple-letter abbreviations can also be formed from a single word that can be long (e.g., Cam stands for calmodulin, which is itself a portmanteau word, Trx for thioredoxin, etc.) as well as short (e.g., Ttn for titin, etc.).

---

log AHSA1; and (7) aminoacyl tRNA synthase complex-interacting multifunctional protein AIMP2, or tRNA synthase complex component JTV1 [608].

3. In general, abbreviations exclude the initials of short function words, such as "and", "or", "of", or "to". However, they are sometimes included in acronyms to make them pronounceable (e.g., radar [originally RADAR] for radio detection and ranging). These letters are often written in lower case. In addition, both cardinal (size, molecular weight, etc.) and ordinal (isoform discovery order) numbers in names are represented by digits.

4. Cytokines are peptidic, proteic, or glycoproteic regulators that are secreted by cells of the immune system. These immunomodulating agents serve as auto- or paracrine signals.

5. The upper case initial P in Prompt is used to avoid confusion with command-line interpreter prompt or prompt book to direct precise timing of actions on theater stage.

In addition, single-letter symbols of amino acids are often used to define a molecule alias (e.g., tyrosine can be abbreviated as Tyr or Y, hence SYK stands for spleen tyrosine kinase).

Aliases use, in general, capital letters and can include hyphens and dots. Yet, as a given protein can represent a proto-oncogene [6] encoded by a gene that can give rise to an oncogene (tumor promoter) after gain- or loss-of-function mutations, [7] the same acronym represents 3 different entities. [8]

Besides, a given abbreviation can designate distinct molecules without necessarily erroneous consequence in a given context (e.g., PAR: poly^{ADP}ribose or protease-activated receptor and GCK: germinal center kinases or glucokinase; in the latter case, the glucokinase abbreviation should be written as GcK or, better, GK).

In addition, a large number of aliases that designate a single molecule results from the fact that molecules have been discovered independently several times with possibly updated functions. Some biochemists uppercase the name of a given molecule, whereas others lowercase (e.g., cell division cycle guanosine triphosphatase of the Rho family CDC42 or Cdc42, adaptor growth factor receptor-bound protein GRB2 or Grb2, chicken tumor virus regulator of kinase CRK or Crk, guanine nucleotide-exchange factor Son of sevenless SOS or Sos, etc.). Acronyms are then not always capitalized. Printing style of aliases should not only avoid confusion, but also help one in remembering alias meaning.

In the present textbook, choice of lower and upper case letters in molecule aliases is dictated by the following criteria.

---

6. In 1911, P. Rous isolated a virus that was capable of generating tumors of connective tissue (sarcomas) in chicken. Proteins were afterward identified, the activity of which, when uncontrolled, can provoke cancer, hence the name oncogene given to genes that encode these proteins. Most of these proteins are enzymes, more precisely kinases. The first oncogene was isolated from the avian Rous virus by D. Stéhelin and called Src (from sarcoma). This investigator demonstrated that the abnormal functioning of the Src protein resulted from mutation of a normal gene, or proto-oncogene, which is involved in cell division.

7. Loss-of-function mutations cause complete or partial loss of function of gene products that operate as tumor suppressors, whereas gain-of-function mutations generate gene products with new or abnormal function that can then act as oncogenes. Typical tumor-inducing agents are enzymes, mostly regulatory kinases and small guanosine triphosphatases, that favor proliferation of cells, which do normally need to be activated to exert their activities. Once their genes are mutated, these enzymes become constitutively active. Other oncogenes include growth factors (a.k.a. mitogens) and transcription factors. Mutations can also disturb signaling axis regulation, thereby raising protein expression. Last, but not least, chromosomal translocation can also provoke the expression of a constitutively active hybrid protein.

8. Like Latin-derived shortened expressions – as well as foreign words – that are currently written in italics, genes can be italicized. However, this usage is not required in scientific textbooks published by Springer. Italic characters are then used to highlight words within a text to easily target them. Proteins are currently romanized (ordinary print), but with a capital initial. Nevertheless, names (not aliases) of chemical species are entirely lowercased in most – if not all – scientific articles, except to avoid confusion with a usual word (e.g., hedgehog animal vs. Hedgehog protein and raptor [bird of prey] vs. Raptor molecule).

(1) An upper case letter is used for initials of words that constitute molecule nouns (e.g., receptor tyrosine kinase RTK). An alias of any compound takes into account added atoms or molecules (e.g., PI: phosphoinositide and PIP: phosphoinositide phosphate) as well as their number (e.g., PIP_2: phosphatidylinositol bisphosphate and DAG: diacylglycerol).

(2) A lower case letter is used when a single letter denotes a subfamily or an isoform when it is preceded by a capital letter (e.g., PTPRe: protein tyrosine phosphatase receptor-like type-E). Nevertheless, an upper case letter is used in an alias after a single or several lower case letters to distinguish the isoform type (e.g., RhoA isoform and DNA-repair protein RecA for recombination protein-A), but OSM stands for oncostatin-M, not osmole Osm [9] to optimize molecule identification.

These criteria enable to use differently written aliases with the same sequence of letters for distinct molecules (e.g., CLIP for corticotropin-like intermediate peptide, CLiP: cytoplasmic CAP-Gly domain-containing linker protein, and iCliP: intramembrane-cleaving protease).

As the exception proves the rule, current aliases, such as PKA and PLA that designate protein kinase-A and phospholipase-A, respectively, have been kept. Preceded by only 2 upper case letters, a lower case letter that should be used to specify an isoform can bring confusion with acronyms of other protein types (e.g., phospholamban alias PLb).

Nouns (e.g., hormone-like fibroblast growth factor [hFGF] and urokinase-type plasminogen activator [uPA]) or adjectives (e.g., intracellular FGF isoform [iFGF]) that categorize a subtype of a given molecule correspond to a lower case letter to emphasize the molecule species. Hence, an upper case letter with a commonly used hyphen (e.g., I[R]-SMAD that stands for inhibitory [receptor-regulated] SMAD; V-ATPase for vacuolar adenosine triphosphatase; MT1-MMP for membrane type-1 matrix metalloproteinase; and T[V]-SNARE for target [vesicle-associated] soluble ^{N}ethylmaleimide-sensitive factor-attachment protein receptor) is then replaced by a lower case letter (e.g., i[r]SMAD, vATPase, mt1MMP, and t[v]SNARE), as is usual for RNA subtypes (mRNA, rRNA, snRNA, and tRNA for messenger, ribosomal, small nuclear, and transfer RNA, respectively). Similarly, membrane-bound and secreted forms of receptors and coreceptors that can derive from alternative mRNA splicing are defined by a lower case letter (e.g., sFGFR for secreted extracellular FGFR form and sFRP for soluble Frizzled-related protein), as well as eukaryotic translation elongation (eEF) and initiation (eIF) factors.

(3) Although `l`, `r`, and `t` can stand for molecule-like, -related, and -type, respectively, when a chemical is related to another one, in general, upper case letters are used for the sake of homogenity and to clearly distinguish between the letter L and numeral 1 (e.g., KLF: Krüppel-like factor, CTK: C-terminal Src kinase (CSK)-type kinase, and SLA: Src-like adaptor).

(4) An upper case letter is most often used for initials of adjectives contained in the molecule name (e.g., AIP: actin-interacting protein; BAX: BCL2-associated

9. Osmole: the amount of osmotically active particles that exerts an osmotic pressure of 1 atm when dissolved in 22.4 l of solvent at 0 °C.

X protein; HIF: hypoxia-inducible factor; KHC: kinesin heavy chain; LAB: linker of activated B lymphocytes; MAPK: mitogen-activated protein kinase; and SNAP: soluble N-ethylmaleimide-sensitive factor-attachment protein);

(5) Lower case letters are used when alias letters do not correspond to initials (e.g., Fox – not fox –: forkhead box), except for portmanteau words that are entirely written in minuscules (e.g., gadkin: γ1-adaptin and kinesin interactor).

This rule applies, whether alias letters do correspond to successive noun letters (e.g., Par: partitioning defective protein and Pax: paxillin, as well as BrK: breast tumor kinase and ChK: checkpoint kinase, whereas CHK denotes C-terminal Src kinase [CSK]-homologous kinase) or not (e.g., Fz: Frizzled and HhIP: Hedgehog-interacting protein),[10] except for composite chemical species (e.g., DAG: diacylglycerol). However, some current usages have been kept for short aliases of chemical species name (e.g., Rho for Ras homolog rather than RHo).

In any case, molecule (super)family (class) aliases as well as those of their members are written in capital letters, such as the IGSF (IGSF*i*: member *i*; immunoglobulin), KIF (KIF*i*; kinesin), SLC (SLC*i*; solute carrier), TNFSF (TNFSF*i*; tumor-necrosis factor), and TNFRSF (TNFRSF*i*; tumor-necrosis factor receptor) superfamily.

Gene names are also written with majuscules when the corresponding protein name contains at least one minuscule, otherwise only the gene name initial is written with an upper case letter that is then followed by lower case letters.

To highlight its function, substrate aliases (e.g., ARF GTPases) contained in a molecule alias are partly written with lower case letters (e.g., ArfRP, ArfGEF, ArfGAP stand for ARF-related protein, ARF guanine-nucleotide exchange factor, and ARF GTPase-activating protein, respectively).

Last, but not least, heavy and pedantic designation of protein isoforms based on roman numerals has been avoided and replaced by usual arabic numerals (e.g., angiotensin-2 rather than angiotensin-II), except for coagulation (or clotting) factors. Moreover, character I can mean either letter I or number 1 without obvious discrimination at first glance (e.g., GAPI that stands for Ras GTPase-activating protein GAP1, but can be used to designate a growth-associated protein inhibitor).

Unnecessary hyphenation in aliases of substances (between an upper case letter, which can define the molecule function, and the chemical alias, or between it and assigned isotype number) has been avoided. In any case, the Notation section serves not only to define aliases, but also, in some instances, as disambiguation pages.

---

10. The Hedgehog gene was originally identified in the fruit fly Drosophila melanogaster. It encodes a protein involved in the determination of segmental polarity and intercellular signaling during morphogenesis. Homologous gene and protein exist in various vertebrate species. The name of the mammal hedgehog comes from hecg and hegge (dense row of shrubs or low trees), as it resides in hedgerows, and hogg and hogge, due to its pig-like, long projecting nose (snout). The word Hedgehog hence is considered as a seamless whole.

A.2 Symbols for Physical Variables

Unlike substances aliases, symbols for physical quantities are most often represented by a single letter of the Latin or Greek alphabet (i: current; J: flux; L: length; m: mass; p: pressure; P: power; T: temperature; t: time; u: displacement; v: velocity; x: space; λ: wavelength; μ: dynamic viscosity; ρ: mass density; etc.). These symbols are specified using sub- and superscripts (c_p and c_v: heat capacity at constant pressure and volume, respectively; $\mathcal{D}_T$: thermal diffusivity; G_h: hydraulic conductivity; G_T: thermal conductivity; α_k: kinetic energy coefficient; α_m: momentum coefficient; etc.).

A physical quantity associated with a given point in space at a given time can be: (1) a scalar uniquely defined by its magnitude; (2) a vector characterized by a magnitude, a support, and a direction represented by an oriented line segment defined by a unit vector; and (3) a tensor specified by a magnitude and a few directions. To ensure a straightforward meaning of symbols used for scalar, vectorial, and tensorial quantities, bold face upper (**T**) and lower (**v**) case letters are used to denote a tensor and a vector, respectively, whereas both roman (plain, upright)-style upper and lower case letters designate a scalar.

List of Currently Used Prefixes and Suffixes

Prefixes (localization)

"ab-" (Latin) and "apo-" (Greek: απο): away from or off (abluminal: endothelial edge opposite to wetted surface; apolipoproteins: lipid carriers that cause egress [also ingress] from cells; aponeurosis (απονευρωσις; νευρον: sinew, tendon) muscle sheath that limits radial motion and enhances axial contraction; and apoptosis: separation ["-ptosis": fall (πτωσις): as leaves fall away from a tree], a type of programmed cell death)

"acr-" (variant "acro-" [ακρος]): top or apex

"ad-" (adfecto: to reach; adfio: to blow toward; adfluo: to flow toward): toward (ad- becomes "ac-" before c, k, or q; "af-" before f [afferent]; "ag-" before g [agglutination]; "al-" before l; "ap-" before p [approximation]; "as-" before s; and "at-" before t)

"cis-", "juxta-", and "para-" (παρα): near, beside, or alongside

"contra-": opposite side; "ipsi-" (ipse): same side; "latero-": side;

"ecto-" (εκτος), "exo-" (εξο), and "extra-": outside, outer, external, or beyond (exogenous chemicals produced by an external source, or xenobiotics ["xeno-": foreigner])

"endo-" (ενδον) and "intra-": inside (endogenous substances synthesized by the body's cells; endomembranes at organelle surfaces within the cell)

"ep-" (variant "eph-", or "epi-" [επι]): upon (epigenetics refers to the inheritance ("-genetic": ability to procreate [γεννητικος]) of variations in gene expression beyond ("epi-": on, upon, above, close to, beside, near, toward, against, among, beyond, and also) change in the DNA sequence.

"front-" and "pre-": anterior or in front of

"post-": behind

"infra-" and "sub-": under or below

"super-" and "supra-": above

"inter-": between or among

"peri-" (περι): around

"tele-" (τελε): remote

"trans-": across

Prefixes (composition)

"an-" and "aniso-" (ανισος): unequal, uneven, heterogeneous

"iso-" (ισος): equal, alike (isomer [μερος: part, portion]

"mono-" (μονος) and "uni-" (unicus): single

"oligo-" (ολιγος): few, little, small

"multi-" (multus), "pluri-" (plus, plures), and "poly-" (πολυς): many, much

"ultra-": in excess.

Prefixes (quantity)

"demi-" (dimidius) and "hemi-" (ημι): half

"sesqui-": one and a half (half more)
"di-" or "dis-" (δυο; δις) as well as "bi-" or "bis-": 2, twice
"tri" (τρεις, τρι-; tres, tria): 3
"tetra-" (τετρα), "quadri-" (variant: "quadr-" and "quadru-"): 4
"penta-" (πεντας; pentas), "quinqu-", and "quint-": 5
"hexa-" (εξ) and "sexa-": 6
"hepta-" (επτα): 7
"octa-" (οκτα): 8
"nona-" (εννεα): 9 (ninth part)
"deca-" (δεκα): 10
"quadra-" (quadragenarius): 40 (elements)
"quinqua-" (quinquagenarius): 50
"sexa-" (sexagenarius [sex: 6]: 60
"septua-" (septuagenarius [septem: 7]): 70
"nona-" (nonagenarius): 90

Prefixes (motion and direction)

"af-": toward the center (single master object); e.g., nerve and vascular afferents (ferre: to carry) to brain and heart, respectively, rather than toward any slave, supplied tissue from the set of the body's organs; also affector, i.e., chemical messenger that brings a signal to the cell considered as the object of interest, this exploration focus being virtually excised from the organism with its central command system, except received signals
"ef-" (effero: to take away): from the center (efferent; effector, i.e., chemical transmitter recruited by the previous mediator of a signaling cascade at a given locus to possibly translocate to another subcellular compartment)
"antero-" (anterior): before, in front of, facing, or forward
"retro-": behind or backward
"tropo-" (τροπος): duct direction; (tropa: rotation; celestial revolution); e.g., tropomyosin (μυς, musculus: muscle; μυο-: refers to muscle [μυοτρωτος: injured at a muscle])

Prefixes (structure and size)

"macro-" (μακρος): large, long, or big
"mega-" (μεγας): great, large
"meso-" (μεσος): middle
"micro-" (μικρος): small
"nano-" (νανος): dwarf, tiny
"homo-" (ομο-): same (ομολογος: agreeing, corroborating; variant: "homeo-" [homeostasis])
"hetero-" (ετερο-): other

Prefixes (timing)

"ana-" (ανα): culminating (anaphase of the cell division cycle), up, above (ανοδος: a way up, anode [positive electrode; οδος; way, path, road, track])
"ante-": before
"circa-": approximately, around (circadian: approximately one day)
"infra-": below, shorter (infradian: rhythm with lower frequency than that of circadian rhythm, not smaller period)
"inter-": among, between, during
"meta-" (μετα): after, beyond, behind, later; in the middle of (metaphase of the cell division cycle); as well as connected to, but with a change of state (metabolism) and about (metadata)
"post-": after
"pre-": earlier
"pro-" (προ): preceding, first, before (prophase of the cell division cycle)
"telo-" (τελος): end, completion
"ultra-": beyond, longer (ultradian: period smaller than that of 24–28-hour cycle, i.e., frequency greater than that of the circadian rhythm)

Prefixes (functioning modality)

"auto-" (αυτος): same, self
"brady-" (βραδυς): slow (decelerate)
"tachy-" (ταχος): rapid (accelerate)
"amphi-" (αμφι): both (amphiphilic substances are both hydrophilic and lipophilic; amphisomes are generated by both autophagosomes and endosomes)

"ana-" : upward (anabolism) or against (anaphylaxis)
"cata-" (κατα): downward (catabolism, cathode [negative electrode; οδος; way, path, road, track])
"anti-" (αντι): against
"pro-": favoring
"co-" (coaccedo: add itself to): together
"contra-": adverse, against, beside, next to, opposite
"de-": remove, reduce, separation after association (Latin de; e.g., deoxy-)
"dys-" (δυς): abnormal (δυσαης): ill-blowing)
"equi-" (æque): equal or alike
"hem-" or "hemat-" (αιμα: blood): related to blood
"hypo-" (υπο): under, beneath, and low
"hyper-" (υπερ): above, beyond, and large
"per-": through (e.g., percutaneous) and during (e.g., peroperative)
"pseudo-" (ψευδο): pretended, false
"re-"; again

Scientific suffixes

"-ase": enzyme (synthase, lipase, etc.)
"-ate": salt of a base
"-cyte" (κυτος): cell (erythro- [ερυθρος: red], leuko- [λευκος: light, bright, clear, white], thrombo- [θρομβος: lump, clot], adipo- [adeps: fat; adipalis, adipatus, adipeus, adipinus: fatty], fibro- [fibra: fiber, filament], myo- [μυς: muscle, mouse, mussel], myocardiocyte [κραδια: heart; cardiacus: related to heart, stomach; to have heart trouble, stomach trouble], etc.);
"-crine" (κρινω): to decide, to separate, and to secrete (e.g., endocrine regulator) (ευκρινεω: keep in order)
"-elle": small (organelle in a cell [like an organ in a body])
"-ium", "-ion", "-isk", and "-iscus": little ("-ium": tissue interface and envelope, such as endothelium and pericardium)
"-phil" (φιλια): attracted (αφιλια: want of friends)
"-phob" (φοβια): repulsed (υδροφοβια, hydrophobia [Latin]: horror of water)
"-phore" (φερω): carrier (αμφερω: to bring up)
"-yl" denotes a radical (molecules with unpaired electrons)
"-ploid" (πλοω): double, fold (diploid, twofold; διπλοω: to double; διαπλοω: unfold)
"-emia": in relation to flow (ανεμια: flatulence; ευηνεμια: fair wind), particularly blood condition
"-genesis" (γενεσις): cause, generation, life source, origin, productive force
"-iasis": for diseased condition
"-itis": inflammation
"-lemma" (λεμμα: skin): sheath
"-ole" and "-ule": small (arteriole and venule; variant "-ula" [blastula] and "-ulum")
"-plasma" (πλασμα): anything molded (plasma: creature generated from silt of earth)
"-plasia" (πλασια): formation, molding
"-podium" (ποδος: foot; podium [Latin]: small knoll, small protuberance): protrusion
"-poiesis" (ποιεω): production
"-soma" (σωμα): body
"-sclerosis" (σκλημα): hardness, induration
"-stasis" (στασις): stabilization (αποκαταστασις: restoration; ανυποστασις: migration)
"-stomosis" (στομα: mouth): equipped with an outlet
"-taxy/tactic" (ταχυ: rapid; τακτικος: to maneuver): related to motion (also prefix, i.e., ταχυκινησις: quick motion; ταχυνω: to accelerate; and ταχυπνοια: short breath; not [δια]ταξις: disposition, arrangement)
"-trophy/trophic" (τροφις: well fed): related to growth
"-oma": tumor of
"-pathy" (παθος, παθεια): disease of
"-tomy" (τομια) and "-ectomy": surgical removal (απλοτομια: simple incision; φαûrhουγγοτομια: laryngotomy)

List of Aliases

A

$\mathcal{A}$: Avogadro number
$\mathcal{A}(p)$: area–pressure relation
$\mathbf{A}$: Almansi strain tensor
A: cross-sectional area
$\mathtt{A}$: actin-binding site
$\mathbf{a}$: acceleration
a: major semi-axis
AA: arachidonic acid
AAA: ATPase associated with diverse cellular activities
AAA: abdominal aortic aneurysm
AAAP: aneurysm-associated antigenic protein
AAK: adaptin-associated kinase
AATK: apoptosis-associated tyrosine kinase
ABC: ATP-binding cassette transporter (transfer ATPase)
AbI: Abelson kinase interactor
Abl: Abelson leukemia viral proto-oncogene product (NRTK)
ABLIM: actin-binding LIM domain-containing protein
ABP: actin-binding protein
AC: atrial contraction
ACAP: ArfGAP with coiled-coil, ankyrin repeat, PH domains
ACase: adenylate cyclase
ACi: adenylate cyclase isoform i
ACAT: acylCoA–cholesterol acyltransferase
ACC: acetyl coenzyme-A carboxylase
ACE: angiotensin-converting enzyme
ACh: acetylcholine
ACK: activated CDC42-associated kinase
ACP1: acid phosphatase-1, soluble (lmwPTP)
ACTH: adrenocorticotropic hormone
^{F}actin: filamentous actin
(Cav–actin: caveolin-associated ^{F}actin)
^{G}actin: monomeric globular actin
AcvR: activin receptor (TGFβ receptor superfamily)
Ad: adrenaline
ADAM: a disintegrin and metallopeptidase (adamalysin)
ADAMTS: a disintegrin and metallopeptidase with thrombospondin
ADAP: adhesion and degranulation-promoting adaptor protein
ADAP: ArfGAP with dual PH domains
ADF: actin-depolymerizing factor (cofilin-related destrin)
ADH: antidiuretic hormone (vasopressin)
ADMA: asymmetric dimethylarginine
ADP: adenosine diphosphate
aDuSP: atypical dual specificity phosphatase
AE: anion exchanger
AEA: N-arachidonoyl ethanolamine (anandamide)
AF: atrial fibrillation
AFAP: ArfGAP with phosphoinositide-binding and PH domains
aFGF: acidic fibroblast growth factor (FGF1)
AGAP: ArfGAP with GTPAse, ankyrin repeat, and PH domains
AGF: autocrine growth factor

AGFG: ArfGAP with FG repeats
Ago: Argonaute protein
AGS: activator of G-protein signaling
AHR: aryl hydrocarbon receptor
AIF: apoptosis-inducing factor
AIP: actin-interacting protein
AIRe: autoimmune regulator
AKAP: A-kinase (PKA)-anchoring protein
ALE: arbitrary Eulerian Lagrangian
ALIX: apoptosis-linked gene-2-interacting protein-X
ALK: anaplastic lymphoma kinase
ALK*i*: type-*i* activin receptor-like kinase (TGFβ receptor superfamily)
ALOx5: arachidonate 5-lipoxygenase
ALOx5AP: arachidonate 5-lipoxygenase activation protein
ALP: actinin-associated LIM protein (PDLIM3)
alsin: amyotrophic lateral sclerosis protein (portmanteau)
ALX: adaptor in lymphocytes of unknown function X
AMAP: A multidomain ArfGAP protein
AMBRA: activating molecule in beclin-1-regulated autophagy protein
AMHR: anti-Müllerian hormone receptor (TGFβ receptor superfamily)
AMIS: apical membrane initiation site (lumenogenesis)
AMP: adenosine monophosphate
AMPAR: α-amino 3-hydroxy 5-methyl 4-isoxazole propionic acid receptor
AMPK: AMP-activated protein kinase
AMSH: associated molecule with SH3 domain (deubiquitinase)
AmyR: amylin receptor
Ang: angiopoietin
AngL: angiopoietin-like molecule
Ank: ankyrin
ANP: atrial natriuretic peptide
ANPR (NP_1): atrial natriuretic peptide receptor (guanylate cyclase)
ARE: activin-response element
ARE: androgen response element
ARE: anti-oxidant response element
ARNT: aryl hydrocarbon nuclear receptor translocator
ANS: autonomic nervous system
ANT: adenine nucleotide transporter
Anx: annexin
AOC: amine oxidase copper-containing protein
AoV: aortic valve
AP: (clathrin-associated) adaptor proteic complex
AP: Activator protein (transcription factor)
AP: activating enhancer-binding protein
AP4A: diadenosine tetraphosphate
APAF: apoptotic peptidase-activating factor
APAP: ArfGAP with PIx- and paxillin-binding domains
APC: antigen-presenting cell
APC: adenomatous polyposis coli protein (Ub ligase)
APC/C: anaphase-promoting complex (or cyclosome; Ub ligase)
APH: anterior pharynx defective phenotype homolog
aPKC: atypical protein kinase C
APl: action potential
Apn: adiponectin
Apo: apolipoprotein
ApoER: apolipoprotein-E receptor
APPL: adaptor containing phospho-Tyr interaction, PH domain, and Leu zipper
APS: adaptor with a PH and SH2 domain
Aqp: aquaporin
AR: adrenergic receptor (adrenoceptor)
AR: androgen receptor (nuclear receptor NR3c4; transcription factor)
AR: area ratio
ARAP: ArfGAP with RhoGAP, ankyrin repeat, PH domains
Areg: amphiregulin (EGF superfamily member)
ARF: ADP-ribosylation factor
ArfRP: ARF-related protein
ARFTS: CKI2A-locus alternate reading frame tumor suppressor (ARF or $p14^{ARF}$)
ARH: autosomal recessive hypercholesterolemia adaptor (low-density lipoprotein receptor adaptor)
ARH: aplysia Ras-related homolog
ArhGEF: RhoGEF
ARL: ADP-ribosylation factor-like protein

ARNO: ARF nucleotide site opener
ARP: absolute refractory period
ARP: actin-related protein
ARPP: cAMP-regulated phosphoprotein
Arr: arrestin
ART: arrestin-related transport adaptor (α-arrestin)
ART: adpribosyltransferase
Artn: artemin
ARVCF: armadillo repeat gene deleted in velocardiofacial syndrome
ARVD: arrythmogenic right ventricular dystrophy
AS: Akt (PKB) substrate
ASAP: artery-specific antigenic protein
ASAP: ArfGAP with SH3, ankyrin repeat, PH domains
ASIC: acid-sensing ion channel
ASK: apoptosis signal-regulating kinase
aSMC: airway smooth muscle cell
ASP: actin-severing protein
AT: antithrombin
ATAA: ascending thoracic aortic aneurysm
ATF: activating transcription factor
AtG: autophagy-related gene product
ATMK: ataxia telangiectasia mutated kinase
ATn: angiotensin
ATng: angiotensinogen
AtOx: anti-oxidant protein (metallochaperone)
ATP: adenosine triphosphate
ATPase: adenosine triphosphatase
ATR ($AT_{1/2}$): angiotensin receptor
ATRK: ataxia telangiectasia and Rad3-related kinase
AVN: atrioventricular node
AVV: atrioventricular valves
AW: analysis window

B

B: Biot-Finger strain tensor
B: bulk modulus
$\mathcal{B}$: bilinear form
b: minor semi-axis
b: body force
$\hat{\mathbf{b}}$: unit binormal
B lymphocyte (B cell): bone marrow lymphocyte
BACE: β-amyloid precursor protein-converting enzyme
BAD: BCL2 antagonist of cell death
BAF: barrier-to-autointegration factor
BAG: BCL2-associated athanogene (chaperone regulator)
BAI: brain-specific angiogenesis inhibitor (adhesion-GPCR)
BAIAP: brain-specific angiogenesis inhibitor-1-associated protein (insulin receptor substrate)
BAK: BCL2-antagonist–killer
(i)BALT: (inducible) bronchus-associated lymphoid tissue
BAMBI: BMP and activin membrane-bound inhibitor homolog
BAnk: B-cell scaffold with ankyrin repeats
Barkor: beclin-1-associated autophagy-related key regulator
BAT: brown adipose tissue
BATF: basic leucine zipper ATF-like transcription factor (B-cell-activating transcription factor)
BAX: BCL2-associated X protein
BBB: blood–brain barrier
BC: boundary condition
bCAM: basal cell adhesion molecule (Lutheran blood group glycoprotein)
BCAP: B-cell adaptor for phosphatidylinositol 3-kinase
BCAR: Breast cancer anti-estrogen resistance docking protein
BCL: B-cell lymphoma (leukemia) protein
BCLxL: B-cell lymphoma extra-large protein
BCR: B-cell receptor
BCR: breakpoint cluster region protein
Bdk: bradykinin
BDNF: brain-derived neurotrophic factor
Becn, beclin: BCL2-interacting protein
BEM: boundary element method
Best: bestrophin
bFGF: basic fibroblast growth factor (FGF2)
BFUe: burst-forming unit erythroid
BFUmeg: burst-forming unit megakaryocyte
BGT: betaine–GABA transporter
BH_4: tetrahydrobiopterin (enzyme cofactor)
BID: BH3-interacting domain death agonist
BIG: brefeldin-A-inhibited GEFs for ARFs

BIK: BCL2-interacting killer
BIM: BH3-containing protein BCL2-like 11 (BCL2L11)
BK: high-conductance, Ca^{++}-activated, voltage-gated K^+ channel
BLK: B-lymphoid tyrosine kinase
Blm: Bloom syndrome, RecQ DNA helicase-like protein
BLnk: B-cell linker protein
BM: basement membrane
BMAL: brain and muscle ARNT-like protein (gene Bmal)
BMAT: bone-marrow adipose tissue
BMF: BCL2 modifying factor
BMP: bone morphogenetic protein (TGFβ superfamily)
BMPR: bone morphogenetic protein receptor
BNIP: BCL2/adenovirus E1B 19-kDa protein-interacting protein
BNP: B-type natriuretic peptide
BMX: bone marrow Tyr kinase gene in chromosome-X product
BOC: brother of CDO
BOK: BCL2-related ovarian killer
BORG: binder of Rho GTPase
BRAG: brefeldin-resistant ArfGEF
BrCa: breast cancer-associated (susceptibility) protein (tumor suppressor; DNA-damage repair; a.k.a. FancD1)
BrD: bromodomain-containing protein
BrK: breast tumor kinase
BrSK: brain-selective kinase
BSEP: bile salt export pump
BTF: basic transcription factor
BTK: Bruton Tyr kinase
BUB: budding uninhibited by benzimidazoles

C

C: stress tensor
C: compliance
C: heat capacity
C: chronotropy
Cx: type-x chemokine C (γ)
C_D: drag coefficient
C_f: friction coefficient
C_L: lift coefficient
C_p: pressure coefficient
c: stress vector
c_τ: shear
c_w: wall shear stress
c: concentration
$c(p)$: wave speed
c_p: isobar heat capacity
c_v: isochor heat capacity
C1P: ceramide 1-phosphate
C-terminus: carboxy (carboxyl group COOH)-terminus
C/EBP: CCAAT/enhancer-binding protein
CA: computed angiography
CAi: carbonic anhydrase isoform i
Ca: calcium
Ca_V: voltage-gated Ca^{++} channel
Ca_V1.x: L-type high-voltage-gated Ca^{++} channel
Ca_V2.x: P/Q/R-type Ca^{++} channel
Ca_V3.x: T-type low-voltage-gated Ca^{++} channel
CAAT: cationic amino acid transporter
CABG: coronary artery bypass grafting
Cables: CDK5 and Abl enzyme substrate
CAK: CDK-activating kinase (pseudokinase)
Cam: calmodulin (calcium-modulated protein)
CamK: calmodulin-dependent kinase
cAMP: cyclic adenosine monophosphate
CAP: adenylate cyclase-associated protein
CAP: carboxyalkylpyrrole protein adduct
CAP: chromosome-associated protein (BrD4)
CAPN: calpain gene
CaPON: carboxy-terminal PDZ ligand of NOS1 (NOS1AP)
CAR: constitutive androstane receptor (NR1i3)
CaR: calcium-sensing receptor
CARP: cell division cycle and apoptosis regulatory protein
CAS: cellular apoptosis susceptibility protein
CAS: CRK-associated substrate (or P130CAS and BCAR1)
CAs: cadherin-associated protein
CASK: calcium–calmodulin-dependent serine kinase (pseudokinase)
CASL: CRK-associated substrate-related protein (CAS2)

CASP: cytohesin-associated scaffold protein
caspase: cysteine-dependent aspartate-specific peptidase
Cav: caveolin
CBF: coronary blood flow
CBF: core-binding factor
CBL: Casitas B-lineage lymphoma adaptor and Ub ligase
CBLb: CBL-related adaptor
CBP: cap-binding protein
CBP: CREB-binding protein
CBP: C-terminal Src kinase-binding protein
CBS: cystathionine β-synthase (H_2S production)
CCDC: coiled-coil domain-containing protein
CCICR: calcium channel-induced Ca^{++} release
CCK4: colon carcinoma kinase 4 (PTK7)
CCL: chemokine CC-motif ligand
CCN: CyR61, CTGF, and NOv (CCN1–CCN3) family
Ccn: cyclin
Ccnx–CDK*i*: type-x cyclin–type-*i* cyclin-dependent kinase dimer
CCPg: cell cycle progression protein
CCS: copper chaperone for superoxide dismutase
CCT: chaperonin containing T-complex protein
CCx: type-x chemokine CC (β)
CCR: chemokine CC motif receptor
CD: cluster determinant protein (cluster of differentiation)
CDase: ceramidase
CDC: cell division cycle protein
cDC: classical dendritic cell
CDH: CDC20 homolog
Cdh: cadherin
CDK: cyclin-dependent kinase
Cdm: caldesmon
CDO: cell adhesion molecule-related/downregulated by oncogenes
CE (CsE): cholesteryl esters
CEC: circulating endothelial cell
CELSR: cadherin, EGF-like, LAG-like, and seven-pass receptor
CenP: centromere protein
CEP: carboxyethylpyrrole
CeP: centrosomal protein
CEPC: circulating endothelial progenitor cell
Cer: ceramide
CerK: ceramide kinase
CerT: ceramide transfer protein
CETP: cholesterol ester transfer protein
CFD: computational fluid dynamics
CFLAR: caspase-8 and FADD-like apoptosis regulator
CFTR: cystic fibrosis transmembrane conductance regulator
CFU: colony-forming unit
CFUb: CFU basophil (basophil-committed stem cells)
CFUc: CFU in culture (granulocyte precursors, i.e., CFUgm)
CFUe: CFU erythroid
CFUeo: CFU eosinophil
CFUg: CFU granulocyte
CFUgm: CFU granulocyte–macrophage
CFUgemm: CFU granulocyte–erythroid–macrophage–megakaryocyte
CFUm: CFU macrophage
CFUmeg: CFU megakaryocyte
CFUs: colony-forming unit spleen (pluripotent stem cells)
CG: chromogranin
cGK: cGMP-dependent protein kinase (protein kinase G)
cGMP: cyclic guanosine monophosphate
CGN: cis-Golgi network
CGRP: calcitonin gene-related peptide
chanzyme: ion channel and enzyme
chemokine: chemoattractant cytokine
CHIP: C-terminus heat shock cognate-70-interacting protein
ChK: checkpoint kinase
CHK: CSK homologous kinase
CHOP: CCAAT/enhancer-binding protein homologous protein
CHREBP: carbohydrate-responsive element-binding protein
ChT: choline transporter
CI: cardiac index
CICR: calcium-induced calcium release
Cin: chronophin
CIP: CDC42-interacting protein

CIP2a: cancerous inhibitor of protein phosphatase-2A
CIPC: CLOCK-interacting protein, circadian
CIS: cytokine-inducible SH2-containing protein
CITED: CBP/P300-interacting transactivator with glutamic (E) and aspartic acid (D)-rich C-terminus-containing protein
CK: creatine kinase
CK: casein kinase
CKI: cyclin-dependent kinase inhibitor
CLAsP: CLiP-associated protein (microtubule binder)
ClASP: clathrin-associated sorting protein
CLC: cardiotrophin-like cytokine
ClC: voltage-gated chloride channel
ClCa: calcium-activated chloride channel
ClIC: chloride intracellular channel
CLINT: clathrin-interacting protein located in the trans-Golgi network
CLIP: corticotropin-like intermediate peptide
CLiP: cytoplasmic CAP-Gly domain-containing linker protein
iCliP: intramembrane-cleaving peptidase (that clips)
CLK: CDC-like kinase
ClNS: Cl^- channel nucleotide-sensitive
CLOCK: circadian locomotor output cycles kaput
CLP: common lymphoid progenitor
CLS: ciliary localization signal
Cmi: chylomicron
CMLP: common myeloid–lymphoid progenitor
CMP: common myeloid progenitor
CMC: cardiomyocyte
Col: collagen
CoLec: collectin
ColF: collagen fiber
CORM: carbon monoxide (CO)-releasing molecule
CNG: cyclic nucleotide-gated channel
CNS: central nervous system
CNT: connecting tubule
CNTi: concentrative nucleoside transporter (SLC28ai)
CNTF: ciliary neurotrophic factor
CntnAP: contactin-associated protein
CO: cardiac output
CoBl: Cordon-bleu homolog (actin nucleator)
COLD: chronic obstructive lung disease
COOL: Cloned out of library (RhoGEF6/7)
coSMAD: common (mediator) SMAD (SMAD4)
COx: cyclooxygenase (prostaglandin endoperoxide synthase)
COx17: cytochrome-C oxidase copper chaperone
CoP: coat protein
CoP: constitutive photomorphogenic protein (Ub ligase)
COPD: chronic obstructive pulmonary disease
COUPTF: chicken ovalbumin upstream promoter transcription factor (NR2f1/2)
CP4H: collagen prolyl 4-hydroxylase
CPC: chromosomal passenger complex
CpG: cytidine-phosphate–guanosine oligodeoxynucleotide (motif)
cPKC: conventional protein kinase C
Cpx: complexin
CR: complement component receptor
Cr: creatine
cRABP: cellular retinoic acid-binding protein
cRBP: cellular retinol-binding protein
CRAC: Ca^{++} release-activated Ca^{++} channel
CRACR: CRAC regulator
Crb: Crumbs homolog polarity complex
CRE: cAMP-responsive element
CREB: cAMP-responsive element-binding protein
CRF: corticotropin-releasing factor (family)
CRH: corticotropin-releasing hormone
CRIB: CDC42/Rac interactive-binding protein
CRIK: citron Rho-interacting, Ser/Thr kinase (STK21)
CRK: CT10 regulator of kinase
CRK: chicken tumor virus regulator of kinase
CRKL: V-CRK avian sarcoma virus CT10 homolog-like

CRL4: cullin-4A RING ubiquitin ligase
CRLR: calcitonin receptor-like receptor
CRP: C-reactive protein
Crt: calreticulin
CRTC: CREB-regulated transcription coactivator
Cry: cryptochrome
Cs: cholesterol
CSBP: cytokine-suppressive anti-inflammatory drug-binding protein
CSE: cystathionine γ-lyase (H_2S production)
CSF: cerebrospinal fluid
CSF: colony-stimulating factor
CSF1: macrophage colony-stimulating factor (mCSF)
CSF2: granulocyte–macrophage colony-stimulating factors (gmCSF and sargramostim)
CSF3: granulocyte colony-stimulating factors (gCSF and filgrastim)
CSK: C-terminal Src kinase
Csk: cytoskeleton
Csq: calsequestrin
CSS: candidate sphingomyelin synthase
CT: cardiotrophin
CT: computed tomography
CTBP: C-terminal-binding protein
CTen: C-terminal tensin-like protein
CTF: C-terminal fragment
CTGF: connective tissue growth factor
CTL: cytotoxic T lymphocyte
CTLA: cytotoxic T-lymphocyte-associated protein
Ctn: catenin
CTr: copper transporter
CtR: calcitonin receptor
CTRC: CREB-regulated transcription coactivator
Cul: cullin
CUT: cryptic unstable transcript
CVI: chronic venous insufficiency
CVLM: caudal ventrolateral medulla
CVP: central venous pressure
CVS: cardiovascular system
Cx: connexin
CXCL*i*: type-*i* CXC (C-X-C motif; α) chemokine ligand
CXCR*i*: type-*i* CXC (C-X-C motif; α) chemokine receptor
CX3CL*i*: type-*i* CX3C (δ) chemokine ligand
CX3CR*i*: type-*i* CX3C (δ) chemokine receptor
cyCK: cytosolic creatine kinase
Cyld: cylindromatosis tumor suppressor protein (deubiquitinase USPL2)
CyP: member of the cytochrome-P450 superfamily
C3G: Crk SH3-binding GEF

D

D: dromotropy
D: vessel distensibility
$\mathcal{D}$: diffusion coefficient
$\mathcal{D}_T$: thermal diffusivity
$\mathbf{D}$: deformation rate tensor
$\mathbf{d}$: displacement vector
D: flexural rigidity
D: demobilization function (from proliferation to quiescence)
d: death, decay, degradation rate
d: duration
Dab: Disabled homolog
DAD: delayed afterdepolarization
DAG: diacylglycerol
DAPC: dystrophin-associated protein complex
DAPK: death-associated protein kinase
DARC: Duffy antigen receptor for chemokine
DAT: dopamine active transporter
DAX: dosage-sensitive sex reversal, adrenal hypoplasia critical region on chromosome X (NR0b1)
DBC: deleted in breast cancer protein
DBP: albumin D-element binding protein (PAR/b–ZIP family)
DC: dendritic cell
DCA: directional coronary atherectomy
DCAF: DDB1- and Cul4-associated factor
DCC: deleted in colorectal carcinoma (netrin receptor)
DCT: distal convoluted tubule
Dctn: dynactin
DDAH: dimethylarginine dimethylaminohydrolase
DDB: damage-specific DNA-binding protein

DDEF: development and differentiation-enhancing factor (ArfGAP)
DDR: discoidin domain receptor
De: Dean number
DEC: differentially expressed in chondrocytes (DEC1 and DEC2 are a.k.a bHLHe40 and bHLHe41, bHLHb2 and bHLHb3, or HRT2 and HRT1)
DEC: deleted in esophageal cancer
DEG: delayed-early gene
deoxyHb: deoxyhemoglobin (deoxygenated hemoglobin)
DETC: dendritic epidermal $\gamma\delta$ T cell
DH: Dbl homology
DHET: dihydroxyeicosatrienoic acid
DHh: desert Hedgehog
Dia: Diaphanous
DICOM: digital imaging and communication for medicine
DICR: depolarization-induced Ca^{++} release
DISC: death-inducing signaling complex
Dkk: Dickkopf
DLg: Disc large homolog
DLL: Delta-like (Notch) ligand
DLx: distal-less homeobox protein
DM: double minute
DMM: DNA methylation modulator
DMPK: myotonic dystrophy-associated protein kinase
DMT: divalent metal transporter
DN1: double-negative-1 cell
DN2: double-negative-2 cell
DN3: double-negative-3 cell
DNA: deoxyribonucleic acid
DNAPK: DNA-dependent protein kinase
DoC2: double C2-like domain-containing protein
DOCK: dedicator of cytokinesis (GEF)
DOK: downstream of Tyr kinase docking protein
DOR: δ-opioid receptor
DPG: diphosphoglyceric acid
DRAM: damage-regulated modulator of autophagy
DRF: Diaphanous-related formin (for GTPase-triggered actin rearrangement)
DRG: dorsal root ganglion
Drl: Derailed
Dsc: desmocollin
Dsg: desmoglein
Dsh: Disheveled (Wnt-signaling mediator)
DSK: dual-specificity kinase
dsRNA: double-stranded RNA
Dst: dystonin
DUb: deubiquitinase
DUS: Doppler ultrasound
DuSP: dual-specificity phosphatase
DV: dead space volume
Dvl: Disheveled (cytoplasmic phosphoprotein; other alias Dsh)
DVT: deep-vein thrombosis
dynactin: dynein activator
DYRK: dual-specificity Tyr (Y) phosphorylation-regulated kinase

E

$\mathbf{E}$: strain tensor
$\mathbf{E}$: electric field
E: elastic modulus
E: elastance
$\mathcal{E}$: energy
$\{\hat{\mathbf{e}}_i\}_{i=1}^3$: basis
$\mathbf{e}$: strain vector
e: specific free energy
E-box: enhancer box sequence of DNA
E2: ubiquitin conjugase
E3: ubiquitin ligase
EAAT: excitatory amino acid (glutamatee–aspartate) transporter
EAD: early afterdepolarization
EAR: V-erbA-related nuclear receptor (NR2f6)
EB: end-binding protein
EBCT: electron beam CT
EBF: early B-cell factor
EC: endothelial cell
ECA: external carotid artery
ECF: extracellular fluid
ECG: electrocardiogram
ECM: extracellular matrix
ED1L: EGF-like repeat- and discoidin-1-like domain-containing protein
EDGR: endothelial differentiation gene receptor
EDHF: endothelial-derived hyperpolarizing factor

EDIL: EGF-like repeats and discoidin-1 (I)-like domain-containing protein
EDV: end-diastolic volume
EEA: early endosomal antigen
eEF: eukaryotic translation elongation factor
EEL: external elastic lamina
EET: epoxyeicosatrienoic acid
EFA6: exchange factor for ARF6 (ArfGEF)
EF-Tu: elongation factor Tu
EGF: epidermal growth factor
EGFL: EGF-like domain-containing protein
EGFR: epidermal growth factor receptor
EGR: early growth response transcription factor
EHD: C-terminal EGFR substrate-15 homology domain-containing protein
eIF: eukaryotic translation initiation factor
EL: endothelial lipase
ELAM: endothelial–leukocyte adhesion molecules
ELCA: excimer laser coronary angioplasty
ELK: ETS-like transcription factor (ternary complex factor [TCF] subfamily)
ElMo: engulfment and cell motility adaptor
Eln: elastin
ElnF: elastin fiber
ELP: early lymphoid progenitor
EMI: early mitotic inhibitor
EMR: EGF-like module-containing, mucin-like, hormone receptor-like protein
EMT: epithelial–mesenchymal transition
ENA–VASP: Enabled homolog and vasoactive (vasodilator)-stimulated phosphoprotein family
ENaC: epithelial Na^+ channel
EnaH: Enabled homolog
endo-siRNA: endogenous small interfering RNA
ENPP: ectonucleotide pyrophosphatase–phosphodiesterase
Ens: endosulfine
ENT: equilibrative nucleoside transporter
ENTPD: ectonucleoside triphosphate diphosphohydrolase
EPAC: exchange protein activated by cAMP
EPAS: endothelial PAS domain protein
EPC: endothelial progenitor cell
EPCR: endothelial protein-C receptor
EPDC: epicardial-derived cell
Epgn: epigen (EGF superfamily member)
EPH: erythropoietin-producing hepatocyte receptor kinase or pseudokinase (EPHa10 and EPHb6)
ephrin: EPH receptor interactor
Epo: erythropoietin
EPS: epidermal growth factor receptor pathway substrate
ER: endoplasmic reticulum
ERx: type-x estrogen receptor (NR3a1/2)
eRas: embryonic stem cell-expressed Ras (or hRas2)
ErbB: erythroblastoma viral gene product B (HER)
ERE: estrogen response element (DNA sequence)
Ereg: epiregulin (EGF superfamily member)
eRF: eukaryotic release factor
ERGIC: endoplasmic reticulum–Golgi intermediate compartment
ERK: extracellular signal-regulated protein kinase
ERK1/2: usually refers to ERK1 and ERK2
ERM: ezrin–radixin–moesin
ERMES: endoplasmic reticulum–mitochondrion encounter structure
ERP: effective refractory period
ERR: estrogen-related receptor (NR3b1–NR3b3)
ESCRT: endosomal sorting complex required for transport
ESL: E-selectin ligand
ESRP: epithelial splicing regulatory protein
ESV: end-systolic volume
ET: endothelin
ETP: early thymocyte progenitor
ETR ($ET_{A/B}$): endothelin receptor
ETS: E-twenty six (transcription factor; erythroblastosis virus E26 proto-oncogene product homolog)
ETV: ETS-related translocation variant
EVAR: endovascular aneurysm repair
Exo: exocyst subunit
Ext: exostosin (glycosyltransferase)

F

F: transformation gradient tensor

F: function fraction of proliferating cells
F: erythrocytic rouleau fragmentation rate
f: surface force
$\hat{\mathbf{f}}$: fiber direction unit vector
f: binding frequency
f_C: cardiac frequency
f_R: breathing frequency
f: friction shape factor
f_v: head loss per unit length
f_i: molar fraction of gas component i
FA: fatty acid
FABP: fatty acid-binding protein
FABP: filamentous actin-binding protein
FACAP: F-actin complex-associated protein
FAD: flavine adenine dinucleotide
FADD: Fas receptor-associated death domain
FAK: focal adhesion kinase
Fanc: Fanconi anemia protein
FAN: Fanconi anemia-associated nuclease
FAPP: phosphatidylinositol four-phosphate adaptor protein
Fas: death receptor (TNFRSF6a)
FasL: death ligand (TNFSF6)
FAST: Forkhead activin signal transducer
FB: fibroblast
Fbln (Fibl): fibulin
Fbn: fibrillin
FBS: F-box, Sec7 protein (ArfGEF)
FBx: F-box only protein (ArfGEF)
FC: fibrocyte
FCHO: FCH domain only protein
FcαR: Fc receptor for IgA
FcγR: Fc receptor for IgG
FcϵR: Fc receptor for IgE
FCP: TF2F-associating C-terminal domain phosphatase
FDM: finite difference method
FEM: finite element method
FERM: four point-1, ezrin–radixin–moesin domain
Fer: Fes-related Tyr kinase
Fes: feline sarcoma kinase
FFA: free fatty acid
FGF: fibroblast growth factor
FGFR: fibroblast growth factor receptor
FGR: viral feline Gardner-Rasheed sarcoma oncogene homolog kinase
FHL: four-and-a-half LIM-only protein
FHoD: formin homology domain-containing protein (FmnL)
FIH: factor inhibiting HIF1α (asparaginyl hydroxylase)
FIP: focal adhesion kinase family-interacting protein
FIT: Fat-inducing transcript
FKBP: FK506-binding protein
FlIP: flice-inhibitory protein
FLK: fetal liver kinase
fMLP: N-formyl methionyl-leucyl-phenylalanine
FN: fibronectin
Fn: fibrin
Fng: fibrinogen
Fos: Finkel Biskis Jinkins murine osteosarcoma virus sarcoma proto-oncogene product
Fox: forkhead box transcription factor
Fpn: ferroportin
FR: flow ratio
FRK: Fyn-related kinase
FrmD: FERM domain-containing adaptor
FRNK: FAK-related non-kinase
FRS: fibroblast growth factor receptor substrate
FSH: follicle-stimulating hormone
FSI: fluid–structure interaction
FVM: finite volume method
FXR: farnesoid X receptor (NR1h4)
Fz: Frizzled (Wnt GPCR)

G

G: Green-Lagrange strain tensor
G: shear modulus
G': storage modulus
G'': loss modulus
$\mathcal{G}$: Gibbs function
G: conductance
G_p: pressure gradient
G_e: electrical conductivity
G_h: hydraulic conductivity
G_T: thermal conductivity
g: gravity acceleration
g: physical quantity
g: detachment frequency
g: free enthalpy

G protein: guanine nucleotide-binding protein ($G\alpha\beta\gamma$ trimer)
$G\alpha$: α subunit (signaling mediator) of G protein
$G\alpha_i$ (Gi): inhibitory $G\alpha$ subunit
$G\alpha_s$ (Gs): stimulatory $G\alpha$ subunit
$G\alpha_t$ (Gt): transducin, $G\alpha$ subunit of rhodopsin
Gs_{XL}: extra-large Gs protein
$G\alpha_{i/o}$ (Gi/o): $G\alpha$ subunit class
$G\alpha_{q/11}$ (Gq/11): $G\alpha$ subunit class
$G\alpha_{12/13}$ (G12/13): $G\alpha$ subunit class
$G\beta\gamma$: dimeric subunit (signaling effector) of G protein
G_{gust}: gustducin, G protein α subunit (Gi/o) of taste receptor
G_{olf}: G protein α subunit (Gs) of olfactory receptor
GAB: GRB2-associated binder
GABA: γ-aminobutyric acid
$GABA_A$: GABA ionotropic receptor (Cl^- channel)
$GABA_B$: GABA metabotropic receptor (GPCR)
GABARAP: $GABA_A$ receptor-associated protein
GaBP: globular actin-binding protein
GADD: growth arrest and DNA-damage-induced protein
gadkin: γ1-adaptin and kinesin interactor
GAG: glycosaminoglycan
GAK: cyclin G-associated kinase
Gal: galanin
GAP: GTPase-activating protein
GAPDH: glyceraldehyde 3-phosphate dehydrogenase
GAS: growth arrest-specific non-coding, single-stranded RNA
GAT: γ-aminobutyric acid transporter
GATA: DNA sequence GATA-binding protein (TF)
GBF: Golgi-associated brefeldin-A-resistant guanine nucleotide-exchange factor
GCAP: guanylate cyclase-activating protein
GCC: Golgi coiled-coil domain-containing protein
GCK: germinal center kinase
GCKR: GCK-related kinase
GCNF: germ cell nuclear factor (NR6a1)
GCN2: general control non-derepressible 2 (pseudokinase)
gCSF: granulocyte colony-stimulating factor (CSF3)
GD: disialoganglioside
GDP: guanosine diphosphate
GDF: growth differentiation factor
GDF: (Rab)GDI displacement (dissociation) factor
GDI: guanine nucleotide-dissociation inhibitor
GDNF: glial cell line-derived neurotrophic factor
GEF: guanine nucleotide (GDP-to-GTP)-exchange factor
GF: growth factor
GFAP: glial fibrillary acidic protein (intermediate filament)
GFL: GDNF family of ligands
GFP: geodesic front propagation
GFR: growth factor receptor
$GFR\alpha i$: type-i GDNF family receptor-α
GGA: Golgi-localized γ-adaptin ear-containing Arf-binding protein
Ggust: (G protein) $G\alpha$ subunit gustducin
GH: growth hormone
GHR: growth hormone receptor
GHRH: growth hormone-releasing hormone
GIP: GPCR-interacting protein
GIRK: $G\beta\gamma$-regulated inwardly rectifying K^+ channel
GIT: GPCR kinase-interacting protein
GKAP: G-kinase-anchoring protein
GKAP: glucokinase-associated phosphatase (DuSP12)
GKAP: guanylate kinase-associated protein
GLK: GCK-like kinase
GluK: ionotropic glutamate receptor (kainate type)
GluN: ionotropic glutamate receptor (NMDA type)
GluR: ionotropic glutamate receptor (AMPA type)
GluT: glucose transporter
GlyCAM: glycosylation-dependent cell adhesion molecule
GlyR: glycine receptor (channel)
GlyT: glycine transporter
GM: monosialoganglioside

gmCSF: granulocyte–monocyte colony-stimulating factor (CSF2)
GMP: granulocyte–monocyte progenitor
GMP: guanosine monophosphate
GnRH: gonadotropin-releasing hormone
GP: glycoprotein
Gpc: glypican
GPI: glycosyl-phosphatidylinositol anchor
gpiAP: GPI-anchored protein
GPCR: G-protein-coupled receptor
GPx: glutathione peroxidase
GQ: quadrisialoganglioside
GR: glucocorticoid receptor (NR3c1)
GRAP: GRB2-related adaptor protein (or GAds)
GRB: growth factor receptor-bound protein
GRC: growth factor-regulated, Ca^{++}-permeable, cation channel (TRPV2)
GRE: glucocorticoid response element (DNA sequence)
GRK: G-protein-coupled receptor kinase
GRP: G-protein-coupled receptor phosphatase
GSH: reduced form of glutathione
GSK: glycogen synthase kinase
GSSG: oxidized form of glutathione (glutathione disulfide)
GT: trisialoganglioside
GTF: general transcription factor
GTP: guanosine triphosphate
GTPase: guanosine triphosphatase
GuCy: guanylate cyclase (CyG)
GWAS: genome-wide association study

H

H: height
$\mathcal{H}$: history function
$\mathtt{H}$: dissipation
$\mathtt{h}$: head loss
h: thickness
h: specific enthalpy
h_T: heat transfer coefficient
h_m: mass transfer coefficient
HA: hyaluronic acid
HAD: haloacid dehalogenase
HAP: huntingtin-associated protein
HAT: histone acetyltransferase
HAAT: heterodimeric amino acid transporter
HAND: heart and neural crest derivatives expressed protein
Hb: hemoglobin
Hb^{SNO}: ^{S}nitrosohemoglobin
HBEGF: heparin-binding EGF-like growth factor
HCK: hematopoietic cell kinase
HCLS: hematopoietic lineage cell-specific Lyn substrate protein
HCN: hyperpolarization-activated, cyclic nucleotide-gated K^+ channel
HCNP: hippocampal cholinergic neurostimulatory peptide
HCT: helical CT
HDAC: histone deacetylase complex
HDL: high-density lipoprotein
HDL–C: HDL–cholesterol
HDL–CE: HDL–cholesteryl ester
HDM: human double minute (Ub ligase)
HEET: hydroxyepoxyeicosatrienoic
hemin: heme oxygenase-1 inducer
HERG: human ether-a-go-go related gene
HER: human epidermal growth factor receptor (HER3: pseudokinase)
HES: Hairy enhancer of split
HETE: hydroxyeicosatrienoic acid
HEV: high endothelial venule
HGF: hepatocyte growth factor
HGFA: hepatocyte growth factor activator (serine peptidase)
HGFR: hepatocyte growth factor receptor
HGS: HGF-regulated Tyr kinase substrate (HRS)
HhIP: Hedgehog-interacting protein
HIF: hypoxia-inducible factor
HIP: huntingtin-interacting protein
HIP1R: HIP1-related protein
His: histamine
Hjv: hemojuvelin
HK: hexokinase
HL: hepatic lipase
HMG: high mobility group protein
HMGB: high mobility group box protein
HMGCoAR: hydroxy methylglutaryl coenzyme-A reductase
HMT: histone methyl transferase
HMWK: high-molecular-weight kininogen
HNF: hepatocyte nuclear factor (NR2a1/2)
HNP: human neutrophil peptide

hnRNP: heterogeneous nuclear ribonucleoprotein
HODE: hydroxy octadecadienoic acid
HOP: HSP70–HSP90 complex-organizing protein
HotAIR: HOX antisense intergenic RNA (large intergenic non-coding RNA)
HOx: heme oxygenase
Hox: homeobox DNA sequence (encodes homeodomain-containing morphogens)
HPK: hematopoietic progenitor kinase (MAP4K)
hpRNA: long hairpin RNA
hRas: Harvey Ras
HRE: hormone response element (DNA sequence)
HRM: hypoxia-regulated microRNA
hRNP: heterogeneous ribonucleoprotein
HRS: hepatocyte growth factor-regulated Tyr kinase substrate
HRT: Hairy and enhancer of Split-related transcription factor
HS: heparan sulfate
HSC: hematopoietic stem cell
HSC: heat shock cognate
HSER: heat stable enterotoxin receptor (guanylate cyclase 2C)
HSP: heat shock protein (chaperone)
HSPG: heparan sulfate proteoglycan
Ht: hematocrit
HTR: high temperature requirement endopeptidase

I

I: identity tensor
i: current
I: inotropy
IAP: inhibitor of apoptosis protein
IBABP: intestinal bile acid-binding protein
IC: isovolumetric contraction
ICA: internal carotid artery
ICAM: intercellular adhesion molecule (IgCAM member)
IgCAM: immunoglobulin-like cell adhesion molecule
ICF: intracellular fluid
ICliP: intramembrane-cleaving peptidase
ID: inhibitor of DNA binding
IDL: intermediate-density lipoprotein
IDmiR: immediately downregulated microRNA
IDOL: inducible degrader of LDL receptor (Ub ligase)
IEG: immediate-early gene
IEL: internal elastic lamina
IEL: intra-epithelial lymphocyte
IfIH: interferon-induced with helicase-C domain-containing protein
Ifn: interferon
IfnAR: interferon-$\alpha/\beta/\omega$ receptor
IFT: intraflagellar transport complex
Ig: immunoglobulin
IGF: insulin-like growth factor
IGFBP: IGF-binding protein
IgHC: immunoglobulin heavy chain
IgLC: immunoglobulin light chain
iGluR: ionotropic glutamate receptor
IH: intimal hyperplasia
IHh: indian Hedgehog
IK: intermediate-conductance Ca^{++}-activated K^{+} channel
IκB: inhibitor of NFκB
IKK: IκB kinase
IL: interleukin
ILC: innate lymphoid cell
iLBP: intracellular lipid-binding protein
ILK: integrin-linked (pseudo)kinase
IMM: inner mitochondrial membrane
IMP: Impedes mitogenic signal propagation
INADl: inactivation no after-potential D protein
InCenP: inner centromere protein
InF: inverted formin
InsIG: insulin-induced gene product (ER anchor)
InsL: insulin-like peptide
InsR (IR): insulin receptor
InsRR: insulin receptor-related receptor
IP: inositol phosphate
IP_3: inositol (1,4,5)-trisphosphate
IP_3R: IP_3 receptor (IP_3-sensitive Ca^{++}-release channel)
IP_4: inositol (1,3,4,5)-tetrakisphosphate
IP_5: inositol pentakisphosphate
IP_6: inositol hexakisphosphate

IPCEF: interaction protein for cytohesin exchange factor
IPOD: (perivacuolar) insoluble protein deposit
IPP: inositol polyphosphate phosphatase
IPP: ILK–PINCH–parvin complex
iPSC: induced pluripotent stem cell
IQGAP: IQ motif-containing GTPase-activating protein (IQ: first 2 amino acids of the motif: isoleucine [I; commonly] and glutamine [Q; invariably]).
IR: isovolumetric relaxation
IRAK: IL1 receptor-associated kinase (IRAK2: pseudokinase)
IRE: irreversible electroporation
IRES: internal ribosome entry site
IRF: interferon-regulatory protein (transcription factor)
IRFF: interferon-regulatory factor family
IRP: iron regulatory protein
IRS: insulin receptor substrate
ISA: intracranial saccular aneurysm
ISG: interferon-stimulated gene product
iSMAD: inhibitory SMAD (SMAD6 or SMAD7)
ITAM: immunoreceptor tyrosine-based activation motif
Itch: Itchy homolog (Ub ligase)
Itg: integrin
ITIM: immunoreceptor tyrosine-based inhibitory motif
ITK: interleukin-2-inducible T-cell kinase
ITPK: inositol trisphosphate kinase
IVC: inferior vena cava
IVP: initial value problem
IVUS: intravascular ultrasound

J

J: flux
J_m: cell surface current density
JAM: junctional adhesion molecule
JaK: Janus (pseudo)kinase
JIP: JNK-interacting protein (MAPK8IP1 and -2)
JMy: junction-mediating and regulatory protein
JNK: Jun N-terminal kinase (MAPK8–MAPK10)
JNKBP: JNK-binding protein;
JNKK: JNK kinase
JSAP: JNK/SAPK-associated protein
Jun: avian sarcoma virus-17 proto-oncogene product (Japanese juunana: seventeen [17]; TF)
JUNQ: juxtanuclear quality-control compartment

K

$\mathbf{K}$: conductivity tensor
K: bending stiffness
$\mathtt{K}$: reflection coefficient
$\mathtt{K}_d$: dissociation constant (index of ligand–target affinity: ([L][T])/[C]; [L], [T], [C]: molar concentrations of the ligand, target, and created complex, respectively)
$\mathtt{K}_M$: Michaelis constant (chemical reaction kinetics)
$\mathtt{K}_m$: material compressibility
k: cross-section ellipticity
k_{ATP}: myosin ATPase rate
k_B: Boltzmann constant (1.38×10^{-23} J/K)
k_c: spring stiffness
k_m: mass-transfer coefficient
k_P: Planck constant
K_R: resistance coefficient
KaP: karyopherin
K_{ATP}: ATP-sensitive K^+ channel
K_{Ca}1.x: BK channel
K_{Ca}2/3/4.x: SK channel
K_{Ca}5.x: IK channel
K_{IR}: inwardly rectifying K^+ channel
K_V: voltage-gated K^+ channel
KAP: kinesin (KIF)-associated protein
Kap: karyopherin
KAT: lysine (K) acetyltransferase
KCC: K^+–Cl^- cotransporter
KChAP: K^+ channel-associated protein
KChIP: K_V channel-interacting protein
KDELR: KDEL (Lys–Asp–Glu–Leu) endoplasmic reticulum retention receptor
KDR: kinase insert domain receptor

KHC: kinesin heavy chain
KIF: kinesin family
KIR: killer cell immunoglobulin-like receptor
KIT: cellular kinase in tyrosine (SCFR)
Kk: kallikrein
KLC: kinesin light chain
KLF: Krüppel-like factor
KLR: killer cell lectin-like receptor
Kn: Knudsen number
KOR: κ-opioid receptor
kRas: Kirsten Ras
Krt: keratin
KSR: kinase suppressor of Ras (adaptor; pseudokinase)

L

L: velocity gradient tensor
L: inertance
L: length
LA: left atrium
LAB: linker of activated B lymphocyte
LAd: LcK-associated adaptor
LANP: long-acting natriuretic peptide
LAP: leucine-rich repeat and PDZ domain-containing protein (4-member family)
LAP: latency-associated peptide (4 isoforms LAP1–LAP4)
LAP: nuclear lamina-associated polypeptide
LAR: leukocyte common-antigen-related receptor (PTPRF)
LAT: linker of activated T lymphocytes
LaTS: large tumor suppressor
LAX: linker of activated X cells (both B and T cells)
LBR: lamin-B receptor
LCA: left coronary artery
LCAT: lysolecithin cholesterol acyltransferase
LCC: left coronary cusp
LCK: leukocyte-specific cytosolic (non-receptor) Tyr kinase
LCP: lymphocyte cytosolic protein (adaptor SLP76)
LDL: low-density lipoprotein
LDLR: low-density lipoprotein receptor
LDV: laser Doppler velocimetry
Le: entry length
LEF: lymphoid enhancer-binding transcription factor
LGalS: lectin, galactoside-binding, soluble cell adhesion molecule
LGIC: ligand-gated ion channel
LGL: lethal giant larva protein
LH: luteinizing hormone
LIF: leukemia-inhibitory factor
LIFR: leukemia-inhibitory factor receptor
LIMA: LIM domain and actin-binding protein
LIME: LcK-interacting molecule
LIMK: Lin1, Isl1, and Mec3 motif-containing kinase
LIMS: LIM and senescent cell antigen-like-containing domain protein
LiNC: linker of nucleoskeleton and cytoskeleton
lincRNA: large intergenic non-coding RNA
LipC: hepatic lipase
LipD: lipoprotein lipase
LipE: hormone-sensitive lipase
LipG: endothelial lipase
LipH: lipase-H
liprin: LAR PTP-interacting protein
LIR: leukocyte immunoglobulin-like receptor
LIS: lissencephaly protein
LKB: liver kinase-B
LKLF: lung Krüppel-like factor
LLTC: large latent TGFβ complex
LMan: lectin, mannose-binding
LMO: LIM domain-only-7 protein
Lmod: leiomodin (actin nucleator)
LMPP: lymphoid-primed multipotent progenitor
LMR: laser myocardial revascularization
Ln: laminin
LOx: lipoxygenase
LP: lipoprotein
LPA: lysophosphatidic acid
LPase: lipoprotein lipase
lpDC: lamina propria dendritic cell
Lphn: latrophilin (adhesion-GPCR)
LPL: lysophospholipid
LPLase: lysophospholipase
LPP: lipid phosphate phosphatase
LPR: lipid phosphatase-related protein

LPS: lipopolysaccharide
LQTS: long-QT syndrome
LRAT: lecithin–retinol acyltransferase
LRH: liver receptor homolog (NR5a2)
LRO: lysosome-related organelle
LRP: LDL receptor-related protein
LRRTM: leucine-rich repeat-containing transmembrane protein
LSK: Lin−, SCA1+, KIT+ cell
LST: lethal with Sec-thirteen
LSV: long saphenous vein
LT (Lkt): leukotriene
LTBP: latent TGFβ-binding protein
LTCC: L-type Ca^{++} channel (Ca_V1)
LTK: leukocyte tyrosine kinase
LUbAC: linear ubiquitin chain assembly complex
LV: left ventricle
LVAD: left ventricular assist device
LX: lipoxin
LXR: liver X receptor (NR1h2/3)
LyVE: lymphatic vessel endothelial hyaluronan receptor

M

M: molar mass
$\mathcal{M}$: moment
m: mass
Ma: Mach number
MACF: microtubule-actin crosslinking factor
mAChR: acetylcholine muscarinic receptor (metabotropic; GPCR)
MAD: mothers against decapentaplegic homolog
MAD: mitotic arrest-deficient protein
MAdCAM: mucosal vascular addressin cell adhesion molecule
MAF: V-maf musculoaponeurotic fibrosarcoma oncogene homolog (TF)
MAGI: membrane-associated guanylate kinase-related protein with inverted domain organization
MAGP: microfibril-associated glycoprotein
MAGuK: membrane-associated guanylate kinase
MAIT: mucosal-associated invariant T lymphocyte
MALT: mucosa-associated lymphoid tissue
MAO: monoamine oxidase
MAP: microtubule-associated protein
MAP1LC3: microtubule-associated protein-1 light chain-3 (LC3)
mAP: mean arterial pressure
MAPK: mitogen-activated protein kinase
MAP2K: MAPK kinase
MAP3K: MA2KP kinase
MAP3K7IP: MAP3K7-interacting protein
MAPKAPK: MAPK-activated protein kinase
MARCKS: myristoylated alanine-rich C kinase substrate
MaRCo: macrophage receptor with collagenous structure (ScaRa2)
MARK: microtubule affinity-regulating kinase
MASTL: microtubule-associated Ser/Thr kinase-like protein
MAT: ménage à trois
MATK: megakaryocyte-associated Tyr kinase
MAVS: mitochondrial antiviral signaling protein
MBP: myosin-binding protein
MBP: myeloid–B-cell progenitor
MBTPS*i*: membrane-bound transcription factor peptidase site *i*
MCAK: mitotic centromere-associated kinesin
MCAM: melanoma cell adhesion molecule
MCL1: BCL2-related myeloid cell leukemia sequence protein-1
MCLC: stretch-gated Mid1-related chloride channel
MCM: minichromosome maintenance protein
MCP: monocyte chemoattractant protein
mCSF: macrophage colony-stimulating factor (CSF1)
MCT: monocarboxylate–proton cotransporter
mDC: myeloid dendritic cell
MDM: mitochondrial distribution and morphology protein

MDR: multiple drug resistance (ABC transporter)
MEF: myocyte enhancer factor
megCSF: megakaryocyte colony-stimulating factor
MEJ: myoendothelial junction
MELK: maternal embryonic leucine zipper kinase
MEP: megakaryocyte erythroid progenitor
MEP: myeloid–erythroid progenitor
MET: mesenchymal–epithelial transition factor (proto-oncogene; HGFR)
METC: mitochondrial electron transport chain
metHb: methemoglobin
MGIC: mechanogated ion channel
mGluR: metabotropic glutamate receptor
MGP: matrix Gla protein
MHC: major histocompatibility complex
MHC: myosin heavy chain
MyHC or MYH: myosin heavy chain gene
miCK: mitochondrial creatine kinase
MiCU: mitochondrial calcium uptake protein
Mid: midline
MinK: misshapen-like kinase
miR: microRNA
miRNP: microribonucleoprotein
MiRP: MinK-related peptide
MIRR: multichain immune-recognition receptor
MIS: Müllerian inhibiting substance
MIS: mini-invasive surgery
MIS: mitochondrial intermembrane space
MIST: mastocyte immunoreceptor signal transducer
MIT: mini-invasive therapy
MiV: mitral valve
MIZ: Myc-interacting zinc finger protein
MJD: Machado-Joseph disease protein domain-containing peptidase (DUb)
MKL: megakaryoblastic leukemia-1 fusion coactivator
MKP: mitogen-activated protein kinase phosphatase
MLC: myosin light chain
MLCK: myosin light-chain kinase
MLCP: myosin light-chain phosphatase
MLK: mixed lineage kinase
MLKL: mixed lineage kinase-like pseudokinase
MLL: mixed lineage [myeloid–lymphoid] leukemia factor
MLLT: mixed lineage leukemia translocated protein
MLP: muscle LIM protein
mmCK: myofibrillar creatine kinase
MME: membrane metalloendopeptidase
MMM: maintenance of mitochondrial morphology protein
MMP: matrix metallopeptidase
MO: mouse protein
Mo: monocyte
MOMP: mitochondrial outer membrane permeabilization
MOR: μ-opioid receptor
MP: MAPK partner
MPF: mitosis (maturation)-promoting factor (CcnB–CDK1 complex)
MPG: N-methylpurine (N-methyladenine)-DNA glycosylase
MPO: median preoptic nucleus
Mpo: myeloperoxidase
MP_P: membrane protein, palmitoylated
MPP: multipotent progenitor
MR: mineralocorticoid receptor (NR3c2)
mRas: muscle Ras (or rRas3)
MRCK: myotonic dystrophy kinase-related CDC42-binding kinase
MRI: (nuclear) magnetic resonance imaging
mRNA: messenger RNA
mRNP: messenger ribonucleoprotein
MRTF: myocardin-related transcription factor
MSC: mesenchymal stem cell
MSH: melanocyte-stimulating hormone
MSIC: mechanosensitive ion channel
MSSCT: multi-slice spiral CT
MST: mammalian sterile-twenty-like kinase
MSt1R: macrophage-stimulating-1 factor receptor (RON)
MT: metallothionein
MTM: myotubularin (myotubular myopathy-associated gene product)
mtMMP: membrane-type MMP (mt*i*MMP: type-*i* mtMMP)
MTMR: myotubularin-related phosphatase
MTOC: microtubule organizing center

MTP: myeloid–T-cell progenitor
MTP: microsomal triglyceride transfer protein
MuRF: muscle-specific RING finger (Ub ligase)
MuSK: muscle-specific kinase
MVB: multivesicular body
MVE: multivesicular endosome (MVB)
MVO2: myocardial oxygen consumption
MWSS: maximal wall shear stress
MyB: V-myb myeloblastosis viral oncogene homolog (TF)
MyC: V-myc myelocytomatosis viral oncogene homolog (TF)
MyHC: myosin heavy chain
MyLC or MYL: myosin light-chain gene
MyPT: myosin phosphatase targeting subunit
MyT: myelin transcription factor

N

N: sarcomere number
$\hat{\mathbf{n}}$: unit normal vector
n: mole number
n: PAM density with elongation x
$\mathtt{n}$: myosin head density
$\mathcal{N}_A$: Avogadro number
N-terminus: amino (amine group NH_2)-terminus
NAADP: nicotinic acid adenine dinucleotide phosphate
nAChR: acetylcholine nicotinic receptor (ionotropic; LGIC)
NAD: nicotine adenine dinucleotide
NADPH: reduced form of nicotinamide adenine dinucleotide phosphate
NAd: noradrenaline
NAF: nutrient-deprivation autophagy factor
NALT: nasal-associated lymphoid tissu
NAmPT: nicotinamide phosphoribosyltransferase
Nanog: ever young (Gaelic)
NAP: NCK-associated protein (NCKAP)
NAT: nucleobase–ascorbate transporter
NAT1: noradrenaline transporter
Na_V voltage-gated Na^+ channel
NBC: Na^+–HCO_3^- cotransporters
NCC: non-coronary cusp
NCC: Na^+–Cl^- cotransporter
Ncdn: neurochondrin
NCK: non-catalytic region of Tyr kinase adaptor
NCoA: nuclear receptor coactivator
NCoR: nuclear receptor corepressor
NCR: natural cytotoxicity-triggering receptor
ncRNA: non-coding RNA
NCS: neuronal calcium sensor
NCKX: Na^+–Ca^{++}–K^+ exchanger
NCLX: Na^+–Ca^{++}–Li^+ exchanger
NCX: Na^+–Ca^{++} exchanger
NDCBE: Na^+-dependent Cl^-–HCO_3^- exchanger
NecL: nectin-like molecule
NEDD: neural precursor cell expressed, developmentally downregulated
NDFIP: NEDD4 family-interacting protein
NeK: never in mitosis gene-A (NIMA)-related kinase
NES: nuclear export signal
NESK: NIK-like embryo-specific kinase
nesprin: nuclear envelope spectrin repeat protein
NeuroD: neurogenic differentiation protein
NF: neurofilament protein (intermediate filament)
NF: neurofibromin (RasGAP)
NFAT: nuclear factor of activated T cells
NFe2: erythroid-derived nuclear factor-2
NFH: neurofilament, heavy polypeptide
NFκB: nuclear factor κ light chain enhancer of activated B cells
NFL: neurofilament, light polypeptide
NFM: neurofilament, medium polypeptide
NGAL: neutrophil gelatinase-associated lipocalin
NGF: nerve growth factor
Ngn: neogenin (netrin receptor)
NHA: Na^+–H^+ antiporter
NHE: Na^+–H^+ exchanger
NHERF: NHE regulatory factor
NHR: nuclear hormone receptor
NIc: nucleoporin-interacting protein
NIK: NFκB-inducing kinase
NIK: NCK-interacting kinase
NIP: neointimal proliferation
NK: natural killer cell
NKCC: Na^+–Ka^+–$2Cl^-$ cotransporter

NKG: NK receptor group
NKT: natural killer T cell
NKx2: NK2 transcription factor-related homeobox protein
NLR: NOD-like receptor (nucleotide-binding oligomerization domain, Leu-rich repeat-containing)
NLS: nuclear localization signal
NMDAR: ^{N}methyl ^{D}aspartate receptor
NmU: neuromedin-U
NO: nitric oxide (nitrogen monoxide)
NonO: non-POU domain-containing octamer-binding protein
NOR: neuron-derived orphan receptor (NR4a3)
NOS: nitric oxide synthase
NOS1: neuronal NOS
NOS1AP: NOS1 adaptor protein
NOS2: inducible NOS
NOS3: endothelial NOS
NOx: NAD(P)H oxidase
Noxa: damage (Latin)
NPAS: neuronal PAS domain-containing transcription factor
NPC: nuclear-pore complex
NPC: Niemann-Pick disease type-C protein
NPC1L: Niemann-Pick protein-C1-like
nPKC: novel protein kinase C
NPY: neuropeptide Y
NR: nuclear receptor
NRAP: nebulin-related actinin-binding protein
nRas: neuroblastoma Ras
NRBP: nuclear receptor-binding protein
NRF: nuclear factor erythroid-derived-2 (NF-E2)-related factor
NRF1: nuclear respiratory factor-1
Nrg: neuregulin (EGF superfamily member)
Nrgn: neuroligin
Nrp: neuropilin (VEGF-binding molecule; VEGFR coreceptor)
NRPTP: non-receptor protein Tyr phosphatase
NRSTK: non-receptor Ser/Thr kinase
NRTK: non-receptor Tyr kinase
Nrxn: neurexin
NSCLC: non-small-cell lung cancer
NSF: N-ethylmaleimide-sensitive factor
NSLTP: non-specific lipid-transfer protein
NST: nucleus of the solitary tract
NT: neurotrophin
NT5E: ecto-5′-nucleotidase
NTCP: sodium–taurocholate cotransporter polypeptide
NTF: N-terminal fragment
NTP: nucleoside triphosphate
NTPase: nucleoside triphosphate hydrolase superfamily member
NTRK: neurotrophic tyrosine receptor kinase (TRK)
NTRKR: neurotrophic Tyr receptor kinase-related protein ($ROR_{(RTK)}$)
NTS: nucleus tractus solitarius
Nu: Nusselt number
NuAK: nuclear AMPK-related kinase
NuP: nucleoporin (nuclear-pore complex protein)
NuRD: nucleosome remodeling and histone deacetylase
NuRR: nuclear receptor-related factor (NR4a2)
nWASP: neuronal WASP

O

^{O}Glc^{N}Ac: β^{N}acetyl ^{D}glucosamine
OCRL: oculocerebrorenal syndrome of Lowe phosphatase
Oct: octamer-binding transcription factor
ODE: ordinary differential equation
OGA: ^{O}Glc^{N}Acase (β^{N}acetylglucosaminidase)
OMM: outer mitochondrial membrane
ORC: origin recognition complex
ORF: open reading frame
ORP: OSBP-related protein
OSA: obstructive sleep apnea
OSBP: oxysterol-binding protein
OSI: oscillatory shear index
OSM: oncostatin M
OSMR: oncostatin M receptor
OSR (OxSR): oxidative stress-responsive kinase
OTK: off-track (pseudo)kinase
OTU: ovarian tumor superfamily peptidase (deubiquitinase)
OTUB: otubain (Ub thioesterase of the OTU superfamily)

OVLT: organum vasculosum lamina terminalis
oxyHb: oxyhemoglobin (oxygenated hemoglobin)

P

$\mathcal{P}$: permeability
P: power
P: cell division rate
p: production rate
p: pressure
p_{i}: partial pressure of gas component i
PA: phosphatidic acid
PAAT: proton–amino acid transporter
PACS: phosphofurin acidic cluster sorting protein
PAF: platelet-activating factor
PAFAH: platelet-activating factor acetylhydrolase
PAG: phosphoprotein associated with glycosphingolipid-enriched microdomains
PAH: polycyclic aromatic hydrocarbon
PAH: pulmonary arterial hypertension
PAI: plasminogen activator inhibitor
PAK: P21-activated kinase
PALR: promoter-associated long RNA
PALS: protein associated with Lin-7
PAMP: pathogen-associated molecular pattern
PAMP: proadrenomedullin peptide
PAR: poly^{ADP}ribose
PAR: promoter-associated, non-coding RNA
PAR*i*: type-*i* peptidase-activated receptor
Par: partitioning defective protein
PARG: polyADPribosyl glycosidase
PARP: polyADPribose polymerase
PASR: promoter-associated short RNA
PATJ: protein (PALS1) associated to tight junctions
Pax: paxillin
Pax*i*: paired box protein-*i* (transcription regulator)
PBC: pre-Bötzinger complex (ventilation frequency)
PBIP: Polo box-interacting protein
PC: polycystin
PC: protein C
PCMRV: phase-contrast MR velocimetry
PCr: phosphocreatine
PCT: proximal convoluted tubule
PCTP: phosphatidylcholine-transfer protein
PD: pharmacodynamics
pDC: plasmacytoid dendritic cell
PdCD: programmed cell death protein
PdCD6IP: PdCD 6-interacting protein
PdCD1Lg: programmed cell death-1 ligand
PDE: phosphodiesterase
PDE: partial differential equation
PDGF: platelet-derived growth factor
PDGFR: platelet-derived growth factor receptor
PDI: protein disulfide isomerase
PDK: phosphoinositide-dependent kinase
Pe: Péclet number
PE: pulmonary embolism
PEBP: phosphatidylethanolamine-binding protein
PECAM: platelet–endothelial cell adhesion molecule
PEDF: pigment epithelium-derived factor (serpin F1)
PEn2: presenilin enhancer-2
PEO: proepicardial organ
Per: Period homolog
PERK: protein kinase-like endoplasmic reticulum kinase
PERP: P53 apoptosis effector related to peripheral myelin protein PMP22
PET: positron emission tomography
Pex: peroxin
PF: platelet factor
PFK: phosphofructokinase
pFRG: parafacial respiratory group
PG: prostaglandin
PGC: PPARγ coactivator
pGC: particulate guanylate cyclase
PGEA: prostaglandin ethanolamide
PGF: paracrine growth factor
PGG: prostaglandin glycerol ester
PGi2: prostacyclin
PGP: permeability glycoprotein
PGx: type-x (D, E, F, H, I) prostaglandin
PGxS: type-x prostaglandin synthase
PH: pleckstrin homology domain
PHD: prolyl hydroxylase
PhK: phosphorylase kinase

PHLPP: PH domain and Leu-rich repeat protein phosphatase
PI: phosphoinositide (phosphorylated phosphatidylinositol)
PI(4)P: phosphatidylinositol 4-phosphate
PI(*i*)P*i*K: phosphatidylinositol *i*-phosphate *i*-kinase
PI(i,j)P_2: phosphatidylinositol (i,j)-bisphosphate (PIP_2)
PI(3,4,5)P_3: phosphatidylinositol (3,4,5)-trisphosphate (PIP_3)
PI3K: phosphatidylinositol 3-kinase
PI3KAP: PI3K adaptor protein
PI*i*K: phosphatidylinositol *i*-kinase
PIAS: protein inhibitor of activated STAT (SUMo ligase)
PIC: pre-initiation complex
PICK: protein that interacts with C-kinase
PIDD: P53-induced protein with a death domain
PIKE: phosphoinositide 3-kinase enhancer (GTPase; ArfGAP)
PIKK: phosphatidylinositol 3-kinase-related kinase (pseudokinase)
PIM: provirus insertion of Molony murine leukemia virus gene product
PIN: peptidyl prolyl isomerase interacting with NIMA
PINCH: particularly interesting new Cys–His protein (or LIMS1)
PInK: PTen-induced kinase
PIP: phosphoinositide monophosphate
PIP*i*K: phosphatidylinositol phosphate *i*-kinase
PIP_2: phosphatidylinositol bisphosphate
PIP_3: phosphatidylinositol triphosphate
PIPP: proline-rich inositol polyphosphate 5-phosphatase
PIR: paired immunoglobulin-like receptor
piRNA: P-element-induced wimpy testis-interacting (PIWI) RNA
PIRT: phosphoinositide-interacting regulator of TRP channels
PITP: phosphatidylinositol-transfer protein
Pitx: pituitary (or paired-like) homeobox transcription factor
PIV: particle image velocimetry
PIX: P21-activated kinase (PAK)-interacting exchange factor (Rho(Arh)GEF6/7)
PK: pharmacokinetics
PK: protein kinase
PKA: protein kinase A
PKB: protein kinase B
PKC: protein kinase C
PKD: protein kinase D
PKG: protein kinase G
PKL: paxillin kinase linker
PKMYT (MYT): membrane-associated Tyr–Thr protein kinase
PKN: protein kinase novel
Pkp: plakophilin
PL: phospholipase
PLA2: phospholipase A2
PLC: phospholipase C
PLD: phospholipase D
PLb: phospholamban
PLd: phospholipid
PlGF: placental growth factor
PLK: Polo-like kinase
PLTP: phospholipid transfer protein
PMCA: plasma membrane Ca^{++} ATPase
PML: promyelocytic leukemia protein
PMR: percutaneous (laser) myocardial revascularization
PMRT: protein arginine methyltransferase
Pn: plasmin
Png: plasminogen
PNS: peripheral nervous system
PoG: proteoglycan
PoM: pore membrane protein
Pon: paraoxonase
POPx: partner of PIX
POSH: scaffold plenty of SH3 domains
PP: protein phosphatase
PP3: protein phosphatase 3 (PP2b or calcineurin)
PPAR: peroxisome proliferator-activated receptor (NR1c1–3)
PPG: photoplethysmography
PPId: peptidyl prolyl isomerase-D
PPIP: monopyrophosphorylated inositol phosphate
$(PP)_2$IP: bispyrophosphorylated inositol phosphate
PPK: PIP kinase
PPM: protein phosphatase (magnesium-dependent)
PPR: pathogen-recognition receptor

PPRE: PPAR response element (DNA sequence)
PR: progesterone receptor (NR3c3)
PRC: protein regulator of cytokinesis
PRC: Polycomb repressive complex
Prdx: peroxiredoxin
pre-cDC: pre-classical dendritic cell
pre-miR: precursor microRNA
preBotC: pre-Bötzinger complex
preKk: prekallikrein
PREx: PIP_3-dependent Rac exchanger (RacGEF)
PRG: plasticity-related gene product
PRH: prolactin-releasing hormone
pri-miR: primary microRNA
PRL: phosphatase of regenerating liver
Prl: prolactin
PrlR: prolactin receptor
PRMT: protein arginine (R) N-methyltransferase
Prompt: promoter upstream transcript
Protor: protein observed with Rictor
PROX: prospero homeobox gene
Prox: PROX gene product (transcription factor)
PrP: processing protein
PRPK: P53-related protein kinase
PRR: pattern recognition receptor
PRR: prorenin and renin receptor
PS: presenilin
PS: protein S
PSC: pluripotent stem cell
PSD: postsynaptic density adaptor
PsD: postsynaptic density
PSEF: pseudo-strain energy function
PSer: phosphatidylserine
PSGL: P-selectin glycoprotein ligand
PSKh: protein serine kinase H
Psm: proteasome subunit
PSTPIP: Pro–Ser–Thr phosphatase-interacting protein
PTA: plasma thromboplastin antecedent
Ptc: Patched receptor (Hedgehog signaling)
PTCA: percutaneous transluminal coronary angioplasty
PtcH: Patched Hedgehog receptor
PTCRA: PTC rotational burr atherectomy
PtdCho (PC): phosphatidylcholine
PtdEtn (PE): phosphatidylethanolamine
PtdSer (PS): phosphatidylserine
PtdIns (PI): phosphatidylinositol
PTen: phosphatase and tensin homolog deleted on chromosome ten (phosphatidylinositol 3-phosphatase)
PTFE: polytetrafluoroethylene
PTH: parathyroid hormone
PTHRP: parathyroid hormone-related protein
PTK: protein Tyr kinase
PTK7: pseudokinase (RTK)
PTP: protein Tyr phosphatase
PTPn*i*: protein Tyr phosphatase non-receptor type *i*
PTPR: protein Tyr phosphatase receptor
PTRF: RNA polymerase-1 and transcript release factor
PUFA: polyunsaturated fatty acid
PUMA: P53-upregulated modulator of apoptosis
PuV: pulmonary valve
PVF: PDGF- and VEGF-related factor
PVNH: paraventricular nucleus of hypothalamus
PVR: pulmonary vascular resistance
PWS: pulse wave speed
Px: pannexin
PXR: pregnane X receptor (NR1i2)
PYK: proline-rich tyrosine kinase
P2X: purinergic ligand-gated channel
P53AIP: P53-regulated apoptosis-inducing protein
p75NtR: pan-neurotrophin receptor

Q

Q: material quantity
Q_e: electric current density
Q_T: thermal energy (heat)
q_T: transfer rate of thermal energy (power)
q: flow rate

R

R: resistance
$\mathcal{R}$: local reaction term
R_h: hydraulic radius
R_g: gas constant
R_R: respiratory quotient

R: recruitment function (from quiescence to proliferation)
r: cell renewal rate
r: radial coordinate
RA: right atrium
RAAS: renin–angiotensin–aldosterone system
Rab: Ras from brain
Rab11FIP: Rab11 family-interacting protein
Rac: Ras-related C3-botulinum toxin substrate
RACC: receptor-activated cation channel
RACK: receptor for activated C-kinase
RAD: recombination protein-A (RecA)-homolog DNA-repair protein
Rad: radiation sensitivity protein
Rag: Ras-related GTP-binding protein
Ral: Ras-related protein
RAlBP: retinaldehyde-binding protein
RalGDS: Ral guanine nucleotide-dissociation stimulator
RAMP: (calcitonin receptor-like) receptor-activity-modifying protein
Ran: Ras-related nuclear protein
RANTES: regulated upon activation, normal T-cell expressed, and secreted product (CCL5)
RAP: receptor-associated protein
Rap: Ras-proximate (Ras-related) protein
Raptor: regulatory associated protein of TOR
RAR: retinoic acid receptor (NR1b2/3)
Ras: rat sarcoma viral oncogene homolog (small GTPase)
RasA: Ras p21 protein activator
rasiRNA: repeat-associated small interfering RNA (PIWI)
RASSF: Ras interaction/interference protein RIN1, afadin, and Ras association domain-containing family protein
RB: retinoblastoma protein
RBC: red blood cell (erythrocyte)
RBP: retinoid-binding protein
RC: ryanodine calcium channel (RyR)
RCA: right coronary artery
RCan: regulator of calcineurin
RCC: right coronary cusp
RCC: regulator of chromosome condensation
Re: Reynolds number
REDD: regulated in development and DNA-damage response gene product
Rel: reticuloendotheliosis proto-oncogene product (TF; member of NFκB)
REP: Rab escort protein
ReR: renin receptor (PRR)
restin: Reed-Steinberg cell-expressed intermediate filament-associated protein (CLiP1)
ReT: rearranged during transfection (receptor Tyr kinase)
RevRE: reverse (Rev)-ErbA (NR1d1/2) response element (DNA sequence)
RFA: radiofrequency ablation
RGL: Ral guanine nucleotide dissociation stimulator-like protein (GEF)
RGS: regulator of G-protein signaling
RHEB: Ras homolog enriched in brain
RHS: equation right-hand side
Rho: Ras homologous
RIAM: Rap1-GTP-interacting adaptor
RIBP: RLK- and ITK-binding protein
RICH: RhoGAP interacting with CIP4 homolog
RICK: receptor for inactive C-kinase
Rictor: rapamycin-insensitive companion of TOR
RIF: Rho in filopodium
RIn: Ras and Rab interactor (RabGEF)
RIN: Ras-like protein expressed in neurons (GTPase)
RIP: regulated intramembrane proteolysis
RIPK: receptor-interacting protein kinase
RISC: RNA-induced silencing complex
RIT: Ras-like protein expressed in many tissues
RKIP: Raf kinase inhibitor protein
RlBP: retinaldehyde-binding protein
RLC: RISC-loading complex
RLK: resting lymphocyte kinase (TXK)
RNA: ribonucleic acid
RNABP: RNA-binding protein
RNase: ribonuclease
RnBP: renin-binding protein
RNF2: RING finger protein-2 (Ub ligase)
RNP: ribonucleoprotein
Robo: roundabout
ROC: receptor-operated channel

RoCK: Rho-associated, coiled-coil-containing protein kinase
ROI: region of interest
ROMK: renal outer medullary potassium channel
ROR: RAR-related orphan receptor (NR1f1–NR1f3)
$ROR_{(RTK)}$: receptor Tyr kinase-like orphan receptor
ROS: reactive oxygen species
Ros: V-ros UR2 sarcoma virus proto-oncogene product (RTK)
RPIP: Rap2-interacting protein
RPS6: ribosomal protein S6
RPTP: receptor protein Tyr phosphatase
rRas: related Ras
rRNA: ribosomal RNA
RSA: respiratory sinus arrhythmia
RSE: rapid systolic ejection
RSK: P90 ribosomal S6 kinase
RSKL: ribosomal protein S6 kinase-like (pseudokinase)
rSMAD: receptor-regulated SMAD (SMAD1–SMAD3, SMAD5, and SMAD9)
RSMCS: robot-supported medical and surgical system
RSpo: R-spondin
RSTK: receptor Ser/Thr kinase
RTK: receptor Tyr kinase
RTN: retrotrapezoid nucleus
Rubicon: RUN domain and Cys-rich domain-containing, beclin-1-interacting protein
Runx: Runt-related transcription factor
RV: right ventricle
RVF: rapid ventricular filling
RVLM: rostral ventrolateral medulla
RVMM: rostral ventromedial medulla
RXR: retinoid X receptor (NR2b1–NR2b3)
RYK: receptor-like Tyr (Y) kinase (pseudokinase)
RyR: ryanodine receptor (ryanodine-sensitive Ca^{++}-release channel)

S

S: Cauchy-Green deformation tensor
s: entropy
s: sarcomere length
s: evolution speed
SAA: serum amyloid A
$SAC_{Cl(K)}$: stretch-activated Cl^- (K^+)-selective channel
SAc: suppressor of actin domain-containing 5-phosphatase
sAC: soluble adenylate cyclase
$SACC_{NS}$: stretch-activated cation non-selective channel
SACM1L: suppressor of actin mutation-1-like
SAH: subarachnoid hemorrhage
SAIC: stretch-activated ion channel
SAN: sinoatrial node
SAP: SLAM-associated protein
SAP: stress-activated protein
SAP*i*: synapse-associated protein *i*
SAPK: stress-activated protein kinase (MAPK)
SAR: secretion-associated and Ras-related protein
SBE: SMAD-binding element
SBF: SET-binding factor
Sc: Schmidt number
SCA: stem cell antigen
SCAMP: secretory carrier membrane protein
SCAP: SREBP cleavage-activating protein (SREBP escort)
SCAR: suppressor of cAMP receptor (WAVe)
ScaR: scavenger receptor
SCF: SKP1–Cul1–F-box Ub-ligase complex
SCF: stem cell factor
SCFR: stem cell factor receptor (KIT)
Scgb: secretoglobin
SCLC: small-cell lung cancer
scLC: squamous-cell lung cancer (NSCLC subtype)
SCN: suprachiasmatic nucleus
SCO: synthesis of cytochrome-C oxidase
SCP (CTDSP): small C-terminal domain (CTD)-containing phosphatase
Scp: stresscopin (urocortin 3)
Scrib: Scribble polarity protein
Sdc: syndecan
SDF: stromal cell-derived factor
SDPR: serum deprivation protein response
SE: systolic ejection

SEF: strain-energy function
SEF: similar expression to FGF genes (inhibitor of RTK signaling)
SEK: SAPK/ERK kinase
Sema: semaphorin (Sema, Ig, transmembrane, and short cytoplasmic domain)
SERCA: sarco(endo)plasmic reticulum calcium ATPase
serpin: serine peptidase inhibitor
SerT: serotonin transporter
SF: steroidogenic factor (NR5a1)
SFK: SRC-family kinase
SFO: subfornical organ
SFPQ: splicing factor proline and glutamine-rich
sFRP: secreted Frizzled-related protein
SftP (SP): surfactant protein
sGC: soluble guanylate cyclase
SGK: serum- and glucocorticoid-regulated kinase
SGlT: Na^+–glucose cotransporter (SLC5a)
Sgo: shugoshin (Japanese: guardian spirit)
SH: Src homology domain
Sh: Sherwood number
SH3P: Src homology-3 domain-containing adaptor protein
Shank: SH3 and multiple ankyrin repeat domain-containing protein
SHAX: SNF7 (VSP32) homolog associated with ALIX
SHB: Src homology-2 domain-containing adaptor
SHC: Src-homologous and collagen-like substrate
SHC: Src homology-2 domain-containing transforming protein
SHh: sonic Hedgehog
SHIP: SH-containing inositol phosphatase
SHP: SH-containing protein Tyr phosphatase (PTPn6/11)
SHP: small heterodimer partner (NR0b2)
shRNA: small (short) hairpin RNA
SIAH: Seven in absentia homolog (Ub ligase)
siglec: sialic acid-binding Ig-like lectin
SIK: salt-inducible kinase
SIn: stress-activated protein kinase-interacting protein
SIP: steroid receptor coactivator-interacting protein
siRNA: small interfering RNA
SiRP: signal-regulatory protein
SIRT: sirtuin (silent information regulator-2 [two]; histone deacetylase)
SIT: SHP2-interacting transmembrane adaptor
SK: small conductance Ca^{++}-activated K^+ channel
SK*i*: sphingosine kinase-*i*
SKIP: sphingosine kinase-1-interacting protein
SKIP: skeletal muscle and kidney-enriched inositol phosphatase
SKP: S-phase kinase-associated protein
SLA: Src-like adaptor
SLAM: signaling lymphocytic activation molecule
SLAMF: SLAM family member
SLAP: Src-like adaptor protein
SLC: solute carrier class member
SLCO: solute carrier organic anion class transporter
SLK: Ste20-like kinase
Sln: sarcolipin
SLPI: secretory leukocyte peptidase inhibitor
SLTC: small latent TGFβ complex
SM: sphingomyelin
SMA: smooth muscle actin
SMAD: small (son of, similar to) mothers against decapentaplegia homolog
SMAP: Small ArfGAP protein, stromal membrane-associated GTPase-activating protein
SMase: sphingomyelinase
SMC: smooth muscle cell
Smo: Smoothened
SMPD: sphingomyelin phosphodiesterase
SMRT: silencing mediator of retinoic acid and thyroid hormone receptor
SMS: sphingomyelin synthase
SMURF: SMAD ubiquitination regulatory factor
SNAAT: sodium-coupled neutral amino acid transporter
SNAP: soluble N-ethylmaleimide-sensitive factor-attachment protein
SnAP: synaptosomal-associated protein

SNARE: SNAP receptor
SNF7: sucrose non-fermenting (VPS32)
SNIP: SMAD nuclear-interacting protein
snoRNA: small nucleolar RNA
snoRNP: small nucleolar ribonucleoprotein
SNP: single-nucleotide polymorphism
snRNA: small nuclear RNA
snRNP: small nuclear ribonucleoprotein
SNx: sorting nexin
SOC: store-operated Ca^{++} channel
SOCE: store-operated Ca^{++} entry
SOCS: suppressor of cytokine signaling protein
SOD: superoxide dismutase
SorbS: sorbin and SH3 domain-containing adaptor
SOS: Son of sevenless (GEF)
Sost: sclerostin
SostDC: sclerostin domain-containing protein
SOX: sex-determining region Y (SRY)-box gene
Sox: SOX gene product (transcription factor)
SP1: specificity protein (transcription factor)
SPARC: secreted protein acidic and rich in cysteine
SPC: sphingosylphosphorylcholine
SPCA: secretory pathway Ca^{++} ATPase
SPECT: single photon emission CT
Sph: sphingosine
SphK: sphingosine kinase
SPI: spleen focus-forming virus (SFFV) proviral integration proto-oncogene product (transcription factor)
SPInt: serine peptidase inhibitor
SPN: supernormal period
SPP: sphingosine phosphate phosphatase
SpRED: Sprouty-related protein with an EVH1 domain
SPURT: secretory protein in upper respiratory tract
SQTS: short-QT syndrome
SR: sarcoplasmic reticulum
SR: Arg/Ser domain-containing protein (alternative splicing)
SRA: steroid receptor RNA activator
SRC: steroid receptor coactivator
Src: sarcoma-associated (Schmidt-Ruppin A2 viral oncogene homolog) kinase
SREBP: sterol regulatory element-binding protein
SRF: serum response factor
SRM/SMRS: Src-related kinase lacking regulatory and myristylation sites
SRP: stresscopin-related peptide (urocortin 2)
SRPK: splicing factor RS domain-containing protein kinase
SRY: sex-determining region Y
SSAC: shear stress-activated channel
SSE: slow systolic ejection
Ssh: slingshot homolog phosphatase
SSI: STAT-induced STAT inhibitor
ssRNA: single-stranded RNA
Sst: somatostatin
SSV: short saphenous vein
St: Strouhal number
STAM: signal-transducing adaptor molecule
STAMBP: STAM-binding protein (Ub isopeptidase)
StAR: steroidogenic acute regulatory protein
StART: StAR-related lipid transfer protein
STAT: signal transducer and activator of transduction
STEAP: six transmembrane epithelial antigen of the prostate
STICK: substrate that interacts with C-kinase
StIM: stromal interaction molecule
STK: protein Ser/Thr kinase
STK1: stem cell protein Tyr kinase receptor
STLK: Ser/Thr kinase-like (pseudo)kinase
Sto: Stokes number
StRAd: STe20-related adaptor
STRAP: Ser/Thr kinase receptor-associated protein
StRAP: stress-responsive activator of P300
Stx: syntaxin ($SNARE^{Q}$)
SUMo: small ubiquitin-related modifier
SUn: Sad1 and Unc84 homology protein
SUR: sulfonylurea receptor
SUT: stable unannotated transcript
SV: stroke volume
SVC: superior vena cava
SVCT: sodium-dependent vitamin-C transporter
SVF: slow ventricular filling
SVP: synaptic vesicle precursor

SVR: systemic vascular resistance
SW: stroke work
SwAP70: 70-kDa switch-associated protein (RacGEF)
SYK: spleen tyrosine kinase
Synj: synaptojanin
Syp: synaptophysin
Syt: synaptotagmin
S1P: sphingosine 1-phosphate
S6K: P70 ribosomal S6 kinase ($P70^{RSK}$)

T

T: extrastress tensor
T: transition rate from a cell cycle phase to the next
T: temperature
T lymphocyte (T cell): thymic lymphocyte
T_C: cytotoxic T lymphocyte (CD8+ effector T cell; CTL)
T_{C1}: type-1 cytotoxic T lymphocyte
T_{C2}: type-2 cytotoxic T lymphocyte
T_{CM}: central memory T lymphocyte
T_{Conv}: conventional T lymphocyte
T_{Eff}: effector T lymphocyte
T_{EM}: effector memory T lymphocyte
T_{FH}: follicular helper T lymphocyte
T_H: helper T lymphocyte (CD4+ effector T cell)
T_{Hi}: type-i helper T lymphocyte ($i = 1/2/9/17/22$)
T_{H3}: TGFβ-secreting T_{Reg} lymphocyte
T_L: lung transfer capacity (alveolocapillary membrane)
T_{R1}: type-1, IL10-secreting, regulatory T lymphocyte
T_{Reg}: regulatory T lymphocyte
aT_{Reg}: CD45RA−, FoxP3^{hi}, activated T_{Reg} cell
iT_{Reg}: inducible T_{Reg} lymphocyte
nT_{Reg}: naturally occurring (natural) T_{Reg} lymphocyte
rT_{Reg}: CD45RA+, FoxP3^{low}, resting T_{Reg} cell
$\hat{\mathbf{t}}$: unit tangent
t: time
TβRi: type-i TGFβ receptor
TAA: thoracic aortic aneurysm
TAB: TAK1-binding protein
TACE: tumor-necrosis factor-α-converting enzyme (ADAM17)
TACE: transarterial chemoembolization
TAF: TBP-associated factor
TAK: TGFβ-activated kinase (MAP3K7)
TALK: TWIK-related alkaline pH-activated K^+ channel
TANK: TRAF family member-associated NFκB activator
TASK: TWIK-related acid-sensitive K^+ channel
TASR: terminus-associated short RNA
TAP: transporter associated with antigen processing (ABC transporter)
Taz: taffazin
TBC1D: Tre2 (or USP6), BUB2, CDC16 domain-containg RabGAP
TBCK: tubulin-binding cofactor kinase (pseudokinase)
TBK: TANK-binding kinase
TBP: TATA box-binding protein (subclass-4F transcription factor)
TBx: T-box transcription factor
TC: thrombocyte (platelet)
TCA: tricarboxylic acid cycle
TCF: T-cell factor
TCF: ternary complex factor
TcFi: type-i transcription factor
TCP: T-complex protein
TCR: T-cell receptor
TEA: transluminal extraction atherectomy
TEC: Tyr kinase expressed in hepatocellular carcinoma
TEF: thyrotroph embryonic factor (PAR/b–ZIP family)
TEK: Tyr endothelial kinase
TEM: transendothelial migration
Ten: tenascin
TF: transcription factor
Tf: transferrin
TFPI: tissue factor pathway inhibitor
TfR: transferrin receptor
TG: triglyceride (triacylglycerol)
TGF: transforming growth factor
TGFBR: TGFβ receptor gene
TGFβRAP: TGFβ receptor-associated protein
TGN: trans-Golgi network
THETE: trihydroxyeicosatrienoic acid

THIK: tandem pore-domain halothane-inhibited K^+ channel
THR: thyroid hormone receptor (NR1a1/2)
TIAM: T-lymphoma invasion and metastasis-inducing protein (RacGEF)
TICE: transintestinal cholesterol efflux
TIE: Tyr kinase with Ig and EGF homology domains (angiopoietin receptor)
TIEG: TGFβ-inducible early gene product
TIGAR: TP53-inducible glycolysis and apoptosis regulator
TIM: T-cell immunoglobulin and mucin domain-containing protein
Tim: timeless homolog
TIMM: translocase of inner mitochondrial membrane
TIMP: tissue inhibitor of metallopeptidase
TIRAP: Toll–IL1R domain-containing adaptor protein
tiRNA: transcription initiation RNA
TJ: tight junction
TKR: Tyr kinase receptor
TLC: total lung capacity
TLR: Toll-like receptor
TLT: TREM-like transcript
TLX: tailless receptor (NR2e1)
TM: thrombomodulin
TM*i*: transmembrane segment-*i* of membrane protein
TMC: twisting magnetocytometry
TMePAI: transmembrane prostate androgen-induced protein
TMy: tropomyosin
Tnn (TN): troponin
Tn: thrombin
TNF: tumor-necrosis factor
TNFαIP: tumor-necrosis factor-α-induced protein
TNFR: tumor-necrosis factor receptor
TNFRSF: tumor-necrosis factor receptor superfamily member
TNFSF: tumor-necrosis factor superfamily member
TNK: Tyr kinase inhitor of NFκB
Tns: tensin
TOR: target of rapamycin
TORC: target of rapamycin complex
TORC: transducer of regulated CREB activity (a.k.a. CRTC)
TP: thromboxane-A2 Gq/11-coupled receptor
TP53I: tumor protein P53-inducible protein
tPA: tissue plasminogen activator
Tpo: thrombopoietin
TPPP: tubulin polymerization-promoting protein
TPST: tyrosylprotein sulftotransferase
TR: testicular receptor (NR2c1/2)
TRAAK: TWIK-related arachidonic acid-stimulated K^+ channel
TRADD: tumor-necrosis factor receptor-associated death domain adaptor
TRAF: tumor-necrosis factor receptor-associated factor
TRAM: TRIF-related adaptor molecule
transceptor: transporter-related receptor
TRAP: TNF receptor-associated protein (HSP75)
TraPP: transport protein particle
TRAT: T-cell receptor-associated transmembrane adaptor
Trb: Tribbles homolog (pseudokinase)
TRE: TPA-response element (AP1/CREB-binding site on promoters)
TRE: trapped in endoderm
TREK: TWIK-related K^+ channel
TREM: triggering receptor expressed on myeloid cells
TRESK: TWIK-related spinal cord K^+ channel
TRF: TBP-related factor
TRH: thyrotropin-releasing hormone
TRIF: Toll–IL1R domain-containing adaptor inducing Ifnβ
TRIM: T-cell receptor-interacting molecule
TRIP: TGFβ receptor-interacting protein (eIF3S2)
TRK: tropomyosin receptor kinase (NTRK)
tRNA: transfer RNA
TRP: transient receptor potential channel
TRPA: ankyrin-like transient receptor potential channel
TRPC: canonical transient receptor potential channel
TRPM: melastatin-related transient receptor potential channel
TRPML: mucolipin-related transient receptor potential channel

TRPN: no mechanoreceptor potential C
TRPP: polycystin-related transient receptor potential channel
TRPV: vanilloid transient receptor potential channel
TrrAP: transactivation (transformation)/transcription domain-associated protein (pseudokinase)
TrV: tricuspid valve
TRx: thioredoxin
TRxIP: thioredoxin-interacting protein
TSC: tuberous sclerosis complex
TSH: thyroid-stimulating hormone
TSLP: thymic stromal lymphopoietin
Tsp: thrombospondin
Tspan: tetraspanin
TsPO: translocator protein of the outer mitochondrial membrane
tSNARE: target SNARE
tsRNA: tRNA-derived small RNA
tssaRNA: transcription start site-associated RNA
Ttn: titin (pseudokinase)
TUT: terminal uridine transferase
TWIK: tandem of P domains in a weak inwardly rectifying K^+ channel
TxA2: thromboxane A2 (thromboxane)
TxB2: thromboxane B2 (thromboxane metabolite)
TXK: Tyr kinase mutated in X-linked agammaglobulinemia
TxaS: thromboxane-A synthase
TyK: tyrosine kinase
T_3: tri-iodothyronine
T_4: thyroxine
$^+$TP: plus-end-tracking proteins

U

U: right stretch tensor
u: displacement vector
u: electrochemical command
u: specific internal energy
Ub: ubiquitin
UbC: ubiquitin-conjugating enzyme
UbE2: E2 ubiquitin conjugase
UbE3: E3 ubiquitin ligase
UbL: ubiquitin-like protein
UCH: ubiquitin C-terminal hydrolase (DUb)
Ucn: urocortin
UCP: uncoupling protein
UDP: uridine diphosphate-glucose
UK: urokinase
ULK: uncoordinated-51-like kinase (pseudokinase)
Unc: uncoordinated receptor
uPA: urokinase-type plasminogen activator (urokinase)
uPAR: uPA receptor
uPARAP: uPAR-associated protein (CLec13e)
UPR: unfolded protein response
UPS: ubiquitin-proteasome system
UP4A: uridine adenosine tetraphosphate
Uro: urodilatin
US: ultrasound
USC: unipotential stem cell
USF: upstream stimulatory factor
USI: ultrasound imaging
USP: ubiquitin-specific peptidase (deubiquitinase)
UTP: uridine triphosphate
UTR: untranslated region
UVRAG: ultraviolet wave resistance-associated gene product

V

V: left stretch tensor
V: volume
V_q: cross-sectional average velocity
V_s: specific volume
v: velocity vector
v: recovery variable
V1(2)R: type-1(2) vomeronasal receptor
$V_{1A/1B/2}$: type-1a/1b/2 arginine vasopressin receptor
VAAC: volume-activated anion channel
$VAC_{Cl(K)}$: volume-activated Cl^- (K^+)-selective channel
VACamKL: vesicle-associated CamK-like (pseudokinase)
$VACC_{NS}$: volume-activated cation non-selective channel
VAChT: vesicular acetylcholine transporter
VAIC: volume-activated ion channel
VAMP: vesicle-associated membrane protein (synaptobrevin)

VanGL: Van Gogh (Strabismus)-like protein
VAP: VAMP-associated protein
VASP: vasoactive stimulatory phosphoprotein
VAT: vesicular amine transporter
vATPase: vesicular-type H^+ ATPase
VAV: ventriculoarterial valve
Vav: GEF named from Hebrew sixth letter
VC: vital capacity
VCAM: vascular cell adhesion molecule
VCt: vasoconstriction
VDAC: voltage-dependent anion channel (porin)
VDACL: plasmalemmal, volume- and voltage-dependent, ATP-conductive, large-conductance, anion channel
VDCC: voltage-dependent calcium channel
VDP: vesicle docking protein
VDt: vasodilation
VEGF: vascular endothelial growth factor
VEGFR: vascular endothelial growth factor receptor
VF: ventricular fibrillation
VF: ventricular filling
VGAT: vesicular GABA transporter
VGC: voltage-gated channel
VgL: Vestigial-like protein
VGluT: vesicular glutamate transporter
VHL: von Hippel-Lindau Ub ligase
VIP: vasoactive intestinal peptide
VLDL: very-low-density lipoprotein
VLDLR: very-low-density lipoprotein receptor
VMAT: vesicular monoamine transporter
VN: vitronectin
VPO: vascular peroxidase
VPS: vacuolar protein sorting-associated kinase
VR: venous return
VRAC: volume-regulated anion channel
VRC: ventral respiratory column
VRK: vaccinia-related kinase
VS: vasostatin
vSMC: vascular smooth muscle cell
vSNARE: vesicular SNAP receptor (SNARE)
VSOR: volume-sensitive outwardly rectifying anion channel
VSP: voltage-sensing phosphatase
VVO: vesiculo-vacuolar organelle
vWF: von Willebrand factor

W

$\mathbf{W}$: vorticity tensor
$\mathcal{W}$: strain energy density
W: work, deformation energy
$\mathtt{w}$: weight
$\mathbf{w}$: grid velocity
WASH: WASP and SCAR homolog
WASP: Wiskott-Aldrich syndrome protein
WAT: white adipose tissue
WAVe: WASP-family verprolin homolog
WBC: white blood cell
WDR: WD repeat-containing protein
Wee: small (Scottish)
WHAMM: WASP homolog associated with actin, membranes, and microtubules
WIP: WASP-interacting protein
WIPF: WASP-interacting protein family protein
WIPI: WD repeat domain-containing phosphoinositide-interacting protein
WNK: with no K (Lys) kinase
Wnt: wingless-type
WPWS: Wolff-Parkinson-White syndrome
WNRRTK: Wnt and neurotrophin receptor-related receptor Tyr kinase ($ROR_{(RTK)}$)
WSB: WD-repeat and SOCS box-containing protein (Ub ligase)
WSS: wall shear stress
WSSTG: WSS transverse gradient
WWTR: WW domain-containing transcription regulator

X

$\mathcal{X}$: trajectory
$\mathtt{X}$: reactance
$\mathbf{X}$: Lagrangian position vector
$\mathbf{x}$: position vector
$\{x, y, z\}$: Cartesian coordinates
XBE: X-factor-binding element
XBP: X-box-binding protein (transcription factor)
XIAP: X-linked inhibitor of apoptosis (Ub ligase)

Y

Y: admittance coefficient
YAP: Yes-associated protein
YBP: Y-box-binding protein (transcription factor)
YY: yin yang (transcriptional repressor)

Z

Z: impedance
ZAP70: ζ-associated protein 70
ZBTB: zinc finger and BTB (Broad complex, Tramtrack, and bric-à-brac) domain-containing transcription factor
ZnF: zinc finger protein
ZO: zonula occludens

Miscellaneous

2-5A: 5′-triphosphorylated, (2′,5′)-phosphodiester-linked oligoadenylate
2AG: 2-arachidonyl glycerol
3DR: three-dimensional reconstruction
3BP2: Abl Src homology-3 domain-binding adaptor
4eBP1: inhibitory eIF4e-binding protein
5HT: serotonin
7TMR: 7-transmembrane receptor (GPCR)

Complementary Lists of Notations

Greek Letters

α: volumic fraction
α: convergence/divergence angle
α: attenuation coefficient
α_{k}: kinetic energy coefficient
α_{m}: momentum coefficient
β: inclination angle
$\{\beta_i\}_1^2$: myocyte parameters
β_T: coefficient of thermal expansion
Γ: domain boundary
Γ_{L}: local reflection coefficient
Γ_{G}: global reflection coefficient
γ: heat capacity ratio
γ: activation factor
γ_{g}: amplitude ratio (modulation rate) of $\mathbf{g}$
γ_s: surface tension
$\dot{\gamma}$: shear rate
δ: boundary layer thickness
ϵ_T: emissivity (thermal energy radiation)
ϵ_e: electric permittivity
ε: strain
ε: small quantity
ζ: singular head loss coefficient
ζ: transmural coordinate
$\{\zeta_j\}_1^3$: local coordinate
η: azimuthal spheroidal coordinate
θ: circumferential polar coordinate
θ: $(\hat{\mathbf{e}}_x, \hat{\mathbf{t}})$ angle
κ: wall curvature
κ_{c}: curvature ratio
κ_{d}: drag coefficient
κ_{h}: hindrance coefficient
κ_{o}: osmotic coefficient
κ_{s}: size ratio
$\{\kappa_k\}_{k=1}^9$: tube law coefficients
κ_{e}: correction factor
Λ: head loss coefficient
λ_{L}: Lamé coefficient
λ: stretch ratio
λ: wavelength
λ_A: area ratio
λ_a: acceleration ratio
λ_L: length ratio
λ_q: flow rate ratio
λ_t: time ratio
λ_v: velocity ratio
μ: dynamic viscosity
μ_{L}: Lamé coefficient
ν: kinematic viscosity
ν_{P}: Poisson ratio
Π: osmotic pressure
ρ: mass density
τ: time constant
Φ: potential
$\phi(t)$: creep function
φ: phase
χ: Lagrangian label
chi_i: molar fraction of species i
χ_i: wetted perimeter
$\psi(t)$: relaxation function
Ψ: porosity
ω: angular frequency
Ω: computational domain

Dual Notations

Bφ: basophil
Eφ: eosinophil
Lφ: lymphocyte
Mφ: macrophage
aaMφ: alternatively activated macrophage
caMφ: classically activated macrophage
Nφ: neutrophil
Σc: sympathetic
pΣc: parasympathetic

Subscripts

$_{A}$: alveolar, atrial
$_{Ao}$: aortic
$_{a}$: arterial
$_{app}$: apparent
$_{atm}$: atmospheric
$_{b}$: blood
$_{\mathbf{c}}$: contractile
$_{c}$: center
$_{c}$: point-contact
$_{D}$: Darcy (filtration)
$_{d}$: diastolic
$_{dyn}$: dynamic
$_{E}$: expiration, Eulerian
$_{e}$: external
$_{e}$: extremum
$_{eff}$: effective
$_{f}$: fluid
$_{g}$: grid
$_{I}$: inspiration
$_{i}$: internal
$_{inc}$: incremental
$_{L}$: Lagrangian
$_{l}$: limit
$_{\ell}$: line-contact
$_{M}$: macroscopic
$_{m}$: mean
$_{max}$: maximum
$_{m}$: muscular, mouth
$_{met}$: metabolic
$_{\mu}$: microscopic
$_{P}$: pulmonary
$_{\mathbf{p}}$: parallel
$_{p}$: particle
$_{q}$: quasi-ovalization
$_{r}$: radial
$_{rel}$: relative
$_{S}$: systemic
$_{\mathbf{s}}$: solute
$_{\mathbf{s}}$: serial
$_{s}$: systolic
$_{t}$: stream division
$_{T}$: total
$_{t}$: turbulence
$_{t}$: time derivative of order 1
$_{tt}$: time derivative of order 2
$_{tis}$: tissue
$_{V}$: ventricular
$_{v}$: venous
$_{W}$: wall
$_{\mathbf{w}}$: water (solvent)
$_{\Gamma}$: boundary
$_{\theta}$: azimuthal
$_{+}$: positive command
$_{-}$: negative command
$_{*}$: at interface
$_{0}$: reference state ($\cdot_0$: unstressed or low shear rate)
$_{\infty}$: high shear rate

Superscripts

^{a}: active state
^{e}: elastic
^{f}: fluid
^{h}: hypertensive
^{n}: normotensive
^{p}: passive state
^{P}: power
^{s}: solid
^{T}: transpose
^{v}: viscoelastic
*: scale
*: complex variable
$\cdot'$: first component of complex elastic and shear moduli
$\cdot''$: second component of complex elastic and shear moduli
$^{\natural}$: static, stationary, steady variable

Mathematical Notations

T: bold face capital letter means tensor
v: bold face lower case letter means vector
S, s: upper or lower case letter means scalar
Δ•: difference
δ•: increment

$d \bullet /dt$: time gradient
∂_t: first-order time partial derivative
∂_{tt}: second-order time partial derivative
∂_i: first-order space partial derivative with respect to spatial coordinate x_i
∇: gradient operator
$\nabla \mathbf{u}$: displacement gradient tensor
$\nabla \mathbf{v}$: velocity gradient tensor
$\nabla \cdot$: divergence operator
∇^2: Laplace operator
$| \, |_+$: positive part
$| \, |_-$: negative part
$\dot{\bullet}$: time derivative
$\bar{\bullet}$: time mean
$\breve{\bullet}$: space averaged
$\langle \bullet \rangle$: ensemble averaged
$\tilde{\bullet}$: dimensionless
$\bullet^+$: normalized ($\in [0,1]$)
$\hat{\bullet}$: peak value
$\bullet_\sim$: modulation amplitude
$\det(\bullet)$: determinant
$\mathrm{cof}(\bullet)$: cofactor
$\mathrm{tr}(\bullet)$: trace

Cranial Nerves

I: olfactory nerve (sensory)
II: optic nerve (sensory)
III: oculomotor nerve (mainly motor)
IV: trochlear nerve (mainly motor)
V: trigeminal nerve (sensory and motor)
VI: abducens nerve (mainly motor)
VII: facial nerve (sensory and motor)
VIII: vestibulocochlear (auditory-vestibular) nerve (mainly sensory)
IX: glossopharyngeal nerve (sensory and motor)
X: vagus nerve (sensory and motor)
XI: cranial accessory nerve (mainly motor)
XII: hypoglossal nerve (mainly motor)

Chemical Notations

[X]: concentration of X species
X (x): upper and lower case letters correspond to gene and corresponding protein or conversely (i.e., Fes, FES, and fes designate protein, a proto-oncogene product that acts as a kinase, and corresponding gene and oncogene product, respectively)
$^\bullet$: radical (unpaired electron[s])
$^{\Delta NT}$: truncated form without the N-terminal domain
$^{\Delta CT}$: truncated form without the C-terminal domain
$^{D(L)}$X: D (L)-stereoisomer of amino acids and carbohydrates (chirality prefixes for dextro- [dexter: right] and levorotation [lævus: left]), i.e., dextro(levo)rotatory enantiomer
^{G}X: globular form of X molecule
$^{F(G)}$actin: polymeric, filamentous (monomeric, globular) actin
$_C$X: carboxy (carboxyl group COOH [C])-terminal cleaved part of molecule X
$_N$X: amino (amine group NH_2 [N])-terminal cleaved part of molecule X
$_c$X: cytosolic molecule
$_m$X: membrane-bound molecule
$_t$X: truncated isoform
X_i: type-*i* isoform of the receptor of ligand X (i: integer)
XRi: receptor isoform i of ligand X (i: integer)
X+: molecule X expressed (X-positive)
X^+: cation; also intermediate product X of oxidation (loss of electron) from a reductant (or reducer) by an oxidant (electron acceptor that removes electrons from a reductant)
X−: molecule X absent (X-negative)
X^-: anion; also intermediate product X of reduction (gain of electron) from an oxidant (or oxidizer) by a reductant (electron donor that transfers electrons to an oxidant)
X^A: activator form of molecule X
X^a: active form of molecule X
X^{ECD}: soluble fragment correponding to the ectodomain of molecule X after extracellular proteolytic cleavage and shedding (possible extracellular messenger or sequestrator)
$X^{(ER)}$: endoplasmic reticulum type of molecule X

small GTPase$^{GTP(GDP)}$: active (inactive) form of small (monomeric), regulatory guanosine triphosphatase
$X^{GTP(GDP)}$: GTP (GDP)-loaded molecule X
X^{ICD}: soluble fragment correponding to intracellular domain of molecule X after intracellular proteolytic cleavage (possible messenger and/or transcription factor; e.g., Notch^{ICD}: intracellular Notch fragment)
X^{M}: methylated molecule X
X^{MT}: mitochondrial type of molecule X
X^{P}: phosphorylated molecule X
pAA: phosphorylated amino acid (pSer, pThr, and pTyr)
X^{PM}: plasmalemmal type of molecule X
X^{R}: repressor form of molecule X
X^{S}: soluble form
X^{SNO}: ^{S}nitrosylated molecule X
X^{U}: ubiquitinated protein X
X_{alt}: alternative splice variant
X_{FL}: full-length protein X
$X_{h(l,m)MW}$: high (low, mid)-molecular-weight isotype
$X_{L(S)}$: long (short) isoform (splice variants)
X_{c}: catalytic subunit
X_{P}: palmitoylated molecule X
X_{i}: number of molecule or atom (i: integer, often 2 or 3)
$(X_1–X_2)_i$: oligomer made of i complexes constituted of molecules X_1 and X_2 (e.g., histones)
a, c, nX: atypical, conventional, novel molecule X (e.g., PKC)
acX: acetylated molecule X (e.g., acLDL)
al, ac, nX: alkaline, acidic, neutral molecule X (e.g., sphingomyelinase)
asX: alternatively spliced molecule X (e.g., asTF)
cX: cellular, cytosolic, constitutive (e.g., cNOS), or cyclic (e.g., cAMP and cGMP) molecule X
caX: cardiomyocyte isoform (e.g., caMLCK)
dX: deoxyX
eX: endothelial isoform (e.g., eNOS and eMLCK)
hX: human form (ortholog); heart type (e.g., hFABP); hormone-like isoform (FGF)
iX: inhibitory mediator (e.g., iSMAD) or intracellular (e.g., iFGF) or inducible (e.g., iNOS) isoform
kX: renal type (kidney) molecule X
ksX: kidney-specific isoform of molecule X
lX: lysosomal molecule X
l,acX: lysosomal, acidic molecule X
mX: mammalian species or membrane-associated molecule X (e.g., mTGFβ)
mtX: mitochondrial type of molecule X
nX: neutral X; neuronal type (e.g., nWASP)
oxX: oxidized molecule X (e.g., oxLDL)
plX: plasmalemmal type of molecule X
rX: receptor-associated mediator or receptor-like enzyme; also regulatory type of molecular species (e.g., rSMAD)
sX: secreted, soluble form of molecule X
s,acX: secreted, acidic molecule X
skX: skeletal myocyte isoform (e.g., skMLCK)
smcX: smooth muscle cell isoform (e.g., smcMLCK)
tX: target type of X (e.g., tSNARE); tissue type (e.g., tPA)
tmX: transmembrane type of X
vX: vesicle-associated (e.g., vSNARE) or vacuolar (e.g., vATPase) type of X
GPX: glycoprotein (X: molecule abbreviation or assigned numeral)
Xx: (x: single letter) splice variants
X1: human form (ortholog)
Xi: isoform type i (paralog or splice variant; i: integer)
Xi/j: (i,j: integers) refers to either both isoforms (i.e., Xi and Xj, such as ERK1/2) or heterodimer (i.e., Xi–Xj, such as ARP2/3)
X1/X2: molecular homologs or commonly used aliases (e.g., contactin-1/F3)
PI(i)P, PI(i,j)P_2, PI(i,j,k)P_3: i,j,k (integers): position(s) of phosphorylated OH groups of the inositol ring of phosphatidylinositol mono-, bis-, and trisphosphates

Amino Acids

Ala (A): alanine
Arg (R): arginine

Asn (N): asparagine
Asp (D): aspartic acid
Asp^{COO^-}: aspartate
CysH (C): cysteine
Cys: cystine
Gln (Q): glutamine
Glu (E): glutamic acid
Glu^{COO^-}: glutamate
Gly (G): glycine
His (H): histidine
Iso, Ile (I): isoleucine
Leu (L): leucine
Lys (K): lysine
Met (M): methionine
Phe (F): phenylalanine
Pro (P): proline
Ser (S): serine
Thr (T): threonine
Trp (W): tryptophan
Tyr (Y): tyrosine
Val (V): valine

Ions

Asp^-: aspartate (carboxylate anion of aspartic acid)
ADP^{3-}: ADP anion
ATP^{4-}: ATP anion
Ca^{++}: calcium cation
Cl^-: chloride anion
Co^{++}: cobalt cation
Cu^+: copper monovalent cation
Cu^{++}: copper divalent cation
Fe^{++}: ferrous iron cation
Fe^{3+}: ferric iron cation
Glu^-: glutamate (carboxylate anion of glutamic acid)
H^+: hydrogen cation (proton)
H_3O^+: hydronium (oxonium or hydroxonium) cation
HCO_3^-: bicarbonate anion
HPO_4^{2-}: hydrogen phosphate anion
K^+: potassium cation
Mg^{++}: magnesium cation
${}^{Mg}ATP^{2-}$: ATP anion
Mn^{++}: manganese cation
Na^+: sodium cation
Ni^{++}: nickel cation (common oxidation state)
OH^-: hydroxide anion
PO_4^{3-}: phosphate anion
SO_4^{2-}: sulfate anion
Zn^{++}: zinc cation (common oxidation state)

Inhaled and Signaling Gas

CO: carbon monoxide (or carbonic oxide; signaling gas and pollutant)
CO_2: carbon dioxide (cell waste)
H_2S: hydrogen sulfide (signaling gas)
He: helium (inert monatomic gas)
N_2: nitrogen (inert diatomic gas)
NO: nitric oxide (or nitrogen monoxide; signaling gas and pollutant)
NO_2: nitrogen dioxide (air pollutant)
O_2: oxygen (cell energy producer)
SO_2: sulfur dioxide (air pollutant)

Nitric Oxide Derivatives

$NO^\bullet$: free radical form
NO^+: nitrosyl or nitrosonium cation
NO^-: nitroxyl or hyponitrite anion (inodilator)
HNO: protonated nitroxyl anion
HNO_2: nitrous acid
NO_2^-: nitrite anion
NO_3^-: nitrate anion

Reactive Oxygen and Nitrogen Species

H_2O_2: hydrogen peroxide
N_2O_3: dinitrogen trioxide
$NO_2^\bullet$: nitrogen dioxide
O_2^-: superoxide
$O{=}C(O^\bullet)O^-$: carbonate radical
$OH^\bullet$: hydroxyl radical (hydroxide ion neutral form)
$ONOO^-$: peroxynitrite

Time Units

d: day
h: hour
mn: minute
s: second
wk: week

SI-Based and Non-SI Units of Quantity

mmol, nmol, μmol: milli-, nano-, micromoles
(amount of a chemical species, one mole containing about $6.02214078 \times 10^{23}$ molecules)
mosm: milliosmole
(osm: number of moles of a osmotically active chemical compound)
kDa: kiloDalton
(Da: atomic or molecular mass unit)
ppm: parts per million
l: liter

Index

D

I

M

N

O

R

T

MIX
Papier aus verantwortungsvollen Quellen
Paper from responsible sources
FSC® C105338

If you have any concerns about our products,
you can contact us on
ProductSafety@springernature.com

In case Publisher is established outside the EU,
the EU authorized representative is:
Springer Nature Customer Service Center GmbH
Europaplatz 3, 69115 Heidelberg, Germany

Printed by Libri Plureos GmbH
in Hamburg, Germany